Symbol	Description
A'	Complement of event A 107
A, B	Events 197
$A \cup B$	Union of events A and B 104
$A \cap B$	Intersection of events A and B 104
α (alpha)	Probability of rejecting H_0 when H_0 is true (Type I error) 291
b	Number of blocks in a randomized block design 429
B_i	Total of observations for block i in an analysis of variance 803
β (beta)	Probability of accepting H_0 when H_0 is false (Type II error) 293
β_0	y-intercept in regression models 476
$\hat{\beta}_0$	Least squares estimator of β_0 480
β_1	Slope of straight-line regression model 476
$\hat{\beta}_1$	Least squares estimator of β_1 480
β_i	Coefficient of independent variable x_i in a multiple regression model 536
$\hat{\beta}_i$	Least squares estimator of β_i 537
c_j	Total count in column j of a contingency table 694
CM	Correction for the mean in an analysis of variance 802
χ^2 (chi-square)	Probability distribution of various test statistics 328
df	Degrees of freedom for the t, χ^2, and F distributions 262
D_0	Hypothesized difference between population means or binomial probabilities 342
$E(n_i)$	Expected number of outcomes of type i under the null hypothesis in a multinomial test 687
$E(n_{ij})$	Expected count in cell (i, j) of a contingency table 695
$\hat{E}(n_{ij})$	Estimated expected count in cell (i, j) of a contingency table 695
$E(x)$	Expected value of the random variable x 163
ϵ (epsilon)	Random error component in regression models 476
$f(x)$	Probability density or frequency function for a continuous random variable x 194
F	Test statistic used to compare two variances, to compare p population means, and to test several terms in a multiple regression 382
F_r	Test statistic for the Friedman nonparametric analysis of variance 745
F_α	Value of F distribution with area α to its right 385
H	Test statistic for the Kruskal–Wallis nonparametric analysis of variance 740
H_a	Alternative (or research) hypothesis 290
H_0	Null hypothesis 290
IQR	Interquartile range 70

Symbol	Description
λ (lambda)	Mean of the Poisson distribution 180
MSB	Mean square for blocks in a randomized block design 435
MSE	Estimator of variance, σ^2, in a multiple regression or analysis of variance model (SSE divided by error degrees of freedom) 409
MST	Mean square for treatments in an analysis of variance 409
μ (mu)	Population mean 45
μ_D	Population mean of differences in a paired difference experiment 360
μ_0	Hypothesized value of a population mean 300
$\mu_{\bar{x}}$	Mean of the sampling distribution of $\bar{x}$ 240
n	(1) Sample size, total number of measurements in a sample 42 (2) Number of trials in a binomial experiment 170
n_D	Number of differences (or pairs) in a paired difference experiment 361
n_i	(1) Number of observations for treatment i in an analysis of variance 408 (2) Count in ith cell in a multinomial sample 686
n_{ij}	Observed count in cell (i, j) in a contingency table 694
$n!$	n factorial; product of first n integers ($0! = 1$) 133
$\binom{N}{n}$	Number of possible combinations when n elements are drawn without replacement from N elements 133
v_1 (nu)	Numerator degrees of freedom for an F statistic 385
v_2	Denominator degrees of freedom for an F statistic 385
p	(1) Observed significance level for a hypothesis test 303 (2) Probability of Success for the binomial distribution 170 (3) Number of treatments being compared in a completely randomized or randomized block design 407
$\hat{p}$	(1) Sample estimator of binomial probability p; proportion of successes in the sample 269 (2) Pooled estimator of $p_1 = p_2 = p$ in test of equality of two population probabilities 373
$p(x)$	Probability distribution for a discrete random variable x 160
p_0	Hypothesized value for binomial probability 314
p_i	Probability of outcome i in a multinomial probability distribution 688
p_{ij}	Probability associated with the (i, j) cell in a contingency table 694
$p_1 - p_2$	Difference between true probabilities of success in two independent binomial experiments 372
$\hat{p}_1 - \hat{p}_2$	Difference between sample probabilities of success in two independent binomial samples; sample estimator of $(p_1 - p_2)$ 372

Continued inside back cover

Statistics

Seventh Edition

James T. McClave
Info Tech, Inc.
University of Florida

Frank H. Dietrich, II
Northern Kentucky University

Terry Sincich
University of South Florida

PRENTICE HALL
Upper Saddle River, New Jersey 07458

Library of Congress Cataloging-in-Publication Data

McClave, James T.
    Statistics / James T. McClave, Frank H. Dietrich II, Terry Sincich. — 7th ed.
      p. cm.
    Includes index.
    ISBN 0-13-471542-X
    ISBN 0-13-471657-4 (Annotated Instructor's Edition)
    1. Statistics. I. Dietrich, Frank H. II. Sincich, Terry.
  III. Title.
  QA276.12.M4   1997
  519.5—dc20   96-12002
                                                      CIP

Executive Editor: Ann Heath
Editorial Director: Tim Bozik
Editor-in-Chief: Jerome Grant
Assistant Vice President of Production and Manufacturing: David W. Riccardi
Development Editor: Millicent Treloar
Text Design/Project Management/Composition: Elm Street Publishing Services, Inc.
Managing Editor: Linda Mihatov Behrens
Executive Managing Editor: Katheen Schiaparelli
Marketing Manager: Evan Girard
Creative Director: Paula Maylahn
Art Director: Amy Rosen
Assistant to the Art Director: Rod Hernandez
Cover Designer: Jeanette Jacobs
Manufacturing Buyer: Alan Fischer
Manufacturing Manager: Trudy Pisciotti
Editorial Assistant: Mindy Ince
Cover Photo: Wynton Marsalis in performance at Tanglewood. Photo by Frank Stewart
Interior Photos:
  Pages ix, 1, 21: Paul Rees/Tony Stone Images
  Pages 91, 155, 193, 227, 685, 715: Frank Stewart
  Pages 253, 289, 339: Mark E. Gibson/Visuals Unlimited
  Pages 401, 473, 535, 605: Mitch Wojnarowicz/The Image Works

© 1997 by Prentice-Hall, Inc.
Simon & Schuster / A Viacom Company
Upper Saddle River, NJ 07458

Printed in the United States of America
10 9 8 7 6 5 4 3 2 1

ISBN 0-13-471542-X

Prentice-Hall International (UK) Limited, *London*
Prentice-Hall of Australia Pty. Limited, *Sydney*
Prentice-Hall Canada, Inc., *Toronto*
Prentice-Hall Hispanoamericano, S.A., *Mexico*
Prentice-Hall of India Private Limited, *New Delhi*
Prentice-Hall of Japan, Inc., *Tokyo*
Simon & Schuster Asia Pte. Ltd., *Singapore*

CONTENTS

LIST OF EXERCISE DATA SETS

Most of the exercise data sets that contain 30 or more measurements are listed below and are available on a computer disk (ASCII format). Case Study data sets available on disk are also listed below. Instructors who adopt this text may obtain a copy of the disk from the publisher.

continued

PREFACE

STATISTICS IN THE 1990's

Most news reports today include results from scientific studies, usually statistical in nature. Consider this one published on page one of the February 12, 1996, *New York Times.* "As Patrons Age, Future of the Arts Is Uncertain." Many of the studies we see reported raise issues of public funding, consumer awareness, or public health and safety, all of which directly affect our daily lives and the decisions we make. In today's world, a solid understanding of the information in statistical reports is as important to artists, actors, and musicians as it is to the sociologists, economists, scientists, and others who produce the reports.

This seventh edition of *Statistics* has been extensively revised to stress the development of statistical thinking, the assessment of credibility and value of the inferences made from data, both by those who consume them and those who produce them. This is an introductory text emphasizing inference, with extensive coverage of data collection and analysis as needed to evaluate the reported results of statistical studies. It assumes a mathematical background of basic algebra.

NEW IN THE SEVENTH EDITION

Major Content Changes

- Chapter 1 has been entirely rewritten to set the groundwork for statistical thinking and to acquaint the student at an early point with the importance of data collection, measurement, types of data, experiments, surveys, and the validity of drawing inferences from data.
- Chapter 2 now more thoroughly covers descriptive analytical tools that are useful in examining data. Pie charts, bar graphs, frequency tables (for qualitative data), and dot plots (for quantitative data) have been added.
- Chapter 4 now includes an optional section on the hypergeometric random variable.
- Chapters 7 through 9 more heavily stress confidence intervals and rely more heavily on computer output than on formulas.
- Chapter 8 now provides an equal balance between applications relying on *p*-values and those relying on critical values in their interpretation.

- Chapter 10 contains a separate section on multiple comparisons with emphasis on computer output rather than formulas. Examples of both Bonferroni's and Tukey's method are presented.
- Chapter 11 now incorporates computer printouts throughout the discussion of simple linear regression.

Pedagogy

- All new, restructured cases, approximately two per chapter, summarize statistical studies on contemporary, controversial issues and include questions to the student prompting them to evaluate findings.
- More than 60 percent of the examples and exercises in the text are new or have been completely revised. Most employ the use of current (post-1990) real data taken from a wide variety of publications.
- New end-of-chapter "Quick Reviews" provide page references to important ideas in the chapter.
- New "Language Lab" feature explains the use of key symbols in formulas and provides a pronunciation guide.
- New "Student Projects" in each chapter emphasize gathering data, analyzing data, and/or report writing.
- Five entirely new, large data bases provide the foundation for the new "Exploring Data with a Computer" feature found in the end-of-chapter exercises, and the data bases are used in examples throughout the text.

TRADITIONAL STRENGTHS

We have maintained the features of *Statistics* that we believe make it unique among introductory statistics texts. These features, which assist the student in achieving an overview of statistics and an understanding of its relevance in the social and life sciences, business, and everyday life, are as follows:

The Use of Examples as a Teaching Device

Almost all new ideas are introduced and illustrated by real data-based applications and examples. We believe that students better understand definitions, generalizations, and abstractions *after* seeing an application.

Many Exercises—Labeled by Type

The text includes more than 1,200 exercises illustrated by applications in almost all areas of research. Many students have trouble learning the mechanics of statistical techniques when all problems are couched in terms of realistic applications. For this reason, all exercise sections are divided into two parts:

Learning the Mechanics. Designed as straightforward applications of new concepts, these exercises allow students to test their ability to comprehend a concept or a definition.

Applying the Concepts. Based on applications taken from a wide variety of journals, newspapers, and other sources, these exercises develop the student's skills at comprehending real-world problems that describe situations to which the techniques may be applied.

A Choice in Level of Coverage of Probability

One of the most troublesome aspects of an introductory statistics course is the study of probability. Probability poses a challenge for instructors because they must decide on the level of presentation, and students find it a difficult subject to comprehend. We believe that one cause for these problems is the mixture of probability and counting rules that occurs in most introductory texts. We have included the counting rules in a separate and optional section at the end of the chapter on probability. In addition, all exercises that require the use of counting rules are marked with an asterisk (*). Thus, the instructor can control the level of coverage of probability.

Extensive Coverage of Multiple Regression Analysis and Model Building

This topic represents one of the most useful statistical tools for the solution of applied problems. Although an entire text could be devoted to regression modeling, we feel that we have presented coverage that is understandable, usable, and much more comprehensive than the presentations in other introductory statistics texts. We devote three chapters to discussing the major types of inferences that can be derived from a regression analysis, showing how these results appear in computer printouts and, most important, selecting multiple regression models to be used in an analysis. Thus, the instructor has the choice of a one-chapter coverage of simple regression, a two-chapter treatment of simple and multiple regression, or a complete three-chapter coverage of simple regression, multiple regression, and model building. This extensive coverage of such useful statistical tools will provide added evidence to the student of the relevance of statistics to the solution of applied problems.

Footnotes

Although the text is designed for students with a non-calculus background, footnotes explain the role of calculus in various derivations. Footnotes are also used to inform the student about some of the theory underlying certain results. The footnotes allow additional flexibility in the mathematical and theoretical level at which the material is presented.

ACKNOWLEDGMENTS

This book reflects the efforts of a great many people over a number of years. First we would like to thank the following professors, whose reviews and comments on this and prior editions have contributed to the seventh edition:

Reviewers of Previous Editions of *Statistics*

David Atkinson
Olivet Nazarene University

William H. Beyer
University of Akron

Patricia M. Buchanan
Pennsylvania State University

Kathryn Chaloner
University of Minnesota

John Dirkse
California State University—
Bakersfield

N. B. Ebrahimi
Northern Illinois University

Dale Everson
University of Idaho

Rudy Gideon
University of Montana

Larry Griffey
Florida Community College

David Groggel
Miami University at Oxford

John E. Groves
California Polytechnic State
University—San Luis Obispo

Jean L. Holton
Virginia Commonwealth University

John H. Kellermeier
State University College at
Plattsburgh

Timothy J. Killeen
University of Connecticut

William G. Koellner
Montclair State University

James R. Lackritz
San Diego State University

Diane Lambert
AT&T/Bell Laboratories

James Lang
Valencia Junior College

Glenn Larson
University of Regina

John J. Lefante, Jr.
University of South Alabama

Pi-Erh Lin
Florida State University

R. Bruce Lind
University of Puget Sound

Rhonda Magel
North Dakota State University

Linda C. Malone
University of Central Florida

Allen E. Martin
California State University—
Los Angeles

Leslie Matekaitis
Cal Genetics

E. Donice McCune
Stephen F. Austin State University

Satya Narayan Mishra
University of South Alabama

A. Mukherjea
Unversity of South Florida

Bernard Ostle
University of Central Florida

William B. Owen
Central Washington University

Won J. Park
Wright State University

John J. Peterson
Smith Kline & French Laboratories

Andrew Rosalsky
University of Florida

C. Bradley Russell
Clemson University

Rita Schillaber
University of Alberta

James R. Schott
University of Central Florida

Susan C. Schott
University of Central Florida

George Schultz
St. Petersburg Junior College

Carl James Schwarz
University of Manitoba

Mike Seyfried
Shippensburg University

Lewis Shoemaker
Millersville University

Charles W. Sinclair
Portland State University

Robert K. Smidt
California Polytechnic State
University—San Luis Obispo

Vasanth B. Solomon
Drake University

W. Robert Stephenson
Iowa State University

Barbara Treadwell
Western Michigan University

Dan Voss
Wright State University

Augustin Vukov
University of Toronto

Dennis D. Wackerly
University of Florida

Reviewers Involved with the Seventh Edition of *Statistics*

Bill Adamson
South Dakota State University

Ibrahim Ahmad
Northern Illinois University

Mary Sue Beersman
Northeast Missouri State University

Hanfeng Chen
Bowling Green State University

Gerardo Chin-Leo
The Evergreen State College

Linda Brant Collins
Iowa State University

John Egenolf
University of Alaska—Anchorage

Christine Franklin
University of Georgia

Victoria Marie Gribshaw
Seton Hill College

Shu-ping Hodgson
Central Michigan University

Soon Hong
Grand Valley State University

Ina Parks S. Howell
Florida International University

Timothy Killeen
University of Connecticut

Edwin G. Landauer
Clackamas Community College

Brenda Masters
Oklahoma State University

Mark M. Meerschaert
University of Nevada—Reno

Christopher Morrell
Loyola College in Maryland

Thomas O'Gorman
Northern Illinois University

Ronald Pierce
Eastern Kentucky University

Betty Rehfuss
North Dakota State University—
Bottineau

Andrew Rosalsky
University of Florida

James R. Schott
University of Central Florida

Arvind K. Shah
University of South Alabama

Thaddeus Tarpey
Wright State University

Kathy Taylor
Clackamas Community College

Special thanks are due to our ancillary authors, Nancy Shafer Boudreau and Mark Dummeldinger, and to typist Brenda Dobson, who have worked with us for many years. Carl Richard Gumina has done an excellent job of accuracy checking the seventh edition and has helped us to ensure a highly accurate, clean text. The Prentice Hall staff of Ann Heath, Millicent Treloar, Mindy Ince, Evan Girard, Bill Paquin, Linda Behrens, and Alan Fischer and Elm Street Publishing Services' Martha Beyerlein, Barb Lange, Sue Langguth, and Cathy Ferguson helped greatly with all phases of the text development, production, and marketing effort. Emily Thompson did an outstanding job as copy editor. Finally, we owe special thanks to Faith Sincich, whose efforts in preparing the manuscript for production and proofreading all stages of the book deserve special recognition.

SUPPLEMENTS FOR THE INSTRUCTOR

The supplements for the seventh edition have been completely revised to reflect the extensive revisions of the text. Each element in the package has been accuracy checked to ensure adherence to the approaches presented in the main text, clarity, and freedom from computational, typographical, and statistical errors.

NEW! *Annotated Instructor's Edition* (AIE) (ISBN 0-13-471657-4)

Marginal notes placed next to discussions of essential teaching concepts include:

- Teaching Tips—suggest alternative presentations or point out common student errors
- Exercises—reference specific section and chapter exercises that reinforce the concept
- ⚡ —identify accompanying PowerPoint slides
- Short Answers—section and chapter exercise answers are provided next to the selected exercises

NEW! *Instructor's Notes* by Mark Dummeldinger (ISBN 0-13-494931-5)

This new printed resource contains suggestions for using the questions at the end of the cases as the basis for class discussion, a complete short answer book with letter of permission to duplicate for student use, and many of the exercises and solutions that were removed from the sixth edition of this text.

Instructor's Solutions Manual by Nancy S. Boudreau (ISBN 0-13-471616-7)

Solutions to all of the even-numbered exercises are given in this manual. Careful attention has been paid to ensure that all methods of solution and notation are consistent with those used in the core text. Solutions to the odd-numbered exercises are found in the *Student's Solution Manual.*

Test Bank by Mark Dummeldinger (ISBN 0-13-472201-9)

Entirely rewritten, the *Test Bank* now includes more than 1,000 problems that correlate to problems presented in the text.

Windows PH Custom Test (ISBN 0-13-471673-6)
Mac PH Custom Test (0-13-472044-X)

Incorporates three levels of test creation: (1) selection of questions from a test bank; (2) addition of new questions with the ability to import test and graphics files from WordPerfect, Microsoft Word, and Wordstar; and (3) algorithmic generation of multiple questions from a single question template. PH Custom Test has a full-featured graphics editor supporting the complex formulas and graphics required by the statistics discipline.

NEW! PowerPoint Presentation Disk (ISBN 0-13-472151-9)

This versatile Windows-based tool may be used by professors in a number of different ways.

- Slide show in an electronic classroom
- Printed and used as transparency masters
- Printed copies may be distributed to students as a convenient note-taking device.

Included on the software disk are: learning objectives, thinking challenges, concept presentation slides, and examples with worked out solutions.

Data Disk (ISBN 0-13-472036-9)

The data sets described in "Appendix B" and the data for all exercises containing twenty or more observations are available on a 3 ½-inch diskette in ASCII format. A list of the exercise data on the disk, with file names, is provided on pages viii–x.

NEW! *New York Times Supplement* (ISBN 0-13-261637-8)

Copies of this supplement may be requested from Prentice Hall by instructors for distribution in their classes. This supplement contains high interest articles published recently in The *New York Times* that relate to topics covered in the text.

NEW! *Computer Software Tutorials: SAS, SPSS, Minitab, ASP* by Terry Sincich (ISBN 0-13-531609-X)

This self-contained manual provides a brief introduction to each of these statistical packages. Keystroke commands and an extensive use of output instructs the student in the use of the chosen statistical software package.

SUPPLEMENTS AVAILABLE FOR PURCHASE BY STUDENTS

Student's Solutions Manual by Nancy S. Boudreau (ISBN 0-13-471666-3)

Fully worked-out solutions to all of the odd-numbered exercises are provided in this manual. Careful attention has been paid to ensure that all methods of solution and notation are consistent with those used in the core text.

Student Versions of SPSS

Student versions of SPSS, the award-winning and market-leading commercial data analysis package, are available for student purchase. They are designed specifically for hands-on classroom teaching and learning of data analysis, statistics, and research methods. Windows, Windows 95, and Power Mac versions of the software allow the user to take full advantage of the of the easy-to-use graphical user interface combined with the traditional power of SPSS. Details on all current products are available from the publisher.

ConStatS by Tufts University (ISBN 0-13-502600-8)

ConStatS is a set of Microsoft Windows-based programs designed to help college students understand concepts taught in a first semester course on probability and statistics. Under development at Tufts University for over eight years, ConStatS helps improve students' conceptual understanding of statistics by engaging them in an active, experimental style of learning. ConStatS is available for individual purchase or to schools on a site license basis. A companion ConStatS workbook (ISBN 0-13-522848-4) that guides students through the labs and ensures they gain the maximum benefit is also available.

For additional information about texts and other materials available from Prentice Hall, visit us on-line at http//.www.prenhall.com.

How to Use This Book

To the Student

The following four pages will demonstrate how to use this text in the most effective way to make studying easier and to understand the connection between statistics and your world.

Chapter Openers Provide a Road Map

- **Where We've Been** quickly reviews how information learned previously applies to the chapter at hand.

- **Where We're Going** highlights how the chapter topics fit into your growing understanding of statistical inference.

Chapter 3

PROBABILITY

Contents

WHERE WE'VE BEEN

We've identified inference, from a sample to a population, as the goal of statistics. And we've seen that to reach this goal, we must be able to describe a set of measurements. Thus, we explored the use of graphical and numerical methods for describing both quantitative and qualitative data sets and for phrasing inferences.

WHERE WE'RE GOING

Now that we know how to phrase an inference about a population, we turn to the problem of making the inference. What is it that permits us to make the inferential jump from sample to population and then to give a measure of reliability for the inference? As you'll see, the answer is *probability*. This chapter is devoted to a study of probability—what it is and some of the basic concepts of the theory behind it.

IQ and the Bell Curve

CASE STUDY 5.1

In their controversial book *The Bell Curve* (Free Press, 1994), Professors Richard J. Herrnstein (a Harvard psychologist) and Charles Murray (a political scientist at MIT) explore, as the subtitle states, "intelligence and class structure in American life." *The Bell Curve* heavily employs statistical analyses in an attempt to support the authors' positions. One of the biggest controversies sparked by the book is the authors' tenet that intelligence levels differ across ethnic groups and that intelligence (or lack thereof) is a cause of a wide range of intractable social problems. Herrnstein and Murray warn, however, that "these group differences do not justify prejudicial assumptions about any member of a given group since his or her intelligence and potential may, in fact, be anywhere under the bell curve." For this case study, we focus on the distribution of intelligence, that is, the **bell curve**.

The measure of intelligence chosen by the authors is the well known Intelligent Quotient (IQ). Numerous tests have been developed to measure IQ: Herrnstein and Murray use the Armed Forces Qualification Test (AFQT), originally designed to measure the cognitive ability of military recruits. Psychologists traditionally treat IQ as a random variable having a normal distribution with mean $\mu = 100$ and standard deviation $\sigma = 15$. This distribution, or *bell curve*, is shown in Figure 5.17.

In their book, Herrnstein and Murray refer to five cognitive classes of people defined by percentiles of the normal distribution. Class I ("very bright") consists of

those with IQs above the 95th percentile; Class II ("bright") are those with IQs between the 75th and 95th percentiles; Class III ("normal") includes IQs between the 25th and 75th percentiles; Class IV ("dull") are those with IQs between the 5th and 25th percentiles; and Class V ("very dull") are IQs below the 5th percentile. These classes are also illustrated in Figure 5.17.

Focus

a. Assuming that the distribution of IQ is accurately represented by the bell curve in Figure 5.17, determine the proportion of people with IQs in each of the five cognitive classes defined by Herrnstein and Murray.

b. Although the cognitive classes above are defined in terms of percentiles, the authors stress that IQ scores should be compared with *z*-scores, not percentiles. In other words, it is more informative to give the difference in *z*-scores for two IQ scores than it is to give the difference in percentiles. To demonstrate this point, calculate the difference in *z*-scores for IQs at the 50th and 55th percentiles. Do the same for IQs at the 94th and 99th percentiles. What do you observe?

c. Researchers have found that scores on many intelligence tests are decidedly nonnormal. Some distributions are skewed toward higher scores, others toward lower scores. How would the proportions in the five cognitive classes differ for an IQ distribution that is skewed right? Skewed left?

Case Studies Explore High Interest Issues

- One to three cases per chapter showcase controversial, contemporary issues.

- Work through the **Focus** questions to help you evaluate the findings.

5.23 The random variable x has a normal distribution with $\mu = 300$ and $\sigma = 30$.

    a. Find the probability that x assumes a value more than 2 standard deviations from its mean. More than 3 standard deviations from μ.

b. Find the probability that x assumes a value within 1 standard deviation of its mean. Within 2 standard deviations of μ.

c. Find the value of x that represents the 80th percentile of this distribution. The 10th percentile.

Colored Boxes Highlight Important Information

- Definitions, Strategies, Key Formulas, and other important information are highlighted.

- Prepare for quizzes and tests by reviewing the highlighted information.

Interesting Examples with Solutions

- Examples, with complete solutions and explanations, illustrate every concept. Work through the solution carefully to prepare for the section exercise set.

- All examples are numbered for easy reference.

- The end of the solution is marked with a ▲ symbol.

Steps for Finding a Probability Corresponding to a Normal Random Variable

1. Sketch the normal distribution and indicate the mean of the random variable x. Then shade the area corresponding to the probability you want to find.
2. Convert the boundaries of the shaded area from x values to standard normal random variable z values using the formula

$$z = \frac{x - \mu}{\sigma}$$

   Show the z values under the corresponding x values on your sketch.
3. Use Table IV in Appendix A (and inside the front cover) to find the areas corresponding to the z values. If necessary, use the symmetry of the normal distribution to find areas corresponding to negative z values and the fact that the total area on each side of the mean equals .5 to convert the areas from Table IV to the probabilities of the event you have shaded.

EXAMPLE 5.7

Suppose an automobile manufacturer introduces a new model that has an advertised mean in-city mileage of 27 miles per gallon. Although such advertisements seldom report any measure of variability, suppose you write the manufacturer for the details of the tests, and you find that the standard deviation is 3 miles per gallon. This information leads you to formulate a probability model for the random variable x, the in-city mileage for this car model. You believe that the probability distribution of x can be approximated by a normal distribution with a mean of 27 and a standard deviation of 3.

a. If you were to buy this model of automobile, what is the probability that you would purchase one that averages less than 20 miles per gallon for in-city driving? In other words, find $P(x < 20)$.
b. Suppose you purchase one of these new models and it does get less than 20 miles per gallon for in-city driving. Should you conclude that your probability model is incorrect?

Solution

a. The probability model proposed for x, the in-city mileage, is shown in Figure 5.13. We are interested in finding the area A to the left of 20 since this area corresponds to the probability that a measurement chosen from this distribution falls below 20. In other words, if this model is correct, the area A represents the fraction of cars that can be expected to get less than 20 miles per gallon for in-city driving. To find A, we first calculate the z value corresponding to $x = 20$. That is,

$$z = \frac{x - \mu}{\sigma} = \frac{20 - 27}{3} = \frac{7}{3} = -2.33$$

FIGURE 5.13

Area under the normal curve for Example 5.7

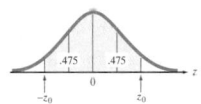

As another example of data from which the central tendency is better described by the median than the mean, consider the household incomes of a community being studied by a sociologist. The presence of just a few households with very high incomes will affect the mean more than the median. Thus, the median will provide a more accurate picture of the typical income for the community. The mean could exceed the vast majority of the sample measurements (household incomes), making it a misleading measure of central tendency.

EXAMPLE 2.6

Calculate the median for the 100 EPA mileages given in Table 2.3. Compare the median to the mean computed in Example 2.4.

Solution

For this large data set, we again resort to a computer analysis. The SPSS printout is reproduced in Figure 2.15, with the median shaded. You can see that the median is 37.0. This value implies that half of the 100 mileages in the data set fall below 37.0 and half lie above 37.0. Note that the median, 37.0, and the mean, 36.994, are almost equal. This fact indicates that the data form an approximately **symmetric distribution**. As indicated in the box on page 47, a comparison of the mean and median gives an indication of the **skewness** (i.e., the tendency of the distribution to have elongated tails) of a data set. ▲

Computer Output Integrated Throughout

- Statistical software packages, such as SPSS, Minitab, SAS, or ASP crunch data quickly so you can spend time analyzing the results. Learning how to interpret statistical output will prove helpful in future classes or on the job.

- When computer output appears in examples, the solution explains how to read and interpret the output.

FIGURE 2.15

SPSS printout of numerical descriptive measures for 100 EPA mileages

Mean	36.994	Std Err	.242	Median	37.000
Mode	37.000	Std Dev	2.418	Variance	5.846
Kurtosis	.770	S E Kurt	.478	Skewness	.051
S E Skew	.241	Range	14.900	Minimum	30.000
Maximum	44.900	Sum	3699.400		
Valid Cases	100	Missing Cases	0		

mode can be used to describe the central tendency of neither... or data (quantitative and qualitative), while the mean and median are primarily useful for quantitative data.

EXERCISES 2.29–2.45

Learning the Mechanics

2.29 Calculate the mode, mean, and median of the following data:

18 10 15 13 17 15 12 15 18 16 11

2.30 Calculate the mean and median of the following grade-point averages:

3.2   2.5   2.1   3.7   .2.8   2.0

2.31 Explain the difference between the calculation of the median for an odd and an even number of measurements. Construct one data set consisting of five measurements and another consisting of six measurements for which the medians are equal.

2.32 Explain how the relationship between the mean and median provides information about the symmetry or skewness of the data's distribution.

2.33 Calculate the mean for samples where
  a. $n = 10, \Sigma x = 85$ **b.** $n = 16, \Sigma x = 400$
  c. $n = 45, \Sigma x = 35$ **d.** $n = 18, \Sigma x = 242$

2.34 Calculate the mean, median, and mode for each of the following samples:
  a. 7, −2, 3, 3, 0, 4
  b. 2, 3, 5, 3, 2, 3, 4, 3, 5, 1, 2, 3, 4
  c. 51, 50, 47, 50, 48, 41, 59, 68, 45, 37

2.35 Describe how the mean compares to the median for a distribution as follows:
  a. Skewed to the left   **b.** Skewed to the right
  c. Symmetric

Applying the Concepts

2.36 *The Condor* (May 1995) published a study of competition for nest holes among collared flycatchers, a bird species. The authors collected the data for the study by periodically inspecting nest boxes located on the island of Gotland in Sweden. The nest boxes were grouped into 14 discrete locations (called plots). The accompanying table gives the number of flycatchers killed and the number of flycatchers breeding at each plot.

Plot Number	Number Killed	Number of Breeders
1	5	30
2	4	28
3	3	38
4	2	34
5	2	26
6	1	124
7	1	68
8	1	86
9	1	32
10	0	30
11	0	46
12	0	132
13	0	100
14	0	6

Source: Merilä, J., and Wiggins, D. A. "Interspecific competition for nest holes causes adult mortality in the collared flycatcher." *The Condor*, Vol. 97, No. 2, May 1995, p. 447 (Table 4) Cooper Ornithological Society.

  a. Calculate the mean, median, and mode for the number of flycatchers killed at the 14 study plots.
  b. Interpret the measures of central tendency, part **a.**
  c. Below is displayed a MINITAB printout of descriptive statistics for the number of breeders at

	N	MEAN	MEDIAN	TRMEAN	STDEV	SEMEAN
BREEDERS	14	55.7	36.0	53.5	39.6	10.6
	MIN	MAX	Q1	Q3		
BREEDERS	6.0	132.0	29.5	89.5		

each plot. Locate the measures of central tendency on the printout and interpret these values.

2.37 The conventional method of measuring the refractive status of an eye involves three quantities: (1) sphere power, (2) cylinder power, and (3) axis. Optometric researchers at a Johannesburg (South Africa) university studied the variation in these three measures of refraction (*Optometry and Vision Science*, June 1995). Twenty-five successive refractive measurements (using a single Topcon RM-A6000 autorefractor) were obtained on the eyes of more than 100 university students. The cylinder power measurements for the left eye of one particular student (ID #11) are listed in the table. [*Note:* All measurements are negative values.] Numerical descriptive measures for the data set are provided in the accompanying SAS printout.

.08 .08 1.07 .09 .16 .04 .07 .17 .11
.06 .12 .17 .20 .12 .17 .09 .07 .16
.15 .16 .09 .06 .10 .21 .06

Source: Rubin, A., and Harris, W. F. "Refractive variation during autorefraction: Multivariate distribution of refractive status." *Optometry and Vision Science*, Vol. 72, No. 6, June 1995, p. 409 (Table 4).

```
                    UNIVARIATE PROCEDURE
Variable=CYLPOWER
                         Moments
N                  25   Sum Wgts            25
Mean          -0.1544   Sum              -3.86
Std Dev      0.196767   Variance      0.038717
Skewness     -4.52208   Kurtosis      21.70196
USS            1.5252   CSS           0.929216
CV           -127.44    Std Mean      0.039353
T:Mean=0     -3.92342   Prob>|T|        0.0006
Sgn Rank     -162.5     Prob>|S|        0.0001
Num ^= 0          25
                    Quantiles(Def=5)
100% Max      -0.04         99%          -0.04
 75% Q3       -0.08         95%          -0.06
 50% Med      -0.11         90%          -0.06
 25% Q1       -0.16         10%           -0.2
  0% Min      -1.07          5%          -0.21
                             1%          -1.07
Range          1.03
Q3-Q1          0.08
Mode          -0.17
                      Extremes
     Lowest   Obs   Highest    Obs
    -1.07(     3)    -0.07(     17)
    -0.21(    24)    -0.06(     10)
     -0.2(    13)    -0.06(     22)
    -0.17(    15)    -0.06(     25)
    -0.17(    12)    -0.04(      6)
```

  a. Locate the measures of central tendency on the printout and interpret their values.
  b. Note that the data contain one unusually large (negative) cylinder power measurement relative to the other measurements in the data set. Find this measurement, called an **outlier**.
  c. Delete the outlier, part **b**, from the data set and recalculate the measures of central tendency. Which measure is most affected by the elimination of the outlier?

2.38 Demographics plays a key role in the recreation industry. According to the *Journal of Leisure Research* (Vol. 23, 1991), difficult times lay ahead for the industry. The article reports that the median age of the population in the United States was 30 in 1980, but will be about 36 by the year 2000.
  a. Interpret the value of the median for both 1980 and 2000 and explain the trend.
  b. If the recreation industry relies on the 18–30 age group for much of its business, what effect will this shift in the median age have on the recreation industry? Explain.

2.39 Applicants for an academic position (e.g., assistant professor) at a college or university are usually required to submit at least three letters of recommendation. A recent study of 148 applicants for an entry-level position in experimental psychology at the University of Alaska Anchorage revealed that many did not meet the three-letter requirement (*American Psychologist*, July 1995). Summary statistics for the number of recommendation letters in each application are given below. Interpret these summary measures.

  Mean = 2.28
  Median = 3
  Mode = 3

2.40 Platelet-activating factor (PAF) is a potent chemical that occurs in patients suffering from shock, inflammation, hypotension, and allergic responses as well as respiratory and cardiovascular disorders. Consequently, drugs that effectively inhibit PAF, keeping it from binding to human cells, may be successful in treating these disorders. A bioassay was undertaken to investigate the potential of 17 traditional Chinese herbal drugs in PAF inhibition (*Progress in Natural Science*, June 1995). The prevention of the PAF binding process, measured as a percentage, for each drug is provided in the accompanying table.
  a. Construct a stem-and-leaf display for the data.
  b. Compute the median inhibition percentage for the 17 herbal drugs. Interpret the result.
  c. Compute the mean inhibition percentage for the 17 herbal drugs. Interpret the result.
  d. Compute the mode of the 17 inhibition percentages. Interpret the result.

Lots of Exercises for Practice

- Every section in the book is followed by an Exercise Set divided into two parts.

- **Learning the Mechanics** has straightforward applicatons of new concepts. Test your mastery of definitons, concepts, and basic computation. Make sure you can answer all of these questions before moving on.

- **Applying the Concepts** tests your understanding of concepts and requires you to apply statistical techniques in solving real world problems. Spending time on these problems will help you develop good problem-solving skills.

Real Data

- Most of the exercises contain data or information taken from newspaper articles, magazines, and journals published since 1990. Statistics are all around you.

Computer Output

- Computer output screens appear in the exercise sets to give you practice in interpretation.

End of Chapter Review

- Each chapter ends with information designed to help you check you understanding of the material, study for tests, and expand your knowledge of statistics.

- **Quick Review** provides a list of key terms and formulas with page number references.

- **Language Lab** helps you learn the language of statistics through pronunciation guides, descriptions of symbols, names, etc.

- **Supplementary Exercises** review all of the important topics covered in the chapter.

Exercises marked with 💾 require a computer for solution.

Data sets for use with the 💾 problems are available on disk.

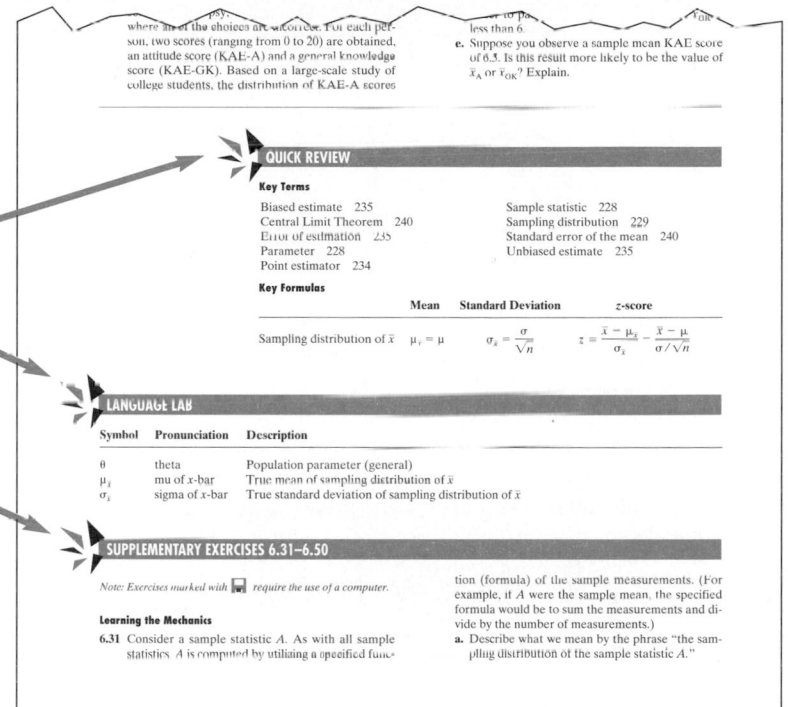

where all of the choices are offered. For each person, two scores (ranging from 0 to 20) are obtained, an attitude score (KAE-A) and a general knowledge score (KAE-GK). Based on a large-scale study of college students, the distribution of KAE-A scores

less than 6.

e. Suppose you observe a sample mean KAE score of 6.5. Is this result more likely to be the value of $\bar{x}_A$ or $\bar{x}_{GK}$? Explain.

QUICK REVIEW

Key Terms

Biased estimate 235
Central Limit Theorem 240
Error of estimation 235
Parameter 228
Point estimator 234

Sample statistic 228
Sampling distribution 229
Standard error of the mean 240
Unbiased estimate 235

Key Formulas

	Mean	Standard Deviation	z-score
Sampling distribution of $\bar{x}$	$\mu_{\bar{x}} = \mu$	$\sigma_{\bar{x}} = \dfrac{\sigma}{\sqrt{n}}$	$z = \dfrac{\bar{x} - \mu_{\bar{x}}}{\sigma_{\bar{x}}} = \dfrac{\bar{x} - \mu}{\sigma/\sqrt{n}}$

LANGUAGE LAB

Symbol	Pronunciation	Description
θ	theta	Population parameter (general)
$\mu_{\bar{x}}$	mu of x-bar	True mean of sampling distribution of $\bar{x}$
$\sigma_{\bar{x}}$	sigma of x-bar	True standard deviation of sampling distribution of $\bar{x}$

SUPPLEMENTARY EXERCISES 6.31–6.50

Note: Exercises marked with 💾 *require the use of a computer.*

Learning the Mechanics

6.31 Consider a sample statistic A. As with all sample statistics, A is computed by utilizing a specified func-

tion (formula) of the sample measurements. (For example, if A were the sample mean, the specified formula would be to sum the measurements and divide by the number of measurements.)
a. Describe what we mean by the phrase "the sampling distribution of the sample statistic A."

STUDENT PROJECTS

To understand the Central Limit Theorem and sampling distribution, consider the following experiment: Toss four identical coins, and record the number of heads observed. Then repeat this experiment four more times, so that you end up with a total of five observations for the random variable x, the number of heads when four coins are tossed.

Now derive and graph the probability distribution for x, assuming the coins are balanced. Note that the mean of this distribution is $\mu = 2$ and the standard deviation is $\sigma = 1$. This probability distribution represents the one from which you are drawing a random sample of five measurements.

Next, calculate the mean $\bar{x}$ of the five measurements—that is, calculate the mean number of heads you observed in five repetitions of the experiment. Although you have

repeated the basic experiment five times, you have only one observed value of $\bar{x}$. To derive the probability distribution or sampling distribution of $\bar{x}$ empirically, you have to repeat the entire process (of tossing four coins five times) many times. Do it 100 times.

The approximate sampling distribution of $\bar{x}$ can be derived theoretically by making use of the Central Limit Theorem. We expect at least an approximate normal probability distribution with a mean $\mu = 2$ and a standard deviation

$$\sigma_{\bar{x}} = \frac{\sigma}{\sqrt{n}} = \frac{1}{\sqrt{5}} = .45$$

Count the number of your 100 $\bar{x}$'s that fall in each of the intervals in the figure on page 251. Use the normal prob-

Student Projects provide challenging projects for further exploration by yourself or with a group of students. These projects give you good practice in gathering and analyzing data and report writing—skills that will be important in future classes and in the workplace.

Exploring Data with a Computer Five large databases available on disk and referenced in Appendix B are the basis for statistical explorations using a statistical software package.

EXPLORING DATA WITH A COMPUTER

The Federal Trade Commission (FTC) annually ranks American cigarette brands in terms of the amount of three hazardous substances—tar, nicotine, and carbon monoxide. The test results are obtained as follows: A sequential smoking machine is used to "smoke" cigarettes to a 23-millimeter butt length. Based on tests of 100 cigarettes per brand, the carbon monoxide, tar, and nicotine concentrations (rounded to the nearest milligram) in the residual "dry" particulate matter are determined.

Appendix B contains the results of the FTC's 1995 tests of 962 domestic cigarette brands. Select one of the three hazardous substances (e.g., tar). Use a statistical software package to obtain the mean and standard deviation of the 962 tar amounts. Assume that these quantities represent the population mean μ and the population standard deviation σ for all domestic cigarette brands.

a. Draw 100 random samples of $n = 20$ observations from the 962 tar amounts. Select the samples with replacement—i.e., replace each measurement before selecting the next.* Calculate the 100 sample means. Generate a stem-and-leaf display or a relative frequency histogram for the 100 means. Then count the number of the 100 sample means that fall in the intervals $\mu \pm \sigma/\sqrt{n}$, $\mu \pm 2\sigma/\sqrt{n}$, and $\mu \pm 3\sigma/\sqrt{n}$. How do the graphical description and the percentage of means falling in the intervals agree with a normal distribution having mean μ and standard deviation $\sigma/\sqrt{n}$?

b. Repeat part a using a sample size of $n = 50$. Is the sampling distribution of the sample means closer to normal for the larger sample size?

Chapter 1

STATISTICS, DATA, AND STATISTICAL THINKING

Contents

Case Studies

WHERE WE'RE GOING

Statistics? Is it a field of study, or is it a group of numbers that summarizes the state of our national economy, the performance of a football team, or the social conditions in a particular locale? Or, as one popular book (Tanur *et al.*, 1989) suggests, is it "a guide to the unknown"? We'll see in Chapter 1 that each of these descriptions is applicable in understanding what statistics is. We'll also see that there are two areas of statistics: *descriptive statistics,* which focuses on developing graphical and numerical summaries that describe some phenomenon, and *inferential statistics,* which uses these numerical summaries to assist in making decisions. The primary theme of this text is inferential statistics. Thus, we'll concentrate on showing how you can use statistics to interpret data and use them to make decisions. Many jobs in industry, government, medicine, and other fields require you to make data-driven decisions so understanding these methods offers you important practical benefits.

1.1 THE SCIENCE OF STATISTICS

What does statistics mean to you? Does it bring to mind batting averages, Gallup polls, unemployment figures, or numerical distortions of facts (lying with statistics!)? Or is it simply a college requirement you have to complete? We hope to persuade you that statistics is a meaningful, useful science whose broad scope of applications to business, government, and the physical and social sciences is almost limitless. We also want to show that statistics can lie only when they are misapplied. Finally, we wish to demonstrate the key role statistics plays in critical thinking—whether in the classroom, on the job, or in everyday life. Our objective is to leave you with the impression that the time you spend studying this subject will repay you in many ways.

The *Random House College Dictionary* defines *statistics* as "the science that deals with the collection, classification, analysis, and interpretation of information or data." Thus, a statistician isn't just someone who calculates batting averages at baseball games or tabulates the results of a Gallup poll. Professional statisticians are trained in *statistical science.* That is, they are trained in collecting numerical information in the form of **data**, evaluating it, and drawing conclusions from it. Furthermore, statisticians determine what information is relevant in a given problem and whether the conclusions drawn from a study are to be trusted.

DEFINITION 1.1

Statistics is the science of data. This involves collecting, classifying, summarizing, organizing, analyzing, and interpreting numerical information.

In the next section, you'll see several real-life examples of statistical applications that involve making decisions and drawing conclusions.

1.2 TYPES OF STATISTICAL APPLICATIONS

Statistics means "numerical descriptions" to most people. Monthly housing starts, the failure rate of liver transplants, and the proportion of African-Americans who feel brutalized by local police all represent statistical descriptions of large sets of data collected on some phenomenon. Often the data are selected from some larger set of data whose characteristics we wish to estimate. We call this selection process *sampling.* For example, you might collect the ages of a sample of customers at a video store to estimate the average age of *all* customers of the store. Then you could use your estimate to target the store's advertisements to the appropriate age group. Notice that statistics involves two different processes: (1) describing large sets of data and (2) drawing conclusions (making estimates, decisions, predictions, etc.) about the sets of data based on sampling. So, the applications of statistics can be divided into two broad areas: *descriptive statistics* and *inferential statistics.*

DEFINITION 1.2

Descriptive statistics utilizes numerical and graphical methods to look for patterns in a data set, to summarize the information revealed in a data set, and to present that information in a convenient form.

DEFINITION 1.3

Inferential statistics utilizes sample data to make estimates, decisions, predictions, or other generalizations about a larger set of data.

Although we'll discuss both descriptive and inferential statistics in the following chapters, the primary theme of the text is **inference**.

Let's begin by examining some studies that illustrate applications of statistics.

Study 1   "Discrimination in the Workplace" (*USA Today,* Aug. 15, 1995).

According to *USA Today,* a record number of job discrimination cases—158,612—were filed with federal or state civil rights agencies in 1994. *USA Today* obtained information on each case from the Equal Employment Opportunity Center (EEOC). Then researchers at this national newspaper determined the basis of each of the 158,612 claims. They reported the results in the graph shown in Figure 1.1. The graph provides an effective summary of the 158,612 claims, clearly depicting that race and sex discrimination are the two most common bases for job discrimination complaints. Thus, Figure 1.1 is an example of *descriptive statistics.*

Study 2   "Who's Number 1 in College Football?... And How Might We Decide?" (*Chance,* Summer 1995).

> College football is perhaps the only major team sport that does not decide a unique champion via competition each year. College football teams are rated during the season by two polls, one a survey of journalists (The Associated Press) and one a survey of coaches (*USA Today*/CNN). [In 1994] both groups preferred undefeated Nebraska to undefeated Penn State in their final polls, but there was substantial debate among experts and fans about whether this was the correct decision.... Statistics is a field concerned with collection, analysis, interpretation, and presentation of numerical data. Certainly, college football provides abundant numerical data that are amenable to statistical analysis. What can statistics say about choosing the number one team?

In this opening paragraph of his article on college football's dilemma, Hal S. Stern clearly describes the problem. Using data on points scored each game by each of the top 25 college football teams in 1994 and their opponents, Stern addressed this problem by formulating several algorithms for ranking the teams. He then evaluated the different ranking methods by assessing their ability to predict the outcome of professional (NFL) football games. An analysis of the NFL predictions led Stern to select a statistical method designed to minimize the effect of "blowouts" on a team's rating.

Penn State and Nebraska didn't play each other in 1994, so we'll never know which team was truly the best. But based on his statistical analysis, Stern *inferred*

FIGURE 1.1

Basis of workplace discrimination claims filed in 1994

[*Note:* The figures add up to more than 100% because charges may involve more than one type.]

Source: *USA Today,* Aug. 15, 1995, p. 10A.

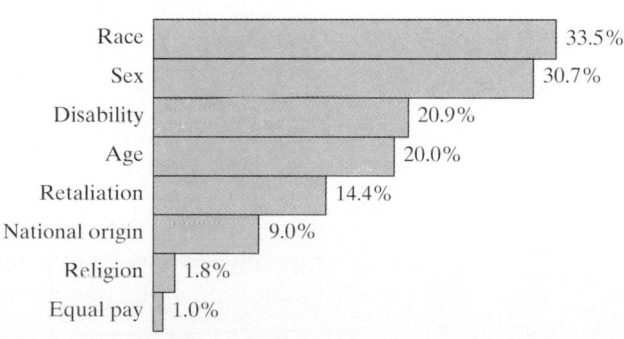

that Penn State was number one in that year. Thus, Stern applied *inferential statistics* to arrive at his conclusion.

Study 3   "The Significance of Belief and Expectancy Within the Spiritual Healing Encounter" (*Social Science & Medicine,* July 1995).

"Spiritual healing techniques have been a fundamental component of the healing rituals of virtually all societies since the advent of man," says researcher Daniel P. Wirth. What effect, if any, do spiritual healing encounters have on the mental and physical health of patients in today's society? And do the effects depend on the patient's expectations? To answer these questions, Wirth conducted a scientific study of 48 individuals residing in a northern California suburb. The mental and physical health of each subject was determined via a battery of standardized tests. Each individual then underwent a spiritual healing encounter that involved a "magnetic laying-on-of-hands approach." One week following the healing session, subjects were retested.

An analysis of the data revealed a significant difference between the pretest and posttest health scores for the subjects. Further analysis led Wirth to conclude that "high healer and patient expectancy [are] important ... predictors as well as facilitators of the healing process."

Like Study 2, this study is an example of the use of *inferential statistics.* Wirth used data from a sample of 48 individuals to make inferences about the beliefs and health improvements of all patients who experience a spiritual healing encounter.

These studies provide three real-life examples of the uses of statistics. Notice that each involves an analysis of data, either for the purpose of describing the data set (Study 1) or for making inferences about a data set (Studies 2 and 3).

1.3   FUNDAMENTAL ELEMENTS OF STATISTICS

Statistical methods are particularly useful for studying, analyzing, and learning about *populations.*

> **DEFINITION 1.4**
>
> A **population** is a set of units (usually, people, objects, transactions, or events) that we are interested in studying.

For example, populations may include (1) *all* employed workers in the United States, (2) *all* registered voters in California, (3) *everyone* who is afflicted with AIDS, (4) *all* the cars produced last year by a particular assembly line, (5) the *entire* stock of spare parts available at United Airlines' maintenance facility, (6) *all* sales made at the drive-in window of a McDonald's restaurant during a given year, and (7) the set of *all* accidents occurring on a particular stretch of interstate highway during a holiday period. Notice that the first three population examples (1–3) are sets (groups) of people, the next two (4–5) are sets of objects, the next (6) is a set of transactions, and the last (7) is a set of events. Also notice that each set includes all the units in the population.

In studying a population, we focus on one or more characteristics or properties of the units in the population. We call such characteristics *variables.* For example, we may be interested in the variables age, gender, and/or the number of years of education of the people currently unemployed in the United States.

FIGURE 1.2

A sample of all
registered voters in
the United States

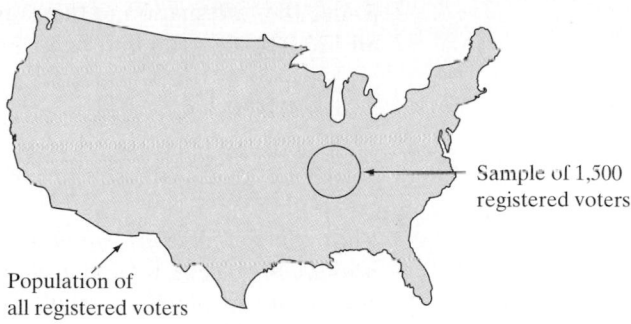

Sample of 1,500
registered voters

Population of
all registered voters

DEFINITION 1.5

A **variable** is a characteristic or property of an individual population unit.

The name "variable" is derived from the fact that any particular characteristic may vary among the units in a population.

In studying a particular variable it is helpful to obtain a numerical representation for it. Often, however, numerical representations are not readily available, so the process of measurement plays an important supporting role in statistical studies. **Measurement** is the process we use to assign numbers to variables of individual population units. We might, for instance, measure the performance of the President by asking a registered voter to rate it on a scale from 1 to 10. Or we might measure workforce age simply by asking each worker how old she is. In other cases, measurement involves the use of instruments such as stop watches, scales, and calipers.

If the population we wish to study is small, it is possible to measure a variable for every unit in the population. For example, if you are measuring the GPA for all incoming first-year students at your university, it is at least feasible to obtain every GPA. When we measure a variable for every unit of a population, the result is a **census** of the population. Typically, however, the populations of interest in most applications are much larger, involving perhaps many thousands or even an infinite number of units. Examples of large populations include those following Definition 1.4, as well as all graduates of your university or college, all potential buyers of a new fax machine, and all pieces of first-class mail handled by the U.S. Post Office. For such populations conducting a census would be prohibitively time-consuming and/or costly. A reasonable alternative would be to select and study a *subset* (or portion) of the units in the population.

DEFINITION 1.6

A **sample** is a subset of the units of a population.

For example, instead of polling all 120,000,000 registered voters in the United States during a presidential election year, a pollster might select and examine a sample of just 1,500 voters (see Figure 1.2). If he is interested in the variable "presidential preference," he would record (measure) the preference of each sampled vote.

After the variable(s) of interest for every unit in the sample (or population) are measured, the data are analyzed, either by descriptive or inferential statistical methods. The pollster, for example, may be interested only in *describing* the voting patterns of the sample of 1,500 voters. More likely, however, he will want to use the information in the sample to make *inferences* about the population of all 120,000,000 voters.

> **DEFINITION 1.7**
>
> A **statistical inference** is an estimate or prediction or some other generalization about a population based on information contained in a sample.

That is, *we use the information contained in the sample to learn about the larger population.** Thus, from the sample of 1,500 voters, the pollster may estimate the percentage of all the voters who would vote for each presidential candidate if the election were held on the day the poll was conducted, or he might use the results to predict the outcome on election day.

EXAMPLE 1.1

A sociologist hypothesizes that the average annual income of households in a particular large city is less than $25,000 per year. To test her hypothesis, she samples 500 households in the city and determines the income of each.

a. Describe the population.
b. Describe the variable of interest.
c. Describe the sample.
d. Describe the inference.

Solution

a. The population is the set of units of interest to the sociologist, which is the set of all households in the city.
b. The total annual income of each household is the variable of interest.
c. The sample must be a subset of the population. In this case, it is the 500 households selected by the sociologist.
d. The inference of interest involves the *generalization* of the information contained in the sample of 500 households to the population of all households in the city. In particular, the sociologist wants to estimate the average income of the households in the city in order to determine whether it is less than $25,000. She might accomplish this by calculating the average income in the sample and using the sample average to estimate the population average.   ▲

EXAMPLE 1.2

"Cola wars" is the popular term for the intense competition between Coca-Cola and Pepsi displayed in their marketing campaigns. Their campaigns have featured movie and television stars, rock videos, athletic endorsements, and claims of consumer preference based on taste tests. Suppose, as part of a Pepsi marketing campaign, 1,000 cola consumers are given a blind taste test (i.e., a taste test in which the two brand names are disguised). Each consumer is asked to state a preference for brand A or brand B.

a. Describe the population.
b. Describe the variable of interest.
c. Describe the sample.
d. Describe the inference.

---

*The terms *population* and *sample* are often used to refer to the sets of measurements themselves, as well as to the units on which the measurements are made. When a single variable of interest is being measured, this usage causes little confusion. But when the terminology is ambiguous, we'll refer to the measurements as *population data sets* and *sample data sets,* respectively.

Solution

a. The population of interest is the collection or set of all consumers.

b. The characteristic that Pepsi wants to measure is the consumer's cola preference as revealed under the conditions of a blind taste test, so cola preference is the variable of interest.

c. The sample is the 1,000 cola consumers selected from the population of all cola consumers.

d. The inference of interest is the *generalization* of the cola preferences of the 1,000 sampled consumers to the population of all cola consumers. In particular, the preferences of the consumers in the sample can be used to *estimate* the percentage of all cola consumers who prefer each brand. ▲

The preceding definitions and examples identify four of the five elements of an inferential statistical problem: a population, one or more variables of interest, a sample, and an inference. But making the inference is only part of the story. We also need to know its **reliability**—that is, how good the inference is. The only way we can be certain that an inference about a population is correct is to include the entire population in our sample. However, because of *resource constraints* (i.e., insufficient time and/or money), we usually can't work with whole populations, so we base our inferences on just a portion of the population (a sample). Thus, we introduce an element of *uncertainty* into our inferences. Consequently, whenever possible, it is important to determine and report the reliability of each inference made. Reliability, then, is the fifth element of inferential statistical problems.

The measure of reliability that accompanies an inference separates the science of statistics from the art of fortune-telling. A palm reader, like a statistician, may examine a sample (your hand) and make inferences about the population (your life). However, unlike statistical inferences, the palm reader's inferences include no measure of reliability.

Suppose, like the sociologist in Example 1.1, we are interested in estimating the average income of a population of households from the average income of a sample of households. Using statistical methods, we can determine a *bound on the estimation error*. This bound is simply a number that our estimation error (the difference between the average income of the sample and the average income of the population of households) is not likely to exceed. We'll see in later chapters that this bound is a measure of the uncertainty of our inference. The reliability of statistical inferences is discussed throughout this text. For now, we simply want you to realize that an inference is incomplete without a measure of its reliability.

DEFINITION 1.8

A **measure of reliability** is a statement (usually quantified) about the degree of uncertainty associated with a statistical inference.

Let's conclude this section with a summary of the elements of both descriptive and inferential statistical problems and an example to illustrate a measure of reliability.

Four Elements of Descriptive Statistical Problems

1. The population or sample of interest
2. One or more variables (characteristics of the population or sample units) that are to be investigated

3. Tables, graphs, or numerical summary tools
4. Identification of patterns in the data

Five Elements of Inferential Statistical Problems

1. The population of interest
2. One or more variables that are to be investigated
3. The sample of population units
4. The inference about the population based on information contained in the sample
5. A measure of reliability for the inference

EXAMPLE 1.3

Refer to Example 1.2, in which the cola preferences of 1,000 consumers were indicated in a taste test. Describe how the reliability of an inference concerning the preferences of all cola consumers in the Pepsi bottler's marketing region could be measured.

Solution

When the preferences of 1,000 consumers are used to estimate those of all consumers in the region, the estimate will not exactly mirror the preferences of the population. For example, if the taste test shows that 56% of the 1,000 consumers chose Pepsi, it does not follow (nor is it likely) that exactly 56% of all cola drinkers in the region prefer Pepsi. Nevertheless, we can use sound statistical reasoning (which we'll explore later in the text) to ensure that our sampling procedure will generate estimates that are almost certainly within a specified limit of the true percentage of all consumers who prefer Pepsi. For example, such reasoning might assure us that the estimate of the preference for Pepsi is almost certainly within 5% of the actual population preference. The implication is that the actual preference for Pepsi is between 51% [i.e., (56 − 5)%] and 61% [i.e., (56 + 5)%]—that is, (56 ± 5)%. This interval represents a measure of reliability for the inference. ▲

1.4 TYPES OF DATA

You have learned that statistics is the science of data and that data are obtained by measuring the values of one or more variables on the units in the sample (or population). All data (and hence the variables we measure) can be classified as one of two general types: *quantitative data* and *qualitative data.*

  Quantitative data are data that are measured on a naturally occurring numerical scale.* The following are examples of quantitative data:

1. The temperature (in degrees Celsius) at which each in a sample of 20 pieces of heat-resistant plastic begins to melt
2. The current unemployment rate (measured as a percentage) for each of the 50 states

---

*Quantitative data can be subclassified as either *interval data* or *ratio data.* For ratio data, the origin (i.e., the value 0) is a meaningful number. But the origin has no meaning with interval data. Consequently, we can add and subtract interval data, but we can't multiply and divide them. Of the four quantitative data sets listed, (1) and (3) are interval data, while (2) and (4) are ratio data.

3. The scores of a sample of 150 law school applicants on the LSAT, a standardized law school entrance exam administered nationwide
4. The number of convicted murderers who receive the death penalty each year over a 10-year period

---

DEFINITION 1.9

Quantitative data are measurements that are recorded on a naturally occurring numerical scale.

In contrast, qualitative data cannot be measured on a natural numerical scale; they can only be classified into categories.* Examples of qualitative data are:

1. The political party affiliation (Democrat, Republican, or Independent) in a sample of 50 voters
2. The defective status (defective or not) of each of 100 computer chips manufactured by Intel
3. The size of a car (subcompact, compact, mid-size, or full-size) rented by each of a sample of 30 business travelers
4. A taste tester's ranking (best, worst, etc.) of four brands of barbecue sauce for a panel of 10 testers

Often, we assign arbitrary numerical values to qualitative data for ease of computer entry and analysis. But these assigned numerical values are simply codes: They cannot be meaningfully added, subtracted, multiplied, or divided. For example, we might code Democrat = 1, Republican = 2, and Independent = 3. Similarly, a taste tester might rank the barbecue sauces from 1 (best) to 4 (worst). These are simply arbitrarily selected numerical codes for the categories and have no utility beyond that.

DEFINITION 1.10

Qualitative data are measurements that cannot be measured on a natural numerical scale; they can only be classified into one of a group of categories.

EXAMPLE 1.4

Chemical and manufacturing plants sometimes discharge toxic-waste materials such as DDT into nearby rivers and streams. These toxins can adversely affect the plants and animals inhabiting the river and the river bank. The U.S. Army Corps of Engineers recently conducted a study of fish in the Tennessee River (in Alabama) and its three tributary creeks: Flint Creek, Limestone Creek, and Spring Creek. A total of 144 fish were captured, and the following variables were measured for each:

1. River/creek where each fish was captured
2. Species (channel catfish, largemouth bass, or smallmouth buffalofish)
3. Length (centimeters)
4. Weight (grams)
5. DDT concentration (parts per million)

Classify each of the five variables measured as quantitative or qualitative.

---

*Qualitative data can be subclassified as either *nominal data* or *ordinal data*. The categories of an ordinal data set can be ranked or meaningfully ordered, but the categories of a nominal data set can't be ordered. Of the four qualitative data sets listed above, (1) and (2) are nominal and (3) and (4) are ordinal.

Solution

The variables length, weight, and DDT concentration are quantitative because each is measured on a numerical scale: length in centimeters, weight in grams, and DDT in parts per million. In contrast, river/creek and species cannot be measured quantitatively: They can only be classified into categories (e.g., channel catfish, largemouth bass, or smallmouth buffalofish for species). Consequently, data on river/creek and species are qualitative.   ▲

As you would expect, the statistical methods for describing, reporting, and analyzing data depend on the type (quantitative or qualitative) of data measured. We demonstrate many useful methods in the remaining chapters of the text. But first we discuss some important ideas on data collection.

1.5   COLLECTING DATA

Once you decide on the type of data—quantitative or qualitative—appropriate for the problem at hand, you'll need to collect the data. Generally, you can obtain data in four different ways:

1. Data from a *published source*
2. Data from a *designed experiment*
3. Data from a *survey*
4. Data collected *observationally*

Sometimes, the data set of interest has already been collected for you and is available in a **published source**, such as a book, journal, or newspaper. For example, you may want to examine and summarize the divorce rates (i.e., number of divorces per 1,000 population) in the 50 states of the United States. You can find this data set (as well as numerous other data sets) at your library in the *Statistical Abstract of the United States,* published annually by the U.S. government. Similarly, someone who is interested in monthly mortgage applications for new home construction would find this data set in the *Survey of Current Business,* another government publication. Other examples of published data sources include *The Wall Street Journal* (financial data), *The Sporting News* (sports information), and America Online (accessed over the Internet).*

A second method of collecting data involves conducting a **designed experiment**, in which the researcher exerts strict control over the units (people, objects, or events) in the study. For example, a recent medical study investigated the potential of aspirin in preventing heart attacks. Volunteer physicians were divided into two groups—the *treatment* group and the *control* group. In the treatment group, each physician took one aspirin tablet a day for one year, while each physician in the control group took an aspirin-free placebo made to look like an aspirin tablet. The researchers, not the physicians under study, controlled who received the aspirin (the treatment) and who received the placebo. As you'll learn in Chapter 10, a properly designed experiment allows you to extract more information from the data than is possible with an uncontrolled study.

---

*With published data, we often make a distinction between the *primary source* and a *secondary source*. If the publisher is the original collector of the data, the source is primary. Otherwise, the data is secondary source data.

Surveys are a third source of data. With a **survey**, the researcher samples a group of people, asks one or more questions, and records the responses. Probably the most familiar type of survey is the political polls conducted by any one of a number of organizations (e.g., Harris, Gallup, Roper, and CNN) and designed to predict the outcome of a political election. Another familiar survey is the Nielsen survey, which provides the major networks with information on the most watched programs on television. Surveys can be conducted through the mail, with telephone interviews, or with in-person interviews. Although in-person surveys are more expensive than mail or telephone surveys, they may be necessary when complex information must be collected.

Finally, observational studies can be employed to collect data. In an **observational study**, the researcher observes the experimental units in their natural setting and records the variable(s) of interest. For example, a child psychologist might observe and record the level of aggressive behavior of a sample of fifth graders playing on a school playground. Similarly, a zoologist may observe and measure the weights of baby elephants born in captivity. Unlike a designed experiment, an observational study is one in which the researcher makes no attempt to control any aspect of the experimental units.

Regardless of the data collection method employed, it is likely that the data will be a sample from some population. And if we wish to apply inferential statistics, we must obtain a *representative sample.*

DEFINITION 1.11

A **representative sample** exhibits characteristics typical of those possessed by the target population.

For example, consider a political poll conducted during a presidential election year. Assume the pollster wants to estimate the percentage of all 120,000,000 registered voters in the United States who favor the incumbent president. The pollster would be unwise to base the estimate on survey data collected for a sample of voters from the incumbent's own state. (See, for example, Figure 1.2.) Such an estimate would almost certainly be *biased* high; consequently, it would not be very reliable.

The most common way to satisfy the representative sample requirement is to select a random sample. A **random sample** ensures that every subset of fixed size in the population has the same chance of being included in the sample. If the pollster samples 1,500 of the 120,000,000 voters in the population so that every subset of 1,500 voters has an equal chance of being selected, he has devised a random sample. The procedure for selecting a random sample is discussed in Chapter 3. Here, however, let's look at two examples involving actual sampling studies.

EXAMPLE 1.5

In the 1970s and early 1980s, state lotteries became commonplace in the United States. In 1985, the *Journal of the Institute for Socioeconomic Studies* (Sept. 1985) reported on a study designed to estimate the proportion of state lottery winners who quit their jobs within one year of striking it rich. Questionnaires were mailed to all 2,000 lottery winners who won at least $50,000 over the 10-year period 1976–1985. Of the 576 who responded, 11% indicated they had quit their jobs.

a. Identify the data collection method.
b. Are the sample data representative of the target population?

The Latest Hite Report—Controversy over the Numbers

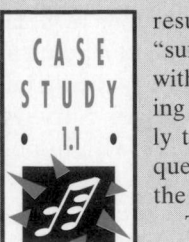

CASE STUDY • 1.1 •

In 1968, researcher Shere Hite shocked conservative America with her now-famous "Hite Report" on the permissive sexual attitudes of American men and women. Twenty years later, Hite engendered controversy again with the publication of *Women and Love: A Cultural Revolution in Progress* (Knopf, 1988). In this book, she revealed some startling statistics describing how women feel about contemporary relationships:

- 84% of women are not emotionally satisfied with their relationships.
- 95% of women report "emotional and psychological harassment" from their partners.
- 70% of women married five years or more are having extramarital affairs.
- Only 13% of women married more than two years are "in love."

Hite conducted the survey by mailing out 100,000 questionnaires to women across the country over a seven-year period. Each questionnaire consisted of 127 open-ended questions, many with numerous subquestions and follow-ups. Hite's instructions read as follows: "It is not necessary to answer every question! Feel free to skip around and answer those questions you choose." Approximately 4,500 completed questionnaires were returned—a response rate of 4.5%. These questionnaires formed the data set from which Hite determined the percentages given above. Claiming that these 4,500 women made up a representative sample of all women in the United States, Hite maintained that her survey results imply that vast numbers of women are "suffering a lot of pain in their love relationships with men." Many people disagreed, however, saying that because only unhappy women were likely to take the time to answer Hite's 127 essay questions, her sample was representative only of the discontented.

The views of several statisticians and expert survey researchers on the validity of Hite's numbers were presented in a subsequent article in *Chance* (Summer 1988). A few of the more critical comments follow.

Hite used a combination of haphazard sampling and volunteer respondents to collect her [data]. First, Hite sent questionnaires to a wide variety of organizations and asked them to circulate the questionnaires to their members. She mentions that they included church groups, women's voting and political groups, women's rights organizations and counseling and walk-in centers for women. These groups would not seem to be representative of women in general; there is an over-representation of feminist groups and of women in troubled circumstances. In addition, the use of groups to distribute the questionnaires meant that gatekeepers had the power of assuring a zero response rate by not distributing the questionnaire, or conversely of greatly stimulating returns by endorsing the study in some fashion. Second, Hite also relied on volunteer respondents who wrote in for copies of the questionnaire. These volunteers seem to have been recruited from readers of her past books and those who saw interviews on television and in the press. This type of volunteer respondent is the exact opposite of the

---

Solution

a. The data collection method is a mail survey since questionnaires were mailed to lottery winners to elicit their job status (quit job within one year or not).

b. Because the data (576 responses to the job status question) clearly make up a subset of the target population (all 2,000 lottery winners), they do form a sample. But whether or not the sample is representative is unclear, since we are given no information on the 576 lottery winners who responded to the survey. However, mail surveys (and surveys in general) often suffer from *nonresponse bias*. The fact that only about one-fourth of the 2,000 lottery winners responded to the survey may indicate apathy on the part of lottery winners. A proportion much larger than 11% may have actually quit their jobs, but they were too busy enjoying their newfound fortune to bother responding to the survey. ▲

EXAMPLE 1.6

Psychologists at the University of Tennessee carried out a study of the susceptibility of people to hypnosis (*Psychological Assessment*, Mar. 1995). In a random sample of 130 undergraduate psychology students at the university, each experienced both traditional hypnosis and computer-assisted hypnosis. Approximately

Case Study 1.1 continued

randomly selected respondent utilized in standard survey research and even more potentially unrepresentative than the group samples cited above. *Source:* Tom Smith, National Opinion Research Center

So few people responded, it's not representative of any group, except the odd group who agreed to respond. Hite has no assurance that even her claimed 4.5% response rate is correct. How do we know how many people passed their hands over these questionnaires? You don't want to fill it out, you give it to your sister, she gives it to a friend. You'll get one response, but that questionnaire may have been turned down by five people. *Source:* Donald Rubin, Professor and Chairman, Department of Statistics, Harvard University

When you get instructions to only answer those questions you wish to, you're likely to skip some. Isn't it more likely that, for example, a woman who feels strongly about affairs would be more likely to answer questions on that subject than a woman who does not feel as strongly? Thus, her finding that 70% of all women married over five years are having affairs is meaningless because she does not report how many people answered each question. I cannot tell whether this means 70% of 1,000 women or 70% of 10 women. *Source:* Judith Tanur, Professor of Sociology and Statistical Specialist in Survey Methodology, State University of New York, Stony Brook

Even in good samples, where you have a 50% or 70% response rate, you usually have some skews—say, with income, race, or region. If she can do a sample like this, she's got the Rosetta Stone, and I'll come study from her. *Source:* Martin Frankel, Professor of Statistics and Computer Information Systems, Baruch

College (commenting on Hite's claim that her sample matches that of the U.S. female population in terms of demographic balance)

According to Hite, whether you're 18 or 71, you're going to answer the questions the same way. Whether white, black, Hispanic, Middle Eastern, or Asian American, you're going to answer the same way. Whether you make $5,000 a year or over $75,000, you'll answer the questions the same way. I've never seen anything like this in my career—and the Kinsey Institute collects data from everybody. *Source:* June Reinisch, Director of the Kinsey Institute, Indiana University, commenting on Hite's numbers showing that no matter what the demographic breakdown of the women married five years or more, about 70% are having extramarital affairs [From Streitfeld, D. "Shere Hite and the trouble with numbers." *Chance: New Directions for Statistics and Computing,* Vol. 1, No. 3, Summer 1988, pp. 26–31. Springer-Verlag, ©1988, the *Washington Post.* Reprinted with permission.]

Focus

a. Identify the population of interest to Shere Hite.
b. Identify the variables of interest to Hite. Are they quantitative or qualitative variables?
c. Describe how Hite obtained her sample.
d. What inferences did Hite make about the population? Comment on the reliability of these inferences.
e. Discuss the difficulty in obtaining a random sample of women across the United States to take part in a survey similar to the one conducted by Shere Hite.

half were randomly assigned to undergo the traditional procedure first, followed by the computer-assisted procedure. The other half were randomly assigned to experience computer-assisted hypnosis first, then traditional hypnosis. Following the hypnosis episodes, all students filled out questionnaires designed to measure a subject's susceptibility to hypnosis. The susceptibility scores of the two groups of students were compared.

a. Identify the data collection method.
b. Are the sample data representative of the target population?

Solution

a. The researchers controlled which type of hypnosis—traditional or computer-assisted—the students experienced first. Consequently, a designed experiment was used to collect the data for the two groups.
b. The sample of 130 psychology students was randomly selected from all psychology students at the University of Tennessee. If the target population is *all University of Tennessee psychology students,* it is likely that the sample is

A "20/20" View of Surveys: Fact or Fiction?

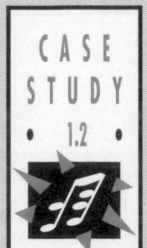

CASE STUDY 1.2

Did you ever notice that, no matter where you stand on popular issues of the day, you can always find statistics or surveys to back up your point of view—whether to take vitamins, whether day care harms kids, or what foods can hurt you or save you? There is an endless flow of information to help you make decisions, but is this information accurate, unbiased? John Stossel decided to check that out, and you may be surprised to learn if the picture you're getting doesn't seem quite right, maybe it isn't.

Barbara Walters gave this introduction to a March 31, 1995 segment of the popular prime-time ABC television program "20/20." The story is titled "Facts or Fiction?—Exposés of So-Called Surveys." One of the surveys investigated by ABC correspondent John Stossel compared the discipline problems experienced by teachers in the 1940s and those experienced today. The results: In the 1940s, teachers worried most about students talking in class, chewing gum, and running in the halls. Today, they worry most about being assaulted! This information was highly publicized in the print media—in daily newspapers, weekly magazines, Ann Landers' column, *The Congressional Quarterly,* and *The Wall Street Journal,* among others—and referenced in speeches by a variety of public figures, including former first lady Barbara Bush and former Education Secretary William Bennett.

"Hearing this made me yearn for the old days when life was so much simpler and gentler, but was life that simple then?" asks Stossel. "Wasn't there juvenile delinquency [in the 1940s]? Is the survey true?" With the help of a Yale School of Management professor, Stossel found the original source of the teacher survey—Texas oilman T. Colin Davis—and discovered it wasn't a survey at all! Davis had simply identified certain disciplinary problems encountered by teachers and published his list in a conservative newsletter—a list he admitted was not obtained from a statistical survey, but from Davis' personal knowledge of the problems in the 1940s. ("I was in school then") and his understanding of the problems today ("I read the papers").

Stossel's commonsense questions of the teacher "survey" is a clear example of the critical thinking process in action. Several more misleading surveys were presented on the ABC program. Listed here, most of these surveys were conducted by businesses or special interest groups with specific objectives in mind.

The "20/20" segment ended with an interview of Cynthia Crossen, author of *Tainted Truth,* an exposé of misleading and biased surveys. Crossen warns:

> If everybody is misusing numbers and scaring us with numbers to get us to do something, however good [that something] is, we've lost the power of numbers. Now, we know certain things from research. For example, we know that smoking cigarettes is hard on your lungs and heart, and because we know that, many people's lives have been extended or saved. We don't want to lose the power of information to help us make decisions, and that's what I worry about.

representative. However, the researchers warn that the sample data should not be used to make inferences about other, more general, populations. ▲

1.6   THE ROLE OF STATISTICS IN CRITICAL THINKING

According to H. G. Wells, author of such science-fiction classics as *The War of the Worlds* and *The Time Machine,* "*Statistical thinking* will one day be as necessary for efficient citizenship as the ability to read and write." Written more than a hundred years ago, Wells' prediction is proving true today.

The growth in data collection associated with scientific phenomena, business operations, and government activities (quality control, statistical auditing, forecasting, etc.) has been remarkable in the past several decades. Every day the media present us with the published results of political, economic, and social surveys. In the increasing government emphasis on drug and product testing, for example, we see vivid evidence of the need to be able to evaluate data sets intelligently. Consequently, each of us has to develop a discerning sense—an ability to

Case Study 1.2 continued

Focus

a. Consider the false March of Dimes report on domestic violence and birth defects. Discuss the type of data required to investigate the impact of domestic violence on birth defects. What data collection method would you recommend?

b. Refer to the American Association of University Women (AAUW) study of self-esteem of high school girls. Explain why the results of the AAUW study are likely to be misleading. What data might be appropriate for assessing the self-esteem of high school girls?

c. Refer to the Food Research and Action Center study of hunger in America. Explain why the results of the study are likely to be misleading. What data would provide insight into the proportion of hungry American children?

Reported Information (Source)	Actual Study Information
Eating oat bran is a cheap and easy way to reduce your cholesterol count. (Quaker Oats)	Diet must consist of nothing but oat bran to achieve a slightly lower cholesterol count.
150,000 women a year die from anorexia. (Feminist group)	Approximately 1,000 women a year die from problems that were likely caused by anorexia.
Domestic violence causes more birth defects than all medical issues combined. (March of Dimes)	No study—false report.
Only 29% of high school girls are happy with themselves, compared to 66% of elementary school girls. (American Association of University Women)	Of 3,000 high school girls 29% responded "Always true" to the statement, "I am happy the way I am." Most answered "Sort of true" and "Sometimes true."
One in four American children under age 12 is hungry or at risk of hunger. (Food Research and Action Center)	Based on responses to questions: "Do you ever cut the size of meals?" "Do you ever eat less than you feel you should?" "Did you ever rely on limited numbers of foods to feed your children because you were running out of money to buy food for a meal?"

use rational thought to interpret the meaning of data. This ability can help you make intelligent decisions, inferences, and generalizations; that is, it helps you *think critically* using statistics.

> **DEFINITION 1.12**
>
> **Statistical thinking** involves applying rational thought to assess data and the inferences made from them critically.

To gain some insight into the role statistics plays in **critical thinking**, i.e., statistical thinking, let's look at a recent *AmStat News* article. This article describes how a group of 27 mathematics and statistics teachers, attending an American Statistical Association course called "Chance," used statistical thinking to evaluate the results of a study. Consider the following excerpt from the article.

> There are few issues in the news that are not in some way statistical. Take one. Should motorcyclists be required by law to wear helmets?... In "The Case for No Helmets" (*New York Times*, June 17, 1995), Dick Teresi, editor of a magazine for Harley-Davidson bikers, argued that helmets may actually kill, since in collisions at

speeds greater than 15 miles an hour the heavy helmet may protect the head but snap the spine. [Teresi], citing a "study," said "nine states without helmet laws had a lower fatality rate (3.05 deaths per 10,000 motorcycles) than those that mandated helmets (3.38)," and "in a survey of 2,500 [at a rally], 98% of the respondents opposed such laws."

[The course instructors] asked: "After reading this [*New York Times*] piece, do you think it is safer to ride a motorcycle without a helmet? Do you think 98% might be a valid estimate of bikers who oppose helmet laws? What further statistical information would you like?" [From Cohn, V. "Chance in college curriculum," *AmStat News,* Aug.–Sept. 1995, No. 223, p. 2.]

You can use several of the key ideas presented in this chapter to help you think statistically about the problem presented in this article. For example, before you can evaluate the validity of the 98% estimate, you would want to know how the data were collected for the study cited by the editor of the biker magazine. If a survey was conducted, it's possible that the 2,500 bikers in the sample were not selected at random from the target population of all bikers, but rather were "self-selected." (Remember, they were all attending a rally—maybe even a rally for bikers who oppose the law.) If the respondents were likely to have strong opinions regarding the helmet law (e.g., strongly oppose the law), the resulting estimate is probably biased high.

You'd also want more information about the study comparing the motorcycle fatality rate of the nine states without a helmet law to those states that mandate helmets. Were the data obtained from a published source? Were all 50 states included in the study? That is, are you seeing sample data or population data? Furthermore, do the helmet laws vary among states? If so, can you really compare the fatality rates?

These questions led the Chance group to the discovery of two scientific and statistically sound studies on helmets. The first, a UCLA study of nonfatal injuries, disputed the charge that helmets shift injuries to the spine. The second study reported a dramatic *decline* in motorcycle crash deaths after California passed its helmet law.

In the remaining chapters of the text, you'll become familiar with the tools essential for building a firm foundation in statistics and statistical thinking.

QUICK REVIEW

Key Terms

Census   5
Critical thinking   15
Data   2
Descriptive statistics   2
Designed experiment   10
Inference   3
Inferential statistics   3
Measure of reliability   7
Measurement   5
Observational study   11
Population   4
Published source   10

Qualitative data   9
Quantitative data   9
Random sample   11
Reliability   7
Representative sample   11
Sample   5
Statistical inference   6
Statistical thinking   15
Statistics   2
Survey   11
Variable   5

EXERCISES 1.1–1.25

Learning the Mechanics

1.1 What is statistics?

1.2 Explain the difference between descriptive and inferential statistics.

1.3 List and define the five elements of an inferential statistical analysis.

1.4 List the four major methods of collecting data and explain their differences.

1.5 Explain the difference between quantitative and qualitative data.

1.6 Explain how populations and variables differ.

1.7 Explain how populations and samples differ.

1.8 What is a representative sample? What is its value?

1.9 Why would a statistician consider an inference incomplete without an accompanying measure of its reliability?

1.10 Define statistical thinking.

1.11 Suppose you're given a data set that classifies each sample unit into one of four categories: A, B, C, or D. You plan to create a computer database consisting of these data, and you decide to code the data and input them as A = 1, B = 2, C = 3, and D = 4. Are the data consisting of the classifications A, B, C, and D qualitative or quantitative? After the data are input as 1, 2, 3, or 4, are they qualitative or quantitative? Explain your answers.

Applying the Concepts

1.12 Consider the set of all students enrolled in your statistics course this term. Suppose you're interested in learning about the current grade point averages (GPAs) of this group.
   a. Define the population and variable of interest.
   b. Is the variable qualitative or quantitative?
   c. Suppose you determine the GPA of every member of the class. Would this represent a census or a sample?
   d. Suppose you determine the GPA of 10 members of the class. Would this represent a census or a sample?
   e. If you determine the GPA of every member of the class and then calculate the average, how much reliability does this calculation have as an "estimate" of the class average GPA?
   f. If you determine the GPA of 10 members of the class and then calculate the average, will the number you get necessarily be the same as the average GPA for the whole class? On what factors would you expect the reliability of the estimate to depend?
   g. What must be true in order for the sample of 10 students you select from your class to be considered a random sample?

1.13 Pollsters regularly conduct opinion polls to determine the popularity rating of the current president. Suppose a poll is to be conducted tomorrow in which 2,000 individuals will be asked whether the president is doing a good or bad job. The 2,000 individuals will be selected by random-digit telephone dialing and asked the question over the phone.
   a. What is the relevant population?
   b. What is the variable of interest? Is it quantitative or qualitative?
   c. What is the sample?
   d. What is the inference of interest to the pollster?
   e. What method of data collection is employed?
   f. How likely is the sample to be representative?

1.14 Colleges and universities are requiring an increasing amount of information about applicants before making acceptance and financial aid decisions. Classify each of the following types of data required on a college application as quantitative or qualitative.
   a. High school GPA
   b. High school class rank
   c. Applicant's score on the SAT or ACT
   d. Gender of applicant
   e. Parents' income
   f. Age of applicant

1.15 Classify the following examples of data as either qualitative or quantitative:
   a. The bacteria count in the water at each of 30 city swimming pools
   b. The occupation of each of 200 shoppers at a supermarket
   c. The marital status of each person living on a city block
   d. The time (in months) between auto maintenance for each of 100 used cars

1.16 A food-products company is considering marketing a new snack food. To see how consumers react to the product, the company conducted a taste test using a sample of 100 randomly selected shoppers at a suburban shopping mall. The shoppers were asked to taste the snack food and then fill out a short questionnaire that requested the following information:
   (1) What is your age?
   (2) Are you the person who typically does the food shopping for your household?
   (3) How many people are in your family?
   (4) How would you rate the taste of the snack food on a scale of 1 to 10, where 1 is least tasty?
   (5) Would you purchase this snack food if it were available on the market?
   (6) If you answered yes to part (5), how often would you purchase the product?

a. Identify the data collection method.

b. Classify the data generated for each question as quantitative or qualitative. Justify your classifications.

1.17 All highway bridges in the United States are inspected periodically for structural deficiency by the Federal Highway Administration (FHWA). Data from the FHWA inspections are compiled into the National Bridge Inventory (NBI). Several of the nearly 100 variables maintained by the NBI are listed below. Classify each variable as quantitative or qualitative.

a. Length of maximum span (feet)

b. Number of vehicle lanes

c. Toll bridge (yes or no)

d. Average daily traffic

e. Condition of deck (good, fair, or poor)

f. Bypass or detour length (miles)

g. Route type (interstate, U.S., state, county, or city)

1.18 Refer to Exercise 1.17. The most recent NBI data were analyzed, and the results were published in the *Journal of Infrastructure Systems* (June 1995). Using the FHWA inspection ratings, each of the 470,515 highway bridges in the United States was categorized as structurally deficient, functionally obsolete, or safe. About 26% of the bridges were found to be structurally deficient, while 19% were functionally obsolete.

a. What is the variable of interest to the researchers?

b. Is the variable of part **a** quantitative or qualitative?

c. Is the data set analyzed a population or a sample? Explain.

d. How did the researchers obtain the data for their study?

1.19 The *Journal of Retailing* (Spring 1988) published a study of the relationship between job satisfaction and the degree of *Machiavellian orientation.* Briefly, the Machiavellian orientation is one in which the executive exerts very strong control, even to the point of deception and cruelty, over the employees he or she supervises. The authors administered a questionnaire to each in a sample of 218 department store executives and obtained both a job-satisfaction score and a Machiavellian rating. They concluded that those with higher job satisfaction scores are likely to have a lower "Mach" rating.

a. What is the population from which the sample was selected?

b. What variables were measured by the authors?

c. Identify the sample.

d. Identify the data collection method used.

e. What inference was made by the authors?

1.20 A Gallup Youth Poll was conducted to determine the topics that teenagers most want to discuss with their parents. The findings show that 46% would like more discussion about the family's financial situation, 37% would like to talk about school, and 30% would like to talk about religion. The survey was based on a national sampling of 505 teenagers, selected at random from all U.S. teenagers.

a. Describe the sample.

b. Describe the population from which the sample was selected.

c. Is the sample representative of the population?

d. What is the variable of interest?

e. How is the inference expressed?

f. Newspaper accounts of most polls usually give a *margin of error* (i.e., plus or minus 3%) for the survey result. What is the purpose of the margin of error and what is its interpretation?

1.21 *USA Today* (Aug. 14, 1995) reported on a study suggesting that "frequently 'heading' the ball in soccer lowers players' IQs." A psychologist tested 60 male soccer players, ages 14–29, who played up to five times a week. Players who averaged 10 or more headers a game had an average IQ of 103, while players who headed one or fewer times per game had an average IQ of 112.

a. Describe the population of interest to the psychologist.

b. Identify the variables of interest.

c. Identify the type (qualitative or quantitative) of the variables, part **b**.

d. Describe the sample.

e. What is the inference made by the psychologist?

f. Discuss possible reasons why the inference, part **e**, may be misleading.

1.22 In order to monitor the quality of care provided to Medicare recipients, 5% of all Medicare surgical cases performed in hospital outpatient departments and ambulatory surgical centers are sampled each year (*Business & Health,* Mar. 1988). Peer-review organizations evaluate the sampled surgeries and assign a quality rating to each.

a. Describe the population being studied.

b. Describe the variable of interest.

c. Describe the sample.

d. Describe the inference of interest.

e. Before any sound statistical conclusions about the quality of Medicare can be drawn, what should accompany the inference?

1.23 A first-year chemistry student conducts an experiment to determine the amount of hydrochloric acid necessary to neutralize 2 milliliters (ml) of a basic solution. The student prepares five 2-ml portions of the solution and adds a known concentration of hydrochloric acid to each. The amount of acid necessary to achieve neutrality of the solution is recorded for each of the five portions.

a. Describe the population of interest to the student.

b. What is the variable of interest? Is it quantitative or qualitative?

c. Describe the sample.

d. Describe the data collection method.

1.24 To evaluate the current status of the dental health of schoolchildren, the American Dental Association wants to estimate the average number of cavities per child in grade school in the United States. One thousand schoolchildren from across the country were randomly selected, and the number of cavities for each was determined.

    a. Describe the population of interest to the American Dental Association.

    b. What is the variable of interest? Is it quantitative or qualitative?

    c. Describe the sample.

    d. Describe the data collection method.

1.25 The Holocaust—the Nazi Germany extermination of the Jews during World War II—has been well documented through films, books, and interviews with the concentration camp survivors. But a recent Roper poll found that one in five U.S. residents doubts the Holocaust really occurred. A national sample of more than 1,000 adults and high school students were asked, "Does it seem possible or does it seem impossible to you that the Nazi extermination of the Jews never happened?" Twenty-two percent of the adults and 20% of the high school students responded "Yes." (*New York Times,* May 16, 1994).

    a. Describe the population of interest to the pollsters.

    b. What is the variable of interest? Is it quantitative or qualitative?

    c. Describe the sample.

    d. Identify the data collection method employed.

    e. What inference was made by the pollsters?

    f. Comment on the reliability of the inference. [*Hint:* Examine, closely, the survey question.]

STUDENT PROJECTS

Scan your daily newspaper or news magazine for articles that contain numerical data. The data might be a summary of the results of a public opinion poll, the results of a vote by the U.S. Senate, or a list of crime rates, birth or death rates, etc. For each article you find, answer the following questions:

1. Do the data constitute a sample or an entire population? If a sample has been taken, clearly identify both the sample and the population; otherwise, identify the population.

2. What type of data (quantitative or qualitative) has been collected?

3. What is the data source?

4. If a sample has been observed, is it likely to be representative of the population?

5. If a sample has been observed, does the article present an explicit (or implied) inference about the population of interest? If so, state the inference made in the article.

6. If an inference has been made, has a measure of reliability been included? What is it?

7. Use your answers to questions 4–6 to critically evaluate the article.

Chapter 2

METHODS FOR DESCRIBING SETS OF DATA

Contents

Case Studies

WHERE WE'VE BEEN

In Chapter 1 we looked at some typical examples of the use of statistics and we discussed the role that statistical thinking plays in supporting decision-making. We examined the difference between descriptive and inferential statistics and described the five elements of inferential statistics: a population, one or more variables, a sample, an inference, and a measure of reliability for the inference. We also learned that data can be of two types—quantitative and qualitative.

WHERE WE'RE GOING

Before we make an inference, we must be able to describe a data set. We can do this by using graphical and/or numerical methods, which we discuss in this chapter. As we'll see in Chapter 7, we use some sample numerical descriptive measures to estimate the values of corresponding population descriptive measures. Therefore, our efforts in this chapter will ultimately lead us to statistical inference.

Suppose you wish to evaluate the mathematical capabilities of a class of 1,000 first-year college students based on their quantitative Scholastic Assessment Test (SAT) scores. How would you describe these 1,000 measurements? Characteristics of interest include the typical or most frequent SAT score, the variability in the scores, the highest and lowest scores, the "shape" of the data, and whether or not the data set contains any unusual scores. Extracting this information isn't easy. The 1,000 scores provide too many bits of information for our minds to comprehend. Clearly we need some method for summarizing and characterizing the information in such a data set. Methods for describing data sets are also essential for statistical inference. Most populations are large data sets. Consequently, we need methods for describing a data set that let us make descriptive statements (inferences) about a population based on information contained in a sample.

Two methods for describing data are presented in this chapter, one *graphical* and the other *numerical*. Both play an important role in statistics. Section 2.1 presents both graphical and numerical methods for describing qualitative data. Graphical methods for describing quantitative data are illustrated in Section 2.2; numerical descriptive methods for quantitative data are presented in Sections 2.3–2.6. We end the chapter with a section on the *misuse* of descriptive techniques.

2.1   DESCRIBING QUALITATIVE DATA

Consider a study of aphasia published in the *Journal of Communication Disorders* (Mar. 1995). Aphasia is the "impairment or loss of the faculty of using or understanding spoken or written language." Three types of aphasia have been identified by researchers: Broca's, conduction, and anomic. They wanted to determine whether one type of aphasia occurs more often than any other, and, if so, how often. Consequently, they measured aphasia type for a sample of 22 adult aphasiacs. Table 2.1 gives the type of aphasia diagnosed for each aphasic in the sample.

For this study, the variable of interest, aphasia type, is qualitative in nature. Qualitative data are nonnumerical in nature; thus, the value of a qualitative variable can only be classified into categories called *classes*. The possible aphasia types—Broca's, conduction, and anomic—represent the classes for this qualitative variable. We can summarize such data numerically in two ways: (1) by computing the *class frequency*—the number of observations in the data set that fall into each class; or (2) by computing the *class relative frequency*—the proportion of the total number of observations falling into each class.

> **DEFINITION 2.1**
>
> A **class** is one of the categories into which qualitative data can be classified.

> **DEFINITION 2.2**
>
> The **class frequency** is the number of observations in the data set falling in a particular class.

> **DEFINITION 2.3**
>
> The **class relative frequency** is the class frequency divided by the total number of observations in the data set.

Examining Table 2.1, we observe that 5 aphasics in the study were diagnosed as suffering from Broca's aphasia, 7 from conduction aphasia, and 10 from anomic

TABLE 2.1   Data on 22 Adult Aphasics

Subject	Type of Aphasia
1	Broca's
2	Anomic
3	Anomic
4	Conduction
5	Broca's
6	Conduction
7	Conduction
8	Anomic
9	Conduction
10	Anomic
11	Conduction
12	Broca's
13	Anomic
14	Broca's
15	Anomic
16	Anomic
17	Anomic
18	Conduction
19	Broca's
20	Anomic
21	Conduction
22	Anomic

Source: Li, E. C., Williams, S. F., and Volpe, A. D., "The effects of topic and listener familiarity on discourse variables in procedural and narrative discourse tasks." *Journal of Communication Disorders,* Vol. 28, No. 1, Mar. 1995, p. 44 (Table 1).

TABLE 2.2   Summary Table for Data on 22 Adult Aphasics

CLASS	FREQUENCY	RELATIVE FREQUENCY
Type of Aphasia	Number of Subjects	Proportion
Broca's	5	.227
Conduction	7	.318
Anomic	10	.455
Totals	22	1.000

aphasia. These numbers—5, 7, and 10—represent the class frequencies for the three classes and are shown in the summary table, Table 2.2.

Table 2.2 also gives the relative frequency of each of the three aphasia classes. From Definition 2.3, we know that we calculate the relative frequency by dividing the class frequency by the total number of observations in the data set. Thus, the relative frequencies for the three types of aphasia are

$$\text{Broca's:} \quad \frac{5}{22} = .227$$

$$\text{Conduction:} \quad \frac{7}{22} = .318$$

$$\text{Anomic:} \quad \frac{10}{22} = .455$$

From these relative frequencies we observe that nearly half (45.5%) of the 22 subjects in the study are suffering from anomic aphasia.

Although the summary table of Table 2.2 adequately describes the data of Table 2.1, we often want a graphical presentation as well. Figures 2.1 and 2.2 show two of the most widely used graphical methods for describing qualitative data—bar graphs and pie charts. Figure 2.1 shows the frequencies of aphasia types in a *bar graph*. Note that the height of the rectangle, or "bar," over each class is equal to the class frequency. (Optionally, the bar heights can be proportional to

FIGURE 2.1

Bar graph for data on 22 aphasics

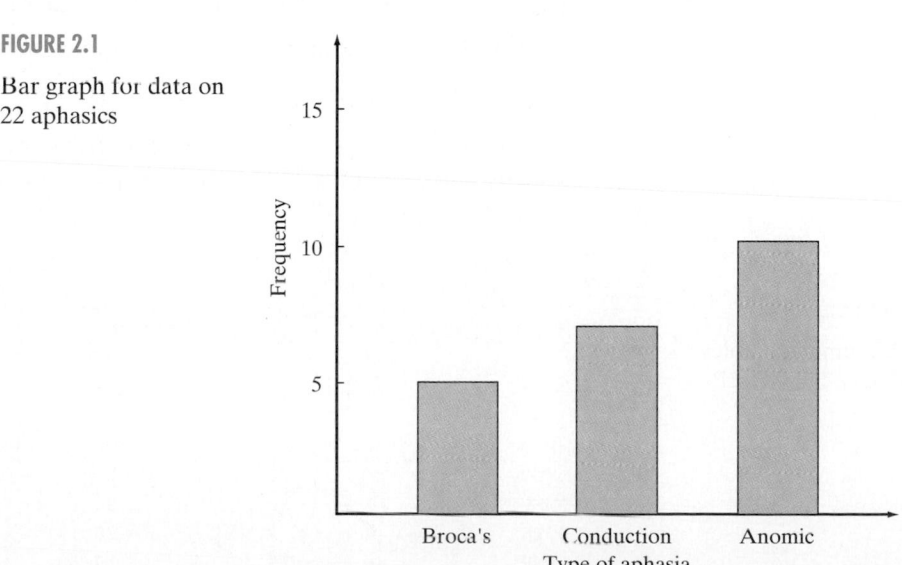

FIGURE 2.2

Pie chart for data on
22 aphasics

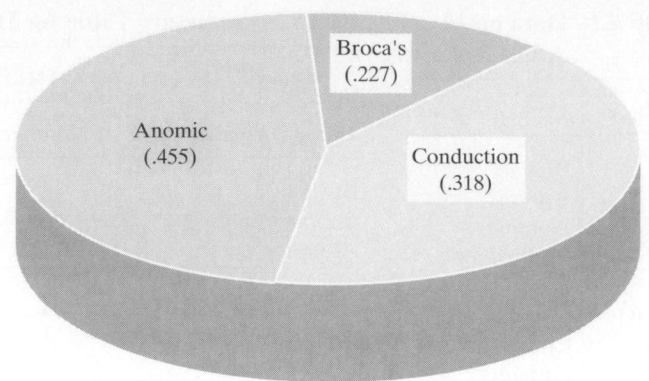

class relative frequencies.) In contrast, Figure 2.2 shows the relative frequencies of
the three types of aphasia in a *pie chart*. Note that the pie is a circle (spanning
360°) and the size (angle) of the "pie slice" assigned to each class is proportional
to the class relative frequency. For example, the slice assigned to anomic aphasia
is 45.5% of 360°, or (.455)(360°) = 163.8°.

Let's look at a practical example that requires interpretation of the graphical results.

EXAMPLE 2.1

A group of cardiac physicians in southwest Florida have been studying a new drug
designed to reduce blood loss in coronary bypass operations. Blood loss data for
114 coronary bypass patients (some who received a dosage of the drug and others
who did not) are provided in Appendix B. Although the drug shows promise in
reducing blood loss, the physicians are concerned about possible side effects and
complications. So their data set includes not only the qualitative variable, DRUG,
which indicates whether or not the patient received the drug, but also the qualitative variable, COMP, which specifies the type (if any) of complication experienced by the patient. The four values of COMP recorded in Appendix B are: (1)
redo surgery, (2) post-op infection, (3) both, or (4) none.

a. Figure 2.3, generated using SAS computer software, shows summary tables
   for the two qualitative variables, DRUG and COMP, in Appendix C. Interpret
   the results.
b. Interpret the SAS graph and summary tables shown in Figure 2.4.

Solution

a. The top table in Figure 2.3 is a summary frequency table for DRUG. Note that
   exactly half (57) of the 114 coronary bypass patients received the drug and half
   did not.

FIGURE 2.3

SAS summary tables for
DRUG and COMP

DRUG	Frequency	Percent	Cumulative Frequency	Cumulative Percent
NO	57	50.0	57	50.0
YES	57	50.0	114	100.0

COMP	Frequency	Percent	Cumulative Frequency	Cumulative Percent
1:REDO	12	10.5	12	10.5
2:INFECT	12	10.5	24	21.0
3:BOTH	4	3.5	28	24.5
4:NONE	86	75.5	114	100.0

FIGURE 2.4

SAS bar graph and
summary tables for COMP
by DRUG

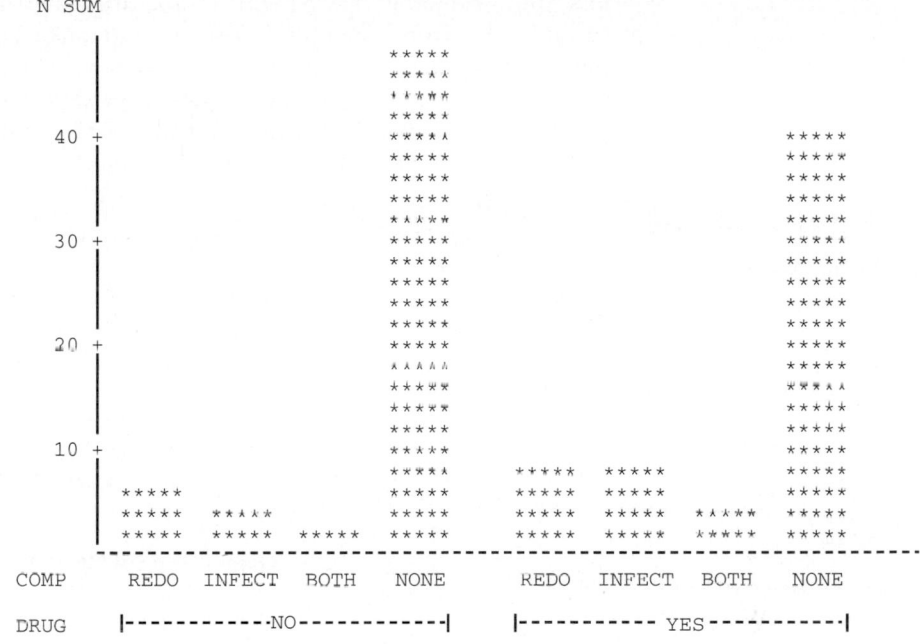

SUM OF N BY COMP GROUPED BY DRUG

| COMP | REDO | INFECT | BOTH | NONE | REDO | INFECT | BOTH | NONE |

DRUG |----------NO----------| |----------YES----------|

```
--------------------------------- DRUG=NO -------------------------------.

                                      Cumulative      Cumulative
      COMP      Frequency    Percent   Frequency       Percent
      --------------------------------------------------------------
      1:REDO         5         8.8          5            8.8
      2:INFECT       4         7.0          9           15.8
      3:BOTH         1         1.8         10           17.5
      4:NONE        47        82.5         57          100.0


------------------------------- DRUG=YES -------------------------------

                                      Cumulative      Cumulative
      COMP      Frequency    Percent   Frequency       Percent
      --------------------------------------------------------------
      1:REDO         7        12.3          7           12.3
      2:INFECT       8        14.0         15           26.3
      3:BOTH         3         5.3          8           31.6
      4:NONE        39        68.4         57          100.0
```

The bottom table in Figure 2.3 is a summary frequency table for COMP. The class relative frequencies are given in the Percent column. We see that 75.5% of the 114 patients had no complications, leaving 24.5% who experienced either a redo surgery, a post-op infection, or both.

b. At the top of Figure 2.4 is a side-by-side bar graph for the data. The first four bars represent the frequencies of COMP for the 57 patients who did not receive the drug; the next four bars represent the frequencies of COMP for the 57 patients who did receive a dosage of the drug. The graph clearly shows that patients who got the drug suffered more complications. The exact percentages are displayed in the summary tables of Figure 2.4. Over 30% of the patients who got the drug had complications, compared to about 17% for the patients who got no drug.

Although these results show that the drug may be effective in reducing blood loss, they also imply that patients on the drug may have a higher risk of complications. But before using this information to make a decision about the drug, the physicians will need to provide a measure of reliability for the inference. That is, the physicians will want to know whether the difference between the percentages of patients with complications observed in this sample of 114 patients is generalizable to the population of all coronary bypass patients.   ▲

EXERCISES 2.1–2.10

Learning the Mechanics

2.1 Complete the following table.

Grade on Statistics Exam	Frequency	Relative Frequency
A: 90–100		.08
B: 80–89	36	
C: 65–79	90	
D: 50–64	30	
F: Below 50	28	
Total	200	1.00

2.2 A qualitative variable with three classes (X, Y, and Z) is measured for each of 20 units randomly sampled from a target population. The data (observed class for each unit) are listed below.

Y X X Z X Y Y Y X X Z X
Y Y X Z Y Y Y X

a. Compute the frequency for each of the three classes.
b. Compute the relative frequency for each of the three classes.
c. Display the results, part **a**, in a frequency bar graph.
d. Display the results, part **b**, in a pie chart.

Applying the Concepts

2.3 How religious are U.S. citizens? To answer this question, researchers at the University of Akron conducted an in-depth random survey of 4,001 people living in the United States. The results, reported in *Newsweek* (Nov. 29, 1993), are summarized in the table below. *Newsweek* illustrated these results with a pie chart. Construct the pie chart and interpret the results.

	Number of U.S. Citizens
Highly committed to religion	761
Modestly religious	880
Nominally religious	1,162
Trace elements of religion	899
Agnostic or atheist	299
Total	4,001

2.4 Audiologists have recently developed a rehabilitation program for hearing-impaired residents in a Canadian home for senior citizens (*Journal of the Academy of Rehabilitative Audiology,* 1994). Each of the 30 residents of the home was diagnosed for degree and type of hearing loss. The results are summarized in the accompanying table. Use a graph to portray the results. Which type of hearing loss appears to be the most prevalent among nursing home residents?

Degree/Type of Hearing Loss	Number of Residents
Hear within normal limits	3
High-frequency sensorineural hearing loss	5
Mild sensorineural hearing loss	1
Mild-to-moderate sensorineural hearing loss	5
Moderate sensorineural hearing loss	7
Moderate-to-severe sensorineural hearing loss	8
Severe-to-profound sensorineural hearing loss	1

Source: Jennings, M. B., and Head, B. G. "Development of an ecological audiologic rehabilitation program in a home-for-the-aged." *Journal of the Academy of Rehabilitative Audiology,* Vol. 27, 1994, p. 77 (Table 1).

2.5 Marine scientists who study dolphin communication have discovered that bottlenose dolphins exhibit an individualized whistle contour known as the *signature whistle.* A study was conducted to categorize the

Whistle Category	Number of Whistles
Type a	97
Type b	15
Type c	9
Type d	7
Type e	7
Type f	2
Type g	2
Type h	2
Type i	2
Type j	4
Type k	13
Other types	25

Source: McCowan, B., and Reiss, D. "Quantitative comparison of whistle repertoires from captive adult bottlenose dolphins (Delphiniae, *Tursiops truncatus*): A re-evaluation of the signature whistle hypothesis." *Ethology,* Vol. 100, No. 3, July 1995, p. 200 (Table 2).

signature whistles emitted by ten captive adult bottlenose dolphins in socially interactive contexts (*Ethology,* July 1995). A total of 185 whistles were recorded during the study period; each whistle contour was analyzed and assigned to a category using a contour similarity (CS) technique. The results are reported in the accompanying table. Use a graphical method to summarize the results. Do you detect any patterns in the data?

2.6 *Choice* magazine, a publication for the academic community, provides new-book reviews each issue. Many librarians rely on these reviews to determine which new books to acquire for their library. A thorough study of the contents of the book reviews published in *Choice* was conducted (*Library Acquisitions: Practice and Theory,* Vol. 19, 1995). A random sample of 375 book reviews in American history, geography, and area studies was selected and the "overall opinion" of the book stated in each review was ascertained. Overall opinion was coded as follows: 1 = would not recommend, 2 = cautious or very little recommendation, 3 = little or no preference, 4 = favorable/recommended, 5 = outstanding/significant contribution. A summary of the data is provided in the accompanying bar graph.

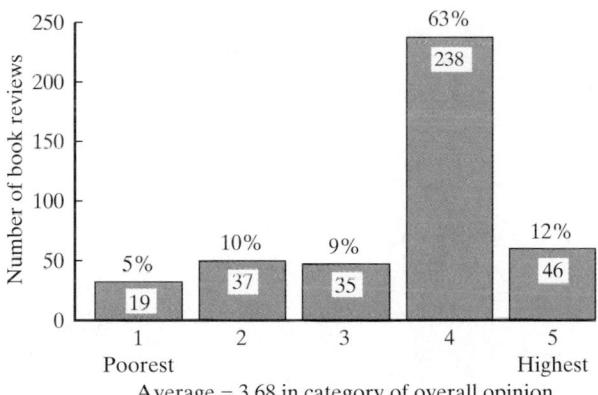

Source: Carlo, P. W., and Natowitz, A. "*Choice* book reviews in American history, geography, and area studies: An analysis for 1988–1993." *Library Acquisitions: Practice & Theory,* Vol. 19, No. 2, 1995, p. 159 (Figure 1).

a. Interpret the bar graph.

b. Comment on the following statement extracted from the study: "A majority (more than 75%) of books reviewed are evaluated favorably and recommended for purchase."

2.7 A survey was conducted to investigate the impact of the mass media on the public's perception of mental illness (*Health Education Journal,* Sept. 1994). The media coverage for each of 562 news items related to mental health in Scotland was classified into one of five categories: violence to others, sympathetic coverage, harm to self, comic images, and criticism of accepted definitions of mental illness. A summary of the results is provided in the accompanying table.

a. Construct a relative frequency table for the data.

b. Display the relative frequencies in a graph.

c. Discuss the findings.

Media Coverage	Number of Items
Violence to others	373
Sympathetic	102
Harm to self	71
Comic images	12
Criticism of definitions	4
Total	562

Source: Philo, G., *et al.* "The impact of the mass media on public images of mental illness: Media content and audience belief." *Health Education Journal,* Vol. 53, No. 3, Sept. 1994, p. 274 (Table 1).

2.8 Each week, *USA Today* reports on how much consumers like a major advertising campaign and how effective they think the ad is in helping the company sell its product. The topic of an August, 1995 report was the "Obey your thirst" ad campaign for Sprite, a lemon-lime soft drink manufactured by Coca-Cola. A *USA Today*/Harris poll of 1,005 adults, selected nationwide, asked, "Do you like the campaign?" and "How effective is the campaign?" The results are shown in the graphs below.

a. What type of graphical method is used to describe the data?

b. Interpret the results for the question "Do you like the campaign?"

c. Interpret the results for the question "How effective is the campaign?"

2.9 Transgenic plants are plants that have been genetically modified using current gene technology. For example,

Do you like the campaign?

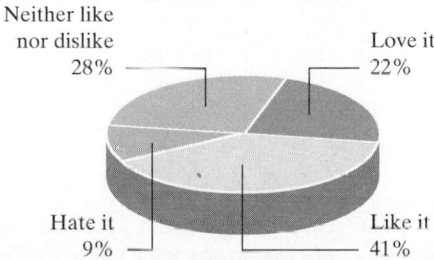

How effective is the campaign?

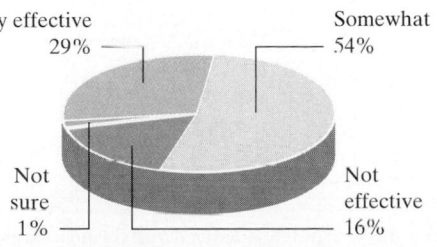

Data for Exercise 2.10

Tanker	Spillage (metric tons, thousands)	CAUSE OF SPILLAGE				
		Collision	Grounding	Fire/Explosion	Hull Failure	Unknown
Atlantic Empress	257	X				
Castillo De Bellver	239			X		
Amoco Cadiz	221				X	
Odyssey	132			X		
Torrey Canyon	124		X			
Sea Star	123	X				
Hawaiian Patriot	101				X	
Independenta	95	X				
Urquiola	91		X			
Irenes Serenade	82			X		
Khark 5	76			X		
Nova	68	X				
Wafra	62		X			
Epic Colocotronis	58		X			
Sinclair Petrolore	57			X		
Yuyo Maru No 10	42	X				
Assimi	50			X		
Andros Patria	48			X		
World Glory	46				X	
British Ambassador	46				X	
Metula	45		X			
Pericles G.C.	44			X		
Mandoil II	41	X				
Jakob Maersk	41		X			
Burmah Agate	41	X				
J. Antonio Lavalleja	38		X			
Napier	37		X			
Exxon Valdez	36		X			
Corinthos	36	X				
Trader	36				X	
St. Peter	33			X		
Gino	32	X				
Golden Drake	32			X		
Ionnis Angelicoussis	32			X		
Chryssi	32				X	
Irenes Challenge	31				X	
Argo Merchant	28		X			
Heimvard	31	X				
Pegasus	25					X
Pacocean	31				X	
Texaco Oklahoma	29				X	
Scorpio	31		X			
Ellen Conway	31		X			
Caribbean Sea	30				X	
Cretan Star	27					X
Grand Zenith	26				X	
Athenian Venture	26			X		
Venoil	26	X				
Aragon	24				X	
Ocean Eagle	21		X			

Source: Daidola, J. C. "Tanker structure behavior during collision and grounding." *Marine Technology,* Vol. 32, No. 1, Jan. 1995, p. 22 (Table 1). Reprinted with permission of The Society of Naval Architects and Marine Engineers (SNAME), Jersey City, N.J.

biologists have recently developed a transgenic toma-to with improved storage properties. The *Journal of Experimental Botany* (May 1995) reported on the current level of experimentation with such genetically modified plants. Each experiment was identified by its trait. The accompanying bar graph describes the number of approved trials of transgenic plants world-wide, by trait, from 1990 to 1992.

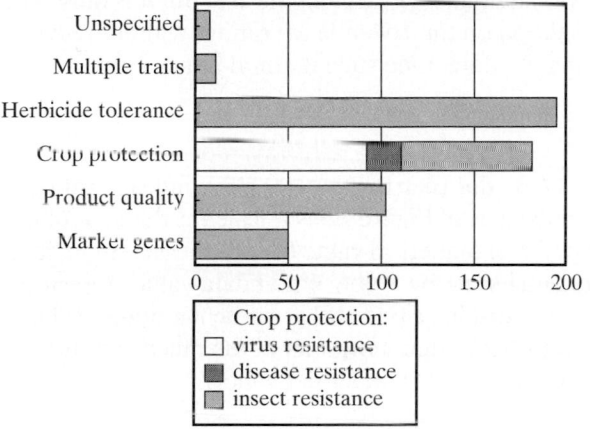

Source: Rogers, H. J., and Parkes, H. C. "Transgenic plants and the environment." *Journal of Experimental Botany*, Vol. 486, No. 286, May 1995, p. 468 (Figure 1.8).

a. Estimate the number of transgenic plant trials approved for herbicide tolerance over this period.
b. Estimate the number of transgenic plant trials approved for developing virus-resistant crops over this period.
c. Modify the bar graph to show relative frequencies rather than frequencies.

2.10 Owing to several major ocean oil spills by tank vessels, Congress passed the 1990 Oil Pollution Act, which requires all tankers to be designed with thicker hulls. Further improvements in the structural design of a tank vessel have been proposed since then, each with the objective of reducing the likelihood of an oil spill and decreasing the amount of outflow in the event of hull puncture. To aid in this development, *Marine Technology* (Jan. 1995) reported on the spillage amount and cause of puncture for 50 recent major oil spills from tankers and carriers. The data are reproduced in the table on page 28.
a. Use a graphical method to describe the cause of oil spillage for the 50 tankers.
b. Does the graph, part **a**, suggest that any one cause is more likely to occur than any other? How is this information of value to the design engineers?

2.2   GRAPHICAL METHODS FOR DESCRIBING QUANTITATIVE DATA

Recall from Section 1.4 that quantitative data sets consist of data that are recorded on a meaningful numerical scale. For describing, summarizing and detecting patterns in such data, we can use three graphical methods: dot plots, stem-and-leaf displays, and histograms.

For example, the Environmental Protection Agency (EPA) performs extensive tests on all new car models to determine their mileage ratings. Suppose that the 100 measurements in Table 2.3 represent the results of such tests on a certain new car model. How can we summarize the information in this rather large sample?

TABLE 2.3   EPA Mileage Ratings on 100 Cars

36.3	41.0	36.9	37.1	44.9	36.8	30.0	37.2	42.1	36.7
32.7	37.3	41.2	36.6	32.9	36.5	33.2	37.4	37.5	33.6
40.5	36.5	37.6	33.9	40.2	36.4	37.7	37.7	40.0	34.2
36.2	37.9	36.0	37.9	35.9	38.2	38.3	35.7	35.6	35.1
38.5	39.0	35.5	34.8	38.6	39.4	35.3	34.4	38.8	39.7
36.3	36.8	32.5	36.4	40.5	36.6	36.1	38.2	38.4	39.3
41.0	31.8	37.3	33.1	37.0	37.6	37.0	38.7	39.0	35.8
37.0	37.2	40.7	37.4	37.1	37.8	35.9	35.6	36.7	34.5
37.1	40.3	36.7	37.0	33.9	40.1	38.0	35.2	34.8	39.5
39.9	36.9	32.9	33.8	39.8	34.0	36.8	35.0	38.1	36.9

FIGURE 2.5

A MINITAB dot plot for
the EPA mileage ratings
on 100 cars

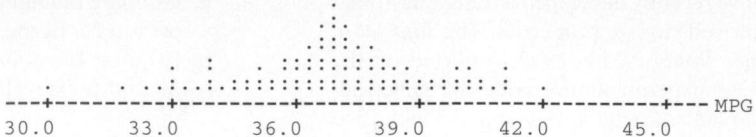

A visual inspection of the data indicates some obvious facts. For example, most of the mileages are in the 30s, with a smaller fraction in the 40s. But it is difficult to provide much additional information on the 100 mileage ratings without resorting to some method of summarizing the data. One such method is a dot plot.

Dot Plots

A computer-generated (MINITAB) **dot plot** for the 100 EPA mileage ratings is shown in Figure 2.5. The horizontal axis of Figure 2.5 is a scale for the quantitative variable in miles per gallon (mpg). The numerical value of each measurement in the data set is located on the horizontal scale by a dot. When data values repeat, the dots are placed above one another, forming a pile at that particular numerical location. As you can see, this dot plot verifies that almost all of the mileage ratings are in the 30s, with most falling between 35 and 39 miles per gallon.

Stem-and-Leaf Display

Another graphical representation of these same data, a **stem-and-leaf display**, is shown in Figure 2.6. In this display the *stem* is the portion of the measurement (mpg) to the left of the decimal point, while the remaining portion to the right of the decimal point is the *leaf*.

The stems for the data set are listed in a column from the smallest (30) to the largest (44). Then the leaf for each observation is recorded in the row of the display corresponding to the observation's stem. For example, the leaf 3 of the first observation (36.3) in Table 2.3 is written in the row corresponding to the stem 36. Similarly, the leaf 7 for the second observation (32.7) in Table 2.3 is recorded in the row corresponding to the stem 32, while the leaf 5 for the third observation (40.5) is recorded in the row corresponding to the stem 40. (The leaves for these

FIGURE 2.6

A stem-and-leaf display for
the EPA mileage ratings
on 100 cars

Stem	Leaf
30	0
31	8
32	5 7 9 9
33	1 2 6 8 9 9
34	0 2 4 5 8 8
35	0 1 2 3 5 6 6 7 8 9 9
36	0 1 2 3 3 4 4 5 5 6 6 7 7 7 8 8 8 9 9 9
37	0 0 0 0 1 1 1 2 2 3 3 4 4 5 6 6 7 7 8 9 9
38	0 1 2 2 3 4 5 6 7 8
39	0 0 3 4 5 7 8 9
40	0 1 2 3 5 5 7
41	0 0 2
42	1
43	
44	9

first three observations are shaded in Figure 2.6.) Typically, the leaves in each row are ordered as shown in Figure 2.6.

The stem-and-leaf display presents another compact picture of the data set. You can see at a glance that the 100 mileage readings are distributed between 30.0 and 44.9, with most of them falling in stem rows 35 to 39. The six leaves in stem row 34 indicate that six of the 100 readings were at least 34.0 but less than 35.0. Similarly, the eleven leaves in stem row 35 indicate that eleven of the 100 readings were at least 35.0 but less than 36.0. Only five cars had readings equal to 41 or larger, and only one was as low as 30.

The definitions of the stem and leaf for a data set can be modified to alter the graphical description. For example, suppose we had defined the stem as the tens digit for the gas mileage data, rather than the ones and tens digits. With this definition, the stems and leaves corresponding to the measurements 36.3 and 32.7 would be as follows:

Stem	Leaf		Stem	Leaf
3	6		3	2

Note that the decimal portion of the numbers has been dropped. Generally, only one digit is displayed in the leaf.

If you look at the data, you'll see why we didn't define the stem this way. All the mileage measurements fall in the 30s and 40s, so all the leaves would fall into just two stem rows in this display. The resulting picture would not be nearly as informative as Figure 2.6.

Histograms

A **relative frequency histogram** for these 100 EPA mileage readings is shown in Figure 2.7. The horizontal axis of Figure 2.7, which gives the miles per gallon for a given automobile, is divided into *intervals* commencing with the interval 29.95–31.45 and proceeding in intervals of equal size to 43.45–44.95 mpg. The vertical axis gives the proportion (or *relative frequency*) of the 100 readings that fall in each interval. Thus, you can see that .33, or 33%, of the owners obtained a mileage between 35.95 and 37.45. This interval contains the highest relative frequency, and the intervals tend to contain a smaller fraction of the measurements as the mileages get smaller or larger.

FIGURE 2.7

Histogram for EPA mileage data

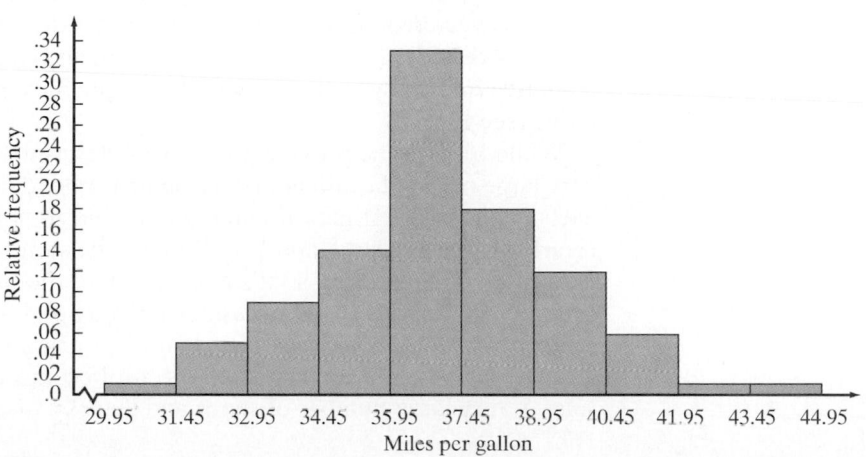

TABLE 2.4   Measurement Classes, Frequencies, and Relative Frequencies for the Car Mileage Data

Measurement Class	Frequency	Relative Frequency
29.95–31.45	1	.01
31.45–32.95	5	.05
32.95–34.45	9	.09
34.45–35.95	14	.14
35.95–37.45	33	.33
37.45–38.95	18	.18
38.95–40.45	12	.12
40.45–41.95	6	.06
41.95–43.45	1	.01
43.45–44.95	1	.01
Totals	100	1.00

By summing the relative frequencies in the intervals 34.45–35.95, 35.95–37.45, and 37.45–38.95, you can see that 65% of the mileages are between 34.45 and 38.95. Similarly, only 2% of the cars obtained a mileage rating over 41.95. Many other summary statements can be made by further study of the histogram.

Dot plots, stem-and-leaf displays, and histograms all provide useful graphical descriptions of quantitative data. Since most statistical software packages can be used to construct these displays, we'll focus here on their interpretation instead of their construction.

Histograms can be used to display either the frequency or the relative frequency of the measurements falling into specified intervals known as **measurement classes**. The measurement classes, frequencies, and relative frequencies for the EPA car mileage data are shown in Table 2.4.

By looking at a histogram (say, the relative frequency histogram in Figure 2.7), you can see two important facts. First, note the total area under the histogram and then note the proportion of the total area that falls over a particular interval of the *x*-axis. You'll see that the proportion of the total area above an interval is equal to the relative frequency of measurements falling in the interval. For example, the relative frequency for the class interval 35.95–37.45 is .33. Consequently, the rectangle above the interval contains .33 of the total area under the histogram.

Second, you can imagine the appearance of the relative frequency histogram for a very large set of data (say, a population). As the number of measurements in a data set is increased, you can obtain a better description of the data by decreasing the width of the class intervals. When the class intervals become small enough, a relative frequency histogram will (for all practical purposes) appear as a smooth curve (see Figure 2.8).

While histograms provide good visual descriptions of data sets—particularly very large ones—they do not let us identify individual measurements. In contrast, each of the original measurements is visible to some extent in a dot plot and clearly visible in a stem-and-leaf display. The stem-and-leaf display arranges the data in ascending order, so it's easy to locate the individual measurements. For example, in Figure 2.6 we can easily see that two of the gas mileage measurements are equal to 36.3, but we can't see that fact by inspecting the histogram in Figure 2.7. However, stem-and-leaf displays can become unwieldy for very large data sets. A very large number of stems and leaves causes the vertical and horizontal

FIGURE 2.8

Effect of the size of a data set on the outline of a histogram

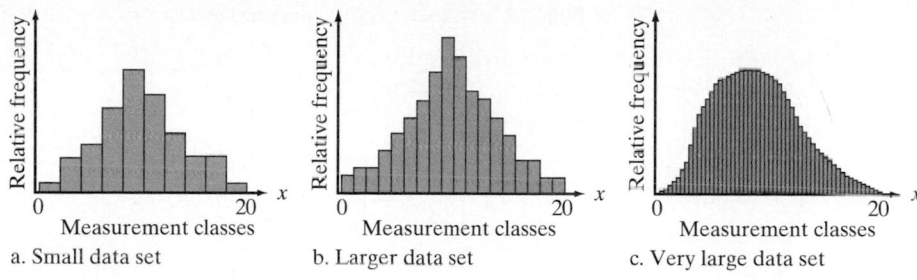

a. Small data set   b. Larger data set   c. Very large data set

dimensions of the display to become cumbersome diminishing the usefulness of the visual display.

EXAMPLE 2.2

The data in Table 2.5 give, by state, the percentages of the total number of college or university student loans that are in default.

a. Use a statistical computer software package to create a relative frequency histogram for these data.
b. Use a statistical computer software package to create a stem-and-leaf display for these data.
c. Compare and interpret the two graphical displays of these data.

Solution

a. We used SAS to generate the relative frequency histogram in Figure 2.9. Note that six classes were formed by the SAS program. The classes are identified by their *midpoints* rather than their endpoints. Thus, the first interval has a midpoint of 3, the second of 6, and so on. The corresponding measurement classes based on these midpoints are therefore 1.5–4.5, 4.5–7.5, etc. Note, too, that the SAS program labels the vertical axis "Percentage" rather than "Relative frequency." Thus, we see that the measurement class with a midpoint of 9 (ranging from 7.5 to 10.5) contains more than 30% of the measurements.

TABLE 2.5   Percentage of Student Loans (per State) in Default

State	%	State	%	State	%	State	%
Ala.	12.0	Ill.	9.3	Mont.	6.4	R.I.	8.8
Alaska	19.7	Ind.	6.7	Nebr.	4.9	S.C.	14.1
Ariz.	12.1	Iowa	6.2	Nev.	10.1	S. Dak.	5.5
Ark.	12.9	Kans.	5.7	N.H.	7.9	Tenn.	12.3
Calif.	11.4	Ky.	10.3	N.J.	12.0	Tex.	15.2
Colo.	9.5	La.	13.5	N. Mex.	7.5	Utah	6.0
Conn.	8.8	Maine	9.7	N.Y.	11.3	Vt.	8.3
Del.	10.9	Md.	16.6	N.C.	15.5	Va.	14.4
D.C.	14.7	Mass.	8.3	N. Dak.	4.8	Wash.	8.4
Fla.	11.8	Mich.	11.4	Ohio	10.4	W. Va.	9.5
Ga.	14.8	Minn.	6.6	Okla.	11.2	Wis.	9.0
Hawaii	12.8	Miss.	15.6	Oreg.	7.9	Wyo.	2.7
Idaho	7.1	Mo.	8.8	Pa.	8.7		

Source: National Direct Student Loan Program.

FIGURE 2.9

SAS relative frequency histogram for student loan default rate

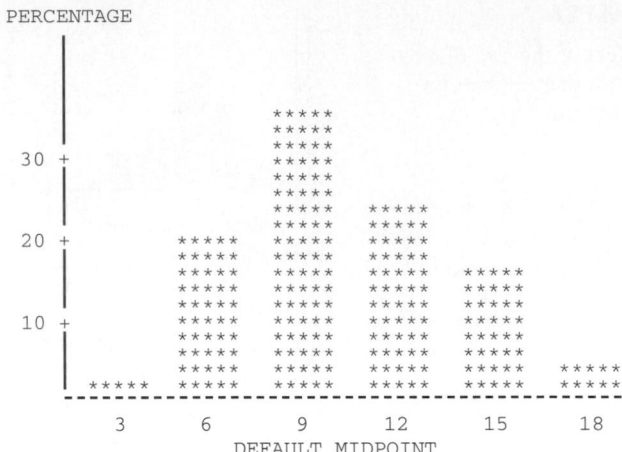

b. We used MINITAB to generate the stem-and-leaf display in Figure 2.10. Note that the stem, which is the *second* column in the printout, has been defined as the number *two* places to the left of the decimal. The leaf is the third column in the printout, and is the number one place to the left of the decimal. (The digit to the right of the decimal is not shown.) MINITAB also indicates the cumulative number of measurements (first column of the printout) from the nearest tail of the distribution to each stem. For the stem that is in the middle of the data set, the number of leaves is indicated in parentheses. The program also prints a key (at the top), giving the units of the leaf.

c. As is usually the case for data sets that are not too large (say, fewer than 100 measurements), the stem-and-leaf display provides more detail than the histogram without being unwieldy. For instance, the stem-and-leaf display in Figure 2.10 clearly indicates that most states' rates are relatively evenly distributed from 6 to 15 (note that the five stem rows in the middle of the display have between 7 and 12 measurements "attached" to each stem). The stem-and-leaf display also clearly indicates the lowest default rate in the first stem row (representing the measurement 2.7) and the highest rate in the last stem row (representing 19.7).

FIGURE 2.10

MINITAB stem-and-leaf display for defaulted student loan data

```
Stem-and-Leaf of DEFAULT     N=51
Leaf Unit = 1.0

   1      0 2
   5      0 4455
  14      0 666667777
 (12)     0 888888899999
  25      1 000011111
  16      1 2222223
   9      1 4444555
   2      1 6
   1      1 9
```

The "Eye Cue" Test: Does Experience Improve Performance?

CASE STUDY • 2.1 •

In 1948, famous child psychologist Jean Piaget devised a test of basic perceptual and conceptual skills dubbed the "water-level task." Subjects were shown a drawing of a glass being held perfectly still (at a 45° angle) by an invisible hand so that any water in it had to be at rest (see Figure 2.11). The task for the subject was to draw a line representing the surface of the water—a line that would touch the black dot pictured on the right side of the glass. Piaget found that young children typically failed the test. Fifty years later, research psychologists still use the water-level task to test the perception of both adults and children. Surprisingly, about 40% of the adult population also fail. In addition, males tend to do better than females, and younger adults tend to do better than older adults.

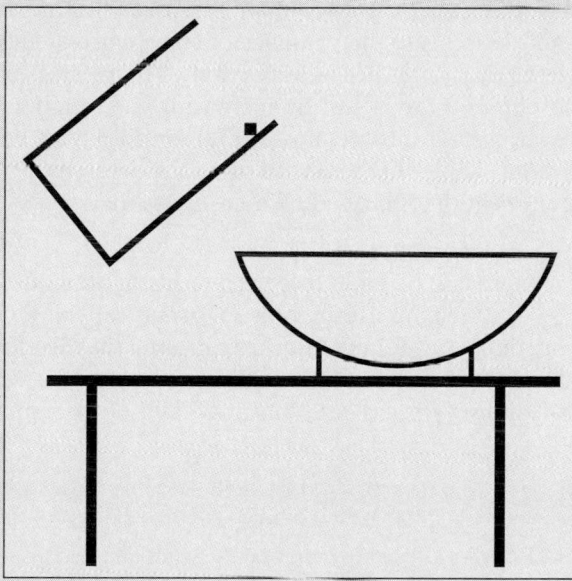

FIGURE 2.11

Drawing of the water-level task

Will people with experience handling liquid-filled containers perform better on this "eye cue" test? This question was the focus of research conducted by psychologists Heiko Hecht (NASA) and Dennis R. Proffitt (University of Virginia) and published in *Psychological Science* (Mar. 1995). The researchers presented the task to each of six different groups: (1) male college students, (2) female college students, (3) professional waitresses, (4) housewives, (5) male bartenders, and (6) male bus drivers. A total of 120 subjects (20 per group) participated in the study. Two of the groups, waitresses and bartenders, were assumed to have considerable experience handling liquid-filled glasses.

After each subject completed the drawing task, the researchers recorded the deviation* (in angle degrees) of the judged line from the true line. If the deviation was within 5° of the true water surface angle, the answer was considered correct. Deviations of more than 5° in either direction were considered incorrect answers. A summary table of the results for each of the six groups is shown in Table 2.6 on page 36. (Numbers in the table represent frequencies of subject responses.)

Focus

a. Use a graphical method to describe the overall results of the study (i.e., the responses for all 120 subjects).

b. Construct graphs that researchers could use to describe the effect of gender on task performance. Do the results tend to support or refute the prevailing theory that males do better than females?

c. Construct graphs that the researchers could use to describe the effect of age on task performance. [*Note:* The college students in the study were much younger than the subjects in the other four groups.] Do the results tend to support or refute the prevailing theory that younger adults do better than older adults?

d. Construct graphs that the researchers could use to describe the effect of experience in handling liquid-filled containers on task performance. Comment on the theory that experience improves task performance.

e. Refer to parts a–d. What are the drawbacks to using graphical displays of sample data to infer the nature of the population?

f. Suppose the researchers had provided the actual deviations in water-line angles (in degrees) for each subject rather than categorizing the answers as "more than 5° below," "more than 5° above," and "within 5°." What graphs would have been appropriate for describing the data? Discuss which graphs—those of parts a–d or those for the data on actual deviations—would be more informative.

*The true surface line is perfectly parallel to the table top.

TABLE 2.6   Results of Water-Level Task (Case Study 2.1)

| | GROUP | | | | | | |
| | FEMALES | | | MALES | | | |
Judged Line	**Students**	**Waitresses**	**Housewives**	**Students**	**Bartenders**	**Bus Drivers**	**Totals**
More than 5° below surface	0	0	1	1	1	1	4
More than 5° above surface	7	15	13	3	11	4	53
Within 5° of surface	13	5	6	16	8	15	63
Totals	20	20	20	20	20	20	120

Source: Hecht, H., and Proffitt, D. R. "The price of experience: Effects of experience on the water-level task." *Psychological Science,* Vol. 6, No. 2, Mar. 1995, p. 93 (Table 1).

The histogram in Figure 2.9 also indicates a relatively even distribution of rates over the four measurement classes with midpoints from 6 to 15. However, it's harder to determine the boundaries since we can't "see" the individual measurements in the display. Similarly, the two small bars over the outside classes with midpoints of 3 and 18 indicate the presence of some extreme measurements, but we can't see the exact number and location of these measurements. Of course, we could obtain more detail by increasing the number of measurement classes. However, histograms are most useful for displaying very large data sets, when the overall shape of the distribution of measurements is more important than the identification of individual measurements.   ▲

Most statistical software packages can be used to generate histograms, stem-and-leaf displays, and dot plots. All three are useful tools for graphically describing data sets. We recommend that you generate and compare the displays whenever you can. You'll find that histograms are generally more useful for very large data sets, while stem-and-leaf displays and dot plots provide useful detail for smaller data sets.

EXERCISES 2.11–2.24

Learning the Mechanics

2.11 Graph the relative frequency histogram for the 500 measurements summarized in the accompanying relative frequency table.

Measurement Class	Relative Frequency
.5–2.5	.10
2.5–4.5	.15
4.5–6.5	.25
6.5–8.5	.20
8.5–10.5	.05
10.5–12.5	.10
12.5–14.5	.10
14.5–16.5	.05

2.12 Refer to Exercise 2.11. Calculate the number of the 500 measurements falling into each of the measurement classes. Then graph a frequency histogram for these data.

2.13 SAS was used to generate the stem-and-leaf display shown here. Note that SAS arranges the stems in descending order.

Stem	*Leaf*
5	1
4	4 5 7
3	0 0 0 3 6
2	1 1 3 4 5 9 9
1	2 2 4 8
0	0 1 2

a. How many observations were in the original data set?

b. In the bottom row of the stem-and-leaf display, identify the stem, the leaves, and the numbers in the original data set represented by this stem and its leaves.

c. Re-create all the numbers in the data set and construct a dot plot.

2.14 The graph summarizes the scores obtained by 100 students on a questionnaire designed to measure aggressiveness. (Scores are integer values that range from 0 to 20. A high score indicates a high level of aggression.)

   a. Which measurement class contains the highest proportion of test scores?

   b. What proportion of the scores lie between 3.5 and 5.5?

   c. What proportion of the scores are higher than 11.5?

   d. How many students scored less than 5.5?

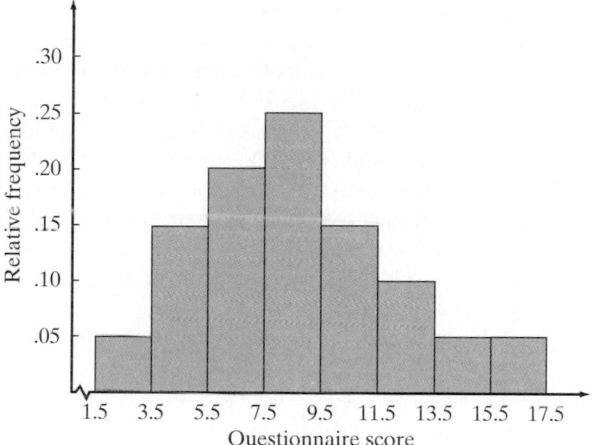

Applying the Concepts

2.15 The role that listener knowledge plays in the perception of dysarthric (imperfectly articulated) speech was recently investigated (*American Journal of Speech–Language Pathology*, Feb. 1995). Thirty female University of Minnesota students, randomly divided into three groups of ten, participated as listeners in the study. All subjects were required to listen to a 48-sentence audiotape of a Korean woman with cerebral palsy and asked to transcribe her entire speech. For the first group of students (the *control group*), the speaker used her normal manner of speaking. For the second group (the *treatment group*), the speaker employed a learned breathing pattern (called breath-group strategy) to improve her speech efficiency. The third group (the *familiarity group*) also listened to the tape with the breath-group strategy, but only after they had practiced by listening twice to another tape in which they were told exactly what the speaker was saying. At the end of the listening/transcribing session, two quantitative variables were measured for each listener: (1) the total number of words transcribed (called the rate of response) and (2) the percentage of words correctly transcribed (called the accuracy score). The data for all 30 subjects are provided in the table below.

   a. MINITAB dot plots of the response rates (RESPRATE) for each listener group are displayed on page 38. Use the information in the plots to describe the differences in the distributions of response rates among the three groups.

   b. Use a graphical method to describe the distribution of accuracy scores for the three listener groups. Construct one graph for each group. Interpret the results.

2.16 The crab spider is a common North American species of arachnid. Male crab spiders, unlike females, rarely feed on prey; rather, they have been observed drinking the nectar of flowers. A group of University of Virginia biologists studied nectivory (nectar drinking) in crab spiders to determine if adult males were feeding on nectar to prevent fluid loss (*Animal Behavior*, June 1995). Nine male spiders were weighed and then placed on the flowers of

CONTROL GROUP		TREATMENT GROUP		FAMILIARIZATION GROUP	
Rate of Response	**Percent Correct**	**Rate of Response**	**Percent Correct**	**Rate of Response**	**Percent Correct**
250	23.6	254	26.0	193	36.0
230	26.0	178	32.6	223	41.0
197	26.0	139	32.6	232	43.0
238	26.7	249	33.0	214	44.0
174	29.2	236	34.4	269	44.0
263	29.5	231	36.5	256	46.0
275	31.3	161	38.5	224	46.0
193	32.9	255	40.0	225	48.0
204	35.4	275	41.7	288	49.0
168	29.2	181	44.8	244	52.0

Source: Tjaden, K., and Liss, J. M. "The influence of familiarity on judgements of treated speech." *American Journal of Speech–Language Pathology*, Vol. 4, No. 1, Feb. 1995, p. 43 (Table 1). Reprinted by permission of the American Speech-Language-Hearing Association.

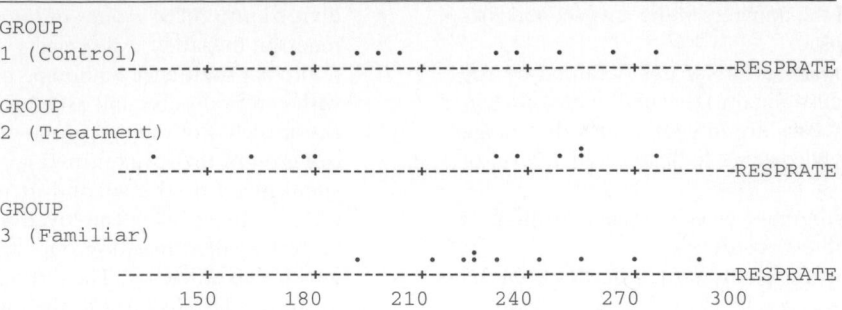

Queen Anne's lace. One hour later, the spiders were removed and reweighed. The evaporative fluid loss (in milligrams) of each of the nine male spiders is given in the accompanying table.

Male Spider	Fluid Loss
A	.018
B	.020
C	.017
D	.024
E	.020
F	.024
G	.003
H	.001
I	.009

Source: Pollard, S. D., *et al.* "Why do male crab spiders drink nectar?" *Animal Behavior*, Vol. 49, No. 6, June 1995, p. 1445 (Table II).

a. Summarize the fluid losses of male crab spiders with a stem-and-leaf display.

b. Of the nine spiders, only three drank any nectar from the flowers of Queen Anne's lace. These three spiders are identified as G, H, and I in the table. Locate and circle these three fluid losses on the stem-and-leaf display. Does the pattern depicted in the graph give you any insight into whether feeding on flower nectar reduces evaporative fluid loss for male crab spiders? Explain.

2.17 While producing many economic benefits to the state of Florida, gypsum and phosphate mines also produce a harmful byproduct: radiation. It has been known for a number of years that the mine tailings (waste) contain radioactive radon 222. In fact, new housing complexes built over the leveled piles of residue have shown disturbing radiation levels within the houses. The radiation levels in waste gypsum and phosphate mounds in Polk County, Florida, are regularly monitored by the Eastern Environmental Radiation Facility (EERF) and by the Polk County Health Department (PCHD), Winter Haven, Florida. Shown in the table at the bottom of the page are measurements of the exhalation rate (a measure of radiation) of soil samples taken on waste piles in Polk County, Florida. They represent part of the data contained in a report by Thomas R. Horton of EERF.

SPSS was used to generate the following stem-and-leaf display for these data. [*Note:* to save space, SPSS places the leaves for two consecutive stems into a single stem row.]

```
Stem-and-leaf display for variable .. EXHLRATE
   0 . 22445668990123455677889
   2 . 013468880088
   4 . 14
   6 . 8
   8 . 1
  10 .
  12 . 0
```

a. Interpret the display. Which digit(s) was used for the stem and which for the leaf? Find the largest measurement in the data set and interpret its representation in the display.

Exhalation Rate of Soil Samples

1,709.79	4,132.28	2,996.49	2,796.42	3,750.83	961.40	1,096.43	1,774.77
357.17	1,489.86	2,367.40	11,968.23	178.99	5,402.35	2,315.52	2,617.57
1,150.94	3,017.48	599.84	2,758.84	3,764.96	1,888.22	2,055.20	205.84
1,572.69	393.55	538.37	1,830.78	878.56	6,815.69	752.89	1,977.97
558.33	880.84	2,770.23	1,426.57	1,322.76	1,480.04	9,139.21	1,698.39

Source: Horton, T. R. "A preliminary radiological assessment of radon exhalation from phosphate gypsum piles and inactive uranium mill tailings piles." EPA–520/5–79–004. Washington, D.C.: Environmental Protection Agency, 1979.

b. Note that the presence of a measurement well removed from the main body of data somewhat distorts the display, since most of the data set is compressed into a small portion of it. The largest measurement was removed from the data set, and a new SPSS stem-and-leaf display was generated, as shown here. Interpret this display, identifying the stem and leaf used. Using both displays, give a verbal description of the data.

```
Stem-and-leaf display for variable .. EXHLRATE
  0 . 2244566899
  1 . 0123455677889
  2 . 01346888
  3 . 0088
  4 . 1
  5 . 4
  6 . 8
  7 .
  8 .
  9 . 1
```

2.18 It's not uncommon for hearing aids to cancel the desired signal. *IEEE Transactions on Speech and Audio Processing* (May 1995) reported on an audio processing system designed to limit the amount of signal cancellation that may occur. The system utilizes a mathematical equation that involves a variable, V, called a *sufficient norm constraint*. A histogram for realizations of V, produced using simulation, is shown below.

a. Estimate the percentage of realizations of V with values ranging from .425 to .675.

b. Cancellation of the desired signal is limited by selecting a norm constraint V so that only 10% of

the realizations have values below the selected level. Find this value of V.

2.19 *Postmortem interval* (PMI) is defined as the elapsed time between death and an autopsy. Knowledge of PMI is considered essential when conducting medical research on human cadavers. The data in the accompanying table are the PMIs of 22 human brain specimens obtained at autopsy in a recent study (*Brain and Language,* June 1995). Graphically describe the PMI data with a dot plot. Based on the plot, make a summary statement about the PMI of the 22 human brain specimens.

Postmortem Intervals for 22 Human Brain Specimens

5.5	14.5	6.0	5.5	5.3	5.8	11.0	6.1
7.0	14.5	10.4	4.6	4.3	7.2	10.5	6.5
3.3	7.0	4.1	6.2	10.4	4.9		

Source: Hayes, T. L., and Lewis, D. A. "Anatomical specialization of the anterior motor speech area: Hemispheric differences in magnopyramidal neurons." *Brain and Language,* Vol. 49, No. 3, June 1995, p. 292 (Table 1).

2.20 Educators are constantly evaluating the efficacy of public schools in the education and training of American students. One quantitative assessment of change over time is the difference in scores on the SAT, which has been used for decades by colleges and universities as one criterion for admission. The following table shows the average SAT scores for each of the 50 states for 1975 and 1990:

State	1975	1990	State	1975	1990
Ala.	883	984	Mont.	1047	987
Alaska	942	914	Nebr.	966	1030
Ariz.	1021	942	Nev.	962	921
Ark.	992	981	N.H.	934	928
Calif.	908	903	N.J.	878	891
Colo.	994	969	N. Mex.	1002	1007
Conn.	913	901	N.Y.	925	882
Del.	915	903	N.C.	827	841
Fla.	915	884	N. Dak.	1064	1069
Ga.	824	844	Ohio	955	949
Hawaii	892	885	Okla.	994	1001
Idaho	1017	968	Oreg.	908	923
Ill.	970	994	Pa.	900	883
Ind.	881	867	R.I.	901	883
Iowa	1091	1088	S.C.	794	834
Kans.	1043	1040	S. Dak.	1084	1061
Ky.	977	994	Tenn.	988	1008
La.	947	993	Tex.	898	874
Maine	908	886	Utah	1069	1031
Md.	907	908	Vt.	915	897
Mass.	903	900	Va.	894	895
Mich.	949	968	Wash.	1011	923
Minn.	1058	1019	W. Va.	964	933
Miss.	980	996	Wis.	1036	1019
Mo.	965	995	Wyo.	1054	977

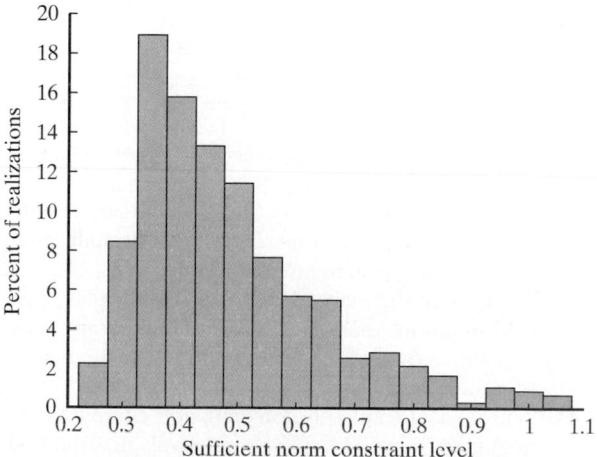

Source: Hoffman, M. W., and Buckley, K. M. "Robust time-domain processing of broadband microphone array data." *IEEE Transactions on Speech and Audio Processing,* Vol. 3, No. 3, May 1995, p. 199 (Figure 4).

We used MINITAB to construct the following stem-and-leaf displays of the two sets of SAT scores:

```
      1975                          1990

                 Leaf Unit = 10

   1    7 9
   1    8
   3    8 22                  1    8 3
   3    8                     3    8 44
   4    8 7                   5    8 67
   9    8 88999              14    8 888888999
  20    9 00000001111        20    9 000001
  22    9 23                 25    9 22223
  (4)   9 4445               25    9 44
  24    9 666677             23    9 6667
  18    9 88999              19    9 88899999
  13   10 011                11   10 00011
  10   10 23                  6   10 33
   8   10 4455                4   10 4
   4   10 66                  3   10 66
   2   10 89                  1   10 8
```

Interpret the two displays, focusing on how the distributions of average state scores changed over this 15-year period.

2.21 Refer to Exercise 2.20. As another method of comparing the 1975 and 1990 average SAT scores, **paired differences** are calculated by subtracting the 1975 score from the 1990 score for each state. The stem-and-leaf display for these differences is shown below:

```
Differences (1990 minus 1975)

Leaf Unit = 10

    1   -0 8
    4   -0 776
    7   -0 444
   16   -0 333322222
  (15)  -0 111111110000000
   19    0 00000111111
    8    0 2223
    4    0 44
    2    0 6
    1    0
    1    1 0
```

a. Interpret the display. How do your conclusions compare to those you reached when comparing the two displays in Exercise 2.20?

b. What is the largest improvement in the average score between 1975 and 1990 as indicated on the display? With which state is this improvement associated in the table of data (Exercise 2.20)?

2.22 Refer to Exercises 2.20 and 2.21. SPSS was used to generate frequency histograms for the average SAT scores for 1975 and 1990. The histograms are shown on page 41. Interpret the histograms, focusing on the difference between the displays for 1975 and 1990. Do you find the stem-and-leaf displays or the histograms more useful for describing these data?

2.23 Saturn has five satellites—Tethys, Mimas, Rhea, Dione, and Enceladus—that rotate around the planet. When Sol (the sun) and Terra (Earth) lie on the same side of the famous Saturn rings, eclipses and occultations of the Saturn satellites are visible. *Astronomy* (Aug. 1995) lists 19 different events involving eclipses or occults of Saturnian satellites during the month of August. For each event, the percentage of light lost by the eclipsed or occulted satellite at midevent is recorded in the accompanying table.

Date		Event	Light Loss (%)
Aug.	2	Eclipse	65
	4	Eclipse	61
	5	Occult	1
	6	Eclipse	56
	8	Eclipse	46
	8	Occult	2
	9	Occult	9
	11	Occult	5
	12	Occult	39
	14	Occult	1
	14	Eclipse	100
	15	Occult	5
	15	Occult	4
	16	Occult	13
	20	Occult	11
	23	Occult	3
	23	Occult	20
	25	Occult	20
	28	Occult	12

Source: Courtesy of *ASTRONOMY* magazine, Aug. 1995, p. 60.

a. Construct a stem-and-leaf display for light loss percentage of the 19 events.

b. Locate on the stem-and-leaf plot, part **a**, the light losses associated with eclipses of Saturnian satellites. (Circle the light losses on the plot.)

c. Based on the marked stem-and-leaf display, part **b**, make an inference about which event type (eclipse or occult) is more likely to lead to a greater light loss.

2.24 Neurological single-photon–emission computed tomography (neuroSPECT) is a relatively new method for detecting abnormalities of the brain that are not apparent on computed tomography (CT) or magnetic resonance imaging (MRI) scans. Fourteen mild

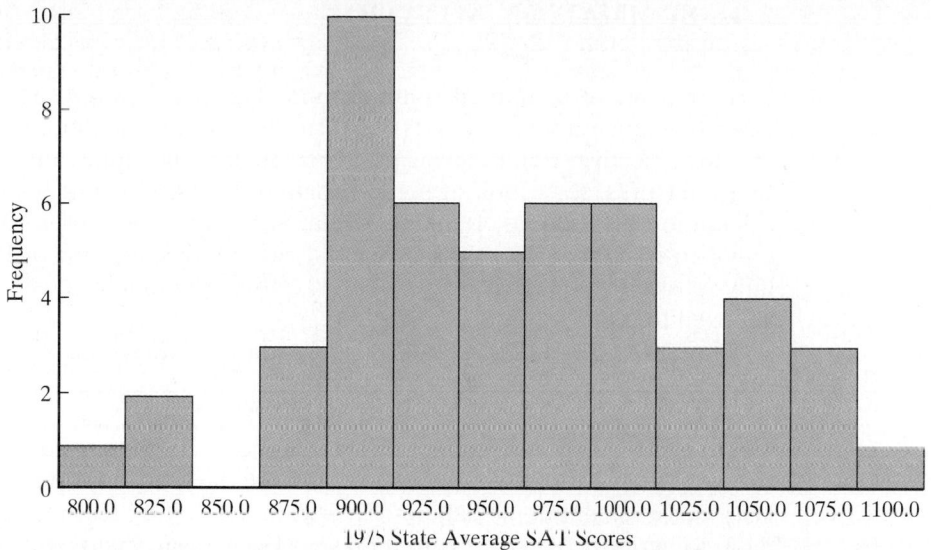

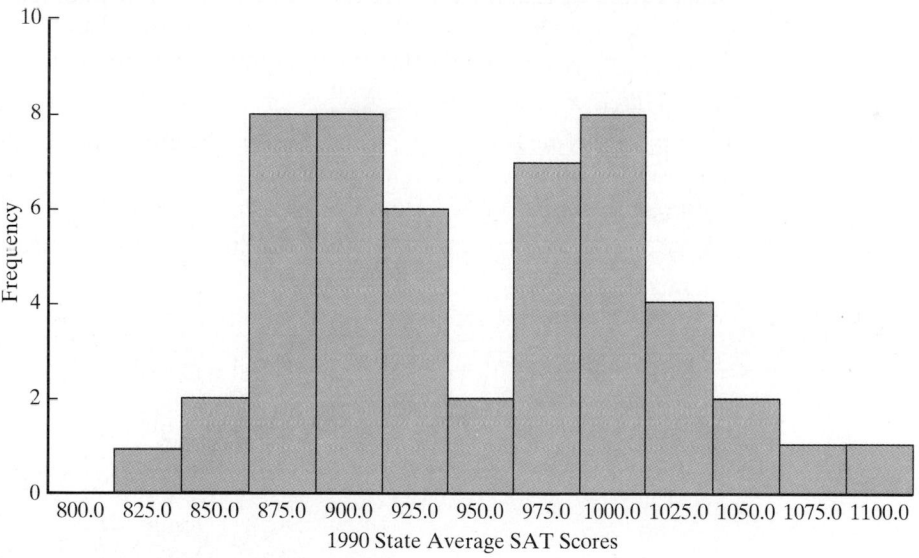

head injury patients with normal CT or MRI scans were examined using neuroSPECT compared to five age-matched hospital employees (controls) who had no history of head injury (*Journal of Head Trauma Rehabilitation,* June 1995). One variable measured by neuroSPECT is mesial-to-lateral HMPAO count density ratio. The ratios for the right temporal lobes of the subjects are shown in the table at right.

a. Construct a dot plot for the neuroSPECT ratios of the 14 head injury patients.

b. Add the neuroSPECT ratios for the five controls to the dot plot, part **a**. Use a different-colored dot for these five measurements.

c. Do you detect a pattern in the dot plot, part **b**? Explain.

Patients		Controls
.852	1.004	1.012
.901	.942	1.111
.849	.977	1.094
.779	.891	1.216
.842	.898	.989
.960	.866	
.782	1.004	

Source: Varney, N. R., *et al.* "NeuroSPECT correlates of disabling mild head injury: Preliminary findings." *Journal of Head Trauma Rehabilitation,* Vol. 10, No. 3, June 1995, p. 24 (Table 2). © 1995 Aspen Publishers, Inc.

2.3   SUMMATION NOTATION

Now that we've examined some graphical techniques for summarizing and describing quantitative data sets, we turn to numerical methods for accomplishing this objective. Before giving the formulas for calculating numerical descriptive measures, let's look at some shorthand notation that will simplify our calculation instructions. Remember that such notation is used for one reason only—to avoid repeating the same verbal descriptions over and over. If you mentally substitute the verbal definition of a symbol each time you read it, you'll soon get used to it.

We denote the measurements of a quantitative data set as:

$$x_1, x_2, x_3, \ldots, x_n$$

where x_1 is the first measurement in the data set, x_2 is the second measurement in the data set, x_3 is the third measurement in the data set, ..., and x_n is the nth (and last) measurement in the data set. Thus, if we have five measurements in a set of data, we write x_1, x_2, x_3, x_4, x_5 to represent the measurements. If the actual numbers are 5, 3, 8, 5, and 4, we have $x_1 = 5$, $x_2 = 3$, $x_3 = 8$, $x_4 = 5$, and $x_5 = 4$.

Most of the formulas we use require a summation of numbers. For example, one sum we'll need to obtain is the sum of all the measurements in the data set, or $x_1 + x_2 + x_3 + \cdots + x_n$. To shorten the notation, we use the symbol Σ for the summation. That is, $x_1 + x_2 + x_3 + \cdots + x_n = \sum_{i=1}^{n} x_i$. Verbally translate $\sum_{i=1}^{5} x_i$ as follows: "The sum of the measurements, whose typical member is x_i, beginning with the member x_1 and ending with the member x_n."

Suppose, as in our earlier example, $x_1 = 5$, $x_2 = 3$, $x_3 = 8$, $x_4 = 5$, and $x_5 = 4$. Then the sum of the five measurements, denoted $\sum_{i=1}^{5} x_i$, is obtained as follows:

$$\sum_{i=1}^{5} x_i = x_1 + x_2 + x_3 + x_4 + x_5$$
$$= 5 + 3 + 8 + 5 + 4 = 25$$

Another important calculation requires that we square each measurement and then sum the squares. The notation for this sum is $\sum_{i=1}^{n} x_i^2$. For the five measurements above, we have

$$\sum_{i=1}^{5} x_i^2 = x_1^2 + x_2^2 + x_3^2 + x_4^2 + x_5^2$$
$$= 5^2 + 3^2 + 8^2 + 5^2 + 4^2$$
$$= 25 + 9 + 64 + 25 + 16 = 139$$

In general, the symbol following the summation sign Σ represents the variable (or function of the variable) that is to be summed.

The Meaning of Summation Notation $\sum_{i=1}^{n} x_i$

Sum the measurements on the variable that appears to the right of the summation symbol, beginning with the 1st measurement and ending with the nth measurement.

EXERCISES 2.25–2.28

Learning the Mechanics

[Note: In all exercises, $\sum$ represents $\sum\limits_{i=1}^{n}$.

2.25 A data set contains the observations 5, 1, 3, 2, 1. Find:
a. $\sum x$ **b.** $\sum x^2$ **c.** $\sum (x - 1)$
d. $\sum (x - 1)^2$ **e.** $(\sum x)^2$

2.26 Suppose a data set contains the observations 3, 8, 4, 5, 3, 4, 6. Find

a. $\sum x$ **b.** $\sum x^2$ **c.** $\sum (x - 5)^2$
d. $\sum (x - 2)^2$ **e.** $(\sum x)^2$

2.27 Refer to Exercise 2.25. Find

a. $\sum x^2 - \dfrac{(\sum x)^2}{5}$ **b.** $\sum (x - 2)^2$ **c.** $\sum x^2 - 10$

2.28 A data set contains the observations 6, 0, -2, -1, 3. Find

a. $\sum x$ **b.** $\sum x^2$ **c.** $\sum x^2 - \dfrac{(\sum x)^2}{5}$

2.4   NUMERICAL MEASURES OF CENTRAL TENDENCY

When we speak of a data set, we refer either to a sample or to a population. If statistical inference is our goal, we'll wish ultimately to use sample numerical descriptive measures to make inferences about the corresponding measures for a population.

As you'll see, a large number of numerical methods are available to describe quantitative data sets. Most of these methods measure one of two data characteristics:

1. The **central tendency** of the set of measurements—that is, the tendency of the data to cluster, or center, about certain numerical values (see Figure 2.12a)
2. The **variability** of the set of measurements—that is, the spread of the data (see Figure 2.12b).

In this section we concentrate on measures of central tendency. In the next section, we discuss measures of variability.

The most popular and best understood measure of central tendency for a quantitative data set is the **arithmetic mean** (or simply the **mean**) of a data set.

> **DEFINITION 2.4**
>
> The **mean** of a set of quantitative data is the sum of the measurements divided by the number of measurements contained in the data set.

In everyday terms, the mean is the average value of the data set and is often used to represent a "typical" value. We denote the **mean of a sample** of measurements by $\bar{x}$ (read "x-bar"), and represent the formula for its calculation as shown in the box on page 44:

FIGURE 2.12

Numerical descriptive measures

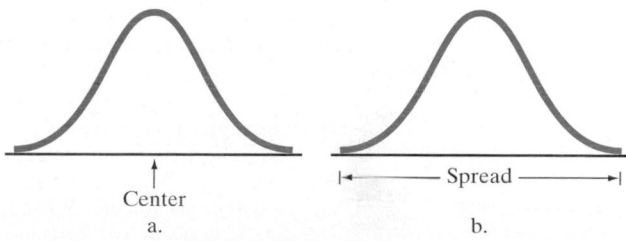

Center
a.

Spread

b.

FIGURE 2.13

SPSS printout of numerical descriptive measures for 100 EPA gas mileages

Mean	36.994	Std Err	.242	Median	37.000
Mode	37.000	Std Dev	2.418	Variance	5.846
Kurtosis	.770	S E Kurt	.478	Skewness	.051
S E Skew	.241	Range	14.900	Minimum	30.000
Maximum	44.900	Sum	3699.400		
Valid Cases	100	Missing Cases	0		

Calculating a Sample Mean

$$\bar{x} = \frac{\sum_{i=1}^{n} x_i}{n}$$

EXAMPLE 2.3   Calculate the mean of the following five sample measurements: 5, 3, 8, 5, 6.

Solution

Using the definition of sample mean and the summation notation, we find

$$\bar{x} = \frac{\sum_{i=1}^{5} x_i}{5} = \frac{5 + 3 + 8 + 5 + 6}{5} = \frac{27}{5} = 5.4$$

Thus, the mean of this sample is 5.4.*   ▲

EXAMPLE 2.4   Calculate the sample mean for the 100 EPA mileages given in Table 2.3.

Solution

The mean gas mileage for the 100 cars is denoted

$$\bar{x} = \frac{\sum_{i=1}^{100} x_i}{100}$$

Rather than compute $\bar{x}$ by hand (or calculator), we entered the data of Table 2.3 into a computer and employed SPSS statistical software to compute the mean. The SPSS printout is shown in Figure 2.13. The sample mean, shaded on the printout, is $\bar{x} = 36.994$.

Given this information, you can visualize a distribution of gas mileage readings centered in the vicinity of $\bar{x} = 36.994$. An examination of the relative frequency histogram (Figure 2.7) confirms that $\bar{x}$ does in fact fall near the center of the distribution.   ▲

The sample mean $\bar{x}$ will play an important role in accomplishing our objective of making inferences about populations based on sample information. For this reason we need to use a different symbol for the **mean of a population**—the mean

---

*In the examples given here, $\bar{x}$ is sometimes rounded to the nearest tenth, sometimes to the nearest hundredth, sometimes to the nearest thousandth, and so on. There is no specific rule for rounding when calculating $\bar{x}$ because $\bar{x}$ is specifically defined to be the sum of all measurements divided by n; that is, it is a specific fraction. When $\bar{x}$ is used for descriptive purposes, it is often convenient to round the calculated value of $\bar{x}$ to the number of significant figures used for the original measurements. When $\bar{x}$ is to be used in other calculations, however, it may be necessary to retain more significant figures.

of the set of measurements on every unit in the population. We use the Greek letter μ (mu) for the population mean.

Symbols for the Sample Mean and the Population Mean

In this text, we adopt a general policy of using Greek letters to represent numerical descriptive measures for the population and Roman letters to represent corresponding descriptive measures for the sample. The symbols for the means are

$$\bar{x} = \text{Sample mean}, \qquad \mu = \text{Population mean}$$

We'll often use the sample mean, $\bar{x}$, to estimate (make an inference about) the population mean, μ. For example, the EPA mileages for the population consisting of *all* cars has a mean equal to some value, μ. Our sample of 100 cars yielded mileages with a mean of $\bar{x} = 36.994$. If, as is usually the case, we don't have access to the measurements for the entire population, we could use $\bar{x}$ as an estimator or approximator for μ. Then we'd need to know something about the reliability of our inference. That is, we'd need to know how accurately we might expect $\bar{x}$ to estimate μ. In Chapter 7, we'll find that this accuracy depends on two factors:

1. The *size of the sample.* The larger the sample, the more accurate the estimate will tend to be.
2. The *variability, or spread, of the data.* All other factors remaining constant, the more variable the data, the less accurate the estimate.

Another important measure of central tendency is the **median.**

DEFINITION 2.5

The **median** of a quantitative data set is the middle number when the measurements are arranged in ascending (or descending) order.

The median is of most value in describing large data sets. If the data set is characterized by a relative frequency histogram (Figure 2.14), the median is the point on the *x*-axis such that half the area under the histogram lies above the median and half lies below. [*Note:* In Section 2.2 we observed that the relative frequency associated with a particular interval on the *x*-axis is proportional to the amount of area under the histogram that lies above the interval.] We denote the *median* of a *sample* by m.

Calculating a Sample Median, m

Arrange the *n* measurements from the smallest to the largest.

1. If *n* is odd, m is the middle number.
2. If *n* is even, m is the mean of the middle two numbers.

FIGURE 2.14

Location of the median

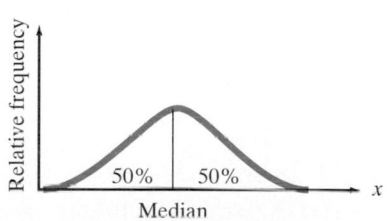

EXAMPLE 2.5   Consider the following sample of $n = 7$ measurements: 5, 7, 4, 5, 20, 6, 2

a. Calculate the median m of this sample.
b. Eliminate the last measurement (the 2) and calculate the median of the remaining $n = 6$ measurements.

Solution

a. The seven measurements in the sample are ranked in ascending order: 2,   4, 5,   5,   6,   7,   20
   Because the number of measurements is odd, the median is the middle measurement. Thus, the median of this sample is m = 5.
b. After removing the 2 from the set of measurements, we rank the sample measurements in ascending order as follows: 4,   5,   5,   6,   7,   20
   Now the number of measurements is even, so we average the middle two measurements. The median is m = (5 + 6)/2 = 5.5.   ▲

In certain situations, the median may be a better measure of central tendency than the mean. In particular, the median is less sensitive than the mean to extremely large or small measurements. Note, for instance, that all but one of the measurements in part a of Example 2.5 center about $x = 5$. The single relatively large measurement, $x = 20$, does not affect the value of the median, 5, but it causes the mean, $\bar{x} = 7$, to lie to the right of most of the measurements.

As another example of data from which the central tendency is better described by the median than the mean, consider the household incomes of a community being studied by a sociologist. The presence of just a few households with very high incomes will affect the mean more than the median. Thus, the median will provide a more accurate picture of the typical income for the community. The mean could exceed the vast majority of the sample measurements (household incomes), making it a misleading measure of central tendency.

EXAMPLE 2.6   Calculate the median for the 100 EPA mileages given in Table 2.3. Compare the median to the mean computed in Example 2.4.

Solution

For this large data set, we again resort to a computer analysis. The SPSS printout is reproduced in Figure 2.15, with the median shaded. You can see that the median is 37.0. This value implies that half of the 100 mileages in the data set fall below 37.0 and half lie above 37.0. Note that the median, 37.0, and the mean, 36.994, are almost equal. This fact indicates that the data form an approximately **symmetric distribution**. As indicated in the box on page 47, a comparison of the mean and median gives an indication of the **skewness** (i.e., the tendency of the distribution to have elongated tails) of a data set.   ▲

FIGURE 2.15

SPSS printout of numerical descriptive measures for 100 EPA mileages

Mean	36.994	Std Err	.242	Median	37.000
Mode	37.000	Std Dev	2.418	Variance	5.846
Kurtosis	.770	S E Kurt	.478	Skewness	.051
S E Skew	.241	Range	14.900	Minimum	30.000
Maximum	44.900	Sum	3699.400		
Valid Cases	100	Missing Cases	0		

Comparing the Mean and the Median

If the data set is skewed to the right, the mean is greater than (to the right of) the median.

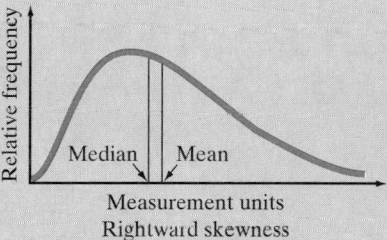

Rightward skewness

If the data set is symmetric, the mean equals the median.

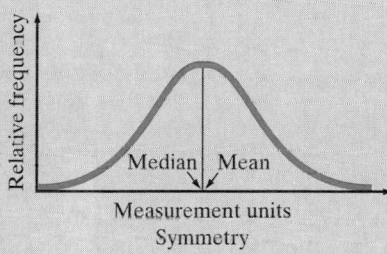

Symmetry

If the data set is skewed to the left, the mean is less than (to the left of) the median.

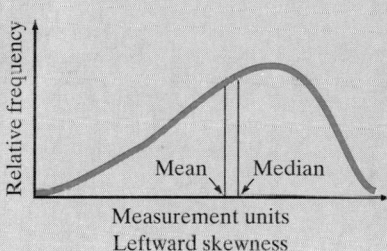

Leftward skewness

A third measure of central tendency is the **mode** of a set of measurements.

DEFINITION 2.6

The **mode** is the measurement that occurs most frequently in the data set.

Therefore, the mode shows where the data tend to concentrate.

EXAMPLE 2.7

Find the mode for the following ten quiz grades:

8, 7, 9, 6, 8, 10, 9, 9, 5, 7

Solution

Since 9 occurs most often, the mode is 9. ▲

The mode is of primary value in describing large data sets. When a large data set has been described using a relative frequency histogram, the mode will be located in the class containing the largest relative frequency. This is called the **modal class**. Several definitions exist for locating the position of the mode within a modal class, but the simplest is to define the mode as the midpoint of the modal class. For example, examine the relative frequency histogram for the EPA mileage readings (Figure 2.7). You can see that the modal class is the interval 35.95–37.45. The mode (the midpoint) is 36.7. Note that this measure of central tendency is very close to the mean, 36.99, and the median, 37.0.

Because it emphasizes data concentration, the mode is often used with large data sets to locate the region in which much of the data are concentrated. A supermarket manager may be interested in the cereal brand with the largest share of the market, that is, the modal brand. The modal income class of the American worker is of interest to the Labor Department. Thus, the mode provides a useful measure of central tendency for various applications.

Note that the cereal brand example involves a qualitative measurement. The mode can be used to describe the central tendency of both types of data (quantitative and qualitative), while the mean and median are primarily useful for quantitative data.

EXERCISES 2.29–2.45

Learning the Mechanics

2.29 Calculate the mode, mean, and median of the following data:

18   10   15   13   17   15   12   15   18   16   11

2.30 Calculate the mean and median of the following grade-point averages:

3.2   2.5   2.1   3.7   2.8   2.0

2.31 Explain the difference between the calculation of the median for an odd and an even number of measurements. Construct one data set consisting of five measurements and another consisting of six measurements for which the medians are equal.

2.32 Explain how the relationship between the mean and median provides information about the symmetry or skewness of the data's distribution.

2.33 Calculate the mean for samples where
 a. $n = 10, \sum x = 85$ **b.** $n = 16, \sum x = 400$
 c. $n = 45, \sum x = 35$ **d.** $n = 18, \sum x = 242$

2.34 Calculate the mean, median, and mode for each of the following samples:
 a. 7, −2, 3, 3, 0, 4
 b. 2, 3, 5, 3, 2, 3, 4, 3, 5, 1, 2, 3, 4
 c. 51, 50, 47, 50, 48, 41, 59, 68, 45, 37

2.35 Describe how the mean compares to the median for a distribution as follows:
 a. Skewed to the left   **b.** Skewed to the right
 c. Symmetric

Applying the Concepts

2.36 *The Condor* (May 1995) published a study of competition for nest holes among collared flycatchers, a bird species. The authors collected the data for the study by periodically inspecting nest boxes located on the island of Gotland in Sweden. The nest boxes were grouped into 14 discrete locations (called plots). The accompanying table gives the number of flycatchers killed and the number of flycatchers breeding at each plot.

Plot Number	Number Killed	Number of Breeders
1	5	30
2	4	28
3	3	38
4	2	34
5	2	26
6	1	124
7	1	68
8	1	86
9	1	32
10	0	30
11	0	46
12	0	132
13	0	100
14	0	6

Source: Merilä, J., and Wiggins, D. A. "Interspecific competition for nest holes causes adult mortality in the collared flycatcher." *The Condor,* Vol. 97, No. 2, May 1995, p. 447 (Table 4) Cooper Ornithological Society.

 a. Calculate the mean, median, and mode for the number of flycatchers killed at the 14 study plots.
 b. Interpret the measures of central tendency, part **a**.
 c. Below is displayed a MINITAB printout of descriptive statistics for the number of breeders at

	N	MEAN	MEDIAN	TRMEAN	STDEV	SEMEAN
BREEDERS	14	55.7	36.0	53.5	39.6	10.6
	MIN	MAX	Q1	Q3		
BREEDERS	6.0	132.0	29.5	89.5		

each plot. Locate the measures of central tendency on the printout and interpret these values.

2.37 The conventional method of measuring the refractive status of an eye involves three quantities: (1) sphere power, (2) cylinder power, and (3) axis. Optometric researchers at a Johannesburg (South Africa) university studied the variation in these three measures of refraction (*Optometry and Vision Science,* June 1995). Twenty-five successive refractive measurements (using a single Topcon RM-A6000 autorefractor) were obtained on the eyes of more than 100 university students. The cylinder power measurements for the left eye of one particular student (ID #11) are listed in the table. [*Note:* All measurements are negative values.] Numerical descriptive measures for the data set are provided in the accompanying SAS printout.

.08	.08	1.07	.09	.16	.04	.07	.17	.11
.06	.12	.17	.20	.12	.17	.09	.07	.16
.15	.16	.09	.06	.10	.21	.06		

Source: Rubin, A., and Harris, W. F. "Refractive variation during autorefraction: Multivariate distribution of refractive status." *Optometry and Vision Science,* Vol. 72, No. 6, June 1995, p. 409 (Table 4).

```
               UNIVARIATE PROCEDURE
Variable=CYLPOWER

                       Moments
N                 25     Sum Wgts              25
Mean          -0.1544   Sum                -3.86
Std Dev      0.196767   Variance        0.038717
Skewness     -4.52208   Kurtosis        21.70196
USS            1.5252   CSS             0.929216
CV            -127.44   Std Mean        0.039353
T:Mean=0     -3.92342   Prob>|T|          0.0006
Sgn Rank       -162.5   Prob>|S|          0.0001
Num ^= 0          25

              Quantiles(Def=5)
  100% Max     -0.04        99%        -0.04
   75% Q3      -0.08        95%        -0.06
   50% Med     -0.11        90%        -0.06
   25% Q1      -0.16        10%         -0.2
    0% Min     -1.07         5%        -0.21
                             1%        -1.07

Range           1.03
Q3-Q1           0.08
Mode           -0.17

                     Extremes
   Lowest      Obs      Highest      Obs
   -1.07(       3)      -0.07(       17)
   -0.21(      24)      -0.06(       10)
    -0.2(      13)      -0.06(       22)
   -0.17(      15)      -0.06(       25)
   -0.17(      12)      -0.04(        6)
```

a. Locate the measures of central tendency on the printout and interpret their values.

b. Note that the data contain one unusually large (negative) cylinder power measurement relative to the other measurements in the data set. Find this measurement, called an **outlier**.

c. Delete the outlier, part **b**, from the data set and recalculate the measures of central tendency. Which measure is most affected by the elimination of the outlier?

2.38 Demographics plays a key role in the recreation industry. According to the *Journal of Leisure Research* (Vol. 23, 1991), difficult times lay ahead for the industry. The article reports that the median age of the population in the United States was 30 in 1980, but will be about 36 by the year 2000.

a. Interpret the value of the median for both 1980 and 2000 and explain the trend.

b. If the recreation industry relies on the 18–30 age group for much of its business, what effect will this shift in the median age have on the recreation industry? Explain.

2.39 Applicants for an academic position (e.g., assistant professor) at a college or university are usually required to submit at least three letters of recommendation. A recent study of 148 applicants for an entry-level position in experimental psychology at the University of Alaska Anchorage revealed that many did not meet the three-letter requirement (*American Psychologist,* July 1995). Summary statistics for the number of recommendation letters in each application are given below. Interpret these summary measures.

Mean = 2.28

Median = 3

Mode = 3

2.40 Platelet-activating factor (PAF) is a potent chemical that occurs in patients suffering from shock, inflammation, hypotension, and allergic responses as well as respiratory and cardiovascular disorders. Consequently, drugs that effectively inhibit PAF, keeping it from binding to human cells, may be successful in treating these disorders. A bioassay was undertaken to investigate the potential of 17 traditional Chinese herbal drugs in PAF inhibition (*Progress in Natural Science,* June 1995). The prevention of the PAF binding process, measured as a percentage, for each drug is provided in the accompanying table.

a. Construct a stem-and-leaf display for the data.

b. Compute the median inhibition percentage for the 17 herbal drugs. Interpret the result.

c. Compute the mean inhibition percentage for the 17 herbal drugs. Interpret the result.

d. Compute the mode of the 17 inhibition percentages. Interpret the result.

Drug	PAF Inhibition (%)
Hai-feng-teng (Fuji)	77
Hai-feng-teng (Japan)	33
Shan-ju	75
Zhang-yiz-hu-jiao	62
Shi-nan-teng	70
Huang-hua-hu-jiao	12
Hua-nan-hu-jiao	0
Xiao-yie-pa-ai-xiang	0
Mao-ju	0
Jia-ju	15
Xie-yie-ju	25
Da-yie-ju	0
Bian-yie-hu-jiao	9
Bi-bo	24
Duo-mai-hu-jiao	40
Yan-sen	0
Jiao-guo-hu-jiao	31

Source: Guiqiu, H. "PAF receptor antagonistic principles from Chinese traditional drugs." *Progress in Natural Science,* Vol. 5, No. 3, June 1995, p. 301 (Table 1).

e. Locate the median, mean, and mode on the stem-and-leaf display, part **a**. Do these measures of central tendency appear to locate the center of the data?

2.41 Would you expect the data sets described below to possess relative frequency distributions that are symmetric, skewed to the right, or skewed to the left? Explain.

a. The salaries of all persons employed by a large university

b. The grades of an easy test

c. The grades on a difficult test

d. The amounts of time students in your class studied last week

e. The ages of automobiles on a used-car lot

f. The amounts of time spent by students on a difficult examination (maximum time is 50 minutes)

2.42 The salaries of superstar professional athletes receive much attention in the media. The multimillion-dollar long-term contract is now commonplace among this elite group. Nevertheless, rarely does a season pass without negotiations between one or more of the players' associations and team owners for additional salary and fringe benefits for *all* players in their particular sports.

a. If a players' association wanted to support its argument for higher "average" salaries, which measure of central tendency do you think it should use? Why?

b. To refute the argument, which measure of central tendency should the owners apply to the players' salaries? Why?

2.43 Ten presumably trained rats were released in a maze. Their times to escape (in seconds) are recorded below. The *N*'s represent two rats that had still not escaped by the end of the experiment.

100   38   *N*   122   95   116   56   135   104   *N*

a. Can you calculate the mean for these data? Explain.

b. Is the median a meaningful measure of central tendency for these data? Explain. Calculate the median.

2.44 Clinical observations suggest that specifically language-impaired (SLI) children have great difficulty with the proper use of pronouns. This phenomenon was investigated and reported in the *Journal of Communication Disorders* (Mar. 1995). Thirty children, all from low-income families, participated in the study. Ten were 5-year-old SLI children, ten

Subject	Gender	Group	DIQ	Pronoun Errors (%)
1	F	YND	110	94.40
2	F	YND	92	19.05
3	F	YND	92	62.50
4	M	YND	100	18.75
5	F	YND	86	0
6	F	YND	105	55.00
7	F	YND	90	100.00
8	M	YND	96	86.67
9	M	YND	90	32.43
10	F	YND	92	0
11	F	SLI	86	60.00
12	M	SLI	86	40.00
13	M	SLI	94	31.58
14	M	SLI	98	66.67
15	F	SLI	89	42.86
16	F	SLI	84	27.27
17	M	SLI	110	33.33
18	F	SLI	107	0
19	F	SLI	87	0
20	M	SLI	95	0
21	M	OND	110	0
22	M	OND	113	0
23	M	OND	113	0
24	F	OND	109	0
25	M	OND	92	0
26	F	OND	108	0
27	M	OND	95	0
28	F	OND	87	0
29	F	OND	94	0
30	F	OND	98	0

Source: Reprinted by permission of the publisher from Moore, M. E. "Error analysis of pronouns by normal and language-impaired children." *Journal of Communication Disorders,* Vol. 28, No. 1, Mar. 1995, p. 62 (Table 2), p. 67 (Table 5). Copyright © 1995 by Elsevier Science, Inc.

```
GROUP EQ 'YND'
PCTERROR
Mean           46.880     Median      43.715    Mode          .000
Std dev        38.214     Variance  1460.347
Valid cases       10     Missing cases      0
-----------------------------------------------------------------
GROUP EQ 'SLI'
PCTERROR
Mean           30.171     Median      32.455    Mode          .000
Std dev        24.108     Variance   581.202
Valid cases       10     Missing cases      0
-----------------------------------------------------------------
GROUP EQ 'OND'
PCTERROR
Mean            .000      Median       .000     Mode          .000
Std dev         .000      Variance     .000
Valid cases       10     Missing cases      0
```

were younger (3-year-old) normally developing (YND) children, and ten were older (5-year-old) normally developing (OND) children. The accompanying table contains the gender, deviation intelligence quotient (DIQ), and percentage of pronoun errors observed for each of the 30 subjects.

a. Identify the variables in the data set as quantitative or qualitative.

b. Why is it nonsensical to compute numerical descriptive measures for qualitative variables?

c. Compute measures of central tendency for DIQ for the ten SLI children.

d. Compute measures of central tendency for DIQ for the ten YND children.

e. Compute measures of central tendency for DIQ for the ten OND children.

f. Use the results, parts **c–e**, to compare the DIQ central tendencies of the three groups of children.

g. An SPSS printout of descriptive statistics for the percentage of pronoun errors, by group, is shown above. Locate the measures of central tendency on the printout and interpret their values.

h. Is it reasonable to use a single number (e.g., mean or median) to describe the center of the percent-age of pronoun error distribution? Or should three "centers" be calculated, one for each of the three groups of children? Explain.

2.45 Refer to Exercises 2.20–2.22, in which we used stem-and-leaf displays and histograms to compare the average SAT scores in 1975 to those in 1990. We now use SPSS to obtain the mean and median for the two sets of average SAT scores:

```
TOTSAT75
Mean        955.300     Median      952.000
Valid cases      50     Missing cases      0
-----------------------------------------------------------
TOTSAT90
Mean        947.460     Median      937.50
Valid cases      50     Missing cases      0
```

a. Use the mean and median to compare the central tendencies of the two data sets.

b. What does the relationship of the mean and median imply about the symmetry or skewness of the data sets? Look at the stem-and-leaf displays and histograms in Exercises 2.20 and 2.22 to assist in your evaluation.

2.5   NUMERICAL MEASURES OF VARIABILITY

Measures of central tendency provide only a partial description of a quantitative data set. The description is incomplete without a measure of the variability, or spread, of the data set. Knowledge of the data's variability along with its center can help us visualize the shape of a data set as well as its extreme values.

If you examine the two histograms in Figure 2.16, you will notice that both hypothetical data sets are symmetric with equal modes, medians, and means.

FIGURE 2.16

Hypothetical data sets

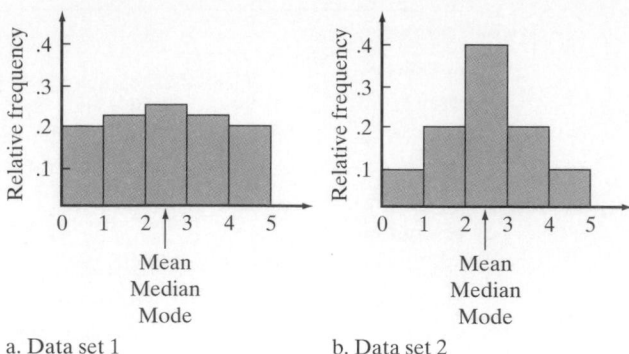

a. Data set 1   b. Data set 2

However, data set 1 (Figure 2.16a) has measurements spread with almost equal relative frequency over the measurement classes, while data set 2 (Figure 2.16b) has most of its measurements clustered about its center. Thus, data set 2 is *less variable* than data set 1. Consequently, you can see that we need a measure of variability as well as a measure of central tendency to describe a data set.

Perhaps the simplest measure of the variability of a quantitative data set is its *range*.

DEFINITION 2.7

The **range** of a quantitative data set is equal to the largest measurement minus the smallest measurement.

The range is easy to compute and easy to understand, but it is a rather insensitive measure of data variation when the data sets are large. This is because two data sets can have the same range and be vastly different with respect to data variation. This phenomenon is demonstrated in Figure 2.16. Both distributions of data shown in the figure have the same range, but most of the measurements in data set 2 tend to concentrate near the center of the distribution. Consequently, the data are much less variable than the data in set 1. Thus, you can see that the range does not always detect differences in data variation for large data sets.

Let's see if we can find a measure of data variation that is more sensitive than the range. Consider the two samples in Table 2.7: Each has five measurements. (We have ordered the numbers for convenience.) Note that both samples have a mean of 3 and that we have also calculated the distance between each measurement and the mean. What information do these distances contain? If they tend to

TABLE 2.7   Two Hypothetical Data Sets

	Sample 1	Sample 2
Measurements	1, 2, 3, 4, 5	2, 3, 3, 3, 4
Mean	$\bar{x} = \dfrac{1 + 2 + 3 + 4 + 5}{5} = \dfrac{15}{5} = 3$	$\bar{x} = \dfrac{2 + 3 + 3 + 3 + 4}{5} = \dfrac{15}{5} = 3$
Distances of measurement values from $\bar{x}$	$(1-3), (2-3), (3-3), (4-3),$ $(5-3)$ or $-2, -1, 0, 1, 2,$	$(2-3), (3-3), (3-3), (3-3),$ $(4-3)$ or $-1, 0, 0, 0, 1$

FIGURE 2.17

Dot plots for
two data sets

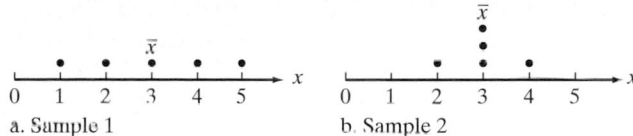

a. Sample 1   b. Sample 2

be large in magnitude, as in sample 1, the data are spread out, or highly variable. If the distances are mostly small, as in sample 2, the data are clustered around the mean, $\bar{x}$, and therefore do not exhibit much variability. You can see that these distances, displayed graphically in Figure 2.17, provide information about the variability of the sample measurements.

The next step is to condense the information in these distances into a single numerical measure of variability. Averaging the distances from $\bar{x}$ won't help because the negative and positive distances cancel; that is, the sum of the deviations (and thus the average deviation) is always equal to zero.

Two methods come to mind for dealing with the fact that positive and negative distances from the mean cancel. The first is to treat all the distances as though they were positive, ignoring the sign of the negative distances. We won't pursue this line of thought because the resulting measure of variability (the mean of the absolute values of the distances) presents analytical difficulties beyond the scope of this text. A second method of eliminating the minus signs associated with the distances is to square them. The quantity we can calculate from the squared distances will provide a meaningful description of the variability of a data set and presents fewer analytical difficulties in inference-making.

To use the squared distances calculated from a data set, we first calculate the *sample variance.*

DEFINITION 2.8

The **sample variance** for a sample of n measurements is equal to the sum of the squared distances from the mean divided by $(n - 1)$. In symbols, using s^2 to represent the sample variance,

$$s^2 = \frac{\sum_{i=1}^{n}(x_i - \bar{x})^2}{n - 1}$$

Note: A shortcut formula for calculating s^2 is

$$s^2 = \frac{\sum_{i=1}^{n}x_i^2 - \dfrac{\left(\sum_{i=1}^{n}x_i\right)^2}{n}}{n - 1}$$

Referring to the two samples in Table 2.7, you can calculate the variance for sample 1 as follows:

$$s^2 = \frac{(1 - 3)^2 + (2 - 3)^2 + (3 - 3)^2 + (4 - 3)^2 + (5 - 3)^2}{5 - 1}$$

$$= \frac{4 + 1 + 0 + 1 + 4}{4} = 2.5$$

The second step in finding a meaningful measure of data variability is to calculate the *standard deviation* of the data set:

DEFINITION 2.9

The **sample standard deviation**, *s*, is defined as the positive square root of the sample variance, s^2. Thus,

$$s = \sqrt{s^2}$$

The population variance, denoted by the symbol σ^2 (sigma squared), is the average of the squared distances of the measurements on *all* units in the population from the mean, μ, and σ (sigma) is the square root of this quantity.

Symbols for Variance and Standard Deviation

s^2 = Sample variance
s = Sample standard deviation
σ^2 = Population variance
σ = Population standard deviation

Notice that, unlike the variance, the standard deviation is expressed in the original units of measurement. For example, if the original measurements are in dollars, the variance is expressed in the peculiar units "dollar squared," but the standard deviation is expressed in dollars.

You may wonder why we use the divisor $(n - 1)$ instead of n when calculating the sample variance. Wouldn't using n be more logical, so that the sample variance would be the average squared distance from the mean? The trouble is, using n tends to produce an underestimate of the population variance, σ^2. So we use $(n - 1)$ in the denominator to provide the appropriate correction for this tendency.* Since sample statistics like s^2 are primarily used to estimate population parameters like σ^2, $(n - 1)$ is preferred to n when defining the sample variance.

EXAMPLE 2.8   Calculate the variance and standard deviation of the following sample: 2, 3, 3, 3, 4.

Solution

As the number of measurements increases, calculating s^2 and s becomes very tedious. Fortunately, as we show in Example 2.9, we can use a statistical software package (or calculator) to find these values. If you must calculate these quantities by hand, it is advantageous to use the shortcut formula provided in Definition 2.8. To do this, we need two summations: $\sum x$ and $\sum x^2$. These can easily be obtained from the following type of tabulation:

x	x^2
2	4
3	9
3	9
3	9
4	16
$\sum x = 15$	$\sum x^2 = 47$

---

*"Appropriate" here means that s^2 with a divisor of $(n - 1)$ is an **unbiased estimator** of σ^2. We define and discuss unbiasedness of estimators in Chapter 6.

FIGURE 2.18

SPSS printout of numerical descriptive measures for 100 EPA mileages

Mean	36.994	Std Err	.242	Median	37.000
Mode	37.000	Std Dev	2.418	Variance	5.846
Kurtosis	.770	S E Kurt	.478	Skewness	.051
S E Skew	.241	Range	14.900	Minimum	30.000
Maximum	44.900	Sum	3699.400		
Valid Cases	100	Missing Cases	0		

Then we use*

$$s^2 = \frac{\sum\limits_{i=1}^{n} x_i^2 - \frac{\left(\sum\limits_{i=1}^{n} x_i\right)^2}{n}}{n-1} = \frac{47 - \frac{(15)^2}{5}}{5-1} = \frac{2}{4} = .5$$

$$s = \sqrt{.5} = .71$$

▲

EXAMPLE 2.9

Use the computer to find the sample variance s^2 and the sample standard deviation s for the 100 gas mileage readings given in Table 2.3.

Solution

The SPSS printout describing the gas mileage data is reproduced in Figure 2.18. The variance and standard deviation, shaded on the printout, are $s^2 = 5.846$ and $s = 2.418$.

▲

You now know that the standard deviation measures the variability of a set of data and how to calculate it. But how can we interpret and use the standard deviation? This is the topic of Section 2.6.

*When calculating s^2, how many decimal places should you carry? Although there are no rules for the rounding procedure, it's reasonable to retain twice as many decimal places in s^2 as you ultimately wish to have in s. If you wish to calculate s to the nearest hundredth (two decimal places), for example, you should calculate s^2 to the nearest ten-thousandth (four decimal places).

EXERCISES 2.46–2.57

Learning the Mechanics

2.46 Answer the following questions about variability of data sets:
   a. What is the primary disadvantage of using the range to compare the variability of data sets?
   b. Describe the sample variance using words rather than a formula. Do the same with the population variance.
   c. Can the variance of a data set ever be negative? Explain. Can the variance ever be smaller than the standard deviation? Explain.

2.47 Calculate the variance and standard deviation for samples where
   a. $n = 10$, $\sum x^2 = 84$, $\sum x = 20$
   b. $n = 40$, $\sum x^2 = 380$, $\sum x = 100$
   c. $n = 20$, $\sum x^2 = 18$, $\sum x = 17$

2.48 Calculate the range, variance, and standard deviation for the following samples:
   a. 4, 2, 1, 0, 1
   b. 1, 6, 2, 2, 3, 0, 3

   c. 8, −2, 1, 3, 5, 4, 4, 1, 3, 3
   d. 0, 2, 0, 0, −1, 1, −2, 1, 0, −1, 1, −1, 0, −3, −2, −1, 0, 1

2.49 Calculate the range, variance, and standard deviation for the following samples:
   a. 39, 42, 40, 37, 41
   b. 100, 4, 7, 96, 80, 3, 1, 10, 2
   c. 100, 4, 7, 30, 80, 30, 42, 2

2.50 Compute $\bar{x}$, s^2, and s for each of the following data sets. If appropriate, specify the units in which your answer is expressed.
   a. 3, 1, 10, 10, 4
   b. 8 feet, 10 feet, 32 feet, 5 feet
   c. −1, −4, −3, 1, −4, −4
   d. ⅕ ounce, ⅕ ounce, ⅕ ounce, ⅖ ounce, ⅕ ounce, ⅘ ounce

2.51 Using only integers between 0 and 10, construct two data sets with at least 10 observations each so that the two sets have the same mean but different variances. Construct dot diagrams for each of your data sets (see Figure 2.17), and mark the mean of each data set on its dot diagram.

2.52 Using only integers between 0 and 10, construct two data sets with at least 10 observations each that have the same range but different means. Construct a dot diagram for each of your data sets (see Figure 2.17), and mark the mean of each data set on its dot diagram.

2.53 Consider the following sample of five measurements: 2, 1, 1, 0, 3
  a. Calculate the range, s^2, and s.
  b. Add 3 to each measurement and repeat part **a**.
  c. Subtract 4 from each measurement and repeat part **a**.
  d. Considering your answers to parts **a**, **b**, and **c**, what seems to be the effect on the variability of a data set by adding the same number to or subtracting the same number from each measurement?

Applying the Concepts

2.54 Improving public education has been an important topic in the United States for many years. A few years ago, *U.S. News & World Report* (Mar. 5, 1990) reported on many factors contributing to the breakdown of public education. One study mentioned in the article found that more than 90% of the nation's school districts reported that their students were scoring "above the national average" on standardized tests. Using your knowledge of measures of central tendency, explain why the schools' reports are incorrect. Does your analysis change if the term "average" refers to the mean? To the median? Explain what effect this misinformation might have on the perception of the nation's schools.

2.55 The final grades given by two professors in introductory statistics courses have been carefully examined. The students in the first professor's class had a grade-point average of 3.0 and a standard deviation of .2. Those in the second professor's class had grade points with an average of 3.0 and a standard deviation of 1.0. If you had a choice, which professor would you take for this course? Explain.

2.56 Consider the following two samples:

Sample 1: 10, 0, 1, 9, 10, 0, 8, 1, 1, 9
Sample 2:   0, 5, 10, 5, 5, 5, 6, 5, 6, 5

  a. Examine both samples and identify the one that you believe has the greater variability.
  b. Calculate the range for each sample. Does the result agree with your answer to part **a**? Explain.
  c. Calculate the standard deviation for each sample. Does the result agree with your answer to part **a**? Explain.
  d. Which of the two, the range or the standard deviation, provides a better measure of variability? Why?

2.57 The U.S. Federal Trade Commission has recently begun assessing fines and other penalties against weight-loss clinics that make unsupported or misleading claims about the effectiveness of their programs. Suppose you have brochures from two weight-loss clinics that both advertise "statistical evidence" about the effectiveness of their programs. Clinic A advertises that the *mean* weight loss during the first month is 15 pounds, while Clinic B advertises a *median* weight loss of 10 pounds.
  a. Assuming the statistics are accurately calculated, which clinic would you recommend if you had no other information? Why?
  b. Upon further research, the median and standard deviation for Clinic A are found to be 10 pounds and 20 pounds, respectively, while the mean and standard deviation for Clinic B are found to be 10 and 5 pounds, respectively. Both are based on samples of more than 100 clients. Describe the two clinics' weight-loss distributions as completely as possible given this additional information. What would you recommend to a prospective client now? Why?
  c. Note that nothing has been said about how the sample of clients upon which the statistics are based was selected. What additional information would be important regarding the sampling techniques employed by the clinics?

2.6 INTERPRETING THE STANDARD DEVIATION

We've seen that if we are comparing the variability of two samples selected from a population, the sample with the larger standard deviation is the more variable of the two. Thus, we know how to interpret the standard deviation on a relative or comparative basis, but we haven't explained how it provides a measure of variability for a single sample.

To understand how the standard deviation provides a measure of variability of a data set, consider a specific data set and answer the following questions: How many measurements are within 1 standard deviation of the mean? How many measurements are within 2 standard deviations? For example, look at the 100 mileage per gallon readings given in Table 2.3. Recall that $\bar{x} = 36.99$ and $s = 2.42$. Then

$$\bar{x} - s = 34.57$$
$$\bar{x} + s = 39.41$$
$$\bar{x} - 2s = 32.15$$
$$\bar{x} + 2s = 41.83$$

If we examine the data, we find that 68 of the 100 measurements, or 68%, are in the interval

$$\bar{x} - s \quad \text{to} \quad \bar{x} + s$$

Similarly, we find that 96, or 96%, of the 100 measurements are in the interval

$$\bar{x} - 2s \quad \text{to} \quad \bar{x} + 2s$$

We usually write these intervals as

$$(\bar{x} - s, \bar{x} + s) \qquad \text{and} \qquad (\bar{x} - 2s, \bar{x} + 2s)$$

Such observations identify criteria for interpreting a standard deviation that apply to *any* set of data, whether a population or a sample. The criteria, expressed as a mathematical theorem and as a rule of thumb, are presented in Tables 2.8 and 2.9. In these tables we give two sets of answers to the questions of how many measurements fall within 1, 2, and 3 standard deviations of the mean. The first, which applies to *any* set of data, is derived from a theorem proved by the Russian mathematician P. L. Chebyshev (1821–1894). The second, which applies to mound-shaped, symmetric distributions of data (where the mean, median, and mode are all about the same), is based upon empirical evidence that has accumulated over many years. However, the percentages given for the intervals in Table 2.8 provide remarkably good approximations even when the distribution of the data is slightly skewed or asymmetric.

EXAMPLE 2.10

Thirty students in an experimental psychology class use various techniques to train a rat to move through a maze. At the end of the course, each student's rat is timed through the maze, with the following results (in minutes):

TABLE 2.8   An Aid to Interpretation of a Standard Deviation: Chebyshev's Rule

Chebyshev's Rule applies to any data set, regardless of the shape of the frequency distribution of the data.

a. No useful information is provided on the fraction of measurements that fall within 1 standard deviation of the mean, i.e., within the interval $(\bar{x} - s, \bar{x} + s)$ for samples and $(\mu - \sigma, \mu + \sigma)$ for populations.

b. At least ¾ will fall within 2 standard deviations of the mean, i.e., within the interval $(x - 2s, \bar{x} + 2s)$ for samples and $(\mu - 2\sigma, \mu + 2\sigma)$ for populations.

c. At least ⅛ of the measurements will fall within 3 standard deviations of the mean, i.e., within the interval $(\bar{x} - 3s, \bar{x} + 3s)$ for samples and $(\mu - 3\sigma, \mu + 3\sigma)$ for populations.

d. Generally, for any number k greater than 1, at least $1 - 1/k^2$ of the measurements will fall within k standard deviations of the mean, i.e., within the interval $(\bar{x} - ks, \bar{x} + ks)$ for samples and $(\mu - k\sigma, \mu + k\sigma)$ for populations.

TABLE 2.9 An Aid to Interpretation of a Standard Deviation:
The Empirical Rule

The **Empirical Rule** is a rule of thumb that applies to data sets with frequency distributions that are mound-shaped and symmetric, as shown below.

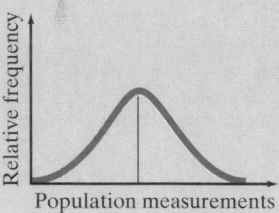

a. Approximately 68% of the measurements will fall within 1 standard deviation of the mean, i.e., within the interval $(\bar{x} - s, \bar{x} + s)$ for samples and $(\mu - \sigma, \mu + \sigma)$ for populations.
b. Approximately 95% of the measurements will fall within 2 standard deviations of the mean, i.e., within the interval $(\bar{x} - 2s, \bar{x} + 2s)$ for samples and $(\mu - 2\sigma, \mu + 2\sigma)$ for populations.
c. Approximately 99.7% (essentially all) the measurements will fall within 3 standard deviations of the mean, i.e., within the interval $(\bar{x} - 3s, \bar{x} + 3s)$ for samples and $(\mu - 3\sigma, \mu + 3\sigma)$ for populations.

1.97	.60	4.02	3.20	1.15	6.06	4.44	2.02	3.37	3.65
1.74	2.75	3.81	9.70	8.29	5.63	5.21	4.55	7.60	3.16
3.77	5.36	1.06	1.71	2.47	4.25	1.93	5.15	2.06	1.65

The mean and standard deviation of these sample data are 3.74 minutes and 2.20 minutes, respectively. Calculate the fraction of the 30 measurements in the intervals $\bar{x} \pm s$, $\bar{x} \pm 2s$, and $\bar{x} \pm 3s$, and compare the results with those predicted in Tables 2.8 and 2.9.

Solution

We first form the interval

$$(\bar{x} - s, \bar{x} + s) = (3.74 - 2.20, 3.74 + 2.20) = (1.54, 5.94)$$

A check of the measurements shows that 23 of the times are within this 1 standard deviation interval around the mean. This number represents $^{23}\!/_{30}$, or $\approx 77\%$ of the sample measurements.

The next interval of interest is

$$(\bar{x} - 2s, \bar{x} + 2s) = (3.74 - 4.40, 3.74 + 4.40) = (-.66, 8.14)$$

All but two of the times are within this interval, so $^{28}\!/_{30}$, or approximately 93%, are within 2 standard deviations of $\bar{x}$.

Finally, the 3 standard deviation interval around $\bar{x}$ is

$$(\bar{x} - 3s, \bar{x} + 3s) = (3.74 - 6.60, 3.74 + 6.60) = (-2.86, 10.34)$$

All of the times fall within 3 standard deviations of the mean.

These 1, 2, and 3 standard deviation percentages (77%, 93%, and 100%) agree fairly well with the approximations of 68%, 95%, and 100% given by the Empirical Rule (Table 2.9) for mound-shaped distributions. If you look at the frequency histogram for this data set in Figure 2.19, you will note that the distribution is not really mound-shaped, nor is it extremely skewed. Thus, we get

FIGURE 2.19

Histogram for times for
rats to move through
the maze

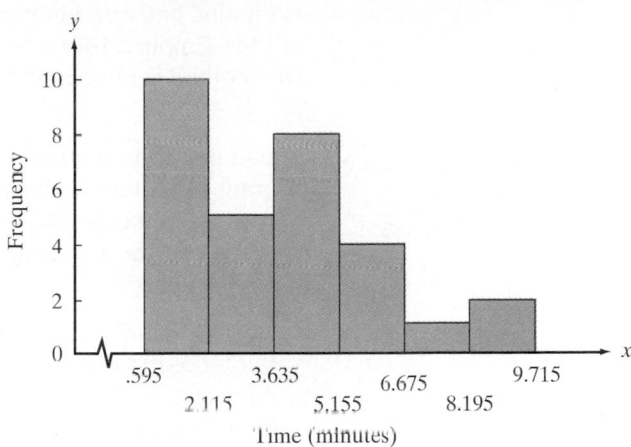

reasonably good results from the mound-shaped approximations. Of course, we
know from Chebyshev's Rule (Table 2.8) that no matter what the shape of the
distribution, we would expect at least 75% and 89% of the measurements to lie
within 2 and 3 standard deviations of $\bar{x}$, respectively.   ▲

EXAMPLE 2.11

Chebyshev's Rule and the Empirical Rule (Tables 2.8 and 2.9) are useful as a check
on the calculation of the standard deviation. For example, suppose we calculated
the standard deviation for the gas mileage data (Table 2.3) to be 5.85. Are there any
"clues" in the data that enable us to judge whether this number is reasonable?

Solution

The range of the mileage data in Table 2.3 is $44.9 - 30.0 = 14.9$. From
Chebyshev's Rule and the Empirical Rule we know that most of the measurements (approximately 95% if the distribution is mound-shaped) will be within 2
standard deviations of the mean. And, regardless of the shape of the distribution
and the number of measurements, almost all of them will fall within 3 standard
deviations of the mean. Consequently, we would expect the range of the measurements to be between 4 (i.e., $\pm 2s$) and 6 (i.e., $\pm 3s$) standard deviations in
length (see Figure 2.20). For the car mileage data, this means that s should fall
between

$$\frac{\text{Range}}{6} = \frac{14.9}{6} = 2.48 \qquad \text{and} \qquad \frac{\text{Range}}{4} = \frac{14.9}{4} = 3.73$$

In particular, the standard deviation should not be much larger than ¼ of the
range, particularly for the data set with 100 measurements. Thus, we have reason
to believe that the calculation of 5.85 is too large. A check of our work reveals
that 5.85 is the variance s^2, not the standard deviation s (see Example 2.9). We
"forgot" to take the square root (a common error); the correct value is $s = 2.42$.
Note that this value is slightly smaller than the range divided by 6 (2.48). The
larger the data set, the greater the tendency for very large or very small measurements (extreme values) to appear, and when they do, the range may exceed 6
standard deviations.   ▲

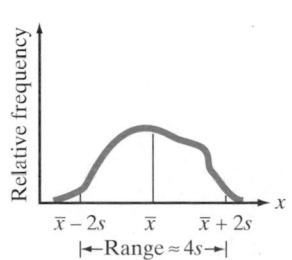

FIGURE 2.20

The relation between the
range and the standard
deviation

In examples and exercises we'll sometimes use $s \approx$ range/4 to obtain a crude,
and usually conservatively large, approximation for s. However, we stress that this
is no substitute for calculating the exact value of s when possible.

Finally, and most importantly, we will use the concepts in Chebyshev's Rule and the Empirical Rule to build the foundation for statistical inference-making. The method is illustrated in Example 2.12.

EXAMPLE 2.12

A manufacturer of automobile batteries claims that the average length of life for its grade A battery is 60 months. However, the guarantee on this brand is for just 36 months. Suppose the standard deviation of the life length is known to be 10 months, and the frequency distribution of the life-length data is known to be mound-shaped.

a. Approximately what percentage of the manufacturer's grade A batteries will last more than 50 months, assuming the manufacturer's claim is true?
b. Approximately what percentage of the manufacturer's batteries will last less than 40 months, assuming the manufacturer's claim is true?
c. Suppose your battery lasts 37 months. What could you infer about the manufacturer's claim?

Solution

If the distribution of life length is assumed to be mound-shaped with a mean of 60 months and a standard deviation of 10 months, it would appear as shown in figure 2.21. Note that we can take advantage of the fact that mound-shaped distributions are (approximately) symmetric about the mean, so that the percentages given by the Empirical Rule can be split equally between the halves of the distribution on each side of the mean. The approximations given in Figure 2.21 are more dependent on the assumption of a mound-shaped distribution than those given by the Empirical Rule (Table 2.9), because the approximations in Figure 2.21 depend on the (approximate) symmetry of the mound-shaped distribution. We saw in Example 2.10 that the Empirical Rule can yield good approximations even for skewed distributions. This will *not* be true of the approximations in Figure 2.21; the distribution *must* be mound-shaped and (approximately) symmetric.

For example, since approximately 68% of the measurements will fall within 1 standard deviation of the mean, the distribution's symmetry implies that approximately 1/2(68%) = 34% of the measurements will fall between the mean and 1 standard deviation on each side. This concept is illustrated in Figure 2.21. The figure also shows that 2.5% of the measurements lie beyond 2 standard deviations in each direction from the mean. This result follows from the fact that if approximately 95% of the measurements fall within 2 standard deviations of the mean, then about 5% fall outside 2 standard deviations; if the distribution is approximately symmetric, then about 2.5% of the measurements fall beyond 2 standard deviations on each side of the mean.

FIGURE 2.21

Battery life-length distribution: manufacturer's claim assumed true

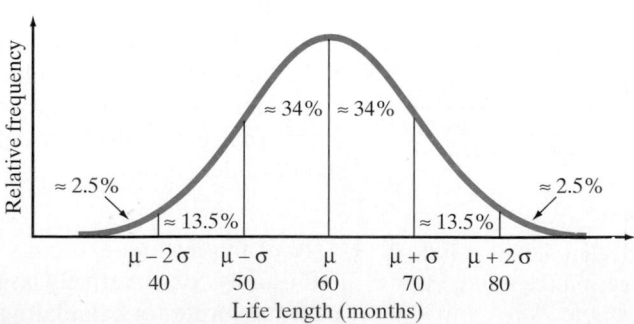

a. It is easy to see in Figure 2.21 that the percentage of batteries lasting more than 50 months is approximately 34% (between 50 and 60 months) plus 50% (greater than 60 months). Thus, approximately 84% of the batteries should have life length exceeding 50 months.

b. The percentage of batteries that last less than 40 months can also be easily determined from Figure 2.21. Approximately 2.5% of the batteries should fail prior to 40 months, assuming the manufacturer's claim is true.

c. If you are so unfortunate that your grade A battery fails at 37 months, you can make one of two inferences: Either your battery was one of the approximately 2.5% that fail prior to 40 months, or something about the manufacturer's claim is not true. Because the chances are so small that a battery fails before 40 months, you would have good reason to have serious doubts about the manufacturer's claim. A mean smaller than 60 months and/or a standard deviation longer than 10 months would both increase the likelihood of failure prior to 40 months.* ▲

Example 2.12 is our initial demonstration of the statistical inference-making process. At this point you should realize that we'll use sample information (in Example 2.12, your battery's failure at 37 months) to make inferences about the population (in Example 2.12, the manufacturer's claims about the life length for the population of all batteries). We'll build on this foundation as we proceed.

*The assumption that the distribution is mound-shaped and symmetric may also be incorrect. However, if the distribution were skewed to the right, as life-length distributions often tend to be, the percentage of measurements more than 2 standard deviations *below* the mean would be even less than 2.5%.

EXERCISES 2.58–2.72

Learning the Mechanics

2.58 To what kind of data sets can Chebyshev's Rule be applied? The Empirical Rule?

2.59 The output from a statistical computer program indicates that the mean and standard deviation of a data set consisting of 200 measurements are $1,500 and $300, respectively.
   a. What are the units of measurement of the variable of interest? Based on the units, what type of data is this: quantitative or qualitative?
   b. What can be said about the number of measurements between $900 and $2,100? Between $600 and $2,400? Between $1,200 and $1,800? Between $1,500 and $2,100?

2.60 For any set of data, what can be said about the percentage of the measurements contained in each of the following intervals?
   a. $\bar{x} - s$ to $\bar{x} + s$ b. $\bar{x} - 2s$ to $\bar{x} + 2s$
   c. $\bar{x} - 3s$ to $x + 3s$

2.61 For a set of data with a mound-shaped relative frequency distribution, what can be said about the percentage of the measurements contained in each of the intervals specified in Exercise 2.60?

2.62 The following is a sample of 25 measurements:

```
7   6   6   11   8   9   11   9   10   8   7   7   5
9   10   7   7   7   7   9   12   10   10   8   6
```

   a. Compute $\bar{x}$, s^2, and s for this sample.
   b. Count the number of measurements in the intervals $\bar{x} \pm s$, $\bar{x} \pm 2s$, $\bar{x} \pm 3s$. Express each count as a percentage of the total number of measurements.
   c. Compare the percentages found in part **b** to the percentages given by the Empirical Rule and Chebyshev's Rule.
   d. Calculate the range and use it to obtain a rough approximation for s. Does the result compare favorably with the actual value for s found in part **a**?

2.63 Given a data set with a largest value of 760 and a smallest value of 135, what would you estimate the standard deviation to be? Explain the logic behind the procedure you used to estimate the standard deviation. Suppose the standard deviation is reported to be 25. Is this feasible? Explain.

Applying the Concepts

2.64 Refer to the *Marine Technology* (Jan. 1995) data on spillage amounts (in thousands of metric tons) for 50 major oil spills, Exercise 2.10. An SPSS histogram for the 50 spillage amounts is shown below.
   a. Interpret the histogram.
   b. Descriptive statistics for the 50 spillage amounts are provided in the SPSS printout. Use this information to form an interval that can be used to predict the spillage amount for the next major oil spill.

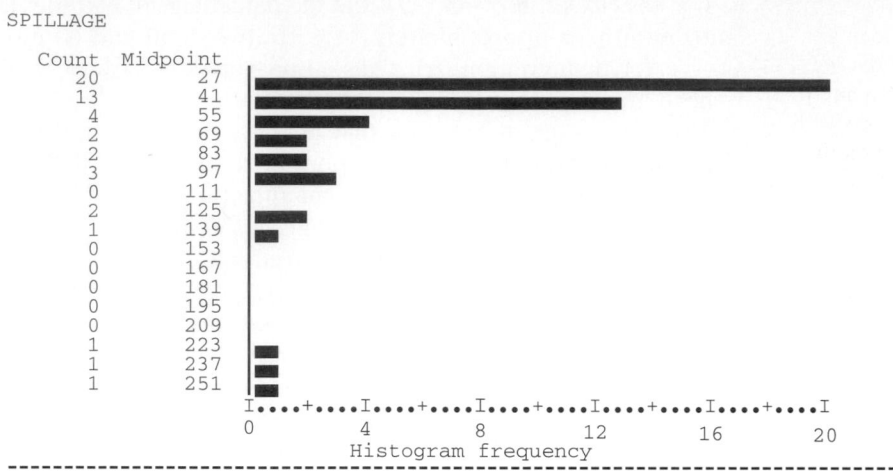

```
SPILLAGE
   Count  Midpoint
      20      27
      13      41
       4      55
       2      69
       2      83
       3      97
       0     111
       2     125
       1     139
       0     153
       0     167
       0     181
       0     195
       0     209
       1     223
       1     237
       1     251
          I....+....I....+....I....+....I....+....I....+....I....+....I
          0         4         8        12        16        20
                         Histogram frequency
```

```
Variable  SPILLAGE

Mean          59.820        S.E. Mean       7.546
Std Dev       53.362        Variance     2847.457
Kurtosis       6.209        S.E. Kurt        .662
Skewness       2.510        S.E. Skew        .337
Range        236.000        Minimum        21.00
Maximum      257.00         Sum          2991.000

Valid Observations -   50      Missing Observations -   0
```

2.65 Refer to the *American Psychologist* (July 1995) study of 148 applicants for a position in experimental psychology, Exercise 2.39. Recall that the mean number of recommendation letters included in each application packet was $\bar{x} = 2.28$. The standard deviation was also reported in the article; its value was $s = 1.48$.

a. Sketch the relative frequency distribution for the number of recommendation letters included in each application for the experimental psychology position.

b. Locate an interval on the distribution, part **a**, that captures approximately 95% of the measurements in the sample.

2.66 Refer to Exercise 2.20, in which we compared the 50 states' average SAT scores in 1975 and 1990. The data are repeated in the table at right. The mean and standard deviation of these data are

1975 SAT Scores	1990 SAT Scores
$\bar{x} = 955.3$	$\bar{x} = 947.5$
$s = 69.5$	$s = 64.0$

Calculate the percentages of measurements in the intervals $\bar{x} \pm s$, $\bar{x} \pm 2s$, and $\bar{x} \pm 3s$ for each data set (1975 and 1990). Check the agreement of these percentages with both Chebyshev's Rule and the Empirical Rule.

2.67 Professor Jon K. Mills of the Illinois School of Professional Psychology measured the Locus of Control (LOC), a measure of one's perception of

State	1975	1990	State	1975	1990
Ala.	883	984	Mont.	1047	987
Alaska	942	914	Nebr.	966	1030
Ariz.	1021	942	Nev.	962	921
Ark.	992	981	N.H.	934	928
Calif.	908	903	N.J.	878	891
Colo.	994	969	N. Mex.	1002	1007
Conn.	913	901	N.Y.	925	882
Del.	915	903	N.C.	827	841
Fla.	915	884	N. Dak.	1064	1069
Ga.	824	844	Ohio	955	949
Hawaii	892	885	Okla.	994	1001
Idaho	1017	968	Oreg.	908	923
Ill.	970	994	Pa.	900	883
Ind.	881	867	R.I.	901	883
Iowa	1091	1088	S.C.	794	834
Kans.	1043	1040	S. Dak.	1084	1061
Ky.	977	994	Tenn.	988	1008
La.	947	993	Tex.	898	874
Maine	908	886	Utah	1069	1031
Md.	907	908	Vt.	915	897
Mass.	903	900	Va.	894	895
Mich.	949	968	Wash.	1011	923
Minn.	1058	1019	W. Va.	964	933
Miss.	980	996	Wis.	1036	1019
Mo.	965	995	Wyo.	1054	977

control over factors affecting one's life, for two groups of individuals undergoing weight-reduction treatment for obesity (*Journal of Psychology,* Mar. 1991). The mean and standard deviation for a sample of 46 adults were 6.45 and 2.89, respectively, while the mean and standard deviation for a sample of 19 adolescents were 10.89 and 2.48, respectively. A lower score on the LOC scale indicates a perception that internal factors are in control, while a higher score indicates a perception that external factors are in control.

a. Calculate the 1- and 2-standard–deviation intervals around the means for each group. Plot these intervals on a line graph using different colors or symbols to represent each group.

b. Assuming that the distributions of LOC scores are approximately mound-shaped, estimate the numbers of individuals within each interval.

c. Based on your answers to parts **a** and **b**, do you think an inference can be made that *all* adults and adolescents undergoing weight-reduction treatment differ with respect to LOC? What factors did you consider in making this inference? [We'll reconsider this exercise in Chapter 9 to show how to measure the reliability of this inference.]

2.68 The *American Rifleman* (June 1993) reported on the velocity of ammunition fired from the FEG P9R pistol, a new 9mm gun manufactured in Hungary. Field tests revealed that Winchester bullets fired from the pistol had a mean velocity (at 15 feet) of 936 feet per second and a standard deviation of 10 feet per second. Tests were also conducted with Uzi and Black Hills ammunition.

a. Describe the velocity distribution of Winchester bullets fired from the FEG P9R pistol.

b. A bullet, brand unknown, is fired from the FEG P9R pistol. Suppose the velocity (at 15 feet) of the bullet is 1,000 feet per second. Is the bullet likely to be manufactured by Winchester? Explain.

2.69 For each day of last year, the number of vehicles passing through a certain intersection was recorded by a city engineer. One objective of this study was to determine the percentage of days that more than 425 vehicles used the intersection. Suppose the mean for the data was 375 vehicles per day and the standard deviation was 25 vehicles.

a. What can you say about the percentage of days that more than 425 vehicles used the intersection? Assume you know nothing about the shape of the relative frequency distribution for the data.

b. What is your answer to part **a** if you know that the relative frequency distribution for the data is mound-shaped?

2.70 A buyer for a lumber company must decide whether to buy a piece of land containing 5,000 pine trees. If 1,000 of the trees are at least 40 feet tall, the buyer will purchase the land; otherwise, he won't. The owner of the land reports that the height of the trees has a mean of 30 feet and a standard deviation of 3 feet. Based on this information, what is the buyer's decision?

2.71 A chemical company produces a substance composed of 98% cracked corn particles and 2% zinc phosphide for use in controlling rat populations in sugarcane fields. Production must be carefully controlled to maintain the 2% zinc phosphide because too much zinc phosphide will cause damage to the sugarcane and too little will be ineffective in controlling the rat population. Records from past production indicate that the distribution of the actual percentage of zinc phosphide present in the substance is approximately mound-shaped, with a mean of 2.0% and a standard deviation of .08%.

a. If the production line is operating correctly, approximately what proportion of batches from a day's production will contain less than 1.84% of zinc phosphide?

b. Suppose one batch chosen randomly actually contains 1.80% zinc phosphide. Does this indicate that there is too little zinc phosphide in that day's production? Explain your reasoning.

2.72 Astronomers theorize that cold dark matter (CDM) caused the formation of galaxies and clusters of galaxies in the universe. The theoretical CDM model requires an estimate of the velocity of light emitted from the galaxy cluster. The *Astronomical Journal* (July 1995) published a study of observed velocities for galaxies in four different galaxy clusters. Galaxy velocity was measured in kilometers per second (km/s) using a spectrograph and a high-power telescope.

a. The observed velocities of 103 galaxies located in the cluster named A2142 are summarized in the accompanying histogram. Comment on whether the Empirical Rule is applicable for describing the velocity distribution for this cluster.

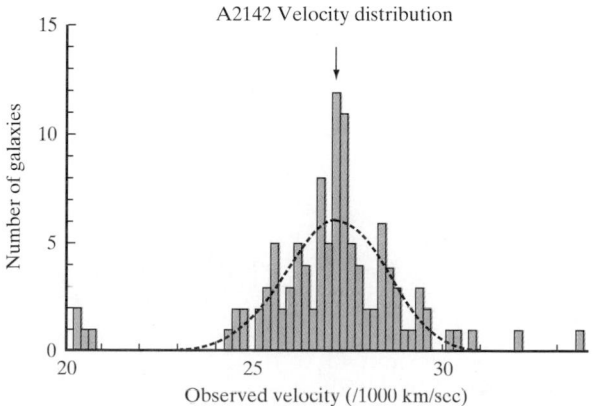

Source: Oegerle, W. R., Hill, J. M., and Fitchett, M. J. "Observations of high dispersion clusters of galaxies: Constraints on cold dark matter." *The Astronomical Journal,* Vol. 110, No. 1, July 1995, p. 37 (Figure 1).

b. The mean and standard deviation of the 103 velocities observed in galaxy cluster A2142 were reported as $\bar{x} = 27{,}117$ km/s and $s = 1{,}280$ km/s, respectively. Use this information to construct an interval that captures approximately 95% of the galaxy velocities in the cluster.

c. Recommend a single velocity value to be used in the CDM model for galaxy cluster A2142. Explain your reasoning.

2.7 NUMERICAL MEASURES OF RELATIVE STANDING

We've seen that numerical measures of central tendency and variability describe the general nature of a quantitative data set (either a sample or a population). In addition, we may be interested in describing the *relative* quantitative location of a particular measurement within a data set. Descriptive measures of the relationship of a measurement to the rest of the data are called **measures of relative standing**.

One measure of the relative standing of a measurement is its **percentile ranking**. For example, suppose you scored an 80 on a test and you want to know how you fared in comparison with others in your class. If the instructor tells you that you scored at the 90th percentile, it means that 90% of the grades were lower than yours and 10% were higher. Thus, if the scores were described by the relative frequency histogram in Figure 2.22, the 90th percentile would be located at a point such that 90% of the total area under the relative frequency histogram lies below the 90th percentile and 10% lies above. If the instructor tells you that you scored in the 50th percentile (the median of the data set), 50% of the test grades would be lower than yours and 50% would be higher.

Percentile rankings are of practical value only for large data sets. Finding them involves a process similar to the one used in finding a median. The measurements are ranked in order and a rule is selected to define the location of each percentile. Since we are primarily interested in interpreting the percentile rankings of measurements (rather than finding particular percentiles for a data set), we define the *pth percentile* of a data set as shown in Definition 2.10.

> **DEFINITION 2.10**
>
> For any set of n measurements (arranged in ascending or descending order), the ***p*th percentile** is a number such that $p\%$ of the measurements fall below the pth percentile and $(100 - p)\%$ fall above it.

FIGURE 2.22

Location of 90th percentile for test grades

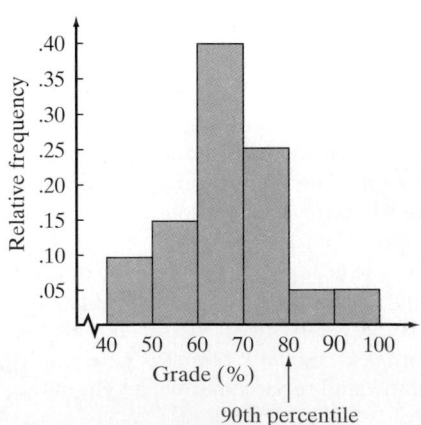

Another measure of relative standing in popular use is the *z-score*. As you can see in Definition 2.11, the *z*-score makes use of the mean and standard deviation of the data set in order to specify the relative location of measurement:

> **DEFINITION 2.11**
>
> The **sample *z*-score** for a measurement *x* is
>
> $$z = \frac{x - \bar{x}}{s}$$
>
> The **population *z*-score** for a measurement *x* is
>
> $$z = \frac{x - \mu}{\sigma}$$

Note that the *z*-score is calculated by subtracting *x* (or μ) from the measurement *x* and then dividing the result by *s* (or σ). The final result, the *z*-score, represents the distance between a given measurement *x* and the mean, expressed in standard deviations.

EXAMPLE 2.13

Suppose a sample of 2,000 high school seniors' verbal SAT scores is selected. The mean and standard deviation are

$$\bar{x} = 550 \qquad s = 75$$

Suppose Joe Smith's score is 475. What is his sample *z*-score?

Solution

You can see that Joe Smith's score lies below the mean score of the 2,000 seniors:

325	475	550	775
$\bar{x} - 3s$	Joe Smith's score	$\bar{x}$	$\bar{x} + 3s$

We compute

$$z = \frac{x - \bar{x}}{s} = \frac{475 - 550}{75} = -1.0$$

which tells us that Joe Smith's score is 1.0 standard deviation *below* the sample mean; in short, his sample *z*-score is −1.0.   ▲

The numerical value of the *z*-score reflects the relative standing of the measurement. A large positive *z*-score implies that the measurement is larger than almost all other measurements, whereas a large negative *z*-score indicates that the measurement is smaller than almost every other measurement. If a *z*-score is 0 or near 0, the measurement is located at or near the mean of the sample or population.

We can be more specific if we know that the frequency distribution of the measurements is mound-shaped. In this case, the following interpretation of the *z*-score can be given:

Interpretation of *z*-Scores for Mound-Shaped Distributions of Data

1. Approximately 68% of the measurements will have a *z*-score between −1 and 1.
2. Approximately 95% of the measurements will have a *z*-score between −2 and 2.
3. Approximately 99.7% (almost all) the measurements will have a *z*-score between −3 and 3.

Note that this interpretation of z-scores is identical to that given by the Empirical Rule for mound-shaped distributions. The statement that a measurement falls in the interval $(\mu - \sigma)$ to $(\mu + \sigma)$ is equivalent to the statement that a measurement has a population z-score between -1 and 1, since all measurements between $(\mu - \sigma)$ and $(\mu + \sigma)$ are within 1 standard deviation of μ.

We end this section with an example that indicates how z-scores may be used to accomplish our primary objective—the use of sample information to make inferences about the population.

EXAMPLE 2.14

Suppose a female bank employee believes that her salary is low as a result of sex discrimination. To substantiate her belief, she collects information on the salaries of her male counterparts in the banking business. She finds that their salaries have a mean of $34,000 and a standard deviation of $2,000. Her salary is $27,000. Does this information support her claim of sex discrimination?

Solution

The analysis might proceed as follows: First, we calculate the z-score for the woman's salary with respect to those of her male counterparts. Thus,

$$z = \frac{\$27{,}000 - \$34{,}000}{\$2{,}000} = -3.5$$

The implication is that the woman's salary is 3.5 standard deviations *below* the mean of the male salary distribution. Furthermore, if a check of the male salary data shows that the frequency distribution is mound-shaped, we can infer that very few salaries in this distribution should have a z-score less than -3, as shown in Figure 2.23. Therefore, a z-score of -3.5 represents either a measurement from a distribution different from the male salary distribution or a very unusual (highly improbable) measurement for the male salary distribution.

Which of the two situations do you think prevails? Do you think the woman's salary is simply unusually low in the distribution of salaries, or do you think her claim of sex discrimination is justified? Most people would probably conclude that her salary does not come from the male salary distribution. However, the careful investigator should require more information before inferring sex discrimination as the cause. We would want to know more about the data collection technique the woman used and more about her competence at her job. Also, perhaps other factors such as length of employment should be considered in the analysis.   ▲

Examples 2.12 and 2.14 exemplify an approach to statistical inference that might be called the **rare-event approach**. An experimenter hypothesizes a specific frequency distribution to describe a population of measurements. Then a sample of measurements is drawn from the population. If the experimenter finds

FIGURE 2.23

Male salary distribution

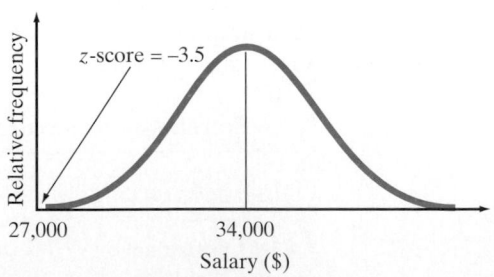

Computer Phobia and Secondary Technical Education Teachers

CASE STUDY 2.2

Prior to, or during, your initial experience with a computer, did you develop computer anxiety? Psychologists define computer anxiety as "the mixture of fear, apprehension, and hope that people feel when planning to interact, or when interacting with a computer." Researchers have found computer anxiety in people at all levels of society, including students, doctors, lawyers, secretaries, managers, and college professors. One profession for which little is known about the level and impact of computer anxiety is secondary technical education (STE), since STE teachers have just recently begun to participate in the hi-tech computer revolution.

To determine the extent of computer anxiety among secondary technical education teachers, Professor Howard Gordon of Marshall University surveyed teachers from West Virginia vo-tech centers during the 1992–1993 school year (*Journal of Studies in Technical Careers,* Vol. 15, 1995). Eight vo-tech centers were randomly selected, and questionnaires were mailed to all 127 STE teachers at the centers. A total of 116 questionnaires were returned, for a response rate of 91%.

Several variables were measured based on the questionnaire responses:

1. Level of computer anxiety, measured by the Computer Anxiety Scale (COMPAS). Scores, ranging from 10 to 50, were categorized as follows: very anxious (37–50); anxious/tense (33–36); some mild anxiety (27–32); generally relaxed/comfortable (20–26); very relaxed/confident (10–19)
2. Sex (male or female)
3. Computer usage (number of hours per week)
4. Computer skill level (novice, intermediate, or expert)
5. Availability of computer in a central office (yes or no)
6. Type of formal computer training completed (in-service workshop, outside training, or none)
7. Amount of formal computer training (total number of hours)
8. Perception of local administrative support of computer use (5-point scale, where 1 = strongly disagree and 5 = strongly agree)
9. Perception of state administrative support of computer use (5-point scale, where 1 = strongly disagree and 5 = strongly agree)

The study had multiple objectives, several of which were descriptive in nature. For this case study, we focus on the following three major objectives.

Objective 1   Summarize the computer anxiety levels of STE teachers in West Virginia.

Objective 2   Compare the computer anxiety levels of male and female STE teachers in West Virginia.

Objective 3   Describe the computer usage, computer skill level, central office computer availability, computer training, and perception of administrative support for STE teachers in West Virginia.

Focus

a. Identify the type, qualitative or quantitative, of each variable in the study.
b. Identify the variables that are relevant to each objective.
c. Some numerical descriptive measures, extracted from the journal article, are provided in Tables 2.10–2.13. [*Source:* Gordon, H. R. D. "Analysis of the computer anxiety levels of secondary technical education teachers in West Virginia." *Journal of Studies in Technical Careers,* Vol. 15, No. 2, 1995, pp. 26–27 (Tables 1, 2, and 3).] Use this information to answer each objective. Whenever possible, support your answer with a graph.

TABLE 2.10   Summary of COMPAS Categories (Case Study 2.2)

Category	Score Range	Frequency	Relative Frequency
Very anxious	37–50	22	.19
Anxious/tense	33–36	8	.07
Some mild anxiety	27–32	23	.20
Generally relaxed/comfortable	20–26	24	.21
Very relaxed/confident	10–19	39	.33
Totals		116	1.00

TABLE 2.11   Comparison of COMPAS Scores by Sex (Case Study 2.2)

	Male Teachers	Female Teachers	All Teachers
n	68	48	116
$\bar{x}$	26.4	24.5	25.6
s	10.6	11.2	10.8

TABLE 2.12   Administrative Support of Computer Use (Case Study 2.2)

	Local Administration	State Administration
n	116	116
$\bar{x}$	3.75	3.63
s	1.19	1.22

TABLE 2.13   Summary of Miscellaneous Variables (Case Study 2.2)

Computer usage: 80% have used; $\bar{x} = 7$ hours/week
Computer skill level: 47% novices
Access to computer in central office: 86% yes
Type of computer training: 5% in-service, 10% outside, 25% none
Amount of formal training: 33% between 25 and 60 hours

it unlikely that the sample came from the hypothesized distribution, the hypothesis is concluded to be false. Thus, in Example 2.14 the woman believes her salary reflects sex discrimination. She hypothesizes that her salary should be just another measurement in the distribution of her male counterparts' salaries if no discrimination exists. However, it is so unlikely that the sample (in this case, her salary) came from the male frequency distribution that she rejects that hypothesis, concluding that the distribution from which her salary was drawn is different from the distribution for the men.

This rare-event approach to inference-making is discussed further in later chapters. Proper application of the approach requires a knowledge of probability, the subject of our next chapter.

EXERCISES 2.73–2.82

Learning the Mechanics

2.73 Compute the z-score corresponding to each of the following values of x:

a. $x = 40$, $s = 5$, $\bar{x} = 30$
b. $x = 90$, $\mu = 89$, $\sigma = 2$
c. $\mu = 50$, $\sigma = 5$, $x = 50$
d. $s = 4$, $x = 20$, $\bar{x} = 30$
e. In parts **a–d**, state whether the z-score locates x within a sample or a population.
f. In parts **a–d**, state whether each value of x lies above or below the mean and by how many standard deviations.

2.74 Give the percentage of measurements in a data set that are above and below each of the following percentiles:

a. 75th percentile   **b.** 50th percentile
c. 20th percentile   **d.** 84th percentile

2.75 What is the 50th percentile of a quantitative data set called?

2.76 Compare the z-scores to decide which of the following x values lie the greatest distance above the mean and the greatest distance below the mean.

a. $x = 100$, $\mu = 50$, $\sigma = 25$
b. $x = 1$, $\mu = 4$, $\sigma = 1$

c. $x = 0$, $\mu = 200$, $\sigma = 100$
d. $x = 10$, $\mu = 5$, $\sigma = 3$

2.77 At one university, the students are given z-scores at the end of each semester rather than the traditional GPAs. The mean and standard deviation of all students' cumulative GPAs, on which the z-scores are based, are 2.7 and .5, respectively.

a. Translate each of the following z-scores to corresponding GPA: $z = 2.0$, $z = -1.0$, $z = .5$, $z = -2.5$.

b. Students with z-scores below -1.6 are put on probation. What is the corresponding probationary GPA?

c. The president of the university wishes to graduate the top 16% of the students with *cum laude* honors and the top 2.5% with *summa cum laude* honors. Where (approximately) should the limits be set in terms of z-scores? In terms of GPAs? What assumption, if any, did you make about the distribution of the GPAs at the university?

2.78 Suppose that 40 and 90 are two elements of a population data set and that their z-scores are -2 and 3, respectively. Using only this information, is it possible to determine the population's mean and standard deviation? If so, find them. If not, explain why it's not possible.

Applying the Concepts

2.79 The distribution of scores on a nationally administered college achievement test has a median of 520 and a mean of 540.

a. Explain why it is possible for the mean to exceed the median for this distribution of measurements.

b. Suppose you are told that the 90th percentile is 660. What does this mean?

c. Suppose you are told that you scored at the 94th percentile. Interpret this statement.

2.80 The U.S. Environmental Protection Agency (EPA) sets a limit on the amount of lead permitted in drinking water. The EPA *Action Level* for lead is .015 milligrams per liter (mg/L) of water. Under EPA guidelines, if 90% of a water system's study samples have a lead concentration less than .015 mg/L, the water is considered safe for drinking. I (co-author Sinich) received a 1994 report on a study of lead levels in the drinking water of homes in my subdivision. The 90th percentile of the study sample had a lead concentration of .00372 mg/L. Are water customers in my subdivision at risk of drinking water with unhealthful lead levels? Explain.

2.81 A city librarian claims that books have been checked out an average of 7 (or more) times in the last year. You suspect he has exaggerated the checkout rate (book usage) and that the mean number of checkouts per book per year is, in fact, less than 7. Using the computerized card catalog, you randomly

select one book and find that it has been checked out 4 times in the last year. Assume that the standard deviation of the number of checkouts per book per year is approximately 1.

a. If the mean number of checkouts per book per year really is 7, what is the z-score corresponding to 4?

b. Considering your answer to part **a**, do you have reason to believe that the librarian's claim is incorrect?

c. If you knew that the distribution of the number of checkouts was mound-shaped, would your answer to part **b** change? Explain.

d. If the standard deviation of the number of checkouts per book per year were 2 (instead of 1), would your answers to parts **b** and **c** change? Explain.

2.82 Refer to the *Optometry and Vision Science* (June 1995) study of refractive variation in eyes, Exercise 2.37. The SAS descriptive statistics printout for the data set consisting of 25 cylinder power measurements is reproduced below.

UNIVARIATE PROCEDURE

Variable=CYLPOWER

Moments

N	25	Sum Wgts	25
Mean	-0.1544	Sum	-3.86
Std Dev	0.196767	Variance	0.038717
Skewness	-4.52208	Kurtosis	21.70196
USS	1.5252	CSS	0.929216
CV	-127.44	Std Mean	0.039353
T:Mean=0	-3.92342	Prob>|T|	0.0006
Sgn Rank	-162.5	Prob>|S|	0.0001
Num ^= 0	25		

Quantiles(Def=5)

100% Max	-0.04		99%	-0.04
75% Q3	-0.08		95%	-0.06
50% Med	-0.11		90%	-0.06
25% Q1	-0.16		10%	-0.2
0% Min	-1.07		5%	-0.21
			1%	-1.07
Range	1.03			
Q3-Q1	0.08			
Mode	-0.17			

Extremes

Lowest	Obs	Highest	Obs
-1.07(	3)	-0.07(	17)
0.21(	24)	-0.06(	10)
-0.2(	13)	-0.06(	22)
-0.17(	15)	-0.06(	25)
-0.17(	12)	-0.04(	6)

a. Find the 10th percentile of cylinder power measurements on the printout. Interpret the result.

b. Find the 95th percentile of cylinder power measurements on the printout. Interpret the result.

c. Use the information on the SAS printout to calculate the z-score for the cylinder power measurement of -1.07.

d. Based on your answer to part **c**, would you classify -1.07 as an extreme cylinder power measurement?

2.8   QUARTILES AND BOX PLOTS (OPTIONAL)

The **box plot**, a relatively recent introduction to the methodology of descriptive measures, is based on the **quartiles** of a data set. Quartiles are values that partition the data set into four groups, each containing 25% of the measurements. The lower quartile Q_L is the 25th percentile, the middle quartile is the median M (the 50th percentile), and the upper quartile Q_U is the 75th percentile (see Figure 2.24).

> **DEFINITION 2.12**
>
> The **lower quartile Q_L** is the 25th percentile of a data set. The **middle quartile M** is the median. The **upper quartile Q_U** is the 75th percentile.

A box plot is based on the *interquartile range (IQR),* the distance between the lower and upper quartiles:

$$IQR = Q_U - Q_L$$

> **DEFINITION 2.13**
>
> The **interquartile range (IQR)** is the distance between the lower and upper quartiles:
>
> $$IQR = Q_U - Q_L$$

The box plot for the gas mileage data (Table 2.3) is given in Figure 2.25. It was generated by the MINITAB statistical software package.* Note that a rectangle (the **box**) is drawn, with the ends of the rectangle (the **hinges**, represented by the "I's" at the ends of the box) drawn at the quartiles Q_L and Q_U. By definition, then, the "middle" 50% of the observations—those between Q_L and Q_U—fall inside the box. For the gas mileage data, these quartiles appear to be at (approximately) 35.5 and 38.5. Thus,

$$IQR = 38.5 - 35.5 = 3.0 \text{ (approximately)}$$

FIGURE 2.24

The quartiles for a data set

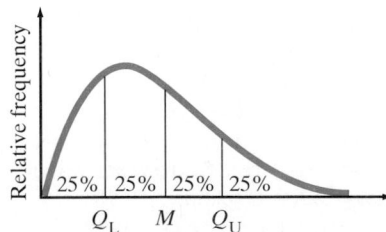

---

*Although box plots can be generated by hand, the amount of detail required makes them particularly well suited for computer generation. We use computer software to generate the box plots in this section.

FIGURE 2.25

MINITAB box plot for gas
mileage data

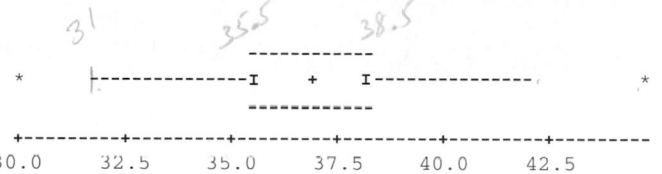

```
          31              35.5        38.5
                       ------------
  *     ┠--------------I   +   I-------------,          *
                       ------------

  +---------+---------+---------+---------+---------+---------
30.0      32.5      35.0      37.5      40.0      42.5
```

Note that the median is shown at about 37 by a plus (+) sign within the box.

   To guide the construction of the "tails" of the box plot, two sets of limits, called **inner fences** and **outer fences**, are used. Neither set of fences actually appears on the box plot. Inner fences are located at a distance of 1.5(IQR) from the hinges. Emanating from the hinges of the box are dashed lines called the **whiskers**. The two whiskers extend to the most extreme observation inside the inner fences. For example, the inner fence on the lower side of the gas mileage box plot is (approximately)

$$\text{Lower inner fence} = \text{Lower hinge} - 1.5(\text{IQR})$$
$$\approx 35.5 - 1.5(3.0)$$
$$= 35.5 - 4.5 = 31.0$$

The smallest measurement *inside* this fence is the second smallest measurement, 31.8. Thus, the lower whisker extends to 31.8. Similarly, the upper whisker extends to 42.1, the largest measurement inside the upper inner fence at about $38.5 + 4.5 = 43.0$.

   Values that are beyond the inner fences receive special attention because they are extreme values that represent relatively rare occurrences. In fact, for mound-shaped distributions, fewer than 1% of the observations are expected to fall outside the inner fences. Two of the 100 gas mileage measurements, 30.0 and 44.9, fall beyond the inner fences, one on each end of the distribution. These measurements are represented by asterisks ($*$).

   The other two imaginary fences, the outer fences, are defined at a distance 3(IQR) from each end of the box. Measurements that fall beyond the outer fences are represented by 0's and are very extreme measurements that require special analysis. Less than one-hundredth of 1% (.01% or .0001) of the measurements from mound-shaped distributions are expected to fall beyond the outer fences. Since no measurement in the gas mileage box plot (Figure 2.25) is represented by a 0, we know that no measurements fall outside the outer fences.

   Generally, any measurements that fall beyond the inner fences—and certainly any that fall beyond the outer fences—are considered potential **outliers**. Outliers are extreme measurements that stand out from the rest of the sample and may be faulty: They may be incorrectly recorded observations, members of a population different from the rest of the sample, or, at the least, very unusual measurements from the same population. For example, the two gas mileage measurements beyond the inner fences may be considered outliers. When we analyze these measurements, we find that they are correctly recorded. Perhaps they represent mileages that correspond to exceptional models of the car being tested or to unusual gas mixtures. Outlier analysis often reveals useful information of this kind and therefore plays an important role in the statistical inference-making process.

   The elements (and nomenclature) of box plots are summarized in the next box. Some aids to the interpretation of box plots are also given.

Elements of a Box Plot

1. A rectangle (the **box**) is drawn with the ends (the **hinges**) drawn at the lower and upper quartiles (Q_L and Q_U). The median of the data is shown in the box, usually by a "+" symbol.
2. The points at distances 1.5(IQR) from each hinge mark the **inner fences** of the data set. Horizontal lines (the **whiskers**) are drawn from each hinge to the most extreme measurement inside the inner fence.
3. A second pair of fences, the **outer fences**, appear at a distance of 3 interquartile ranges, 3(IQR), from the hinges. One symbol (usually "*") is used to represent measurements falling between the inner and outer fences, and another (usually "0") is used to represent measurements beyond the outer fences. Thus, outer fences are not shown unless one or more measurements lie beyond them.
4. The symbols used to represent the median and the extreme data points (those beyond the fences) will vary depending on the software you use to construct the box plot. (You may use your own symbols if you are constructing a box plot by hand.) You should consult the program's documentation to determine exactly which symbols are used.

Aids to the Interpretation of Box Plots

1. Examine the length of the box. The IQR is a measure of the sample's variability and is especially useful for the comparison of two samples (see Example 2.16).
2. Visually compare the lengths of the whiskers. If one is clearly longer, the distribution of the data is probably skewed in the direction of the longer whisker.
3. Analyze any measurements that lie beyond the fences. Fewer than 5% should fall beyond the inner fences, even for very skewed distributions. Measurements beyond the outer fences are probably outliers, with one of the following explanations:
   a. The measurement is incorrect. It may have been observed, recorded, or entered into the computer incorrectly.
   b. The measurement belongs to a population different from the population that the rest of the sample was drawn from (see Example 2.16).
   c. The measurement is correct *and* from the same population as the rest. Generally, we accept this explanation only after carefully ruling out all others.

EXAMPLE 2.15

Use a statistical software package to draw a box plot for the student loan default data, Table 2.5.

Solution

The MINITAB box plot for the student loan default rates is shown in Figure 2.26. Note that the median appears to be about 9.5, and, with the exception of a single extreme observation, the distribution appears to be symmetrically distributed between approximately 3% and 17%. The single outlier is beyond the inner fence but inside the outer fence. Examination of the data reveals that this observation corresponds to Alaska's default rate of 19.7%.   ▲

EXAMPLE 2.16

A Ph.D. student in psychology conducted a stimulus reaction experiment as a part of her dissertation research. She subjected 50 students to a threatening stimulus and 50 to a nonthreatening stimulus. The reaction times of all 100 students were recorded electronically to the nearest tenth of a second. Box plots of the two

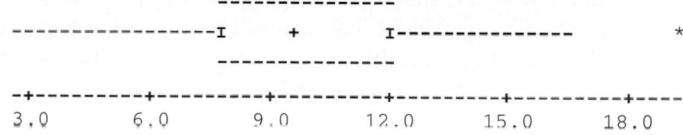

```
                                    ---------------
   ------------------I      +      I---------------          *
                                    ---------------
   -+---------+---------+---------+---------+---------+----
   3.0       6.0       9.0      12.0      15.0      18.0
```

resulting samples of reaction times, generated using SAS, are shown in Figure 2.27. Interpret the box plots.

Solution

Perhaps the first thing you notice about the two box plots is that they are arranged vertically rather than horizontally. Some statistical software packages, including SAS used here, use this arrangement. Also, note that the median is represented by a dashed line through the box. The plus (+) symbol represents the mean in the SAS box plot. Analysis of the box plots on the same numerical scale reveals that the distribution of times corresponding to the threatening stimulus lies below that of the nonthreatening stimulus. The implication is that the reaction times tend to be faster to the threatening stimulus. Note, too, that the upper whiskers of both samples are longer than the lower whiskers, indicating that the reaction times are positively skewed. The box length corresponding to the threatening stimulus is smaller than that for the nonthreatening stimulus, indicating less variability in the reaction times to the threatening stimulus.

No observations in the two samples fall between the inner and outer fences (denoted by 0 in SAS). However, note that one of the observations corresponding to the threatening stimulus is beyond the outer fence (denoted by *). When the researcher examined her notes from the experiments, she found that the subject whose time was beyond the outer fence had mistakenly been given the nonthreatening stimulus. You can see in Figure 2.27 that his time would have been within the upper whisker if moved to the box plot corresponding to the nonthreatening stimulus. The box plots should be reconstructed since they will both change slightly when this misclassified reaction time is moved from one sample to the other.

The researcher concluded that the reactions to the threatening stimulus were faster and more predictable (less variable) than those to the nonthreatening stim-

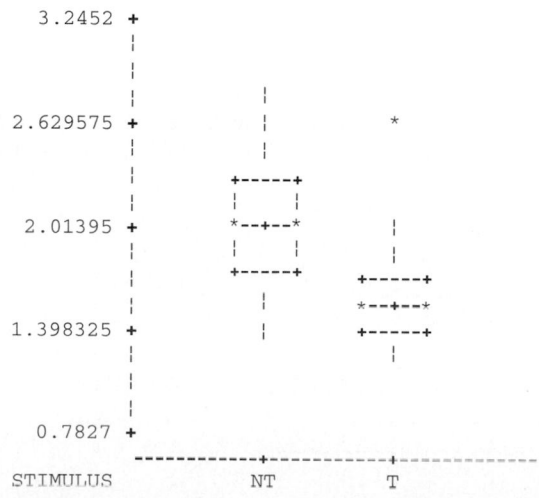

```
Variable=TIME

 3.2452 +
        |
        |
        |
        |                       |
2.629575 +                      |              *
        |                       |
        |                    +-----+
        |                    |     |
2.01395 +                    *--+--*          |
        |                    |     |           |
        |                    +-----+        +-----+
        |                       |           *--+--*
1.398325 +                      |           +-----+
        |                                     |
        |
        |
 0.7827 +
        --------------+-----------+-----------------
STIMULUS             NT           T
```

ulus. However, she was asked by her Ph.D. committee whether the results were *statistically significant.* Their question addresses the issue of whether the observed difference between the samples might be attributable to chance or sampling variation rather than to real differences between the populations. To answer their question, the researcher must use inferential statistics rather than graphical descriptions. We discuss how to compare two samples using inferential statistics in Chapter 9.   ▲

EXERCISES 2.83–2.92

Note: Exercises marked with 💾 *require the use of a computer.*

Learning the Mechanics

2.83 Define the 25th, 50th, and 75th percentiles of a data set. Explain how they provide a description of the data.

2.84 Suppose a data set consisting of exam scores has a lower quartile $Q_L = 60$, a median $M = 75$, and an upper quartile $Q_U = 85$. The scores on the exam range from 18 to 100. Without having the actual scores available to you, construct as much of the box plot as possible.

2.85 MINITAB was used to generate the following box plot:

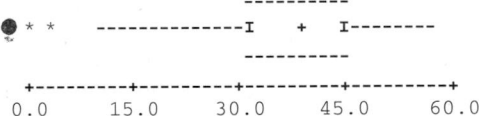

```
         ----------
● * *    --------------I  +  I--------
         ----------

+---------+---------+---------+---------+
0.0     15.0      30.0      45.0      60.0
```

a. What is the median of the data set (approximately)?

b. What are the upper and lower quartiles of the data set (approximately)?

c. What is the interquartile range of the data set (approximately)?

d. Is the data set skewed to the left, skewed to the right, or symmetric?

e. What percentage of the measurements in the data set lie to the right of the median? To the left of the upper quartile?

2.86 MINITAB was used to generate the accompanying box plots. Compare and contrast the frequency distributions of the two data sets. Your answer should

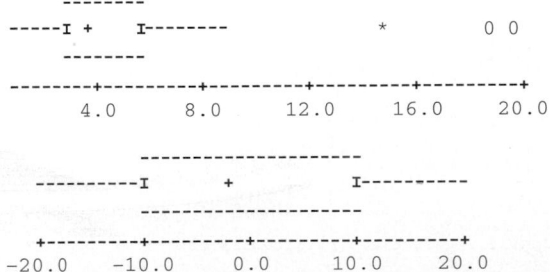

```
      --------
-----I +    I--------              *        0 0
      --------

--------+---------+---------+---------+---------+
       4.0       8.0       12.0      16.0      20.0
```

```
          --------------------
----------I     +        I----------
          --------------------

+---------+---------+---------+---------+
-20.0    -10.0      0.0      10.0      20.0
```

include comparisons of the following characteristics: central tendency, variation, skewness, and outliers.

💾 **2.87** Consider the following two sample data sets:

Sample A			Sample B		
121	171	158	171	152	170
173	184	163	168	169	171
157	85	145	190	183	185
165	172	196	140	173	206
170	159	172	172	174	169
161	187	100	199	151	180
142	166	171	167	170	188

a. Use a statistical software package to construct a box plot for each data set.

b. Using information reflected in your box plots, describe the similarities and differences in the two data sets.

c. Identify any outliers that may exist in the two data sets.

Applying the Concepts

2.88 Refer to the *American Journal of Speech–Language Pathology* (Feb. 1995) study, Exercise 2.15. Recall that three groups of ten college students listened to an audio tape of a woman with dysarthric speech. The first group (*control*) listened to the speaker's normal manner of speaking, while the second group (*treatment*) listened to the speaker while she employed a learned breathing pattern strategy. The third group (*familiarity*) also listened to the speaker while she used the learned breathing pattern strategy, but did so only after they had practiced listening to her speech on another tape. At the end of the session, each student transcribed the spoken words. Box plots, constructed using SAS, for the percentage of words correctly transcribed (i.e., accuracy rate) by each group are shown below.

a. How do the median accuracy rates compare for the three groups? [*Hint:* Recall that SAS uses a dashed line through the box to represent the median.]

b. How do the variabilities of the accuracy rates compare for the three groups?

c. The standard deviations of the accuracy rates are 3.56 for the control group, 5.45 for the treatment group, and 4.46 for the familiarity group. Do the standard deviations agree with the interquartile

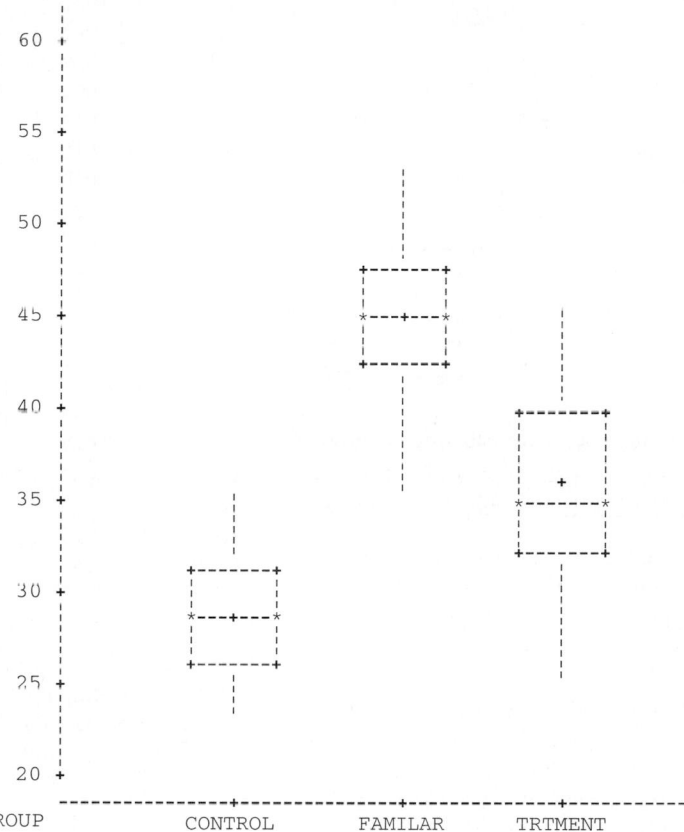

UNIVARIATE PROCEDURE
Schematic Plots

Variable=ACCRATE

ranges (part **b**) with regard to the comparison of the variabilities of the accuracy rates?

d. Is there evidence of outliers in any of the three distributions?

2.89 A manufacturer of minicomputer systems is interested in improving its customer support services. As a first step, its marketing department has been charged with the responsibility of summarizing the extent of customer problems in terms of system down time. The 40 most recent customers were surveyed to determine the amount of down time (in hours) they had experienced during the previous month. These data are listed in the table.

a. Use a statistical software package to construct a box plot for these data. Use the information reflected in the box plot to describe the frequency distribution of the data set. Your description should address central tendency, variation, and skewness.

b. Use your box plot to determine which customers are having unusually lengthy down times.

c. Find and interpret the z-scores associated with customers you identified in part **b**.

Customer Number	Down Time	Customer Number	Down Time	Customer Number	Down Time
230	12	244	2	258	28
231	16	245	11	259	19
232	5	246	22	260	34
233	16	247	17	261	26
234	21	248	31	262	17
235	29	249	10	263	11
236	38	250	4	264	64
237	14	251	10	265	19
238	47	252	15	266	18
239	0	253	7	267	24
240	24	254	20	268	49
241	15	255	9	269	50
242	13	256	22		
243	8	257	18		

2.90 In Exercise 2.17 we constructed a stem-and-leaf display for 40 radiation (exhalation rate) measurements on waste gypsum and phosphate mounds in Florida. The accompanying figure, constructed using SAS, is a box plot for the same data.

```
Variable=EXHLRATE

   12312.5 +              *
           |
           |
           |
    9234.375 +            *
           |
           |
           |              0
    6156.25 +
           |              |
           |              |
           |           +-----+
    3078.125 +         |     |
           |           |  +  |
           |           *-----*
           |           +-----+
         0 +              |
           +-------------+------------
```

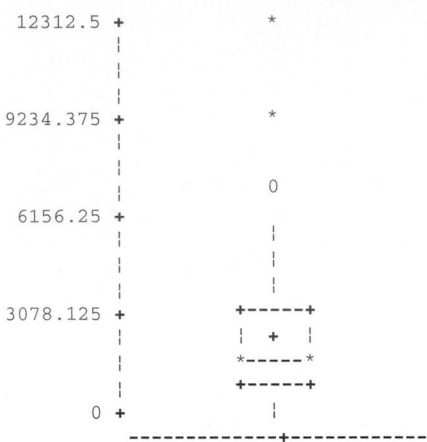

a. Use the box plot to estimate the lower quartile, median, and upper quartile of these data.

b. Does the distribution appear to be skewed? Explain.

c. Are there any outliers among these data? Explain.

d. Examine the stem-and-leaf display for these data in Exercise 2.17, and the box plot here. Which do you prefer as a description of these radiation measurements?

2.91 Refer to Exercises 2.20 and 2.66, in which we compared the 50 states' average SAT scores in 1975 and 1990. The data are repeated in the following table. SPSS was used to generate the accompanying box plots for the 50 states' SAT scores for 1975 and 1990.

State	1975	1990	State	1975	1990
Ala.	883	984	Mont.	1047	987
Alaska	942	914	Nebr.	966	1030
Ariz.	1021	942	Nev.	962	921
Ark.	992	981	N.H.	934	928
Calif.	908	903	N.J.	878	891
Colo.	994	969	N. Mex.	1002	1007
Conn.	913	901	N.Y.	925	882
Del.	915	903	N.C.	827	841
Fla.	915	884	N. Dak.	1064	1069
Ga.	824	844	Ohio	955	949
Hawaii	892	885	Okla.	994	1001
Idaho	1017	968	Oreg.	908	923
Ill.	970	994	Pa.	900	883
Ind.	881	867	R.I.	901	883
Iowa	1091	1088	S.C.	794	834
Kans.	1043	1040	S. Dak.	1084	1061
Ky.	977	994	Tenn.	988	1008
La.	947	993	Tex.	898	874
Maine	908	886	Utah	1069	1031
Md.	907	908	Vt.	915	897
Mass.	903	900	Va.	894	895
Mich.	949	968	Wash.	1011	923
Minn.	1058	1019	W. Va.	964	933
Miss.	980	996	Wis.	1036	1019
Mo.	965	995	Wyo.	1054	977

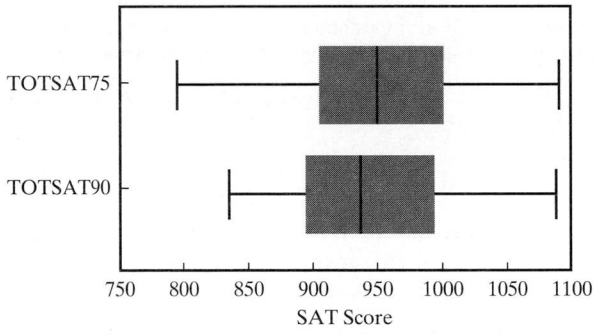

a. Compare the central tendency of the SAT scores for the 2 years. [*Note:* SPSS box plots show only the median, represented by the line within each box.]

b. Compare the variability of the SAT scores for the 2 years.

c. Are any states' SAT scores outliers in either year? If so, identify them.

2.92 The accompanying MINITAB-generated box plots describe the U.S. Environmental Protection Agency's 1986 automobile mileage estimates for all models manufactured by Ford and Honda.

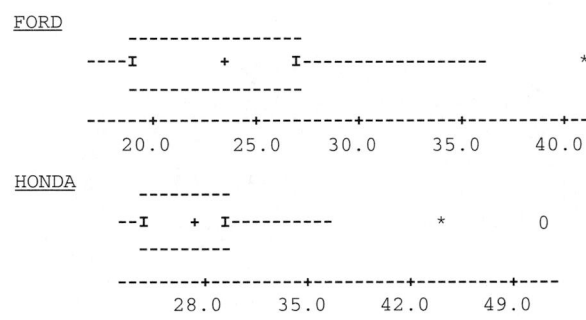

a. Which manufacturer has the higher median mileage estimate?

b. Which manufacturer's mileage estimates have the greater range?

c. Which manufacturer's mileage estimates have the greater interquartile range?

d. Which manufacturer has the model with the highest mileage estimate? Approximately what is that mileage?

Suicide in Urban Jails

Suicide is the leading cause of death of Americans incarcerated in correctional facilities. Moreover, the rate of completed suicide among jailed inmates who have made previous attempts is more than 100 times the rate in the general population. What factors increase the risk of suicide in urban jails?

To answer this question, a group of researchers (with backgrounds in political science, psychology, psychiatry, and correctional facilitation) collected data on all suicides that occurred from 1967 to 1992 in the Wayne County Jail, Detroit, Michigan (*American Journal of Psychiatry,* July 1995). A total of 37 suicides occurred during this period, all by hanging. For each suicide victim, the following variables were measured:

> Number of days in jail before suicide
> Marital status (married, single, widowed, or divorced)
> Race (white or nonwhite)
> Charge (murder/manslaughter or other)
> Shift on which suicide occurred: day (7 AM–3 PM), afternoon (3 PM–11 PM), or night (11 PM–7 AM)

The complete data set is provided in Table 2.14.

CASE STUDY • 2.3 •

Use graphs and numerical descriptive measures to summarize and describe the data in Table 2.14. Next, use these descriptive techniques to answer the following questions raised by the researchers. Then use your answers to the questions to make inferences about the general population of suicidal inmates. (We'll show you how to attach measures of reliability to these inferences in Chapters 7 and 8.)

Focus

a. Are suicides at the jail more likely to be committed by inmates charged with murder/manslaughter or with lesser crimes?

b. Are suicides at the jail more likely to be committed at night?

c. What is the typical length of time an inmate is in jail before committing suicide?

d. Are more suicides committed by white or nonwhite inmates?

e. Have suicides at the jail declined over the years?

f. Which are more likely to commit suicide earlier in their length of stay at the jail, white or nonwhite inmates? Inmates charged with murder/manslaughter or other inmates? Married or nonmarried inmates?

2.9 DISTORTING THE TRUTH WITH DESCRIPTIVE TECHNIQUES

A picture may be "worth a thousand words," but pictures can also color messages or distort them. In fact, the pictures in statistics—relative frequency histograms, charts, and other graphical descriptions—are susceptible to distortion, so we have to examine each of them with care.

We will mention a few of the pitfalls to watch for when interpreting a chart or graph. But first we should mention the **time series graph**, which is often the object of distortion. This type of graph records the behavior of some variable of interest recorded over time. Examples of variables commonly graphed as time series abound: economic indices, the U.S. food surplus, defense spending, presidential popularity index, etc. Since time series graphs often appear in newspapers or magazines, we'll use some of them to demonstrate several ways in which pictures are commonly distorted.

One common way to change the impression conveyed by a graph is to change the scale on the vertical axis, the horizontal axis, or both. For example, Figure 2.28 on page 79 is a bar graph that shows the market share of sales for a company for each of the years 1990 to 1995. If you want to show that the change in firm A's market share over time is moderate, you should pack in a large number of units per inch on the vertical axis. That is, make the distance between successive units

TABLE 2.14   Data on 37 Suicides in Wayne County Jail (Case Study 2.3)

Victim	Days in Jail Before Suicide	Marital Status	Race	Murder/ Manslaughter Charge	Time of Suicide	Year
1	3	Married	W	Yes	Night	1972
2	4	Single	W	Yes	Night	1987
3	5	Single	NW	Yes	Afternoon	1975
4	7	Widowed	NW	Yes	Night	1981
5	10	Single	NW	Yes	Afternoon	1982
6	15	Married	NW	Yes	Night	1970
7	15	Divorced	NW	Yes	Night	1985
8	19	Married	NW	Yes	Day	1980
9	22	Single	NW	Yes	Night	1983
10	29	Married	W	Yes	Night	1981
11	31	Single	NW	Yes	Night	1970
12	85	Single	NW	Yes	Afternoon	1985
13	126	Married	NW	Yes	Night	1981
14	221	Divorced	W	Yes	Night	1970
15	14	Single	NW	No	Night	1980
16	22	Single	W	No	Afternoon	1970
17	42	Single	W	No	Day	1976
18	122	Single	NW	No	Night	1983
19	309	Married	NW	No	Night	1970
20	69	Married	NW	No	Night	1979
21	143	Single	NW	No	Night	1985
22	6	Married	NW	No	Night	1971
23	1	Single	NW	No	Day	1968
24	1	Single	NW	No	Night	1973
25	1	Single	W	No	Night	1989
26	2	Single	W	No	Day	1987
27	3	Single	NW	No	Night	1985
28	4	Married	W	No	Afternoon	1969
29	4	Single	W	No	Afternoon	1974
30	4	Single	NW	No	Night	1980
31	6	Single	W	No	Day	1968
32	6	Married	NW	No	Night	1975
33	10	Single	W	No	Night	1967
34	18	Single	NW	No	Night	1979
35	26	Married	W	No	Night	1976
36	41	Single	NW	No	Night	1985
37	86	Married	W	No	Night	1968

Source: DuRand, C. J., *et al.* "A quarter century of suicide in a major urban jail: Implications for community psychiatry." *American Journal of Psychiatry,* Vol. 152, No. 7, July 1995, p. 1078 (Table 1). Copyright 1995, the American Psychiatric Association. Reprinted by permission.

on the vertical scale small, as shown in Figure 2.28. You can see that a change in the firm's market share over time is barely apparent.

If you want to use the same data to make the changes in firm A's market share appear large, you should increase the distance between successive units on the vertical axis. That is, stretch the vertical axis by graphing only a few units per inch as in Figure 2.29. A telltale sign of stretching is a long vertical axis, but this is often hidden by starting the vertical axis at some point above 0, as shown in Figure 2.30a. The same effect can be achieved by using a broken line—called a *scale break*—for the vertical axis, as shown in Figure 2.30b.

FIGURE 2.28

Firm Λ's market share from 1990 to 1995—packed vertical axis

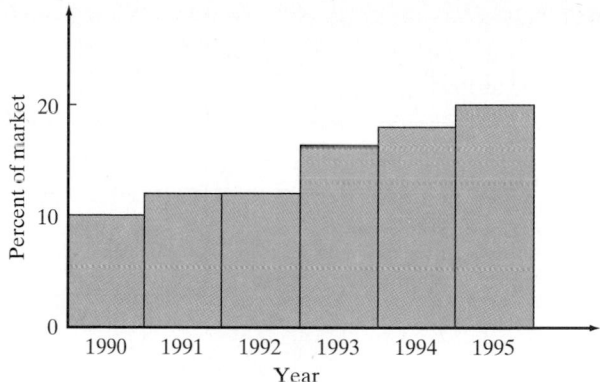

FIGURE 2.29

Firm A's market share from 1990 to 1995—stretched vertical axis

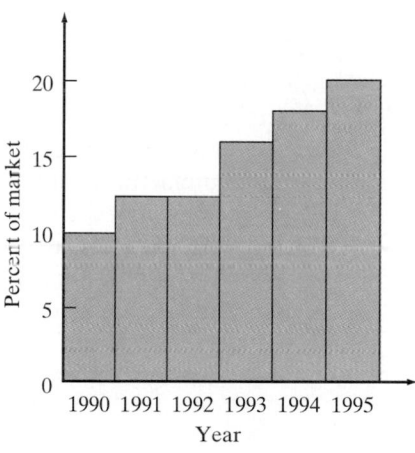

FIGURE 2.30

Changes in money supply from January to June

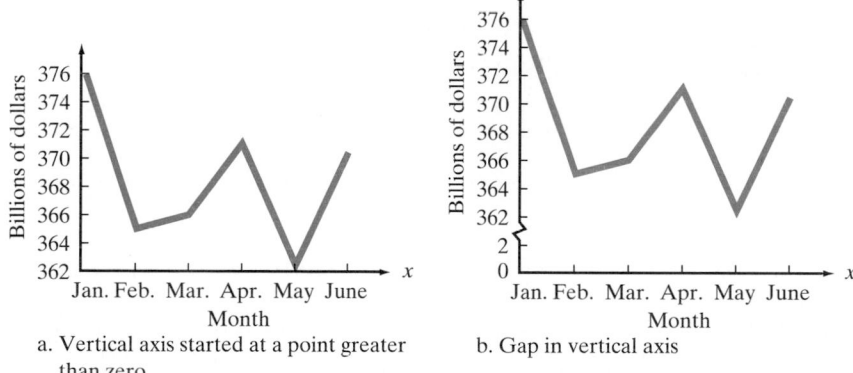

a. Vertical axis started at a point greater than zero

b. Gap in vertical axis

Stretching the horizontal axis (increasing the distance between successive units) may also lead you to incorrect conclusions. For example, Figure 2.31a depicts the change in the Gross Domestic Product (GDP) from the first quarter of 1993 to the last quarter of 1994. If you increase the size of the horizontal axis, as in Figure 2.31b, the change in GDP over time seems less pronounced.

The changes in categories indicated by a bar graph can also be emphasized or deemphasized by stretching or shrinking the vertical axis. Another method of achieving visual distortion with bar graphs is by making the width of the bars

FIGURE 2.31

Gross Domestic Product
from 1993 to 1994

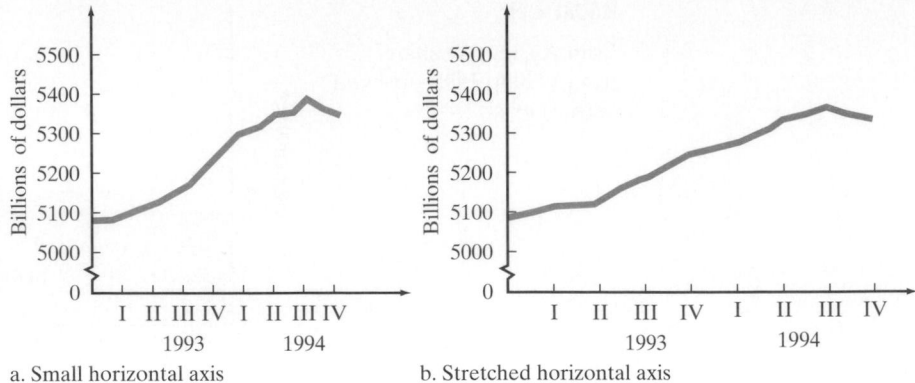

a. Small horizontal axis   b. Stretched horizontal axis

proportional to the height. For example, look at the bar chart in Figure 2.32a, which depicts the percentage of a year's total automobile sales attributable to each of the four major manufacturers. Now suppose we make both the width and the height grow as the market share grows. This change is shown in Figure 2.32b. The reader may tend to equate the *area* of the bars with the relative market share of each manufacturer. But in fact, the true relative market share is proportional only to the *height* of the bars.

   Although we've discussed only a few of the ways that graphs can be used to convey misleading pictures of phenomena, the lesson is clear. Look at all graphical descriptions of data with a critical eye. Particularly, check the axes and the size of the units on each axis. Ignore the visual changes and concentrate on the actual numerical changes indicated by the graph or chart.

   The information in a data set can also be distorted by using numerical descriptive measures, as Example 2.17 indicates.

EXAMPLE 2.17

Suppose you're considering working for a small law firm—one that currently has a senior member and three junior members. You inquire about the salary you could expect to earn if you join the firm. Unfortunately, you receive two answers:

> *Answer A:*   The senior member tells you that an "average employee" earns $57,500.

> *Answer B:*   One of the junior members later tells you that an "average employee" earns $45,000.

Which answer can you believe?

FIGURE 2.32

Relative share of the
automobile market for
each of four major
manufacturers

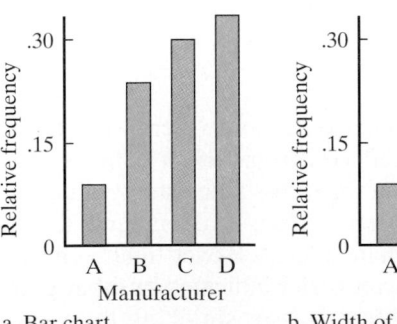
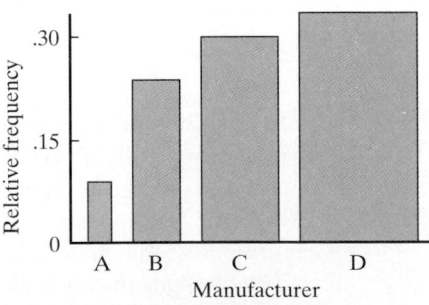

a. Bar chart   b. Width of bars grows with height

Solution

The confusion exists because the phrase "average employee" has not been clearly defined. Suppose the four salaries paid are $45,000 for each of the three junior members and $95,000 for the senior member. Thus,

$$\bar{x} = \frac{3(\$45,000) + \$95,000}{4} = \frac{\$230,000}{4} = \$57,500$$

$$\text{Median} = \$45,000$$

You can now see how the two answers were obtained. The senior member reported the mean of the four salaries, and the junior member reported the median. The information you received was distorted because neither person stated which measure of central tendency was being used. ▲

Another distortion of information in a sample occurs when *only* a measure of central tendency is reported. Both a measure of central tendency and a measure of variability are needed to obtain an accurate mental image of a data set.

Suppose you want to buy a new car and are trying to decide which of two models to purchase. Since energy and economy are both important issues, you decide to purchase model A because its EPA mileage rating is 32 miles per gallon in the city, whereas the mileage rating for model B is only 30 miles per gallon in the city.

However, you may have acted too quickly. How much variability is associated with the ratings? As an extreme example, suppose that further investigation reveals that the standard deviation for model A mileages is 5 miles per gallon, whereas that for model B is only 1 mile per gallon. If the mileages form a mound-shaped distribution, they might appear as shown in Figure 2.33. Note that the larger amount of variability associated with model A implies that more risk is involved in purchasing model A. That is, the particular car you purchase is more likely to have a mileage rating that will greatly differ from the EPA rating of 32 miles per gallon if you purchase model A, while a model B car is not likely to vary from the 30 miles per gallon rating by more than 2 miles per gallon.

We conclude this section with another example on distorting the truth with numerical descriptive measures.

EXAMPLE 2.18

Children Out of School in America is a report on delinquency of school-age children prepared by the Children's Defense Fund (CDF). Consider the following three reported results of the CDF survey.

FIGURE 2.33

Mileage distributions for two car models

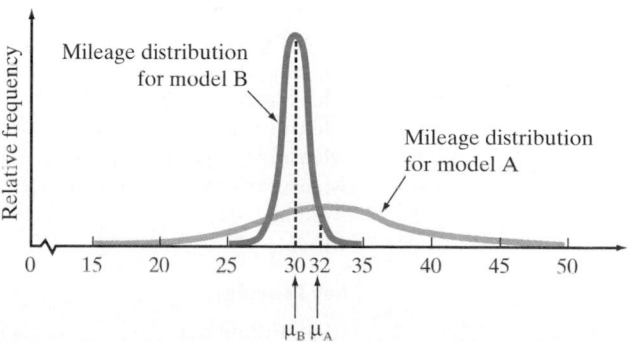

Reported result: 25% of the 16- and 17-year-olds in the Portland, Maine, Bayside East Housing Project were out of school. Actual data: *Only eight children were surveyed; two were found to be out of school.*

Reported result: Of all the secondary school students who had been suspended more than once in census tract 22 in Columbia, South Carolina, 33% had been suspended two times and 67% had been suspended three or more times. Actual data: *CDF found only three children in that entire census tract who had been suspended; one child was suspended twice and the other two children, three or more times.*

Reported result: In the Portland Bayside East Housing Project, 50% of all the secondary school children who had been suspended more than once had been suspended three or more times. Actual data: *The survey found two secondary school children who had been suspended in that area; one of them had been suspended three or more times.*

Identify the potential distortions in the results reported by the CDF.

Solution

In each of these examples the reporting of percentages (i.e., relative frequencies) instead of the numbers themselves is misleading. No inference we might draw from the cited examples would be reliable. (We'll see how to measure the reliability of estimated percentages in Chapter 7.) In short, either the report should state the numbers alone instead of percentages, or, better yet, it should state that the numbers were too small to report by region. If several regions were combined, the numbers (and percentages) would be more meaningful. ▲

QUICK REVIEW

Key Terms

Note: Starred terms (*) refer to the optional section in this chapter.

Bar graph 23	Mound-shaped distribution 58
Box plots* 70	Numerical descriptive measures 42
Chebyshev's Rule 57	Outer fences* 71
Class frequency 22	Outliers* 71
Class relative frequency 22	Percentile 64
Classes 22	Pie chart 24
Dot plot 30	Quartiles* 70
Empirical Rule 58	Range 52
Hinges* 70	Rare-event approach 66
Histogram 31	Relative frequency histogram 31
Inner fences* 71	Skewness 46
Interquartile range* 70	Standard deviation 54
Lower quartile* 70	Stem-and-leaf display 30
Mean 43	Symmetric distribution 46
Measurement classes 32	Time series graph 77
Measures of central tendency 43	Upper quartile* 70
Measures of relative standing 64	Variance 53
Measures of variation or spread 43	Whiskers* 71
Median 45	z-score 65
Mode 47	

Key Formulas

$$\frac{\text{Class Frequency}}{n}$$

Class relative frequency   22

$$\bar{x} = \frac{\sum\limits_{i=1}^{n} x_i}{n}$$ Sample mean   44

$$s^2 = \frac{\sum\limits_{i=1}^{n}(x_i - \bar{x})^2}{n-1} = \frac{\sum\limits_{i=1}^{n} x_i^2 - \dfrac{\left(\sum\limits_{i=1}^{n} x_i\right)^2}{n}}{n-1}$$ Sample variance   53

$$s = \sqrt{s^2}$$ Sample standard deviation   54

$$z = \frac{x - \bar{x}}{s}$$ Sample z-score   65

$$z = \frac{x - \mu}{\sigma}$$ Population z-score   65

$$IQR = Q_U - Q_L$$ Interquartile range   70

LANGUAGE LAB

Symbol	Pronunciation	Description
Σ	sum of	Summation notation; $\sum\limits_{i=1}^{n} x_i$ represents the sum of the measurements $x_1, x_2, \ldots, x_n$
μ	mu	Population mean
$\bar{x}$	x-bar	Sample mean
σ^2	sigma squared	Population variance
σ	sigma	Population standard deviation
s^2		Sample variance
s		Sample standard deviation
z		z-score for a measurement
m		Median (middle quartile) of a sample data set
Q_L		Lower quartile (25th percentile)
Q_U		Upper quartile (75th percentile)
IQR		Interquartile range

SUPPLEMENTARY EXERCISES 2.93–2.116

Note: Exercises marked with  *require the use of a computer.*

Learning the Mechanics

2.93 Construct a relative frequency histogram for the data summarized in the accompanying table.

Measurement Class	Relative Frequency	Measurement Class	Relative Frequency
.00– .75	.02	5.25–6.00	.15
.75–1.50	.01	6.00–6.75	.12
1.50–2.25	.03	6.75–7.50	.09
2.25–3.00	.05	7.50–8.25	.05
3.00–3.75	.10	8.25–9.00	.04
3.75–4.50	.14	9.00–9.75	.01
4.50–5.25	.19		

2.94 Discuss the conditions under which the median is preferred to the mean as a measure of central tendency.

2.95 Consider the following three measurements: 50, 70, 80. Find the z-score for each measurement if they are from a population with a mean and standard deviation equal to
a. $\mu = 60, \sigma = 10$ **b.** $\mu = 60, \sigma = 5$
c. $\mu = 40, \sigma = 10$ **d.** $\mu = 40, \sigma = 100$

2.96 If the range of a set of data is 20, find a rough approximation to the standard deviation of the data set.

2.97 Compute s^2 for data sets with the following characteristics:

a. $\sum\limits_{i=1}^{n} x_i^2 = 246,\ \sum\limits_{i=1}^{n} x_i = 63,\ n = 22$

b. $\sum\limits_{i=1}^{n} x_i^2 = 666,\ \sum\limits_{i=1}^{n} x_i = 106,\ n = 25$

c. $\sum\limits_{i=1}^{n} x_i^2 = 76,\ \sum\limits_{i=1}^{n} x_i = 11,\ n = 7$

2.98 For each of the following data sets, compute $\bar{x}$, s^2, and s:

    a. 13, 1, 10, 3, 3   **b.** 13, 6, 6, 0

    c. 1, 0, 1, 10, 11, 11, 15   **d.** 3, 3, 3, 3

    e. For each data set in parts **a–d**, form the interval $\bar{x} \pm 2s$ and calculate the percentage of the measurements that fall in the interval.

2.99 For each of the following data sets, compute $\bar{x}$, s^2, and s. If appropriate, specify the units in which your answers are expressed.

    a. 4, 6, 6, 5, 6, 7   **b.** −$1, $4, −$3, $0, −$3, −$6

    c. ⅗%, ⅘%, ⅖%, ⅕%, ¹⁄₁₆%

    d. Calculate the range of each data set in parts **a–c**.

2.100 Explain why we generally prefer the standard deviation to the range as a measure of variability for quantitative data.

Applying the Concepts

2.101 The following data sets have been invented to demonstrate that the lower bounds given by Chebyshev's Rule are appropriate. Notice that the data are contrived and would not be encountered in a real-life problem.

    a. Consider a data set that contains ten 0's, two 1's, and ten 2's. Calculate $\bar{x}$, s^2, and s. What percentage of the measurements are in the interval $\bar{x} \pm s$? Compare this result to Chebyshev's Rule.

    b. Consider a data set that contains five 0's, thirty-two 1's, and five 2's. Calculate $\bar{x}$, s^2, and s. What percentage of the measurements are in the interval $\bar{x} \pm 2s$? Compare this result to Chebyshev's Rule.

    c. Consider a data set that contains three 0's, fifty 1's, and three 2's. Calculate $\bar{x}$, s^2, and s. What percentage of the measurements are in the interval $\bar{x} \pm 3s$? Compare this result to Chebyshev's Rule.

    d. Draw a histogram for each of the data sets in parts **a**, **b**, and **c**. What do you conclude from these graphs and the answers to parts **a**, **b**, and **c**?

2.102 The pie charts shown below, extracted from *Christianity Today* (May 15, 1995), describe the ethnic background of legal immigrants into the United States during two periods, 1951–1960 and 1981–1990. What trends, if any, are depicted in the pie charts?

2.103 The environmental impact of predator control and regulated killing of predator species by the U.S. Fish and Wildlife Service was the topic of a recent study (*UCLA Journal of Environmental Law & Policy,* Vol. 13, 1994/95). One of the methods used to control, trap, and kill predators is the steel-jaw leghold trap, invented more than 300 years ago. Unfortunately, the traps often end up capturing and killing animals other than the intended predators. For example, the study revealed that in one year, 44,982 animals were killed by steel traps intended for coyotes; but, only 25,026 of these animals were actually coyotes. The following table lists the species of all animals captured by steel traps during that year and the number of each killed. Use a graphical method to summarize the data. Interpret the graph.

Species	Number Killed by Steel Traps
Coyotes	25,026
Opossums	2,698
Porcupines	1,367
Raccoons	3,345
Skunks	6,348
Beavers	682
Dogs	273
Cats	73
Goats	49
Foxes	114
Rabbits	98
Deer	20
Muskrats	52
Other	4,837

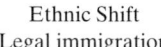

Ethnic Shift
Legal immigration

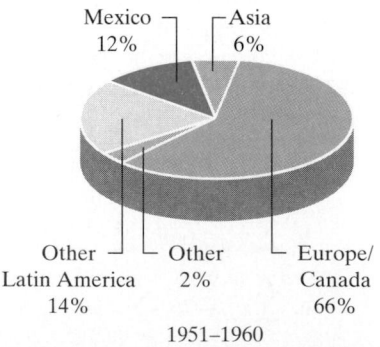

1951–1960

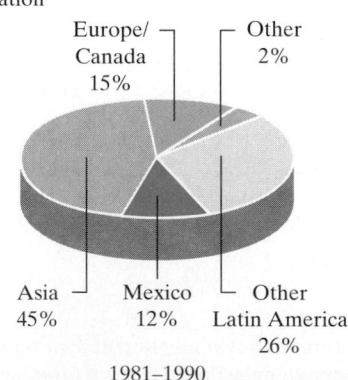

1981–1990

2.104 A psychologist has developed a new technique intended to improve rote memory. To test the method against other standard methods, 20 high school students are selected at random, and each is taught the new technique. The students are then asked to memorize a list of 100 word phrases using the technique. The following are the number of word phrases memorized correctly by the students:

91	64	98	66	83	87	83	86	80	93
83	75	72	79	90	80	90	71	84	68

a. Define the terms *mean, median,* and *mode* in the context of this problem.

b. Construct a relative frequency histogram for the data.

c. Compute the mean, median, and mode for the data set and locate them on the histogram. Do these measures of central tendency appear to locate the center of distribution of data?

2.105 In some locations, radiation levels in homes are measured at well above normal background levels in the environment. As a result, many architects and builders are making design changes to ensure adequate air exchange so that radiation will not be "trapped" in homes. In one such location, 50 homes' levels were measured, and the mean level was 10 parts per billion (ppb), the median was 8 ppb, and the standard deviation was 3 ppb. Background levels in this location are at about 4 ppb.

a. Based on these results, is the distribution of the 50 homes' radiation levels symmetric, skewed to the left, or skewed to the right? Why?

b. Use both Chebyshev's Rule and the Empirical Rule to describe the distribution of radiation levels. Which do you think is most appropriate in this case? Why?

c. Use the results from part **b** to approximate the number of homes in this sample that have radiation levels above the background level.

d. Suppose another home is measured at a location 10 miles from the one sampled, and has a level of 20 ppb. What is the z-score for this measurement relative to the 50 homes sampled in the other location? Is it likely that this new measurement comes from the same distribution of radiation levels as the other 50? Why? How would you go about confirming your conclusion?

2.106 A study in the *Journal of Leisure Research* (Vol. 24, 1992) investigated the relationship between academic performance and leisure activities. One hundred fifty-nine high school students were given a list of 43 leisure activities (sports, fishing, music, drama, photography, writing, watching TV, etc.). Each was asked to state how many activities they participated in each week. From this list, activities that involved reading, writing, or arithmetic were selected to form another variable, "academic leisure activities." Some of the results of the study are presented below:

	$\bar{x}$	s
GPA	2.96	.71
Number of leisure activities	12.38	5.07
Number of academic leisure activities	2.77	1.97

a. For GPA, calculate the intervals $\bar{x} \pm s$, $\bar{x} \pm 2s$, and $\bar{x} \pm 3s$. Based on these intervals, comment on the skewness or symmetry you would expect in these data. (Remember, the range of GPAs is 0 to 4.) Approximately what percentage of the students would you expect to find in each interval?

b. For number of leisure activities, calculate the intervals $x \pm s$, $x \pm 2s$, and $x \pm 3s$. Based on these intervals, comment on the skewness or symmetry you would expect in these data. (Remember, the number of leisure activities cannot be negative.) Approximately what percentage of students would you expect to find in each interval?

c. For number of academic leisure activities, calculate the intervals $\bar{x} \pm s$, $\bar{x} \pm 2s$, and $\bar{x} \pm 3s$. Based on these intervals, comment on the skewness or symmetry you would expect in these data. Approximately what percentage of students would you expect to find in each interval?

d. Based on your answers, which of the variables' distributions would you expect to be most skewed? Why?

2.107 Refer to *The Astronomical Journal* study of galaxy velocities, Exercise 2.72. A second cluster of galaxies, named A1775, is thought to be a *double cluster,* i.e., two clusters of galaxies in close proximity. Fifty-one velocity observations (in kilometers per second, km/s) from cluster A1775 are listed in the accompanying table.

22922	20210	21911	19225	18792	21993	23059
20785	22781	23303	22192	19462	19057	23017
20186	23292	19408	24909	19866	22891	23121
19673	23261	22796	22355	19807	23432	22625
22744	22426	19111	18933	22417	19595	23408
22809	19619	22738	18499	19130	23220	22647
22718	22779	19026	22513	19740	22682	19179
19404	22193					

Source: Oegerle, W. R., Hill, J. M., and Fitchett, M. J. "Observations of high dispersion clusters of galaxies: Constraints on cold dark matter." *The Astronomical Journal,* Vol. 110, No. 1, July 1995, p. 34 (Table 1), p. 37 (Figure 1).

a. Use a graphical method to describe the velocity distribution of galaxy cluster A1775.

b. Examine the graph, part **a**. Is there evidence to support the double cluster theory? Explain.

c. Calculate numerical descriptive measures (e.g., mean and standard deviation) for galaxy velocities in cluster A1775. Depending on your answer to part **b**, you may need to calculate two sets of numerical descriptive measures, one for each of the clusters (say, A1775A and A1775B) within the double cluster.

d. Suppose you observe a galaxy velocity of 20,000 km/s. Is this galaxy likely to belong to cluster A1775A or A1775B? Explain.

2.108 A study was conducted to compare the effects of fructose and glucose on the high endurance of women athletes (*Research Quarterly for Exercise and Sport,* Vol. 54, 1983). Six women athletes received 300 milliliters each of a certain drink (say, water and glucose) and then ran until exhausted. Various measurements were then taken on each athlete. The mean and standard deviation of the performance time (in minutes) for the six athletes after receiving the glucose drink were 61.9 and 20.3, respectively. What do the mean and standard deviation tell you about this small data set?

2.109 A tornado is defined as a violently rotating column of air extending from a thunderstorm to the ground. The most violent tornadoes, with wind speeds of 250 mph or more, are capable of tremendous destruction. The accompanying table lists, for each state, the reported number of tornadoes per year and the month of peak tornado occurrence.

a. Identify the variables measured, number of tornadoes per year and month of peak occurrence, as quantitative or qualitative.

b. Use a graphical technique to summarize the month of peak tornado occurrence in the 50 states. Interpret the graph.

2.110 Refer to Exercise 2.109. The descriptive statistics and stem-and-leaf display reproduced on page 87 were obtained by using SPSS to analyze the reported number of tornadoes per year of the 50 states.

a. Use the stem-and-leaf display to give a verbal description of the data set. Do any of the measurements appear to be outliers?

b. Examine the computer output to determine the value of $\bar{x}$, the median, s^2, and s for the number of tornadoes per year. [*Note:* SPSS also produces some descriptive statistics we haven't covered yet; ignore these for now.]

c. Based on the stem-and-leaf display and the relationship of the mean and median, is the distribution of these data approximately mound-shaped or is it skewed to the right or left?

State	Number of Tornadoes per Year	Month of Peak Occurrence	State	Number of Tornadoes per Year	Month of Peak Occurrence
AL	23	April	MT	5	June
AK	0	—	NE	36	June
AZ	4	August	NV	1	June
AR	21	April	NH	2	July
CA	4	February	NJ	3	July
CO	24	June	NM	9	June
CT	1	July	NY	5	July
DE	1	July	NC	14	April
FL	52	June	ND	20	July
GA	21	April	OH	16	May
HI	1	January	OK	47	May
ID	2	July	OR	1	May
IL	27	May	PA	10	June
IN	23	May	RI	0.23	August
IA	35	May	SC	10	April
KS	36	May	SD	28	June
KY	10	May	TN	12	April
LA	27	April	TX	137	May
ME	2	July	UT	2	June
MD	3	May	VT	1	July
MA	3	August	VA	6	May
MI	21	June	WA	1	May
MN	19	June	WV	2	May
MS	26	April	WI	18	July
MO	27	May	WY	11	June

Source: "Tornadoes ... nature's most violent storms." U.S. Dept. of Commerce, National Oceanic and Atmospheric Administration, Sept. 1992.

```
        NUMBER

Valid cases:          50.0   Missing cases:        .0   Percent missing:      .0

Mean        16.2046  Std Err      3.0858  Min          .0000  Skewness      3.6931
Median      10.0000  Variance   476.0939  Max       137.0000  S E Skew       .3366
5% Trim     13.2667  Std Dev     21.8196  Range     137.0000  Kurtosis     18.8773
                                          IQR        21.2500  S E Kurt       .6619
-----------------------------------------------------------------------------------

        NUMBER

Frequency   Stem &  Leaf

     23.00     0  .  00111111122222333445569
      9.00     1  .  000124689
     12.00     2  .  011133467778
      3.00     3  .  566
      1.00     4  .  7
      1.00     5  .  2
      1.00  Extremes    (137)

Stem width:        10.00
Each leaf:          1 case(s)
```

d. Based on your answer to part **c**, what percentage of the measurements do you expect to find in the intervals $\bar{x} \pm s$, $\bar{x} \pm 2s$, and $\bar{x} \pm 3s$?

e. Count the number of measurements that actually fall in each interval of part **d** and express each interval count as a percentage of the total number of measurements. Compare these results to your estimates from part **d**.

2.111 The Community Attitude Assessment Scale (CAAS) measures citizens' attitudes toward 15 life areas (e.g., education, employment, and health) on four dimensions—importance, influence, equality of opportunity, and satisfaction. In order to develop the CAAS, a number of households in each of 25 communities were randomly selected and sent questionnaires. Because relatively low response rates suggest the possibility of a substantial but unknown opinion bias in the reported data, the percentage of the sample responding to the survey was determined in each community. The results are given here (in percent):

21 14 18 20 14 16 6 22 28 16 26 14 13
15 25 21 14 7 12 8 15 14 21 22 10

a. Use statistical software to construct a stem-and-leaf display for the data. Use this display to give a verbal description of the data set.

b. Use the software to construct a relative frequency histogram for the data given, locating the mean, median, and mode.

c. Find the range for the data and use it to calculate an approximate value for s. Use this value to check your answer to part **d**.

d. Calculate the variance and standard deviation for the data.

e. Find the proportion of the measurements that fall in the interval $\bar{x} \pm 2s$.

2.112 Refer to Exercise 2.111. Use a statistical software package to construct a box plot for the percent responses. Use the box plot to describe the distribution of responses.

2.113 A professor believes that if a class is allowed to work on an examination as long as desired, the times spent by the students would be approximately mound-shaped with mean 40 minutes and standard deviation 6 minutes. Approximately how long should be allotted for the examination if the professor wants almost all (say, 97.5%) of the class to finish?

2.114 Polychlorinated biphenyls (PCBs), considered to be extremely hazardous to humans, are often used in the insulation of large electric transformers. The *Gainesville Sun* (Mar. 24, 1984) reported on the discovery of a particularly high PCB count at a salvage company in Clay County, Florida. The company, which salvaged the copper in electrical transformers, allowed oil contaminated with PCBs to seep into the soil in and around the salvage site. One soil sample in the vicinity registered 200 parts per million (ppm) of PCBs, four times the safe limit established by the Florida Department of Environmental Regulation. Suppose that the PCB count in samples of soil in the vicinity of the salvage operation has a distribution with mean equal to 25 ppm and standard deviation equal to 5 ppm of PCBs. Would a soil sample showing 200 ppm be classified as an extreme observation? Explain.

2.115 Most people living in metropolitan areas receive impressions of what is happening in their area primarily through their major newspapers. A study was conducted to determine whether the *Uniform Crime Report,* compiled by the Federal Bureau of Investigation, and the daily newspaper gave consistent information about the trend and distribution of crime in a metropolitan area. An attention score, based on the amount of space devoted to a story, was calculated for each paper's coverage of murders, assaults, robberies, etc. Suppose μ, the average murder attention score of metropolitan newspapers across the country in 1995, was 60, with σ = 4.5. One metropolitan newspaper in the midwest had a 1995 murder attention score of 69.

a. Approximately what percentage of the newspapers had a murder attention score higher than 69 in 1995? (Make no assumptions about the nature of the distribution of scores.)

b. Repeat part **a**, assuming attention scores were mound-shaped.

2.116 Systematic inflammatory response syndrome (SIRS) is a major problem for patients with dys-

functioning organs. A study of the incidence of SIRS in 22 patients with cirrhosis of the liver was conducted by medical researchers at the University of Pittsburgh (*Journal of the American Medical Association,* July 5, 1995). The clinical characteristics of the 22 patients—all awaiting liver transplants in the intensive care unit (ICU)—are reported in the accompanying table.

a. Identify each variable measured as quantitative or qualitative.

b. Use a graphical method to describe each qualitative variable. Interpret the graphs.

c. Use a computer to describe each quantitative variable with a graph. Interpret the graphs.

d. Use a computer to generate numerical descriptive measures for each quantitative variable. Interpret the results.

e. Combine the information from parts **b–d** to make a general statement about a typical patient who awaits a liver transplant in ICU.

Patient	Organ Dysfunction Group	Bile Blood Level (mg/dL)	Age (years)	ORGAN FAILURE SCORES* Apache Scale	Goris Scale	SIRS	Blood Culture	Infection	Outcome
1	Mild	1.22	64	21	1	No	Negative	No	Stabilized, left ICU
2	Mild	1.89	57	21	5	No	Negative	Yes	Stable, left ICU
3	Mild	.58	69	19	5	No	Negative	No	Stabilized, died
4	Mild	.13	54	16	2	No	Negative	No	Stable, left ICU
5	Mild	2.21	19	7	3	No	Negative	No	Stable, transplant
6	Mild	.96	55	10	2	No	Negative	No	Stable, transplant
7	Moderate	.54	40	23	9	Yes	Negative	No	Stabilized, transplant
8	Moderate	1.11	44	27	8	Yes	Negative	No	Stabilized, left ICU
9	Moderate	.81	51	19	4	Yes	Negative	No	Stabilized, transplant
10	Moderate	1.11	28	18	7	Yes	Negative	No	Stabilized, left ICU
11	Moderate	.58	65	21	4	Yes	Negative	No	Died
12	Moderate	.38	57	15	5	No	Negative	No	Stabilized, transplant
13	Severe	.88	61	20	5	Yes	Positive	Yes	Died
14	Severe	2.40	47	15	5	Yes	Negative	Yes	Died
15	Severe	2.12	40	21	8	Yes	Negative	Yes	Died
16	Severe	.68	48	29	11	Yes	Positive	Yes	Died
17	Severe	.33	47	29	9	Yes	Positive	Yes	Died
18	Severe	.10	60	30	7	Yes	Negative	Yes	Died
19	Severe	.76	67	30	9	Yes	Negative	Yes	Died
20	Severe	1.05	48	23	7	Yes	Negative	Yes	Died
21	Severe	.44	58	25	8	Yes	Negative	Yes	Died
22	Severe	.39	65	27	9	Yes	Negative	Yes	Died

*The higher the organ failure score, the more serious the disease.

Source: Rosenbloom, A. J., *et al.* "Leukocyte activation in the peripheral blood of patients with cirrhosis of the liver and SIRS." *Journal of the American Medical Association,* Vol. 274, No. 1, July 5, 1995, p. 60. Copyright 1995, American Medical Association.

STUDENT PROJECTS

We list here several sources of real-life data sets (many in the list have been obtained from Wasserman and Bernero's *Statistics Sources*). This index of data sources is very complete and is a useful reference for anyone interested in finding almost any type of data. First we list some almanacs:

> *CBS News Almanac*
>
> *Information Please Almanac*
>
> *World Almanac and Book of Facts*

United States Government publications are also rich sources of data:

> *Agricultural Statistics*
>
> *Digest of Educational Statistics*
>
> *Handbook of Labor Statistics*
>
> *Housing and Urban Development Yearbook*
>
> *Social Indicators*
>
> *Uniform Crime Reports for the United States*
>
> *Vital Statistics of the United States*
>
> *Business Conditions Digest*
>
> *Economic Indicators*
>
> *Monthly Labor Review*
>
> *Survey of Current Business*
>
> *Bureau of the Census Catalog*
>
> *Statistical Abstract of the United States*

Main data sources are published on an annual basis:

> *Commodity Yearbook*
>
> *Facts and Figures on Government Finance*
>
> *Municipal Yearbook*
>
> *Standard and Poor's Corporation, Trade and Securities: Statistics*

Some sources contain data that are international in scope:

> *Compendium of Social Statistics*
>
> *Demographic Yearbook*
>
> *United Nations Statistical Yearbook*
>
> *World Handbook of Political and Social Indicators*

The Internet or America Online are useful sources of data if you have access to a modem and a computer.

Utilizing the data sources listed, sources suggested by your instructor, or your own resourcefulness, find one real-life quantitative data set that stems from an area of particular interest to you.

a. Describe the data set by using a relative frequency histogram, stem-and-leaf display, or dot plot.
b. Find the mean, median, variance, standard deviation, and range of the data set.
c. Use Chebyshev's Rule and the Empirical Rule to describe the distribution of this data set. Count the actual number of observations that fall within 1, 2, and 3 standard deviations of the mean of the data set and compare these counts with the description of the data set you developed in part b.

EXPLORING DATA WITH A COMPUTER

Note: The appendix of this text contains several large data sets obtained from a variety of real-world studies. These data sets are the focus of this and all subsequent "Exploring Data with a Computer" sections. Consequently, they are made available (from the publisher) on a 3½" DOS diskette in ASCII format so that you can readily access and analyze them on the computer with a statistical software package.

Periodically, *Car and Driver* magazine conducts comprehensive road tests on all new car models. The results of the tests are reported in *Car and Driver* "Road Test Digest." The "Road Test Digest" for the August 1995 issue is provided in Appendix B. For each of the 139 new cars tested, the appendix includes the following variables:

1. List price ($)
2. Elapsed time from 0 to 60 mph (seconds)
3. Elapsed time for ¼ mile at full throttle (seconds)
4. Maximum speed (mph)
5. Braking distance from 70 to 0 mph (feet)
6. EPA-estimated city fuel economy (mpg)
7. Road-holding (grip) during cornering (gravitational force, in *g*'s)

a. Note that each variable is quantitative in nature. Select one of the variables and perform the following analyses on the computer: (i) stem-and-leaf display, (ii) relative frequency histogram, and (iii) box plot.

b. Compare the graphical descriptions, part a, and discuss what each reveals about the distribution of the selected variable for the 139 new cars.

c. Find and interpret the following numerical descriptive measures for the variable: (i) mean, (ii) median, and (iii) standard deviation.

d. Use the computer to count the number of measurements in the intervals $\bar{x} \pm s$, $\bar{x} \pm 2s$, and $\bar{x} \pm 3s$. Compare the results with those given by Chebyshev's Rule and the Empirical Rule.

e. Find and interpret the following measures of relative standing for the variable selected: (i) 10th percentile, (ii) 25th percentile, (iii) 75th percentile, (iv) 90th percentile, and (v) *z*-score for the measurement for Dodge Viper.

f. Do you detect any unusual measurements in the data set?

Chapter 3

PROBABILITY

Contents

Case Studies

WHERE WE'VE BEEN

We've identified inference, from a sample to a population, as the goal of statistics. And we've seen that to reach this goal, we must be able to describe a set of measurements. Thus, we explored the use of graphical and numerical methods for describing both quantitative and qualitative data sets and for phrasing inferences.

WHERE WE'RE GOING

Now that we know how to phrase an inference about a population, we turn to the problem of making the inference. What is it that permits us to make the inferential jump from sample to population and then to give a measure of reliability for the inference? As you'll see, the answer is *probability*. This chapter is devoted to a study of probability—what it is and some of the basic concepts of the theory behind it.

Recall that one branch of statistics is concerned with decisions about a population based on sample information. You can see how this is accomplished more easily if you understand the relationship between population and sample—a relationship that becomes clearer if we reverse the statistical procedure of making inferences from sample to population. In this chapter, then, we assume that the population is *known* and calculate the chances of obtaining various samples from the population. Thus, we show that probability is the reverse of statistics: In probability, we use the population information to infer the probable nature of the sample.

Probability plays an important role in inference making. Suppose, for example, you have an opportunity to invest in an oil exploration company. Past records show that out of ten previous oil drillings (a sample of the company's experiences), all ten came up dry. What do you conclude? Do you think the chances are better than 50:50 that the company will hit a gusher? Should you invest in this company? Chances are, your answer to these questions will be an emphatic No. If the company's exploratory prowess is sufficient to hit a producing well 50% of the time, a record of ten dry wells out of ten drilled is an event that is just too improbable.

Or suppose you're playing poker with what your opponents assure you is a well-shuffled deck of cards. In three consecutive five-card hands, the person on your right is dealt four aces. Based on this sample of three deals, do you think the cards are being adequately shuffled? Again, your answer is likely to be No because dealing three hands of four aces is just too improbable if the cards were properly shuffled.

Note that the decisions concerning the potential success of the oil drilling company and the adequacy of card shuffling both involve knowing the chance—or probability—of a certain sample result. Both situations were contrived so that you could easily conclude that the probabilities of the sample results were small. Unfortunately, the probabilities of many observed sample results aren't so easy to evaluate intuitively. For these cases we need the assistance of a theory of probability.

3.1   EVENTS, SAMPLE SPACES, AND PROBABILITY

Let's begin our treatment of probability with straightforward examples that are easily described. With the aid of these simple examples, we can introduce important definitions that will help us develop the notion of probability more easily.

Suppose a coin is tossed once and the up face is recorded. The result we see and record is called an *observation,* or *measurement,* and the process of making an observation is called an *experiment.* Notice that our definition of experiment is broader than the one used in the physical sciences, where you tend to picture test tubes, microscopes, and other laboratory equipment. Among other things, statistical experiments may include recording a customer's preference for one of two computer operating systems (DOS or Macintosh), recording a voter's opinion on an important political issue, measuring the amount of dissolved oxygen in a polluted river, observing the closing price of a stock, counting the number of errors in an inventory, and observing the fraction of insects killed by a new insecticide. The point is that a statistical experiment can be almost any act of observation as long as the outcome is uncertain.

> **DEFINITION 3.1**
>
> An **experiment** is an act or process of observation that leads to a single outcome that cannot be predicted with certainty.

Consider the simple experiment of tossing a die and observing the number on the up face. The six possible outcomes to this experiment are:

1. Observe a 1
2. Observe a 2
3. Observe a 3
4. Observe a 4
5. Observe a 5
6. Observe a 6

Note that if this experiment is conducted once, *you can observe one and only one of these six basic outcomes, and the outcome cannot be predicted with certainty.* Also, these possibilities cannot be decomposed into more basic outcomes. Because observing the outcome of an experiment is similar to selecting a sample from a population, the basic possible outcomes to an experiment are called *sample points.**

> **DEFINITION 3.2**
>
> A **sample point** is the most basic outcome of an experiment.

EXAMPLE 3.1

Two coins are tossed, and their up faces are recorded. List all the sample points for this experiment.

Solution

Even for a seemingly trivial experiment, we must be careful when listing the sample points. At first glance, we might expect three basic outcomes: Observe two heads, Observe two tails, or Observe one head and one tail. However, further reflection reveals that the last of these, Observe one head and one tail, can be decomposed into two outcomes: Head on coin 1, Tail on coin 2; and Tail on coin 1, Head on coin 2.[†] Thus, we have four sample points:

1. Observe *HH*
2. Observe *HT*
3. Observe *TH*
4. Observe *TT*

where *H* in the first position means "Head on coin 1," *H* in the second position means "Head on coin 2," and so on.   ▲

We often wish to refer to the collection of all the sample points of an experiment. This collection is called the *sample space* of the experiment. For example, there are six sample points in the sample space associated with the die-toss experiment. The sample spaces for the experiments discussed thus far are shown in Table 3.1.

*Alternatively, the term "simple event" can be used.
[†]Even if the coins are identical in appearance, there are, in fact, two distinct coins. Thus, the designation of one coin as coin 1 and the other as coin 2 is legitimate in any case.

TABLE 3.1   Experiments and Their Sample Spaces

Experiment: Observe the up face on a coin.
Sample space: 1. Observe a head
                           2. Observe a tail
This sample space can be represented in set notation as a set containing two sample points:

$$S: \{H, T\}$$

where H represents the sample point Observe a head and T represents the sample point Observe a tail.

Experiment: Observe the up face on a die.
Sample space: 1. Observe a 1
                           2. Observe a 2
                           3. Observe a 3
                           4. Observe a 4
                           5. Observe a 5
                           6. Observe a 6
This sample space can be represented in set notation as a set of six sample points:

$$S: \{1, 2, 3, 4, 5, 6\}$$

Experiment: Observe the up faces on two coins.
Sample space: 1. Observe *HH*
                           2. Observe *HT*
                           3. Observe *TH*
                           4. Observe *TT*
This sample space can be represented in set notation as a set of four sample points:

$$S: \{HH, HT, TH, TT\}$$

a. Experiment: Observe the up face on a coin

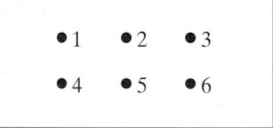

b. Experiment: Observe the up face on a die

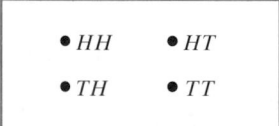

c. Experiment: Observe the up faces on two coins

FIGURE 3.1

Venn diagrams for the three experiments from Table 3.1

DEFINITION 3.3

The **sample space** of an experiment is the collection of all its sample points.

Just as graphs are useful in describing sets of data, a pictorial method for presenting the sample space can often be useful. Figure 3.1 shows such a representation for each of the experiments in Table 3.1. In each case, the sample space is shown as a closed figure, labeled *S,* containing all possible sample points. Each sample point is represented by a solid dot (i.e., a "point") and labeled accordingly. Such graphical representations are known as **Venn diagrams**.

Now that we know that an experiment will result in *only one* basic outcome—called a sample point—and that the sample space is the collection of all possible sample points, we're ready to discuss the probabilities of the sample points. You have undoubtedly used the term *probability* and have some intuitive idea about its meaning. Probability is generally used synonymously with "chance," "odds," and similar concepts. For example, if a fair coin is tossed, we might reason that both the sample points, Observe a head and Observe a tail, have the same *chance* of occurring. Thus, we might state that "the probability of observing a head is 50%" or "the *odds* of seeing a head are 50:50." Both these statements are based on an informal knowledge of probability. We'll begin our treatment of probability by using such informal concepts and then solidify what we mean later.

FIGURE 3.2

Proportion of heads in *N* tosses of a coin

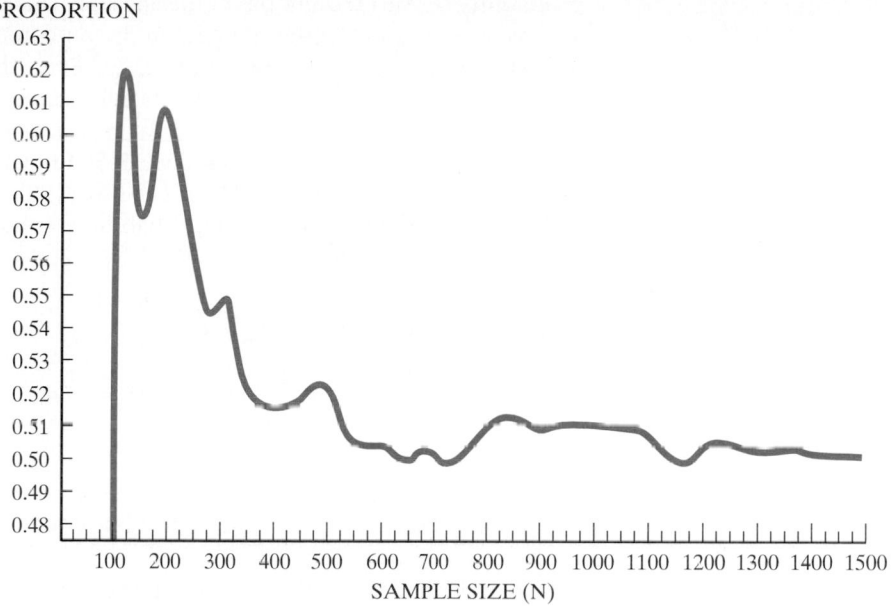

PROPORTION

The probability of a sample point is a number between 0 and 1 that measures the likelihood that the outcome will occur when the experiment is performed. This number is usually taken to be the relative frequency of the occurrence of a sample point in a very long series of repetitions of an experiment.* For example, if we are assigning probabilities to the two sample points in the coin-toss experiment (Observe a head and Observe a tail), we might reason that if we toss a balanced coin a very large number of times, the sample points Observe a head and Observe a tail will occur with the same relative frequency of .5.

Our reasoning is supported by Figure 3.2. The figure plots the relative frequency of the number of times that a head occurs when simulating (by computer) the toss of a coin *N* times, where *N* ranges from as few as 25 tosses to as many as 1,500 tosses of the coin. You can see that when *N* is large (i.e., *N* = 1,500), the relative frequency is converging to .5. Thus, the probability of each sample point in the coin-tossing experiment is .5.

For some experiments, we may have little or no information on the relative frequency of occurrence of the sample points; consequently, we must assign probabilities to the sample points based on general information about the experiment. For example, if the experiment is to invest in a business venture and to observe whether it succeeds or fails, the sample space would appear as in Figure 3.3. We are unlikely to be able to assign probabilities to the sample points of this experiment based on a long series of repetitions since unique factors govern each performance of this kind of experiment. Instead, we may consider factors such as the personnel managing the venture, the general state of the economy at the time, the rate of success of similar ventures, and any other pertinent information. If we finally decide that the venture has an 80% chance of succeeding, we assign a

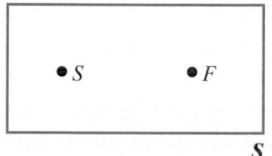

FIGURE 3.3

Experiment: Invest in a business venture and observe whether it succeeds (*S*) or fails (*F*)

---

*The result derives from an axiom in probability theory called the **Law of Large Numbers**. Phrased informally, this law states that the relative frequency of the number of times that an outcome occurs when an experiment is replicated over and over again (i.e., a large number of times) approaches the true (or theoretical) probability of the outcome.

probability of .8 to the sample point Success. This probability can be interpreted as a measure of our degree of belief in the outcome of the business venture; that is, it is a subjective probability. Notice, however, that such probabilities must be based on expert information that is carefully assessed. If not, we may be misled on any decisions based on these probabilities or based on any calculations in which they appear. [*Note:* For a text that deals in detail with the subjective evaluation of probabilities, see Winkler (1972) or Lindley (1985).]

No matter how you assign the probabilities to sample points, the probabilities assigned must obey two rules:

Probability Rules for Sample Points

1. All sample point probabilities *must* lie between 0 and 1.
2. The probabilities of all the sample points within a sample space *must* sum to 1.

Assigning probabilities to sample points is easy for some experiments. For example, if the experiment is to toss a fair coin and observe the face, we would probably all agree to assign a probability of ½ to the two sample points, Observe a head and Observe a tail. However, many experiments have sample points whose probabilities are more difficult to assign.

EXAMPLE 3.2

A retail computer store sells two basic types of personal computers (PCs): standard desktop units and laptop units. Thus the owner must decide how many of each type of PC to stock. An important factor affecting the solution is the proportion of customers who purchase each type of PC. Show how this problem might be formulated in the framework of an experiment with sample points and a sample space. Indicate how probabilities might be assigned to the sample points.

Solution

If we use the term *customer* to refer to a person who purchases one of the two types of PCs, the experiment can be defined as the entrance of a customer and the observation of which type of PC is purchased. There are two sample points in the sample space corresponding to this experiment:

> D: {The customer purchases a standard desktop unit}
>
> L: {The customer purchases a laptop unit}

The difference between this and the coin-toss experiment becomes apparent when we attempt to assign probabilities to the two sample points. What probability should we assign to the sample point D? If you answer .5, you are assuming that the events D and L should occur with equal likelihood, just like the sample points Heads and Tails in the coin-toss experiment. But assignment of sample point probabilities for the PC purchase experiment is not so easy. Suppose a check of the store's records indicates that 80% of its customers purchase desktop units. Then it might be reasonable to approximate the probability of the sample point D as .8 and that of the sample point L as .2. Here we see that sample points are not always equally likely so assigning probabilities to them can be complicated—particularly for experiments that represent real applications (as opposed to coin- and die-toss experiments). ▲

Although the probabilities of sample points are often of interest in their own right, it is usually probabilities of collections of sample points that are important. Example 3.3 demonstrates this point.

EXAMPLE 3.3

A fair die is tossed, and the up face is observed. If the face is even, you win $1. Otherwise, you lose $1. What is the probability that you win?

Solution

Recall that the sample space for this experiment contains six sample points:

$$S: \{1, 2, 3, 4, 5, 6\}$$

Since the die is balanced, we assign a probability of ⅙ to each of the sample points in this sample space. An even number will occur if one of the sample points, Observe a 2, Observe a 4, or Observe a 6, occurs. A collection of sample points such as this is called an *event*, which we denote by the letter A. Since the event A contains three sample points—each with probability ⅙—and since no sample points can occur simultaneously, we reason that the probability of A is the sum of the probabilities of the sample points in A. Thus, the probability of A is ⅙ + ⅙ + ⅙ = ½. This probability implies that, *in the long run*, you will win $1 half the time and lose $1 half the time. ▲

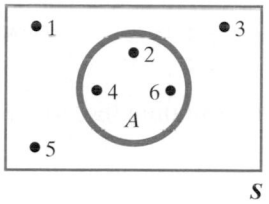

FIGURE 3.4

Die-toss experiment with event *A:* Observe an even number

Figure 3.4 is a Venn diagram depicting the sample space associated with a die-toss experiment and the event A, Observe an even number. The event A is represented by the closed figure inside the sample space S. This closed figure A contains all the sample points that comprise it.

To decide which sample points belong to the set associated with an event A, test each sample point in the sample space S. If event A occurs, then that sample point is in the event A. For example, the event A, Observe an even number, in the die-toss experiment will occur if the sample point Observe a 2 occurs. By the same reasoning, the sample points Observe a 4 and Observe a 6 are also in event A.

To summarize, we have demonstrated that an event can be defined in words or it can be defined as a specific set of sample points. This leads us to the following general definition of an event:

DEFINITION 3.4

An **event** is a specific collection of sample points.

EXAMPLE 3.4

Consider the experiment of tossing two coins. Suppose the coins are *not* balanced and the correct probabilities associated with the sample points are given in the accompanying table. [*Note:* The necessary properties for assigning probabilities to sample points are satisfied.]

Consider the events

$$A: \{\text{Observe exactly one head}\}$$
$$B: \{\text{Observe at least one head}\}$$

Calculate the probability of A and the probability of B.

Sample Point	Probability
HH	$4/9$
HT	$2/9$
TH	$2/9$
TT	$1/9$

Solution

Event *A* contains the sample points *HT* and *TH*. Since two or more sample points cannot occur at the same time, we can easily calculate the probability of event *A* by summing the probabilities of the two sample points. Thus, the probability of observing exactly one head (event *A*), denoted by the symbol $P(A)$, is

$$P(A) = P(\text{Observe } HT) + P(\text{Observe } TH) = 2/9 + 2/9 = 4/9$$

Similarly, since *B* contains the sample points *HH, HT,* and *TH,*

$$P(B) = 4/9 + 2/9 + 2/9 = 8/9 \qquad \blacktriangle$$

The preceding example leads us to a general procedure for finding the probability of an event *A:*

> The probability of an event *A* is calculated by summing the probabilities of the sample points in the sample space for *A*.

Thus, we can summarize the steps for calculating the probability of any event as indicated in the next box.

Steps for Calculating the Probabilities of Events
___

1. Define the experiment; that is, describe the process used to make an observation and the type of observation that will be recorded.
2. List the sample points.
3. Assign probabilities to the sample points.
4. Determine the collection of sample points contained in the event of interest.
5. Sum the sample point probabilities to get the event probability.

EXAMPLE 3.5

Ohio's Learning, Earning, and Parenting (LEAP) program is designed to encourage school attendance among pregnant and parenting teens on welfare. Eligible teens who provide evidence of school enrollment receive a bonus payment, while those who do not attend school have money deducted from their grant (i.e., the teens are "sanctioned"). In a survey of LEAP teens, published in *Children and Youth Services Review* (Vol. 17, 1995), the bonus and/or sanction requests were recorded for each teen. The results are summarized in Table 3.2.

a. Define the experiment that generated the data in Table 3.2 and list the sample points.
b. Assign probabilities to the sample points.
c. What is the probability that both bonuses and sanctions are requested for a LEAP teen?

TABLE 3.2   Bonus and Sanction Requests for Ohio LEAP Teens

Request	Percentage
No bonuses or sanctions (*N*)	7
Only bonuses (*OB*)	37
Only sanctions (*OS*)	18
Both, more bonuses than sanctions (*BB*)	14
Both, more sanctions than bonuses (*BS*)	18
Both, equal number of bonuses and sanctions (*BE*)	6
Total	100%

Source: Reprinted from *Children and Youth Services Review*, Vol. 17, Nos. 1/2, Wood, R. G., *et al.* "Encouraging school enrollment and attendance among teenage parents on welfare: Early impacts of Ohio's LEAP program." p. 302, Figure 1, with kind permission from Elsevier Science Ltd., The Boulevard, Langford Lane, Kidlington OX5 1GB, UK.

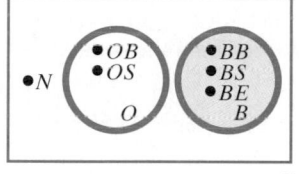

FIGURE 3.5

Venn diagram for LEAP teen survey

d. What is the probability that only one type of request (either a bonus or a sanction, but not both) is made?

Solution

a. The experiment is the act of surveying an Ohio LEAP teen. The sample points, the simplest outcomes of the experiment, are the six response categories listed in Table 3.2. These sample points are shown in the Venn diagram in Figure 3.5.

b. If, as in Example 3.1 we were to assign equal probabilities in this case, each of the response categories would have a probability of one-sixth (⅙), or .167. But, by examining Table 3.2 you can see that equal probabilities are not reasonable here because the response percentages were not even approximately the same in the six classifications. It is more reasonable to assign a probability equal to the response percentage in each class, as shown in Table 3.3.*

c. The event *B* that both bonuses and sanctions are requested is not a sample point because it consists of more than one of the response classifications (the sample points). In fact, as shown in Figure 3.5, *B* consists of three sample points. The probability of *B* is defined to be the sum of the probabilities of the sample points in *B:*

$$P(B) = P(BB) + P(BS) + P(BE) = .14 + .18 + .06 = .38$$

d. The event *O* that only bonuses or only sanctions are requested consists of two sample points, and the probability is the sum of the corresponding sample point probabilities:

$$P(O) = P(OB) + P(OS) = .37 + .18 = .55 \qquad \blacktriangle$$

TABLE 3.3   Sample Point Probabilities for LEAP Teen Survey

Sample Point	Probability
N	.07
OB	.37
OS	.18
BB	.14
BS	.18
BE	.06

For the experiments discussed so far, listing the sample points has been easy. For more complex experiments, the number of sample points may be so large that listing them is impractical. In solving probability problems for experiments with many sample points, we employ the same principles that we use for experiments

*Since the response percentages were based on a sample of LEAP teens, these assigned probabilities are estimates of the true population response percentages. You'll learn how to measure the reliability of probability estimates in Chapter 7.

Game Show Strategy: To Switch or Not to Switch?

Marilyn vos Savant, who is listed in the *Guinness Book of World Records Hall of Fame* for "Highest IQ," writes a monthly column in the Sunday newspaper supplement, *Parade Magazine*. Her column, "Ask Marilyn," is devoted to games of skill, puzzles, and mind-bending riddles. In one issue (*Parade Magazine,* Feb. 17, 1991), vos Savant posed the following question:

CASE STUDY 3.1

> Suppose you're on a game show, and you're given a choice of three doors. Behind one door is a car; behind the others, goats. You pick a door—say, #1—and the host, who knows what's behind the doors, opens another door—say #3—which has a goat. He then says to you, "Do you want to pick door #2?" Is it to your advantage to switch your choice?

Marilyn's answer: "Yes, you should switch. The first door has a ⅓ chance of winning [the car], but the second has a ⅔ chance [of winning the car]."Predictably, vos Savant's surprising answer elicited thousands of critical letters, many of them from Ph.D. mathematicians, who disagreed with her. Some of the more interesting and critical letters, which were printed in her next column (*Parade Magazine,* Feb. 24, 1991) are condensed below:

- "May I suggest you obtain and refer to a standard textbook on probability before you try to answer a question of this type again?" (University of Florida)
- "Your logic is in error, and I am sure you will receive many letters on this topic from high school and college students. Perhaps you should keep a few addresses for help with future columns." (Georgia State University)
- "You are utterly incorrect about the game-show question, and I hope this controversy will call some public attention to the serious national crisis in mathematical education. If you can admit your error you will have contributed constructively toward the solution of a deplorable situation. How many irate mathematicians are needed to get you to change your mind?" (Georgetown University)
- "I am in shock that after being corrected by at least three mathematicians, you still do not see your mistake." (Dickinson State University)

- "You are the goat!" (Western State University)
- "You're wrong, but look on the positive side. If all the Ph.D.'s were wrong, the country would be in serious trouble." (U.S. Army Research Institute)

The logic employed by those who disagree with vos Savant is as follows: Once the host shows you door #3 (a goat), only two doors remain. The probability of the car being behind door #1 (your door) is ½; similarly, the probability is ½ for door #2. Therefore, in the long run (i.e., over a long series of trials) it doesn't matter whether you switch to door #2 or keep door #1. Approximately 50% of the time you'll win the car, and 50% of the time you'll get the goat.

Who is correct, the Ph.D.'s or Marilyn? By answering the following series of questions, you'll arrive at the correct solution.

Focus

a. Before the show is taped, the host randomly decides the door behind which to put the car; then the goats go behind the remaining two doors. List the sample points for this experiment.

b. Suppose you choose at random door #1. Now, for each sample point in part a, circle door #1 and put an X through one of the remaining two doors that hides a goat. (This is the door that the host shows—always a goat.)

c. Refer to the altered sample points in part b. Assume your strategy is to keep door #1. Count the number of sample points for which this is a "winning" strategy (i.e., you win the car). Assuming equally likely sample points, what is the probability that you win the car?

d. Repeat part c, but assume your strategy is to always switch doors.

e. Based on the probabilities of parts c and d, is it to your advantage to switch your choice?

with few sample points. The only difference is that we need *counting rules* for determining the number of sample points without actually enumerating all of them. In Section 3.9 (an optional section), we present several of the more useful counting rules.

EXERCISES 3.1–3.19

Learning the Mechanics

3.1 An experiment results in one of the following sample points: E_1, E_2, E_3, E_4, or E_5.
   a. Find $P(E_3)$ if $P(E_1) = .1$, $P(E_2) = .2$, $P(E_4) = .1$, and $P(E_5) = .1$.
   b. Find $P(E_3)$ if $P(E_1) = P(E_3)$, $P(E_2) = .1$, $P(E_4) = .2$, and $P(E_5) = .1$.
   c. Find $P(E_3)$ if $P(E_1) = P(E_2) = P(E_4) = P(E_5) = .1$.

3.2 The accompanying diagram describes the sample space of a particular experiment and events A and B.
   a. What is this type of diagram called?
   b. Suppose the sample points are equally likely. Find $P(A)$ and $P(B)$.
   c. Suppose $P(1) = P(2) = P(3) = P(4) = P(5) = \frac{1}{20}$ and $P(6) = P(7) = P(8) = P(9) = P(10) = \frac{3}{20}$. Find $P(A)$ and $P(B)$.

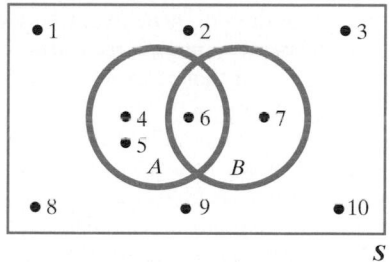

S

3.3 The sample space for an experiment contains five sample points with probabilities as shown in the table. Find the probability of each of the following events:

   A: {Either 1, 2, or 3 occurs}

   B: {Either 1, 3, or 5 occurs}

   C: {4 does not occur}

Sample Points	Probabilities
1	.05
2	.20
3	.30
4	.30
5	.15

3.4 Consider the experiment of tossing a die and observing the up face.
   a. Draw a Venn diagram for the experiment. On your diagram indicate the event Observe a number greater than 4. Call this event A. Also indicate the event Observe an even number. Call this event B.
   b. We would all agree that the probability of observing a 3 on the toss of a "fair" die is $\frac{1}{6}$. Explain what it means for a die to be fair (or balanced) and explain how knowing that a die is fair leads us to $P(3) = \frac{1}{6}$.

   c. For your Venn diagram of part **a**, assume the die is fair and find $P(A)$ and $P(B)$.
   d. If you knew that a particular die was unfair (i.e., "loaded"), how would you determine the probability of observing a 3?

3.5 The accompanying Venn diagram depicts an experiment with six sample points. The events A and B are also shown. The probabilities of the sample points are:

$$P(1) = P(2) = P(4) = \frac{2}{9}$$
$$P(3) = P(5) = P(6) = \frac{1}{9}$$

   a. Find $P(A)$.
   b. Find $P(B)$.
   c. Find the probability that the events A and B occur *simultaneously*.

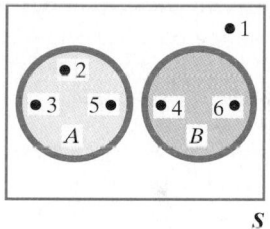

S

3.6 Two fair dice are tossed, and the up face on each die is recorded.
   a. List the 36 sample points contained in the sample space.
   b. Find the probability of observing each of the following events:

   A: {A 3 appears on each of the two dice}

   B: {The sum of the numbers is even}

   C: {The sum of the numbers is equal to 7}

   D: {A 5 appears on at least one of the dice}

   E: {The sum of the numbers is 10 or more}

3.7 Consider the experiment composed of one roll of a fair die followed by one toss of a fair coin. List the sample points. Assign a probability to each sample point. Determine the probability of observing each of the following events:

   A: {6 on the die; H on the coin}

   B: {Even number on the die; T on the coin}

   C: {Even number on the die}

   D: {T on the coin}

3.8 Two marbles are drawn at random and without replacement from a box containing two blue marbles and three red marbles. Determine the probability of observing each of the following events:

A: {Two blue marbles are drawn}

B: {A red and a blue marble are drawn}

C: {Two red marbles are drawn}

3.9 Simulate the experiment described in Exercise 3.8 using any five identically shaped objects, two of which are one color and three, another. Mix the objects, draw two, record the results, and then replace the objects. Repeat the experiment a large number of times (at least 100). Calculate the proportion of time events *A, B,* and *C* occur. How do these proportions compare with the probabilities you calculated in Exercise 3.8? Should these proportions equal the probabilities? Explain.

Applying the Concepts

3.10 Diversity training of employees is the latest trend in U.S. business. *USA Today* (Aug. 15, 1995) reported on the primary reasons businesses give for making diversity training part of their strategic planning process. The reasons are summarized in the pie chart reproduced here. Assume that one business is selected at random from all U.S. businesses that use diversity training and the primary reason is determined.
   a. List the sample points for this experiment.
   b. Assign reasonable probabilities to the sample points. Verify that the probabilities sum to 1.
   c. What is the probability that increasing productivity or staying competitive is the primary reason for diversity training?
   d. What is the probability that social responsibility is not the primary reason for diversity training?

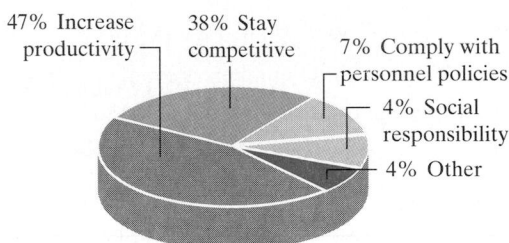

47% Increase productivity

38% Stay competitive

7% Comply with personnel policies

4% Social responsibility

4% Other

3.11 In Exercise 2.7 you read about a *Health Education Journal* (Sept. 1994) study of media coverage of stories involving mental illness in Scotland. The media coverage of each of 562 stories was classified by type, with the results shown in the accompanying table. Assume that one of the 562 stories on mental illness is selected and the type of media coverage is noted.
   a. List the sample points for this experiment.
   b. Explain why the sample points, part **a**, are not equally likely to occur.
   c. Assign reasonable probabilities to the sample points.
   d. Find the probability that a randomly selected story on mental health is portrayed sympathetically or comically by the Scottish media.

Media Coverage	Number of Items
Violence to others	373
Sympathetic	102
Harm to self	71
Comic images	12
Criticism of definitions	4
Total	562

Source: Philo, G. *et al.* "The impact of the mass media on public images of mental illness: Media content and audience belief." *Health Education Journal,* Vol. 53, No. 3, Sept. 1994, p. 274 (Table 1).

3.12 An individual's genetic makeup is determined by the genes obtained from each parent. For every genetic trait, each parent possesses a gene pair; and each contributes one-half of this gene pair, with equal probability, to their offspring, forming a new gene pair. The offspring's traits (eye color, baldness, etc.) come from this new gene pair, where each gene in this pair possesses some characteristic.

For the gene pair that determines eye color, each gene trait may be one of two types: dominant brown (*B*) or recessive blue (*b*). A person possessing the gene pair *BB* or *Bb* has brown eyes, whereas the gene pair *bb* produces blue eyes.
   a. Suppose both parents of an individual are brown-eyed, each with a gene pair of the type *Bb*. What is the probability that a randomly selected child of this couple will have blue eyes? [*Hint:* Construct the sample space for the experiment.]
   b. If one parent has brown eyes, type *Bb,* and the other has blue eyes, what is the probability that a randomly selected child of this couple will have blue eyes?
   c. Suppose one parent is brown-eyed, type *BB.* What is the probability that a child has blue eyes?

3.13 Scientists have already discovered two genes (on chromosome 19 and 21) that mutate to cause the early onset of Alzheimer's disease. *Science News* (July 8, 1995) reported on a search for a third gene that causes Alzheimer's. An international team of scientists gathered genetic information from 21 families afflicted by the early-onset form of the disease. In six of these families, mutations in the S182 gene on chromosome 14 accounted for early-onset Alzheimer's.
   a. Consider a family afflicted by early-onset Alzheimer's disease. Find the approximate probability that the researchers will find mutations in the S182 gene on chromosome 14 for this family.
   b. Discuss the reliability of the probability, part **a**. How could a more accurate estimate of the probability be obtained?

3.14 Consider the following question posed to Marilyn vos Savant in her weekly newspaper column "Ask Marilyn":

I have two pairs of argyle socks, and they look nearly identical—one navy blue and the other black. [When doing the laundry] my wife matches the socks incorrectly much more often than she does correctly. ... If all four socks are in front of her, it seems to me that her chances are 50% for a wrong match and 50% for a right match. What do you think? *Source: Parade Magazine,* Feb. 27, 1994.

Use your knowledge of probability to answer this question. [*Hint:* List the sample points in the experiment.]

3.15 Three people play a game called "Odd Man Out." In this game, each player flips a fair coin until the outcome (heads or tails) for one of the players is not the same as the other two players'. This player is then "the odd man out" and loses the game. Find the probability that the game ends (i.e., either exactly one of the coins will fall heads or exactly one of the coins will fall tails) after only one toss by each player. Suppose one of the players, hoping to reduce the chances of being the odd man, uses a two-headed coin. Will this ploy be successful? Solve by listing the sample points in the sample space.

3.16 Carbon monoxide (CO) is an odorless, colorless, highly toxic gas which is produced by fires as well as by motor vehicles and appliances that burn carbon-based fuels. The *American Journal of Public Health* (July 1995) published a study on unintentional CO poisoning of Colorado residents for the years 1986–1991. A total of 981 cases of CO poisoning were reported during the six-year period. Each case was classified as fatal or nonfatal and by source of exposure. The number of cases occurring in each of the categories is shown in the accompanying table. Assume that one of the 981 cases of unintentional CO poisoning is randomly selected.

Source of Exposure	Fatal	Nonfatal	Total
Fire	63	53	116
Auto exhaust	60	178	238
Furnace	18	345	363
Kerosene or spaceheater	9	18	27
Appliance	9	63	72
Other gas-powered motor	3	73	76
Fireplace	0	16	16
Other	3	19	22
Unknown	9	42	51
Total	174	807	981

Source: Cook, M. C., Simon, P. A., and Hoffman, R. E. "Unintentional carbon monoxide poisoning in Colorado, 1986 through 1991." *American Journal of Public Health,* Vol. 85, No. 7, July 1995, p. 989 (Table 1). © 1995 American Public Health Association.

   a. List all sample points for this experiment.
   b. What is the set of all sample points called?
   c. Let A be the event that the CO poisoning is caused by fire. Find $P(A)$.
   d. Let B be the event that the CO poisoning is fatal. Find $P(B)$.
   e. Let C be the event that the CO poisoning is caused by auto exhaust. Find $P(C)$.
   f. Let D be the event that the CO poisoning is caused by auto exhaust and is fatal. Find $P(D)$.
   g. Let E be the event that the CO poisoning is caused by fire but is nonfatal. Find $P(E)$.

3.17 Often, probabilities are expressed in terms of **odds**, especially in gambling settings. For example, handicappers for horse races express their belief about the probabilities that each horse will win a race in terms of odds. If the probability of event E is $P(E)$, then the *odds in favor of E* are $P(E)$ to $1 - P(E)$. Thus, if a handicapper assesses a probability of .25 that Snow Chief will win the Belmont Stakes, the odds in favor of Snow Chief are $^{25}/_{100}$ to $^{75}/_{100}$, or 1 to 3. It follows that the *odds against E* are $1 - P(E)$ to $P(E)$, or 3 to 1 against a win by Snow Chief. In general, if the odds in favor of event E are a to b, then $P(E) = a/(a + b)$.

   a. A second handicapper assesses the probability of a win by Snow Chief to be $\frac{1}{3}$. According to the second handicapper, what are the odds in favor of a Snow Chief win?
   b. A third handicapper assesses the odds in favor of Snow Chief to be 1 to 1. According to the third handicapper, what is the probability of a Snow Chief win?
   c. A fourth handicapper assesses the odds against Snow Chief's winning to be 3 to 2. Find this handicapper's assessment of the probability that Snow Chief will win.

3.18 Before placing a person in a highly skilled position, a company gives the applicants a series of three examinations. The first is a physical examination, and each applicant is classified as satisfactory or unsatisfactory. The other two are verbal and quantitative examinations, and the scores are used to classify each applicant as high, medium, or low in each area. Thus, each individual will receive a health score, a verbal score, and a quantitative score.

   a. List the different sets of classifications that can result from this battery of examinations.
   b. If all applicants who take the examinations are equally qualified, and all the variation in test scores is random, what is the probability that an applicant will receive the lowest classification on all three examinations?
   c. If an applicant scores in the highest category on at least two of the three examinations, the applicant will get a position. What is the probability that a randomly selected applicant will get a position? (Make the same assumption as in part **b.**)

3.19 If you earn a Ph.D. in one of the fields in science or engineering, how likely are you to get a full-time job in your specialty? According to a recent survey of doctorate-granting institutions and educational institutions that employ people with doctorates, "Universities in the U.S. are producing about 25 percent more doctorates in science and engineering fields than the U.S. economy can afford" (*New York Times,* July 4, 1995). Suppose there are 1,000 jobs available for new Ph.D.'s in science and engineering in a given year. Then, according to the survey, 1,250 new Ph.D.'s will be eligible for these jobs.

a. Assuming that all new Ph.D.'s actually apply for these jobs and that each is equally qualified, find the probability that a new earned Ph.D. in science or engineering will obtain one of the available jobs.
b. Which of the assumptions in **a** is not likely to be true? Explain.

3.2 UNIONS AND INTERSECTIONS

An event can often be viewed as a composition of two or more other events. Such events, which are called **compound events**, can be formed (composed) in two ways, as defined and illustrated here.

> **DEFINITION 3.5**
>
> The **union** of two events A and B is the event that occurs if either A or B or both occur on a single performance of the experiment. We denote the union of events A and B by the symbol $A \cup B$. $A \cup B$ consists of all the sample points that belong to A or B or *both*. (See Figure 3.6a.)

> **DEFINITION 3.6**
>
> The **intersection** of two events A and B is the event that occurs if both A and B occur on a single performance of the experiment. We write $A \cap B$ for the intersection of A and B. $A \cap B$ consists of all the sample points belonging to *both* A and B. (See Figure 3.6b.)

EXAMPLE 3.6

Consider the die-toss experiment. Define the following events:

A: {Toss an even number}
B: {Toss a number less than or equal to 3}

a. Describe $A \cup B$ for this experiment.
b. Describe $A \cap B$ for this experiment.
c. Calculate $P(A \cup B)$ and $P(A \cap B)$ assuming the die is fair.

Solution

Draw the Venn diagram as shown in Figure 3.7.

FIGURE 3.6

Venn diagrams for union and intersection

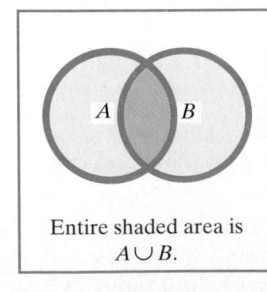

Entire shaded area is
$A \cup B$.

a. Union

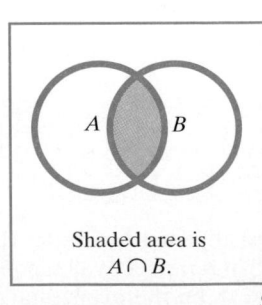

Shaded area is
$A \cap B$.

b. Intersection

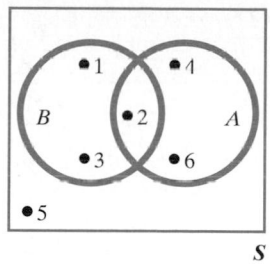

FIGURE 3.7

Venn diagram for die toss

a. The union of *A* and *B* is the event that occurs if we observe either an even number, a number less than or equal to 3, or both on a single throw of the die. Consequently, the sample points in the event $A \cup B$ are those for which *A* occurs, *B* occurs, or both *A* and *B* occur. Checking the sample points in the entire sample space, we find that the collection of sample points in the union of *A* and *B* is

$$A \cup B = \{1, 2, 3, 4, 6\}$$

b. The intersection of *A* and *B* is the event that occurs if we observe *both* an even number and a number less than or equal to 3 on a single throw of the die. Checking the sample points to see which imply the occurrence of *both* events *A* and *B*, we see that the intersection contains only one sample point:

$$A \cap B = \{2\}$$

In other words, the intersection of *A* and *B* is the sample point Observe a 2.

c. Recalling that the probability of an event is the sum of the probabilities of the sample points of which the event is composed, we have

$$P(A \cup B) = P(1) + P(2) + P(3) + P(4) + P(6)$$
$$= \tfrac{1}{6} + \tfrac{1}{6} + \tfrac{1}{6} + \tfrac{1}{6} + \tfrac{1}{6} = \tfrac{5}{6}$$

and

$$P(A \cap B) = P(2) = \tfrac{1}{6} \qquad \blacktriangle$$

Unions and intersections can be defined for more than two events. For example, the event $A \cup B \cup C$ represents the union of three events, *A*, *B*, and *C*. This event, which includes the set of sample points in *A*, *B*, or *C*, will occur if any one or more of the events *A*, *B*, or *C* occurs. Similarly, the intersection $A \cap B \cap C$ is the event that all three of the events *A*, *B*, and *C* occur. Therefore, $A \cap B \cap C$ is the set of sample points that are in all three of the events, *A*, *B*, and *C*.

EXAMPLE 3.7   Refer to Example 3.6 and define the event

$$C: \{\text{Toss a number greater than 1}\}$$

Find the sample points in

a. $A \cup B \cup C$
b. $A \cap B \cap C$

where

$$A: \{\text{Toss an even number}\}$$
$$B: \{\text{Toss a number less than or equal to 3}\}$$

Solution

a. Event *C* contains the sample points corresponding to tossing a 2, 3, 4, 5, or 6; event *A* contains the sample points 2, 4, and 6; and event *B* contains the sample points 1, 2, and 3. Therefore, the event that either *A*, *B*, or *C* occurs contains all six sample points in *S*—that is, those corresponding to tossing a 1, 2, 3, 4, 5, or 6.

b. You will observe all of the events *A*, *B*, and *C* only if you observe a 2. Therefore, the intersection $A \cap B \cap C$ contains the single sample point Toss a 2. $\qquad \blacktriangle$

TABLE 3.4   Percentage of Respondents in Age–Income Classes

Age	INCOME		
	<$25,000	**$25,000–$50,000**	**>$50,000**
<30 yrs	5%	12%	10%
30–50 yrs	14%	22%	16%
>50 yrs	8%	10%	3%

EXAMPLE 3.8

Many firms undertake direct marketing campaigns to promote their products. The campaigns typically involve mailing information to millions of households. The response rates are carefully monitored to determine the demographic characteristics of respondents. By studying tendencies to respond, the firms can better target future mailings to those segments of the population most likely to purchase their products.

Suppose a distributor of mail-order tools is analyzing the results of a recent mailing. The probability of response is believed to be related to income and age. The percentages of the total number of respondents to the mailing are given by income and age classification in Table 3.4.

Define the following events:

> *A:* {A respondent's income is more than $50,000}
>
> *B:* {A respondent's age is 30 or more}

a. Find $P(A)$ and $P(B)$.
b. Find $P(A \cup B)$.
c. Find $P(A \cap B)$.

Solution

Following the steps for calculating probabilities of events, we first note that the objective is to characterize the income and age distribution of respondents to the mailing. To accomplish this, we define the experiment to consist of selecting a respondent from the collection of all respondents and observing which income and age class he or she occupies. The sample points are the nine different age–income classifications:

$$E_1: \{<30 \text{ yrs}, <\$25,000\}$$
$$E_2: \{30\text{–}50 \text{ yrs}, <\$25,000\}$$
$$\vdots \qquad\qquad \vdots$$
$$E_9: \{>50 \text{ yrs}, >\$50,000\}$$

Next, we assign probabilities to the sample points. If we blindly select one of the respondents, the probability that he or she will occupy a particular age–income classification is just the proportion, or relative frequency, of respondents in the classification. These proportions are given (as percentages) in Table 3.4. Thus,

$$P(E_1) = \text{Relative frequency of respondents in}$$
$$\text{age–income class } \{<30 \text{ yrs}, <\$25,000\}$$
$$= .05$$
$$P(E_2) = .14$$

and so forth. You may verify that the sample points probabilities add to 1.

a. To find $P(A)$, we first determine the collection of sample points contained in event A. Since A is defined as {>$50,000}, we see from Table 3.4 that A contains the three sample points represented by the last column of the table. In words, the event A consists of the income classification {>$50,000} in all three age classifications. The probability of A is the sum of the probabilities of the sample points in A:

$$P(A) = .10 + .16 + .03 = .29$$

Similarly, B consists of the six sample points in the second and third rows of Table 3.4:

$$P(B) = .14 + .22 + .16 + .08 + .10 + .03 = .73$$

b. The union of events A and B, $A \cup B$, consists of all the sample points in *either A or B or both.* That is, the union of A and B consists of all respondents whose income exceeds $50,000 *or* whose age is 30 or more. In Table 3.4 this is any sample point found in the third column *or* the last two rows. Thus,

$$P(A \cup B) = .10 + .14 + .22 + .16 + .08 + .10 + .03 = .83$$

c. The intersection of events A and B, $A \cap B$, consists of all sample points in *both A and B.* That is, the intersection of A and B consists of all respondents whose income exceeds $50,000 *and* whose age is 30 or more. In Table 3.4 this is any sample point found in the third column *and* the last two rows. Thus,

$$P(A \cap B) = .16 + .03 = .19 \qquad \blacktriangle$$

3.3 COMPLEMENTARY EVENTS

A very useful concept in the calculation of event probabilities is the notion of **complementary events**:

> **DEFINITION 3.7**
>
> The **complement** of an event A is the event that A does *not* occur—that is, the event consisting of all sample points that are not in event A. We denote the complement of A by A'.

An event A is a collection of sample points, and the sample points included in A' are those not in A. Figure 3.8 demonstrates this idea. Note from the figure that all sample points in S are included in *either* A or A' and that *no* sample point is in both A and A'. This leads us to conclude that the probabilities of an event and its complement *must sum to 1:*

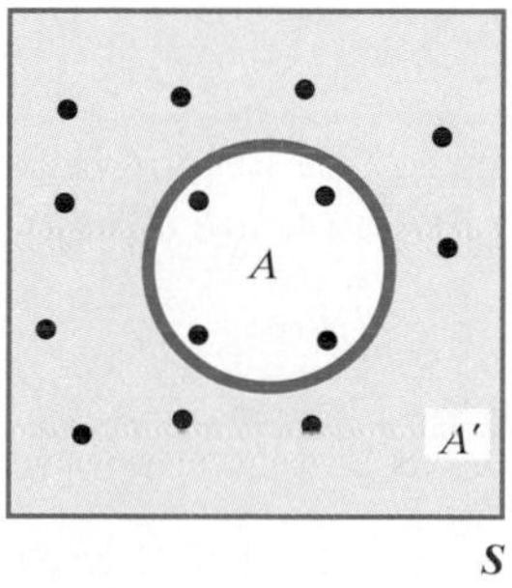

> The sum of the probabilities of complementary events equals 1; that is,
>
> $$P(A) + P(A') = 1$$

In many probability problems calculating the probability of the complement of the event of interest is easier than calculating the event itself. Then, because

$$P(A) + P(A') = 1$$

FIGURE 3.8

Venn diagram of complementary events

we can calculate $P(A)$ by using the relationship

$$P(A) = 1 - P(A')$$

EXAMPLE 3.9

Consider the experiment of tossing two fair coins. Use the complementary relationship to calculate the probability of event A: {Observing at least one head}.

Solution

We know that the event A: {Observing at least one head} consists of the sample points

$$A: \{HH, HT, TH\}$$

The complement of A is defined as the event that occurs when A does not occur. Therefore,

$$A': \{\text{Observe no heads}\} = \{TT\}$$

This complementary relationship is shown in Figure 3.9. Assuming the coins are balanced,

$$P(A') = P(TT) = \tfrac{1}{4}$$

and

$$P(A) = 1 - P(A') = 1 - \tfrac{1}{4} = \tfrac{3}{4}. \qquad \blacktriangle$$

EXAMPLE 3.10

A fair coin is tossed ten times, and the up face is recorded after each toss. What is the probability of event A: {Observe at least one head}?

Solution

We solve this problem by following the five steps for calculating probabilities of events (see Section 3.1).

Step 1 Define the experiment. The experiment is to record the results of the ten tosses of the coin.

Step 2 List the sample points. A sample point consists of a particular sequence of ten heads and tails. Thus, one sample point is *HHTTTHTHTT,* which denotes head on first toss, head on second toss, tail on third toss, etc. Others are *HTHHHTTTTT* and *THHTHTHTTH*. Obviously, the number of sample points is very large—too many to list. It can be shown (see Section 3.9) that there are $2^{10} = 1{,}024$ sample points for this experiment.

Step 3 Assign probabilities. Since the coin is fair, each sequence of heads and tails has the same chance of occurring, and therefore all the sample points are equally likely. Then

$$P(\text{Each sample point}) = \frac{1}{1{,}024}$$

Step 4 Determine the sample points in event A. A sample point is in A if at least one H appears in the sequence of ten tosses. However, if we consider the complement of A, we find that

$$A' = \{\text{No heads are observed in 10 tosses}\}$$

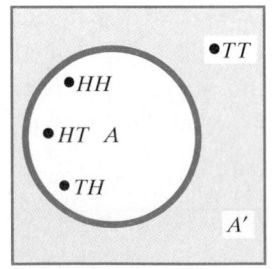

FIGURE 3.9

Complementary events in the toss of two coins

Thus, A' contains only one sample point:

$$A': \{TTTTTTTTTT\}$$

and $P(A') = \dfrac{1}{1,024}.$

Step 5   Now we use the relationship for complementary events to find $P(A)$:

$$P(A) = 1 - P(A') = 1 - \frac{1}{1,024} = \frac{1,023}{1,024} = .999$$

That is, we are virtually certain of observing at least one head in ten tosses of the coin.   ▲

3.4   THE ADDITIVE RULE AND MUTUALLY EXCLUSIVE EVENTS

In Section 3.2 we saw how to determine which sample points are contained in a union and how to calculate the probability of the union by adding the probabilities of the sample points in the union. It is also possible to obtain the probability of the union of two events by using the additive rule, as illustrated in the following example.

EXAMPLE 3.11

A loaded die is tossed and the up face is observed. The following two events are defined:

$A:$ {Observe an even number}

$B:$ {Observe a number less than 3}

Suppose we know that $P(A) = .4$, $P(B) = .2$, and $P(A \cap B) = .1$. Find $P(A \cup B)$.

Solution

By studying the Venn diagram in Figure 3.10, we can obtain information that will help us find $P(A \cup B)$. We can see that

$$P(A \cup B) = P(1) + P(2) + P(4) + P(6)$$

Also, we know that

$$P(A) = P(2) + P(4) + P(6) = .4$$
$$P(B) = P(1) + P(2) = .2$$
$$P(A \cap B) = P(2) = .1$$

If we add the probabilities of the sample points that comprise events A and B, we find

$$P(A) + P(B) = \overbrace{P(2) + P(4) + P(6)}^{P(A)} + \overbrace{P(1) + P(2)}^{P(B)}$$

$$= \overbrace{P(1) + P(2) + P(4) + P(6)}^{P(A \cup B)} + \overbrace{P(2)}^{P(A \cap B)}$$

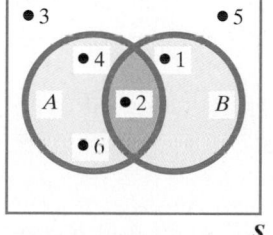

FIGURE 3.10

Venn diagram for die toss

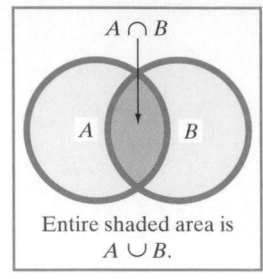

$A \cap B$

A B

Entire shaded area is
$A \cup B$.

S

FIGURE 3.11

Venn diagram of union

Thus, by subtraction, we have

$$P(A \cup B) = P(A) + P(B) - P(A \cap B) = .4 + .2 - .1 = .5 \quad \blacktriangle$$

By studying the Venn diagram in Figure 3.11, you can see that the method used in Example 3.10 can be generalized to find the union of two events for any experiment. The probability of the union of two events, *A* and *B*, can always be obtained by summing $P(A)$ and $P(B)$ and subtracting $P(A \cap B)$. We must subtract $P(A \cap B)$ because the sample point probabilities in $A \cap B$ have been included twice—once in $P(A)$ and once in $P(B)$.

The formula for calculating the probability of the union of two events, often called the *additive rule of probability,* is given in the next box.

Additive Rule of Probability

The probability of the union of events *A* and *B* is the sum of the probability of events *A* and *B* minus the probability of the intersection of events *A* and *B*, that is,

$$P(A \cup B) = P(A) + P(B) - P(A \cap B)$$

EXAMPLE 3.12

Hospital records show that 12% of all patients are admitted for surgical treatment, 16% are admitted for obstetrics, and 2% receive both obstetrics and surgical treatment. If a new patient is admitted to the hospital, what is the probability that the patient will be admitted either for surgery, obstetrics, or both?

Solution

Consider the following events:

  A: {A patient admitted to the hospital receives surgical treatment}
  B: {A patient admitted to the hospital receives obstetrics treatment}

Then, from the given information,

$$P(A) = .12 \qquad P(B) = .16$$

and the probability of the event that a patient receives both obstetrics and surgical treatment is

$$P(A \cap B) = .02$$

The event that a patient admitted to the hospital receives either surgical treatment, obstetrics treatment, or both is the union $A \cup B$. The probability of $A \cup B$ is given by the additive rule of probability:

$$P(A \cup B) = P(A) + P(B) - P(A \cap B)$$
$$= .12 + .16 - .02 = .26$$

Thus, 26% of all patients admitted to the hospital receive either surgical treatment, obstetrics treatment, or both. $\blacktriangle$

A very special relationship exists between events *A* and *B* when $A \cap B$ contains no sample points. In this case we call the events *A* and *B* *mutually exclusive events.*

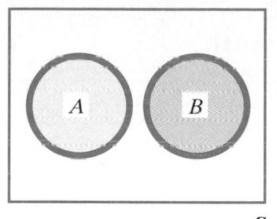

FIGURE 3.12

Venn diagram of mutually exclusive events

DEFINITION 3.8

Events A and B are **mutually exclusive events** if $A \cap B$ contains no sample points, that is, if A and B have no sample points in common.

Figure 3.12 shows a Venn diagram of two mutually exclusive events. The events A and B have no sample points in common; that is, A and B cannot occur simultaneously and $P(A \cap B) = 0$. Thus, we have the important relationship given in the following box.

If two events A and B are *mutually exclusive*, the probability of the union of A and B equals the sum of the probabilities of A and B; that is, $P(A \cup B) = P(A) + P(B)$

Caution: The formula shown above is *false* if the events are *not* mutually exclusive. In this case (i.e., two nonmutually exclusive events), you must apply the general additive rule of probability.

EXAMPLE 3.13

Consider the experiment of tossing two balanced coins. Find the probability of observing *at least* one head.

Solution

Define the events

$A:$ {Observe at least one head}

$B:$ {Observe exactly one head}

$C:$ {Observe exactly two heads}

Note that

$$A = B \cup C$$

and that $B \cap C$ contains no sample points (see Figure 3.13). Thus, B and C are mutually exclusive, so that

$$P(A) = P(B \cup C) = P(B) + P(C) = \tfrac{1}{2} + \tfrac{1}{4} = \tfrac{3}{4} \quad \blacktriangle$$

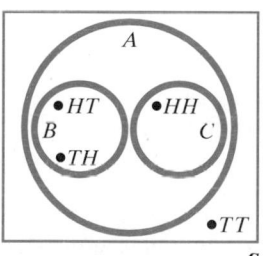

FIGURE 3.13

Venn diagram for coin toss experiment

Although Example 3.13 is very simple, it shows us that writing events with verbal descriptions that include the phrases "at least" or "at most" as unions of mutually exclusive events is very useful. This practice enables us to find the probability of the event by adding the probabilities of the mutually exclusive events.

EXERCISES 3.20–3.33

Learning the Mechanics

3.20 A fair coin is tossed three times and the events A and B are defined as follows:

$A:$ {At least one head is observed}

$B:$ {The number of heads observed is odd}

a. Identify the sample points in the events A, B, $A \cup B$, A', and $A \cap B$.

b. Find $P(A)$, $P(B)$, $P(A \cup B)$, $P(A')$, and $P(A \cap B)$ by summing the probabilities of the appropriate sample points.

c. Find $P(A \cup B)$ using the additive rule. Compare your answer to the one you obtained in part **b**.

d. Are the events A and B mutually exclusive? Why?

3.21 What are mutually exclusive events? Give a verbal description, then draw a Venn diagram.

3.22 A pair of fair dice is tossed. Define the following events:

> *A:* {You will roll a 7 (i.e., the sum of the dots on the up faces of the two dice is equal to 7)}
>
> *B:* {At least one of the two dice shows a 4}

  a. Identify the sample points in the events A, B, $A \cap B$, $A \cup B$, and A'.
  b. Find $P(A)$, $P(B)$, $P(A \cap B)$, $P(A \cup B)$, and $P(A')$ by summing the probabilities of the appropriate sample points.
  c. Find $P(A \cup B)$ using the additive rule. Compare your answer to that for the same event in part **b**.
  d. Are A and B mutually exclusive? Why?

3.23 Consider the accompanying Venn diagram, where $P(E_1) = P(E_2) = P(E_3) = \frac{1}{5}$, $P(E_4) = P(E_5) = \frac{1}{20}$, $P(E_6) = \frac{1}{10}$, and $P(E_7) = \frac{1}{5}$. Find each of the following probabilities:
  a. $P(A)$ **b.** $P(B)$ **c.** $P(A \cup B)$ **d.** $P(A \cap B)$
  e. $P(A')$ **f.** $P(B')$ **g.** $P(A \cup A')$ **h.** $P(A' \cap B)$

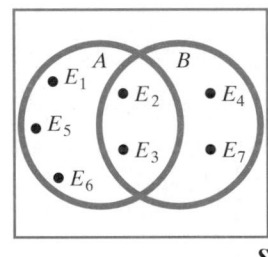

3.24 Consider the accompanying Venn diagram, where $P(E_1) = .10$, $P(E_2) = .05$, $P(E_3) = P(E_4) = .2$, $P(E_5) = .06$, $P(E_6) = .3$, $P(E_7) = .06$, and $P(E_8) = .03$. Find the following probabilities:
  a. $P(A')$ **b.** $P(B')$ **c.** $P(A' \cap B)$
  d. $P(A \cup B)$ **e.** $P(A \cap B)$ **f.** $P(A' \cup B')$
  g. Are events A and B mutually exclusive? Why?

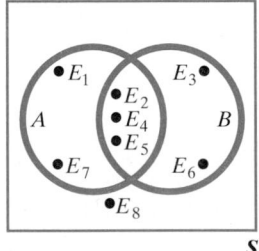

3.25 The following table describes the adult population of a small suburb of a large southern city. A marketing research firm plans to randomly select one adult from the suburb to evaluate a new food product. For this experiment the nine age–income categories are the sample points. Consider the following events:

| | | INCOME | |
Age	<$20,000	$20,000–$50,000	>$50,000
<25	950	1,000	50
25–45	450	2,050	1,500
>45	50	950	1,000

> *A:* {Person is under 25}
>
> *B:* {Person is between 25 and 45}
>
> *C:* {Person is over 45}
>
> *D:* {Person has income under $20,000}
>
> *E:* {Person has income of $20,000–$50,000}
>
> *F:* {Person has income over $50,000}

Convert the frequencies in the table to relative frequencies and use them to calculate the following probabilities:
  a. $P(B)$ **b.** $P(F)$ **c.** $P(C \cap F)$ **d.** $P(B \cup C)$
  e. $P(A')$ **f.** $P(A' \cap F)$
  g. Consider each pair of events (A and B, A and C, etc.) and list the pairs of events that are mutually exclusive. Justify your choices.

3.26 Refer to Exercise 3.25. Use the same event definitions to do the following exercises.
  a. Write the event that the person selected is under 25 with an income over $50,000 as an intersection of two events.
  b. Write the event that the person selected is age 25 or older as the union of two mutually exclusive events and as the complement of an event.

3.27 Three fair coins are tossed. We wish to find the probability of the event A: {Observe at least one head}.
  a. Express A as the union of three mutually exclusive events. Find the probability of A using this expression.
  b. Express A as the complement of an event. Find the probability of A using this expression.

Applying the Concepts

3.28 A buyer for a large metropolitan department store must choose two firms from the four available to supply the store's fall line of men's slacks. The buyer has not dealt with any of the four firms before and considers their products equally attractive. Unknown to the buyer, two of the four firms are having serious financial problems that may result in their not being able to deliver the fall line of slacks as soon as promised. The four firms are identified as G_1 and G_2 (firms in good financial conditions) and P_1 and P_2 (firms in poor financial condition). Sample points identify the pairs of firms selected. If the probability of the buyer's selecting a particular firm from among the four is the same for each firm, the sample points and their probabilities for this buying experiment are those listed in the following table.

Sample Points	Probability
$G_1 G_2$	$\frac{1}{6}$
$G_1 P_1$	$\frac{1}{6}$
$G_1 P_2$	$\frac{1}{6}$
$G_2 P_1$	$\frac{1}{6}$
$G_2 P_2$	$\frac{1}{6}$
$P_1 P_2$	$\frac{1}{6}$

We define the following events:

A: {At least one of the selected firms is in good financial condition}

B: {Firm P_1 is selected}

a. Define the event $A \cap B$ as a specific collection of sample points.
b. Define the event $A \cup B$ as a specific collection of sample points.
c. Define the event A' as a specific collection of sample points.
d. Find $P(A)$, $P(B)$, $P(A \cap B)$, $P(A \cup B)$, and $P(A')$ by summing the probabilities of the appropriate sample points.
e. Find $P(A \cup B)$ using the additive rule. Are events A and B mutually exclusive? Why?

3.29 A study of binge alcohol drinking by college students was published in the *American Journal of Public Health* (July 1995). A national sample of undergraduate college students received a questionnaire that requested detailed information about drinking behavior and several other variables, such as race, marital status, sex, religion, GPA, major, and housing. One hundred forty questionnaires were returned. Suppose an experiment consists of randomly selecting one of the returned questionnaires. Consider the following events:

A: {The student is a binge drinker}

B: {The student is a male}

C: {The student lives in a coed dorm}

Describe each of the following events in terms of unions, intersections, and complements ($A \cup B$, $A \cap B$, A', etc.):
a. The student is male and a binge drinker.
b. The student is not a binge drinker.
c. The student is male or lives in a coed dorm.
d. The student is female and not a binge drinker.

3.30 *Roulette* is a very popular game in many American casinos. In Roulette, a ball spins on a circular wheel that is divided into 38 arcs of equal length, bearing the numbers 00, 0, 1, 2, ..., 35, 36. The number of the arc on which the ball stops is the outcome of one play of the game. The numbers are also colored in the manner shown at the bottom of the page.

Players may place bets on the table in a variety of ways, including bets on odd, even, red, black, high, low, etc. Define the following events:

A: {Outcome is an odd number (00 and 0 are considered neither odd nor even)}

B: {Outcome is a black number}

C: {Outcome is a low number (1–18)}

a. Define the event $A \cap B$ as a specific set of sample points.
b. Define the event $A \cup B$ as a specific set of sample points.
c. Find $P(A)$, $P(B)$, $P(A \cap B)$, $P(A \cup B)$, and $P(C)$ by summing the probabilities of the appropriate sample points.
d. Define the event $A \cap B \cap C$ as a specific set of sample points.
e. Find $P(A \cup B)$ using the additive rule. Are events A and B mutually exclusive? Why?
f. Find $P(A \cap B \cap C)$ by summing the probabilities of the sample points given in part **d**.
g. Define the event $(A \cup B \cup C)$ as a specific set of sample points.
h. Find $P(A \cup B \cup C)$ by summing the probabilities of the sample points given in part **g**.

3.31 *Ear & Hearing* (Apr. 1995) reported on a study of elderly hearing-impaired subjects who wear conventional analog hearing aids. Two of the variables measured for each hearing aid wearer were duration of experience with hearing aids and daily hearing aid use. The results are summarized in the table on page 114, where the numbers in the table represent the percentages of hearing aid wearers who fall into the respective categories. (Note that the percentages add to 100%.) Consider a randomly selected elderly hearing-impaired subject who wears a conventional analog hearing aid.

a. List all the sample points for this experiment.
b. What is the set of all sample points called?
c. Let C be the event that an elderly hearing-impaired subject has more than 10 years of experience wearing hearing aids. Find $P(C)$ by summing the probabilities of the sample points in C.
d. Let G be the event that an elderly hearing-impaired subject uses hearing aids 8–16 hours per day. Find $P(G)$.

Red:	1,	3,	5,	7,	9,	12,	14,	16,	18,	19,	21,	23,	25,	27,	30,	32,	34,	36
Black:	2,	4,	6,	8,	10,	11,	13,	15,	17,	20,	22,	24,	26,	28,	29,	31,	33,	35
Green:	00,	0																

Hearing Aid Experience	DAILY USE (HOURS)				Total
	Less than 1	1–4	4–8	8–16	
Less than 1 year	4	6	7	15	42
1–10 years	4	9	6	28	47
More than 10 years	0	3	6	12	21
Total	8	18	19	55	110

Source: Cox, R. M. and Alexander, G. C. "The abbreviated profile of hearing aid benefit." *Ear & Hearing*, Vol. 16. No. 2, Apr. 1995, p. 178 (Table 2).

e. Let A be the event that an elderly hearing-impaired subject has less than 1 year of experience wearing hearing aids. Find $P(A)$.

f. Let D be the event that an elderly hearing-impaired subject uses hearing aids less than 1 hour per day. Find $P(D)$.

g. Let E be the event that an elderly hearing-impaired subject uses hearing aids between 1 and 4 hours per day. Find $P(E)$.

3.32 Refer to Exercise 3.31. Define the characteristics of an elderly hearing-impaired person portrayed by the following events, then find the probability of each. For each union, use the additive rule to find the probability. Also, determine whether the events are mutually exclusive.
a. $A \cap G$ **b.** $C \cup E$ **c.** $C \cap D$
d. $A \cup G$ **e.** $A \cup D$

3.33 Identifying managerial prospects who are both talented and motivated is difficult. A human-resources director constructed the following two-way table to define nine combination of talent–motivation levels. The number in a cell is the director's estimate of the probability that a managerial prospect will fall in that category. Suppose the director has decided to hire a new manager. Define the following events:

A: {Prospect places in high motivation category}

B: {Prospect places in high talent category}

C: {Prospect is average or better in both categories}

D: {Prospect places low in at least one category}

E: {Prospect places highest in both categories}

Motivation	TALENT		
	High	Medium	Low
High	.05	.16	.05
Medium	.19	.32	.05
Low	.11	.05	.02

a. Does the sum of the cell probabilities equal 1?

b. List the sample points in each of the events described above and find their probabilities.

c. Find $P(A \cup B)$, $P(A \cap B)$, and $P(A \cup C)$.

d. Find $P(A')$ and explain what this means from a practical point of view.

e. Consider each pair of events (A and B, A and C, etc.). Which of the pairs are mutually exclusive? Why?

3.5   CONDITIONAL PROBABILITY

The event probabilities we've been discussing give the relative frequencies of the occurrences of the events when the experiment is repeated a very large number of times. Such probabilities are often called **unconditional probabilities** because no special conditions are assumed, other than those that define the experiment.

Often, however, we have additional knowledge that might affect the outcome of an experiment, so we need to alter the probability of an event of interest. A probability that reflects such additional knowledge is called the **conditional probability** of the event. For example, we've seen that the probability of observing an even number (event A) on a toss of a fair die is $\frac{1}{2}$. But suppose we're given the information that on a particular throw of the die the result was a number less than or equal to 3 (event B). Would the probability of observing an even number

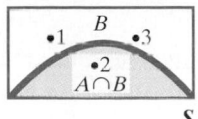

FIGURE 3.14

Reduced sample space for the die-toss experiment given that event *B* has occurred

on that throw of the die still be equal to ½? It can't be, because making the assumption that *B* has occurred reduces the sample space from six sample points to three sample points (namely, those contained in event *B*). This reduced sample space is shown in Figure 3.14. Because the sample points for the die-toss experiment are equally likely, each of the three sample points in the reduced sample space is assigned an equal *conditional probability* of ⅓. Since the only even number of the three in the reduced sample space *B* is the number 2 and the die is fair, we conclude that the probability that *A* occurs *given that B occurs* is ⅓. We use the symbol $P(A \mid B)$ to represent the probability of event *A* given that event *B* occurs. For the die-toss example

$$P(A \mid B) = \frac{1}{3}$$

To get the probability of event *A* given that event *B* occurs, we proceed as follows. We divide the probability of the part of *A* that falls within the reduced sample space *B*, namely $P(A \cap B)$, by the total probability of the reduced sample space, namely, $P(B)$. Thus, for the die-toss example with event *A*: {Observe an even number} and event *B*: {Observe a number less than or equal to 3}, we find

$$P(A \mid B) = \frac{P(A \cap B)}{P(B)} = \frac{P(2)}{P(1) + P(2) + P(3)} = \frac{\frac{1}{6}}{\frac{3}{6}} = \frac{1}{3}$$

The formula for $P(A \mid B)$ is true in general:

> To find the *conditional probability that event A occurs given that event B occurs*, divide the probability that *both A* and *B* occur by the probability that *B* occurs, that is,
>
> $$P(A \mid B) = \frac{P(A \cap B)}{P(B)} \qquad \text{[We assume that } P(B) \neq 0.]$$

This formula adjusts the probability of $A \cap B$ from its original value in the complete sample space *S* to a conditional probability in the reduced sample space *B*. If the sample points in the complete sample space are equally likely, then the formula will assign equal probabilities to the sample points in the reduced sample space, as in the die-toss experiment. If, on the other hand, the sample points have unequal probabilities, the formula will assign conditional probabilities proportional to the probabilities in the complete sample space. This is illustrated by the following examples.

EXAMPLE 3.14

Many medical researchers have conducted experiments to examine the relationship between cigarette smoking and cancer. Consider an individual randomly selected from the adult male population. Let *A* represent the event that the individual smokes, and let *C* represent the event that the individual develops cancer. Therefore, $A \cap C$ is the sample point that an adult male smokes and develops cancer; $A \cap C'$ is the sample point that an adult male smokes and does not develop cancer, etc. Assume that the probabilities associated with the four sample points are as shown in the following table for a certain section of the United States. How can these sample point probabilities be used to examine the relationship between smoking and cancer?

Sample Points	Probabilities
$A \cap C$	.05
$A \cap C'$	.20
$A' \cap C$	.03
$A' \cap C'$	.72

Solution

One method of determining whether these probabilities indicate that smoking and cancer are related is to compare the conditional probability that an adult male acquires cancer given that he smokes with the conditional probability that an adult male acquires cancer given that he does not smoke.

First, we consider the reduced sample space A corresponding to adult male smokers. The two sample points $A \cap C$ and $A \cap C'$ are contained in this reduced sample space, and the adjusted probabilities of these two sample points are the two conditional probabilities:

$$P(C|A) = \frac{P(A \cap C)}{P(A)} \quad \text{and} \quad P(C'|A) = \frac{P(A \cap C')}{P(A)}$$

The probability of event A is the sum of the probabilities of the sample points in A:

$$P(A) = P(A \cap C) + P(A \cap C') = .05 + .20 = .25$$

Then the values of the two conditional probabilities in the reduced sample space A are:

$$P(C|A) = \frac{.05}{.20} = .20 \quad \text{and} \quad P(C'|A) = \frac{.20}{.25} = .80$$

These two numbers represent the probabilities that an adult male smoker develops cancer and does not develop cancer, respectively. Notice that the conditional probabilities .80 and .20 are in the same 4 to 1 ratio as the original (unconditional) probabilities, .20 and .05. The conditional probability formula simply adjusts the unconditional probabilities so that they add to 1 in the reduced sample space, A, of adult male smokers.

In a like manner, the conditional probabilities of an adult male nonsmoker developing cancer and not developing cancer are

$$P(C|A') = \frac{P(A' \cap C)}{P(A')} = \frac{.03}{.75} = .04$$

$$P(C'|A') = \frac{P(A' \cap C')}{P(A')} = \frac{.72}{.75} = .96$$

Observe that the conditional probabilities .96 and .04 are in the same 24 to 1 ratio as the unconditional probabilities .72 and .03.

Two of the conditional probabilities give some insight into the relationship between cancer and smoking: the probability of developing cancer given that the adult male is a smoker, and the probability of developing cancer given that the adult male is not a smoker. The conditional probability that an adult male smoker develops cancer (.20) is the probability that a nonsmoker develops cancer (.04). This does not imply that smoking *causes* cancer, but it does suggest a pronounced link between smoking and cancer.   ▲

EXAMPLE 3.15

The investigation of consumer product complaints by the Federal Trade Commission (FTC) has generated much interest by manufacturers in the quality of their products. A manufacturer of an electromechanical kitchen utensil conducted an analysis of a large number of consumer complaints and found that they fell into the six categories shown in Table 3.5. If a consumer complaint is received, what is the probability that the cause of the complaint was product appearance given that the complaint originated during the guarantee period?

TABLE 3.5   Distribution of Product Complaints

	REASON FOR COMPLAINT			
	Electrical	Mechanical	Appearance	Totals
During Guarantee Period	18%	13%	32%	63%
After Guarantee Period	12%	22%	3%	37%
Totals	30%	35%	35%	100%

Solution

Let *A* represent the event that the cause of a particular complaint is product appearance, and let *B* represent the event that the complaint occurred during the guarantee period. Checking Table 3.5, you can see that $(18 + 13 + 32)\% = 63\%$ of the complaints occur during the guarantee period. Hence, $P(B) = .63$. The percentage of complaints that were caused by appearance and occurred during the guarantee period (the event $A \cap B$) is 32%. Therefore, $P(A \cap B) = .32$.

Using these probability values, we can calculate the conditional probability $P(A \mid B)$ that the cause of a complaint is appearance given that the complaint occurred during the guarantee time:

$$P(A \mid B) = \frac{P(A \cap B)}{P(B)} = \frac{.32}{.63} = .51$$

Consequently, we can see that slightly more than half the complaints that occurred during the guarantee period were due to scratches, dents, or other imperfections in the surface of the kitchen devices.   ▲

EXERCISES 3.34–3.45

Learning the Mechanics

3.34 Consider the experiment defined by the accompanying Venn diagram, with the sample space *S* containing five sample points. The sample points are assigned the following probabilities: $P(E_1) = .1$, $P(E_2) = .1$, $P(E_3) = .2$, $P(E_4) = .5$, $P(E_5) = .1$.

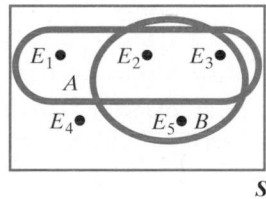

S

a. Calculate $P(A)$, $P(B)$, and $P(A \cap B)$.
b. Suppose we know event *A* has occurred, so the reduced sample space consists of the three sample points in *A*: E_1, E_2, and E_3. Use the formula for conditional probability to determine the probabilities of these three sample points given that *A* has occurred. Verify that the conditional probabilities are in the same ratio to one another as the original sample point probabilities.

c. Calculate the conditional probability $P(B \mid A)$ in two ways: First, add the adjusted (conditional) probabilities of the sample points in the intersection $A \cap B$, since these represent the event that *B* occurs given that *A* has occurred. Second, use the formula for conditional probability:

$$P(B \mid A) = \frac{P(A \cap B)}{P(A)}$$

Verify that the two methods yield the same result.

3.35 Given that $P(A) = .3$, $P(B) = .6$, and $P(A \cap B) = .15$, find $P(A \mid B)$ and $P(B \mid A)$.

3.36 A sample space contains six sample points and events *A*, *B*, and *C* as shown in the accompanying Venn diagram. The probabilities of the sample points are $P(1) = .20$, $P(2) = .05$, $P(3) = .25$, $P(4) = .10$, $P(5) = .15$, $P(6) = .25$.

a. Find $P(A)$, $P(B)$, and $P(C)$.
b. Find $P(A \cap B)$, $P(A \cap C)$, and $P(B \cap C)$.
c. Suppose you know that event *A* has occurred. Assign conditional probabilities to the three sample points contained in *A*. Verify that they add to 1 and are in the same ratio as the original (unconditional) probabilities.

d. Use the conditional probabilities from part **c** to calculate $P(B|A)$. Use the formula for $P(B|A)$ to verify your answer.

e. Use the formula for conditional probability to calculate $P(C|A)$ and $P(C|A')$. Verify the results by inspection of the Venn diagram, remembering that the "given event" is a reduced sample space for a conditional probability.

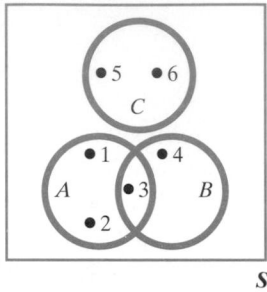

3.37 Two fair coins are tossed and the events A and B are defined as follows:

$A:$ {At least one head appears}

$B:$ {Exactly one head appears}

a. Draw a Venn diagram for the experiment, labeling each sample point and showing events A and B. Assign probabilities to the sample points.

b. Find $P(A)$, $P(B)$, and $P(A \cap B)$.

c. Use the formula for conditional probability to find $P(A|B)$ and $P(B|A)$. Verify your answers by inspecting the Venn diagram and using the concept of reduced sample spaces.

3.38 A box contains two white, two red, and two blue poker chips. Two chips are randomly chosen without replacement and their colors are noted. Define the following events:

$A:$ {Both chips are of the same color}

$B:$ {Both chips are red}

$C:$ {At least one chip is red or white}

Find $P(B|A)$, $P(B|A')$, $P(B|C)$, $P(A|C)$, and $P(C|A')$.

Applying the Concepts

3.39 In "Crime, race and reporting to the police," R. Shah and K. Pease examined the relationship between the race of the attacker, the race of the victim, and the degree of the injury sustained in reported crimes in a British crime survey (*The Howard Journal of Criminal Justice,* Aug. 1992). The table at the bottom of the page is cited in the article.

a. What is the probability that a randomly selected reported crime involved a white attacker and a white victim?

b. What is the probability that a randomly selected reported crime involved serious injuries?

c. Given that both the attacker and victim were white, what is the probability that a randomly selected reported crime involved fatalities?

d. If no injury was reported, what is the probability that a randomly selected reported crime involved a nonwhite attacker and a white victim?

3.40 The National Center for Health Statistics compiles data on U.S. suicide rates by race and gender. These data were used to estimate the probability of committing suicide during a lifetime in *Geriatric Psychiatry* (Vol. 27, 1994). The probabilities of lifetime suicide for four race/gender cohorts are given in the accompanying table. For example, a black female selected at random from her race/gender cohort has a $1/535$ chance of committing suicide during her lifetime.

a. Are the probabilities listed in the table conditional or unconditional probabilities? Explain.

b. Consider the following statement: "The probability that a randomly selected male commits suicide is obtained by summing the probabilities for black male and white male: $1/133 + 1/66$." Defend or refute this argument.

Cohort	Probability of Suicide
Black female	$1/535$
White female	$1/196$
Black male	$1/133$
White male	$1/66$

3.41 "Go" is one of the oldest and most popular strategic board games in the world, especially in Japan and

Degree of Injury	ATTACKER/VICTIM				
	White/White	**White/Nonwhite**	**Nonwhite/Nonwhite**	**Nonwhite/White**	**Totals**
Fatal	183	18	18	18	237
Serious	580	49	111	141	881
Slight	4,336	440	594	1,656	7,026
None	1,422	136	214	1,801	3,573
Totals	6,521	643	937	3,616	11,717

Source: Shah, R. and Pease, K. "Crime, race and reporting to the police," *The Howard Journal of Criminal Justice,* Volume 31, No. 3, Aug. 1992.

Korea. This two-player game is played on a flat surface marked with 19 vertical and 19 horizontal lines. The objective is to control territory by placing pieces called "stones" on vacant points on the board. Players alternate placing their stones. The player using black stones goes first, followed by the player using white stones. [*Note:* The University of Virginia requires MBA students to learn Go to understand how the Japanese conduct business.] *Chance* (Summer 1995) published an article that investigated the advantage of playing first (i.e., using the black stones) in Go. The results of 577 games recently played by professional Go players were analyzed.

a. In the 577 games, the player with the black stones won 319 times and the player with white stones won 258 times. Use this information to estimate the probability of winning when you play first in Go.

b. Professional Go players are classified by level. Group C includes the top-level players, followed by Group B (middle-level) and Group A (low-level) players. The table below describes the number of games won by the player with the black stones, categorized by level of the black player and level of the opponent. Estimate the probability of winning when you play first in Go for each combination of player and opponent level.

c. If the player with the black stones is ranked higher than the player with the white stones, what is the probability that black wins?

d. Given that the players are of the same level, what is the probability that the player with the black stones wins?

Black Player Level	Opponent Level	Number of Wins	Number of Games
C	A	34	34
C	B	69	79
C	C	66	118
B	A	40	54
B	B	52	95
B	C	27	79
A	A	15	28
A	B	11	51
A	C	5	39
	Totals	319	577

Source: J. Kim, and H. J. Kim, "The advantage of playing first in Go." *Chance*, Vol. 8, No. 3, Summer 1995, p. 26 (Table 3). Reprinted with permission.

3.42 A soap manufacturer has decided to market two new brands. After an analysis of current market conditions and a review of the firm's past successes and failures with new brands, the manufacturer believes that the sample points and probabilities of their occurrence in this marketing experiment are as shown in the table below (where *S* means the brand succeeds and *F* means the brand fails in the first year). Define the following events:

A: {Both new brands are successful in the first year}

B: {At least one new brand is successful in the first year}

Sample Points	Probabilities
SS	.09
SF	.21
FS	.21
FF	.49

a. Find $P(A)$, $P(B)$, and $P(A \cap B)$.

b. Find $P(A \mid B)$ and $P(B \mid A)$.

3.43 Refer to the *American Journal of Public Health* study of unintentional carbon monoxide (CO) poisonings in Colorado, Exercise 3.16. The 981 cases were classified in a table, which is reproduced below. A case of unintentional CO poisoning is chosen at random from the 981 cases.

a. Given that the source of the poisoning is fire, what is the probability that the case is fatal?

b. Given that the case is nonfatal, what is the probability that it is caused by auto exhaust?

c. If the case is fatal, what is the probability that the source is unknown?

d. If the case is nonfatal, what is the probability that the source is not fire or a fireplace?

Source of Exposure	Fatal	Nonfatal	Total
Fire	63	53	116
Auto exhaust	60	178	238
Furnace	18	345	363
Kerosene or spaceheater	9	18	27
Appliance	9	63	72
Other gas-powered motor	3	73	76
Fireplace	0	16	16
Other	3	19	22
Unknown	9	42	51
Total	174	807	981

Source: Cook, M. C., Simon, P. A., and Hoffman, R. E. "Unintentional carbon monoxide poisoning in Colorado, 1986 through 1991." *American Journal of Public Health*, Vol. 85, No. 7, July 1995, p. 989 (Table 1). © 1995 American Public Health Association.

3.44 Physicians and pharmacists sometimes fail to inform patients adequately about the proper application of prescription drugs and about the precautions to take in order to avoid potential side effects. This failure is an ongoing problem in the United States. One method of increasing patients' awareness of the problem is for physicians to provide Patient Medication Instruction (PMI) sheets. The American Medical Association, however, has found that only 20% of the doctors who prescribe drugs frequently distribute PMI sheets to their patients. Assume that 20% of all patients receive the PMI sheet with their prescriptions and that 12% receive the PMI sheet and are hospitalized because of a drug-related problem. What is the probability that a person will be hospitalized for a drug-related problem given that the person has received the PMI sheet?

3.45 There are several methods of typing, or classifying, human blood. The most common procedure types blood into the general classifications of A, B, O, or AB. A method that is not as well known examines phosphoglucomutase (PGM) and classifies the blood into one of three main categories, 1-1, 2-1, or 2-2. Suppose a certain geographic region of the United States has the PGM percentages shown in the accompanying table. A person is to be chosen at random from this region.

Race	1-1	2-1	2-2
White	46.3%	39.2%	4.0%
Black	6.7%	3.4%	.4%

a. What is the probability that a black person is chosen?

b. Given that a black person is chosen, what is the probability he or she is PGM type 1-1?

c. Given that a white person is chosen, what is the probability he or she is PGM type 1-1?

3.6 THE MULTIPLICATIVE RULE AND INDEPENDENT EVENTS

The probability of an intersection of two events can be calculated using the *multiplicative rule*, which employs the conditional probabilities we defined in the previous section, as shown in the following example.

EXAMPLE 3.16

An agriculturist, who is interested in planting wheat next year, is concerned with the following events:

> *B:* {The production of wheat will be profitable}
>
> *A:* {A serious drought will occur}

Based on available information, the agriculturist believes that the probability is .01 that production of wheat will be profitable *assuming* a serious drought will occur in the same year and that the probability is .05 that a serious drought will occur. That is,

$$P(B|A) = .01 \quad \text{and} \quad P(A) = .05$$

Based on the information provided, what is the probability that a serious drought will occur *and* that a profit will be made? That is, find $P(A \cap B)$, the probability of the intersection of events A and B.

Solution

As you will see, we've already developed a formula for finding the probability of an intersection of two events. Recall that the conditional probability of B given A is

$$P(B|A) = \frac{P(A \cap B)}{P(A)}$$

Multiplying both sides of this equation by $P(A)$, we obtain a formula for the probability of the intersection of events A and B. This is often called the **multiplicative rule of probability** and is given by

$$P(A \cap B) = P(A)P(B|A)$$

Thus,

$$P(A \cap B) = (.05)(.01)$$
$$= .0005$$

The probability that a serious drought occurs *and* the production of wheat is profitable is only .0005. As we might expect, this intersection is a very rare event.

▲

Multiplicative Rule of Probability

$$P(A \cap B) = P(A)P(B \mid A) = P(B)P(A \mid B)$$

Intersections often contain only a few sample points. In this case, the probability of an intersection is easy to calculate by summing the appropriate sample point probabilities. However, the formula for calculating intersection probabilities is invaluable when the intersection contains numerous sample points, as the next example illustrates.*

EXAMPLE 3.17

A county welfare agency employs ten welfare workers who interview prospective food stamp recipients. Periodically the supervisor selects, at random, the forms completed by two workers to audit for illegal deductions, which give recipients extra food stamps. Unknown to the supervisor, three of the workers have regularly been giving illegal deductions to applicants. What is the probability that both of the two workers chosen have been giving illegal deductions?

Solution

Define the following two events:

> *A:* {First worker selected gives illegal deductions}
>
> *B:* {Second worker selected gives illegal deductions}

We want to find the probability of the event that both selected workers have been giving illegal deductions. This event can be restated as: {First worker gives illegal deductions *and* second worker gives illegal deductions}. Thus, we want to find the probability of the intersection, $A \cap B$. Applying the multiplicative rule, we have

$$P(A \cap B) = P(A)P(B \mid A)$$

To find $P(A)$ it is helpful to consider the experiment as selecting one worker from the ten. Then the sample space for the experiment contains ten sample points (representing the ten welfare workers), where the three workers giving illegal deductions are denoted by the symbol I (I_1, I_2, I_3), and the seven workers not giving illegal deductions are denoted by the symbol N ($N_1, ..., N_7$). The resulting Venn diagram is shown in Figure 3.15.

Since the first worker is selected at random from the ten, it is reasonable to assign equal probabilities to the 10 sample points. Thus, each sample point has a probability of $\frac{1}{10}$. The sample points in event A are $\{I_1, I_2, I_3\}$—the three workers who are giving illegal deductions. Thus,

$$P(A) = P(I_1) + P(I_2) + P(I_3) = \tfrac{1}{10} + \tfrac{1}{10} + \tfrac{1}{10} = \tfrac{3}{10}$$

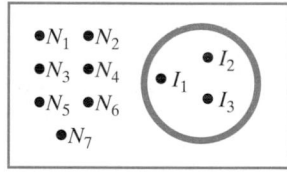

FIGURE 3.15

Venn diagram for finding $P(A)$

---

*The multiplicative rule of probability also plays an important role in an area of statistics known as **Bayesian statistics**. Consult the references at the end of the chapter for detailed discussions of Bayesian statistics.

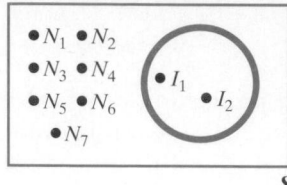

FIGURE 3.16

Venn diagram for finding $P(B \mid A)$

To find the conditional probability, $P(B \mid A)$, we need to alter the sample space S. Since we know A has occurred—i.e., the first worker selected is giving illegal deductions—only two of the nine remaining workers in the sample space are giving illegal deductions. The Venn diagram for this new sample space is shown in Figure 3.16. Each of these 9 sample points is equally likely, so each is assigned a probability of $\frac{1}{9}$. Since the event $B \mid A$ contains the sample points $\{I_1, I_2\}$, we have

$$P(B \mid A) = P(I_1) + P(I_2) = \frac{1}{9} + \frac{1}{9} = \frac{2}{9}$$

Substituting $P(A) = \frac{3}{10}$ and $P(B \mid A) = \frac{2}{9}$ into the formula for the multiplicative rule, we find

$$P(A \cap B) = P(A)P(B \mid A) = (\tfrac{3}{10})(\tfrac{2}{9}) = \tfrac{6}{90} = \tfrac{1}{15}$$

Thus, there is a 1 in 15 chance that both workers chosen by the supervisor have been giving illegal deductions to food stamp recipients. ▲

The sample space approach is only one way to solve the problem posed in Example 3.17. An alternative method employs the concept of a **tree diagram**. Tree diagrams are helpful for calculating the probability of an intersection.

To illustrate, a tree diagram for Example 3.17 is displayed in Figure 3.17. The tree begins at the far left with two branches. These branches represent the two possible outcomes N (no illegal deductions) and I (illegal deductions) for the first worker selected. The unconditional probability of each outcome is given (in parentheses) on the appropriate branch. That is, for the first worker selected, $P(N) = \frac{7}{10}$ and $P(I) = \frac{3}{10}$. (These can be obtained by summing sample point probabilities as in Example 3.17.)

The next level of the tree diagram (moving to the right) represents the outcomes for the second worker selected. The probabilities shown here are conditional probabilities since the outcome for the first worker is assumed to be known. For example, if the first worker is giving illegal deductions (I), the probability that the second worker is also giving illegal deductions (I) is $\frac{2}{9}$ since of the nine workers left to be selected, only two remain who are giving illegal deductions. This conditional probability, $\frac{2}{9}$, is shown in parentheses on the bottom branch of Figure 3.17.

FIGURE 3.17

Tree diagram for Example 3.17

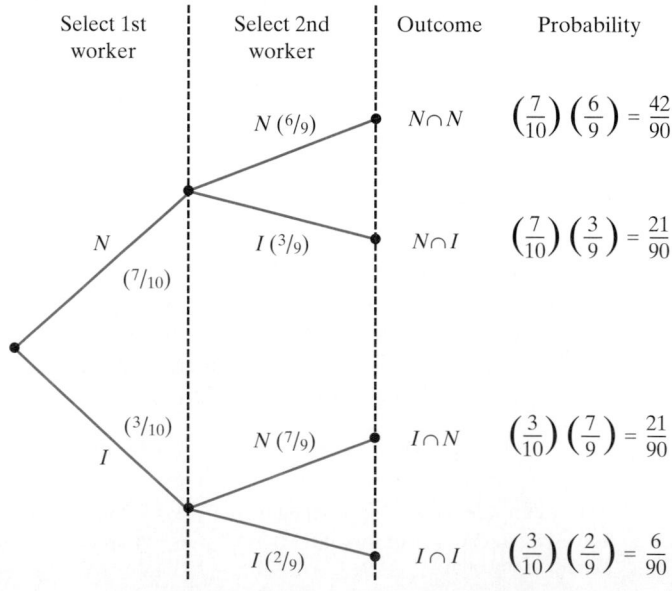

Finally, the four possible outcomes of the experiment are shown at the end of each of the four tree branches. These events are intersections of two events (outcome of first worker *and* outcome of second worker). Consequently, the multiplicative rule is applied to calculate each probability, as shown in Figure 3.17. You can see that the intersection $\{I \cap I\}$— the event that both workers selected are giving illegal deductions—has probability $6/90 = 1/15$, which is the same value we obtained in Example 3.17.

The next example demonstrates a special application of the multiplicative rule.

EXAMPLE 3.18

Consider the experiment of tossing a fair coin twice and recording the up face on each toss. The following events are defined:

$$A: \{\text{First toss is a head}\}$$
$$B: \{\text{Second toss is a head}\}$$

Does *knowing* that event A has occurred affect the probability that B will occur?

Solution

Intuitively, the answer should be No, since what occurs on the first toss should in no way affect what occurs on the second toss. Let's check our intuition. Recall the sample space for this experiment:

1. Observe *HH*
2. Observe *HT*
3. Observe *TH*
4. Observe *TT*

Each of these sample points has a probability of ¼. Thus,

$$P(B) = P(HH) + P(TH) \quad \text{and} \quad P(A) = P(HH) + P(HT)$$
$$= ¼ + ¼ = ½ \qquad\qquad\qquad = ¼ + ¼ = ½$$

Now, what is $P(B \,|\, A)$?

$$P(B \,|\, A) = \frac{P(A \cap B)}{P(A)} = \frac{P(HH)}{P(A)}$$
$$= \frac{¼}{½} = \frac{1}{2}$$

We can now see that $P(B) = ½$ and $P(B \,|\, A) = ½$. Knowing that the first toss resulted in a head does not affect the probability that the second toss will be a head. The probability is ½ whether or not we know the result of the first toss. When this occurs, we say that the two events A and B are *independent events*. ▲

DEFINITION 3.9

Events A and B are **independent events** if the occurrence of B does not alter the probability that A has occurred; that is, events A and B are independent if

$$P(A \,|\, B) = P(A)$$

When events A and B are independent, it is also true that

$$P(B \,|\, A) = P(B)$$

Events that are not independent are said to be **dependent**.

EXAMPLE 3.19

Consider the experiment of tossing a fair die and let

> *A:* {Observe an even number}
>
> *B:* {Observe a number less than or equal to 4}

Are events *A* and *B* independent?

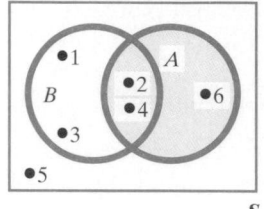

FIGURE 3.18

Venn diagram for die-toss experiment

Solution

The Venn diagram for this experiment is shown in Figure 3.18. We first calculate

$$P(A) = P(2) + P(4) + P(6) = \tfrac{1}{2}$$
$$P(B) = P(1) + P(2) + P(3) + P(4) = \tfrac{4}{6} = \tfrac{2}{3}$$
$$P(A \cap B) = P(2) + P(4) = \tfrac{2}{6} = \tfrac{1}{3}$$

Now assuming *B* has occurred, the conditional probability of *A* given *B* is

$$P(A \mid B) = \frac{P(A \cap B)}{P(B)} = \frac{\tfrac{1}{3}}{\tfrac{2}{3}} = \frac{1}{2} = P(A)$$

Thus, assuming that event *B* occurs does not alter the probability of observing an even number—it remains $\tfrac{1}{2}$. Therefore, the events *A* and *B* are independent. Note that if we calculate the conditional probability of *B* given *A*, our conclusion is the same:

$$P(B \mid A) = \frac{P(A \cap B)}{P(A)} = \frac{\tfrac{1}{3}}{\tfrac{1}{2}} = \frac{2}{3} = P(B) \qquad \blacktriangle$$

EXAMPLE 3.20

Refer to the consumer product complaint study in Example 3.15. The percentages of complaints of various types during and after the guarantee period are shown in Table 3.5. Define the following events:

> *A:* {Cause of complaint is product appearance}
>
> *B:* {Complaint occurred during the guarantee term}

Are *A* and *B* independent events?

Solution

Events *A* and *B* are independent if $P(A \mid B) = P(A)$. We calculated $P(A \mid B)$ in Example 3.15 to be .51, and from Table 3.5 we see that

$$P(A) = .32 + .03 = .35$$

Therefore, $P(A \mid B)$ is not equal to $P(A)$, and *A* and *B* are dependent events. $\blacktriangle$

To gain an intuitive understanding of independence, think of situations in which the occurrence of one event does not alter the probability that a second event will occur. For example, new medical procedures are often tested on laboratory animals. The scientists conducting the tests generally try to perform the procedures on the animals so the results for one animal do not affect the results for the others. That is, the event that the procedure is successful on one animal is *independent* of the result for another. In this way, the scientists can get a more accurate idea of the efficacy of the procedure than if the results were dependent, with the success or failure for one animal affecting the results for other animals.

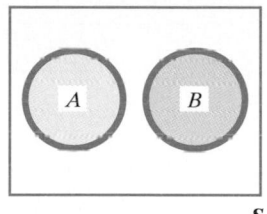

S

FIGURE 3.19

Mutually exclusive events are dependent events

As a second example, consider an election poll in which 1,000 registered voters are asked their preference between two candidates. Pollsters try to use procedures for selecting a sample of voters so that the responses are independent. That is, the objective of the pollster is to select the sample so the event that one polled voter prefers candidate A does not alter the probability that a second polled voter prefers candidate A.

Now consider the world of sports. Do you think the results of a batter's successive trips to the plate in baseball, or of a basketball player's successive shots at the basket, are independent? If a basketball player makes two successive shots, is the probability of making the next shot altered from its value if the result of the first shot is not known? If a player makes two shots in a row, the probability of a third successful shot is likely to be different from what we would assign if we knew nothing about the first two shots. Why should this be so? Research has shown that many such results in sports tend to be *dependent* because players (and even teams) tend to get on "hot" and "cold" streaks, during which their probabilities of success may increase or decrease significantly.

We will make three final points about independence. The first is that the property of independence, unlike the mutually exclusive property, cannot be shown on or gleaned from a Venn diagram. This means *you can't trust your intuition.* In general, the only way to check for independence is by performing the calculations of the probabilities in the definition.

The second point concerns the relationship between the mutually exclusive and independence properties. Suppose that events A and B are mutually exclusive, as shown in Figure 3.19, and both events have nonzero probabilities. Are these events independent or dependent? That is, does the assumption that B occurs alter the probability of the occurrence of A? It certainly does, because if we assume that B has occurred, it is impossible for A to have occurred simultaneously. That is, $P(A \mid B) = 0$. Thus, *mutually exclusive events are dependent events* since $P(A) \neq P(A \mid B)$.

The third point is that the probability of the intersection of independent events is very easy to calculate. Referring to the formula for calculating the probability of an intersection, we find

$$P(A \cap B) = P(A)P(B \mid A)$$

Thus, since $P(B \mid A) = P(B)$ when A and B are independent, we have the following useful rule:

> *If events A and B are independent,* the probability of the intersection of A and B equals the product of the probabilities of A and $B;$ that is
>
> $$P(A \cap B) = P(A)P(B)$$
>
> The converse is also true: If $P(A \cap B) = P(A)P(B)$, then events A and B are independent.

In the die-toss experiment, we showed in Example 3.19 that the events $A:$ {Observe an even number} and $B:$ {Observe a number less than or equal to 4} are independent if the die is fair. Thus,

$$P(A \cap B) = P(A)P(B) = (\tfrac{1}{2})(\tfrac{2}{3}) = \tfrac{1}{3}$$

This agrees with the result that we obtained in the example:

$$P(A \cap B) = P(2) + P(4) = \tfrac{2}{6} = \tfrac{1}{3}$$

O.J., Spousal Abuse, and Murder

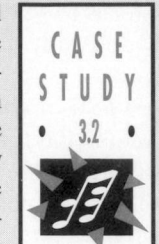

CASE
STUDY
• 3.2 •

In September 1995, a Los Angeles jury acquitted O.J. Simpson, a former NFL star turned movie actor and TV celebrity, of murdering his ex-wife, Nicole Brown, and her companion Ron Goldman. The "trial of the century," as some journalists called it, was watched on television by a captivated nationwide audience for an entire year. During the trial, Judge Lance Ito made several critical and difficult decisions with respect to the admission of evidence. One of these concerned Simpson's history of spousal abuse.

Simpson had had several violent encounters with Brown while they were married, at least two of which were documented with audiotapes of 911 calls that Nicole Brown had made to police. Attorney Alan Dershowitz, a consultant for O.J.'s team of defense lawyers, argued in the *L.A. Times* (Jan. 15, 1995) that such evidence should be excluded owing to the tenuous relationship between spousal abuse and spousal murder. In support of his position, Dershowitz noted that about 2,000 women are murdered by a current or former mate and about 2 million spousal assaults occur annually. He inferred from this that a woman in an abusive relationship has only about a 1 in 1,000 chance of being murdered by her mate each

year. Although Judge Ito ultimately admitted some of the spousal abuse evidence, several jurors apparently agreed with Dershowitz's point of view. After the trial, at least one stated publicly that the prosecution "wasted its time" presenting the evidence of spousal abuse since it had no direct connection to the murders.

Prior to the acquittal, two analysts at RAND Corporation, Jon Mertz and Jonathan Caulkins, commented on Dershowitz's argument in *Chance* (Spring 1995). Dershowitz had stated in the *L.A. Times* article that "the issue is whether a history of abuse is necessarily a prelude to murder." According to the RAND analysts,

He's wrong. In a murder trial we have another crucial piece of evidence: A woman is dead. The issue is not whether abuse leads to murder but whether a history of abuse helps identify the murderer. Of the 4,936 women who were murdered in 1992, about 1,430 were killed by their current or former husband or boyfriend. Thus, if it is simply known that a woman has been killed, there is a 29% probability that she was killed by a current or former mate. It's no wonder spouses are considered suspects in murder cases.

EXAMPLE 3.21

Almost every retail business has the problem of determining how much inventory to purchase. Insufficient inventory may result in lost business, and excess inventory may have a detrimental effect on profits. Suppose a retail computer store owner is planning to place an order for personal computers (PCs). She is trying to decide how many IBM PCs and how many IBM compatibles to order.

The owner's records indicate that 80% of the previous PC customers purchased IBM PCs and 20% purchased compatibles.

a. What is the probability that the next two customers will purchase compatibles?
b. What is the probability that the next ten customers will purchase compatibles?

Solution

a. Let C_1 represent the event that customer 1 will purchase a compatible and C_2 represent the event that customer 2 will purchase a compatible. The event that *both* customers purchase compatibles is the intersection of the two events, $C_1 \cap C_2$. From the records the store owner could reasonably conclude that $P(C_1) = .2$ (based on the fact that 20% of past customers have purchased compatibles), and the same reasoning would apply to C_2. However, in order to compute the probability of $C_1 \cap C_2$, we need more information. Either the records must be examined for the occurrence of consecutive purchases of compatibles, or some assumption must be made to allow the calculation of $P(C_1 \cap C_2)$ from the multiplicative rule. It seems reasonable to make the assumption that the two events are independent, since the decision of the first

Case Study 3.2 continued

Focus

a. Show how Dershowitz arrived at his "1/1,000" probability value.
b. Show how Mertz and Caulkins arrived at their "29%" probability value.
c. Which probability is more relevant to the O.J. Simpson case? Why?

Mertz and Caulkins provide additional support for their position by employing the formula for a conditional probability. Assuming a woman has been murdered, define the following events:

   A: {Current or former mate has a history of spousal abuse}

   M: {Current or former mate is the murderer}

The probability that Mertz and Caulkins seek is $P(M|A)$, i.e., the probability that the mate is the murderer, given that the mate has a history of abuse. Using the conditional probability rule and multiplicative rule,

$$P(M|A) = \frac{P(A \cap M)}{P(A)} = \frac{P(A|M)P(M)}{P(A)}$$

Note that the event A occurs if either the two events $\{A \cap M\}$ or $\{A \cap M'\}$ occur. That is, the mate may be an abuser **and** the murderer, **or**, the mate may be an abuser

and not the murderer. Since "and" represents intersection and "or" represents union, the denominator of the above equation, $P(A)$, can be written as follows:

$$P(A) = P(A \cap M) = P(A \cap M')$$
$$= P(A|M)P(M) + P(A|M')P(M')$$

Substituting this expression into the desired probability, we obtain

$$P(M|A) = \frac{P(A|M)P(M)}{P(A)}$$
$$= \frac{P(A|M)P(M)}{P(A|M)P(M) + P(A|M')P(M')}$$

[This formula involving conditional probabilities is known as **Bayes' Rule**.]

Focus

d. Compute $P(M|A)$ using the above formula, the probability of part b, and the following information (provided by the RAND analysts): 5% of all women killed by someone other than a current or former mate have been abused; 50% of all women killed by their current or former mate have been abused.
e. Interpret the probability, part d, in the context of the O.J. Simpson trial.

customer is not likely to affect the decision of the second customer. Assuming independence, we have

$$P(C_1 \cap C_2) = P(C_1)P(C_2) = (.2)(.2) = .04$$

b. To see how to compute the probability that ten consecutive purchases will be compatibles, first consider the event that three consecutive customers purchase compatibles. If C_3 represents the event that the third customer purchases a compatible, then we want to compute the probability of the intersection $C_1 \cap C_2$ with C_3. Again assuming independence of the purchasing decisions, we have

$$P(C_1 \cap C_2 \cap C_3) = P(C_1 \cap C_2)P(C_3) = (.2)^2(.2) = .008$$

Similar reasoning leads to the conclusion that the intersection of ten such events can be calculated as follows:

$$P(C_1 \cap C_2 \cap \cdots \cap C_{10}) = P(C_1)P(C_2) \cdot \cdots \cdot P(C_{10})$$
$$= (.2)^{10} = .0000001024$$

Thus, the probability that ten consecutive customers purchase IBM compatibles is about 1 in 10 million, assuming the probability of each customer's purchase of a compatible is .2 and the purchase decisions are independent.   ▲

EXERCISES 3.46–3.62

Learning the Mechanics

3.46 Three fair coins are tossed and the following events are defined:

> *A:* {Observe at least one head}
>
> *B:* {Observe exactly two heads}
>
> *C:* {Observe exactly two tails}
>
> *D:* {Observe at most one head}

a. Sum the probabilities of the appropriate sample points to find $P(A)$, $P(B)$, $P(C)$, $P(D)$, $P(A \cap B)$, $P(A \cap D)$, $P(B \cap C)$, and $P(B \cap D)$.

b. Use your answers to part **a** to calculate $P(B \mid A)$, $P(A \mid D)$, and $P(C \mid B)$.

c. Which pairs of events, if any, are independent? Why?

3.47 An experiment results in one of five sample points with the following probabilities: $P(E_1) = .22$, $P(E_2) = .31$, $P(E_3) = .15$, $P(E_4) = .22$, and $P(E_5) = .1$. The following events have been defined:

> *A:* $\{E_1, E_3\}$
>
> *B:* $\{E_2, E_3, E_4\}$
>
> *C:* $\{E_1, E_5\}$

Find each of the following probabilities:

a. $P(A)$ **b.** $P(B)$ **c.** $P(A \cap B)$
d. $P(A \mid B)$ **e.** $P(B \cap C)$ **f.** $P(C \mid B)$
g. Consider each pair of events: A and B, A and C, and B and C. Are any of the pairs of events independent? Why?

3.48 Two fair dice are tossed, and the following events are defined:

> *A:* {Sum of the numbers showing is odd}
>
> *B:* {Sum of the numbers showing is 9, 11, or 12}

Are events A and B independent? Why?

3.49 A sample space contains six sample points and events A, B, and C as shown in the accompanying Venn diagram. The probabilities of the sample points are $P(1) = .20$, $P(2) = .05$, $P(3) = .30$, $P(4) = .10$, $P(5) = .10$, $P(6) = .25$.

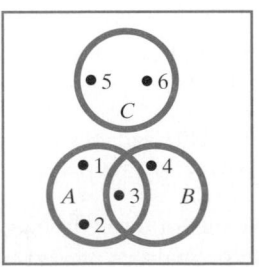

a. Which pairs of events, if any, are mutually exclusive? Why?

b. Which pairs of events, if any, are independent? Why?

c. Find $P(A \cup B)$ by adding the probabilities of the sample points and then by using the additive rule. Verify that the answers agree. Repeat for $P(A \cup C)$.

3.50 Defend or refute each of the following statements:
a. Dependent events are always mutually exclusive.
b. Mutually exclusive events are always dependent.
c. Independent events are always mutually exclusive.

3.51 For two events, A and B, $P(A) = .4$ and $P(B) = .2$.
a. If A and B are independent, find $P(A \cap B)$, $P(A \mid B)$, and $P(A \cup B)$.
b. If A and B are dependent, with $P(A \mid B) = .6$, find $P(A \cap B)$ and $P(B \mid A)$.

Applying the Concepts

3.52 Each of 50 people in a random sample was asked to name his or her favorite soft drink. The responses are shown below:

Pepsi-Cola	18	Sprite	4
Coca-Cola	16	Mountain Dew	1
Seven-Up	6	Dr. Pepper	1
Mr. Pibb	4		

Suppose a person is selected at random from the survey. Let A be the event that the person preferred a soft drink bottled by the Coca-Cola Company (Coca-Cola, Mr. Pibb, or Sprite). Let B be the event that the person did *not* choose a cola (either Pepsi-Cola or Coca-Cola).
a. Find $P(A)$. **b.** Find $P(B)$.
c. Describe the event $A \cap B$, and find its probability.
d. Describe the event $A \cup B$ and use the additive rule to find its probability.
e. Use the multiplicative rule to find $P(A \mid B)$.
f. Use the multiplicative rule to find $P(A \mid B')$.

3.53 The probability that a certain computer microchip fails upon first use is .10. If it does not fail immediately, the probability that it lasts 1 year is .99. Define the experiment as observing whether a microchip fails or not during its first year, and define the following events:

> *A:* {Microchip lasts through first use}
>
> *B:* {Microchip lasts from first use through end of first year}

a. List the sample points of the experiment in terms of the events A and B. [*Hint:* The event that the chip fails during the first year is *not* a sample point because it can be decomposed into two more basic events. There are three sample points.]

b. Assign probabilities to the sample points. [*Hint:* You'll need to use the multiplicative rule for some of the sample points.]

c. Use the sample point probabilities to find the probability that the microchip does not fail during the first year.

3.54 Enterococci are a family of bacteria that cause blood infections in hospitalized patients. The antibiotic vancomycin has been used to successfully battle enterococci. However, strains of the bacteria resistant to the antibiotic have recently evolved. A study by the Centers for Disease Control and Prevention in Atlanta, Georgia, revealed that 8% of all enterococci isolated in hospitals nationwide were resistant to vancomycin (*New York Times,* Sept. 12, 1995). Consider a random sample of three patients with blood infections caused by the enterococci bacteria. Assume that all three patients are treated with the antibiotic vancomycin.

a. What is the probability that all three patients are successfully treated? What assumption did you make concerning the patients?

b. What is the probability that the bacteria resist the antibiotic for at least one patient?

3.55 Scoring a hole-in-one is the single greatest shot a golfer can make. In the 1989 U.S. Open, four players made holes-in-one on the sixth hole at Oak Hill Country Club. Lois Hains, in an article in *Golf Digest* (1990), reports that the estimated probability of making a hole-in-one is approximately $1/2{,}780$ for male professional golfers and $1/2{,}920$ for female professional golfers. Suppose we randomly select four professional male golfers.

a. What is the probability that each of these four golfers makes a hole-in-one on the sixth hole during the same round? What assumption did you make to calculate this probability?

b. What is the probability that none of these golfers makes a hole-in-one on the sixth hole?

c. What is the probability that at least one of these golfers makes a hole-in-one on the sixth hole?

d. Do you think the probability of a hole-in-one on the sixth hole at Oak Hill Country Club is the same as the number reported in *Golf Digest*? Explain.

3.56 On Apr. 25, 1980, the *Bakersfield Californian* reported on a record number of payouts in the Pennsylvania lottery. On that day, the winning number, with options from 000 to 999, was 666. Each digit of this winning number was determined by allowing a blast of air to propel one of ten Ping-Pong balls, numbered 0, 1, 2, ..., 9, into a basket. Harried state officials were denying the possibility of tampering with the lottery, but savvy Pittsburgh bookies knew something was up: They were refusing bets on any numbers involving 4's or 6's. Later, it

was found that a TV host (eventually indicted and convicted) and some friends had injected liquid into all the balls except those numbered 4 and 6. Thus, only numbers involving the 4 and the 6 digits could be winners.

a. If the balls had not been fixed, what is the probability that a number involving only 4's and 6's would win? What assumption about the determination of the three digits did you make to solve this problem?

b. Given the conditions of part **a**, what is the probability that the number 666 would win?

c. Since the TV announcer and collaborators knew that the winning number would involve only 4's and 6's, what is the probability that the winning number would be 666?

3.57 "Channel One" is an education television network that is available to all secondary schools in the United States. Participating schools are equipped with TV sets in every classroom in order to receive the Channel One broadcasts. According to *Educational Technology* (May–June 1995), 40% of all U.S. secondary schools subscribe to the Channel One Communications Network (CCN). Of these subscribers, only 5% never use the CCN broadcasts, while 20% use CCN more than five times per week.

a. Find the probability that a randomly selected U.S. secondary school subscribes to CCN but never uses the CCN broadcasts.

b. Find the probability that a randomly selected U.S. secondary school subscribes to CCN and uses the broadcasts more than five times per week.

3.58 In October 1994, a flaw was discovered in the Pentium chip installed in many new personal computers. The chip produced an incorrect result when dividing two numbers. Intel, the manufacturer of the Pentium chip, initially announced that such an error would occur once in 9 billion divides, or "once in every 27,000 years" for a typical user; consequently, it did not immediately offer to replace the chip. Assume that the probability of a divide error with the Pentium chip is, in fact, $1/9{,}000{,}000{,}000$.

a. For a division performed using the flawed Pentium chip, what is the probability that no error will occur?

b. Consider two successive divisions performed using the flawed chip. What is the probability that neither result will be in error? (Assume that any one division has no impact on the result of any other division performed by the chip.)

c. Depending on the procedure, statistical software packages may perform an extremely large number of divisions to produce the required output. For heavy users of the software, 1 billion divisions over a short time frame is not unusual. Calculate the probability that 1 billion divisions performed using the flawed Pentium chip will result in no errors.

d. Use the result, part **c**, to compute the probability of at least one error in the 1 billion divisions. [*Note:* Two months after the flaw was discovered, Intel agreed to replace all Pentium chips free of charge.]

3.59 A version of the dice game "craps" is played in the following manner: A player starts by rolling two dice. If the result is a 7 or 11, the player wins. For most other sums appearing on the dice, the player continues to roll the dice until that sum recurs (in which case the player wins) or until a 7 or 11 occurs (in which case the player loses). But if on any roll the outcome is 2 ("snake-eyes"), the game is over, and the player loses.

a. What is the probability that a player wins the game on the first roll of the dice? (Assume the dice are balanced.)

b. What is the probability that a player loses the game on the first roll of the dice?

c. If the player throws a total of 3 on the first roll, what is the probability that the game ends on the next roll?

3.60 The genetic origin and properties of maize (modern-day corn), a domestic plant developed 8,000 years ago in Mexico, was investigated in *Economic Botany* (Jan.–Mar. 1995). Seeds from maize ears carry either single spikelets or paired spikelets, but not both. Progeny tests on approximately 600 maize ears revealed the following information. Forty percent of all seeds carry single spikelets, while 60% carry paired spikelets. A seed with single spikelets will produce maize ears with single spikelets 29% of the time and paired spikelets 71% of the time. A seed with paired spikelets will produce maize ears with single spikelets 26% of the time and paired spikelets 74% of the time.

a. Find the probability that a randomly selected maize ear seed carries a single spikelet and produces ears with single spikelets.

b. Find the probability that a randomly selected maize ear seed produces ears with paired spikelets.

3.61 One in every 500 African-Americans in the United States is reported to have sickle-cell anemia. One in ten is a carrier of the trait, and carriers have a 1 in 4 chance of passing the disease to their children. Suppose a carrier has three children and the transmission of the disease from the carrier to any one child is independent of whether or not the carrier transmits to another. Find the probability that

a. None of the children acquires the disease

b. All three acquire the disease

c. Exactly one acquires the disease

3.62 One of the problems encountered in organ transplants is the body's rejection of the transplanted tissue; that is, the body's white blood cells tend to attack the foreign tissue. The key to reception or rejection is the nature of the antigens attached to the tissue cells. If the antigens of the donor and receiver match, the body will accept the transplanted tissue. While the antigens in identical twins always match, the probability of a match in other siblings is .25 and that of a match in two people from the population at large is .001. Suppose you need a kidney, and you have two brothers and a sister.

a. If one of your three siblings offers a kidney, what is the probability that the antigens will match?

b. If all three siblings offer a kidney, what is the probability that all three antigens will match?

c. If all three siblings offer a kidney, what is the probability that none of the antigens will match?

d. Repeat parts **b** and **c**, this time assuming that the three donors were obtained from the population at large.

3.7   PROBABILITY AND STATISTICS: AN EXAMPLE

We've encountered several new concepts in the preceding sections, the sheer number of which can make the study of probability seem a particularly arduous task. But it's essential to understand the connection between probability and statistics. We'll establish this connection clearly in the remaining chapters. For now, let's look at one brief example here so that you can begin to grasp why some knowledge of probability is important in the study of statistics.

Suppose a research psychologist is studying the hypothesis that trained rats will pass on at least part of their training to their offspring. To test the hypothesis, three offspring (no two with the same parents) of trained rats are randomly selected and subjected to a training test. From many previous experiments the psychologist knows that the relative frequency distribution of the scores for untrained rats is mound-shaped with a mean (μ) of 60 and a standard deviation (σ) of 10.

FIGURE 3.20

Relative frequency
distribution of training
test scores

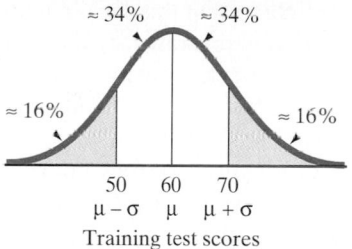

Suppose all three of the trained rats' offspring score more than 70 on the test. What can the researcher conclude?

The relative frequency distribution of the scores for untrained rats is shown in Figure 3.20. If the distribution is mound-shaped and approximately symmetric about the mean, we can conclude that approximately 16% of untrained rats will score more than 70 on the test (see also Table 2.9).

Now define the events

A_1: {Offspring 1 scores more than 70}

A_2: {Offspring 2 scores more than 70}

A_3: {Offspring 3 scores more than 70}

We want to find $P(A_1 \cap A_2 \cap A_3)$, the probability that all three offspring score more than 70 on the training test.

Since the offspring are selected so that they have different parents, it may be plausible to assume that the events $A_1, A_2,$ and A_3 are independent. That is,

$$P(A_2 \mid A_1) = P(A_2)$$

In words, knowing that the first offspring scores more than 70 on the test does not affect the probability that the second offspring scores more than 70. With the assumption of independence, we can calculate the probability of the intersection by multiplying the individual probabilities:

$$P(A_1 \cap A_2 \cap A_3) = P(A_1)P(A_2)P(A_3)$$
$$\approx (.16)(.16)(.16) = .004096$$

Thus, the probability that the research psychologist will observe all three offspring scoring more than 70 is only about .004 *if the offspring are untrained.* If this event were to occur, the psychologist might conclude that it lends credence to the theory that the offspring inherit some of the parents' training *since it is so unlikely to occur if they are untrained.* Such a conclusion would be an application of the rare-event approach to statistical inference. You can see that the basic principles of probability play an important role.

3.8   RANDOM SAMPLING

How a sample is selected from a population is of vital importance in statistical inference because the probability of an observed sample will be used to infer the characteristics of the sampled population. To illustrate, suppose you deal yourself four cards from a deck of 52 cards and all four cards are aces. Do you conclude that your deck is an ordinary bridge deck, containing only four aces, or do you conclude that the deck is stacked with more than four aces? It depends on how the cards were drawn. If the four aces were always placed at the top of a standard bridge deck, drawing four aces is not unusual—it is certain. On the other hand, if the cards are thoroughly mixed, drawing four aces in a sample of four cards is highly improbable. The point, of course, is that in order to use the observed sample of four cards to draw inferences about the population (the deck of 52 cards), you need to know how the sample was selected from the deck.

One of the simplest and most frequently employed sampling procedures is implied in the previous examples and exercises. It produces what is known as a *random sample*.

DEFINITION 3.10

If n elements are selected from a population in such a way that every set of n elements in the population has an equal probability of being selected, the n elements are said to be a **random sample**.*

EXAMPLE 3.22

Suppose a lottery consists of 10 tickets. (This number is small to simplify our example.) One ticket stub is to be chosen, and the corresponding ticket holder will receive a generous prize. How would you select this ticket stub so that the prize will be awarded fairly?

Solution

If the prize is to be awarded fairly, it seems reasonable to require that each ticket stub have the same probability of being drawn. That is, each stub should have a probability of $\frac{1}{10}$ of being selected. A method to achieve the objective of equal selection probabilities is to *mix* the 10 stubs thoroughly and *blindly* pick one of the stubs. If this procedure were repeatedly used, each time replacing the selected stub, a particular stub should be chosen approximately $\frac{1}{10}$ of the time in a long series of draws. This method of sampling is known as **random sampling**.

If a population is not too large and the elements can be numbered on slips of paper, poker chips, etc., you can physically mix the slips of paper or chips and remove n elements from the total. The numbers that appear on the chips selected would indicate the population elements to be included in the sample. Since it is often difficult to achieve a thorough mix, such a procedure provides only an approximation to random sampling. Most researchers rely on **random number generators** to automatically generate the random sample. Random number generators are available in table form and they are built into most statistical software packages.   ▲

You can think of sampling as an experiment consisting of drawing n elements from a population of N elements, with each different sample representing a sample point of the experiment. Thus, in Example 3.22, in which a lottery consisted of drawing one of 10 tickets, there are 10 different sample points. If the drawing is held so that each sample point is equally likely with probability $\frac{1}{10}$, then the result of the experiment is a *random sample*.

Of course, for most applications the population will consist of more than $N = 10$ elements, and the sample will consist of more than $n = 1$ element. Often, the

---

*Strictly speaking, this is a **simple random sample**. There are many different types of random samples. The simple random sample is the most common.

TABLE 3.6   Listing of All Possible Samples of Two Tickets Drawn from 10 Tickets

T_1, T_2	T_2, T_3	T_3, T_4	T_4, T_5	T_5, T_6	T_6, T_7	T_7, T_8	T_8, T_9	T_9, T_{10}
T_1, T_3	T_2, T_4	T_3, T_5	T_4, T_6	T_5, T_7	T_6, T_8	T_7, T_9	T_8, T_{10}	
T_1, T_4	T_2, T_5	T_3, T_6	T_4, T_7	T_5, T_8	T_6, T_9	T_7, T_{10}		
T_1, T_5	T_2, T_6	T_3, T_7	T_4, T_8	T_5, T_9	T_6, T_{10}			
T_1, T_6	T_2, T_7	T_3, T_8	T_4, T_9	T_5, T_{10}				
T_1, T_7	T_2, T_8	T_3, T_9	T_4, T_{10}					
T_1, T_8	T_2, T_9	T_3, T_{10}						
T_1, T_9	T_2, T_{10}							
T_1, T_{10}								

total number of possible samples will not be easy to visualize, so a method for counting the number of samples is needed. For example, suppose the lottery consists of drawing two tickets from ten. We can list the possible samples as in Table 3.6, where T_1 represents ticket 1, T_2 ticket 2, ..., and T_{10} ticket 10. The systematic listing in Table 3.6 shows the 45 possible sample points of the experiment of sampling two elements from 10. However, the listing is tedious and only gets more so as the values of the population size N and the sample size n are increased.

A second method of determining the number of samples is to use **combinatorial mathematics**. The combinatorial symbol for the number of different ways of selecting n elements from N elements is $\binom{N}{n}$, which is read "the number of combinations of N elements taken n at a time." The formula* for calculating the number is

$$\binom{N}{n} = \frac{N!}{n!(N-n)!}$$

where "!" is the factorial symbol and is a shorthand for the following multiplication:

$$n! = n(n-1)(n-2) \cdot \cdots \cdot (3)(2)(1)$$

Thus, for example, $5! = 5 \cdot 4 \cdot 3 \cdot 2 \cdot 1 = 120$. (The quantity $0!$ is defined to be 1.)

EXAMPLE 3.23

a. Use the combinatorial formula to count the number of different ways of drawing one lottery ticket from a total of 10 tickets.
b. Use the combinatorial formula to count the number of ways of drawing two lottery tickets from a total of 10 tickets.

Solution

a. Substituting $n = 1$ and $N = 10$ into the formula, we find

$$\binom{N}{n} = \binom{10}{1} = \frac{10!}{1!(10-1)!} = \frac{10!}{1!9!}$$

$$= \frac{10 \cdot 9 \cdot 8 \cdot 7 \cdot 6 \cdot 5 \cdot 4 \cdot 3 \cdot 2 \cdot 1}{(1)(9 \cdot 8 \cdot 7 \cdot 6 \cdot 5 \cdot 4 \cdot 3 \cdot 2 \cdot 1)} = 10$$

Thus, as we know intuitively, there are 10 different samples that can be selected when drawing one ticket from 10.

---

*For a more thorough discussion of the reasoning behind this and other counting rules, see (optional) Section 3.9.

b. Substituting $n = 2$ and $N = 10$ into the formula, we find

$$\binom{N}{n} = \binom{10}{2} = \frac{10!}{2!(10-2)!} = \frac{10!}{2!8!}$$

$$= \frac{10 \cdot 9 \cdot 8 \cdot 7 \cdot 6 \cdot 5 \cdot 4 \cdot 3 \cdot 2 \cdot 1}{(2 \cdot 1)(8 \cdot 7 \cdot 6 \cdot 5 \cdot 4 \cdot 3 \cdot 2 \cdot 1)} = \frac{10 \cdot 9}{2 \cdot 1} = 45$$

This agrees with the number obtained by listing all possible samples in Table 3.6 but requires much less effort. And, as the next example illustrates, the combinatorial formula works long after listing has ceased to be practical. ▲

EXAMPLE 3.24

Suppose you wish to randomly sample five households from a population of 100,000 households.

a. How many different samples can be selected?
b. Use a random number generator to select a random sample.

Solution

a. Using the combinatorial rule, we find

$$\binom{100,000}{5} = \frac{100,000!}{5!99,995!}$$

$$= \frac{100,000 \cdot 99,999 \cdot 99,998 \cdot 99,997 \cdot 99,996}{5 \cdot 4 \cdot 3 \cdot 2 \cdot 1}$$

$$= 8.33 \times 10^{22}$$

Thus, there are 83.3 billion trillion different samples of five households that can be selected from 100,000.

b. To ensure that each of the possible samples has an equal chance of being selected, as required for random sampling, we can employ a **random number table**, as provided in Table I of Appendix A. Random number tables are constructed in such a way that every number occurs with (approximately) equal probability. Furthermore, the occurrence of any one number in a position is independent of any of the other numbers that appear in the table. To use a table of random numbers, number the N elements in the population from 1 to N. Then turn to Table I and select a starting number in the table. Proceeding from this number either across the row or down the column, remove and record n numbers from the table.

To illustrate, first we number the households in the population from 1 to 100,000. Then, we turn to a page of Table I, say the first page. (A partial reproduction of the first page of Table I is shown in Table 3.7.) Now, we arbitrarily select a starting number, say the random number appearing in the third row, second column. This number is 48,360. Then we proceed down the second column to obtain the remaining four random numbers. In this case we have selected five random numbers, which are shaded in Table 3.7. Using the first five digits to represent households from 1 to 99,999 and the number 00000 to represent household 100,000, we can see that the households numbered

48,360   93,093   39,975   6,907   72,905

should be included in our sample.

TABLE 3.7   Partial Reproduction of Table I in Appendix A

Row	Column 1	2	3	4	5	6
1	10480	15011	01536	02011	81647	91646
2	22368	46573	25595	85393	30995	89198
3	24130	48360	22527	97265	76393	64809
4	42167	93093	06243	61680	07856	16376
5	37570	39975	81837	16656	06121	91782
6	77921	06907	11008	42751	27756	53498
7	99562	72905	56420	69994	98872	31016
8	96301	91977	05463	07972	18876	20922
9	89579	14342	63661	10281	17453	18103
10	85475	36857	53342	53988	53060	59533
11	28918	69578	88231	33276	70997	79936
12	63553	40961	48235	03427	49626	69445
13	09429	93969	52636	92737	88974	33488

Note: Use only the necessary number of digits in each random number to identify the element to be included in the sample. If, in the course of recording the *n* numbers from the table, you select a number that has already been selected, simply discard the duplicate and select a replacement at the end of the sequence. Thus, you may have to record more than *n* numbers from the table to obtain a sample of *n* unique numbers.

Can we be perfectly sure that all 83.3 billion trillion samples have an equal chance of being selected? That fact is, we can't; but to the extent that the random number table contains truly random sequences of digits, the sample should be very close to random.   ▲

Table I in Appendix A is just one example of a random number generator. For most scientific studies that require a large random sample, computers are used to generate the random sample. The SAS, MINITAB, and ASP statistical software packages all have easy-to-use random number generators.

For example, suppose we required a random sample of $n = 50$ households from the population of 100,000 households in Example 3.24. Here, we might employ the SAS random number generator. Figure 3.21 shows a SAS printout

FIGURE 3.21

SAS-generated random sample of 50 households

OBS	HOUSENUM	OBS	HOUSENUM	OBS	HOUSENUM	OBS	HOUSENUM
1	47122	14	47271	27	17098	40	4260
2	94231	15	3642	28	23259	41	58140
3	95531	16	7611	29	30512	42	22903
4	41445	17	81646	30	91548	43	65959
5	80287	18	92158	31	7673	44	13962
6	11731	19	36667	32	68549	45	25819
7	47523	20	71811	33	85433	46	66497
8	84847	21	78988	34	5231	47	79559
9	69822	22	3819	35	13455	48	87017
10	18270	23	21873	36	71666	49	28483
11	52636	24	74938	37	66280	50	91806
12	21750	25	23635	38	66210		
13	63363	26	35807	39	21998		

listing 50 random numbers (from a population of 100,000). The households with these identification numbers would be included in the random sample.

EXERCISES 3.63–3.67

Learning the Mechanics

3.63 Suppose you wish to sample $n = 3$ elements from a total of $N = 6$ elements.

    a. Count the number of different samples that can be drawn, first by listing them, and then by using combinatorial mathematics.

    b. If random sampling is to be employed, what is the probability that any particular sample will be selected?

    c. Show how to use the random number table, Table I in Appendix A, to select a random sample of three elements from a population of six elements. Perform the sampling procedure 20 times. Do any two of the samples contain the same three elements? Given your answer to part **b**, did you expect repeated samples?

3.64 Suppose you wish to sample $n = 3$ elements from a total of $N = 600$ elements.

    a. Count the number of different samples by using combinatorial mathematics.

    b. If random sampling is to be employed, what is the probability that any particular sample will be selected?

    c. Show how to use the random number table, Table I in Appendix A, to select a random sample of 3 elements from a population of 600 elements. Perform the sampling procedure 20 times. Do any two of the samples contain the same three elements? Given your answer to part **b**, did you expect repeated samples?

    d. Use a computer to generate a random sample of 3 from the population of 600 elements.

Applying the Concepts

3.65 Archaeologists plan to perform test digs at a location they believe was inhabited several thousand years ago. The site is approximately 10,000 meters long and 5,000 meters wide. They first draw rectangular grids over the area, consisting of lines every 100 meters, creating a total of $100 \cdot 50 = 5,000$ intersections (not counting one of the outer boundaries).

The plan is to randomly sample 50 intersection points and dig at the sampled intersections. Explain how you could use a random number generator to obtain a random sample of 50 intersections. Develop at least two plans: one that numbers the intersections from 1 to 5,000 prior to selection and another that selects the row and column of each sampled intersection (from the total of 100 rows and 50 columns).

3.66 To ascertain the effectiveness of their advertising campaigns, firms frequently conduct telephone interviews with consumers. They may use random samples of telephone numbers that are arbitrarily or systematically selected from telephone directories, or they may employ an innovation called *random-digit dialing*. In this approach, a random number generator mechanically creates the sample of phone numbers to be called. An advantage of random-digit dialing is that it can obtain a representative sample from the population of *all* households with telephones, whereas telephone-directory sampling obtains a sample only from the population of households with *listed* telephone numbers.

    a. Explain how the random number table (Table I of Appendix A, or a computer) could be used to generate a sample of 7-digit telephone numbers.

    b. Use the procedure you described in part **a** to generate a sample of ten 7-digit telephone numbers.

    c. Use the procedure you described in part **a** to generate five 7-digit telephone numbers whose first three digits are 373.

3.67 In addition to its decennial enumeration of the population, the U.S. Bureau of the Census regularly samples the population to estimate level of and changes in a number of other attributes, such as income, family size, employment, and marital status. Suppose the Bureau plans to sample 1,000 households in a city that has a total of 534,322 households. Show how the Bureau could use the random number table in Appendix A or a computer to generate the sample. Select the first 10 households to be included in the sample.

3.9 SOME COUNTING RULES (OPTIONAL)

In Section 3.1 we pointed out that experiments sometimes have so many sample points that it is impractical to list them all. However, many of these experiments possess sample points with identical characteristics. If you can develop a **counting rule** to count the number of sample points for such an experiment, it can be used to aid in the solution of the problems.

EXAMPLE 3.25

A product can be shipped by four airlines and each airline can ship via three different routes. How many distinct ways exist to ship the product?

Solution

A pictorial representation of the different ways to ship the product will aid in counting them. The tree diagram (see Section 3.6) is useful for this purpose. Consider the tree diagram shown in Figure 3.22. At the starting point (stage 1), there are four choices—the different airlines—to begin the journey. Once we have chosen an airline (stage 2), there are three choices—the different routes—to complete the shipment and reach the final destination. Thus, the tree diagram clearly shows that there are $(4)(3) = 12$ distinct ways to ship the product. ▲

The method of solving this example can be generalized to any number of stages with sets of different elements. This method is given by the *multiplicative rule*.

> **The Multiplicative Rule**
>
> You have k sets of elements, n_1 in the first set, n_2 in the second set, ..., and n_k in the kth set. Suppose you wish to form a sample of k elements by *taking one element from each* of the k sets. The number of different samples that can be formed is the product
>
> $$n_1 n_2 n_3 \cdots n_k$$

EXAMPLE 3.26

There are 20 candidates for three different executive positions, E_1, E_2, and E_3. How many different ways could you fill the positions?

Solution

For this example, there are $k = 3$ sets of elements corresponding to:

Set 1: Candidates available to fill position E_1

Set 2: Candidates remaining (after filling E_1) that are available to fill E_2

Set 3: Candidates remaining (after filling E_1 and E_2) that are available to fill E_3

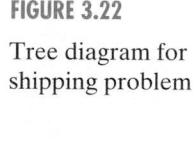

FIGURE 3.22

Tree diagram for shipping problem

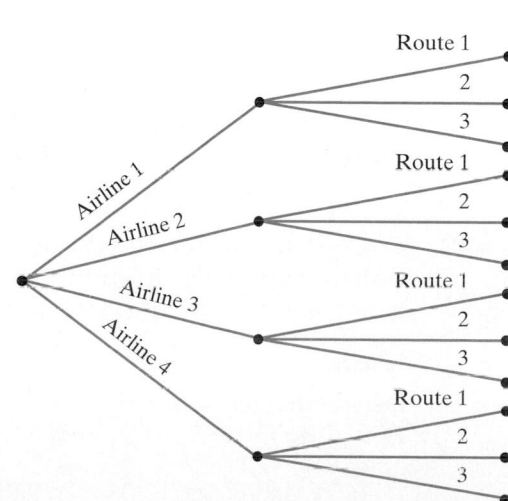

The numbers of elements in the sets are $n_1 = 20, n_2 = 19, n_3 = 18$. Therefore, the number of different ways of filling the three positions is given by the multiplicative rule as $n_1 n_2 n_3 = (20)(19)(18) = 6,840$.   ▲

EXAMPLE 3.27

Consider the experiment discussed in Example 3.10 of tossing a coin 10 times. Show how we found that there were $2^{10} = 1,024$ sample points for this experiment.

Solution

There are $k = 10$ sets of elements for this experiment. Each set contains two elements: a head and a tail. Thus, there are

$$(2)(2)(2)(2)(2)(2)(2)(2)(2)(2) = 2^{10} = 1,024$$

different outcomes (sample points) of this experiment.   ▲

EXAMPLE 3.28

Suppose there are five dangerous military missions, each requiring one soldier. In how many different ways can five soldiers from a squadron of 100 be assigned to these five missions?

Solution

We can solve this problem by using the multiplicative rule. The entire set of 100 soldiers is available for the first mission, and after the selection of one soldier for that mission, 99 are available for the second mission, etc. Thus, the total number of different ways of choosing five soldiers for the five missions is

$$n_1 n_2 n_3 n_4 n_5 = (100)(99)(98)(97)(96) = 9,034,502,400$$ ▲

The arrangement of elements in a distinct order is called a **permutation**. Thus, in Example 3.28 there are more than 9 billion different *permutations* of 5 elements (soldiers) drawn from a set of 100 elements!

Permutations Rule

Given a *single set* of N distinctly different elements, you wish to select n elements from the N and *arrange* them within n positions. The number of different **permutations** of the N elements taken n at a time is denoted by P_n^N and is equal to

$$P_n^N = N(N-1)(N-2) \cdot \cdots \cdot (N-n+1) = \frac{N!}{(N-n)!}$$

where $n! = n(n-1)(n-2) \cdot \cdots \cdot (3)(2)(1)$ and is called n **factorial**. (Thus, for example, $5! = 5 \cdot 4 \cdot 3 \cdot 2 \cdot 1 = 120$.) The quantity $0!$ is defined to be 1.

EXAMPLE 3.29

You wish to drive, in sequence, from a starting point to each of five cities, and you wish to compare the distances—and ultimately the costs—of the different routes. How many different routes would you have to compare?

Solution

Denote the cities as $C_1, C_2, \ldots, C_5$. Then a route moving from the starting point to C_2 to C_1 to C_3 to C_4 to C_5 would be represented as $C_2 C_1 C_3 C_4 C_5$. The total number

TABLE 3.8

Group 1	Group 2	Group 3	Group 4
ABCD	ABDC	ACDB	BCDA
ACBD	ADBC	ADCB	BDCA
BACD	BADC	CADB	CBDA
BCAD	BDAC	CDAB	CDBA
CABD	DABC	DACB	DBCA
CBAD	DBAC	DCAB	DCBA

TABLE 3.9

Job 1	Job 2
ABC	D
ABD	C
ACD	B
BCD	A

of routes would equal the number of ways you could rearrange the $N = 5$ cities in $n = 5$ positions. This number is

$$P_n^N = P_5^5 = \frac{5!}{(5-5)!} = \frac{5!}{0!} = \frac{5 \cdot 4 \cdot 3 \cdot 2 \cdot 1}{1} = 120$$

(Recall that $0! = 1$.)   ▲

EXAMPLE 3.30

You are supervising four construction workers. You must assign three to job 1 and one to job 2. In how many different ways can you make this assignment?

Solution

To begin, suppose each worker is to be assigned to a distinct job. Then, using the multiplicative rule, you obtain $(4)(3)(2)(1) = 24$ ways of assigning the workers to four distinct jobs. The 24 ways are listed in four groups in Table 3.8 (where ABCD signifies that worker A was assigned the first distinct job, worker B the second, etc.).

Now suppose the first three positions represent job 1 and the last position represents job 2. You can now see that all the listings in group 1 represent the same outcome of the experiment of interest. That is, workers A, B, and C are assigned to job 1 and worker D to job 2. Similarly, group 2 listings are equivalent, as are group 3 and group 4 listings. Thus, there are only four different assignments of four workers to the two jobs. These are shown in Table 3.9.

To generalize this result, we point out that the final result can be found by

$$\frac{(4)(3)(2)(1)}{(3)(2)(1)(1)} = 4$$

The $(4)(3)(2)(1)$ is the number of different ways *(permutations)* the workers could be assigned four distinct jobs. The division by $(3)(2)(1)$ is to remove the duplicated permutations resulting from the fact that three workers are assigned the same jobs. And the division by 1 is associated with the worker assigned to job 2.   ▲

Partitions Rule

There exists a *single* set of N distinctly different elements. You wish to partition them into k sets, with the first set containing n_1 elements, the second containing n_2 elements, ..., and the kth set containing n_k elements. The number of different partitions is

$$\frac{N!}{n_1! n_2! \cdot \cdots \cdot n_k!} \qquad \text{where} \quad n_1 + n_2 + n_3 + \cdots + n_k = N$$

EXAMPLE 3.31

You have 12 constructions workers. You wish to assign three to job site 1, four to job site 2, and five to job site 3. In how many different ways can you make this assignment?

Solution

For this example, $k = 3$ (corresponding to the $k = 3$ different job sites), $N = 12$, and $n_1 = 3$, $n_2 = 4$, and $n_3 = 5$. Then the number of different ways to assign the workers to the job sites is

$$\frac{N!}{n_1!n_2!n_3!} = \frac{12!}{3!4!5!} = \frac{12 \cdot 11 \cdot 10 \cdot \cdots \cdot 3 \cdot 2 \cdot 1}{(3 \cdot 2 \cdot 1)(4 \cdot 3 \cdot 2 \cdot 1)(5 \cdot 4 \cdot 3 \cdot 2 \cdot 1)} = 27{,}720 \quad \blacktriangle$$

EXAMPLE 3.32

How many samples of four courses can you select from a list of 25 courses in a university catalog?

Solution

For this example, $k = 2$ (corresponding to the $n_1 = 4$ classes you *do* choose and the $n_2 = 21$ classes you *do not* choose) and $N = 25$. Then, the number of different ways to choose the four classes from 25 is

$$\frac{N!}{n_1!n_2!} = \frac{25!}{(4!)(21!)} = \frac{25 \cdot 24 \cdot 23 \cdot \cdots \cdot 3 \cdot 2 \cdot 1}{(4 \cdot 3 \cdot 2 \cdot 1)(21 \cdot 20 \cdot \cdots \cdot 2 \cdot 1)} = 12{,}650$$

(Now you know why there's so much confusion surrounding registration time at many schools.) $\blacktriangle$

This special application of the partitions rule—partitioning a set of N elements into $k = 2$ groups (the elements that appear in a sample and those that do not)—is very common. Therefore, we give a different name to the rule for counting the number of different ways of partitioning a set of elements into two parts: the **combinations rule**.

Combinations Rule

A sample of n elements is to be chosen from a set of N elements. Then the number of different samples of n elements that can be selected from N is denoted by $\binom{N}{n}$ and is equal to

$$\binom{N}{n} = \frac{N!}{n!(N-n)!}$$

Note that the order in which the n elements are drawn is not important.

EXAMPLE 3.33

Five soldiers from a squadron of 100 are to be chosen for a dangerous mission. In how many ways can groups of five be formed?

Solution

This problem is equivalent to sampling $n = 5$ elements from a set of $N = 100$ elements. Thus, the number of ways is the number of possible combinations of five soldiers selected from 100, or

$$\binom{100}{5} = \frac{100!}{(5!)(95!)} = \frac{100 \cdot 99 \cdot 98 \cdot 97 \cdot 96 \cdot 95 \cdot 94 \cdot \cdots \cdot 2 \cdot 1}{(5 \cdot 4 \cdot 3 \cdot 2 \cdot 1)(95 \cdot 94 \cdot \cdots \cdot 2 \cdot 1)}$$

$$= \frac{100 \cdot 99 \cdot 98 \cdot 97 \cdot 96}{5 \cdot 4 \cdot 3 \cdot 2 \cdot 1} = 75{,}287{,}520$$

Compare this result with that of Example 3.28, where we found that the number of permutations of five elements drawn from 100 was more than 9 billion. Because the order of the elements does not affect combinations, there are *fewer* combinations than permutations.   ▲

Summary of Counting Rules

1. *Multiplicative rule.* If you are drawing *one element from each of k sets of elements,* with the sizes of the sets $n_1, n_2, ..., n_k$, the number of different results is

$$n_1 n_2 n_3 \cdots n_k$$

2. *Permutations rule.* If you are drawing *n elements from a set of N elements and arranging the n elements in a distinct order,* the number of different results is

$$P_n^N = \frac{N!}{(N-n)!}$$

3. *Partitions rule.* If you are partitioning the *elements of a set of N elements into k groups consisting of $n_1 n_2, ..., n_k$ elements* $(n_1 + \cdots + n_k = N)$, the number of different results is

$$\frac{N!}{n_1! n_2! \cdots n_k!}$$

4. *Combinations rule.* If you are drawing *n elements from a set of N elements without regard to the order of the n elements,* the number of different results is

$$\binom{N}{n} = \frac{N!}{n!(N-n)!}$$

[*Note:* The combinations rule is a special case of the partitions rule when $k = 2$.]

When working a probability problem, you should carefully examine the experiment to determine whether you can use one or more of the rules we have discussed in this section. Let's see how these rules can help solve a probability problem.

EXAMPLE 3.34

A consumer testing service is commissioned to rank the top three brands of laundry detergent. A total of 10 brands are to be included in the study.

a. In how many different ways can the consumer testing service arrive at the final ranking?

b. If the testing service can distinguish no difference among the brands and therefore arrives at the final ranking by random selection, what is the probability that company Z's brand is ranked first? In the top three?

Solution

a. Since the testing service is drawing three elements (brands) from a set of 10 elements and arranging the three elements in a distinct order, we use the permutations rule to find the number of different results:

$$P_3^{10} = \frac{10!}{(10-3)!} = 10 \cdot 9 \cdot 8 = 720$$

b. The steps for calculating the probability of interest are as follows:

Step 1   The experiment is to select and rank three brands of detergent from 10 brands.

Step 2   There are too many sample points to list. However, we know from part a that there are 720 different outcomes (i.e., sample points) of this experiment.

Step 3   If we assume the testing service determines the rankings at random, each of the 720 sample points should have an equal probability of occurrence. Thus,

$$P(\text{Each sample point}) = \tfrac{1}{720}$$

Step 4   One event of interest to company Z is that its brand receives top ranking. We will call this event A. The list of sample points that result in the occurrence of event A is long, but the *number* of sample points contained in event A is determined by breaking event A into two parts:

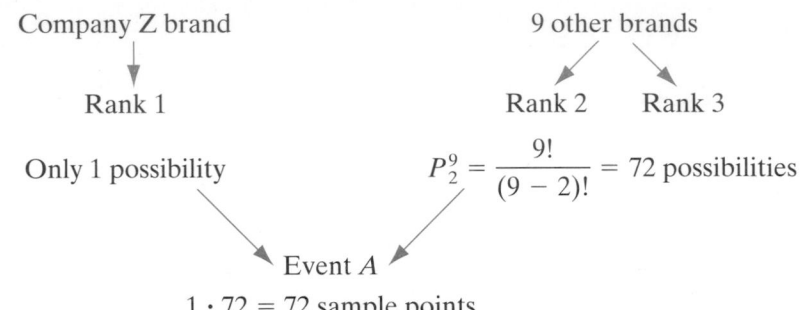

Company Z brand

Rank 1

Only 1 possibility

9 other brands

Rank 2   Rank 3

$$P_2^9 = \frac{9!}{(9-2)!} = 72 \text{ possibilities}$$

Event A

$1 \cdot 72 = 72$ sample points

Thus, event A can occur in 72 different ways.

Now define B as the event that company Z's brand is ranked in the top three. Since event B specifies only that brand Z appear in the top three, we repeat the calculations above, fixing brand Z in position 2 and then in position 3. We conclude that the number of sample points contained in event B is $3(72) = 216$.

Step 5   The final step is to calculate the probabilities of events A and B. Since the 720 sample points are equally likely to occur, we find

$$P(A) = \frac{\text{Number of sample points in } A}{\text{Total number of sample points}} = \frac{72}{720} = \frac{1}{10}$$

Similarly,

$$P(B) = \tfrac{216}{720} = \tfrac{3}{10} \qquad \blacktriangle$$

EXAMPLE 3.35

Refer to Example 3.34. Suppose the consumer testing service is to choose the top three laundry detergents from the group of 10 *but is not to rank the three.*

a. In how many different ways can the testing service choose the three to be designated as top detergents?
b. Assuming that the testing service makes its choice by a random selection and that company X has two brands in the group of 10, what is the probability that exactly one of the company X brands is selected in the top three? At least one?

Solution

a. The testing service is selecting three elements (brands) from a set of 10 elements *without regard to order,* so we can apply the combinations rule to determine the number of different results:

$$\binom{10}{3} = \frac{10!}{3!(10-3)!} = \frac{10 \cdot 9 \cdot 8}{3 \cdot 2 \cdot 1} = 120$$

b. The steps for calculating the probability of interest are as follows:

Step 1   The experiment is to select *(but not rank)* three brands from 10.

Step 2   There are 120 sample points for this experiment.

Step 3   Since the selection is made at random,

$$P(\text{Each sample point}) = \tfrac{1}{120}$$

Step 4   Define events A and B as follows:

A: {Exactly one company X brand is selected}

B: {At least one company X brand is selected}

Since each of the sample points is equally likely to occur, we need only know the number of sample points in A and B to determine their probabilities.

For event A to occur, exactly one company X brand must be selected, along with two of the remaining eight brands. We thus break A into two parts:

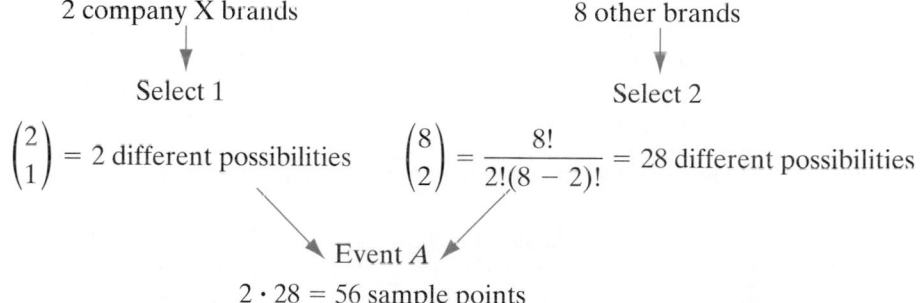

2 company X brands

8 other brands

Select 1

Select 2

$\binom{2}{1} = 2$ different possibilities $\qquad \binom{8}{2} = \dfrac{8!}{2!(8-2)!} = 28$ different possibilities

Event A

$2 \cdot 28 = 56$ sample points

Note that the one company X brand can be selected in two ways, whereas the two other brands can be selected in 28 ways. (We use the combinations rule because the order of selection is not important.) Then, we use the multiplicative rule to combine one of the two ways to select a company X brand with one of the 28 ways to select two other brands, yielding a total of 56 sample points for event A.

To count the sample points contained in B, we first note that

$$B = A \cup C$$

where A is defined as before and

C: {Both company X brands are selected}

We have determined that A contains 56 sample points. Using the same method for event C, we find

2 company X brands

8 other brands

Select 2

Select 1

$\binom{2}{2} = 1$ possibility $\qquad \binom{8}{1} = 8$ different possibilities

Event C

$1 \cdot 8 = 8$ sample points

Thus, event C contains eight sample points and $P(C) = \tfrac{8}{120}$.

Step 5   Since all the sample points are equally likely,

$$P(A) = \frac{\text{Number of sample points in } A}{\text{Total number of sample points}} = \frac{56}{120} = \frac{7}{15}$$

$$P(B) = P(A \cup C) = P(A) + P(C) - P(A \cap C)$$

Lottery Buster!

CASE STUDY 3.3

Welcome to the Wonderful World of Lottery Bu$ters." So begins the premier issue of *Lottery Buster,* a monthly publication for players of the state lottery games. *Lottery Buster* provides interesting facts and figures on the nearly 40 state lotteries currently operating in the United States and purported "tips" on how to increase a player's odds of winning the lottery.

New Hampshire, in 1963, was the first state in modern times to authorize a state lottery as an alternative to increasing taxes. (Prior to this time, beginning in 1895, lotteries were banned in America for fear of corruption.) Since then, lotteries have become immensely popular for two reasons. First, they lure you with the opportunity to win millions of dollars with a $1 investment; second, when you lose, at least you know your money is going to a good cause.

The popularity of the state lottery has brought with it an avalanche of self-proclaimed "experts" and "mathematical wizards" (such as the editors of *Lottery Buster*) who provide advice on how to win the lottery—for a fee, of course! These experts—the legitimate ones, anyway—base their "systems" of winning on their knowledge of probability and statistics.

For example, more experts would agree that the "golden rule" or "first rule" in winning lotteries is *game selection.* State lotteries generally offer three types of games: Instant (scratch-off) tickets, Daily Numbers (Pick-3 and Pick-4), and the weekly Pick-6 Lotto game.

The Instant game involves scratching off the thin, opaque covering on a ticket to determine whether you have won or lost. The cost of a ticket is usually 50¢, and the amount to be won ranges from $1 to $100,000 in most states, while it reaches $1 million in others. *Lottery Buster* advises against playing the Instant game because it is "a pure chance play, and you can win only by dumb luck. No skill can be applied to this game."

The Daily Numbers game permits you to choose either a three-digit (Pick-3) or four-digit (Pick-4) number at a cost of $1 per ticket. Each night, the winning number is drawn. If your number matches the winning number, you win a large sum of money, usually $100,000. You do have some control over the Daily Numbers game (since you pick the numbers that you play) and, consequently, there are strategies available to increase your chances of winning. However, the Daily Numbers game, like the Instant game, is not available for out-of-state play. For this reason, and because the payoffs are relatively small, lottery experts prefer the weekly Pick-6 Lotto game.

To play Pick-6 Lotto, you select six numbers of your choice from a field of numbers ranging from 1 to N, where N depends on which state's game you are playing.

For example, Florida's Lotto game involves picking six numbers ranging from 1 to 49 (denoted 6/49) as shown on the Florida Lotto ticket, Figure 3.23. Delaware's Lotto is a 6/30 game, and Pennsylvania's is a 6/40 game. The cost of a ticket is $1 and the payoff, if your six numbers match the winning numbers drawn at the end of each week, is $6 million or more, depending on the number of tickets purchased. (To date, Pennsylvania has had the largest weekly payoff of $97 million.) In addition to the grand prize, you can win second-, third-, and fourth-prize payoffs by matching five, four, and three of the six numbers drawn, respectively. And you don't have to be a resident of the state to play the state's Lotto game. Anyone can play by calling a toll-free "hotline" number.

Focus

a. Consider Florida's 6/49 Lotto game. Calculate the number of possible ways in which you can choose the six numbers from the 49 available. If you purchase a single $1 ticket, what is the probability that you win the grand prize (i.e., match all six numbers)?

b. Repeat part a for Delaware's 6/30 game.

c. Repeat part a for Pennsylvania's 6/40 game.

d. Since you can play any state's Lotto game, which of the three, Florida, Delaware, or Pennsylvania, would you choose to play? Why?

e. One strategy used to increase your odds of winning a Lotto is to employ a *wheeling system.* In a complete wheeling system, you select more than six numbers, say, seven, and play every combination of six of those seven numbers. Suppose you choose to "wheel" the following seven numbers in a 6/40 game: 2, 7, 18, 23, 30, 32, 39. How many tickets would you need to purchase to have every possible combination of the seven numbers? List the six numbers on each of these tickets.

f. Refer to part e. What is the probability of winning the 6/40 Lotto when you wheel seven numbers? Does the strategy, in fact, increase your odds of winning?

g. Another strategy is to play neighboring pairs. Neighboring pairs are two consecutive numbers that come up together on the winning ticket. In one state lottery, for example, 79% of the winning tickets had at least one neighboring pair. Thus, some "experts" feel that you have a better chance of winning if you include at least one neighboring pair in your number selection. Calculate the probability of winning the 6/40 Lotto with the six numbers: 2, 15, 19, 20, 27, 37. [*Note:* 19, 20 is a neighboring pair.] Compare this probability to the one in part c. Comment on the neighboring pairs strategy.

FIGURE 3.23

Reproduction of Florida's 6/49 Lotto ticket (Case Study 3.3)

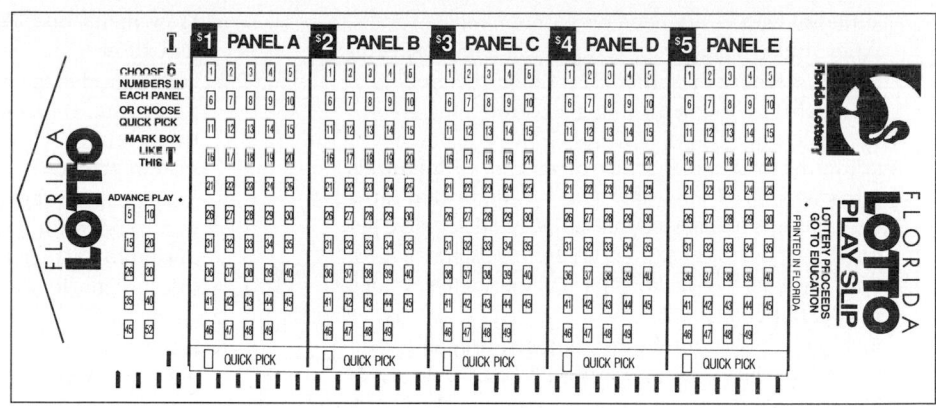

where $P(A \cap C) = 0$ because the testing service cannot select exactly one *and* exactly two of the company X brands; that is, A and C are mutually exclusive events. Thus,

$$P(B) = {}^{56}\!/_{120} + {}^{8}\!/_{120} = {}^{64}\!/_{120} = {}^{8}\!/_{15} \qquad \blacktriangle$$

Learning how to decide whether a particular counting rule applies to an experiment takes patience and practice. If you wish to develop this skill, attempt to use the rules to solve the following exercises and especially the optional exercises (marked with an asterisk) at the end of this chapter. Proofs of the counting rules can be found in the text by W. Feller (1968, Chapter 3).

EXERCISES 3.68–3.86

Learning the Mechanics

3.68 Use the multiplicative rule to determine the number of sample points in the sample space corresponding to the experiment of tossing a coin the following number of times:
a. 2 times   **b.** 3 times   **c.** 5 times   **d.** n times

3.69 Find the numerical values of
a. $\binom{6}{3}$ **b.** P_2^5 **c.** P_2^4 **d.** $\binom{100}{98}$
e. $\binom{50}{50}$ **f.** $\binom{50}{0}$ **g.** P_3^5 **h.** P_0^{10}

3.70 An experiment consists of choosing objects without regard to order. How many sample points are there if you choose the following?
a. 3 objects from 7   **b.** 2 objects from 6
c. 2 objects from 30   **d.** 8 objects from 10
e. r objects from q

3.71 Determine the number of sample points contained in the sample space when you toss the following:
a. 1 die   **b.** 2 dice   **c.** 4 dice   **d.** n dice

3.72 Suppose you are managing 10 employees, and you need to form three teams to work on different projects. Assume that each employee may serve on any team. In how many different ways can the teams be formed if the number of members on each project team are as follows:
a. 3, 3, 4   **b.** 2, 3, 5   **c.** 1, 4, 5   **d.** 2, 4, 4

Applying the Concepts

3.73 Flying into New York from a certain city, you can choose one of three airlines and can travel either first class or economy. How many travel options do you have?

3.74 A computer retail store has nine personal computers (PCs) in stock. A university buyer plans to purchase three of them. Unknown to either buyer or seller, two of the PCs in stock have defective disk drives. The three computers are selected at random from the nine available.
a. In how many different ways can the three computers be selected?
b. What is the probability that none of the computers selected will have a defective disk drive?
c. What is the probability that at least one of the computers selected will have a defective disk drive?

3.75 The National Resident Matching Program (NRMP) is a service provided by the Association of American Medical Colleges to match graduating medical students with residency appointments at hospitals. After students and hospitals have evaluated each other, they submit rank-order lists of their preferences to the NRMP. Using a matching algorithm, the NRMP then generates final, nonnegotiable assignments of students to the residency programs of hospitals (*Academic Medicine,* June 1995). Assume that three graduating medical students (#1, #2, and #3) have applied for positions at three different hospitals (A, B, and C), where each hospital has one and only one resident opening.

a. How many different assignments of medical students to hospitals are possible? List them.

b. Suppose student #1 prefers hospital B. If the NRMP algorithm is entirely random, what is the probability that the student is assigned to hospital B?

3.76 Suppose an automobile license plate is designed to begin with a number from 1 to 26, corresponding to the 26 counties in a state. If this number is followed by another five-digit number, how many different license plates can the state issue?

3.77 How many different five-card hands can be dealt from a 52-card bridge deck?

3.78 A company sells midlevel models of automobiles in five different styles. A buyer can get an automobile in one of eight colors with either standard or automatic transmission. Would it be reasonable to expect a dealer to stock at least one automobile in every style, color, and transmission combination? At a minimum, how many automobiles would the dealer have to stock?

3.79 The Republican governor of a state with no income tax is appointing a committee of 5 members to consider changes in the income tax law. There are 15 state representatives available for appointment to the committee, 7 Democrats and 8 Republicans. Assume that the governor selects the committee of 5 members randomly from the 15 representatives.

a. In how many different ways can the committee members be selected?

b. What is the probability that no Democrat is appointed to the committee? If this were to occur, what would you conclude about the assumption that the appointments were made at random? Why?

c. What is the probability that the majority of the committee members are Republican? If this were to occur, would you have reason to doubt that the governor made the selections randomly? Why?

3.80 In the mid-1980s the state of Florida ran out of combinations of letters and numbers for its license plates. Then, each license plate contained three letters of the alphabet followed by three digits selected from the 10 digits, 0, 1, 2, ..., 9.

a. How many different license plates did this system allow?

b. New Florida tags were obtained by reversing the procedure, starting with three digits followed by three letters. How many new tags did the new system provide?

c. Since the new tag numbers were added to the old numbers, what is the total number of licenses available for registration in Florida?

3.81 A salesperson living in city A wishes to visit four cities, B, C, D, and E.

a. If the cities are all connected by airlines, how many different travel plans could be constructed to visit each city exactly once and then return home?

b. Suppose all the cities are connected except that B and C are not directly connected. How many different flight plans would be available to the salesperson?

3.82 The new nine-digit Zip code has become an integral part of the U.S. postal system. How many different Zip codes are available for use by the postal service? (Assume any nine-digit number can be used as a Zip code.) The first three digits of a Chicago Zip code are 606. If no other city in the United States has these first three digits as part of its Zip, how many different Zip codes may possibly exist in Chicago?

3.83 Suppose you are to choose a basketball team (five players) from eight available athletes.

a. How many ways can you choose a team (ignoring positions)?

b. How many ways can you choose a team composed of two guards, two forwards, and a center?

c. How many ways can you choose a team composed of point guard, shooting guard, power forward, small forward, and center?

3.84 Intercollegiate volleyball rules require that after the opposing team has lost its serve, each of the six members of the serving team must rotate into new positions on the court. Hence, each player must be able to play all six different positions. How many different team combinations by player and position are possible during a volleyball game? If players are initially assigned to the positions in a random manner, find the probability that the best server on the team is in the serving position.

3.85 A college professor hands out a list of 10 questions, five of which will appear on the final examination for the course. One of the students taking the course is pressed for time and can prepare for only seven of the 10 questions. Suppose the professor chooses the five questions at random from the 10.

a. What is the probability that the student will be prepared for all five questions that appear on the final examination?

b. What is the probability that the student will be prepared for fewer than three questions?

c. What is the probability that the student will be prepared for exactly four questions?

3.86 Consider five-card poker hands dealt from a standard 52-card bridge deck. Two important events are

> *A:* {You draw a flush}
>
> *B:* {You draw a straight}

[*Note:* A *flush* consists of any five cards of the same suit. A *straight* consists of any five cards with values in sequence. In a straight, the cards may be of any suit and an ace may have a value of 1 or a value higher than a king.]

a. Find $P(A)$. **b.** Find $P(B)$.

c. The event that both A and B occur—that is, $A \cap B$—is called a *straight flush*. Find $P(A \cap B)$.

QUICK REVIEW

Key Terms

Note: Starred (*) terms refer to the optional section in this chapter.

Additive rule of probability   110
Bayes' Rule   127
Combinations rule*   140
Combinatorial mathematics   133
Complementary events   107
Compound event   104
Conditional probability   114
Counting rule*   136
Event   97
Experiment   93
Independent events   123
Intersection   104
Multiplicative rule*   137

Multiplicative rule of probability   120
Mutually exclusive events   111
Partitions rule*   139
Permutation*   138
Permutations rule*   138
Probability rules   96
Random number generator   132
Random sample   132
Sample point   93
Sample space   94
Tree diagram   122
Union   104
Venn diagram   94

Key Formulas

$P(A) + P(A') = 1$ ⟶ Complementary events   107

$P(A \cup B) = P(A) + P(B) - P(A \cap B)$ ⟶ Additive rule   110

$P(A \cap B) = 0$ ⟶ Mutually exclusive events   111

$P(A \cup B) = P(A) + P(B)$ ⟶ Additive rule for mutually exclusive events   111

$P(A \mid B) = \dfrac{P(A \cap B)}{P(B)}$ ⟶ Conditional probability   115

$P(A \cap B) = P(A)P(B \mid A) = P(B)P(A \mid B)$ ⟶ Multiplicative rule   121

$P(A \mid B) = P(A)$ ⟶ Independent events   123

$P(A \cap B) = P(A)P(B)$ ⟶ Multiplicative rule for independent events   125

$\dbinom{N}{n} = \dfrac{N!}{n!(N - n)!}$ ⟶ Combinatorial rule   133

where $N! = N(N - 1)(N - 2) \cdots (2)(1)$

[*Note:* For a summary of formulas for counting rules, see page 141.]

LANGUAGE LAB

SYMBOL	PRONUNCIATION	DESCRIPTION
S		Sample space
S: $\{1, 2, 3, 4, 5\}$		Set of sample points, 1, 2, 3, 4, 5, in sample space
A: $\{1, 2\}$		Set of sample points, 1, 2, in event A
$P(A)$	Probability of A	Probability that event A occurrs
$A \cup B$	A union B	Union of events A and B (either A or B or both occur)
$A \cap B$	A intersect B	Intersection of events A and B (both A and B occur)
A'	A complement	Complement of event A (the event that A does not occur)
$P(A \mid B)$	Probability of A given B	Conditional probability that event A occurs given that event B occurs
$\binom{N}{n}$	N chose n	Number of combinations of N elements taken n at a time
$N!$	N factorial	Multiply $N(N-1)\,(N-2)\cdots(2)(1)$
P_n^N	N permute n	Number of different permutations of N elements taken n at a time

SUPPLEMENTARY EXERCISES 3.87–3.118

Note: Starred (*) exercises refer to the optional section in this chapter.

Learning the Mechanics

3.87 A fair die is tossed and the up face is noted. If the number is even, the die is tossed again; if the number is odd, a fair coin is tossed. Define the events

> A: {A head appears on the coin}
>
> B: {The die is tossed only one time}

  a. List the sample points in the sample space.
  b. Give the probability for each of the sample points.
  c. Find $P(A)$ and $P(B)$.
  d. Identify the sample points in A', B', $A \cap B$, and $A \cup B$.
  e. Find $P(A')$, $P(B')$, $P(A \cap B)$, $P(A \cup B)$, $P(A \mid B)$, and $P(B \mid A)$.
  f. Are A and B mutually exclusive events? Independent events? Why?

3.88 The accompanying Venn diagram illustrates a sample space containing six sample points and three events, A, B, and C. The probabilities of the sample points are: $P(1) = .3$, $P(2) = .2$, $P(3) = .1$, $P(4) = .1$, $P(5) = .1$, and $P(6) = .2$.

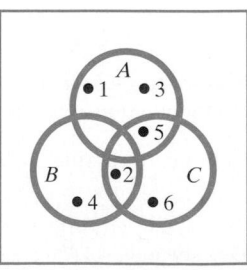

S

  a. Find $P(A \cap B)$, $P(B \cap C)$, $P(A \cup C)$, $P(A \cup B \cup C)$, $P(B')$, $P(A' \cap B)$, $P(B \mid C)$, and $P(B \mid A)$.
  b. Are A and B independent? Mutually exclusive? Why?
  c. Are B and C independent? Mutually exclusive? Why?

3.89 Which of the following pairs of events are mutually exclusive? Justify your response.
  a. Cleveland Indians win the World Series.
      Albert Belle, an Indians outfielder, hits 60 home runs.
  b. An IBM microcomputer is purchased.
      An Apple microcomputer is purchased.
  c. Subject A in a psychology experiment responds to a stimulus within 5 seconds.
      Subject A in a psychology experiment records the fastest response to a stimulus (2.3 seconds).

3.90 A balanced die is thrown once. If a 4 appears, a ball is drawn from urn 1; otherwise, a ball is drawn from urn 2. Urn 1 contains four red, three white, and three black balls. Urn 2 contains six red and four white balls.
  a. Find the probability that a red ball is drawn.
  b. Find the probability that urn 1 was used given that a red ball was drawn.

3.91 Two events, A and B, are independent, with $P(A) = .3$ and $P(B) = .1$.
  a. Are A and B mutually exclusive? Why?
  b. Find $P(A \mid B)$ and $P(B \mid A)$.
  c. Find $P(A \cup B)$.

*3.92 A random sample of five students is to be selected from 50 sociology majors for participation in a special program.

a. In how many different ways can the sample be drawn?

b. Show how the random number table, Table 1 of Appendix A, can be used to select the sample of students.

3.93 Find the numerical value of:

a. $6!$ **b.** $\binom{10}{9}$ **c.** $\binom{10}{1}$ **d.** P_2^6

e. $\binom{6}{3}$ **f.** $0!$ **g.** P_4^{10} **h.** P_2^{50}

Applying the Concepts

3.94 In Exercise 2.3, you read about the University of Akron survey of 4,001 U.S. residents and their commitment to religion. The survey results are reproduced in the accompanying table.

a. Find the probability that a randomly selected U.S. resident is highly committed to religion.

b. Find the probability that a randomly selected U.S. resident is agnostic or atheist.

c. Find the probability that a randomly selected U.S. resident is religious to any degree.

d. Explain why the probabilities you calculated in parts **a**–**c** are only approximations to the true probabilities.

	Number of U.S. Residents
Highly committed to religion	761
Modestly religious	880
Nominally religious	1,162
Trace elements of religion	899
Agnostic or atheist	299
Total	4,001

3.95 In college basketball games a player may be afforded the opportunity to shoot two consecutive foul shots (free throws).

a. Suppose a player who scores on 80% of his foul shots has been awarded two free throws. If the two throws are considered independent, what is the probability that the player scores on both shots? Exactly one? Neither shot?

b. Suppose a player who scores on 80% of his first attempted foul shots has been awarded two free throws, and the outcome on the second shot is dependent on the outcome of the first shot. In fact, if this player makes the first shot, he makes 90% of the second shots; and if he misses the first shot, he makes 70% of the second shots. In this case, what is the probability that the player scores on both shots? Exactly one? Neither shot?

c. In parts **a** and **b**, we considered two ways of *modeling* the probability that a basketball player scores on two consecutive foul shots. Which mod-

el do you think is a more realistic attempt to explain the outcome of shooting foul shots; that is, do you think two consecutive foul shots are independent or dependent? Explain.

3.96 Psychologists tend to believe that there is a relationship between aggressiveness and order of birth. To test this belief, a psychologist chose 500 elementary school students at random and administered each a test designed to measure the student's aggressiveness. Each student was classified according to one of four categories. The percentages of students falling in the four categories are shown here.

	Firstborn	Not Firstborn
Aggressive	15%	15%
Not Aggressive	25%	45%

a. If one student is chosen at random from the 500, what is the probability that the student is firstborn?

b. What is the probability that the student is aggressive?

c. What is the probability that the student is aggressive, given the student was firstborn?

d. If

  A: {Student chosen is aggressive}

  B: {Student chosen is firstborn}

are A and B independent events? Explain

3.97 The corporations in the highly competitive razor blade industry do a tremendous amount of advertising each year. Corporation G gave a supply of three top name brands, G, S, and W, to a consumer and asked her to use them and rank them in order of preference. The corporation was, of course, hoping the consumer would prefer its brand and rank it first, thereby giving them some material for a consumer interview advertising campaign. If the consumer did not prefer one blade over any other, but was still required to rank the blades, what is the probability that

a. The consumer ranked brand G first?

b. The consumer ranked brand G last?

c. The consumer ranked brand G last and brand W second?

d. The consumer ranked brand W first, brand G second, and brand S third?

3.98 Despite penicillin and other antibiotics, bacterial pneumonia kills thousands of U.S. residents every year. The U.S. Food and Drug Administration has approved the use of a new antipneumonia vaccine called Pneumovax. It is designed especially for elderly or debilitated patients who are usually the most vulnerable to bacterial pneumonia. Field trials proved the new vaccine to be 90% effective in stimulating the production of antibodies to pneumonia-producing bacteria (i.e., it was 90% successful in preventing persons exposed to pneumonia-producing bacteria from acquiring the disease). Suppose the probability that an elderly or debilitated person is exposed to these bacteria is .40 (whether innoculated or not) and, after being exposed, the probability that the person will contract bacterial pneumonia if not inoculated with the vaccine is .95. Find the probability that an elderly or debilitated person inoculated with this new vaccine acquires pneumonia. What is the probability if this person has not been inoculated?

3.99 Use your intuitive understanding of independence to form an opinion about whether each of the following scenarios represents an independent event.

a. The results of consecutive tosses of a coin

b. The opinions of randomly selected individuals in a pre-election poll

c. A major league baseball player's results in two consecutive at-bats

d. The amount of gain or loss associated with investments in different stocks if these stocks are bought on the same day and sold on the same day 1 month later

e. The amount of gain or loss associated with investments in different stocks if these stocks are bought and sold in different time periods, 5 years apart

f. The responses of two different subjects to the same stimulus in a psychology experiment

3.100 What are the characteristics of families with young children (under age 6)? This was one of several questions posed by a University of Michigan researcher in *Children and Youth Services Review* (Vol. 17, 1995). Using data obtained from the National Child Care Survey, the income distribution and employment status of these families were summarized as indicated in the accompanying table.

Income Characteristic	Percentage
No parent	1
Below poverty line; not employed	7
Below poverty line; employed	7
Above poverty line, but less than $25,000; not employed	2
Above poverty line, but less than $25,000; employed	22
$25,000 or more	61
Total	100

a. Find the probability that a randomly selected family with young children has an income above the poverty line, but less than $25,000.

b. Find the probability that a randomly selected family with young children has unemployed parents or no parents.

c. Find the probability that a randomly selected family with young children has an income below the poverty line.

3.101 In the game of Parcheesi each player rolls a pair of dice on each turn. In order to begin the game, you must throw a 5 on at least one of the dice, or a total of 5 on the two dice. What is the probability that you can begin the game on your first turn? The second turn? The third turn? The nth turn?

3.102 Entomologists are often interested in studying the effect of chemical sex attractants (pheromones) on insects. One common technique is to release several insects to a site equidistant from the pheromone under study and a control substance. If the pheromone has an effect, more insects will travel toward it than toward the control. Otherwise, the insects are equally likely to travel in either direction. Suppose the pheromone under study has no effect, so that it is equally likely that an insect will move toward either the pheromone or the control. If five insects are released, what is the probability that

a. All five travel toward the pheromone?

b. Exactly four?

c. What inference would you make if the event in part **b** actually occurs? Explain.

3.103 All-terrain vehicles (ATVs) came under fire in the 1980s owing to the high number of injuries and deaths attributed to these machines. In response, manufacturers agreed to provide extensive safety warnings to owners, to develop a media safety-awareness program, and to implement a nationwide training program. The *Journal of Risk and Uncertainty* (May 1992) published an article investigating the relationship of injury rate to a variety of factors. One of the more interesting factors studied, age of the driver, was found to have a strong relationship to injury rate. The article reports that prior to the safety-awareness program, 14% of the ATV drivers were under age 12; another 13% were 12–15, and 48% were under age 25. Suppose an ATV driver is selected at random prior to the implementation of the safety-awareness program.

a. Find the probability that the ATV driver is 15 years old or younger.

b. Find the probability that the ATV driver is 25 years old or older.

c. Given that the ATV driver is under age 25, what is the probability the driver is under age 12?

d. Are the events Under age 25 and Under age 12 mutually exclusive? Why or why not?

e. Are the events Under age 25 and Under age 12 independent? Why or why not?

3.104 The performance of quality inspectors affects both the quality of outgoing products and the cost of the products. A product that passes inspection is assumed to meet quality standards; a product that fails inspection may be reworked, scrapped, or reinspected. Quality engineers at Westinghouse Electric Corporation evaluated performances of inspectors in judging the quality of solder joints by comparing each inspector's classifications of a set of 153 joints with the consensus evaluation of a panel of experts. The results for a particular inspector are shown in the accompanying table.

	INSPECTOR'S JUDGMENT	
Committee's Judgment	**Joint Acceptable**	**Joint Rejectable**
Joint acceptable	101	10
Joint rejectable	23	19

Source: Meagher, J. J., and Scazzero, J. A. "Measuring inspector variability," *39th Annual Quality Congress Transactions,* May 1985, pp. 75–81. © 1985 American Society for Quality Control. Reprinted with permission.

One of the 153 solder joints is to be selected at random.
a. What is the probability that the inspector judges the joint to be acceptable? That the committee judges the joint to be acceptable?
b. What is the probability that both the inspector and the committee judge the joint to be acceptable? That neither judge the joint to be acceptable?
c. What is the probability that the inspector and the committee disagree? Agree?

3.105 Sandoz, a pharmaceutical firm, has developed a drug, cyclosporine, that appears to retard a body's immune response to transplanted organs, thus slowing rejection. Sandoz reports that kidney transplant patients who receive the drug have an 80% chance of living through the first year, and that survival rates for other types of transplants have also improved. Suppose a hospital performs four kidney transplants, and all patients receive the drug cyclosporine. Assuming one patient's survival is independent of another's, what is the probability that
a. All four patients are alive at the end of 1 year?
b. None of the four patients is alive at the end of 1 year?
c. At least one of the patients is alive at the end of 1 year?

3.106 The probability that an Avon salesperson sells beauty products to a prospective customer on the first visit to the customer is .4. If the salesperson fails to make the sale on the first visit, the probability that the sale will be made on the second visit is .65. The salesperson never visits a prospective customer more than twice. What is the probability that the salesperson will make a sale to a particular customer?

3.107 Seventy-five percent of all women who submit to pregnancy tests are really pregnant. A certain pregnancy test gives a *false positive* result with probability .02 and a *valid positive result* with probability .99. If a particular woman's test is positive, what is the probability that she really is pregnant? [*Hint:* If *A* is the event that a woman is pregnant and *B* is the event that the pregnancy test is positive, then *B* is the union of the two mutually exclusive events $A \cap B$ and $A' \cap B$. Also, the probability of a false positive result may be written as $P(B \mid A') = .02$.]

3.108 Blackjack, a favorite game of gamblers, is played by a dealer and at least one opponent. At the outset of the game, two cards of a 52-card bridge deck are dealt to the player and two cards to the dealer. Drawing an ace and a face card is called a *blackjack*. If the dealer does not draw a blackjack and the player does, the player wins. If both the dealer and player draw blackjack, a "push" (i.e., a tie) occurs.
a. What is the probability that the dealer will draw a blackjack?
b. What is the probability that the player wins with a blackjack?

3.109 The figure shown at the bottom of the page is a schematic representation of a system comprised of three components. The system operates properly only if all three components operate properly. The three components are said to operate *in series*. The components could be mechanical or electrical; they could be work stations in an assembly process; or they could represent the functions of three different departments in an organization. The probability of failure for each component is listed in the accompanying table. Assume the components operate independently of each other.

Component	Probability of Failure
1	.12
2	.09
3	.11

A System Comprised of Three Components in Series

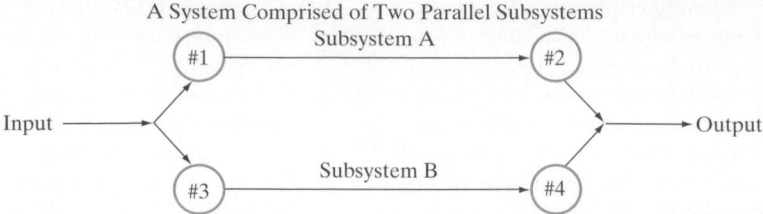

A System Comprised of Two Parallel Subsystems

a. Find the probability that the system operates properly.

b. What is the probability that at least one of the components will fail and therefore that the system will fail?

3.110 The figure above is a representation of a system that comprises two subsystems said to operate *in parallel*. Each subsystem has two components that operate in series (refer to Exercise 3.109). The system will operate properly as long as at least one of the subsystems functions properly. The probability of failure for each component in the system is .1. Assume the components operate independently of each other.

a. Find the probability that the system operates properly.

b. Find the probability that exactly one subsystem fails.

c. Find the probability that the system fails to operate properly.

d. How many parallel subsystems like the two shown here would be required to guarantee that the system would operate properly at least 99% of the time?

3.111 Refer to *The Howard Journal of Criminal Justice* study of the relationship between the race of the attacker, the race of the victim, and the degree of injury sustained in reported crimes, Exercise 3.39. The information is reproduced in the table below.

a. What is the probability that a randomly selected reported crime involved a nonwhite attacker?

b. What is the probability that a randomly selected reported crime involved no injury?

c. What is the probability that a randomly selected reported crime involved both a nonwhite attacker and no injury?

d. Are the race of the attacker and the degree of injury independent?

3.112 A small brewery has two bottling machines. Machine A produces 75% of the bottles and machine B produces 25%. One out of every 20 bottles filled by A is rejected for some reason, while one out of every 30 bottles from B is rejected. What proportion of bottles is rejected? What is the probability that a randomly selected bottle comes from machine A, given that it is accepted?

3.113 A clinical psychologist is asked to view tapes in which each of six experimental subjects is discussing his or her recent dreams. Three of the six subjects have previously been classified as "high-anxiety" individuals and the other three as "low-anxiety." The psychologist is told only that there are three of each type and is asked to select the three high-anxiety subjects.

a. List all possible outcomes (sample points) for this experiment.

b. Assuming that the psychologist guesses at the classifications of the subjects, assign probabilities to the sample points.

c. Find the probability that the psychologist guesses all classifications correctly.

d. Find the probability that the psychologist guesses at least two of the three high-anxiety subjects correctly.

*3.114 Refer to Exercise 3.113. Suppose the clinical psychologist is given tapes of 20 subjects and is told

	ATTACKER/VICTIM				
Degree of Injury	**White/White**	**White/Nonwhite**	**Nonwhite/Nonwhite**	**Nonwhite/White**	**Totals**
Fatal	183	18	18	18	237
Serious	580	49	111	141	881
Slight	4,336	440	594	1,656	7,026
None	1,422	136	214	1,801	3,573
Totals	6,521	643	937	3,616	11,717

Source: Shah, R. and Pease, K. "Crime, race and reporting to the police," *The Howard Journal of Criminal Justice,* Volume 31, No. 3, Aug. 1992.

only that 10 are high-anxiety individuals and 10 are low-anxiety individuals. The psychologist is asked to select the 10 high-anxiety subjects.

a. Using the appropriate counting rule from Section 3.9, count the number of sample points for this experiment.

b. Assuming that the psychologist is guessing, assign probabilities to each of the sample points.

c. Find the probability that the psychologist guesses all classifications correctly.

d. Find the probability that the psychologist guesses at least 9 of the 10 high-anxiety subjects correctly.

*3.115 Suppose there are 500 applicants for five equivalent positions at a factory and the company is able to narrow the field to 30 equally qualified applicants. Seven of the finalists are minority candidates. Assume that the five who are chosen are selected at random from this final group of 30.

a. In how many different ways can the selection be made?

b. What is the probability that none of the minority candidates is hired?

c. What is the probability that no more than one minority candidate is hired?

*3.116 Suppose a manufacturer of big-screen television sets is prepared to send 20 new sets to three retail dealers. Dealer A is to get ten of the sets, dealer B six of the sets, and dealer C four of the sets.

a. In how many different ways can the 20 sets be divided among the dealers in the prescribed numbers?

b. Assuming the sets are randomly divided among the dealers and there are three defective sets among the 20, what is the probability that dealer A gets all three defective sets?

c. Making the assumptions given in part **b**, what is the probability that dealer A gets two defective sets and dealer B, the other defective set?

*3.117 According to a morning news program, a very rare event recently occurred in Dubuque, Iowa. Each of four women playing bridge was astounded to note that she had been dealt a perfect bridge hand. That is, one woman was dealt all 13 spades, another all 13 hearts, another all diamonds, and another all the clubs. What is the probability of this rare event?

*3.118 A poll is taken in an attempt to determine the top three prime-time television shows. Each person interviewed is asked to rank his or her first, second, and third favorite show from a list of 40 shows (10 from each of the four major broadcast networks, ABC, CBS, NBC, and Fox).

a. In how many different ways can the ranking be made by an interviewee?

b. If an interviewee ranks the top three shows by making a random selection, what is the probability that the first two programs selected are ABC programs while the third is an NBC program?

c. If the ranking is random, what is the probability that all three program selections are from the same network?

STUDENT PROJECTS

Obtain a standard deck of 52 playing cards (the kind commonly used for bridge, poker, or solitaire). An experiment will consist of drawing one card at random from the deck of cards and recording which card was observed. This random drawing will be simulated by shuffling the deck thoroughly and observing the top card. Consider the following two events:

A: {Card observed is a heart}

B: {Card observed is an ace, king, queen, or jack}

a. Find $P(A)$, $P(B)$, $P(A \cap B)$, and $P(A \cup B)$.

b. Conduct the experiment 10 times and record the observed card each time. Be sure to return the observed card each time and thoroughly shuffle the deck before making the draw. After you've observed 10 cards, calculate the proportion of observations that satisfy event *A*, event *B*, event *A* ∩ *B*, and event *A* ∪ *B*. Compare the observed proportions with the true probabilities calculated in part a.

c. Conduct the experiment 40 more times to obtain a total of 50 observed cards. Now calculate the proportion of observations that satisfy event *A*, event *B*, event *A* ∩ *B*, and event *A* ∪ *B*. Compare these proportions with those found in part b and the true probabilities found in part a.

d. Conduct the experiment 50 more times to obtain a total of 100 observations. Compare the observed proportions for the 100 trials with those found previously. What comments do you have concerning the different proportions found in parts b, c, and d as compared to the true probabilities found in part a? How do you think the observed proportions and true probabilities would compare if the experiment were conducted 1,000 times? 1 million times?

EXPLORING DATA WITH A COMPUTER

Most universities provide job placement services for graduating students through a Career Resource Center (CRC). The CRC at the University of South Florida (USF) conducts a survey of students who have found jobs prior to graduation. Information requested on the survey questionnaire includes graduation date, degree earned, major, gender, job title, and starting salary. The data for 170 USF students graduating between August 1992 and May 1995 are described in Appendix B and available in an ASCII file on a 3½" DOS disk.

Suppose the director of the CRC is studying starting salary patterns of USF students who have secured employment at the time of graduation. She will use the data set of Appendix B to determine factors that affect the starting salary of USF graduates.

a. If one of the 170 students surveyed were to be randomly selected, what is the probability that the USF graduate's starting salary exceeds \$35,000?

b. Suppose the graduate were to be selected from those with business majors. What is the probability that the graduate's starting salary exceeds \$35,000?

c. Are the events described in parts a and b independent? Why or why not? What are the practical implications of the events' independence, or lack thereof?

d. Determine whether the events {Starting salary exceeds \$35,000} and {Male graduate} are independent. What are the practical implications of your answer?

Chapter 4

DISCRETE RANDOM VARIABLES

Contents

Case Studies

*W*HERE WE'VE BEEN

We saw by illustration in Chapter 3 how probability would be used to make an inference about a population from data contained in an observed sample. We also noted that probability would be used to measure the reliability of the inference.

*W*HERE WE'RE GOING

Most of the experimental events we encountered in Chapter 3 were events described in words and denoted by capital letters. In real life, most sample observations are numerical—in other words, they are numerical data. In this chapter, we learn that data are observed values of random variables. We study three important random variables and learn how to find the probabilities of specific numerical outcomes.

| HH ● (2) | HT ● (1) |
| TT ● (0) | TH ● (1) |

S

FIGURE 4.1

Venn diagram for coin-tossing experiment

You may have noticed that many of the examples of experiments in Chapter 3 generated quantitative (numerical) observations. The unemployment rate, the percentage of voters favoring a particular candidate, the cost of textbooks for a school term, and the amount of pesticide in the discharge waters of a chemical plant are all examples of numerical measurements of some phenomenon. Thus, most experiments have sample points that correspond to values of some numerical variable.

To illustrate, consider the coin-tossing experiment of Chapter 3. Figure 4.1 is a Venn diagram showing the sample points when two coins are tossed and the up faces (heads or tails) of the coins are observed. One possible numerical outcome is the total number of heads observed. These values (0, 1, or 2) are shown in parentheses on the Venn diagram, one numerical value associated with each sample point. In the jargon of probability, the variable "total number of heads observed in two tosses of a coin" is called a *random variable*.

> **DEFINITION 4.1**
>
> A **random variable** is a variable that assumes numerical values associated with the random outcomes of an experiment, where one (and only one) numerical value is assigned to each sample point.

The term *random variable* is more meaningful than the term *variable* alone because the adjective *random* indicates that the coin-tossing experiment may result in one of the several possible values of the variable—0, 1, and 2—according to the *random* outcome of the experiment, *HH, HT, TH,* and *TT.* Similarly, if the experiment is to count the number of customers who use the drive-up window of a bank each day, the random variable (the number of customers) will vary from day to day, partly because of the random phenomena that influence whether customers use the drive-up window. Thus, the possible values of this random variable range from 0 to the maximum number of customers the window could possibly serve in a day.

We define two different types of random variables, *discrete* and *continuous,* in Section 4.1. Then we spend the remainder of this chapter discussing specific types of discrete random variables and the aspects that make them important to the statistician. We discuss continuous random variables in Chapter 5.

4.1   TWO TYPES OF RANDOM VARIABLES

Recall that the sample point probabilities corresponding to an experiment must sum to 1. Dividing one unit of probability among the sample points in a sample space and consequently assigning probabilities to the values of a random variable is not always as easy as the examples in Chapter 3 might lead you to believe. If the number of sample points is finite, that is, if they can be completely listed, the job is relatively easy. However, some experiments result in an infinite number of sample points, in which case assignment of probabilities is more difficult. In fact, we have to use different probability models depending on the number of values that a random variable can assume.

EXAMPLE 4.1

A panel of 10 experts for the *Wine Spectator* (a national publication) is asked to taste a new white wine and assign a rating of 0, 1, 2, or 3. A score is then obtained by adding together the ratings of the 10 experts. How many values can this random variable assume?

Solution

A sample point is a sequence of 10 numbers associated with the rating of each expert. For example, one sample point is

$$\{1, 0, 0, 1, 2, 0, 0, 3, 1, 0\}$$

The random variable assigns a score to each one of these sample points by adding the 10 numbers together. Thus, the smallest score is 0 (if all 10 ratings are 0) and the largest score is 30 (if all 10 ratings are 3). Since every integer between 0 and 30 is a possible score, the random variable denoted by the symbol x can assume 31 values. Note that the value of the random variable for the sample point above is $x = 8.$*

This is an example of a **discrete random variable**, since there is a finite number of distinct possible values. Whenever all the possible values a random variable can assume can be listed (or *counted*), the random variable is *discrete*.   ▲

EXAMPLE 4.2

Suppose the Environmental Protection Agency (EPA) takes readings once a month on the amount of pesticide in the discharge water of a chemical company. If the amount of pesticide exceeds the maximum level set by the EPA, the company is forced to take corrective action and may be subject to penalty. Consider the following random variable:

> Number, x, of months before the company's discharge
> exceeds the EPA's maximum level

What values can x assume?

Solution

The company's discharge of pesticide may exceed the maximum allowable level on the first month of testing, the second month of testing, etc. It is possible that the company's discharge will *never* exceed the maximum level. Thus, the set of possible values for the number of months until the level is first exceeded is the set of all positive integers

$$1, 2, 3, 4, \ldots$$

If we can list the values of a random variable x, even though the list is never-ending, we call the list **countable** and the corresponding random variable *discrete*. Thus, the number of months until the company's discharge first exceeds the limit is a *discrete random variable*.   ▲

EXAMPLE 4.3

Refer to Example 4.2. A second random variable of interest is the amount x of pesticide (in milligrams per liter) found in the monthly sample of discharge waters from the chemical company. What values can this random variable assume?

Solution

Unlike the *number* of months before the company's discharge exceeds the EPA's maximum level, the set of all possible values for the *amount* of discharge *cannot*

---

*The standard mathematical convention is to use a capital letter (e.g., X) to denote the theoretical random variable. The possible values (or realizations) of the random variable are typically denoted with a lowercase letter (e.g., x). Thus, in Example 4.1, the random variable X can take on the values $x = 0, 1, 2, \ldots, 30$. Since this notation can be confusing for introductory statistics students, we simplify the notation by using the lowercase x to represent the random variable throughout.

be listed—i.e., is not countable. The possible values for the amount x of pesticide would correspond to the points on the interval between 0 and the largest possible value the amount of the discharge could attain, the maximum number of milligrams that could occupy 1 liter of volume. (Practically, the interval would be much smaller, say, between 0 and 500 milligrams per liter.) When the values of a random variable are not countable but instead correspond to the points on some interval, we call it a **continuous random variable**. Thus, the *amount* of pesticide in the chemical plant's discharge waters is a *continuous random variable*.   ▲

DEFINITION 4.2

Random variables that can assume a *countable* number of values are called **discrete**.

DEFINITION 4.3

Random variables that can assume values corresponding to any of the points contained in one or more intervals are called **continuous**.

The following are examples of discrete random variables:

1. The number of sales made by a salesperson in a given week: $x = 0, 1, 2, \ldots$
2. The number of students in a sample of 500 who favor increasing student activities and, correspondingly, an increase in student activity fees: $x = 0, 1, 2, \ldots, 500$
3. The number of students applying to medical schools this year: $x = 0, 1, 2, \ldots$
4. The number of errors on a page of an accountant's ledger: $x = 0, 1, 2, \ldots$
5. The number of customers waiting to be served in a restaurant at a particular time: $x = 0, 1, 2, \ldots$

Note that each of the examples of discrete random variables begins with the words "The number of …." This wording is very common, since the discrete random variables most frequently observed are counts. The following are examples of continuous random variables:

1. The length of time between arrivals at a hospital clinic: $0 \leq x < \infty$ (infinity)
2. For a new apartment complex, the length of time from completion until a specified number of apartments are rented: $0 \leq x < \infty$
3. The amount of carbonated beverage loaded into a 12-ounce can in a can-filling operation: $0 \leq x \leq 12$
4. The depth at which a successful oil drilling venture first strikes oil: $0 \leq x \leq c$, where c is the maximum depth obtainable
5. The weight of a food item bought in a supermarket: $0 \leq x \leq 500$
   [*Note:* Theoretically, there is no upper limit on x, but it is unlikely that it would exceed 500 pounds.]

Discrete random variables and their probability distributions are discussed in this chapter. Continuous random variables and their probability distributions are the topic of Chapter 5.

EXERCISES 4.1–4.5

Applying the Concepts

4.1 What is a random variable?

4.2 How do discrete and continuous random variables differ?

4.3 Classify the following random variables according to whether they are discrete or continuous:

   a. The number of words spelled correctly by a student on a spelling test

b. The amount of water flowing through the Hoover Dam in a day

c. The length of time an employee is late for work

d. The number of bacteria per cubic centimeter of drinking water

e. The amount of carbon monoxide produced per gallon of unleaded gas

f. Your weight

4.4 Identify the following random variables as discrete or continuous:

a. The amount of flu vaccine in a syringe

b. The heart rate (number of beats per minute) of an American male

c. The time it takes a student to complete an examination

d. The barometric pressure at a given location

e. The number of registered voters who vote in a national election

f. Your score on the Scholastic Assessment Test (SAT)

4.5 Identify the following variables as discrete or continuous:

a. The reaction time difference to the same stimulus before and after training

b. The number of violent crimes committed per month in your community

c. The number of commercial aircraft near-misses per month

d. The number of winners each week in a state lottery

e. The number of free throws made per game by a basketball team

f. The distance traveled by a school bus each day

4.2   PROBABILITY DISTRIBUTIONS FOR DISCRETE RANDOM VARIABLES

A complete description of a discrete random variable requires that we *specify the possible values the random variable can assume and the probability associated with each value.* To illustrate, consider Example 4.4.

EXAMPLE 4.4

Recall the experiment of tossing two coins (Section 4.1), and let x be the number of heads observed. Find the probability associated with each value of the random variable x, assuming the two coins are fair.

Solution

The sample space and sample points for this experiment are reproduced in Figure 4.2. Note that the random variable x can assume values 0, 1, 2. Recall (from Chapter 3) that the probability associated with each of the four sample points is $\frac{1}{4}$. Then, identifying the probabilities of the sample points associated with each of these values of x, we have

$$P(x = 0) = P(TT) = \frac{1}{4}$$
$$P(x = 1) = P(TH) + P(HT) = \frac{1}{4} + \frac{1}{4} = \frac{1}{2}$$
$$P(x = 2) = P(HH) = \frac{1}{4}$$

Thus, we now know the values the random variable can assume (0, 1, 2) and how the probability is *distributed over* these values ($\frac{1}{4}$, $\frac{1}{2}$, $\frac{1}{4}$). This completely describes the random variable and is referred to as the *probability distribution,* denoted by the symbol $p(x)$.* The probability distribution for the coin-toss example is shown in tabular form in Table 4.1 and in graphical form in Figure 4.3. Since the probability distribution for a discrete random variable is concentrated at specific points (values of x), the graph in Figure 4.3a represents the probabilities

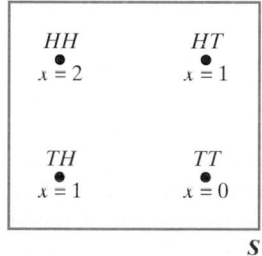

FIGURE 4.2

Venn diagram for the two-coin–toss experiment

*In standard mathematical notation, the probability that a random variable X takes on a value x is denoted $P(X = x) = p(x)$. Thus, $P(X = 0) = p(0)$, $P(X = 1) = p(1)$, etc. In this text, we adopt the simpler $p(x)$ notation.

FIGURE 4.3

Probability distribution for coin-toss experiment: Graphical form

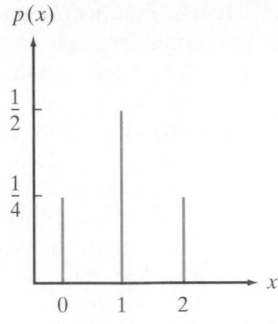

a. Point representation of $p(x)$

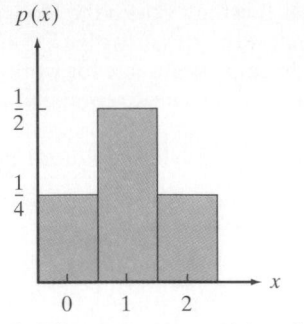

b. Histogram representation of $p(x)$

TABLE 4.1   **Probability Distribution for Coin-Toss Experiment: Tabular Form**

x	$p(x)$
0	¼
1	½
2	¼

as the heights of vertical lines over the corresponding values of x. Although the representation of the probability distribution as a histogram, as in Figure 4.3b, is less precise (since the probability is spread over a unit interval), the histogram representation will prove useful when we approximate probabilities of certain discrete random variables in Section 4.4.

We could also present the probability distribution for x as a formula, but this would unnecessarily complicate a very simple example. We give the formulas for the probability distributions of some common discrete random variables later in this chapter.   ▲

DEFINITION 4.4

The **probability distribution** of a discrete random variable is a graph, table, or formula that specifies the probability associated with each possible value the random variable can assume.

Two requirements must be satisfied by all probability distributions for discrete random variables.

Requirements for the Probability Distribution of a Discrete Random Variable x

$$p(x) \geq 0 \qquad \text{for all values of } x$$
$$\sum p(x) = 1$$

where the summation of $p(x)$ is over all possible values of x.*

Example 4.4 illustrates how the probability distribution for a discrete random variable can be derived, but for many practical situations the task is much more difficult. Fortunately, many experiments and associated discrete random variables observed in nature possess identical characteristics. Thus, you might observe a random variable in a psychology experiment that would possess the same probability distribution as a random variable observed in an engineering experiment or a social sample survey. We classify random variables according to type of experiment, derive the probability distribution for each of the different types, and then use the appropriate probability distribution when a particular type of random variable is observed in a practical situation. The probability distributions

---

*Unless otherwise indicated, summations will always be over all possible values of x.

for most commonly occurring discrete random variables have already been derived. This fact simplifies the problem of finding the probability distributions for random variables.

In Sections 4.4 and 4.5, we describe two important types of discrete random variables, give their probability distributions, and explain where and how they can be applied in practice. (Mathematical derivations of the probability distributions will be omitted, but these details can be found in the references at the back of the book.)

But first, in Section 4.3, we discuss some descriptive measures of these sometimes complex probability distributions. Since probability distributions are analogous to the relative frequency distributions of Chapter 2, it should be no surprise that the mean and standard deviation are useful descriptive measures.

EXERCISES 4.6–4.17

Learning the Mechanics

4.6 Consider the following probability distribution:

x	−4	0	1	3
$p(x)$	.1	.2	.4	.3

a. List the values that x may assume.
b. What value of x is most probable?
c. What is the probability that x is greater than 0?
d. What is the probability that $x = -2$?

4.7 A discrete random variable x can assume five possible values: 2, 3, 5, 8, and 10. Its probability distribution is shown here:

x	2	3	5	8	10
$p(x)$	.15	.10		.25	.25

a. What is $p(5)$?
b. What is the probability that x equals 2 or 10?
c. What is $P(x \le 8)$?

4.8 Explain why each of the following is or is not a valid probability distribution for a discrete random variable x:

a.

x	0	1	2	3
$p(x)$	.2	.3	.3	.2

b.

x	−2	−1	0
$p(x)$	.25	.50	.20

c.

x	4	9	20
$p(x)$	−.3	1.0	.3

d.

x	2	3	5	6
$p(x)$	.15	.20	.40	.35

4.9 The random variable x has the following discrete probability distribution:

x	10	11	12	13	14
$p(x)$	.2	.3	.2	.1	.2

Since the values that x can assume are mutually exclusive events, the event $\{x \le 12\}$ is the union of three mutually exclusive events:

$$\{x = 10\} \cup \{x = 11\} \cup \{x = 12\}$$

a. Find $P(x \le 12)$. **b.** Find $P(x > 12)$.
c. Find $P(x \le 14)$. **d.** Find $P(x = 14)$.
e. Find $P(x \le 11$ or $x > 12)$.

4.10 The random variable x has the discrete probability distribution shown here.

x	−2	−1	0	1	2
$p(x)$	.10	.15	.40	.30	.05

a. Find $P(x \le 0)$. **b.** Find $P(x > -1)$.
c. Find $P(-1 \le x \le 1)$. **d.** Find $P(x < 2)$.
e. Find $P(-1 < x < 2)$. **f.** Find $P(x < 1)$.

4.11 Toss three fair coins and let x equal the number of heads observed.
a. Identify the sample points associated with this experiment and assign a value of x to each sample point.
b. Calculate $p(x)$ for each value of x.
c. Construct a probability histogram for $p(x)$.
d. What is $P(x = 2$ or $x = 3)$?

Applying the Concepts

4.12 Nitrous oxide, more commonly known as "laughing gas," is used extensively in dental procedures. The sweet-smelling gas induces a state of conscious sedation that suppresses a patient's anxiety level. According to the American Dental Association, 60% of all dentists use nitrous oxide in their practice (*New York Times,* June 20, 1995). Suppose x equals the number of dentists in a random sample of five dentists who use laughing gas in practice. If $p = .60$ is the probability of any one dentist's using laughing gas (and the dentists operate independently), then the probability distribution of x (we show how to calculate these probabilities in Section 4.4) is

x	0	1	2	3	4	5
$p(x)$	.0102	.0768	.2304	.3456	.2592	.0778

Find the probability that the number of dentists using laughing gas in the sample of five is
a. 4   **b.** less than 2   **c.** greater than or equal to 3

4.13 In Exercise 2.6 you read about a study of book reviews in American history, geography, and area studies published in *Choice* magazine (*Library Acquisitions: Practice and Theory,* Vol. 19, 1995). The overall rating stated in each review was ascertained and recorded as follows: 1 = would not recommend, 2 = cautious or very little recommendation, 3 = little or no preference, 4 = favorable/recommended, 5 = outstanding/significant contribution. Based on a sample of 375 book reviews, the probability distribution of rating, x, is

Book Rating x	$p(x)$
1	.051
2	.099
3	.093
4	.635
5	.122

a. Is this a valid probability distribution?
b. What is the probability that a book reviewed in *Choice* has a rating of 1?
c. What is the probability that a book reviewed in *Choice* has a rating of at least 4?
d. What is the probability that a book reviewed in *Choice* has a rating of 2 or 3?

4.14 In Exercise 3.61 we noted that 1 in every 500 African-Americans in the United States has sickle-cell anemia. One in 10 is a carrier of the trait, and those who marry have a 1 in 4 chance of passing the disease to their children. Suppose a carrier has three children and let x equal the number who acquire the disease.
a. Find $p(x)$ for $x = 0, 1, 2, 3$.
b. Graph $p(x)$.   **c.** Find $p(x \le 1)$.

4.15 Every human possesses two sex chromosomes. A copy of one or the other (equally likely) is contributed to an offspring. Males have one X chromosome and one Y chromosome. Females have two X chromosomes. If a couple has three children, what is the probability that they have at least one boy? [*Hint:* Define the random variable z as the number of male offspring, and find the probability distribution of z.]

4.16 In 1993, President Clinton encountered difficulties in appointing an attorney general. One troublesome issue involved the nominees' hiring of illegal aliens and/or failing to pay Social Security taxes for their domestic help. Both of these practices are against the law. Suppose that a study reveals that 10% of households with incomes exceeding $50,000 annually have hired an illegal alien and/or failed to pay Social Security taxes. Let x be the number of households with incomes in that range who are contacted before one that has not broken either law is found.
a. What is the range of possible values for x?
b. Find $P(x < 3)$.   **c.** Find $P(x > 2)$.
d. Find the probability distribution for values of x from 1 to 10 and graph it over that domain. Can x exceed 10?

4.17 Many real-world systems (e.g., electric power transmission, transportation, telecommunications, and manufacturing systems) can be regarded as capacitated-flow networks, whose arcs have independent, but random, capacities. A team of Chinese university professors investigated the reliability of several flow networks in the journal *Networks* (May 1995). One

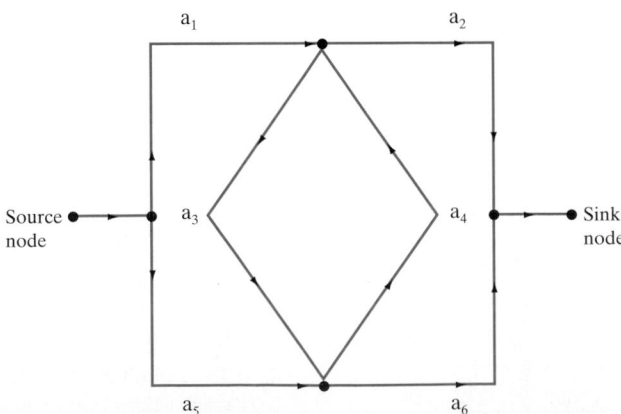

Arc	Capacity (x)	$p(x)$
a_1	3	.60
	2	.25
	1	.10
	0	.05
a_2	2	.60
	1	.30
	0	.10
a_3	1	.90
	0	.10

Arc	Capacity (x)	$p(x)$
a_4	1	.90
	0	.10
a_5	1	.90
	0	.10
a_6	2	.70
	1	.25
	0	.05

Source: Lin, J., *et al.* "On reliability evaluation of capacitated-flow network in terms of minimal pathsets." *Networks*, Vol. 25, No. 3, May 1995, p. 135 (Table I). Copyright 1995 John Wiley and Sons. Reprinted by permission of John Wiley and Sons, Inc.

network examined in the article (illustrated on page 162) is a bridge network with arcs a_1, a_2, a_3, a_4, a_5, and a_6. The probability distribution of the capacity x for each of the six arcs is provided at the left.

a. Verify that the properties of discrete probability distributions are satisfied for each arc capacity distribution.

b. Find the probability that the capacity for arc a_1 will exceed 1.

c. Repeat part **b** for each of the remaining five arcs.

d. One path from the source node to the sink node is through arcs a_1 and a_2. Find the probability that the system maintains a capacity of more than 1 through the a_1–a_2 path. (Recall that the arc capacities are independent.)

4.3   EXPECTED VALUES OF DISCRETE RANDOM VARIABLES

If a discrete random variable x were observed a very large number of times and the data generated were arranged in a relative frequency distribution, the relative frequency distribution would be indistinguishable from the probability distribution for the random variable. Thus, the probability distribution for a random variable is a theoretical model for the relative frequency distribution of a population. To the extent that the two distributions are equivalent (and we will assume they are), the probability distribution for x possesses a mean μ and a variance σ^2 that are identical to the corresponding descriptive measures for the population. This section explains how you can find the mean value for a random variable. We illustrate the procedure with an example.

Examine the probability distribution for x (the number of heads observed in the toss of two fair coins) in Figure 4.4. Try to locate the mean of the distribution intuitively. We may reason that the mean μ of this distribution is equal to 1 as follows: In a large number of experiments, $\frac{1}{4}$ should result in $x = 0$, $\frac{1}{2}$ in $x = 1$, and $\frac{1}{4}$ in $x = 2$ heads. Therefore, the average number of heads is

$$\mu = 0(\tfrac{1}{4}) + 1(\tfrac{1}{2}) + 2(\tfrac{1}{4}) = 0 + \tfrac{1}{2} + \tfrac{1}{2} = 1$$

Note that to get the population mean of the random variable x, we multiply each possible value of x by its probability $p(x)$, and then we sum this product over all possible values of x. The *mean of x* is also referred to as the *expected value of x*, denoted $E(x)$.

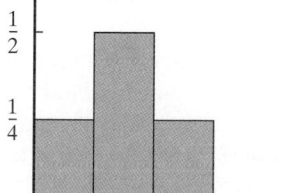

FIGURE 4.4

Probability distribution for a two-coin toss

DEFINITION 4.5

The **mean**, or **expected value**, of a discrete random variable x is

$$\mu = E(x) = \sum x p(x)$$

The term *expected* is a mathematical term and should not be interpreted as it is typically used. Specifically, a random variable might never be equal to its "expected value." Rather, the expected value is the mean of the probability distribution, or a measure of its central tendency. You can think of μ as the mean value of x in a *very large* (actually, *infinite*) number of repetitions of the experiment.

EXAMPLE 4.5

Suppose you work for an insurance company, and you sell a $10,000 whole-life insurance policy at an annual premium of $290. Actuarial tables show that the probability of death during the next year for a person of your customer's age, sex, health, etc., is .001. What is the expected gain (amount of money made by the company) for a policy of this type?

Solution

The experiment is to observe whether the customer survives the upcoming year. The probabilities associated with the two sample points, Live and Die, are .999 and .001, respectively. The random variable you are interested in is the gain x, which can assume the values shown in the following table.

Gain x	Sample Point	Probability
$290	Customer lives	.999
$290−$10,000	Customer dies	.001

If the customer lives, the company gains the $290 premium as profit. If the customer dies, the gain is negative because the company must pay $10,000, for a net "gain" of $(290 − 10,000)$. The expected gain is therefore

$$\mu = E(x) = \sum xp(x)$$
$$= (290)(.999) + (290 - 10,000)(.001) = \$280$$

In other words, if the company were to sell a very large number of 1-year $10,000 policies to customers possessing the characteristics described above, it would (on the average) net $280 per sale in the next year.   ▲

Example 4.5 illustrates that the expected value of a random variable x need not equal a possible value of x. That is, the expected value is $280, but x will equal either $290 or −$9,710 each time the experiment is performed (a policy is sold and a year elapses). The expected value is a measure of central tendency—and in this case represents the average over a very large number of 1-year policies—but is not a possible value of x.

We learned in Chapter 2 that the mean and other measures of central tendency tell only part of the story about a set of data. The same is true about probability distributions. We need to measure variability as well. Since a probability distribution can be viewed as a representation of a population, we will use the population variance to measure its variability.

The **population variance** σ^2 is defined as the average of the squared distance of x from the population mean μ. Since x is a random variable, the squared distance, $(x - \mu)^2$, is also a random variable. Using the same logic used to find the mean value of x, we find the mean value of $(x - \mu)^2$ by multiplying all possible values of $(x - \mu)^2$ by $p(x)$ and then summing over all possible x values.* This quantity,

$$E\big[(x - \mu)^2\big] = \sum_{\text{all } x}(x - \mu)^2 p(x)$$

*It can be shown that $E[(x - \mu)^2] = E(x^2) - \mu^2$, where $E(x^2) = \sum x^2 p(x)$. Note the similarity between this expression and the shortcut formula $\sum (x - \bar{x})^2 = \sum x^2 - (\sum x)^2/n$ given in Chapter 2.

is also called the **expected value of the squared distance from the mean**; that is, $\sigma^2 = E[(x - \mu)^2]$. The standard deviation of x is defined as the square root of the variance σ^2.

DEFINITION 4.6

The **variance** of a random variable x is

$$\sigma^2 = E[(x - \mu)^2] = \sum(x - \mu)^2 p(x)$$

DEFINITION 4.7

The **standard deviation** of a discrete random variable is equal to the square root of the variance, i.e., to $\sigma = \sqrt{\sigma^2}$.

Knowing the mean μ and standard deviation σ of the probability distribution of x, in conjunction with Chebyshev's Rule (Table 2.8) and the Empirical Rule (Table 2.9), we can make statements about the likelihood that values of x will fall within the intervals $\mu \pm \sigma$, $\mu \pm 2\sigma$, and $\mu \pm 3\sigma$. These probabilities are given in the next box.

Chebyshev's Rule and Empirical Rule for a Discrete Random Variable

Let x be a discrete random variable with probability distribution $p(x)$, mean μ, and standard deviation σ. Then, depending on the shape of $p(x)$, the following probability statements can be made:

	CHEBYSHEV'S RULE	**EMPIRICAL RULE**
	Applies to any probability distribution (see Figure 4.5a)	Applies to probability distributions that are mound-shaped and symmetric (see Figure 4.5b)
$P(\mu - \sigma < x < \mu + \sigma)$	≥ 0	$\approx .68$
$P(\mu - 2\sigma < x < \mu + 2\sigma)$	$\geq \tfrac{3}{4}$	$\approx .95$
$P(\mu - 3\sigma < x < \mu + 3\sigma)$	$\geq \tfrac{8}{9}$	≈ 1.00

FIGURE 4.5

Shapes of two probability distributions for a discrete random variable x

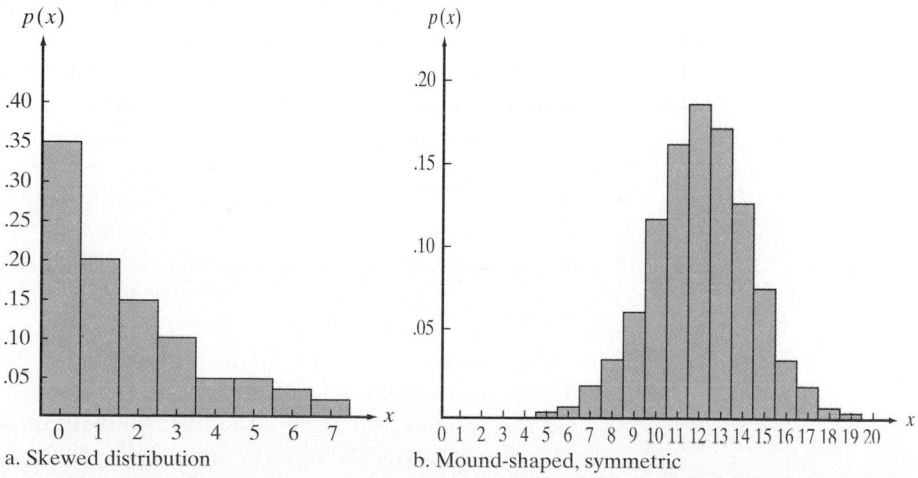

a. Skewed distribution

b. Mound-shaped, symmetric

"The Showcase Showdown"

The popular television game show *The Price is Right*, hosted by Bob Barker, employs a number of games to test those contestants lucky and skilled enough to make it onto the main stage. A few of these are games of skill (e.g., contestants are required to estimate the prices of items placed on display), but many are games of chance. "The Showcase Showdown" is a game of chance played twice on every show.

CASE STUDY • 4.1 •

"The Showcase Showdown" involves a large wheel with twenty nickel values, 5, 10, 15, 20, ..., 95, 100, marked on it. Contestants (usually three) spin the wheel once or twice, with the objective of obtaining the highest total score *without going over a dollar (100)*. The player who spins first is at something of a disadvantage since he or she must decide whether to stop after the initial spin or to spin a second time. The player who spins last has the advantage of knowing the first players' scores. Researchers writing in the *American Statistician* (Aug. 1995) used probability theory to show that the optimal strategy for the first player in a three-player game is to spin a second time only if the value of the initial spin is 65 or less.

For this case study, consider a version of "The Showcase Showdown" involving only a single player. Let *x* represent the total score obtained by the player, assuming a "fair" wheel, i.e., a wheel with equally likely outcomes. If the total of the player's spins exceeds 100, the total score is set to 0.

Focus

a. If the player is permitted only one spin of the wheel, find the probability distribution for *x*.

b. Refer to the probability distribution, part a. Find $E(x)$ and interpret this value.

c. Refer to the probability distribution, part a. Give a range of values within which x is likely to fall.

d. Suppose the player will spin the wheel twice, no matter what the outcome of the first spin. Find the probability distribution for *x*.

e. What assumption did you make to obtain the probability distribution, part d? Is it a reasonable assumption?

f. Find μ and σ for the probability distribution, part d, and interpret the results.

g. Refer to part d. What is the probability that in two spins the player's total score exceeds a dollar (i.e., is set to 0)?

h. Suppose the player obtains a 20 on the first spin and decides to spin again. Find the probability distribution for *x*.

i. Refer to part h. What is the probability that the player's total score exceeds a dollar?

j. Given that the player obtains a 65 on the first spin and decides to spin again, find the probability that the player's total score exceeds a dollar.

k. Repeat part j for different first-spin outcomes. (Optionally, each student in the class can be assigned a different outcome and the results of the class pooled.) Use this information to suggest a strategy for the one-player game.

EXAMPLE 4.6

Medical research has shown that a certain type of chemotherapy is successful 70% of the time when used to treat skin cancer. Suppose five skin cancer patients are treated with this type of chemotherapy and let *x* equal the number of successful cures out of the five. The probability distribution for the number *x* of successful cures out of five is given in the table:

x	0	1	2	3	4	5
$p(x)$	.002	.029	.132	.309	.360	.168

a. Find $\mu = E(x)$. Interpret the result.

b. Find $\sigma = \sqrt{E[(x - \mu)^2]}$. Interpret the result.

c. Graph $p(x)$. Locate μ and the interval $\mu \pm 2\sigma$ on the graph. Use either Chebyshev's Rule or the Empirical Rule to approximate the probability that *x* falls in this interval. Compare this result with the actual probability.

d. Would you expect to observe fewer than two successful cures out of five?

Solution

a. Applying the formula,

$$\mu = E(x) = \sum xp(x)$$
$$= 0(.002) + 1(.029) + 2(.132) + 3(3.09) + 4(.360) + 5(.168) = 3.50$$

If five skin cancer patients receive the chemotherapy treatment, we expect the number x who are cured to be near 3.5. Remember that this expected value has meaning only when the experiment—treating five skin cancer patients with chemotherapy—is repeated a large number of times.

b. Now we calculate the variance of x:

$$\sigma^2 = E\big[(x - \mu)^2\big] = \sum (x - \mu)^2 p(x)$$
$$= (0 - 3.5)^2(.002) + (1 - 3.5)^2(.029) + (2 - 3.5)^2(.132)$$
$$+ (3 - 3.5)^2(.309) + (4 - 3.5)^2(.360) + (5 - 3.5)^2(.168)$$
$$= 1.05$$

Thus, the standard deviation is

$$\sigma = \sqrt{\sigma^2} = \sqrt{1.05} = 1.02$$

This value measures the spread of the probability distribution of x, the number of successful cures out of five.

c. The graph of $p(x)$ is shown in Figure 4.6 with the mean μ and the interval $\mu \pm 2\sigma = 3.50 \pm 2(1.02) = 3.50 \pm 2.04 = (1.46, 5.54)$ shown on the graph. Note particularly that $\mu = 3.5$ locates the center of the probability distribution. Since this distribution is a theoretical relative frequency distribution that is moderately mound-shaped (see Figure 4.6), we expect (from Chebyshev's Rule) at least 75% and, more likely (from the Empirical Rule), approximately 95% of observed x values to fall in the interval $\mu \pm 2\sigma$—that is, between 1.46 and 5.54. You can see from Figure 4.6 that the actual probability that x falls in the interval $\mu \pm 2\sigma$ includes the sum of $p(x)$ for the values $x = 2$, $x = 3$, $x = 4$, and $x = 5$. This probability is $p(2) + p(3) + p(4) + p(5) = .132 + .309 + .360 + .168 = .969$. Therefore, 96.9% of the probability distribution lies within 2

FIGURE 4.6

Graph of $p(x)$ for
Example 4.6

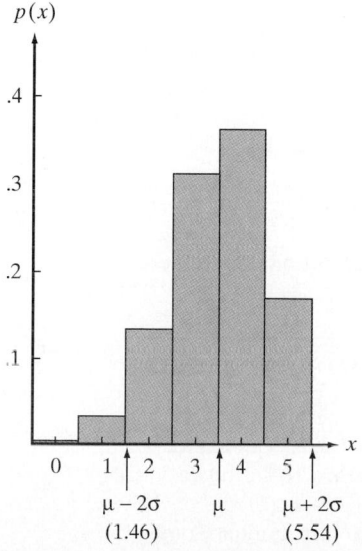

standard deviations of the mean. This percentage is consistent with both Chebyshev's Rule and the Empirical Rule.

d. Fewer than two successful cures out of five implies that $x = 0$ or $x = 1$. Since both these values of x lie outside the interval $\mu \pm 2\sigma$, we know from the Empirical Rule that such a result is unlikely (approximate probability of .05). The exact probability, $P(x \le 1)$, is $p(0) + p(1) = .002 + .029 = .031$. Consequently, in a single experiment where five skin cancer patients are treated with chemotherapy, we would not expect to observe fewer than two successful cures.   ▲

EXERCISES 4.18–4.29

Learning the Mechanics

4.18 Consider the probability distribution for the random variable x shown here.

x	10	20	30	40	50	60
$p(x)$	.05	.20	.30	.25	.10	.10

a. Find μ, σ^2, and σ. b. Graph $p(x)$.
c. Locate μ and the interval $\mu \pm 2\sigma$ on your graph. What is the probability that x will fall within the interval $\mu \pm 2\sigma$?

4.19 Consider the probability distribution shown here.

x	1	2	4	10
$p(x)$	.2	.4	.2	.2

a. Find $\mu = E(x)$. b. Find $\sigma^2 = E[(x - \mu)^2]$.
c. Find σ. d. Interpret the value you obtained for μ.
e. In this case, can the random variable x ever assume the value μ? Explain.
f. In general, can a random variable ever assume a value equal to its expected value? Explain.

4.20 Consider the probability distribution shown here.

x	−4	−3	−2	−1	0	1	2	3	4
$p(x)$	.02	.07	.10	.15	.30	.18	.10	.06	.02

a. Calculate μ, σ^2, and σ.
b. Graph $p(x)$. Locate μ, $\mu - 2\sigma$, and $\mu + 2\sigma$ on the graph.
c. What is the probability that x is in the interval $\mu \pm 2\sigma$?

4.21 Consider the probability distributions shown here.

x	0	1	2		y	0	1	2
$p(x)$	.3	.4	.3		$p(y)$	.1	.8	.1

a. Use your intuition to find the mean for each distribution. How did you arrive at your choice?
b. Which distribution appears to be more variable? Why?
c. Calculate μ and σ^2 for each distribution. Compare these answers to your answers in parts **a** and **b**.

Applying the Concepts

4.22 A patient complaining of severe stomach pains checked into a local hospital. After a series of tests the doctors narrowed their diagnosis to four possible ailments. They believe there is a 40% chance that the patient has hepatitis; a 10% chance that she has cirrhosis; a 45% chance of gallstones; and a 5% chance of cancer of the pancreas. The doctors are certain that the patient has only one of the diseases, but will not know which disease until further tests are performed. The cost associated with treating each disease is given in the table:

Disease	Hepatitis	Cirrhosis	Gallstones	Pancreatic Cancer
Cost	$700	$1,100	$3,320	$16,450

a. Construct the probability distribution for the cost of treating the patient.
b. Calculate the mean of the probability distribution you constructed in part **a**. What does this number represent?
c. Further testing reveals that the patient has either hepatitis or cirrhosis. Given this information, construct the new probability distribution for the cost of treating the patient.
d. Calculate the mean of the probability distribution you constructed in part **c**. What does this number represent?

4.23 Exercise 4.13 gives the probability distribution for the rating x of books reviewed in *Choice* magazine. The probability distribution is reproduced here.

Book Rating x	$p(x)$
1	.051
2	.099
3	.093
4	.635
5	.122

a. Find $E(x)$. Give a meaningful interpretation of the result.
b. Find σ.

c. Find the exact probability that x is in the interval $\mu \pm 2\sigma$. Compare to Chebyshev's Rule and the Empirical Rule.

4.24 Odds makers try to predict which professional and college football teams will win and by how much (the *spread*). If the odds makers do this accurately, adding the spread to the underdog's score should make the final score a tie. Suppose a bookie will give you $6 for every $1 you risk if you pick the winners in three ballgames (adjusted by the spread) on a "parlay" card. Thus, for every $1 bet you will either lose $1 or gain $5. What is the bookie's expected earnings per dollar wagered?

4.25 The number of training units that must be passed before a complex computer software program is mastered varies from one to five, depending on the student. After much experience, the software manufacturer has determined the probability distribution that describes the fraction of users mastering the software after each number of training units:

Number of Units	1	2	3	4	5
Probability of Mastery	.1	.25	.4	.15	.1

a. Calculate the mean number of training units necessary to master the program. Calculate the median. Interpret each.

b. If the firm wants to ensure that at least 75% of the students master the program, what is the minimum number of training units that must be administered? At least 90%?

c. Suppose the firm develops a new training program that increases the probability that only one unit of training is needed from .1 to .25, increases the probability that only two units are needed to .35, leaves the probability that three units are needed at .4, and completely eliminates the need for four or five units. How do your answers to parts **a** and **b** change for this new program?

4.26 A capacitated-bridge network with six arcs (*Network*, May 1995) was presented in Exercise 4.17. The probability distributions for the capacities of the six arcs are reproduced at right. Compute the mean capacity of each arc and interpret its value.

4.27 A rock concert producer has scheduled an outdoor concert for Saturday, May 24. If it does not rain, the producer expects to make $20,000 profit from the concert. If it does rain, the producer will be forced to cancel the concert and will lose $12,000 (rock star's

fee, advertising costs, stadium rental, administrative costs, etc.). The producer has learned from the National Weather Service that the probability of rain on May 24 is .4.

a. Find the producer's expected profit from the concert.

b. For a fee of $1,000, an insurance company has offered to insure the producer against all losses resulting from a rained-out concert. If the producer buys the insurance, what is her expected profit from the concert?

c. Assuming the forecast is accurate, do you believe the insurance company has charged too much or too little for the policy? Explain.

4.28 Exercise 2.3 presented the results of a *Newsweek* (Nov. 28, 1993) survey of U.S. residents' commitment to religion. The responses, coded from 0 to 4, are summarized in tabular form at the bottom of the page.

a. Can *Newsweek* report that more U.S. residents are religious than not?

b. Treating the response categories as probabilities associated with the response codes, what is the mean response? Interpret the mean.

4.29 Most states offer weekly lotteries to generate revenue for the state. Despite the long odds of winning, residents continue to gamble on the lottery each week (see Case Study 3.3). The chance of winning Florida's Pick-6 Lotto game is 1 in approximately 14 million. Suppose you buy a $1 Lotto ticket in anticipation of winning the $7-million grand prize. Calculate your expected net winnings. Interpret the result.

Arc	Capacity (x)	$p(x)$	Arc	Capacity (x)	$p(x)$
a_1	3	.60	a_4	1	.90
	2	.25		0	.10
	1	.10			
	0	.05			
a_2	2	.60	a_5	1	.90
	1	.30		0	.10
	0	.10			
a_3	1	.90	a_6	2	.70
	0	.10		1	.25
				0	.05

Source: Lin, J., *et al.* "On reliability evaluation of capacitated-flow network in terms of minimal pathsets." *Networks,* Vol. 25, No. 3, May 1995, p. 135 (Table I). Copyright 1995 John Wiley and Sons. Reprinted by permission of John Wiley and Sons, Inc.

	Agnostic/ Atheist	Trace Elements of Religion	Normally Religious	Modestly Religious	Highly Committed
Response, x	0	1	2	3	4
Percent	8	22	29	22	19

4.4 THE BINOMIAL RANDOM VARIABLE

Many experiments result in *dichotomous* responses—i.e., responses for which there exist two possible alternatives, such as Yes–No, Pass–Fail, Defective–Nondefective, or Male–Female. A simple example of such an experiment is the coin-toss experiment. A coin is tossed a number of times, say 10. Each toss results in one of two outcomes, Head or Tail, and the probability of observing each of these two outcomes remains the same for each of the 10 tosses. Ultimately, we are interested in the probability distribution of x, the number of heads observed. Many other experiments are equivalent to tossing a coin (either balanced or unbalanced) a fixed number n of times and observing the number x of times that one of the two possible outcomes occurs. Random variables that possess these characteristics are called **binomial random variables**.

Public opinion and consumer preference polls (e.g., the Gallup and Harris polls) frequently yield observations on binomial random variables. For example, suppose a sample of 100 students is selected from a large student body and each person is asked whether he or she favors (a Head) or opposes (a Tail) a certain campus issue. Ultimately, we are interested in x, the number of people in the sample who favor the issue. If each student is randomly selected from the student body and if the (unknown) proportion of students favoring the issue is p, then observing whether a student favors or is opposed to the issue is analogous to tossing an unbalanced coin. The chance that any randomly selected student favors the issue is p; the probability that he or she opposes the issue is $(1 - p)$. Sampling 100 students is analogous to tossing the coin 100 times. Thus, you can see that opinion polls which record the number of people who favor a certain issue are real-life equivalents of coin-toss experiments. We have been describing a **binomial experiment**; it is identified by the following characteristics.

Characteristics of a Binomial Random Variable

---

1. The experiment consists of n identical trials.
2. There are only two possible outcomes on each trial. We will denote one outcome by S (for Success) and the other by F (for Failure).
3. The probability of S remains the same from trial to trial. This probability is denoted by p, and the probability of F is denoted by q. Note that $q = 1 - p$.
4. The trials are independent.
5. The binomial random variable x is the number of S's in n trials.

EXAMPLE 4.7   For the following examples, decide whether x is a binomial random variable.

a. Suppose a university scholarship committee must select two students to receive a scholarship for the next academic year. The committee receives 10 applications for the scholarships—six from male students and four from female students. Suppose the applicants are all equally qualified, so that the selections are randomly made. Let x be the number of female students who receive a scholarship.

b. Before marketing a new product on a large scale, many companies will conduct a consumer-preference survey to determine whether the product is likely to be successful. Suppose a company develops a new diet soda and then conducts a taste-preference survey in which 100 randomly chosen consumers state their preferences among the new soda and the two leading sellers. Let x be the number of the 100 who choose the new brand over the two others.

c. Some surveys are conducted by using a method of sampling other than simple random sampling (defined in Chapter 3). For example, suppose a television cable company plans to conduct a survey to determine the fraction of households in the city that would use the cable television service. The sampling method is to choose a city block at random and then survey every household on that block. This sampling technique is called **cluster sampling**. Suppose 10 blocks are so sampled, producing a total of 124 household responses. Let *x* be the number of the 124 households that would use the television cable service.

Solution

a. In checking the binomial characteristics, a problem arises with independence (characteristic 4 in the preceding box). Given that the first student selected is female, the probability that the second chosen is female is ³⁄₉. On the other hand, given that the first selection is a male student, the probability that the second is female is ⁴⁄₉. Thus, the conditional probability of a Success (choosing a female student to receive a scholarship) on the second trial (selection) depends on the outcome of the first trial, and the trials are therefore dependent. Since the trials are *not independent*, this is not a binomial random variable.
b. Surveys that produce dichotomous responses and use random sampling techniques are classic examples of binomial experiments. In our example, each randomly selected consumer either states a preference for the new diet soda or does not. The sample of 100 consumers is a very small proportion of the totality of potential consumers, so the response of one would be, for all practical purposes, independent of another.* Thus, *x* is a binomial random variable.
c. This example is a survey with dichotomous responses (Yes or No to the cable service), but the sampling method is not simple random sampling. Again, the binomial characteristic of independent trials would probably not be satisfied. The responses of households within a particular block would be dependent, since households within a block tend to be similar with respect to income, level of education, and general interests. Thus, the binomial model would not be satisfactory for *x* if the cluster sampling technique were employed.   ▲

EXAMPLE 4.8

The Heart Association claims that only 10% of U.S. adults over 30 can pass the President's Physical Fitness Commission's minimum requirements. Suppose four adults are randomly selected, and each is given the fitness test. Let *x* be the number of the four who pass the minimum requirements. Find the probability distribution for *x*, assuming that the Heart Association's claim is true.

Solution

Recall that a probability distribution describes a discrete random variable by assigning probabilities to each of its values. In this example, the possible values of *x*, the number of four adults who pass the minimum requirements, are 0, 1, 2, 3, 4. Furthermore, there are four identical trials (the sampling and observation of the four adults), each with two possible outcomes (pass or fail). We are assuming that the probability that each adult passes is .1, and the result for each adult will be (at least approximately) independent of that for the others. Thus, the number, *x*, of adults who pass the minimum requirements is a binomial random variable.

---

*In most real-life applications of the binomial distribution, the population of interest has a finite number of elements (trials), denoted *N*. When *N* is large and the sample size *n* is small relative to *N*, say $n/N \leq .05$, the sampling procedure, for all practical purposes, satisfies the conditions of a binomial experiment.

Let us first consider the event $x = 0$, that is, the event that none of the four tested adults passes the test. You can see that the event $x = 0$ is equivalent to the sample point

$$FFFF$$

where F in the first position implies that adult 1 fails, F in the second position implies that adult 2 fails, etc. Since the trials are independent in this binomial experiment (knowing whether adult 1 passes should not affect the probability that adult 2 passes), we can find the probability of an intersection by multiplying the probabilities of the events. Thus,

$$P(x = 0) = P(FFFF) = P(F)P(F)P(F)p(F) = (.9)(.9)(.9)(.9) = (.9)^4 = .6561$$

The event $x = 1$ implies that one of the four adults passes the physical fitness test and three fail it. The following list of sample points contains all sample points that imply $x = 1$:

$$SFFF \qquad FSFF \qquad FFSF \qquad FFFS$$

where S in the first position corresponds to adult 1 passing the test, S in the second position corresponds to adult 2 passing the test, etc. Note that each of these sample points will have the same probability, $(.1)(.9)^3$, where .1 corresponds to the one adult who passes the test and $(.9)^3$ corresponds to the three who fail it. Remembering from Chapter 3 that we obtain the probability of an event by summing the probabilities of the sample points of which it is composed, we get

$$P(x = 1) = 4[(.1)(.9)^3] = .2916$$

The event $x = 2$ implies that two adults pass the test and two fail it; this event consists of the following six sample points:

$$SSFF \qquad SFSF \qquad SFFS \qquad FSSF \qquad FSFS \qquad FFSS$$

Each of these sample points has probability $(.1)^2(.9)^2$, so that

$$P(x = 2) = 6[(.1)^2(.9)^2] = .0486$$

Similarly,

$$P(x = 3) = 4[(.1)^3(.9)] = .0036$$
$$P(x = 4) = (.1)^4 = .0001$$

The complete probability distribution is shown in Figure 4.7 and listed in Table 4.2.

▲

TABLE 4.2 Probability Distribution for Physical Fitness Example: Tabular Form

x	$p(x)$
0	.6561
1	.2916
2	.0486
3	.0036
4	.0001

FIGURE 4.7

Probability distribution for physical fitness example: Graphical form

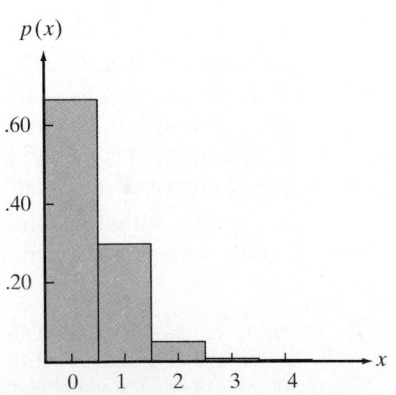

Before we give a formula for $p(x)$, we will refresh your memory on factorial notation. Particularly, the symbol $n!$ is to be read "n factorial" and is calculated by

$$n! = n(n-1)(n-2) \cdots \cdot 3 \cdot 2 \cdot 1$$

We define $0! = 1$. Thus, for example, $4! = 4 \cdot 3 \cdot 2 \cdot 1 = 24$.

Using factorial notation, we write the formula for a binomial probability distribution with $n = 4$ and $p = .1$:

$$p(x) = \frac{4!}{x!(4-x)!} (.1)^x (.9)^{4-x}$$

Then, for $x = 2$, we have

$$p(2) = \frac{4!}{2!(4-2)!} (.1)^2 (.9)^{4-2} = \frac{4 \cdot 3 \cdot 2 \cdot 1}{(2 \cdot 1)(2 \cdot 1)} (.1)^2 (.9)^2$$
$$= 6(.1)^2(.9)^2 = .0486$$

which agrees with our sample point calculation. Note that the first part of the formula, $4!/[x!(4-x)!]$, counts the number of sample points that result in x adults passing the physical fitness test. The second part of the formula, $(.1)^x(.9)^{4-x}$, is the probability assigned to each sample point that has x adults passing and $(4-x)$ failing. When we multiply the *number* of sample points by the *probability* assigned to each sample point, we get the probability that x adults pass the test. We will see that this formula can be generalized to give the probability distribution of any binomial random variable.

Note that $\binom{n}{x}$, shorthand for $n!/[x!(n-x)!]$, is the number of sample points that have x successes and $(n-x)$ failures (see Section 3.9 for a discussion of the combinations rule for counting sample points), and $p^x q^{n-x}$ is the probability assigned to each sample point that has x successes and $(n-x)$ failures. The product of these two quantities, $\binom{n}{x} p^x q^{n-x}$, is the probability that x successes and $(n-x)$ failures are observed.

The Binomial Probability Distribution

$$p(x) = \binom{n}{x} p^x q^{n-x} \qquad (x = 0, 1, 2, \ldots, n)$$

where

   p = Probability of a success on a single trial

   $q = 1 - p$

   n = Number of trials

   x = Number of successes in n trials

$\binom{n}{x} = \dfrac{n!}{x!(n-x)!}$

The binomial probability distribution is so named because the probabilities, $p(x)$, $x = 0, 1, \ldots, n$, are terms of the binomial expansion, $(q + p)^n$.

EXAMPLE 4.9   Refer to Example 4.8. Calculate μ and σ, the mean and standard deviation, respectively, of the number of the four adults who pass the test.

Solution

From Section 4.3 we know that the mean of a discrete probability distribution is

$$\mu = \sum xp(x)$$

Referring to Table 4.2, the probability distribution for the number x who pass the fitness test, we find

$$\mu = 0(.6561) + 1(.2916) + 2(.0486) + 3(.0036) + 4(.0001) = .4 = 4(.1) = np$$

The relationship $\mu = np$ holds in general for a binomial random variable.
   The variance is

$$
\begin{aligned}
\sigma^2 &= \sum (x - \mu)^2 p(x) = \sum (x - .4)^2 p(x) \\
&= (0 - .4)^2(.6561) + (1 - .4)^2(.2916) + (2 - .4)^2(.0486) \\
&\quad + (3 - .4)^2(.0036) + (4 - .4)^2(.0001) \\
&= .104976 + .104976 + .124416 + .024336 + .001296 \\
&= .36 = 4(.1)(.9) = npq
\end{aligned}
$$

The relationship $\sigma^2 = npq$ holds in general for a binomial random variable.
   Finally, the standard deviation of the number who pass the fitness test is

$$\sigma = \sqrt{\sigma^2} = \sqrt{.36} = .6 \qquad \blacktriangle$$

We emphasize that you need not use the expectation summation rules to calculate μ and σ^2 for a binomial random variable. You can find them easily using the formulas $\mu = np$ and $\sigma^2 = npq$.

Mean, Variance, and Standard Deviation for a Binomial Random Variable

Mean: $\mu = np$

Variance: $\sigma^2 = npq$

Standard deviation: $\sigma = \sqrt{npq}$

As we demonstrated in Chapter 2, the mean and standard deviation provide measures of the central tendency and variability, respectively, of a distribution. Thus, we can use μ and σ to obtain a rough visualization of the probability distribution for x when the calculation of the probabilities is too tedious. To see the binomial probability distribution in action, consider Example 4.10.

EXAMPLE 4.10

A poll of 20 voters is taken in a large city. The purpose is to determine x, the number in favor of a certain candidate for mayor. Suppose that (unknown to us) 60% of all the city's voters favor this candidate.

   a. Find the mean and standard deviation of x. Interpret the results.
   b. Find the probability that x is less than or equal to 10 ($x \leq 10$).
   c. Find the probability that x exceeds 12 ($x > 12$).
   d. Find the probability that x equals 11 ($x = 11$).
   e. Graph the probability distribution of x and locate the interval $\mu - 2\sigma$ to $\mu + 2\sigma$ on the graph.

Solution

   a. Given that the sample of 20 was randomly selected from a large number of voters, x, the number of the 20 who favor the candidate, is (approximately) a

TABLE 4.3   Reproduction of Part of Table II in Appendix A

k \ p	.01	.05	.10	.20	.30	.40	.50	.60	.70	.80	.90	.95	.99
0	.818	.358	.122	.012	.001	.000	.000	.000	.000	.000	.000	.000	.000
1	.983	.736	.392	.069	.008	.001	.000	.000	.000	.000	.000	.000	.000
2	.999	.925	.677	.206	.035	.004	.000	.000	.000	.000	.000	.000	.000
3	1.000	.984	.867	.411	.107	.016	.001	.000	.000	.000	.000	.000	.000
4	1.000	.997	.957	.630	.238	.051	.006	.000	.000	.000	.000	.000	.000
5	1.000	1.000	.989	.804	.416	.126	.021	.002	.000	.000	.000	.000	.000
6	1.000	1.000	.998	.913	.608	.250	.058	.006	.000	.000	.000	.000	.000
7	1.000	1.000	1.000	.968	.772	.416	.132	.021	.001	.000	.000	.000	.000
8	1.000	1.000	1.000	.990	.887	.596	.252	.057	.005	.000	.000	.000	.000
9	1.000	1.000	1.000	.997	.952	.755	.412	.128	.017	.001	.000	.000	.000
10	1.000	1.000	1.000	.999	.983	.872	.588	.245	.048	.003	.000	.000	.000
11	1.000	1.000	1.000	1.000	.995	.943	.748	.404	.113	.010	.000	.000	.000
12	1.000	1.000	1.000	1.000	.999	.979	.868	.584	.228	.032	.000	.000	.000
13	1.000	1.000	1.000	1.000	1.000	.994	.942	.750	.392	.087	.002	.000	.000
14	1.000	1.000	1.000	1.000	1.000	.998	.979	.874	.584	.196	.011	.000	.000
15	1.000	1.000	1.000	1.000	1.000	1.000	.994	.949	.762	.370	.043	.003	.000
16	1.000	1.000	1.000	1.000	1.000	1.000	1.000	.984	.893	.589	.133	.016	.000
17	1.000	1.000	1.000	1.000	1.000	1.000	1.000	.996	.965	.794	.323	.075	.001
18	1.000	1.000	1.000	1.000	1.000	1.000	1.000	.999	.992	.931	.608	.264	.017
19	1.000	1.000	1.000	1.000	1.000	1.000	1.000	1.000	.999	.988	.878	.642	.182

binomial random variable. The value of p is the fraction of the total voters who favor the candidate; that is $p = .6$. Therefore, we calculate the mean and variance:

$$\mu = np = 20(.6) = 12$$
$$\sigma^2 = npq = 20(.6)(.4) = 4.8$$

The standard deviation is then

$$\sigma = \sqrt{4.8} = 2.2$$

If we were to repeatedly sample 20 voters, we'd find that $\mu = 12$, which implies that, on average, 12 of the 20 voters will favor the candidate. Thus, we expect about 12 voters to favor the candidate in this sample of 20. The value $\sigma = 2.2$ is a measure of the spread of the distribution of x, which we illustrate in detail in part e.

b. Calculating binomial probabilities when n is large is a formidable task. For example, to find the probability that $x \le 10$, we would calculate

$$P(x \le 10) = p(0) + p(1) + p(2) + \cdots + p(10)$$

$$= \sum_{x=0}^{10} p(x) = \sum_{x=0}^{10} \binom{20}{x}(.6)^x(.4)^{20-x}$$

We can avoid these tedious calculations by making use of cumulative binomial probability tables (Table II in Appendix A). Part of Table II is shown in Table 4.3. The columns correspond to values of p, and the rows correspond to values of the random variable x.

FIGURE 4.8

Graph of binomial probability distribution for $n = 20$ and $p = .6$

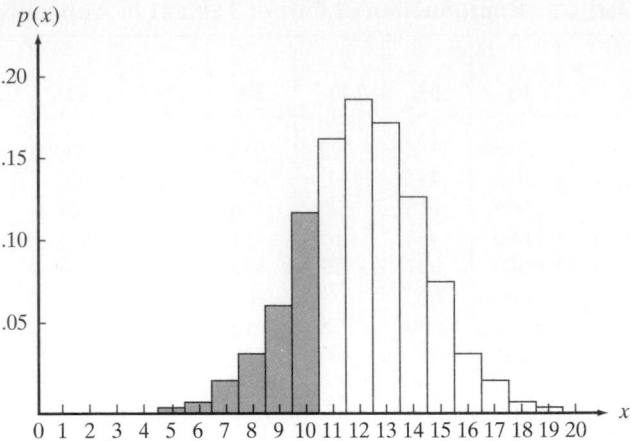

The entries in Table II are the cumulative sums

$$P(x \le k) = p(0) + p(1) + p(2) + \cdots + p(k)$$

for values of $k = 0, 1, 2, \ldots, (n - 1)$. Observe that the bottom row of the table, the one corresponding to $k = n$, is omitted. This is because the sum of $p(x)$ from $x = 0$ to $x = n$ is always equal to 1; that is, $P(x \le n) = 1$ for any binomial random variable.

To find $P(x \le 10)$ for $n = 20$ and $p = .6$, we first find the column corresponding to $p = .6$ and then the row corresponding to $k = 10$. The recorded value, shaded in Table 4.3 and in Figure 4.8, is

$$P(x \le 10) = .245$$

c. To find the probability

$$P(x > 12) = p(13) + p(14) + \cdots + p(19) + p(20) = \sum_{x=13}^{20} p(x)$$

we use the fact that for all probability distributions, $\sum p(x) = 1$. Therefore, using the complementary event, we have

$$P(x > 12) = 1 - [p(0) + p(1) + \cdots + p(12)]$$

$$= 1 - P(x \le 12) = 1 - \sum_{x=0}^{12} p(x)$$

Consulting Table II, we find the entry in row $k = 12$, column $p = .6$ to be .584. Thus,

$$P(x > 12) = 1 - .584 = .416$$

d. To find the probability that exactly 11 voters favor the candidate, recall that the entries in Table II are cumulative probabilities and use the relationship

$$P(x = 11) = [p(0) + p(1) + \cdots + p(10) + p(11)]$$
$$- [p(0) + p(1) + \cdots + p(9) + p(10)]$$
$$= P(x \le 11) - P(x \le 10)$$

Then

$$P(x = 11) = .404 - .245 = .159$$

The Space Shuttle *Challenger:* Catastrophe in Space

On January 28, 1986, at 11:39.13 AM (EST), while traveling at mach 1.92 at an altitude of 46,000 feet, the space shuttle *Challenger* was totally enveloped in an explosive burn that destroyed the shuttle and resulted in the deaths of all seven astronauts aboard. What happened? What was the cause of this catastrophe? This was the 25th shuttle mission. The preceding 24 missions had all been successful.

According to *Discover* (Apr. 1986), the report of the Presidential Commission assigned to investigate the accident concluded that the explosion was caused by the failure of the O-ring seal in the joint between the two lower segments of the right solid-fuel rocket booster. The seal is supposed to prevent superhot gases from leaking through the joint during the propellant burn of the booster rocket. The failure of the seal permitted a jet of white-hot gases to escape and to ignite the liquid fuel of the external fuel tank. The fuel-tank fireburst destroyed the *Challenger.*

What were the chances that this event would occur? In a report made one year prior to the catastrophe, the

CASE STUDY • 4.2 •

National Aeronautics and Space Administration (NASA) claimed that the probability of such a failure was about $\frac{1}{60,000}$, or about once in every 60,000 flights. But a risk-assessment study conducted for the Air Force at about the same time assessed the probability of shuttle catastrophe due to booster rocket "burn-through" to be $\frac{1}{35}$, or about once in every 35 missions.

Focus

a. Assuming NASA's failure-rate estimate was accurate, compute the probability that no disasters would have occurred during 25 shuttle missions.

b. Repeat part a, but use the Air Force's failure-rate estimate.

c. What conditions must exist for the probabilities, parts a and b, to be valid?

d. Given the events of January 28, 1986, which risk assessment—NASA's or the Air Force's—appears to be more appropriate? [*Hint:* Consider the complement of the events, parts a and b.]

e. The probability distribution for x is shown in Figure 4.9. Note that

$$\mu - 2\sigma = 12 - 2(2.2) = 7.6 \qquad \mu + 2\sigma = 12 + 2(2.2) = 16.4$$

The interval $\mu - 2\sigma$ to $\mu + 2\sigma$ is shown in Figure 4.9. The probability that x falls in the interval $\mu \pm 2\sigma$ is $P(x = 8, 9, 10, \ldots, 16) = P(x \le 16) - P(x \le 7) = .984 - .021 = .963$. Note that this probability is very close to the .95 given by the Empirical Rule.   ▲

FIGURE 4.9

The binomial probability distribution for x in Example 4.10: $n = 20$ and $p = .6$

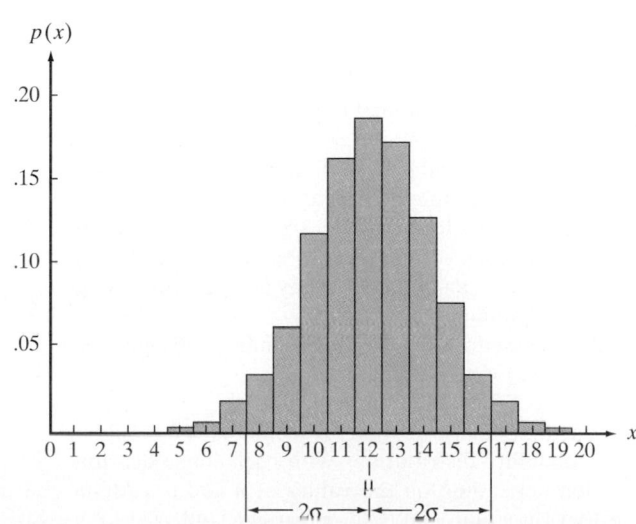

EXERCISES 4.30–4.49

Learning the Mechanics

4.30 Compute the following:

  a. $\dfrac{6!}{2!(6-2)!}$ **b.** $\dbinom{5}{2}$ **c.** $\dbinom{7}{0}$ **d.** $\dbinom{6}{6}$ **e.** $\dbinom{4}{3}$

4.31 Consider the following probability distribution:

$$p(x) = \binom{5}{x}(.7)^x(.3)^{5-x} \qquad (x = 0, 1, 2, \ldots, 5)$$

  a. Is x a discrete or a continuous random variable? Explain.
  b. What is the name of this probability distribution?
  c. Graph the probability distribution.
  d. Find the mean and standard deviation of x.
  e. Show the mean and the 2-standard-deviation interval on each side of the mean on the graph you drew in part **c.**

4.32 If x is a binomial random variable, compute $p(x)$ for each of the following cases:

  a. $n = 5, x = 1, p = .2$ **b.** $n = 4, x = 2, q = .4$
  c. $n = 3, x = 0, p = .7$ **d.** $n = 5, x = 3, p = .1$
  e. $n = 4, x = 2, q = .6$ **f.** $n = 3, x = 1, p = .9$

4.33 Suppose x is a binomial random variable with $n = 3$ and $p = .3$.

  a. Calculate the value of $p(x)$, $x = 0, 1, 2, 3$, using the formula for a binomial probability distribution.
  b. Using your answers to part **a**, give the probability distribution for x in tabular form.

4.34 If x is a binomial random variable, calculate μ, σ^2, and σ for each of the following:

  a. $n = 25, p = .5$ **b.** $n = 80, p = .2$
  c. $n = 100, p = .6$ **d.** $n = 70, p = .9$
  e. $n = 60, p = .8$ **f.** $n = 1{,}000, p = .04$

4.35 If x is a binomial random variable, use Table II in Appendix A to find the following probabilities:

  a. $P(x = 2)$ for $n = 10, p = .4$
  b. $P(x \le 5)$ for $n = 15, p = .6$
  c. $P(x > 1)$ for $n = 5, p = .1$
  d. $P(x < 10)$ for $n = 25, p = .7$
  e. $P(x \ge 10)$ for $n = 15, p = .9$
  f. $P(x = 2)$ for $n = 20, p = .2$

4.36 Suppose x is a binomial random variable with $n = 5$ and $p = .5$. Compute $p(x)$ for $x = 0, 1, 2, 3, 4$, and 5 using the following three methods:

  a. List the sample points (using S for a Success and F for a Failure on each trial) corresponding to each value of x, assign probabilities to each sample point, and obtain $p(x)$ by adding sample point probabilities.
  b. Use the formula for the binomial probability distribution to obtain $p(x)$.
  c. Use Table II to obtain $p(x)$.

4.37 The binomial probability distribution is a family of probability distributions with each single distribution depending on the values of n and p. Assume that x is a binomial random variable with $n = 4$.

  a. Determine a value of p such that the probability distribution of x is symmetric.
  b. Determine a value of p such that the probability distribution of x is skewed to the right.
  c. Determine a value of p such that the probability distribution of x is skewed to the left.
  d. Graph each of the binomial distributions you obtained in parts **a**, **b**, and **c**. Locate the mean for each distribution on its graph.
  e. In general, for what values of p will a binomial distribution be symmetric? Skewed to the right? Skewed to the left?

Applying the Concepts

4.38 Would most wives marry the same man again if given the chance? According to a poll of 608 married women conducted by *Ladies Home Journal* (June 1988), 80% would, in fact, marry their current husbands. Assume the women in the sample were randomly selected from among all married women in the United States. Does the number x in the sample who would marry their husbands again possess (approximately) a binomial probability distribution? Explain.

4.39 Zoologists have discovered that animals spend a great deal of time resting, although this rest time can have functional importance (e.g., predators lying in wait for their prey). One University of Vermont researcher estimated the percentage of time various species spend at rest, discounting time spent in deep sleep (*National Wildlife,* Aug.–Sept. 1993). For example, the probability that a female fence lizard will be resting at any given time is .97.

  a. In a random sample of 20 female fence lizards, what is the probability that at least 15 will be resting at any given time?
  b. In a random sample of 20 female fence lizards, what is the probability that fewer than 10 will be resting at any given time?
  c. In a random sample of 200 female fence lizards, would you expect to observe fewer than 190 at rest at any given time? Explain.

4.40 In Exercise 3.55, you learned that the probability that a male professional golfer makes a hole-in-one is $^1\!/_{2,780}$. Suppose 36 professional male golfers play the sixth hole during a round of golf. Let x be the number of the 36 who make a hole-in-one.

  a. Can x be reasonably treated as a binomial random variable? (Check the characteristics.)
  b. Calculate the probability that none of the 36 golfers makes a hole-in-one on the sixth hole.
  c. Calculate the probability that exactly four of the 36 golfers make a hole-in-one on the sixth hole—as actually happened during the 1989 U.S. Open.

4.41 A large southern university has determined from past records that the probability a student who registers for fall classes will have his or her schedule rejected (due to overfilled classrooms, clerical error, etc.) is .2.

a. Suppose 25,000 students register for fall classes, and x is the number of students whose schedules are rejected. Is x a binomial random variable? Explain.

b. Suppose a random sample of 20 students is selected from the total of 25,000, and x is the number of these students whose schedules are rejected. Is x a binomial random variable? Explain.

c. Suppose you sample the results of the first 1,000 students who register next fall and record x, the number of rejected registrations. Is x a binomial random variable? Explain.

d. For the random variables in parts **a**, **b**, and **c** that you identified as being binomial, find μ, σ^2, and σ.

4.42 Ataxia-telangiectasia (A-T) is a neurological disorder that weakens immune systems and causes premature aging. Most A-T patients die in their teens or early twenties. According to *Science News* (June 24, 1995), when both members of a couple carry the A-T gene, their children have a one in four chance of developing the disease.

a. Consider 15 couples in which both members of each couple carry the A-T gene. What is the probability that more than 10 of 15 couples have children that develop the neurological disorder?

b. Consider 10,000 couples in which both members of each couple carry the A-T gene. What is the probability that fewer than 3,000 will have children that develop the disease? Set up the solution, but don't perform the calculations. [*Note:* In Section 5.4, we discuss a procedure you can use to obtain an approximate answer to this question without having to perform the tedious calculations required by the binomial distribution.] Calculate the mean and standard deviation of the distribution, and give an "intuitive" approximation to this probability.

4.43 According to researchers at Johns Hopkins University School of Medicine, one in every three women has been a victim of domestic abuse (*Annals of Internal Medicine,* Nov. 1995). This probability was obtained from a survey of nearly 2,000 adult women residing in Baltimore, Maryland. Suppose we randomly sample 15 women and find that 4 have been abused.

a. What is the probability of observing 4 or more abused women in a sample of 15 if the proportion p of women who are victims of domestic abuse is really $p = \frac{1}{3}$.

b. Many experts on domestic violence believe that the proportion of women who are domestically abused is closer to $p = .10$. Calculate the probability of observing 4 or more abused women in a sample of 15 if $p = .10$.

c. Why might your answers to parts **a** and **b** lead you to believe that $p = \frac{1}{3}$?

4.44 Suppose you are a purchasing officer for a large company. You have purchased 5 million electrical switches and your supplier has guaranteed that the shipment will contain no more than .1% defectives. To check the shipment, you randomly sample 500 switches, test them, and find that four are defective. If the switches are as represented, calculate μ and σ for this sample of 500. Based on this evidence, do you think the supplier has complied with the guarantee? Explain. [*Hint:* Calculate μ and σ for this binomial random variable with $p = .001$ to see if a value of x as large as 4 is probable.]

4.45 For a space vehicle to gain reentry into the earth's atmosphere, a particular system must work properly. One component of the system operates successfully only 85% of the time. To increase the reliability of the system, four of these components are installed in such a way that the system will operate successfully if at least one component is working. What is the probability that the system will fail? Assume the components operate independently.

4.46 According to the Centers for Disease Control and Prevention (CDCP), 60% of all HIV-infected males in the United States are homosexual (*Social Science & Medicine,* July 1995). In a random sample of 20 HIV-infected males, you determine that 11 are homosexual.

a. Assuming that the probability p that an HIV-infected male is homosexual is, in fact, .6, find the probability of observing 11 or more homosexuals in the sample.

b. Does the probability, part **a**, tend to support or refute the CDCP's claim of $p = .6$? Explain.

4.47 An experiment is to be conducted to see whether an acclaimed psychic has extrasensory perception (ESP). Five different cards are to be shuffled, and one chosen at random. The psychic will then try to identify which card was drawn without seeing it. The experiment is to be repeated 20 times, and x, the number of correct decisions, will be recorded. (Assume that the 20 trials are independent.)

a. If the psychic is guessing—i.e., if the psychic does *not* possess ESP—what is the value of p, the probability of a correct decision on each trial?

b. If the psychic is guessing, what is the expected number of correct decisions in 20 trials?

c. If the psychic is guessing, what is the probability of six or more correct decisions in 20 trials?

d. Suppose that the psychic makes six correct decisions in 20 trials. Is there evidence to indicate that the psychic is *not* guessing and actually has ESP? Explain.

4.48 A literature professor decides to give a 20-question true–false quiz to determine who has read an assigned novel. She wants to choose the passing grade such that the probability of passing a student who guesses on every question is less than .05. What score should she set as the lowest passing grade?

4.49 Every quarter, the Food and Drug Administration (FDA) produces a report called the *Total Diet Study*. The FDA's report covers more than 200 food items, each of which is analyzed for potentially harmful chemical compounds. A recent *Total Diet Study* reported that no pesticides at all were found in 65% of the domestically produced food samples (*Consumer's Research,* June 1995). Consider a random sample of 800 food items analyzed for the presence of pesticides.

a. Compute μ and σ for the random variable *x*, the number of food items found that showed no trace of pesticide.

b. Based on a sample of 800 food items, is it likely you would observe less than half without any traces of pesticide? Explain.

4.5   THE POISSON RANDOM VARIABLE (OPTIONAL)

A type of probability distribution that is often useful in describing the number of events that will occur in a specific period of time or in a specific area or volume is the **Poisson distribution** (named after the 18th-century physicist and mathematician, Siméon Poisson). Typical examples of random variables for which the Poisson probability distribution provides a good model are

1. The number of traffic accidents per month at a busy intersection
2. The number of noticeable surface defects (scratches, dents, etc.) found by quality inspectors on a new automobile
3. The parts per million of some toxin found in the water or air emission from a manufacturing plant
4. The number of diseased trees per acre of a certain woodland
5. The number of death claims received per day by an insurance company
6. The number of unscheduled admissions per day to a hospital

Characteristics of a Poisson Random Variable

1. The experiment consists of counting the number of times a certain event occurs during a given unit of time or in a given area or volume (or weight, distance, or any other unit of measurement).
2. The probability that an event occurs in a given unit of time, area, or volume is the same for all the units.
3. The number of events that occur in one unit of time, area, or volume is independent of the number that occur in other units.
4. The mean (or expected) number of events in each unit is denoted by the Greek letter lambda, λ.

The characteristics of the Poisson random variable are usually difficult to verify for practical examples. The examples given satisfy them well enough that the Poisson distribution provides a good model in many instances. As with all probability models, the real test of the adequacy of the Poisson model is in whether it provides a reasonable approximation to reality—that is, whether empirical data support it.

The Poisson probability distribution also provides a good approximation to a binomial probability distribution with mean

$$\lambda = np$$

when n is large, p is small, and $np \leq 7$. To illustrate with a relatively small value of n, if $n = 25$ and $p = .05$, then the exact value of the binomial probability $p(2)$ is .231. The Poisson approximation is .224. The approximations are better for $n \geq 100$.

The probability distribution, mean, and variance for a Poisson random variable are shown in the next box.

Probability Distribution, Mean, and Variance for a Poisson Random Variable

$$p(x) = \frac{\lambda^x e^{-\lambda}}{x!} \qquad (x = 0, 1, 2, \dots)$$

$$\mu = \lambda \qquad \sigma^2 = \lambda$$

where

  λ = Mean number of events during given unit of time, area, volume, etc.

  $e = 2.71828\dots$

The calculation of Poisson probabilities is made easier by the use of Table III in Appendix A, which gives the cumulative probabilities $P(x \leq k)$ for various values of λ. The use of Table III is illustrated in Example 4.11.

EXAMPLE 4.11

Ecologists often use the number of reported sightings of a rare species of animal to estimate the remaining population size. For example, suppose the number, x, of reported sightings per week of blue whales is recorded. Assume that x has (approximately) a Poisson probability distribution. Furthermore, assume that the average number of weekly sightings is 2.6.

a. Find the mean and standard deviation of x, the number of blue whale sightings per week.
b. Use Table III to find the probability that fewer than two sightings are made during a given week.
c. Use Table III to find the probability that more than five sightings are made during a given week.
d. Use Table III to find the probability that exactly five sightings are made during a given week.

Solution

a. The mean and variance of a Poisson random variable are both equal to λ. Thus, for this example,

$$\mu = \lambda = 2.6 \qquad \sigma^2 = \lambda = 2.6$$

Then the standard deviation of x is

$$\sigma = \sqrt{2.6} = 1.61$$

Remember that the mean measures the central tendency of the distribution and does not necessarily equal a possible value of x. In this example, the mean is 2.6 sightings, and although there cannot be 2.6 sightings during a given week, the average number of weekly sightings is 2.6. Similarly, the standard deviation of 1.61 measures the variability of the number of sightings per week. Perhaps a more helpful measure is the interval $\mu \pm 2\sigma$, which in this case stretches from $-.62$ to 5.82. We expect the number of sightings to fall in this interval most of the time—with at least 75% relative frequency (according to Chebyshev's

FIGURE 4.10

Probability distribution for number of blue whale sightings

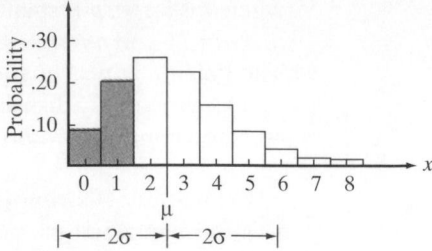

Rule) and probably with more than 90% relative frequency (the Empirical Rule). The mean and the 2-standard-deviation interval around it are shown in Figure 4.10.

b. A partial reproduction of Table III is shown in Table 4.4. The rows of the table correspond to different values of λ, and the columns correspond to different values of the Poisson random variable x. The entries in the table are cumulative probabilities (much like the binomial probabilities in Table II). To find the probability that fewer than two sightings are made during a given week, we first note that

$$P(x < 2) = P(x \le 1)$$

This probability is a cumulative probability and therefore is the entry in Table III in the row corresponding to λ = 2.6 and the column corresponding to $x = 1$. The entry is .267, shown shaded in Table 4.4. This probability corresponds to the shaded area in Figure 4.10 and may be interpreted as meaning that there is a 26.7% chance that fewer than two sightings will be made during a given week.

TABLE 4.4   Reproduction of Part of Table III in Appendix A

λ \ x	0	1	2	3	4	5	6	7	8	9
2.2	.111	.355	.623	.819	.928	.975	.993	.998	1.000	1.000
2.4	.091	.308	.570	.779	.904	.964	.988	.997	.999	1.000
2.6	.074	.267	.518	.736	.877	.951	.983	.995	.999	1.000
2.8	.061	.231	.469	.692	.848	.935	.976	.992	.998	.999
3.0	.050	.199	.423	.647	.815	.916	.966	.988	.996	.999
3.2	.041	.171	.380	.603	.781	.895	.955	.983	.994	.998
3.4	.033	.147	.340	.558	.744	.871	.942	.977	.992	.997
3.6	.027	.126	.303	.515	.706	.844	.927	.969	.988	.996
3.8	.022	.107	.269	.473	.668	.816	.909	.960	.984	.994
4.0	.018	.092	.238	.433	.629	.785	.889	.949	.979	.992
4.2	.015	.078	.210	.395	.590	.753	.867	.936	.972	.989
4.4	.012	.066	.185	.359	.551	.720	.844	.921	.964	.985
4.6	.010	.056	.163	.326	.513	.686	.818	.905	.955	.980
4.8	.008	.048	.143	.294	.476	.651	.791	.887	.944	.975
5.0	.007	.040	.125	.265	.440	.616	.762	.867	.932	.968
5.2	.006	.034	.109	.238	.406	.581	.732	.845	.918	.960
5.4	.005	.029	.095	.213	.373	.546	.702	.822	.903	.951
5.6	.004	.024	.082	.191	.342	.512	.670	.797	.886	.941
5.8	.003	.021	.072	.170	.313	.478	.638	.771	.867	.929
6.0	.002	.017	.062	.151	.285	.446	.606	.744	.847	.916

c. To find the probability that more than five sightings are made during a given week, we consider the complementary event

$$P(x > 5) = 1 - P(x \le 5) = 1 - .951 = .049$$

where .951 is the entry in Table III corresponding to $\lambda = 2.6$ and $x = 5$ (see Table 4.4). Note from Figure 4.10 that this is the area in the interval $\mu \pm 2\sigma$, or $-.62$ to 5.82. Then the number of sightings should exceed 5—or, equivalently, should be more than 2 standard deviations from the mean—during only about 4.9% of all weeks. Note that this percentage agrees remarkably well with that given by the Empirical Rule for mound-shaped distributions, which tells us to expect approximately 5% of the measurements (values of the random variable) to lie farther than 2 standard deviations from the mean.

d. To use Table III to find the probability that *exactly* five sightings are made during a given week, we must write the probability as the difference between two cumulative probabilities:

$$P(x = 5) = P(x \le 5) - P(x \le 4) = .951 - .877 = .074 \qquad \blacktriangle$$

Note that the probabilities in Table III are all rounded to three decimal places. Thus, although in theory a Poisson random variable can assume infinitely large values, the values of x in Table III are extended only until the cumulative probability is 1.000. This does not mean that x *cannot* assume larger values, but only that the likelihood is less than .001 (in fact, less than .0005) that it will do so.

Finally, you may need to calculate Poisson probabilities for values of λ not found in Table III. You may be able to obtain an adequate approximation by interpolation, but if not, consult more extensive tables for the Poisson distribution.

EXERCISES 4.50–4.63

Learning the Mechanics

4.50 Consider the probability distribution shown here:

$$p(x) = \frac{3^x e^{-3}}{x!} \qquad (x = 0, 1, 2, \ldots)$$

   a. Is x a discrete or continuous random variable? Explain.
   b. What is the name of this probability distribution?
   c. Graph the probability distribution.
   d. Find the mean and standard deviation of x.
   e. Find the mean and standard deviation of the probability distribution.

4.51 Given that x is a random variable for which a Poisson probability distribution provides a good approximation, use Table III to compute the following:
   a. $P(x \le 2)$ when $\lambda = 1$ **b.** $P(x \le 2)$ when $\lambda = 2$
   c. $P(x \le 2)$ when $\lambda = 3$
   d. What happens to the probability of the event $\{x \le 2\}$ as λ increases from 1 to 3? Is this intuitively reasonable?

4.52 Assume that x is a random variable having a Poisson probability distribution with a mean of 1.5. Use Table III to find the following probabilities:

   a. $P(x \le 3)$ **b.** $P(x \ge 3)$ **c.** $P(x = 3)$
   d. $P(x = 0)$ **e.** $P(x > 0)$ **f.** $P(x > 6)$

4.53 Suppose x is a random variable for which a Poisson probability distribution with $\lambda = 1$ provides a good characterization.
   a. Graph $p(x)$ for $x = 0, 1, 2, \ldots, 9$.
   b. Find μ and σ for x, and locate μ and the interval $\mu \pm 2\sigma$ on the graph.
   c. What is the probability that x will fall within the interval $\mu \pm 2\sigma$?

4.54 Suppose x is a random variable for which a Poisson probability distribution with $\lambda = 3$ provides a good characterization.
   a. Graph $p(x)$ for $x = 0, 1, 2, \ldots, 9$.
   b. Find μ and σ for x, and locate μ and the interval $\mu \pm 2\sigma$ on the graph.
   c. What is the probability that x will fall within the interval $\mu \pm 2\sigma$?

4.55 As mentioned in Section 4.5, when n is large, p is small, and $np \le 7$, the Poisson probability distribution provides a good approximation to the binomial probability distribution. Since we provide exact binomial probabilities (Table II in Appendix A) for relatively small values of *n*, you can investigate the adequacy of the approximation for *n* = 25. Use Table II to find $p(0)$, $p(1)$, and $p(2)$ for *n* = 25 and *p* = .05. Calculate the corresponding Poisson approximations using $\lambda = \mu = np$. [*Note:* These approximations are reasonably good for *n* as small as 25, but to use the approximation in a practical situation we would prefer to have $n \geq 100$.]

Applying the Concepts

4.56 The mean number of patients admitted per day to the emergency room of a small hospital is 2.5. If, on a given day, there are only four beds available for new patients, what is the probability the hospital will not have enough beds to accommodate its newly admitted patients?

4.57 The Environmental Protection Agency (EPA) requires manufacturers of vinyl chloride and similar compounds to limit the amount of these chemicals in plant air emissions to no more than 10 parts per million. Suppose the mean emission of vinyl chloride for a particular plant is 4 parts per million. Assume that the number of parts per million of vinyl chloride in air samples, *x*, follows a Poisson probability distribution.

  a. What is the standard deviation of *x* for the plant?

  b. Is it likely that a sample of air from the plant would yield a value of *x* that would exceed the EPA limit? Explain.

  c. Discuss conditions that would make the Poisson assumption plausible.

4.58 A can company reports that the number of breakdowns per 8-hour shift on its machine-operated assembly line follows a Poisson distribution with a mean of 1.5.

  a. What is the probability of exactly two breakdowns on the midnight shift?

  b. What is the probability of fewer than two breakdowns on the afternoon shift?

  c. What is the probability that more than two breakdowns occur on the midnight shift?

  d. What is the probability of no breakdowns during three consecutive 8-hour shifts? (Assume that the machine operates independently across shifts.)

4.59 A certain automatic car wash takes exactly 5 minutes to wash a car. On the average, 10 cars per hour arrive at the car wash. Suppose that, 30 minutes before closing time, five cars are in line. If the car wash is in continuous use until closing time, what is the probability that no one will be in line at closing time?

4.60 The safety supervisor at a large manufacturing plant believes the expected number of industrial accidents per month is 3.4.

  a. What is the probability of exactly two accidents occurring next month?

  b. What is the probability of three or more accidents occurring next month?

  c. What assumptions do you need to make to solve this problem using the methodology of this chapter?

4.61 Flying approximately 26 billion passenger-miles per month, U.S. airlines average about 11.8 fatalities per month (*Statistical Abstract of the United States*). Assume the probability distribution for *x*, the number of fatalities per month, can be approximated by a Poisson probability distribution.

  a. What is the probability that no fatalities will occur during any given month? [*Hint:* Either use Table III of Appendix A and interpolate to approximate the probability, or use a calculator or computer to calculate the probability exactly.]

  b. Find $E(x)$ and the standard deviation of *x*.

  c. Use your answers to part **b** to describe the probability that as many as 20 fatalities will occur in any given month.

  d. Discuss conditions that would make the Poisson assumption plausible.

4.62 The number *x* of people who arrive at a cashier's counter in a bank during a specified period of time often exhibits (approximately) a Poisson probability distribution. If we know the mean arrival rate λ, the Poisson probability distribution can be used to aid in the design of the customer service facility. Suppose you estimate that the mean number of arrivals per minute for cashier service at a bank is one person per minute.

  a. What is the probability that in a given minute the number of arrivals will equal three or more?

  b. Can you tell the bank manager that the number of arrivals will rarely exceed two per minute?

4.63 In many cities, neighborhood Crime Watch groups are formed in an attempt to reduce the amount of criminal activity. Suppose one neighborhood that has experienced an average of 10 crimes per year organizes such a group. During the first year following the creation of the group, three crimes are committed in the neighborhood.

  a. Use the Poisson distribution to calculate the probability that three or fewer crimes are committed in a year assuming the average number is still 10 crimes per year.

  b. Do you think this event provides some evidence that the Crime Watch group has been effective in this neighborhood?

4.6   THE HYPERGEOMETRIC RANDOM VARIABLE (OPTIONAL)

The **hypergeometric probability distribution** provides a realistic model for some types of enumerative (countable) data. The characteristics of the hypergeometric distribution are listed in the box.

> **Characteristics of a Hypergeometric Random Variable**
>
> 1. The experiment consists of randomly drawing n elements without replacement from a set of N elements, r of which are S's (for Success) and $(N - r)$ of which are F's (for Failure).
> 2. The hypergeometric random variable x is the number of S's in the draw of n elements.

Note that both the hypergeometric and binomial characteristics stipulate that each draw, or trial, results in one of two outcomes. The basic difference between these random variables is that the hypergeometric trials are dependent, while the binomial trials are independent. The draws are dependent because the probability of drawing an S (or an F) is dependent on what occurred on preceding draws.

To illustrate the dependence between trials, we note that the probability of drawing an S on the first draw is r/N. Then, the probability of drawing an S on the second draw depends on the outcome of the first. It will be either $(r - 1)/(N - 1)$, or $r/(N - 1)$, depending on whether the first draw was an S or an F. Consequently, the results of the draws represent dependent events.

For example, suppose we define x as the number of women hired in a random selection of three applicants from a total of six men and four women. This random variable satisfies the characteristics of a hypergeometric random variable with $N = 10$ and $n = 3$. The possible outcomes on each trial are either selection of a female (S) or selection of a male (F). Another example of a hypergeometric random variable is the number, x, of defective large-screen television picture tubes in a random selection of $n = 4$ from a shipment of $N = 8$ tubes. And, as a third example, suppose $n = 5$ stocks are randomly selected from a list of $N = 15$ stocks. Then, the number x of the five selected companies that pay regular dividends to stockholders is a hypergeometric random variable.

> **Probability Distribution, Mean, and Variance of the Hypergeometric Random Variable**
>
> $$p(x) = \frac{\binom{r}{x}\binom{N - r}{n - x}}{\binom{N}{n}} \qquad [x = \text{Maximum}[0, n - (N - r)], \ldots, \text{Minimum}(r, n)]$$
>
> $$\mu = \frac{nr}{N} \qquad \sigma^2 = \frac{r(N - r)n(N - n)}{N^2(N - 1)}$$
>
> where
>
> N = Total number of elements
> r = Number of S's in the N elements
> n = Number of elements drawn
> x = Number of S's drawn in the n elements

EXAMPLE 4.12

Suppose, as we mentioned earlier, an employer randomly selects three new employees from a total of ten applicants, six men and four women. Let x be the number of women who are hired.

a. Find the mean and standard deviation of x.
b. Find the probability that no women are hired.

Solution

a. Since x is a hypergeometric random variable with $N = 10$, $n = 3$, and $r = 4$, the mean and variance are

$$\mu = \frac{nr}{N} = \frac{(3)(4)}{10} = 1.2$$

$$\sigma^2 = \frac{r(N - r)n(N - n)}{N^2(N - 1)} = \frac{4(10 - 4)3(10 - 3)}{(10)^2(10 - 1)}$$

$$= \frac{(4)(6)(3)(7)}{(100)(9)} = .56$$

The standard deviation is

$$\sigma = \sqrt{.56} = .75$$

b. The probability that no women are hired by the employer, assuming the selection is truly random, is

$$P(x = 0) = p(0) = \frac{\binom{4}{0}\binom{10 - 4}{3 - 0}}{\binom{10}{3}}$$

$$= \frac{\dfrac{4!}{0!(4 - 0)!} \dfrac{6!}{3!(6 - 3)!}}{\dfrac{10!}{3!(10 - 3)!}} = \frac{(1)(20)}{120} = \frac{1}{6}$$

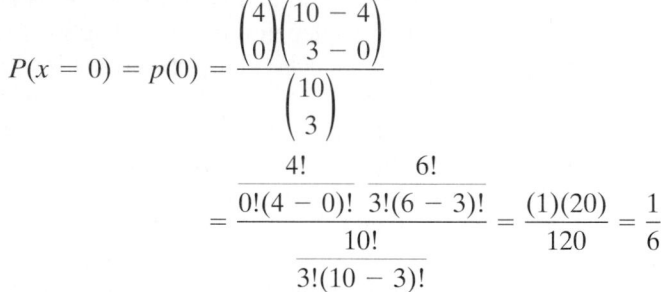

FIGURE 4.11

Probability distribution for x in Example 4.12

The entire probability distribution for x is shown in Figure 4.11. The mean $\mu = 1.2$ and the interval $\mu \pm 2\sigma = (-.3, 2.7)$ are indicated. You can see that if this random variable were to be observed over and over again a large number of times, most of the values of x would fall within the interval $\mu \pm 2\sigma$. ▲

EXERCISES 4.64–4.77

Learning the Mechanics

4.64 Explain the difference between sampling with replacement and sampling without replacement.

4.65 How do binomial and hypergeometric random variables differ? In what respects are they similar?

4.66 Given that x is a hypergeometric random variable with $N = 8$, $n = 3$, and $r = 5$, compute the following:
a. $P(x = 1)$ b. $P(x = 0)$ c. $P(x = 3)$ d. $P(x \geq 4)$

4.67 Given that x is a hypergeometric random variable, compute $p(x)$ for each of the following cases:
a. $N = 5$, $n = 3$, $r = 3$, $x = 1$
b. $N = 9$, $n = 5$, $r = 3$, $x = 3$
c. $N = 4$, $n = 2$, $r = 2$, $x = 2$
d. $N = 4$, $n = 2$, $r = 2$, $x = 0$

4.68 Given that x is a hypergeometric random variable with $N = 10$, $n = 5$, and $r = 7$:
a. Display the probability distribution for x in tabular form.
b. Compute the mean and variance of x.
c. Graph $p(x)$ and locate μ and the interval $\mu \pm 2\sigma$ on the graph.
d. What is the probability that x will fall within the interval $\mu \pm 2\sigma$?

4.69 Given that x is a hypergeometric random variable with $N = 12$, $n = 8$, and $r = 6$:
a. Display the probability distribution for x in tabular form.
b. Compute μ and σ for x.

Probability in a Reverse Cocaine Sting

CASE STUDY · 4.3 ·

he American Statistician (May 1991) described an interesting application of a discrete probability distribution in a case involving illegal drugs. It all started with a "bust" in a mid-sized Florida city. During the bust, police seized 496 foil packets of a white, powdery substance, presumably cocaine. However, it is not a crime to buy or sell nonnarcotic cocaine lookalikes (e.g., inert powders), so detectives had to prove that the packets contained genuine cocaine in order to convict their suspects of drug trafficking. When the police laboratory randomly selected and chemically tested four of the packets, all four tested positive for cocaine. This finding led to the conviction of the traffickers.

Focus

a. Of the 496 packets confiscated, suppose 331 contained genuine cocaine and 165 contained an inert (legal) powder. Find the probability that four randomly selected packets will test positive for cocaine.

After the conviction, the police decided to use the 492 remaining foil packets (i.e., those not tested) in reverse sting operations. Two of the 492 packets were randomly selected and sold by undercover officers to a buyer. Between the sale and the arrest, however, the buyer disposed of the evidence. The key question is: Beyond a reasonable doubt, did the defendant really purchase cocaine?

In court, the defendant's attorney argued that his client should not be convicted because the police could not prove that the missing foil packets contained cocaine. The police contended, however, that since four of the original 496 packets tested positive for cocaine, the two packets sold in the reverse sting were also highly likely to contain cocaine.

Focus

b. Given that four of the original packets tested positive for cocaine, what is the probability that two packets sold in the reverse sting did not contain cocaine? Assume the information provided in part a is correct.

c. Use the results, parts a and b, to compute the probability that of the original 496 foil packets, the first four selected (at random) would test positive for cocaine and the next two selected (at random) would test negative.

A statistician hired as an expert witness, Professor Jonathan Shuster of the University of Florida, showed that the probability of the event described in part c is maximized when 331 of the 496 packets contain cocaine and 165 do not. Consequently, the probability, part c, represents the best possible case for the defendant. Nevertheless, the defense was satisfied that the probability of part c was large enough to establish reasonable doubt in a criminal trial.

Focus

d. Recalculate the probabilities, parts a–c, assuming that 400 of the original 496 packets contained cocaine. Comment on the reasonable-doubt claim using this probability.

To the surprise of the defense, the prosecution revealed that the remaining 490 packets had not been used in any other reverse sting operations and offered to test a sample of them. On the advice of the statistician, the defense requested that an additional 20 packets be tested. All 20 tested positive for cocaine! As a consequence, the defendant was ultimately convicted. The case represented Florida's first cocaine-possession conviction without the actual physical evidence.

Focus

e. Assuming the best possible case for the defendant (i.e., 331 of the original 496 packets contain cocaine), find the probability that the first four test positive, the next two test negative, and the next 20 test positive. [*Hint:* $P(ABC) = P(C|AB) \cdot P(B|A) \cdot P(A)$.] Comment on the reasonable-doubt claim.

---

c. Graph $p(x)$ and locate μ and the interval $\mu \pm 2\sigma$ on the graph.

d. What is the probability that x will fall within the interval $\mu \pm 2\sigma$?

4.70 Use the results of Exercise 4.69 to find the following probabilities:

    a. $P(x = 1)$ **b.** $P(x = 4)$ **c.** $P(x \leq 4)$

    d. $P(x \geq 5)$ **e.** $P(x < 3)$ **f.** $P(x \geq 8)$

4.71 Suppose you plan to sample 10 items from a population of 100 items and would like to determine the probability of observing 4 defective items in the sample. Which probability distribution should you use to compute this probability under the following conditions? Justify your answers.

    a. The sample is drawn without replacement.

    b. The sample is drawn with replacement.

Applying the Concepts

4.72 *Hotspots* are species-rich geographical areas. A *Nature* (Sept. 1993) study estimated the probability that a bird species in Great Britain will inhabit a butterfly hotspot at .70. Consider a random sample of 4 British bird species selected from a total of 10 tagged species. Assume that 7 of the 10 tagged species inhabit a butterfly hotspot.

a. What is the probability that exactly half of the 4 bird species sampled inhabit a butterfly hotspot?

b. What is the probability that at least 1 of the 4 bird species sampled inhabit a butterfly hotspot?

4.73 Suppose you are purchasing cases of wine (twelve bottles per case) and that, periodically, you select a test case to determine the adequacy of the bottles' seals. To do this, you randomly select and test three bottles in the case. If a case contains one spoiled bottle of wine, what is the probability that this bottle will turn up in your sample?

4.74 Imagine you are purchasing small lots of a manufactured product. If it is very costly to test a single item, it may be desirable to test a sample of items from the lot instead of testing every item in the lot. Such a sampling plan would be based on a hypergeometric probability distribution. For example, suppose each lot contains ten items. You decide to sample four items per lot and reject the lot if you observe one or more defectives. If the lot contains one defective item, what is the probability that you will accept the lot? What is the probability that you will accept the lot if it contains two defective items? Three? Four?

4.75 The cornrake is a European bird species in danger of worldwide extinction. A census revealed that 12 cornrakes inhabit mainland Scotland (*Journal of Applied Ecology,* 1993). Suppose that two of these Scottish cornrakes are captured for mating purposes. Let *x* be the number of these captured cornrakes that are capable of mating. If exactly four of the original 12 cornrakes inhabiting Scotland are infertile and hence unable to mate, find the probability distribution for *x*.

4.76 A nursery advertises that it has ten elm trees for sale. Unknown to the nursery, three of the trees have already been infected with Dutch elm disease and will die within a year. If a buyer purchases two trees, what is the probability that both trees will be healthy? What is the probability that at least one of the trees is infected?

4.77 A curious event was described in the *Minneapolis Star and Tribune.* The Minneapolis Community Development Agency (MCDA) makes home improvement grants each year to homeowners in depressed city neighborhoods. Of the $708,000 granted one year, $233,000 was awarded by the city council using a "random selection" of 140 homeowners' applications from among a total of 743 applications: 601 from the north side and 142 from the south side of Minneapolis. Oddly, all 140 grants awarded were from the north side—clearly a highly improbable outcome if, in fact, the 140 winners were randomly selected from among the 743 applicants.

a. Suppose the 140 winning applications were randomly selected from among the total of 743, and let *x* equal the number in the sample from the north side. Find the mean and standard deviation of *x*.

b. Use the results of part **a** to support a contention that the grant winners were not randomly selected.

QUICK REVIEW

Key Terms

Note: Starred (*) items refer to optional sections in this chapter.

Binomial experiment 170	Expected value 163
Binomial random variable 170	Hypergeometric random variable* 185
Continuous random variable 158	Poisson random variable* 180
Cumulative binomial probabilities 175	Probability distribution 160
Discrete random variable 157	Random variable 156

Key Formulas

	Probability Distribution	Mean (μ)	Variance (σ^2)
General discrete random variable	$p(x)$	$\displaystyle\sum_{\text{all } x} xp(x)$	$\displaystyle\sum_{\text{all } x}(x-\mu)^2 p(x)$
Binomial random variable	$\binom{n}{x}p^x q^{n-x}$	np	npq
*Poisson random variable	$\dfrac{\lambda^x e^{-\lambda}}{x!}$	λ	λ
*Hypergeometric random variable	$\dfrac{\binom{r}{x}\binom{N-r}{n-x}}{\binom{N}{n}}$	$\dfrac{nr}{N}$	$\dfrac{r(N-r)n(N-n)}{N^2(N-1)}$

LANGUAGE LAB

Symbol	Pronunciation	Description
$p(x)$		Probability distribution of the random variable x
S		The outcome of a binomial trial denoted a "success"
F		The outcome of a binomial trial denoted a "failure"
p		The probability of success (S) in a binomial trial
q		The probability of failure (F) in a binomial trial, where $q = 1 - p$
λ	lambda	The mean (or expected) number of events for a Poisson random variable
e		A constant used in the Poisson probability distribution, where $e = 2.71828 \ldots$

SUPPLEMENTARY EXERCISES 4.78–4.98

Note: Starred (*) exercises refer to optional sections in this chapter.

Learning the Mechanics

4.78 Which of the following describe discrete random variables and which describe continuous random variables?

  a. The length of time that an exercise physiologist's program takes to elevate her client's heart rate to 140 beats per minute

  b. The number of crimes committed on a college campus per year

  c. The number of square feet of vacant office space in a large city

  d. The number of voters who favor a new tax proposal

*4.79 Identify the type of random variable—binomial, Poisson, or hypergeometric—described by each of the following probability distributions:

  a. $p(x) = \dfrac{.5^x e^{-.5}}{x!}$ $(x = 0, 1, 2, \ldots)$

  b. $p(x) = \binom{6}{x}(.2)^x(.8)^{6-x}$ $(x = 0, 1, 2, \ldots, 6)$

  c. $p(x) = \dfrac{10!}{x!(10-x)!}(.9)^x(.1)^{10-x}$
  $(x = 0, 1, 2, \ldots, 10)$

4.80 For each of the following examples, decide whether x is a binomial random variable and explain your decision:

  a. A manufacturer of computer chips randomly selects 100 chips from each hour's production in order to estimate the proportion of defectives. Let x represent the number of defectives in the 100 sampled chips.

  b. Of five applicants for a job, two will be selected. Although all applicants appear to be equally qualified, only three have the ability to fulfill the expectations of the company. Suppose that the two selections are made at random from the five applicants, and let x be the number of qualified applicants selected.

c. A software developer establishes a support hot-line for customers to call in with questions about the use of the software. Let x represent the number of calls received on the support hotline during a specified workday.

d. Florida is one of a minority of states with no state income tax. A poll of 1,000 registered voters is conducted to determine how many would favor a state income tax in light of the state's current fiscal condition. Let x be the number in the sample who would favor the tax.

4.81 Suppose x is a binomial random variable. Find $p(x)$ for each of the following combinations of x, n, and p:
a. $x = 1, n = 3, p = .1$ b. $x = 4, n = 20, p = .3$
c. $x = 0, n = 2, p = .4$ d. $x = 4, n = 5, p = .5$
e. $n = 15, x = 12, p = .9$ f. $n = 10, x = 8, p = .6$

*4.82 Given that x is a hypergeometric random variable, compute $p(x)$ for each of the following cases:
a. $N = 8, n = 5, r = 3, x = 2$
b. $N = 6, n = 2, r = 2, x = 2$
c. $N = 5, n = 4, r = 4, x = 3$

4.83 Consider the discrete probability distribution shown here.

x	10	12	18	20
$p(x)$	.2	.3	.1	.4

a. Calculate μ, σ^2, and σ. b. What is $P(x < 15)$?
c. Calculate $\mu \pm 2\sigma$.
d. What is the probability that x is in the interval $\mu \pm 2\sigma$?

4.84 Suppose x is a binomial random variable with $n = 20$ and $p = .7$.
a. Find $P(x = 14)$. b. Find $P(x \leq 12)$.
c. Find $P(x > 12)$. d. Find $P(9 \leq x \leq 18)$.
e. Find $P(8 < x < 18)$. f. Find μ, σ^2, and σ.
g. What is the probability that x is in the interval $\mu \pm 2\sigma$?

*4.85 Suppose x is a Poisson random variable. Compute $p(x)$ for each of the following cases:
a. $\lambda = 2, x = 3$ b. $\lambda = 1, x = 4$ c. $\lambda = .5, x = 2$

Applying the Concepts

4.86 A recent study in *The International Journal of Sports Psychology* (Jan.–Mar. 1990) investigated the role that handedness plays in the sport of golf. *Handedness* refers to whether a player is right- or left-handed. The researchers report that approximately 10.4% of the population of golfers are left-handed.

a. Suppose a random sample of 20 professional golfers is selected. Find the probability that none of the 20 professional golfers is left-handed.

b. The article reports that none of the top 100 professional golfers on the American tour in 1985 was left-handed. Assuming $p = .104$, find the probability of this occurrence.

c. Based on the probability found in part **b**, can you reach any conclusion about the proportion of all professional golfers who are left-handed? Be sure to state any assumptions you make in reaching your conclusion.

4.87 In Exercise 3.105 we noted that the drug cyclosporine appears to retard the body's immune response, slowing the rejection of transplanted organs. The drug's developer, Sandoz, reports that kidney transplant patients who receive the drug have an 80% chance of living through the first year. Suppose four kidney transplant patients are given cyclosporine, and let x equal the number living through the first year.
a. Calculate $p(x)$ for $x = 0, 1, 2, 3, 4$.
b. Graph $p(x)$. c. Find $P(x < 2)$.

4.88 Many minor operations at a hospital can be performed the same day the patient is admitted. A hospital serving a large metropolitan area has found that 20% of newly admitted patients needing an operation are scheduled for same-day surgery. Suppose that 10 patients are randomly selected from those admitted to the hospital for surgery over the past year. Let x, the number in the sample of 10 who receive same-day surgery, have a binomial probability distribution.

a. What is the probability that exactly five of these patients have same-day surgery?

b. What is the probability that at most one has same-day surgery?

c. If the 10 patients are selected from the admissions on a single given day, is it reasonable to expect x to possess the characteristics of a binomial random variable? That is, is this a binomial experiment? Explain.

4.89 Anticipating a substantial growth in sales over the next 5 years, a printing company is planning today for the warehouse space it will need 5 years hence. It obviously cannot be certain exactly how many square feet of storage space, x, it will need in 5 years, but the company can project its needs by using a probability distribution such as the following:

x	10,000	15,000	20,000	25,000	30,000	35,000
$p(x)$	.05	.15	.35	.25	.15	.05

What is the expected number of square feet of storage space the printing company will need in 5 years?

4.90 The probability that a person responds to a mailed questionnaire is .4.

a. What is the probability that of 20 questionnaires, more than 12 will be returned?

b. How many questionnaires should be mailed if you want to be reasonably certain that at least 100 will be returned?

4.91 The *Journal of Applied Physics* (Vol. 71, 1986) reported the results of an extensive survey conducted to determine the extent of whistle blowing among federal employees and to study the factors associated with retaliation against whistle blowers. *Whistle blowing* refers to an employee's reporting of wrongdoing by co-workers. Among other things, the survey found that about 5% of employees contacted had reported wrongdoing during the past 12 months. Assume that a sample of 25 employees in one agency are contacted, and let x be the number who have observed and reported wrongdoing in the last 12 months. Assume that the probability of whistle blowing is .05 for any federal employee over the past 12 months.

a. Find the mean and standard deviation of x. Can x be equal to its expected value? Explain.

b. Write the event that at least five of the employees are whistle blowers in terms of x. Find the probability of the event.

c. If five of the 25 contacted have been whistle blowers over the past 12 months, what would you conclude about the applicability of the 5% assumption to this agency? Use your answer to part **b** to justify your conclusion.

*4.92 By mistake, a manufacturer of CD mini-rack systems includes three defective systems in a shipment of ten going out to a small retailer. The retailer has decided to accept the shipment of CD systems only if none are found to be defective. Upon receipt of the shipment, the retailer examines only five of the CD systems. What is the probability that the shipment will be rejected? If the retailer inspects six of the CD systems, what is the probability the shipment will be accepted?

4.93 The state highway patrol has determined that one out of every six calls for help originating from roadside call boxes is a hoax. Five calls for help have come in and five tow trucks have been dispatched.

a. What is the probability that none of the calls was a hoax?

b. What is the probability that only three of the callers really needed assistance?

c. What assumptions do you have to make in order to solve this problem?

d. If the highway patrol answers 10,000 calls for help next year and each call costs the patrol about $30 (labor, gas, etc.), approximately how much money will be wasted answering false alarms?

4.94 The owner of construction company A makes bids on jobs so that if awarded the job, company A will make a $10,000 profit. The owner of construction company B makes bids on jobs so that if awarded the job, company B will make a $15,000 profit. Each company describes the probability distribution of the number of jobs the company is awarded per year as shown in the table.

Company A		Company B	
2	.05	2	.15
3	.15	3	.30
4	.20	4	.30
5	.35	5	.20
6	.25	6	.05

a. Find the expected number of jobs each will be awarded in a year.

b. What is the expected profit for each company?

c. Find the variance and standard deviation of the distribution of number of jobs awarded per year for each company.

d. Graph $p(x)$ for both companies A and B. For each company, what proportion of the time will x fall in the interval $\mu \pm 2\sigma$?

4.95 The efficacy of insecticides is often measured by the dose necessary to kill a certain percentage of insects. Suppose a certain dose of a new insecticide is supposed to kill 80% of the insects that receive it. To test the claim, 25 insects are exposed to the insecticide.

a. If the insecticide really kills 80% of the exposed insects, what is the probability that fewer than 15 die?

b. If you observed such a result, what would you conclude about the new insecticide? Explain your logic.

4.96 A physical fitness specialist claims that the probability is greater than .5 that an average adult male can improve his physical condition by spending 5 minutes per day on a certain exercise program. To test the claim, 15 randomly selected adult males follow the program for a specified amount of time. Maximum oxygen uptake is measured before and after the program for each male and serves as the criterion for assessing physical condition.

a. If the program is not really beneficial (i.e., the probability of improvement is only .5), what is the probability that 11 or more of the 15 men have improved maximum oxygen uptake?

b. Suppose 11 or more showed increased maximum oxygen uptake. Assuming that the specialist's exercise program is ineffective, would you regard $x \geq 11$ as a rare event, or would you conclude that, in actuality, the probability of improvement exceeds $p = \frac{1}{2}$ and that the program is effective?

*4.97 An emergency rescue vehicle is used an average of 1.3 times daily.
  a. What is the probability that the vehicle will be used exactly twice tomorrow?
  b. What is the probability that it will be used more than twice?
  c. Exactly three times?
*4.98 Large bakeries typically have fleets of delivery trucks. One such bakery determined that the ex-pected number of delivery truck breakdowns per day is 1.5. Assume that the number of breakdowns is independent from day to day.
  a. What is the probability that exactly two break-downs will occur today and exactly three to-morrow?
  b. Fewer than two today and more than two to-morrow?

STUDENT PROJECTS

Consider the following random variables:

1. The number x of people who recover from a certain disease out of a sample of five patients
2. The number x of voters in a sample of five who favor a new method of tax reform
3. The number x of hits a baseball player gets in five offi-cial times at bat

In each case, x is a binomial random variable (or approx-imately so) with $n = 5$ trials. Assume that in each case the probability of Success is .3. (What is a Success in each of the examples?) If this were true, the probability distribu-tion for x would be the same for each of the three exam-ples. To obtain a relative frequency histogram for x, conduct the following experiment. Place 10 poker chips (pennies, marbles, or any 10 *identical* items) in a bowl and mark three of the 10 Success—the remaining seven will represent Failure. Randomly select a chip from the 10, observing whether it was a Success or Failure. Then re-turn the chip and randomly select a second chip from the 10 available chips, and record this outcome. Repeat this process until a total of five trials has been conducted. Count the number x of Successes observed in the five tri-als. Repeat the entire process 100 times to obtain 100 ob-served values of x.

  a. Use the 100 values of x obtained from the simulation to construct a relative frequency histogram for x. Note that this histogram is an approximation to $p(x)$.
  b. Calculate the exact values of $p(x)$ for $n = 5$ and $p = .3$ and compare these values with the approximations found in part a.
  c. If you were to repeat the simulation an extremely large number of times (say 100,000), how do you think the relative frequency histogram and true probability dis-tribution would compare?

EXPLORING DATA WITH A COMPUTER

Refer to the data on 114 coronary bypass patients in Appendix B. Recall (from Example 2.1) that some of the patients received a new drug designed to reduce blood loss prior to surgery and some did not. Use the computer to calculate the percentage of drugged patients who expe-rience some type of complication during or after surgery. Round this percentage to the nearest integer and use the result to estimate the probability of complications for a coronary bypass patient who receives a dosage of the drug prior to surgery.

  a. Suppose 20 coronary bypass patients are given the new drug prior to surgery. If these people represent a random sample of all coronary bypass patients, what is the probability that more than a third of them experi-ence a complication during or after surgery?
  b. If 2,000 coronary bypass patients are given the drug, find the mean μ and the standard deviation σ of the number of them who have complications. Calculate $\mu \pm 2\sigma$ and $\mu \pm 3\sigma$, and use Chebyshev's Rule and the Empirical Rule to estimate the probability that the number of patients with complications falls in each of the intervals. Based on your answers, assess the likeli-hood that more than a third of the 2,000 patients will have complications from surgery.

Chapter 5

CONTINUOUS RANDOM VARIABLES

Contents

Case Studies

*W*HERE WE'VE BEEN

We've seen that because numerical data represent observed values of random variables, we need to find the probabilities associated with specific sample observations. The probability theory of Chapter 3 provided the mechanism for finding the probabilities associated with discrete random variables. Finding and describing this set of probabilities—the probability distribution for a discrete random variable—was the subject of Chapter 4.

*W*HERE WE'RE GOING

Data may be derived from observations on continuous as well as discrete random variables. Thus, we need to know about probability distributions associated with continuous random variables and also how to use the mean and standard deviation to describe these distributions. Chapter 5 addresses this problem and, in particular, introduces the *normal probability distribution.* As you'll see, the normal probability distribution is one of the most useful distributions in statistics.

In this chapter we'll consider some continuous random variables that are commonly encountered. Recall that a continuous random variable is one that can assume any value within some interval or intervals. For example, the length of time between a person's visits to a doctor, the thickness of sheets of steel produced in a rolling mill, and the yield of wheat per acre of farmland are all continuous random variables. The methodology we employ to describe continuous random variables will necessarily be somewhat different from that used to describe discrete random variables. We first discuss the general form of *continuous probability distributions,* and then we explore three specific types that are used in making statistical decisions. The *normal probability distribution,* which plays a basic and important role in both the theory and application of statistics, is essential to the study of most of the subsequent chapters in this book. The other types have practical applications, but a study of these topics is optional.

5.1   CONTINUOUS PROBABILITY DISTRIBUTIONS

The graphical form of the probability distribution for a continuous random variable *x* is a smooth curve that might appear as shown in Figure 5.1. This curve, a function of *x,* is denoted by the symbol $f(x)$ and is variously called a **probability density function**, a **frequency function**, or a **probability distribution**.

The areas under a probability distribution correspond to probabilities for *x.* For example, the area *A* beneath the curve between the two points *a* and *b,* as shown in Figure 5.1, is the probability that *x* assumes a value between *a* and *b* $(a < x < b)$. Because there is no area over a point, say $x = a$, it follows that (according to our model) the probability associated with a particular value of *x* is equal to 0; that is, $P(x = a) = 0$ and hence $P(a < x < b) = P(a \leq x \leq b)$. In other words, the probability is the same whether or not you include the endpoints of the interval. Also, because areas over intervals represent probabilities, it follows that the total area under a probability distribution, the probability assigned to all values of *x,* should equal 1. Note that probability distributions for continuous random variables possess different shapes depending on the relative frequency distributions of real data that the probability distributions are supposed to model.

The areas under most probability distributions are obtained by using calculus or numerical methods.* Because these methods often involve difficult procedures, we will give the areas for some of the most common probability distributions in

FIGURE 5.1

A probability distribution $f(x)$ for a continuous random variable *x*

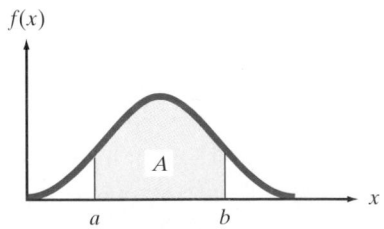

*Students with knowledge of calculus should note that the probability that *x* assumes a value in the interval $a < x < b$ is $P(a < x < b) = \int_a^b f(x)\, dx$, assuming the integral exists. Similar to the requirements for a discrete probability distribution, we require $f(x) \geq 0$ and $\int_{-\infty}^{\infty} f(x)\, dx = 1$.

tabular form in Appendix A. Then, to find the area between two values of x, say $x = a$ and $x = b$, you simply have to consult the appropriate table.

For each of the continuous random variables presented in this chapter, we will give the formula for the probability distribution along with its mean μ and standard deviation σ. These two numbers will enable you to make some approximate probability statements about a random variable even when you do not have access to a table of areas under the probability distribution.

5.2   THE UNIFORM DISTRIBUTION

All the probability problems discussed in Chapter 3 had sample spaces that contained a finite number of sample points. In many of these problems, the sample points were assigned equal probabilities—for example, the die toss or the coin toss. For continuous random variables, there is an infinite number of values in the sample space, but in some cases the values may appear to be equally likely. For example, if a short exists in a 5-meter stretch of electrical wire, it may have an equal probability of being in any particular 1-centimeter segment along the line. Or if a safety inspector plans to choose a time at random during the 4 afternoon work-hours to pay a surprise visit to a certain area of a plant, then each 1-minute time interval in this 4-work-hour period will have an equally likely chance of being selected for the visit.

Continuous random variables that appear to have equally likely outcomes over their range of possible values possess a **uniform probability distribution**, perhaps the simplest of all continuous probability distributions. Suppose the random variable x can assume values only in an interval $c \leq x \leq d$. Then the uniform frequency function has a rectangular shape, as shown in Figure 5.2. Note that the possible values of x consist of all points in the interval between point c and point d. The height of $f(x)$ is constant in that interval and equals $1/(d - c)$. Therefore, the total area under $f(x)$ is given by

$$\text{Total area of rectangle } = (\text{Base})(\text{Height}) = (d - c)\left(\frac{1}{d - c}\right) = 1$$

The uniform probability distribution provides a model for continuous random variables that are *evenly distributed* over a certain interval. That is, a uniform random variable is one that is just as likely to assume a value in one interval as it is to assume a value in any other interval of equal size. There is no clustering of values around any value; instead, there is an even spread over the entire region of possible values.

The uniform distribution is sometimes referred to as the **randomness distribution**, since one way of generating a uniform random variable is to perform an

FIGURE 5.2

The uniform probability distribution

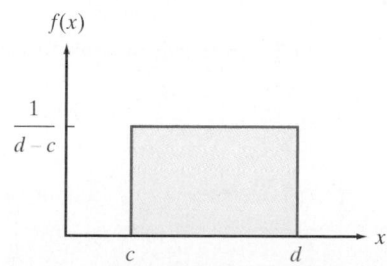

experiment in which a point is *randomly selected* on the horizontal axis between the points c and d. If we were to repeat this experiment infinitely often, we would create a uniform probability distribution like that shown in Figure 5.2. The random selection of points in an interval can also be used to generate random numbers such as those in Table I of Appendix A. Recall that random numbers are selected in such a way that every number would have an equal probability of selection. Therefore, random numbers are realizations of a uniform random variable. (Random numbers were used to draw random samples in Section 3.8.) The formulas for the uniform probability distribution, its mean, and standard deviation are shown in the box.

Probability Distribution, Mean, and Standard Deviation of a Uniform Random Variable x

$$f(x) = \frac{1}{d - c} \qquad (c \le x \le d)$$

$$\mu = \frac{c + d}{2} \qquad\qquad \sigma = \frac{d - c}{\sqrt{12}}$$

Suppose the interval $a < x < b$ lies within the domain of x; that is, it falls within the larger interval $c \le x \le d$. Then the probability that x assumes a value within the interval $a < x < b$ is equal to the area of the rectangle over the interval, namely, $(b - a)/(d - c)$.*

EXAMPLE 5.1

An unprincipled used car dealer sells a car to an unsuspecting buyer, even though the dealer knows that the car will have a major breakdown within the next 6 months. The dealer provides a warranty of 45 days on all cars sold. Let x represent the length of time until the breakdown occurs. Assume that x is a uniform random variable with values between 0 and 6 months.

a. Calculate the mean and standard deviation of x. Graph the probability distribution of x, and show the mean on the horizontal axis. Also show 1- and 2-standard-deviation intervals around the mean.
b. Calculate the probability that the breakdown occurs while the car is still under warranty.

Solution

a. To calculate the mean and standard deviation for x, we substitute 0 and 6 months for c and d, respectively, in the formulas for uniform random variables. Thus,

$$\mu = \frac{c + d}{2} = \frac{0 + 6}{2} = 3 \text{ months}$$

and

$$\sigma = \frac{d - c}{\sqrt{12}} = \frac{6 - 0}{\sqrt{12}} = \frac{6}{3.464} = 1.73 \text{ months}$$

The uniform probability distribution is

$$f(x) = \frac{1}{d - c} = \frac{1}{6 - 0} = \frac{1}{6} \qquad (0 \le x \le 6)$$

---

*The student with knowledge of calculus should note that

$$P(a < x < b) = \int_a^b f(x)\, d(x) = \int_a^b 1/(d - c)\, dx = (b - a)/(d - c)$$

FIGURE 5.3

Distribution for x in Example 5.1

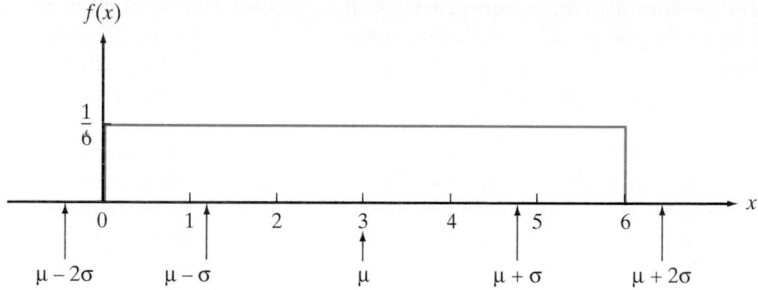

FIGURE 5.4

Probability that car breaks down within 1.5 months of purchase

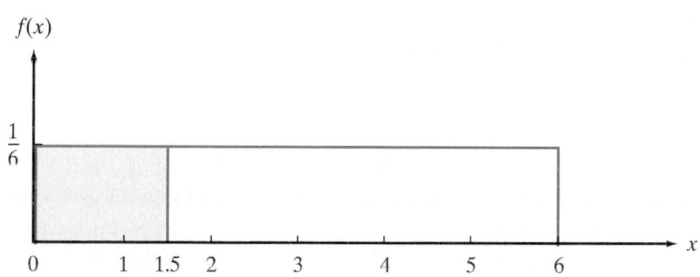

The graph of this function is shown in Figure 5.3. The mean and 1- and 2-standard-deviation intervals around the mean are shown on the horizontal axis.

b. To find the probability that the car is still under warranty when it breaks down, we must find the probability that x, the time until the breakdown occurs, is less than 45 days or (about) 1.5 months. As indicated in Figure 5.4, we need to calculate the area under the frequency function $f(x)$ between the points $x = 0$ and $x = 1.5$. This is the area of a rectangle with base $1.5 - 0 = 1.5$ and height $\frac{1}{6}$. The probability that the unsuspecting buyer will be able to have the car repaired at the dealer's expense is then

$$P(0 < x < 1.5) = (\text{Base})(\text{Height}) = (1.5)(\tfrac{1}{6}) = .25 \qquad \blacktriangle$$

EXERCISES 5.1–5.12

Learning the Mechanics

5.1 Suppose x is a random variable best described by a uniform probability distribution with $c = 10$ and $d = 30$.
  a. Find $f(x)$.
  b. Find the mean and standard deviation of x.
  c. Graph $f(x)$ and locate μ and the interval $\mu \pm 2\sigma$ on the graph. Note that the probability that x assumes a value within the interval $\mu \pm 2\sigma$ is equal to 1.

5.2 Refer to Exercise 5.1. Find the following probabilities:
  a. $P(10 \leq x \leq 25)$ **b.** $P(20 < x < 30)$
  c. $P(x \geq 25)$ **d.** $P(x < 10)$ **e.** $P(x \leq 25)$
  f. $P(10 \leq x \leq 25)$ **g.** $P(20.5 \leq x \leq 25.5)$

5.3 Suppose x is a random variable best described by a uniform probability distribution with $c = 2$ and $d = 4$.
  a. Find $f(x)$.
  b. Find the mean and standard deviation of x.
  c. Find $P(\mu - \sigma \leq x \leq \mu + \sigma)$.   **d.** Find $P(x > 2.78)$.
  e. Find $P(2.4 \leq x \leq 3.7)$.   **f.** Find $P(x < 2)$.

5.4 Refer to Exercise 5.3. Find the value of a that makes each of the following probability statements true.
  a. $P(x \geq a) = .5$ **b.** $P(x \leq a) = .2$ **c.** $P(x \leq a) = 0$
  d. $P(2.5 \leq x \leq a) = .5$

5.5 The random variable x is best described by a uniform probability distribution with $c = 100$ and $d = 200$. Find the probability that x assumes a value
  a. More than 2 standard deviations from μ.
  b. Less than 3 standard deviations from μ.
  c. Within 2 standard deviations of μ.

5.6 The random variable x is best described by a uniform probability distribution with mean 50 and standard deviation 5. Find c, d, and $f(x)$. Graph the probability distribution.

Applying the Concepts

5.7 The manager of a local soft-drink bottling company believes that when a new beverage-dispensing machine is set to dispense 7 ounces, it in fact dispenses an amount x at random anywhere between 6.5 and 7.5 ounces inclusive. Suppose x has a uniform probability distribution.

a. Is the amount dispensed by the beverage machine a discrete or a continuous random variable? Explain.

b. Graph the frequency function for x, the amount of beverage the manager believes is dispensed by the new machine when it is set to dispense 7 ounces.

c. Find the mean and standard deviation for the distribution graphed in part **b**, and locate the mean and the interval $\mu \pm 2\sigma$ on the graph.

d. Find $P(x \geq 7)$.   **e.** Find $P(x < 6)$.

f. Find $P(6.5 \leq x \leq 7.25)$.

g. What is the probability that each of the next six bottles filled by the new machine will contain more than 7.25 ounces of beverage? Assume that the amount of beverage dispensed in one bottle is independent of the amount dispensed in another bottle.

5.8 Researchers at the University of California-Berkeley have designed, built, and tested a switched-capacitor circuit for generating random signals (*International Journal of Circuit Theory and Applications,* May–June 1990). The circuit's trajectory was shown to be uniformly distributed on the interval $(0, 1)$.

a. Give the mean and variance of the circuit's trajectory.

b. Compute the probability that the trajectory falls between .2 and .4.

c. Would you expect to observe a trajectory that exceeds .995? Explain.

5.9 The manager of a large department store with three floors reports that the time a customer on the second floor must wait for an elevator has a uniform distribution ranging from 0 to 4 minutes. Find the mean and standard deviation of x, the time a customer on the second floor waits for an elevator. If it takes the elevator 15 seconds to go from floor to floor, find the probability that a hurried customer can reach the first floor in less than 1.5 minutes after pushing the second-floor elevator button.

5.10 The weather on a tropical island in January is fairly constant. Records indicate that the high temperatures for each day of the month tend to have a uniform distribution over the interval from 75 to 90°F. A tourist arrives on the island on a randomly selected day in January.

a. What is the probability that the temperature will be above 80°F?

b. What is the probability that the temperature will be between 80 and 85°F?

c. What is the expected temperature?

5.11 A tool-and-die machine shop produces extremely high-tolerance spindles. The spindles are 18-inch slender rods used in a variety of military equipment. A piece of equipment used in the manufacture of the spindles malfunctions on occasion and places a single gouge somewhere on the spindle. However, if the spindle can be cut so that it has 14 consecutive inches without a gouge, the spindle can be salvaged for other purposes. Assuming that the location of the gouge along the spindle is best described by a uniform distribution, what is the probability that a defective spindle can be salvaged?

5.12 A bus is scheduled to stop at a certain bus stop every half hour on the hour and the half hour. At the end of the day, buses still stop after every 30 minutes, but because delays often occur earlier in the day, the bus is never early and likely to be late. The director of the bus line claims that the length of time a bus is late is uniformly distributed and the maximum time that a bus is late is 20 minutes.

a. If the director's claim is true, what is the expected number of minutes a bus will be late?

b. If the director's claim is true, what is the probability that the last bus on a given day will be more than 19 minutes late?

c. If you arrive at the bus stop at the end of a day at exactly half-past the hour and must wait more than 19 minutes for the bus, what would you conclude about the director's claim? Why?

5.3   THE NORMAL DISTRIBUTION

One of the most commonly observed continuous random variables has a **bell-shaped** probability distribution as shown in Figure 5.5. It is known as a **normal random variable** and its probability distribution is called a **normal distribution**.

The normal distribution plays a very important role in the science of statistical inference. Moreover, many phenomena generate random variables with probability distributions that are very well approximated by a normal distribution. For example, the error made in measuring a person's blood pressure may be a normal random variable, and the probability distribution for the yearly rainfall in a certain region might be approximated by a normal probability distribution. The normal distribution might also provide an accurate model for the normal

FIGURE 5.5

A normal probability distribution

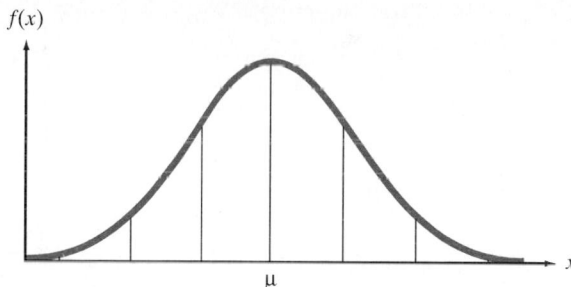

approximation to an existing population of data by comparing the relative frequency distribution of a large sample of the data to the normal probability distribution. Tests to detect disagreement between a set of data and the assumption of normality are available, but they are beyond the scope of this book.

The normal distribution is perfectly symmetric about its mean μ, as can be seen in the examples in Figure 5.6. Its spread is determined by the value of its standard deviation σ.

The formula for the normal probability distribution is shown in the box. When plotted, this formula yields a curve like that shown in Figure 5.5.

Probability Distribution for a Normal Random Variable x

$$f(x) = \frac{1}{\sigma\sqrt{2\pi}} e^{-(1/2)[(x-\mu)/\sigma]^2}$$

where

μ = Mean of the normal random variable x

σ = Standard deviation

π = 3.1416 ...

e = 2.71828 ...

Note that the mean μ and standard deviation σ appear in this formula, so that no separate formulas for μ and σ are necessary. To graph the normal curve we have to know the numerical values of μ and σ.

Computing the area over intervals under the normal probability distribution is a difficult task.* Consequently, we will use the computed areas listed in Table IV

FIGURE 5.6

Several normal distributions with different means and standard deviations

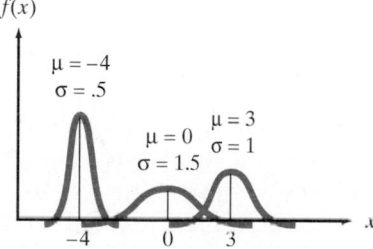

---

*The student with knowledge of calculus should note that there is not a closed-form expression for $P(a < x < b) = \int_{a}^{b} f(x)\, dx$ for the normal probability distribution. The value of this definite integral can be obtained to any desired degree of accuracy by numerical approximation procedures. For this reason, it is tabulated for the user.

FIGURE 5.7

Standard normal
distribution: $\mu = 0$, $\sigma = 1$

$f(z)$

z

$-3 \quad -2 \quad -1 \quad 0 \quad 1 \quad 2 \quad 3$

of Appendix A (and inside the front cover). Although there are an infinitely
large number of normal curves—one for each pair of values for μ and σ—we
have formed a single table that will apply to any normal curve.

Table IV is based on a normal distribution with mean $\mu = 0$ and standard
deviation $\sigma = 1$, called a *standard normal distribution*. A random variable with a
standard normal distribution is typically denoted by the symbol z. The formula for
the probability distribution of z is given by $f(z) = \dfrac{1}{\sqrt{2\pi}} e^{-(1/2)z^2}$. Figure 5.7 shows
the graph of a standard normal distribution.

Since we will ultimately convert all normal random variables to standard
normal in order to use Table IV to find probabilities, it is important that you learn
to use Table IV well. A partial reproduction of Table IV is shown in Table 5.1.
Note that the values of the standard normal random variable z are listed in the

TABLE 5.1   Reproduction of Part of Table IV in Appendix A

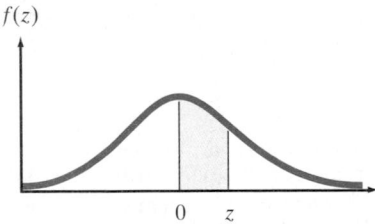

$f(z)$

$0 \quad z$

z	.00	.01	.02	.03	.04	.05	.06	.07	.08	.09
.0	.0000	.0040	.0080	.0120	.0160	.0199	.0239	.0279	.0319	.0359
.1	.0398	.0438	.0478	.0517	.0557	.0596	.0636	.0675	.0714	.0753
.2	.0793	.0832	.0871	.0910	.0948	.0987	.1026	.1064	.1103	.1141
.3	.1179	.1217	.1255	.1293	.1331	.1368	.1406	.1443	.1480	.1517
.4	.1554	.1591	.1628	.1664	.1700	.1736	.1772	.1808	.1844	.1879
.5	.1915	.1950	.1985	.2019	.2054	.2088	.2123	.2157	.2190	.2224
.6	.2257	.2291	.2324	.2357	.2389	.2422	.2454	.2486	.2517	.2549
.7	.2580	.2611	.2642	.2673	.2704	.2734	.2764	.2794	.2823	.2852
.8	.2881	.2910	.2939	.2967	.2995	.3023	.3051	.3078	.3106	.3133
.9	.3159	.3186	.3212	.3238	.3264	.3289	.3315	.3340	.3365	.3389
1.0	.3413	.3438	.3461	.3485	.3508	.3531	.3554	.3577	.3599	.3621
1.1	.3643	.3665	.3686	.3708	.3729	.3749	.3770	.3790	.3810	.3830
1.2	.3849	.3869	.3888	.3907	.3925	.3944	.3962	.3980	.3997	.4015
1.3	.4032	.4049	.4066	.4082	.4099	.4115	.4131	.4147	.4162	.4177
1.4	.4192	.4207	.4222	.4236	.4251	.4265	.4279	.4292	.4306	.4319
1.5	.4332	.4345	.4357	.4370	.4382	.4394	.4406	.4418	.4429	.4441

FIGURE 5.8

Areas under the standard
normal curve for
Example 5.2

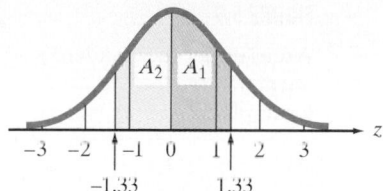

left-hand column. The entries in the body of the table give the area (probability)
between 0 and z. Examples 5.2–5.5 illustrate the use of the table.

EXAMPLE 5.2

Find the probability that the standard normal random variable z falls between
-1.33 and $+1.33$.

Solution

The standard normal distribution is shown again in Figure 5.8. Since all probabilities associated with standard normal random variables can be depicted as areas
under the standard normal curve, you should always draw the curve and then
equate the desired probability to an area.

In this example we want to find the probability that z falls between -1.33 and
$+1.33$, which is equivalent to the area between -1.33 and $+1.33$, shown shaded in
Figure 5.8. Table IV provides the area between $z = 0$ and any value of z, so that if
we look up $z = 1.33$, we find that the area between $z = 0$ and $z = 1.33$ is .4082.
This is the area labeled A_1 in Figure 5.8. To find the area A_2 between $z = 0$ and
$z = -1.33$, we note that the symmetry of the normal distribution implies that the
area between $z = 0$ and any point to the left is equal to the area between $z = 0$
and the point equidistant to the right. Thus, in this example the area between
$z = 0$ and $z = -1.33$ is equal to the area between $z = 0$ and $z = +1.33$. That is,

$$A_1 = A_2 = .4082$$

The probability that z falls between -1.33 and $+1.33$ is the sum of the areas of A_1
and A_2. We summarize in probabilistic notation:

$$P(-1.33 < z < +1.33) = P(-1.33 < z < 0) + P(0 \leq z < 1.33)$$
$$= A_1 + A_2$$
$$= .4082 + .4082 = .8164$$

Remember that "$<$" and "$\leq$" are equivalent in events involving z, because the
inclusion (or exclusion) of a single point does not alter the probability of an event
involving a continuous random variable. ▲

EXAMPLE 5.3

Find the probability that a standard normal random variable exceeds 1.64; that is,
find $P(z > 1.64)$.

Solution

The area under the standard normal distribution to the right of 1.64 is the shaded
area labeled A_1 in Figure 5.9. This area represents the desired probability that z
exceeds 1.64. However, when we look up $z = 1.64$ in Table IV, we must remember that the probability given in the table corresponds to the area between $z = 0$
and $z = 1.64$ (the area labeled A_2 in Figure 5.9). From Table IV we find that
$A_2 = .4495$. To find the area A_1 to the right of 1.64, we make use of two facts:

FIGURE 5.9

Areas under the standard
normal curve for
Example 5.3

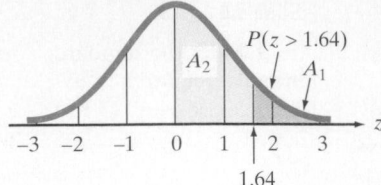

1. The standard normal distribution is symmetric about its mean, $z = 0$.
2. The total area under the standard normal probability distribution equals 1.

Taken together, these two facts imply that the areas on either side of the mean $z = 0$ equal .5; thus, the area to the right of $z = 0$ in Figure 5.9 is $A_1 + A_2 = .5$. Then

$$P(z > 1.64) = A_1 = .5 - A_2 = .5 - .4495 = .0505$$

To attach some practical significance to this probability, note that the implication is that the chance of a standard normal random variable exceeding 1.64 is approximately .05. Or, since z represents the number of standard deviations between *any* normal random variable and its mean, a normal random variable will exceed its mean by more than 1.64 standard deviations only about 5% of the time. ▲

EXAMPLE 5.4 Find the probability that a normal random variable lies to the right of a point $-.74$ standard deviation from its mean.

Solution

We first must interpret the event in terms of a standard normal random variable, so that we can use Table IV to find the event's probability. Since the standard normal random variable z is simply the number of standard deviations between an observed value of a normal random variable and its mean, the event that a normal random variable lies to the right of a point $-.74$ standard deviation from the mean is equivalent to the event that the standard normal random variable z exceeds $-.74$. The event is shown as the shaded area in Figure 5.10, and we want to find $P(z > -.74)$.

We divide the shaded area into two parts: the area A_1 between $z = -.74$ and $z = 0$, and the area A_2 to the right of $z = 0$. We must always make such a division when the desired area lies on both sides of the mean ($z = 0$) because Table IV contains areas between $z = 0$ and the point you look up. To find A_1, we remember that the sign of z is unimportant when determining the area, because the standard normal distribution is symmetric about its mean. We look up $z = .74$ in Table IV to find that $A_1 = .2704$. The symmetry also implies that half the distribution lies on each side of the mean, so the area A_2 to the right of $z = 0$ is .5. Then,

$$P(z > -.74) = A_1 + A_2 = .2704 + .5 = .7704$$ ▲

FIGURE 5.10

Areas under the standard
normal curve for
Example 5.4

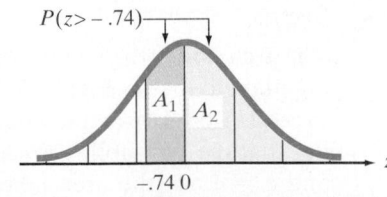

EXAMPLE 5.5

Find the probability that a normal random variable lies more than 1.96 standard deviations from its mean *in either direction.* Compare the result to the Empirical Rule.

Solution

The event that a normal random variable exceeds 1.96 standard deviations in either direction is equivalent to the event that the standard normal random variable z exceeds 1.96 in absolute value. That is, we want to find

$$P(z > |1.96|) = P(z < -1.96 \text{ or } z > 1.96)$$

This probability is the shaded area in Figure 5.11. Note that the total shaded area is the sum of two areas, A_1 and A_2—areas that are equal because of the symmetry of the normal distribution.

We look up $z - 1.96$ and find the area between $z = 0$ and $z - 1.96$ to be .4750. Then the area to the right of 1.96, A_2, is $.5 - .4750 = .0250$, so that

$$P(z > |1.96|) = A_1 + A_2 = .0250 + .0250 = .05$$

The implication is that any normal random variable lies more than 1.96 standard deviations from its mean 5% of the time. Recall from Chapter 2 that the Empirical Rule tells us that about 5% of the measurements in mound-shaped distributions will lie beyond 2 standard deviations from the mean; the normal distribution, which is certainly mound-shaped, has 5% of its area beyond 1.96 standard deviations. In fact, the normal distribution provides the model on which the Empirical Rule is based, along with much "empirical" experience with real data that often approximately obey the rule, whether drawn from a normal distribution or not.   ▲

To apply Table IV to a normal random variable x with any mean μ and any standard deviation σ, we must first convert the value of x to a z-score. The population z-score for a measurement was defined (in Section 2.6) as the *distance* between the measurement and the population mean, divided by the population standard deviation. Thus, the z-score gives the distance between a measurement and the mean in units equal to the standard deviation. In symbolic form, the z-score for the measurement x is

$$z = \frac{x - \mu}{\sigma}$$

Note that when $x = \mu$, we obtain $z = 0$.

An important property of the normal distribution is that if x is normally distributed with any mean and any standard deviation, z is *always* normally distributed with mean 0 and standard deviation 1. That is, z is a standard normal random variable. This property (stated in the box) is applied in the next several examples.

FIGURE 5.11

Areas under the standard normal curve for Example 5.5

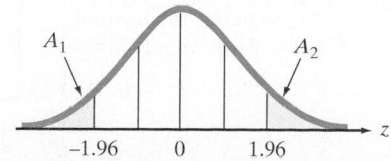

DEFINITION 5.1

The **standard normal distribution** is a normal distribution with $\mu = 0$ and $\sigma = 1$. A random variable with a standard normal distribution, denoted by the symbol z, is called a **standard normal random variable**.

Property of Normal Distributions

If x is a normal random variable with mean μ and standard deviation σ, then the random variable z defined by the formula

$$z = \frac{x - \mu}{\sigma}$$

has a standard normal distribution. The value z describes the number of standard deviations between x and μ.

EXAMPLE 5.6

Assume that the length of time, x, between charges of a pocket calculator is normally distributed with a mean of 100 hours and a standard deviation of 15 hours. Find the probability that the calculator will last between 80 and 120 hours between charges.

Solution

The normal distribution with mean $\mu = 100$ and $\sigma = 15$ is shown in Figure 5.12. The desired probability that the calculator lasts between 80 and 120 hours is shaded. In order to find the probability, we must first convert the distribution to standard normal, which we do by calculating the z-score:

$$z = \frac{x - \mu}{\sigma}$$

The z-scores corresponding to the important values of x are shown beneath the x values on the horizontal axis in Figure 5.12. Note that $z = 0$ corresponds to the mean of $\mu = 100$ hours, whereas the x values 80 and 120 yield z-scores of -1.33 and $+1.33$, respectively. Thus, the event that the calculator lasts between 80 and 120 hours is equivalent to the event that a standard normal random variable lies between -1.33 and $+1.33$. We found this probability in Example 5.2 (see Figure 5.8) by doubling the area corresponding to $z = 1.33$ in Table IV. That is,

$$P(80 \le x \le 120) = P(-1.33 \le z \le 1.33) = 2(.4082) = .8164 \qquad \blacktriangle$$

The steps to follow when calculating a probability corresponding to a normal random variable are shown in the box.

FIGURE 5.12

Areas under the normal curve for Example 5.6

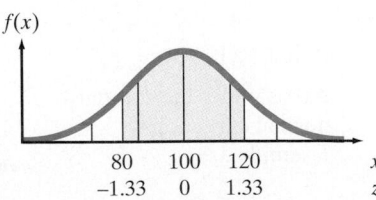

**Steps for Finding a Probability Corresponding to
a Normal Random Variable**

---

1. Sketch the normal distribution and indicate the mean of the random variable x.
   Then shade the area corresponding to the probability you want to find.
2. Convert the boundaries of the shaded area from x values to standard normal
   random variable z values using the formula

$$z = \frac{x - \mu}{\sigma}$$

   Show the z values under the corresponding x values on your sketch.
3. Use Table IV in Appendix A (and inside the front cover) to find the areas corresponding to the z values. If necessary, use the symmetry of the normal distribution to find areas corresponding to negative z values and the fact that the total
   area on each side of the mean equals .5 to convert the areas from Table IV to the
   probabilities of the event you have shaded.

EXAMPLE 5.7

Suppose an automobile manufacturer introduces a new model that has an advertised mean in-city mileage of 27 miles per gallon. Although such advertisements
seldom report any measure of variability, suppose you write the manufacturer for
the details of the tests, and you find that the standard deviation is 3 miles per
gallon. This information leads you to formulate a probability model for the
random variable x, the in-city mileage for this car model. You believe that the
probability distribution of x can be approximated by a normal distribution with a
mean of 27 and a standard deviation of 3.

a. If you were to buy this model of automobile, what is the probability that you
   would purchase one that averages less than 20 miles per gallon for in-city driving? In other words, find $P(x < 20)$.
b. Suppose you purchase one of these new models and it does get less than 20
   miles per gallon for in-city driving. Should you conclude that your probability
   model is incorrect?

Solution

a. The probability model proposed for x, the in-city mileage, is shown in Figure
   5.13. We are interested in finding the area A to the left of 20 since this area corresponds to the probability that a measurement chosen from this distribution
   falls below 20. In other words, if this model is correct, the area A represents the
   fraction of cars that can be expected to get less than 20 miles per gallon for in-city
   driving. To find A, we first calculate the z value corresponding to $x = 20$. That is,

$$z = \frac{x - \mu}{\sigma} = \frac{20 - 27}{3} = -\frac{7}{3} = -2.33$$

FIGURE 5.13

Area under the normal
curve for Example 5.7

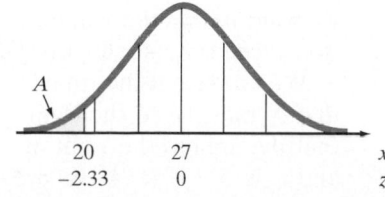

Then

$$P(x < 20) = P(z < -2.33)$$

as indicated by the shaded area in Figure 5.13. Since Table IV gives only areas to the right of the mean (and because the normal distribution is symmetric about its mean), we look up 2.33 in Table IV and find that the corresponding area is .4901. This is equal to the area between $z = 0$ and $z = -2.33$, so we find

$$P(x < 20) = A = .5 - .4901 = .0099 \approx .01$$

According to this probability model, you should have only about a 1% chance of purchasing a car of this make with an in-city mileage under 20 miles per gallon.

b. Now you are asked to make an inference based on a sample—the car you purchased. You are getting less than 20 miles per gallon for in-city driving. What do you infer? We think you will agree that one of two possibilities is true:

1. The probability model is correct. You simply were unfortunate to have purchased one of the cars in the 1% that get less than 20 miles per gallon in the city.

2. The probability model is incorrect. Perhaps the assumption of a normal distribution is unwarranted, or the mean of 27 is an overestimate, or the standard deviation of 3 is an underestimate, or some combination of these errors was made. At any rate, the form of the actual probability model certainly merits further investigation.

You have no way of knowing with certainty which possibility is correct, but the evidence points to the second one. We are again relying on the rare-event approach to statistical inference that we introduced earlier. The sample (one measurement in this case) was so unlikely to have been drawn from the proposed probability model that it casts serious doubt on the model. We would be inclined to believe that the model is somehow in error.   ▲

Occasionally you will be given a probability and will want to find the values of the normal random variable that correspond to the probability. For example, suppose the scores on a college entrance examination are known to be normally distributed, and a certain prestigious university will consider for admission only those applicants whose scores exceed the 90th percentile of the test score distribution. To determine the minimum score for admission consideration, you will need to be able to use Table IV in reverse, as demonstrated in the following example.

EXAMPLE 5.8   Find the value of z, call it z_0, in the standard normal distribution that will be exceeded only 10% of the time. That is, find z_0 such that $P(z \geq z_0) = .10$.

Solution

In this case we are given a probability, or an area, and asked to find the value of the standard normal random variable that corresponds to the area. Specifically, we want to find the value z_0 such that only 10% of the standard normal distribution exceeds z_0 (see Figure 5.14).

We know that the total area to the right of the mean $z = 0$ is .5, which implies that z_0 must lie to the right of (above) 0. To pinpoint the value, we use the fact that the area to the right of z_0 is .10, which implies that the area between $z = 0$ and z_0 is $.5 - .1 = .4$. But areas between $z = 0$ and some other z values are exactly the types given in Table IV. Therefore, we look up the area .4000 in the body of

FIGURE 5.14

Areas under the
standard normal curve
for Example 5.8

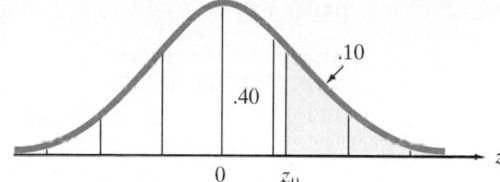

Table IV and find that the corresponding z value is (to the closest approximation) $z_0 = 1.28$. The implication is that the point 1.28 standard deviations above the mean is the 90th percentile of a normal distribution.   ▲

EXAMPLE 5.9

Find the value of z_0 such that 95% of the standard normal z values lie between $-z_0$ and $+z_0$—that is, $P(-z_0 \leq z \leq z_0) = .95$.

Solution

Here we wish to move an equal distance z_0 in the positive and negative directions from the mean $z = 0$ until 95% of the standard normal distribution is enclosed. This means that the area on each side of the mean will be equal to $\frac{1}{2}(.95) = .475$, as shown in Figure 5.15. Since the area between $z = 0$ and z_0 is .475, we look up .475 in the body of Table IV to find the value $z_0 = 1.96$. Thus, as we found in the reverse order in Example 5.5, 95% of a normal distribution lies between +1.96 and −1.96 standard deviations of the mean.   ▲

Now that you have learned to use Table IV to find a standard normal z value that corresponds to a specified probability, we demonstrate a practical application in Example 5.10.

EXAMPLE 5.10

Suppose the scores, x, on a college entrance examination are normally distributed with a mean of 550 and a standard deviation of 100. A certain prestigious university will consider for admission only those applicants whose scores exceed the 90th percentile of the distribution. Find the minimum score an applicant must achieve in order to receive consideration for admission to the university.

Solution

In this example, we want to find a score, x_0, such that 90% of the scores (x-values) in the distribution fall below x_0 and only 10% fall above x_0. That is,

$$P(x \leq x_0) = .90$$

Converting x to a standard normal random variable, where $\mu = 550$ and $\sigma = 100$, we have

FIGURE 5.15

Areas under the standard
normal curve for
Example 5.9

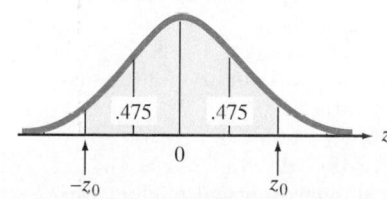

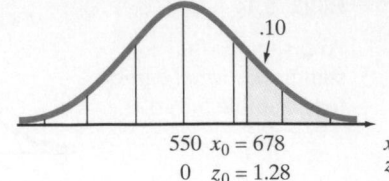

FIGURE 5.16

Area under the normal
curve for Example 5.10

$$P(x \le x_0) = P\left(z \le \frac{x_0 - \mu}{\sigma}\right)$$

$$= P\left(z \le \frac{x_0 - 550}{100}\right) = .90$$

In Example 5.8 (see Figure 5.14) we found the 90th percentile of the standard normal distribution to be $z_0 = 1.28$. That is, we found $P(z \le 1.28) = .90$. Consequently, we know the minimum test score x_0 corresponds to a z-score of 1.28; that is,

$$\frac{x_0 - 550}{100} = 1.28$$

If we solve this equation for x_0, we find

$$x_0 = 550 + 1.28(100) = 550 + 128 = 678$$

This x value is shown in Figure 5.16. Thus, the 90th percentile of the test score distribution is 678. An applicant must score at least 678 on the entrance exam to be considered for admission to the university. ▲

EXERCISES 5.13–5.41

Learning the Mechanics

5.13 Find the area under the standard normal probability distribution between the following pairs of z-scores:
a. $z = 0$ and $z = 2.00$ **b.** $z = 0$ and $z = 1.00$
c. $z = 0$ and $z = 3$ **d.** $z = 0$ and $z = .58$
e. $z = -2.00$ and $z = 0$ **f.** $z = -1.00$ and $z = 0$
g. $z = -1.69$ and $z = 0$ **h.** $z = -.58$ and $z = 0$

5.14 Find the following probabilities for the standard normal random variable z:
a. $P(-1 \le z \le 1)$ **b.** $P(-2 \le z \le 2)$
c. $P(-2.16 < z \le .55)$ **d.** $P(-.42 < z < 1.96)$
e. $P(z \ge -2.33)$ **f.** $P(z < 2.33)$

5.15 Find the following probabilities for the standard normal random variable z:
a. $P(z > 1.46)$ **b.** $P(z < -1.56)$
c. $P(.67 \le z \le 2.41)$ **d.** $P(-1.96 \le z < -.33)$
e. $P(z \ge 0)$ **f.** $P(-2.33 < z < 1.50)$

5.16 Find each of the following probabilities for a standard normal random variable z:
a. $P(z = 1)$ **b.** $P(z \le 1)$ **c.** $P(z < 1)$ **d.** $P(z > 1)$

5.17 Find each of the following probabilities for the standard normal random variable z:
a. $P(-1 \le z \le 1)$ **b.** $P(-1.96 \le z \le 1.96)$
c. $P(-1.645 \le z \le 1.645)$ **d.** $P(-2 \le z \le 2)$

5.18 Find a value of the standard normal random variable z, call it z_0, such that

a. $P(z \ge z_0) = .05$ **b.** $P(z \ge z_0) = .025$
c. $P(z \le z_0) = .025$ **d.** $P(z \ge z_0) = .10$
e. $P(z > z_0) = .10$

5.19 Find a value of the standard normal random variable z, call it z_0, such that
a. $P(z \le z_0) = .0401$ **b.** $P(-z_0 \le z \le z_0) = .95$
c. $P(-z_0 \le z < z_0) = .90$
d. $P(-z_0 \le z \le z_0) = .8740$
e. $P(z_0 \le z \le 0) = .2967$
f. $P(-2 < z < z_0) = .9710$

5.20 Find a z-score, call it z_0, such that
a. $P(z \ge z_0) = .5$ **b.** $P(z \ge z_0) = .0057$
c. $P(0 \le z \le z_0) = .4713$ **d.** $P(z < z_0) = .0392$

5.21 Give the z-score for a measurement from a normal distribution for the following:
a. 1 standard deviation above the mean
b. 1 standard deviation below the mean
c. Equal to the mean
d. 2.5 standard deviations below the mean
e. 3 standard deviations above the mean

5.22 Suppose the random variable x is best described by a normal distribution with $\mu = 25$ and $\sigma = 5$. Find the z-score that corresponds to each of the following x values:
a. $x = 25$ **b.** $x = 30$ **c.** $x = 37.5$
d. $x = 10$ **e.** $x = 50$ **f.** $x = 32$

IQ and the Bell Curve

In their controversial book *The Bell Curve* (Free Press, 1994), Professors Richard J. Herrnstein (a Harvard psychologist) and Charles Murray (a political scientist at MIT) explore, as the subtitle states, "intelligence and class structure in American life." *The Bell Curve* heavily employs statistical analyses in an attempt to support the authors' positions. One of the biggest controversies sparked by the book is the authors' tenet that intelligence levels differ across ethnic groups and that intelligence (or lack thereof) is a cause of a wide range of intractable social problems. Herrnstein and Murray warn, however, that "these group differences do not justify prejudicial assumptions about any member of a given group since his or her intelligence and potential may, in fact, be anywhere under the bell curve." For this case study, we focus on the distribution of intelligence, that is, the **bell curve**.

The measure of intelligence chosen by the authors is the well known Intelligent Quotient (IQ). Numerous tests have been developed to measure IQ; Herrnstein and Murray use the Armed Forces Qualification Test (AFQT), originally designed to measure the cognitive ability of military recruits. Psychologists traditionally treat IQ as a random variable having a normal distribution with mean $\mu = 100$ and standard deviation $\sigma = 15$. This distribution, or *bell curve,* is shown in Figure 5.17.

In their book, Herrnstein and Murray refer to five cognitive classes of people defined by percentiles of the normal distribution. Class I ("very bright") consists of those with IQs above the 95th percentile; Class II ("bright") are those with IQs between the 75th and 95th percentiles; Class III ("normal") includes IQs between the 25th and 75th percentiles; Class IV ("dull") are those with IQs between the 5th and 25th percentiles; and Class V ("very dull") are IQs below the 5th percentile. These classes are also illustrated in Figure 5.17.

Focus

a. Assuming that the distribution of IQ is accurately represented by the bell curve in Figure 5.17, determine the proportion of people with IQs in each of the five cognitive classes defined by Herrnstein and Murray.

b. Although the cognitive classes above are defined in terms of percentiles, the authors stress that IQ scores should be compared with z-scores, not percentiles. In other words, it is more informative to give the difference in z-scores for two IQ scores than it is to give the difference in percentiles. To demonstrate this point, calculate the difference in z-scores for IQs at the 50th and 55th percentiles. Do the same for IQs at the 94th and 99th percentiles. What do you observe?

c. Researchers have found that scores on many intelligence tests are decidedly nonnormal. Some distributions are skewed toward higher scores, others toward lower scores. How would the proportions in the five cognitive classes differ for an IQ distribution that is skewed right? Skewed left?

5.23 The random variable x has a normal distribution with $\mu = 300$ and $\sigma = 30$.

a. Find the probability that x assumes a value more than 2 standard deviations from its mean. More than 3 standard deviations from μ.

b. Find the probability that x assumes a value within 1 standard deviation of its mean. Within 2 standard deviations of μ.

c. Find the value of x that represents the 80th percentile of this distribution. The 10th percentile.

FIGURE 5.17

The distribution of IQ

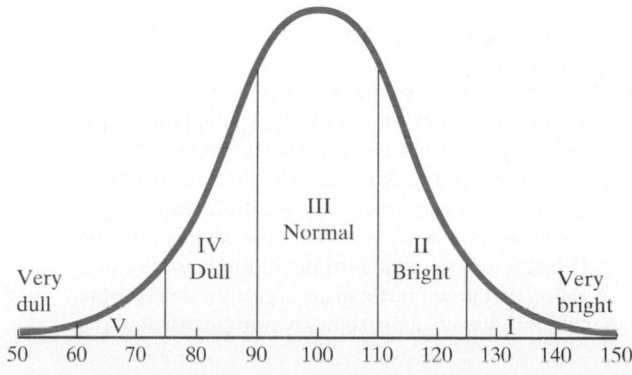

5.24 Suppose x is a normally distributed random variable with $\mu = 11$ and $\sigma = 2$. Find each of the following:
   a. $P(10 \leq x \leq 12)$ **b.** $P(6 \leq x \leq 10)$
   c. $P(13 \leq x \leq 16)$ **d.** $P(7.8 \leq x \leq 12.6)$
   e. $P(x \geq 13.24)$ **f.** $P(x \geq 7.62)$

5.25 Suppose x is a normally distributed random variable with $\mu = 40$ and $\sigma = 10$. Find each of the following:
   a. $P(x \leq 50)$ **b.** $P(x \leq 35.6)$ **c.** $P(35 \leq x \leq 56.8)$
   d. $P(22.9 \leq x \leq 33.2)$
   e. $P(x \geq 25.3)$ **f.** $P(x \leq 25.3)$

5.26 Suppose x is a normally distributed random variable with $\mu = 30$ and $\sigma = 8$. Find a value of the random variable, call it x_0, such that
   a. $P(x \geq x_0) = .5$ **b.** $P(x < x_0) = .025$
   c. $P(x > x_0) = .10$ **d.** $P(x > x_0) = .95$
   e. 10% of the values of x are less than x_0
   f. 80% of the values of x are less than x_0
   g. 1% of the values of x are greater than x_0

5.27 Suppose x is a normally distributed random variable with mean 100 and standard deviation 8. Draw a rough graph of the distribution of x. Locate μ and the interval $\mu \pm 2\sigma$ on the graph. Find the following probabilities:
   a. $P(\mu - 2\sigma \leq x \leq \mu + 2\sigma)$ **b.** $P(x \geq \mu + 2\sigma)$
   c. $P(x \leq 92)$ **d.** $P(92 \leq x \leq 116)$
   e. $P(92 \leq x \leq 96)$ **f.** $P(76 \leq x \leq 124)$

5.28 The random variable x has a normal distribution with standard deviation 25. It is known that the probability that x exceeds 150 is .90. Find the mean μ of the probability distribution.

Applying the Concepts

5.29 In studying the dynamics of fish populations, knowing the length of a species at different ages is critical. *Fisheries Science* (Feb. 1995) published a study of the length distributions of sardines inhabiting Japanese waters. At two years of age, fish have a length distribution that is approximately normal with $\mu = 20.20$ centimeters (cm) and $\sigma = .65$ cm.
   a. Find the probability that a two-year-old sardine inhabiting Japanese waters is between 20 and 21 cm long.
   b. A sardine captured in Japanese waters has a length of 19.84 cm. Is this sardine likely to be two years old?
   c. Repeat part **b** for a sardine with a length of 22.01 cm.

5.30 In a laboratory experiment, researchers at Barry University (Miami Shores, Florida) studied the rate at which sea urchins ingested turtle grass (*Florida Scientist,* Summer/Autumn 1991). The urchins, starved for 48 hours, were fed 5-cm blades of green turtle grass. The mean ingestion time was found to be 2.83 hours and the standard deviation was .79 hour. Assume that green turtle grass ingestion time for the sea urchins has an approximately normal distribution.

   a. Find the probability that a sea urchin will require 4 or more hours to ingest a 5-cm blade of green turtle grass.
   b. Find the probability that a sea urchin will require between 2 and 3 hours to ingest a 5-cm blade of green turtle grass.

5.31 Personnel tests are designed to test a job applicant's cognitive and/or physical abilities. An IQ test is an example of the former; a speed test involving the arrangement of pegs on a peg board is an example of the latter. A particular dexterity test is administered nationwide by a private testing service. It is known that for all tests administered last year the distribution of scores was approximately normal with mean 75 and standard deviation 7.5.
   a. A particular employer requires job candidates to score at least 80 on the dexterity test. Approximately what percentage of the test scores during the past year exceeded 80?
   b. The testing service reported to a particular employer that one of its job candidate's scores fell at the 98th percentile of the distribution (i.e., approximately 98% of the scores were lower than the candidate's, and only 2% were higher). What was the candidate's score?

5.32 The ability of unsaturated field soils to retain water is critical to the soil ecosystem. A team of soil scientists investigated the water retention properties of soil cores sampled from an uncropped field consisting of silt loam (*Soil Science,* Jan. 1995). At a pressure of .1 megapascal (MPa), the water content of the soil (measured in cubic meters of water per cubic meter of soil, m^3/m^3) was determined to be approximately normally distributed with $\mu = .27$ and $\sigma = .04$.
   a. For a soil core at a pressure of .1 MPa, what is the probability its water content is less than .30 m^3/m^3?
   b. In addition to water content readings at a pressure of .1 MPa, measurements were obtained at pressures of 0, .005, .01, .03, and 1.5 MPa. Consider a soil core with a water content reading of .14. Is it likely that this reading was obtained at a pressure of .1 MPa? Explain.

5.33 The average salary for a major league baseball player has risen steadily from $19,000 per year in 1967 to $3,500,000 in 1995.
   a. If the 1995 distribution of salaries is normally distributed with a standard deviation equal to $800,000, what percentage of major league baseball players are making $5,000,000 per year or more?
   b. Can you give a reason why it is unlikely that the distribution of major league baseball salaries is normally distributed?

5.34 In baseball, a "no-hitter" is a regulation 9-inning game in which the pitcher yields no hits to the opposing batters. *Chance* (Summer 1994) reported on a study of no-hitters in major league baseball (MLB). The initial analysis focused on the total number of hits yielded per game per team for all 9-inning MLB games played between 1989 and 1993. The distribution of hits/9-innings is approximately normal with mean 8.72 and standard deviation 1.10.

a. What percentage of 9-inning MLB games result in fewer than 6 hits?

b. Demonstrate, statistically, why a no-hitter is considered an extremely rare occurrence in MLB.

5.35 A study of serum cholesterol levels in psychiatric patients of a maximum-security forensic hospital revealed that cholesterol level is approximately normally distributed with a mean of 208 milligrams per deciliter (mg/dL) and a standard deviation of 25 mg/dL (*Journal of Behavioral Medicine*, Feb. 1995). Prior research has shown that patients who exhibit violent behavior have cholesterol levels below 200 mg/dL.

a. What is the probability of observing a psychiatric patient with a cholesterol level below 200 mg/dL?

b. For three randomly selected patients, what is the probability that at least one will have a cholesterol level below 200 mg/dL?

5.36 The amount of oxygen dissolved in rivers and streams depends on the water temperature and on the amounts of decaying organic matter from natural processes or human disturbances that are present in the water. The Council on Environmental Quality (CEQ) considers a dissolved oxygen content of less than 5 milligrams per liter (mg/L) of water to be undesirable because it is unlikely to support aquatic life. Suppose an industrial plant discharges its waste into a river and the downstream daily oxygen content measurements are normally distributed with a mean equal to 6.3 mg/L and a standard deviation of .6 mg/L.

a. What percentage of the days would the dissolved oxygen content in the river be considered undesirable by the CEQ?

b. Within what limits would we expect the dissolved oxygen content to fall?

5.37 A machine used to regulate the amount of dye dispensed for mixing shades of paint can be set so that it discharges an average of μ milliliters (mL) of dye per can of paint. The amount of dye discharged is known to have a normal distribution with a standard deviation of .4 mL. If more than 6 mL of dye is discharged when making a certain shade of blue paint, the shade is unacceptable. Determine the setting for μ so that only 1% of the cans of paint will be unacceptable.

5.38 A physical-fitness association is including the mile run in its secondary-school fitness test for boys. The time for this event for boys in secondary school is approximately normally distributed with a mean of 450 seconds and a standard deviation of 40 seconds. If the association wants to designate the fastest 10% as "excellent," what time should the association set for this criterion?

5.39 *Ecological Applications* (May 1995) published a study on the development of forests following wildfires in the Pacific Northwest. One variable of interest to the researcher was tree diameter at breast height 110 years after the fire. The population of Douglas fir trees was shown to have an approximately normal diameter distribution, with $\mu = 50$ centimeters (cm) and $\sigma = 12$ cm. Find the diameter, d, such that 30% of the Douglas fir trees in the population have diameters that exceed d.

5.40 Geologists have successfully used statistical models to evaluate the nature of sedimentary deposits (called *facies*) in reservoirs. One of the model's key parameters is the proportion P of facies bodies in a reservoir. An article in *Mathematical Geology* (Apr. 1995) demonstrated that the number of facies bodies that must be sampled to satisfactorily estimate P is approximately normally distributed with $\mu = 99$ and $\sigma = 4.3$. How many facies bodies are required to satisfactorily estimate P for 99% of the reservoirs evaluated?

5.41 What relationship exists between the standard normal distribution and the box-plot methodology (optional Section 2.8) for describing distributions of data using quartiles? The answer depends on the true underlying probability distribution of the data. Assume for the remainder of this exercise that the distribution is normal.

a. Calculate the values of the standard normal random variable z, call them z_L and z_U, that correspond to the hinges of the box plot—i.e., the lower and upper quartiles, Q_L and Q_U—of the probability distribution.

b. Calculate the z values that correspond to the inner fences of the box plot for a normal probability distribution.

c. Calculate the z values that correspond to the outer fences of the box plot for a normal probability distribution.

d. What is the probability that an observation lies beyond the inner fences of a normal probability distribution? The outer fences?

e. Can you better understand why the inner and outer fences of a box plot are used to detect outliers in a distribution? Explain.

5.4   APPROXIMATING A BINOMIAL DISTRIBUTION WITH A NORMAL DISTRIBUTION

When a binomial random variable can assume a large number of values, the calculation of its probabilities may become very tedious. To contend with this problem, we provide tables in Appendix A to give the probabilities for some values of n and p, but these tables are by necessity incomplete. In particular, the binomial probability table (Table II) can be used only for $n = 5, 6, 7, 8, 9, 10, 15, 20,$ or 25. To deal with this limitation, we seek approximation procedures for calculating the probabilities associated with a binomial probability distribution.

When n is large, a normal probability distribution may be used to provide a good approximation to the probability distribution of a binomial random variable. To show how this approximation works, we refer to Example 4.10, in which we used the binomial distribution to model the number x of 20 voters who favor a candidate. We assumed that 60% of all the eligible voters favored the candidate. The mean and standard deviation of x were found to be $\mu = 12$ and $\sigma = 2.2$. The binomial distribution for $n = 20$ and $p = .6$ is shown in Figure 5.18, and the approximating normal distribution with mean $\mu = 12$ and standard deviation $\sigma = 2.2$ is superimposed.

As part of Example 4.10, we used Table II to find the probability that $x \leq 10$. This probability, which is equal to the sum of the areas contained in the rectangles (shown in Figure 5.18) that correspond to $p(0), p(1), p(2), \ldots, p(10)$, was found to equal .245. The portion of the approximating normal curve that would be used to approximate the area $p(0) + p(1) + \cdots + p(10)$ is shaded in Figure 5.18. Note that this shaded area lies to the left of 10.5 (not 10), so we may include all of the probability in the rectangle corresponding to $p(10)$. Because we are approximating a discrete distribution (the binomial) with a continuous distribution (the normal), we call the use of 10.5 (instead of 10 or 11) a **correction for continuity**. That is, we are correcting the discrete distribution so that it can be approximated by the continuous one. The use of the correction for continuity leads to the calculation of the following standard normal z value:

$$z = \frac{x - \mu}{\sigma} = \frac{10.5 - 12}{2.2} = -.68$$

FIGURE 5.18

Binomial distribution for $n = 20, p = .6$ and normal distribution with $\mu = 12, \sigma = 2.2$

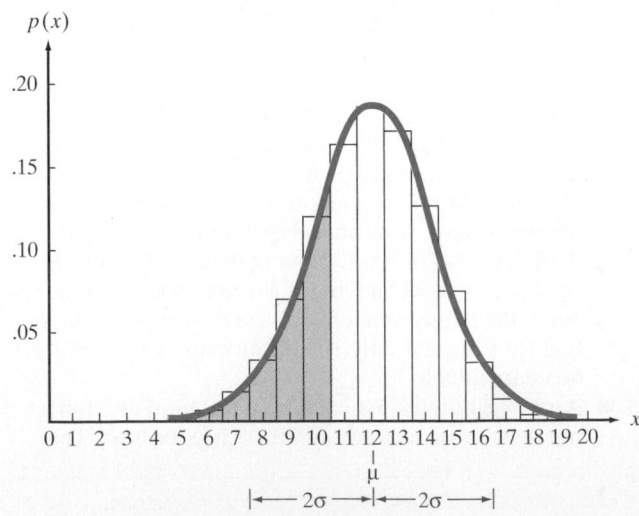

Using Table IV, we find the area between $z = 0$ and $z = .68$ to be .2517. Then the probability that x is less than or equal to 10 is approximated by the area under the normal distribution to the left of 10.5, shown shaded in Figure 5.18. That is,

$$P(x \leq 10) \approx P(z \leq -.68) = .5 - P(-.68 < z \leq 0)$$
$$= .5 - .2517 = .2438$$

The approximation differs only slightly from the exact binomial probability, .245. Of course, when tables of exact binomial probabilities are available, we will use the exact value rather than a normal approximation.

Use of the normal distribution will not always provide a good approximation for binomial probabilities. The following is a useful rule of thumb to determine when n is large enough for the approximation to be effective: The interval $\mu \pm 3\sigma$ should lie within the range of the binomial random variable x (i.e., 0 to n) in order for the normal approximation to be adequate. The rule works well because almost all of the normal distribution falls within 3 standard deviations of the mean, so if this interval is contained within the range of x values, there is "room" for the normal approximation to work.

As shown in Figure 5.19a for the preceding example with $n = 20$ and $p = .6$, the interval $\mu \pm 3\sigma = 12 \pm 3(2.19) = (5.43, 18.57)$ lies within the range 0 to 20. However, if we were to try to use the normal approximation with $n = 10$ and $p = .1$, the interval $\mu \pm 3\sigma$ is $1 \pm 3(.95)$, or $(-1.85, 3.85)$. As shown in Figure 5.19b, this interval is not contained within the range of x since $x = 0$ is the lower bound for a binomial random variable. Note in Figure 5.19b that the normal distribution

FIGURE 5.19

Rule of thumb for normal approximation to binomial probabilities

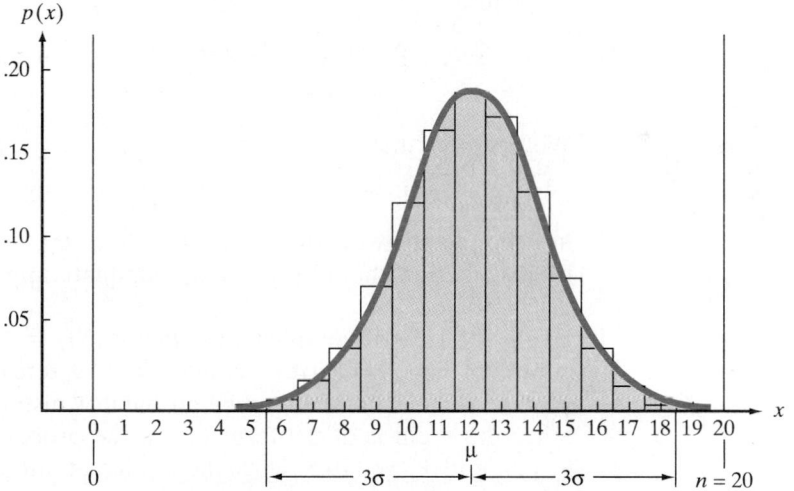

a. $n = 20, p = .6$: Normal approximation is good

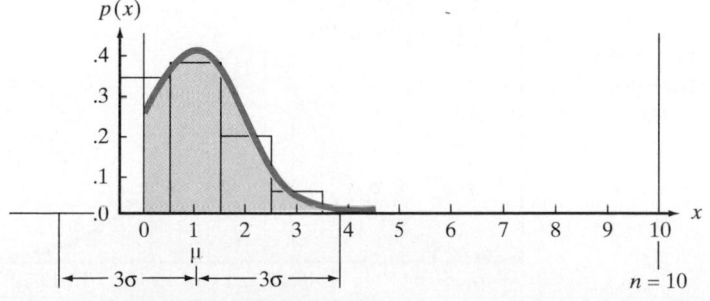

b. $n = 10, p = .1$: Normal approximation is poor

will not "fit" in the range of *x*, and therefore it will not provide a good approximation to the binomial probabilities.

EXAMPLE 5.11 The pocket calculator has become relatively inexpensive because its solid-state circuitry is stamped by machine, thus making mass production feasible. One problem with anything that is mass-produced is quality control. The process must somehow be monitored or audited to be sure the output of the process conforms to requirements.

One method of dealing with this problem is **lot acceptance sampling**, in which items being produced are sampled at various stages of the production process and are carefully inspected. The lot of items from which the sample is drawn is then accepted or rejected, based on the number of defectives in the sample. Lots that are accepted may be sent forward for further processing or may be shipped to customers; lots that are rejected may be reworked or scrapped. For example, suppose a manufacturer of calculators chooses 200 stamped circuits from the day's production and determines *x*, the number of defective circuits in the sample. Suppose that up to a 6% rate of defectives is considered acceptable for the process.

a. Find the mean and standard deviation of *x*, assuming the defective rate is 6%.
b. Use the normal approximation to determine the probability that 20 or more defectives are observed in the sample of 200 circuits (i.e., find the approximate probability that $x \geq 20$).

Solution

a. The random variable *x* is binomial with $n = 200$ and the fraction defective $p = .06$. Thus,

$$\mu = np = 200(.06) = 12$$
$$\sigma = \sqrt{npq} = \sqrt{200(.06)(.94)} = \sqrt{11.28} = 3.36$$

We first note that

$$\mu \pm 3\sigma = 12 \pm 3(3.36) = 12 \pm 10.08 = (1.92, 22.08)$$

lies completely within the range from 0 to 200. Therefore, a normal probability distribution should provide an adequate approximation to this binomial distribution.

b. Using the rule of complements, $P(x \geq 20) = 1 - P(x \leq 19)$. To find the approximating area corresponding to $x \leq 19$, refer to Figure 5.20. Note that we want to include all the binomial probability histogram from 0 to 19, inclusive. Since the event is of the form $x \leq a$, the proper correction for continuity is $a + .5 = 19 + .5 = 19.5$. Thus, the *z* value of interest is

FIGURE 5.20

Normal approximation to the binomial distribution with $n = 200, p = .06$

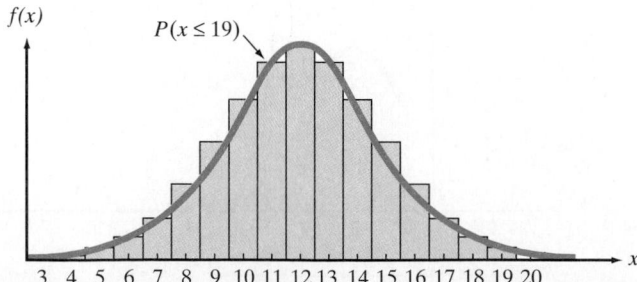

FIGURE 5.21

Standard normal distribution

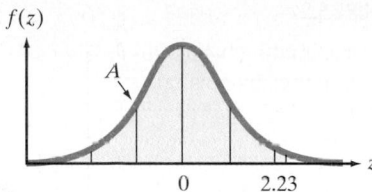

$$z = \frac{(a + .5) - \mu}{\sigma} = \frac{19.5 - 12}{3.36} = \frac{7.5}{3.36} = 2.23$$

Referring to Table IV in Appendix A, we find that the area to the right of the mean 0 corresponding to $z = 2.23$ (see Figure 5.21) is .4871. So the area $A = P(z \leq 2.23)$ is

$$A = .5 + .4871 = .9871$$

Thus, the normal approximation to the binomial probability we seek is

$$P(x \geq 20) = 1 - P(x \leq 19) \approx 1 - .9871 = .0129$$

In other words, the probability is extremely small that 20 or more defectives will be observed in a sample of 200 circuits—*if in fact the true fraction of defectives is* .06. If the manufacturer observes $x \geq 20$, the likely reason is that the process is producing more than the acceptable 6% defectives. The lot acceptance sampling procedure is another example of using the rare-event approach to make inferences.   ▲

The steps for approximating a binomial probability by a normal probability are given in the accompanying box.

Using a Normal Distribution to Approximate Binomial Probabilities

1. After you have determined n and p for the binomial distribution, calculate the interval

$$\mu \pm 3\sigma = np \pm 3\sqrt{npq}$$

   If the interval lies in the range 0 to n, the normal distribution will provide a reasonable approximation to the probabilities of most binomial events.

2. Express the binomial probability to be approximated in the form $P(x \leq a)$ or $P(x \leq b) - P(x \leq a)$. For example,

$$P(x < 3) = P(x \leq 2)$$
$$P(x \geq 5) = 1 - P(x \leq 4)$$
$$P(7 \leq x \leq 10) = P(x \leq 10) - P(x \leq 6)$$

3. For each value of interest a, the correction for continuity is $(a + .5)$, and the corresponding standard normal z value is

$$z = \frac{(a + .5) - \mu}{\sigma} \qquad \text{(See Figure 5.22)}$$

4. Sketch the approximating normal distribution and shade the area corresponding to the probability of the event of interest, as in Figure 5.22. Verify that the rectangles you have included in the shaded area correspond to the event probability you wish to approximate. Using Table IV and the z value(s) you calculated in step 3, find the shaded area. This is the approximate probability of the binomial event.

FIGURE 5.22

Approximating binomial probabilities by normal probabilities

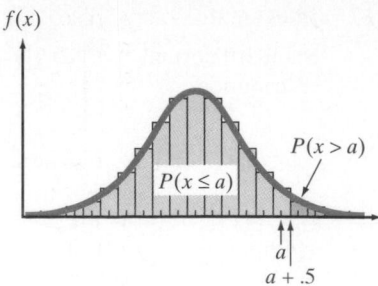

EXERCISES 5.42–5.59

Learning the Mechanics

5.42 Why might you want to use a normal distribution to approximate a binomial distribution?

5.43 What conditions must be satisfied in order for the normal distribution to provide a good approximation to the binomial probability distribution?

5.44 Assume that x is a binomial random variable with n and p as specified in parts **a–f**. For which cases would it be appropriate to use a normal distribution to approximate the binomial distribution?
a. $n = 100, p = .01$ **b.** $n = 20, p = .6$
c. $n = 10, p = .4$ **d.** $n = 1,000, p = .05$
e. $n = 100, p = .8$ **f.** $n = 35, p = .7$

5.45 Suppose x is a binomial random variable with $p = .4$ and $n = 25$.
a. Would it be appropriate to approximate the probability distribution of x with a normal distribution? Explain.
b. Assuming that a normal distribution provides an adequate approximation to the distribution of x, what are the mean and variance of the approximating normal distribution?
c. Use Table II of Appendix A to find the exact value of $P(x \geq 9)$.
d. Use the normal approximation to find $P(x \geq 9)$.

5.46 Assume that x is a binomial random variable with $n = 25$ and $p = .5$. Use Table II of Appendix A and the normal approximation to find the exact and approximate values, respectively, for the following probabilities:
a. $P(x \leq 11)$ **b.** $P(x \geq 16)$
c. $P(8 \leq x \leq 16)$

5.47 Assume that x is a binomial random variable with $n = 100$ and $p = .40$. Use a normal approximation to find the following:
a. $P(x \leq 35)$ **b.** $P(40 \leq x \leq 50)$ **c.** $P(x \geq 38)$

5.48 Assume that x is a binomial random variable with $n = 100$ and $p = .50$. Find each of the following probabilities:
a. $P(x > 500)$ **b.** $P(490 \leq x < 500)$
c. $P(x > 550)$

Applying the Concepts

5.49 In Exercise 4.43 you learned that some researchers believe that one in every three women is a victim of domestic abuse (*Annals of Internal Medicine,* Nov. 1995).
a. For a random sample of 150 women, what is the approximate probability that more than half are victims of domestic abuse?
b. For a random sample of 150 women, what is the approximate probability that fewer than 50 are victims of domestic abuse?
c. Would you expect to observe fewer than 30 domestically abused women in a sample of 150? Explain.

5.50 Refer to the *National Wildlife* study of resting habits of fence lizards, Exercise 4.39. Recall that the probability that a female fence lizard will be resting at any given time is .97.
a. Suppose 1,000 female fence lizards are under observation in a laboratory. For a certain experiment you want to identify the 10 most active lizards at any given time. What is the approximate probability that you observe at least 10 active (nonresting) lizards?
b. Suppose 50 female fence lizards are under observation and you are interested in the probability that fewer than 40 are resting at any given time. Explain why the approximate method of part **a** may not yield an accurate estimate of the probability.

5.51 An article in *The International Journal of Sports Psychology* (July–Sept. 1990) evaluated the relationship between physical fitness and stress. The research revealed that white-collar workers in good physical condition have only a 10% probability of developing a stress-related health problem. What is the probability that more than 60 in a random sample of 400 white-collar employees in good physical condition will develop stress-related illnesses?

5.52 The *Statistical Abstract of the United States* reports that 25% of the country's households are composed of one person. If 1,000 homes are randomly selected to participate in a Nielsen survey to determine television ratings, find the approximate probability that no more than 240 of these homes are one-person households.

5.53 In Case Study 4.2, the number of shuttle catastrophes caused by booster failure in n missions was treated as a binomial random variable. Using the binomial distribution and the probability of catastrophe determined by the Air Force's risk assessment study ($1/35$), you determined the probability of at least 1 shuttle catastrophe in 25 missions to be .5155.

   a. Based on the guidelines presented in this section, would it have been advisable to approximate this probability using the normal approximation to the binomial distribution? Explain.

   b. Regardless of your answer to part **a**, use the normal distribution to approximate the binomial probability. Comment on the difference between the exact and approximate probabilities.

   c. Refer to part **a**. Would the normal approximation be advisable if $n = 100$? If $n = 500$? If $n - 1,000$?

   d. Approximate the probability that more than 25 catastrophes occur in 1,000 flights, assuming that the probability of a catastrophe in any given flight remains $1/35$.

5.54 If you are attracted to lotteries, you should know that one of the best is run by the U.S. government. Furthermore, with some research you can improve your chances of winning. We refer to the Interior Department's lottery for oil and gas leases. The *Wall Street Journal* reported on abuses in the lottery that occurred in the past—particularly the tendency of the Interior Department to include valuable oil leases, some worth millions of dollars, among the many included in the lottery. The Interior Department claims that only 5 to 10% of the leases included in the $75 per ticket lottery should be salable to an oil company. Yet 184 of 328 winners in the July 1980 lottery of Wyoming lands were able to sell their leases to oil companies. Suppose as many as 10% of all leases awarded by the Interior Department are salable to oil companies to answer the following questions:

   a. What is the probability that as many as 50 leases (i.e., 50 or more) would be salable to oil companies?

   b. If 184 of the 328 leases awarded in the July 1980 Wyoming lottery were in fact salable to oil companies, would you regard this as a rare event? Explain.

   c. Given the outcome of the Wyoming lottery and considering your answer to part **b**, do you believe the Interior Department's claim? Explain.

5.55 According to the *American Cancer Society*, melanoma, a form of skin cancer, kills 60% of

Americans who suffer from the disease each year. Consider a sample of 10,000 melanoma patients.

   a. What are the expected value and variance of x, the number of the 10,000 melanoma patients who die of the affliction this year?

   b. Find the probability that x will exceed 6,100 patients per year.

   c. Would you expect x, the number of patients dying of melanoma, to exceed 6,500 in any single year? Explain.

5.56 A recent study involving attrition rates at a major university has shown that 43% of all incoming first-year students do not graduate within 4 years of entrance.

   a. If 200 first-year students are randomly sampled this year and their progress through college is followed, what is the approximate probability that no less than half will graduate within the next 4 years?

   b. What is the approximate probability that the number of sampled first-year students graduating within 4 years will be between 40 and 80?

5.57 A manufacturer of 3.5-inch computer disks claims that 99.4% of its disks are defect-free. A large software company that buys and uses large numbers of the disks wants to verify this claim, so it selects 1,600 disks to be tested. The tests reveal 12 disks to be defective. Assuming that the disk manufacturer's claim is correct, what is the probability of finding 12 or more defective disks in a sample of 1,600? Does your answer cast doubt on the manufacturer's claim? Explain.

5.58 The median time a patient waits to see a doctor in a large clinic is 20 minutes. On a day when 150 patients visit the clinic, what is the approximate probability that

   a. More than half will wait more than 20 minutes?

   b. More than 85 will wait more than 20 minutes?

   c. More than 60 but fewer than 90 will wait more than 20 minutes?

5.59 The percentage of fat in the bodies of American men is an approximately normal random variable with mean equal to 15% and standard deviation equal to 2%.

   a. If these values were used to describe the body fat of men in the United States Army and if 20% or more body fat is characterized as obese, what is the approximate probability that a random sample of 10,000 soldiers will contain fewer than 50 who would be characterized as obese?

   b. If the army actually were to check the percentage of body fat for a random sample of 10,000 men and if only 30 contained 20% (or higher) body fat, would you conclude that the army was successful in reducing the percentage of obese men below the percentage in the general population? Explain your reasoning.

5.5   THE EXPONENTIAL DISTRIBUTION (OPTIONAL)

The length of time between emergency arrivals at a hospital, the length of time between breakdowns of manufacturing equipment, the length of time between catastrophic events (floods, earthquakes, etc.), and the distance traveled by a wildlife ecologist between sightings of an endangered species are all random phenomena that we might want to describe probabilistically. The amount of time or distance between occurrences of random events like these can often be described by the **exponential probability distribution**. For this reason, the exponential distribution is sometimes called the **waiting time distribution**. The formula for the exponential probability distribution is shown in the box along with its mean and standard deviation.

Probability Distribution, Mean, and Standard Deviation for an Exponential Random Variable x

$$f(x) = \frac{1}{\theta} e^{-x/\theta} \qquad (x > 0)$$

$$\mu = \theta \qquad\qquad \sigma = \theta$$

Unlike the normal distribution, which has a shape and location determined by the values of the two quantities μ and σ, the shape of the exponential distribution is governed by a single quantity, θ. Further, it is a probability distribution with the property that its mean equals its standard deviation. Exponential distributions corresponding to $\theta = .5$, 1, and 2 are shown in Figure 5.23.

To calculate probabilities for exponential random variables, we need to be able to find areas under the exponential probability distribution. Suppose we want to find the area A to the right of some number a, as shown in Figure 5.24. This area can be calculated by using the following formula:

Finding the Area A **to the Right of a Number** a **for an Exponential Distribution***

$$A = P(x \geq a) = e^{-a/\theta}$$

FIGURE 5.23

Exponential distributions

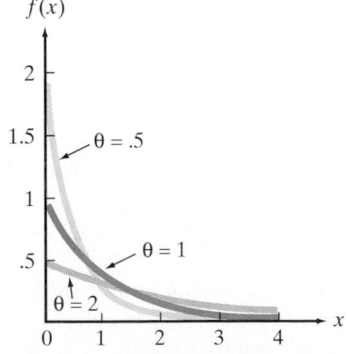

*For students with a knowledge of calculus, the shaded area in Figure 5.24 corresponds to the integral
$$\int_{a}^{\infty} \frac{1}{\theta} e^{-x/\theta}\, dx = -e^{-x/\theta}\Big|_{a}^{\infty} = e^{-a/\theta}$$

FIGURE 5.24

The area A to the right of a number a for an exponential distribution

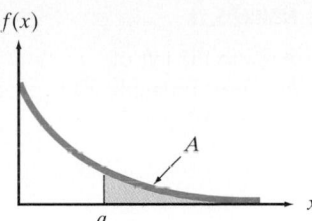

Use Table V in Appendix A to find the value of $e^{-a/\theta}$ after substituting the appropriate numerical values for θ and a.

EXAMPLE 5.12

Suppose the length of time (in hours) between emergency arrivals at a certain hospital is modeled as an exponential distribution with $\theta = 2$. What is the probability that more than 5 hours pass without an emergency arrival?

Solution

The probability we want is the area A to the right of $a = 5$ in Figure 5.25. To find this probability, use the formula given for area:

$$A = e^{-a/\theta} = e^{-(5/2)} = e^{-2.5}$$

Referring to Table V, we find

$$A = e^{-2.5} = .082085$$

Our exponential model indicates that the probability that more than 5 hours pass between emergency arrivals is about .08 for this hospital. ▲

EXAMPLE 5.13

A microwave oven manufacturer is trying to determine the length of warranty period it should attach to its magnetron tube, the most critical component in the oven. Preliminary testing has shown that the length of life (in years), x, of a magnetron tube has an exponential probability distribution with $\theta = 6.25$.

a. Find the mean and standard deviation of x.
b. Suppose a warranty period of 5 years is attached to the magnetron tube. What fraction of tubes must the manufacturer plan to replace, assuming that the exponential model with $\theta = 6.25$ is correct?
c. Find the probability that the length of life of a magnetron tube will fall within the interval $\mu - 2\sigma$ to $\mu + 2\sigma$.

FIGURE 5.25

Area to the right of $a = 5$ for Example 5.12

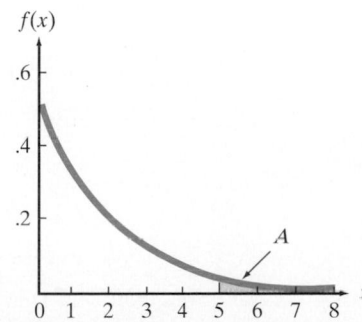

FIGURE 5.26

Area to the left of
$a = 5$ for Example 5.13

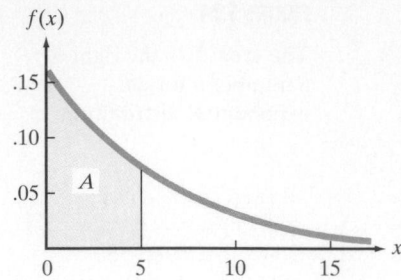

Solution

a. Since $\theta = \mu = \sigma$, both μ and σ equal 6.25.
b. To find the fraction of tubes that will have to be replaced before the 5-year warranty period expires, we need to find the area between 0 and 5 under the distribution. This area, *A*, is shown in Figure 5.26. To find the required probability, we recall the formula

$$P(x > a) = e^{-a/\theta}$$

Using this formula, we can find

$$P(x > 5) = e^{-a/\theta} = e^{-5/6.25} = e^{-.80} = .449329$$

(see Table V). To find the area *A*, we use the complementary relationship:

$$P(x \leq 5) = 1 - P(x > 5) = 1 - .449329 = .550671$$

So approximately 55% of the magnetron tubes will have to be replaced during the 5-year warranty period.

c. We would expect the probability that the life of a magnetron tube, *x*, falls within the interval $\mu - 2\sigma$ to $\mu + 2\sigma$ to be quite large. A graph of the exponential distribution showing the interval $\mu - 2\sigma$ to $\mu + 2\sigma$ is given in Figure 5.27. Since the point $\mu - 2\sigma$ lies below $x = 0$, we need to find only the area between $x = 0$ and $x = \mu + 2\sigma = 6.25 + 2(6.25) = 18.75$. This area, *P*, which is shaded in Figure 5.27, is

$$P = 1 - P(x > 18.75)$$
$$= 1 - e^{-18.75/\theta} = 1 - e^{-18.75/6.25} = 1 - e^{-3}$$

FIGURE 5.27

Area in the interval $\mu \pm 2\sigma$
for Example 5.13

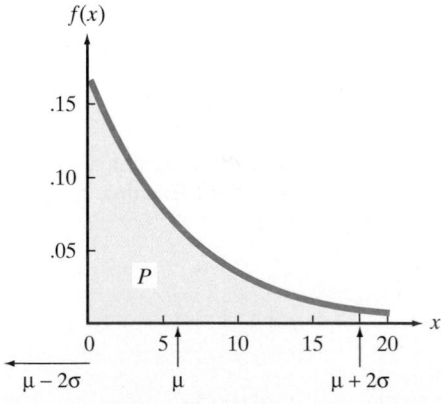

Is New Always Better Than Used?

CASE STUDY • 5.2 •

Continuous probability distributions provide useful theoretical models for the lifelength of a component (e.g., a computer chip, halogen light bulb, compact disk player, and so on). Often, it is important to know whether or not it is better to periodically replace an old component with a new component. For example, for certain types of light bulbs, an old bulb that has been in use for a while tends to have a longer lifelength than a new bulb.

Let the continuous random variable x represent the lifelength (in hours) of some component—say a new halogen light bulb. Assume that x has an exponential distribution with mean $\theta = 250$ hours.

Focus

a. Find the probability that the lifelength of the new halogen bulb exceeds 500 hours.

b. For two independently selected new halogen bulbs, find the probability that one has a lifelength exceeding 300 hours and the other has a lifelength exceeding 200 hours.

According to *Microelectronics and Reliability* (Jan. 1986), the "life" distribution of x is considered *new better than used* (NBU) if

$$P(x > a + b) \le P(x > a)P(x > b)$$

Alternatively, a "life" distribution is considered *new worse than used* (NWU) if

$$P(x > a + b) \ge P(x > a)P(x > b)$$

Focus

c. Use the probabilities, parts a–b, to show that when $a = 300$ and $b = 200$ the exponential distribution is both NBU and NWU.

d. Choose any two positive numbers a and b, where $a > 0$ and $b > 0$, and repeat part **c**. (Have each member of your class choose different values of a and b.)

e. Show that, in general, for any a and b ($a > 0$ and $b > 0$), the exponential distribution with mean θ is both NBU and NWU. Such a "life" distribution is said to be *new same as used* or *memoryless*. Explain why.

Checking Table V for the value of e^{-3}, we find $e^{-3} = .049787$. Therefore, the probability that the life x of a magnetron tube will fall within the interval $\mu - 2\sigma$ to $\mu + 2\sigma$ is

$$P = 1 - e^{-3} = 1 - .049787 = .950213$$

You can see that this probability agrees very well with the interpretation of a standard deviation given by the Empirical Rule in Table 2.9 even though this probability distribution is not mound-shaped. (It is strongly skewed to the right.)   ▲

EXERCISES 5.60–5.74

Learning the Mechanics

5.60 The random variables x and y have exponential distributions with $\theta = 3$ and $\theta = 1$, respectively. Using Table V in Appendix A, carefully plot both distributions on the same set of axes.

5.61 Use Table V in Appendix A to determine the value of $e^{-a/\theta}$ for each of the following cases.

   a. $\theta = 1, a = 1$ **b.** $\theta = 1, a = 2.5$
   c. $\theta = .4, a = 3$ **d.** $\theta = .2, a = .3$

5.62 Suppose x has an exponential distribution with $\theta = 1$. Find the following probabilities:

   a. $P(x > 1)$ **b.** $P(x \le 3)$
   c. $P(x > 1.5)$ **d.** $P(x \le 5)$

5.63 Suppose x has an exponential distribution with $\theta = 2.5$. Find the following probabilities:

   a. $P(x \le 4)$ **b.** $P(x > .5)$
   c. $P(x \le 2)$ **d.** $P(x > 3)$

5.64 Suppose the random variable x has an exponential probability distribution with $\theta = 1$. Find the mean and standard deviation of x. Find the probability that x will assume a value within the interval $\mu \pm 2\sigma$.

5.65 The random variable x can be adequately approximated by an exponential probability distribution with $\theta = 2$. Find the probability that x assumes a value

   a. More than 3 standard deviations from μ

   b. Less than 2 standard deviations from μ

   c. Within .5 standard deviation of μ

Applying the Concepts

5.66 In the National Hockey League (NHL), games that are tied at the end of three periods are sent to "sudden-death" overtime. In overtime, the team to score the first goal wins. An analysis of all NHL overtime games played between 1970 and 1993 showed that the length of time elapsed before the winning goal is scored has an approximate exponential distribution with mean 9.15 minutes (*Chance,* Winter 1995).

  a. For a randomly selected overtime NHL game, find the probability that the winning goal is scored in 3 minutes or less.

  b. In the NHL, each period (including overtime) lasts 20 minutes. If neither team scores a goal in overtime, the game is considered a tie. What is the probability of an NHL game ending in a tie?

5.67 After bacteria are subjected to a certain drug, the length of time until the bacteria die follows an exponential distribution with a mean of $\frac{1}{2}$ hour.

  a. How long will it take for half the bacteria to die after the drug is administered?

  b. What proportion of bacteria will die between .2 and 1 hour?

5.68 A part processed in a flexible manufacturing system (FMS) is routed through a set of operations, some of which are sequential and some of which are parallel. In addition, an FMS operation can be processed by alternative machines. An article in *IEEE Transactions* (Mar. 1990) gave an example of an FMS with four machines operating independently. The repair rates for the machines (i.e., the time, in hours, it takes to repair a failed machine) are exponentially distributed with means $\mu_1 = 1$, $\mu_2 = 2$, $\mu_3 = .5$, and $\mu_4 = .5$, respectively.

  a. Find the probability that the repair time for machine 1 exceeds 1 hour.

  b. Repeat part **a** for machine 2.

  c. Repeat part **a** for machines 3 and 4.

  d. If all four machines fail simultaneously, find the probability that the repair time for the entire system exceeds 1 hour.

5.69 A taxi service based at an airport can be characterized as a transportation system with one source terminal and a fleet of vehicles. Each vehicle takes passengers from the terminal to different destinations; then it returns to the terminal after some random trip time and makes another trip. To improve the vehicle-dispatching decisions involved in such a system (e.g., How many passengers should be allocated to a waiting taxi?), a study was conducted and published in the *European Journal of Operational Research* (Vol. 21, 1985). In modeling the system, the authors assumed travel times of successive trips to be independent exponential random variables. Assume $\theta = 20$ minutes.

  a. What is the mean trip time for the taxi service?

  b. What is the probability that a particular trip will take more than 30 minutes?

  c. Two taxis have just been dispatched. What is the probability that both will be gone for more than 30 minutes? That at least one of the taxis will return within 30 minutes?

5.70 Refer to the *Ecological Applications* (May 1995) study of forest development 110 years following a wildfire, Exercise 5.39. Another species of tree, western hemlock, was found to have a breast height diameter distribution that resembled an exponential distribution with $\theta = 30$ centimeters. Find the probability that a western hemlock tree growing in the forest damaged by wildfire 110 years ago has a diameter that exceeds 25 centimeters.

5.71 An infestation of a certain species of caterpillar, the spruce budworm, can cause extensive damage to the timberlands of the northern United States. It is known that an outbreak of this type of infestation occurs, on the average, every 30 years. Assuming that this phenomenon obeys an exponential probability law, what is the probability that catastrophic outbreaks of spruce budworm infestation will occur within 6 years of each other?

5.72 An article in the Jacksonville, Florida, *Times Union* (Mar. 11, 1984) reported on the unexplained crash of a small plane on takeoff and the resulting injury to its 24-year-old student pilot. The article noted that Shields Aviation, the renter of the plane, inspects (and presumably performs maintenance on) the aircraft every 100 flight hours. The plane had flown 30 hours since its last inspection. Suppose *x,* the time between malfunctions for this particular plane, has an exponential distribution with mean equal to 300 hours. What is the probability that a plane of this type will malfunction within 30 hours after the last inspection?

5.73 The importance of modeling machine downtime correctly in simulation studies was discussed in *Industrial Engineering* (Aug. 1990). The paper presented simulation results for a single-machine-tool system with the following properties:

  • The interarrival times of jobs are exponentially distributed with a mean of 1.25 minutes

  • The amount of time the machine operates before breaking down is exponentially distributed with a mean of 540 minutes

  a. Find the probability that two jobs arrive for processing at most 1 minute apart.

  b. Find the probability that the machine operates for at least 720 minutes (12 hours) before breaking down.

5.74 The probability distribution of the length of service time is important in the design of service facilities, and it has a definite effect on sales. Suppose the time an individual has to wait in line to be served at a fast-food franchise is assumed to have an exponential distribution with mean equal to 1 minute.

a. What is the probability that an individual would have to wait more than 2 minutes before being served?

b. What is the probability that an individual will be served within 30 seconds after arriving in line?

QUICK REVIEW

Key Terms

Note: Starred (*) items refer to the optional section in this chapter.

Bell curve   209
Continuous probability distribution   194
Correction for continuity   212
Exponential distribution*   218
Frequency function   194

Normal distribution   198
Probability density function   194
Standard normal distribution   204
Uniform distribution   195
Waiting time distribution*   218

Key Formulas

Note: Starred (*) items refer to the optional section in this chapter.

Random Variable	Density Function	Mean	Standard Deviation	
Uniform, x	$f(x) = \dfrac{1}{d-c}$ $(c \leq x \leq d)$	$\mu = \dfrac{c+d}{2}$	$\sigma = \dfrac{d-c}{\sqrt{12}}$	196
Normal, x	$f(x) = \dfrac{1}{\sigma\sqrt{2\pi}}\, e^{-(1/2)[(x-\mu)/\sigma]^2}$	μ	σ	199
Standard Normal, $z = \left(\dfrac{x-\mu}{\sigma}\right)$	$f(z) = \dfrac{1}{\sqrt{2\pi}}\, e^{-(1/2)z^2}$	$\mu = 0$	$\sigma = 1$	200
*Exponential, x	$f(x) = \dfrac{1}{\theta}\, e^{-x/\theta}$ $(x > 0)$	$\mu = \theta$	$\sigma = \theta$	218

Normal Approximation to Binomial

$$P(x \leq a) = P\left[z \leq \frac{(a+.5) - \mu}{\sigma} \right]$$

215

LANGUAGE LAB

Symbol	Pronunciation	Description
$f(x)$	f of x	Probability density function for a continuous random variable x
θ	theta	Mean of exponential random variable

SUPPLEMENTARY EXERCISES 5.75–5.98

Note: Starred (*) exercises refer to the optional section in this chapter.

Learning the Mechanics

5.75 Assume that x is a random variable best described by a uniform distribution with $c = 40$ and $d = 70$.
  a. Find $f(x)$.
  b. Find the mean and standard deviation of x.
  c. Graph the probability distribution for x and locate its mean and the interval $\mu \pm 2\sigma$ on the graph.
  d. Find $P(x \le 45)$.   **e.** Find $P(x \ge 58)$.
  f. Find $P(x \le 100)$.
  g. Find $P(\mu - \sigma \le x \le \mu + \sigma)$.   **h.** Find $P(x > 60)$.

5.76 Find the following probabilities for the standard normal random variable z:
  a. $P(z \le 2.1)$ **b.** $P(z \ge 2.1)$ **c.** $P(z \ge -1.65)$
  d. $P(-2.13 \le z \le -.41)$ **e.** $P(-1.45 \le z \le 2.15)$
  f. $P(z \le -1.43)$

5.77 Find a z-score, call it z_0, such that
  a. $P(z \le z_0) = .8708$ **b.** $P(z \ge z_0) = .0526$
  c. $P(z \le z_0) = .5$ **d.** $P(-z_0 \le z \le z_0) = .8164$
  e. $P(z \ge z_0) = .8023$ **f.** $P(z \ge z_0) = .0041$

5.78 The random variable x has a normal distribution with $\mu = 70$ and $\sigma = 10$. Find the following probabilities:
  a. $P(x \le 75)$ **b.** $P(x \ge 90)$ **c.** $P(60 \le x \le 75)$
  d. $P(x > 75)$ **e.** $P(x = 75)$ **f.** $P(x \le 95)$

5.79 The random variable x has a normal distribution with $\mu = 40$ and $\sigma^2 = 36$. Find a value of x, call it x_0, such that
  a. $P(x \ge x_0) = .5$ **b.** $P(x \le x_0) = .9911$
  c. $P(x \le x_0) = .0028$ **d.** $P(x \ge x_0) = .0228$
  e. $P(x \le x_0) = .1003$ **f.** $P(x \ge x_0) = .7995$

5.80 Assume that x is a binomial random variable with $n = 100$ and $p = .5$. Use the normal probability distribution to approximate the following probabilities:
  a. $P(x \le 48)$ **b.** $P(50 \le x \le 65)$ **c.** $P(x \ge 70)$
  d. $P(55 \le x \le 58)$ **e.** $P(x = 62)$
  f. $P(x \le 49 \text{ or } x \ge 72)$

****5.81** Assume that x has an exponential distribution with $\theta = 3.0$. Find
  a. $P(x \le 1)$ **b.** $P(x > 1)$ **c.** $P(x = 1)$
  d. $P(x \le 6)$ **e.** $P(2 \le x \le 10)$

Applying the Concepts

5.82 The *Sociology of Sport Journal* (Vol. 9, 1992) examined the SAT scores of male student-athletes at NCAA Division I institutions. The SAT score is used to determine whether athletes are eligible to participate in athletics during their first year, with the NCAA requiring a minimum score of 700. Suppose that SAT scores of athletes on scholarship have an average of 950 and a standard deviation of 200. Assuming the distribution of SAT scores is approximately normal, what percentage of the athletes will not be eligible their first year?

5.83 Recall the *Astronomical Journal* (July 1995) study of galaxy velocity, Exercise 2.72. The observed velocity of a galaxy located in galaxy cluster A2142 was found to have a normal distribution with mean 27,117 kilometers per second (km/s) and standard deviation 1,280 km/s. A galaxy with a velocity of 24,350 km/s is observed. Comment on the likelihood that this galaxy will be located in cluster A2142.

5.84 The distribution of the demand (in number of units per unit time) for a product can often be approximated by a normal probability distribution. For example, a bakery has determined that the number of loaves of its white bread demanded daily has a normal distribution with mean 7,200 loaves and standard deviation 300 loaves. Based on cost considerations, the company has decided that its best strategy is to produce a sufficient number of loaves so that it will fully supply demand on 94% of all days.
  a. How many loaves of bread should the company produce?
  b. Based on the production in part **a**, on what percentage of days will the company be left with more than 500 loaves of unsold bread?

5.85 For a student to graduate, a high school requires that each student demonstrate competence in mathematics by scoring 70% or above on a mathematics achievement test. The scores of those students taking the test for the first time are approximately normally distributed with a mean of 77% and a standard deviation of 7.3%. What percentage of students who take the test for the first time will pass the test?

****5.86** The length of time between arrivals at a hospital clinic and the length of clinical service time are two random variables that play important roles in designing a clinic and deciding how many physicians and nurses are needed for its operation. The probability distributions of both the length of time between arrivals and the length of service time are often approximately exponential. Suppose the mean time between arrivals for patients at a clinic is 4 minutes.
  a. What is the probability that a particular interarrival time (the time between the arrival of two patients) is less than 1 minute?
  b. What is the probability that the next four interarrival times are all less than 1 minute?
  c. What is the probability that an interarrival time will exceed 10 minutes?

5.87 On the average, the main chute fails in one of every 1,000 parachutes. Suppose during a lifetime a professional parachutist makes 4,000 jumps, and let x equal the number of times the main chute fails.
  a. Show that x has a binomial distribution.
  b. What is the approximate probability that the parachutist's main chute fails on at least one jump?

c. Because n is so large and p is so small, the Poisson probability distribution will also provide a good approximation to the probability, part **b**. (See Exercise 4.55.) If you covered Section 4.5, find the Poisson approximation to $P(x > 0)$.

5.88 The net weight per package of a certain brand of corn chips is listed as 10 ounces. The weight actually delivered to each package by an automated machine is a normal random variable with mean 10.5 ounces and standard deviation .2 ounce. Suppose 1,500 packages are chosen at random and the net weights are ascertained. Let x be the number of the 1,500 selected packages that contain at least 10 ounces of corn chips. Then x is a binomial random variable with $n - 1,500$ and $p -$ probability that a randomly selected package contains at least 10 ounces. What is the probability that they all contain at least 10 ounces of corn chips? What is the probability that at least 90% of the packages contain 10 ounces or more?

5.89 A local track club has decided to sponsor a 10,000-meter road race. From past results of races across the state, it is known that the length of time to complete the race has an approximately normal distribution with a mean of 49 minutes and a standard deviation of 8 minutes. The club has decided that everyone who completes the race will receive a T-shirt. Those who run between 34 and 60 minutes will also receive a medal, and those finishing under 34 minutes will receive a plaque.
a. What proportion of racers would you expect to receive a medal and a T-shirt?
b. What proportion of racers would you expect to receive a plaque and a T-shirt?

5.90 A company has a lump-sum incentive plan for salespeople that is dependent on the level of their sales. If they sell less than $100,000 per year, they receive a $1,000 bonus; from $100,000 to $200,000, they receive $5,000; and above $200,000, they receive $10,000. Suppose the annual sales per salesperson has approximately a normal distribution with $\mu = \$180,000$ and $\sigma = \$50,000$.
a. Find p_1, the proportion of salespeople who receive a $1,000 bonus.
b. Find p_2, the proportion of salespeople who receive a $5,000 bonus.
c. Find p_3, the proportion of salespeople who receive a $10,000 bonus.
d. What is the mean value of the bonus payout for the company? [*Hint:* See the definition for the expected value of a random variable in Chapter 4.]

5.91 Sixteen percent of the African-American population is known to suffer from sickle-cell anemia. If 1,000 African-Americans are sampled at random, what is the approximate probability that
a. More than 175 have the disease?
b. Fewer than 140 have the disease?

c. The number of people in the sample with the disease is between 130 and 180 inclusive?

5.92 An admissions officer for a law school indicates that 35% of the applicants meet all 10 requirements and 95% meet at least eight of the 10 requirements.
a. If a random sample of 300 applicants is taken, what is the approximate probability that fewer than 250 will fail to meet all 10 requirements?
b. What is the approximate probability that more than 280 will meet at least eight of the 10 requirements?

5.93 A. K. Shah published a simple approximation for areas under the normal curve in *The American Statistician* (Feb., 1985). Shah showed that the area A under the standard normal curve between 0 and z is

$$A \approx \begin{cases} z(4.4 - z)/10 & \text{for } 0 \le z \le 2.2 \\ .49 & \text{for } 2.2 < z < 2.6 \\ .50 & \text{for } z \ge 2.6 \end{cases}$$

a. Use the approximation to find
(i) $P(0 < z < 1.2)$ (ii) $P(0 < z < 2.5)$
(iii) $P(z > .8)$ (iv) $P(z < 1.0)$
b. Find the exact probabilities in part **a**.
c. Shah showed that the approximation has a maximum absolute error of .0052. Verify this for the approximations in part **a**.

*5.94 Based on sample data collected in the Denver area, a study found that in some cases the exponential distribution is an adequate approximation for the distribution of the time (in weeks) an individual is unemployed (*Journal of Economics,* Vol. 28, 1985). In particular, the author found the exponential distribution to be appropriate for white and African-American workers but not for Hispanics. Use $\theta = 13$ to answer the following questions.
a. What is the mean time workers are unemployed according to the exponential distribution?
b. Find the probability that a white worker who just lost her job will be unemployed for at least 2 weeks. For more than 6 weeks.
c. What is the probability that an unemployed worker will find a new job within 12 weeks?

5.95 Contrary to our intuition, very reliable decisions concerning the proportion of a large group of consumers favoring a certain product or a certain social issue can be based on relatively small samples. For example, suppose the target population of consumers contains 50 million people and we wish to decide whether the proportion of consumers, p, in the population that favor some product (or issue) is as large as some value, say .2. Suppose you randomly select a sample as small as 1,600 from the 50 million and you observe the number, x, of consumers in the sample who favor the new product. Assuming that $p = .2$, find the mean and standard deviation of x. Suppose that 400 or (25%) of the sample of 1,600 consumers favor the new product. Why might this sample result lead you to conclude that p (the proportion of consumers favoring the product in the population of 50 million) is at least as large as .2? [*Hint:* Find the values of μ and σ for $p = .2$, and use them to decide whether the observed value of x is unusually large.]

5.96 The median income of residents in a certain community is $20,000. If 1,000 people are randomly sampled from the population of residents, let x equal the number whose incomes are less than $20,000. Compute approximate values for the following:
   a. $P(x > 500)$ **b.** $P(x \le 480)$ **c.** $P(475 \le x \le 525)$

*5.97 The number of serious accidents in a manufacturing plant has (approximately) a Poisson probability distribution with a mean of two serious accidents per month. If x, the number of events per unit time, has a Poisson distribution with mean λ, then it can be shown that the time between two successive events has an exponential probability distribution with mean $\theta = 1/\lambda$.
   a. If an accident occurs today, what is the probability that the next serious accident will not occur within the next month?

b. What is the probability that more than one serious accident will occur within the next month?

*5.98 The Poisson probability distribution, like the binomial, can be approximated by a normal probability distribution. This approximation, using $\mu = \lambda$ and $\sigma = \sqrt{\lambda}$, will be good when λ is large (large enough so that the distance between $x = 0$ and λ is at least $3\sigma = 3\sqrt{\lambda}$, or $\lambda \ge 9$). The number of union complaints per month at a certain manufacturing plant has a Poisson probability distribution with $\lambda = 40$ complaints per month. Use the normal approximation to the Poisson probability distribution.
   a. Approximate the probability that the number of complaints in a given month will be less than 35.
   b. Approximate the probability that the number of complaints in a given month exceeds 40.
   c. If the mean number of complaints per month remains constant and if the number of complaints in 1 month is independent of the number in any other, what is the probability that in each of 3 successive months, the number of complaints exceeds 40?

STUDENT PROJECTS

For large values of n the computational effort involved in working with the binomial probability distribution is considerable. Fortunately, in many instances the normal distribution provides a good approximation to the binomial distribution. This project was designed to enable you to demonstrate to yourself how well the normal distribution approximates the binomial distribution.

a. Let the random variable x have a binomial probability with $n = 10$ and $p = .5$. Using the binomial distribution, find the probability that x takes on a value in each of the following intervals: $\mu \pm \sigma$, $\mu \pm 2\sigma$, and $\mu \pm 3\sigma$.

b. Find the probabilities requested in part a by using a normal approximation to the given binomial distribution.

c. Determine the magnitude of the difference between each of the three probabilities as determined by the binomial distribution and by the normal approximation.

d. Letting x have a binomial distribution with $n = 20$ and $p = .5$, repeat parts a, b, and c. Notice that the probability estimates provided by the normal distribution are more accurate for $n = 20$ than for $n = 10$.

e. Letting x have a binomial distribution with $n = 20$ and $p = .01$, repeat parts a, b, and c. Notice that the probability estimates provided by the normal distribution are very poor in this case. Explain why this occurs.

EXPLORING DATA WITH A COMPUTER

Refer to Exploring Data with a Computer in Chapter 4. Again use the computer-generated percentage of drugged patients that experience some type of complication during coronary bypass surgery as an estimate of the true percentage.

a. Use the normal approximation to the binomial to estimate the probability that fewer than $\frac{1}{3}$ of 20 coronary bypass patients (still assuming they represent a random sample) have complications during surgery.

Compare your answer with the exact binomial probability of the same event.

b. Use the normal approximation to the binomial to estimate the probability that fewer than $\frac{1}{3}$ of 200 patients experience complications. If your statistical software package has a function that calculates exact binomial probabilities, use it to calculate the same probability you approximated, and compare the results.

Chapter 6

SAMPLING DISTRIBUTIONS

Contents

Case Studies

*W*HERE WE'VE BEEN

We've learned that the objective of most statistical investigations is inference—that is, making decisions or predictions about a population based on information in a sample. To actually make the decision, we use the sample data to compute sample statistics, such as the sample mean or variance. The knowledge of random variables and their probability distributions enables us to construct theoretical models of populations.

*W*HERE WE'RE GOING

Because sample measurements are observed values of random variables, the value for a sample statistic will vary in a random manner from sample to sample. In other words, since sample statistics are random variables, they possess probability distributions that are either discrete or continuous. In this chapter, we'll explore these probability distributions, which are called *sampling distributions*. You'll learn why many sampling distributions tend to be approximately normal, and you'll see how sampling distributions can be used to evaluate the reliability of inferences made using the statistics.

In Chapters 4 and 5 we assumed that we knew the probability distribution of a random variable, and using this knowledge we were able to compute the mean, variance, and probabilities associated with the random variable. However, in most practical applications, this information is not available. To illustrate, in Example 4.10 we calculated the probability that the binomial random variable *x*, the number of 20 polled voters who favor a certain mayoral candidate, assumed specific values. To do this, it was necessary to assume some value for *p*, the proportion of all voters who favor the candidate. Thus, for the purposes of illustration, we assumed $p = .6$ when, in all likelihood, the exact value of *p* would be unknown. In fact, the probable purpose of taking the poll is to estimate *p*. Similarly, when we modeled the in-city gas mileage of a certain automobile model, we used the normal probability distribution with an *assumed* mean and standard deviation of 27 and 3 miles per gallon, respectively. In most situations, the true mean and standard deviation are unknown quantities that would have to be estimated. Numerical quantities that describe probability distributions are called *parameters*. Thus, *p*, the probability of a success in a binomial experiment, and μ and σ, the mean and standard deviation of a normal distribution, are examples of parameters.

> **DEFINITION 6.1**
>
> A **parameter** is a numerical descriptive measure of a population. Because it is based on the observations in the population, its value is almost always unknown.

We have also discussed the sample mean $\bar{x}$, sample variance s^2, sample standard deviation *s*, etc., which are numerical descriptive measures calculated from the sample. We will often use the information contained in these *sample statistics* to make inferences about the parameters of a population.

> **DEFINITION 6.2**
>
> A **sample statistic** is a numerical descriptive measure of a sample. It is calculated from the observations in the sample.

Note that the term *statistic* refers to a *sample* quantity and the term *parameter* refers to a *population* quantity.

Before we can show you how to use sample statistics to make inferences about population parameters, we need to be able to evaluate their properties. Does one sample statistic contain more information than another about a population parameter? On what basis should we choose the "best" statistic for making inferences about a parameter? The purpose of this chapter is to answer these questions.

6.1 WHAT IS A SAMPLING DISTRIBUTION?

If we want to estimate a parameter of a population—say, the population mean μ—we could use a number of sample statistics for our estimate. Two possibilities are the sample mean $\bar{x}$ and the sample median *m*. Which of these do you think will provide a better estimate of μ?

Before answering this question, consider the following example: Toss a fair die, and let *x* equal the number of dots showing on the up face. Suppose the die is tossed three times, producing the sample measurements 2, 2, 6. The sample mean

FIGURE 6.1

Comparing the sample mean ($\bar{x}$) and sample median (m) as estimators of the population mean (μ)

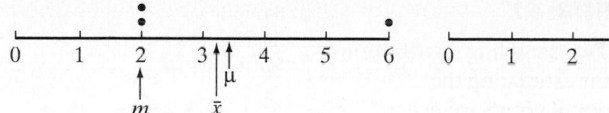

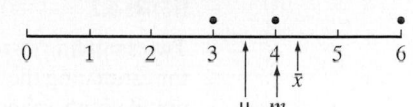

a. Sample 1: $\bar{x}$ is closer than m to μ

b. Sample 2: m is closer than $\bar{x}$ to μ

is $\bar{x} = 3.33$ and the sample median is $m = 2$. Since the population mean of x is $\mu = 3.5$, you can see that for this sample of three measurements, the sample mean $\bar{x}$ provides an estimate that falls closer to μ than does the sample median (see Figure 6.1a). Now suppose we toss the die three more times and obtain the sample measurements 3, 4, 6. The mean and median of this sample are $\bar{x} = 4.33$ and $m = 4$, respectively. This time m is closer to μ (see Figure 6.1b).

This simple example illustrates an important point: Neither the sample mean nor the sample median will *always* fall closer to the population mean. Consequently, we cannot compare these two sample statistics, or, in general, any two sample statistics, on the basis of their performance for a single sample. Instead, we need to recognize that sample statistics are themselves random variables, because different samples can lead to different values for the sample statistics. As random variables, sample statistics must be judged and compared on the basis of their probability distributions, i.e., the *collection* of values and associated probabilities of each statistic that would be obtained if the sampling experiment were repeated a *very large number of times.* We will illustrate this concept with another example.

Suppose it is known that in a certain part of Canada the daily high temperature recorded for all past months of January has a mean of $\mu = 10°F$ and a standard deviation of $\sigma = 5°F$. Consider an experiment consisting of randomly selecting 25 daily high temperatures from the records of past months of January and calculating the sample mean $\bar{x}$. If this experiment were repeated a very large number of times, the value of $\bar{x}$ would vary from sample to sample. For example, the first sample of 25 temperature measurements might have a mean $\bar{x} = 9.8$, the second sample a mean $\bar{x} = 11.4$, the third sample a mean $\bar{x} = 10.5$, etc. If the sampling experiment were repeated a very large number of times, the resulting histogram of sample means would be approximately the probability distribution of $\bar{x}$. If $\bar{x}$ is a good estimator of μ, we would expect the values of $\bar{x}$ to cluster around μ as shown in Figure 6.2. This probability distribution is called a *sampling distribution* because it is generated by repeating a sampling experiment a very large number of times.

DEFINITION 6.3

The **sampling distribution** of a sample statistic calculated from a sample of n measurements is the probability distribution of the statistic.

FIGURE 6.2

Sampling distribution for $\bar{x}$ based on a sample of $n = 25$ measurements

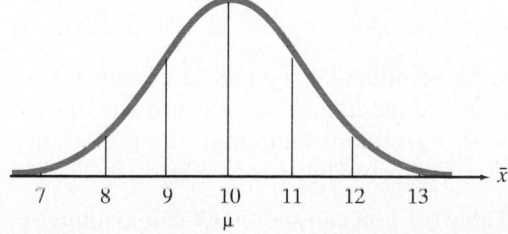

FIGURE 6.3

Two sampling distributions for estimating the population variance, σ^2

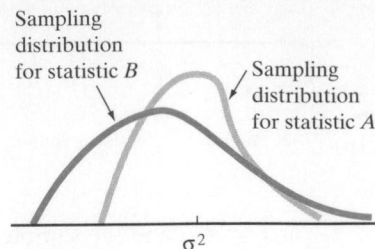

In actual practice, the sampling distribution of a statistic is obtained mathematically or (at least approximately) by simulating the sample on a computer using a procedure similar to that just described.

If $\bar{x}$ has been calculated from a sample of $n = 25$ measurements selected from a population with mean $\mu = 10$ and standard deviation $\sigma = 5$, the sampling distribution (Figure 6.2) provides information about the behavior of $\bar{x}$ in repeated sampling. For example, the probability that you will draw a sample of 25 measurements and obtain a value of $\bar{x}$ in the interval $9 \leq \bar{x} \leq 10$ will be the area under the sampling distribution over that interval.

Since the properties of a statistic are typified by its sampling distribution, it follows that to compare two sample statistics you compare their sampling distributions. For example, if you have two statistics, A and B, for estimating the same parameter (for purposes of illustration, suppose the parameter is the population variance σ^2) and if their sampling distributions are as shown in Figure 6.3, you would choose statistic A in preference to statistic B. You would make this choice because the sampling distribution for statistic A centers over σ^2 and has less spread (variation) than the sampling distribution for statistic B. When you draw a single sample in a practical sampling situation, the probability is higher that statistic A will fall nearer σ^2.

Remember that in practice we will not know the numerical value of the unknown parameter σ^2, so we will not know whether statistic A or statistic B is closer to σ^2 for a sample. We have to rely on our knowledge of the theoretical sampling distributions to choose the best sample statistic and then use it sample after sample. The procedure for finding the sampling distribution for a statistic is demonstrated in Example 6.1.

EXAMPLE 6.1

Consider a population consisting of the measurements 0, 3, and 12 and described by the probability distribution shown here. A random sample of $n = 3$ measurements is selected from the population.

x	0	3	12
$p(x)$	$\frac{1}{3}$	$\frac{1}{3}$	$\frac{1}{3}$

a. Find the sampling distribution of the sample mean $\bar{x}$.
b. Find the sampling distribution of the sample median m.

Solution

Every possible sample of $n = 3$ measurements is listed in Table 6.1 along with the sample mean and median. Also, because any one sample is as likely to be selected as any other (random sampling), the probability of observing any particular sample is $\frac{1}{27}$. The probability is also listed in Table 6.1.

a. From Table 6.1 you can see that $\bar{x}$ can assume the values 0, 1, 2, 3, 4, 5, 6, 8, 9, and 12. Because $\bar{x} = 0$ occurs in only one sample, $P(\bar{x} = 0) = \frac{1}{27}$. Similarly,

TABLE 6.1

Possible Samples	$\bar{x}$	m	Probability
0, 0, 0	0	0	$1/27$
0, 0, 3	1	0	$1/27$
0, 0, 12	4	0	$1/27$
0, 3, 0	1	0	$1/27$
0, 3, 3	2	3	$1/27$
0, 3, 12	5	3	$1/27$
0, 12, 0	4	0	$1/27$
0, 12, 3	5	3	$1/27$
0, 12, 12	8	12	$1/27$
3, 0, 0	1	0	$1/27$
3, 0, 3	2	3	$1/27$
3, 0, 12	5	3	$1/27$
3, 3, 0	2	3	$1/27$
3, 3, 3	3	3	$1/27$
3, 3, 12	6	3	$1/27$
3, 12, 0	5	3	$1/27$
3, 12, 3	6	3	$1/27$
3, 12, 12	9	12	$1/27$
12, 0, 0	4	0	$1/27$
12, 0, 3	5	3	$1/27$
12, 0, 12	8	12	$1/27$
12, 3, 0	5	3	$1/27$
12, 3, 3	6	3	$1/27$
12, 3, 12	9	12	$1/27$
12, 12, 0	8	12	$1/27$
12, 12, 3	9	12	$1/27$
12, 12, 12	12	12	$1/27$

$\bar{x} = 1$ occurs in three samples: $(0, 0, 3)$, $(0, 3, 0)$, and $(3, 0, 0)$. Therefore, $P(\bar{x} = 1) = 3/27 = 1/9$. Calculating the probabilities of the remaining values of $\bar{x}$ and arranging them in a table, we obtain the probability distribution shown here.

$\bar{x}$	0	1	2	3	4	5	6	8	9	12
$p(\bar{x})$	$1/27$	$3/27$	$3/27$	$1/27$	$3/27$	$6/27$	$3/27$	$3/27$	$3/27$	$1/27$

This is the sampling distribution for $\bar{x}$ because it specifies the probability associated with each possible value of $\bar{x}$.

b. In Table 6.1 you can see that the median m can assume one of the three values 0, 3, or 12. The value $m = 0$ occurs in seven different samples. Therefore, $P(m = 0) = 7/27$. Similarly, $m = 3$ occurs in 13 samples and $m = 12$ occurs in seven samples. Therefore, the probability distribution (i.e., the sampling distribution) for the median m is as shown below.

m	0	3	12
$p(m)$	$7/27$	$13/27$	$7/27$

▲

Example 6.1 demonstrates the procedure for finding the exact sampling distribution of a statistic when the number of different samples that could be selected from the population is relatively small. In the real world, populations often consist of a large number of different values, making samples difficult (or impossible) to enumerate. When this situation occurs, we may choose to obtain the approximate sampling distribution for a statistic by simulating the sampling over and over again and recording the proportion of times different values of the statistic occur. Example 6.2 illustrates this procedure.

EXAMPLE 6.2

Suppose we perform the following experiment over and over again: Take a sample of 11 measurements from the uniform distribution shown in Figure 6.4. Calculate the two sample statistics

$$\bar{x} = \text{Sample mean} = \frac{\sum x}{11}$$

m = Median = Sixth sample measurement when the 11 measurements are arranged in ascending order

Obtain approximations to the sampling distributions of $\bar{x}$ and m.

Solution

We use a computer to generate 1,000 samples, each with $n = 11$ observations. Then we compute $\bar{x}$ and m for each sample. Our goal is to obtain approximations to the sampling distributions of $\bar{x}$ and m to find out which sample statistic ($\bar{x}$ or m) contains more information about μ. (Note that, in this particular example, we *know* the population mean is $\mu = .5$) The first 10 of the 1,000 samples generated are presented in Table 6.2. For instance, the first computer-generated sample from the uniform distribution (arranged in ascending order) contained the following measurements: .125, .138, .139, .217, .419, .506, .516, .757, .771, .786, and .919. The sample mean $\bar{x}$ and median m computed for this sample are

$$x = \frac{.125 + .138 + \cdots + .919}{11} = .481$$

m = Sixth ordered measurement = .506

The relative frequency histograms for $\bar{x}$ and m for the 1,000 samples of size $n = 11$ are shown in Figure 6.5.

You can see that the values of $\bar{x}$ tend to cluster around μ to a greater extent than do the values of m. Thus, on the basis of the observed sampling distributions, we conclude that $\bar{x}$ contains more information about μ than m does—at least for samples of $n = 11$ measurements from the uniform distribution.   ▲

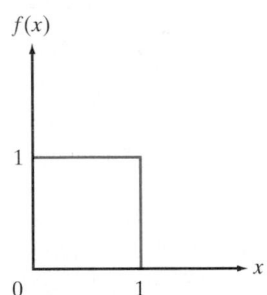

$f(x)$

1

0   1 x

FIGURE 6.4

Uniform distribution from 0 to 1

TABLE 6.2   First 10 Samples of $n = 11$ Measurements from a Uniform Distribution

Sample	Measurements										
1	.217	.786	.757	.125	.139	.919	.506	.771	.138	.516	.419
2	.303	.703	.812	.650	.848	.392	.988	.469	.632	.012	.065
3	.383	.547	.383	.584	.098	.676	.091	.535	.256	.163	.390
4	.218	.376	.248	.606	.610	.055	.095	.311	.086	.165	.665
5	.144	.069	.485	.739	.491	.054	.953	.179	.865	.429	.648
6	.426	.563	.186	.896	.628	.075	.283	.549	.295	.522	.674
7	.643	.828	.465	.672	.074	.300	.319	.254	.708	.384	.534
8	.616	.049	.324	.700	.803	.399	.557	.975	.569	.023	.072
9	.093	.835	.534	.212	.201	.041	.889	.728	.466	.142	.574
10	.957	.253	.983	.904	.696	.766	.880	.485	.035	.881	.732

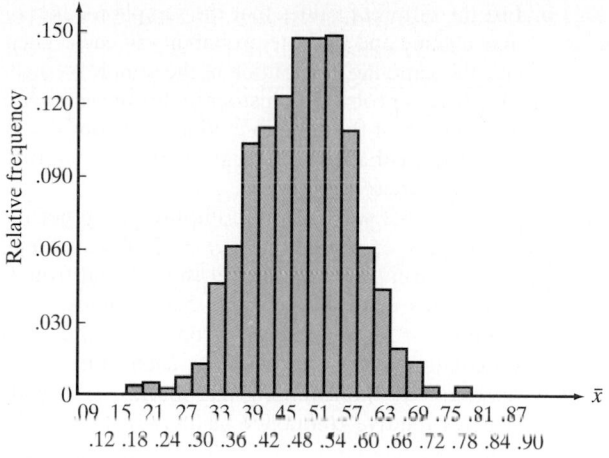

a. Sampling distribution for $\bar{x}$ (based on 1,000 samples of $n = 11$ measurements)

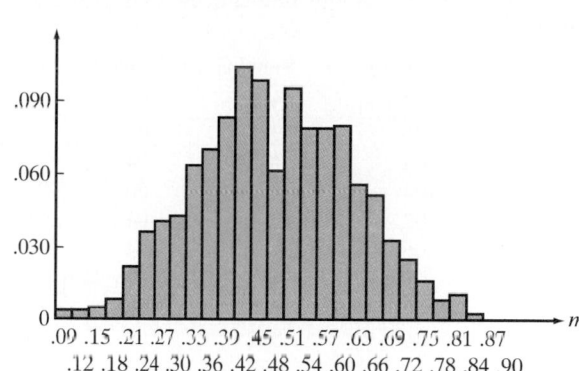

b. Sampling distribution for m (based on 1,000 samples of $n = 11$ measurements)

FIGURE 6.5   Relative frequency histograms for $\bar{x}$ and m, Example 6.2

As noted earlier, many sampling distributions can be derived mathematically, but the theory necessary to do this is beyond the scope of this text. Consequently, when we need to know the properties of a statistic, we will present its sampling distribution and simply describe its properties. Several of the important properties we look for in sampling distributions are discussed in the next section.

EXERCISES 6.1–6.7

Note: Exercises marked with 💾 *require the use of a computer.*

Learning the Mechanics

6.1 The probability distribution shown here describes a population of measurements that can assume values of 0, 2, 4, and 6, each of which occurs with the same relative frequency:

x	0	2	4	6
$p(x)$	¼	¼	¼	¼

  a. List all the different samples of $n = 2$ measurements that can be selected from this population.
  b. Calculate the mean of each different sample listed in part **a**.
  c. If a sample of $n = 2$ measurements is randomly selected from the population, what is the probability that a specific sample will be selected?
  d. Assume that a random sample of $n = 2$ measurements is selected from the population. List the different values of $\bar{x}$ found in part **b**, and find the probability of each. Then give the sampling distribution of the sample mean $\bar{x}$ in tabular form.
  e. Construct a probability histogram for the sampling distribution of $\bar{x}$.

6.2 Simulate sampling from the population described in Exercise 6.1 by marking the values of x, one on each

of four identical coins (or poker chips, etc.). Place the coins (marked 0, 2, 4, and 6) into a bag, randomly select one, and observe its value. Replace this coin, draw a second coin, and observe its value. Finally, calculate the mean $\bar{x}$ for this sample of $n = 2$ observations randomly selected from the population (Exercise 6.1, part **b**). Replace the coins, mix, and using the same procedure, select a sample of $n = 2$ observations from the population. Record the numbers and calculate $\bar{x}$ for this sample. Repeat this sampling process until you acquire 100 values of $\bar{x}$. Construct a relative frequency distribution for these 100 sample means. Compare this distribution to the exact sampling distribution of $\bar{x}$ found in part **e** of Exercise 6.1. *[Note: the distribution obtained in this exercise is an approximation to the exact sampling distribution. But, if you were to repeat the sampling procedure, drawing two coins not 100 times but 10,000 times, the relative frequency distribution for the 10,000 sample means would be almost identical to the sampling distribution of $\bar{x}$ found in Exercise 6.1, part **e**.]*

6.3 Consider the population described by the probability distribution shown here.

x	1	2	3	4	5
$p(x)$	.2	.3	.2	.2	.1

The random variable x is observed twice. If these observations are independent, verify that the different samples of size 2 and their probabilities are as shown here.

Sample	Probability	Sample	Probability
1, 1	.04	3, 4	.04
1, 2	.06	3, 5	.02
1, 3	.04	4, 1	.04
1, 4	.04	4, 2	.06
1, 5	.02	4, 3	.04
2, 1	.06	4, 4	.04
2, 2	.09	4, 5	.02
2, 3	.06	5, 1	.02
2, 4	.06	5, 2	.03
2, 5	.03	5, 3	.02
3, 1	.04	5, 4	.02
3, 2	.06	5, 5	.01
3, 3	.04		

a. Find the sampling distribution of the sample mean $\bar{x}$.

b. Construct a probability histogram for the sampling distribution of $\bar{x}$.

c. What is the probability that $\bar{x}$ is 4.5 or larger?

d. Would you expect to observe a value of $\bar{x}$ equal to 4.5 or larger? Explain.

6.4 Refer to Exercise 6.3 and find $E(x) = \mu$. Then use the sampling distribution of $\bar{x}$ found in Exercise 6.3 to find the expected value of $\bar{x}$. Note that $E(\bar{x}) = \mu$.

6.5 Refer to Exercise 6.3. Assume that a random sample of $n = 2$ measurements is randomly selected from the population.

a. List the different values that the sample median m may assume and find the probability of each. Then give the sampling distribution of the sample median.

b. Construct a probability histogram for the sampling distribution of the sample median and compare it with the probability histogram for the sample mean (Exercise 6.3, part **b**).

6.6 In Example 6.2 we use the computer to generate 1,000 samples, each containing $n = 11$ observations, from a uniform distribution over the interval from 0 to 1. For this exercise, generate 500 samples, each containing $n = 15$ observations, from this population.

a. Calculate the sample mean for each sample. To approximate the sampling distribution of $\bar{x}$, construct a relative frequency histogram for the 500 values of $\bar{x}$.

b. Repeat part **a** for the sample median. Compare this approximate sampling distribution with the approximate sampling distribution of $\bar{x}$ found in part **a**.

6.7 Consider a population that contains values of x equal to 00, 01, 02, 03, ..., 96, 97, 98, 99. Assume that these values of x occur with equal probability. Generate 500 samples, each containing $n = 25$ measurements, from this population. Calculate the sample mean $\bar{x}$ and sample variance s^2 for each of the 500 samples.

a. To approximate the sampling distribution of $\bar{x}$, construct a relative frequency histogram for the 500 values of $\bar{x}$.

b. Repeat part **a** for the 500 values of s^2.

6.2   PROPERTIES OF SAMPLING DISTRIBUTIONS: UNBIASEDNESS AND MINIMUM VARIANCE

The simplest type of statistic used to make inferences about a population parameter is a *point estimator*. A point estimator is a rule or formula that tells us how to use the sample data to calculate a single number that is intended to estimate the value of some population parameter. For example, the sample mean $\bar{x}$ is a point estimator of the population mean μ. Similarly, the sample variance s^2 is a point estimator of the population variance σ^2.

> **DEFINITION 6.4**
>
> A **point estimator** of a population parameter is a rule or formula that tells us how to use the sample data to calculate a single number that can be used as an *estimate* of the population parameter.

Often, many different point estimators can be found to estimate the same parameter. Each will have a sampling distribution that provides information about the point estimator. By examining the sampling distribution, we can determine how large the difference between an estimate and the true value of the

FIGURE 6.6

Sampling distributions of unbiased and biased estimators

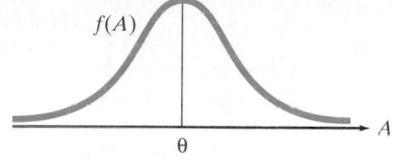

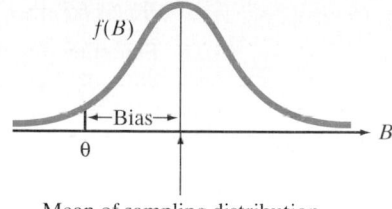

Mean of sampling distribution

a. Unbiased sample statistic
   for the parameter θ

b. Biased sample statistic
   for the parameter θ

parameter (called the **error of estimation**) is likely to be. We can also tell whether an estimator is more likely to overestimate or to underestimate a parameter.

EXAMPLE 6.3

Suppose two statistics, *A* and *B*, exist to estimate the same population parameter, θ (theta). (Note that θ could be any parameter, μ, $σ^2$, σ, etc.) Suppose the two statistics have sampling distributions as shown in Figure 6.6. Based on these sampling distributions, which statistic is more attractive as an estimator of θ?

Solution

As a first consideration, we would like the sampling distribution to center over the value of the parameter we wish to estimate. One way to characterize this property is in terms of the mean of the sampling distribution. Consequently, we say that a statistic is *unbiased* if the mean of the sampling distribution is equal to the parameter it is intended to estimate. This situation is shown in Figure 6.6a, where the mean $μ_A$ of statistic *A* is equal to θ. If the mean of a sampling distribution is not equal to the parameter it is intended to estimate, the statistic is said to be *biased*. The sampling distribution for a biased statistic is shown in Figure 6.6b. The mean $μ_B$ of the sampling distribution for statistic *B* is not equal to θ; in fact, it is shifted to the right of θ. ▲

You can see that biased statistics tend either to overestimate or to underestimate a parameter. Consequently, when other properties of statistics tend to be equivalent, we will choose an unbiased statistic to estimate a parameter of interest.*

DEFINITION 6.5

If the sampling distribution of a sample statistic has a mean equal to the population parameter the statistic is intended to estimate, the statistic is said to be an **unbiased estimate** of the parameter.

If the mean of the sampling distribution is not equal to the parameter, the statistic is said to be a **biased estimate** of the parameter.

The standard deviation of a sampling distribution measures another important property of statistics—the spread of these estimates generated by repeated sampling. Suppose two statistics, *A* and *B*, are both unbiased estimators of the population parameter. Since the means of the two sampling distributions are the same, we turn to their standard deviations in order to decide which will provide estimates that fall closer to the unknown population parameter we are estimating.

*Unbiased statistics do not exist for all parameters of interest, but they do exist for the parameters considered in this text.

FIGURE 6.7

Sampling distributions for two unbiased estimators

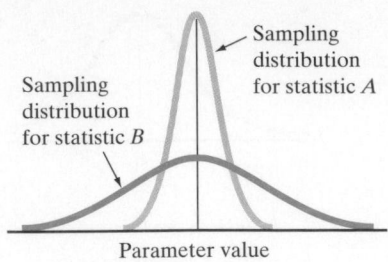

Parameter value

Naturally, we will choose the sample statistic that has the smaller standard deviation. Figure 6.7 depicts sampling distributions for A and B. Note that the standard deviation of the distribution of A is smaller than the standard deviation for B, indicating that over a large number of samples, the values of A cluster more closely around the unknown population parameter than do the values of B. Stated differently, the probability that A is close to the parameter value is higher than the probability that B is close to the parameter value.

In summary, to make an inference about a population parameter, we use the sample statistic with a sampling distribution that is unbiased and has a small standard deviation (usually smaller than the standard deviation of other unbiased sample statistics). The derivation of this sample statistic will not concern us, because the "best" statistic for estimating specific parameters is a matter of record. We will simply present an unbiased estimator with its standard deviation for each population parameter we consider. [*Note:* The standard deviation of the sampling distribution of a statistic is also called the **standard error of the statistic**.]

EXAMPLE 6.4

In Example 6.1, we found the sampling distributions of the sample mean $\bar{x}$ and the sample median m for random samples of $n = 3$ measurements from a population defined by the probability distribution shown here.

x	0	3	12
$p(x)$	$\frac{1}{3}$	$\frac{1}{3}$	$\frac{1}{3}$

The sampling distributions of $\bar{x}$ and m were found to be as shown here.

$\bar{x}$	0	1	2	3	4	5	6	8	9	12
$p(\bar{x})$	$\frac{1}{27}$	$\frac{3}{27}$	$\frac{3}{27}$	$\frac{1}{27}$	$\frac{3}{27}$	$\frac{6}{27}$	$\frac{3}{27}$	$\frac{3}{27}$	$\frac{3}{27}$	$\frac{1}{27}$

m	0	3	12
$p(m)$	$\frac{7}{27}$	$\frac{13}{27}$	$\frac{7}{27}$

a. Show that $\bar{x}$ is an unbiased estimator of μ in this situation.
b. Show that m is a biased estimator of μ in this situation.

Solution

a. The expected value of a discrete random variable x (see Section 4.3) is defined to be $E(x) = \sum xp(x)$, where the summation is over all values of x. Then

$$E(x) = \mu = \sum xp(x) = (0)(\tfrac{1}{3}) + (3)(\tfrac{1}{3}) + (12)(\tfrac{1}{3}) = 5$$

The expected value of the discrete random variable $\bar{x}$ is

$$E(\bar{x}) - \sum(\bar{x})p(\bar{x})$$

summed over all values of $\bar{x}$. Or

$$F(\bar{x}) = (0)(\tfrac{1}{27}) + (1)(\tfrac{3}{27}) + 2(\tfrac{3}{27}) + \cdots + (12)(\tfrac{1}{27}) = 5$$

Since $E(\bar{x}) = \mu$, we see that $\bar{x}$ is an unbiased estimator of μ.

b. The expected value of the sample median m is

$$E(m) = \sum mp(m) = (0)(\tfrac{7}{27}) + (3)(\tfrac{13}{27}) + (12)(\tfrac{7}{27}) = 4.56$$

Since the expected value of m is not equal to μ ($\mu = 5$), the sample median m is a biased estimator of μ.   ▲

EXAMPLE 6.5

Refer to Example 6.4 and find the standard deviations of the sampling distributions of $\bar{x}$ and m. Which statistic would appear to be a better estimator of μ?

Solution

The variance of the sampling distribution of $\bar{x}$ (we denote it by the symbol $\sigma_{\bar{x}}^2$) is found to be

$$\sigma_{\bar{x}}^2 = E\left\{[\bar{x} - E(\bar{x})]^2\right\} = \sum(\bar{x} - \mu)^2 p(\bar{x})$$

where, from Example 6.4,

$$E(\bar{x}) = \mu = 5$$

Then

$$\sigma_{\bar{x}}^2 = (0 - 5)^2(\tfrac{1}{27}) + (1 - 5)^2(\tfrac{3}{27}) + (2 - 5)^2(\tfrac{3}{27}) + \cdots + (12 - 5)^2(\tfrac{1}{27}) = 8.6667$$

and

$$\sigma_{\bar{x}} = \sqrt{8.6667} = 2.94$$

Similarly, the variance of the sampling distribution of m (we denote it by σ_m^2) is

$$\sigma_m^2 = E\left\{[m - E(m)]^2\right\}$$

where, from Example 6.4, the expected value of m is $E(m) = 4.56$. Then

$$\sigma_m^2 = E\left\{[m - E(m)]^2\right\} = \sum[m - E(m)]^2 p(m)$$
$$= (0 - 4.56)^2(\tfrac{7}{27}) + (3 - 4.56)^2(\tfrac{13}{27}) + (12 - 4.56)^2(\tfrac{7}{27}) = 20.9136$$

and

$$\sigma_m = \sqrt{20.9136} = 4.57$$

Which statistic appears to be the better estimator for the population mean μ: the sample mean $\bar{x}$ or the median m? To answer this question, we compare the sampling distributions of the two statistics. The sampling distribution of the sample median m is biased (i.e., it is located to the left of the mean μ) and its standard deviation $\sigma_m = 4.57$ is much larger than the standard deviation of the sampling distribution of $\bar{x}$, $\sigma_{\bar{x}} = 2.94$. Consequently, the sample mean $\bar{x}$ would be a better estimator of the population mean μ, for the population in question, than would the sample median m.   ▲

EXERCISES 6.8–6.14

Note: Exercises marked with ▢ *require the use of a computer.*

Learning the Mechanics

6.8 Consider the probability distribution shown here.

x	0	1	4
$p(x)$	⅓	⅓	⅓

a. Find μ and σ^2.
b. Find the sampling distribution of the sample mean $\bar{x}$ for a random sample of $n = 2$ measurements from this distribution.
c. Show that $\bar{x}$ is an unbiased estimator for μ. [*Hint:* Show that $E(\bar{x}) = \sum \bar{x} p(\bar{x}) = \mu$.]
d. Find the sampling distribution of the sample variance s^2 for a random sample of $n = 2$ measurements from this distribution.
e. Show that s^2 is an unbiased estimator for σ^2.

6.9 Consider the probability distribution shown here.

x	2	4	9
$p(x)$	⅓	⅓	⅓

a. Calculate μ for this distribution.
b. Find the sampling distribution of the sample mean $\bar{x}$ for a random sample of $n = 3$ measurements from this distribution, and show that $\bar{x}$ is an unbiased estimator of μ.
c. Find the sampling distribution of the sample median m for a random sample of $n = 3$ measurements from this distribution, and show that the median is a biased estimator of μ.
d. If you wanted to estimate μ using a sample of three measurements from this population, which estimator would you use? Why?

6.10 Consider the probability distribution shown here.

x	0	1	2
$p(x)$	⅓	⅓	⅓

a. Find μ.
b. For a random sample of $n = 3$ observations from this distribution, find the sampling distribution of the sample mean.
c. Find the sampling distribution of the median of a sample of $n = 3$ observations from this population.
d. Refer to parts **b** and **c** and show that both the mean and median are unbiased estimators of μ for this population.
e. Find the variances of the sampling distributions of the sample mean and the sample median.
f. Which estimator would you use to estimate μ? Why?

▢ **6.11** Generate 500 samples, each containing $n = 25$ measurements, from a population that contains values of x equal to 1, 2, ..., 48, 49, 50. Assume that these values of x are equally likely. Calculate the sample mean $\bar{x}$ and median m for each sample. Construct relative frequency histograms for the 500 values of $\bar{x}$ and the 500 values of m. Use these approximations to the sampling distributions of $\bar{x}$ and m to answer the following questions:
a. Does it appear that $\bar{x}$ and m are unbiased estimators of the population mean? [*Note:* $\mu = 25.5$.]
b. Which sampling distribution displays greater variation?

6.12 Refer to Exercise 6.3.
a. Show that $\bar{x}$ is an unbiased estimator of μ.
b. Find $\sigma_{\bar{x}}^2$.
c. Find the probability that $\bar{x}$ will fall within $2\sigma_{\bar{x}}$ of μ.

6.13 Refer to Exercise 6.3.
a. Find the sampling distribution of s^2.
b. Find the population variance σ^2.
c. Show that s^2 is an unbiased estimator of σ^2.
d. Find the sampling distribution of the sample standard deviation s.
e. Show that s is a biased estimator of σ.

6.14 Refer to Exercise 6.5, where we found the sampling distribution of the sample median. Is the median an unbiased estimator of the population mean μ?

6.3 THE CENTRAL LIMIT THEOREM

Estimating the mean useful life of automobiles, the mean number of crimes per month in a large city, and the mean yield per acre of a new soybean hybrid are practical problems with something in common. In each case we are interested in making an inference about the mean μ of some population. As we mentioned in Chapter 2, the sample mean $\bar{x}$ is, in general, a good estimator of μ. We now develop pertinent information about the sampling distribution for this useful statistic.

FIGURE 6.8

Sampled uniform population

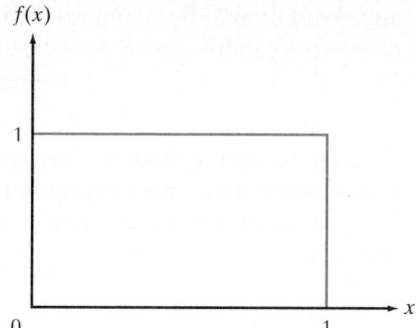

EXAMPLE 6.6

Suppose a population has the uniform probability distribution given in Figure 6.8. The mean and standard deviation of this probability distribution are $\mu = .5$ and $\sigma = .29$. (See Section 5.2 for the formulas for μ and σ.) Now suppose a sample of 11 measurements is selected from this population. Describe the sampling distribution of the sample mean $\bar{x}$ based on the 1,000 sampling experiments discussed in Example 6.2.

Solution

You will recall that in Example 6.2 we generated 1,000 samples of $n = 11$ measurements each. The relative frequency histogram for the 1,000 sample means is shown in Figure 6.9 with a normal probability distribution superimposed. You can see that this normal probability distribution approximates the computer-generated sampling distribution very well.

To fully describe a normal probability distribution, it is necessary to know its mean and standard deviation. Inspection of Figure 6.9 indicates that the mean of the distribution of $\bar{x}$, $\mu_{\bar{x}}$, appears to be very close to .5, the mean of the sampled uniform population. Furthermore, for a mound-shaped distribution such as that shown in Figure 6.9, almost all the measurements should fall within 3 standard deviations of the mean. Since the number of values of $\bar{x}$ is very large (1,000), the range of the observed $\bar{x}$'s divided by 6 (rather than 4) should give a reasonable

FIGURE 6.9

Relative frequency histogram for $\bar{x}$ in 1,000 samples of $n = 11$ measurements with normal distribution superimposed

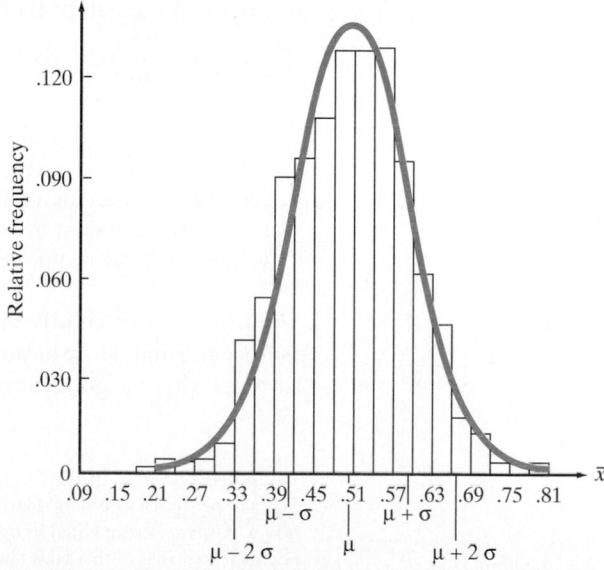

approximation to the standard deviation of the sample means, $\sigma_{\bar{x}}$. The values of $\bar{x}$ range from about .2 to .8, so we calculate

$$\sigma_{\bar{x}} \approx \frac{\text{Range of } \bar{x}\text{'s}}{6} = \frac{.8 - .2}{6} = .1$$

To summarize our findings based on 1,000 samples, each consisting of 11 measurements from a uniform population, the sampling distribution of $\bar{x}$ appears to be approximately normal with a mean of about .5 and a standard deviation of about .1.   ▲

The sampling distribution of $\bar{x}$ has the properties given in the next box, assuming only that a random sample of n observations has been selected from *any* population.

Properties of the Sampling Distribution of $\bar{x}$

1. Mean of sampling distribution equals mean of sampled population. That is, $\mu_{\bar{x}} = E(\bar{x}) = \mu$.
2. Standard deviation of sampling distribution equals

$$\frac{\text{Standard deviation of sampled population}}{\text{Square root of sample size}}$$

That is, $\sigma_{\bar{x}} = \sigma / \sqrt{n}$.
The standard deviation $\sigma_{\bar{x}}$ is often referred to as the **standard error of the mean**.

You can see that our approximation to $\mu_{\bar{x}}$ in Example 6.6 was precise, since property 1 assures us that the mean is the same as that of the sampled population: .5. Property 2 tells us how to calculate the standard deviation of the sampling distribution of $\bar{x}$. Substituting $\sigma = .29$, the standard deviation of the sampled uniform distribution, and the sample size $n = 11$ into the formula for $\sigma_{\bar{x}}$, we find

$$\sigma_{\bar{x}} = \frac{\sigma}{\sqrt{n}} = \frac{.29}{\sqrt{11}} = .09$$

Thus, the approximation we obtained in Example 6.6, $\sigma_{\bar{x}} \approx .1$, is very close to the exact value, $\sigma_{\bar{x}} = .09$.

A third property, applicable when the sample size n is large, is contained in one of the most important theoretical results in statistics: the *Central Limit Theorem*.

Central Limit Theorem

Consider a random sample of n observations selected from a population (*any* population) with mean μ and standard deviation σ. Then, when n is sufficiently large, the sampling distribution of $\bar{x}$ will be approximately a normal distribution with mean $\mu_{\bar{x}} = \mu$ and standard deviation $\sigma_{\bar{x}} = \sigma / \sqrt{n}$. The larger the sample size, the better will be the normal approximation to the sampling distribution of $\bar{x}$.*

Thus, for sufficiently large samples the sampling distribution of $\bar{x}$ is approximately normal. How large must the sample size n be so that the normal distribution provides a good approximation for the sampling distribution of $\bar{x}$? The

---

*Moreover, because of the Central Limit Theorem, the sum of a random sample of n observations, Σx, will possess a sampling distribution that is approximately normal for large samples. This distribution will have a mean equal to $n\mu$ and a variance equal to $n\sigma^2$. Proof of the Central Limit Theorem is beyond the scope of this book, but it can be found in many mathematical statistics texts.

FIGURE 6.10

Sampling distributions of *x̄* for different populations and different sample sizes

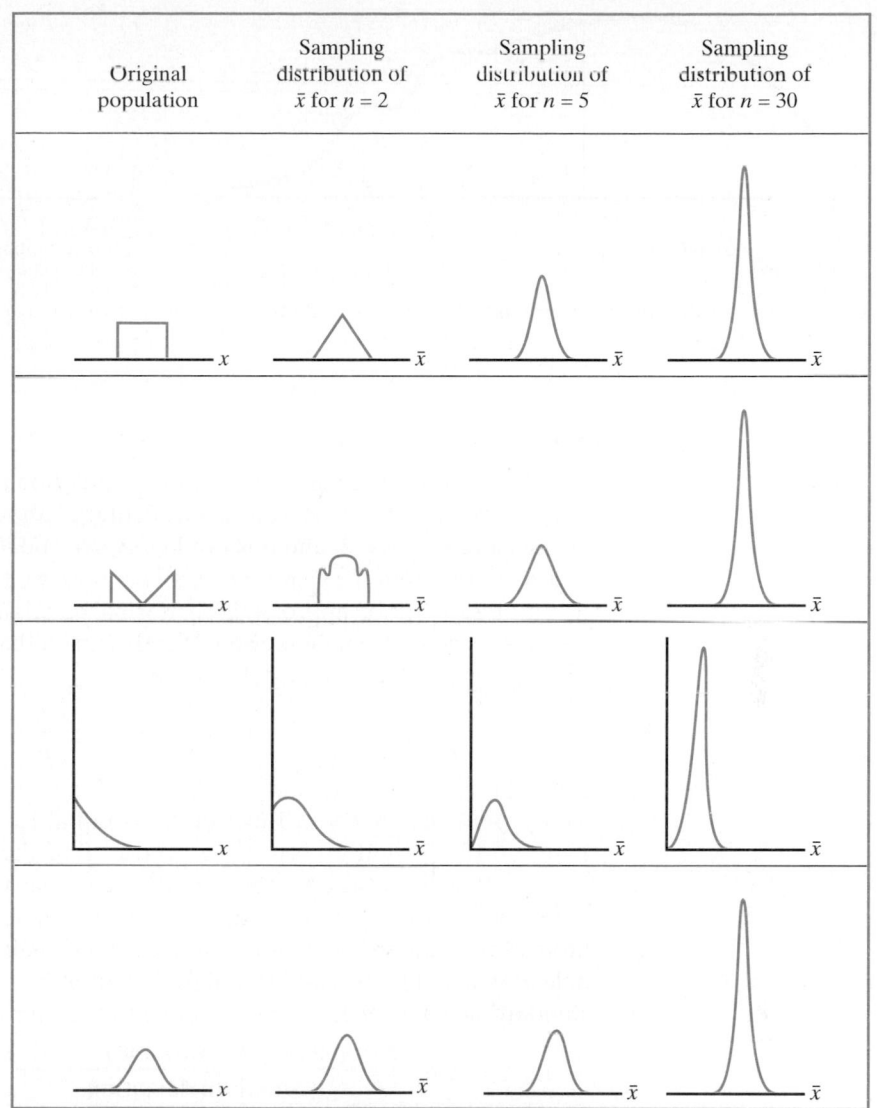

Original population	Sampling distribution of $\bar{x}$ for $n = 2$	Sampling distribution of $\bar{x}$ for $n = 5$	Sampling distribution of $\bar{x}$ for $n = 30$

answer depends on the shape of the distribution of the sampled population, as shown by Figure 6.10. Generally speaking, the greater the skewness of the sampled population distribution, the larger the sample size must be before the normal distribution is an adequate approximation for the sampling distribution of *x̄*. For most sampled populations, sample sizes of $n \geq 30$ will suffice for the normal approximation to be reasonable. We will use the normal approximation for the sampling distribution of *x̄* when the sample size is at least 30.

EXAMPLE 6.7

Suppose we have selected a random sample of $n = 25$ observations from a population with mean equal to 80 and standard deviation equal to 5. It is known that the population is not extremely skewed.

a. Sketch the relative frequency distributions for the population and for the sampling distribution of the sample mean, *x̄*.

b. Find the probability that *x̄* will be larger than 82.

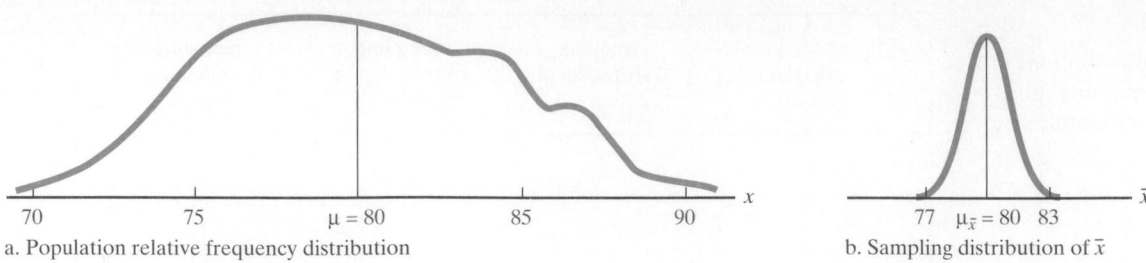

a. Population relative frequency distribution

b. Sampling distribution of $\bar{x}$

FIGURE 6.11   A population relative frequency distribution and the sampling distribution for $\bar{x}$

Solution

a. We do not know the exact shape of the population relative frequency distribution, but we do know that it should be centered about $\mu = 80$, its spread should be measured by $\sigma = 5$, and it is not highly skewed. One possibility is shown in Figure 6.11a. From the Central Limit Theorem, we know that the sampling distribution of $\bar{x}$ will be approximately normal since the sampled population distribution is not extremely skewed. We also know that the sampling distribution will have mean and standard deviation

$$\mu_{\bar{x}} = \mu = 80 \qquad \text{and} \qquad \sigma_{\bar{x}} = \frac{\sigma}{\sqrt{n}} = \frac{5}{\sqrt{25}} = 1$$

The sampling distribution of $\bar{x}$ is shown in Figure 6.11b.

b. The probability that $\bar{x}$ will exceed 82 is equal to the lightly shaded area in Figure 6.12. To find this area, we need to find the z value corresponding to $\bar{x} = 82$. Recall that the standard normal random variable z is the difference between any normally distributed random variable and its mean, expressed in units of its standard deviation. Since $\bar{x}$ is a normally distributed random variable with mean $\mu_{\bar{x}} = \mu$ and standard deviation $\sigma_{\bar{x}} = \sigma/\sqrt{n}$, it follows that the standard normal z value corresponding to the sample mean, $\bar{x}$, is

$$z = \frac{(\text{Normal random variable}) - (\text{Mean})}{\text{Standard deviation}} = \frac{\bar{x} - \mu_{\bar{x}}}{\sigma_{\bar{x}}}$$

Therefore, for $\bar{x} = 8$, we have

$$z = \frac{\bar{x} - \mu_{\bar{x}}}{\sigma_{\bar{x}}} = \frac{82 - 80}{1} = 2$$

The area A in Figure 6.12 corresponding to $z = 2$ is given in the table of areas under the normal curve (see Table IV of Appendix A) as .4772. Therefore, the tail area corresponding to the probability that $\bar{x}$ exceeds 82 is

$$P(\bar{x} > 82) = P(z > 2) = .5 - .4772 = .0228 \qquad \blacktriangle$$

FIGURE 6.12

The sampling distribution
of $\bar{x}$

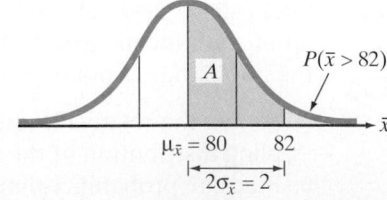

EXAMPLE 6.8

A manufacturer of automobile batteries claims that the distribution of the lengths of life of its best battery has a mean of 54 months and a standard deviation of 6 months. Suppose a consumer group decides to check the claim by purchasing a sample of 50 of these batteries and subjecting them to tests that determine battery life.

a. Assuming that the manufacturer's claim is true, describe the sampling distribution of the mean lifetime of a sample of 50 batteries.
b. Assuming that the manufacturer's claim is true, what is the probability the consumer group's sample has a mean life of 52 or fewer months?

Solution

a. Even though we have no information about the shape of the probability distribution of the lives of the batteries, we can use the Central Limit Theorem to deduce that the sampling distribution for a sample mean lifetime of 50 batteries is approximately normally distributed. Furthermore, the mean of this sampling distribution is the same as the mean of the sampled population, which is $\mu = 54$ months according to the manufacturer's claim. Finally, the standard deviation of the sampling distribution is given by

$$\sigma_{\bar{x}} = \frac{\sigma}{\sqrt{n}} = \frac{6}{\sqrt{50}} = .85 \text{ month}$$

Note that we used the claimed standard deviation of the sampled population, $\sigma = 6$ months. Thus, if we assume that the claim is true, the sampling distribution for the mean life of the 50 batteries sampled is as shown in Figure 6.13.

b. If the manufacturer's claim is true, the probability that the consumer group observes a mean battery life of 52 or fewer months for their sample of 50 batteries, $P(\bar{x} \le 52)$, is equivalent to the lightly shaded area in Figure 6.13. Since the sampling distribution is approximately normal, we can find this area by computing the standard normal z value:

$$z = \frac{\bar{x} - \mu_{\bar{x}}}{\sigma_{\bar{x}}} = \frac{\bar{x} - \mu}{\sigma_{\bar{x}}} = \frac{52 - 54}{.85} = -2.35$$

where $\mu_{\bar{x}}$, the mean of the sampling distribution of $\bar{x}$, is equal to μ, the mean of the lives of the sampled population, and $\sigma_{\bar{x}}$ is the standard deviation of the sampling distribution of $\bar{x}$. Note that z is the familiar standardized distance (z-score) of Section 2.6 and, since $\bar{x}$ is approximately normally distributed, it will possess the standard normal distribution of Section 5.3.

The area A shown in Figure 6.13 between $\bar{x} = 52$ and $\bar{x} = 54$ (corresponding to $z = -2.35$) is found in Table IV of Appendix A to be .4906. Therefore, the area to the left of $\bar{x} = 52$ is

$$P(\bar{x} \le 52) = .5 - A = .5 - .4906 = .0094$$

Thus, the probability the consumer group will observe a sample mean of 52 or less is only .0094 if the manufacturer's claim is true. If the 50 tested batteries do exhibit a mean of 52 or fewer months, the consumer group will have strong

FIGURE 6.13

Sampling distribution of $\bar{x}$
in Example 6.8 for $n = 50$

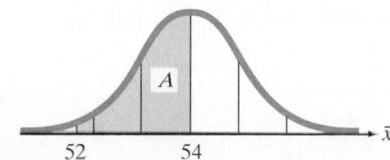

evidence that the manufacturer's claim is untrue, because such an event is very unlikely to occur if the claim is true. (This is still another application of the *rare-event approach* to statistical inference.) ▲

We conclude this section with two final comments on the sampling distribution of $\bar{x}$. First, from the formula $\sigma_{\bar{x}} = \sigma/\sqrt{n}$, we see that the standard deviation of the sampling distribution of $\bar{x}$ gets smaller as the sample size n gets larger. For example, we computed $\sigma_{\bar{x}} = .85$ when $n = 50$ in Example 6.8. However, for $n = 100$ we obtain $\sigma_{\bar{x}} = \sigma/\sqrt{n} = 6/\sqrt{100} = .60$. This relationship will hold true for most of the sample statistics encountered in this text. That is: *The standard deviation of the sampling distribution decreases as the sample size increases.* Consequently, the larger the sample size, the more accurate the sample statistic (e.g., $\bar{x}$) is in estimating a population parameter (e.g., μ). We will use this result in Chapter 7 to help us determine the sample size needed to obtain a specified accuracy of estimation.

Our second comment concerns the Central Limit Theorem. In addition to providing a very useful approximation for the sampling distribution of a sample mean, the Central Limit Theorem offers an explanation for the fact that many relative frequency distributions of data possess mound-shaped distributions. Many of the measurements we take in various areas of research are really means or sums of a large number of small phenomena. For example, a year's growth of a pine seedling is the total of the many individual components that affect the plant's growth. Similarly, we can view the length of time a construction company takes to build a house as the total of the times taken to complete a multitude of distinct jobs, and we can regard the monthly demand for blood at a hospital as the total of the many individual patients' needs. Whether or not the observations entering into these sums satisfy the assumptions basic to the Central Limit Theorem is open to question. However, it is a fact that many distributions of data in nature are mound-shaped and possess the appearance of normal distributions.

EXERCISES 6.15–6.30

Note: Exercises marked with 💾 *require the use of a computer.*

Learning the Mechanics

6.15 Suppose a random sample of n measurements is selected from a population with mean $\mu = 100$ and variance $\sigma^2 = 100$. For each of the following values of n, give the mean and standard deviation of the sampling distribution of the sample mean $\bar{x}$.
  a. $n = 4$ **b.** $n = 25$ **c.** $n = 100$
  d. $n = 50$ **e.** $n = 500$ **f.** $n = 1,000$

6.16 Suppose a random sample of $n = 25$ measurements is selected from a population with mean μ and standard deviation σ. For each of the following values of μ and σ, give the values of $\mu_{\bar{x}}$ and $\sigma_{\bar{x}}$.
  a. $\mu = 10, \sigma = 3$ **b.** $\mu = 100, \sigma = 25$
  c. $\mu = 20, \sigma = 40$ **d.** $\mu = 10, \sigma = 100$

6.17 Consider the probability distribution shown here.

x	1	2	3	8
$p(x)$	.1	.4	.4	.1

a. Find μ, σ^2, and σ.

b. Find the sampling distribution of $\bar{x}$ for random samples of $n = 2$ measurements from this distribution by listing all possible values of $\bar{x}$, and find the probability associated with each.

c. Use the results of part **b** to calculate $\mu_{\bar{x}}$ and $\sigma_{\bar{x}}$. Confirm that $\mu_{\bar{x}} = \mu$ and that $\sigma_{\bar{x}} = \sigma/\sqrt{n} = \sigma/\sqrt{2}$.

6.18 Will the sampling distribution of $\bar{x}$ always be approximately normally distributed? Explain.

6.19 A random sample of $n = 64$ observations is drawn from a population with a mean equal to 20 and standard deviation equal to 16.

a. Give the mean and standard deviation of the (repeated) sampling distribution of $\bar{x}$.

b. Describe the shape of the sampling distribution of $\bar{x}$. Does your answer depend on the sample size?

c. Calculate the standard normal z-score corresponding to a value of $\bar{x} = 15.5$.

d. Calculate the standard normal z-score corresponding to $\bar{x} = 23$.

6.20 Refer to Exercise 6.19. Find the probability that
a. $\bar{x}$ is less than 16   **b.** $\bar{x}$ is greater than 23
c. $\bar{x}$ is greater than 25
d. $\bar{x}$ falls between 16 and 22   **e.** $\bar{x}$ is less than 14

6.21 A random sample of $n = 100$ observations is selected from a population with $\mu = 30$ and $\sigma = 16$. Approximate the following probabilities:
a. $P(\bar{x} \geq 28)$ **b.** $P(22.1 \leq \bar{x} \leq 26.8)$
c. $P(\bar{x} \leq 28.2)$ **d.** $P(\bar{x} \geq 27.0)$

6.22 A random sample of $n = 900$ observations is selected from a population with $\mu = 100$ and $\sigma = 10$.
a. What are the largest and smallest values of $\bar{x}$ that you would expect to see?
b. How far, at the most, would you expect $\bar{x}$ to deviate from μ?
c. Did you have to know μ to answer part **b**? Explain.

6.23 Consider a population that contains values of x equal to 0, 1, 2, ..., 97, 98, 99. Assume that the values of x are equally likely. For each of the following values of n, generate 500 random samples and calculate $\bar{x}$ for each sample. For each sample size, construct a relative frequency histogram of the 500 values of $\bar{x}$. What changes occur in the histograms as the value of n increases? What similarities exist? Use $n = 2$, $n = 5$, $n = 10$, $n = 30$, and $n = 50$.

Applying the Concepts

6.24 The *College Student Journal* (Dec. 1992) investigated differences in traditional and nontraditional students, where nontraditional students are generally defined as those 25 years old or older. Based on the study results, we can assume that the population mean and standard deviation for the GPA of all nontraditional students is $\mu = 3.5$ and $\sigma = .5$. Suppose that a random sample of $n = 100$ nontraditional students is selected from the population of all nontraditional students, and the GPA of each student is determined. Then $\bar{x}$, the sample mean, will be approximately normally distributed (because of the Central Limit Theorem).
a. Calculate $\mu_{\bar{x}}$ and $\sigma_{\bar{x}}$.
b. What is the approximate probability that the nontraditional student sample has a mean GPA between 3.40 and 3.60?
c. What is the approximate probability that the sample of 100 nontraditional students has a mean GPA that exceeds 3.62?
d. How would the sampling distribution of $\bar{x}$ change if the sample size n were doubled from 100 to 200? How do your answers to parts **b** and **c** change when the sample size is doubled?

6.25 Interpersonal violence (e.g., rape) generally leads to psychological stress for the victim. *Clinical Psychology Review* (Vol. 15, 1995) reported on the results of all recently published studies of the relationship between interpersonal violence and psychological stress. The distribution of the time elapsed between the violent incident and the initial sign of stress has a mean of 5.1 years and a standard deviation of 6.1 years. Consider a random sample of $n = 150$ victims of interpersonal violence. Let $\bar{x}$ represent the mean time elapsed between the violent act and the first sign of stress for the sampled victims.
a. Give the mean and standard deviation of the sampling distribution of $\bar{x}$.
b. Will the sampling distribution of $\bar{x}$ be approximately normal? Explain.
c. Find $P(\bar{x} > 5.5)$.   **d.** Find $P(4 < \bar{x} < 5)$.

6.26 The Computer-Assisted Hypnosis Scale (CAHS) is designed to measure a person's susceptibility to hypnosis. In computer-assisted hypnosis, the computer serves as a facilitator of hypnosis by using digitized speech processing coupled with interactive involvement with the hypnotic subject. CAHS scores range from 0 (no susceptibility) to 12 (extremely high susceptibility). A study in *Psychological Assessment* (Mar. 1995) reported a mean CAHS score of 4.59 and a standard deviation of 2.95 for University of Tennessee undergraduates. Assume that $\mu = 4.59$ and $\sigma = 2.95$ for this population. Suppose a psychologist uses CAHS to test a random sample of 50 subjects.
a. Would you expect to observe a sample mean CAHS score of $\bar{x} = 6$ or higher? Explain.
b. Suppose the psychologist actually observes $\bar{x} = 6.2$. Based on your answer to part **a**, make an inference about the population from which the sample was selected.

6.27 The ocean quahog is a type of clam found in the coastal waters of New England and the mid-Atlantic states. Extensive beds of ocean quahogs along the New Jersey shore gave rise to the development of the largest U.S. shellfish harvesting program. A federal survey of offshore ocean quahog harvesting in New Jersey, conducted from 1980 to 1992, revealed an average catch per unit effort (CPUE) of 89.34 clams. The CPUE standard deviation was 7.74 (*Journal of Shellfish Research,* June 1995). Let $\bar{x}$ represent the mean CPUE for a sample of 35 attempts to catch ocean quahogs off the New Jersey shore.
a. Compute $\mu_{\bar{x}}$ and $\sigma_{\bar{x}}$. Interpret their values.
b. Sketch the sampling distribution of $\bar{x}$.
c. Find $P(\bar{x} > 88)$.   **d.** Find $P(\bar{x} < 87)$.

6.28 Last year a company began a program to compensate its employees for unused sick days, paying each employee a bonus of one-half the usual wage earned for each unused sick day. The question that naturally arises is: "Did this policy motivate employees to use fewer sick days?" *Before* last year,

The Insomnia Pill

CASE STUDY 6.1

A research report published in the *Proceedings of the National Academy of Sciences* (Mar. 1994) brought encouraging news to insomniacs and long-distance travelers. Neuroscientists at the Massachusetts Institute of Technology (MIT) have been experimenting with melatonin—a hormone secreted by the pineal gland in the brain—as a sleep-inducing hormone. Since the hormone is naturally produced, it is considered nonaddictive. The researchers believe melatonin may also be effective in treating jet lag—the body's response to rapid travel across many time zones so that a daylight–darkness change disrupts sleep patterns.

In the MIT study, young male volunteers were given various doses of melatonin or a placebo (no dosage of melatonin). Then they were placed in a dark room at midday and told to close their eyes for 30 minutes. The variable of interest was the time (in minutes) elapsed before each volunteer fell asleep.

According to the lead investigator, Professor Richard Wurtman, "Our volunteers fall asleep in five or six min-

utes on melatonin, while those on placebo take about 15 minutes." Wurtman warns, however, that uncontrolled doses of melatonin could cause serious mood-altering side effects. (Melatonin is sold in some health food stores. However, sales are unregulated, and the purity and strength of the hormone are often uncertain.)

Focus

With the placebo (i.e., no hormone) the researchers found that the mean time to fall asleep was 15 minutes. Assume that with the placebo treatment $\mu = 15$ and $\sigma = 5$. Now, consider a random sample of 40 young males, each of whom is given a dose of the sleep-inducing hormone, melatonin. The times (in minutes) to fall asleep for these 40 males are listed in Table 6.3.* Use the data to make an inference about the true value of μ for those taking the melatonin. Does melatonin appear to be an effective drug against insomnia?

---

the number of sick days used by employees had a distribution with a mean of 7 days and a standard deviation of 2 days.

a. Assuming that these parameters did not change last year, find the approximate probability that the sample mean number of sick days used by 100 employees chosen at random was less than or equal to 6.4 last year.

b. How would you interpret the result if the sample mean for the 100 employees was 6.4.?

6.29 A soft drink bottler purchases glass bottles from a vendor. The bottles are required to have an internal pressure of at least 150 pounds per square inch (psi). A prospective bottle vendor claims that its produc-

tion process yields bottles with a mean internal pressure of 157 psi and a standard deviation of 3 psi. The bottler strikes an agreement with the vendor that permits the bottler to sample from the vendor's production process to verify the vendor's claim. The bottler randomly selects 40 bottles from the last 10,000 produced, measures the internal pressure of each, and finds the mean pressure for the sample to be 1.3 psi below the process mean cited by the vendor.

a. Assuming the vendor's claim to be true, what is the probability of obtaining a sample mean this far or farther below the process mean? What does your answer suggest about the validity of the vendor's claim?

TABLE 6.3   Times (in Minutes) for 40 Male Volunteers to Fall Asleep (Case Study 6.1)

6.4	6.0	3.2	4.4	6.2	1.7	5.1
5.9	1.6	4.4	16.2	4.8	8.3	7.5
4.8	3.3	4.0	6.2	6.3	5.0	6.3
5.1	6.4	15.6	3.4	3.1	6.1	6.0
5.0	1.8	6.1	4.5	4.5	1.5	4.7
7.6	8.2	4.9	6.1	3.9		

---

*These are simulated sleep times based on summary information provided in the MIT study.

b. If the process standard deviation were 3 psi as claimed by the vendor, but the mean were 156 psi, would the observed sample result be more or less likely than in part **a**? What if the mean were 158 psi?

c. If the process mean were 157 psi as claimed, but the process standard deviation were 2 psi, would the sample result be more or less likely than in part **a**? What if instead the standard deviation were 6 psi?

6.30 The Test of Knowledge About Epilepsy (KAE), which is designed to measure attitudes toward persons with epilepsy, uses 20 multiple-choice items where all of the choices are incorrect. For each person, two scores (ranging from 0 to 20) are obtained, an attitude score (KAE-A) and a general knowledge score (KAE-GK). Based on a large-scale study of college students, the distribution of KAE-A scores

has a mean of $\mu = 11.92$ and a standard deviation of $\sigma = 2.95$ while the distribution of KAE-GK scores has a mean of $\mu = 6.35$ and a standard deviation of $\sigma = 2.12$ (*Rehabilitative Psychology*, Spring 1995).

a. Let $\bar{x}_A$ represent the mean KAE-A score for a random sample of 100 college students. Describe the sampling distribution of $\bar{x}_A$.

b. Refer to part **a**. Find the probability that x_A falls between 11.5 and 12.5.

c. Let $\bar{x}_{GK}$ represent the mean KAE-GK score for a random sample of 100 college students. Describe the sampling distribution of $\bar{x}_{GK}$.

d. Refer to part **c**. Find the probability that $\bar{x}_{GK}$ is less than 6.

e. Suppose you observe a sample mean KAE score of 6.5. Is this result more likely to be the value of $\bar{x}_A$ or $\bar{x}_{GK}$? Explain.

 **QUICK REVIEW**

Key Terms

Biased estimate   235
Central Limit Theorem   240
Error of estimation   235
Parameter   228
Point estimator   234

Sample statistic   228
Sampling distribution   229
Standard error of the mean   240
Unbiased estimate   235

Key Formulas

	Mean	**Standard Deviation**	**z-score**
Sampling distribution of $\bar{x}$	$\mu_{\bar{x}} = \mu$	$\sigma_{\bar{x}} = \dfrac{\sigma}{\sqrt{n}}$	$z = \dfrac{\bar{x} - \mu_{\bar{x}}}{\sigma_{\bar{x}}} = \dfrac{\bar{x} - \mu}{\sigma/\sqrt{n}}$

 **LANGUAGE LAB**

Symbol	Pronunciation	Description
θ	theta	Population parameter (general)
$\mu_{\bar{x}}$	mu of x-bar	True mean of sampling distribution of $\bar{x}$
$\sigma_{\bar{x}}$	sigma of x-bar	True standard deviation of sampling distribution of $\bar{x}$

 **SUPPLEMENTARY EXERCISES 6.31–6.50**

Note: Exercises marked with 💾 *require the use of a computer.*

Learning the Mechanics

6.31 Consider a sample statistic A. As with all sample statistics, A is computed by utilizing a specified func-

tion (formula) of the sample measurements. (For example, if A were the sample mean, the specified formula would be to sum the measurements and divide by the number of measurements.)

a. Describe what we mean by the phrase "the sampling distribution of the sample statistic A."

b. Suppose A is to be used to estimate a population parameter α. What is meant by the assertion that A is an unbiased estimator of α?

c. Consider another sample statistic, B. Assume that B is also an unbiased estimator of the population parameter α. How can we use the sampling distributions of A and B to decide which is the better estimator of α?

d. If the sample sizes on which A and B are based are large, can we apply the Central Limit Theorem and assert that the sampling distributions of A and B are approximately normal? Why or why not?

6.32 The standard deviation (or, as it is usually called, the *standard error*) of the sampling distribution for the sample mean, $\bar{x}$, is equal to the standard deviation of the population from which the sample was selected divided by the square root of the sample size. That is,

$$\sigma_{\bar{x}} = \frac{\sigma}{\sqrt{n}}$$

a. As the sample size is increased, what happens to the standard error of $\bar{x}$? Why is this property considered important?

b. Suppose that a sample statistic has a standard error that is not a function of the sample size. In other words, the standard error remains constant as n changes. What would this imply about the statistic as an estimator of a population parameter?

c. Suppose another unbiased estimator (call it A) of the population mean is a sample statistic with a standard error equal to

$$\sigma_A = \frac{\sigma}{\sqrt[3]{n}}$$

Which of the sample statistics, $\bar{x}$ or A, is preferable as an estimator of the population mean? Why?

d. Suppose that the population standard deviation σ is equal to 10 and that the sample size is 64. Calculate the standard errors of $\bar{x}$ and A. Assuming that the sampling distribution of A is approximately normal, interpret the standard errors. Why is the assumption of (approximate) normality unnecessary for the sampling distribution of $\bar{x}$?

6.33 A random sample of $n = 68$ observations is selected from a population with $\mu = 19.6$ and $\sigma = 3.2$. Approximate each of the following probabilities.
a. $P(\bar{x} \leq 19.6)$ **b.** $P(\bar{x} \leq 19)$
c. $P(\bar{x} \geq 20.1)$ **d.** $P(19.2 \leq \bar{x} \leq 20.6)$

6.34 A random sample of 40 observations is to be drawn from a large population of measurements. It is known that 30% of the measurements in the population are 1's, 20% are 2's, 20% are 3's, and 30% are 4's.
a. Give the mean and standard deviation of the (repeated) sampling distribution of $\bar{x}$, the sample mean of the 40 observations.

b. Describe the shape of the sampling distribution of $\bar{x}$. Does your answer depend on the sample size?

6.35 Use a statistical software package to generate 100 random samples of size $n = 2$ from a population characterized by a uniform probability distribution (Section 5.2) and $c = 0$ and $d = 10$. Compute $\bar{x}$ for each sample, and plot a frequency distribution for the 100 $\bar{x}$ values. Repeat this process for $n = 5, 10, 30$, and 50. Explain how your plots illustrate the Central Limit Theorem.

6.36 Use a statistical software package to generate 100 random samples of size $n = 2$ from a population characterized by a normal probability distribution with mean 100 and standard deviation 10. Compute $\bar{x}$ for each sample and plot a frequency distribution for the 100 values of $\bar{x}$. Repeat this process for $n = 5$, 10, 30, and 50. How does the fact that the sampled population is normal affect the sampling distribution of $\bar{x}$?

6.37 A random sample of size n is to be drawn from a large population with mean 100 and standard deviation 10, and the sample mean $\bar{x}$ is to be calculated. To see the effect of different sample sizes on the standard deviation of the sampling distribution of $\bar{x}$, plot $\sigma / \sqrt{n}$ against n for $n = 1, 5, 10, 20, 30, 40$, and 50.

6.38 Suppose x equals the number of heads observed when a single coin is tossed; that is, $x = 0$ or $x = 1$. The population corresponding to x is the set of 0's and 1's generated when the coin is tossed repeatedly a large number of times. Suppose we select $n = 2$ observations from this population. (That is, we toss the coin twice and observe two values of x.)
a. List the three different samples (combinations of 0's and 1's) that could be obtained.
b. Calculate the value of $\bar{x}$ for each of the samples.
c. List the values that $\bar{x}$ can assume, and find the probabilities of observing these values.
d. Construct a graph of the sampling distribution of $\bar{x}$.

Applying the Concepts

6.39 The fourth *Annual Report: Florida Employer Opinion Survey* (1992) gives the results of an extensive survey of employer opinions in Florida. Each employer was asked to rate his or her satisfaction with the preparation of employees by the public education system. Responses were 1, 1.5, or 2, representing very dissatisfied, neither satisfied nor dissatisfied, and very satisfied, respectively. A sample of 651 employers was selected. Assume that the mean for all employers in Florida is 1.50 (the "dividing line" between satisfied and dissatisfied) and the standard deviation is .45.
a. Which type of distribution describes the individual survey responses, continuous or discrete?
b. Describe the distribution that best approximates the sample mean response of 651 employers.

What are the mean and standard deviation of this distribution? What assumptions, if any, are necessary to ensure the validity of your answers?

c. What is the approximate probability that the sample mean will be 1.45 or less?

d. The mean of the sample of 651 employers surveyed in 1992 was 1.36. Given this result, do you think it is likely that all Florida employers' opinions were evenly divided on the effectiveness of public education? That is, do you think the assumption that the population mean is 1.50 is correct? Why or why not?

6.40 Many species of terrestrial tree frogs that hibernate at or near the ground surface can survive prolonged exposure to low winter temperatures. In freezing conditions, the frog's body temperature, called its *supercooling temperature,* remains relatively higher because of an accumulation of glycerol in its body fluids. Studies have shown that the supercooling temperature of terrestrial frogs frozen at $-6°C$ has a relative frequency distribution with a mean of $-2.18°C$ and a standard deviation of .32°C (*Science,* May 1983). Consider the mean supercooling temperature, $\bar{x}$, of a random sample of $n = 42$ terrestrial frogs frozen at $-6°C$.

a. Find the probability that $\bar{x}$ exceeds $-2.05°C$.

b. Find the probability that $\bar{x}$ falls between $-2.20°C$ and $-2.10°C$.

6.41 The relationship among socioeconomic status, IQ, and juvenile delinquency has long been the subject of psychological research. In one study (*Journal of Abnormal Psychology,* 1981), for example, researchers examined the relationship between delinquent behavior and poor verbal abilities. Assume that scores on the verbal IQ test administered during this research have a population mean $\mu = 107$ and a population standard deviation $\sigma = 15$.

a. What shape would you expect the (repeated) sampling distribution of $\bar{x}$ for $n = 84$ juveniles to have? Does your answer depend on the shape of the distribution of verbal IQ scores for all juveniles?

b. Assuming that the population mean and standard deviation for juveniles with no record of delinquency are the same as those for all juveniles, approximate the probability that the sample mean verbal IQ for $n = 84$ juveniles will be 110 or more. State any assumptions you make.

c. As part of the cited study, the researchers found that a sample of $n = 84$ juveniles with no record of delinquency had a mean verbal IQ of $\bar{x} = 110$. Considering your answer to part **b**, do you think that the population mean and standard deviation for nondelinquent juveniles are the same as those for all juveniles? Explain.

6.42 The distribution of the number of barrels of oil produced by a certain oil well each day for the past 3 years has a mean of 400 and a standard deviation of 75.

a. Describe the sampling distribution of the mean number of barrels produced per day for samples of 40 production days drawn from the past 3 years.

b. What is the approximate probability that the sample mean will be greater than 425?

c. What is the approximate probability that the sample mean will be less than 400?

d. What assumptions did you have to make in order to answer parts **a–c**? Justify the assumptions.

6.43 The distribution of violent crimes per day in a certain city possesses a mean equal to 1.3 and a standard deviation equal to 1.7. A random sample of 50 days is observed, and the daily mean number of crimes for this sample, $\bar{x}$, is calculated.

a. Give the mean and standard deviation of the sampling distribution of $\bar{x}$.

b. Will the sampling distribution of $\bar{x}$ be approximately normal? Explain.

c. Find an approximate value of $P(\bar{x} < 1)$.

d. Find an approximate value of $P(\bar{x} > 1.9)$.

6.44 To determine whether a metal lathe that produces machine bearings is properly adjusted, a random sample of 25 bearings is collected and the diameter of each is measured.

a. If the standard deviation of the diameters of the bearings measured over a long period of time is .001 inch, what is the approximate probability that the mean diameter $\bar{x}$ of the sample of 25 bearings will lie within .0001 inch of the population mean diameter of the bearings?

b. If the population of diameters has an extremely skewed distribution, how will your approximation in part **a** be affected?

6.45 Refer to Exercise 6.44. The mean diameter of the bearings produced by the machine is supposed to be .5 inch. The company decides to use the sample mean (from Exercise 6.44) to decide whether the process is in control; i.e., whether it is producing bearings with a mean diameter of .5 inch. The machine will be considered out of control if the mean of the sample of $n = 25$ diameters is less than .4994 inch or larger than .5006 inch. If the true mean diameter of the bearings produced by the machine is .501 inch, what is the approximate probability that the test will imply that the process is out of control?

6.46 This past year, an elementary school began using a new method to teach arithmetic to first graders. A standardized test, administered at the end of the year, was used to measure the effectiveness of the new method. The distribution of past scores on the standardized test produced a mean of 75 and a standard deviation of 10.

a. If the new method is no different from the old method, what is the approximate probability that the mean score $\bar{x}$ of a random sample of 36 students will be greater than 79?

b. What assumptions must be satisfied to make your answer valid?

6.47 In the last decade preventative health behavior began to receive much attention. In one study at a health fair (*Psychological Reports,* 1985), 100 attendees were administered a questionnaire on preventative health behaviors. Assume that the population mean and standard deviation of the questionnaire scores in the area of preventing heart disease are 38 and 5, respectively, when administered to the general public.

a. Describe the sampling distribution of the sample mean questionnaire score of 100 health-fair attendees, assuming that they are a random sample of the general public.

b. What is the probability that the sample mean score of the 100 health-fair attendees will exceed 39.1, if they represent a random sample of the general public?

c. The scores of the 100 health-fair attendees on the preventative behavior questionnaire had a mean of 39.1. Considering your answer to part **b**, do you think the attendees represent a random sample of the general public with regard to the heart disease questionnaire?

6.48 Refer to Exercise 6.47. The 100 health-fair attendees were also administered a questionnaire on malignancies. Assume that the mean and standard deviation of scores of the general public are 42 and 6.5, respectively.

a. Describe the sampling distribution of the sample mean score of 100 attendees, assuming that they are a random sample of the general public.

b. What is the probability that the sample mean questionnaire score of the 100 attendees will exceed 43.6, if they represent a random sample of the general public?

c. The scores of the 100 health-fair attendees on the preventative behavior questionnaire related to malignancies had a mean of 43.6. Considering your answer to part **b**, do you think the attendees represent a random sample of the general public with regard to the malignancy questionnaire?

6.49 [*Note:* This exercise refers to an optional section in Chapter 4.] A building contractor has decided to purchase a load of factory-reject aluminum siding as long as the average number of flaws per piece of siding in a sample of size 35 from the factory's reject pile is 2.1 or less. If it is known that the number of flaws per piece of siding in the factory's reject pile has a Poisson probability distribution with a mean of 2.5, find the approximate probability that the contractor will not purchase a load of siding. [*Hint:* If x is a Poisson random variable with mean λ, then σ_x^2 also equals λ.]

6.50 [*Note:* This exercise refers to an optional section in Chapter 5.] An article in *Industrial Engineering* (Aug. 1990) discussed the importance of modeling machine downtime correctly in simulation studies. As an illustration, the researcher considered a single-machine-tool system with repair times (in minutes) that can be modeled by an exponential distribution with $\theta = 60$ (see optional Section 5.5). Of interest is the mean repair time, $\bar{x}$, of a sample of 100 machine breakdowns.

a. Find $E(\bar{x})$ and the variance of $\bar{x}$.

b. What probability distribution provides the best model of the sampling distribution of $\bar{x}$? Why?

c. Calculate the probability that the mean repair time, $\bar{x}$, is no longer than 30 minutes.

STUDENT PROJECTS

To understand the Central Limit Theorem and sampling distribution, consider the following experiment: Toss four identical coins, and record the number of heads observed. Then repeat this experiment four more times, so that you end up with a total of five observations for the random variable x, the number of heads when four coins are tossed.

Now derive and graph the probability distribution for x, assuming the coins are balanced. Note that the mean of this distribution is $\mu = 2$ and the standard deviation is $\sigma = 1$. This probability distribution represents the one from which you are drawing a random sample of five measurements.

Next, calculate the mean $\bar{x}$ of the five measurements—that is, calculate the mean number of heads you observed in five repetitions of the experiment. Although you have

repeated the basic experiment five times, you have only one observed value of $\bar{x}$. To derive the probability distribution or sampling distribution of $\bar{x}$ empirically, you have to repeat the entire process (of tossing four coins five times) many times. Do it 100 times.

The approximate sampling distribution of $\bar{x}$ can be derived theoretically by making use of the Central Limit Theorem. We expect at least an approximate normal probability distribution with a mean $\mu = 2$ and a standard deviation

$$\sigma_{\bar{x}} = \frac{\sigma}{\sqrt{n}} = \frac{1}{\sqrt{5}} = .45$$

Count the number of your 100 $\bar{x}$'s that fall in each of the intervals in the figure on page 251. Use the normal probability distribution with $\mu = 2$ and $\sigma_{\bar{x}} = .45$ to calculate the expected number of the 100 $\bar{x}$'s in each of the intervals. How closely does the theory describe your experimental results?

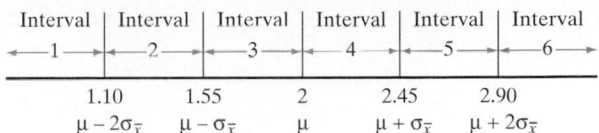

Interval 1	Interval 2	Interval 3	Interval 4	Interval 5	Interval 6

1.10	1.55	2	2.45	2.90
$\mu - 2\sigma_{\bar{x}}$	$\mu - \sigma_{\bar{x}}$	μ	$\mu + \sigma_{\bar{x}}$	$\mu + 2\sigma_{\bar{x}}$

EXPLORING DATA WITH A COMPUTER

The Federal Trade Commission (FTC) annually ranks American cigarette brands in terms of the amount of three hazardous substances—tar, nicotine, and carbon monoxide. The test results are obtained as follows: A sequential smoking machine is used to "smoke" cigarettes to a 23-millimeter butt length. Based on tests of 100 cigarettes per brand, the carbon monoxide, tar, and nicotine concentrations (rounded to the nearest milligram) in the residual "dry" particulate matter are determined.

Appendix B contains the results of the FTC's 1995 tests of 962 domestic cigarette brands. Select one of the three hazardous substances (e.g., tar). Use a statistical software package to obtain the mean and standard deviation of the 962 tar amounts. Assume that these quantities represent the population mean μ and the population standard deviation σ for all domestic cigarette brands.

a. Draw 100 random samples of $n = 20$ observations from the 962 tar amounts. Select the samples with replacement—i.e., replace each measurement before selecting the next.* Calculate the 100 sample means. Generate a stem-and-leaf display or a relative frequency histogram for the 100 means. Then count the number of the 100 sample means that fall in the intervals $\mu \pm \sigma/\sqrt{n}$, $\mu + 2\sigma\sqrt{n}$, and $\mu \pm 3\sigma\sqrt{n}$. How do the graphical description and the percentage of means falling in the intervals agree with a normal distribution having mean μ and standard deviation $\sigma/\sqrt{n}$?

b. Repeat part a using a sample size of $n = 50$. Is the sampling distribution of the sample means closer to normal for the larger sample size?

---

*Here and in the future we will specify sampling with replacement to simulate the sampling from very large or infinite populations.

Chapter 7

INFERENCES BASED ON A SINGLE SAMPLE

Estimation with Confidence Intervals

Contents

Case Studies

*W*HERE WE'VE BEEN

We've learned that populations are characterized by numerical descriptive measures (called *parameters*) and that decisions about their values are based on sample statistics computed from sample data. Since statistics vary in a random manner from sample to sample, inferences based on them are subject to uncertainty. This property is reflected in the sampling (probability) distribution of a statistic.

*W*HERE WE'RE GOING

In this chapter, we'll put all the preceding material into practice; that is, we'll estimate population means and proportions based on a single sample selected from the population of interest. Most important, we use the sampling distribution of a sample statistic to assess the reliability of an estimate.

The estimation of the mean gas mileage for a new car model, the estimation of the expected life of a computer monitor, and the estimation of the mean yearly sales for companies in the steel industry are problems with a common element. In each case, we're interested in estimating the mean of a population of measurements. This important problem constitutes the primary topic of this chapter.

You'll see that different techniques are used for estimating a mean, depending on whether a sample contains a large or small number of measurements. Nevertheless, our objectives remain the same: We want to use the sample information to estimate the mean and to assess the reliability of the estimate.

First, we consider a method of estimating a population mean using a large sample (Section 7.1) and a small sample (Section 7.2). Then, we consider estimation of population proportions (Section 7.3). Finally, we see how to determine the sample sizes necessary for reliable estimates (Section 7.4).

7.1   LARGE-SAMPLE CONFIDENCE INTERVAL FOR A POPULATION MEAN

We illustrate the **large-sample method** of estimating a population mean with an example. Suppose a large hospital wants to estimate the average length of time patients remain in the hospital. To accomplish this objective, the hospital administrators plan to sample 100 of all previous patients' records and to use the sample mean, $\bar{x}$, of the lengths of stay to estimate the mean stay, μ, of *all* patients' visits. The sample mean $\bar{x}$ represents a *point estimator* of the population mean μ (Definition 6.4). How can we assess the accuracy of this point estimator?

According to the Central Limit Theorem, the sampling distribution of the sample mean is approximately normal for large samples, as shown in Figure 7.1. Let us calculate the interval

$$\bar{x} \pm 2\sigma_{\bar{x}} = \bar{x} \pm \frac{2\sigma}{\sqrt{n}}$$

That is, we form an interval 4 standard deviations wide—from 2 standard deviations below the sample mean to 2 standard deviations above the mean. What are the chances (answer before we have drawn a sample) that this interval will enclose μ, the population mean?

To answer this question, refer to Figure 7.1. If the 100 measurements yield a value of $\bar{x}$ that falls between the two lines on either side of μ—i.e., within 2 standard deviations of μ—then the interval $\bar{x} \pm 2\sigma_{\bar{x}}$ will contain μ; if $\bar{x}$ falls outside these boundaries, the interval $\bar{x} \pm 2\sigma_{\bar{x}}$ will not contain μ. Since the area under the normal curve (the sampling distribution of $\bar{x}$) between these boundaries is about .95 (more precisely, from Table IV in Appendix A the area is .9544), we

FIGURE 7.1

Sampling distribution of $\bar{x}$

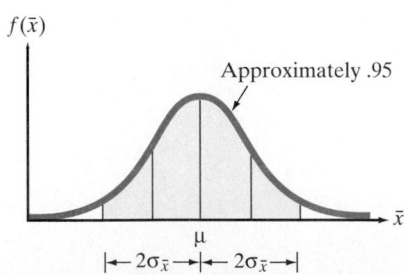

TABLE 7.1   Lengths of Stay (in Days) for 100 Patients

2	3	8	6	4	4	6	4	2	5
8	10	4	4	4	2	1	3	2	10
1	3	2	3	4	3	5	2	4	1
2	9	1	7	17	9	9	9	4	4
1	1	1	3	1	6	3	3	2	5
1	3	3	14	2	3	9	6	6	3
5	1	4	6	11	22	1	9	6	5
2	2	5	4	3	6	1	5	1	6
17	1	2	4	5	4	4	3	2	3
3	5	2	3	3	2	10	2	4	2

know that the interval $x \pm 2\sigma_{\bar{x}}$ will contain μ with a probability approximately equal to .95.

For instance, consider the lengths of time spent in the hospital for 100 patients shown in Table 7.1. A SAS printout of summary statistics for the sample of 100 lengths of stay is shown in Figure 7.2.

From the printout, we find $\bar{x} = 4.53$ days and $s = 3.68$ days. To achieve our objective, we must construct the interval

$$\bar{x} \pm 2\sigma_{\bar{x}} = 4.53 \pm 2\frac{\sigma}{\sqrt{100}}$$

But now we face a problem. You can see that without knowing the standard deviation σ of the original population—that is, the standard deviation of the lengths of stay of *all* patients—we cannot calculate this interval. However, since we have a large sample ($n = 100$ measurements), we can approximate the interval by using the sample standard deviation s to approximate σ. Thus,

$$\bar{x} \pm 2\frac{\sigma}{\sqrt{100}} \approx \bar{x} \pm 2\frac{s}{\sqrt{100}} = 4.53 \pm 2\left(\frac{3.68}{10}\right) = 4.53 \pm .74$$

That is, we estimate the mean length of stay in the hospital for all patients to fall in the interval 3.79 to 5.27 days.

Can we be sure that μ, the true mean, is in the interval 3.79 to 5.27? We cannot be certain, but we can be reasonably confident that it is. This confidence is derived from the knowledge that if we were to draw repeated random samples of 100 measurements from this population and form the interval $\bar{x} \pm 2\sigma_{\bar{x}}$ each time, approximately 95% of the intervals would contain μ. We have no way of knowing (without looking at all the patients' records) whether our sample interval is one of the 95% that contain μ or one of the 5% that do not, but the odds certainly favor its containing μ. Consequently, the interval 3.79 to 5.27 provides a reliable estimate of the mean length of patient stay in the hospital.

The formula that tells us how to calculate an interval estimate based on sample data is called an *interval estimator,* or *confidence interval.* The probability, .95, that measures the confidence we can place in the interval estimate is called a *confidence coefficient.* The percentage, 95%, is called the *confidence level* for the interval estimate. It is not usually possible to assess precisely the reliability of point

FIGURE 7.2

SAS summary statistics for the lengths of 100 hospital stays

```
Analysis Variable : LOS

N Obs     N        Minimum       Maximum          Mean          Std Dev
 100     100     1.0000000    22.0000000     4.5300000       3.6775458
```

estimators because they are single points rather than intervals. So, because we prefer to use estimators for which a measure of reliability can be calculated, we will generally use interval estimators.

DEFINITION 7.1

An **interval estimator** (or **confidence interval**) is a formula that tells us how to use sample data to calculate an interval that estimates a population parameter.

DEFINITION 7.2

The **confidence coefficient** is the probability that an interval estimator encloses the population parameter—that is, the relative frequency with which the interval estimator encloses the population parameter when the estimator is used repeatedly a very large number of times. The **confidence level** is the confidence coefficient expressed as a percentage.

Now we have seen how an interval can be used to estimate a population mean. When we use an interval estimator, we can usually calculate the probability that the estimation *process* will result in an interval that contains the true value of the population mean. That is, the probability that the interval contains the parameter in repeated usage is usually known. Figure 7.3 shows what happens when 10 different samples are drawn from a population, and a confidence interval for μ is calculated from each. The location of μ is indicated by the vertical line in the figure. Ten confidence intervals, each based on one of 10 samples, are shown as horizontal line segments. Note that the confidence intervals move from sample to sample—sometimes containing μ and other times missing μ. If our confidence level is 95%, in the long run, 95% of our sample confidence intervals will contain μ.

Suppose you wish to choose a confidence coefficient other than .95. Notice in Figure 7.1 that the confidence coefficient .95 is equal to the total area under the sampling distribution, less .05 of the area, which is divided equally between the two tails. Using this idea, we can construct a confidence interval with any desired confidence coefficient by increasing or decreasing the area (call it α) assigned to the tails of the sampling distribution (see Figure 7.4). For example, if we place area $\alpha/2$ in each tail and if $z_{\alpha/2}$ is the z value such that the area $\alpha/2$ lies to its right, the confidence interval with confidence coefficient $(1 - \alpha)$ is

$$\bar{x} \pm z_{\alpha/2}\sigma_{\bar{x}}$$

To illustrate, for a confidence coefficient of .90 we have $(1 - \alpha) = .90$, $\alpha = .10$, and $\alpha/2 = .05$; $z_{.05}$ is the z value that locates area .05 in the upper tail of the sampling distribution. Recall that Table IV in Appendix A gives the areas between the mean and a specified z value. Since the total area to the right of the mean is .5, we find that $z_{.05}$ will be the z value corresponding to an area of $.5 - .05 = .45$ to the right of the mean (see Figure 7.5). This z value is $z_{.05} = 1.645$.

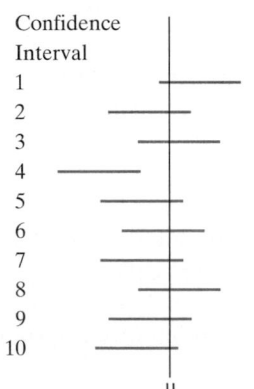

Confidence Interval

1
2
3
4
5
6
7
8
9
10

μ

FIGURE 7.3

Confidence intervals for μ: 10 samples

FIGURE 7.4

Locating $z_{\alpha/2}$ on the standard normal curve

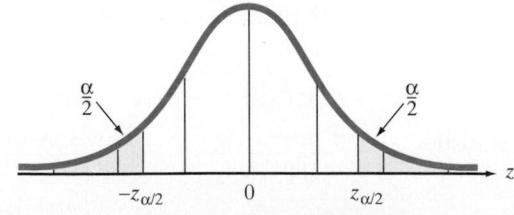

FIGURE 7.5

The z value ($z_{.05}$) corresponding to an area equal to .05 in the upper tail of the z distribution

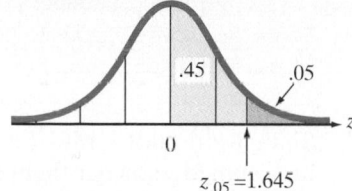

Confidence coefficients used in practice usually range from .90 to .99. The most commonly used confidence coefficients with corresponding values of α and $z_{\alpha/2}$ are shown in Table 7.2.

Large-Sample 100$(1 - \alpha)$% Confidence Interval for μ

$$\bar{x} \pm z_{\alpha/2}\sigma_{\bar{x}} = \bar{x} \pm z_{\alpha/2}\frac{\sigma}{\sqrt{n}}$$

where $z_{\alpha/2}$ is the z value with an area $\alpha/2$ to its right (see Figure 7.4) and $\sigma_{\bar{x}} = \sigma/\sqrt{n}$. The parameter σ is the standard deviation of the sampled population and n is the sample size.

When n is equal to 30 or more, the confidence interval is approximately equal to

$$\bar{x} \pm z_{\alpha/2}\left(\frac{s}{\sqrt{n}}\right)$$

where s is the sample standard deviation.

EXAMPLE 7.1

Unoccupied seats on flights cause airlines to lose revenue. Suppose a large airline wants to estimate its average number of unoccupied seats per flight over the past year. To accomplish this, the records of 225 flights are randomly selected, and the number of unoccupied seats is noted for each of the sampled flights. The sample mean and standard deviation are

$$\bar{x} = 11.6 \text{ seats} \qquad s = 4.1 \text{ seats}$$

Estimate μ, the mean number of unoccupied seats per flight during the past year, using a 90% confidence interval.

Solution

The general form of the 90% confidence interval for a population mean is

$$\bar{x} \pm z_{\alpha/2}\sigma_{\bar{x}} = \bar{x} \pm z_{.05}\sigma_{\bar{x}} = \bar{x} \pm 1.645\left(\frac{\sigma}{\sqrt{n}}\right)$$

For the 225 records sampled, we have

$$11.6 \pm 1.645\left(\frac{\sigma}{\sqrt{225}}\right)$$

TABLE 7.2   Commonly Used Values of $z_{\alpha/2}$

CONFIDENCE LEVEL 100$(1 - \alpha)$	α	$\alpha/2$	$z_{\alpha/2}$
90%	.10	.05	1.645
95%	.05	.025	1.96
99%	.01	.005	2.575

FIGURE 7.6

MINITAB printout for the confidence intervals in Example 7.1

```
                N    MEAN   STDEV   SE MEAN    90.0 PERCENT C.I.
NoSeats        225    11.6    4.1     0.273   (  11.15,   12.05)
```

Since we do not know the value of σ (the standard deviation of the number of unoccupied seats per flight for all flights of the year), we use our best approximation—the sample standard deviation s. Then the 90% confidence interval is, approximately,

$$11.6 \pm 1.645\left(\frac{4.1}{\sqrt{225}}\right) = 11.6 \pm .45$$

or from 11.15 to 12.05. That is, at the 90% confidence level, we estimate the mean number of unoccupied seats per flight to be between 11.15 and 12.05 during the sampled year. This result is verified on the MINITAB printout of the analysis shown in Figure 7.6.

We stress that the confidence level for this example, 90%, refers to the procedure used. If we were to apply this procedure repeatedly to different samples, approximately 90% of the intervals would contain μ. We do not know whether this particular interval (11.15, 12.05) is one of the 90% that contain μ or one of the 10% that do not.   ▲

The interpretation of confidence intervals for a population mean is summarized in the accompanying box.

Interpretation of a Confidence Interval for a Population Mean

---

When we form a $100(1 - \alpha)\%$ confidence interval for μ, we usually express our confidence in the interval with a statement such as, "We can be $100(1 - \alpha)\%$ confident that μ lies between the lower and upper bounds of the confidence interval," where for a particular application, we substitute the appropriate numerical values for the confidence and for the lower and upper bounds. *The statement reflects our confidence in the estimation process rather than in the particular interval that is calculated from the sample data.* We know that repeated application of the same procedure will result in different lower and upper bounds on the interval. Furthermore, we know that $100(1 - \alpha)\%$ of the resulting intervals will contain μ. There is (usually) no way to determine whether any particular interval is one of those that contain μ, or one that does not. However, unlike point estimators, confidence intervals have some measure of reliability, the confidence coefficient, associated with them. For that reason they are generally preferred to point estimators.

Sometimes, the estimation procedure yields a confidence interval that is too wide for our purposes. In this case, we will want to reduce the width of the interval to obtain a more precise estimate of μ. One way to accomplish this is to decrease the confidence coefficient, $1 - \alpha$. For example, consider the problem of estimating the mean length of stay, μ, for hospital patients. Recall that for a sample of 100 patients, $\bar{x} = 4.53$ days and $s = 3.68$ days. A 90% confidence interval for μ is

$$\bar{x} \pm 1.645\sigma/\sqrt{n} \approx 4.53 \pm (1.645)(3.68/\sqrt{100}) = 4.53 \pm .61,$$

or (3.92, 5.14). You can see that this interval is narrower than the previously calculated 95% confidence interval, (3.79, 5.27). Unfortunately, we also have "less confidence" in the 90% confidence interval. An alternative method used to decrease the width of an interval without sacrificing "confidence" is to increase the sample size n. We demonstrate this method in Section 7.4.

EXERCISES 7.1–7.20

Learning the Mechanics

7.1 Find $z_{\alpha/2}$ for each of the following:
   a. $\alpha = .10$ **b.** $\alpha = .01$ **c.** $\alpha = .05$ **d.** $\alpha = .20$

7.2 What is the confidence level of each of the following confidence intervals for μ?

   a. $\bar{x} \pm 1.96\left(\dfrac{\sigma}{\sqrt{n}}\right)$ **b.** $\bar{x} \pm 1.645\left(\dfrac{\sigma}{\sqrt{n}}\right)$

   c. $\bar{x} \pm 2.575\left(\dfrac{\sigma}{\sqrt{n}}\right)$ **d.** $\bar{x} \pm 1.282\left(\dfrac{\sigma}{\sqrt{n}}\right)$

   e. $\bar{x} \pm .99\left(\dfrac{\sigma}{\sqrt{n}}\right)$

7.3 A random sample of 64 observations from a population produced the following summary statistics:

$$\sum x = 700 \qquad \sum (x_i - \bar{x})^2 - 4{,}238$$

   a. Find a 95% confidence interval for μ.
   b. Interpret the confidence interval you found in part **a**.

7.4 A random sample of 90 observations produced a mean $\bar{x} = 25.9$ and a standard deviation $s = 2.7$.
   a. Find a 95% confidence interval for the population mean μ.
   b. Find a 90% confidence interval for μ.
   c. Find a 99% confidence interval for μ.

7.5 A random sample of 100 observations from a normally distributed population possesses a mean equal to 83.2 and a standard deviation equal to 6.4.
   a. Find a 95% confidence interval for μ.
   b. What do you mean when you say that a confidence coefficient is .95?
   c. Find a 99% confidence interval for μ.
   d. What happens to the width of a confidence interval as the value of the confidence coefficient is increased while the sample size is held fixed?
   e. Would your confidence intervals of parts **a** and **c** be valid if the distribution of the original population was not normal? Explain.

7.6 A random sample of n measurements was selected from a population with unknown mean μ and standard deviation σ. Calculate a 95% confidence interval for μ for each of the following situations:
   a. $n = 75, \bar{x} = 28, s^2 = 12$
   b. $n = 200, \bar{x} = 102, s^2 = 22$
   c. $n = 100, \bar{x} = 15, s = .3$
   d. $n = 100, \bar{x} = 4.05, s = .83$
   e. Is the assumption that the underlying population of measurements is normally distributed necessary to ensure the validity of the confidence intervals in parts **a–d**? Explain.

7.7 Explain the difference between an interval estimator and a point estimator for μ.

7.8 Explain what is meant by the statement, "We are 95% confident that an interval estimate contains μ."

7.9 Will a large sample confidence interval be valid if the population from which the sample is taken is not normally distributed? Explain.

7.10 The mean and standard deviation of a random sample of n measurements are equal to 33.9 and 3.3, respectively.
   a. Find a 95% confidence interval for μ if $n = 100$.
   b. Find a 95% confidence interval for μ if $n = 400$.
   c. Find the widths of the confidence intervals found in parts **a** and **b**. What is the effect on the width of a confidence interval of quadrupling the sample size while holding the confidence coefficient fixed?

Applying the Concepts

7.11 The *Journal of the American Medical Association* (Apr. 21, 1993) reported on the results of a National Health Interview Survey designed to determine the prevalence of smoking among U.S. adults. More than 40,000 adults responded to questions such as "Have you smoked at least 100 cigarettes in your lifetime?" and "Do you smoke cigarettes now?" Current smokers (more than 11,000 adults in the survey) were also asked: "On the average, how many cigarettes do you now smoke a day?" The results yielded a mean of 20.0 cigarettes per day with an associated 95% confidence interval of (19.7, 20.3).
   a. Interpret the 95% confidence interval.
   b. State any assumptions about the target population of current cigarette smokers that must be satisfied for inferences derived from the interval to be valid.
   c. A tobacco industry researcher claims that the mean number of cigarettes smoked per day by regular cigarette smokers is less than 15. Comment on this claim.

7.12 Refer to *The Astronomical Journal* (July 1995) study of the velocity of light emitted from a galaxy in the universe, Exercise 2.72. A sample of 103 galaxies located in the galaxy cluster A2142 had a mean velocity of $\bar{x} = 27{,}117$ kilometers per second (km/s) and a standard deviation of $s = 1{,}280$ km/s. Suppose your goal is to make an inference about the population mean light velocity of galaxies in cluster A2142.
   a. In part **b** of Exercise 2.72, you constructed an interval that captured approximately 95% of the galaxy velocities in the cluster. Explain why this interval is inappropriate for this inference.
   b. Construct a 95% confidence interval for the mean light velocity emitted from all galaxies in cluster A2142. Interpret the result.

7.13 Research indicates that bicycle helmets save lives. A study reported in *Public Health Reports* (May–June

N Obs	N	Minimum	Maximum	Mean	Std Dev
50	50	0.8000000	48.2000000	20.8480000	13.4138253

1992) was intended to identify ways of encouraging helmet use in children. One of the variables measured was the children's perception of the risk involved in bicycling. A 4-point scale was used, with scores ranging from 1 (no risk) to 4 (very high risk). A sample of 797 children in grades 4–6 yielded the following results on the perception of risk variable: $\bar{x} = 3.39$, $s = .80$.

  a. Calculate a 90% confidence interval for the average perception of risk for all students in grades 4–6. What assumptions did you make to ensure the validity of the confidence interval?

  b. If the population mean perception of risk exceeds 2.50, the researchers will conclude that students in these grades exhibit an awareness of the risk involved with bicycling. Interpret the confidence interval constructed in part **a** in this context.

7.14 A fact long known but little understood is that twins, in their early years, tend to have lower IQs and pick up language more slowly than do nontwins. Psychologists have found that the slower intellectual growth of most twins may be caused by benign parental neglect. Suppose it is desired to estimate the mean attention time given to twins per week by their parents. A sample of 50 sets of 2½-year-old twin boys is taken, and at the end of 1 week the attention time given to each pair is recorded. The data (in hours) are listed in the accompanying table. A SAS printout at the top of the page gives summary statistics. Use this information to find a 90% confidence interval for the mean attention time given to all twin boys by their parents. Interpret the confidence interval.

20.7	16.7	22.5	12.1	2.9
23.5	6.4	1.3	39.6	35.6
10.9	7.1	46.0	23.4	29.4
44.1	13.8	24.3	9.3	3.4
15.7	46.6	10.6	6.7	5.4
14.0	20.7	48.2	7.7	22.2
20.3	34.0	44.5	23.8	20.0
43.1	14.3	21.9	17.5	9.6
36.4	0.8	1.1	19.3	14.6
32.5	19.1	36.9	27.9	14.0

7.15 Automotive engineers are continually improving their products. Suppose a new type of brake light has been developed by General Motors. As part of a product safety evaluation program, General Motors' engineers wish to estimate the mean driver response time to the new brake light. (Response time is the

length of time from the point that the brake is applied until the driver in the following car takes some corrective action.) Fifty drivers are selected at random and the response time (in seconds) for each driver is recorded, yielding the following results: $\bar{x} = .72$, $s^2 = .022$. Estimate the mean driver response time to the new brake light using a 99% confidence interval. Interpret the confidence interval.

7.16 Named for the section of the 1978 Internal Revenue Code that authorized them, 401(k) plans permit employees to shift part of their before-tax salaries into investments such as mutual funds. Employers typically match 50% of an employee's contribution up to about 6% of salary (*Fortune,* Dec. 28, 1992). One company, concerned with what it believed was a low employee participation rate in its 401(k) plan, sampled 30 other companies with similar plans and asked for their 401(k) participation rates. The rates (in percentages) at the bottom of the page were obtained. Descriptive statistics for the data are given in the SPSS printout on page 261.

  a. Use the information on the SPSS printout to construct a 95% confidence interval for the mean participation rate for all companies that have 401(k) plans.

  b. Interpret the interval in the context of this problem.

  c. What assumption is necessary to ensure the validity of this confidence interval?

  d. If the company that conducted the sample has a 71% participation rate, can it safely conclude that its rate is below the population mean rate for all companies with 401(k) plans? Explain.

  e. If the 60% in the data set had been 80%, how would the center and width of the confidence interval you constructed in part **a** be affected?

7.17 Tropical swarm-founding wasps, like ants and bees, rely on workers to raise their offspring. Interestingly, the workers of this species of wasp are mostly fully developed females, capable of producing offspring of their own. However, many do not do so. Instead, they rear the young of a few others in the brood. One possible factor in this strange behavior is inbreeding, which increases relatedness among the wasps, presumably making it easier for the workers to pick out their closest relatives as propagators of their own genetic material. To test this theory, 197 swarm-founding wasps were captured in Venezuela, frozen at $-70°C$, and then subjected to a series of genetic tests (*Science,* Nov. 1988). The data were used

80	76	81	77	82	80
88	85	80	79	83	75
90	84	82	77	75	86

85	60	80	79	82	70
87	78	80	84	72	75

```
Number of Valid Observations (Listwise) =      30.00
Variable     Mean    Std Dev  Minimum   Maximum    N Label
PARTRATE    79.73      5.96    60.00     90.00     30
```

to generate an inbreeding coefficient, x, for each wasp specimen, with the following results: $\bar{x} = .044$ and $s = .884$.

a. Construct a 99% confidence interval for the mean inbreeding coefficient of this species of wasp.

b. A coefficient of 0 implies that the wasp has no tendency to inbreed. Use the confidence interval, part **a**, to make an inference about the tendency for this species of wasp to inbreed.

7.18 A process has been developed that can transform ordinary iron into a kind of super-iron called *metallic glass*. Metallic glass is three to four times stronger than the toughest steel alloys, but it becomes brittle at very high temperatures. To estimate the mean temperature, μ, at which a particular type of metallic glass becomes brittle, 36 pieces of this metallic glass were randomly sampled from a recent production run. Each piece was independently subjected to higher and higher temperatures until it became brittle. The temperature at which brittleness first appeared was recorded for each piece in the sample. The following results were obtained: $\bar{x} = 480°F$, $s = 11°F$. Use a 90% confidence interval to estimate μ. Interpret your confidence interval.

7.19 Research reported in the *Journal of Psychology and Aging* (May 1992) studied the role that the age of workers has in determining their level of job satisfaction. The researcher hypothesized that both younger and older workers would have a higher job satisfaction rating than middle-age workers. Each of a sample of 1,686 adults was given a job satisfaction score based on answers to a series of questions. Higher job satisfaction scores indicate higher levels of job satisfaction. The data are given in the accompanying table, arranged by age group.

a. Construct 95% confidence intervals for the mean job satisfaction scores of each age group. Carefully interpret each interval.

b. In the construction of three 95% confidence intervals, is it more or less likely that at least one of them will *not* contain the population mean it is

	AGE GROUP		
	Younger 18–24	**Middle-Age 25–44**	**Older 45–64**
$\bar{x}$	4.17	4.04	4.31
s	.75	.81	.82
n	241	768	677

intended to estimate than it is for a single confidence interval to miss the population mean? [*Hint:* Assume the three intervals are independent and calculate the probability that at least one of them will not contain the population mean it estimates. Compare this probability to the probability that a single interval fails to enclose the mean.]

c. Based on these intervals, does it appear that the researcher's hypothesis is supported? [*Caution:* We'll learn how to use sample information to compare population means in Chapter 9, and we'll return to this exercise at that time. Here, simply base your opinion on the individual confidence intervals you constructed in part **a**.]

7.20 According to scientists, the cockroach has had 300 million years to develop a resistance to destruction. That they are "reproductively brilliant" is evidenced by a study conducted by researchers for S.C. Johnson & Son, Inc. (manufacturers of Raid and Off). Five thousand roaches, the expected number in a roach-infested house, were released in the Raid test kitchen. One week later the kitchen was fumigated and 16,298 dead roaches were counted, a gain of 11,298 roaches for the 1-week period. Assume that none of the original roaches died during the 1-week period and that the standard deviation of x, the number of roaches produced per roach in a 1-week period, is 1.5. Use the number of roaches produced by the sample of 5,000 roaches to find a 95% confidence interval for the mean number of roaches produced per week for each roach in a typical roach-infested house.

7.2   SMALL-SAMPLE CONFIDENCE INTERVAL FOR A POPULATION MEAN

Federal legislation requires pharmaceutical companies to perform extensive tests on new drugs before they can be marketed. Initially, a new drug is tested on animals. If the drug is deemed safe after this first phase of testing, the pharmaceutical company is then permitted to begin human testing on a limited basis. During this second phase, inferences must be made about the safety of the drug based on information in very small samples.

Suppose a pharmaceutical company must estimate the average increase in blood pressure of patients who take a certain new drug. Assume that only six patients can be used in the initial phase of human testing. The use of a **small sample** in making an inference about μ presents two immediate problems when we attempt to use the standard normal z as a test statistic.

PROBLEM 1

The shape of the sampling distribution of the sample mean $\bar{x}$ (and the z statistic) now depends on the shape of the population that is sampled. We can no longer assume that the sampling distribution of $\bar{x}$ is approximately normal, because the Central Limit Theorem ensures normality only for samples that are sufficiently large.

Solution to Problem 1

The sampling distribution of $\bar{x}$ (and z) is exactly normal even for relatively small samples if the sampled population is normal. It is approximately normal if the sampled population is approximately normal.   ▲

PROBLEM 2

The population standard deviation σ is almost always unknown. Although it is still true that $\sigma_{\bar{x}} = \sigma/\sqrt{n}$, the sample standard deviation s may provide a poor approximation for σ when the sample size is small.

Solution to Problem 2

Instead of using the standard normal statistic

$$z = \frac{\bar{x} - \mu}{\sigma_{\bar{x}}} = \frac{\bar{x} - \mu}{\sigma/\sqrt{n}}$$

which requires knowledge of or a good approximation to σ, we define and use the statistic

$$t = \frac{\bar{x} - \mu}{s/\sqrt{n}}$$

in which the sample standard deviation, s, replaces the population standard deviation, σ.   ▲

The distribution of the *t* **statistic** in repeated sampling was discovered by W. S. Gosset, a chemist in the Guinness brewery in Ireland, who published his discovery in 1908 under the pen name of Student. The main result of Gosset's work is that if we are sampling from a normal distribution, the t statistic has a sampling distribution very much like that of the z statistic: mound-shaped, symmetric, with mean 0. The primary difference between the sampling distributions of t and z is that the t statistic is more variable than the z, which follows intuitively when you realize that t contains two random quantities ($\bar{x}$ and s), whereas z contains only one ($\bar{x}$).

The actual amount of variability in the sampling distribution of t depends on the sample size n. A convenient way of expressing this dependence is to say that the t statistic has $(n-1)$ **degrees of freedom (df)**. Recall that the quantity $(n-1)$ is the divisor that appears in the formula for s^2. This number plays a key role in the sampling distribution of s^2 and appears in discussions of other statistics in later chapters. Particularly, the smaller the number of degrees of freedom associated with the t statistic, the more variable will be its sampling distribution.

FIGURE 7.7

Standard normal (z) distribution and t-distribution with 4 df

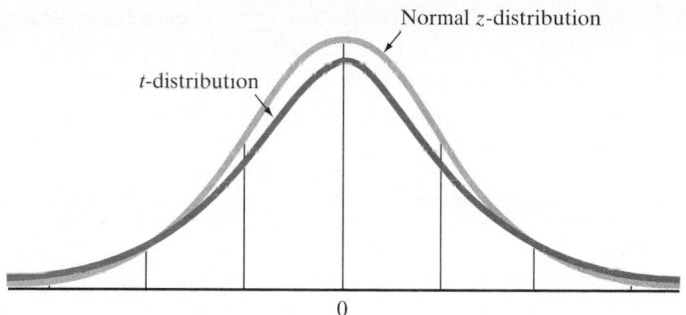

In Figure 7.7 we show both the sampling distribution of z and the sampling distribution of a t statistic with 4 df. You can see that the increased variability of the t statistic means that the t value, t_α, that locates an area α in the upper tail of the t-distribution is larger than the corresponding value z_α. For any given value of α, the t value t_α increases as the number of degrees of freedom (df) decreases. Values of t that will be used in forming small-sample confidence intervals of μ are given in Table VI of Appendix A and inside the back cover of the text. A partial reproduction of this table is shown in Table 7.3.

Note that t_α values are listed for degrees of freedom from 1 to 29, where α refers to the tail area under the t-distribution to the right of t_α. For example, if we

TABLE 7.3   Reproduction of Part of Table VI in Appendix A

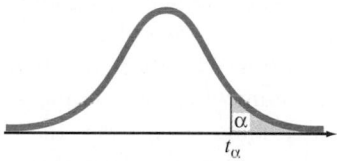

Degrees of Freedom	$t_{.100}$	$t_{.050}$	$t_{.025}$	$t_{.010}$	$t_{.005}$	$t_{.001}$	$t_{.0005}$
1	3.078	6.314	12.706	31.821	63.657	318.31	636.62
2	1.886	2.920	4.303	6.965	9.925	22.326	31.598
3	1.638	2.353	3.182	4.541	5.841	10.213	12.924
4	1.533	2.132	2.776	3.747	4.604	7.173	8.610
5	1.476	2.015	2.571	3.365	4.032	5.893	6.869
6	1.440	1.943	2.447	3.143	3.707	5.208	5.959
7	1.415	1.895	2.365	2.998	3.499	4.785	5.408
8	1.397	1.860	2.306	2.896	3.355	4.501	5.041
9	1.383	1.833	2.262	2.821	3.250	4.297	4.781
10	1.372	1.812	2.228	2.764	3.169	4.144	4.587
11	1.363	1.796	2.201	2.718	3.106	4.025	4.437
12	1.356	1.782	2.179	2.681	3.055	3.930	4.318
13	1.350	1.771	2.160	2.650	3.012	3.852	4.221
14	1.345	1.761	2.145	2.624	2.977	3.787	4.140
15	1.341	1.753	2.131	2.602	2.947	3.733	4.073
⋮	⋮	⋮	⋮	⋮	⋮	⋮	⋮
∞	1.282	1.645	1.960	2.326	2.576	3.090	3.291

FIGURE 7.8

The $t_{.025}$ value in a t-distribution with 4 df and the corresponding $z_{.025}$ value

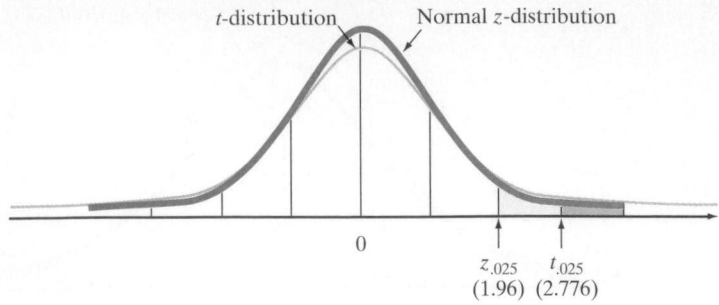

want the t value with an area of .025 to its right and 4 df, we look in the table under the column $t_{.025}$ for the entry in the row corresponding to 4 df. This entry is $t_{.025} = 2.776$, as shown in Figure 7.8. The corresponding standard normal z-score is $z_{.025} = 1.96$.

Note that the last row of Table VI, where df = ∞ (infinity), contains the standard normal z values. This follows from the fact that as the sample size n grows very large, s becomes closer to σ and thus t becomes closer in distribution to z. In fact, when df = 29, there is little difference between corresponding tabulated values of z and t. Thus, we choose the arbitrary cutoff of $n = 30$ (df = 29) to distinguish between the large-sample and small-sample inferential techniques.

Returning to the example of testing a new drug, suppose that the six test patients have blood pressure increases of 1.7, 3.0, .8, 3.4, 2.7, and 2.1 points. How can we use this information to construct a 95% confidence interval for μ, the mean increase in blood pressure associated with the new drug for all patients in the population?

First, we know that we are dealing with a sample too small to assume that the sample mean $\bar{x}$ is approximately normally distributed by the Central Limit Theorem. That is, we do not get the normal distribution of $\bar{x}$ "automatically" from the Central Limit Theorem when the sample size is small. Instead, we must assume that the measured variable, in this case the increase in blood pressure, is normally distributed in order for the distribution of $\bar{x}$ to be normal.

Second, unless we are fortunate enough to know the population standard deviation σ, which in this case represents the standard deviation of *all* the patients' increases in blood pressure when they take the new drug, we cannot use the standard normal z statistic to form our confidence interval for μ. Instead, we must use the t-distribution, with $(n - 1)$ degrees of freedom.

In this case, $n - 1 = 5$ df, and the t value is found in Table 7.3 to be

$$t_{.025} = 2.571 \qquad \text{with 5 df}$$

Recall that the large-sample confidence interval would have been of the form

$$\bar{x} \pm z_{\alpha/2}\sigma_{\bar{x}} = \bar{x} \pm z_{\alpha/2}\frac{\sigma}{\sqrt{n}} = \bar{x} \pm z_{.025}\frac{\sigma}{\sqrt{n}}$$

where 95% is the desired confidence level. To form the interval for a small sample from *a normal distribution, we simply substitute t for z and s for σ in the preceding formula:*

$$\bar{x} \pm t_{\alpha/2}\frac{s}{\sqrt{n}}$$

A MINITAB printout showing descriptive statistics for the six blood pressure increases is displayed in Figure 7.9. Note that $\bar{x} = 2.283$ and $s = .950$.

FIGURE 7.9

MINITAB analysis of six
blood pressure increases

```
        N    MEAN   STDEV   SE MEAN   95.0 PERCENT C.I.
BPIncr  6   2.283   0.950    0.388   (  1.287,   3.280)
```

Substituting these numerical values into the confidence interval formula, we get

$$2.283 \pm (2.571)\left(\frac{.950}{\sqrt{6}}\right) = 2.283 \pm .997$$

or 1.286 to 3.280 points. Note that this interval agrees (except for rounding) with the confidence interval generated by MINITAB in Figure 7.9.

We interpret the interval as follows: We can be 95% confident that the mean increase in blood pressure associated with taking this new drug is between 1.286 and 3.28 points. As with our large-sample interval estimates, our confidence is in the process, not in this particular interval. We know that if we were to use this estimation procedure repeatedly, 95% of the confidence intervals produced would contain the true mean μ, *assuming that the probability distribution of changes in blood pressure from which our sample was selected is normal.* The latter assumption is necessary for the small-sample interval to be valid.

What price did we pay for having to utilize a small sample to make the inference? First, we had to assume the underlying population is normally distributed, and if the assumption is invalid, our interval might also be invalid.* Second, we had to form the interval using a *t* value of 2.571 rather than a *z* value of 1.96, resulting in a wider interval to achieve the same 95% level of confidence. If the interval from 1.286 to 3.28 is too wide to be of much use, we know how to remedy the situation: increase the number of patients sampled in order to decrease the interval width (on average).

The procedure for forming a small-sample confidence interval is summarized in the accompanying box.

Small-Sample Confidence Interval† for μ

$$\bar{x} \pm t_{\alpha/2}\left(\frac{s}{\sqrt{n}}\right)$$

where $t_{\alpha/2}$ is based on $(n-1)$ degrees of freedom

Assumption: A random sample is selected from a population with a relative frequency distribution that is approximately normal.

EXAMPLE 7.2

Some quality control experiments require *destructive sampling* (i.e., the test to determine whether the item is defective destroys the item) in order to measure some particular characteristic of the product. For example, suppose a manufacturer of printers for personal computers wishes to estimate the mean number of characters printed before the printhead fails. The cost of destructive sampling

---

*By *invalid,* we mean that the probability that the procedure will yield an interval that contains μ is not equal to $(1 - \alpha)$. Generally, if the underlying population is approximately normal, the confidence coefficient will approximate the probability that the interval contains μ.

†The procedure given in the box assumes that the population standard deviation σ is unknown, which is almost always the case. If σ is known, we can form the small-sample confidence interval just as we would a large-sample confidence interval using a standard normal *z* value instead of *t*. However, we must still assume that the underlying population is approximately normal.

often dictates small samples. Suppose the printer manufacturer tests $n = 15$ printheads and calculates the following statistics:

$$\bar{x} = 1.23 \text{ million characters} \qquad s = .27 \text{ million characters}$$

Form a 99% confidence interval for the mean number of characters printed before the printhead fails.

Solution

If we assume that the number of characters printed before printhead failure is normally distributed, we can use the t statistic to form the confidence interval. We use a confidence coefficient of .99 and $n - 1 = 14$ degrees of freedom to find $t_{\alpha/2}$ in Table VI:

$$t_{\alpha/2} = t_{.005} = 2.977$$

Thus, the small sample forces us to assume normality and extend the interval almost 3 standard deviations (of $\bar{x}$) on each side of the sample mean in order to form the 99% confidence interval. For these data, the interval is

$$\bar{x} \pm t_{.005}\left(\frac{s}{\sqrt{n}}\right) = 1.23 \pm 2.977\left(\frac{.27}{\sqrt{15}}\right)$$

$$= 1.23 \pm .21 \qquad \text{or} \qquad (1.02, 1.44)$$

Thus, the manufacturer can be 99% confident that the printhead has a mean life of between 1.02 and 1.44 million characters. If the manufacturer were to advertise that the mean life of its printheads is (at least) 1 million characters, the interval would support such a claim. Our confidence is derived from the fact that 99% of the intervals formed in repeated applications of this procedure will contain μ. ▲

We have emphasized throughout this section that an assumption that the population is approximately normally distributed is necessary for making small-sample inferences about μ when using the t statistic. Although many phenomena do have approximately normal distributions, it is also true that many random phenomena have distributions that are not normal or even mound-shaped. Empirical evidence acquired over the years has shown that the t-distribution is rather insensitive to moderate departures from normality. That is, use of the t statistic when sampling from mound-shaped populations generally produces credible results; however, for cases in which the distribution is distinctly nonnormal, we must either take a large sample or use a *nonparametric method* (the topic of Chapter 15).

What Do You Do When the Population Relative Frequency Distribution Departs Greatly from Normality?

Answer: Use the nonparametric statistical methods of Chapter 15.

EXERCISES 7.21–7.34

Learning the Mechanics

7.21 Explain the differences in the sampling distributions of $\bar{x}$ for large and small samples under the following assumptions.

 a. The variable of interest, x, is normally distributed.

 b. Nothing is known about the distribution of the variable x.

7.22 Suppose you have selected a random sample of $n = 7$ measurements from a normal distribution. Compare the standard normal z values with the corresponding

t values if you were forming the following confidence intervals.
  a. 80% confidence interval
  b. 90% confidence interval
  c. 95% confidence interval
  d. 98% confidence interval
  e. 99% confidence interval
  f. Use the table values you obtained in parts **a–e** to sketch the *z*- and *t*-distributions. What are the similarities and differences?

7.23 Let t_0 be a specific value of *t*. Use Table VI in Appendix A to find t_0 values such that the following statements are true.
  a. $P(t \geq t_0) = .025$ where df = 10
  b. $P(t \geq t_0) = .01$ where df = 17
  c. $P(t \leq t_0) = .005$ where df = 6
  d. $P(t \leq t_0) = .05$ where df = 13

7.24 Let t_0 be a particular value of *t*. Use Table VI of Appendix A to find t_0 values such that the following statements are true.
  a. $P(-t_0 < t < t_0) = .95$ where df = 16
  b. $P(t \leq -t_0 \text{ or } t \geq t_0) = .05$ where df = 16
  c. $P(t \leq t_0) = .05$ where df = 16
  d. $P(t \leq -t_0 \text{ or } t \geq t_0) = .10$ where df = 12
  e. $P(t \leq -t_0 \text{ or } t \geq t_0) = .01$ where df = 8

7.25 The following random sample was selected from a normal distribution: 4, 6, 3, 5, 9, 3.
  a. Construct a 90% confidence interval for the population mean μ.
  b. Construct a 95% confidence interval for the population mean μ.
  c. Construct a 99% confidence interval for the population mean μ.
  d. Assume that the sample mean $\bar{x}$ and sample standard deviation *s* remain exactly the same as those you just calculated but that they are based on a sample of *n* = 25 observations rather than *n* = 6 observations. Repeat parts **a–c**. What is the effect of increasing the sample size on the width of the confidence intervals?

7.26 The following sample of 16 measurements was selected from a population that is approximately normally distributed:

| 91 | 80 | 99 | 110 | 95 | 106 | 78 | 121 |
| 106 | 100 | 97 | 82 | 100 | 83 | 115 | 104 |

  a. Construct an 80% confidence interval for the population mean.
  b. Construct a 95% confidence interval for the population mean and compare the width of this interval with that of part **a**.

  c. Carefully interpret each of the confidence intervals and explain why the 80% confidence interval is narrower.

Applying the Concepts

7.27 Spinifex pigeons are one of the few bird species that inhabit the desert of Western Australia. The pigeons rely almost entirely on seeds for food. The *Australian Journal of Zoology* (Vol. 43, 1995) reported on a study of the diets and water requirements of these spinifex pigeons. Sixteen pigeons were captured in the desert and the crop (i.e., stomach) contents of each were examined. The following table reports the weight (in grams) of dry seed in the crop of each pigeon. Use the descriptive statistics in the SAS printout at the bottom of the page to construct a 99% confidence interval for the average weight of dry seeds in the crops of spinifex pigeons inhabiting the Western Australian desert. Interpret the result.

| .457 | 3.751 | .238 | 2.967 | 2.509 | 1.384 | 1.454 | .818 |
| .335 | 1.436 | 1.603 | 1.309 | .201 | .530 | 2.144 | .834 |

Source: Excerpted from Williams, J. B., Bradshaw, D., and Schmidt, L. "Field metabolism and water requirements of spinifex pigeons (*Geophaps plumifera*) in Western Australia." *Australian Journal of Zoology*, Vol. 43, No. 1, 1995, p. 7 (Table 2).

7.28 Patients with normal hearing in one ear and unaidable sensorineural hearing loss in the other are characterized as suffering from unilateral hearing loss. In a study reported in the *American Journal of Audiology* (Mar. 1995), eight patients with unilateral hearing loss were fitted with a special wireless hearing aid in the "bad" ear. The absolute sound pressure level (SPL) was then measured near the eardrum of that ear when noise was produced at a frequency of 500 hertz. The SPLs of the eight patients, recorded in decibels, are listed below:

| 73.0 | 80.1 | 82.8 | 76.8 | 73.5 | 74.3 | 76.0 | 68.1 |

Construct and interpret a 90% confidence interval for the true mean SPL of unilateral hearing loss in these patients when noise is produced at a frequency of 500 hertz.

7.29 Health insurers and the federal government are both putting pressure on hospitals to shorten the average length of stay (LOS) of their patients. A random sample of 20 hospitals in one state had a mean LOS in 1995 of 3.8 days and a standard deviation of 1.2 days.
  a. Use a 90% confidence interval to estimate the population mean of the LOS for the state's hospitals in 1995.

```
Analysis Variable : SEED_WT

N Obs    N      Minimum      Maximum        Mean      Std Dev
  16    16    0.2010000    3.7510000    1.3731250    1.0339506
```

b. Interpret the interval in terms of this application.

c. What is meant by the phrase "90% confidence interval"?

7.30 Pulse rate is an important measure of the fitness of a person's cardiovascular system. The mean pulse rate for all U.S. adult males is approximately 72 heart beats per minute. A random sample of 21 U.S. adult males who jog at least 15 miles per week had a mean pulse rate of 52.6 beats per minute and a standard deviation of 3.22 beats per minute.

a. Find a 95% confidence interval for the mean pulse rate of all U.S. adult males who jog at least 15 miles per week.

b. Interpret the interval found in part **a**. Does it appear that jogging at least 15 miles per week reduces the mean pulse rate for adult males?

c. What assumptions are required to ensure the validity of the confidence interval and the inference, part **b**?

7.31 Accidental spillage and misguided disposal of petroleum wastes have resulted in extensive contamination of soils across the country. A common hazardous compound found in the contaminated soil is benzo(a)pyrene [B(a)p]. An experiment was conducted to determine the effectiveness of a method designed to remove B(a)p from soil (*Journal of Hazardous Materials,* June 1995). Three soil specimens contaminated with a known amount of B(a)p were treated with a toxin that inhibits microbial growth. After 95 days of incubation, the percentage of B(a)p removed from each soil specimen was measured. The experiment produced the following summary statistics: $\bar{x} = 49.3$ and $s = 1.5$.

a. Use a 99% confidence interval to estimate the mean percentage of B(a)p removed from a soil specimen in which the toxin was used.

b. Interpret the interval in terms of this application.

c. What assumption is necessary to ensure the validity of this confidence interval?

7.32 The cost of hiring an employee (excluding salary) can range from about $1,500 for a secretary to more than $40,000 for a manager. To estimate its mean cost of hiring an entry-level secretary, a large corporation randomly selected eight of the entry-level secretaries it had hired during the last two years and determined the costs (in dollars) involved in hiring each. The following data were obtained:

2,100   1,650   1,315   2,035   2,245   1,980   1,700   2,190

Assume that the population from which these data were sampled is approximately normally distributed.

a. Describe the population from which the corporation collected the sample data.

b. Use a 90% confidence interval to estimate the mean of interest to the corporation.

c. How wide is the confidence interval you constructed in part **b**? Would a 95% confidence interval be wider or narrower? Explain.

7.33 A *mortgage* is a type of loan that is secured by a designated piece of property. If the borrower defaults on the loan, the lender can sell the property to recover the outstanding debt. A federal bank examiner is interested in estimating the mean outstanding principal balance of all home mortgages foreclosed by the bank due to default by the borrower during the last 3 years. A random sample of 12 foreclosed mortgages yielded the following data (in dollars):

95,982	81,422	39,888	46,836	66,899	69,110
59,200	62,331	105,812	55,545	56,635	72,123

a. Describe the population from which the bank examiner collected the sample data. What characteristic must this population possess to enable us to construct a confidence interval for the mean outstanding principal balance using the method described in this section?

b. Construct a 90% confidence interval for the mean of interest.

c. Carefully interpret your confidence interval in the context of the problem.

7.34 In Exercise 2.19 you learned that postmortem interval (PMI) is the elapsed time between the death of an individual and the performance of an autopsy on the cadaver. *Brain and Language* (June 1995) reported on the PMIs of 22 randomly selected human brain specimens obtained at autopsy. The data are reproduced in the accompanying table, followed by a MINITAB printout of the analysis of the data.

a. Locate a 95% confidence interval for the true mean PMI of human brain specimens obtained at autopsy.

b. Interpret the interval, part **a**.

c. What assumption is required for the interval, part **a**, to be valid? Is this assumption satisfied? Explain.

d. What is meant by the phrase "95% confidence interval"?

**POSTMORTEM INTERVALS FOR 22
HUMAN BRAIN SPECIMENS**

5.5	14.5	6.0	5.5	5.3	5.8	11.0	6.1
7.0	14.5	10.4	4.6	4.3	7.2	10.5	6.5
3.3	7.0	4.1	6.2	10.4	4.9		

Source: Hayes, T. L., and Lewis, D. A. "Anatomical specialization of the anterior motor speech area: Hemispheric differences in magnopyramidal neurons." *Brain and Language,* Vol. 49, No. 3, June 1995, p. 292 (Table 1).

```
         N    MEAN   STDEV   SE MEAN   95.0 PERCENT C.I.
PMI     22   7.300   3.185    0.679 (   5.888,   8.712)
```

7.3   LARGE-SAMPLE CONFIDENCE INTERVAL FOR A POPULATION PROPORTION

The number of public opinion polls has grown at an astounding rate in recent years. Almost daily, the news media report the results of some poll. Pollsters regularly determine the percentage of people in favor of the president's welfare-reform program, the fraction of voters in favor of a certain candidate, the fraction of customers who favor a particular brand of wine, and the proportion of people who smoke cigarettes. In each case, we are interested in estimating the percentage (or proportion) of some group with a certain characteristic. In this section we consider methods for making inferences about population proportions when the sample is large.

EXAMPLE 7.3

Public opinion polls are conducted regularly to estimate the fraction of U.S. citizens who trust the president. Suppose 1,000 people are randomly chosen and 637 answer that they trust the president. How would you estimate the true fraction of *all* U.S. citizens who trust the president?

Solution

What we have really asked is how you would estimate the probability p of success in a binomial experiment, where p is the probability that a person chosen trusts the president. One logical method of estimating p for the population is to use the proportion of successes in the sample. That is, we can estimate p by calculating

$$\hat{p} = \frac{\text{Number of people sampled who trust the president}}{\text{Number of people sampled}}$$

where $\hat{p}$ is read "p hat." Thus, in this case,

$$\hat{p} = \frac{637}{1,000} = .637 \qquad \blacktriangle$$

To determine the reliability of the estimator $\hat{p}$, we need to know its sampling distribution. That is, if we were to draw samples of 1,000 people over and over again, each time calculating a new estimate $\hat{p}$, what would be the frequency distribution of all the $\hat{p}$ values? The answer lies in viewing $\hat{p}$ as the average, or mean, number of successes per trial over the n trials. If each success is assigned a value equal to 1 and a failure is assigned a value of 0, then the sum of all n sample observations is x, the total number of successes, and $\hat{p} = x/n$ is the average, or mean, number of successes per trial in the n trials. The Central Limit Theorem tells us that the relative frequency distribution of the sample mean for any population is approximately normal for sufficiently large samples.

The repeated sampling distribution of $\hat{p}$ has the characteristics listed in the next box and shown in Figure 7.10.

FIGURE 7.10

Sampling distribution of $\hat{p}$

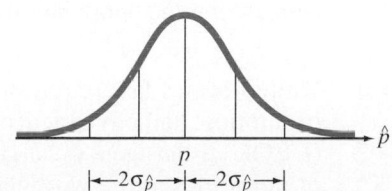

Sampling Distribution of $\hat{p}$

1. The mean of the sampling distribution of $\hat{p}$ is p; that is, $\hat{p}$ is an unbiased estimator of p.
2. The standard deviation of the sampling distribution of $\hat{p}$ is $\sqrt{pq/n}$; that is, $\sigma_{\hat{p}} = \sqrt{pq/n}$, where $q = 1 - p$.
3. For large samples, the sampling distribution of $\hat{p}$ is approximately normal. A sample size is considered large if the interval $\hat{p} \pm 3\sigma_{\hat{p}}$ does not include 0 or 1. [*Note:* This requirement is almost equivalent to that given in Section 5.4 for approximating a binomial distribution with a normal one. The difference is that we assumed p to be known in Section 5.4; now we are trying to make inferences about an unknown p, so we use $\hat{p}$ to estimate p in checking the adequacy of the normal approximation.]

The fact that $\hat{p}$ is a "sample mean fraction of successes" allows us to form confidence intervals about p in a manner that is completely analogous to that used for large-sample estimation of μ.

Large-Sample Confidence Interval for p

$$\hat{p} \pm z_{\alpha/2}\sigma_{\hat{p}} = \hat{p} \pm z_{\alpha/2}\sqrt{\frac{pq}{n}} \approx \hat{p} \pm z_{\alpha/2}\sqrt{\frac{\hat{p}\hat{q}}{n}}$$

where $\hat{p} = \dfrac{x}{n}$ and $\hat{q} = 1 - \hat{p}$

Note: When n is large, $\hat{p}$ can approximate the value of p in the formula for $\sigma_{\hat{p}}$.

TABLE 7.4 Values of pq for Several Different Values of p

p	pq	$\sqrt{pq}$
.5	.25	.50
.6 or .4	.24	.49
.7 or .3	.21	.46
.8 or .2	.16	.40
.9 or .1	.09	.30

Thus, if 637 of 1,000 U.S. citizens say they trust the president, a 95% confidence interval for the proportion of *all* U.S. citizens who trust the president is

$$\hat{p} \pm z_{\alpha/2}\sigma_{\hat{p}} = .637 \pm 1.96\sqrt{\frac{pq}{1,000}}$$

where $q = 1 - p$. Just as we needed an approximation for σ in calculating a large-sample confidence interval for μ, we now need an approximation for p. As Table 7.4 shows, the approximation for p does not have to be especially accurate, because the value of $\sqrt{pq}$ needed for the confidence interval is relatively insensitive to changes in p. Therefore, we can use $\hat{p}$ to approximate p. Keeping in mind that $\hat{q} = 1 - \hat{p}$, we substitute these values into the formula for the confidence interval:

$$\hat{p} \pm 1.96\sqrt{pq/1,000} \approx \hat{p} \pm 1.96\sqrt{\hat{p}\hat{q}/1,000}$$
$$= .637 \pm 1.96\sqrt{(.637)(.363)/1,000} = .637 \pm .030$$
$$= (.607, .667)$$

Then we can be 95% confident that the interval from 60.7% to 66.7% contains the true percentage of *all* U.S. citizens who trust the president. That is, in repeated construction of confidence intervals, approximately 95% of all samples would produce confidence intervals that enclose p. Note that the guidelines for interpreting a confidence interval about μ also apply to interpreting a confidence interval for p because p is the "population mean fraction of successes" in a binomial experiment.

EXAMPLE 7.4

United States law requires companies to treat minorities fairly in their hiring, promotion, and pay practices. The Equal Employment Opportunity Commission (EEOC) is the agency charged with the responsibility of monitoring the treatment of minorities in the workplace. Suppose the EEOC samples 135 recent hires of one large company and determines that 12 are minorities. Use a 90% confidence interval to estimate the proportion of all new hires of the company that are minorities.

Solution

The number, x, of the 135 sampled hires who are minorities is a binomial random variable if we can assume that the sample was randomly selected from all the company's hires and that the process by which the company hires is relatively stable during the period of interest (so that the probability of a minority hire remains constant over the period).

Then the point estimate of the proportion of minority hires for this company is

$$\hat{p} = \frac{x}{n} = \frac{12}{135} = .089$$

We first check to be sure that the sample size is sufficiently large that the normal distribution provides a reasonable approximation for this binomial proportion. We check the 3-standard-deviation interval around $\hat{p}$:

$$\hat{p} \pm 3\sigma_{\hat{p}} \approx \hat{p} \pm 3\sqrt{\frac{\hat{p}\hat{q}}{n}}$$

$$= .089 \pm 3\sqrt{\frac{(.089)(.911)}{135}} = .089 \pm .074 = (.015, .163)$$

Since this interval is wholly contained in the interval $(0, 1)$, we may conclude that the normal approximation is reasonable.

We now form the 90% confidence interval for p, the true proportion of minority hires for the company:

$$\hat{p} \pm z_{\alpha/2}\sigma_{\hat{p}} = \hat{p} \pm z_{\alpha/2}\sqrt{\frac{pq}{n}} \approx \hat{p} \pm z_{\alpha/2}\sqrt{\frac{\hat{p}\hat{q}}{n}}$$

$$= .089 \pm 1.645\sqrt{\frac{(.089)(.911)}{135}} = .089 \pm .040 = (.049, .129)$$

Thus, we can be 90% confident that the proportion of all the company's hires (under the current policy) who are minorities is between .049 and .129. As always, our confidence stems from the fact that 90% of all similarly formed intervals will contain the true proportion p and not from any knowledge about whether this particular interval does.

Suppose the EEOC knows that the minority proportion of the qualified work force for jobs in the industry represented by this company is 15%. Can we conclude that the company is hiring minorities at a rate below the 15% qualification rate? Since the 90% confidence interval lies completely below .15 (the upper confidence limit is only .129), we would conclude based on this interval that the company's minority hire rate is below the 15% qualification rate. In fact, we would conclude, with 90% confidence, that the minority hire rate is between 4.9% and 12.9%.   ▲

In Example 7.4 we used the confidence interval to make an inference about whether the true value of p is less than .15. That is, we used the sample to *test* whether p is less than .15. When we want to use sample information to test the value of a population parameter, we usually conduct a *test of hypothesis,* the subject of Chapter 8. But first, we'll conclude Chapter 7 with a section showing how to determine the sample size necessary to estimate either a population mean or a population proportion.

Suicide in Urban Jails—Revisited

CASE STUDY
• 7.1 •

I n Case Study 2.3 we presented the results of a study on suicide in correctional facilities, published in the *American Journal of Psychiatry* (July 1995). The researchers wanted to know what factors increase the risk of suicide by those incarcerated in urban jails. Recall that they collected data on 37 suicides that occurred from 1967 to 1992 in Wayne County Jail, Detroit, Michigan. (All 37 suicides were committed by hanging.) The data are reproduced in Table 7.5.

You answered several questions posed in Case Study 2.3 using descriptive statistics, then used your answers to make inferences about the general population of suicidal inmates of the jail. The concepts developed in this chapter will allow you to attach measures of reliability to several of these inferences. Answer each of these questions again (as posed below), using a 95% confidence interval for the target parameter implied in the question. Fully interpret each confidence interval.

Focus

a. Are suicides at the jail more likely to be committed by inmates charged with murder/manslaughter or with lesser crimes?

b. Are suicides at the jail more likely to be committed at night?

c. What is the typical length of time an inmate is in jail before committing suicide?

d. Are more suicides committed by white or nonwhite inmates?

EXERCISES 7.35–7.50

Learning the Mechanics

7.35 Describe the sampling distribution of $\hat{p}$ based on large samples of size n. That is, give the mean, the standard deviation, and the (approximate) shape of the distribution of $\hat{p}$ when large samples of size n are (repeatedly) selected from the binomial distribution with probability of success p.

7.36 Explain the meaning of the phrase "$\hat{p}$ is an unbiased estimator of p."

7.37 A random sample of size $n = 196$ yielded $\hat{p} = .64$.
   a. Is the sample size large enough to use the methods of this section to construct a confidence interval for p? Explain.
   b. Construct a 95% confidence interval for p.
   c. Interpret the 95% confidence interval.
   d. Explain what is meant by the phrase "95% confidence interval."

7.38 A random sample size $n = 144$ yielded $\hat{p} = .76$.
   a. Is the sample size large enough to use the methods of this section to construct a confidence interval for p? Explain.
   b. Construct a 90% confidence interval for p.
   c. What assumption is necessary to ensure the validity of this confidence interval?

7.39 For the binomial sample information summarized in each part, indicate whether the sample size is large enough to use the methods of this chapter to construct a confidence interval for p.
   a. $n = 500$, $\hat{p} = .05$.
   b. $n = 100$, $\hat{p} = .05$
   c. $n = 10$, $\hat{p} = .5$
   d. $n = 10$, $\hat{p} = .3$

7.40 A random sample of 50 consumers taste-tested a new snack food. Their responses were coded (0: do not like; 1: like; 2: indifferent) and recorded as follows:

1	0	0	1	2	0	1	1	0	0
0	1	0	2	0	2	2	0	0	1
1	0	0	0	0	1	0	2	0	0
0	1	0	0	1	0	0	1	0	1
0	2	0	0	1	1	0	0	0	1

   a. Use an 80% confidence interval to estimate the proportion of consumers who like the snack food.
   b. Provide a statistical interpretation for the confidence interval you constructed in part **a**.

7.41 A poll is taken in which 278 of 500 randomly sampled voters indicate their preference for a certain candidate. Use a 95% confidence interval to estimate the true fraction, p, of voters who prefer the candidate.

Applying the Concepts

7.42 The *American Journal of Orthopsychiatry* (July 1995) published an article on the prevalence of homelessness in the United States. A sample of 487 U.S. adults was taken, and the members were asked to respond to the question: "Was there ever a time in your life when you did not have a place to live?" A 95% confidence interval for the true proportion who were/are homeless, based on the survey data, was determined to be (.045, .091). Fully interpret the result.

7.43 Obstructive sleep apnea is a sleep disorder that causes a person to stop breathing momentarily and then awaken briefly. These sleep interruptions,

TABLE 7.5 **Data on 37 Suicides in Wayne County Jail (Case Study 7.1)**

Victim	Days in Jail Before Suicide	Marital Status	Race	Murder/ Manslaughter Charge	Time of Suicide
1	3	Married	W	Yes	11 pm–7 am
2	4	Single	W	Yes	11 pm–7 am
3	5	Single	NW	Yes	3 pm–11 pm
4	7	Widowed	NW	Yes	11 pm–7 am
5	10	Single	NW	Yes	3 pm–11 pm
6	15	Married	NW	Yes	11 pm–7 am
7	15	Divorced	NW	Yes	11 pm–7 am
8	19	Married	NW	Yes	7 am–3 pm
9	22	Single	NW	Yes	11 pm–7 am
10	29	Married	W	Yes	11 pm–7 am
11	31	Single	NW	Yes	11 pm–7 am
12	85	Single	NW	Yes	3 pm–11 pm
13	126	Married	NW	Yes	11 pm–7 am
14	221	Divorced	W	Yes	11 pm–7 am
15	14	Single	NW	No	11 pm–7 am
16	22	Single	W	No	3 pm–11 pm
17	42	Single	W	No	7 am–3 pm
18	122	Single	NW	No	11 pm–7 am
19	309	Married	NW	No	11 pm–7 am
20	69	Married	NW	No	11 pm–7 am
21	143	Single	NW	No	11 pm–7 am
22	6	Married	NW	No	11 pm–7 am
23	1	Single	NW	No	7 am–3 pm
24	1	Single	NW	No	11 pm–7 am
25	1	Single	W	No	11 pm–7 am
26	2	Single	W	No	7 am–3 pm
27	3	Single	NW	No	11 pm–7 am
28	4	Married	W	No	3 pm–11 pm
29	4	Single	W	No	3 pm–11 pm
30	4	Single	NW	No	11 pm–7 am
31	6	Single	W	No	7 am–3 pm
32	6	Married	NW	No	11 pm–7 am
33	10	Single	W	No	11 pm–7 am
34	18	Single	NW	No	11 pm–7 am
35	26	Married	W	No	11 pm–7 am
36	41	Single	NW	No	11 pm–7 am
37	86	Married	W	No	11 pm–7 am

Source: DuRand, C. J., *et al.* "A quarter century of suicide in a major urban jail: Implications for community psychiatry." *American Journal of Psychiatry*, Vol. 152, No. 7, July 1995, p. 1078 (Table 1). Copyright 1995 American Psychiatric Association. Reprinted by permission.

which may occur hundreds of times in a night, can drastically reduce the quality of rest and cause fatigue during waking hours. Researchers at Stanford University studied 159 commercial truck drivers and found that 124 of them suffered from obstructive sleep apnea (*Chest*, May 1995).

a. Use the study results to estimate, with 90% confidence, the fraction of truck drivers who suffer from the sleep disorder.

b. Sleep researchers believe that about 25% of the general population suffer from obstructive sleep apnea. Comment on whether or not this value represents the true percentage of truck drivers who suffer from the sleep disorder.

7.44 Past research has clearly indicated that the stress produced by today's lifestyles results in health problems for a large proportion of society. An article in *The International Journal of Sports Psychology*

Data for Exercise 7.48

Tanker	Spillage (metric tons, thousands)	CAUSE OF SPILLAGE				
		Collision	Grounding	Fire/Explosion	Hull Failure	Unknown
Atlantic Empress	257	X				
Castillo De Bellver	239			X		
Amoco Cadiz	221				X	
Odyssey	132			X		
Torrey Canyon	124		X			
Sea Star	123	X				
Hawaiian Patriot	101				X	
Independenta	95	X				
Urquiola	91		X			
Irenes Serenade	82			X		
Khark 5	76			X		
Nova	68	X				
Wafra	62		X			
Epic Colocotronis	58		X			
Sinclair Petrolore	57			X		
Yuyo Maru No 10	42	X				
Assimi	50			X		
Andros Patria	48			X		
World Glory	46				X	
British Ambassador	46				X	
Metula	45		X			
Pericles G.C.	44			X		
Mandoil II	41	X				
Jakob Maersk	41		X			
Burmah Agate	41	X				
J. Antonio Lavalleja	38		X			
Napier	37		X			
Exxon Valdez	36		X			
Corinthos	36	X				
Trader	36				X	
St. Peter	33			X		
Gino	32	X				
Golden Drake	32			X		
Ionnis Angelicoussis	32			X		
Chryssi	32				X	
Irenes Challenge	31				X	
Argo Merchant	28		X			
Heimvard	31	X				
Pegasus	25					X
Pacocean	31				X	
Texaco Oklahoma	29				X	
Scorpio	31		X			
Ellen Conway	31		X			
Caribbean Sea	30				X	
Cretan Star	27					X
Grand Zenith	26				X	
Athenian Venture	26			X		
Venoil	26	X				
Aragon	24				X	
Ocean Eagle	21		X			

Source: Daidola, J. C. "Tanker structure behavior during collision and grounding." *Marine Technology*, Vol. 32, No. 1, Jan. 1995, p. 22 (Table 1). Reprinted with permission of The Society of Naval Architects and Marine Engineers (SNAME), Jersey City, N.J.

(Jul.–Sept. 1992) evaluates the relationship between physical fitness and stress. Employees of companies that participate in the Health Examination Program offered by Health Advancement Services (HAS) were classified into three fitness levels: poor, average, and good. Each person was tested for signs of stress. The results for the three groups are reported in the accompanying table.

Fitness Level	Sample Size	Proportion with Signs of Stress
Poor	242	.155
Average	212	.133
Good	95	.108

a. Check to see whether each of these samples is large enough to construct a confidence interval for the true proportion of all employees at each fitness level exhibiting signs of stress.

b. Assuming each sample represents a random sample from its corresponding population, calculate and interpret a 95% confidence interval for the proportion of people with signs of stress within each of the three fitness levels.

c. Interpret each of the confidence intervals constructed in part **b** using the terminology of this exercise.

7.45 Astronauts often report episodes of disorientation as they move around the zero-gravity spacecraft. To compensate, crew members rely heavily on visual information to establish up–down orientation. An empirical study was conducted to assess the potential of using color brightness as a body orientation cue (*Human Factors,* Dec. 1988). Ninety college students, reclining on their backs in the dark, were disoriented when positioned on a rotating platform under a slowly rotating disk that filled their entire field of vision. Half the disk was painted with a brighter level of color than the other half. The students were asked to say "stop" when they believed they were right-side up, and the brightness level of the disk was recorded. Of the 90 students, 58 selected the brighter color level as their cue to stop.

a. Use this information to estimate the true proportion of subjects who use the bright color level as a cue to being right-side up. Construct a 95% confidence interval for the true proportion.

b. Can you infer from the result, part **a**, that a majority of subjects would select bright color levels over dark color levels as a cue to being right-side up?

7.46 For the last five years, the accounting firm Price Waterhouse has monitored the U.S. Postal Service's performance. One parameter of interest is the percentage of mail delivered on time. In a sample of 332,000 items mailed between Dec. 10, 1994 and Mar. 3, 1995—the most difficult delivery season owing to bad weather and holidays—Price Waterhouse determined that 282,200 items were delivered on time (*Tampa Tribune,* Mar. 26, 1995). Use this information to estimate with 99% confidence the true percentage of items delivered on time by the U.S. Postal Service. Interpret the result.

7.47 In Exercise 2.5 you learned about the signature whistles of bottlenose dolphins. Marine scientists categorize signature whistles by type—type a, type b, type c, etc. For one study of a sample of 185 whistles emitted by bottlenose dolphins in captivity, 97 were categorized as type a whistles (*Ethology,* July 1995).

a. Estimate the true proportion of bottlenose dolphin signature whistles that are type a whistles. Use a 99% confidence interval.

b. Interpret the interval, part **a**.

7.48 Refer to the *Marine Technology* (Jan. 1995) study of the causes of fifty recent major oil spills from tankers and carriers, Exercise 2.10. The data are reproduced in the table on page 274 .

a. Give a point estimate for the proportion of major oil spills that are caused by hull failure.

b. Form a 95% confidence interval for the estimate, part **a**. Interpret the result.

7.49 What is the greatest problem facing adolescents today? According to their peers, it's drugs—not crime, social pressure, grades, or sex. A telephone survey of 400 12–17-year-old youths, conducted by the Center on Addiction and Substance Abuse at Columbia University, found 128 who ranked drugs as their greatest problem (*New York Times,* July 17, 1995). The next most common worry, crime in school, was ranked first by 62 of the 400 youths. What "margin of error" (using the language of the media) would you attach to these findings? Explain.

7.50 Refer to the *American Psychologist* (July 1995) study of applications for university positions in experimental psychology, Exercise 2.39. One of the objectives of the analysis was to identify problems and errors (e.g., not submitting at least three letters of recommendation) in the application packages. In a sample of 148 applications for the positions, 30 applicants failed to provide *any* letters of recommendation. Estimate, with 90% confidence, the true fraction of applicants for a university position in experimental psychology that fail to provide letters of recommendation. Interpret the result.

7.4 DETERMINING THE SAMPLE SIZE

Recall (Section 1.5) that one way to collect the relevant data for a study used to make inferences about the population is to implement a designed (planned) experiment. Perhaps the most important design decision faced by the analyst is to determine the size of the sample. We show in this section that the appropriate sample size for making an inference about a population mean or proportion depends on the desired reliability.

Estimating a Population Mean

Consider the example from Section 7.1 in which we estimated the mean length of stay for patients in a large hospital. A sample of 100 patients' records produced the 95% confidence interval: $\bar{x} \pm 2\sigma_{\bar{x}} \approx 4.53 \pm .74$. Consequently, our estimate $\bar{x}$ was within .74 day of the true mean length of stay, μ, for all the hospital's patients at the 95% confidence level. That is, the 95% confidence interval for μ was $2(.74) = 1.48$ days wide when 100 accounts were sampled. This is illustrated in Figure 7.11a.

Now suppose we want to estimate μ to within .25 day with 95% confidence. That is, we want to narrow the width of the confidence interval from 1.48 days to .50 day, as shown in Figure 7.11b. How much will the sample size have to be increased to accomplish this? If we want the estimator $\bar{x}$ to be within .25 day of μ, we must have

$$2\sigma_{\bar{x}} = .25 \quad \text{or, equivalently,} \quad 2\left(\frac{\sigma}{\sqrt{n}}\right) = .25$$

The necessary sample size is obtained by solving this equation for n. To do this, we need an approximation for σ. We have an approximation from the initial sample of 100 patients' records—namely, the sample standard deviation, $s = 3.68$. Thus,

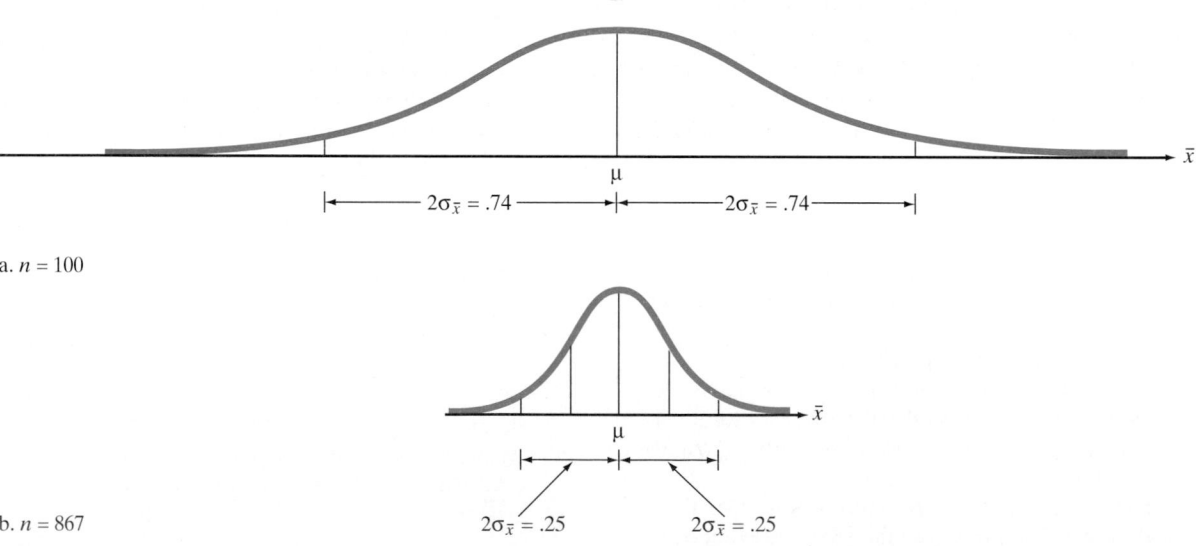

a. $n = 100$

b. $n = 867$

FIGURE 7.11

Relationship between sample size and width of confidence interval: Hospital stay example

$$2\left(\frac{\sigma}{\sqrt{n}}\right) \approx 2\left(\frac{s}{\sqrt{n}}\right) = 2\left(\frac{3.68}{\sqrt{n}}\right) = .25$$

$$\sqrt{n} = \frac{2(3.68)}{.25} = 29.44$$

$$n = (29.44)^2 = 866.71 \approx 867$$

Approximately 867 patients' records will have to be sampled to estimate the mean length of stay μ to within .25 day with (approximately) 95% confidence. The confidence interval resulting from a sample of this size will be approximately .50 day wide (see Figure 7.11b).

In general, we express the reliability associated with a confidence interval for the population mean μ by specifying the **bound, *B*,** within which we want to estimate μ with $100(1 - \alpha)\%$ confidence. The bound *B* then is equal to the half width of the confidence interval, as shown in Figure 7.12.

The procedure for finding the sample size necessary to estimate μ to within a given bound *B* is given in the next box.

Sample Size Determination for $100(1 - \alpha)\%$ Confidence Intervals for μ

In order to estimate μ to within a **bound *B*** with $100(1 - \alpha)\%$ confidence, the required sample size is found as follows:

$$z_{\alpha/2}\left(\frac{\sigma}{\sqrt{n}}\right) = B$$

The solution can be written in terms of *B* as follows:

$$n = \frac{(z_{\alpha/2})^2 \sigma^2}{B^2}$$

The value of σ is usually unknown. It can be estimated by the standard deviation, *s*, from a prior sample. Alternatively, we may approximate the range R of observations in the population, and (conservatively) estimate $\sigma \approx R/4$. In any case, you should round the value of *n* obtained *upward* to ensure that the sample size will be sufficient to achieve the specified reliability.

EXAMPLE 7.5

Suppose the manufacturer of official NFL footballs uses a machine to inflate the new balls to a pressure of 13.5 pounds. When the machine is properly calibrated, the mean inflation pressure is 13.5 pounds, but uncontrollable factors cause the pressures of individual footballs to vary randomly from about 13.3 to 13.7 pounds. For quality control purposes, the manufacturer wishes to estimate the mean inflation pressure to within .025 pound of its true value with a 99% confidence interval. What sample size should be specified for the experiment?

FIGURE 7.12

Specifying the bound *B* as the half-width of a confidence interval

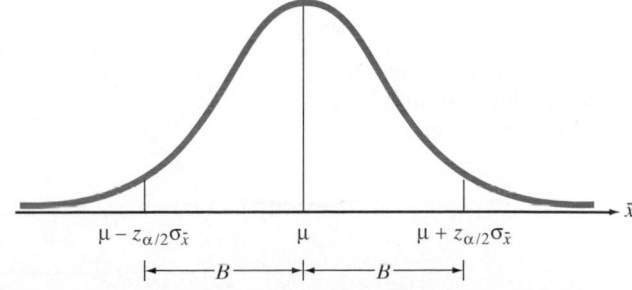

Solution

We desire a 99% confidence interval that estimates μ to within $B = .025$ pound of its true value. For a 99% confidence interval, we have $z_{\alpha/2} = z_{.005} = 2.575$. To estimate σ, we note that the range of observations is $R = 13.7 - 13.3 = .4$ and use $\sigma \approx R/4 = .1$. Now we use the formula derived in the box to find the sample size *n*:

$$n = \frac{(z_{\alpha/2})^2 \sigma^2}{B^2} \approx \frac{(2.575)^2 (.1)^2}{(.025)^2} = 106.09$$

We round this up to $n = 107$. Realizing that σ was approximated by $R/4$, we might even advise that the sample size be specified as $n = 110$ to be more certain of attaining the objective of a 99% confidence interval with bound $B = .025$ pound or less.   ▲

Sometimes the formula will lead to a solution that indicates a small sample size is sufficient to achieve the confidence interval goal. Unfortunately, the procedures and assumptions for small samples differ from those for large samples, as we discovered in Section 7.2. Therefore, if the formulas yield a small sample size ($n < 30$), one simple strategy is to select a sample size $n \geq 30$.

Estimating a Population Proportion

The method outlined above is easily applied to a population proportion *p*. For example, in Section 7.3 a pollster used a sample of 1,000 U.S. citizens to calculate a 95% confidence interval for the proportion who trust the president, obtaining the interval $.637 \pm .03$. Suppose the pollster wishes to estimate more precisely the proportion who trust the president, say to within .015 with a 95% confidence interval.

The pollster wants a confidence interval with a bound *B* on the estimate of *p* of $B = .015$. The sample size required to generate such an interval is found by solving the following equation for *n*:

$$z_{\alpha/2} \sigma_{\hat{p}} = B \quad \text{or} \quad z_{\alpha/2}\sqrt{\frac{pq}{n}} = .015 \qquad \text{(see Figure 7.13)}$$

Since a 95% confidence interval is desired, the appropriate *z* value is $z_{\alpha/2} = z_{.025} = 1.96$. We must approximate the value of the product *pq* before we can solve the equation for *n*. As shown in Table 7.4, the closer the values of *p* and *q* to .5, the larger the product *pq*. Thus, to find a conservatively large sample size that will generate a confidence interval with the specified reliability, we generally choose an approximation of *p* close to .5. In the case of the proportion of U.S. citizens who trust the president, however, we have an initial sample estimate of $\hat{p} = .637$. A conservatively large estimate of *pq* can therefore be obtained by using $p = .60$. We now substitute into the equation and solve for *n*:

FIGURE 7.13

Specifying the bound *B* of a confidence interval for a population proportion *p*

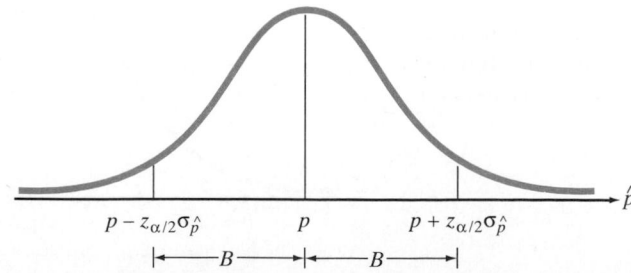

$$1.96\sqrt{\frac{(.60)(.40)}{n}} = .015$$

$$n = \frac{(1.96)^2(.60)(.40)}{(.015)^2} = 4,097.7 \approx 4,098$$

The pollster must sample about 4,098 U.S. citizens to estimate the percentage who trust the president with a confidence interval of width .03.

The procedure for finding the sample size necessary to estimate a population proportion p to within a given bound B is given in the next box.

Sample Size Determination for $100(1 - \alpha)\%$ Confidence Interval for p

In order to estimate a binomial probability p to within a bound B with $100(1 - \alpha)\%$ confidence, the required sample size is found by solving the following equation for n:

$$z_{\alpha/2}\sqrt{\frac{pq}{n}} = B$$

The solution can be written in terms of B:

$$n = \frac{(z_{\alpha/2})^2(pq)}{B^2}$$

Since the value of the product pq is unknown, it can be estimated by using the sample fraction of successes, $\hat{p}$, from a prior sample. Remember (Table 7.4) that the value of pq is at its maximum when p equals .5, so that you can obtain conservatively large values of n by approximating p by .5 or values close to .5. In any case, you should round the value of n obtained *upward* to ensure that the sample size will be sufficient to achieve the specified reliability.

EXAMPLE 7.6

Suppose a large telephone manufacturer that entered the post-regulation market quickly has an initial problem with excessive customer complaints and consequent returns of the phones for repair or replacement. The manufacturer wants to estimate the magnitude of the problem in order to design a quality control program. How many telephones should be sampled and checked in order to estimate the fraction defective, p, to within .01 with 90% confidence?

Solution

In order to estimate p to within a bound of .01, we set the half-width of the confidence interval equal to $B = .01$, as shown in Figure 7.14.

The equation for the sample size n requires an estimate of the product pq. We could most conservatively estimate $pq = .25$ (i.e., use $p = .5$), but this may be overly conservative when estimating a fraction defective. A value of .1, corresponding to 10% defective, will probably be conservatively large for this application. The solution is therefore

FIGURE 7.14

Specified reliability for estimate of fraction defective in Example 7.6

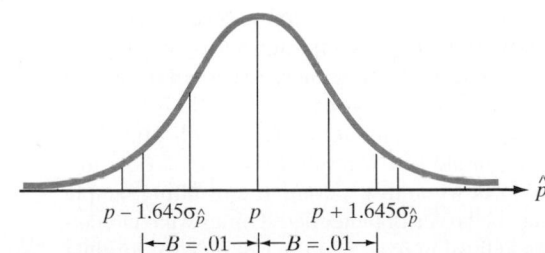

Is Caffeine Addictive?

CASE STUDY • 7.2 •

Scientific research has established that certain substances, when taken in excess, can become addictive. These substances range from heavy narcotic drugs such as heroin and cocaine to alcohol in beer and wine and nicotine in cigarettes. Media blitzes have made drug users, smokers, and drinkers well aware of the dangers of these addictions. But what about the millions of Americans who regularly drink coffee, tea, or cola products each day? Does the caffeine in coffee, tea, and cola induce an addiction similar to that induced by alcohol, tobacco, heroin, and cocaine?

In an attempt to answer this question, researchers at Johns Hopkins University recently examined 27 caffeine drinkers and found 25 who displayed some type of withdrawal symptoms when abstaining from caffeine. [*Note:* The 27 caffeine drinkers volunteered for the study.] Furthermore, of 11 caffeine drinkers who were diag-

nosed as caffeine-dependent, 8 displayed dramatic withdrawal symptoms (including impairment in normal functioning) when they consumed a caffeine-free diet in a controlled setting.

The National Coffee Association claimed, however, that the study group was too small to draw conclusions. "It is inappropriate to overgeneralize the results of this ... study, since caffeine as normally consumed poses no health risk to the average consumer and cannot be described as a substance of dependence" (*Los Angeles Times,* Oct. 5, 1994).

Focus

Give your (supported) opinion on the issue. If the sample size is deemed satisfactory, establish confidence intervals for the parameters of interest. If not, how large a sample should be selected?

$$n = \frac{(z_{\alpha/2})^2(pq)}{B^2} = \frac{(1.645)^2(.1)(.9)}{(.01)^2} = 2{,}435.4 \approx 2{,}436$$

Thus, the manufacturer should sample 2,436 telephones in order to estimate the fraction defective, *p,* to within .01 with 90% confidence. Remember that this answer depends on our approximation for *pq,* where we used .09. If the fraction defective is closer to .05 than .10, we can use a sample of 1,286 telephones (check this) to estimate *p* to within .01 with 90% confidence. ▲

The cost of sampling will also play an important role in the final determination of the sample size to be selected to estimate either μ or *p.* Although more complex formulas can be derived to balance the reliability and cost considerations, we will solve for the necessary sample size and note that the sampling budget may be a limiting factor. Consult the references at the back of the book for a more complete treatment of this problem.

EXERCISES 7.51–7.67

Learning the Mechanics

7.51 If you wish to estimate a population mean to within a bound $B = .2$ using a 95% confidence interval and you know from prior sampling that σ^2 is approximately equal to 5.4, how many observations would you have to include in your sample?

7.52 If nothing is known about *p,* .5 can be substituted for *p* in the sample-size formula for a population proportion. But when this is done, the resulting sample size may be larger than needed. Under what circumstances will using $p = .5$ in the sample-size formula yield a sample size larger than needed to construct a

confidence interval for *p* with a specified bound and a specified confidence level?

7.53 Suppose you wish to estimate a population mean correct to within a bound $B = .15$ with probability equal to .90. You do not know σ^2, but you know that the observations will range in value between 31 and 39.

   a. Find the approximate sample size that will produce the desired accuracy of the estimate. You wish to be conservative to ensure that the sample size will be ample to achieve the desired accuracy of the estimate. [*Hint:* Using your knowledge of data variation from Section 2.5, assume that the range of the observations will equal 4σ.]

b. Calculate the approximate sample size, making the less conservative assumption that the range of the observations is equal to 6σ.

7.54 In each case, find the approximate sample size required to construct a 95% confidence interval for p that has bound $B = .06$.

a. Assume p is near .3.

b. Assume you have no prior knowledge about p, but you wish to be certain that your sample is large enough to achieve the specified accuracy for the estimate.

7.55 The following is a 90% confidence interval for p: (.26, .54). How large was the sample used to construct this interval?

7.56 It costs you $10 to draw a sample of size $n = 1$ and measure the attribute of interest. You have a budget of $1,200.

a. Do you have sufficient funds to estimate the population mean for the attribute of interest with a 95% confidence interval 4 units in width? Assume $\sigma = 12$.

b. If you used a 90% confidence level, would your answer to part **a** change? Explain.

7.57 Suppose you wish to estimate the mean of a normal population using a 95% confidence interval, and you know from prior information that $\sigma^2 \approx 1$.

a. To see the effect of the sample size on the width of the confidence interval, calculate the width of the confidence interval for $n = 16, 25, 49, 100,$ and 400.

b. Plot the width as a function of sample size n on graph paper. Connect the points by a smooth curve and note how the width decreases as n increases.

Applying the Concepts

7.58 The EPA standard on the amount of suspended solids that can be discharged into rivers and streams is a maximum of 60 milligrams per liter (mg/L) daily, with a maximum monthly average of 30 mg/L. Suppose you want to test a randomly selected sample of n water specimens and estimate the mean daily rate of pollution produced by a mining operation. If you want a 95% confidence interval estimate with a bound on the error of 1 mg/L, how many water specimens would you have to include in your sample? Assume prior knowledge indicates that pollution readings in water samples taken during a day are approximately normally distributed with a standard deviation equal to 5 mg/L.

7.59 Although corporate executives are probably not as highly stressed as air-traffic controllers or inner-city police, research has indicated that they are among the more highly pressured work groups. In order to estimate p, the proportion of managers who perceive themselves to be frequently under stress, one study sampled 532 managers in western Australian corporations (*Harvard Business Review*, Jan.–Feb. 1986). One hundred ninety of these managers fell into the high-stress group. Assume that random sampling was used in this study. Was the sample size large enough to estimate p to within .03 with 95% confidence? Explain.

7.60 According to estimates made by the General Accounting Office, the Internal Revenue Service (IRS) answered 18.3 million telephone inquiries during a recent tax season, and 17% of the IRS offices provided answers that were wrong. These estimates were based on data collected from sample calls to numerous IRS offices. How many IRS offices should be randomly selected and contacted in order to estimate the proportion of IRS offices that fail to correctly answer questions about gift taxes with a 90% confidence interval of width .06?

7.61 Refer to Exercise 7.19, in which workers' level of job satisfaction was related to age. Suppose that we want to estimate the mean level of job satisfaction for younger workers (age 18–24) to within .04 with 95% confidence. How large a sample should be selected? Recall that the standard deviation for this age group was .75.

7.62 According to a Food and Drug Administration (FDA) study, a cup of coffee contains an average of 115 milligrams (mg) of caffeine, with the amount per cup ranging from 60 to 180 mg. In contrast, sugar-free Mr. Pibb tested at 58.8 mg of caffeine per 12-ounce serving, Coca-Cola and Diet Coke at 45.6 mg, and Pepsi at 38.4 mg. Suppose you want to repeat the FDA experiment to obtain an estimate of the mean caffeine content in a cup of coffee correct to within 5 mg with 95% confidence. How many cups of coffee would have to be included in your sample?

7.63 The United States Golf Association (USGA) tests all new brands of golf balls to ensure that they meet USGA specifications. One test conducted is intended to measure the average distance traveled when the ball is hit by a machine called "Iron Byron," a name inspired by the swing of the famous golfer Byron Nelson. Suppose the USGA wishes to estimate the mean distance for a new brand to within 1 yard with 90% confidence. Assume that past tests have indicated that the standard deviation of the distances Iron Byron hits golf balls is approximately 10 yards. How many golf balls should be hit by Iron Byron to achieve the desired accuracy in estimating the mean?

7.64 If you want to estimate the proportion of operating automobiles that are equipped with both driver-side and passenger-side air bags, approximately how large a sample would be required to estimate p to within .02 with probability equal to .95?

7.65 Destructive sampling, in which the test to determine whether an item is defective destroys the item, is generally expensive, and the high costs involved often prohibit large sample sizes. For example, suppose the National Highway Safety Administration (NHSA) wishes to determine the proportion of new tires that will fail when subjected to hard braking at a speed of 60 miles per hour. NHSA can obtain the tires for $25 (wholesale price) each. Suppose the budget for the experiment is $10,000, and NHSA wishes to estimate the percentage that will fail to within .02 with 95% confidence. Assuming that the entire $10,000 can be spent on tires (ignoring other costs) and that the true fraction that will fail is approximately .05, can NHSA attain its goal while staying within the budget? Explain.

7.66 It costs more to produce defective items—since they must be scrapped or reworked—than it does to produce nondefective items. This simple fact suggests that manufacturers should ensure the quality of their products by perfecting their production processes instead of depending on inspection of finished products (Deming, 1986). In order to better understand a particular metal stamping process, a manufacturer wishes to estimate the mean length of parts produced by this stamping process during the past 24 hours.

a. How many parts should be sampled in order to estimate the population mean to within .1 millimeter (mm) with 90% confidence? Previous studies of this machine have indicated that the standard deviation of lengths produced by the stamping operation is about 2 mm.

b. Time permits the use of a sample size no larger than 100. If a 90% confidence interval for μ is constructed using $n = 100$, will it be wider or narrower than would have been obtained using the sample size determined in part **a**? Explain.

c. If management requires that μ be estimated to within .1 mm and that a sample size of no more than 100 be used, what is (approximately) the maximum confidence level that could be attained for a confidence interval that meets management's specifications?

7.67 Refer to the *International Journal of Sports Psychology* study, Exercise 7.44, where a sample of 95 workers in good physical condition was used to estimate the proportion of all workers in good condition who showed signs of stress. Recall that the point estimate of the proportion was .108. How many workers from this category must be sampled to estimate the true proportion who are stressed to within .01 with 95% confidence?

QUICK REVIEW

Key Terms

Bound on the error of estimation 277
Confidence coefficient 256
Confidence interval 256
Confidence level 256

Degrees of freedom 262
Interval estimator 256
t Statistic 262

Key Formulas

$\hat{\theta} \pm (z_{\alpha/2})\sigma_{\hat{\theta}}$

Large-sample confidence interval for population parameter θ

where $\hat{\theta}$ and and $\sigma_{\hat{\theta}}$ are obtained from the table below

Parameter θ	Estimator $\hat{\theta}$	Standard Error $\sigma_{\hat{\theta}}$	
Mean, μ	$\bar{x}$	$\sigma/\sqrt{n}$	257
Proportion, p	$\hat{p}$	$\sqrt{\dfrac{pq}{n}}$	270

$\bar{x} \pm t_{\alpha/2}\left(\dfrac{s}{\sqrt{n}}\right)$

Small-sample confidence interval for population mean μ 265

$n = \dfrac{(z_{\alpha/2})^2\sigma^2}{B^2}$

Determining the sample size n for estimating μ 277

$n = \dfrac{(z_{\alpha/2})^2(pq)}{B^2}$

Determining the sample size n for estimating p 279

LANGUAGE LAB

Symbol	Pronunciation	Description
θ	theta	General population parameter
μ	mu	Population mean
p		Population proportion
B		Bound on error of estimation
α	alpha	$(1 - \alpha)$ represents the confidence coefficient
$z_{\alpha/2}$	z of alpha over 2	z value used in a $100(1 - \alpha)\%$ large-sample confidence interval
$t_{\alpha/2}$	t of alpha over 2	t value used in a $100(1 - \alpha)\%$ small-sample confidence interval
$\bar{x}$	x-bar	Sample mean; point estimate of μ
$\hat{p}$	p-hat	Sample proportion; point estimate of p
σ	sigma	Population standard deviation
s		Sample standard deviation; point estimate of σ
$\sigma_{\bar{x}}$	sigma of $\bar{x}$	Standard deviation of sampling distribution of $\bar{x}$
$\sigma_{\hat{p}}$	sigma of $\hat{p}$	Standard deviation of sampling distribution of $\hat{p}$

SUPPLEMENTARY EXERCISES 7.68–7.92

Note: List the assumptions necessary for the valid implementation of the statistical procedures you use in solving all these exercises.

Learning the Mechanics

7.68 Let t_0 represent a particular value of t from Table VI of Appendix A. Find the table values such that the following statements are true:
  a. $P(t \leq t_0) = .05$ where df = 17
  b. $P(t \geq t_0) = .005$ where df = 14
  c. $P(t \leq -t_0 \text{ or } t \geq t_0) = .10$ where df = 6
  d. $P(t \leq -t_0 \text{ or } t \geq t_0) = .01$ where df = 22

7.69 In each of the following instances, determine whether you would use a z or t statistic (or neither) to form a 95% confidence interval; then look up the appropriate z or t value.
  a. Random sample of size $n = 21$ from a normal distribution with unknown mean μ and standard deviation σ
  b. Random sample of size $n = 175$ from a normal distribution with unknown mean μ and standard deviation σ
  c. Random sample of size $n = 12$ from a normal distribution with unknown mean μ and standard deviation $\sigma = 5$
  d. Random sample of size $n = 65$ from a distribution about which nothing is known
  e. Random sample of size $n = 8$ from a distribution about which nothing is known

7.70 A random sample of 225 measurements is selected from a population, and the sample mean and standard deviation are $\bar{x} = 32.5$ and $s = 30.0$, respectively.
  a. Use a 99% confidence interval to estimate the mean of the population, μ.
  b. How large a sample would be needed to estimate μ to within .5 with 99% confidence?
  c. What is meant by the phrase "99% confidence" as it is used in this exercise?

Applying the Concepts

7.71 A large New York City bank is interested in estimating (1) the proportion of weeks in which it processes more than 100,000 checks and (2) the average number of checks it processes per week. The bank maintains records of x, the number of checks processed each week. Suppose the bank records the number of checks, x, processed per week for 50 weeks randomly sampled from among the past 6 years. Define each of the following *in the context of the problem.*
  a. $\bar{x}$ **b.** $\hat{p}$ **c.** σ_x **d.** μ_x **e.** n **f.** $\sigma_{\bar{x}}$ **g.** p **h.** s_x

7.72 A random sample of 900 college students is selected, and 670 are found to receive some form of financial aid.
  a. Use a 90% confidence interval to estimate the proportion of all college students, p, who receive some form of financial aid.
  b. How large a sample would be needed to estimate p to within .01 with 90% confidence?
  c. What is meant by the phrase "90% confidence" as it is used in this exercise?

7.73 The Centers for Disease Control and Prevention (CDCP) in Atlanta, Georgia, conducts an annual survey of the general health of the U.S. population as part of its Behavioral Risk Factor Surveillance System (*New York Times*, Mar. 29, 1995). Using random-digit dialing, the CDCP telephones U.S. citizens over 18 years of age and asks them the following four questions:

(1) Is your health generally excellent, very good, good, fair, or poor?

(2) How many days during the previous 30 days was your physical health not good because of injury or illness?

(3) How many days during the previous 30 days was your mental health not good because of stress, depression, or emotional problems?

(4) How many days during the previous 30 days did your physical or mental health prevent you from performing your usual activities?

Identify the parameter of interest for each question.

7.74 Refer to Exercise 7.73. According to the CDCP, 89,582 of 102,263 adults interviewed stated that their health was good, very good, or excellent.

  a. Use a 99% confidence interval to estimate the true proportion of U.S. adults who believe their health to be good to excellent. Interpret the interval.

  b. Why might the estimate, part **a**, be overly optimistic (i.e., biased high)?

7.75 W.S. Snyder and J.W. Chrissis (1990) presented a hybrid algorithm for solving a polynomial zero–one mathematical programming problem. The algorithm incorporates a mixture of pseudo-Boolean concepts and time-proven implicit enumeration procedures. Fifty-two random problems were solved using the hybrid algorithm; the times to solution (CPU time in seconds) are listed in the table below. A stem-and-leaf display and descriptive statistics for the data set are provided in the SAS printout on page 285. Use this information to estimate, with 95% confidence, the mean solution time for the hybrid algorithm. Interpret the result.

7.76 Research reported in *The Professional Geographer* (May 1992) investigates the hypothesis that the disproportionate housework responsibility of women in two-income households is a major factor in determining the proximity of a woman's place of employment. The researcher studied the distance (in miles) to work for both men and women in two-income households. Random samples of men and women yielded the results shown in the accompanying table.

	CENTRAL CITY RESIDENCE		SUBURBAN RESIDENCE	
	Men	**Women**	**Men**	**Women**
Sample Size	159	119	138	93
Mean	7.4	4.5	9.3	6.6
Std. Deviation	6.3	4.2	7.1	5.6

  a. For central city residences, calculate a 95% confidence interval for the average distance to work for men and women in two-income households. Interpret the intervals.

  b. Repeat part **a** for suburban residences.

  [*Note:* We will show how to use statistical techniques to compare two population means in Chapter 9.]

7.77 Refer to the *Journal of the American Medical Association* (Apr. 21, 1993) report on the prevalence of cigarette smoking among U.S. adults, Exercise 7.11. Of the 43,732 survey respondents, 11,239 indicated that they were current smokers and 10,539 indicated they were former smokers.

  a. Construct and interpret a 90% confidence interval for the percentage of U.S. adults who currently smoke cigarettes.

  b. Construct and interpret a 90% confidence interval for the percentage of U.S. adults who are former cigarette smokers.

7.78 A company is interested in estimating μ, the mean number of days of sick leave taken by all its employees. The firm's statistician selects at random 100 personnel files and notes the number of sick days taken by each employee. The following sample statistics are computed: $\bar{x} = 12.2$ days, $s = 10$ days.

  a. Estimate μ using a 90% confidence interval.

  b. How many personnel files would the statistician have to select in order to estimate μ to within 2 days with a 99% confidence interval?

7.79 In the United States, people over age 50 represent 25% of the population, yet they control 70% of the wealth. Research indicates the highest priority of retirees is travel. A study in the *Annals of Tourism Research* (Vol. 19, 1992) investigates the relationship of retirement status (pre- and post-retirement)

.045	1.055	.136	1.894	.379	.136	.336	.258	1.070
.506	.088	.242	1.639	.912	.412	.361	8.788	.579
1.267	.567	.182	.036	.394	.209	.445	.179	.118
.333	.554	.258	.182	.070	3.985	.670	3.888	.136
.091	.600	.291	.327	.130	.145	4.170	.227	.064
.194	.209	.258	3.046	.045	.049	.079		

Source: Snyder, W. S., and Chrissis, J. W. "A hybrid algorithm for solving zero–one mathematical programming problems." *IEEE Transactions,* Vol. 22, No. 2, June 1990, p. 166 (Table 1). Reprinted with the permission of the Institute of Industrial Engineers, 25 Technology Park/Atlanta, Norcross, GA 30092, (770) 449-0461. Copyright © 1990.

UNIVARIATE PROCEDURE

Variable=SOLTIME

Moments

N	52	Sum Wgts	52
Mean	0.812192	Sum	42.234
Std Dev	1.50476	Variance	2.264303
Skewness	3.65356	Kurtosis	15.73264
USS	149.7816	CSS	115.4795
CV	185.2714	Std Mean	0.208673
T:Mean=0	3.892183	Prob>$|$T$|$	0.0003
Sgn Rank	689	Prob>$|$S$|$	0.0001
Num ^= 0	52		
W:Normal	0.530623	Prob<W	0.0

Quantiles(Def=5)

100%	Max	8.788	99%	8.788
75%	Q3	0.5895	95%	3.985
50%	Med	0.2745	90%	1.894
25%	Q1	0.136	10%	0.07
0%	Min	0.036	5%	0.045
			1%	0.036

Range	8.752
Q3-Q1	0.4535
Mode	0.136

Extremes

Lowest	Obs	Highest	Obs
0.036(	22)	3.046(	49)
0.045(	50)	3.888(	35)
0.045(	1)	3.985(	33)
0.049(	51)	4.17(	43)
0.064(	45)	8.788(	17)

```
Stem Leaf                                                      #          Boxplot
   8 8                                                         1             *
   8
   7
   7
   6
   6
   5
   5
   4
   4 02                                                        2             *
   3 9                                                         1             *
   3 0                                                         1             *
   2
   2
   1 69                                                        2             0
   1 113                                                       3             |
   0 5666679                                                   7          +--+--+
   0 00001111111111112222222222333333334444                   35         *-----*
     ----+----+----+----+----+----+----+----+
```

to various items of interest in the travel industry. As one part of the study, a sample of 323 post-retirees was selected, and the number of nights each typically stayed away from home on trips was determined. One hundred seventy-two (172) responded that their typical stays ranged from 4 to 7 nights. Use a 90% confidence interval to estimate the true proportion of post-retirement travelers who stay between 4 and 7 nights on a typical trip. Interpret the interval.

7.80 A meteorologist wishes to estimate the mean amount of snowfall per year in Spokane, Washington. A random sample of the recorded snowfalls for 20 years produces a sample mean equal to 54 inches and a standard deviation of 9.59 inches.

   a. Estimate the true mean amount of snowfall in Spokane using a 99% confidence interval.

   b. If you were purchasing snow-removal equipment for a city, what numerical descriptive measure of the distribution of depth of snowfall would be of most interest to you? Would it be the mean?

7.81 A company purchases large quantities of naphtha in 50-gallon drums. Because the purchases are ongoing, small shortages in the drums can represent a sizable loss to the company. The weights of the drums vary slightly from drum to drum, so the weight of the naphtha is determined by removing it from the drums and measuring it. Suppose the company samples the contents of 20 drums, measures the naphtha in each, and calculates $\bar{x} = 49.70$ gallons and $s = .32$ gallon. Find a 95% confidence interval for the mean number of gallons of naphtha per drum. What assumptions are necessary to ensure the validity of the confidence interval?

7.82 Before approval is given for the use of a chemical as an insecticide, the United States Department of Agriculture (USDA) requires that several tests be performed to see how the substance will affect wildlife. In one case, the USDA wants to know the proportion of starlings that will die after being exposed to a new insecticide. A random sample of 80 starlings was caught and fed a diet of regular food that had been treated with the substance. After 10 days, 10 starlings died. Use a 99% confidence interval to estimate the true proportion of starlings that will be killed by the substance.

7.83 A random sample of 122 Illinois law firms was selected to determine their degree of computer usage (*Legal Assistant Today,* Mar.–Apr. 1988). Seventy-six of the firms used microcomputers (PCs).

   a. Use a 95% confidence interval to estimate the proportion of all Illinois law firms that used microcomputers at the time of the survey.

   b. Interpret the interval in the terms of this application.

   c. What is meant by the phrase "95% confidence interval"?

   d. Do you think this interval provides an estimate for the proportion of all U.S. law firms that were using microcomputers at the time of the survey? Why or why not?

7.84 Refer to Exercise 7.83. In a similar sample survey of 71 Illinois banks, 55 used microcomputers (*Banking Administration,* Jan. 1988).

   a. Use a 95% confidence interval to estimate the proportion of all Illinois banks that used microcomputers at the time of the survey.

   b. Interpret the interval in the terms of this application.

   c. How does this interval compare to the one you constructed for the proportion of Illinois law firms that used microcomputers at the time of the survey?

7.85 A health researcher wishes to estimate the mean number of cavities per child for children under the age of 12 who live in a specified environment. The number of cavities per child for a random sample of 35 children under the age of 12 has a mean of 2 and a standard deviation of 1.7. Construct a 90% confidence interval for the mean number of cavities per child under the age of 12 who lives in the sampled environment.

7.86 Recycling is receiving increased emphasis among environmental agencies as a means of dealing with the significant growth in the trash and garbage of our "throw-away" society. To estimate the degree of awareness about recycling in one major city, a random sample of 346 households was selected, and 212 were found to use available recycling facilities. Use a 95% confidence interval to estimate the proportion of households in the city that are using available recycling facilities.

7.87 In 1988 tire sales in the United States reached a historic high (*Chemical Week,* 1989). Tires made of synthetic rubber accounted for much of this growth. Suppose one manufacturer was testing a new synthetic rubber design. Twenty tires of the new design were produced and subjected to wear tests. The results indicate that the mean wear for the test tires was 42,250 miles, and the standard deviation was 4,355 miles.

   a. What is the point estimate of the true mean wear for the new design?

   b. Construct a 90% confidence interval estimate for the mean wear associated with the new design.

   c. Which method of estimation is better, point estimation or interval estimation? Why?

7.88 Refer to Exercise 7.87. Suppose that 200 tires rather than 20 were tested, and assume that the sample mean and standard deviation for the 200 tires remain the same: $\bar{x}$ = 42,250 miles, s = 4,355 miles. Repeat parts **a–c** of Exercise 7.87 and comment on the similarities and differences in your answers.

7.89 Recently, an outbreak of salmonella (bacterial) poisoning was traced to a particular brand of ice cream bar, and the manufacturer removed the bars from the market. Despite this response, many consumers refused to purchase *any* brand of ice cream bars for some period of time after the event (McClave, personal consulting). One manufacturer conducted a survey of consumers 6 months after the outbreak. A sample of 244 ice cream bar consumers was contacted, and 23 respondents indicated that they would not purchase ice cream bars because of the potential for food poisoning.

a. What is the point estimate of the true fraction of the entire market who refuse to purchase ice cream bars 6 months after the outbreak?

b. Is the sample size large enough to use the normal approximation for the sampling distribution of the estimator of the binomial probability? Justify your response.

c. Construct a 95% confidence interval for the true proportion of the market who still refuse to purchase ice cream bars 6 months after the event.

d. Interpret both the point estimate and confidence interval in terms of this application.

7.90 Refer to Exercise 7.89. Suppose it is now 1 year after the outbreak of food poisoning was traced to ice cream bars. The manufacturer wishes to estimate the proportion who still will not purchase ice cream bars to within .02 using a 95% confidence interval. How many consumers should be sampled?

7.91 Medicaid health-assistance programs are administered by the individual states, even though part of the funding is federal. The federal government requires that the states perform regular audits in order to ensure that payments are accurate. One Florida hospital was audited by the Florida Department of Health and Rehabilitative Services (HRS), and a random sample of 25 Medicaid claims was selected. The sample mean of the claims was $34.76 and the standard deviation was $11.34.

a. Use a 99% confidence interval to estimate the mean of all claims submitted by this hospital.

b. What assumptions are necessary to ensure the validity of this confidence interval?

7.92 Refer to Exercise 7.91.

a. How many claims must be sampled if the HRS wants to estimate the mean size of the hospital's claims to within $1.00 using a 99% confidence interval?

b. If a sample of this size were to be selected and a 99% confidence interval constructed, what assumptions would be necessary to ensure the validity of the interval?

STUDENT PROJECTS

Choose a population pertinent to your major area of interest—a population that has an unknown mean or, if the population is binomial, that has an unknown probability of success. For example, a marketing major may be interested in the proportion of consumers who prefer a certain product. A sociology major may be interested in estimating the proportion of people in a particular socioeconomic group or the mean income of people living in a particular part of a city. A political science major may wish to estimate the proportion of an electorate in favor of a certain candidate, a certain amendment, or a certain presidential policy. A pre-med student might want to find the average length of time patients stay in the hospital or the average number of people treated daily in the emergency room. We could continue with examples, but the point should be clear—choose something of interest to you.

Define the parameter you want to estimate and conduct a *pilot study* to obtain an initial estimate of the parameter of interest and, more importantly, an estimate of the variability associated with the estimator. A pilot study is a small experiment (perhaps 20 to 30 observations) used to gain some information about the population of interest. The purpose of the study is to help plan more elaborate future experiments. Using the results of your pilot study, determine the sample size necessary to estimate the parameter to within a reasonable bound (of your choice) with a 95% confidence interval.

EXPLORING DATA WITH A COMPUTER

Refer to Exploring Data with a Computer in Chapter 6. Recall the values of the "population" mean μ and standard deviation σ for the 962 FTC measurements on tar, nicotine, or carbon monoxide in cigarette smoke. Suppose our objective is to sample from this population and to estimate the mean μ using a 95% confidence interval.

a. Determine the sample size n_1 necessary to estimate μ to within 1 milligram with 95% confidence. Then generate one hundred 95% confidence intervals by repeatedly drawing samples of size n_1 (with replacement) from the 962 measurements and using the sample statistics to form a confidence interval. Treat σ as unknown when forming the confidence intervals. What percentage of confidence intervals will contain μ?

b. Determine the sample size n_2 necessary to estimate μ to within .5 milligram with 95% confidence. Then generate one hundred 95% confidence intervals by repeatedly drawing samples of size n_2 (with replacement) from the 962 measurements and using the sample statistics to form a confidence interval. Treat σ as unknown when forming the confidence intervals. What percentage of confidence intervals will contain μ?

c. Repeat part a, but this time use an 80% confidence interval.

Chapter 8

INFERENCES BASED ON A SINGLE SAMPLE

Tests of Hypothesis

Contents

Case Studies

*W*HERE WE'VE BEEN

We saw how to use sample information to estimate population parameters in Chapter 7. The sampling distribution of a statistic is used to assess the reliability of an estimate, which we express in terms of a confidence interval.

*W*HERE WE'RE GOING

We'll see how to utilize sample information to test what the value of a population parameter may be. This type of inference is called a *test of hypothesis*. We'll also see how to conduct a test of hypothesis about a population mean μ and a population proportion *p*. And, just as with estimation, we'll stress the measurement of the reliability of the inference. An inference without a measure of reliability is little more than a guess.

Suppose you wanted to determine whether the mean level of a driver's blood alcohol exceeds the legal limit after two drinks, or whether the mean breaking strength of sewer pipe exceeds 2,400 pounds per foot, or whether the majority of registered voters approve of the president's performance. In each case you are interested in making an inference about how the value of a parameter relates to a specific numerical value. Is it less than, equal to, or greater than the specified number? This type of inference, called a **test of hypothesis**, is the subject of this chapter.

We introduce the elements of a test of hypothesis in Section 8.1. We then show how to conduct a large-sample test of hypothesis about a population mean in Sections 8.2 and 8.3. In Section 8.4 we utilize small samples to conduct tests about means. Large-sample tests about binomial probabilities are the subject of Section 8.5, and some advanced methods for determining the reliability of a test are covered in optional Section 8.6. Finally, we show how to make inferences about a population variance in optional Section 8.7.

8.1   THE ELEMENTS OF A TEST OF HYPOTHESIS

Suppose building specifications in a certain city require that the average breaking strength of residential sewer pipe be more than 2,400 pounds per foot of length (i.e., per lineal foot). Each manufacturer who wants to sell pipe in this city must demonstrate that its product meets the specification. Note that we are again interested in making an inference about the mean μ of a population. However, in this example we are less interested in estimating the value of μ than we are in testing a *hypothesis* about its value. That is, we want to decide whether the mean breaking strength of the pipe exceeds 2,400 pounds per lineal foot.

The method used to reach a decision is based on the rare-event concept explained in earlier chapters. We define two hypotheses: (1) The **null hypothesis** is that which represents the status quo to the party performing the sampling experiment—the hypothesis that will be accepted unless the data provide convincing evidence that it is false. (2) The **alternative**, or **research**, **hypothesis** is that which will be accepted only if the data provide convincing evidence of its truth. From the point of view of the city conducting the tests, the null hypothesis is that the manufacturer's pipe does *not* meet specifications unless the tests provide convincing evidence otherwise. The null and alternative hypotheses are therefore

> *Null hypothesis (H_0):* $\mu \leq 2{,}400$
> (i.e., the manufacturer's pipe does not meet specifications)
>
> *Alternative (research) hypothesis (H_a):* $\mu > 2{,}400$
> (i.e., the manufacturer's pipe meets specifications)

How can the city decide when enough evidence exists to conclude that the manufacturer's pipe meets specifications? Since the hypotheses concern the value of the population mean μ, it is reasonable to use the sample mean $\bar{x}$ to make the inference, just as we did when forming confidence intervals for μ in Sections 7.1 and 7.2. The city will conclude that the pipe meets specifications only when the sample mean $\bar{x}$ convincingly indicates that the population mean exceeds 2,400 pounds per lineal foot.

"Convincing" evidence in favor of the alternative hypothesis will exist when the value of $\bar{x}$ exceeds 2,400 by an amount that cannot be readily attributed to sampling variability. To decide, we compute a **test statistic**, which is the z value

FIGURE 8.1

The sampling distribution of $\bar{x}$, assuming $\mu = 2,400$

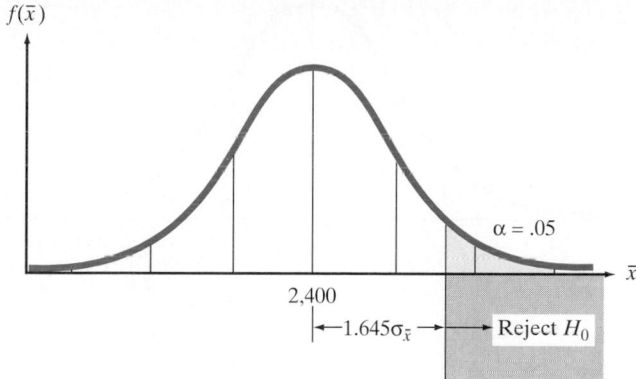

that measures the distance between the value of $\bar{x}$ and the value of μ specified in the null hypothesis. When the null hypothesis contains more than one value of μ, as in this case (H_0: $\mu \le 2,400$), we use the value of μ closest to the values specified in the alternative hypothesis. The idea is that if the hypothesis that μ *equals* 2,400 can be rejected in favor of $\mu > 2,400$, then μ *less than or equal to* 2,400 can certainly be rejected. Thus, the test statistic is

$$z = \frac{\bar{x} - 2,400}{\sigma_x} = \frac{\bar{x} - 2,400}{\sigma / \sqrt{n}}$$

Note that a value of $z = 1$ means that $\bar{x}$ is 1 standard deviation above $\mu = 2,400$; a value of $z = 1.5$ means that $\bar{x}$ is 1.5 standard deviations above $\mu = 2,400$, etc. How large must z be before the city can be convinced that the null hypothesis can be rejected in favor of the alternative and conclude that the pipe meets specifications?

If you examine Figure 8.1, you will note that the chance of observing $\bar{x}$ more than 1.645 standard deviations above 2,400 is only .05—*if in fact the true mean μ is* 2,400. Thus, if the sample mean is more than 1.645 standard deviations above 2,400, either H_0 is true and a relatively rare event has occurred (.05 probability) or H_a is true and the population mean exceeds 2,400. Since we would most likely reject the notion that a rare event has occurred, we would reject the null hypothesis ($\mu \le 2,400$) and conclude that the alternative hypothesis ($\mu > 2,400$) is true. What is the probability that this procedure will lead us to an incorrect decision?

Such an incorrect decision—deciding that the null hypothesis is false when in fact it is true—is called a **Type I decision error**. As indicated in Figure 8.1, the risk of making a Type I error is denoted by the symbol α. That is,

$\alpha = P(\text{Type I error})$

$= P(\text{Rejecting the null hypothesis when in fact the null hypothesis is true})$

In our example

$$\alpha = P(z > 1.645 \text{ when in fact } \mu = 2,400) = .05$$

We now summarize the elements of the test:

H_0: $\mu \le 2,400$

H_a: $\mu > 2,400$

Test statistic: $z = \dfrac{\bar{x} - 2,400}{\sigma_{\bar{x}}}$

Rejection region: $z > 1.645$, which corresponds to $\alpha = .05$

FIGURE 8.2

Location of the test statistic
for a test of the hypothesis
H_0: $\mu = 2,400$

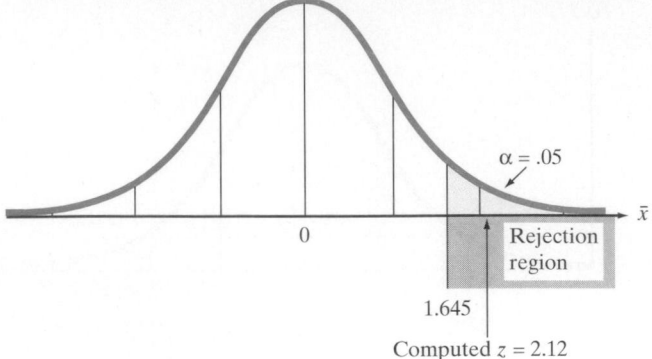

Note that the **rejection region** refers to the values of the test statistic for which we will *reject the null hypothesis.*

To illustrate the use of the test, suppose we test 50 sections of sewer pipe and find the mean and standard deviation for these 50 measurements to be

$$\bar{x} = 2,460 \text{ pounds per linear foot} \qquad s = 200 \text{ pounds per linear foot}$$

As in the case of estimation, we can use s to approximate σ when s is calculated from a large set of sample measurements.

The test statistic is

$$z = \frac{\bar{x} - 2,400}{\sigma_{\bar{x}}} = \frac{\bar{x} - 2,400}{\sigma/\sqrt{n}} \approx \frac{\bar{x} - 2,400}{s/\sqrt{n}}$$

Substituting $\bar{x} = 2,460$, $n = 50$, and $s = 200$, we have

$$z \approx \frac{2,460 - 2,400}{200/\sqrt{50}} = \frac{60}{28.28} = 2.12$$

Therefore, the sample mean lies $2.12\sigma_{\bar{x}}$ above the hypothesized value of μ, 2,400, as shown in Figure 8.2. Since this value of z exceeds 1.645, it falls in the rejection region. That is, we reject the null hypothesis that $\mu = 2,400$ and conclude that $\mu > 2,400$. Thus, it appears that the company's pipe has a mean strength that exceeds 2,400 pounds per linear foot.

How much faith can be placed in this conclusion? What is the probability that our statistical test could lead us to reject the null hypothesis (and conclude that the company's pipe meets the city's specifications) when in fact the null hypothesis is true? The answer is $\alpha = .05$. That is, we selected the level of risk, α, of making a Type I error when we constructed the test. Thus, the chance is only 1 in 20 that our test would lead us to conclude the manufacturer's pipe satisfies the city's specifications when in fact the pipe does *not* meet specifications.

Now, suppose the sample mean breaking strength for the 50 sections of sewer pipe turned out to be $\bar{x} = 2,430$ pounds per linear foot. Assuming that the sample standard deviation is still $s = 200$, the test statistic is

$$z = \frac{2,430 - 2,400}{200/\sqrt{50}} = \frac{30}{28.28} = 1.06$$

Therefore, the sample mean $\bar{x} = 2,430$ is only 1.06 standard deviations above the null hypothesized value of $\mu = 2,400$. As shown in Figure 8.3, this value does not fall into the rejection region ($z > 1.645$). Therefore, we know that we cannot reject H_0 using $\alpha = .05$. Even though the sample mean exceeds the city's specification of

FIGURE 8.3

Location of test statistic
when $\bar{x} = 2,430$

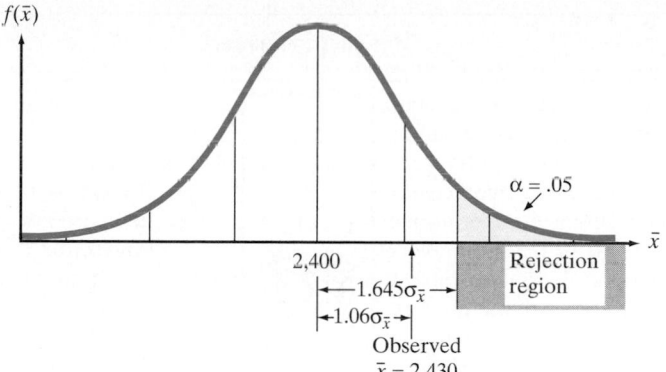

2,400 by 30 pounds per linear foot, it does not exceed the specification by enough to provide *convincing* evidence that the *population mean* exceeds 2,400.

Should we accept the null hypothesis $H_0: \mu \le 2,400$ and conclude that the manufacturer's pipe does not meet specifications? To do so would be to risk a **Type II error**—that of concluding that the null hypothesis is true (the pipe does not meet specifications) when in fact it is false (the pipe does meet specifications). We denote the probability of committing a Type II error by β, and we show in optional Section 8.6 that β is often difficult to determine precisely. Rather than make a decision (accept H_0) for which the probability of error (β) is unknown, we avoid the potential Type II error by avoiding the conclusion that the null hypothesis is true. Instead, we will simply state that *the sample evidence is insufficient to reject H_0 at $\alpha = .05$*. Since the null hypothesis is the "status-quo" hypothesis, the effect of not rejecting H_0 is to maintain the status quo. In our pipe-testing example, the effect of having insufficient evidence to reject the null hypothesis that the pipe does not meet specifications is probably to prohibit the utilization of the manufacturer's pipe unless and until there is sufficient evidence that the pipe does meet specifications. That is, until the data indicate convincingly that the null hypothesis is false, we usually maintain the status quo implied by its truth.

Table 8.1 summarizes the four possible outcomes of a test of hypothesis. The "true state of nature" columns in Table 8.1 refer to the fact that either the null hypothesis H_0 is true or the alternative hypothesis H_a is true. Note that the true state of nature is unknown to the researcher conducting the test. The "decision" rows in Table 8.1 refer to the action of the researcher, assuming that he or she will either conclude that H_0 is true or that H_a is true, based on the results of the sampling experiment. Note that a Type I error can be made *only* when the alternative hypothesis is accepted (equivalently, when the null hypothesis is rejected), and a Type II error can be made *only* when the null hypothesis is accepted. Our policy

TABLE 8.1   Conclusions and Consequences for a Test of Hypothesis

	TRUE STATE OF NATURE	
Conclusion	H_0 **True**	H_a **True**
H_0 *True*	Correct decision	Type II error (probability β)
H_a *True*	Type I error (probability α)	Correct decision

Statistics Is Murder!

CASE STUDY • 8.1 •

We've already seen how statistics and probability can play a key role in establishing credible evidence in a trial by jury (see Case Study 3.2). Sometimes, the outcome of a jury trial defies the "commonsense" expectations of the general public (e.g., the O.J. Simpson verdict in the "Trial of the Century"). Such a verdict is more acceptable if we understand that the jury trial of an accused murderer is analogous to the statistical hypothesis-testing process. Each of the elements of a test of hypothesis applies to the jury system of deciding the guilt or innocence of the accused:

1. *Null hypothesis* (H_0): The null hypothesis in a jury trial is that the accused is innocent. The status-quo hypothesis in the U.S. system of justice is innocence, which is assumed to be true until proven false *beyond a reasonable doubt*.

2. *Alternative hypothesis* (H_a): The alternative hypothesis is guilt, which is accepted only when sufficient evidence exists to establish its truth.

3. *Test statistic:* The test statistic in a trial is the final vote of the jury—that is, the number of the jury members who vote "guilty."

4. *Rejection region:* In a murder trial the jury vote must be unanimous in favor of guilt before the null hypothesis of innocence is rejected in favor of the alternative hypothesis of guilt. Thus, for a 12-member jury trial, the rejection region is $x = 12$, where x is the number of "guilty" votes.

5. *Assumption:* The primary assumption made in trials concerns the method of selecting the jury. The jury is assumed to represent a random sample of citizens who have no prejudice concerning the case.

6. *Experiment and calculation of the test statistic:* The sampling experiment is analogous to the jury selection, the trial, and the jury deliberations. The final vote of the jury is analogous to the calculation of the test statistic.

7. *Conclusion:*
   (a) If the vote of the jury is unanimous in favor of guilt, the null hypothesis of innocence is rejected and the court concludes that the accused murderer is guilty.

   (b) Any vote other than a unanimous one for guilt results in the court's reserving judgment about the hypotheses, either by declaring the accused "not guilty," or by declaring a mistrial and repeating the "test" with a new jury. (The latter is analogous to collecting more data and repeating a statistical test of hypothesis.) The court never accepts the null hypothesis by declaring the accused "innocent."

Focus

a. Define Type I and Type II errors in a murder trial.

b. Which of the two errors is the more serious? Explain.

c. The court does not, in general, know the values of α and β; but ideally, both should be small. One of these probabilities is assumed to be smaller than the other in a jury trial. Which one, and why?

d. The court system relies on the belief that the value of α is made very small by requiring a unanimous vote before guilt is concluded. Explain why this is so.

e. For a jury prejudiced against a guilty verdict as the trial begins, will the value of α increase or decrease? Explain.

f. For a jury prejudiced against a guilty verdict as the trial begins, will the value of β increase or decrease? Explain.

---

will be to make a decision only when we know the probability of making the error that corresponds to that decision. Since α is usually specified by the analyst, we will generally be able to reject H_0 (accept H_a) when the sample evidence supports that decision. However, since β is usually not specified, we will generally avoid the decision to accept H_0, preferring instead to state that the sample evidence is insufficient to reject H_0 when the test statistic is not in the rejection region.

The elements of a test of hypothesis are summarized in the following box. Note that the first four elements are all specified *before* the sampling experiment is performed. In no case will the results of the sample be used to determine the hypotheses—the data are collected to test the predetermined hypotheses, not to formulate them.

Elements of a Test of Hypothesis

1. *Null hypothesis* (H_0): A theory about the values of one or more population parameters. The theory generally represents the status quo, which we accept until it is proven false.
2. *Alternative (research) hypothesis* (H_a): A theory that contradicts the null hypothesis. The theory generally represents that which we will accept only when sufficient evidence exists to establish its truth.
3. *Test statistic:* A sample statistic used to decide whether to reject the null hypothesis.
4. *Rejection region:* The numerical values of the test statistic for which the null hypothesis will be rejected. The rejection region is chosen so that the probability is α that it will contain the test statistic when the null hypothesis is true, thereby leading to a Type I error. The value of α is usually chosen to be small (e.g., .01, .05, or .10), and is referred to as the **level of significance** of the test.
5. *Assumptions:* Clear statement(s) of any assumptions made about the population(s) being sampled.
6. *Experiment and calculation of test statistic:* Performance of the sampling experiment and determination of the numerical value of the test statistic.
7. *Conclusion:*
   a. If the numerical value of the test statistic falls in the rejection region, we reject the null hypothesis and conclude that the alternative hypothesis is true. We know that the hypothesis-testing process will lead to this conclusion incorrectly (Type I error) only $100\alpha\%$ of the time when H_0 is true.
   b. If the test statistic does not fall in the rejection region, we do not reject H_0. Thus, we reserve judgment about which hypothesis is true. We do not conclude that the null hypothesis is true because we do not (in general) know the probability β that our test procedure will lead to an incorrect acceptance of H_0 (Type II error).

EXERCISES 8.1–8.10

Learning the Mechanics

8.1 Which hypothesis, the null or the alternative, is the status-quo hypothesis? Which is the research hypothesis?

8.2 Which element of a test of hypothesis is used to decide whether to reject the null hypothesis in favor of the alternative hypothesis?

8.3 What is the level of significance of a test of hypothesis?

8.4 What is the difference between Type I and Type II errors in hypothesis testing? How do α and β relate to Type I and Type II errors?

8.5 List the four possible results of the combinations of decisions and true states of nature for a test of hypothesis.

8.6 We (generally) reject the null hypothesis when the test statistic falls in the rejection region, but we do not accept the null hypothesis when the test statistic does not fall in the rejection region. Why?

8.7 If you test a hypothesis and reject the null hypothesis in favor of the alternative hypothesis, does your test prove that the alternative hypothesis is correct? Explain.

Applying the Concepts

8.8 In 1895 an Italian criminologist, Cesare Lombroso, proposed that blood pressure be used to test for truthfulness. In the 1930s, William Marston added the measurements of respiration and perspiration to the process, built a machine to do the measuring, and called his invention the *polygraph*—or *lie detector.* Today, the federal court system will not consider polygraph results as evidence, but nearly half the state courts do permit polygraph tests under certain circumstances. In addition, its use in screening job applicants is on the rise. Physicians Michael Phillips, Allan Brett, and John Beary subjected the polygraph to the same careful testing given to medical diagnostic tests. They found that if 1,000 people were subjected to the polygraph and 500 told the truth and 500 lied, the polygraph would indicate that approximately 185 of the truth-tellers were liars and that approximately 120 of the liars were truth-tellers ("Lie Detectors Can Make a Liar of You," *Discover,* June 1986).

a. In the application of a polygraph test, an individual is presumed to be a truth-teller (H_0) until "proven" a liar (H_a). In this context, what is a Type I error? A Type II error?

b. According to Phillips, Brett, and Beary, what is the probability (approximately) that a polygraph test will result in a Type I error? A Type II error?

8.9 When a new drug is formulated, the pharmaceutical company must subject it to lengthy and involved testing before receiving the necessary permission from the Food and Drug Administration (FDA) to market the drug. The FDA's policy is that the pharmaceutical company must provide substantial evidence that a new drug is safe prior to receiving FDA approval, so that the FDA can confidently certify the safety of the drug to potential consumers.

  a. If the new drug testing were to be placed in a test of hypothesis framework, would the null hypothesis be that the drug is safe or unsafe? The alternative hypothesis?

  b. Given the choice of null and alternative hypotheses in part a, describe Type I and Type II errors in terms of this application. Define α and β in terms of this application.

  c. If the FDA wants to be very confident that the drug is safe before permitting it to be marketed, is it more important that α or β be small? Explain.

8.10 One of the most pressing problems in high-technology industries is computer security. Computer security is typically achieved by use of a *password*—a collection of symbols (usually letters and numbers) that must be supplied by the user before the computer permits access to the account. The problem is that persistent hackers can create programs that enter millions of combinations of symbols into a target system until the correct password is found. The newest systems solve this problem by requiring authorized users to identify themselves by unique body characteristics. For example, a system developed by Palmguard, Inc., tests the hypothesis

  H_0: The proposed user is authorized

versus

  H_a: The proposed user is unauthorized

by checking characteristics of the proposed user's palm against those stored in the authorized users' data bank (*Omni*, 1984).

  a. Define a Type I error and Type II error for this test. Which is the more serious error? Why?

  b. Palmguard reports that the Type I error rate for its system is less than 1%, whereas the Type II error rate is .00025%. Interpret these error rates.

  c. Another successful security system, the EyeDentifyer, "spots authorized computer users by reading the one-of-a-kind patterns formed by the network of minute blood vessels across the retina at the back of the eye." The EyeDentifyer reports Type I and II error rates of .01% (1 in 10,000) and .005% (5 in 100,000), respectively. Interpret these rates.

8.2 LARGE-SAMPLE TEST OF HYPOTHESIS ABOUT A POPULATION MEAN

In Section 8.1 we learned that the null and alternative hypotheses form the basis for a test of hypothesis inference. The null and alternative hypotheses may take one of several forms. In the sewer pipe example we tested the null hypothesis that the population mean strength of the pipe is less than or equal to 2,400 pounds per lineal foot against the alternative hypothesis that the mean strength exceeds 2,400. That is, we tested

$$H_0:\ \mu \leq 2{,}400$$
$$H_a:\ \mu > 2{,}400$$

This is a **one-tailed** (or **one-sided**) **statistical test** because the alternative hypothesis specifies that the population parameter (the population mean μ, in this example) is strictly greater than a specified value (2,400, in this example). If the null hypothesis had been $H_0:\ \mu \geq 2{,}400$ and the alternative hypothesis had been $H_a:\ \mu < 2{,}400$, the test would still be one-sided, because the parameter is still specified to be on "one side" of the null hypothesis value. Some statistical investigations seek to show that the population parameter is *either larger or smaller* than some specified value. Such an alternative hypothesis is called a **two-tailed** (or **two-sided**) **hypothesis**.

FIGURE 8.4

Rejection regions corresponding to one- and two-tailed tests

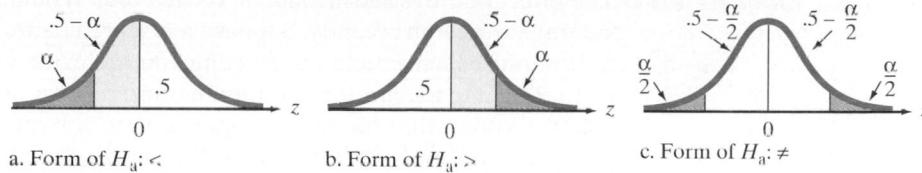

a. Form of H_a: $<$ b. Form of H_a: $>$ c. Form of H_a: $\neq$

While alternative hypotheses are always specified as strict inequalities, such as $\mu < 2{,}400$, $\mu > 2{,}400$, or $\mu \neq 2{,}400$, null hypotheses are usually specified as equalities, such as $\mu = 2{,}400$. Even when the null hypothesis is an inequality, such as $\mu \leq 2{,}400$, we specify H_0: $\mu = 2{,}400$, reasoning that if sufficient evidence exists to show that H_a: $\mu > 2{,}400$ is true when tested against H_0: $\mu = 2{,}400$, then surely sufficient evidence exists to reject $\mu < 2{,}400$ as well. Therefore, the null hypothesis is specified as the value of μ closest to a one-sided alternative hypothesis and as the only value *not* specified in a two-tailed alternative hypothesis. The steps for selecting the null and alternative hypotheses are summarized in the accompanying box.

Steps for Selecting the Null and Alternative Hypotheses

---

1. Select the *alternative hypothesis* as that which the sampling experiment is intended to establish. The alternative hypothesis will assume one of three forms:
   a. One-tailed, upper-tailed   *Example:* H_a: $\mu > 2{,}400$
   b. One-tailed, lower-tailed   *Example:* H_a: $\mu < 2{,}400$
   c. Two-tailed   *Example:* H_a: $\mu \neq 2{,}400$
2. Select the *null hypothesis* as the status quo, that which will be presumed true unless the sampling experiment conclusively establishes the alternative hypothesis. The null hypothesis will be specified as that parameter value closest to the alternative in one-tailed tests, and as the complementary (or only unspecified) value in two-tailed tests.

Example: H_0: $\mu = 2{,}400$

The rejection region for a two-tailed test differs from that for a one-tailed test. When we are trying to detect departure from the null hypothesis in *either* direction, we must establish a rejection region in both tails of the sampling distribution of the test statistic. Figures 8.4a and 8.4b show the one-tailed rejection regions for lower- and upper-tailed tests, respectively. The two-tailed rejection region is illustrated in Figure 8.4c. Note that a rejection region is established in each tail of the sampling distribution for a two-tailed test.

The rejection regions corresponding to typical values selected for α are shown in Table 8.2 for one- and two-tailed tests. Note that the smaller α you select, the more evidence (the larger z) you will need before you can reject H_0.

TABLE 8.2   Rejection Regions for Common Values of α

	ALTERNATIVE HYPOTHESES		
	Lower-Tailed	**Upper-Tailed**	**Two-Tailed**
$\alpha = .10$	$z < -1.28$	$z > 1.28$	$z < -1.645$ or $z > 1.645$
$\alpha = .05$	$z < -1.645$	$z > 1.645$	$z < -1.96$ or $z > 1.96$
$\alpha = .01$	$z < -2.33$	$z > 2.33$	$z < -2.575$ or $z > 2.575$

EXAMPLE 8.1

The effect of drugs and alcohol on the nervous system has been the subject of considerable research recently. Suppose a research neurologist is testing the effect of a drug on response time by injecting 100 rats with a unit dose of the drug, subjecting each to a neurological stimulus, and recording its response time. The neurologist knows that the mean response time for rats not injected with the drug (the "control" mean) is 1.2 seconds. She wishes to test whether the mean response time for drug-injected rats differs from 1.2 seconds. Set up the test of hypothesis for this experiment, using $\alpha = .01$.

Solution

Since the neurologist wishes to detect whether the mean response time, μ, for drug-injected rats differs from the control mean of 1.2 seconds in *either* direction—that is, $\mu < 1.2$ or $\mu > 1.2$—we conduct a two-tailed statistical test. Following the procedure for selecting the null and alternative hypotheses, we specify as the alternative hypothesis that the mean differs from 1.2 seconds, since determining whether the drug-injected mean differs from the control mean is the purpose of the experiment. The null hypothesis is the presumption that drug-injected rats have the same mean response time as control rats unless the research indicates otherwise. Thus,

$$H_0: \mu = 1.2$$
$$H_a: \mu \neq 1.2 \qquad (\text{i.e., } \mu < 1.2 \text{ or } \mu > 1.2)$$

The test statistic measures the number of standard deviations between the observed value of $\bar{x}$ and the null hypothesized value $\mu = 1.2$:

$$\text{Test statistic:} \quad z = \frac{\bar{x} - 1.2}{\sigma_{\bar{x}}}$$

The rejection region must be designated to detect a departure from $\mu = 1.2$ in *either* direction, so we will reject H_0 for values of z that are either too small (negative) or too large (positive). To determine the precise values of z that comprise the rejection region, we first select α, the probability that the test will lead to incorrect rejection of the null hypothesis. Then we divide α equally between the lower and upper tail of the distribution of z, as shown in Figure 8.5. In this example, $\alpha = .01$, so $\alpha/2 = .005$ is placed in each tail. The areas in the tails correspond to $z = -2.575$ and $z = 2.575$, respectively (from Table 8.2):

$$\text{Rejection region:} \quad z < -2.575 \text{ or } z > 2.575 \qquad (\text{see Figure 8.5})$$

Assumptions: Since the sample size of the experiment is large enough ($n > 30$), the Central Limit Theorem will apply, and no assumptions need be made about the population of response time measurements. The sampling distribution of the sample mean response of 100 rats will be approximately normal regardless of the distribution of the individual rats' response times. ▲

FIGURE 8.5

Two-tailed rejection region: $\alpha = .01$

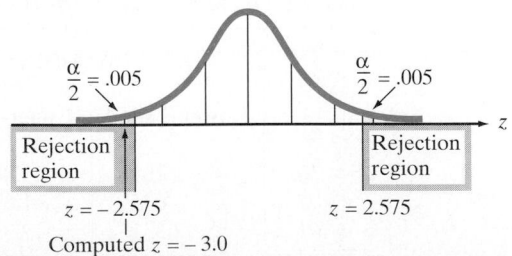

Note that the test in Example 8.1 is set up *before* the sampling experiment is conducted. The data are not used to develop the test. Evidently, the neurologist wants to conclude that the mean response time for the drug-injected rats differs from the control mean only when the evidence is very convincing, because the value of α has been set quite low at .01. If the experiment results in the rejection of H_0, she can be 99% confident that the mean response time of the drug-injected rats differs from the control mean.

Once the test is set up, she is ready to perform the sampling experiment and conduct the test. The test is performed in Example 8.2.

EXAMPLE 8.2

Refer to the neurological response-time test set up in Example 8.1. Suppose the sampling experiment is conducted with the following results:

$$n = 100 \text{ response times for drug-injected rats}$$

$$\bar{x} = 1.05 \text{ seconds} \qquad s = .5 \text{ second}$$

Use the results of the sampling experiment to conduct the test of hypothesis.

Solution

Since the test is completely specified in Example 8.1, we simply substitute the sample statistics into the test statistic:

$$z = \frac{\bar{x} - 1.2}{\sigma_{\bar{x}}} = \frac{\bar{x} - 1.2}{\sigma / \sqrt{n}} = \frac{1.05 - 1.2}{\sigma / \sqrt{100}}$$

$$\approx \frac{1.05 - 1.2}{s/10} = \frac{-.15}{.5/10} = -3.0$$

The implication is that the sample mean, 1.05, is (approximately) 3 standard deviations below the null hypothesized value of 1.2 in the sampling distribution of $\bar{x}$. You can see in Figure 8.5 that this value of z is in the lower-tail rejection region, which consists of all values of $z < -2.575$. This sampling experiment provides sufficient evidence to reject H_0 and conclude, at the $\alpha = .01$ level of significance, that the mean response time for drug-injected rats differs from the control mean of 1.2 seconds. It appears that the rats receiving an injection of this drug have a mean response time that is less than 1.2 seconds. ▲

Two final points about the test of hypothesis in Example 8.2 apply to all statistical tests:

1. Since z is less than -2.575, it is tempting to state our conclusion at a significance level lower than $\alpha = .01$. We resist the temptation because the level of α is determined *before* the sampling experiment is performed. If we decide that we are willing to tolerate a 1% Type I error rate, the result of the sampling experiment should have no effect on that decision. *In general, the same data should not be used both to set up and to conduct the test.*

2. When we state our conclusion at the .01 level of significance, we are referring to the failure rate of the *procedure,* not the result of this particular test. We know that the test procedure will lead to the rejection of the null hypothesis only 1% of the time when in fact $\mu = 1.2$. *Therefore, when the test statistic falls in the rejection region, we infer that the alternative $\mu \neq 1.2$ is true and express our confidence in the procedure by quoting the α level of significance, or the $100(1 - \alpha)\%$ confidence level.*

The setup of a large-sample test of hypothesis about a population mean is summarized in the following box. Both the one- and two-tailed tests are shown.

Large-Sample Test of Hypothesis About μ

ONE-TAILED TEST	TWO-TAILED TEST
H_0: $\mu = \mu_0$*	H_0: $\mu = \mu_0$*
H_a: $\mu < \mu_0$	H_a: $\mu \neq \mu_0$
(or H_a: $\mu > \mu_0$)	

Test statistic: $z = \dfrac{\bar{x} - \mu_0}{\sigma_{\bar{x}}}$ Test statistic: $z = \dfrac{\bar{x} - \mu_0}{\sigma_{\bar{x}}}$

Rejection region: $z < -z_\alpha$ Rejection region: $z < -z_{\alpha/2}$
(or $z > z_\alpha$ when H_a: $\mu > \mu_0$)   or $z > z_{\alpha/2}$

where z_α is chosen so that   where $z_{\alpha/2}$ is chosen so that
$P(z > z_\alpha) = \alpha$ $P(z > z_{\alpha/2}) = \alpha/2$

Assumptions: No assumptions need to be made about the probability distribution of the population because the Central Limit Theorem assures us that, for large samples, the test statistic will be approximately normally distributed regardless of the shape of the underlying probability distribution of the population.

*Note: μ_0 is the symbol for the numerical value assigned to μ under the null hypothesis.

Once the test has been set up, the sampling experiment is performed and the test statistic is calculated. The next box contains possible conclusions for a test of hypothesis, depending on the result of the sampling experiment.

Possible Conclusions for a Test of Hypothesis

1. If the calculated test statistic falls in the rejection region, reject H_0 and conclude that the alternative hypothesis H_a is true. State that you are rejecting H_0 at the α level of significance. Remember that the confidence is in the testing *process*, not the particular result of a single test.
2. If the test statistic does not fall in the rejection region, conclude that the sampling experiment does not provide sufficient evidence to reject H_0 at the α level of significance. [Generally, we will not "accept" the null hypothesis unless the probability β of a Type II error has been calculated (see optional Section 8.6).]

EXERCISES 8.11–8.23

Learning the Mechanics

8.11 For each of the following rejection regions, sketch the sampling distribution for z and indicate the location of the rejection region.
  a. $z > 1.96$ **b.** $z > 1.645$
  c. $z > 2.575$ **d.** $z < -1.28$
  e. $z < -1.645$ or $z > 1.645$
  f. $z < -2.575$ or $z > 2.575$
  g. For each of the rejection regions specified in parts **a–f**, what is the probability that a Type I error will be made?

8.12 Suppose you are interested in conducting the statistical test of H_0: $\mu = 200$ against H_a: $\mu > 200$, and you have decided to use the following decision rule: Reject H_0 if the sample mean of a random sample of 100 items is more than 215. Assume that the standard deviation of the population is 80.
  a. Express the decision rule in terms of z.
  b. Find α, the probability of making a Type I error, by using this decision rule.

8.13 A random sample of 100 observations from a population with standard deviation 60 yielded a sample mean of 110.
  a. Test the null hypothesis that $\mu = 100$ against the alternative hypothesis that $\mu > 100$ using α = .05. Interpret the results of the test.

b. Test the null hypothesis that μ = 100 against the alternative hypothesis that μ ≠ 100 using α = .05. Interpret the results of the test.

c. Compare the results of the two tests you conducted. Explain why the results differ.

8.14 A random sample of 64 observations produced the following summary statistics: $\bar{x}$ = .323 and s^2 = .034.

a. Test the null hypothesis that μ = .36 against the alternative hypothesis that μ < .36 using α = .10.

b. Test the null hypothesis that μ = .36 against the alternative hypothesis that μ ≠ .36 using α = .10. Interpret the result.

Applying the Concepts

8.15 Psychologists studying post-traumatic stress disorder (PTSD) often use as subjects Vietnam veterans who were held as prisoners of war (POWs). Very little attention has been focused on PTSD in World War II combat veterans. Recently, however, *Psychological Assessment* (Mar. 1995) published the results of a study of World War II aviators who were captured by German forces after they were shot down. Having located a total of 239 World War II aviator POW survivors, the researchers asked each veteran to participate in the study; 33 responded to the letter of invitation. Each of the 33 POW survivors was administered the Minnesota Multiphasic Personality Inventory, one component of which measures level of PTSD. [*Note:* The higher the score, the higher the level of PTSD.] The aviators produced a mean PTSD score of $\bar{x}$ = 9.00 and a standard deviation of s = 9.32.

a. Set up the null and alternative hypotheses for determining whether the true mean PTSD score of all World War II aviator POWs is less than 16. [*Note:* The value 16 represents the mean PTSD score established for Vietnam POWs.]

b. Conduct the test, part **a**, using α = .10. What are the practical implications of the test?

c. Discuss the representativeness of the sample used in the study and its ramifications.

8.16 A study reported in the *Journal of Occupational and Organizational Psychology* (Dec. 1992) investigated the relationship of employment status to mental health. Each of a sample of 49 unemployed men was given a mental health examination using the General Health Questionnaire (GHQ). The GHQ is a widely recognized measure of present mental health, with lower values indicating better mental health. The mean and standard deviation of the GHQ scores were $\bar{x}$ = 10.94 and s = 5.10, respectively.

a. Specify the appropriate null and alternative hypotheses if we wish to test the research hypothesis that the mean GHQ score for all unemployed men exceeds 10. Is the test one-tailed or two-tailed? Why?

b. If we specify α = .05, what is the appropriate rejection region for this test?

c. Conduct the test, and state your conclusion clearly in the language of this exercise.

8.17 Humerus bones from the same species of animal tend to have approximately the same length-to-width ratios. When fossils of humerus bones are discovered, archeologists can often determine the species of animal by examining the length-to-width ratios of the bones. It is known that species A exhibits a mean ratio of 8.5. Suppose 41 fossils of humerus bones were unearthed at an archeological site in East Africa, where species A is believed to have lived. (Assume that the unearthed bones were all from the same unknown species.) The length-to-width ratios of the bones were calculated and listed, as shown in the following table. Summary statistics for the data set are shown in the SAS printout at the bottom of the page.

a. Test whether the population mean ratio of all bones of this particular species differs from 8.5. Use α = .01.

b. What are the practical implications of the test, part **a**?

Length-to-Width Ratios of a Sample of Humerus Bones

10.73	8.89	9.07	9.20	10.33	9.98	9.84	9.59
8.48	8.71	9.57	9.29	9.94	8.07	8.37	6.85
8.52	8.87	6.23	9.41	6.66	9.35	8.86	9.93
8.91	11.77	10.48	10.39	9.39	9.17	9.89	8.17
8.93	8.80	10.02	8.38	11.67	8.30	9.17	12.00
9.38

8.18 What factors inhibit the learning process in the classroom? To answer this question, researchers at Murray State University surveyed 40 students from a senior-level marketing class (*Marketing Education Review*, Fall 1994). Each student was given a list of factors and asked to rate the extent to which each factor inhibited the learning process in courses offered in their department. A 7-point rating scale was used, where 1 = "not at all" and 7 = "to a great extent." The factor with the highest rating was instructor-related: "Professors who place too much emphasis on a single right answer rather than overall thinking

```
Analysis Variable : LWRATIO

N Obs      Minimum       Maximum         Mean      Std Dev
----------------------------------------------------------
   41     6.2300000    12.0000000    9.2575610    1.2035651
----------------------------------------------------------
```

and creative ideas." Summary statistics for the student ratings of this factor are $\bar{x} = 4.70$ and $s = 1.62$.

a. Conduct a test to determine if the true mean rating for this instructor-related factor exceeds 4. Use $\alpha = .05$. Interpret the test results.

b. Because the variable of interest, rating, is measured on a 7-point scale, it is unlikely that the population of ratings will be normally distributed. Consequently, some analysts may perceive the test, part **a**, to be invalid and search for alternative methods of analysis. Defend or refute this position.

8.19 The pain reliever currently used in a hospital is known to bring relief to patients in a mean time of 3.5 minutes. To compare a new pain reliever with the current one, the new drug is administered to a random sample of 50 patients. The mean time to relief for the sample of patients is 2.8 minutes and the standard deviation is 1.14 minutes. Do the data provide sufficient evidence to conclude that the new drug was effective in reducing the mean time until a patient receives relief from pain? Test using $\alpha = .10$.

8.20 The introduction of printed circuit boards (PCBs) in the 1950s revolutionized the electronics industry. However, solder-joint defects on PCBs have plagued electronics manufacturers since the introduction of the PCB. A single PCB may contain thousands of solder joints. Current technology uses X-rays and lasers for inspection (*Quality Congress Transactions,* 1986). A particular manufacturer of laser-based inspection equipment claims that its product can inspect on average at least 10 solder joints per second when the joints are spaced .1 inch apart. The equipment was tested by a potential buyer on 48 different PCBs. In each case, the equipment was operated for exactly 1 second. The number of solder joints inspected on each run follows:

```
10   9  10  10  11   9  12   8   8   9   6  10
 7  10  11   9   9  13   9  10  11  10  12   8
 9   9   9   7  12   6   9  10  10   8   7   9
11  12  10   0  10  11  12   9   7   9   9  10
```

a. The potential buyer wants to know whether the sample data refute the manufacturer's claim. Specify the null and alternative hypotheses that the buyer should test.

b. In the context of this exercise, what is a Type I error? A Type II error?

c. Conduct the hypothesis test you described in part **a**, and interpret the test's results in the context of this exercise. Use $\alpha = .05$ and the SPSS descriptive statistics printout at the bottom of the page.

8.21 *Environmental Science & Technology* (Oct. 1993) reported on a study of contaminated soil in The Netherlands. Seventy-two 400-gram soil specimens were sampled, dried, and analyzed for the contaminant cyanide. The cyanide concentration [in milligrams per kilogram (mg/kg) of soil] of each soil specimen was determined using an infrared spectroscopic method. The sample resulted in a mean cyanide level of $\bar{x} = 84$ mg/kg and a standard deviation of $s = 80$ mg/kg. Use this information to test the hypothesis that the true mean cyanide level in soil in The Netherlands exceeds 100 mg/kg. Test at $\alpha = .10$.

8.22 Nutritionists stress that weight control generally requires significant reductions in the intake of fat. A random sample of 64 middle-aged men on weight-control programs is selected to determine whether their mean intake of fat exceeds the recommended 30 grams per day. The sample mean and standard deviation are $\bar{x} = 37$ and $s = 32$, respectively.

a. Considering the sample mean and standard deviation, would you expect the distribution for fat intake per day to be symmetric or skewed? Explain.

b. Do the sample results indicate that the mean intake for middle-aged men on weight-control programs exceeds 30 grams? Test using $\alpha = .10$.

c. Would you reach the same conclusion as in part **b** using $\alpha = .05$? Using $\alpha = .01$? Why can the conclusion of a test change when the value of α is changed?

8.23 Refer to the *Science* (Nov. 1988) study of inbreeding in tropical swarm-founding wasps, Exercise 7.17. A sample of 197 wasps, captured, frozen, and subjected to a series of genetic tests, yielded a sample mean inbreeding coefficient of $\bar{x} = .044$ with a standard deviation of $s = .884$. Recall that if the wasp has no tendency to inbreed, the true mean inbreeding coefficient μ for the species will equal 0.

a. Test the hypothesis that the true mean inbreeding coefficient μ for this species of wasp exceeds 0. Use $\alpha = .05$.

b. Compare the inference, part **a**, to the inference obtained in Exercise 7.17 using a confidence interval. Do the inferences agree? Explain.

Variable	Mean	Std Dev	Minimum	Maximum	N	Label
NUMBER	9.29	2.10	.00	13.00	48	

8.3   OBSERVED SIGNIFICANCE LEVELS: *p*-VALUES

According to the statistical test procedure described in Section 8.2, the rejection region and, correspondingly, the value of α are selected prior to conducting the test, and the conclusions are stated in terms of rejecting or not rejecting the null hypothesis. A second method of presenting the results of a statistical test is one that reports the extent to which the test statistic disagrees with the null hypothesis and leaves to the reader the task of deciding whether to reject the null hypothesis. This measure of disagreement is called the *observed significance level* (or *p-value*) for the test.

> **DEFINITION 8.1**
>
> The **observed significance level**, or *p* **value**, for a specific statistical test is the probability (assuming H_0 is true) of observing a value of the test statistic that is at least as contradictory to the null hypothesis, and supportive of the alternative hypothesis, as the actual one computed from the sample data.

For example, the value of the test statistic computed for the sample of $n = 50$ sections of sewer pipe was $z = 2.12$. Since the test is one-tailed—i.e., the alternative (research) hypothesis of interest is H_a: $\mu > 2,400$—values of the test statistic even more contradictory to H_0 than the one observed would be values larger than $z = 2.12$. Therefore, the observed significance level (*p*-value) for this test is

$$p\text{-value} = P(z \geq 2.12)$$

or, equivalently, the area under the standard normal curve to the right of $z = 2.12$ (see Figure 8.6).

The area A in Figure 8.6 is given in Table IV in Appendix A as .4830. Therefore, the upper-tail area corresponding to $z = 2.12$ is

$$p\text{-value} = .5 - .4830 = .0170$$

Consequently, we say that these test results are "very significant"; i.e., they disagree rather strongly with the null hypothesis, H_0: $\mu = 2,400$, and favor H_a: $\mu > 2,400$. The probability of observing a z value as large as 2.12 is only .0170, if in fact the true value of μ is 2,400.

If you are inclined to select $\alpha = .05$ for this test, then you would reject the null hypothesis because the *p*-value for the test, .0170, is less than .05. In contrast, if you choose $\alpha = .01$, you would not reject the null hypothesis because the *p*-value for the test is larger than .01. Thus, the use of the observed significance level is identical to the test procedure described in the preceding sections except that the choice of α is left to you.

The steps for calculating the *p*-value corresponding to a test statistic for a population mean are given in the next box.

FIGURE 8.6

Finding the *p*-value for an upper-tailed test when $z = 2.12$

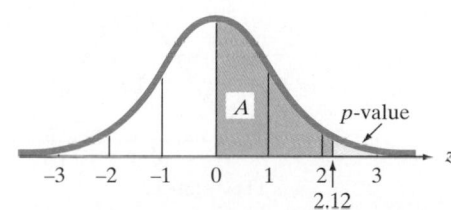

FIGURE 8.7

Finding the *p*-value for a
one-tailed test

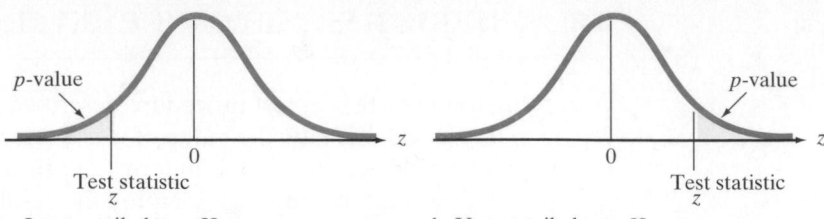

a. Lower–tailed test, H_a: $\mu < \mu_0$ b. Upper–tailed test, H_a: $\mu > \mu_0$

Steps for Calculating the *p*-Value for a Test of Hypothesis

1. Determine the value of the test statistic z corresponding to the result of the sampling experiment.
2. a. If the test is one-tailed, the *p*-value is equal to the tail area beyond z in the same direction as the alternative hypothesis. Thus, if the alternative hypothesis is of the form $>$, the *p*-value is the area to the right of, or above, the observed z value. Conversely, if the alternative is of the form $<$, the *p*-value is the area to the left of, or below, the observed z value. (See Figure 8.7.)
   b. If the test is two-tailed, the *p*-value is equal to twice the tail area beyond the observed z value in the direction of the sign of z. That is, if z is positive, the *p*-value is twice the area to the right of, or above, the observed z value. Conversely, if z is negative, the *p*-value is twice the area to the left of, or below, the observed z value. (See Figure 8.8.)

EXAMPLE 8.3   Find the observed significance level for the test of the mean response time for drug-injected rats in Examples 8.1 and 8.2.

Solution

Example 8.1 presented a two-tailed test of the hypothesis

$$H_0: \mu = 1.2 \text{ seconds}$$

against the alternative hypothesis

$$H_a: \mu \neq 1.2 \text{ seconds}$$

The observed value of the test statistic in Example 8.2 was $z = -3.0$, and any value of z less than -3.0 or greater than $+3.0$ (because this is a two-tailed test) would be even more contradictory to H_0. Therefore, the observed significance level for the test is

$$p\text{-value} = P(z < -3.0 \text{ or } z > +3.0)$$

FIGURE 8.8

Finding the *p*-value for a
two-tailed test: *p*-value =
$2(p/2)$

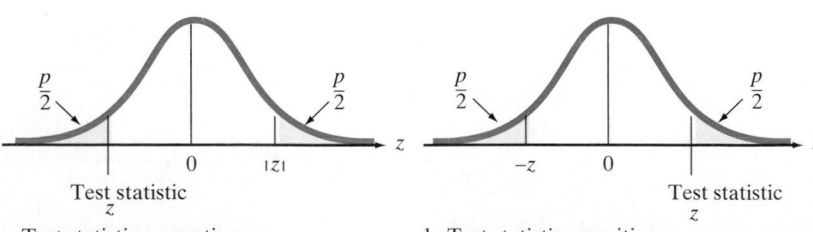

a. Test statistic z negative   b. Test statistic z positive

Thus, we calculate the area below the observed z value, $z = -3.0$, and double it. Consulting Table IV in Appendix A, we find that $P(z < -3.0) = .5 - .4987 = .0013$. Therefore, the p-value for this two-tailed test is

$$2P(z < -3.0) = 2(.0013) = .0026$$

We can interpret this p-value as a strong indication that the mean reaction time of drug-injected rats differs from the control mean ($\mu \neq 1.2$), since we would observe a test statistic this extreme or more extreme only 26 in 10,000 times if the drug-injected mean were equal to the control mean ($\mu = 1.2$). The extent to which the mean differs from 1.2 could be better determined by calculating a confidence interval for μ.   ▲

When publishing the results of a statistical test of hypothesis in journals, case studies, reports, etc., many researchers make use of p values. Instead of selecting α beforehand and then conducting a test, as outlined in this chapter, the researcher computes (usually with the aid of a statistical software package) and reports the value of the appropriate test statistic and its associated p-value. It is left to the reader of the report to judge the significance of the result—i.e., the reader must determine whether to reject the null hypothesis in favor of the alternative hypothesis, based on the reported p-value. Usually, the null hypothesis is rejected if the observed significance level is *less than* the fixed significance level, α, chosen by the reader. The inherent advantage of reporting test results in this manner is twofold: (1) Readers are permitted to select the maximum value of α that they would be willing to tolerate if they actually carried out a standard test of hypothesis in the manner outlined in this chapter, and (2) a measure of the degree of significance of the result (i.e., the p-value) is provided.

Reporting Test Results as *p*-Values: How to Decide Whether to Reject H_0

1. Choose the maximum value of α that you are willing to tolerate.
2. If the observed significance level (p-value) of the test is less than the chosen value of α, reject the null hypothesis. Otherwise, do not reject the null hypothesis.

EXAMPLE 8.4

The lengths of stay (in days) for 100 randomly selected hospital patients, first presented in Table 7.1, are reproduced in Table 8.3. Suppose we want to test the hypothesis that the true mean length of stay (LOS) at the hospital is less than 5 days, i.e.,

TABLE 8.3   Lengths of Stay for 100 Hospital Patients

2	3	8	6	4	4	6	4	2	5
8	10	4	4	4	2	1	3	2	10
1	3	2	3	4	3	5	2	4	1
2	9	1	7	17	9	9	9	4	4
1	1	1	3	1	6	3	3	2	5
1	3	3	14	2	3	9	6	6	3
5	1	4	6	11	22	1	9	6	5
2	2	5	4	3	6	1	5	1	6
17	1	2	4	5	4	4	3	2	3
3	5	2	3	3	2	10	2	4	2

FIGURE 8.9

MINITAB printout for the lower-tailed test in Example 8.4

```
TEST OF MU = 5.000 VS MU L.T. 5.000
THE ASSUMED SIGMA = 3.68

              N      MEAN    STDEV    SE MEAN       Z    P VALUE
LOS         100     4.530    3.678     0.368    -1.28       0.10
```

$$H_0: \mu = 5$$
$$H_a: \mu < 5$$

Use the data in the table to conduct the test at $\alpha = .05$.

Solution

Instead of conducting the test by hand, we will use a statistical software package. The data were entered into a computer and MINITAB was used to conduct the analysis. The MINITAB printout for the lower-tailed test is displayed in Figure 8.9. Both the test statistic, $z = 1.28$, and p-value of the test, $p = .10$, are shaded on the MINITAB printout. Since the p-value exceeds our selected α value, $\alpha = .05$, we cannot reject the null hypothesis. Hence, there is insufficient evidence (at $\alpha = .05$) to conclude that the true mean LOS at the hospital is less than 5 days.   ▲

Note: MINITAB provides an option for selecting one-tailed or two-tailed tests and reports the appropriate p-value. Some statistical software packages such as SAS and SPSS will conduct only two-tailed tests of hypothesis. For these packages, you obtain the p-value for a one-tailed test as follows:

$$p = \frac{\text{Reported } p\text{-value}}{2} \quad \text{if} \begin{cases} H_a \text{ is of form} > \text{ and } z \text{ is positive} \\ H_a \text{ is of form} < \text{ and } z \text{ is negative} \end{cases}$$

$$p = 1 - \left(\frac{\text{Reported } p\text{-value}}{2}\right) \quad \text{if} \begin{cases} H_a \text{ is of form} > \text{ and } z \text{ is negative} \\ H_a \text{ is of form} < \text{ and } z \text{ is positive} \end{cases}$$

EXERCISES 8.24–8.38

Learning the Mechanics

8.24 If a hypothesis test were conducted using $\alpha = .05$, for which of the following p-values would the null hypothesis be rejected?
a. .06 **b.** .10 **c.** .01 **d.** .001 **e.** .251 **f.** .042

8.25 For each α and observed significance level (p-value) pair, indicate whether the null hypothesis would be rejected.
a. $\alpha = .05, p\text{-value} = .10$
b. $\alpha = .10, p\text{-value} = .05$
c. $\alpha = .01, p\text{-value} = .001$
d. $\alpha = .025, p\text{-value} = .05$
e. $\alpha = .10, p\text{-value} = .45$

8.26 An analyst tested the null hypothesis $\mu \geq 20$ against the alternative hypothesis that $\mu < 20$. The analyst reported a p-value of .06. What is the smallest value of α for which the null hypothesis would be rejected?

8.27 In a test of $H_0: \mu = 100$ against $H_a: \mu > 100$, the sample data yielded the test statistic $z = 2.17$. Find the p-value for the test.

8.28 In a test of $H_0: \mu = 100$ against $H_a: \mu \neq 100$, the sample data yielded the test statistic $z = 2.17$. Find the p-value for the test.

8.29 In a test of the hypothesis $H_0: \mu = 50$ versus $H_a: \mu > 50$, a sample of $n = 100$ observations possessed mean $\bar{x} = 49.4$ and standard deviation $s = 4.1$. Find and interpret the p-value for this test.

8.30 In a test of the hypothesis $H_0: \mu = 10$ versus $H_a: \mu \neq 10$, a sample of $n = 50$ observations possessed mean $\bar{x} = 10.7$ and standard deviation $s = 3.1$. Find and interpret the p-value for this test.

8.31 Consider a test of $H_0: \mu = 75$ performed using the computer. SAS reports a two-tailed p-value of .1032. Make the appropriate conclusion for each of the following situations:

a. H_a: $\mu < 75$, $z = -1.63$, $\alpha = .05$
b. H_a: $\mu < 75$, $z = 1.63$, $\alpha = .10$
c. H_a: $\mu > 75$, $z = 1.63$, $\alpha = .10$
d. H_a: $\mu \neq 75$, $z = -1.63$, $\alpha = .01$

Applying the Concepts

8.32 Refer to the *Psychological Assessment* study of World War II aviator POWs, Exercise 8.15. You tested whether the true mean post-traumatic stress disorder score of World War II aviator POWs is less than 16. Recall that $\bar{x} = 9.00$ and $s = 9.32$ for a sample of $n = 33$ POWs. Compute the *p*-value of the test and interpret the result.

8.33 The manufacturer of an over-the-counter analgesic claims that its product brings pain relief to headache sufferers in less than 3.5 minutes, on average. In order to be able to make this claim in its television advertisements, the manufacturer was required by a particular television network to present statistical evidence in support of the claim. The manufacturer reported that for a random sample of 50 headache sufferers, the mean time to relief was 3.3 minutes and the standard deviation was 1.1 minutes.
  a. Do these data support the manufacturer's claim? Test using $\alpha = .05$.
  b. Report the *p*-value of the test.
  c. In general, do large *p*-values or small *p*-values support the manufacturer's claim? Explain.

8.34 Refer to Exercise 8.16, in which a random sample of 49 unemployed men were administered the General Health Questionnaire (GHQ). The sample mean and standard deviation were 10.94 and 5.10, respectively. Denoting the population mean GHQ for unemployed workers by μ, we wish to test the null hypothesis H_0: $\mu = 10$ versus the one-tailed alternative H_a: $\mu > 10$.
  a. When the data are run through MINITAB, the results (in part) are as shown below. Check the program's results for accuracy.

```
TEST OF MU = 10.00 VS MU G.T.  10.00
THE ASSUMED SIGMA = 5.10

       N    MEAN   STDEV   SE MEAN     Z    P VALUE
GHQ   49   10.94   5.10     0.73     1.29    .0985
```

  b. What conclusion would you reach about the test based on the computer analysis?

8.35 In Exercise 7.13 we examined research about bicycle helmets reported in *Public Health Reports* (May–

June 1992). One of the variables measured was the children's perception of the risk involved in bicycling. A random sample of 797 children in grades 4–6 were asked to rate their perception of bicycle risk without wearing a helmet, ranging from 1 (no risk) to 4 (very high risk). The mean and standard deviation of the sample were $x = 3.39$ and $s = .80$, respectively.
  a. Assume that a mean score, μ, of 2.5 is indicative of indifference to risk, and values of μ exceeding 2.5 indicate a perception that a risk exists. What are the appropriate null and alternative hypotheses for testing the research hypothesis that children in this age group perceive a risk associated with failure to wear helmets?
  b. Calculate the *p*-value for the data collected in this study.
  c. Interpret the *p*-value in the context of this research.

8.36 In Exercise 8.17 you tested H_0: $\mu = 8.5$ versus H_a: $\mu \neq 8.5$, where μ is the population mean length-to-width ratio of humerus bones of a particular species of animal. A SAS printout for the hypothesis test is shown at the bottom of the page. Locate the *p*-value on the printout and interpret its value.

8.37 Golf-course designers are concerned that old courses are becoming obsolete because new equipment enables golfers to hit the ball so far: In effect, golf courses are "shrinking." One designer believes that new courses need to be built with the expectation that players will be able to hit the ball more than 250 yards (with their drivers), on average. Suppose a sample of 135 golfers is tested, and their mean driving distance is 256.3 yards, with a standard deviation of 43.4 yards.
  a. What are appropriate null and alternative hypotheses to test the designer's research hypothesis?
  b. Calculate and interpret the *p*-value for this test.
  c. If you were to make this a two-tailed test, how would your answer to part **b** change?

8.38 In Exercise 8.20 you tested H_0: $\mu \geq 10$ versus H_a: $\mu < 10$, where μ is the average number of solder joints inspected per second when the joints are spaced .1 inch apart. An SPSS printout of the hypothesis test is shown on page 308.
  a. Locate the two-tailed *p*-value of the test shown on the printout.
  b. Adjust the *p*-value for the one-tailed test (if necessary) and interpret its value.

Analysis Variable : RATIO_85				
N Obs	Mean	Std Dev	T	Prob>|T|
41	0.7575610	1.2035651	4.0303238	0.0002

Variable	Number of Cases	Mean	Standard Deviation	Standard Error			
NUMBER	48	9.2917	2.103	.304			
MU	48	10.0000	.	.			

(Difference) Mean	Standard Deviation	Standard Error	t Value	Degrees of Freedom	2-Tail Prob.
−.7083	2.103	.304	−2.33	47	.024

8.4   SMALL-SAMPLE TEST OF HYPOTHESIS ABOUT A POPULATION MEAN

Most water-treatment facilities monitor the quality of their drinking water on an hourly basis. One variable monitored is pH, which measures the degree of alkalinity or acidity in the water. A pH below 7.0 is acidic, one above 7.0 is alkaline, and a pH of 7.0 is neutral. One water-treatment plant has a target pH of 8.5 (most try to maintain a slightly alkaline level). The mean and standard deviation of 1 hour's test results, based on 17 water samples at this plant, are

$$\bar{x} = 8.42 \qquad s = .16$$

Does this sample provide sufficient evidence that the mean pH level in the water differs from 8.5?

This inference can be placed in a test of hypothesis framework. We establish the target pH as the null hypothesized value and then utilize a two-tailed alternative that the true mean pH differs from the target:

$$H_0: \mu = 8.5$$
$$H_a: \mu \neq 8.5$$

Recall from Section 7.3 that when we are faced with making inferences about a population mean using the information in a small sample, two problems emerge:

1. The normality of the sampling distribution for $\bar{x}$ does not follow from the Central Limit Theorem when the sample size is small. We must assume that the distribution of measurements from which the sample was selected is approximately normally distributed in order to ensure the approximate normality of the sampling distribution of $\bar{x}$.

2. If the population standard deviation σ is unknown, as is usually the case, then we cannot assume that s will provide a good approximation for σ when the sample size is small. Instead, we must use the t-distribution rather than the standard normal z-distribution to make inferences about the population mean μ.

Therefore, as the test statistic of a small-sample test of a population mean, we use the t statistic:

$$\textit{Test statistic:} \quad t = \frac{\bar{x} - \mu_0}{s/\sqrt{n}} = \frac{\bar{x} - 8.5}{s/\sqrt{n}}$$

where μ_0 is the null hypothesized value of the population mean, μ. In our example, $\mu_0 = 8.5$.

To find the rejection region, we must specify the value of α, the probability that the test will lead to rejection of the null hypothesis when it is true, and then

FIGURE 8.10

Two-tailed rejection region for small-sample *t*-test

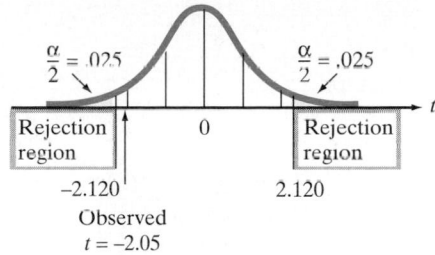

consult the *t* table (Table VI of Appendix A). Using $\alpha = .05$, the two-tailed rejection region is

> *Rejection region:* $t_{\alpha/2} = t_{.025} = 2.120$ with $n - 1 = 16$ degrees of freedom
> Reject H_0 if $t < -2.120$ or $t > 2.120$

The rejection region is shown in Figure 8.10.

We are now prepared to calculate the test statistic and reach a conclusion:

$$t = \frac{\bar{x} - \mu_0}{s / \sqrt{n}} = \frac{8.42 - 8.50}{.16 / \sqrt{17}} = \frac{-.08}{.039} = -.025$$

Since the calculated value of *t* does not fall in the rejection region (Figure 8.10), we cannot reject H_0 at the $\alpha = .05$ level of significance. Thus, the water-treatment plant should not conclude that the mean pH differs from the 8.5 target based on the sample evidence.

It is interesting to note that the calculated *t* value, -2.05, is *less than* the .05 level *z* value, -1.96. The implication is that if we had *incorrectly* used a *z* statistic for this test, we would have rejected the null hypothesis at the .05 level, concluding that the mean pH level differs from 8.5. The important point is that the statistical procedure to be used must always be closely scrutinized and all the assumptions understood. Many statistical lies are the result of misapplications of otherwise valid procedures.

The technique for conducting a small-sample test of hypothesis about a population mean is summarized in the following box.

Small-Sample Test of Hypothesis About μ

ONE-TAILED TEST	TWO-TAILED TEST
H_0: $\mu = \mu_0$	H_0: $\mu = \mu_0$
H_a: $\mu < \mu_0$	H_a: $\mu \neq \mu_0$
(or H_a: $\mu > \mu_0$)	
Test statistic: $t = \dfrac{\bar{x} - \mu_0}{s / \sqrt{n}}$	*Test statistic:* $t = \dfrac{\bar{x} - \mu_0}{s / \sqrt{n}}$
Rejection region: $t < -t_\alpha$	*Rejection region:* $t < -t_{\alpha/2}$
(or $t > t_\alpha$ when H_a: $\mu > \mu_0$)	or $t > t_{\alpha/2}$

where t_α and $t_{\alpha/2}$ are based on $(n - 1)$ degrees of freedom

Assumption: A random sample is selected from a population with a relative frequency distribution that is approximately normal.

FIGURE 8.11

A *t*-distribution with 9 df
and the rejection region for
Example 8.5

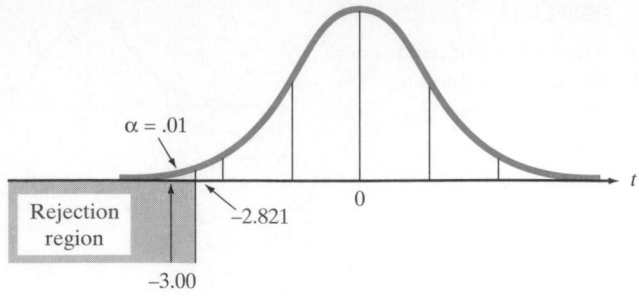

EXAMPLE 8.5

A major car manufacturer wants to test a new engine to determine whether it meets new air-pollution standards. The mean emission μ of all engines of this type must be less than 20 parts per million of carbon. Ten engines are manufactured for testing purposes, and the emission level of each are determined. The data (in parts per million) are listed below:

| 15.6 | 16.2 | 22.5 | 20.5 | 16.4 | 19.4 | 16.6 | 17.9 | 12.7 | 13.9 |

Do the data supply sufficient evidence to allow the manufacturer to conclude that this type of engine meets the pollution standard? Assume that the manufacturer is willing to risk a Type I error with probability α = .01.

Solution

The manufacturer wants to support the research hypothesis that the mean emission level μ for all engines of this type is less than 20 parts per million. The elements of this small-sample one-tailed test are

$$H_0: \mu = 20$$
$$H_a: \mu < 20$$
$$\text{Test statistic:} \quad t = \frac{\bar{x} - 20}{s/\sqrt{n}}$$

Assumption: The relative frequency distribution of the population of emission levels for all engines of this type is approximately normal.

Rejection region: For α = .01 and df = $n - 1 = 9$, the one-tailed rejection region (see Figure 8.11) is $t < -t_{.01} = -2.821$.

To calculate the test statistic, we entered the data into a computer and analyzed it using SAS. The SAS descriptive statistics printout is shown in Figure 8.12. From the printout, we obtain $\bar{x} = 17.17$, $s = 2.98$. Substituting these values into the test statistic formula, we get

$$t = \frac{\bar{x} - 20}{s/\sqrt{n}} = \frac{17.17 - 20}{2.98/\sqrt{10}} = -3.00$$

Since the calculated *t* falls in the rejection region (see Figure 8.11), the manufacturer concludes that μ < 20 parts per million and the new engine type meets the pollution standard. Are you satisfied with the reliability associated with this inference? The probability is only α = .01 that the test would support the research hypothesis if in fact it were false. ▲

FIGURE 8.12

SAS descriptive statistics
for 10 emission levels

```
Analysis Variable : EMIT

N Obs   N      Minimum       Maximum          Mean       Std Dev
-------------------------------------------------------------------
   10  10    12.7000000    22.5000000    17.1700000     2.9814426
-------------------------------------------------------------------
```

FIGURE 8.13

The observed significance level for the test of Example 8.5

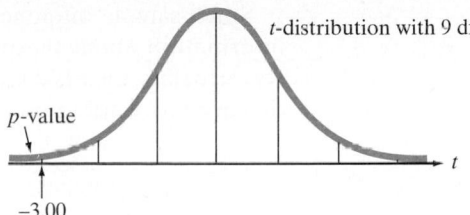

p-value

t-distribution with 9 df

−3.00

t

EXAMPLE 8.6

Find the observed significance level for the test described in Example 8.5. Interpret the result.

Solution

The test of Example 8.5 was a lower-tailed test: H_0: $\mu = 20$ versus H_a: $\mu < 20$. Since the value of t computed from the sample data was $t = -3.00$, the observed significance level (or *p*-value) for the test is equal to the probability that t would assume a value less than or equal to -3.00 if in fact H_0 were true. This is equal to the area in the lower tail of the *t*-distribution (shaded in Figure 8.13).

One way to find this area—i.e., the *p*-value for the test—is to consult the *t*-table (Table VI in Appendix A). Unlike the table of areas under the normal curve, Table VI gives only the *t* values corresponding to the areas .100, .050, .025, .010, .005, .001, and .0005. Therefore, we can only approximate the *p*-value for the test. Since the observed *t* value was based on 9 degrees of freedom, we use the df = 9 row in Table VI and move across the row until we reach the *t* values that are closest to the observed $t = -3.00$. [*Note:* We ignore the minus sign.] The *t* values corresponding to *p*-values of .010 and .005 are 2.821 and 3.250, respectively. Since the observed *t* value falls between $t_{.010}$ and $t_{.005}$, the *p*-value for the test lies between .005 and .010. In other words, $.005 < p\text{-value} < .01$. Thus, we would reject the null hypothesis, H_0: $\mu = 20$ parts per million, for any value of α larger than .01 (the upper bound of the *p*-value).

A second, more accurate, way to obtain the *p*-value is to use a statistical software package to conduct the test of hypothesis. The SAS printout for the test of H_0: $\mu = 20$ is displayed in Figure 8.14. Both the test statistic (-3.00) and *p*-value (.0149) are shaded in Figure 8.14. Recall (from Section 8.4) that SAS conducts, by default, a two-tailed test. That is, SAS tests the alternative H_a: $\mu \neq 20$. Thus, the *p*-value reported on the printout must be adjusted to obtain the appropriate *p*-value for our lower-tailed test. Since the value of the test statistic is negative and H_a is of the form $<$ (i.e., the value of t agrees with the direction specified in H_a), the *p*-value is obtained by dividing the printout value in half:

$$\text{One-tailed } p\text{-value} = \frac{\text{Reported } p\text{-value}}{2} = \frac{.0149}{2} = .00745$$

You can see that the actual *p*-value of the test falls within the bounds obtained from Table VI. Thus, the two methods agree; we will reject H_0: $\mu = 20$ in favor of H_a: $\mu < 20$ for any α level larger than .01.   ▲

FIGURE 8.14

SAS test of H_0: $\mu = 20$ for Example 8.6

```
Analysis Variable : EMIT_20

N Obs          Mean         Std Dev               T   Prob>|T|
   10    -2.8300000       2.9814426      -3.0016495     0.0149
```

Small-sample inferences typically require more assumptions and provide less information about the population parameter than do large-sample inferences. Nevertheless, the *t*-test is a method of testing a hypothesis about a population mean of a normal distribution when only a small number of observations is available.

EXERCISES 8.39–8.53

Learning the Mechanics

8.39 Under what circumstances should you use the *t*-distribution in testing a hypothesis about a population mean?

8.40 In what ways are the distributions of the *z*-statistic and *t*-test statistic alike? How do they differ?

8.41 For each of the following rejection regions, sketch the sampling distribution of *t*, and indicate the location of the rejection region on your sketch:
  a. $t > 1.440$ where df = 6
  b. $t < -1.782$ where df = 12
  c. $t < -2.060$ or $t > 2.060$ where df = 25

8.42 For each of the rejection regions defined in Exercise 8.41, what is the probability that a Type I error will be made?

8.43 A random sample of *n* observations is selected from a normal population to test the null hypothesis that $\mu = 10$. Specify the rejection region for each of the following combinations of H_a, α, and *n:*
  a. H_a: $\mu \neq 10$; $\alpha = .05$; $n = 14$
  b. H_a: $\mu > 10$; $\alpha = .01$; $n = 24$
  c. H_a: $\mu > 10$; $\alpha = .10$; $n = 9$
  d. H_a: $\mu < 10$; $\alpha = .01$; $n = 12$
  e. H_a: $\mu \neq 10$; $\alpha = .10$; $n = 20$
  f. H_a: $\mu < 10$; $\alpha = .05$; $n = 4$

8.44 The following sample of six measurements was randomly selected from a normally distributed population: 1, 3, −1, 5, 1, 2.
  a. Test the null hypothesis that the mean of the population is 3 against the alternative hypothesis, $\mu < 3$. Use $\alpha = .05$.
  b. Test the null hypothesis that the mean of the population is 3 against the alternative hypothesis, $\mu \neq 3$. Use $\alpha = .05$.
  c. Find the observed significance level for each test.

8.45 A sample of five measurements, randomly selected from a normally distributed population, resulted in the following summary statistics: $\bar{x} = 4.8$, $s = 1.3$.
  a. Test the null hypothesis that the mean of the population is 6 against the alternative hypothesis, $\mu < 6$. Use $\alpha = .05$.
  b. Test the null hypothesis that the mean of the population is 6 against the alternative hypothesis, $\mu \neq 6$. Use $\alpha = .05$.
  c. Find the observed significance level for each test.

8.46 MINITAB is used to conduct a *t*-test for the null hypothesis H_0: $\mu = 1,000$ versus the alternative hypothesis H_a: $\mu > 1,000$ based on a sample of 17 observations. The software's output is

```
TEST OF MU = 1,000 VS MU G.T. 1,000

      N   MEAN   STDEV   SE MEAN       T   P VALUE
GHQ  17   1020   43.54     10.56   1.894     .0382
```

  a. What assumptions are necessary for the validity of this procedure?
  b. Interpret the results of the test.
  c. Suppose the alternative hypothesis had been the two-tailed H_a: $\mu \neq 1,000$. If the *t* statistic were unchanged, what would the *p*-value be for this test? Interpret the *p*-value for the two-tailed test.

Applying the Concepts

8.47 Organochlorine pesticides (OCPs), like polychlorinated biphenyls (PCBs), are highly toxic organic compounds that are often found in fish. By law, the levels of OCPs and PCBs in fish must be constantly monitored, so it is important to be able to accurately measure the amounts of these compounds in fish specimens. A new technique, called matrix solid-phase dispersion (MSPD), has been developed for chemically extracting trace organic compounds from solids (*Chromatographia*, Mar. 1995). The MSPD method was tested as follows. Uncontaminated fish fillets were injected with a known amount of OCP or PCB. The MSPD method was then used to extract the contaminant and the percentage of the toxic compound recovered was measured. The recovery percentages for $n = 5$ fish fillets injected with the OCP Aldrin are listed below:

99   102   94   99   95

Do the data provide sufficient evidence to indicate that the mean recovery percentage of Aldrin exceeds 85% using the new MSPD method? Test using $\alpha = .05$.

8.48 In Exercise 7.27 you analyzed data from a study of the diets of spinifex pigeons. The data, extracted from the *Australian Journal of Zoology*, are reproduced in the table below. Each measurement in the data set represents the weight (in grams) of the crop content of a spinifex pigeon. MINITAB was used to conduct a test of the null hypothesis H_0: $\mu = 1$, where μ represents the mean weight in the crops of all spinifex pigeons. The resulting printout accompanies the list of data on page 313.
  a. Identify the alternative hypothesis specified on the printout.

| .457 | 3.751 | .238 | 2.967 | 2.509 | 1.384 | 1.454 | .818 |
| .335 | 1.436 | 1.603 | 1.309 | .201 | .530 | 2.144 | .834 |

Source: Excerpted from Williams, J. B., Bradshaw, D., and Schmidt, L. "Field metabolism and water requirements of spinifex pigeons (*Geophaps plumifera*) in Western Australia." *Australian Journal of Zoology,* Vol. 43, No. 1, 1995, p. 7 (Table 2).

```
TEST OF MU = 1.000 VS MU N.E. 1.000

              N     MEAN   STDEV   SE MEAN      T   P VALUE
WEIGHT       16    1.373   1.034    0.258    1.44      0.17
```

b. Locate the value of the test statistic and give its value.

c. Find and interpret the *p*-value of the test shown on the printout.

8.49 The changing ecology of the Everglades National Park in Florida, considered a national treasure by many, has been the subject of much environmental research. One water-quality parameter of concern in the park is the total phosphorus level. Suppose that the EPA makes 12 measurements in one section of the park, yielding a mean level of total phosphorus at 12.3 parts per billion (ppb) and a standard deviation of 5.4 ppb. The EPA wants to test whether the data support the conclusion that the mean level is less than 15 ppb.

a. What are the null and alternative hypotheses appropriate for the EPA's test?

b. The EPA statistician analyzed the data using SAS, with the results shown below. Find and adjust (if necessary) the *p*-value of the test.

c. Interpret the results of the test.

```
Analysis Variable : PHOS_15

 N Obs        T      Prob>|T|
----------------------------------
    12     -1.732       0.1112
----------------------------------
```

8.50 One of the most feared predators in the ocean is the great white shark. It is known that the white shark grows to a mean length of 21 feet; however, one marine biologist believes that great white sharks off the Bermuda coast grow much longer owing to unusual feeding habits. To test this claim, some full-grown great white sharks were captured off the Bermuda coast, measured, and then set free. However, because the capture of sharks is difficult, costly, and very dangerous, only three specimens were sampled. Their lengths were 24, 20, and 22 feet.

a. Do the data provide sufficient evidence to support the marine biologist's claim? Use $\alpha = .10$.

b. Give the approximate observed significance level for the test in part **a**, and interpret its value.

c. What assumptions must be made in order to carry out the test?

d. Do you think these assumptions are likely to be satisfied in this sampling situation?

8.51 Periodic assessment of stress in paved highways is important to maintaining safe roads. The Mississippi Department of Transportation recently collected data on the number of cracks (called *crack intensity*) in an undivided two-lane highway using van-mounted state-of-the-art video technology (*Journal of Infrastructure Systems,* Mar. 1995). The mean number of cracks found in a sample of 8 fifty-meter sections of the highway was $\bar{x} = .210$, with a variance of $s^2 = .011$. Suppose the American Association of State Highway and Transportation Officials (AASHTO) recommends a maximum mean crack intensity of .100 for safety purposes. Test the hypothesis that the true mean crack intensity of the Mississippi highway exceeds the AASHTO recommended maximum. Use $\alpha = .01$.

8.52 The SCL-90-R is a 90-item symptom inventory checklist designed to reflect the psychological status of an individual. Each symptom (e.g., obsessive-compulsive behavior) is scored on a scale of 0 (none) to 4 (extreme). The total of these scores yields an individual's Positive Symptom Total (PST). "Normal" individuals are known to have a mean PST of about 40. The *Journal of Head Trauma Rehabilitation* (Apr. 1995) reported that a sample of 23 patients diagnosed with mild to moderate traumatic brain injury had a mean PST score of $\bar{x} = 48.43$ and a standard deviation of $s = 20.76$. Is there sufficient evidence to claim that the true mean PST score of all patients with mild to moderate traumatic brain injury exceeds the "normal" value of 40? Test using $\alpha = .05$.

8.53 Research reported in the *Journal of Psychology* (Mar. 1991) studied the personality characteristics of obese individuals. One variable, the locus of control (LOC), measures the individual's degree of belief that he or she has control over situations. High scores on the LOC scale indicate less perceived control. For one sample of 19 obese adolescents, the mean LOC score was 10.89 with a standard deviation of 2.48. Suppose we wish to test whether the mean

LOC score for all obese adolescents exceeds 10, the average for "normal" individuals.

a. Specify the null and alternative hypotheses for this test.

b. The data are analyzed by computer, and the following MINITAB output is obtained. Check the calculation of the t statistic, and determine whether the p-value is in the correct range according to Table VI of Appendix A.

	N	MEAN	STDEV	SE MEAN	T	P VALUE
TEST OF MU = 10.000 VS MU G.T. 10.000						
LOC	19	10.89	2.48	.569	1.564	.0676

8.5 LARGE-SAMPLE TEST OF HYPOTHESIS ABOUT A POPULATION PROPORTION

Inferences about population proportions (or percentages) are often made in the context of the probability, p, of "success" for a binomial distribution. We saw how to use large samples from binomial distributions to form confidence intervals for p in Section 7.3. We now consider tests of hypotheses about p.

Consider, for example, a method currently used by doctors to screen women for possible breast cancer. The method fails to detect cancer in 20% of the women who actually have the disease. Suppose a new method has been developed that researchers hope will detect cancer more accurately. This new method was used to screen a random sample of 140 women known to have breast cancer. Of these, the new method failed to detect cancer in 12 women. Does this sample provide evidence that the failure rate of the new method differs from the one currently in use?

We first view this as a binomial experiment with 140 screened women as the trials and failure to detect breast cancer as "Success" (in binomial terminology). Let p represent the probability that the new method fails to detect the breast cancer. If the new method is no better than the current one, then the failure rate is $p = .2$. On the other hand, if the new method is either better or worse than the current method, then the failure rate is either smaller or larger than 20%; i.e., $p \neq .2$.

We can now place the problem in the context of a test of hypothesis:

$$H_0: \ p = .2$$
$$H_a: \ p \neq .2$$

Recall that the sample proportion, $\hat{p}$, is really just the sample mean of the outcomes of the individual binomial trials and, as such, is approximately normally distributed (for large samples) according to the Central Limit Theorem. Thus, for large samples we can use the standard normal z as the test statistic:

Test statistic: $\quad z = \dfrac{\text{Sample proportion} - \text{Null hypothesized proportion}}{\text{Standard deviation of sample proportion}}$

$$= \dfrac{\hat{p} - p_0}{\sigma_{\hat{p}}}$$

where we use the symbol p_0 to represent the null hypothesized value of p.

Rejection region: We use the standard normal distribution to find the appropriate rejection region for the specified value of α. Using $\alpha = .05$, the two-tailed rejection region is

$$z < -z_{\alpha/2} = -z_{.025} = -1.96 \qquad \text{or} \qquad z > z_{\alpha/2} = z_{.025} = 1.96$$

FIGURE 8.15

Rejection region for breast cancer example

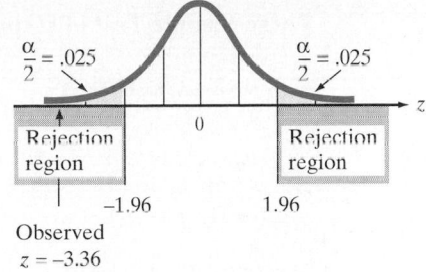

See Figure 8.15.

We are now prepared to calculate the value of the test statistic. Before doing so, we want to be sure that the sample size is large enough to ensure that the normal approximation for the sampling distribution of $\hat{p}$ is reasonable. To check this, we calculate a 3-standard-deviation interval around the null hypothesized value, p_0, which is assumed to be the true value of p until our test procedure proves otherwise. Recall that $\sigma_{\hat{p}} = \sqrt{pq/n}$ and that we need an estimate of the product pq in order to calculate a numerical value of the test statistic z. Since the null hypothesized value is generally the accepted-until-proven-otherwise value, we use the value of $p_0 q_0$ (where $q_0 = 1 - p_0$) to estimate pq in the calculation of z. Thus,

$$\sigma_{\hat{p}} = \sqrt{\frac{pq}{n}} \approx \sqrt{\frac{p_0 q_0}{n}} = \sqrt{\frac{(.2)(.8)}{140}} = .034$$

and the 3-standard-deviation interval around p_0 is

$$p_0 \pm 3\sigma_{\hat{p}} \approx .2 \pm 3(.034) = (.166, .234)$$

As long as this interval does not contain 0 or 1 (i.e., is completely contained in the interval 0 to 1), as is the case here, the normal distribution will provide a reasonable approximation for the sampling distribution of $\hat{p}$.

Returning to the hypothesis test at hand, the proportion of the screenings that failed to detect breast cancer is

$$\hat{p} = \frac{12}{140} = .086$$

Finally, we calculate the number of standard deviations (the z value) between the sampled and hypothesized value of the binomial proportion:

$$z = \frac{\hat{p} - p_0}{\sigma_{\hat{p}}} = \frac{\hat{p} - p_0}{\sqrt{p_0 q_0/n}} = \frac{.086 - .2}{.034} = \frac{-.114}{.034} = -3.36$$

The implication is that the observed sample proportion is (approximately) 3.36 standard deviations below the null hypothesized proportion .2 (Figure 8.15). Therefore, we reject the null hypothesis, concluding at the .05 level of significance that the true failure rate of the new method for detecting breast cancer differs from .20. Since $\hat{p} = .086$, it appears that the new method is better (i.e., has a smaller failure rate) than the method currently in use. (To estimate the magnitude of the failure rate for the new method, a confidence interval can be constructed.)

The test of hypothesis about a population proportion p is summarized in the next box. Note that the procedure is entirely analogous to that used for conducting large-sample tests about a population mean.

Large-Sample Test of Hypothesis About p

ONE-TAILED TEST	TWO-TAILED TEST

$H_0: p = p_0$ 　　　　　　　　　　　　$H_0: p = p_0$
$\quad$ (p_0 = hypothesized value of p)

$H_a: p < p_0$ 　　　　　　　　　　　　$H_a: p \neq p_0$
$\quad$ (or $H_a: p > p_0$)

Test statistic: $\quad z = \dfrac{\hat{p} - p_0}{\sigma_{\hat{p}}}$ 　　　　Test statistic: $\quad z = \dfrac{\hat{p} - p_0}{\sigma_{\hat{p}}}$

where, according to H_0, $\sigma_{\hat{p}} = \sqrt{p_0 q_0 / n}$ and $q_0 = 1 - p_0$

Rejection region: $\quad z < -z_\alpha$ 　　　　Rejection region: $\quad z < -z_{\alpha/2}$
$\quad$ (or $z > z_\alpha$ when $H_a: p > p_0$) 　　　　　　　or $z > z_{\alpha/2}$

Assumption:　The experiment is binomial, and the sample size is large enough that the interval $p_0 \pm 3\sigma_{\hat{p}}$ does not include 0 or 1.

EXAMPLE 8.7

The reputations (and hence sales) of many businesses can be severely damaged by shipments of manufactured items that contain a large percentage of defectives. For example, a manufacturer of alkaline batteries may want to be reasonably certain that fewer than 5% of its batteries are defective. Suppose 300 batteries are randomly selected from a very large shipment; each is tested and 10 defective batteries are found. Does this provide sufficient evidence for the manufacturer to conclude that the fraction defective in the entire shipment is less than .05? Use $\alpha = .01$.

Solution

Before conducting the test of hypothesis, we check to determine whether the sample size is large enough to use the normal approximation for the sampling distribution of $\hat{p}$. The criterion is tested by the interval

$$p_0 \pm 3\sigma_{\hat{p}} = p_0 \pm 3\sqrt{\frac{p_0 q_0}{n}} = .05 \pm 3\sqrt{\frac{(.05)(.95)}{300}}$$

$$= .05 \pm .04 \quad \text{or } (.01, .09)$$

Since the interval lies within the interval (0, 1), the normal approximation will be adequate.

The objective of the sampling is to determine whether there is sufficient evidence to indicate that the fraction defective, p, is less than .05. Consequently, we will test the null hypothesis that $p = .05$ against the alternative hypothesis that $p < .05$. The elements of the test are

$H_0: p = .05$

$H_a: p < .05$

Test statistic: $\quad z = \dfrac{\hat{p} - p_0}{\sigma_{\hat{p}}}$

Rejection region: $\quad z < -z_{.01} = -2.33 \qquad$ (see Figure 8.16)

We now calculate the test statistic:

$$z = \frac{\hat{p} - .05}{\sigma_{\hat{p}}} = \frac{(10/300) - .05}{\sqrt{p_0 q_0 / n}} = \frac{.033 - .05}{\sqrt{p_0 q_0 / 300}}$$

FIGURE 8.16

Rejection region for
Example 8.7

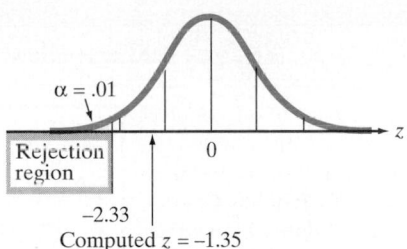

Notice that we use p_0 to calculate $\sigma_{\hat{p}}$ because, in contrast to calculating $\sigma_{\hat{p}}$ for a confidence interval, the test statistic is computed on the assumption that the null hypothesis is true—that is, $p = p_0$. Therefore, substituting the values for $\hat{p}$ and p_0 into the z statistic, we obtain

$$z \approx \frac{-.017}{\sqrt{(.05)(.95)/300}} = \frac{-.017}{.0126} = -1.35$$

As shown in Figure 8.16, the calculated z value does not fall in the rejection region. Therefore, there is insufficient evidence at the .01 level of significance to indicate that the shipment contains fewer than 5% defective batteries.   ▲

EXAMPLE 8.8

In Example 8.7 we found that we did not have sufficient evidence, at the $\alpha = .01$ level of significance, to indicate that the fraction defective p of alkaline batteries was less than $p = .05$. How strong was the weight of evidence favoring the alternative hypothesis ($H_a: p < .05$)? Find the observed significance level for the test.

Solution

The computed value of the test statistic z was $z = -1.35$. Therefore, for this lower-tailed test, the observed significance level is

$$\text{Observed significance level} = P(z \leq -1.35)$$

This lower-tail area is shown in Figure 8.17. The area between $z = 0$ and $z = 1.35$ is given in Table IV in Appendix A as .4115. Therefore, the observed significance level is $.5 - .4115 = .0885$. Note that this probability is quite small. Although we did not reject $H_0: p = .05$ at $\alpha = .01$, the probability of observing a z value as small as or smaller than -1.35 is only .0885 if in fact H_0 is true. Therefore, we would reject H_0 if we choose $\alpha = .10$ (since the observed significance level is less than .10), and we would not reject H_0 (the conclusion of Example 8.7) if we choose $\alpha = .05$ or $\alpha = .01$.   ▲

Small-sample test procedures are also available for p. These are omitted from our discussion because most surveys use samples that are large enough to employ the large-sample tests presented in this section.

FIGURE 8.17

The observed significance
level for Example 8.8

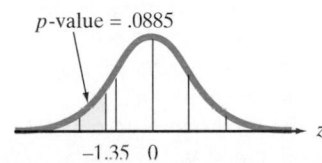

Verifying Petitions—How Many to Check?

CASE
STUDY
• 8.2 •

To get their names on the ballot of a local election, political candidates often must obtain petitions bearing the signatures of a minimum number of registered voters. In Pinellas County, Florida, a certain political candidate obtained petitions with 18,200 signatures. To verify that the names on the petitions were signed by actual registered voters, election officials randomly sampled 100 of the names and checked each for authenticity. This sampling strategy was questioned by several readers of the *St. Petersburg Times*.

Consider the opening paragraphs in a "Letter to the Editor" of the *St. Petersburg Times* (Apr. 7, 1992), written by H. Burstein of Tarpon Springs, Florida:

> Relevant to a petition on behalf of [the political candidate], a recent letter writer asked: "How is it mathematically possible to verify 18,200 signatures by verifying 100? That is ... an insult to common sense, let alone common mathematics. I would suggest at least 100 per 1,000 of those signatures be verified."
>
> Despite its seeming assault on common sense, the correct answer is that a sample of 100 *could* be sufficient, depending on the required accuracy. The key role that scientific (mathematical) sampling plays in

our society—in political polls, medical research, product evaluation, industrial quality control, etc.—justifies an expansion of my brief answer.

The letter writer's "expansion" involves a discussion of the following hypothetical (but realistic) situation. Suppose 17,000 valid signatures are required to place the candidate on the ballot. Since 18,200 signatures were obtained (call this the population), as many as 1,200 of them can be invalid for the candidate to get on the ballot. Now suppose that of the 100 names checked, only 2 were invalid signatures. The question is whether or not 98 out of 100 verified signatures is sufficient to believe that more than 17,000 of the 18,200 signatures are valid.

Focus

a. Use the hypothesis testing methodology presented in this chapter to solve the problem. Compute the approximate *p*-value of the test and interpret its value.

b. Repeat part a if only 16,000 valid signatures are required.

c. Write up the results of the analysis in a report for city election officials.

EXERCISES 8.54–8.65

Learning the Mechanics

8.54 For the binomial sample sizes and null hypothesized values of p in each part, determine whether the sample size is large enough to use the normal approximation methodology presented in this section to conduct a test of the null hypothesis $H_0: p = p_0$.
   a. $n = 500, p_0 = .05$ **b.** $n = 100, p_0 = .99$
   c. $n = 50, p_0 = .2$ **d.** $n = 20, p_0 = .2$
   e. $n = 10, p_0 = .4$

8.55 Suppose a random sample of 100 observations from a binomial population gives a value of $\hat{p} = .69$ and you wish to test the null hypothesis that the population parameter p is equal to .75 against the alternative hypothesis that p is less than .75.
   a. Noting that $\hat{p} = .69$, what does your intuition tell you? Does the value of $\hat{p}$ appear to contradict the null hypothesis?
   b. Use the large-sample z-test to test $H_0: p = .75$ against the alternative hypothesis, $H_a: p < .75$. Use $\alpha = .05$. How do the test results compare with your intuitive decision from part **a**?
   c. Find and interpret the observed significance level of the test you conducted in part **b**.

8.56 Suppose the sample in Exercise 8.55 has produced $\hat{p} = .84$ and we wish to test $H_0: p = .9$ against the alternative $H_a: p < .9$.
   a. Calculate the value of the z statistic for this test.
   b. Note that the numerator of the z statistic ($\hat{p} - p_0$ = .84 − .90 = −.06) is the same as for Exercise 8.55. Considering this, why is the absolute value of z for this exercise larger than that calculated in Exercise 8.55?
   c. Complete the test using $\alpha = .05$ and interpret the result.
   d. Find the observed significance level for the test and interpret its value.

8.57 A random sample of 100 observations is selected from a binomial population with unknown probability of success p. The computed value of $\hat{p}$ is equal to .74.
   a. Test $H_0: p = .65$ against $H_a: p > .65$. Use $\alpha = .01$.
   b. Test $H_0: p = .65$ against $H_a: p > .65$. Use $\alpha = .10$.
   c. Test $H_0: p = .90$ against $H_a: p \neq .90$. Use $\alpha = .05$.
   d. Form a 95% confidence interval for p.
   e. Form a 99% confidence interval for p.

8.58 Refer to Exercise 7.40, in which 50 consumers taste-tested a new snack food.

  a. Test H_0: $p = .5$ against H_a: $p > .5$, where p is the proportion of customers who do not like the snack food. Use $\alpha = .10$.

  b. Report the observed significance level of your test.

8.59 A statistics student will use a computer program to test the null hypothesis H_0: $p = .5$ against the one-tailed alternative, H_a: $p > .5$. A sample of 500 observations are input into the computer, which returns the following result: $z = .44$, one-tailed p-value = .3300.

  a. The student concludes, based on the p-value, that there is a 33% chance that the alternative hypothesis is true. Do you agree? If not, correct the interpretation.

  b. How would the p-value change if the alternative hypothesis were two-tailed, H_a: $p \ne .5$? Interpret this p-value.

Applying the Concepts

8.60 Earthquakes are not uncommon in California. An article in the *Annals of the Association of American Geographers* (June 1992) investigated many factors that California residents consider when purchasing earthquake insurance. The survey revealed that only 133 of 337 randomly selected residences in Los Angeles County were protected by earthquake insurance.

  a. What are the appropriate null and alternative hypotheses to test the research hypothesis that less than 40% of the residents of Los Angeles County were protected by earthquake insurance?

  b. Do the data provide sufficient evidence to support the research hypothesis? Use $\alpha = .10$.

  c. Calculate and interpret the p-value for the test.

8.61 A placebo is a pill that looks and tastes real but contains no medically active chemicals. The *placebo effect* describes the phenomenon of improvement in the condition of a patient taking placebos. One of the keys is that both doctors and patients must believe that the placebo being administered is really a drug. Such was the case at a clinic in La Jolla, California. Physicians gave what they thought were drugs to 7,000 asthma, ulcer, and herpes patients. Although the doctors later learned that the drugs were really placebos, 70% of the patients reported an improved condition (*Forbes,* May 22, 1995). Use this information to test (at $\alpha = .05$) if there is a placebo effect at the clinic. Assume that if the placebo is ineffective, the probability of a patient's condition improving is .5.

8.62 "Take the Pepsi Challenge" was a marketing campaign used by the Pepsi-Cola Company. Coca-Cola drinkers participated in a blind taste test in which they tasted unmarked cups of Pepsi and Coke and were asked to select their favorite. In one Pepsi television commercial, an announcer states that "in recent blind taste tests, more than half the Diet Coke drinkers surveyed said they preferred the taste of Diet Pepsi" (*Consumer's Research,* May 1993). Suppose 100 Diet Coke drinkers took the Pepsi Challenge and 56 preferred the taste of Diet Pepsi. Test the hypothesis that more than half of all Diet Coke drinkers will select Diet Pepsi in a blind taste test. Use $\alpha = .05$. What are the consequences of the test results from Coca-Cola's perspective?

8.63 Refer to the *Chest* (May 1995) study of obstructive sleep apnea, Exercise 7.43. Recall that the disorder causes a person to stop breathing and awaken briefly during a sleep cycle. Stanford University researchers found that 124 of 159 commercial truck drivers suffered from obstructive sleep apnea.

  a. Sleep researchers theorize that 25% of the general population suffers from obstructive sleep apnea. Use a test of hypothesis (at $\alpha = .10$) to determine whether this percentage differs for commercial truck drivers.

  b. Find the observed significance level of the test and interpret its value.

  c. In part **b** of Exercise 7.43, you used a 90% confidence interval to make the inference of part **a**. Explain why these two inferences must necessarily agree.

8.64 *Meteoritics* (Mar. 1995) reported the results of a study of lunar soil evolution. Data were obtained from the Apollo 16 mission to the moon, during which a 62-cm core was extracted from the soil near the landing site. Monomineralic grains of lunar soil were separated out and examined for coating with dust and glass fragments. Each grain was then classified as coated or uncoated. Of interest is the "coat index," i.e., the proportion of grains that are coated. According to soil evolution theory, the coat index will exceed .5 at the top of the core, equal .5 in the middle of the core, and fall below .5 at the bottom of the core. Use the summary data in the accompanying table to test each part of the 3-part theory. Use $\alpha = .05$ for each test.

	LOCATION (DEPTH)		
	Top (4.25 cm)	**Middle (28.1 cm)**	**Bottom (54.5 cm)**
Number of grains sampled	84	73	81
Number coated	64	35	29

Source: Basu, A. and McKay, D. S. "Lunar soil evolution processes and Apollo 16 core 60013/60014." *Meteoritics,* Vol. 30, No. 2, Mar. 1995, p. 166 (Table 2).

8.65 The National Science Foundation, in a survey of 2,237 engineering graduate students who earned their Ph.D. degrees, found that 607 were U.S. citizens; the majority (1,630) of the Ph.D. degrees were awarded to foreign nationals (*Science,* Sept. 24, 1993). Conduct a test to determine whether the true percentage of engineering Ph.D. degrees awarded to foreign nationals exceeds 50%. Use $\alpha = .01$.

8.6   CALCULATING TYPE II ERROR PROBABILITIES: MORE ABOUT β (OPTIONAL)

In our introduction to hypothesis testing in Section 8.1, we showed that the probability of committing a Type I error, α, can be controlled by the selection of the rejection region for the test. Thus, when the test statistic falls in the rejection region and we make the decision to reject the null hypothesis, we do so knowing the error rate for incorrect rejections of H_0. The situation corresponding to accepting the null hypothesis, and thereby risking a Type II error, is not generally as controllable. For that reason, we adopted a policy of nonrejection of H_0 when the test statistic does not fall in the rejection region, rather than risking an error of unknown magnitude.

To see how β, the probability of a Type II error, can be calculated for a test of hypothesis, recall the example in Section 8.1 in which a city tests a manufacturer's pipe to see whether it meets the requirement that the mean strength exceed 2,400 pounds per linear foot. The setup for the test is as follows:

$$H_0: \mu = 2,400$$
$$H_a: \mu > 2,400$$

Test statistic: $z = \dfrac{\bar{x} - 2,400}{\sigma / \sqrt{n}}$

Rejection region: $z > 1.645$ for $\alpha = .05$

Figure 8.18a shows the rejection region for the **null distribution**—that is, the distribution of the test statistic assuming the null hypothesis is true. The area in the rejection region is .05, and this area represents α, the probability that the test statistic leads to rejection of H_0 when in fact H_0 is true.

The Type II error probability β is calculated assuming that the null hypothesis is false, because it is defined as the *probability of accepting H_0 when it is false.* Since H_0 is false for any value of μ exceeding 2,400, one value of β exists for each possible value of μ greater than 2,400 (an infinite number of possibilities). Figures 8.18b, 8.18c, and 8.18d show three of the possibilities, corresponding to alternative hypothesis values of μ equal to 2,425, 2,450, and 2,475, respectively. Note that β is the area in the *nonrejection* (or *acceptance) region* in each of these distributions and that β decreases as the true value of μ moves farther from the null hypothesized value of $\mu = 2,400$. This is sensible because the probability of incorrectly accepting the null hypothesis should decrease as the distance between the null and alternative values of μ increases.

In order to calculate the value of β for a specific value of μ in H_a, we proceed as follows:

1. Calculate the value of $\bar{x}$ that corresponds to the border between the acceptance and rejection regions. For the sewer pipe example, this is the value of $\bar{x}$ that lies 1.645 standard deviations above $\mu = 2,400$ in the sampling distribution of $\bar{x}$.

FIGURE 8.18

Values of α and β for
various values of μ

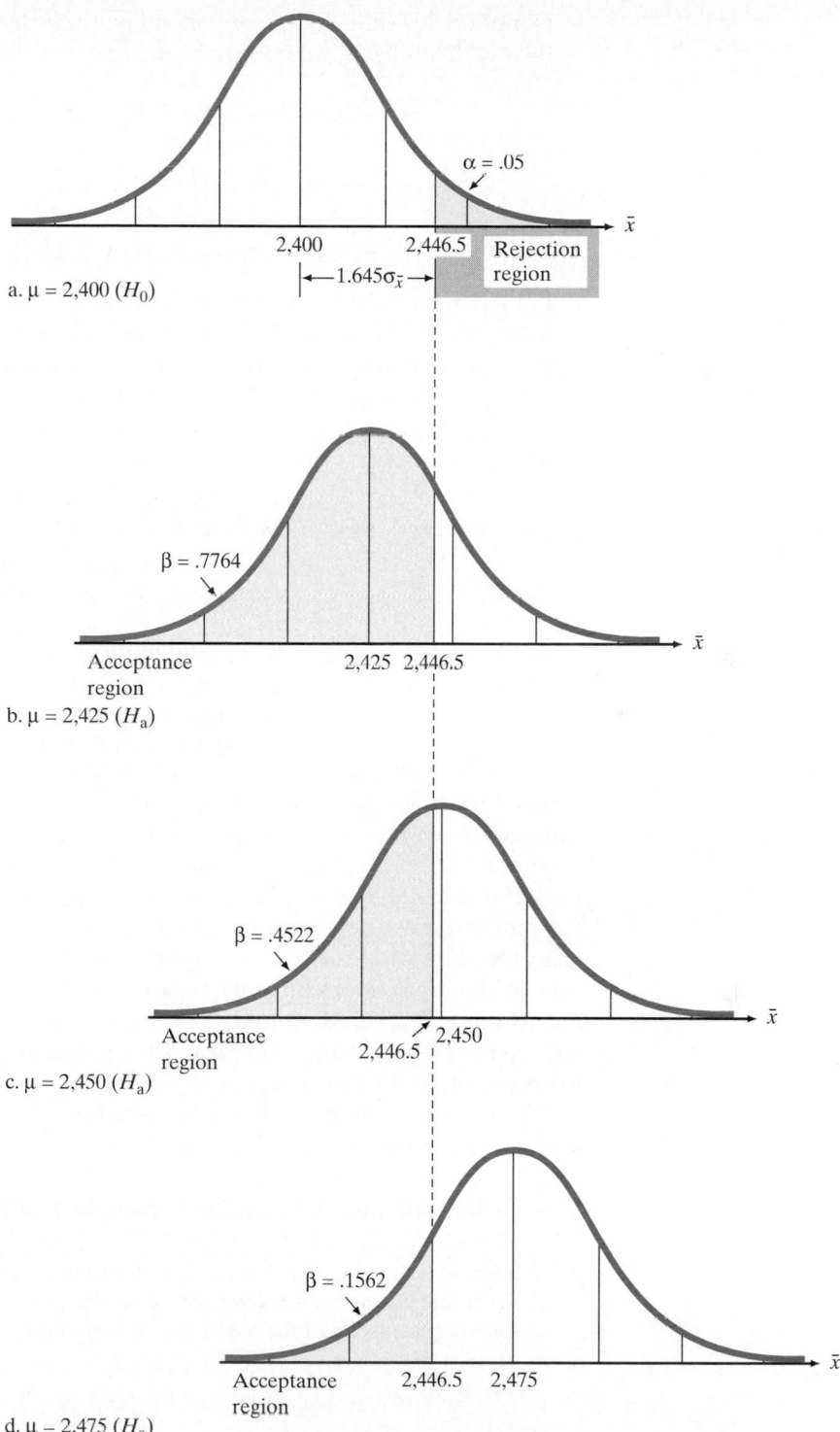

a. μ = 2,400 (H_0)

b. μ = 2,425 (H_a)

c. μ = 2,450 (H_a)

d. μ – 2,475 (H_a)

Denoting this value by $\bar{x}_0$, corresponding to the largest value of $\bar{x}$ that supports the null hypothesis, we find (recalling that $s = 200$ and $n = 50$)

$$\bar{x}_0 = \mu_0 + 1.645\sigma_{\bar{x}} = 2{,}400 + 1.645\left(\frac{\sigma}{\sqrt{n}}\right)$$

$$\approx 2{,}400 + 1.645\left(\frac{s}{\sqrt{n}}\right) = 2{,}400 + 1.645\left(\frac{200}{\sqrt{50}}\right)$$

$$= 2{,}400 + 1.645(28.28) = 2{,}446.5$$

2. For a particular alternative distribution corresponding to a value of μ, denoted by μ_a, we calculate the z value corresponding to $\bar{x}_0$, the border between the rejection and acceptance regions. We then use this z value and Table IV of Appendix A to determine the area in the *acceptance region* under the alternative distribution. This area is the value of β corresponding to the particular alternative μ_a. For example, for the alternative $\mu_a = 2{,}425$, we calculate

$$z = \frac{\bar{x}_0 - 2{,}425}{\sigma_{\bar{x}}} = \frac{\bar{x}_0 - 2{,}425}{\sigma / \sqrt{n}}$$

$$\approx \frac{\bar{x}_0 - 2{,}425}{s / \sqrt{n}} = \frac{2{,}446.5 - 2{,}425}{28.28} = .76$$

Note in Figure 8.18b that the area in the acceptance region is the area to the left of $z = .76$. This area is

$$\beta = .5 + .2764 = .7764$$

Thus, the probability that the test procedure will lead to an incorrect acceptance of the null hypothesis $\mu = 2{,}400$ when in fact $\mu = 2{,}425$ is about .78. As the average strength of the pipe increases to 2,450, the value of β decreases to .4522 (Figure 8.18c). If the mean strength is further increased to 2,475, the value of β is further decreased to .1562 (Figure 8.18d). Thus, even if the true mean strength of the pipe exceeds the minimum specification by 75 pounds per linear foot, the test procedure will lead to an incorrect acceptance of the null hypothesis (rejection of the pipe) approximately 16% of the time. The upshot is that the pipe must be manufactured so that the mean strength well exceeds the minimum requirement if the manufacturer wants the probability of its acceptance by the city to be large (i.e., β to be small).

The steps for calculating β for a large-sample test about a population mean are summarized in the next box.

Steps for Calculating β for a Large-Sample Test About μ

1. Calculate the value(s) of $\bar{x}$ corresponding to the border(s) of the rejection region. There will be one border value for a one-tailed test, and two for a two-tailed test. The formula is one of the following, corresponding to a test with level of significance α:

   Upper-tailed test: $\bar{x}_0 = \mu_0 + z_\alpha\sigma_{\bar{x}} \approx \mu_0 + z_\alpha\left(\dfrac{s}{\sqrt{n}}\right)$

   Lower-tailed test: $\bar{x}_0 = \mu_0 - z_\alpha\sigma_{\bar{x}} \approx \mu_0 - z_\alpha\left(\dfrac{s}{\sqrt{n}}\right)$

   Two-tailed test: $\bar{x}_{0,L} = \mu_0 - z_{\alpha/2}\sigma_{\bar{x}} \approx \mu_0 - z_{\alpha/2}\left(\dfrac{s}{\sqrt{n}}\right)$

   $\bar{x}_{0,U} = \mu_0 + z_{\alpha/2}\sigma_{\bar{x}} \approx \mu_0 + z_{\alpha/2}\left(\dfrac{s}{\sqrt{n}}\right)$

2. Specify the value of μ_a in the alternative hypothesis for which the value of β is to be calculated. Then convert the border value(s) of x_0 to z value(s) using the alternative distribution with mean μ_a. The general formula for the z value is

$$z = \frac{\bar{x}_0 - \mu_a}{\sigma_{\bar{x}}}$$

Sketch the alternative distribution (centered at μ_a), and shade the area in the acceptance (nonrejection) region. Use the z statistic(s) and Table IV of Appendix A to find the shaded area, which is β.

Following the calculation of β for a particular value of μ_a, you should interpret the value in the context of the hypothesis testing application. It is often useful to interpret the value of $1 - β$, which is known as the *power of the test* corresponding to a particular alternative, μ_a. Since β is the probability of accepting the null hypothesis when the alternative hypothesis is true with $\mu = \mu_a$, $1 - β$ is the probability of the complementary event, or the probability of rejecting the null hypothesis when the alternative H_a: $\mu = \mu_a$ is true. That is, the power $1 - β$ measures the likelihood that the test procedure will lead to the correct decision (reject H_0) for a particular value of the mean in the alternative hypothesis.

DEFINITION 8.2

The **power of a test** is the probability that the test will correctly lead to the rejection of the null hypothesis for a particular value of μ in the alternative hypothesis. The power is equal to $1 - β$ for the particular alternative considered.

For example, in the sewer pipe example we found that $β = .7764$ when $\mu = 2,425$. This is the probability that the test leads to the (incorrect) acceptance of the null hypothesis when $\mu = 2,425$. Or, equivalently, the power of the test is $1 - .7764 = .2236$, which means that the test will lead to the (correct) rejection of the null hypothesis only 22% of the time when the pipe exceeds specifications by 25 pounds per linear foot. When the manufacturer's pipe has a mean strength of 2,475 (that is, 75 pounds per linear foot in excess of specifications), the power of the test increases to $1 - .1562 = .8438$. That is, the test will lead to the acceptance of the manufacturer's pipe 84% of the time if $\mu = 2,475$.

EXAMPLE 8.9

Recall the drug experiment in Examples 8.1 and 8.2, in which we tested to determine whether the mean response time for rats injected with a drug differs from the control mean response time of $\mu = 1.2$ seconds. The test setup is repeated here:

H_0: $\mu = 1.2$

H_a: $\mu \neq 1.2$ (i.e., $\mu < 1.2$ or $\mu > 1.2$)

Test statistic: $z = \dfrac{\bar{x} - 1.2}{\sigma_{\bar{x}}}$

Rejection region: $z < -1.96$ or $z > 1.96$ for $\alpha = .05$

$z < -2.575$ or $z > 2.575$ for $\alpha = .01$

Note that two rejection regions have been specified corresponding to values of $\alpha = .05$ and $\alpha = .01$, respectively. Assume that $n = 100$ and $s = .5$.

a. Suppose drug-injected rats have a mean response time of 1.1 seconds, that is, $\mu = 1.1$. Calculate the values of β corresponding to the two rejection regions. Discuss the relationship between the values of α and β.

b. Calculate the power of the test for each of the rejection regions when $\mu = 1.1$.

Solution

a. We first consider the rejection region corresponding to $\alpha = .05$. The first step is to calculate the border values of $\bar{x}$ corresponding to the two-tailed rejection region, $z < -1.96$ or $z > 1.96$:

$$\bar{x}_{0,L} = \mu_0 - 1.96\sigma_{\bar{x}} \approx \mu_0 - 1.96\left(\frac{s}{\sqrt{n}}\right) = 1.2 - 1.96\left(\frac{.5}{10}\right) = 1.102$$

$$\bar{x}_{0,U} = \mu_0 + 1.96\sigma_{\bar{x}} \approx \mu_0 + 1.96\left(\frac{s}{\sqrt{n}}\right) = 1.2 + 1.96\left(\frac{.5}{10}\right) = 1.298$$

These border values are shown in Figure 8.19.

Next, we convert these values to z values in the alternative distribution with $\mu_a = 1.1$:

$$z_L = \frac{\bar{x}_{0,L} - \mu_a}{\sigma_{\bar{x}}} \approx \frac{1.102 - 1.1}{.05} = .04$$

$$z_U = \frac{\bar{x}_{0,U} - \mu_a}{\sigma_{\bar{x}}} \approx \frac{1.298 - 1.1}{.05} = 3.96$$

These z values are shown in Figure 8.19b. You can see that the acceptance (or nonrejection) region is the area between them. Using Table IV of Appendix A, we find that the area between $z = 0$ and $z = .04$ is .0160, and the area between $z = 0$ and $z = 3.96$ is (approximately) .5 (since $z = 3.96$ is off the scale of Table IV). Then the area between $z = .04$ and $z = 3.96$ is, approximately,

$$\beta = .5 - .0160 = .4840$$

Thus, the test with $\alpha = .05$ will lead to a Type II error about 48% of the time when the mean reaction time for drug-injected rats is .1 second less than the control mean response time.

For the rejection region corresponding to $\alpha = .01$, $z < -2.575$ or $z > 2.575$, we find

$$\bar{x}_{0,L} = 1.2 - 2.575\left(\frac{.5}{10}\right) = 1.0712$$

$$\bar{x}_{0,U} = 1.2 + 2.575\left(\frac{.5}{10}\right) = 1.3288$$

These border values of the rejection region are shown in Figure 8.19c.

Converting these to z values in the alternative distribution with $\mu_a = 1.1$, we find $z_L = -.58$ and $z_U = 4.58$. The area between these values is, approximately,

$$\beta = .2190 + .5 = .7190$$

Thus, the chance that the test procedure with $\alpha = .01$ will lead to an incorrect acceptance of H_0 is about 72%.

Note that the value of β increases from .4840 to .7190 when we decrease the value of α from .05 to .01. This is a general property of the relationship between α and β: *as α is decreased (increased), β is increased (decreased).*

b. The power is defined to be the probability of (correctly) rejecting the null hypothesis when the alternative is true. When $\mu = 1.1$ and $\alpha = .05$, we find

$$\text{Power} = 1 - \beta = 1 - .4840 = .5160$$

When $\mu = 1.1$ and $\alpha = .01$, we find

$$\text{Power} = 1 - \beta = 1 - .7190 = .2810$$

FIGURE 8.19

Calculation of β for drug-injected rats (Example 8.9)

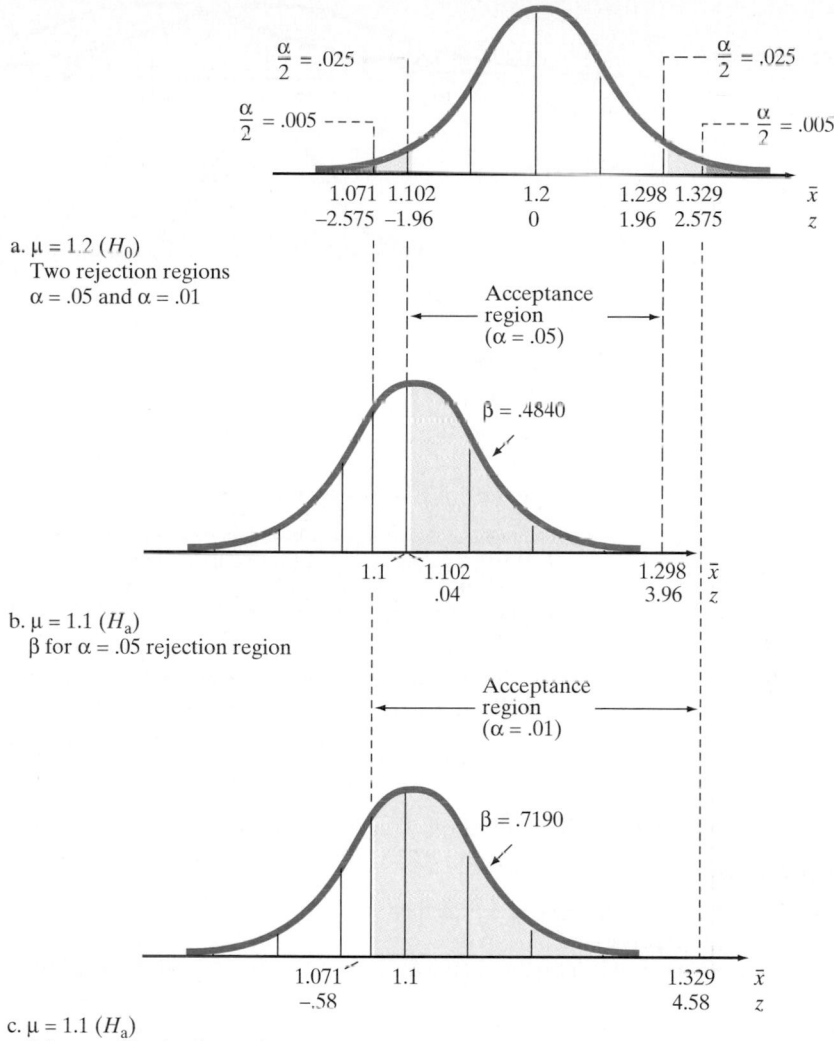

a. $\mu = 1.2$ (H_0)
  Two rejection regions
  $\alpha = .05$ and $\alpha = .01$

b. $\mu = 1.1$ (H_a)
  β for $\alpha = .05$ rejection region

c. $\mu = 1.1$ (H_a)
  β for $\alpha = .01$ rejection region

You can see that the power of the test is decreased as the level of α is decreased. This means that as the probability of incorrectly rejecting the null hypothesis is decreased, the probability of correctly accepting the null hypothesis for a given alternative is also decreased. Thus, the value of α must be selected carefully, with the realization that a test is made less capable of detecting departures from the null hypothesis when the value of α is decreased.   ▲

We have shown that the probability of committing a Type II error, β, is inversely related to α (Example 8.9), and that the value of β decreases as the value of μ_a moves farther from the null hypothesis value (sewer pipe example). The sample size n also affects β. Remember that the standard deviation of the sampling distribution of $\bar{x}$ is inversely proportional to the square root of the sample size ($\sigma_{\bar{x}} = \sigma / \sqrt{n}$). Thus, as illustrated in Figure 8.20, the variability of both the null and alternative sampling distributions is decreased as n is increased. If the value of α is specified and remains fixed, the value of β decreases as n

FIGURE 8.20

Relationship between α, β, and n

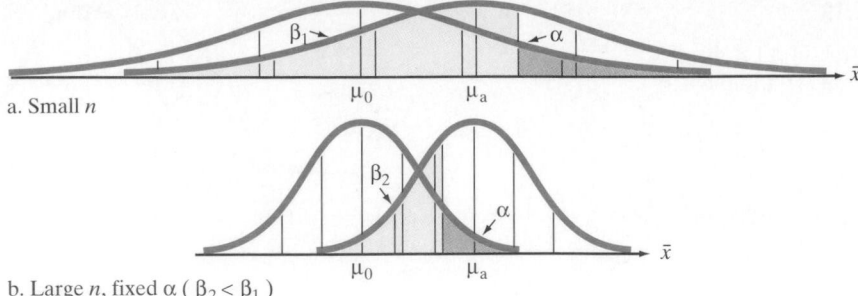

a. Small n

b. Large n, fixed α ($\beta_2 < \beta_1$)

increases, as illustrated in Figure 8.20. Conversely, the power of the test for a given alternative hypothesis is increased as the sample size is increased.

The properties of β and power are summarized in the following box.

Properties of β and Power

1. For fixed n and α, the value of β decreases and the power increases as the distance between the specified null value μ_0 and the specified alternative value μ_a increases (see Figure 8.18).
2. For fixed n and values of μ_0 and μ_a, the value of β increases and the power decreases as the value of α is decreased (see Figure 8.19).
3. For fixed α and values of μ_0 and μ_a, the value of β decreases and the power increases as the sample size n is increased (see Figure 8.20).

EXERCISES 8.66–8.76

Learning the Mechanics

8.66 What is the relationship between β, the probability of committing a Type II error, and the power of a test?

8.67 List three factors that will increase the power of a test.

8.68 Suppose you want to test H_0: $\mu = 1{,}000$ against H_a: $\mu > 1{,}000$ using α = .05. The population in question is normally distributed with standard deviation 120. A random sample of size $n = 36$ will be used.
   a. Sketch the sampling distribution of $\bar{x}$ assuming that H_0 is true.
   b. Find the value of $\bar{x}_0$, that value of $\bar{x}$ above which the null hypothesis will be rejected. Indicate the rejection region on your graph of part **a**. Shade the area above the rejection region and label it α.
   c. On your graph of part **a**, sketch the sampling distribution of $\bar{x}$ if $\mu = 1{,}020$. Shade the area under this distribution that corresponds to the probability that $\bar{x}$ falls in the nonrejection region when $\mu = 1{,}020$. Label this area β.
   d. Find β.
   e. Compute the power of this test for detecting the alternative H_a: $\mu = 1{,}020$.

8.69 Refer to Exercise 8.68.
   a. If $\mu = 1{,}040$ instead of 1,020, what is the probability that the hypothesis test will incorrectly fail to reject H_0? That is, what is β?

   b. If $\mu = 1{,}040$, what is the probability that the test will correctly reject the null hypothesis? That is, what is the power of the test?
   c. Compare β and the power of the test when $\mu = 1{,}040$ to the values you obtained in Exercise 8.68 for $\mu = 1{,}020$. Explain the differences.

8.70 It is desired to test H_0: $\mu = 50$ against H_a: $\mu < 50$ using α = .10. The population in question is uniformly distributed with standard deviation 20. A random sample of size 64 will be drawn from the population.
   a. Describe the (approximate) sampling distribution $\bar{x}$ under the assumption that H_0 is true.
   b. Describe the (approximate) sampling distribution $\bar{x}$ under the assumption that the population mean is 45.
   c. If μ were really equal to 45, what is the probability that the hypothesis test would lead the investigator to commit a Type II error?
   d. What is the power of this test for detecting the alternative H_a: $\mu = 45$?

8.71 Refer to Exercise 8.70.
   a. Find β for each of the following values of the population mean: 49, 47, 45, 43, and 41.
   b. Plot each value of β you obtained in part **a** against its associated population mean. Show β on the vertical axis and μ on the horizontal axis. Draw a curve through the five points on your graph.

c. Use your graph of part **b** to find the approximate probability that the hypothesis test will lead to a Type II error when $\mu = 48$.

d. Convert each of the β values you calculated in part **a** to the power of the test at the specified value of μ. Plot the power on the vertical axis against μ on the horizontal axis. Compare the graph of part **b** to the *power curve* of this part.

e. Examine the graphs of parts **b** and **d**. Explain what they reveal about the relationships among the distance between the true mean μ and the null hypothesized mean μ_0, the value of β, and the power.

8.72 Suppose you want to conduct the two-tailed test of H_0: $\mu = 10$ against H_a: $\mu \neq 10$ using $\alpha = .05$. A random sample of size 100 will be drawn from the population in question. Assume the population has a standard deviation equal to 1.0.

a. Describe the sampling distribution of $\bar{x}$ under the assumption that H_0 is true.

b. Describe the sampling distribution of $\bar{x}$ under the assumption that $\mu = 9.9$.

c. If μ were really equal to 9.9, find the value of β associated with the test.

d. Find the value of β for the alternative H_a: $\mu = 10.1$.

Applying the Concepts

8.73 Refer to Exercise 8.15, in which the null hypothesis that the mean PTSD score of World War II POWs is 16 is tested against the alternative hypothesis that the mean is less than 16. Recall that 33 POWs were included in the sample. Assume that the resulting standard deviation of $s = 9.32$ is a good estimate of the true standard deviation.

a. Calculate the power of the test for the mean values of 15.5, 15.0, 14.5, 14.0, and 13.5.

b. Plot the power of the test on the vertical axis against the mean on the horizontal axis. Draw a curve through the points.

c. Use the power curve of part **b** to estimate the power for the mean value $\mu = 14.25$. Calculate the power for this value of μ, and compare it to your approximation.

d. Use the power curve to approximate the power of the test when $\mu = 12.10$. If the true value of the mean PTSD score for WWII POWs is really 10, what (approximately) are the chances that the test will fail to reject the null hypothesis that the mean is 16?

8.74 Refer to Exercise 8.73. Show what happens to the power curve when the sample size is increased from $n = 33$ to $n = 100$. Assume that the standard deviation is $\sigma = 9.32$.

8.75 If a manufacturer (the vendee) buys all items of a particular type from a particular vendor, the manufacturer is practicing *sole sourcing*. Sole sourcing is a purchasing policy that is generally recognized as an important component of a firm's quality system. One of the major benefits of sole sourcing for the vendee is the improved communication that results from the closer vendee/vendor relationship (*Quality Congress Transactions,* 1986). As part of a sole sourcing arrangement, a vendor agrees to periodically supply its vendee with sample data from its production process. The vendee uses the data to investigate whether the mean length of rods produced by the vendor's production process is truly 5.0 millimeters (mm) or more, as claimed by the vendor and desired by the vendee.

a. If the production process has a standard deviation of .01 mm, the vendor supplies $n = 100$ items to the vendee, and the vendee uses $\alpha = .05$ in testing H_0: $\mu = 5.0$ mm against H_a: $\mu < 5.0$ mm, what is the probability that the vendee's test will fail to reject the null hypothesis when in fact $\mu = 4.9975$ mm? What is the name given to this type of error?

b. Refer to part **a**. What is the probability that the vendee's test will reject the null hypothesis when in fact $\mu = 5.0$? What is the name given to this type of error?

c. What is the power of the test to detect a departure of .0025 mm below the specified mean rod length of 5.0 mm?

8.76 Refer to Exercise 8.20, in which the performance of a particular type of laser-based inspection equipment was investigated. Assume that the standard deviation of the number of solder joints inspected on each run is 1.2. If $\alpha = .05$ is used in conducting the hypothesis test of interest using a sample of 48 circuit boards, and if the true mean number of solder joints that can be inspected is really equal to 9.5, what is the probability that the test will result in a Type II error?

8.7 INFERENCES ABOUT A POPULATION VARIANCE (OPTIONAL)

Although many practical problems involve inferences about a population mean (or proportion), it is sometimes of interest to make an inference about a population variance, σ^2. To illustrate, a quality control supervisor in a cannery knows

FIGURE 8.21

Several χ^2 probability distributions

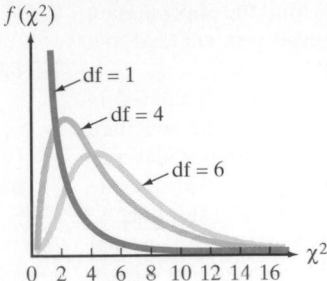

that the exact amount each can contains will vary since there are certain uncontrollable factors that affect the amount of fill. The mean fill per can is important, but equally important is the variation of fill. If σ^2, the variance of the fill, is large, some cans will contain too little and others too much. Suppose regulatory agencies specify that the standard deviation of the amount of fill should be less than .1 ounce. To determine whether the process is meeting this specification, the supervisor randomly selects 10 cans, weighs the contents of each, and finds that the sample standard deviation of these measurements is .04. Do these data provide sufficient evidence to indicate that the variability is as small as desired? To answer this question, we need a procedure for testing a hypothesis about σ^2.

Intuitively it seems that we should compare the sample variance s^2 to the hypothesized value of σ^2 (or s to σ) in order to make a decision about the population's variability. The quantity

$$\frac{(n-1)s^2}{\sigma^2}$$

has been shown to have a sampling distribution called a **chi-square (χ^2) distribution** when the population from which the sample is taken is *normally distributed*. Several chi-square distributions are shown in Figure 8.21.

The upper-tail areas for this distribution have been tabulated and are given in Table VII of Appendix A, a portion of which is reproduced in Table 8.4. The table gives the values of χ^2, denoted as χ^2_α, that locate an area of α in the upper tail of the chi-square distribution; that is, $P(\chi^2 > \chi^2_\alpha) = \alpha$. In this case, as with the t statistic, the shape of the chi-square distribution depends on the degrees of freedom associated with s^2, namely $(n-1)$. Thus, for $n = 10$ and an upper-tail value $\alpha = .05$, you will have $n - 1 = 9$ df and $\chi^2_{.05} = 16.9190$ (shaded area in Table 8.4). To further illustrate the use of Table VII, we return to the can-filling example.

EXAMPLE 8.10

According to the previous discussion, the quality control supervisor sampled $n = 10$ cans and calculated $s = .04$. Does this value of s provide sufficient evidence to indicate that the standard deviation σ of the fill measurements is less than .1 ounce?

Solution

Since the null and alternative hypotheses must be stated in terms of σ^2 (rather than σ), we want to test the null hypothesis that $\sigma^2 = .01$ against the alternative that $\sigma^2 < .01$. Therefore, the elements of the test are

$$H_0: \ \sigma^2 = .01$$
$$H_a: \ \sigma^2 < .01$$

$$\textit{Test statistic:} \quad \chi^2 = \frac{(n-1)s^2}{\sigma^2}$$

TABLE 8.4   Reproduction of Part of Table VII in Appendix A: Critical Values of χ^2

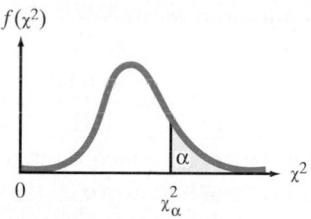

Degrees of Freedom	$\chi^2_{.100}$	$\chi^2_{.050}$	$\chi^2_{.025}$	$\chi^2_{.010}$	$\chi^2_{.005}$
1	2.70554	3.84146	5.02389	6.63490	7.87944
2	4.60517	5.99147	7.37776	9.21034	10.5966
3	6.25139	7.81473	9.34840	11.3449	12.8381
4	7.77944	9.48773	11.1433	13.2767	14.8602
5	9.23635	11.0705	12.8325	15.0863	16.7496
6	10.6446	12.5916	14.4494	16.8119	18.5476
7	12.0170	14.0671	16.0128	18.4753	20.2777
8	13.3616	15.5073	17.5346	20.0902	21.9550
9	14.6837	16.9190	19.0228	21.6660	23.5893
10	15.9871	18.3070	20.4831	23.2093	25.1882
11	17.2750	19.6751	21.9200	24.7250	26.7569
12	18.5494	21.0261	23.3367	26.2170	28.2995
13	19.8119	22.3621	24.7356	27.6883	29.8194
14	21.0642	23.6848	26.1190	29.1413	31.3193
15	22.3072	24.9958	27.4884	30.5779	32.8013
16	23.5418	26.2962	28.8454	31.9999	34.2672
17	24.7690	27.5871	30.1910	33.4087	35.7185
18	25.9894	28.8693	31.5264	34.8053	37.1564
19	27.2036	30.1435	32.8523	36.1908	38.5822

Assumption: The distribution of the amounts of fill is approximately normal.

Rejection region: The smaller the value of s^2 we observe, the stronger the evidence in favor of H_a. Thus, we reject H_0 for "small values" of the test statistic. With $\alpha = .05$ and 9 df, the χ^2 value for rejection is found in Table VII and pictured in Figure 8.22. We will reject H_0 if $\chi^2 < 3.32511$.

FIGURE 8.22

Rejection region for
Example 8.10

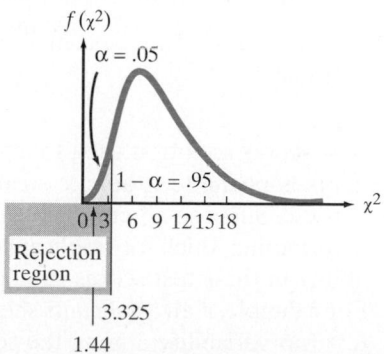

1.44

Remember that the area given in Table VII is the area to the *right* of the numerical value in the table. Thus, to determine the lower-tail value, which has $\alpha = .05$ to its *left*, we used the $\chi^2_{.95}$ column in Table VII.

Since

$$\chi^2 = \frac{(n-1)s^2}{\sigma^2} = \frac{9(.04)^2}{.01} = 1.44$$

is less than 3.32511, the supervisor can conclude that the variance σ^2 of the population of all amounts of fill is less than .01 ($\sigma < .1$) with probability of a Type I error equal to $\alpha = .05$. If this procedure is repeatedly used, it will incorrectly reject H_0 only 5% of the time. Thus, the quality control supervisor is confident in the decision that the cannery is operating within the desired limits of variability.

▲

It is also possible to form a confidence interval for a population variance through the manipulation of the χ^2 distribution. This confidence interval, as well as the one-tailed and two-tailed tests of hypothesis for σ^2, are given in the following boxes.

Test of a Hypothesis About σ^2

| **ONE-TAILED TEST** | **TWO-TAILED TEST** |

H_0: $\sigma^2 = \sigma_0^2$
    (σ_0^2 = hypothesized variance)

H_a: $\sigma^2 < \sigma_0^2$
    (or H_a: $\sigma^2 > \sigma_0^2$)

Test statistic: $\chi^2 = \dfrac{(n-1)s^2}{\sigma_0^2}$

Rejection region: $\chi^2 < \chi^2_{(1-\alpha)}$
  (or $\chi^2 > \chi^2_\alpha$ when H_a: $\sigma^2 > \sigma_0^2$)

H_0: $\sigma^2 = \sigma_0^2$

H_a: $^2 \neq \sigma_0^2$

Test statistic: $\chi^2 = \dfrac{(n-1)s^2}{\sigma_0^2}$

Rejection region: $\chi^2 < \chi^2_{(1-\alpha/2)}$
  or $\chi^2 > \chi^2_{\alpha/2}$

---

where the distribution of χ^2 is based on $(n-1)$ degrees of freedom.

Assumption: The population from which the sample is drawn is approximately normal.

Confidence Interval for σ^2

$$\frac{(n-1)s^2}{\chi^2_{\alpha/2}} < \sigma^2 < \frac{(n-1)s^2}{\chi^2_{(1-\alpha/2)}}$$

where the distribution of χ^2 is based on $(n-1)$ degrees of freedom.

Assumption: The population from which the sample is drawn is approximately normal.

EXAMPLE 8.11

Test scores are often used to evaluate individuals applying for the same job, students applying to graduate school, attorneys trying to become members of the bar, etc. Suppose an employment agency uses a 500-point examination to help in determining which job applicants are best qualified for certain positions. The variability in these test scores should be considered when evaluating the test results. For example, if all applicants should somehow score exactly the same score on the test (no variability among the scores), the test would be of no value in deciding which applicants should be employed. A large amount of variability among the test scores would be desirable in order to differentiate the relative merits of the applicants. To evaluate the variability of the test scores, the employment agency randomly selects 100 test scores and calculates $s^2 = 127$. Use this information to form a 95% confidence interval for σ^2, the variability for *all* test scores.

Solution

The general form of the confidence interval is

$$\frac{(n-1)s^2}{\chi^2_{\alpha/2}} < \sigma^2 < \frac{(n-1)s^2}{\chi^2_{(1-\alpha/2)}}$$

where we assume the test scores are approximately normally distributed. Using Table VII in Appendix A, we obtain the tabulated values of chi-square corresponding to $\alpha/2 = .025$ and $1 - \alpha/2 = .975$, based on $n - 1 = 100 - 1 = 99$ degrees of freedom:

$$\chi^2_{.025} \approx 129.56 \qquad \text{and} \qquad \chi^2_{.975} \approx 74.22$$

The interval of interest is therefore

$$\frac{99(127)}{129.56} < \sigma^2 < \frac{99(127)}{74.22}$$

$$97.04 < \sigma^2 < 169.40$$

Thus, the employment agency can be 95% confident that the variance of all test scores is (approximately) between 97 and 170 or, equivalently, that the standard deviation is between 10 and 13, rounding to the nearest whole number.

This information can be used to determine whether the amount of variability is sufficient. If the standard deviation of 13 is deemed too small to allow the company to discriminate among the applicants, the structure of the test should be changed. ▲

EXERCISES 8.77–8.89

Learning the Mechanics

8.77 Let χ^2_0 be a particular value of χ^2. Find the value of χ^2_0 such that:
  a. $P(\chi^2 > \chi^2_0) = .10$ for $n = 12$
  b. $P(\chi^2 > \chi^2_0) = .05$ for $n = 9$
  c. $P(\chi^2 > \chi^2_0) = .025$ for $n = 5$

8.78 A random sample of n observations is selected from a normal population to test the null hypothesis that $\sigma^2 = 25$. Specify the rejection region for each of the following combinations of H_a, α, and n:
  a. H_a: $\sigma^2 \neq 25$; $\alpha = .05$; $n = 16$
  b. H_a: $\sigma^2 > 25$; $\alpha = .01$; $n = 23$
  c. H_a: $\sigma^2 > 25$; $\alpha = .10$; $n = 15$
  d. H_a: $\sigma^2 < 25$; $\alpha = .01$; $n = 13$
  e. H_a: $\sigma^2 \neq 25$; $\alpha = .10$; $n = 7$
  f. H_a: $\sigma^2 < 25$; $\alpha = .05$; $n = 25$

8.79 A random sample of 17 observations is selected from a normal population with variance σ^2. Give the values of $\chi^2_{\alpha/2}$ and $\chi^2_{(1-\alpha/2)}$ that would be used to form a confidence interval for σ^2 for each of the following levels of confidence:

  a. 95% **b.** 90% **c.** 99%

8.80 A random sample of seven measurements gave $\bar{x} = 9.4$ and $s^2 = 4.84$.
  a. What assumptions must you make concerning the population in order to test a hypothesis about (or estimate) σ^2?
  b. Suppose the assumptions in part **a** are satisfied. Test the null hypothesis, $\sigma^2 = 1$, against the alternative hypothesis, $\sigma^2 > 1$. Use $\alpha = .05$.
  c. Test the null hypothesis that $\sigma^2 = 1$ against the alternative hypothesis that $\sigma^2 \neq 1$. Use $\alpha = .05$.
  d. Find a 90% confidence interval for σ^2.

8.81 Refer to Exercise 8.80. Suppose we had $n = 100$, $\bar{x} = 9.4$, and $s^2 = 4.84$.
  a. Test the null hypothesis, H_0: $\sigma^2 = 1$, against the alternative hypothesis, H_a: $\sigma^2 > 1$.
  b. Compare your test result with that of Exercise 8.80.
  c. Find a 90% confidence interval for σ^2. Compare this confidence interval with the confidence interval obtained in Exercise 8.80 and note the effect of an increase in sample size on the width of the interval.

8.82 A random sample of $n = 7$ observations from a normal population produced the following measurements: 4, 0, 6, 3, 3, 5, 9.

    a. Do the data provide sufficient evidence to indicate that $\sigma^2 < 1$? Test using $\alpha = .05$.

    b. Find a 90% confidence interval for σ^2.

Applying the Concepts

8.83 Recording electrical activity of the brain is important in clinical problems as well as neurophysiological research. To improve the signal-to-noise ratio (SNR) in the electrical activity, it is necessary to repeatedly stimulate subjects and average the responses—a procedure that assumes that single responses are homogeneous. A study was conducted to test the homogeneous signal theory (*IEEE Engineering in Medicine and Biology Magazine,* Mar. 1990). The null hypothesis is that the variance of the SNR readings of subjects equals the "expected" level under the homogeneous signal theory. For this study, the "expected" level was assumed to be .54. If the SNR variance exceeds this level, the researchers will conclude that the signals are nonhomogeneous.

    a. Set up the null and alternative hypotheses for the researchers.

    b. Signal-to-noise ratios recorded for a sample of 41 normal children ranged from .03 to 3.0. Use this information to obtain an estimate of the sample standard deviation. [*Hint:* Assume that the distribution of SNRs is normal, and that most the SNRs in the population will fall within $\mu \pm 2\sigma$, i.e., from $\mu - 2\sigma$ to $\mu + 2\sigma$. Note that the range of the interval equals 4σ.]

    c. Use the estimate of s in part **b** to conduct the test of part **a**. Test using $\alpha = .10$.

8.84 A marine biologist wishes to use male angelfish for experimental purposes because she believes that their weight is fairly stable (i.e., the variability in weights among male angelfish is small). The biologist randomly samples 16 male angelfish and finds that their mean weight is 4.1 pounds and the standard deviation is 1.73 pounds. Find a 95% confidence interval for the variability in weights of all male angelfish. What assumptions must you make in order to form the interval?

8.85 Refer to Exercise 8.84. It is suggested that the marine biologist use parrotfish instead of male angelfish in the experiment. Since these are more difficult to obtain, the biologist decides to use parrotfish only if there is evidence that the variance of their weights is less than 4. A random sample of 10 parrotfish produces a mean of 4.3 pounds and a variance of 2. Is there sufficient evidence for the biologist to claim that the variability in weights among parrotfish is small enough to justify their use in the experiment? Test at $\alpha = .05$, and state any assumptions that are needed.

8.86 Refer to the *Chromatographia* (Mar. 1995) study of a new technique for extracting toxic organic compounds from solids, Exercise 8.47. Recall that uncontaminated fish fillets were injected with a toxic substance and the method was used to extract the toxicant. To test the precision of the new method, 7 measurements were obtained on a single fish fillet injected with the contaminant Aldrin. Summary statistics on percent of Aldrin recovered are given as follows: $\bar{x} = 99$, $s = 9$.

    a. Construct a 90% confidence interval for the variation in the percent recovery of Aldrin using the new method.

    b. Interpret the interval, part **a**.

8.87 It is essential in the manufacture of machinery to utilize parts that conform to specifications. In the past, diameters of the ball-bearings produced by a certain manufacturer had a variance of .00156. To cut costs, the manufacturer instituted a less expensive production method. The variance of the diameters of 100 randomly sampled bearings produced by the new process was .00211. Do the data provide sufficient evidence to indicate that diameters of ball-bearings produced by the new process are more variable than those produced by the old process?

8.88 Refer to the *IEEE Transactions* (June 1990) study of a new hybrid algorithm for solving polynomial 0/1 mathematical problems, Exercise 7.75. A SAS printout giving descriptive statistics for the sample of 52 solution times is reproduced below. Use this information to compute an approximate 95% confidence interval for the variance of the solution times. Interpret the result.

```
Analysis Variable : CPU

N Obs    N         Mean      Variance     Std Dev
------------------------------------------------------
  52     52   0.8121923   2.2643035   1.5047603
------------------------------------------------------
```

8.89 Geologists analyze *fluid inclusions* (pockets of gas or liquid) in rock to infer the concentrations of exogenous fluids present when the rocks crystallized. To do this they use a technique called laser Raman microprobe (LRM) spectroscopy. An experiment was conducted to estimate the precision of the LRM technique (*Applied Spectroscopy,* Feb. 1986). A chip of natural Brazilian quartz was artificially injected with several fluid inclusions of liquid carbon dioxide (CO_2) and then subjected to LRM spectroscopy. The amount of CO_2 present in the inclusion was recorded for the same inclusion on four different days. The data (in mole percent) follow:

    86.6   84.6   85.5   85.9

a. Obtain an estimate of the precision of the LRM technique by constructing a 99% confidence interval for the variation in the CO_2 concentration measurements.

b. What assumption is required for the interval estimate to be valid?

QUICK REVIEW

Key Terms

Alternative (research) hypothesis   290
Chi-square distribution   328
Conclusion   295
Level of significance   295
Lower-tailed test   297
Null hypothesis   290
Observed significance level
  (*p*-value)   303

One-tailed test   296
Power of the test   323
Rejection region   292
Test statistic   290
Two-tailed test   296
Type I error   291
Type II error   293
Upper-tailed test   297

Key Formulas

Note: Starred (*) formulas are from the optional sections in this chapter.

For testing H_0: $\theta = \theta_0$, the **large-sample test statistic** is

$$z = \frac{\hat{\theta} - \theta_0}{\sigma_{\hat{\theta}}}$$

where $\hat{\theta}$, θ_0, and $\sigma_{\hat{\theta}}$ are obtained from the table below:

Parameter, θ	Hypothesized Parameter Value, θ_0	Estimator, $\hat{\theta}$	Standard Error of Estimator, $\sigma_{\hat{\theta}}$	
μ	μ_0	$\bar{x}$	$\dfrac{\sigma}{\sqrt{n}}$	300
p	p_0	$\hat{p}$	$\sqrt{\dfrac{p_0 q_0}{n}}$	316

For testing H_0: $\mu = \mu_0$, the **small-sample test statistic** is   309

$$t = \frac{\bar{x} - \mu_0}{s/\sqrt{n}}$$

*For testing H_0: $\sigma^2 = \sigma_0^2$, the test statistic is   330

$$\chi^2 = \frac{(n-1)s^2}{\sigma_0^2}$$

*A $(1 - \alpha)$ 100% confidence interval for σ^2 is   330

$$\frac{(n-1)s^2}{\chi_{\alpha/2}^2} < \sigma^2 < \frac{(n-1)s^2}{\chi_{(1-\alpha/2)}^2}$$

*$\beta = P(\hat{\theta}$ falls in acceptance region $\mid \theta = \theta_a)$ 320

*Power = $1 - \beta$ 323

LANGUAGE LAB

Symbol	Pronunciation	Description
H_0	H-oh	Null hypothesis
H_a	H-a	Alternative hypothesis
α	alpha	Probability of Type I error
β	beta	Probability of Type II error
χ^2	chi-square	Sampling distribution of s^2 when data are from a normal population

SUPPLEMENTARY EXERCISES 8.90–8.121

Note: List the assumptions necessary for the valid implementation of the statistical procedures you use in solving all these exercises. Starred (*) exercises refer to the optional sections in this chapter.

Learning the Mechanics

8.90 *Complete the following statement:* The smaller the *p*-value associated with a test of hypothesis, the stronger the support for the _____ hypothesis. Explain your answer.

8.91 Specify the differences between a large-sample and small-sample test of hypothesis about a population mean μ. Focus on the assumptions and test statistics.

8.92 Which of the elements of a test of hypothesis can and should be specified *prior* to analyzing the data that are to be utilized to conduct the test?

8.93 If the rejection of the null hypothesis of a particular test would cause your firm to go out of business, would you want α to be small or large? Explain.

8.94 *Complete the following statement:* The larger the *p*-value associated with a test of hypothesis, the stronger the support for the _____ hypothesis. Explain your answer.

8.95 A random sample of 20 observations selected from a normal population produced $\bar{x} = 72.6$ and $s^2 = 19.4$.
  a. Form a 90% confidence interval for the population mean.
  b. Test H_0: $\mu = 80$ against H_a: $\mu < 80$. Use $\alpha = .05$.
  c. Test H_0: $\mu = 80$ against H_a: $\mu \neq 80$. Use $\alpha = .01$.
  d. Form a 99% confidence interval for μ.
  e. How large a sample would be required to estimate μ to within 1 unit with 95% confidence?

8.96 A random sample of $n = 200$ observations from a binomial population yields $\hat{p} = .29$.
  a. Test H_0: $p = .35$ against H_a: $p < .35$. Use $\alpha = .05$.
  b. Test H_0: $p = .35$ against H_a: $p \neq .35$. Use $\alpha = .05$.
  c. Form a 95% confidence interval for p.
  d. Form a 99% confidence interval for p.

  e. How large a sample would be required to estimate p to within .05 with 99% confidence?

8.97 A random sample of 175 measurements possessed a mean $\bar{x} = 8.2$ and a standard deviation $s = .79$.
  a. Form a 95% confidence interval for μ.
  b. Test H_0: $\mu = 8.3$ against H_a: $\mu \neq 8.3$. Use $\alpha = .05$.
  c. Test H_0: $\mu = 8.4$ against H_a: $\mu \neq 8.4$. Use $\alpha = .05$.

*8.98 A random sample of 41 observations from a normal population possessed a mean $\bar{x} = 88$ and a standard deviation $s = 6.9$.
  a. Form a 90% confidence interval for σ^2.
  b. Form a 99% confidence interval for σ^2.
  c. Test H_0: $\sigma^2 = 30$ against H_a: $\sigma^2 > 30$. Use $\alpha = .05$.
  d. Test H_0: $\sigma^2 = 30$ against H_a: $\sigma^2 \neq 30$. Use $\alpha = .05$.

8.99 A *t*-test is conducted for the null hypothesis H_0: $\mu = 10$ versus the alternative H_a: $\mu > 10$ for a random sample of $n = 17$ observations. The data are analyzed using MINITAB, with the results shown at the bottom of the page.
  a. Interpret the *p*-value.
  b. What assumptions are necessary for the validity of this test?
  c. Calculate and interpret the *p*-value assuming the alternative hypothesis was instead H_a: $\mu \neq 10$.

Applying the Concepts

8.100 Medical tests have been developed to detect many serious diseases. A medical test is designed to minimize the probability that it will produce a "false positive" or a "false negative." A false positive is a positive test result for an individual who does not have the disease, whereas a false negative is a negative test result for an individual who does have the disease.
  a. If we treat a medical test for a disease as a statistical test of hypothesis, what are the null and alternative hypotheses for the medical test?

```
TEST OF MU = 10.000 VS MU G.T. 10.000

        N     MEAN    STDEV   SE MEAN      T    P VALUE
X      17    12.50     8.78      2.13   1.174     .1288
```

b. What are the Type I and Type II errors for the test? Relate each to false positives and false negatives.

c. Which of these errors has graver consequences? Considering this error, is it more important to minimize α or β? Explain.

8.101 How does lack of sleep affect one's creative ability? One British study, believed to be the first to measure systematically the effect of sleep loss on divergent thinking, found that loss of sleep sabotages creative faculties and compromises the ability to deal with unfamiliar situations (*Sleep,* Jan. 1989). In the study, 12 healthy college students, deprived of one night's sleep, received an array of tests intended to measure thinking time, fluency, flexibility, and originality of thought. The overall test scores of the sleep-deprived students were compared to the average score expected from students who received their accustomed sleep. Suppose the overall scores of the 12 sleep-deprived students had a mean of $\bar{x} = 63$ and a standard deviation of 17. (Lower scores are associated with a decreased ability to think creatively.)

a. Test the hypothesis that the true mean score of sleep-deprived subjects is less than 80, the mean score of subjects who received sleep prior to taking the test. Use $\alpha = .05$.

b. What assumption is required for the hypothesis test of part **a** to be valid?

8.102 Failure to meet payments on student loans guaranteed by the U.S. government has been a major problem both for banks and for the government. Approximately 50% of all student loans guaranteed by the government are in default. A random sample of 350 loans to college students in one region of the United States indicates that 147 loans are in default.

a. Do the data indicate that the proportion of student loans in default in this area of the country differs from the proportion of all student loans in the United States that are in default? Use $\alpha = .01$.

b. Find the observed significance level for the test and interpret its value.

8.103 In order to be effective, a certain mechanical component used in a spacecraft must have a mean length of life greater than 1,100 hours. Owing to the prohibitive cost of this component, only three were tested under simulated space conditions. The lifetimes (hours) of the components were recorded and the computed statistics were: $\bar{x} = 1,173.6$ and

$s = 36.3$. These data were analyzed using MINITAB (shown at the bottom of the page).

a. Verify that the software has correctly calculated the t statistic, and use Table VI of Appendix A to determine whether the p-value is in the appropriate range.

b. Interpret the p-value.

c. What assumptions are necessary for the validity of this test?

d. Which type of error, I or II, is of greater concern for this test? Explain.

e. Would you recommend that this component be passed as meeting specifications?

8.104 A consumer protection group is concerned that a ketchup manufacturer is filling its 20-ounce family-size containers with less than 20 ounces of ketchup. The group purchases 10 family-size bottles of this ketchup, weighs the contents of each, and finds that the mean weight is equal to 19.86 ounces, and the standard deviation is equal to .22 ounce.

a. Do the data provide sufficient evidence for the consumer group to conclude that the mean fill per family-size bottle is less than 20 ounces? Test using $\alpha = .05$.

b. If the test in part **a** were conducted on a periodic basis by the company's quality control department, is the consumer group more concerned about making a Type I error or a Type II error? (The probability of making this type of error is called the *consumer's risk.*)

c. The ketchup company is also interested in the mean amount of ketchup per bottle. It does not wish to overfill them. For the test conducted in part **a**, which type of error is more serious from the company's point of view—a Type I error or a Type II error? (The probability of making this type of error is called the *producer's risk.*)

8.105 A sporting goods manufacturer who produces both white and yellow golf balls claims that more than 75% of all golf balls sold are white. A marketing study of the purchases of white and yellow golf balls at a number of stores showed that of 470 balls sold, 410 were white and 60 were yellow.

a. Is there sufficient evidence to support the manufacturer's claim? Test using $\alpha = .01$.

b. Find the observed significance level for the test and interpret its value.

*c. Calculate the probability, β, of a Type II error if in fact 80% of the golf balls sold are white.

*8.106 To perform an experiment, a chemist has to use a substance that contains 50% by weight of sodium

```
TEST OF MU = 1100 VS MU G.T. 1100

            N      MEAN    STDEV   SE MEAN      T    P VALUE
COMP        3    1,173.6    36.3    20.96    3.512     .0362
```

nitrate. The chemist suspects that a particular batch of the substance has not been mixed thoroughly, thus causing the amount of sodium nitrate to vary from one portion of the batch to another. The results of 20 randomly selected 10-gram samples yield a sample standard deviation equal to .05 gram. Estimate the true variance of the amount of sodium nitrate in 10-gram samples selected from the batch. Use a 95% confidence interval.

8.107 The EPA sets a limit of 5 parts per million (ppm) on PCB (a dangerous substance) in water. A major manufacturing firm producing PCB for electrical insulation discharges small amounts from the plant. The company management, attempting to control the PCB in its discharge, has given instructions to halt production if the mean amount of PCB in the effluent exceeds 3 ppm. A random sample of 50 water specimens produced the following statistics: $\bar{x} = 3.1$ ppm and $s = .5$ ppm.

    a. Do these statistics provide sufficient evidence to halt the production process? Use $\alpha = .01$.

    b. If you were the plant manager, would you want to use a large or a small value for α for the test in part **a**?

*8.108 Refer to Exercise 8.107.

    a. In the context of the problem, define a Type II error.

    b. Calculate β for the test described in part **a** of Exercise 8.107 assuming that the true mean is $\mu = 3.1$ ppm.

    c. What is the power of the test to detect the effluent's departure from the standard of 3.0 ppm when the mean is 3.1 ppm?

    d. Repeat parts **b** and **c** assuming that the true mean is 3.2 ppm. What happens to the power of the test as the plant's mean PCB departs further from the standard?

*8.109 Refer to Exercises 8.107 and 8.108.

    a. Suppose an α value of .05 is used to conduct the test. Does this change favor the manufacturer? Explain.

    b. Determine the value of β and the power for the test when $\alpha = .05$ and $\mu = 3.1$.

    c. What happens to the power of the test when α is increased?

8.110 One study (*Journal of Political Economy*, Feb. 1988) of gambling newsletters that purport to improve a bettor's odds of winning bets on NFL football games indicates that the newsletters' betting schemes were not profitable. Suppose a random sample of 50 games is selected to test one gambling newsletter. Following the newsletter's recommendations, 30 of the 50 games produced winning wagers.

    a. Test whether the newsletter can be said to significantly increase the odds of winning over what one could expect by selecting the winner at random. Use $\alpha = .05$.

    b. Calculate and interpret the *p*-value for the test.

*8.111 Refer to Exercise 8.110.

    a. Describe a Type II error in terms of this application.

    b. Calculate the probability β of a Type II error for this test assuming that the newsletter really does increase the probability of winning a wager on an NFL game to $p = .55$.

    c. Suppose the number of games sampled is increased from 50 to 100. How does this affect the probability of a Type II error for $p = .55$?

8.112 A large mail-order company has placed an order for 5,000 electric can openers with a supplier on the condition that no more than 2% of the devices will be defective. To check the shipment, the company tests a random sample of 400 of the can openers and finds that 11 are defective. Does this provide sufficient evidence to indicate that the proportion of defective can openers in the shipment exceeds 2%? Test using $\alpha = .05$.

8.113 Refer to Exercise 8.112. Find and interpret the observed significance level for the hypothesis test.

8.114 A psychologist was interested in knowing whether male heroin addicts' assessments of self-worth differ from those of the general male population. On a test designed to measure assessment of self-worth, the mean score for males from the general population is 48.6. A random sample of 25 scores achieved by heroin addicts yielded a mean of 44.1 and a standard deviation of 6.2.

    a. Do the data indicate a difference in assessment of self-worth between male heroin addicts and the general male population? Test using $\alpha = .01$.

    b. Give the approximate observed significance level for the test and interpret its value.

8.115 The "beta coefficient" of a stock is a measure of the stock's volatility (or risk) relative to the market as a whole. Stocks with beta coefficients greater than 1 generally bear greater risk (more volatility) than the market, whereas stocks with beta coefficients less than 1 are less risky (less volatile) than the overall market. A random sample of 15 high-technology stocks was selected at the end of 1995, and the mean and standard deviation of the beta coefficients were calculated: $\bar{x} = 1.23$, $s = .37$.

    a. Set up the appropriate null and alternative hypotheses to test whether the average high-technology stock is riskier than the market as a whole.

b. Establish the appropriate test statistic and rejection region for the test. Use $\alpha = .10$.

c. What assumptions are necessary to ensure the validity of the test?

d. Calculate the test statistic and state your conclusion.

e. What is the approximate p-value associated with this test? Interpret it.

8.116 Refer to Exercise 8.115. SAS was used to analyze the data, and the output was as shown below.

```
Analysis Variable : BETA_1

  N Obs              T    Prob>|T|
     15     2.4080000      0.0304
```

a. Interpret the p-value on the computer output. (**Prob**$>$|**T**| on the SAS printout corresponds to a two-tailed test of the null hypothesis $\mu = 1$.)

b. If the alternative hypothesis of interest is $\mu > 1$, what is the appropriate p-value of the test?

8.117 Officials from a high school claim that at least 85% of the students who have graduated from the school have received a college degree or are enrolled in a college degree program. A random sample of 60 former graduates indicates that 47 have received or are enrolled in a program to receive a college degree. Do the data contradict the school official's claim? (Use $\alpha = .05$.)

8.118 In the past, a chemical company produced 880 pounds of a certain type of plastic per day. Now, using a newly developed and less expensive process, the mean daily yield of plastic for the first 50 days of production is 871 pounds; the standard deviation is 21 pounds.

a. Do the data provide sufficient evidence to indicate that the mean daily yield for the new process is less than that for the old procedure? (Test using $\alpha = .01$.)

b. What assumptions must you make in order to use the statistical test you employed?

*8.119 Refer to Exercise 8.118. Calculate the probability β that the test fails to reject the null hypothesis that the new process has a mean daily yield of 880 when in fact the true mean is 875. What is the power of the test to determine the $\mu < 880$ when $\mu = 875$?

*8.120 Ophthalmologists require an instrument that can rapidly measure interocular pressure for glaucoma patients. The device now in general use is known to yield readings of this pressure with a variance of 10.3. The variance of five pressure readings on the same eye by a newly developed instrument is equal to 9.8. Does this sample variance provide sufficient evidence to indicate that the new instrument is more reliable than the instrument currently in use? (Use $\alpha = .05$.)

*8.121 The standard deviation of the diameters of screw-top lids must not be larger than .6 millimeter (mm) to ensure that the lids will fit properly on glass jars. A random sample of 15 lids yields a sample standard deviation of .78 mm. Do the data provide sufficient evidence to conclude that the standard deviation of the lid diameters is larger than .6? (Use $\alpha = .01$.)

STUDENT PROJECTS

The "efficient market" theory postulates that the best predictor of a stock's price at some point in the future is the current price of the stock (with some adjustments for inflation and transaction costs, which we shall assume to be negligible for the purpose of this exercise). To test this theory, select a random sample of 25 stocks on the New York Stock Exchange and record the closing prices on the last days of two recent consecutive months. Calculate the increase or decrease in the stock price over the 1-month period.

a. Define μ as the mean change in price of all stocks over a 1-month period. Set up the appropriate null and alternative hypotheses in terms of μ.

b. Use the sample of the 25 stock price differences to conduct the test of hypothesis established in part a. Use $\alpha = .05$.

c. Tabulate the number of rejections and nonrejections of the null hypothesis in your class. If the null hypothesis were true, how many rejections of the null hypothesis would you expect among those in your class? How does this expectation compare with the actual number of nonrejections? What does the result of this exercise indicate about the efficient market theory?

EXPLORING DATA WITH A COMPUTER

Refer to Exploring Data with a Computer in Chapters 6 and 7. Select one of the variables—tar, nicotine, or carbon monoxide content—measured in Appendix B. Let μ_0 be the mean value of this variable for the population of 972 cigarettes in Appendix B—i.e., μ_0 is the "true" value of the population mean. In each of the following scenarios you will be testing the null hypothesis H_0: $\mu = \mu_0$ versus the two-tailed alternative hypothesis H_a: $\mu \neq \mu_0$. Thus, in this exercise you know that the null hypothesis is true.

a. Select 100 samples of size 50 (with replacement) from the 972 cigarettes, and conduct the test for each sample using $\alpha = .05$. In how many of the tests does the sample lead to an incorrect rejection of the null hypothesis? How does this compare with what you expected to occur? Repeat the exercise using $\alpha = .10$.

b. Select 100 samples of size 20 (with replacement) from the 972 cigarettes, and conduct the test for each sample using $\alpha = .05$. In how many of the tests does the sample lead to an incorrect rejection of the null hypothesis? How does this compare with what you expected to occur? Repeat the exercise using $\alpha = .10$.

Chapter 9

INFERENCES BASED ON TWO SAMPLES

Confidence Intervals and Tests of Hypotheses

Contents

Case Studies

*W*HERE WE'VE BEEN

We explored two methods for making statistical inferences, confidence intervals and tests of hypotheses based on single samples, in Chapters 7 and 8. In particular, we gave confidence intervals and tests of hypotheses concerning a population mean μ and a population proportion p, and we learned how to select the sample size necessary to obtain a specified amount of information concerning a parameter. (Inferences about a population variance σ^2 were included in an optional section.)

*W*HERE WE'RE GOING

Now that we've learned to make inferences about a single population, we'll learn how to compare two populations. For example, we may wish to compare the mean gas mileages for two models of automobiles, or the mean reaction times of men and women to a visual stimulus, or the proportions of subjects in a psychological experiment that respond to two different stimuli. In this chapter we'll see how to decide whether differences exist and how to estimate the differences between population means and proportions.

Many experiments involve a comparison of two populations. For instance, a sociologist may want to estimate the difference in mean life expectancy between inner-city and suburban residents. A consumer group may want to test whether two major brands of food freezers differ in the mean amount of electricity they use. A political candidate may wish to estimate the difference in the proportions of voters in two districts who favor his or her candidacy. A professional golfer may wish to compare the variability in the distance that two competing brands of golf balls travel when struck with the same club. In this chapter we consider techniques for using two samples to compare the populations from which they were selected.

9.1 COMPARING TWO POPULATION MEANS: INDEPENDENT SAMPLING

Many of the same procedures that are used to estimate and test hypotheses about a single parameter can be modified to make inferences about two parameters. Both the z and t statistics may be adapted to make inferences about the difference between two population means.

In this section we develop both large-sample and small-sample methodologies for comparing two population means. In the large-sample case we use the z statistic, while in the small-sample case we use the t statistic.

Large Samples

EXAMPLE 9.1

A dietitian has developed a diet that is low in fats, carbohydrates, and cholesterol. Although the diet was initially intended to be used by people with heart disease, the dietitian wishes to examine the effect this diet has on the weights of obese people. Two random samples of 100 obese people each are selected, and one group of 100 is placed on the low-fat diet. The other 100 are placed on a diet that contains approximately the same quantity of food but is not as low in fats, carbohydrates, and cholesterol. For each person, the amount of weight lost (or gained) in a 3-week period is recorded. Using the summary data given in Table 9.1, form a 95% confidence interval for the difference between the population mean weight losses for the two diets. Interpret the result.

Solution

Recall that the general form of a large-sample confidence interval for a single mean μ is $\bar{x} \pm z_{\alpha/2}\sigma_{\bar{x}}$. That is, we add and subtract $z_{\alpha/2}$ standard deviations of the sample estimate, $\bar{x}$, to the value of the estimate. We employ a similar procedure to form the confidence interval for the difference between two population means.

Let μ_1 represent the mean of the conceptual population of weight losses for all obese people who could be placed on the low-fat diet. Let μ_2 be similarly defined

TABLE 9.1 Summary Statistics for Diet Study

	Low-Fat Diet	**Other Diet**
Sample size	100	100
Sample mean weight loss	9.3 pounds	7.4 pounds
Sample variance	22.4	16.3

for the other diet. We wish to form a confidence interval for $(\mu_1 - \mu_2)$. An intuitively appealing estimator for $(\mu_1 - \mu_2)$ is the difference between the sample means, $(\bar{x}_1 - \bar{x}_2)$. Thus, we will form the confidence interval of interest by

$$(\bar{x}_1 - \bar{x}_2) \pm z_{\alpha/2}\sigma_{(\bar{x}_1 - \bar{x}_2)}$$

Assuming the two samples are independent, the standard deviation of the difference between the sample means is

$$\sigma_{(\bar{x}_1 - \bar{x}_2)} = \sqrt{\frac{\sigma_1^2}{n_1} + \frac{\sigma_2^2}{n_2}} \approx \sqrt{\frac{s_1^2}{n_1} + \frac{s_2^2}{n_2}}$$

Using the sample data and noting that $\alpha = .05$ and $z_{.025} = 1.96$, we find that the 95% confidence interval is, approximately,

$$(9.3 - 7.4) \pm 1.96\sqrt{\frac{22.4}{100} + \frac{16.3}{100}} = 1.9 \pm (1.96)(.62) = 1.9 \pm 1.22$$

or (.68, 3.12). Using this estimation procedure over and over again for different samples, we know that approximately 95% of the confidence intervals formed in this manner will enclose the difference in population means $(\mu_1 - \mu_2)$. Therefore, we are highly confident that the mean weight loss for the low-fat diet is between .68 and 3.12 pounds more than the mean weight loss for the other diet. With this information, the dietitian better understands the potential of the low-fat diet as a weight-reducing diet.   ▲

The justification for the procedure used in Example 9.1 to estimate $(\mu_1 - \mu_2)$ relies on the properties of the sampling distribution of $(\bar{x}_1 - \bar{x}_2)$. The performance of the estimator in repeated sampling is pictured in Figure 9.1, and its properties are summarized in the next box.

Properties of the Sampling Distribution of $(\bar{x}_1 - \bar{x}_2)$

1. The mean of the sampling distribution of $(\bar{x}_1 - \bar{x}_2)$ is $(\mu_1 - \mu_2)$.
2. If the two samples are independent, the standard deviation of the sampling distribution is

$$\sigma_{(\bar{x}_1 - \bar{x}_2)} = \sqrt{\frac{\sigma_1^2}{n_1} + \frac{\sigma_2^2}{n_2}}$$

where σ_1^2 and σ_2^2 are the variances of the two populations being sampled and n_1 and n_2 are the respective sample sizes. We also refer to $\sigma_{(\bar{x}_1 - \bar{x}_2)}$ as the **standard error** of the statistic $(\bar{x}_1 - \bar{x}_2)$.
3. The sampling distribution of $(\bar{x}_1 - \bar{x}_2)$ is approximately normal for *large samples* by the Central Limit Theorem.

FIGURE 9.1

Sampling distribution of $(\bar{x}_1 - \bar{x}_2)$

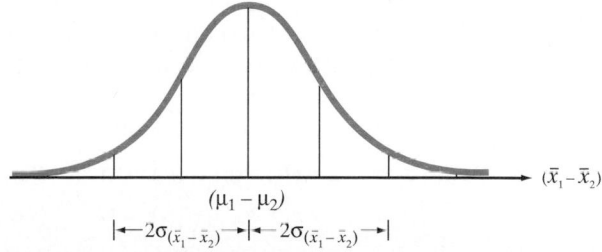

In Example 9.1, we noted the similarity in the procedures for forming a large-sample confidence interval for one population mean and a large-sample confidence interval for the difference between two population means. When we are testing hypotheses, the procedures are again very similar. The general large-sample procedures for forming confidence intervals and testing hypotheses about $(\mu_1 - \mu_2)$ are summarized in the next two boxes.

Large Sample Confidence Interval for $(\mu_1 - \mu_2)$

$$(\bar{x}_1 - \bar{x}_2) \pm z_{\alpha/2}\sigma_{(\bar{x}_1 - \bar{x}_2)} = (\bar{x}_1 - \bar{x}_2) \pm z_{\alpha/2}\sqrt{\frac{\sigma_1^2}{n_1} + \frac{\sigma_2^2}{n_2}}$$

Assumptions: The two samples are randomly selected in an independent manner from the two populations. The sample sizes, n_1 and n_2, are large enough so that $\bar{x}_1$ and $\bar{x}_2$ both have approximately normal sampling distributions and so that s_1^2 and s_2^2 provide good approximations to σ_1^2 and σ_2^2. This will be true if $n_1 \geq 30$ and $n_2 \geq 30$.

Large-Sample Test of Hypothesis for $(\mu_1 - \mu_2)$

        ONE-TAILED TEST   **TWO-TAILED TEST**

H_0: $(\mu_1 - \mu_2) = D_0$ H_0: $(\mu_1 - \mu_2) = D_0$
H_a: $(\mu_1 - \mu_2) < D_0$ H_a: $(\mu_1 - \mu_2) \neq D_0$
    [or H_a: $(\mu_1 - \mu_2) > D_0$]

where D_0 = Hypothesized difference between the means (this difference is often hypothesized to be equal to 0)

Test statistic:

$$z = \frac{(\bar{x}_1 - \bar{x}_2) - D_0}{\sigma_{(\bar{x}_1 - \bar{x}_2)}} \quad \text{where} \quad \sigma_{(\bar{x}_1 - \bar{x}_2)} = \sqrt{\frac{\sigma_1^2}{n_1} + \frac{\sigma_2^2}{n_2}}$$

Rejection region: $z < -z_{\alpha}$ *Rejection region:* $|z| > z_{\alpha/2}$
    [or $z > z_{\alpha}$ when
    H_a: $(\mu_1 - \mu_2) > D_0$]

Assumptions: Same as for the large-sample confidence interval.

EXAMPLE 9.2

Refer to the study of obese people on a low-fat diet and a regular diet, Example 9.1. Another way to compare the mean weight losses for the two different diets is to conduct a test of hypothesis. Use the summary data in Table 9.1 to conduct the test. Use $\alpha = .05$.

Solution

Again, we let μ_1 and μ_2 represent the population mean weight losses of obese people on the low-fat diet and regular diet, respectively. If one diet is more effective in reducing the weights of obese people, then either $\mu_1 < \mu_2$ or $\mu_2 < \mu_1$; that is, $\mu_1 \neq \mu_2$.

Thus, the elements of the test are as follows:

H_0: $(\mu_1 - \mu_2) = 0$ (i.e., $\mu_1 = \mu_2$; note that $D_0 = 0$ for this hypothesis test)
H_a: $(\mu_1 - \mu_2) \neq 0$ (i.e., $\mu_1 \neq \mu_2$)

FIGURE 9.2

Rejection region for
Example 9.2

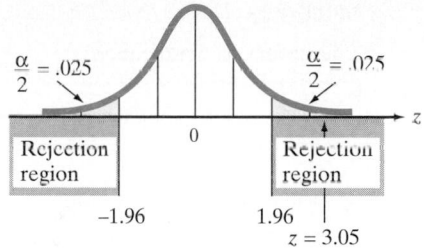

Test statistic: $z = \dfrac{(\bar{x}_1 - \bar{x}_2) - D_0}{\sigma_{(\bar{x}_1 - \bar{x}_2)}} = \dfrac{\bar{x}_1 - \bar{x}_2 - 0}{\sigma_{(\bar{x}_1 - \bar{x}_2)}}$

Rejection region: $z < -z_{w2} = -1.96$ or $z > z_{w2} = 1.96$ (see Figure 9.2)

Substituting the summary statistics given in Table 9.1 into the test statistic,
we obtain

$$z = \frac{(\bar{x}_1 - \bar{x}_2) - 0}{\sigma_{(\bar{x}_1 - \bar{x}_2)}} = \frac{9.3 - 7.4}{\sqrt{\dfrac{\sigma_1^2}{n_1} + \dfrac{\sigma_2^2}{n_2}}}$$

$$\approx \frac{1.9}{\sqrt{\dfrac{s_1^2}{n_1} + \dfrac{s_2^2}{n_2}}} = \frac{1.9}{\sqrt{\dfrac{22.4}{100} + \dfrac{16.3}{100}}} = \frac{1.9}{.622} = 3.05$$

As you can see in Figure 9.2, the calculated z value clearly falls in the rejection
region. Therefore, the samples provide sufficient evidence, at $\alpha = .05$, for the
dietician to conclude that the mean weight losses for the two diets differ.

This conclusion agrees with the inference drawn from the 95% confidence
interval in Example 9.1. However, the confidence interval provides more information on the mean weight losses. From the hypothesis test, we know only that
the two means differ; that is, $\mu_1 \neq \mu_2$. From the confidence interval in Example
9.1, we found that the mean weight loss μ_1 of the low-fat diet was between .68 and
3.12 pounds more than the mean weight loss μ_2 of the regular diet. In other words,
the test tells us that the means differ, but the confidence interval tells us how large
the difference is. Both inferences are made with the same degree of reliability—
namely, with 95% confidence (or, at $\alpha = .05$). ▲

EXAMPLE 9.3   Find the observed significance level for the test in Example 9.2. Interpret the result.

 Solution

The alternative hypothesis in Example 9.2, H_a: $\mu_1 - \mu_2 \neq 0$, required a two-tailed
test using

$$z = \frac{\bar{x}_1 - \bar{x}_2}{\sigma_{(\bar{x}_1 - \bar{x}_2)}}$$

as a test statistic. Since the z value calculated from the sample data was 3.05, the
observed significance level (p-value) for the two-tailed test is the probability of
observing a value of z at least as contradictory to the null hypothesis as $z =
3.05$; that is,

$$p\text{-value} = 2 \cdot P(z \geq 3.05)$$

FIGURE 9.3

The observed significance level for Example 9.2

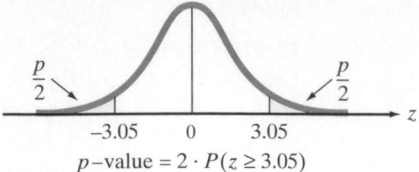

$$p\text{-value} = 2 \cdot P(z \geq 3.05)$$

This probability is computed assuming H_0 is true and is equal to the shaded area shown in Figure 9.3.

The tabulated area corresponding to $z = 3.05$ in Table IV of Appendix A is .4989. Therefore,

$$P(z \geq 3.05) = .5 - .4989 = .0011$$

and the observed significance level for the test is

$$p\text{-value} = 2(.0011) = .0022$$

Since our selected α value, .05, exceeds this p-value, we have sufficient evidence to reject $H_0: \mu_1 - \mu_2 = 0$.

The p-value of the test is more easily obtained from a statistical software package. An ASP printout for the hypothesis test is displayed in Figure 9.4. The two-tailed p-value, shaded on the printout, is .002256.* This value agrees (except for rounding) with our calculated p-value.   ▲

Small Samples

When comparing two population means with small samples (say, $n_1 < 30$ and $n_2 < 30$), the methodology of the previous three examples is invalid. The reason? When the sample sizes are small, estimates of σ_1^2 and σ_2^2 are unreliable and the Central Limit Theorem (which guarantees that the z statistic is normal) can no

FIGURE 9.4

ASP printout for the hypothesis test of Example 9.2

```
         HYPOTHESIS: MEAN LOWFAT (X) = MEAN REGULAR (Y)

        SAMPLE MEAN OF X   =     9.3
    SAMPLE VARIANCE OF X   =    22.4
        SAMPLE SIZE OF X   =   100
        SAMPLE MEAN OF Y   =     7.4
    SAMPLE VARIANCE OF Y   =    16.3
        SAMPLE SIZE OF Y   =   100

        MEAN X - MEAN Y    =     1.9
                     Z     =     3.0542
               P-VALUE     =     2.25658E-3
             P-VALUE/2     =     1.12829E-3
             SD. ERROR     =     0.622093
                     t     =     3.0542
                  D. F.    =   193
               P-VALUE     =     2.57535E-3
             P-VALUE/2     =     1.28768E-3
             SD. ERROR     =     0.622093
```

*ASP uses E-notation where necessary for extremely large or small numbers. E-3 instructs the user to move the decimal point of the preceding number 3 places to the left. Similarly, E+5 instructs the user to move the decimal 5 places to the right.

FIGURE 9.5

Assumptions for the
two-sample t:
(1) Normal populations,
(2) equal variances

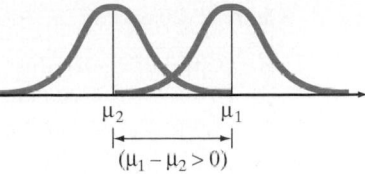

μ_2 μ_1

$(\mu_1 - \mu_2 > 0)$

longer be applied. But as in the case of a single mean (Section 8.4), we use the familiar Student's t-distribution described in Chapter 7.

To use the t-distribution, both sampled populations must be approximately normally distributed with equal population variances, and the random samples must be selected independently of each other.

The normality and equal variances assumptions imply relative frequency distributions for the populations that would appear as shown in Figure 9.5.

Since we assume the two populations have equal variances ($\sigma_1^2 = \sigma_2^2 = \sigma^2$), it is reasonable to use the information contained in both samples to construct a **pooled sample estimator** of σ^2 for use in confidence intervals and test statistics. Thus, if s_1^2 and s_2^2 are the two sample variances (both estimating the variance σ^2 common to both populations), the pooled estimator of σ^2, denoted as s_p^2, is

$$s_p^2 = \frac{(n_1 - 1)s_1^2 + (n_2 - 1)s_2^2}{(n_1 - 1) + (n_2 - 1)} = \frac{(n_1 - 1)s_1^2 + (n_2 - 1)s_2^2}{n_1 + n_2 - 2}$$

or

$$s_p^2 = \frac{\overbrace{\sum(x_1 - \bar{x}_1)^2}^{\text{From sample 1}} + \overbrace{\sum(x_2 - \bar{x}_2)^2}^{\text{From sample 2}}}{n_1 + n_2 - 2}$$

where x_1 represents a measurement from sample 1 and x_2 represents a measurement from sample 2. Recall that the term *degrees of freedom* was defined in Section 7.2 as 1 less than the sample size. Thus, in this case, we have $(n_1 - 1)$ degrees of freedom for sample 1 and $(n_2 - 1)$ degrees of freedom for sample 2. Since we are pooling the information on σ^2 obtained from both samples, the degrees of freedom associated with the pooled variance s_p^2 is equal to the sum of the degrees of freedom for the two samples, namely, the denominator of s_p^2; that is, $(n_1 - 1) + (n_2 - 1) = n_1 + n_2 - 2$.

Note that the second formula given for s_p^2 shows that the pooled variance is simply a **weighted average** of the two sample variances, s_1^2 and s_2^2. The weight given each variance is proportional to its degrees of freedom. If the two variances have the same number of degrees of freedom (i.e., if the sample sizes are equal), then the pooled variance is a simple average of the two sample variances. The result is an average or "pooled" variance that is a better estimate of σ^2 than either s_1^2 or s_2^2 alone.

Both the confidence interval and the test of hypothesis procedures for comparing two population means with small samples are summarized in the accompanying boxes.

Small-Sample Confidence Interval for ($\mu_1 - \mu_2$) (Independent Samples)

$$(\bar{x}_1 - \bar{x}_2) \pm t_{\alpha/2}\sqrt{s_p^2\left(\frac{1}{n_1} + \frac{1}{n_2}\right)}$$

where $s_p^2 = \dfrac{(n_1 - 1)s_1^2 + (n_2 - 1)s_2^2}{n_1 + n_2 - 2}$

continued

and $t_{\alpha/2}$ is based on $(n_1 + n_2 - 2)$ degrees of freedom.

Assumptions:
1. Both sampled populations have relative frequency distributions that are approximately normal.
2. The population variances are equal.
3. The samples are randomly and independently selected from the populations.

Small-Sample Test of Hypothesis for $(\mu_1 - \mu_2)$ (Independent Samples)

ONE-TAILED TEST	TWO-TAILED TEST

H_0: $(\mu_1 - \mu_2) = D_0$ $\qquad\qquad\qquad$ H_0: $(\mu_1 - \mu_2) = D_0$

H_a: $(\mu_1 - \mu_2) < D_0$ $\qquad\qquad\quad$ H_a: $(\mu_1 - \mu_2) \neq D_0$

$\qquad$ [or H_a: $(\mu_1 - \mu_2) > D_0$]

Test statistic:

$$t = \frac{(\bar{x}_1 - \bar{x}_2) - D_0}{\sqrt{s_p^2\left(\dfrac{1}{n_1} + \dfrac{1}{n_2}\right)}}$$

Rejection region: $\quad t < -t_\alpha$ $\qquad\qquad$ *Rejection region:* $\quad |t| > t_{\alpha/2}$

$\qquad$ [or $t > t_\alpha$ when

$\qquad$ H_a: $(\mu_1 - \mu_2) > D_0$]

where t_α and $t_{\alpha/2}$ are based on $(n_1 + n_2 - 2)$ degrees of freedom.

Assumptions: Same as for the small-sample confidence interval for $(\mu_1 - \mu_2)$ in the previous box.

EXAMPLE 9.4

Suppose you wish to compare a new method of teaching reading to "slow learners" to the current standard method. You decide to base this comparison on the results of a reading test given at the end of a learning period of 6 months. Of a random sample of 20 slow learners, 8 are taught by the new method and 12 are taught by the standard method. All 20 children are taught by qualified instructors under similar conditions for a 6-month period. The results of the reading test at the end of this period are given in Table 9.2. Use the data in the table to estimate the true mean difference between the test scores for the new method and the standard method. Use a 95% confidence interval, and interpret the interval. What assumptions must be made in order that the estimate be valid?

Solution

For this experiment, let μ_1 and μ_2 represent the mean reading test scores of slow learners taught with the new and standard methods, respectively. Then, the objective is to obtain a 95% confidence interval for $(\mu_1 - \mu_2)$. To use the small-sample confidence interval for $(\mu_1 - \mu_2)$, the following assumptions must be satisfied:

TABLE 9.2 Reading Test Scores for Slow Learners

New Method				Standard Method			
80	80	79	81	79	62	70	68
76	66	79	76	73	76	86	73
				72	68	75	66

FIGURE 9.6

SAS printout for Example 9.4

```
Analysis Variable : Score

-------------------------- METHOD=New --------------------------
 N Obs    N     Minimum       Maximum           Mean       Std Dev
    8     8   66.0000000   81.0000000    77.1250000    4.8532022


-------------------------- METHOD=Std --------------------------
 N Obs    N     Minimum       Maximum           Mean       Std Dev
   12    12   62.0000000   86.0000000    72.3333333    6.3436917
```

1. We assume that the populations of test scores are normally distributed for both the new and standard methods of instruction. Since a test score can be viewed as a sum of the results on the various components of the test, the Central Limit Theorem lends credence to this assumption.

2. The variance of the test scores is assumed to be the same for the two populations. Under the circumstances, we might expect the variation in test scores to be approximately the same for both methods.

3. The samples are randomly and independently selected from the two populations. We have randomly chosen 20 different slow learners for the two samples in such a way that the test score for one child is not dependent on the test score for any other child. Therefore, this assumption would probably be valid.

The first step in constructing the confidence interval is to obtain summary statistics (e.g., $\bar{x}$ and s) on reading test scores for each method. The data of Table 9.2 were entered into a computer, and SAS used to obtain these descriptive statistics. The SAS printout appears in Figure 9.6. Note that $\bar{x}_1 = 77.125$, $s_1 = 4.853$, $\bar{x}_2 = 72.333$, and $s_2 = 6.344$.

Next, we calculate the pooled estimate of variance:

$$s_p^2 = \frac{(n_1 - 1)s_1^2 + (n_2 - 1)s_2^2}{n_1 + n_2 - 2}$$

$$= \frac{(8 - 1)(4.853)^2 + (12 - 1)(6.344)^2}{8 + 12 - 2} = 33.754$$

where s_p^2 is based on $(n_1 + n_2 - 2) = (8 + 12 - 2) = 18$ degrees of freedom. Also, we find $t_{\alpha/2} = t_{.025} = 2.101$ (based on 18 degrees of freedom) from Table VI of Appendix A.

Finally, the 95% confidence interval for $(\mu_1 - \mu_2)$, the difference between mean test scores for the two methods, is

$$(\bar{x}_1 - \bar{x}_2) \pm t_{\alpha/2}\sqrt{s_p^2\left(\frac{1}{n_1} + \frac{1}{n_2}\right)} = (77.12 - 72.33) \pm t_{.025}\sqrt{33.754\left(\frac{1}{8} + \frac{1}{12}\right)}$$

$$= 4.79 \pm (2.101)(2.652)$$

$$= 4.79 \pm 5.57$$

or $(-.78, 10.36)$. Therefore, with a confidence coefficient equal to .95, we estimate the difference in mean test scores between using the new method of teaching and the standard method to fall in the interval $-.78$ to 10.36. In other words, we estimate the mean test score for the new method to be anywhere from .78 points less to 10.36 points more than the mean test score for the standard method. Although the sample means seem to suggest that the new method is associated with a higher mean test score, there is insufficient evidence to indicate that $(\mu_1 - \mu_2)$ differs from 0 because the interval includes 0 as a possible value for $(\mu_1 - \mu_2)$. To show a difference in mean test scores (if it exists), you could

Detection of Rigged Milk Prices

Many products and services are purchased by government agencies, cities, states, and businesses on the basis of sealed bids, and contracts are awarded to the lowest bidders. This process works extremely well in competitive markets, but it has the potential to increase the cost of purchasing if the markets are noncompetitive or if bidders are engaged in collusive practices. The latter may have occurred recently in the Kentucky milk market.

CASE STUDY • 9.1 •

Each year, the state of Kentucky invites bids from dairies to supply half-pint containers of fluid milk products for its school districts. In several districts in northern Kentucky, the suppliers (dairies) were accused of "price-fixing," i.e., conspiring to set the price of half-pint milk containers above the "fair," or competitive, price. A market analysis disclosed that two dairies— Meyer Dairy and Trauth Dairy—were the only two bidders on the milk contracts in the school districts in the "tri-county market" of Boone, Campbell, and Kenton Counties between 1983 and 1991. Consequently, these two companies were awarded all the milk contracts in the market. Speculation is that Meyer and Trauth conspired to rig their bids in the tri-county market.

Numerous economic and statistical methods exist for detecting the possibility of collusive practices among bidders. These typically involve (1) market-shares analyses, (2) an analysis of incumbency rates (i.e., the percentage of markets that are won by the same vendor who won the previous year), (3) an examination of bid levels and their dispersion, (4) price-distance analyses, (5) bid-sequence analyses, (6) comparison of mean winning prices, and (7) econometric modeling of the winning price. For this case, we'll focus on a comparison of the mean winning bid prices; however, a complete examination would consist of all seven types of analyses.

Consider two similar markets, one in which the bids are rigged and the other in which bids are made competitively. In theory, the mean winning price in the "rigged" market will be higher than the mean winning price in the competitive market. Recall that in this case, the speculation is that from 1983 to 1991 the tri-county milk market (where Meyer and Trauth dairies supplied most of the milk) was rigged while the school districts in northern Kentucky that surround the tri-county market (the "surrounding market") was competitive. Consequently, you will want to compare the mean winning price in the tri-county market with that of the surrounding market for each of the years in question.

The state of Kentucky has granted permission for you to analyze its milk data. The data, described in Appendix B, consist of the prices (in dollars) of all winning bids in the tri-county and surrounding markets from 1983 to 1991 for each of three products: whole white milk, low-fat white milk, and low-fat chocolate milk.

---

increase the sample size, thereby narrowing the width of the confidence interval for $(\mu_1 - \mu_2)$. An alternative is to design the experiment differently. This possibility is discussed in the next section. ▲

The two-sample t statistic is a powerful tool for comparing population means when the assumptions are satisfied. It has also been shown to retain its usefulness when the sampled populations are only approximately normally distributed. And when the sample sizes are equal, the assumption of equal population variances can be relaxed. That is, if $n_1 = n_2$, then σ_1^2 and σ_2^2 can be quite different and the test statistic will still possess, approximately, a Student's t-distribution. When the assumptions are not satisfied, you can select larger samples from the populations or you can use other available statistical tests (nonparametric statistical tests, which are described in Chapter 15).

What Should You Do If the Assumptions Are Not Satisfied?

Answer: If you are concerned that the assumptions are not satisfied, use the Wilcoxon rank sum test for independent samples to test for a shift in population distributions. See Chapter 15.

Case Study 9.1 continued

Focus

a. The mean winning bid prices for half-pint containers of low-fat white milk in the two markets are plotted by year on the same graph in Figure 9.7. Do you detect a pattern in the graph that supports the bid-collusion theory? Explain.

b. Why is Figure 9.7 insufficient to claim, with a high degree of confidence, that the true population mean winning price of low-fat white milk in the tri-county market exceeds the population mean in the surrounding market?

c. Figure 9.8 is a series of Student *t*-tests, performed using SAS, that compare the means by year. Use the information on the printouts to test the bid-collusion theory. Give your conclusions and recommendations in a written report intended for the state of Kentucky. [*Note:* Two *p*-values are reported in Figure 9.8 for each *t*-test performed. The *p*-value in the **Equal** row is appropriate when the assumption of equal vari-

ances is satisfied. (This is the *t*-test discussed in Section 9.1.) The *p*-value in the **Unequal** row is appropriate when the assumption of equal variances is not satisfied, i.e., when the two population variances are unequal.* For this question, use the *p*-value in the **Equal** row.]

d. (Optional question for those covering Section 9.5.) For each test conducted in part **c**, test whether the assumption about equal population variances is satisfied. [*Note:* The *p*-values for those tests can be found on the SAS printout, Figure 9.8.] After performing these tests, adjust (if necessary) the conclusions you made in part **c**.

e. The entire milk data set described in Appendix B is available in ASCII format on a 3.5" diskette from the publisher. Repeat the analysis in parts a–d for both whole milk and low-fat chocolate milk. Summarize your conclusions and make recommendations in a written report.

FIGURE 9.7

Plot of average winning low-fat white milk price

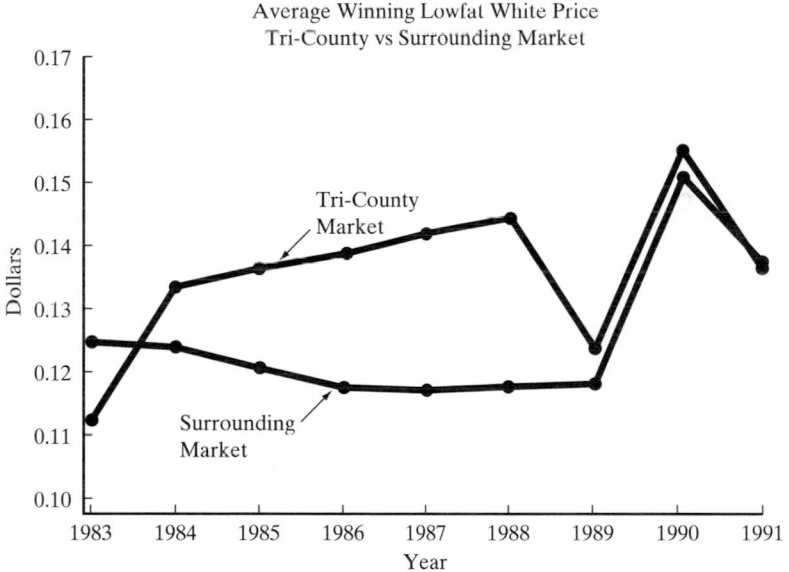

Average Winning Lowfat White Price
Tri-County vs Surrounding Market

*The *t*-test used in the unequal variances case is an approximate test based on the test statistic

$$t = \frac{(\bar{x}_1 - \bar{x}_2)}{\sqrt{\dfrac{s_1^2}{n_1} + \dfrac{s_2^2}{n_2}}}$$

with degrees of freedom

$$\nu \approx \left[\frac{\left(\dfrac{s_1^2}{n_1} + \dfrac{s_2^2}{n_2} \right)}{\dfrac{\left(\dfrac{s_1^2}{n_1} \right)^2}{n_1 - 1} + \dfrac{\left(\dfrac{s_2^2}{n_2} \right)^2}{n_2 - 1}} \right]$$

The test is based on Satterthwaite's (1946) approximation.

FIGURE 9.8 SAS printout of *t*-tests for low-fat white milk (Case Study 9.1)

```
                             TTEST PROCEDURE

*********************************YEAR=1983*********************************

Variable: LFWBID

Market          N            Mean           Std Dev          Std Error
---------------------------------------------------------------------------
Surround        26         0.12468077       0.01610881       0.00315920
Tri-County      11         0.11211818       0.01222545       0.00368611

Variances       T      DF      Prob>|T|
---------------------------------------
Unequal       2.5877   24.7     0.0159
Equal         2.3128   35.0     0.0267

For H0: Variances are equal, F' = 1.74   DF = (25,10)   Prob>F' = 0.3644


*********************************YEAR=1984*********************************

Variable: LFWBID

Market          N            Mean           Std Dev          Std Error
---------------------------------------------------------------------------
Surround        27         0.12380741       0.01446619       0.00278402
Tri-County      12         0.13384167       0.00635717       0.00183516

Variances       T      DF      Prob>|T|
---------------------------------------
Unequal      -3.0093   37.0     0.0047
Equal        -2.2931   37.0     0.0276

For H0: Variances are equal, F' = 5.18   DF = (26,11)   Prob>F' = 0.0067


*********************************YEAR=1985*********************************

Variable: LFWBID

Market          N            Mean           Std Dev          Std Error
---------------------------------------------------------------------------
Surround        30         0.12065667       0.01473012       0.00268934
Tri-County      13         0.13658462       0.00537445       0.00149061

Variances       T      DF      Prob>|T|
---------------------------------------
Unequal      -5.1801   40.4     0.0001
Equal        -3.7696   41.0     0.0005

For H0: Variances are equal, F' = 7.51   DF = (29,12)   Prob>F' = 0.0007


*********************************YEAR=1986*********************************

Variable: LFWBID

Market          N            Mean           Std Dev          Std Error
---------------------------------------------------------------------------
Surround        33         0.11782121       0.01180640       0.00205523
Tri-County      13         0.13909231       0.00533205       0.00147884

Variances       T      DF      Prob>|T|
---------------------------------------
Unequal      -8.4010   43.0     0.0001
Equal        -6.2183   44.0     0.0000

For H0: Variances are equal, F' = 4.90   DF = (32,12)   Prob>F' = 0.0055


*********************************YEAR=1987*********************************

Variable: LFWBID

Market          N            Mean           Std Dev          Std Error
---------------------------------------------------------------------------
Surround        35         0.11730857       0.01235100       0.00208770
Tri-County      13         0.14242308       0.00701738       0.00194627

Variances       T      DF      Prob>|T|
---------------------------------------
Unequal      -8.7991   37.8     0.0001
Equal        -6.8995   46.0     0.0000

For H0: Variances are equal, F' = 3.10   DF = (34,12)   Prob>F' = 0.0404
```

FIGURE 9.8   Continued

```
*****************************YEAR=1988************************************

Variable: LFWBID

Market          N               Mean            Std Dev          Std Error
----------------------------------------------------------------------------
Surround        35          0.11818571        0.01285522        0.00217923
Tri-County      13          0.14481538        0.00618019        0.00171408

Variances       T       DF      Prob>|T|
----------------------------------------
Unequal     -9.6219    42.7      0.0001
Equal       -7.1332    46.0      0.0000

For H0: Variances are equal, F' = 4.33   DF = (34,12)     Prob>F' = 0.0095


*****************************YEAR=1989************************************

Variable: LFWBID

Market          N               Mean            Std Dev          Std Error
----------------------------------------------------------------------------
Surround        36          0.11847222        0.00655951        0.00109325
Tri-County      13          0.12401538        0.00828350        0.00229743

Variances       T       DF      Prob>|T|
----------------------------------------
Unequal     -2.1787    17.7      0.0431
Equal       -2.4334    47.0      0.0188

For H0: Variances are equal, F' = 1.59   DF = (12,35)     Prob>F' = 0.2771


*****************************YEAR=1990************************************

Variable: LFWBID

Market          N               Mean            Std Dev          Std Error
----------------------------------------------------------------------------
Surround        34          0.15185588        0.00954524        0.00163700
Tri-County      14          0.15599286        0.00615236        0.00164429

Variances       T       DF      Prob>|T|
----------------------------------------
Unequal     -1.7830    37.2      0.0828
Equal       -1.4938    46.0      0.1421

For H0: Variances are equal, F' = 2.41   DF = (33,13)     Prob>F' = 0.0936


*****************************YEAR=1991************************************

Variable: LFWBID

Market          N               Mean            Std Dev          Std Error
----------------------------------------------------------------------------
Surround         5          0.13638000        0.00718485        0.00321316
Tri-County      12          0.13538333        0.00585768        0.00169097

Variances       T       DF      Prob>|T|
----------------------------------------
Unequal      0.2745     6.3      0.7925
Equal        0.3001    15.0      0.7682

For H0: Variances are equal, F' = 1.50   DF = (4,11)      Prob>F' = 0.5343
```

EXERCISES 9.1–9.28

Learning the Mechanics

9.1 In order to compare the means of two populations, independent random samples of 400 observations are selected from each population, with the following results:

Sample 1	Sample 2
$\bar{x}_1 = 5{,}275$	$\bar{x}_2 = 5{,}240$
$s_1 = 150$	$s_2 = 200$

a. Use a 95% confidence interval to estimate the difference between the population means ($\mu_1 - \mu_2$). Interpret the confidence interval.

b. Test the null hypothesis H_0: ($\mu_1 - \mu_2$) = 0 versus the alternative hypothesis H_a: ($\mu_1 - \mu_2$) ≠ 0. Give the significance level of the test, and interpret the result.

c. Suppose the test in part **b** was conducted with the alternative hypothesis H_a: $(\mu_1 - \mu_2) > 0$. How would your answer to part **b** change?

d. Test the null hypothesis H_0: $(\mu_1 - \mu_2) = 25$ versus the alternative H_a: $(\mu_1 - \mu_2) \neq 25$. Give the significance level, and interpret the result. Compare your answer to the test conducted in part **b**.

e. What assumptions are necessary to ensure the validity of the inferential procedures applied in parts **a–d**?

9.2 The purpose of this exercise is to compare the variability of $\bar{x}_1$ and $\bar{x}_2$ with the variability of $(\bar{x}_1 - \bar{x}_2)$.

a. Suppose the first sample is selected from a population with mean $\mu_1 = 150$ and variance $\sigma_1^2 = 900$. Within what range should the sample mean vary about 95% of the time in repeated samples of 100 measurements from this distribution? That is, construct an interval extending 2 standard deviations of $\bar{x}_1$ on each side of μ_1.

b. Suppose the second sample is selected independently of the first from a second population with mean $\mu_2 = 150$ and variance $\sigma_2^2 = 1,600$. Within what range should the sample mean vary about 95% of the time in repeated samples of 100 measurements from this distribution? That is, construct an interval extending 2 standard deviations of $\bar{x}_2$ on each side of μ_2.

c. Now consider the difference between the two sample means $(\bar{x}_1 - \bar{x}_2)$. What are the mean and standard deviation of the sampling distribution of $(\bar{x}_1 - \bar{x}_2)$?

d. Within what range should the difference in sample means vary about 95% of the time in repeated independent samples of 100 measurements each from the two populations?

e. What, in general, can be said about the variability of the difference between independent sample means relative to the variability of the individual sample means?

9.3 Independent random samples of 100 observations each are chosen from two normal populations with the following means and standard deviations:

Population 1	Population 2
$\mu_1 = 14$	$\mu_2 = 10$
$\sigma_1 = 4$	$\sigma_2 = 3$

Let $\bar{x}_1$ and $\bar{x}_2$ denote the two sample means.

a. Give the mean and standard deviation of the sampling distribution of $\bar{x}_1$.

b. Give the mean and standard deviation of the sampling distribution of $\bar{x}_2$.

c. Suppose you were to calculate the difference $(\bar{x}_1 - \bar{x}_2)$ between the sample means. Find the mean and standard deviation of the sampling distribution of $(\bar{x}_1 - \bar{x}_2)$.

d. Will the statistic $(\bar{x}_1 - \bar{x}_2)$ be normally distributed? Explain.

9.4 To use the t statistic to test for a difference between the means of two populations, what assumptions must be made about the two populations? About the two samples?

9.5 Two populations are described in each of the following cases. In which cases would it be appropriate to apply the small-sample t-test to investigate the difference between the population means?

a. Population 1: Normal distribution with variance σ_1^2
Population 2: Skewed to the right with variance $\sigma_2^2 = \sigma_1^2$

b. Population 1: Normal distribution with variance σ_1^2
Population 2: Normal distribution with variance $\sigma_2^2 \neq \sigma_1^2$

c. Population 1: Skewed to the left with variance σ_1^2
Population 2: Skewed to the left with variance $\sigma_2^2 = \sigma_1^2$

d. Population 1: Normal distribution with variance σ_1^2
Population 2: Normal distribution with variance $\sigma_2^2 = \sigma_1^2$

e. Population 1: Uniform distribution with variance σ_1^2
Population 2: Uniform distribution with variance $\sigma_2^2 = \sigma_1^2$

9.6 In the t-tests of this section, σ_1^2 and σ_2^2 are assumed to be equal. Thus, we say $\sigma_1^2 = \sigma_2^2 = \sigma^2$. Why is a pooled estimator of σ^2 used instead of either s_1^2 or s_2^2?

9.7 Assume that $\sigma_1^2 = \sigma_2^2 = \sigma^2$. Calculate the pooled estimator of σ^2 for each of the following cases:

a. $s_1^2 = 200, s_2^2 = 180, n_1 = n_2 = 25$

b. $s_1^2 = 25, s_2^2 = 40, n_1 = 20, n_2 = 10$

c. $s_1^2 = .20, s_2^2 = .30, n_1 = 8, n_2 = 12$

d. $s_1^2 = 2,500, s_2^2 = 1,800, n_1 = 16, n_2 = 17$

e. Note that the pooled estimate is a weighted average of the sample variances. To which of the variances does the pooled estimate fall nearer in each of the above cases?

9.8 Independent random samples from normal populations produced the results shown below:

Sample 1	Sample 2
1.2	4.2
3.1	2.7
1.7	3.6
2.8	3.9
3.0	

a. Calculate the pooled estimate of σ^2.

b. Do the data provide sufficient evidence to indicate that $\mu_2 > \mu_1$? Test using $\alpha = .10$.

c. Find a 90% confidence interval for $(\mu_1 - \mu_2)$.

d. Which of the two inferential procedures, the test of hypothesis in part **b** or the confidence interval in part **c**, provides more information about $(\mu_1 - \mu_2)$?

9.9 Two independent random samples have been selected, 100 observations from population 1 and 100 from population 2. Sample means $\bar{x}_1 = 70$ and $\bar{x}_2 = 50$ were obtained. From previous experience with these populations, it is known that the variances are $\sigma_1^2 = 100$ and $\sigma_2^2 = 64$.

a. Find $\sigma_{(\bar{x}_1 - \bar{x}_2)}$.

b. Sketch the approximate sampling distribution for $(\bar{x}_1 - \bar{x}_2)$ assuming $(\mu_1 - \mu_2) = 5$.

c. Locate the observed value of $(\bar{x}_1 - \bar{x}_2)$ on the graph you drew in part **b**. Does it appear that this value contradicts the null hypothesis H_0: $(\mu_1 - \mu_2) = 5$?

d. Use the z table on the inside of the front cover to determine the rejection region for the test of H_0: $(\mu_1 - \mu_2) = 5$ against H_a: $(\mu_1 - \mu_2) \neq 5$. Use $\alpha = .05$.

e. Conduct the hypothesis test of part **d** and interpret your result.

9.10 Refer to Exercise 9.9. Construct a 95% confidence interval for $(\mu_1 - \mu_2)$. Interpret the interval. Which inference provides more information about the value of $(\mu_1 - \mu_2)$—the test of hypothesis in Exercise 9.9 or the confidence interval in this exercise?

9.11 Independent random samples are selected from two populations and used to test the hypothesis H_0: $(\mu_1 - \mu_2) = 0$ against the alternative H_a: $(\mu_1 - \mu_2) \neq 0$. A total of 233 observations from population 1 and 312 from population 2 are analyzed by a statistical software package, with the following result:

$\bar{x}_1 = 473$	$\bar{x}_2 = 485$
$s_1 = 84$	$s_2 = 93$
Z = -1.576	P-VALUE = .1150

a. Interpret the results of the computer analysis.

b. If the alternative hypothesis had been H_a: $(\mu_1 - \mu_2) < 0$, how would the p-value change? Interpret the p-value for this one-tailed test.

9.12 Independent random samples from approximately normal populations produced the results shown below:

Sample 1				Sample 2			
52	33	42	44	52	43	47	56
41	50	44	51	62	53	61	50
45	38	37	40	56	52	53	60
44	50	43		50	48	60	55

a. Do the data provide sufficient evidence to conclude that $(\mu_2 - \mu_1) > 10$? Test using $\alpha = .01$.

b. Construct a 98% confidence interval for $(\mu_2 - \mu_1)$. Interpret your result.

9.13 Independent random samples selected from two normal populations produced the sample means and standard deviations shown at top right:

Sample 1	Sample 2
$n_1 = 17$	$n_2 = 12$
$\bar{x}_1 = 5.4$	$\bar{x}_2 = 7.9$
$s_1 = 3.4$	$s_2 = 4.8$

a. The test H_0: $(\mu_1 - \mu_2) = 0$ against H_a: $(\mu_1 - \mu_2) \neq 0$ was conducted using statistical software, with the results shown. Check and interpret the results.

T = -1.646   DF = 27   P-VALUE(2-TAILED) = .1114

b. Estimate $(\mu_1 - \mu_2)$ using a 95% confidence interval.

Applying the Concepts

9.14 Some college professors make bound lecture notes available to their classes in an effort to improve teaching effectiveness. Because students pay the additional cost, educators want to know whether the students consider the lecture notes to be a good educational value. *Marketing Educational Review* (Fall 1994) published a study of business students' opinions of lecture notes. Two groups of students were surveyed—86 students enrolled in a promotional strategy class that required the purchase of lecture notes, and 35 students enrolled in a sales/retailing elective that did not offer lecture notes. In both courses, the instructor used lectures as the main method of delivery. At the end of the semester, the students were asked to respond to the statement: "Having a copy of the lecture notes was [would be] helpful in understanding the material." Responses were measured on a 9-point semantic difference scale, where 1 = "strongly disagree" and 9 = "strongly agree." A summary of the results is shown in the accompanying table.

Classes Buying Lecture Notes	Classes Not Buying Lecture Notes
$n_1 = 86$	$n_2 = 35$
$\bar{x}_1 = 8.48$	$\bar{x}_2 = 7.80$
$s_1^2 = 0.94$	$s_2^2 = 2.99$

Source: Gray, J. I., and Abernathy, A. M. "Pros and cons of lecture notes and handout packages: Faculty and student opinions." *Marketing Education Review*, Vol. 4, No. 3, Fall 1984, p. 25 (Table 4). Reprinted with permission of the American Marketing Association.

a. Describe the two populations involved in the comparison.

b. Do the samples provide sufficient evidence to conclude that there is a difference in the mean responses of the two groups of students? Test using $\alpha = .01$.

c. Construct a 99% confidence interval for $(\mu_1 - \mu_2)$. Interpret the result.

d. Would a 95% confidence interval for $(\mu_1 - \mu_2)$ be narrower or wider than the one you found in part **c**? Why?

9.15 Bear gallbladder has been used in Chinese medicine for more than 1,300 years. Today it is used to treat a wide variety of ailments, including gallstones and other liver diseases. Due to the difficulty of obtaining bear gallbladder, Chinese medical researchers are searching for a more readily available source of animal bile. A study in the *Journal of Ethnopharmacology* (June 1995) examined pig gallbladder as an effective substitute for bear gallbladder. Twenty male mice were divided randomly into two groups: 10 were given a dosage of bear bile and 10 were given a dosage of pig bile. All the mice then received an injection of croton oil in the left ear lobe to induce inflammation. Four hours later, both the left and right ear lobes were weighed, with the difference (in milligrams) representing the degree of swelling. Summary statistics on the degree of swelling are provided in the following table.

Bear Bile	Pig Bile
$n_1 = 10$	$n_2 = 10$
$\bar{x}_1 = 9.19$	$\bar{x}_2 = 9.71$
$s_1 = 4.17$	$s_2 = 3.33$

a. Compare the mean degree of swelling for mice treated with bear and pig bile with a 95% confidence interval. Interpret the result.

b. What assumptions are necessary for the inference, part **a**, to be valid?

c. Another group of 10 mice (the control group) was given normal saline instead of bear or pig bile. The degree of swelling for this control group produced the following summary statistics: $\bar{x} = 18.30, s = 3.38$. Compare the mean degree of swelling for the control group to the mice treated with pig bile using a 95% confidence interval. Interpret the results.

9.16 Excess postexercise oxygen consumption (EPOC) describes energy expended during the body's recovery period immediately following aerobic exercise. *The Journal of Sports Medicine and Physical Fitness* (Dec. 1994) published a study designed to investigate the effect of fitness level on the magnitude and duration of EPOC. Ten healthy young adult males volunteered for the study. Five of these were endurance-trained and comprised the fit group; the other five were not engaged in any systematic training and comprised the sedentary group. Each volunteer engaged in a weight-supported exercise on a cycle ergometer until 300 kilocalories were expended. The magnitude (in kilocalories) and duration (in minutes) of the EPOC of each exerciser was measured. The study results are summarized in the table at the bottom of the page.

a. Conduct a test of hypothesis to determine whether the true mean magnitude of EPOC differs for fit and sedentary young adult males. Use $\alpha = .10$.

b. The *p*-value for the test, part **a**, is given in the table. Interpret this value.

c. Conduct a test of hypothesis to determine whether the true mean duration of EPOC differs for fit and sedentary young adult males. Use $\alpha = .10$.

d. The *p*-value for the test, part **c**, is given in the table. Interpret this value.

9.17 Researchers at Rochester Institute of Technology investigated the use of isolation timeout as a behavioral management technique (*Exceptional Children*, Feb. 1995). Subjects for the study were 155 emotionally disturbed students enrolled in a special education facility. The students were randomly assigned to one of two types of classrooms—Option II classrooms (one teacher, one paraprofessional, and a maximum of 12 students) and Option III classrooms (one teacher, one paraprofessional, and a maximum of 6 students). Over the academic year the number of behavioral incidents resulting in an isolation timeout was recorded for each student. Summary statistics for the two groups of students are shown in the following table.

	Option II	Option III
Number of students	100	55
Mean number of timeout incidents	78.67	102.87
Standard deviation	59.08	69.33

Source: Costenbader, V., and Reading-Brown, M. "Isolation timeout used with students with emotional disturbance." *Exceptional Children*, Vol. 61, No. 4, Feb. 1995, p. 359 (Table 3). Copyright 1995 by The Council for Exceptional Children. Reprinted with permission.

Variable		Fit ($n = 5$)	Sedentary ($n = 5$)	*p*-Value
Magnitude (kcal)	Mean	12.2	12.2	.998
	Std. Dev.	3.1	4.3	
Duration (min)	Mean	16.6	20.4	.344
	Std. Dev.	3.1	7.8	

Source: Sedlock, D. A. "Fitness levels and postexercise energy expenditure." *The Journal of Sports Medicine and Physical Fitness*, Vol. 34, No. 4, Dec. 1994, p. 339 (Table III).

Snail Mating System	Sample Size	EFFECTIVE POPULATION SIZE	
		Mean	**Standard Deviation**
Outcrossing	17	4,894	1,932
Selfing	5	4,133	1,890

Source: Jarne, P. "Mating system, bottlenecks, and genetic polymorphism in hermaphroditic animals." *Genetical Research,* Vol. 65, No. 3, June 1995, p. 197 (Table 4). Copyright 1995 Genetical Research. Reprinted with the permission of Cambridge University Press.

Do you agree with the following statement: "On average, students in Option III classrooms had significantly more timeout incidents than students in Option II classrooms?" Use the statistical methodology presented in this section to arrive at your decision.

9.18 Hermaphrodites are animals that possess the reproductive organs of both sexes. *Genetical Research* (June 1995) published a study of the mating systems of hermaphroditic snail species. The mating habits of the snails were classified into two groups: (1) self-fertilizing (selfing) snails that mate with snails of the same sex and (2) cross-fertilizing (outcrossing) snails that mate with snails of the opposite sex. One variable of interest in the study was the effective population size of the snail species. The means and standard deviations of the effective population size for independent random samples of 17 outcrossing snail species and 5 selfing snail species are given in the table at the top of the page. Compare the mean effective population sizes of the two types of snail species with a 90% confidence interval. Interpret the result.

9.19 Do cocaine abusers have radically different personalities than nonabusing college students? This was one of the questions researched in *Psychological Assessment* (June 1995). The Zuckerman–Kuhlman Personality Questionnaire (ZKPQ) was administered to a sample of 450 cocaine abusers and a sample of 589 college students. The ZKPQ yields scores (measured on a 20-point scale) on each of five dimensions: impulsive–sensation seeking, sociability, neuroticism–anxiety, aggression–hostility, and activity. The results are summarized in the table at the bottom of the page. Compare the mean ZKPQ scores of the two groups on each dimension using a statistical test of hypothesis. Interpret the results.

9.20 Marine biochemists at the University of Tokyo studied the properties of crustacean striated muscles (*The Journal of Experimental Zoology*, Sept. 1993). It is well known that certain muscles contract faster than others. The main purpose of the experiment was to compare the biochemical properties of fast and slow muscles of crayfish. Using crayfish obtained from a local supplier, the researchers excised twelve fast-muscle fiber bundles and tested each fiber bundle for uptake of calcium. Twelve slow-muscle fiber bundles were excised from a second sample of crayfish, and calcium uptake was measured. The results of the experiment are summarized below. (All calcium measurements are in moles per milligram.) Analyze the data using a 95% confidence interval. Make an inference about the difference between the calcium uptake means of fast and slow muscles.

Fast Muscle	Slow Muscle
$n_1 = 12$	$n_2 = 12$
$\bar{x}_1 = .57$	$\bar{x}_2 = .37$
$s_1 = .104$	$s_2 = .035$

Source: Ushio, H., and Watabe, S. "Ultrastructural and biochemical analysis of the sarcoplasmic reticulum from crayfish fast and slow striated muscles." *The Journal of Experimental Zoology*, Vol. 267, Sept. 1993, p. 16 (Table 1). Copyright © 1993 John Wiley & Sons, Inc. Reprinted with permission.

ZKPQ Dimension	COCAINE ABUSERS (*n* = 450)		COLLEGE STUDENTS (*n* = 589)	
	Mean	**Std. Dev.**	**Mean**	**Std. Dev.**
Impulsive–sensation seeking	9.4	4.4	9.5	4.4
Sociability	10.4	4.3	12.5	4.0
Neuroticism–anxiety	8.6	5.1	9.1	4.6
Aggression–hostility	8.6	3.9	7.3	4.1
Activity	11.1	3.4	8.0	4.1

Source: Ball, S. A. "The validity of an alternative five-factor measure of personality in cocaine abusers." *Psychological Assessment,* Vol. 7, No. 2, June 1995, p. 150 (Table 1). Copyright © 1995 by the American Psychological Association. Reprinted with permission.

9.21 Recent research in nursing education has focused on teaching strategies that link scientific theory with practice. One study compared a traditional approach to teaching basic nursing skills with an innovative approach (*Journal of Nursing Education*, Jan. 1992). The innovative approach utilizes two strategies (Vee heuristics and concept maps) that consciously link theoretical concepts with practical skills. Forty-two students enrolled in an upper-division nursing course participated in the study. Half (21) were randomly assigned to labs that utilized the innovative approach. After completing the course, all students were given short-answer questions about scientific principles underlying each of ten nursing skills. The objective of the research is to compare the mean scores of the two groups of students.

   a. What is the appropriate test to use to compare the two groups?

   b. Are any assumptions required for the test?

   c. One question dealt with the use of clean/sterile gloves. The mean scores for this question were 3.28 (traditional) and 3.40 (innovative). Is there sufficient information to perform the test?

   d. Refer to part **c**. The *p*-value for the test was reported as $p = .79$. Interpret this result.

   e. Another question concerned the choice of a stethoscope. The mean scores of the two groups were 2.55 (traditional) and 3.60 (innovative) with an associated *p*-value of .02. Interpret these results.

9.22 The *Florida Scientist* (Summer/Autumn 1991) reported on a study of the feeding habits of sea urchins. A sample of 20 urchins were captured from Biscayne Bay (Miami), placed in marine aquaria, then starved for 48 hours. Each sea urchin was then fed a 5-cm blade of turtle grass. Ten of the urchins received only green blades, while the other half received only decayed blades. (Assume that the two samples of sea urchins—10 urchins per sample—were randomly and independently selected.) The ingestion time, measured from the time the blade first made contact with the urchin's mouth to the time the urchin had finished ingesting the blade, was recorded. A summary of the results is provided in the table below.

	Green Blades	Decayed Blades
Number of sea urchins	10	10
Mean ingestion time (hours)	3.35	2.36
Standard deviation (hours)	.79	.47

Source: Montague, J. R., *et al.* "Laboratory measurement of ingestion rate for the sea urchin *Lytechinus variegatus*." *Florida Scientist*, Vol. 54, Nos. 3/4, Summer/Autumn, 1991 (Table 1).

   a. Construct a 90% confidence interval for the difference between the mean ingestion times of sea urchins feeding on green and decayed turtle grass.

   b. According to the researchers, "the difference in rates at which the urchins ingested the blades suggest that green, unblemished turtle grass may not be a particularly palatable food compared with decayed turtle grass. If so, urchins in the field may find it more profitable to selectively graze on decayed portions of the leaves." Does the result, part **a**, support this conclusion?

9.23 Helping smokers kick the habit is big business in today's no-smoking environment. One of the more commonly used treatments according to an article in the *Journal of Imagination, Cognition and Personality* (Vol. 12, 1992/93) is Spiegel's three-point message:

1. For your body, smoking is a poison.
2. You need your body to live.
3. You owe your body this respect and protection.

To determine the effectiveness of this treatment, the authors conducted a study consisting of a sample of 52 smokers placed in two groups, a Spiegel treatment group or a control group (no treatment). Each participant was asked to record the number of cigarettes he or she smoked each week. The results for the study are shown below for the beginning period and four follow-up time periods.

	NUMBER OF CIGARETTES SMOKED IN WEEK		
	n	$\bar{x}$	*s*
Beginning			
Treatment	35	165.09	71.20
Control	17	159.00	67.45
First follow-up (2 wks)			
Treatment	35	105.00	69.08
Control	17	157.24	66.80
Second follow-up (4 wks)			
Treatment	35	111.11	69.08
Control	17	159.52	65.73
Third follow-up (8 wks)			
Treatment	35	120.20	67.59
Control	17	157.88	64.41
Fourth follow-up (12 wks)			
Treatment	35	123.63	74.09
Control	17	162.17	67.01

   a. Create 95% confidence intervals for the difference in the average number of cigarettes smoked per week for the two groups for the beginning and each follow-up period. Interpret the results.

   b. What assumptions are necessary for the validity of these confidence intervals?

9.24 The relationship between handedness and motor competence in preschool children was investigated in *Child Development* (Vol. 56). Random samples of 41 right-handers and 41 left-handers were administered several tests of motor skills, yielding the means and standard deviations shown in the accompanying table.

	Left-Handed	**Right-Handed**
n	41	41
$\bar{x}$	97.5	98.1
s	17.5	19.2

Source: Tan, L. E. "Laterality and motor skills in four-year-olds." *Child Development,* 56.

a. Is there evidence of a difference between the motor skills of right- and left-handed preschool children based on this experiment? Use $\alpha = .10$.

b. Use a 90% confidence interval to estimate the true difference in mean motor skill scores between left- and right-handed preschoolers. Does the confidence interval support the result of the test you conducted in part **a**?

c. What assumptions about the distributions of the populations of test scores are necessary to ensure the validity of the inferences you made in parts **a** and **b**?

d. What is the observed significance level of the test you conducted in part **a**?

e. The researcher concluded that "… tests indicated no difference between left-handers and right-handers…." Does your analysis support this conclusion?

9.25 Suppose you manage a plant that purifies its liquid waste and discharges the water into a local river. An EPA inspector has collected water specimens of the discharge of your plant and also water specimens in the river upstream from your plant. Each water specimen is divided into five parts, the bacteria count is read on each, and the mean count for each specimen is reported. The average bacteria counts for each of six specimens are reported in the accompanying table for the two locations.

a. Why might the bacteria counts shown here tend to be approximately normally distributed?

Plant Discharge			**Upstream**		
30.1	36.2	33.4	29.7	30.3	26.4
28.2	29.8	34.9	27.3	31.7	32.3

b. What are the appropriate null and alternative hypotheses to test whether the mean bacteria count for the plant discharge exceeds that for the upstream location? Be sure to define any symbols you use.

c. When the data are submitted to SPSS analysis, part of the output is shown at the bottom of the page. Carefully interpret this output.

d. What assumptions are necessary to ensure the validity of this test?

9.26 How are physical and mental health related? Research published in the *Journal of Psychology* (Mar. 1991) investigates the relationship between age and perception of control among obese females. Perception of control was measured by the Locus of Control (LOC), a test instrument widely used for determining whether individuals perceive themselves as being in control ("internal" LOC), or as being controlled by others ("external" LOC). Forty-six obese female adults and 19 obese female adolescents were sampled and an LOC score for each was determined. Higher scores represent external LOC and lower scores represent internal LOC. The data reported in the article are shown below:

	Adults	**Adolescents**
$\bar{x}$	6.45	10.89
s	2.89	2.48
n	46	19

a. What are the appropriate null and alternative hypotheses for determining whether a difference exists between the mean LOC score for obese female adults and adolescents? Define any symbols you use.

```
Independent samples of  LOCATION

Group 1:  LOCATION EQ      1.00        Group 2:  LOCATION EQ      2.00
t-test for:  BACOUNT

                   Number                Standard    Standard
                   of Cases    Mean      Deviation   Error
        Group 1       6      32.1000      3.189      1.302
        Group 2       6      29.6167      2.355       .961

              | Pooled Variance Estimate | Separate Variance Estimate
      F    2-Tail |    t    Degrees of 2-Tail |    t    Degrees of  2-Tail
   Value  Prob.   | Value   Freedom    Prob.  | Value   Freedom     Prob.
   1.03    .522   | 1.53      10        .156  | 1.53      9.20        .159
```

b. The data were analyzed using ASP, with the results shown here. Interpret these results.

```
HYPOTHESIS: MEAN ADULTS = MEAN ADOLESCENTS
    SAMPLE MEAN OF X     =     6.45
SAMPLE VARIANCE OF X     =     8.3521
    SAMPLE SIZE OF X     =    46
    SAMPLE MEAN OF Y     =    10.89
SAMPLE VARIANCE OF Y     =     6.1504
    SAMPLE SIZE OF Y     =    19

    MEAN X - MEAN Y      =    -4.44
                   t     =    -5.85852
               D. F.     =    38
             P-VALUE     =     8.9011E-7
           P-VALUE/2     =     4.45055E-7
           SD. ERROR     =     0.757871
```

c. Another part of the ASP output is shown at the bottom of the page. Interpret the interval shown on the bottom of the printout.

d. Which do you find more informative, the test of hypothesis or the confidence interval? Explain.

9.27 Do the early questions in an exam affect your performance on later questions? Research reported in the *Journal of Psychology* (Jan. 1991) investigated this relationship by conducting the following experiment: 140 college students were randomly assigned to two groups. Group A took an exam that contained 15 difficult questions followed by five moderate questions. Group B took an exam that contained 15 easy questions followed by the same five moderate questions. One of the goals of the study was to compare the average score on the moderate questions for the two groups.

a. The researchers hypothesized that the students who had the easy questions would score better than the students who had the difficult questions. Set up the appropriate null and alternative hypotheses to test their research hypothesis, defining any symbols you use.

b. Suppose the *p*-value for this test was reported as .3248. What conclusion would you reach based on this *p*-value?

c. What assumptions are necessary to ensure the validity of this conclusion?

9.28 Research reported in *The Professional Geographer* (May 1992) examines the hypothesis that the disproportionate housework responsibility of women in two-income households is a major factor in determining the proximity of a woman's place of employment. The distance to work for both men and women in two-income households was reported for random samples of both central city and suburban residences:

	CENTRAL CITY RESIDENCE		SUBURBAN RESIDENCE	
	Men	**Women**	**Men**	**Women**
n	159	119	138	93
$\bar{x}$	7.4	4.5	9.3	6.6
s	6.3	4.2	7.1	5.6

a. For central city residences, calculate a 99% confidence interval for the difference in average distance to work for men and women in two-income households. Interpret the interval.

b. Repeat part **a** for suburban residences.

c. Interpret the confidence intervals. Do they indicate that women tend to work closer to home than men?

d. What assumptions have you made to ensure the validity of the confidence intervals constructed in parts **a** and **b**?

```
        95% CONFD INTERVAL  MEAN ADULTS - MEAN ADOLESCENTS
        SAMPLE MEAN OF X     =     6.45
    SAMPLE VARIANCE OF X     =     8.3521
        SAMPLE SIZE OF X     =    46
        SAMPLE MEAN OF Y     =    10.89
    SAMPLE VARIANCE OF Y     =     6.1504
        SAMPLE SIZE OF Y     =    19
             CONFIDENCE      =    95

          % CONFIDENCE       =    95
                  D. F.      =    38
                      t      =     2.02439
              SD. ERROR      =     0.757871
          t*SD. ERROR        =     1.53423
          UPPER LIMIT        =    -2.90577
        MEAN X - MEAN Y      =    -4.44
          LOWER LIMIT        =    -5.97423
```

9.2   COMPARING TWO POPULATION MEANS: PAIRED DIFFERENCE EXPERIMENTS

In Example 9.4, we compared two methods of teaching reading to slow learners by means of a 95% confidence interval. Suppose it is possible to measure the slow learners' "reading IQs" *before* they are subjected to a teaching method. Eight pairs of slow learners with similar reading IQs are found, and one member of each pair is randomly assigned to the standard teaching method while the other is assigned to the new method. The data are given in Table 9.3, followed by an SPSS printout (Figure 9.9) of descriptive statistics for each method. Do the data support the hypothesis that the population mean reading test score for slow learners taught by the new method is greater than the mean reading test score for those taught by the standard method?

We want to test

$$H_0: (\mu_1 - \mu_2) = 0$$
$$H_a: (\mu_1 - \mu_2) > 0$$

If we employ the t statistic for independent samples (Section 9.1), we first calculate s_p^2 using the values of s_1 and s_2 shaded on the SPSS printout:

$$s_p^2 = \frac{(n_1 - 1)s_1^2 + (n_2 - 1)s_2^2}{n_1 + n_2 - 2} = \frac{(8 - 1)(6.93)^2 + (8 - 1)(7.01)^2}{8 + 8 - 2} = 48.55$$

Then we substitute the values of $\bar{x}_1$ and $\bar{x}_2$, also shaded on the printout, to form the test statistic

$$t = \frac{(\bar{x}_1 - \bar{x}_2) - 0}{\sqrt{s_p^2\left(\frac{1}{n_1} + \frac{1}{n_2}\right)}} = \frac{76.0 - 71.625}{\sqrt{48.55\left(\frac{1}{8} + \frac{1}{8}\right)}} = \frac{4.375}{3.485} = 1.26$$

TABLE 9.3   Reading Test Scores for Eight Pairs of Slow Learners

Pair	New Method (1)	Standard Method (2)
1	77	72
2	74	68
3	82	76
4	73	68
5	87	84
6	69	68
7	66	61
8	80	76

FIGURE 9.9

SPSS descriptive statistics for data in Table 9.3

```
Summaries of    SCORE
By levels of    METHOD

Variable        Value   Label              Mean     Std Dev    Cases

For Entire Population                      73.8125   7.1013      16

METHOD          NEW                        76.0000   6.9282       8
METHOD          STD                        71.6250   7.0089       8

    Total Cases =        16
```

TABLE 9.4

Pair	New Method	Standard Method	Difference (New Method − Standard Method)
1	77	72	5
2	74	68	6
3	82	76	6
4	73	68	5
5	87	84	3
6	69	68	1
7	66	61	5
8	80	76	4

This small t value will not lead to rejection of H_0 when compared to the t-distribution with $n_1 + n_2 - 2 = 14$ df, even if α is chosen as large as .10 ($t_{.10} = 1.345$). Thus, from *this* analysis we might conclude that we do not have sufficient evidence to infer a difference in the mean test scores for the two methods.

If you carefully examine the data in Table 9.3, however, you will find this result difficult to accept. The test score of the new method is larger than the corresponding test score for the standard method *for every one of the eight pairs of slow learners.* This, in itself, seems to provide strong evidence to indicate that μ_1 exceeds μ_2. Why, then, did the t-test fail to detect this difference? The answer is: *The independent samples t-test is not a valid procedure to use with this set of data.*

The t-test is inappropriate because the assumption of independent samples is invalid. We have randomly chosen *pairs of test scores,* and thus, once we have chosen the sample for the new method, we have *not* independently chosen the sample for the standard method. The dependence between observations within pairs can be seen by examining the pairs of test scores, which tend to rise and fall together as we go from pair to pair. This pattern provides strong visual evidence of a violation of the assumption of independence required for the two-sample t-test of Section 9.1. In this situation, you will note the *large variation within samples* (reflected by the large value of s_p^2) in comparison to the relatively *small difference between the sample means.* Because s_p^2 is so large, the t-test of Section 9.1 is unable to detect a difference between μ_1 and μ_2.

We now consider a valid method of analyzing the data of Table 9.3. In Table 9.4 we add the column of differences between the test scores of the pairs of slow learners. We can regard these differences in test scores as a random sample of differences for all pairs (matched on reading IQ) of slow learners, past and present. Then we can use this sample to make inferences about the mean of the population of differences, μ_D, which is equal to the difference ($\mu_1 - \mu_2$). That is, the mean of the population (and sample) of differences equals the difference between the population (and sample) means. Thus, our test becomes

$$H_0: \mu_D = 0 \qquad (\mu_1 - \mu_2 = 0)$$
$$H_a: \mu_D > 0 \qquad (\mu_1 - \mu_2 > 0)$$

The test statistic is a one-sample t (Section 8.4), since we are now analyzing a single sample of differences for small n:

$$\textit{Test statistic:} \quad t = \frac{\bar{x}_D - 0}{s_D / \sqrt{n_D}}$$

FIGURE 9.10

Rejection region for Example 9.4

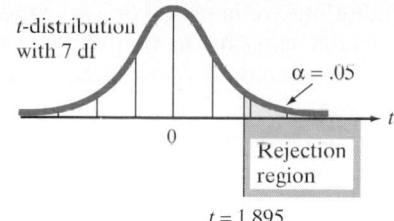

$t = 1.895$

where $\bar{x}_D$ = Sample mean difference
s_D = Sample standard deviation of differences
n_D = Number of differences = Number of pairs

Assumptions: The population of differences in test scores is approximately normally distributed. The sample differences are randomly selected from the population differences. [*Note:* We do not need to make the assumption that $\sigma_1^2 = \sigma_2^2$.]

Rejection region: At significance level $\alpha = .05$, we will reject H_0 if $t > t_{.05}$, where $t_{.05}$ is based on $(n_D - 1)$ degrees of freedom.

Referring to Table IV in Appendix A, we find the t value corresponding to $\alpha = .05$ and $n_D - 1 = 8 - 1 = 7$ df to be $t_{.05} = 1.895$. Then we will reject the null hypothesis if $t > 1.895$ (see Figure 9.10). Note that the number of degrees of freedom decreases from $n_1 + n_2 - 2 = 14$ to 7 when we use the paired difference experiment rather than the two independent random samples design.

Summary statistics for the $n = 8$ differences are shown on the MINITAB printout, Figure 9.11. Note that $\bar{x}_D = 4.375$ and $s_D = 1.685$. Substituting these values into the formula for the test statistic, we have

$$t = \frac{\bar{x}_D - 0}{s_D / \sqrt{n_D}} = \frac{4.375}{1.685 / \sqrt{8}} = 7.34$$

Because this value of t falls in the rejection region, we conclude (at $\alpha = .05$) that the population mean test score for slow learners taught by the new method exceeds the population mean score for those taught by the standard method. We can reach the same conclusion by noting that the p-value of the test, shaded in Figure 9.11, is much smaller than $\alpha = .05$.

This kind of experiment, in which observations are paired and the differences are analyzed, is called a **paired difference experiment**. In many cases, a paired difference experiment can provide more information about the difference between population means than an independent samples experiment. The idea is to compare population means by comparing the differences between pairs of experimental units (objects, people, etc.) that were very similar prior to the experiment. The differencing removes sources of variation that tend to inflate σ^2. For example, when two children are taught to read by two different methods, the observed difference in achievement may be due to a difference in the effectiveness of the two teaching methods *or* it may be due to differences in the initial reading levels and IQs of the two children (random error). To reduce the effect of differences in the children on the observed differences in reading achievement, the two methods of

FIGURE 9.11

MINITAB analysis of differences in Table 9.4

```
TEST OF MU = 0.000 VS MU G.T. 0.000

          N    MEAN    STDEV   SE MEAN      T    P VALUE
DIFF      8    4.375   1.685   0.596     7.34    0.0000
```

reading are imposed on two children who are more likely to possess similar intellectual capacity, namely children with nearly equal IQs. The effect of this pairing is to remove the larger source of variation that would be present if children with different abilities were randomly assigned to the two samples. Making comparisons within groups of similar experimental units is called **blocking**, and the paired difference experiment is a simple example of a **randomized block experiment**. In our example, pairs of children with matching IQ scores represent the blocks.

Some other examples for which the paired difference experiment might be appropriate are the following:

1. Suppose you want to estimate the difference $(\mu_1 - \mu_2)$ in mean price per gallon between two major brands of premium gasoline. If you choose two independent random samples of stations for each brand, the variability in price due to geographic location may be large. To eliminate this source of variability you could choose pairs of stations of similar size, one station for each brand, in close geographic proximity and use the sample of differences between the prices of the brands to make an inference about $(\mu_1 - \mu_2)$.

2. Suppose a college placement center wants to estimate the difference $(\mu_1 - \mu_2)$ in mean starting salaries for men and women graduates who seek jobs through the center. If it independently samples men and women, the starting salaries may vary because of their different college majors and differences in grade-point averages. To eliminate these sources of variability, the placement center could match male and female job-seekers according to their majors and grade-point averages. Then the differences between the starting salaries of each pair in the sample could be used to make an inference about $(\mu_1 - \mu_2)$.

3. Suppose you wish to estimate the difference $(\mu_1 - \mu_2)$ in mean absorption rate into the bloodstream for two drugs that relieve pain. If you independently sample people, the absorption rates might vary because of age, weight, sex, blood pressure, etc. In fact, there are many possible sources of nuisance variability, and pairing individuals who are similar in all the possible sources would be quite difficult. However, it may be possible to obtain two measurements *on the same person*. First, we administer one of the two drugs and record the time until absorption. After a sufficient amount of time, the other drug is administered and a second measurement on absorption time is obtained. The differences between the measurements for each person in the sample could then be used to estimate $(\mu_1 - \mu_2)$. This procedure would be advisable only if the amount of time allotted between drugs is sufficient to guarantee little or no carryover effect. Otherwise, it would be better to use different people matched as closely as possible on the factors thought to be most important.

The hypothesis-testing procedures and the method of forming confidence intervals for the difference between two means using a paired difference experiment are summarized in the accompanying boxes for both large and small *n*.

Paired Difference Confidence Interval for $\mu_D = \mu_1 - \mu_2$

LARGE SAMPLE

$$\bar{x}_D \pm z_{\alpha/2}\frac{\sigma_D}{\sqrt{n_D}} \approx \bar{x}_D \pm z_{\alpha/2}\frac{s_D}{\sqrt{n_D}}$$

Assumption: The sample differences are randomly selected from the population of differences.

SMALL SAMPLE

$$\bar{x}_D \pm t_{\alpha/2} \frac{s_D}{\sqrt{n_D}}$$

where $t_{\alpha/2}$ is based on $(n_D - 1)$ degrees of freedom

Assumptions: 1. The relative frequency distribution of the population of differences is normal.
    2. The sample differences are randomly selected from the population of differences.

Paired Difference Test of Hypothesis for $\mu_D = \mu_1 - \mu_2$

ONE-TAILED TEST	**TWO-TAILED TEST**
H_0: $\mu_D = D_0$	H_0: $\mu_D = D_0$
H_a: $\mu_D < D_0$	H_a: $\mu_D \neq D_0$
[or H_a: $\mu_D > D_0$]	

LARGE SAMPLE

Test statistic:

$$z = \frac{\bar{x}_D - D_0}{\sigma_D / \sqrt{n_D}} \approx \frac{\bar{x}_D - D_0}{s_D / \sqrt{n_D}}$$

Rejection region: $z < -z_\alpha$ *Rejection region:* $|z| > z_{\alpha/2}$
  [or $z > z_\alpha$ when H_a: $\mu_D > D_0$]

Assumption: The differences are randomly selected from the population of differences.

SMALL SAMPLE

Test statistic:

$$t = \frac{\bar{x}_D - D_0}{s_D / \sqrt{n_D}}$$

Rejection region: $t < -t_\alpha$ *Rejection region:* $|t| > t_{\alpha/2}$
  [or $t > t_\alpha$ when H_a: $\mu_D > D_0$]

where t_α and $t_{\alpha/2}$ are based on $(n_D - 1)$ degrees of freedom

Assumptions: 1. The relative frequency distribution of the population of differences is normal.
    2. The differences are randomly selected from the population of differences.

EXAMPLE 9.5

An experiment is conducted to compare the starting salaries of male and female college graduates who find jobs. Pairs are formed by choosing a male and a female with the same major and similar grade-point averages (GPA). Suppose a random sample of 10 pairs is formed in this manner and the starting annual salary of each person is recorded. The results are shown in Table 9.5. Compare the mean starting salary, μ_1, for males to the mean starting salary, μ_2, for females using a 95% confidence interval. Interpret the results.

TABLE 9.5 Data on Annual Salaries for Matched Pairs of College Graduates

Pair	Male	Female	Difference Male – Female	Pair	Male	Female	Difference Male – Female
1	$29,300	$28,800	$ 500	6	$27,800	$28,000	$ −200
2	31,500	31,600	−100	7	29,500	29,200	300
3	30,400	29,800	600	8	31,200	30,100	1,100
4	28,500	28,500	0	9	28,400	28,200	200
5	33,500	32,600	900	10	29,200	28,500	700

Solution

Since the data on annual salary are collected in pairs of males and females matched on GPA and major, a paired difference experiment is performed. To conduct the analysis, we first compute the differences between the salaries, as shown in Table 9.5. Summary statistics for these $n = 10$ differences are displayed in the MINITAB printout, Figure 9.12.

The 95% confidence interval for $\mu_D = (\mu_1 - \mu_2)$ for this small sample is

$$\bar{x}_D \pm t_{\alpha/2} \frac{s_D}{\sqrt{n_D}}$$

where $t_{\alpha/2} = t_{.025} = 2.262$ (obtained from Table VI, Appendix A) is based on $n - 2 = 8$ degrees of freedom. Substituting the values of $\bar{x}_D$ and s_D shown on the printout, we obtain

$$\bar{x}_D \pm 2.262 \frac{s_D}{\sqrt{n_D}} = 400 \pm 2.262 \left(\frac{434.613}{\sqrt{10}} \right)$$
$$= 400 \pm 310.88 \approx 400 \pm 311 = (\$89, \$711)$$

[*Note:* This interval is also shown shaded on the MINITAB printout, Figure 9.12.] Our interpretation is that the true mean difference between the starting salaries of males and females falls between $89 and $711, with 95% confidence. Since the interval falls above 0, we infer that $\mu_1 - \mu_2 > 0$; that is, the mean salary for males exceeds the mean salary for females. ▲

To measure the amount of information about $(\mu_1 - \mu_2)$ gained by using a paired difference experiment in Example 9.5 rather than an independent samples experiment, we can compare the relative widths of the confidence intervals obtained by the two methods. A 95% confidence interval for $(\mu_1 - \mu_2)$ using the paired difference experiment is, from Example 9.5, ($89, $711). If we analyzed the same data as though this were an independent samples experiment,* we would first obtain the descriptive statistics shown in the MINITAB printout, Figure 9.13.

FIGURE 9.12

MINITAB analysis of differences in Table 9.5

```
              N       MEAN      STDEV    SE MEAN    95.0 PERCENT C.I.
DIFF         10    400.000    434.613   137.437   ( 89.013, 710.987)
```

*This is done only to provide a measure of the increase in the amount of information obtained by a paired design in comparison to an unpaired design. Actually, if an experiment is designed using pairing, an unpaired analysis would be invalid because the assumption of independent samples would not be satisfied.

FIGURE 9.13

MINITAB analysis of data in Table 9.5, assuming independent samples

```
TWOSAMPLE T FOR C1 VS C2
        N       MEAN     STDEV    SE MEAN
C1    10       29930      1735        549
C2    10       29530      1527        483

95 PCT CI FOR MU C1 - MU C2:   (-1136, 1936)

TTEST MU C1 = MU C2 (VS NE): T= 0.55   P=0.59   DF=   18

POOLED STDEV =          1634
```

Then we substitute the sample means and standard deviations shown on the printout into the formula for a 95% confidence interval for $(\mu_1 - \mu_2)$ using independent samples:

$$(\bar{x}_1 - \bar{x}_2) \pm t_{.025}\sqrt{s_p^2\left(\frac{1}{n_1} + \frac{1}{n_2}\right)}$$

where

$$s_p^2 = \frac{(n_1 - 1)s_1^2 + (n_2 - 1)s_2^2}{n_1 + n_2 - 2}$$

MINITAB performed these calculations and obtained the interval $(-\$1,136, \$1,936)$. This interval is shaded on Figure 9.13.

Notice that the independent samples interval includes 0. Consequently, if we were to use this interval to make an inference about $(\mu_1 - \mu_2)$, we would incorrectly conclude that the mean starting salaries of males and females do not differ! You can see that the confidence interval for the independent sampling experiment is about five times wider than for the corresponding paired difference confidence interval. Blocking out the variability due to differences in majors and grade-point averages significantly increases the information about the difference in male and female mean starting salaries by providing a much more accurate (smaller confidence interval for the same confidence coefficient) estimate of $(\mu_1 - \mu_2)$.

You may wonder whether conducting a paired difference experiment is always superior to an independent samples experiment. The answer is: Most of the time, but not always. We sacrifice half the degrees of freedom in the t statistic when a paired difference design is used instead of an independent samples design. This is a loss of information, and unless this loss is more than compensated for by the reduction in variability obtained by blocking (pairing), the paired difference experiment will result in a net loss of information about $(\mu_1 - \mu_2)$. Thus, we should be convinced that the pairing will significantly reduce variability before performing the paired difference experiment. Most of the time this will happen.

One final note: The pairing of the observations is determined *before* the experiment is performed (that is, by the *design* of the experiment). A paired difference experiment is *never* obtained by pairing the sample observations after the measurements have been acquired.

What Do You Do When the Assumption of a Normal Distribution for the Population of Differences Is Not Satisfied?

Answer: Use the Wilcoxon signed rank test for the paired difference design (Chapter 15).

An IQ Comparison of Identical Twins Reared Apart

CASE STUDY • 9.2 •

How much of our personality, our likes and dislikes, our individuality, is predetermined by our genes? And which of our traits are shaped and changed by our environment? Twins, because they share an identical genotype, make ideal subjects for investigating the degree to which various environmental conditions may instigate change. The classical method of studying this phenomenon and the subject of an interesting book by Susan Farber (*Identical Twins Reared Apart*, New York: Basic Books, 1981) is the study of identical twins separated early in life and reared apart.

Identical twins, genetically called *monozygotic (MZ) twins,* are formed when a single egg, fertilized by a single sperm, splits into two parts, each of which develops into a separate embryo. In contrast, fraternal twins, or *dizygotic (DZ) twins,* develop when two eggs are released from one or both ovaries and each is fertilized by a different sperm. Although they have been the subjects in many studies of twins, DZ twins are no more genetically similar than any two siblings. Therefore, a study of DZ twins would leave unanswered the question of whether a given individual trait is due to hereditary or environmental differences. Likewise, MZ twins reared in the same family share almost identical environments, making it difficult to separate the two factors. Thus, claims Farber, "in theory at least, the clearest demarcation of heredity and environment is found when identical twins have been separated early in life and reared apart in different homes, by different parents, and often in widely varying socioeconomic and geographic circumstances."

Several studies of MZ twins have been conducted. Farber's book contains a chronicle and reanalysis of 95 pairs of identical twins reared apart. Much of her discussion focuses on a comparison of IQ scores.

The question is, "Are there significant differences between the IQ scores of identical twins, where one member of the pair is reared by the natural parents and the other member of the pair is not?" The data for this analysis (extracted from Table E6 of Farber's book) appear in Table 9.6. One member (A) of each of the $n = 32$ pairs of twins was reared by a natural parent; the other member (B) was reared by a relative or some other person.

Focus

Analyze the data in Table 9.6 using the methodology presented in this chapter. Interpret the results and discuss the implications.

TABLE 9.6 IQ Scores of Twins Reared Apart

Pair ID	Twin A	Twin B	Pair ID	Twin A	Twin B
112	113	109	228	100	88
114	94	100	232	100	104
126	99	86	236	93	84
132	77	80	306	99	95
136	81	95	308	109	98
148	91	106	312	95	100
170	111	117	314	75	86
172	104	107	324	104	103
174	85	85	328	73	78
180	66	84	330	88	99
184	111	125	338	92	111
186	51	66	342	108	110
202	109	108	344	88	83
216	122	121	350	90	82
218	97	98	352	79	76
220	82	94	416	97	98

Source: Adapted from *Identical Twins Reared Apart,* by Susan L. Farber. Copyright © 1981 by Basic Books, Inc. Reprinted by permission of Basic Books, Inc., Publishers, New York.

EXERCISES 9.29–9.43

Learning the Mechanics

9.29 Suppose a paired difference experiment is to be performed in order to test the null hypothesis H_0: $\mu_D - 2$ against H_a: $\mu_D > 2$. Assume that the second observation in each pair will be subtracted from the first. Define μ_D, and explain what is meant by the hypotheses $\mu_D = 2$ and $\mu_D > 2$.

9.30 A paired difference experiment yielded n_D pairs of observations. In each case, what is the rejection region for testing H_0: $\mu_D = 2$ against H_a: $\mu_D > 2$?
a. $n_D = 10, \alpha = .05$ **b.** $n_D = 20, \alpha = .10$
c. $n_D = 5, \alpha = .025$ **d.** $n_D = 9, \alpha = .01$

9.31 The data for a random sample of six paired observations are shown in the table below.

Pair	Sample from Population 1 Observation 1	Sample from Population 2 Observation 2
1	7	4
2	3	1
3	9	7
4	6	2
5	4	4
6	8	7

a. Calculate the difference between each pair of observations by subtracting observation 2 from observation 1. Use the differences to calculate $\bar{x}_D$ and s_D^2.
b. If μ_1 and μ_2 are the means of populations 1 and 2, respectively, express μ_D in terms of μ_1 and μ_2.
c. Form a 95% confidence interval for μ_D.
d. Test the null hypothesis H_0: $\mu_D = 0$ against the alternative hypothesis H_a: $\mu_D \neq 0$. Use $\alpha = .05$.

9.32 The data for a random sample of 10 paired observations are shown in the accompanying table
a. If you wish to test whether these data are sufficient to indicate that the mean for population 2 is larger than that for population 1, what are the appropriate null and alternative hypotheses? Define any symbols you use.
b. The data are analyzed using MINITAB, with the results shown below. Interpret these results.

Pair	Sample from Population 1	Sample from Population 2
1	19	24
2	25	27
3	31	36
4	52	53
5	49	55
6	34	34
7	59	66
8	47	51
9	17	20
10	51	55

c. The output of MINITAB also included a confidence interval. Interpret this output.
d. What assumptions are necessary to ensure the validity of this analysis?

9.33 A paired difference experiment produced the following data:

$$n_D - 16 \quad \bar{x}_1 - 143 \quad \bar{x}_2 - 150 \quad \bar{x}_D = -7 \quad s_D^2 = 64$$

a. Determine the values of t for which the null hypothesis, $\mu_1 - \mu_2 = 0$, would be rejected in favor of the alternative hypothesis, $\mu_1 - \mu_2 < 0$. Use $\alpha = .10$.
b. Conduct the paired difference test described in part **a**. Draw the appropriate conclusions.
c. What assumptions are necessary so that the paired difference test will be valid?
d. Find a 90% confidence interval for the mean difference μ_D.
e. Which of the two inferential procedures, the confidence interval of part **d** or the test of hypothesis of part **b**, provides more information about the difference between the population means?

9.34 A paired difference experiment yielded the data shown in the next table.

Pair	x	y	Pair	x	y
1	55	44	5	75	62
2	68	55	6	52	38
3	40	25	7	49	31
4	55	56			

```
TEST OF MU = 0.000 VS MU L.T. 0.000

           N      MEAN   STDEV  SE MEAN      T   P VALUE
DIFF      10    -3.700   2.214    0.700   -5.29   0.0002



           N      MEAN   STDEV  SE MEAN   95.0 PERCENT C.I.
DIFF      10    -3.700   2.214    0.700  ( -5.284,  -2.116)
```

a. Test H_0: $\mu_D = 10$ against H_a: $\mu_D \neq 10$, where $\mu_D = (\mu_1 - \mu_2)$. Use $\alpha = .05$.

b. Report the *p*-value for the test you conducted in part **a**. Interpret the *p*-value.

Applying the Concepts

9.35 Dr. Philip Lieberman, a neuroscientist at Brown University, recently conducted a field experiment to gauge the effect of high altitude on a person's ability to think critically (*New York Times,* Aug. 23, 1995). The subjects of the experiment were five male members of an American expedition climbing Mount Everest. At the base camp, Lieberman read sentences to the climbers while they looked at simple pictures in a book. The length of time (in seconds) it took for each climber to match the picture with a sentence was recorded. Using a radio, Lieberman repeated the task when the climbers reached a camp 5 miles above sea level. At this altitude, the researcher noted that the climbers took 50% longer to complete the task.

a. What is the variable measured for this experiment?

b. What are the experimental units?

c. Discuss how the data should be analyzed.

9.36 Hypersexual behavior caused by traumatic brain injury (TBI) is often treated with the drug Depo-Provera. In one clinical study, eight young male TBI patients who exhibited hypersexual behavior were treated weekly with 400 milligrams of Depo-Provera for six months (*Journal of Head Trauma Rehabilitation,* June 1995). The testosterone levels [in nanograms per deciliter (ng/dL)] of the patients both prior to treatment and at the end of the 6-month treatment period are given in the table below.

	TESTOSTERONE LEVELS	
Patient	Pretreatment	After 6 Months on Depo-Provera
1	849	96
2	903	41
3	890	31
4	1,092	124
5	362	46
6	900	53
7	1,006	113
8	672	174

Source: Emory, L. E., Cole, C. M., and Meyer, W. J. "Use of Depo-Provera to control sexual aggression in persons with traumatic brain injury." *Journal of Head Trauma Rehabilitation,* Vol. 10, No. 3, June 1995, p. 52 (Table 2). © 1995 Aspen Publishing, Inc.

a. Construct a 99% confidence interval for the true mean difference between the pretreatment and after-treatment testosterone levels of young male TBI patients.

b. Use the interval, part **a**, to make an inference about the effectiveness of Depo-Provera in reducing testosterone levels of young male TBI patients.

9.37 A *homophone* is a word whose pronunciation is the same as that of another word having a different meaning and spelling (e.g., *nun* and *none, doe* and *dough,* etc.). *Brain and Language* (Apr. 1995) reported on a study of homophone spelling in patients with Alzheimer's disease. Twenty Alzheimer's patients were asked to spell 24 homophone pairs given in random order, then the number of homophone confusions (e.g., spelling *doe* given the context, *bake bread dough*) was recorded for each patient. One year later, the same test was given to the same patients. The data for the study are provided in the table below. The researchers posed the following question: "Do Alzheimer's patients show a significant increase in mean homophone confusion errors over time?" Perform an analysis of the data to answer the researchers' question. Use the relevant information in the SAS printout on page 369. What assumptions are necessary for the procedure used to be valid? Are they satisfied?

	NUMBER OF HOMOPHONE CONFUSIONS	
Patient	Time 1	Time 2
1	5	5
2	1	3
3	0	0
4	1	1
5	0	1
6	2	1
7	5	6
8	1	2
9	0	9
10	5	8
11	7	10
12	0	3
13	3	9
14	5	8
15	7	12
16	10	16
17	5	5
18	6	3
19	9	6
20	11	8

Source: Neils, J., Roeltgen, D. P., and Constantinidou, F. "Decline in homophone spelling associated with loss of semantic influence on spelling in Alzheimer's disease." *Brain and Language,* Vol. 49, No. 1, Apr. 1995, p. 36 (Table 3).

```
N Obs   Variable  N      Minimum      Maximum        Mean      Std Dev
---------------------------------------------------------------------------
  20    TIME1     20            0   11.0000000   4.1500000    3.4984959
        TIME2     20            0   16.0000000   5.8000000    4.2127626
        DIFF      20   -9.0000000    3.0000000  -1.6500000    3.1999178
```

9.38 A study published in *Clinical Kinesiology* (Spring 1995) was designed to examine the metabolic and cardiopulmonary responses during exercise of persons diagnosed with multiple sclerosis (MS). Leg cycling and arm cranking exercises were performed by ten MS patients and ten healthy (non-MS) control subjects. Each member of the control group was selected based on gender, age, height, and weight to match (as closely as possible) with one member of the MS group. Consequently, the researchers compared the MS and non-MS groups by matched-pairs *t*-tests on such outcome variables as oxygen uptake, carbon dioxide output, and peak aerobic power. The data on the matching variables used in the experiment are shown in the table below.

a. Visually examine the data table. Does it appear that the researchers have successfully matched the MS and non-MS subjects?

b. Conduct a test to determine whether the mean ages of the MS and non-MS groups are significantly different. Use α = .05.

c. MINITAB printouts for comparing the mean heights and weights of the two groups are shown below the table. Interpret the results.

d. Based on the results, parts **b** and **c**, comment on the validity of the *t*-tests performed by the researchers on the outcome variables.

9.39 Inositol is a complex cyclic alcohol found to be effective against clinical depression. Medical researchers believe inositol may also be used to treat panic disorder. To test this theory, a double-blind, placebo-controlled study of 21 patients diagnosed with panic disorder was conducted (*American Journal of Psychiatry,* July 1995). Patients completed daily panic diaries recording the occurrence of their panic attacks. The data for a week in which patients received a glucose placebo and for a week when they were treated with inositol are provided in the table on page 370. [*Note:* Neither the patients nor the treating physicians knew which week the placebo was given.] Analyze the data and interpret the results. Comment on the validity of the assumptions.

Matched Pair	MS SUBJECTS				NON-MS SUBJECTS			
	Gender	Age (years)	Height (cm)	Weight (kg)	Gender	Age (years)	Height (cm)	Weight (kg)
1	M	48	171.0	80.8	M	45	173.0	76.3
2	F	34	158.5	75.0	F	34	158.0	75.6
3	F	34	167.6	55.5	F	34	164.5	57.7
4	M	38	167.0	71.3	M	34	161.3	70.0
5	M	45	182.5	90.9	M	39	179.0	96.0
6	F	42	166.0	72.4	F	42	167.0	77.8
7	M	32	172.0	70.5	M	34	165.8	74.7
8	F	35	166.5	55.3	F	43	165.1	71.4
9	F	33	166.5	57.9	F	31	170.1	60.4
10	F	46	175.0	79.9	F	43	175.0	77.9

Source: Ponichtera-Mulcare, J. A., *et al.* "Maximal aerobic exercise of individuals with multiple sclerosis using three modes of ergometry." *Clinical Kinesiology,* Vol. 49, No. 1, Spring 1995, p. 7 (Table 1). Reprinted by permission.

```
TEST OF MU = 0.000 VS MU N.E. 0.000

           N     MEAN    STDEV   SE MEAN       T   P VALUE
HTDiff    10    1.380    3.230     1.022    1.35      0.21


TEST OF MU = 0.000 VS MU N.E. 0.000

           N     MEAN    STDEV   SE MEAN       T   P VALUE
WTDiff    10   -2.830    5.670     1.793   -1.58      0.15
```

	PANIC ATTACKS PER WEEK	
Patient	**Placebo**	**Inositol**
1	0	0
2	2	1
3	0	3
4	1	2
5	0	0
6	10	5
7	2	0
8	6	4
9	1	1
10	1	0
11	1	3
12	0	2
13	3	1
14	3	1
15	3	4
16	4	2
17	6	4
18	15	21
19	28	8
20	30	0
21	13	0

Source: Benjamin, J., *et al.* "Double-blind, placebo-controlled, crossover trial of inositol treatment for panic disorder." *American Journal of Psychiatry,* Vol. 152, No. 7, July 1995, p. 1085 (Table 1). Copyright 1995, the American Psychiatric Association. Reprinted by permission.

9.40 *Arctic and Alpine Research* (Vol. 14, 1982) investigated the relationship between the mean daily air temperature and the cocoon temperature of wooly-bear caterpillars of the High Arctic. The data in the table below indicate that the caterpillar's body temperature (inside the cocoon) is higher than the outside air temperature. Estimate the mean difference in temperature between the cocoon and the outside air. Use a 95% confidence interval. Interpret the result.

9.41 A study reported in the *Journal of Psychology* (Mar. 1991) measures the change in female students' self-concepts as they move from high school to college. A sample of 133 Boston College first-year female students were selected for the study. Each was asked to evaluate several aspects of her life at two points in time: at the end of her senior year of high school, and during her sophomore year of college. Each student was asked to evaluate where she believed she stood on a scale that ranged from top 10% of class (1) to lowest 10% of class (5). The results for three of the traits evaluated are reported in the table below.

Trait	n	**SENIOR YEAR OF HIGH SCHOOL** $\bar{x}$	**SOPHOMORE YEAR OF COLLEGE** $\bar{x}$
Leadership	133	2.09	2.33
Popularity	133	2.48	2.69
Intellectual self-confidence	133	2.29	2.55

a. What null and alternative hypotheses would you test to determine whether the mean self-concept of females decreases between the senior year of high school and the sophomore year of college as measured by each of these three traits?

b. Are these traits more appropriately analyzed using an independent samples test or a paired difference test? Explain.

c. Noting the size of the sample, what assumptions are necessary to ensure the validity of the tests?

d. The article reports that the leadership test results in a *p*-value greater than .05, while the tests for popularity and intellectual self-confidence result in *p*-values less than .05. Interpret these results.

	TEMPERATURE (°C)			**TEMPERATURE (°C)**	
Day	**Air**	**Cocoon**[a]	**Day**	**Air**	**Cocoon**[a]
1	10.4	15.1	7	1.7	3.6
2	9.2	14.6	8	2.0	5.3
3	2.2	6.8	9	3.0	7.0
4	2.6	6.8	10	3.5	7.1
5	4.1	8.0	11	4.5	9.6
6	3.7	8.7	12	4.4	9.5

[a]Each cocoon temperature is the average of the temperatures of two cocoons.

Source: Kevan, P. G., Jensen, T. S., and Shorthouse, J. D. "Body temperatures and behavioral thermoregulation of High Arctic wooly-bear caterpillars and pupae (*Gynaephora rossii,* Lymantridae: Lepidoptera) and the importance of sunshine," *Arctic and Alpine Research,* 1982, 14. Reproduced with permission of the Regents of the University of Colorado.

9.42 Merck Research Labs conducted an experiment to evaluate the effect of a new drug using the single-T swim maze. Nineteen impregnated dam rats were allocated a dosage of 12.5 milligrams of the drug. One male and one female rat pup were randomly selected from each resulting litter to perform in the swim maze. Each pup was placed in the water at one end of the maze and allowed to swim until it escaped at the opposite end. If the pup failed to escape after a certain period of time, it was placed at the beginning of the maze and given another chance. The experiment was repeated until each pup accomplished three successful escapes. The accompanying table reports the number of swims required by each pup to perform three successful escapes. Is there sufficient evidence of a difference between the mean number of swims required by male and female pups? Use the accompanying MINITAB printout to conduct the test (at $\alpha = .10$). Comment on the assumptions required for the test to be valid.

9.43 The data shown in the table below are part of a study conducted to compare the abilities of men and women to perform the strenuous tasks required of a shipboard firefighter (*Human Factors*, Vol. 24, 1982). They represent the pulling force (in newtons) that a firefighter was able to exert in pulling the starter cord of a P-250 water pump. Firefighters were matched in pairs according to weight, thus producing data for a matched pairs (or paired difference) experiment.

Pair	Female	Male
1	40.03	84.51
2	75.62	80.06
3	53.38	102.30
4	62.27	88.96

Source: M. D. Phillips and R. L. Pepper, "Shipboard Firefighting Performance of Females and Males." Reprinted with permission from *Human Factors*, Vol. 24, No. 3, 1982. Copyright 1982 by The Human Factors Society. All rights reserved.

a. Do the data provide sufficient evidence to indicate a difference in mean pulling force between female and male firefighters? Test using $\alpha = .05$.

b. Find a 95% confidence interval for the difference in mean pulling force between female and male firefighters. Interpret the interval.

Litter	Male	Female	Litter	Male	Female
1	8	5	11	6	5
2	8	4	12	6	3
3	6	7	13	12	5
4	6	3	14	3	8
5	6	5	15	3	4
6	6	3	16	8	12
7	3	8	17	3	6
8	5	10	18	6	4
9	4	4	19	9	5
10	4	4			

Source: Thomas E. Bradstreet, Merck Research Labs, BL 3-2, West Point, PA 19486

```
TEST OF MU = 0.000 VS MU N.E.  0.000

              N     MEAN    STDEV   SE MEAN      T    P VALUE
SwimDiff     19    0.368    3.515    0.806     0.46      0.65
```

9.3 COMPARING TWO POPULATION PROPORTIONS: INDEPENDENT SAMPLING

Suppose a presidential candidate wants to compare the preference of registered voters in the northeastern United States (NE) to those in the southeastern United States (SE). Such a comparison would help determine where to concentrate campaign efforts. The candidate hires a professional pollster to randomly choose 1,000 registered voters in the northeast and 1,000 in the southeast and interview each to learn her or his voting preference. The objective is to use this sample information to make an inference about the difference $(p_1 - p_2)$ between the proportion p_1 of *all* registered voters in the northeast and the proportion p_2 of *all* registered voters in the southeast who plan to vote for the presidential candidate.

The two samples represent independent binomial experiments. (See Section 4.4 for the characteristics of binomial experiments.) The binomial random variables are the numbers x_1 and x_2 of the 1,000 sampled voters in each area who indicate they will vote for the candidate. The results are summarized below.

NE	SE
$n_1 = 1{,}000$	$n_2 = 1{,}000$
$x_1 = 546$	$x_2 = 475$

We can now calculate the sample proportions $\hat{p}_1$ and $\hat{p}_2$ of the voters in favor of the candidate in the northeast and southeast, respectively:

$$\hat{p}_1 = \frac{x_1}{n_1} = \frac{546}{1{,}000} = .546 \qquad \hat{p}_2 = \frac{x_2}{n_2} = \frac{475}{1{,}000} = .475$$

The difference between the sample proportions $(\hat{p}_1 - \hat{p}_2)$ makes an intuitively appealing point estimator of the difference between the population parameters $(p_1 - p_2)$. For our example, the estimate is

$$(\hat{p}_1 - \hat{p}_2) = .546 - .475 = .071$$

To judge the reliability of the estimator $(\hat{p}_1 - \hat{p}_2)$, we must observe its performance in repeated sampling from the two populations. That is, we need to know the sampling distribution of $(\hat{p}_1 - \hat{p}_2)$. The properties of the sampling distribution are given in the following box. Remember that $\hat{p}_1$ and $\hat{p}_2$ can be viewed as means of the number of successes per trial in the respective samples, so the Central Limit Theorem applies when the sample sizes are large.

Properties of the Sampling Distribution of $(\hat{p}_1 - \hat{p}_2)$

1. The mean of the sampling distribution of $(\hat{p}_1 - \hat{p}_2)$ is $(p_1 - p_2)$; that is,

$$E(\hat{p}_1 - \hat{p}_2) = p_1 - p_2$$

   Thus, $(\hat{p}_1 - \hat{p}_2)$ is an unbiased estimator of $(p_1 - p_2)$.
2. The standard deviation of the sampling distribution of $(\hat{p}_1 - \hat{p}_2)$ is

$$\sigma_{(\hat{p}_1 - \hat{p}_2)} = \sqrt{\frac{p_1 q_1}{n_1} + \frac{p_2 q_2}{n_2}}$$

3. If the sample sizes n_1 and n_2 are large (see Section 7.3 for a guideline), the sampling distribution of $(\hat{p}_1 - \hat{p}_2)$ is approximately normal.

Since the distribution of $(\hat{p}_1 - \hat{p}_2)$ in repeated sampling is approximately normal, we can use the z statistic to derive confidence intervals for $(p_1 - p_2)$ or to test a hypothesis about $(p_1 - p_2)$.

For the voter example, a 95% confidence interval for the difference $(p_1 - p_2)$ is

$$(\hat{p}_1 - \hat{p}_2) \pm 1.96\sigma_{(\hat{p}_1 - \hat{p}_2)} \quad \text{or} \quad (\hat{p}_1 - \hat{p}_2) \pm 1.96\sqrt{\frac{p_1 q_1}{n_1} + \frac{p_2 q_2}{n_2}}$$

The quantities $p_1 q_1$ and $p_2 q_2$ must be estimated in order to complete the calculation of the standard deviation, $\sigma_{(\hat{p}_1 - \hat{p}_2)}$ and hence the calculation of the confidence interval. In Section 7.3 we showed that the value of pq is relatively insensitive to the value chosen to approximate p. Therefore, $\hat{p}_1\hat{q}_1$ and $\hat{p}_2\hat{q}_2$ will provide satisfactory estimates to approximate $p_1 q_1$ and $p_2 q_2$, respectively. Then

$$\sqrt{\frac{p_1 q_1}{n_1} + \frac{p_2 q_2}{n_2}} \approx \sqrt{\frac{\hat{p}_1\hat{q}_1}{n_1} + \frac{\hat{p}_2\hat{q}_2}{n_2}}$$

and we will approximate the 95% confidence interval by

$$(\hat{p}_1 - \hat{p}_2) \pm 1.96 \sqrt{\frac{\hat{p}_1 \hat{q}_1}{n_1} + \frac{\hat{p}_2 \hat{q}_2}{n_2}}$$

Substituting the sample quantities yields

$$(.546 - .475) \pm 1.96 \sqrt{\frac{(.546)(.454)}{1,000} + \frac{(.475)(.525)}{1,000}}$$

or $.071 \pm .044$. Thus, we are 95% confident that the interval from .027 to .115 contains $(p_1 - p_2)$.

We infer that there are between 2.7% and 11.5% more registered voters in the northeast than in the southeast who plan to vote for the presidential candidate. It seems that the candidate should direct a greater campaign effort in the southeast compared to the northeast.

The general form of a confidence interval for the difference $(p_1 - p_2)$ between population proportions is given in the following box.

Large-Sample $100(1 - \alpha)\%$ Confidence Interval for $(p_1 - p_2)$

$$(\hat{p}_1 - \hat{p}_2) \pm z_{\alpha/2} \sigma_{(\hat{p}_1 - \hat{p}_2)} = (\hat{p}_1 - \hat{p}_2) \pm z_{\alpha/2} \sqrt{\frac{p_1 q_1}{n_1} + \frac{p_2 q_2}{n_2}}$$

$$\approx (\hat{p}_1 - \hat{p}_2) \pm z_{\alpha/2} \sqrt{\frac{\hat{p}_1 \hat{q}_1}{n_1} + \frac{\hat{p}_2 \hat{q}_2}{n_2}}$$

Assumptions: The two samples are independent random samples. Both samples should be large enough that the normal distribution provides an adequate approximation to the sampling distribution of $\hat{p}_1$ and $\hat{p}_2$ (see Section 8.5).

The z statistic,

$$z = \frac{(\hat{p}_1 - \hat{p}_2) - (p_1 - p_2)}{\sigma_{(\hat{p}_1 - \hat{p}_2)}}$$

is used to test the null hypothesis that $(p_1 - p_2)$ equals some specified difference, say D_0. For the special case where $D_0 = 0$, that is, where we want to test the null hypothesis $H_0: (p_1 - p_2) = 0$ (or, equivalently, $H_0: p_1 = p_2$), the best estimate of $p_1 = p_2 = p$ is obtained by dividing the total number of successes $(x_1 + x_2)$ for the two samples by the total number of observations $(n_1 + n_2)$; that is,

$$\hat{p} = \frac{x_1 + x_2}{n_1 + n_2} \quad \text{or} \quad \hat{p} = \frac{n_1 \hat{p}_1 + n_2 \hat{p}_2}{n_1 + n_2}$$

The second equation shows that $\hat{p}$ is a weighted average of $\hat{p}_1$ and $\hat{p}_2$, with the larger sample receiving more weight. If the sample sizes are equal, then $\hat{p}$ is a simple average of the two sample proportions of successes.

We now substitute the weighted average $\hat{p}$ for both p_1 and p_2 in the formula for the standard deviation of $(\hat{p}_1 - \hat{p}_2)$:

$$\sigma_{(\hat{p}_1 - \hat{p}_2)} = \sqrt{\frac{p_1 q_1}{n_1} + \frac{p_2 q_2}{n_2}} \approx \sqrt{\frac{\hat{p}\hat{q}}{n_1} + \frac{\hat{p}\hat{q}}{n_2}} = \sqrt{\hat{p}\hat{q} \left(\frac{1}{n_1} + \frac{1}{n_2} \right)}$$

The test is summarized in the next box.

Large-Sample Test of Hypothesis About ($p_1 - p_2$)

ONE-TAILED TEST	TWO-TAILED TEST
H_0: $(p_1 - p_2) = 0*$	H_0: $(p_1 - p_2) = 0$
H_a: $(p_1 - p_2) < 0$	H_a: $(p_1 - p_2) \neq 0$
[or H_a: $(p_1 - p_2) > 0$]	

Test statistic:

$$z = \frac{(\hat{p}_1 - \hat{p}_2)}{\sigma_{(\hat{p}_1 - \hat{p}_2)}}$$

Rejection region: $z < -z_\alpha$ *Rejection region:* $|z| > z_{\alpha/2}$
  [or $z > z_\alpha$ when
  H_a: $(p_1 - p_2) > 0$]

Note: $\sigma_{(\hat{p}_1 - \hat{p}_2)} = \sqrt{\dfrac{p_1 q_1}{n_1} + \dfrac{p_2 q_2}{n_2}} \approx \sqrt{\hat{p}\hat{q}\left(\dfrac{1}{n_1} + \dfrac{1}{n_2}\right)}$ where $\hat{p} = \dfrac{x_1 + x_2}{n_1 + n_2}$

Assumption: Same as for large-sample confidence interval for $(p_1 - p_2)$.
(See previous box.)

EXAMPLE 9.6

In the past decade intensive antismoking campaigns have been sponsored by both federal and private agencies. Suppose the American Cancer Society randomly sampled 1,500 adults in 1985 and then sampled 2,000 adults in 1995 to determine whether there was evidence that the percentage of smokers had decreased. The results of the two sample surveys are shown in the table below, where x_1 and x_2 represent the numbers of smokers in the 1985 and 1995 samples, respectively. Do these data indicate that the fraction of smokers decreased over this 10-year period? Use $\alpha = .05$.

1985	1995
$n_1 = 1,500$	$n_2 = 2,000$
$x_1 = 576$	$x_2 = 652$

Solution

If we define p_1 and p_2 as the true proportions of adult smokers in 1985 and 1995, the elements of our test are

$$H_0: (p_1 - p_2) = 0$$
$$H_a: (p_1 - p_2) > 0$$

(The test is one-tailed since we are interested only in determining whether the proportion of smokers *decreased*.)

Test statistic: $z = \dfrac{(\hat{p}_1 - \hat{p}_2) - 0}{\sigma_{(\hat{p}_1 - \hat{p}_2)}}$

Rejection region: $\alpha = .05$

$z > z_\alpha = z_{.05} = 1.645$ (see Figure 9.14)

We now calculate the sample proportions of smokers

$$\hat{p}_1 = \frac{576}{1,500} = .384 \qquad \hat{p}_2 = \frac{652}{2,000} = .326$$

---

*The test can be adapted to test for a difference $D_0 \neq 0$. Because most applications call for a comparison of p_1 and p_2, implying $D_0 = 0$, we will confine our attention to this case.

FIGURE 9.14

Rejection region for Example 9.6

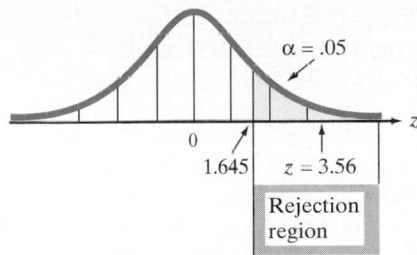

Then

$$z = \frac{(\hat{p}_1 - \hat{p}_2) - 0}{\sigma_{(\hat{p}_1 - \hat{p}_2)}} \approx \frac{(\hat{p}_1 - \hat{p}_2)}{\sqrt{\hat{p}\hat{q}\left(\dfrac{1}{n_1} + \dfrac{1}{n_2}\right)}}$$

where

$$\hat{p} = \frac{x_1 + x_2}{n_1 + n_2} = \frac{576 + 652}{1{,}500 + 2{,}000} = .351$$

Note that $\hat{p}$ is a weighted average of $\hat{p}_1$ and $\hat{p}_2$, with more weight given to the larger (1995) sample.

Thus, the computed value of the test statistic is

$$z = \frac{.384 - .326}{\sqrt{(.351)(.649)\left(\dfrac{1}{1{,}500} + \dfrac{1}{2{,}000}\right)}} = \frac{.058}{.0164} - 3.56$$

There is sufficient evidence at the $\alpha = .05$ level to conclude that the proportion of adults who smoke has decreased over the 1985–1995 period. We could place a confidence interval on $(p_1 - p_2)$ if we were interested in estimating the extent of the decrease.   ▲

EXERCISES 9.44–9.62

Learning the Mechanics

9.44 What are the characteristics of a binomial experiment?

9.45 Explain why the Central Limit Theorem is important in finding an approximate distribution for $(\hat{p}_1 - \hat{p}_2)$.

9.46 In each case, determine whether the sample sizes are large enough to conclude that the sampling distribution of $(\hat{p}_1 - \hat{p}_2)$ is approximately normal.
  a. $n_1 = 10, n_2 = 12, \hat{p}_1 = .50, \hat{p}_2 = .50$
  b. $n_1 = 10, n_2 = 12, \hat{p}_1 = .10, \hat{p}_2 = .08$
  c. $n_1 = n_2 = 30, \hat{p}_1 = .20, \hat{p}_2 = .30$
  d. $n_1 = 100, n_2 = 200, \hat{p}_1 = .05, \hat{p}_2 = .09$
  e. $n_1 = 100, n_2 = 200, \hat{p}_1 = .95, \hat{p}_2 = .91$

9.47 For each of the following values of α, find the values of z for which $H_0: (p_1 - p_2) = 0$ would be rejected in favor of $H_a: (p_1 - p_2) < 0$.
  a. $\alpha = .01$ **b.** $\alpha = .025$ **c.** $\alpha = .05$ **d.** $\alpha = .10$

9.48 Independent random samples, each containing 800 observations, were selected from two binomial populations. The samples from populations 1 and 2 produced 320 and 400 successes, respectively.

a. Test $H_0: (p_1 - p_2) = 0$ against $H_a: (p_1 - p_2) \neq 0$. Use $\alpha = .05$.
b. Test $H_0: (p_1 - p_2) = 0$ against $H_a: (p_1 - p_2) \neq 0$. Use $\alpha = .01$.
c. Test $H_0: (p_1 - p_2) = 0$ against $H_a: (p_1 - p_2) < 0$. Use $\alpha = .01$.
d. Form a 90% confidence interval for $(p_1 - p_2)$.

9.49 Construct a 95% confidence interval for $(p_1 - p_2)$ in each of the following situations:
  a. $n_1 = 400, \hat{p}_1 = .65; n_2 = 400, \hat{p}_2 = .58$
  b. $n_1 = 180, \hat{p}_1 = .31; n_2 = 250, \hat{p}_2 = .25$
  c. $n_1 = 100, \hat{p}_1 = .46; n_2 = 120, \hat{p}_2 = .61$

9.50 Sketch the sampling distribution of $(\hat{p}_1 - \hat{p}_2)$ based on independent random samples of $n_1 = 100$ and $n_2 = 200$ observations from two binomial populations with success probabilities $p_1 = .1$ and $p_2 = .5$, respectively.

9.51 Random samples of size $n_1 = 50$ and $n_2 = 60$ were drawn from populations 1 and 2, respectively. The samples yielded $\hat{p}_1 = .4$ and $\hat{p}_2 = .2$. Test $H_0: (p_1 - p_2) = .1$ against $H_a: (p_1 - p_2) > .1$ using $\alpha = .05$.

Applying the Concepts

9.52 Stroke is the nation's third leading cause of death after heart attacks and cancer. Doctors have theorized that renegade antibodies in blood may be a major underlying cause of stroke. Until recently, no clear evidence existed to support this theory. To investigate the causal link between antibodies and stroke, medical researchers examined two groups of patients for renegade antibodies (*Tampa Tribune,* Feb. 2, 1992). One group of 255 patients had suffered a first stroke while the other group of 257 patients were hospitalized for other reasons. Twenty-five of the stroke patients had antibodies in their blood compared to 11 of the other patients. Analyze the data for the medical researchers. What conclusions can you draw from the analysis?

9.53 Do you have an insatiable craving for chocolate or some other food? Since many North Americans apparently do, psychologists are designing scientific studies to examine the phenomenon. According to the *New York Times* (Feb. 22, 1995), one of the largest studies of food cravings involved a survey of 1,000 McMaster University (Canada) students. The survey revealed that 97% of the women in the study acknowledged specific food cravings while only 67% of the men did. Assume that 600 of the respondents were women and 400 were men.

a. Is there sufficient evidence to claim that the true proportion of women who acknowledge having food cravings exceeds the corresponding proportion of men? Test using $\alpha = .01$.

b. Why is it dangerous to conclude from the study that women have a higher incidence of food cravings than men?

9.54 Many female undergraduates at four-year colleges and universities switch from science, mathematics, and engineering (SME) majors into disciplines that are not science-based, thereby contributing to the underrepresentation of women in SME fields. When female undergraduates switch majors, are their reasons different from those of their male counterparts? This question was investigated in *Science Education* (July 1995). A sample of 335 junior/senior undergraduates—172 females and 163 males—at two large research universities were identified as "switchers", that is, they left a declared SME major for a non-SME major. Each student listed one or more factors that contributed to the switching decision.

a. Of the 172 females in the sample, 74 listed lack or loss of interest in SME (i.e., "turned off" by science) as a major factor, compared to 72 of the 163 males. Conduct a test (at $\alpha = .10$) to determine whether the proportion of female switchers who give "lack of interest in SME" as a major

reason for switching differs from the corresponding proportion of males.

b. Thirty-three of the 172 females in the sample indicated that they were discouraged or lost confidence because of low grades in SME during their early years, compared to 44 of 163 males. Construct a 90% confidence interval for the difference between the proportions of female and male switchers who lost confidence due to low grades in SME. Interpret the result.

9.55 The *Journal of Fish Biology* (Aug. 1990) reported on a study to compare the incidence of parasites (tapeworms) in species of Mediterranean and Atlantic fish. In the Mediterranean Sea, 588 brill were captured and dissected, and 211 were found to be infected by the parasite. In the Atlantic Ocean, 123 brill were captured and dissected, and 26 were found to be infected. Compare the proportions of infected brill at the two capture sites using a 90% confidence interval. Interpret the interval.

9.56 Women are filling managerial positions in increasing numbers, although there is much dissension about whether progress has been fast enough. Does marriage hinder the career progression of women to a greater degree than that of men? An article in *The Journal of Applied Psychology* (Vol. 77, 1992) investigates this and other questions relating to managerial careers of men and women in today's workforce. In a random sample of 795 male managers and 223 female managers from 20 Fortune 500 corporations, 86% of the male managers and 45% of the female managers were married.

a. Use a 95% confidence interval to estimate the difference between the proportions of men and women managers who are married.

b. Interpret the interval. State your conclusion in terms of the populations from which the samples were drawn.

9.57 Does cohabitation prior to marriage change the probability of the marriage's success? A recent article in *The Journal of Marriage and the Family* (Feb. 1992) compares the divorce rate for couples who cohabited prior to marriage against that for couples who did not cohabit. Suppose that a random sample of 500 couples are selected from each group.

a. What are the appropriate null and alternative hypotheses to be tested? Define any symbols you use.

b. What is the appropriate test statistic? Sketch the probability distribution of this test statistic, assuming the null hypothesis is true.

c. The authors report that cohabitation prior to marriage is associated with a greater risk of divorce, with a *p*-value $< .05$. In which tail of the distribution you sketched in part **b** did the observed test statistic fall? Show its approximate location on the sketch of the distribution.

d. Based on the observed significance level, what conclusion would you reach about the test you set up in part **a**?

9.58 It is estimated that more than half the votes cast in upcoming elections will be cast by women. Of concern to the dominant political parties is the possibility of a "gender gap." According to the gender gap hypothesis, there is a difference between male and female perceptions of which political party best addresses the nation's problems. Particularly, those who favor this hypothesis believe that the Democratic platform may appear to be more in line with philosophies of the majority of women. Suppose a Republican campaign manager wishes to test this theory. Assume that 300 registered voters (150 men, 150 women) were given the platforms of both the Republicans and Democrats for five major issues and then were asked to state a preference for one of the parties. Suppose 81 men and 70 women preferred the Republican views.

a. Does this result provide sufficient evidence to indicate that a gender gap favoring the Democratic party exists? Test at $\alpha = .05$.

b. If there were no undecided responses, can we say that women prefer the Democrats' responses to issues? Test at $\alpha = .01$.

9.59 According to new research, inositol—a complex alcohol nutrient found in breast milk—has been found to reduce the risk of lung and eye damage in premature infants. The study, published in the *New England Journal of Medicine* (May 7, 1992), involved 220 premature infants. The infants were randomly divided into two groups; half (110) of the infants received an intravenous feeding of inositol, while the other half received a standard diet. The researchers found that 14 of the inositol-fed infants suffered retinopathy of prematurity, an eye injury that results from the high oxygen levels needed to compensate for poorly developed lungs. In contrast, 29 of the infants on the standard diet developed this disease. Analyze the results with a test of hypothesis. Use $\alpha = .01$.

9.60 Geneticists at Duke University Medical Center have identified the E2F1 transcription factor as an important component of cell proliferation control (*Nature*, Sept. 23, 1993). The researchers induced DNA synthesis in two batches of serum-starved cells. Each cell in one batch was microinjected with the E2F1 gene, while the cells in the second batch (the controls) were not exposed to E2F1. After 30 hours, the number of cells in each batch that exhibited altered growth was determined. The results of the experiment are summarized in the next table.

	Control	**E2F1-Treated Cells**
Total number of cells	158	92
Number of growth-altered cells	15	41

Source: Johnson, D. G., *et al.* "Expression of transcription factor E2F1 induces quiescent cells to enter S phase." *Nature,* Vol. 365, No. 6444, Sept. 23, 1993, p. 351 (Table 1). Reprinted with permission from *Nature.* Copyright © 1993 Macmillan Magazines Limited.

a. Compare the percentages of cells exhibiting altered growth in the two batches with a 90% confidence interval.

b. Use the interval, part **a**, to make an inference about the ability of the E2F1 transcription factor to induce cell growth.

9.61 Scientists have linked a catastrophic decline in the number of frogs inhabiting the world to ultraviolet radiation from the Earth's tattered ozone layer (*Tampa Tribune,* Mar. 1, 1994). The Pacific tree frog, however, is not believed to be in decline, possibly because it manufactures an enzyme that protects its eggs from ultraviolet radiation. Researchers at Oregon State University compared the hatching rates of two groups of Pacific tree frog eggs. One group of eggs was shielded with ultraviolet-blocking sun shades, while the second group was not. The number of eggs successfully hatched in each group is provided in the following table. Compare the hatching rates of the two groups of Pacific tree frog eggs with a test of hypothesis. Use $\alpha = .01$.

	Sun-Shaded Eggs	**Unshaded Eggs**
Total number	70	80
Number hatched	34	31

9.62 Refer to Exercise 7.44, in which we examined data from an article in the *International Journal of Sports Psychology* (1990) studying the relationship between levels of fitness and stress among employees of companies offering health and fitness programs. Random samples of employees were selected from each of three fitness level categories, and each was evaluated for signs of stress. The data are repeated here.

Fitness Level	Sample Size	Proportion with Signs of Stress
Poor	242	.155
Average	212	.133
Good	95	.108

a. What are the appropriate null and alternative hypotheses to test whether a greater proportion of employees in the poor fitness category show signs of stress than those in the average fitness category? Define any symbols you use.

b. Conduct the test constructed in part **a** using α = .10. Interpret the result.

c. How would your null and alternative hypotheses change if you wanted to compare the proportions showing signs of stress in the poor and good fitness categories?

d. To conduct the test in part **c**, the data are analyzed by a statistical software package with the results shown here. Based on these results, state whether you would be willing to support the following statement: "Fitness level has no bearing on whether an individual shows signs of stress." Explain.

```
Z = 1.11    P-VALUE = .1335
```

9.4 DETERMINING THE SAMPLE SIZE

You can find the appropriate sample size to estimate the difference between a pair of parameters with a specified degree of reliability by using the method described in Section 7.4. That is, to estimate the difference between a pair of parameters correct to within B units with probability $(1 - \alpha)$, let $z_{\alpha/2}$ standard deviations of the sampling distribution of the estimator equal B. Then solve for the sample size. To do this, you have to solve the problem for a specific ratio between n_1 and n_2. Most often, you will want to have equal sample sizes, that is, $n_1 = n_2 = n$. We will illustrate the procedure with two examples.

EXAMPLE 9.7

New fertilizer compounds are often advertised with the promise of increased crop yields. Suppose we want to compare the mean yield μ_1 of wheat when a new fertilizer is used to the mean yield μ_2 with a fertilizer in common use. The estimate of the difference in mean yield per acre is to be correct to within .25 bushel with a confidence coefficient of .95. If the sample sizes are to be equal, find $n_1 = n_2 = n$, the number of 1-acre plots of wheat assigned to each fertilizer.

Solution

To solve the problem, you need to know something about the variation in the bushels of yield per acre. Suppose from past records you know the yields of wheat possess a range of approximately 10 bushels per acre. You could then approximate $\sigma_1 = \sigma_2 = \sigma$ by letting the range equal 4σ. Thus,

$$4\sigma \approx 10 \text{ bushels}$$

$$\sigma \approx 2.5 \text{ bushels}$$

The next step is to solve the equation

$$z_{\alpha/2}\sigma_{(\bar{x}_1 - \bar{x}_2)} = B \quad \text{or} \quad z_{\alpha/2}\sqrt{\frac{\sigma_1^2}{n_1} + \frac{\sigma_2^2}{n_2}} = B$$

for n, where $n = n_1 = n_2$. Since we want the estimate to lie within B = .25 of $(\mu_1 - \mu_2)$ with confidence coefficient equal to .95, we have $z_{\alpha/2} = z_{.025} = 1.96$. Then, letting $\sigma_1 = \sigma_2 = 2.5$ and solving for n, we have

$$1.96\sqrt{\frac{(2.5)^2}{n} + \frac{(2.5)^2}{n}} = .25$$

$$1.96\sqrt{\frac{2(2.5)^2}{n}} = .25$$

$$n = 768.32 \approx 769 \text{ (rounding up)}$$

Consequently, you will have to sample 769 acres of wheat for each fertilizer to estimate the difference in mean yield per acre to within .25 bushel. Since this would necessitate extensive and costly experimentation, you might decide to allow a larger bound (say, $B = .50$ or $B = 1$) in order to reduce the sample size, or you might decrease the confidence coefficient. The point is that we can obtain an idea of the experimental effort necessary to achieve a specified precision in our final estimate by determining the approximate sample size *before* the experiment is begun.   ▲

EXAMPLE 9.8

A production supervisor suspects that a difference exists between the proportions p_1 and p_2 of defective items produced by two different machines. Experience has shown that the proportion defective for each of the two machines is in the neighborhood of .03. If the supervisor wants to estimate the difference in the proportions to within .005 using a 95% confidence interval, how many items must be randomly sampled from the production of each machine? (Assume that the supervisor wants $n_1 = n_2 = n$.)

Solution

In this sampling problem, $B = .005$, and for the specified level of reliability, $z_{\alpha/2} = z_{.025} = 1.96$. Then, letting $p_1 = p_2 = .03$ and $n_1 = n_2 = n$, we find the required sample size per machine by solving the following equation for n:

$$z_{\alpha/2}\sigma_{(\hat{p}_1 - \hat{p}_2)} = B$$

or

$$z_{\alpha/2}\sqrt{\frac{p_1 q_1}{n_1} + \frac{p_2 q_2}{n_2}} = B$$

$$1.96\sqrt{\frac{(.03)(.97)}{n} + \frac{(.03)(.97)}{n}} = .005$$

$$1.96\sqrt{\frac{2(.03)(.97)}{n}} = .005$$

$$n = 8{,}943.2$$

You can see that this may be a tedious sampling procedure. If the supervisor insists on estimating $(p_1 - p_2)$ correct to within .005 with 95% confidence, approximately 9,000 items will have to be inspected for each machine.   ▲

You can see from the calculations in Example 9.8 that $\sigma_{(\hat{p}_1 - \hat{p}_2)}$ (and hence the solution, $n_1 = n_2 = n$) depends on the actual (but unknown) values of p_1 and p_2. In fact, the required sample size $n_1 = n_2 = n$ is largest when $p_1 = p_2 = .5$. Therefore, if you have no prior information on the approximate values of p_1 and p_2, use $p_1 = p_2 = .5$ in the formula for $\sigma_{(\hat{p}_1 - \hat{p}_2)}$. If p_1 and p_2 are in fact close to .5, then the values of n_1 and n_2 that you have calculated will be correct. If p_1 and p_2 differ substantially from .5, then your solutions for n_1 and n_2 will be larger than needed. Consequently, using $p_1 = p_2 = .5$ when solving for n_1 and n_2 is a conservative procedure because the sample sizes n_1 and n_2 will be at least as large as (and probably larger than) needed.

The procedures for determining sample sizes necessary for estimating $(\mu_1 - \mu_2)$ or $(p_1 - p_2)$ for the case $n_1 = n_2$ are given in the following box.

Determination of Sample Size for Two-Sample Procedures

1. To estimate $(\mu_1 - \mu_2)$ to within a given bound B with probability $(1 - \alpha)$, use the following formula to solve for equal sample sizes that will achieve the desired reliability:

$$n_1 = n_2 = \frac{(z_{\alpha/2})^2(\sigma_1^2 + \sigma_2^2)}{B^2}$$

You will need to substitute estimates for the values of σ_1^2 and σ_2^2 before solving for the sample size. These estimates might be sample variances s_1^2 and s_2^2 from prior sampling (e.g., a pilot sample), or from an educated (and conservatively large) guess based on the range—that is, $s \approx R/4$.

2. To estimate $(p_1 - p_2)$ to within a given bound B with probability $(1 - \alpha)$, use the following formula to solve for equal sample sizes that will achieve the desired reliability:

$$n_1 = n_2 = \frac{(z_{\alpha/2})^2(p_1 q_1 + p_2 q_2)}{B^2}$$

You will need to substitute estimates for the values of p_1 and p_2 before solving for the sample size. These estimates might be based on prior samples, obtained from educated guesses or, most conservatively, specified as $p_1 = p_2 = .5$.

EXERCISES 9.63–9.74

Learning the Mechanics

9.63 Suppose you want to estimate the difference between two population means correct to within 2.2 with probability .95. If prior information suggests that the population variances are approximately equal to $\sigma_1^2 = \sigma_2^2 = 15$ and you want to select independent random samples of equal size from the populations, how large should the sample sizes, n_1 and n_2, be?

9.64 A pollster wants to estimate the difference between the proportions of men and women who favor a particular national candidate using a 90% confidence interval of width .04. Suppose the pollster has no prior information about the proportions. If equal numbers of men and women are to be polled, how large should the sample sizes be?

9.65 Find the appropriate values of n_1 and n_2 (assume $n_1 = n_2$) needed to estimate $(\mu_1 - \mu_2)$ with:
  a. A bound on the error of estimation equal to 3.2 with 95% confidence. From prior experience it is known that $\sigma_1 \approx 15$ and $\sigma_2 \approx 17$.
  b. A bound on the error of estimation equal to 8 with 99% confidence. The range of each population is 60.
  c. A 90% confidence interval of width 1.0. Assume that $\sigma_1^2 \approx 5.8$ and $\sigma_2^2 \approx 7.5$.

9.66 Assuming that $n_1 = n_2$, find the sample sizes needed to estimate $(p_1 - p_2)$ for each of the following situations:
  a. Bound = .01 with 99% confidence. Assume that $p_1 \approx .4$ and $p_2 \approx .7$.

  b. A 90% confidence interval of width .05. Assume there is no prior information available to obtain approximate values of p_1 and p_2.
  c. Bound = .03 with 90% confidence. Assume that $p_1 \approx .2$ and $p_2 \approx .3$.

9.67 Enough money has been budgeted to collect independent random samples of size $n_1 = n_2 = 100$ from populations 1 and 2 in order to estimate $(\mu_1 - \mu_2)$. Prior information indicates that $\sigma_1 = \sigma_2 = 12$. Have sufficient funds been allocated to construct a 90% confidence interval for $(\mu_1 - \mu_2)$ of width 5 or less? Justify your answer.

Applying the Concepts

9.68 Is housework hazardous to your health? A study in the *Public Health Reports* (July–Aug. 1992) compares the life expectancies of 25-year-old white women in the labor force to those who are housewives. How large a sample would have to be taken from each group in order to be 95% confident that the estimate of difference in life expectancies for the two groups is within 1 year of the true difference in life expectancies? Assume that equal sample sizes will be selected from the two groups, and that the standard deviation for both groups is approximately 15 years.

9.69 Nationally televised home shopping was introduced in 1985. Overnight it became the hottest craze in television programming. Today, nearly all cable companies carry at least one home shopping channel. Who uses these home shopping services? Are the shoppers primarily men or women? Suppose you want to estimate the difference in the proportions of men and women who say they have used or expect to use televised home shopping using an 80% confidence interval of width .06 or less.

a. Approximately how many people should be included in your samples?

b. Suppose you want to obtain individual estimates for the two proportions of interest. Will the sample size found in part **a** be large enough to provide estimates of each proportion correct to within .02 with probability equal to .90? Justify your response.

9.70 One reason high school seniors are encouraged to attend college is that the job opportunities are much better for those with college degrees than for those without. A high school counselor wants to estimate the difference in mean income per day between high school graduates who have a college education and those who have not gone on to college. Suppose it is decided to compare the daily incomes of 30-year-olds, and the range of daily incomes for both groups is approximately $200 per day. How many people from each group should be sampled in order to estimate the true difference between mean daily incomes correct to within $10 per day with probability .9? Assume that $n_1 = n_2$.

9.71 Rat damage creates a large financial loss in the production of sugarcane. One aspect of the problem that has been investigated by the U.S. Department of Agriculture concerns the optimal place to locate rat poison. To be most effective in reducing rat damage, should the poison be located in the middle of the field or on the outer perimeter? One way to answer this question is to determine where the greater amount of damage occurs. If damage is measured by the proportion of cane stalks that have been damaged by rats, how many stalks from each section of the field should be sampled in order to estimate the true difference between proportions of stalks damaged in the two sections to within .02 with 95% confidence?

9.72 A manufacturer of large-screen televisions wants to compare the proportions of its best sets that need repair within 1 year with those of a competitor. If it is desired to estimate the difference between proportions to within .05 with 90% confidence, and if the

manufacturer plans to sample twice as many buyers (n_1) of its sets as buyers (n_2) of the competitor's sets, how many buyers of each brand must be sampled? Assume that the proportion of sets that need repair will be about .2 for both brands.

9.73 In seeking a good professional football running back, a coach is looking for a player with high mean yards gained per carry and a small standard deviation. Suppose the coach wishes to compare the mean yards gained per carry for two major prospects based on independent random samples of their yards gained per carry in the early part of the coming pro football season. Suppose data from last year indicate that $\sigma_1 = \sigma_2 \approx 5$ yards. If the coach wants to estimate the difference in means correct to within 1 yard with probability equal to .9, how many runs would have to be observed for each player? (Assume equal sample sizes.)

9.74 Refer to Exercise 9.62, where we examined data relating levels of fitness and stress (*International Journal of Sports Psychology,* July–Sept. 1990). Essentially, we failed to reject the null hypothesis that those in poor physical condition exhibit signs of stress in the same proportion as those in good physical condition, even though the sample proportions differed by nearly .05. (That is, of those sampled, almost 5% more of those in poor condition exhibited signs of stress than those in good condition.)

a. How large would the samples have to be to estimate the difference in the proportions showing signs of stress to within .04 with 95% confidence? Assume the sample sizes for the two groups will be equal, and remember that the first samples selected resulted in proportions of .155 and .108 for the poor and good fitness levels, respectively.

b. Suppose samples of the size you calculated in part **a** were selected from each group, and the sample proportions again turned out to be .155 and .108. Test the null hypothesis $H_0: (p_1 - p_2) = 0$ against the alternative $H_a: (p_1 - p_2) > 0$, where p_1 is the proportion of all employees at the poor fitness level who show signs of stress and p_2 is the proportion of all employees at the good fitness level who show signs of stress. Calculate and interpret the observed significance level.

9.5 COMPARING TWO POPULATION VARIANCES: INDEPENDENT SAMPLING (OPTIONAL)

Many times it is of practical interest to use the techniques developed in this chapter to compare the means or proportions of two populations. However, there are also important instances when we wish to compare two population variances. For example, when two devices are available for producing precision measurements (scales, calipers, thermometers, etc.), we might want to compare the variability of the measurements of the devices before deciding which one to purchase. Or when two standardized tests can be used to rate job applicants, the variability of the scores for both tests should be taken into consideration before deciding which test to use.

For problems like these we need to develop a statistical procedure to compare population variances. The common statistical procedure for comparing population variances, σ_1^2 and σ_2^2, makes an inference about the ratio σ_1^2/σ_2^2. In this section, we will show how to test the null hypothesis that the ratio σ_1^2/σ_2^2 equals 1 (the variances are equal) against the alternative hypothesis that the ratio differs from 1 (the variances differ):

$$H_0: \frac{\sigma_1^2}{\sigma_2^2} = 1 \qquad (\sigma_1^2 = \sigma_2^2)$$

$$H_a: \frac{\sigma_1^2}{\sigma_2^2} \neq 1 \qquad (\sigma_1^2 \neq \sigma_2^2)$$

To make an inference about the ratio σ_1^2/σ_2^2, it seems reasonable to collect sample data and use the ratio of the sample variances, s_1^2/s_2^2. We will use the test statistic

$$F = \frac{s_1^2}{s_2^2}$$

To establish a rejection region for the test statistic, we need to know the sampling distribution of s_1^2/s_2^2. As you will subsequently see, the sampling distribution of s_1^2/s_2^2 is based on two of the assumptions already required for the *t*-test:

1. The two sampled populations are normally distributed.
2. The samples are randomly and independently selected from their respective populations.

When these assumptions are satisfied and when the null hypothesis is true (that is, $\sigma_1^2 = \sigma_2^2$), the sampling distribution of $F = s_1^2/s_2^2$ is the **F-distribution** with $(n_1 - 1)$ numerator degrees of freedom and $(n_2 - 1)$ denominator degrees of freedom, respectively. The shape of the F-distribution depends on the degrees of freedom associated with s_1^2 and s_2^2—that is, on $(n_1 - 1)$ and $(n_2 - 1)$. An F-distribution with 7 and 9 df is shown in Figure 9.15. As you can see, the distribution is skewed to the right, since s_1^2/s_2^2 cannot be less than 0 but can increase without bound.

We need to be able to find F values corresponding to the tail areas of this distribution in order to establish the rejection region for our test of hypothesis because we expect the ratio F of the sample variances to be either very large or very small when the population variances are unequal. The upper-tail F values for $\alpha = .10, .05, .025,$ and $.01$ can be found in Tables VIII, IX, X, and XI of Appendix A. Table IX is partially reproduced in Table 9.7. It gives F values that correspond to $\alpha = .05$ upper-tail areas for different degrees of freedom ν_1 for the

FIGURE 9.15

An *F*-distribution with 7 numerator and 9 denominator degrees of freedom

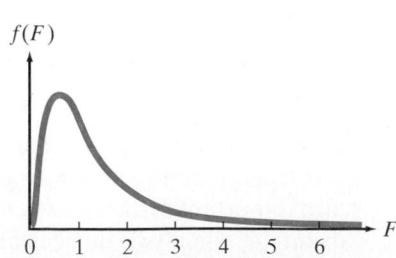

TABLE 9.7 **Reproduction of Part of Table IX in Appendix A:**
Percentage Points of the *F*-distribution, α = .05

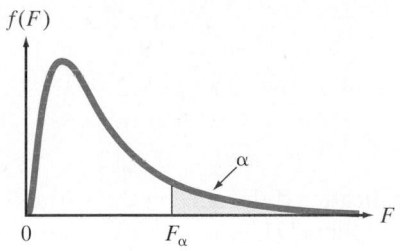

ν_2 \ ν_1	NUMERATOR DEGREES OF FREEDOM								
	1	**2**	**3**	**4**	**5**	**6**	**7**	**8**	**9**
1	161.4	199.5	215.7	224.6	230.2	234.0	236.8	238.9	240.5
2	18.51	19.00	19.16	19.25	19.30	19.33	19.35	19.37	19.38
3	10.13	9.55	9.28	9.12	9.01	8.94	8.89	8.85	8.81
4	7.71	6.94	6.59	6.39	6.26	6.16	6.09	6.04	6.00
5	6.61	5.79	5.41	5.19	5.05	4.95	4.88	4.82	4.77
6	5.99	5.14	4.76	4.53	4.39	4.28	4.21	4.15	4.10
7	5.59	4.74	4.35	4.12	3.97	3.87	3.79	3.73	3.68
8	5.32	4.46	4.07	3.84	3.69	3.58	3.50	3.44	3.39
9	5.12	4.26	3.86	3.63	3.48	3.37	3.29	3.23	3.18
10	4.96	4.10	3.71	3.48	3.33	3.22	3.14	3.07	3.02
11	4.84	3.98	3.59	3.36	3.20	3.09	3.01	2.95	2.90
12	4.75	3.89	3.49	3.25	3.11	3.00	2.91	2.85	2.80
13	4.67	3.81	3.41	3.18	3.03	2.92	2.83	2.77	2.71
14	4.60	3.74	3.34	3.11	2.96	2.85	2.76	2.70	2.65

Denominator Degrees of Freedom

numerator sample variance, s_1^2, whereas the rows correspond to the degrees of freedom ν_2 for the denominator sample variance, s_2^2. Thus, if the numerator degrees of freedom is $\nu_1 = 7$ and the denominator degrees of freedom is $\nu_2 = 9$, we look in the seventh column and ninth row to find $F_{.05} = 3.29$. As shown in Figure 9.16, α = .05 is the tail area to the right of 3.29 in the *F*-distribution with 7 and 9 df. That is, if $\sigma_1^2 = \sigma_2^2$, then the probability that the *F* statistic will exceed 3.29 is α = .05.

EXAMPLE 9.9

An experimenter wants to compare the metabolic rates of white mice subjected to different drugs. The weights of the mice may affect their metabolic rates, and thus the experimenter wishes to obtain mice that are relatively homogeneous with

FIGURE 9.16

An *F*-distribution for $\nu_1 = 7$ and $\nu_2 = 9$ df; α = :05

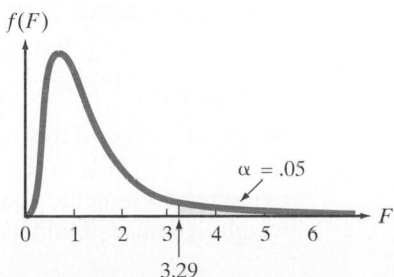

respect to weight. Five hundred mice will be needed to complete the study. Currently, 18 mice from supplier 1 and another 13 mice from supplier 2 are available for comparison. The experimenter weighs these mice and obtains the following summary information:

Supplier 1	Supplier 2
$n_1 = 18$	$n_2 = 13$
$\bar{x}_1 = 4.21$ ounces	$\bar{x}_2 = 4.18$ ounces
$s_1^2 = .019$	$s_2^2 = .049$

Do these data provide sufficient evidence to indicate a difference in the variability of weights of mice obtained from the two suppliers? (Use $\alpha = .10$.) Using the results of this analysis, what would you suggest to the experimenter?

Solution

Let

$$\sigma_1^2 = \text{Population variance of weights of white mice from supplier 1}$$
$$\sigma_2^2 = \text{Population variance of weights of white mice from supplier 2}$$

The hypotheses of interest are then

$$H_0: \frac{\sigma_1^2}{\sigma_2^2} = 1 \qquad (\sigma_1^2 = \sigma_2^2)$$

$$H_a: \frac{\sigma_1^2}{\sigma_2^2} \neq 1 \qquad (\sigma_1^2 \neq \sigma_2^2)$$

The nature of the F-tables given in Appendix A affects the form of the test statistic. To form the rejection region for a two-tailed F-test, we want to make certain that the upper tail is used, because only the upper-tail values of F are shown in Tables VIII, IX, X, and XI. To accomplish this, *we will always place the larger sample variance in the numerator of the F-test statistic.* This has the effect of doubling the tabulated value for α, since we double the probability that the F-ratio will fall in the upper tail by always placing the larger sample variance in the numerator. That is, we establish a one-tailed rejection region by putting the larger variance in the numerator rather than establishing rejection regions in both tails.

Thus, for our example, we have a denominator s_1^2 with df $= v_2 = n_1 - 1 = 17$ and a numerator s_2^2 with df $= v_1 = n_2 - 1 = 12$. Therefore, the test statistic will be

$$F = \frac{\text{Larger sample variance}}{\text{Smaller sample variance}} = \frac{s_2^2}{s_1^2}$$

and we will reject $H_0: \sigma_1^2 = \sigma_2^2$ for $\alpha = .10$ when the calculated value of F exceeds the tabulated value:

$$F_{\alpha/2} = F_{.05} = 2.38 \qquad \text{(see Figure 9.17)}$$

We can now calculate the value of the test statistic and complete the analysis:

$$F = \frac{s_2^2}{s_1^2} = \frac{.049}{.019} = 2.58$$

When we compare this result to the rejection region shown in Figure 9.17, we see that $F = 2.58$ falls in the rejection region. Therefore, the data provide sufficient evidence to indicate that the population variances differ. It appears that the weights of mice obtained from supplier 1 tend to be more homogeneous than the

FIGURE 9.17

Rejection region for
Example 9.9

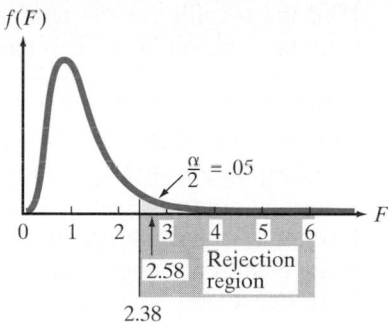

weights of mice obtained from supplier 2. On the basis of this evidence, we would
advise the experimenter to purchase the mice from supplier 1.   ▲

What would you have concluded if the value of F calculated from the samples
had not fallen in the rejection region? Would you conclude that the null hypothesis of equal variances is true? No, because then you risk the possibility of a Type
II error (accepting H_0 if H_a is true) without knowing the value of β, the probability of accepting H_0: $\sigma_1^2 = \sigma_2^2$ if in fact it is false. Since we will not consider the
calculation of β for specific alternatives in this text, when the F statistic does not
fall in the rejection region we simply conclude that insufficient sample evidence
exists to refute the null hypothesis that $\sigma_1^2 = \sigma_2^2$.

The F-test for equal population variances is summarized in the next box.

F-Test for Equal Population Variances

ONE-TAILED TEST	**TWO-TAILED TEST**

H_0: $\sigma_1^2 = \sigma_2^2$ H_0: $\sigma_1^2 = \sigma_2^2$

H_a: $\sigma_1^2 < \sigma_2^2$ H_a: $\sigma_1^2 \neq \sigma_2^2$

  (or H_a: $\sigma_1^2 > \sigma_2^2$)

Test statistic: *Test statistic:*

$$F = \frac{s_2^2}{s_1^2}$$ $$F = \frac{\text{Larger sample variance}}{\text{Smaller sample variance}}$$

$$\left(\text{or } F = \frac{s_1^2}{s_2^2} \quad \text{when } H_a\text{: } \sigma_1^2 > \sigma_2^2\right)$$ $$= \frac{s_1^2}{s_2^2} \quad \text{when } s_1^2 > s_2^2$$

$$\left(\text{or } \frac{s_2^2}{s_1^2} \quad \text{when } s_2^2 > s_1^2\right)$$

Rejection region: *Rejection region:*

$F > F_\alpha$ $F > F_{\alpha/2}$

where F_α and $F_{\alpha/2}$ are based on v_1 = numerator degrees of freedom and v_2 =
denominator degrees of freedom; v_1 and v_2 are the degrees of freedom for the
numerator and denominator sample variances, respectively.

Assumptions: 1. Both sampled populations are normally distributed.
             2. The samples are random and independent.

EXAMPLE 9.10

Find the *p*-value for the test in Example 9.9 using the *F*-tables in Appendix A. Compare this to the exact *p*-value obtained from a computer printout.

Solution

Since the observed value of the *F* statistic in Example 9.9 was 2.58, the observed significance level of the test would equal the probability of observing a value of *F* at least as contradictory to H_0: $\sigma_1^2 = \sigma_2^2$ as $F = 2.58$, if in fact H_0 is true. Since we give the *F*-tables in Appendix A only for values of α equal to .10, .05, .025, and .01, we can only approximate the observed significance level. Checking Tables VIII, IX, X, and XI, we find $F_{.05} = 2.38$ and $F_{.025} = 2.82$. Since the observed value of *F* exceeds $F_{.05}$ but is less than $F_{.025}$, the observed significance level for the test is less than 2(.05) = .10, but greater than 2(.025) = .05; that is,

$$.05 < p\text{-value} < .10$$

(Note that we double the α value shown in Table IX because this is a two-tailed test.)

An ASP printout of the *F*-test is displayed in Figure 9.18. The exact *p*-value, shaded on the printout, is .073. Clearly this value falls between .05 and .10 as stated above. Our interpretation is that we will reject H_0 and conclude that the population variances differ for any α value that exceeds .073.   ▲

EXAMPLE 9.11

An investor believes that although the price of stock 1 usually exceeds that of stock 2, stock 1 represents a riskier investment, where the risk of a given stock is measured by the variation in daily price changes. Suppose we obtain a random sample of 25 daily price changes for stock 1 and 25 for stock 2. The sample results are summarized in the table below. Compare the risks associated with the two stocks by testing the null hypothesis that the variances of the price changes for the stocks are equal against the alternative that the price variance of stock 1 exceeds that of stock 2. Use $\alpha = .05$.

Stock 1	Stock 2
$n_1 = 25$	$n_2 = 25$
$\bar{x}_1 = .250$	$\bar{x}_2 = .125$
$s_1 = .76$	$s_2 = .46$

Solution

$$H_0:\ \sigma_1^2 = \sigma_2^2$$
$$H_a:\ \sigma_1^2 > \sigma_2^2$$

FIGURE 9.18

ASP *F*-test of Example 9.9

```
            HYPOTHESIS: VAR SUPPLIER 1 (X) = VAR SUPPLIER 2 (Y)

SAMPLE VARIANCE OF X   =      0.019
        SAMPLE SIZE OF X   =    18
SAMPLE VARIANCE OF Y   =      0.049
        SAMPLE SIZE OF Y   =    13

                    F   =      2.57895
   NUMERATOR D. F.   =     12
 DENOMINATOR D. F.   =     17
          P-VALUE    =      0.0730396
        P-VALUE/2    =      0.0365198
```

Test statistic: $F = \dfrac{s_1^2}{s_2^2}$

Assumptions: 1. The changes in daily stock prices have relative frequency distributions that are approximately normal.
2. The stock samples are randomly and independently selected from a set of daily stock reports.

Rejection region: $F > F_\alpha = F_{.05} = 1.98$, where $F_{.05}$ is based on $v_1 = 24$ df and $v_2 = 24$ df.

We calculate

$$F = \frac{s_1^2}{s_2^2} = \frac{(.76)^2}{(.46)^2} = 2.73$$

The calculated F exceeds the rejection value of 1.98. Therefore, we conclude that the variance of the daily price change for stock 1 exceeds that for stock 2. It appears that stock 1 is a riskier investment than stock 2. How much reliability can we place in this inference? Only one time in 20 (since $\alpha = .05$), on the average, would this statistical test lead us to conclude erroneously that σ_1^2 exceeds σ_2^2 if in fact they are equal.

Since this is a one-tailed test, the *p*-value is equal to the probability that F exceeds the computed value, 2.73; that is,

$$p\text{-value} = P(F > 2.73)$$

Checking Tables X and XI, we find that $F_{.025} = 2.27$ and $F_{.01} = 2.66$. Since the computed value of the F statistic exceeds 2.66, the observed significance level for the test is slightly less:

$$\text{Approximate } p\text{-value} = .01 \qquad \blacktriangle$$

As a final example of an application, consider the comparison of population variances as a check of the assumption $\sigma_1^2 = \sigma_2^2$ needed for the two-sample *t*-test. Rejection of the null hypothesis $\sigma_1^2 = \sigma_2^2$ would indicate that the assumption is invalid. [*Note:* Nonrejection of the null hypothesis does *not* imply that the assumption is valid.] We illustrate with an example.

EXAMPLE 9.12

In Example 9.4 (Section 9.1) we used the two-sample *t* statistic to compare the mean reading scores of two groups of slow learners who had been taught to read using two different methods. The data are repeated in Table 9.8 for convenience. The use of the *t* statistic was based on the assumption that the population variances of the test scores were equal for the two methods. Use the computer to check this assumption at $\alpha = .10$.

TABLE 9.8   Reading Test Scores for Slow Learners

New Method				Standard Method			
80	80	79	81	79	62	70	68
76	66	79	76	73	76	86	73
				72	68	75	66

FIGURE 9.19

SAS *F*-test for the data in
Table 9.8

```
                              TTEST PROCEDURE
Variable: SCORE

METHOD        N                 Mean          Std Dev        Std Error
-------------------------------------------------------------------------
New           8           77.12500000        4.85320218      1.71586609
Std          12           72.33333333        6.34369169      1.83126605

Variances         T      DF     Prob>|T|
-------------------------------------------
Unequal        1.9094   17.5     0.0727
Equal          1.8070   18.0     0.0875

For H0: Variances are equal,  F' = 1.71   DF = (11,7)   Prob>F' = 0.4891
```

Solution

We want to test

$$H_0: \sigma_1^2 = \sigma_2^2$$

$$H_a: \sigma_1^2 \neq \sigma_2^2$$

This *F*-test is shown on the SAS printout, Figure 9.19. Both the test statistic, $F = 1.71$, and two-tailed *p*-value, .4891, are shaded on the printout. Since $\alpha = .10$ is less than the *p*-value, we do not reject the null hypothesis that the population variances of the reading test scores are equal. It is here that the temptation to misuse the *F*-test is strongest. *We cannot conclude that the data justify the use of the t statistic.* This is equivalent to accepting H_0, and we have repeatedly warned against this conclusion because the probability of a Type II error, β, is unknown. The α level of .10 protects us only against rejecting H_0 if it is true. This use of the *F*-test may prevent us from abusing the *t* procedure when we obtain a value of *F* that leads to a rejection of the assumption that $\sigma_1^2 = \sigma_2^2$. But when the *F* statistic does not fall in the rejection region, we know little more about the validity of the assumption than before we conducted the test. ▲

We have presented the *F*-test as a test of a hypothesis of equality of variances, that is, $\sigma_1^2 = \sigma_2^2$. Although this is the most common application of the test, it can also be used to test a hypothesis that the ratio between the population variances is equal to some specified value, $H_0: \sigma_1^2 / \sigma_2^2 = k$. The test would be conducted in exactly the same way as a test of a hypothesis concerning the equality of variances except that we would use the test statistic

$$F = \frac{s_1^2}{s_2^2}\left(\frac{1}{k}\right)$$

What Do You Do If the Assumption of Normal Population Distributions Is Not Satisfied?

Answer: The *F*-test is much less robust (i.e., much more sensitive) to departures from normality than the *t*-test for comparing the population means (Section 9.1). If you have doubts about the normality of the population frequency distributions, use a **nonparametric method** for comparing the two population variances. A method can be found in the nonparametric statistics texts listed in the references.

Rather than test a hypothesis concerning σ_1^2 / σ_2^2, you may wish to gain insight into the relative magnitudes of σ_1^2 and σ_2^2 by estimating their ratio. A confidence interval for the ratio of two population variances can be obtained by using the formula shown in the next box.

Confidence Interval for σ_1^2/σ_2^2

$$\left(\frac{s_1^2}{s_2^2}\right)\left(\frac{1}{F_{L,\alpha/2}}\right) < \frac{\sigma_1^2}{\sigma_2^2} < \left(\frac{s_1^2}{s_2^2}\right)F_{U,\alpha/2}$$

where $F_{L,\alpha/2}$ is the tabulated value of F that places an area $\alpha/2$ in the upper tail of the F-distribution and that is based on $v_1 = (n_1 - 1)$ numerator and $v_2 = (n_2 - 1)$ denominator degrees of freedom, and $F_{U,\alpha/2}$ is the tabulated upper-tail value of F based on $v_1 = (n_2 - 1)$ numerator and $v_2 = (n_1 - 1)$ denominator degrees of freedom.

Assumptions: The samples were randomly and independently selected from populations that possess approximately normal distributions.

EXAMPLE 9.13   Refer to Example 9.9 and find a 90% confidence interval for the ratio of the variances of the weights of mice obtained from the two suppliers.

Solution

In Example 9.9, we were given the following information:

Supplier 1	**Supplier 2**
$n_1 = 18$	$n_2 = 13$
$\bar{x}_1 = 4.21$ ounces	$\bar{x}_2 = 4.18$ ounces
$s_1^2 = .019$	$s_2^2 = .049$

Then

$$n_1 - 1 = 18 - 1 = 17 \qquad n_2 - 1 = 13 - 1 = 12$$

and, from Table IX of Appendix A,

$$F_{L,\alpha/2} = F_{L,.05} \approx 2.59 \qquad \text{(where } v_1 = 17 \text{ and } v_2 = 12)$$
$$F_{U,\alpha/2} = F_{U,.05} = 2.38 \qquad \text{(where } v_1 = 12 \text{ and } v_2 = 17)$$

Substituting these values into the formula for the confidence interval, we obtain

$$\left(\frac{s_1^2}{s_2^2}\right)\left(\frac{1}{F_{L,\alpha/2}}\right) < \frac{\sigma_1^2}{\sigma_2^2} < \left(\frac{s_1^2}{s_2^2}\right)F_{U,\alpha/2}$$

$$\left(\frac{.019}{.049}\right)\left(\frac{1}{2.59}\right) < \frac{\sigma_1^2}{\sigma_2^2} < \left(\frac{.019}{.049}\right)2.38$$

$$.150 < \frac{\sigma_1^2}{\sigma_2^2} < .923$$

According to this confidence interval, we estimate that σ_1^2, the variance in the weights of mice obtained from supplier 1, could be as small as .150 or as large as .923 times the size of σ_2^2, the variance in the weights of mice obtained from supplier 2.   ▲

EXERCISES 9.75–9.88

Learning the Mechanics

9.75 Under what conditions is the sampling distribution of s_1^2/s_2^2 an F-distribution?

9.76 Use Tables VIII, IX, X, and XI of Appendix A to find each of the following F values:
   a. $F_{.05}$ where $v_1 = 8$ and $v_2 = 5$
   b. $F_{.01}$ where $v_1 = 20$ and $v_2 = 14$
   c. $F_{.025}$ where $v_1 = 10$ and $v_2 = 5$
   d. $F_{.10}$ where $v_1 = 20$ and $v_2 = 5$

9.77 Given v_1 and v_2, find the following probabilities:
   a. $v_1 = 2$, $v_2 = 30$, $P(F \geq 4.18)$
   b. $v_1 = 24$, $v_2 = 14$, $P(F < 1.94)$
   c. $v_1 = 9$, $v_2 = 1$, $P(F \leq 6,022.0)$
   d. $v_1 = 30$, $v_2 = 30$, $P(F > 1.84)$

9.78 For each of the following cases, identify the rejection region that should be used to test H_0: $\sigma_1^2 = \sigma_2^2$ against H_a: $\sigma_1^2 > \sigma_2^2$. Assume $v_1 = 20$ and $v_2 = 30$.
a. $\alpha = .10$ **b.** $\alpha = .05$ **c.** $\alpha = .025$ **d.** $\alpha = .01$

9.79 For each of the following cases, identify the rejection region that should be used to test H_0: $\sigma_1^2 = \sigma_2^2$ against H_a: $\sigma_1^2 \neq \sigma_2^2$. Assume $v_1 = 8$ and $v_2 = 40$.
a. $\alpha = .20$ **b.** $\alpha = .10$ **c.** $\alpha = .05$ **d.** $\alpha = .02$

9.80 Specify the appropriate rejection region for testing H_0: $\sigma_1^2 = \sigma_2^2$ in each of the following situations:
a. H_a: $\sigma_1^2 > \sigma_2^2$; $\alpha = .05$, $n_1 = 25$, $n_2 = 20$
b. H_a: $\sigma_1^2 < \sigma_2^2$; $\alpha = .05$, $n_1 = 10$, $n_2 = 15$
c. H_a: $\sigma_1^2 \neq \sigma_2^2$; $\alpha = .10$, $n_1 = 21$, $n_2 = 31$
d. H_a: $\sigma_1^2 < \sigma_2^2$; $\alpha = .01$, $n_1 = 31$, $n_2 = 41$
e. H_a: $\sigma_1^2 \neq \sigma_2^2$; $\alpha = .05$, $n_1 = 7$, $n_2 = 16$

9.81 Give the appropriate tabled F values to form a 95% confidence interval for σ_1^2 / σ_2^2 for each of the following combinations of n_1 and n_2:
a. $n_1 = 16$, $n_2 = 10$ **b.** $n_1 = 13$, $n_2 = 13$
c. $n_1 = 21$, $n_2 = 41$ **d.** $n_1 = 16$, $n_2 = 20$

9.82 Independent random samples were selected from each of two normally distributed populations, $n_1 = 16$ from population 1 and $n_2 = 25$ from population 2. The means and variances for the two samples are shown in the table.

Sample 1	Sample 2
$n_1 = 16$	$n_2 = 25$
$\bar{x}_1 = 22.5$	$\bar{x}_2 = 28.2$
$s_1^2 = 2.87$	$s_2^2 = 9.85$

a. Test the null hypothesis H_0: $\sigma_1^2 = \sigma_2^2$ against the alternative hypothesis H_a: $\sigma_1^2 \neq \sigma_2^2$. Use $\alpha = .05$.
b. Form a 95% confidence interval for σ_1^2 / σ_2^2.
c. Test H_0: $\sigma_1^2 = \sigma_2^2$ against H_a: $\sigma_1^2 < \sigma_2^2$. Use $\alpha = .05$.

9.83 Independent random samples were selected from each of two normally distributed populations, $n_1 = 6$ from population 1 and $n_2 = 4$ from population 2. The data are shown in the table.

Sample 1	Sample 2
3.1	2.3
4.3	1.4
1.2	3.7
1.7	8.9
.6	
3.4	

a. Form a 90% confidence interval for σ_1^2 / σ_2^2.
b. Test H_0: $\sigma_1^2 = \sigma_2^2$ against H_a: $\sigma_1^2 < \sigma_2^2$. Use $\alpha = .01$.
c. Test H_0: $\sigma_1^2 = \sigma_2^2$ against H_a: $\sigma_1^2 \neq \sigma_2^2$. Use $\alpha = .10$.

Applying the Concepts

9.84 A recent study in the *Journal of Occupational and Organizational Psychology* (Dec. 1992) investigated the relationship of employment status and mental health. A sample of working and unemployed people was selected, and each person was given a mental health examination using the General Health Questionnaire (GHQ), a widely recognized measure of mental health. Although the article focused on comparing the mean GHQ levels, a comparison of the variability of GHQ scores for employed and unemployed men and women is of interest as well.
a. In general terms, what does the amount of variability in GHQ scores tell us about the group?
b. What are the appropriate null and alternative hypotheses to compare the variability of the mental health scores of the employed and unemployed groups? Define any symbols you use.
c. The standard deviation for a sample of 142 employed men was 3.26, while the standard deviation for 49 unemployed men was 5.10. Conduct the test you set up in part **b** using $\alpha = .05$. Interpret the results.
d. What assumptions are necessary to ensure the validity of the test?

9.85 Tests of product quality can be completely automated or they can be conducted using human inspectors or human inspectors aided by mechanical devices. Although human inspection is frequently the most economical alternative, it can lead to serious inspection error problems (*Journal of Quality Technology,* Apr. 1986). Numerous studies have demonstrated that inspectors often cannot detect as many as 85% of the defective items that they inspect; moreover, performance varies from inspector to inspector. To evaluate the performance of inspectors in a new company, a quality manager had a sample of 12 novice inspectors evaluate 200 finished products. The same 200 items were evaluated by 12 experienced inspectors. The quality of each item—whether defective or nondefective—was known to the manager. The following table lists the number of inspection errors (classifying a defective item as nondefective or vice versa) made by each inspector. A SAS printout with descriptive statistics for the two types of inspectors is on page 391.

Novice Inspectors				Experienced Inspectors			
30	35	26	40	31	15	25	19
36	20	45	31	28	17	19	18
33	29	21	48	24	10	20	21

```
Analysis Variable : ERRORS

--------------------------------INSPECT=EXPER------------------------

N Obs   N     Minimum      Maximum           Mean        Std Dev
-------------------------------------------------------------------
   12  12   10.0000000   31.0000000    20.5833333      5.7439032

------------------------------INSPECT=NOVICE ----------------------

N Obs   N     Minimum      Maximum           Mean        Std Dev
-------------------------------------------------------------------
   12  12   20.0000000   48.0000000    32.8333333      8.6427409
```

a. Prior to conducting this experiment, the manager believed the variance in inspection errors was lower for experienced inspectors than for novice inspectors. Do the sample data support her belief? Test using $\alpha = .05$.

b. What is the appropriate *p*-value of the test you conducted in part **a**?

9.86 A series of experiments has been conducted to compare the quantity of hemoglobin in the blood of men and women who are between the ages of 20 and 30. One phase of the study deals with a comparison of the variability in the hemoglobin measurements between the two groups. In random samples of 25 women and 20 men, all between the ages of 20 and 30, the researcher recorded the amount of hemoglobin in the blood of each. (The quantity of the hemoglobin is measured as a percentage of the total volume of blood.) The results are summarized in the table below. Form a 90% confidence interval for the ratio of the variance of the amount of hemoglobin in women's blood to the corresponding variance for men.

Women	Men
$n_1 = 25$	$n_2 = 20$
$\bar{x}_1 = 42.7$	$\bar{x}_2 = 41.8$
$s_1^2 = 18.3$	$s_2^2 = 8.5$

9.87 Refer to the *Genetical Research* (June 1995) study of the mating habits of hermaphroditic snails, Exercise 9.18. Recall that the snails were identified as either selfers or outcrossers and that the mean effective population sizes of the two groups were compared. The data for the study are reproduced in the next table. Geneticists are often more interested in comparing the variation in population size of the two types of mating systems. Conduct this analysis for the researcher. Interpret the result.

9.88 The quality control department of a paper company measures the brightness (a measure of reflectance)

Snail Mating System	Sample Size	EFFECTIVE POPULATION SIZE	
		Mean	**Standard Deviation**
Outcrossing	17	4,894	1,932
Selfing	5	4,133	1,890

Source: Jarne, P. "Mating system, bottlenecks, and genetic polymorphism in hermaphroditic animals." *Genetical Research*, Vol. 65, No. 3, June 1995, p. 197 (Table 4). Copyright 1995 Genetical Research. Reprinted with the permission of Cambridge University Press.

of finished paper on a periodic basis throughout the day. Two instruments that are available to measure the paper specimens are subject to error, but they can be adjusted so that the mean readings for a control paper specimen are the same for both instruments. Suppose you are concerned about the precision of the two instruments and want to compare the variability in the readings of instrument 1 to those of instrument 2. Five brightness measurements were made on a single paper specimen using each of the two instruments. The data are shown in the table.

a. Form a 95% confidence interval for the ratio of the variance of the measurements obtained by instrument 1 to the variance of the measurements obtained using instrument 2.

b. Interpret the interval you found in part **a** in terms of the precision of the two instruments.

c. What assumptions must be satisfied for the confidence interval in part **a** to be valid?

Instrument 1	Instrument 2
29	26
28	34
30	30
28	32
30	28

Note: Starred (*) items are from the optional section in this chapter.

Key Terms

Blocking   362
F-distribution*   382
Nonparametric method*   388
Paired difference experiment   361

Pooled sample estimator   345
Randomized block experiment   362
Standard error   341
Weighted average   345

Key Formulas

$(1 - \alpha)100\%$ confidence interval for θ: (see table)
Large samples: $\hat{\theta} \pm z_{\alpha/2}\sigma_{\hat{\theta}}$
Small samples: $\hat{\theta} \pm t_{\alpha/2}\sigma_{\hat{\theta}}$

For testing H_0: $\theta = D_0$: (see table)

Large samples: $z = \dfrac{\hat{\theta} - D_0}{\sigma_{\hat{\theta}}}$

Small samples: $t = \dfrac{\hat{\theta} - D_0}{\sigma_{\hat{\theta}}}$

Parameter, θ	Estimator, $\hat{\theta}$	Standard Error of Estimator, $\sigma_{\hat{\theta}}$	Estimated Standard Error
$\mu_1 - \mu_2$ (independent samples)	$\bar{x}_1 - \bar{x}_2$	$\sqrt{\dfrac{\sigma_1^2}{n_1} + \dfrac{\sigma_2^2}{n_2}}$	Large n: $\sqrt{\dfrac{s_1^2}{n_1} + \dfrac{s_2^2}{n_2}}$ Small n: $\sqrt{s_p^2\left(\dfrac{1}{n_1} + \dfrac{1}{n_2}\right)}$
μ_D (paired sample)	$\bar{x}_D$	$\dfrac{\sigma_D}{\sqrt{n_D}}$	$\dfrac{s_D}{\sqrt{n_D}}$
$p_1 - p_2$ (large, independent samples)	$\hat{p}_1 - \hat{p}_2$	$\sqrt{\dfrac{p_1 q_1}{n_1} + \dfrac{p_2 q_2}{n_2}}$	Hypotheses tests: $\sqrt{\hat{p}\hat{q}\left(\dfrac{1}{n_1} + \dfrac{1}{n_2}\right)}$ Confidence intervals: $\sqrt{\dfrac{\hat{p}_1\hat{q}_1}{n_1} + \dfrac{\hat{p}_2\hat{q}_2}{n_2}}$

Pooled sample variance:

$$s_p^2 = \frac{(n_1 - 1)s_1^2 + (n_2 - 1)s_2^2}{n_1 + n_2 - 2}$$

Pooled sample proportion:

$$\hat{p} = \frac{x_1 + x_2}{n_1 + n_2}, \qquad \hat{q} = 1 - \hat{p}$$

Determining the sample size for estimating $\mu_1 - \mu_2$:

$$n_1 = n_2 = \frac{(z_{\alpha/2})^2(\sigma_1^2 + \sigma_2^2)}{B^2}$$

Determing the sample size for estimating $p_1 - p_2$:

$$n_1 = n_2 = \frac{(z_{\alpha/2})^2(p_1 q_1 + p_2 q_2)}{B^2}$$

*Test statistic for testing H_0: $\dfrac{\sigma_1^2}{\sigma_2^2}$

$$F = \frac{\text{larger } s^2}{\text{smaller } s^2} \quad \text{if} \quad H_a: \frac{\sigma_1^2}{\sigma_2^2} \neq 1$$

$$F = \frac{s_1^2}{s_2^2} \quad \text{if} \quad H_a: \frac{\sigma_1^2}{\sigma_2^2} > 1$$

*$(1 - \alpha)100\%$ confidence interval for $\dfrac{\sigma_1^2}{\sigma_2^2}$:

$$\left(\frac{s_1^2}{s_2^2}\right)\left(\frac{1}{F_{L,\alpha/2}}\right) < \frac{\sigma_1^2}{\sigma_2^2} < \left(\frac{s_1^2}{s_2^2}\right)\left(F_{U,\alpha/2}\right)$$

LANGUAGE LAB

Symbol	Pronunciation	Description
$\mu_1 - \mu_2$	mu-1 minus mu-2	Difference between population means
$\bar{x}_1 - \bar{x}_2$	x-bar-1 minus x-bar-2	Difference between sample means
$\sigma_{(\bar{x}_1 - \bar{x}_2)}$	sigma of x-bar-1 minus x-bar-2	Standard deviation of the sampling distribution of $(\bar{x}_1 - \bar{x}_2)$
s_p^2	s-p squared	Pooled sample variance
D_0	D naught	Hypothesized value of difference
μ_D	mu D	Difference between population means, paired data
$\bar{x}_D$	x-bar D	Mean of sample differences
s_D	s-D	Standard deviation of sample differences
n_D	n-D	Number of differences in sample
$p_1 - p_2$	p-1 minus p-2	Difference between population proportions
$\hat{p}_1 - \hat{p}_2$	p-1 hat minus p-2 hat	Difference between sample proportions
$\sigma_{(\hat{p}_1 - \hat{p}_2)}$	sigma of p-1 hat minus p-2 hat	Standard deviation of the sampling distribution of $(\hat{p}_1 - \hat{p}_2)$
F_α	F-alpha	Critical value of F associated with tail area α
v_1	nu-1	Numerator degrees of freedom for F statistic
v_2	nu-2	Denominator degrees of freedom for F statistic
$\dfrac{\sigma_1^2}{\sigma_2^2}$	sigma-1 squared over sigma-2 squared	Ratio of two population variances

SUPPLEMENTARY EXERCISES 9.89–9.119

Note: List the assumptions necessary to ensure the validity of the statistical procedures you use to work these exercises. Starred (*) exercises are from the optional section in this chapter.

Learning the Mechanics

9.89 Independent random samples were selected from two normally distributed populations with means μ_1 and μ_2, respectively. The sample sizes, means, and variances are shown in the following table.

Sample 1	Sample 2
$n_1 = 12$	$n_2 = 14$
$\bar{x}_1 = 17.8$	$\bar{x}_2 = 15.3$
$s_1^2 = 74.2$	$s_2^2 = 60.5$

a. Test $H_0: (\mu_1 - \mu_2) = 0$ against $H_a: (\mu_1 - \mu_2) > 0$. Use $\alpha = .05$.

b. Form a 99% confidence interval for $(\mu_1 - \mu_2)$.

c. How large must n_1 and n_2 be if you wish to estimate $(\mu_1 - \mu_2)$ to within 2 units with 99% confidence? Assume that $n_1 = n_2$.

9.90 Two independent random samples were selected from normally distributed populations with means and variances (μ_1, σ_1^2) and (μ_2, σ_2^2), respectively. The sample sizes, means, and variances are shown in the table below.

Sample 1	Sample 2
$n_1 = 20$	$n_2 = 15$
$\bar{x}_1 = 123$	$\bar{x}_2 = 116$
$s_1^2 = 31.3$	$s_2^2 = 120.1$

a. Form a 95% confidence interval for σ_1^2 / σ_2^2.

*b. Test $H_0: \sigma_1^2 = \sigma_2^2$ against $H_a: \sigma_1^2 \neq \sigma_2^2$. Use $\alpha = .05$.

c. Would you be willing to use a t-test to test the null hypothesis $H_0: (\mu_1 - \mu_2) = 0$ against the alternative hypothesis $H_a: (\mu_1 - \mu_2) \neq 0$? Why?

9.91 Two independent random samples are taken from two populations. The results of these samples are summarized in the following table.

Sample 1	Sample 2
$n_1 = 135$	$n_2 = 148$
$\bar{x}_1 = 12.2$	$\bar{x}_2 = 8.3$
$s_1^2 = 2.1$	$s_2^2 = 3.0$

a. Form a 90% confidence interval for $(\mu_1 - \mu_2)$.

b. Test $H_0: (\mu_1 - \mu_2) = 0$ against $H_a: (\mu_1 - \mu_2) \neq 0$. Use $\alpha = .01$.

c. What sample sizes would be required if you wish to estimate $(\mu_1 - \mu_2)$ to within .2 with 90% confidence? Assume that $n_1 = n_2$.

9.92 Independent random samples were selected from two binomial populations. The sizes and number of observed successes for each sample are shown in the table below.

Sample 1	Sample 2
$n_1 = 200$	$n_2 = 200$
$x_1 = 110$	$x_2 = 130$

a. Test H_0: $(p_1 - p_2) = 0$ against H_a: $(p_1 - p_2) < 0$. Use $\alpha = .10$.

b. Form a 95% confidence interval for $(p_1 - p_2)$.

c. What sample sizes would be required if we wish to use a 95% confidence interval of width .01 to estimate $(p_1 - p_2)$?

9.93 A random sample of five pairs of observations were selected, one of each pair from a population with mean μ_1, the other from a population with mean μ_2. The data are shown in the accompanying table.

Pair	Value from Population 1	Value from Population 2
1	28	22
2	31	27
3	24	20
4	30	27
5	22	20

a. Test the null hypothesis H_0: $\mu_D = 0$ against H_a: $\mu_D \neq 0$, where $\mu_D = \mu_1 - \mu_2$. Use $\alpha = .05$.

b. Form a 95% confidence interval for μ_D.

c. When are the procedures you used in parts **a** and **b** valid?

9.94 List the assumptions necessary for each of the following inferential techniques:

a. Large-sample inferences about the difference $(\mu_1 - \mu_2)$ between population means using a two-sample z statistic

b. Small-sample inferences about $(\mu_1 - \mu_2)$ using an independent samples design and a two-sample t statistic

c. Small-sample inferences about $(\mu_1 - \mu_2)$ using a paired difference design and a single-sample t statistic to analyze the differences

d. Large-sample inferences about the differences $(p_1 - p_2)$ between binomial proportions using a two-sample z statistic

***e.** Inferences about the ratio σ_1^2 / σ_2^2 of two population variances using an F test.

Applying the Concepts

9.95 Nontraditional university students, generally defined as those at least 25 years old, comprise an increasingly large proportion of undergraduate student bodies at most universities. A study reported in the *College Student Journal* (Dec. 1992) compared traditional and nontraditional students on a number of factors, including grade-point average (GPA). The table below summarizes the information from the sample.

GPA	Traditional Students	Nontraditional Students
n	94	73
$\bar{x}$	2.90	3.50
s	.50	.50

a. What are the appropriate null and alternative hypotheses if we want to test whether the mean GPAs of traditional and nontraditional students differ?

b. Conduct the test using $\alpha = .01$, and interpret the result.

c. What assumptions are necessary to ensure the validity of the test?

9.96 A pupillometer is a device used to observe changes in an individual's pupil dilations as he or she is exposed to different visual stimuli. Since there is a direct correlation between the amount an individual's pupil dilates and his or her interest in the stimuli, marketing organizations sometimes use pupillometers to help them evaluate potential consumer interest in new products, alternative package designs, and other factors. The Design and Market Research Laboratories of the Container Corporation of America used a pupillometer to evaluate consumer reaction to different silverware patterns for one of its clients. Suppose 15 consumers were chosen at random, and each was shown two different silverware patterns. The pupillometer readings (in millimeters) for each consumer, with the means and standard deviations of each sample of observations and their differences, are shown in the tables below.

Consumer	Pattern 1	Pattern 2
1	1.00	.80
2	.97	.66
3	1.45	1.22
4	1.21	1.00
5	.77	.81
6	1.32	1.11
7	1.81	1.30
8	.91	.32
9	.98	.91
10	1.46	1.10
11	1.85	1.60
12	.33	.21
13	1.77	1.50
14	.85	.65
15	.15	.05

	$\bar{x}$	s
Pattern 1	1.12	.50
Pattern 2	.88	.45
Difference (1 − 2)	.24	.16

a. Which type of experiment does this represent—independent samples or paired difference? Explain.

b. Use a 90% confidence interval to estimate the difference in mean pupil dilation per consumer for silverware patterns 1 and 2. Interpret the confidence interval, assuming that the pupillometer indeed measures consumer interest.

c. Test the hypothesis that the mean dilation differs for the two patterns. Use $\alpha = .10$. Does the test conclusion support your interpretation of the confidence interval in part **b**?

d. What assumptions are necessary to ensure the validity of the inferences in parts **b** and **c**?

9.97 An experiment was conducted to measure the effects of fructose and glucose on high-endurance performance of athletes (*Research Quarterly for Exercise and Sport,* Vol. 54, 1983). Six trained female runners were used in the experiment. Each was given 300 milliliters of a liquid 45 minutes prior to running for 85 minutes or until she reached a state of exhaustion, whichever occurred first. Various measures of endurance, such as performance time (time to exhaustion), rating of perceived exertion, etc., were recorded at the end of each run. Four liquids (treatments) were used in the experiment. The first contained fructose, the second contained glucose, the third contained water sweetened with a calcium saccharine solution (a placebo designed to suggest the presence of fructose or glucose), and the fourth contained water alone. Each of the six subjects performed the run for each of the four liquids, which were arranged in random order. The table below gives the averages of the six runners' times (in minutes) to exhaustion for only two of the mixtures, glucose and the placebo. The table also gives the sample sizes and standard deviations for the two samples.

	Glucose	**Placebo**
n	6	6
$\bar{x}$	63.9	52.2
s	20.3	13.5

Source: McMurry, R. G., Wilson, J. R., and Kitchell, B. S. "The effects of fructose and glucose on high-endurance performance." *Research Quarterly for Exercise and Sport,* 1983, Vol. 54. Reprinted by permission of the American Alliance for Health, Physical Education, Recreation, and Dance, Reston, VA 22091.

a. Describe the experiment and explain how and why it does or does not satisfy the assumptions of independent random sampling.

b. Suppose that the data were based on independent random samples. Would the data provide sufficient evidence to indicate a difference in the mean time to exhaustion between runners given the glucose mixture and those given the placebo? Test using $\alpha = .05$. Interpret your result.

c. Consider the manner in which the experiment was actually conducted. What do you gain or lose by analyzing the data using the method of part **b**?

9.98 It has been known for a number of years that the tailings (waste) of gypsum and phosphate mines in Florida contain radioactive radon 222. The radiation levels in waste gypsum and phosphate mounds in Polk County, Florida, are regularly monitored by the Eastern Environmental Radiation Facility (EERF) and by the Polk County Health Department (PCHD), Winter Haven, Florida. The following table shows measurements of the exhalation rate (a measure of radiation) for 15 soil samples obtained from waste mounds in Polk County, Florida. The exhalation rate was measured for each soil sample by both the PCHD and the EERF. The objective of selecting the paired measurements was to determine whether there is a bias—a difference in the mean readings—between PCHD and EERF. The data in the table represent part of the data contained in a report by Thomas R. Horton of EERF.

Charcoal Canister No.	PCHD	EERF
71	1,709.79	1,479.0
58	357.17	257.8
84	1,150.94	1,287.0
91	1,572.69	1,395.0
44	558.33	416.5
43	4,132.28	3,993.0
79	1,489.86	1,351.0
61	3,017.48	1,813.0
85	393.55	187.7
46	880.84	630.4
4	2,996.49	3,707.0
20	2,367.40	2,791.0
36	599.84	706.8
42	538.37	618.5
55	2,770.23	2,639.0

Source: Horton, T. R. "Preliminary radiological assessment of radon exhalation from phosphate gypsum piles and inactive uranium mill tailings piles." EPA-520/5-79-004. Washington, D.C.: Environmental Protection Agency, 1979.

a. Considering the relative size of the measurements from canister to canister, explain why a paired difference experiment was conducted rather than an independent samples experiment.

b. Given that the mean difference (PCHD − EERF) for the 15 sampled canisters is 84.17 with standard deviation 408.92, do the data provide sufficient evidence to indicate a difference in the mean exhalation rates between PCHD and EERF? Test using $\alpha = .05$.

c. Find a 95% confidence interval for the difference in mean measurements between PCHD and EERF. Interpret the interval. Does it support the result of the test in part **a**?

9.99 A random sample of 400 female attorneys and 200 male attorneys, from a total of approximately

606,000 attorneys in the United States, revealed that 25% of the women finished in the top 10% of their classes versus 18% of the males (*American Bar Association Journal,* Oct. 1983). Based on the independent random samples of 400 female and 200 male attorneys selected from all attorneys in the United States, do the data provide sufficient evidence to indicate that the probabilities of finishing in the upper 10% of their law school class differ between male and female lawyers?

9.100 One theory regarding the mobility of college and university faculty members is that those who publish the most scholarly articles are also the most mobile. The logic behind this theory is that good researchers who publish frequently receive more job offers and are therefore more likely to move from one university to another. The *Academy of Management Journal* (Vol. 25, 1982) examined this relationship for persons employed in industry. Using the personnel records of a large national oil company, the researchers obtained the early career performance records for 529 of the company's employees. Of these, 174 were classified as *stayers,* those who stayed with the company; the other 355, who left the company at varying points during a 15-year period, were classified as *leavers.* Summary statistics on three variables—initial performance, rate of career advancement (number of promotions per year), and final performance appraisals—for both stayers and leavers are shown in the following table. For each variable, compare the means of stayers and leavers using an appropriate statistical method. Interpret the results.

Variable	STAYERS ($n_1 = 174$)		LEAVERS ($n_2 = 355$)	
	$\bar{x}_1$	s_1	$\bar{x}_2$	s_2
Initial performance	3.51	.51	3.24	.52
Rate of career advancement	.43	.20	.31	.31
Final performance appraisal	3.78	.62	3.15	.68

9.101 A study in the *Journal of Psychology & Marketing* (Jan. 1992) investigates the degree to which American consumers are concerned about product tampering. Large random samples of male and female consumers were asked to rate their concern about product tampering on a scale of 1 (little or no concern) to 9 (very concerned).

a. What are the appropriate null and alternative hypotheses to determine whether a difference exists in the mean level of concern about product tampering between men and women? Define any symbols you use.

b. The statistics reported include those shown in the MINITAB printout at the bottom of the page. Interpret these results.

c. What assumptions are necessary to ensure the validity of this test?

9.102 A new insect spray, type A, is to be compared with a spray, type B, that is currently in use. Two rooms of equal size are sprayed with the same amount of spray, one room with A, the other with B. Two hundred insects are released into each room, and after 1 hour the numbers of dead insects are counted. The results are given in the table below.

	Spray A	Spray B
Number of insects	200	200
Number of dead insects	120	90

a. Do the data provide sufficient evidence to indicate that spray A is more effective than spray B in controlling the insects? Test using $\alpha = .05$.

b. Find a 90% confidence interval for $(p_1 - p_2)$, the difference in the rates of kill for the two sprays. Interpret this interval.

9.103 In the past, many bodily functions were thought to be beyond conscious control. However, recent experimentation suggests that a person may be able to control certain body functions if that person is trained in a program of *biofeedback* exercises. An experiment is conducted to show that blood pressure levels can be consciously reduced in people trained in this program. The blood pressure measurements (in millimeters of mercury) listed in the table below represent readings before and after the biofeedback training of six subjects.

Subject	Before	After
1	136.9	130.2
2	201.4	180.7
3	166.8	149.6
4	150.0	153.2
5	173.2	162.6
6	169.3	160.1

```
TWOSAMPLE T FOR MSCORE VS FSCORE
             N     MEAN     STDEV    SE MEAN
MSCORE      200    3.209    2.33     0.165
FSCORE      200    3.923    2.94     0.208

TTEST MU MSCORE = MU FSCORE (VS NE): T= -2.69 P=0.0072 DF=398
```

```
TEST OF MU = 0.000 VS MU G.T. 0.000

            N      MEAN    STDEV    SE MEAN     T    P VALUE
BminusA     6    10.200    8.393     3.426    2.98    0.015


            N      MEAN    STDEV    SE MEAN    95.0 PERCENT C.I.
BminusA     6    10.20     8.39      3.43   (    1.39,    19.01)
```

a. If we want to test whether the mean blood pressure decreases after the training, what are the appropriate null and alternative hypotheses? Define any symbols you use.

b. When the test is conducted using MINITAB, the results are as shown in the first printout above. Interpret these results.

c. The output of MINITAB also includes a confidence interval. Interpret this interval.

9.104 Students enrolled in music classes at the University of Texas (Austin) participated in a study to compare the observations and teacher evaluations of music education majors and non-music majors (*Journal of Research in Music Education,* Winter 1991). Independent random samples of 100 music majors and 100 nonmajors rated the overall performance of their teacher using a 6-point scale, where 1 = lowest rating and 6 = highest rating. Use a 95% confidence interval to compare the mean teacher ratings of the two groups of music students. Interpret the result.

	Music majors	**Non-music majors**
Sample size	100	100
Mean "overall" rating	4.26	4.59
Standard deviation	.81	.78

Source: Duke, R. A., and Blackman, M. D. "The relationship between observers' recorded teacher behavior and evaluation of music instruction." *Journal of Research in Music Education,* Vol. 39, No. 4, Winter 1991 (Table 2).

9.105 Some power plants are located near rivers or oceans so that the available water can be used for cooling the condensers. Suppose that, as part of an environmental impact study, a power company wants to estimate the difference in mean water temperature between the discharge of its plant and the offshore waters. How many sample measurements must be taken at each site in order to estimate the true difference between means to within .2°C with 95% confidence? Assume that the range in readings will be about 4°C at each site and the same number of readings will be taken at each site.

9.106 Consider the following experimental situation proposed more than 20 years ago in the *Training School Bulletin* (May 1973). Suppose it is desired to form two groups of mentally impaired subjects to compare two methods of educational therapy. Subjects are randomly selected from an existing (large) group of subjects and ordered (from lowest to highest) based on the scores of an appropriate matching variable such as IQ. The two highest-ranking subjects on the matching variable would then form pair 1, the next two pair 2, etc., and one member from each pair would be randomly assigned to each therapy group. Explain the advantages of this experiment over one that utilizes independent random samples of mentally impaired subjects.

9.107 Does the time of day during which one works affect job satisfaction? A study in *The Journal of Occupational Psychology* (Sept. 1991) examined differences in job satisfaction between day-shift and night-shift nurses. Nurses' satisfaction with their hours of work, free time away from work, and breaks during work were measured. The following table shows the mean scores for each measure of job satisfaction (higher scores indicate greater satisfaction), along with the observed significance level comparing the means for the day-shift and night-shift samples:

MEAN SATISFACTION

	Day Shift	**Night Shift**	*p*-**Value**
Satisfaction with:			
Hours of work	3.91	3.56	.813
Free time	2.55	1.72	.047
Breaks	2.53	3.75	.0073

a. Specify the null and alternative hypotheses if we wish to test whether a difference in job satisfaction exists between day-shift and night-shift nurses on each of the three measures. Define any symbols you use.

b. Interpret the *p*-value for each of the tests. (Each of the *p*-values in the table is two-tailed.)

c. Assume that each of the tests is based on small samples of nurses from each group. What assumptions are necessary for the tests to be valid?

9.108 When new instruments are developed to perform chemical analyses of products (food, medicine, etc.), they are usually evaluated with respect to two criteria: accuracy and precision. *Accuracy* refers to the ability of the instrument to identify correctly the nature and amounts of a product's components. *Precision* refers to the consistency with which the instrument will identify the components of the same material. Thus, a large variability in the identification of a single batch of a product indicates a lack of precision. Suppose a pharmaceutical firm is considering two brands of an instrument designed to identify the components of certain drugs. As part of a comparison of precision, 10 test-tube samples of a well-mixed batch of a drug are selected and then five are analyzed by instrument A and five by instrument B. The data shown below are the percentages of the primary component of the drug given by the instruments. Do these data provide evidence of a difference in the precision of the two machines? Use $\alpha = .10$.

Instrument A	Instrument B
43	46
48	49
37	43
52	41
45	48

9.109 A large shipment of produce contains Valencia and navel oranges. To determine whether there is a difference in the proportions of nonmarketable fruit between the two varieties, random samples of 850 Valencia oranges and 1,500 navel oranges are independently selected and the number of nonmarketable oranges of each type is counted. It is found that 30 Valencia and 90 navel oranges from these samples are nonmarketable. Do these data provide sufficient evidence to indicate a difference between the proportions of nonmarketable Valencia and navel oranges? Test at the $\alpha = .05$ level of significance.

9.110 The threat of earthquakes is a part of life for homeowners in California. Scientists have been warning about "the big one" for decades. An article in the *Annals of the Association of American Geographers* (June 1992) explored some factors that are considered when California homeowners purchase earthquake insurance, including the proximity to a major earthquake fault. Surveys were mailed to residents in four California counties. The data collected are shown in the table below.

   a. Los Angeles County is the closest of the four to a major earthquake fault. Calculate 95% confidence intervals for the difference in the proportions of earthquake-insured residents in Los Angeles County and each of the other counties.

   b. Do these results support the contention that closer proximities to major earthquake faults result in higher proportions of earthquake-insured residents?

9.111 Refer to Exercise 9.110. How large would the samples from Los Angeles and San Bernardino counties have to be in order to estimate the difference between earthquake-insured proportions to within .03 with 95% confidence?

9.112 The state of Florida requires all high school students to pass a literacy test before they receive a high school diploma. A student who fails the test can enroll in a refresher course and retake the test at a later date. To evaluate the effectiveness of the refresher course, eight students' test scores were compared, before and after, with the results shown in the following table. Do the data provide sufficient evidence to conclude that the mean test score has increased? (Use $\alpha = .05$.)

Student	Before	After
1	45	49
2	52	50
3	63	70
4	68	71
5	57	53
6	55	61
7	60	62
8	59	67

9.113 Suppose you use either the SAS or SPSS statistical program package to test the null hypothesis H_0: $(\mu_1 - \mu_2) = 0$ against H_a: $(\mu_1 - \mu_2) \neq 0$ with independent samples of size 12 and 10, respectively. Assuming you are using $\alpha = .05$, what conclusion would you reach in each of the following instances of observed significance level reported by the program? (Recall that both SAS and SPSS report the two-tailed *p*-value.)

   a. P-VALUE = .0429   **b.** P-VALUE = .1984
   c. P-VALUE = .0001   **d.** P-VALUE = .0344
   e. P-VALUE = .0545   **f.** P-VALUE = .9633
   g. What assumptions are necessary to ensure the validity of this test?

9.114 The National Football League (NFL) rules committee in 1974 changed the way in which the football was turned over following a missed field goal. Before 1974, the ball was always placed on the 20-yard line for the start of the next possession. Since 1974, the ball has been placed at the line of scrimmage from which the missed field goal was attempted. An article in the *Sociology of Sport Journal* (Dec. 1992) hypothesizes that this rule change made accurate placekickers more valuable to NFL teams. To test the theory, salary data were collected for the five statistically most accurate placekickers in the league 6 years prior to the rule

	Contra Costa	Santa Clara	Los Angeles	San Bernardino
Sample size	521	556	337	372
Number with earthquake insurance	117	222	133	109

change and 13 years after the rule change. The data in the table below present the ratio of the average salary of the five most accurate placekickers divided by the entire league mean salary for each year. Descriptive statistics for the ratios are provided in the SAS printout at the bottom of the page.

Before Rule Change	After Rule Change	
.710	.718	.719
.721	.697	.718
.701	.722	.696
.714	.707	.711
.729	.721	.708
.709	.704	.704
	.703	

a. What are the appropriate null and alternative hypotheses for testing that accurate placekickers became more valuable (relative to players in other positions) after the 1974 rule change took effect? Define any symbols you use.

b. Conduct the test and interpret the result. Use $\alpha = .10$.

c. What assumption must be made about the population variances to ensure the validity of the test?

*d. Test the assumption, part **c**. Use $\alpha = .05$.

9.115 How does gender affect the type of advertising that proves to be most effective? An article in the *Journal of Advertising Research* (May/June 1990) makes reference to numerous studies that conclude males tend to be more competitive with others than with themselves. To apply this conclusion to advertising, the author creates two ads promoting a new brand of soft drink:

Ad 1: Four men are shown competing in racquetball

Ad 2: One man is shown competing against himself in racquetball

The author hypothesized that the first ad will be more effective when shown to males. To test this hypothesis, 43 males were shown both ads and asked to measure their attitude toward the advertisement (Aad), their attitude toward the brand of soft-drink (Ab), and their intention to purchase the soft drink (Intention). Each variable was measured using a 7-point scale, with higher scores indicating a more favorable attitude. The results are shown below:

	SAMPLE MEANS		
	Aad	**Ab**	**Intention**
Ad 1	4.465	3.311	4.366
Ad 2	4.150	2.902	3.813
Level of significance	$p = .091$	$p = .032$	$p = .050$

a. What are the appropriate null and alternative hypotheses to test the author's research hypothesis? Define any symbols you use.

b. Based on the information provided about this experiment, do you think this is an independent samples experiment or a paired difference experiment? Explain.

c. Interpret the *p*-value for each test.

d. What assumptions are necessary for the validity of the tests?

9.116 As part of a study of participative management, workers from two types of New Zealand sociocultural backgrounds were sampled: those who believed in the existence of a class system and those who believed that they lived and worked in a classless society (*Academy of Management Journal*, Vol. 17, 1974). Each worker in the sample was selected from a work environment with participatory management. Do workers who consider themselves to be social equals with their management superiors possess different levels of job satisfaction than those workers who see themselves as socially different from management? Each worker in the independent random samples was asked to answer this question by rating his or her job satisfaction on a scale of 1 (poor) to 7 (excellent). Using the results shown in the table on page 400, what can you say about the differences in job satisfaction for the two different sociocultural types of workers?

9.117 An article in *Psychological Reports* (Vol. 56, 1985) examined the relationship between the intensity of

```
Analysis Variable : RATIO

-------------------------PERIOD=After ------------------------

N Obs    N     Minimum       Maximum          Mean        Std Dev
    13   13   0.6960000     0.7220000     0.7098462      0.0090078

-------------------------PERIOD=Before-----------------------

N Obs    N     Minimum       Maximum          Mean        Std Dev
     6    6   0.7010000     0.7290000     0.7140000      0.0098387
```

	BELIEF IN EXISTENCE OF A CLASS SYSTEM	
	Yes	**No**
Sample size	175	277
Mean	5.42	5.19
Standard deviation	1.24	1.17

Source: Excerpted from Hines, G. H. "Influences on employee expectancy and participative management." *Academy of Management Journal,* 1974, Vol. 17.

parental punishment practices and various demographic characteristics of mothers. The intensity score descriptive statistics are shown in the following table, with higher scores showing a greater intensity of parental punishments.

	Single	**Married**
n	8	6
$\bar{x}$	70.75	77.33
s	14.80	13.69

Source: Excerpted from Shorkey, C. T., McRoy, R. G., and Armendariz, J. "Intensity of parental punishments and problem-solving attitudes and behaviors." *Psychological Reports,* 1985, Vol. 56.

a. Use a 95% confidence interval to estimate the difference between the mean intensity of parental punishment scores for single and married mothers. Interpret the interval in terms of this application.

b. Does the confidence interval in part **a** support an inference that the true means for single and married mothers differ?

c. Conduct a test of hypothesis comparing the two means using $\alpha = .05$. Does your conclusion agree with that in part **b**?

d. What assumptions are necessary to ensure the validity of the inferences in parts **a–c**? State them in terms of this application.

*9.118 Refer to Exercise 9.117. Conduct a test to determine whether the variation in the intensity of punishment scores differs for single and married mothers. Use $\alpha = .05$.

*9.119 In Exercise 9.73 we planned to compare the mean yards gained per run for two professional football running backs. Suppose a record of early season runs shows that the standard deviations of yards gained per run, based on samples of 50 runs per player, are $s_1 = 3.2$ and $s_2 = 5.7$ yards, respectively. Do these standard deviations indicate that the variation in the distributions of yards gained per carry differs for the two players? Test using $\alpha = .05$.

STUDENT PROJECTS

We have now discussed two methods of collecting data to compare two population means. In many experimental situations a decision must be made either to collect two independent samples or to conduct a paired difference experiment. The importance of this decision cannot be overemphasized, since the amount of information obtained and the cost of the experiment are both directly related to the method of experimentation that is chosen.

Choose two populations (pertinent to your major area) which have unknown means and for which you could both collect two independent samples and collect paired observations. Before conducting the experiment, state which method of sampling you think will provide more information (and why). Compare the two methods, first performing the independent sampling procedure by collecting 10 observations from each population (a total of 20 measurements), then performing the paired difference experiment by collecting 10 pairs of observations.

Construct two 95% confidence intervals, one for each experiment you conduct. Which method provides the narrower confidence interval and hence more information on this performance of the experiment? Does this result agree with your preliminary expectations?

EXPLORING DATA WITH A COMPUTER

Consider the blood loss data for cardiac surgery patients given in Appendix B. A group of physicians wants to compare the mean blood loss of patients who take a new drug prior to surgery to the mean blood loss of patients who do not take the drug.

a. Treat the measurements for the two patient groups as independent random samples of all cardiac surgery patients. Test the null hypothesis that the population means are equal using $\alpha = .01$, and place a 99% confidence interval on the true difference between the mean blood loss for the two patient groups.

b. Repeat part a using the following pairs of α and confidence levels for the tests and confidence intervals: (.05, 95%), (.10, 90%), and (.20, 80%). Describe what happens to the tests and confidence intervals as α is increased and the confidence level is decreased. Which do you think is more informative—the tests or the confidence intervals?

Chapter 10

ANALYSIS OF VARIANCE

Comparing More Than Two Means

Contents

Case Studies

*W*HERE WE'VE BEEN

As we've seen in preceding chapters, the solutions of many practical problems are based on inferences about population means. Methods for estimating and testing hypotheses about a single mean and the comparison of two means were presented in Chapters 7–9.

*W*HERE WE'RE GOING

In this chapter we extend the methodology of Chapters 7–9 in two important ways. First, we discuss the critical elements in the *design* of a sampling experiment. Then we see how to *analyze* the experiment in order to compare more than two populations. We'll look at several of the more popular experimental designs, and we'll see how to use the computer to analyze designed experiments using analysis of variance programs.

Most of the data analyzed in previous chapters were collected in *observational* sampling experiments rather than *designed* sampling experiments. In *observational experiments* the analyst has little or no control over the variables under study and merely observes their values. In contrast, *designed experiments* are those in which the analyst attempts to control the levels of one or more variables to determine their effect on a variable of interest. Although many practical situations do not present the opportunity for such control, it is instructive, even for observational experiments, to have a working knowledge of the analysis and interpretation of data that result from designed experiments and to know the basics of how to design experiments when the opportunity arises.

We first present the basic elements of an experimental design in Section 10.1. We then discuss two of the simpler, and more popular, experimental designs in Sections 10.2 and 10.4. Slightly more complex experiments are discussed in Section 10.5.

10.1 ELEMENTS OF A DESIGNED EXPERIMENT

Certain elements are common to almost all designed experiments, regardless of the specific area of application. For example, the *response* is the variable of interest in the experiment. The response might be the SAT scores of a high school senior, the total sales of a firm last year, or the total income of a particular household this year. We will also refer to the response as the *dependent variable*.

> **DEFINITION 10.1**
>
> The **response variable** is the variable of interest to be measured in the experiment. We also refer to the response as the **dependent variable**.

The intent of most statistical experiments is to determine the effect of one or more variables on the response. These variables are usually referred to as the *factors* in a designed experiment. Factors are either *quantitative* or *qualitative,* depending on whether the variable is measured on a numerical scale or not. For example, we might want to explore the effect of the qualitative factor Gender on the response SAT score. In other words, we want to compare the SAT scores of male and female high school seniors. Or, we might wish to determine the effect of the quantitative factor Number of salespeople on the response Total sales for retail firms. Often two or more factors are of interest. For example, we might want to determine the effect of the quantitative factor Number of wage earners and the qualitative factor Location on the response Household income.

> **DEFINITION 10.2**
>
> **Factors** are those variables whose effect on the response is of interest to the experimenter. **Quantitative factors** are measured on a numerical scale, whereas **qualitative factors** are those that are not (naturally) measured on a numerical scale.

Levels are the values of the factors that are utilized in the experiment. The levels of qualitative factors are usually nonnumerical. For example, the levels of Gender are Male and Female, and the levels of Location might be North, East,

South, and West.* The levels of quantitative factors are the numerical values of the variable utilized in the experiment. The Number of salespeople for each of a set of companies, the Number of wage earners in each of a set of households, and the GPAs for a set of high school seniors all represent levels of the respective quantitative factors.

DEFINITION 10.3

Factor levels are the values of the factor utilized in the experiment.

When a *single factor* is employed in an experiment, the *treatments* of the experiment are the levels of the factor. For example, if the effect of the factor Gender on the response SAT score is being investigated, the treatments of the experiment are the two levels of Gender—Female and Male. Or, if the effect of the Number of wage earners on Household income is the subject of the experiment, the numerical values assumed by the quantitative factor Number of wage earners are the treatments. If *two or more factors* are utilized in an experiment, the treatments are the factor-level combinations used. For example, if the effects of the factors Gender and GPA on the response SAT score are being investigated, the treatments are the combinations of the levels of Gender and GPA used; thus (Female, 2.61), (Male, 3.43), and (Female, 3.82) would all be treatments.

DEFINITION 10.4

The **treatments** of an experiment are the factor-level combinations utilized.

The objects on which the response variable and factors are observed are the *experimental units.* For example, SAT score, High school GPA, and Gender are all variables that can be observed on the same experimental unit—a high school senior. Or, the Total sales, the Earnings per share, and the Number of salespeople can be measured on a particular firm in a particular year, and the firm–year combination is the experimental unit. The Total income, the Number of female wage-earners, and the Location can be observed for a household at a particular point in time, and the household–time combination is the experimental unit. Every experiment, whether observational or designed, has experimental units on which the variables are observed. However, the identification of the experimental units is more important in designed experiments, when the experimenter must actually sample the experimental units and measure the variables.

DEFINITION 10.5

An **experimental unit** is the object on which the response and factors are observed or measured.[†]

When the specification of the treatments and the method of assigning the experimental units to each of the treatments is controlled by the analyst, the experiment is said to be *designed.* In contrast, if the analyst is just an observer of the treatments on a sample of experimental units, the experiment is *observational.*

---

*The levels of a qualitative variable may bear numerical labels. For example, the Locations could be numbered 1, 2, 3, and 4. However, in such cases the numerical labels for a qualitative variable will usually be codes representing nonnumerical levels.
[†]Recall (Chapter 1) that the set of all experimental units is the population.

FIGURE 10.1

Sampling experiment:
Process and terminology

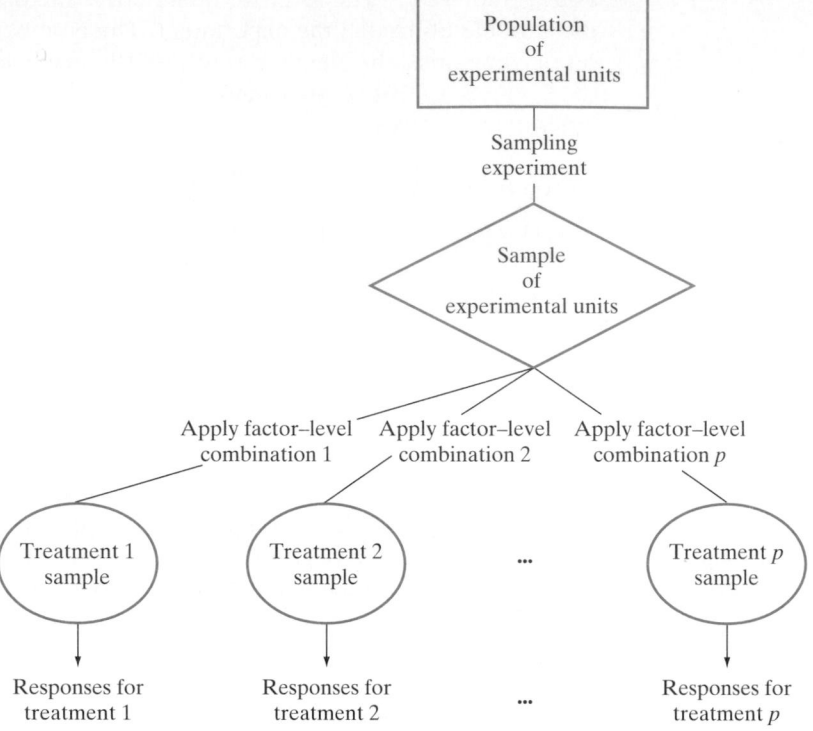

For example, if you specify the number of Female and Male high school students within each GPA range to be randomly selected in order to evaluate the effect of Gender and GPA on SAT scores, you are designing the experiment. If, on the other hand, you simply observe the SAT scores, Gender, and GPA for all students who took the SAT test last month at a particular high school, the experiment is observational.

> **DEFINITION 10.6**
>
> A **designed experiment** is one for which the analyst controls the specification of the treatments and the method of assigning the experimental units to each treatment. An **observational experiment** is one for which the analyst simply observes the treatments and the response on a sample of experimental units.

The diagram in Figure 10.1 provides an overview of the experimental process and a summary of the terminology introduced in this section. Note that the experimental unit is at the core of the process. The method by which the sample of experimental units is selected from the population determines the type of experiment. The level of every factor (the treatment) and the response are all variables that are observed or measured on each experimental unit.

EXAMPLE 10.1

The USGA (United States Golf Association) regularly tests golf equipment to ensure that it conforms to USGA standards. Suppose it wishes to compare the mean distance traveled by four different brands of golf balls when struck by a driver (the club used to maximize distance). The following experiment is conducted:

10 balls of each brand are randomly selected. Each is struck by "Iron Byron" (the USGA's golf robot named for the famous golfer, Byron Nelson) using a driver, and the distance traveled is recorded. Identify each of the following elements in this experiment: response, factors, factor types, levels, treatments, and experimental units.

Solution

The response is the variable of interest, Distance traveled. The only factor being investigated is Brand of golf ball, and it is nonnumerical and therefore qualitative. The four brands (say A, B, C, and D) represent the levels of this factor. Since only one factor is utilized, the treatments are the four levels of this factor—that is, the four brands. The experimental unit is a golf ball; more specifically, it is a golf ball at a particular position in the striking sequence, since the distance traveled can be recorded only when the ball is struck, and we would expect the distance to be different (due to random factors such as wind resistance, landing place, and so forth) if the same ball is struck a second time. Note that 10 experimental units are sampled for each treatment, generating a total of 40 observations.

This experiment, like many real applications, is a blend of designed and observational: The analyst cannot control the assignment of the brand to each golf ball (observational), but he or she can control the assignment of each ball to the position in the striking sequence (designed). ▲

EXAMPLE 10.2

Suppose the USGA is also interested in comparing the mean distances the four brands of golf balls travel when struck by a five-iron and by a driver. Ten balls of each brand are randomly selected, five to be struck by the driver, and five by the five-iron. Identify the elements of the experiment, and construct a schematic diagram similar to Figure 10.1 to provide an overview of this experiment.

Solution

The response is the same as in Example 10.1—Distance traveled. The experiment now has two factors, Brand of golf ball and Club utilized. There are four levels of Brand (A, B, C, and D) and two of Club (driver and five-iron, or 1 and 5). Treatments are factor-level combinations, so there are $4 \times 2 = 8$ treatments in this experiment: (A, 1), (A, 5), (B, 1), (B, 5), (C, 1), (C, 5), (D, 1), and (D, 5). The experimental units are still the combinations of golf ball and hitting position. Note that five experimental units are sampled per treatment, generating 40 observations. The experiment is summarized in Figure 10.2. ▲

Our objective in designing an experiment is usually to maximize the amount of information obtained about the relationship between the treatments and the response. Of course, we are almost always subject to constraints on budget, time, and even the availability of experimental units. Nevertheless, designed experiments are generally preferred to observational experiments. Not only do we have better control of the amount and quality of the information collected, but we also avoid the biases inherent in observational experiments in the selection of the experimental units representing each treatment. Inferences based on observational experiments always carry the implicit assumption that the sample has no hidden bias that was not considered in the statistical analysis. Better understanding of the potential problems with observational experiments is a by-product of our study of experimental design in the remainder of this chapter.

FIGURE 10.2

Two-factor golf experiment
summary: Example 10.2

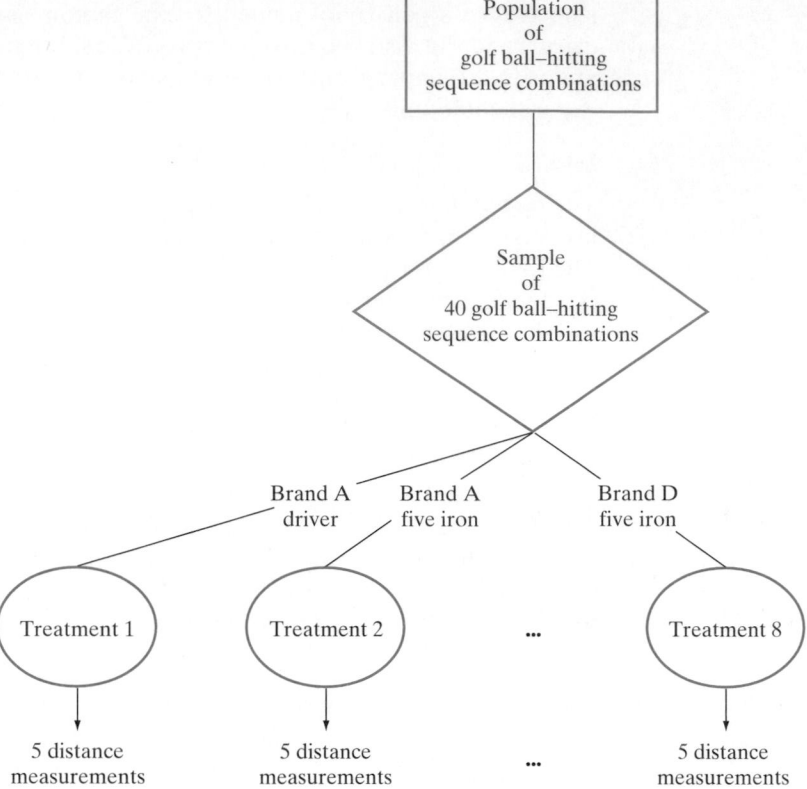

EXERCISES 10.1–10.7

Learning the Mechanics

10.1 What are the treatments for a designed experiment that utilizes one qualitative factor with four levels—A, B, C, and D?

10.2 What are the treatments for a designed experiment with two factors, one qualitative with two levels (A and B) and one quantitative with five levels (50, 60, 70, 80, and 90)?

10.3 What are the experimental units on which each of the following responses are observed?
a. College GPA
b. Household income
c. Your time in running the 100-yard dash
d. A patient's reaction to a new drug

10.4 What is the difference between an observational and a designed experiment?

Applying the Concepts

10.5 Suppose the mean cholesterol levels are to be compared for adults in four socioeconomic classes—poverty, low income, middle income, and high income. If random samples of each socioeconomic class are selected to make the comparison, identify each of the following elements of the experiment.

a. Response **b.** Factor(s) and factor type(s)
c. Treatments **d.** Experimental units

10.6 A quality control supervisor measures the quality of a steel ingot on a scale from 0 to 10. He designs an experiment in which three different temperatures (ranging from 1,100 to 1,200°F) and five different pressures (ranging from 500 to 600 psi) are utilized, with 20 ingots produced at each Temperature–Pressure combination. Identify the following elements of the experiment:
a. Response **b.** Factor(s) and factor type(s)
c. Treatments **d.** Experimental units

10.7 Brief descriptions of a number of experiments are given next. Determine whether each is observational or designed, and explain your reasoning.
a. An economist obtains the unemployment rate and gross state product for a sample of states over the past 10 years, with the objective of examining the relationship between the unemployment rate and the gross state product by census region.
b. A psychologist tests the effects of three different feedback programs by randomly assigning five rats to each program and recording their response times at specified intervals during the program.

c. A marketer of computers runs ads in each of four national publications for one quarter, and keeps track of the number of sales that are attributable to each publication's ad.

d. An electric utility engages a consultant to monitor the discharge from its smokestack on a monthly basis over a 1-year period in order to relate the level of sulfur dioxide in the discharge to the load on the facility's generators.

e. Intrastate trucking rates are compared before and after governmental deregulation of prices charged, with the comparison also taking into account distance of haul, goods hauled, and the price of diesel fuel.

f. An agriculture student compares the amount of rainfall in four different states over the past 5 years.

10.2   THE COMPLETELY RANDOMIZED DESIGN

The simplest experimental design, a *completely randomized design,* consists of the *independent random selection* of experimental units representing each treatment. For example, we could independently select random samples of 20 female and 15 male high school seniors to compare their mean SAT scores. Or, we could independently select random samples of 30 households from each of four census districts to compare the mean income per household among the districts. In both examples our objective is to compare treatment means by selecting random, independent samples for each treatment.

> **DEFINITION 10.7**
>
> A **completely randomized design** is a design for which independent random samples of experimental units are selected for each treatment.*

The objective of a completely randomized design is usually to compare the treatment means. If we denote the true, or population, means of the p treatments as $\mu_1, \mu_2, ..., \mu_p$, then we will test the null hypothesis that the treatment means are all equal against the alternative that at least two of the treatment means differ:

$$H_0:\ \mu_1 = \mu_2 = \cdots = \mu_p$$
$$H_a:\ \text{At least two of the } p \text{ treatment means differ}$$

The μ's might represent the means of *all* female and male high school seniors' SAT scores or the means of *all* households' income in each of four census regions.

To conduct a statistical test of these hypotheses, we will use the means of the independent random samples selected from the treatment populations using the completely randomized design. That is, we compare the p sample means $\bar{x}_1, \bar{x}_2, ..., \bar{x}_p$.

For example, suppose you select independent random samples of five female and five male high school seniors and obtain sample mean SAT scores of 550 and 590, respectively. Can we conclude that males score 40 points higher, on average, than females? To answer this question, we must consider the amount of sampling variability among the experimental units (students). If the scores are as depicted in the dot plot shown in Figure 10.3, then the difference between the means is small relative to the sampling variability of the scores within the treatments, Female

---

*We use *completely randomized design* to refer to both designed and observational experiments. Thus, the only requirement is that the experimental units to which treatments are applied (designed) or on which treatments are observed (observational) are independently selected for each treatment.

FIGURE 10.3

Dot plot of SAT scores: Difference between means dominated by sampling variability

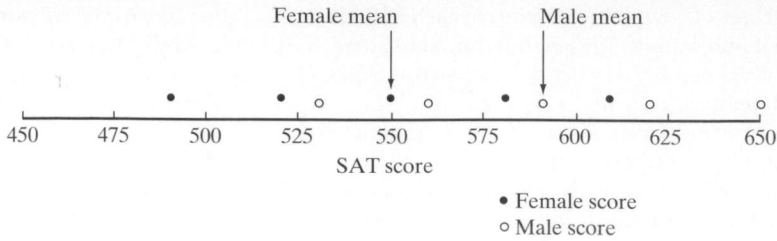

and Male. We would be inclined not to reject the null hypothesis of equal population means in this case.

In contrast, if the data are as depicted in the dot plot of Figure 10.4, then the sampling variability is small relative to the difference between the two means. We would be inclined to favor the alternative hypothesis that the population means differ in this case.

You can see that the key is to compare the difference between the treatment means to the amount of sampling variability. To conduct a formal statistical test of the hypotheses requires numerical measures of the difference between the treatment means and the sampling variability within each treatment. The variation between the treatment means is measured by the **Sum of Squares for Treatments** (SST), which is calculated by squaring the distance between each treatment mean and the overall mean of *all* sample measurements, multiplying each squared distance by the number of sample measurements for the treatment, and adding the results over all treatments:

$$\text{SST} = \sum_{i=1}^{p} n_i(\bar{x}_i - \bar{x})^2 = 5(550 - 570)^2 + 5(590 - 570)^2 = 4{,}000$$

where we use $\bar{x}$ to represent the overall mean response of all sample measurements, that is, the mean of the combined samples. The symbol n_i is used to denote the sample size for the ith treatment. You can see that the value of SST is 4,000 for the two samples of five female and five male SAT scores depicted in Figures 10.3 and 10.4.

Next, we must measure the sampling variability within the treatments. We call this the **Sum of Squares for Error** (SSE) because it measures the variability around the treatment means that is attributed to sampling error. Suppose the 10 measurements in the first dot plot (Figure 10.3) are 490, 520, 550, 580, and 610 for females and 530, 560, 590, 620, and 650 for males. Then the value of SSE is computed by summing the squared distance between each response measurement and the corresponding treatment mean, and then adding the squared differences over all measurements in the entire sample:

$$\text{SSE} = \sum_{j=1}^{n_1}(x_{1j} - \bar{x}_1)^2 + \sum_{j=1}^{n_2}(x_{2j} - \bar{x}_2)^2 + \cdots + \sum_{j=1}^{n_p}(x_{pj} - \bar{x}_p)^2$$

FIGURE 10.4

Dot plot of SAT scores: Difference between means large relative to sampling variability

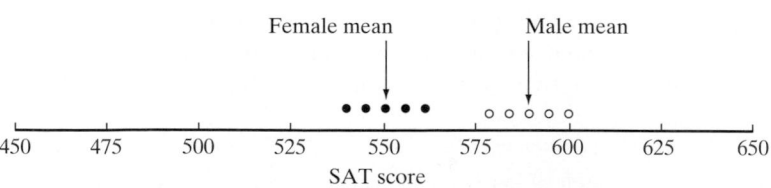

where the symbol x_{1j} is the jth measurement in sample 1, x_{2j} is the jth measurement in sample 2, and so on. This rather complex-looking formula can be simplified by recalling the formula for the sample variance, s^2, given in Chapter 3:

$$s^2 = \sum_{i=1}^{n} \frac{(x_i - \bar{x})^2}{n-1}$$

Note that each sum in SSE is simply the numerator of s^2 for that particular treatment. Consequently, we can rewrite SSE as

$$\text{SSE} = (n_1 - 1)s_1^2 + (n_2 - 1)s_2^2 + \cdots + (n_p - 1)s_p^2$$

where $s_1^2, s_2^2, \ldots, s_p^2$ are the sample variances for the p treatments. For our samples of SAT scores, we find $s_1^2 = 2{,}250$ (for females) and $s_2^2 = 2{,}250$ (for males); then we have

$$\text{SSE} = (5 - 1)(2{,}250) + (5 - 1)(2{,}250) = 18{,}000$$

To make the two measurements of variability comparable, we divide each by the degrees of freedom to convert the sums of squares to mean squares. First, the **Mean Square for Treatments** (MST), which measures the variability *among* the treatment means, is equal to

$$\text{MST} = \frac{\text{SST}}{p-1} = \frac{4{,}000}{2-1} = 4{,}000$$

where the number of degrees of freedom for the p treatments is $(p - 1)$. Next, the **Mean Square for Error** (MSE), which measures the sampling variability *within* the treatments, is

$$\text{MSE} = \frac{\text{SSE}}{n-p} = \frac{18{,}000}{10-2} = 2{,}250$$

Finally, we calculate the ratio of MST to MSE—an ***F* statistic**:

$$F = \frac{\text{MST}}{\text{MSE}} = \frac{4{,}000}{2{,}250} = 1.78$$

Values of the F statistic near 1 indicate that the two sources of variation, between treatment means and within treatments, are approximately equal. In this case, the difference between the treatment means may well be attributable to sampling error, which provides little support for the alternative hypothesis that the population treatment means differ. Values of F well in excess of 1 indicate that the variation among treatment means well exceeds that within means and therefore support the alternative hypothesis that the population treatment means differ.

When does F exceed 1 by enough to reject the null hypothesis that the means are equal? This depends on the degrees of freedom for treatments and for error, and on the value of α selected for the test. We compare the calculated F value to a table F value (Tables VIII–XI of Appendix A) with $v_1 = (p - 1)$ degrees of freedom in the numerator and $v_2 = (n - p)$ degrees of freedom in the denominator and corresponding to a Type I error probability of α. For the SAT score example, the F statistic has $v_1 = (2 - 1) = 1$ numerator degree of freedom and $v_2 = (10 - 2) = 8$ denominator degrees of freedom. Thus, for $\alpha = .05$ we find (Table IX of Appendix A)

$$F_{.05} = 5.32$$

The implication is that MST would have to be 5.32 times greater than MSE before we could conclude at the .05 level of significance that the two population treatment means differ. Since the data yielded $F = 1.78$, our initial impressions for the dot plot in Figure 10.3 are confirmed—there is insufficient information to

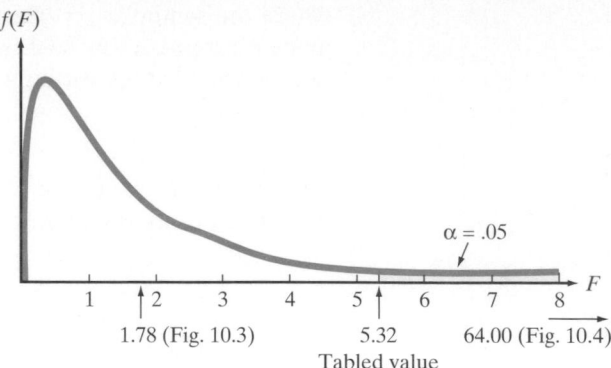

FIGURE 10.5

Rejection region and
calculated *F* values for
SAT score samples

conclude that the mean SAT scores differ for the populations of female and male high school seniors. The rejection region and the calculated *F* value are shown in Figure 10.5.

In contrast, consider the dot plot in Figure 10.4. Since the means are the same as in the first example, 550 and 590, respectively, the variation between the means is the same, MST = 4,000. But the variation within the two treatments appears to be considerably smaller. The observed SAT scores are 540, 545, 550, 555, and 560 for females and 580, 585, 590, 595, and 600 for males. These values yield $s_1^2 = 62.5$ and $s_2^2 = 62.5$. Thus, the variation within the treatments is measured by

$$\text{SSE} = (5 - 1)(62.5) + (5 - 1)(62.5)$$
$$= 500$$
$$\text{MSE} = \frac{\text{SSE}}{n - p} = \frac{500}{8} = 62.5$$

Then the *F*-ratio is

$$F = \frac{\text{MST}}{\text{MSE}} = \frac{4,000}{62.5} = 64.0$$

Again, our visual analysis of the dot plot is confirmed statistically: $F = 64.0$ well exceeds the tabled *F* value, 5.32, corresponding to the .05 level of significance. We would therefore reject the null hypothesis at that level and conclude that the SAT mean score of Males differs from that of Females.

Recall that we performed a hypothesis test for the difference between two means in Section 9.1 using a two-sample *t* statistic for two independent samples. When two independent samples are being compared, the *t*- and *F*-tests are equivalent. To see this, recall the formula

$$t = \frac{\bar{x}_1 - \bar{x}_2}{\sqrt{s_p^2 \left(\frac{1}{n_1} + \frac{1}{n_2} \right)}} = \frac{590 - 550}{\sqrt{(62.5)\left(\frac{1}{5} + \frac{1}{5} \right)}} = \frac{40}{5} = 8$$

where we used the fact that $s_p^2 = \text{MSE}$, which you can verify by comparing the formulas. Note that the calculated *F* for these samples ($F = 64$) equals the square of the calculated *t* for the same samples ($t = 8$). Likewise, the tabled *F* value (5.32) equals the square of the tabled *t* value at the two-sided .05 level of significance ($t_{.025} = 2.306$ with 8 df). Since both the rejection region and the calculated values are related in the same way, the tests are equivalent. Moreover, the assumptions that must be met to ensure the validity of the *t*- and *F*-tests are the same:

1. The probability distributions of the populations of responses associated with each treatment must all be normal.
2. The probability distributions of the populations of responses associated with each treatment must have equal variances.
3. The samples of experimental units selected for the treatments must be random and independent.

In fact, the only real difference between the tests is that the *F*-test can be used to compare *more than two* treatment means, whereas the *t*-test is applicable to two samples only. The **F-test** is summarized in accompanying the box.

Test to Compare *p* Treatment Means for a Completely Randomized Design

$$H_0: \mu_1 = \mu_2 = \cdots = \mu_p$$

H_a: At least two treatment means differ

Test statistic: $F = \dfrac{\text{MST}}{\text{MSE}}$

Assumptions: 1. Samples are selected randomly and independently from the respective populations.
2. All *p* population probability distributions are normal.
3. The *p* population variances are equal.

Rejection region: $F > F_\alpha$, where F_α is based on $(p - 1)$ numerator degrees of freedom (associated with MST) and $(n - p)$ denominator degrees of freedom (associated with MSE).

Computational formulas for MST and MSE are given in Appendix C. We will rely on some of the many statistical software packages available to compute the *F* statistic, concentrating on the interpretation of the results rather than their calculations.

EXAMPLE 10.3

Suppose the USGA wants to compare the mean distances associated with four different brands of golf balls when struck with a driver. A completely randomized design is employed, with Iron Byron, the USGA's robotic golfer, using a driver to hit a random sample of 10 balls of each brand in a random sequence. The distance is recorded for each hit, and the results are shown in Table 10.1, organized by brand.

a. Set up the test to compare the mean distances for the four brands. Use $\alpha = .10$.
b. Use the SAS Analysis of Variance program to obtain the test statistic and *p*-value. Interpret the results.

Solution

a. To compare the mean distances of the four brands, we first specify the hypotheses to be tested. Denoting the population mean of the *i*th brand by μ_i, we test

$$H_0: \mu_1 = \mu_2 = \mu_3 = \mu_4$$

H_a: The mean distances differ for at least two of the brands

The test statistic compares the variation among the four treatment (Brand) means to the sampling variability within each of the treatments.

TABLE 10.1   Results of Completely Randomized Design: Iron Byron Driver

	Brand A	Brand B	Brand C	Brand D
	251.2	263.2	269.7	251.6
	245.1	262.9	263.2	248.6
	248.0	265.0	277.5	249.4
	251.1	254.5	267.4	242.0
	260.5	264.3	270.5	246.5
	250.0	257.0	265.5	251.3
	253.9	262.8	270.7	261.8
	244.6	264.4	272.9	249.0
	254.6	260.6	275.6	247.1
	248.8	255.9	266.5	245.9
Sample Means	250.8	261.1	270.0	249.3

Test statistic: $F = \dfrac{MST}{MSE}$

Rejection region: $F > F_\alpha = F_{.10}$

with $v_1 = (p - 1) = 3$ df   and $v_2 = (n - p) = 36$ df

From Table VIII of Appendix A, we find $F_{.10} \approx 2.25$ for 3 and 36 df. Thus, we will reject H_0 if $F > 2.25$. (See Figure 10.7.)

The assumptions necessary to ensure the validity of the test are as follows:

1. The samples of 10 golf balls for each brand are selected randomly and independently.
2. The probability distributions of the distances for each brand are normal.
3. The variances of the distance probability distributions for each brand are equal.

b. The SAS printout for the data in Table 10.1 resulting from this completely randomized design is given in Figure 10.6. The Total Sum of Squares is designated the **Corrected Total**, and it is partitioned into the **Model** and **Error Sums of Squares**. The bottom part of the printout further partitions the **Model** component into the factors that comprise the model. In this single-factor experiment, the Model and Brand sums of squares are the same. The **Sum of Squares** column is headed **Anova SS**.

FIGURE 10.6

SAS analysis of variance printout for golf ball distance data: Completely randomized design

```
                              Analysis of Variance Procedure

Dependent Variable: DISTANCE

                                    Sum of          Mean
Source                 DF          Squares        Square    F Value    Pr > F

Model                   3      2794.388750    931.462917      43.99    0.0001
Error                  36       762.301000     21.175028
Corrected Total        39      3556.689750

                 R-Square          C.V.       Root MSE       DISTANCE Mean

                 0.785671      1.785118       4.601633          257.777500


Source                 DF        Anova SS   Mean Square    F Value    Pr > F

BRAND                   3      2794.388750    931.462917      43.99    0.0001
```

FIGURE 10.7

F-Test for completely randomized design: Golf ball experiment

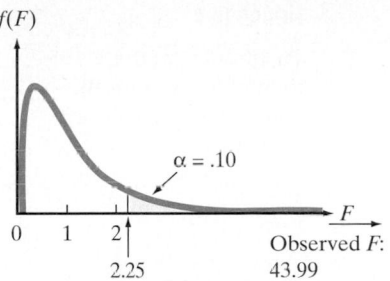

The values of the mean squares, MST and MSE (shaded on the printout), are 931.46 and 21.18, respectively. The *F*-ratio, 43.99, also shaded on the printout, exceeds the tabled value of 2.25. We therefore reject the null hypothesis at the .10 level of significance, concluding that at least two of the brands differ with respect to mean distance traveled when struck by the driver.

The observed significance level of the *F*-test is also shaded on the printout: .0001. This is the area to the right of the calculated *F* value and it implies that we would reject the null hypothesis that the means are equal at any α level greater than .0001. The result of the test is displayed in Figure 10.7.   ▲

The results of an **analysis of variance** (ANOVA) can be summarized in a simple tabular format similar to that obtained from the SAS program in Example 10.3. The general form of the table is shown in Table 10.2, where the symbols df, SS, and MS stand for degrees of freedom, Sum of Squares, and Mean Square, respectively. Note that the two sources of variation, Treatments and Error, add to the Total Sum of Squares, SS(Total). The ANOVA summary table for Example 10.3 is given in Table 10.3, and the partitioning of the Total Sum of Squares into its two components is illustrated in Figure 10.8.

Suppose the *F*-test results in a rejection of the null hypothesis that the treatment means are equal. Is the analysis complete? Usually, the conclusion that at

TABLE 10.2   General ANOVA Summary Table for a Completely Randomized Design

Source	df	SS	MS	F
Treatments	$p - 1$	SST	$MST = \dfrac{SST}{p - 1}$	$\dfrac{MST}{MSE}$
Error	$n - p$	SSE	$MSE = \dfrac{SSE}{n - p}$	
Total	$n - 1$	SS(Total)		

TABLE 10.3   ANOVA Summary Table for Example 10.3

Source	df	SS	MS	F	p-Value
Brands	3	2,794.39	931.46	43.99	.0001
Error	36	762.30	21.18		
Total	39	3,556.69			

FIGURE 10.8

Partitioning of the Total Sum of Squares for the completely randomized design

Total sum of squares
SS(Total)

Sum of squares for treatments
SST

Sum of squares for error
SSE

least two of the treatment means differ leads to other questions. Which of the means differ, and by how much? For example, the *F*-test in Example 10.3 leads to the conclusion that at least two of the brands of golf balls have different mean distances traveled when struck with a driver. Now the question is, which of the brands differ? How are the brands ranked with respect to mean distance?

One way to obtain this information is to construct a confidence interval for the difference between the means of any pair of treatments using the method of Section 9.1. For example, if a 95% confidence interval for $\mu_A - \mu_C$ in Example 10.3 is found to be $(-24, -13)$, we are confident that the mean distance for Brand C exceeds the mean for Brand A (since all differences in the interval are negative). Constructing these confidence intervals for all possible brand pairs allows you to rank the brand means. A method for conducting these *multiple comparisons*—one that controls for Type I errors—is presented in Section 10.3.

EXAMPLE 10.4

Refer to the completely randomized design ANOVA conducted in Example 10.3. Are the assumptions required for the test approximately satisfied?

Solution

The assumptions for the test are repeated below.

1. The samples of golf balls for each brand are selected randomly and independently.
2. The probability distributions of the distances for each brand are normal.
3. The variances of the distance probability distributions for each brand are equal.

Since the sample consisted of 10 randomly selected balls of each brand and the robotic golfer Iron Byron was used to drive all the balls, the first assumption of independent random samples is satisfied. To check the next two assumptions, we will employ two graphical methods presented in Chapter 2: stem-and-leaf displays and dot plots. A MINITAB stem-and-leaf display for the sample distances of each brand of golf ball is shown in Figure 10.9, followed by a MINITAB dot plot in Figure 10.10.

The normality assumption can be checked by examining the stem-and-leaf displays in Figure 10.9. With only 10 sample measurements for each brand, however, the displays are not very informative. More data would need to be collected for each brand before we could assess whether the distances come from normal distributions. Fortunately, analysis of variance has been shown to be a very **robust method** when the assumption of normality is not satisfied exactly: That is, moderate departures from normality do not have much effect on the significance level of the ANOVA *F*-test or on confidence coefficients. Rather than spend the time, energy,

FIGURE 10.9

MINITAB stem-and-leaf displays for distance data: Completely randomized design

```
Stem-and-leaf of BrandA   N = 10
Leaf Unit = 1.0

   2    24  45
   2    24
   4    24  88
  (3)   25  011
   3    25  3
   2    25  4
   1    25
   1    25
   1    26  0

Stem-and-leaf of BrandB   N = 10
Leaf Unit = 1.0

   2    25  45
   3    25  7
   3    25
   4    26  0
  (3)   26  223
   3    26  445

Stem-and-leaf of BrandC   N = 10
Leaf Unit = 1.0

   1    26  3
   2    26  5
   4    26  67
   5    26  9
   5    27  00
   3    27  2
   2    27  5
   1    27  7

Stem-and-leaf of BrandD   N = 10
Leaf Unit = 1.0

   1    24  2
   2    24  5
   4    24  67
  (3)   24  899
   3    25  11
   1    25
   1    25
   1    25
   1    25
   1    26  1
```

or money to collect additional data for this experiment in order to verify the normality assumption, we will rely on the robustness of the ANOVA methodology.

Dot plots are a convenient way to obtain a rough check on the assumption of equal variances. With the exception of a possible outlier for Brand D, the dot plots in Figure 10.10 show that the spread of the distance measurements is about the same for each brand. Since the sample variances appear to be the same, the assumption of equal population variances for the brands is probably satisfied. Although robust with respect to the normality assumption, ANOVA is *not robust* with respect to the equal variances assumption. Departures from the assumption

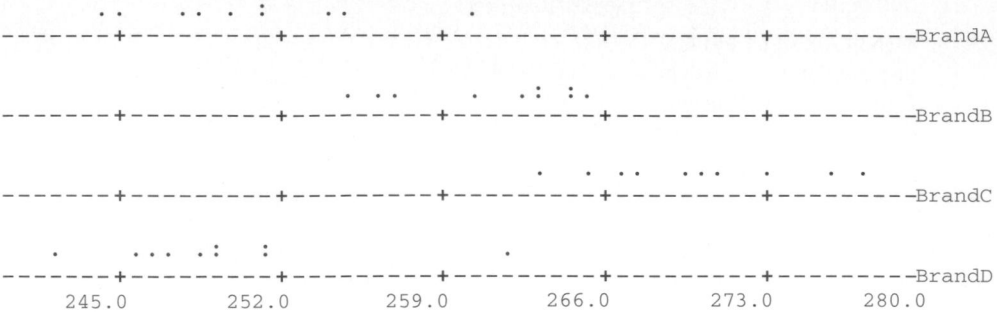

FIGURE 10.10 MINITAB dot plots for distance data: Completely randomized design

of equal population variances can affect the associated measures of reliability (e.g., *p*-values and confidence levels). Fortunately, the effect is slight when the sample sizes are equal, as in this experiment. ▲

Although graphs can be used to check the ANOVA assumptions, as in Example 10.4, no measures of reliability can be attached to these graphs. When you have a plot that is unclear as to whether or not an assumption is satisfied, you can use formal statistical tests which are beyond the scope of this text. Consult the references for Chapter 10 at the end of the book for information on these tests. When the validity of the ANOVA assumptions is in doubt, nonparametric statistical methods are useful.

> **What Do You Do When the Assumptions Are Not Satisfied for the Analysis of Variance for a Completely Randomized Design?**
>
> *Answer:* Use a nonparametric statistical method such as the Kruskal-Wallis *H*-Test of Section 15.4.

The procedure for conducting an analysis of variance for a completely randomized design is summarized in the following box. Remember that the hallmark of this design is independent random samples of experimental units associated with each treatment. We discuss a design with dependent samples in the next Section 10.4.

> **Steps for Conducting an ANOVA for a Completely Randomized Design**
>
> 1. Be sure the design is truly completely randomized, with independent random samples for each treatment.
> 2. Check the assumptions of normality and equal variances.
> 3. Create an ANOVA summary table that specifies the variability attributable to treatments and error, making sure that it leads to the calculation of the *F* statistic for testing the null hypothesis that the treatment means are equal in the population. Use a statistical software program to obtain the numerical results. If no such package is available, use the calculation formulas in Appendix C.
> 4. If the *F*-test leads to the conclusion that the means differ:
>   a. Conduct a multiple comparisons procedure for as many of the pairs of means as you wish to compare (see Section 10.3). Use the results to summarize the statistically significant differences among the treatment means.

b. If desired, form confidence intervals for one or more individual treatment means.

5. If the *F*-test leads to the nonrejection of the null hypothesis that the treatment means are equal, consider the following possibilities:

a. The treatment means are equal—that is, the null hypothesis is true.

b. The treatment means really differ, but other important factors affecting the response are not accounted for by the completely randomized design. These factors inflate the sampling variability, as measured by MSE, resulting in smaller values of the *F* statistic. Either increase the sample size for each treatment or use a different experimental design (as in Section 10.4) that accounts for the other factors affecting the response.

[*Note:* Be careful not to automatically conclude that the treatment means are equal since the possibility of a Type II error must be considered if you accept H_0.]

EXERCISES 10.8–10.26

Learning the Mechanics

10.8 Use Tables VIII, IX, X, and XI of Appendix A to find each of the following *F* values:
   a. $F_{.05}, v_1 = 3, v_2 = 4$ **b.** $F_{.01}, v_1 = 3, v_2 = 4$
   c. $F_{.10}, v_1 = 20, v_2 = 40$ **d.** $F_{.025}, v_1 = 12, v_2 = 9$

10.9 Find the following probabilities:
   a. $P(F < 3.48)$ for $v_1 = 5, v_2 = 9$
   b. $P(F > 3.09)$ for $v_1 = 15, v_2 = 20$
   c. $P(F > 2.40)$ for $v_1 = 15, v_2 = 15$
   d. $P(F \leq 1.83)$ for $v_1 = 8, v_2 = 40$

10.10 In which dot plot is the difference between the sample means small relative to the variability within the sample observations? Justify your answer. (See **a.** and **b.** at the bottom of the page.)

10.11 Refer to Exercise 10.10. Assume that the two samples represent independent, random samples corresponding to two treatments in a completely randomized design.
   a. Calculate the treatment means, i.e., the means of samples 1 and 2, for both dot plots.
   b. Use the means to calculate the Sum of Squares for Treatments (SST) for each dot plot.
   c. Calculate the sample variance for each sample and use these values to obtain the Sum of Squares for Error (SSE) for each dot plot.
   d. Calculate the Total Sum of Squares [SS(Total)] for the two dot plots by adding the Sums of Squares for Treatment and Error. What percentage of SS(Total) is accounted for by the

treatments—that is, what percentage of the Total Sum of Squares is the Sum of Squares for Treatment in each case?

   e. Convert the Sum of Squares for Treatment and Error to mean squares by dividing each by the appropriate number of degrees of freedom. Calculate the *F*-ratio of the Mean Square for Treatment (MST) to the Mean Square for Error (MSE) for each dot plot.
   f. Use the *F*-ratios to test the null hypothesis that the two samples are drawn from populations with equal means. Use $\alpha = .05$.
   g. What assumptions must be made about the probability distributions corresponding to the responses for each treatment in order to ensure the validity of the *F*-tests conducted in part **f**?

10.12 Refer to Exercises 10.10 and 10.11. Conduct a two-sample *t*-test (Section 9.1) of the null hypothesis that the two treatment means are equal for each dot plot. Use $\alpha = .05$ and two-tailed tests. In the course of the test, compare each of the following with the *F*-tests in Exercise 10.11:
   a. The pooled variances and the MSEs
   b. The *t*- and the *F*-test statistics
   c. The tabled values of *t* and *F* that determine the rejection regions
   d. The conclusions of the *t*- and *F*-tests
   e. The assumptions that must be made in order to ensure the validity of the *t*- and *F*-tests

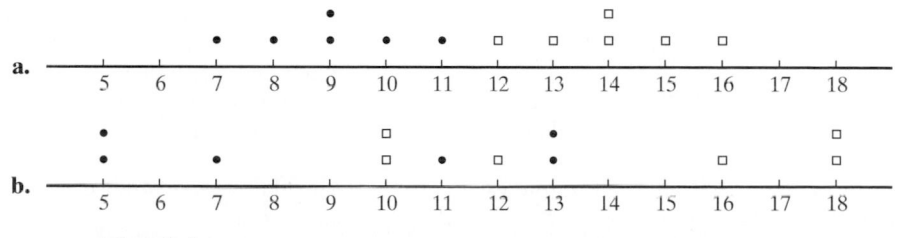

• Sample 1
□ Sample 2

10.13 Refer to Exercises 10.10 and 10.11. Complete the following ANOVA table for each of the two dot plots:

Source	df	SS	MS	F
Treatments				
Error				
Total				

10.14 Suppose the Total Sum of Squares for a completely randomized design with $p = 5$ treatments and $n = 30$ total measurements (six per treatment) is equal to 500. In each of the following cases, conduct an *F*-test of the null hypothesis that the mean responses for the five treatments are the same. Use $\alpha = .10$.
   a. Sum of Squares for Treatment (SST) is 20% of SS(Total)
   b. SST is 50% of SS(Total)
   c. SST is 80% of SS(Total)
   d. What happens to the *F*-ratio as the percentage of the Total Sum of Squares attributable to treatments is increased?

10.15 A partially completed ANOVA table for a completely randomized design is shown here:

Source	df	SS	MS	F
Treatments	6	18.4		
Error				
Total	41	45.2		

   a. Complete the ANOVA table.
   b. How many treatments are involved in the experiment?
   c. Do the data provide sufficient evidence to indicate a difference among the population means? Test using $\alpha = .10$.
   d. Find the approximate observed significance level for the test in part **c**, and interpret it.
   e. Suppose that $\bar{x}_1 = 3.7$ and $\bar{x}_2 = 4.1$. Do the data provide sufficient evidence to indicate a difference between μ_1 and μ_2? Assume that there are six observations for each treatment. Test using $\alpha = .10$.
   f. Refer to part **e**. Find a 90% confidence interval for $(\mu_1 - \mu_2)$.
   g. Refer to part **e**. Find a 90% confidence interval for μ_1.

10.16 See the accompanying MINITAB printout for an experiment utilizing a completely randomized design.
   a. How many treatments are involved in the experiment? What is the total sample size?
   b. Conduct a test of the null hypothesis that the treatment means are equal. Use $\alpha = .01$.

```
ANALYSIS OF VARIANCE
SOURCE     DF       SS       MS       F
FACTOR      3    57258    19086    14.80
ERROR      34    43836     1289
TOTAL      37   101094
```

Note: MINITAB uses "FACTOR" instead of "Treatments" in this printout.

   c. What additional information is needed in order to be able to compare specific pairs of treatment means?

10.17 The data in the table below resulted from an experiment that utilized a completely randomized design.

Treatment 1	Treatment 2	Treatment 3
3.9	5.4	1.3
1.4	2.0	.7
4.1	4.8	2.2
5.5	3.8	
2.3	3.5	

   a. Use the appropriate calculation formulas in Appendix C to complete the following ANOVA table:

Source	df	SS	MS	F
Treatments				
Error				
Total				

   b. Test the null hypothesis that $\mu_1 = \mu_2 = \mu_3$, where μ_i represents the true mean for treatment *i*, against the alternative that at least two of the means differ. Use $\alpha = .01$.

Applying the Concepts

10.18 Speech recognition technology has advanced to the point that it is now possible to communicate with a computer through verbal commands. A study was conducted to evaluate the value of speech recognition in human interactions with computer systems (*Special Interest Group on Computer–Human Interaction Bulletin*, July 1993.) A sample of 45 subjects was randomly divided into three groups (15 subjects per group), and each subject was asked to perform tasks on a basic voice-mail system. A different interface was employed in each group: (1) touchtone, (2) human operator, or (3) simulated speech recognition. One of the variables measured was overall time (in seconds) to perform the assigned tasks. An analysis was conducted to compare the mean overall performance times of the three groups.
   a. Identify the experimental design employed in this study.

b. The sample mean performance times for the three groups are given below. Although the sample means are different, the null hypothesis of the ANOVA could not be rejected at $\alpha = .05$. Explain how this is possible.

Group	Mean Performance Time (seconds)
Touchtone	1,400
Human operator	1,030
Speech recognition	1,040

10.19 The Minnesota Multiphasic Personality Inventory (MMPI) is a questionnaire used to gauge personality type. Several scales are built into the MMPI to assess response distortion; these include the Infrequency (I), Obvious (O), Subtle (S), Obvious-subtle (O-S), and Dissimulation (D) scales. *Psychological Assessment* (Mar. 1995) published a study that investigated the effectiveness of these MMPI scales in detecting deliberately distorted responses. A completely randomized design with four treatments was employed. The treatments consisted of independent random samples of females in the following four groups: nonforensic psychiatric patients ($n_1 = 65$), forensic psychiatric patients ($n_2 = 28$), college students who were requested to respond honestly ($n_3 = 140$), and college students who were instructed to provide "fake bad" responses ($n_4 = 45$). All 278 participants were given the MMPI and the I, O, S, O-S, and D scores were recorded for each. Each scale was treated as a response variable, and an analysis of variance was conducted. The ANOVA *F*-values are reported in the following table.

Response Variable	ANOVA *F*-Value
Infrequency (I)	155.8
Obvious (O)	49.7
Subtle (S)	10.3
Obvious-subtle (O-S)	45.4
Dissimulation (D)	39.1

a. For each response variable, determine whether the mean scores of the four groups completing the MMPI differ significantly. Use $\alpha = .05$ for each test.

b. If the MMPI is effective in detecting distorted responses, then the mean score for the "fake bad" treatment group will be largest. Based on the information provided, can the researchers make an inference about the effectiveness of the MMPI? Explain.

10.20 Organic chemical solvents are used for cleaning fabricated metal parts in industries such as aero-space, electronics, and automobiles. These solvents, when disposed of, have the potential to become hazardous waste. The *Journal of Hazardous Materials* (July 1995) published the results of a study of the chemical properties of three different types of hazardous organic solvents used to clean metal parts: aromatics, chloroalkanes, and esters. One variable studied was sorption rate, measured as mole percentage. Independent samples of solvents from each type were tested and their sorption rates were recorded, as shown in the table below. A MINITAB analysis of variance of the data is shown in the accompanying printout.

a. Construct an ANOVA table from the MINITAB printout.

b. Is there evidence of differences among the mean sorption rates of the three organic solvent types? Test using $\alpha = .10$.

Aromatics		Chloroalkanes		Esters		
1.06	.95	1.58	1.12	.29	.43	.06
.79	.65	1.45	.91	.06	.51	.09
.82	1.15	.57	.83	.44	.10	.17
.89	1.12	1.16	.43	.61	.34	.60
1.05				.55	.53	.17

Source: Reprinted from *Journal of Hazardous Materials,* Vol. 42, No. 2, J. D. Ortego *et al,* "A review of polymeric geosynthetics used in hazardous waste facilities." p. 142 (Table 9), July 1995, with kind permission of Elsevier Science-NL, Sara Burgerhartstraat 25, 1055 KV Amsterdam, The Netherlands.

```
Analysis of Variance for SORPRATE

Source    DF      SS       MS       F       P

SOLVENT    2    3.3054   1.6527   24.51   0.000
Error     29    1.9553   0.0674
Total     31    5.2607
```

10.21 Studies conducted at the University of Melbourne (Australia) indicate that there may be a difference between the pain thresholds of blonds and brunettes. Men and women of various ages were divided into four categories according to hair color: light blond, dark blond, light brunette, and dark brunette. The purpose of the experiment was to determine whether hair color is related to the amount of pain evoked by common types of mishaps and assorted types of trauma. Each person in the experiment was given a pain threshold score based on his or her performance in a pain sensitivity test (the higher the score, the higher the person's pain tolerance). SAS was used to conduct the analysis of variance calculations. The table of results and SAS printout are on page 420.

```
                       Analysis of Variance Procedure

Dependent Variable: PAIN

                                Sum of          Mean
Source                 DF       Squares         Square    F Value    Pr > F

Model                   3    1360.7263158    453.5754386    6.79     0.0041
Error                  15    1001.8000000     66.7866667
Corrected Total        18    2362.5263158

             R-Square           C.V.        Root MSE            PAIN Mean
             0.575962      17.081838       8.1723110           47.84210526

Source                 DF       Anova SS    Mean Square    F Value    Pr > F

COLOR                   3    1360.726316    453.575439       6.79     0.0041
```

HAIR COLOR

Light Blond	Dark Blond	Light Brunette	Dark Brunette
62	63	42	32
60	57	50	39
71	52	41	51
55	41	37	30
48	43		35

a. Based on the given information, what type of experimental design appears to have been employed?

b. Using the SAS printout, conduct a test to determine whether the mean pain thresholds differ among people possessing the four types of hair color. Use $\alpha = .05$.

c. What is the observed significance level for the test in part **b**? Interpret it.

d. What assumptions must be met in order to ensure the validity of the inferences you made in part **b**?

10.22 A researcher at St. Louis University conducted a study to determine whether entrepreneurs, newly hired managers, and newly promoted managers differ in their risk-taking propensities. For the purpose of this study, entrepreneurs were defined as individuals who, within 3 months prior to the study, had ceased working for their employers in order to manage their own business ventures. Thirty-one individuals of each type were selected to partici-pate in the study. Each was asked to complete a questionnaire which required the respondent to choose between a safe alternative and a more attractive but risky one. Test scores were designed to measure risk-taking propensity. (Lower scores are associated with greater conservatism in risk-taking situations.) Summary statistics for the test scores are given in the table at the bottom of the page.

a. The following partial ANOVA table is calculated from the data. Complete the table.

Source	df	SS	MS	F
Treatments		509.87		
Error		12,259.96		
Total				

b. Do the data provide sufficient evidence to indicate differences in the mean risk-taking propensities among the three groups? Test using $\alpha = .05$.

c. What assumptions must be satisfied in order for the test of part **b** to be valid?

d. Would you advise conducting tests to compare the individual pairs of means? Explain.

e. Would you classify this experiment as observational or designed? Explain.

10.23 Two major eating disorders that affect women are anorexia nervosa and bulimia. Researchers have found that patients with eating disorders have lower self-esteem than normal eaters. However, it is unclear whether the low self-esteem results directly

Group	Sample Size	Sample Mean	Standard Deviation	Group Totals
Entrepreneurs	31	71.00	11.94	2,201
Newly hired managers	31	72.52	12.19	2,248
Promoted managers	31	66.97	10.84	2,076
	93			6,525

from the eating disorder, from possible depression, or from both the eating disorder and depression. To clarify this issue, a designed experiment was conducted in which patients with eating disorders were compared to depressed female psychiatric patients having no history of eating disorder (*Journal of Abnormal Psychology,* Vol. 97, 1987). The study included three age-matched groups of women: 37 eating-disorder patients, 12 depressed patients, and 34 normal eaters. All 83 participants were administered the Bern Sex Role Inventory (BSRI), a questionnaire designed to measure self-esteem. The scores were subjected to an analysis of variance to compare the mean self-esteem ratings of the three groups of women.

a. Identify the treatments in this experiment.

b. Give both the null and alternative hypotheses of interest to the researchers.

c. Give the degrees of freedom associated with the sources of variation in the ANOVA.

d. The researchers reported the test statistic for testing the hypothesis, part **b**, as $F = 11.5$. Interpret the result. (Use $\alpha = .01$.)

10.24 A recent study in the *Journal of Psychology and Marketing* (Jan. 1992) investigates consumer attitudes toward product tampering. One variable considered was the education level of the consumer. Consumers were divided into five educational classifications and asked to rate their concern about product tampering on a scale of 1 (little or no concern) to 9 (very concerned). The education levels and the means are shown in the following table.

Education Level	Mean	Sample Size
Non–high school graduate	3.731	26
High school graduate	3.224	49
Some college completed	3.330	94
College graduate	3.167	60
Some postgraduate work	4.341	86

a. Identify the type of ANOVA design used in this experiment. Identify the treatments in this experiment.

b. The article compared the mean concern ratings for the five education levels. The F statistic for this test was reported to be 3.298. Conduct a test of hypothesis to determine whether the mean concern ratings differ for at least two of the education levels. [*Hint:* Calculate the degrees of freedom for Treatment and Error from the information given.]

10.25 An experiment is conducted to determine whether there is a difference among the mean increases in growth produced by five inoculins of growth hormones for plants. The experimental material consists of 20 cuttings of a shrub (all of equal weight), with four cuttings randomly assigned to each of the five different inoculins. The results of the experiment are given in the following table; all measurements represent an increase in weight (in grams).

INOCULIN

A	B	C	D	E
15	21	22	10	6
18	13	19	14	11
9	20	24	21	15
16	17	21	13	8

a. Given that SST = 292.80 and SS(Total) = 500.55, complete an ANOVA table for this experiment.

b. Is there evidence of a difference among the mean increases in weight for the five inoculins of growth hormone? Test using $\alpha = .05$.

c. Find the approximate observed significance level for the test in part **a**, and interpret its value.

d. What assumptions are necessary to ensure the validity of the inferences in parts **a** and **b**?

10.26 Probably the most difficult time in a child's life comes in the adjustment years of adolescence (age 13 to 17). It is during this period that the adolescent–parent relationships suffer the most. An article in the *Journal of Family Violence* (Mar. 1992) investigated the differences in the opinions of adolescents and their parents when determining who should make decisions about a variety of issues. Thirty-five adolescents and their parents were independently given a list of 44 issues common to adolescent children (e.g., curfew, homework, chores, etc.). Each was asked to state how many of the 44 issues the adolescent should make decisions about on his or her own. The results are presented below.

	WHO COMPLETED SURVEYS		
	Mother	**Father**	**Adolescent**
Mean	7.57	7.51	15.94
Std. Dev.	6.07	5.93	9.11

a. State the hypotheses necessary to determine whether parents and adolescents disagree on who should make decisions about issues common to adolescent children.

b. The article reported an F statistic of $F = 15.92$. Conduct the test using $\alpha = .05$. [*Hint:* A total of 105 people were sampled.]

10.3 MULTIPLE COMPARISONS OF MEANS

Consider a completely randomized design with three treatments, A, B, and C. Suppose we determine that the treatment means are statistically different via the ANOVA F-test of Section 10.2. To complete the analysis, we want to rank the three treatment means. As mentioned in Section 10.2, we start by placing confidence intervals on the difference between various pairs of treatment means in the experiment. In the three-treatment experiment, for example, we would construct confidence intervals for the following differences: $\mu_A - \mu_B, \mu_A - \mu_C$, and $\mu_B - \mu_C$.

In general, if there are p treatment means, there are

$$c = p(p - 1)/2$$

pairs of means that can be compared. If we want to have $100(1 - \alpha)\%$ confidence that each of the c confidence intervals contains the true difference it is intended to estimate, we must use a smaller value of α for each individual confidence interval than we would use for a single interval. For example, suppose we want to rank the means of the three treatments, A, B, and C, with 95% confidence that all three confidence intervals comparing the means contain the true differences between the treatment means. Then each individual confidence interval will need to be constructed using a level of significance smaller than $\alpha = .05$ in order to have 95% confidence that the three intervals collectively include the true differences.*

To make **multiple comparisons** of a set of treatment means we can use a number of procedures which, under various assumptions, ensure that the overall confidence level associated with all the comparisons remains at or above the specified $100(1 - \alpha)\%$ level. Three widely used techniques are the Bonferroni, Scheffé, and Tukey methods. For each of these procedures, the risk of making a Type I error applies to the comparisons of the treatment means in the experiment; thus, the value of α selected is called an **experimentwise error rate** (in contrast to a **comparisonwise error rate**).

The choice of a multiple comparisons method in ANOVA will depend on the type of experimental design used and the comparisons of interest to the analyst. For example, Tukey (1949) developed his procedure specifically for pairwise comparisons when the sample sizes of the treatments are equal. The Bonferroni method (see Miller, 1981), like the Tukey procedure, can be applied when pairwise comparisons are of interest; however, Bonferroni's method does not require equal sample sizes. Scheffé (1953) developed a more general procedure for comparing all possible linear combinations of treatment means (called *contrasts*). Consequently, when making pairwise comparisons, the confidence intervals produced by Scheffé's method will generally be wider than the Tukey or Bonferroni confidence intervals.

The formulas for constructing confidence intervals for differences between treatment means using the Tukey, Bonferroni, or Scheffé method are beyond the scope of this text. However, these procedures (and many others) are available

---

*The reason each interval must be formed at a higher confidence level than that specified for the collection of intervals can be demonstrated as follows:

$$P\{\text{At least one of } c \text{ intervals fails to contain the true difference}\}$$
$$= 1 - P\{\text{All } c \text{ intervals contain the true differences}\}$$
$$= 1 - (1 - \alpha)^c \geq \alpha$$

Thus, to make this probability of at least one failure equal to α, we must specify the individual levels of significance to be less than α.

in the ANOVA programs of most statistical software packages. The programs generate a confidence interval for the difference between two treatment means for all possible pairs of treatments, based on the experimentwise error rate (α) selected by the analyst.

EXAMPLE 10.5

Refer to the completely randomized design of Example 10.3, in which we concluded that at least two of the four brands of golf balls are associated with different mean distances traveled when struck with a driver.

a. Use Tukey's multiple comparisons procedure to rank the treatment means with an overall confidence level of 95%.
b. Estimate the mean distance traveled for balls manufactured by the brand with the highest rank.

Solution

a. To rank the treatment means with an overall confidence level of .95, we require the experimentwise error rate of $\alpha = .05$. The confidence intervals generated by Tukey's method appear at the bottom of the SAS ANOVA printout, Figure 10.11. Note that for any pair of means, μ_i and μ_j, SAS computes two confidence intervals—one for $(\mu_i - \mu_j)$ and one for $(\mu_j - \mu_i)$. Only one of these intervals is necessary to decide whether the means differ significantly.

In this example, we have $p - 4$ brand means to compare. Consequently, the number of relevant pairwise comparisons—that is, the number of nonredundant confidence intervals—is $c = 4(3)/2 = 6$. These six intervals, shaded in Figure 10.11, are given below.

Brand Comparison	Confidence Interval
$(\mu_A - \mu_B)$	$(-15.822, -4.738)$
$(\mu_A - \mu_C)$	$(-24.712, -13.628)$
$(\mu_A - \mu_D)$	$(-4.082, 7.002)$
$(\mu_B - \mu_C)$	$(-14.432, -3.348)$
$(\mu_B - \mu_D)$	$(6.198, 17.282)$
$(\mu_C - \mu_D)$	$(15.088, 26.172)$

We are 95% confident that the intervals *collectively* contain all the differences between the true brand mean distances. Note that intervals that contain 0, such as the (Brand A–Brand D) interval from −4.082 to 7.002, do not support a conclusion that the true brand mean distances differ. If both endpoints of the interval are positive, as with the (Brand B–Brand D) interval from 6.198 to 17.282, the implication is that the first brand (B) mean distance exceeds the second (D). Conversely, if both endpoints of the interval are negative, as with the (Brand A–Brand C) interval from −24.712 to −13.628, the implication is that the second brand (C) mean distance exceeds the first brand (A) mean distance.

A convenient summary of the results of the Tukey multiple comparisons is a listing of the brand means from highest to lowest, with a solid line connecting those that are *not* significantly different. This summary is shown in Figure 10.12. The interpretation is that brand C's mean distance exceeds all others; brand B's mean exceeds that of brands A and D; and the means of brands A and D do not differ significantly. All these inferences are made with 95% confidence, the overall confidence level of the Tukey multiple comparisons.

b. Brand C is ranked highest; thus, we want a confidence interval for μ_C. Since the samples were selected independently in a completely randomized design, a

FIGURE 10.11

SAS printout for
Example 10.5

```
                              Analysis of Variance Procedure
Dependent Variable: DISTANCE

                                    Sum of          Mean
Source                   DF        Squares         Square    F Value    Pr > F
Model                     3    2794.388750     931.462917      43.99    0.0001
Error                    36     762.301000      21.175028
Corrected Total          39    3556.689750

              R-Square            C.V.      Root MSE      DISTANCE Mean
              0.785671        1.785118      4.601633           257.777500


Source                   DF      Anova SS  Mean Square    F Value    Pr > F
BRAND                     3    2794.388750     931.462917     43.99    0.0001
```

```
        Tukey's Studentized Range (HSD) Test for variable: DISTANCE

       NOTE: This test controls the type I experimentwise error rate.

           Alpha= 0.05 Confidence= 0.95 df= 36 MSE= 21.17503
                Critical Value of Studentized Range= 3.809
                  Minimum Significant Difference= 5.5424

      Comparisons significant at the 0.05 level are indicated by '***'.

                                 Simultaneous              Simultaneous
                                     Lower   Difference          Upper
                       BRAND    Confidence      Between     Confidence
                  Comparison         Limit        Means          Limit

              C    - B              3.348        8.890         14.432     ***
              C    - A             13.628       19.170         24.712     ***
              C    - D             15.088       20.630         26.172     ***

              B    - C            -14.432       -8.890         -3.348     ***
              B    - A              4.738       10.280         15.822     ***
              B    - D              6.198       11.740         17.282     ***

              A    - C            -24.712      -19.170        -13.628     ***
              A    - B            -15.822      -10.280         -4.738     ***
              A    - D             -4.082        1.460          7.002

              D    - C            -26.172      -20.630        -15.088     ***
              D    - B            -17.282      -11.740         -6.198     ***
              D    - A             -7.002       -1.460          4.082
```

confidence interval for an individual treatment mean is obtained with the one-sample t confidence interval of Section 7.2, using the standard deviation, $s = \sqrt{\text{MSE}}$, as the measure of sampling variability for the experiment. A 95% confidence interval on the mean distance traveled by brand C (apparently the "longest ball" of those tested), is

$$\bar{x}_C \pm t_{.025} s \sqrt{1/n}$$

BRAND	MEAN
C	270.0
B	261.1
A	250.8
D	249.3

FIGURE 10.12

Summary of Tukey
Multiple Comparisons

where $n = 10$, $t_{.025} \approx 2$ (based on 36 degrees of freedom), and $s = 4.601633$ (obtained from the SAS printout, Figure 10.11). Substituting, we obtain

$$270.0 \pm (2)(4.60)(\sqrt{.1})$$

$$270.0 \pm 2.9 \text{ or } (267.1, 272.9)$$

Thus, we are 95% confident that the true mean distance traveled for brand C is between 267.1 and 272.9 yards, when hit with a driver by Iron Byron. ▲

FIGURE 10.13

SAS Bonferroni Analysis
for four golf ball brands

```
                        Analysis of Variance Procedure

                  Bonferroni (Dunn) T tests for variable: DISTANCE

        NOTE: This test controls the type I experimentwise error rate, but
              generally has a higher type II error rate than REGWQ.

                     Alpha= 0.05 df= 36 MSE= 21.17503
                          Critical Value of T= 2.79
                     Minimum Significant Difference= 5.7456

              Means with the same letter are not significantly different.

                    Bon Grouping        Mean      N   BRAND
                            A         269.950     10   C

                            B         261.060     10   B

                            C         250.780     10   A
                            C
                            C         249.320     10   D
```

Many of the available statistical software packages that have multiple comparisons routines will also produce the rankings shown in Figure 10.12. For example, the SAS printout in Figure 10.13 displays the ranking of the mean distances for the four golf ball brands using Bonferroni's method.

The experimentwise error rate (.05), MSE value (21.175), and critical t value (2.79) used in the analysis are shaded on the printout. The Bonferroni rankings of the means are displayed at the bottom of Figure 10.13. Instead of a solid line, note that SAS uses a **Bon Grouping** letter to connect means that are not significantly different. You can see that the mean for Brand C is ranked highest, followed by the mean for Brand B. These two brand means are significantly different since they are associated with a different Bon Grouping letter. Brands A and D are ranked lowest and their means are not significantly different (since they have the same Bon Grouping letter).

Remember that the Tukey and Bonferroni methods are just two of numerous multiple comparisons procedures available. Another technique may be more appropriate for the experimental design you employ. Consult the references for details on these other methods and when they should be applied.

Can Therapy Help Binge Eaters?

Do you experience episodes of excessive eating accompanied by being overweight? If so, you may suffer from binge eating disorder (BED). Clinical psychologists estimate that between 25% and 35% of overweight patients are binge eaters. Cognitive-behavioral therapy (CBT), in which patients are taught how to make changes in specific behavior patterns, can be effective in treating BED. For example, CBT is used to teach binge eaters to weigh themselves each week, adopt an exercise program, eat only low-fat foods, and monitor their thoughts and moods before binge eating. However, only

CASE
STUDY
• 10.1 •

about half of BED patients treated with CBT actually stop binge eating. Consequently, binge eaters who do not respond to CBT need a second level of treatment.

A group of Stanford University researchers investigated the effectiveness of interpersonal therapy (IPT) as a second level of treatment for binge eaters. In IPT, patients meet in a group format to discuss the source of their binge eating (e.g., depression, personal problems) and how to deal with the problem. The 24-week study, published in the *Journal of Consulting and Clinical Psychology* (June

continued

Case Study 10.1 continued

1995), used as subjects 41 overweight individuals diagnosed with BED.

The subjects were selected from a pool of 260 individuals who either were referred to the researchers or responded to advertisements placed in the local media. Some were excluded from the study because they lacked interest or time, others were excluded because they did not meet the criteria for being both overweight and a binge eater. Sixty-four subjects were deemed eligible for the study, and all agreed to participate; however, 14 did not complete a baseline assessment and 9 dropped out in the middle of the study, leaving a final sample size of $n = 41$.

The researchers employed a design that randomly assigned subjects to either a treatment group (30 subjects) or a control group (11 subjects). Subjects in the treatment group received 12 weeks of cognitive-behavior therapy. At the end of the 12 weeks, the 30 subjects who underwent CBT were subdivided into two groups. Those who responded successfully to CBT by losing weight, limiting binge eating, and establishing a minimum exercise program (17 subjects) were assigned to a weight loss therapy (WLT) program for the next 12 weeks. Those CBT subjects who did not respond to treatment (13 subjects), received 12 weeks of IPT. The 11 subjects in the control group received no therapy of any type, but were assessed at the end of the 12-week and 24-week periods. Thus, the study ultimately consisted of three groups of overweight BED subjects: the CBT-WLT group, the CBT-IPT group, and the control group.

The researchers measured several outcome (response) variables for each subject, including the number of binge eating episodes per week and quantity of weight lost (in kilograms). Summary statistics for each of the three groups at the end of the 24-week period are shown in Table 10.4.

One way to analyze the data is to treat the design as completely randomized with three treatments (CBT-WLT, CBT-IPT, and Control), then conduct an analysis of variance for each response variable. Although the ANOVA tables were not provided in the article, sufficient information is provided in Table 10.4 to reconstruct them.

Focus

a. Consider number of binges per week as the response variable, x. Compute SST for the ANOVA on number of binges per week using the formula on page 408:

$$SST = \sum_{i=1}^{3} n_i(\bar{x}_i - \bar{x})^2$$

where $\bar{x}$ is the overall mean number of binges per week of all 41 subjects. [*Hint:* $\bar{x} = \left(\sum_{i=1}^{3} n_i\bar{x}_i \right) / 41.$]

b. Again, consider the number of binges per week as the response variable. Recall that SSE for the ANOVA can be written as

$$SSE = (n_1 - 1)s_1^2 + (n_2 - 1)s_2^2 + (n_3 - 1)s_3^2$$

where s_1^2, s_2^2, and s_3^2 are the sample variances associated with the three treatments. Compute SSE for the ANOVA.

c. Refer to parts **a** and **b**. Complete the ANOVA table below.

Source	df	SS	MS	F
Treatments	—	—	—	—
Error	—	—	—	
Total	—	—		

d. Is there sufficient evidence (at $\alpha = .01$) of differences among the mean number of binges experienced per week by the subjects in the three groups?

e. Refer to part **d**. If appropriate, conduct a Bonferroni analysis to rank the three treatment means. Use an experimentwise error rate of $\alpha = .03$. Interpret the results for the researchers. [*Hint:* The Bonferroni formula for a confidence interval for the difference $(\mu_i - \mu_j)$ is:

$$(\bar{x}_i - \bar{x}_j) \pm t_{\alpha*/2}(s)\sqrt{(1/n_i) + (1/n_j)}$$

where $\alpha* = 2\alpha/[(p)(p-1)]$, α is the experimentwise error rate, and p is the total number of treatment means compared.]

f. Repeat parts a–e for the response variable, weight.

g. Comment on the validity of the ANOVA assumptions. How might this affect the results of the study?

h. Comment on the use of a completely randomized design for this study. Have subjects been randomly and independently assigned to each group? How might this impact the results of the study?

i. Comment on the manner in which subjects were selected for the study. Will the sample of 41 subjects adequately represent the target population? Explain the impact of this on the inferences derived from the study.

TABLE 10.4   Summary Data Table for Three Groups of Binge Eaters

	CBT-WLT	CBT-IPT	Control
Sample size	17	13	11
Mean number of binges per week	0.2	1.9	2.9
(Standard deviation)	(0.4)	(1.7)	(2.0)
Mean weight	98.8	119.2	110.2
(Standard deviation)	(22.3)	(32.1)	(22.8)

Source: Agras, W. S., *et al.* "Does interpersonal therapy help patients with binge eating disorder who fail to respond to cognitive-behavioral therapy?" *Journal of Consulting and Clinical Psychology,* Vol. 63, No. 3, June 1995, p. 358 (Tables 1 and 2).

EXERCISES 10.27–10.57

Learning the Mechanics

10.27 Consider a completely randomized design with p treatments. Assume all pairwise comparisons of treatment means are to be made using a multiple comparisons procedure. Determine the total number of treatment means to be compared for the following values of p.
  a. $p = 3$ **b.** $p - 5$ **c.** $p - 4$ **d.** $p = 10$

10.28 Define an experimentwise error rate.

10.29 Define a comparisonwise error rate.

10.30 Consider a completely randomized design with 5 treatments, A, B, C, D, and E. The ANOVA F-test revealed significant differences among the means. A multiple comparisons procedure was used to compare all possible pairs of treatment means at $\alpha = .05$. The ranking of the 5 treatment means is summarized below. Identify which pairs of means are significantly different.

  a. $\overline{A \quad C} \quad \overline{E \quad B} \quad D$

  b. $\overline{A \quad C \quad E} \quad B \quad D$

  c. $A \quad \overline{C \quad E} \quad B \quad D$

  d. $\overline{A \quad C} \quad \overline{E \quad B} \quad D$

Applying the Concepts

10.31 Refer to the *Journal of Communication Disorders* (Mar. 1995) study of pronoun misusage by specifically language-impaired (SLI) children, Exercise 2.44. Recall that percentage of pronoun errors were recorded for three groups of children: ten 5-year-old SLI children, ten younger (3-year old) normally developing (YND) children, and ten older (5-year old) normally developing (OND) children. A completely randomized design ANOVA with multiple comparisons of means was conducted on the data, producing the results shown below. Interpret these results.

Source	df	SS	MS	F	p-value
Groups	2	11,292.45	5,646.22	8.295	.0016
Error	27	18,377.65	680.65		
Total	29	29,670.10			

Mean percentage of errors:	0.00	30.17	46.88
Group:	OND	SLI	YND

Source: Reprinted by permission of the publisher from Moore, M. E. "Error analysis of pronouns by normal and language-impaired children." *Journal of Communication Disorders,* Vol. 28, No. 1, Mar. 1995, p. 67 (Table 6). Copyright 1995 by Elsevier Science, Inc.

10.32 Refer to the *Psychological Assessment* (Mar. 1995) study of the effectiveness of the Minnesota Multiphasic Personality Inventory (MMPI) in detecting distorted responses, Exercise 10.19. You conducted an ANOVA on each of five MMPI scales and determined that the mean scores of four groups of females differed significantly for each scale. Recall that the four groups were nonforensic psychiatric patients (NFP), forensic psychiatric patients (FP), college students who responded honestly (CSH), and college students who "faked bad" (CSFB) responses. The Bonferroni method was used to rank the means of the four groups using an experimentwise error rate of $\alpha = .05$. The results for each response variable are shown in the next table. [*Note:* Overbars connect means that are not significantly different.] Interpret the results for each response variable. Recall that the MMPI will be deemed effective in detecting distorted responses if the mean score of the "fake bad" group is largest.

10.33 Refer to the *Journal of Hazardous Materials* (July 1995) study of mean sorption rates for three types of organic solvents, Exercise 10.20. SAS was used to produce Bonferroni confidence intervals for all

Response Variable

Infrequency (F)	Mean	7.1	11.3	14.6	33.6
	Group	CSH	NFP	FP	CSFB
Obvious (O)	Mean	240.9	270.7	287.5	341.9
	Group	CSH	FP	NFP	CSFB
Subtle (S)	Mean	231.4	244.1	244.5	259.7
	Group	CSFB	CSH	FP	NFP
Obvious-subtle (O-S)	Mean	51.3	88.2	93.4	198.8
	Group	CSH	NFP	FP	CSFB
Dissimulation (D)	Mean	12.0	12.1	20.8	21.0
	Group	FP	NFP	CSH	CSFB

Source: Bagby, R. M., Burs, T., and Nicholson, R. A. "Relative effectiveness of the standard validity scales in detecting fake-bad and fake-good responding: Replication and extension." *Psychological Assessment,* Vol. 7, No. 1, Mar. 1995, p. 86 (Table 1). Copyright 1995 by the American Psychological Association. Reprinted with permission.

pairs of treatment means. The SAS printout is shown below.

a. Find and interpret the experimentwise error rate shown on the printout.

b. Locate and interpret the confidence interval for the difference between the mean sorption rates of aromatics and esters.

c. Use the confidence intervals to determine which pairs of treatment means are significantly different.

10.34 Refer to the *Journal of Abnormal Psychology* study of self-esteem, eating disorders, and depression in women described in Exercise 10.23. Tukey's multiple comparisons procedure was conducted on the three group means. The ranked means are listed, with overbars connecting the means that are not significantly different at $\alpha = .01$. Interpret the results.

Group	Depression	Eating Disorders	Normal
Mean self-esteem rating	3.6	3.9	4.9

10.35 *Research Quarterly* published a study that investigated the strength and endurance of women athletes. A completely randomized design was used to compare the maximum grip strength (in kilograms) of women athletes who engaged in eight different sports. An analysis of variance, with the eight sports as the treatments, yielded the results in the ANOVA table on page 429.

a. Is there sufficient evidence (at $\alpha = .05$) to indicate a difference in mean maximum grip strength among the eight sports?

```
                Analysis of Variance Procedure

           Bonferroni (Dunn) T tests for variable: SORPRATE

NOTE: This test controls the type I experimentwise error rate but
      generally has a higher type II error rate than Tukey's for all
      pairwise comparisons.

      Alpha= 0.05  Confidence= 0.95  df= 29  MSE= 0.067426
               Critical Value of T= 2.54091
   Comparisons significant at the 0.05 level are indicated by '***'.

                         Simultaneous              Simultaneous
                            Lower    Difference       Upper
              SOLVENT     Confidence   Between      Confidence
            Comparison      Limit       Means         Limit

         CHLOR - AROMA    -0.2566      0.0640        0.3846
         CHLOR - ESTER     0.3874      0.6763        0.9651     ***

         AROMA - CHLOR    -0.3846     -0.0640        0.2566
         AROMA - ESTER     0.3340      0.6122        0.8904     ***

         ESTER - CHLOR    -0.9651     -0.6763       -0.3874     ***
         ESTER - AROMA    -0.8904     -0.6122       -0.3340     ***
```

Mean Education Level	3.167 College graduate	3.224 High school graduate	3.330 Some college	3.731 Non–high school graduate	4.3441 Post-graduate

Source	df	SS	MS	F
Treatments	7	511.50	73.07	3.15
Error	42	974.88	23.21	

b. The sample mean grip strengths for the eight sports are given below. Note that the largest (strongest) mean grip strengths were associated with women in tennis and golf, whereas the

Sport	Mean Maximum Strength (kilograms)
Tennis ($n_1 = 5$)	45.44
Golf ($n_2 = 6$)	42.74
Basketball ($n_3 = 7$)	40.68
Field hockey ($n_4 = 7$)	39.16
Gymnastics ($n_5 = 5$)	39.14
Volleyball ($n_6 = 6$)	38.37
Track and field ($n_7 = 8$)	37.26
Swimming ($n_8 = 6$)	33.38

Source: Heyward, V. and McCreary, L. "Analysis of the static strength and relative endurance of women athletes." *Research Quarterly,* Vol. 48, 1977, pp. 703–709.

smallest was associated with women in swimming. (The researchers attribute this result to the extensive use of forearm and hand muscles in tennis and golf.) The means are ranked statistically at $\alpha = .15$ with a multiple comparisons procedure. Interpret the results.

10.36 Refer to the *Journal of Psychology and Marketing* study, Exercise 10.24. Using $\alpha = .05$, a Bonferroni analysis yielded the result shown at the top of the page at $\alpha = .05$. Interpret the result.

10.37 Refer to the *Journal of Family Violence* study, Exercise 10.26. Recall that the mean number of decisions the adolescent should make on his/her own differed for surveys completed by mothers, fathers, and adolescents. The Bonferroni technique was used to compare the pairs of means at an overall significance level of $\alpha = .06$. The resulting confidence intervals are shown below. Interpret the intervals.

Comparison	Confidence Interval
(Mother − Father)	$.06 \pm 4.06$
(Mother − Adolescent)	-8.37 ± 4.06
(Father − Adolescent)	-8.43 ± 4.06

10.4   THE RANDOMIZED BLOCK DESIGN

If the completely randomized design results in nonrejection of the null hypothesis that the treatment means differ because the sampling variability (as measured by MSE) is large, we may want to consider an experimental design that better controls the variability. In contrast to the selection of independent samples of experimental units specified by the completely randomized design, the *randomized block design* utilizes experimental units that are *matched sets,* assigning one from each set to each treatment. The matched sets of experimental units are called *blocks.* The theory behind the randomized block design is that the sampling variability of the experimental units in each block will be reduced, in turn reducing the measure of error, MSE.

DEFINITION 10.8

The **randomized block design** consists of a two-step procedure:

1. Matched sets of experimental units, called **blocks**, are formed, each block consisting of p experimental units (where p is the number of treatments). The b blocks should consist of experimental units that are as similar as possible.
2. One experimental unit from each block is randomly assigned to each treatment, resulting in a total of $n = bp$ responses.

For example, if we wish to compare SAT scores of female and male high school seniors, we could select independent random samples of five females and five males, and analyze the results of the completely randomized design as outlined in Section 10.2. Or, we could select matched pairs of females and males according to their scholastic records, and analyze the SAT scores of the pairs. For instance, we could select pairs of students with approximately the same GPAs from the same high school. Five such pairs (blocks) are depicted in Table 10.5. Note that this is just a *paired difference experiment*, first discussed in Section 9.2.

As before, the variation between the treatment means is measured by squaring the distance between each treatment mean and the overall mean, multiplying each squared distance by the number of measurements for the treatment, and summing over treatments:

$$
\begin{aligned}
\text{SST} &= \sum_{i=1}^{p} b(\bar{x}_{T_i} - \bar{x})^2 \\
&= 5(606 - 600)^2 + 5(594 - 600)^2 = 360
\end{aligned}
$$

where $\bar{x}_{T_i}$ represents the sample mean for the ith treatment, b (the number of blocks) is the number of measurements for each treatment, and p is the number of treatments.

The blocks also account for some of the variation among the different responses. That is, just as SST measures the variation between the female and male means, we can calculate a measure of variation among the five block means representing different schools and scholastic abilities. Analogous to the computation of SST, we sum the squares of the differences between each block mean and the overall mean, multiplying each squared difference by the number of measurements for each block, and sum over blocks to calculate the **Sum of Squares for Blocks** (SSB):

$$
\begin{aligned}
\text{SSB} &= \sum_{i=1}^{b} p(\bar{x}_{B_i} - \bar{x})^2 \\
&= 2(535 - 600)^2 + 2(560 - 600)^2 + 2(585 - 600)^2 + \\
&\quad\ 2(630 - 600)^2 + 2(690 - 600)^2 \\
&= 30{,}100
\end{aligned}
$$

where $\bar{x}_{B_i}$ represents the sample mean for the ith block and p (the number of treatments) is the number of measurements in each block. As we expect, the variation in SAT scores attributable to Schools and Levels of scholastic achievement is apparently large.

As before, we want to compare the variability attributed to treatments with that which is attributed to sampling variability. In a randomized block design, the sampling variability is measured by subtracting that portion attributed to treatments

TABLE 10.5 Randomized Block Design: SAT Score Comparison

Block	Female SAT Score	Male SAT Score	Block Mean
1 (School A, 2.75 GPA)	540	530	535
2 (School B, 3.00 GPA)	570	550	560
3 (School C, 3.25 GPA)	590	580	585
4 (School D, 3.50 GPA)	640	620	630
5 (School E, 3.75 GPA)	690	690	690
Treatment Mean	606	594	

FIGURE 10.14

Partitioning of the Total Sum of Squares for the randomized block design

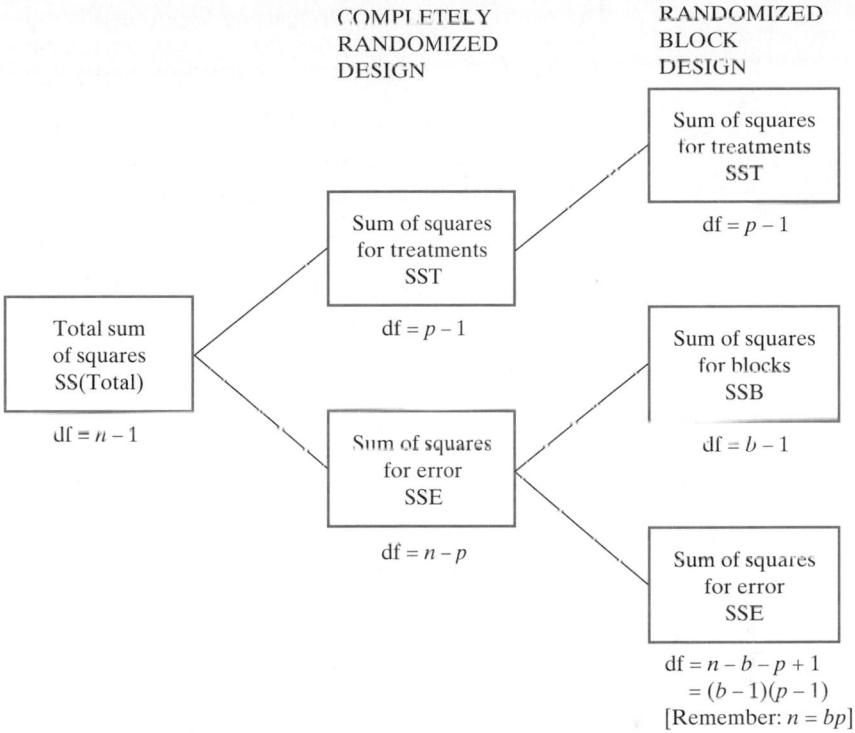

and blocks from the Total Sum of Squares, SS(Total). The total variation is the sum of squared differences of each measurement from the overall mean:

$$SS(Total) = \sum_{i=1}^{n}(x_i - \bar{x})^2$$
$$= (540 - 600)^2 + (530 - 600)^2 + (570 - 600)^2 + (550 - 600)^2 + \cdots + (690 - 600)^2$$
$$= 30,600$$

Then the variation attributable to sampling error is found by subtraction:

$$SSE = SS(Total) - SST - SSB = 30,600 - 360 - 30,100 = 140$$

In summary, the Total Sum of Squares, 30,600, is divided into three components: 360 attributed to treatments (Gender), 30,100 attributed to blocks (Scholastic ability), and 140 attributed to sampling error.

The mean squares associated with each source of variability are obtained by dividing the sum of squares by the appropriate number of degrees of freedom. The partitioning of the Total Sum of Squares and the total degrees of freedom for a randomized block experiment is summarized in Figure 10.14.

To determine whether we can reject the null hypothesis that the treatment means are equal in favor of the alternative that at least two of them differ, we calculate

$$MST = \frac{SST}{p - 1} = \frac{360}{2 - 1} = 360$$

$$MSE = \frac{SSE}{n - b - p + 1} = \frac{140}{10 - 5 - 2 + 1} = 35$$

The F-ratio that is used to test the hypothesis is

$$F = \frac{360}{35} = 10.29$$

Comparing this ratio to the tabled F value corresponding to $\alpha = .05$, $v_1 = (p - 1) = 1$ degree of freedom in the numerator, and $v_2 = (n - b - p + 1) = 4$ degrees of freedom in the denominator, we find that

$$F = 10.29 > F_{.05} = 7.71$$

which indicates that we should reject the null hypothesis and conclude that the mean SAT scores differ for females and males.

If you review Section 9.2, you will find that the analysis of a paired difference experiment results in a one-sample t-test on the differences between the treatment responses within each block. Applying the procedure to the differences between female and male scores in Table 10.5, we find

$$t = \frac{\bar{x}_D}{s_D / \sqrt{n_D}} = \frac{12}{\sqrt{70} / \sqrt{5}} = 3.207$$

At the .05 level of significance with $(n_D - 1) = 4$ degrees of freedom,

$$t = 3.21 > t_{.025} = 2.776$$

Since $t^2 = (3.207)^2 = 10.29$ and $t^2_{.025} = (2.776)^2 = 7.71$, we find that the paired difference t-test and the ANOVA F-test are equivalent, with both the calculated test statistics and the rejection region related by the formula $F = t^2$. The difference between the tests is that the paired difference t-test can be used to compare only two treatments in a randomized block design, whereas the F-test can be applied to *two or more* treatments in a randomized block design. The F-test is summarized in the following box.

Test to Compare p Treatment Means: Randomized Block Design

$$H_0: \quad \mu_1 = \mu_2 = \cdots = \mu_p$$
$$H_a: \quad \text{At least two treatment means differ}$$

Test statistic: $\quad F = \dfrac{\text{MST}}{\text{MSE}}$

Rejection region: $\quad F > F_\alpha$, where F_α is based on $(p - 1)$ numerator degrees of freedom and $(n - b - p + 1)$ denominator degrees of freedom.

Assumptions: 1. The probability distributions of observations corresponding to all the block–treatment combinations are normal.
2. The variances of all the probability distributions are equal.

Note that the assumptions concern the probability distributions associated with each block–treatment combination. The experimental unit selected for each combination is assumed to have been randomly selected from all possible experimental units for that combination, and the response is assumed to be normally distributed with the same variance for each of the block–treatment combinations. For example, the F-test comparing female and male SAT score means requires the scores for each combination of gender and scholastic ability (e.g., females with 3.25 GPA) to be normally distributed with the same variance as the other combinations employed in the experiment.

The calculation formulas for randomized block designs are given in Appendix C. We will rely on statistical software packages to analyze randomized block

FIGURE 10.15

Illustration of completely
randomized design and
randomized block design:
Comparison of four golf
ball brands

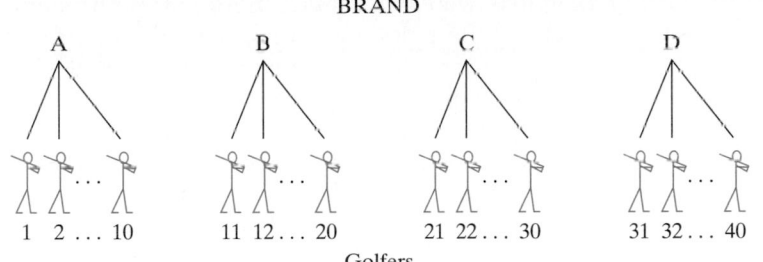

BRAND

a. Completely randomized design

b. Randomized block design

designs and to obtain the necessary ingredients for testing the null hypothesis
that the treatment means are equal.

EXAMPLE 10.6

Refer to Examples 10.3–10.5. Suppose the USGA wants to compare the mean distances associated with the four brands of golf balls when struck by a driver, but wishes to employ human golfers rather than the robot, Iron Byron. Assume that 10 balls of each brand are to be utilized in the experiment.

a. Explain how a completely randomized design could be employed.
b. Explain how a randomized block design could be employed.
c. Which design is likely to provide more information about the differences among the brand mean distances?

Solution

a. Since the completely randomized design calls for independent samples, we can employ such a design by randomly selecting 40 golfers and then randomly assigning 10 golfers to each of the four brands. Finally, each golfer will strike the ball of the assigned brand, and the distance will be recorded. The design is illustrated in Figure 10.15a.
b. The randomized block design employs blocks of relatively homogeneous experimental units. For example, we could randomly select 10 golfers, and permit each golfer to hit four balls, one of each brand, in a random sequence. Then each golfer is a block, with each treatment (brand) assigned to each block (golfer). The design is summarized in Figure 10.15b.

c. Because we expect much more variability among distances generated by "real" golfers than by Iron Byron, we would expect the randomized block design to control the variability better than the completely randomized design. That is, with 40 different golfers, we would expect the sampling variability among the measured distances within each brand to be greater than that among the four distances generated by each of 10 golfers hitting one ball of each brand. ▲

EXAMPLE 10.7 Refer to Example 10.6. Suppose the randomized block design of part **b** is employed, utilizing a random sample of 10 golfers, with each golfer using a driver to hit four balls, one of each brand, in a random sequence.

a. Set up a test of the research hypothesis that the brand mean distances differ. Use $\alpha = .05$.
b. The data for the experiment are given in Table 10.6. Use statistical software to analyze the data, and conduct the test set up in part **a**.

Solution

a. We want to test whether the data in Table 10.6 provide sufficient evidence to conclude that the brand mean distances differ. Denoting the population mean of the ith brand by μ_i, we test

$$H_0: \mu_1 = \mu_2 = \mu_3 = \mu_4$$

$$H_a: \text{The mean distances differ for at least two of the brands}$$

The test statistic compares the variation among the four treatment (brand) means to the sampling variability within each of the treatments.

Test statistic: $\quad F = \dfrac{\text{MST}}{\text{MSE}}$

Rejection region: $\quad F > F_\alpha = F_{.05}$, with $v_1 = (p - 1) = 3$ numerator degrees of freedom and $v_2 = (n - p - b + 1) = 27$ denominator degrees of freedom. From Table IX of Appendix A, we find $F_{.05} = 2.96$. Thus, we will reject H_0 if $F > 2.96$.

The assumptions necessary to ensure the validity of the test are as follows: (1) The probability distributions of the distances for each brand–golfer combination

TABLE 10.6 Distance Data for Randomized Block Design

Golfer (Block)	Brand A	Brand B	Brand C	Brand D
1	202.4	203.2	223.7	203.6
2	242.0	248.7	259.8	240.7
3	220.4	227.3	240.0	207.4
4	230.0	243.1	247.7	226.9
5	191.6	211.4	218.7	200.1
6	247.7	253.0	268.1	244.0
7	214.8	214.8	233.9	195.8
8	245.4	243.6	257.8	227.9
9	224.0	231.5	238.2	215.7
10	252.2	255.2	265.4	245.2
Sample Means	227.0	233.2	245.3	220.7

FIGURE 10.16

MINITAB printout for randomized block design: Golf ball brand comparison

```
Analysis of Variance for DISTANCE

Source    DF       SS       MS       F       P
BRAND      3    3298.7   1099.6   54.31   0.000
GOLFER     9   12073.9   1341.5   66.26   0.000
Error     27     546.6     20.2
Total     39   15919.2
```

are normal. (2) The variances of the distance probability distributions for each brand–golfer combination are equal.

b. MINITAB was used to analyze the data in Table 10.6, and the result is shown in Figure 10.16.

The values of MST and MSE (shaded on the printout) are 1,099.6 and 20.2, respectively. The F-ratio for Brand (also shaded on the printout) is $F = 54.31$, which exceeds the tabled value of 2.96. We therefore reject the null hypothesis at the $\alpha = .05$ level of significance, concluding that at least two of the brands differ with respect to mean distance traveled when struck by the driver. (This result is confirmed by noting that the observed significance level of the test, shaded on the printout, is $p \approx 0$.) ▲

The results of an ANOVA can be summarized in a simple tabular format, similar to that utilized for the completely randomized design in Section 10.2. The general form of the table is shown in Table 10.7, and that for Example 10.7 is given in Table 10.8. Note that the randomized block design is characterized by three sources of variation—Treatments, Blocks, and Error—which sum to the Total Sum of Squares. We hope that employing blocks of experimental units will reduce the error variability, thereby making the test for comparing treatment means more powerful.

When the F-test results in the rejection of the null hypothesis that the treatment means are equal, we will usually want to compare the various pairs of treatment means to determine which specific pairs differ. We can employ a multiple comparisons procedure as in Section 10.3. The number of pairs of means to be

TABLE 10.7 General ANOVA Summary Table for a Randomized Block Design

Source	df	SS	MS	F
Treatment	$p - 1$	SST	MST	MST/MSE
Block	$b - 1$	SSB	MSB	
Error	$n - p - b + 1$	SSE	MSE	
Total	$n - 1$	SS(Total)		

TABLE 10.8 ANOVA Table for Example 10.7

Source	df	SS	MS	F	p
Treatment (Brand)	3	3,298.7	1,099.6	54.31	.000
Block (Golfer)	9	12,073.9	1,341.5		
Error	27	546.6	20.2		
Total	39	15,919.2			

FIGURE 10.17

SAS printout of Bonferroni
confidence intervals for
randomized block design

```
                          Analysis of Variance Procedure

                  Bonferroni (Dunn) T tests for variable: DISTANCE

NOTE: This test controls the type I experimentwise error rate but
      generally has a higher type II error rate than Tukey's for all
      pairwise comparisons.

          Alpha= 0.05  Confidence= 0.95  df= 27  MSE= 20.24521
                      Critical Value of T= 2.84690
                  Minimum Significant Difference= 5.7286

     Comparisons significant at the 0.05 level are indicated by '***'.

                             Simultaneous             Simultaneous
                                Lower     Difference      Upper
                    BRAND     Confidence   Between     Confidence
                  Comparison    Limit       Means        Limit

                 C   - B        6.421      12.150       17.879     ***
                 C   - A       12.551      18.280       24.009     ***
                 C   - D       18.871      24.600       30.329     ***

                 B   - C      -17.879     -12.150       -6.421     ***
                 B   - A        0.401       6.130       11.859     ***
                 B   - D        6.721      12.450       18.179     ***

                 A   - C      -24.009     -18.280      -12.551     ***
                 A   - B      -11.859      -6.130       -0.401     ***
                 A   - D        0.591       6.320       12.049     ***

                 D   - C      -30.329     -24.600      -18.871     ***
                 D   - B      -18.179     -12.450       -6.721     ***
                 D   - A      -12.049      -6.320       -0.591     ***
```

compared will again be $c = p(p - 1)/2$, where p is the number of treatment means. In Example 10.7, $c = 4(3)/2 = 6$; that is, there are six pairs of golf ball brand means to be compared.

EXAMPLE 10.8

Bonferroni's procedure is used to compare the mean distances of the four golf ball brands in Example 10.7. The resulting confidence intervals, with an experiment-wise error rate of $\alpha = .05$, are shown in the SAS printout, Figure 10.17. Interpret the results.

Solution

Note that 12 confidence intervals are shown at the bottom of Figure 10.17 rather than 6. As we saw in Example 10.5, SAS computes intervals for both $\mu_i - \mu_j$ and $\mu_j - \mu_i$, $i \neq j$. Only half of these are necessary to conduct the analysis, and these are shaded on the printout. The intervals are summarized below:

$$(A - B): \quad (-11.8, -.4)$$
$$(A - C): \quad (-24.0, -12.6)$$
$$(A - D): \quad (.6, 12.0)$$
$$(B - C): \quad (-17.9, -6.4)$$
$$(B - D): \quad (6.7, 18.2)$$
$$(C - D): \quad (18.9, 30.3)$$

Note that we are 95% confident that all the brand means differ because none of the intervals contains 0. The listing of the brand means in Figure 10.18 has no

Brand	Mean
C	245.3
B	233.2
A	227.0
D	220.7

FIGURE 10.18

Listing of brand means for randomized block design

[*Note:* All differences are statistically significant.]

lines connecting them because there are no nonsignificant differences at the .05 level.　▲

Unlike the completely randomized design, the randomized block design cannot, in general, be used to estimate individual treatment means. Whereas the completely randomized design employs a random sample for each treatment, the randomized block design does not necessarily employ a random sample of experimental units for each treatment. The experimental units within the blocks are assumed to be randomly selected, but the blocks themselves may not be randomly selected.

We can, however, test the hypothesis that the block means are significantly different. We simply compare the variability attributable to differences among the block means to that associated with sampling variability. The ratio of MSB to MSE is an F-ratio similar to that formed in testing treatment means. The F statistic is compared to a tabled value for a specific value of α, with numerator degrees of freedom $(b - 1)$ and denominator degrees of freedom $(n - p - b + 1)$. The test is usually given on the same printout as the test for treatment means. Refer to the MINITAB printout in Figure 10.16, and note that the test statistic for comparing the block means is

$$F = \frac{\text{MSB}}{\text{MSE}} = \frac{\text{MS(Golfers)}}{\text{MS(Error)}} = \frac{1{,}341.5}{20.2} = 66.26$$

with a p-value of .000. Since $\alpha = .05$ exceeds this p-value, we conclude that the block means are different. The results of the test are summarized in Table 10.9.

In the golf example, the test for block means confirms our suspicion that the golfers vary significantly; therefore, the use of the block design was a good decision. However, be careful not to conclude that the block design was a mistake if the F-test for blocks does not result in rejection of the null hypothesis that the block means are the same. Remember that the possibility of a Type II error exists, and we are not controlling its probability as we are the probability α of a Type I error. If the experimenter believes that the experimental units are more homogeneous within blocks than between blocks, then he or she should use the randomized block design regardless of the results of a single test comparing the block means.

The procedure for conducting an analysis of variance for a randomized block design is summarized in the next box. Remember that the hallmark of this design is the utilization of blocks of homogeneous experimental units in which each treatment is represented.

TABLE 10.9　ANOVA Table for Randomized Block Design: Test for Blocks Included

Source	df	SS	MS	*F*	*p*
Treatments (Brands)	3	3,298.7	1,099.6	54.31	.000
Blocks (Golfers)	9	12,073.9	1,341.5	66.26	.000
Error	27	546.6	20.2		
Total	39	15,919.2			

Steps for Conducting an ANOVA for a Randomized Block Design

1. Be sure the design consists of blocks (preferably, blocks of homogeneous experimental units) and that each treatment is randomly assigned to one experimental unit in each block.
2. If possible, check the assumptions of normality and equal variances for all block–treatment combinations. [*Note:* This may be difficult to do since the design will likely have only one observation for each block–treatment combination.]
3. Create an ANOVA summary table that specifies the variability attributable to Treatments, Blocks, and Error, and that leads to the calculation of the F statistic to test the null hypothesis that the treatment means are equal in the population. Use a statistical software package or the calculation formulas in Appendix C to obtain the necessary numerical ingredients.
4. If the F-test leads to the conclusion that the means differ, use the Bonferroni, Tukey, or similar procedure to conduct multiple comparisons of as many of the pairs of means as you wish. Use the results to summarize the statistically significant differences among the treatment means. Remember that, in general, the randomized block design cannot be used to form confidence intervals for individual treatment means.
5. If the F-test leads to the nonrejection of the null hypothesis that the treatment means are equal, several possibilities exist:
    a. The treatment means are equal—that is, the null hypothesis is true.
    b. The treatment means really differ, but other important factors affecting the response are not accounted for by the randomized block design. These factors inflate the sampling variability, as measured by MSE, resulting in smaller values of the F statistic. Either increase the sample size for each treatment, or conduct an experiment that accounts for the other factors affecting the response (as in Section 10.5). Do not automatically reach the former conclusion, since the possibility of a Type II error must be considered if you accept H_0.
6. If desired, conduct the F-test of the null hypothesis that the block means are equal. Rejection of this hypothesis lends statistical support to the utilization of the randomized block design.

Note: As stated in the box, it is often difficult to check whether the assumptions for a randomized block design are satisfied. When you feel these assumptions are likely to be violated, a nonparametric procedure is advisable.

What Do You Do When Assumptions Are Not Satisfied for the Analysis of Variance for a Randomized Block Design?

Answer: Use a nonparametric statistical method such as the Friedman F_r test of Section 15.5.

EXERCISES 10.38–10.49

Learning the Mechanics

10.38 A randomized block design yielded the ANOVA table at right:
  a. How many blocks and treatments were used in the experiment?
  b. How many observations were collected in the experiment?
  c. Specify the null and alternative hypotheses you would use to compare the treatment means.

Source	df	SS	MS	F
Treatments	4	501	125.25	9.109
Blocks	2	225	112.50	8.182
Error	8	110	13.75	
Total	14	836		

d. What test statistic should be used to conduct the hypothesis test of part **c**?

e. Specify the rejection region for the test of parts **c** and **d**. Use $\alpha = .01$.

f. Conduct the test of parts **c–e**, and state the proper conclusion.

g. What assumptions are necessary to ensure the validity of the test you conducted in part **f**?

10.39 An experiment was conducted using a randomized block design. The data from the experiment are displayed in the following table.

Treatment	BLOCK 1	2	3
1	2	3	5
2	8	6	7
3	7	6	5

a. Fill in the missing entries in the ANOVA table.

Source	df	SS	MS	F
Treatments		21.5555		
Blocks				
Error				
Total		30.2222		

b. Specify the null and alternative hypotheses you would use to investigate whether a difference exists among the treatment means.

c. What test statistic should be used in conducting the test of part **b**?

d. Describe the Type I and Type II errors associated with the hypothesis test of part **b**.

e. Conduct the hypothesis test of part **b** using $\alpha = .05$.

10.40 A randomized block design was used to compare the mean responses for three treatments. Four blocks of three homogeneous experimental units were selected, and each treatment was randomly assigned to one experimental unit within each block. The data are shown in the table, followed by the MINITAB ANOVA printout for this experiment.

Treatment	BLOCK 1	2	3	4
A	3.4	5.5	7.9	1.3
B	4.4	5.8	9.6	2.8
C	2.2	3.4	6.9	.3

```
ANALYSIS OF VARIANCE ON RESPONSE
SOURCE      DF       SS        MS
TREATMNT     2    12.032     6.016
BLOCK        3    71.749    23.916
ERROR        6     0.708     0.118
TOTAL       11    84.489
```

a. Use the printout to fill in the entries in the following ANOVA table.

Source	df	SS	MS	F
Treatments				
Blocks				
Error				
Total				

b. Do the data provide sufficient evidence to indicate that the treatment means differ? Use $\alpha = .05$.

c. Do the data provide sufficient evidence to indicate that blocking was effective in reducing the experimental error? Use $\alpha = .05$.

d. What assumptions are necessary to ensure the validity of the inferences made in parts **b** and **c**?

10.41 The analysis of variance for a randomized block design produced the ANOVA table entries shown here.

Source	df	SS	MS	F
Treatments	3	28.2		
Blocks	5		13.80	
Error		34.1		
Total				

a. Complete the ANOVA table.

b. Do the data provide sufficient evidence to indicate a difference among the treatment means? Test using $\alpha = .01$.

c. Do the data provide sufficient evidence to indicate that blocking was a useful design strategy for this experiment? Explain.

d. If the sample means for treatments A and B are $\bar{x}_A = 9.7$ and $\bar{x}_B = 12.1$, respectively, find a 90% confidence interval for $(\mu_A - \mu_B)$. Interpret the interval.

10.42 Suppose an experiment utilizing a randomized block design has four treatments and nine blocks, for a total of $4 \times 9 = 36$ observations. Assume that the Total Sum of Squares for the response is SS(Total) = 500. For each of the following partitions of SS(Total), test the null hypothesis that the treatment means are equal, and the null hypothesis that the block means are equal. Use $\alpha = .05$ for each test.

a. The Sum of Squares for Treatments (SST) is 20% of SS(Total), and the Sum of Squares for Blocks (SSB) is 30% of SS(Total).

b. SST is 50% of SS(Total), and SSB is 20% of SS(Total).

c. SST is 20% of SS(Total), and SSB is 50% of SS(Total).

d. SST is 40% of SS(Total), and SSB is 40% of SS(Total).

e. SST is 20% of SS(Total), and SSB is 20% of SS(Total).

Applying the Concepts

10.43 *Physical Therapy* (Aug. 1986) reported on a study to "determine whether the medial rotation that accompanies flexion of the shoulder took place during the performance of the flexion–abduction–lateral-rotation proprioceptive neuromuscular facilitation pattern (D_2F)." Ten college students, who exhibited no evidence of disease or limitation of movement in their shoulders, served as the subjects for the study. For each subject, the medial rotation was measured (in degrees) at each of three positions in the D_2F pattern: (1) beginning position, (2) point at which rotation changed directions, and (3) ending position. The goal of the analysis is to compare the mean medial rotation measurements of the three positions.

a. Identify the treatments in this experiment.

b. Identify the blocks in this experiment.

c. Identify the response variable.

d. Explain why a randomized block design is appropriate for this experiment.

10.44 A study was conducted to investigate the effect of prompting in a walking program designed to meet the American College of Sports Medicine's cardiovascular exercise goals (*Health Psychology,* Mar. 1995). Five groups of walkers—27 in each group—agreed to participate by walking for 20 minutes at least one day per week over a 24-week period. The participants were prompted to walk each week via telephone calls from research assistants, but different prompting schemes were used for each group. Walkers in the control group received no prompting phone calls; walkers in the "frequent/low" group received a call once a week with low structure (i.e., "just touching base"); walkers in the "frequent/high" group received a call once a week with high structure (i.e., goals are set); walkers in the "infrequent/low" group received a call once every 3 weeks with low structure; and walkers in the "infrequent/high" group received a call once every 3 weeks with high structure. The table below lists the number of participants in each group who actually walked the minimum requirement each week for weeks 1, 4, 8, 12, 16, and 24.

The data were subjected to an analysis of variance for a randomized block design, with the five walker groups representing the treatments and the six time periods (weeks) representing the blocks. A SAS printout of the ANOVA is on page 441:

a. What is the purpose of blocking on weeks in this study?

b. Construct an ANOVA summary table using the printout.

c. Is there sufficient evidence of a difference in the mean number of walkers per week among the five walker groups? Use $\alpha = .05$.

d. Tukey's technique was used to compare all pairs of treatment means with an experiment-wise error rate of $\alpha = .05$. The rankings are shown at the bottom of the SAS printout on the next page. Interpret these results.

e. What assumptions must hold to ensure the validity of the inferences in parts **c** and **d**?

10.45 Plant therapists believe that plants can reduce the stress levels of humans. A Kansas State University study was conducted to investigate this phenomenon. Two weeks prior to final exams, ten undergraduate students took part in an experiment to determine what effect the presence of a live plant, a photo of a plant, or absence of a plant has on the student's ability to relax while isolated in a dimly lit room. Each student participated in three sessions—one with a live plant, one with a plant photo, and one with no plant (control).* During each session, finger temperature was measured at one-minute intervals for 20 minutes. Since increasing finger temperature indicates an increased level of relaxation,

	TREATMENT CONDITION				
Week	**Control**	**Frequent/Low**	**Frequent/High**	**Infrequent/Low**	**Infrequent/High**
1	7	23	25	21	19
4	2	19	25	10	12
8	2	18	19	9	9
12	2	7	20	8	2
16	2	18	18	8	7
24	1	17	17	7	6

Source: Lombard, D. N., *et al.* "Walking to meet health guidelines: The effect of prompting frequency and prompt structure." *Health Psychology,* Vol. 14, No. 2, Mar. 1995, p. 167 (Table 2). Copyright 1995 American Psychological Association. Reprinted with permission.

*The experiment is simplified for this exercise. The actual experiment involved 30 students who participated in 12 sessions.

```
                    Analysis of Variance Procedure
Dependent Variable: WALKERS

                              Sum of          Mean
Source             DF         Squares        Square      F Value    Pr > F

Model               9      1571.400000     174.600000     23.50     0.0001
Error              20       148.600000       7.430000
Corrected Total    29      1720.000000

               R-Square           C.V.        Root MSE        WALKERS Mean

               0.913605        22.71502       2.725803        12.0000000


Source             DF       Anova SS    Mean Square    F Value    Pr > F

PROMPT              4     1185.000000    296.250000      39.87     0.0001
WEEK                5      386.400000     77.280000      10.40     0.0001
```

Tukey's Studentized Range (HSD) Test for variable: WALKERS

Note: This test controls the type I experimentwise error rate, but generally has a higher type II error rate than REGWQ.

Alpha= 0.05 df= 20 MSE= 7.43
Critical Value of Studentized Range= 4.232
Minimum Significant Difference= 4.7092

Means with the same letter are not significantly different.

```
      Tukey Grouping        Mean       N    PROMPT

             A            20.667       6    FREQ/HI
             A
             A            17.000       6    FREQ/LO

             B            10.500       6    INFR/LO
             B
             B             9.167       6    INFR/HI

             C             2.667       6    CONTROL
```

Student	Live Plant	Plant Photo	No Plant (control)
1	91.4	93.5	96.6
2	94.9	96.6	90.5
3	97.0	95.8	95.4
4	93.7	96.2	96.7
5	96.0	96.6	93.5
6	96.7	95.5	94.8
7	95.2	94.6	95.7
8	96.0	97.2	96.2
9	95.6	94.8	96.0
10	95.6	92.6	96.6

Source: Elizabeth Schreiber, Department of Statistics, Kansas State University, Manhattan, Kansas

the maximum temperature (in degrees) was used as the response variable. The data for the experiment, provided in the table above, were analyzed using the ANOVA procedure of SPSS. Use the accompanying SPSS printout at the top of page 442 to make the proper inferences.

10.46 Using decoys is a common method of hunting waterfowl. A study in the *Journal of Wildlife Management* (July 1995) compared the effectiveness of three different decoy types—taxidermy-mounted decoys, plastic shell decoys, and full-bodied plastic decoys—in attracting Canada geese to sunken pit blinds. In order to account for an extraneous source of variation, three pit blinds were used as blocks in the experiment. Thus, a randomized block design with three treatments (decoy types) and three blocks (pit blinds) was employed. The response variable was the percentage of a goose flock to approach within 46 meters of the pit blind on a given day. The data are given in the next table*, followed by a MINITAB printout of the analysis. Fully interpret the results.

*The actual design employed in the study was more complex than the randomized block design shown here. In the actual study, each number in the table represents the mean daily percentage of goose flocks attracted to the blind, averaged over 13–17 days.

```
* * * A N A L Y S I S   O F   V A R I A N C E * * *

          TEMP
    By    PLANT
          STUDENT

                              Sum of              Mean              Signif
Source of Variation           Squares     DF     Square      F      of F

Main Effects                   18.537     11     1.685      .523     .863
    PLANT                        .122      2      .061      .019     .981
    STUDENT                    18.415      9     2.046      .635     .754

Explained                      18.537     11     1.685      .523     .863

Residual                       58.038     18     3.224

Total                          76.575     29     2.641

        30 Cases were processed.
        0 Cases (    .0 PCT) were missing.
```

	DECOY TYPE		
Blind	**Shell**	**Full-Bodied**	**Taxidermy-Mounted**
1	7.3	13.6	17.8
2	12.6	10.4	17.0
3	16.4	23.4	13.6

Source: Harrey, W. F., Hindman, L. J., and Rhodes, W. E. "Vulnerability of Canada geese to taxidermy-mounted decoys." *Journal of Wildlife Management,* Vol. 59, No. 3, July 1995, p. 475 (Table 1).

```
Analysis of Variance for PERCENT

Source     DF     SS        MS       F       P

DECOY       2    30.07    15.03    0.61    0.589
BLIND       2    44.15    22.07    0.89    0.479
Error       4    99.34    24.83
Total       8   173.56
```

10.47 A simulation study was conducted to investigate the machine performance of several new algorithms for functions in the FORTRAN computer program library (*IBM Journal of Research and Development,* Mar. 1986). The next table gives the time per call (in microseconds) for several randomly selected scalar functions (averaged over 10,000 random arguments) on each of three different IBM System/370 machines. Treating the functions as blocks, the data were subjected to an ANOVA for a randomized block design using SAS. Use the SAS printout on page 443 to answer the following questions.

a. Is there sufficient evidence to indicate that the mean function call times differ for the three IBM System/370 machines? Test using $\alpha = .10$.

b. Conduct a test to determine whether blocking on functions was effective in removing an extraneous source of variation. Use $\alpha = .10$.

Function	IBM 4331	IBM 4361	IBM 4341
EDUM	9.90	3.07	4.88
ACOS CIRC(0,PI)	179.62	33.28	33.23
SIN LINEAR(−PI,PI)	105.72	24.13	27.08
EXP LINEAR(−16,16)	254.82	39.14	37.46
D2DUM	13.47	4.63	5.72

Source: Agarwal, R. C., *et al.* "New scalar and vector elementary functions for the IBM System/370." *IBM Journal of Research and Development,* Vol. 30, No. 2, Mar. 1986, p. 139 (Table 4). Copyright 1986 by International Business Machines Corporation, reprinted with permission.

10.48 Two drugs, A and B, used for the treatment of glaucoma (an eye disease) were tested for effectiveness on 10 diseased dogs. Drug A was administered to one eye (chosen randomly) of each dog and drug B to the other eye. Pressure measurements were taken 1 hour later on both eyeballs of each dog. The 10 diseased dogs serve as the blocks for comparing the two treatments, drugs A and B. Pressure measurements are given in the table. (The smaller the measurement, the less serious the eye disease.)

	TREATMENT	
Dog	**Drug A**	**Drug B**
1	.17	.15
2	.20	.18
3	.14	.13
4	.18	.18
5	.23	.19
6	.19	.12
7	.12	.07
8	.10	.09
9	.16	.14
10	.13	.08

a. Perform an analysis of variance for these data. Do the data provide sufficient evidence to indicate a

```
                    Analysis of Variance Procedure

Dependent Variable: TIME

                              Sum of         Mean
Source              DF        Squares       Square    F Value    Pr > F

Model               6      53025.20409    8837.53402    3.21     0.0653
Error               8      22016.91424    2752.11428
Corrected Total    14      75042.11833

                  R-Square          C.V.      Root MSE            TIME Mean

                  0.706606       101.3862     52.46060          51.7433333

Source              DF      Anova SS    Mean Square    F Value    Pr > F

MACHINE             2     27875.04789   13937.52395      5.06     0.0379
FUNCTION            4     25150.15620    6287.53905      2.28     0.1487
```

difference in mean pressure readings for the two treatments (i.e., is one of the glaucoma drugs better than the other)? Use $\alpha = .05$.

b. What is the purpose of using the dogs as blocks in this experiment?

c. Recall that a randomized block design with $p = 2$ treatments is a paired difference experiment (Chapter 9). Analyze the data as a paired difference experiment using a t-test to compare the treatment means. Use $\alpha = .05$.

d. Compare the computed F and t values from parts **a** and **c**, and verify that $F = t^2$. Also verify that for the rejection region values of F and t, $F_\alpha = t^2_{\alpha/2}$.

e. Find the approximate observed significance level for the test in part **a** and interpret its value.

10.49 Refer to Exercise 9.97 *(Research Quarterly for Exercise and Sport)*. The table below presents the mean and the estimated standard deviation of the mean (also called standard error of the mean,

SEM) for each of the four treatments and for seven response variables (performance time, rating of perceived exertion, etc.) observed in the experiment. Since each mean is based on six observations, the standard error of a sample mean is $s/\sqrt{6}$—that is, the sample standard deviation divided by $\sqrt{6}$. Therefore, the standard deviation for each of the samples can be calculated by multiplying the standard error by $\sqrt{6} = 2.45$. Focus your attention on any one of the response variables in the table, say performance time, and ignore the other response variables.

a. What experimental design was used to collect the data?

b. Is it possible to perform an analysis of variance for the performance data based on the information in the table?

c. Suppose the information on performance times given in the table is derived from a completely randomized design. Perform the analysis of

Metabolic and Performance Responses of Six Female Subjects Running Until Exhaustion After the Ingestion of Fructose, Glucose, Water, and a Placebo (Mean ± SEM)

	Fructose	Glucose	Placebo	Control (H_2O)
Performance time (min)	61.9 ± 8.3	63.9 ± 8.5	52.2 ± 5.5[a]	65.6 ± 7.6[a]
Rating of perceived exertion[b]	11.6 ± 1.3	13.1 ± 1.1	14.6 ± 1.3	11.9 ± 1.3
Heart rate (beats/min)	173.4 ± 7.8	172.0 ± 5.9	175.4 ± 6.1	172.1 ± 6.5
Resting oxygen uptake (l/min)	$.31 \pm .06$	$.26 \pm .08$	$.30 \pm .03$	$.31 \pm .05$
Exercise oxygen uptake (l/min)	$3.15 \pm .14$	$2.96 \pm .07$[a]	$3.46 \pm .11$[a]	$3.19 \pm .14$
Resting R value[c]	$.94 \pm .02$	$.99 \pm .02$	$.73 \pm .02$	$.71 \pm .02$
Exercise R value[c]	$.93 \pm .02$	$.99 \pm .02$	$.81 \pm .03$	$.85 \pm .03$

[a]Significant difference ($p < .05$) from each other.

[b]Significant difference ($p < .05$) placebo vs. fructose or control trials.

[c]Significant difference ($p < .05$) fructose and glucose vs. control and placebo.

Source: McMurray, R. G., Wilson, J. R., and Kitchell, B. S. "The effects of fructose and glucose on high-endurance performance." *Research Quarterly for Exercise and Sport,* 1983, 54. Reprinted by permission of the American Alliance for Health, Physical Education, Recreation, and Dance, Reston, VA 22091.

variance, and present your results in an ANOVA table. Do the data provide sufficient evidence to indicate differences in mean performance times among the four treatments? Test using $\alpha = .05$ and explain the practical consequences of the test results.

d. Refer to part **c**. What is gained or lost by analyzing data as if they had been collected according to a completely randomized design?

10.5 FACTORIAL EXPERIMENTS

All the experiments discussed in Sections 10.2 and 10.4 were **single-factor experiments**. The treatments were levels of a single factor, with the sampling of experimental units performed using either a completely randomized or randomized block design. However, most responses are affected by more than one factor, and we will therefore often wish to design experiments involving more than one factor.

Consider an experiment in which the effects of two factors on the response are being investigated. Assume that factor A is to be investigated at a levels, and factor B at b levels. Recalling that treatments are factor-level combinations, you can see that the experiment has, potentially, ab treatments that could be included in the experiment. A *complete factorial experiment* is one in which all possible ab treatments are utilized.

> **DEFINITION 10.9**
>
> A **complete factorial experiment** is one in which every factor-level combination is utilized. That is, the number of treatments in the experiment equals the total number of factor-level combinations.

For example, suppose the USGA wants to determine not only the relationship between distance and brand of golf ball, but also between distance and the club used to hit the ball. If they decide to use four brands and two clubs (say, driver and five-iron) in the experiment, then a complete factorial would call for utilizing all $4 \times 2 = 8$ Brand–Club combinations. This experiment is referred to more specifically as a **complete 4×2 factorial**. A layout for a two-factor factorial experiment (we are henceforth referring to a *complete factorial* when we use the term *factorial*) is given in Table 10.10. The factorial experiment is also referred to as a **two-way classification** because it can be arranged in the row–column format exhibited in Table 10.10.

TABLE 10.10 **Schematic Layout of Two-Factor Factorial Experiment**

| | Level | **FACTOR *B* AT *b* LEVELS** | | | | |
		1	**2**	**3**	$\cdots$	***b***
	1	Trt. 1	Trt. 2	Trt. 3	$\cdots$	Trt. b
	2	Trt. $b + 1$	Trt. $b + 2$	Trt. $b + 3$	$\cdots$	Trt. $2b$
Factor A	*3*	Trt. $2b + 1$	Trt. $2b + 2$	Trt. $2b + 3$	$\cdots$	Trt. $3b$
at a Levels	$\vdots$	$\vdots$	$\vdots$	$\vdots$	$\cdots$	$\vdots$
	a	Trt. $(a - 1)b + 1$	Trt. $(a - 1)b + 2$	Trt. $(a - 1)b + 3$	$\cdots$	Trt. ab

FIGURE 10.19

Illustration of possible treatment effects: Factorial experiment

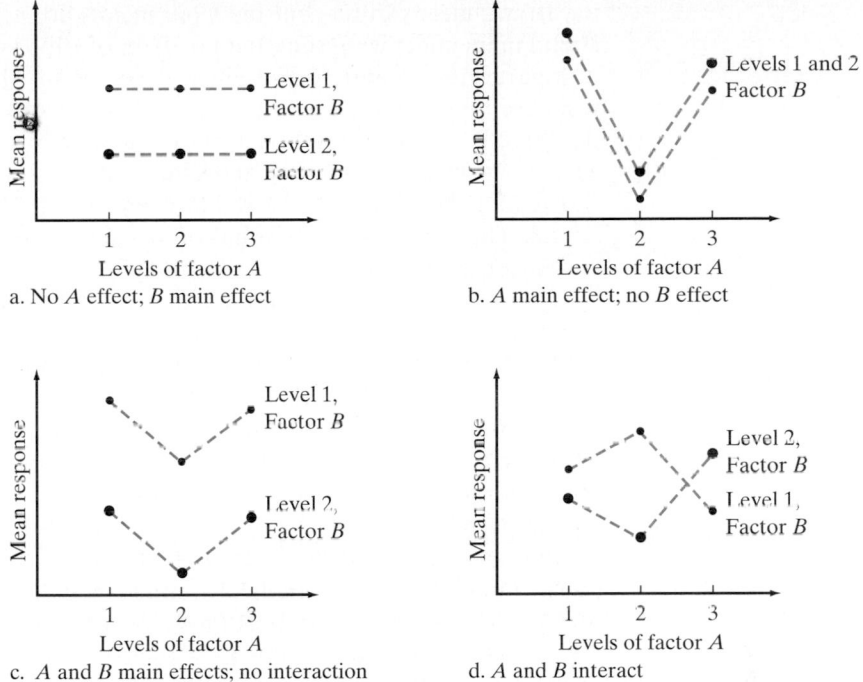

a. No A effect; B main effect

b. A main effect; no B effect

c. A and B main effects; no interaction

d. A and B interact

In order to complete the specification of the experimental design, the treatments must be assigned to the experimental units. If the assignment of the ab treatments in the factorial experiment is random and independent, the design is completely randomized. For example, if the machine Iron Byron is used to hit 80 golf balls, 10 for each of the eight Brand–Club combinations, in a random sequence, the design would be completely randomized. On the other hand, if the assignment is made within homogeneous blocks of experimental units, then the design is a randomized block. For example, if ten golfers are employed to hit each of the eight golf balls, and each golfer hits all eight Brand–Club combinations in a random sequence, then the design is randomized block, with the golfers serving as blocks. In the remainder of this section, we confine our attention to factorial experiments utilizing completely randomized designs.

If we utilize a completely randomized design to conduct a factorial experiment with ab treatments, we can proceed with the analysis in exactly the same way as we did in Section 10.2. That is, we calculate (or let the computer calculate) the measure of treatment mean variability (MST) and the measure of sampling variability (MSE) and use the F-ratio of these two quantities to test the null hypothesis that the treatment means are equal. However, if this hypothesis is rejected, so that we conclude some differences exist among the treatment means, important questions remain. Are both factors affecting the response, or only one? If both, do they affect the response independently, or do they interact to affect the response?

For example, suppose the distance data indicate that at least two of the eight treatment (Brand–Club combinations) means differ in the golf experiment. Does the brand of ball (factor A) or the club utilized (factor B) affect mean distance, or do both affect it? Several possibilities are shown in Figure 10.19. In Figure 10.19a, the brand means are equal (only three are shown for the purpose of illustration), but the distances differ for the two levels of factor B (Club). Thus, there is no effect of Brand on distance, but a Club main effect is present. In Figure 10.19b, the Brand means differ, but the Club means are equal for each Brand. Here a Brand main effect is present, but no effect of Club is present.

Figures 10.19c and 10.19d illustrate cases in which both factors affect the response. In Figure 10.19c, the mean distance between clubs does not change for the three Brands, so that the effect of Brand on distance is independent of Club. That is, the two factors Brand and Club *do not interact.* In contrast, Figure 10.19d shows that the difference between mean distances between Clubs varies with Brand. Thus, the effect of Brand on distance depends on Club; therefore, the two factors *do interact.*

In order to determine the nature of the treatment effect, if any, on the response in a factorial experiment, we need to break the treatment variability into three components: Interaction Between Factors *A* and *B*, Main Effect of Factor *A*, and Main Effect of Factor *B*. The **factor Interaction** component is used to test whether the factors combine to affect the response, while the **factor Main Effect** components are used to determine whether the factors separately affect the response.

The partitioning of the Total Sum of Squares into its various components is illustrated in Figure 10.20. Notice that at stage 1 the components are identical to those in the one-factor, completely randomized designs of Section 10.2; the Sums of Squares for Treatment and Error sum to the Total Sum of Squares. The degrees of freedom for treatments is equal to $(ab - 1)$, one less than the number of treatments. The degrees of freedom for error is equal to $(n - ab)$, the total sample size minus the number of treatments. Only at stage 2 of the partitioning does the factorial experiment differ from those previously discussed. Here we divide the Treatment Sum of Squares into its three components: Interaction and

FIGURE 10.20

Partitioning the Total Sum of Squares for a two-factor factorial

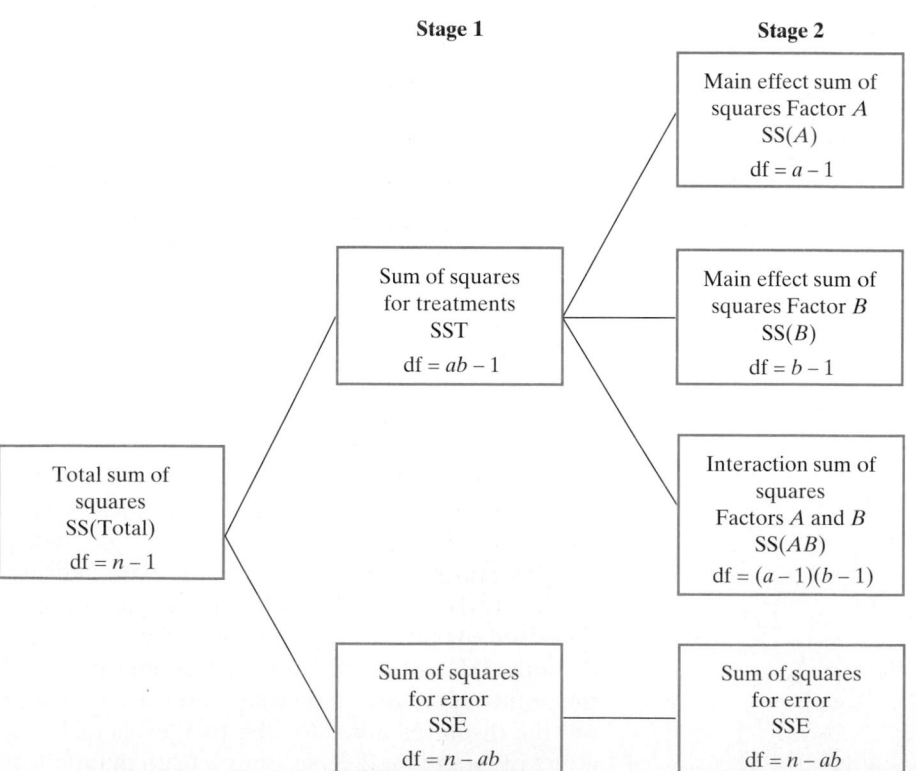

the two Main Effects. These components can then be used to test the nature of the differences, if any, among the treatment means.

There are a number of ways to proceed in the testing and estimation of factors in a factorial experiment. We present one approach in the next box.

Procedure for Analysis of Two-Factor Factorial Experiment

1. Partition the Total Sum of Squares into the Treatment and Error components (stage 1 of Figure 10.20). Use either a statistical software package or the calculation formulas in Appendix C to accomplish the partitioning.
2. Use the F-ratio of Mean Square for Treatments to Mean Square for Error to test the null hypothesis that the treatment means are equal.*
   a. If the test results in nonrejection of the null hypothesis, consider refining the experiment by increasing the number of replications or introducing other factors. Also consider the possibility that the response is unrelated to the two factors.
   b. If the test results in rejection of the null hypothesis, then proceed to step 3.
3. Partition the Treatment Sum of Squares into the Main Effect and Interaction Sum of Squares (stage 2 of Figure 10.20). Use either a statistical software package or the calculation formulas in Appendix C to accomplish the partitioning.
4. Test the null hypothesis that factors A and B do not interact to affect the response by computing the F-ratio of the Mean Square for Interaction to the Mean Square for Error.
   a. If the test results in nonrejection of the null hypothesis, proceed to step 5.
   b. If the test results in rejection of the null hypothesis, conclude that the two factors interact to affect the mean response. Then proceed to step 6a.
5. Conduct tests of two null hypotheses that the mean response is the same at each level of factor A and factor B. Compute two F-ratios by comparing the Mean Square for each Factor Main Effect to the Mean Square for Error.
   a. If one or both tests result in rejection of the null hypothesis, conclude that the factor affects the mean response. Proceed to step 6b.
   b. If both tests result in nonrejection, an apparent contradiction has occurred. Although the treatment means apparently differ (step 2 test), the interaction (step 4) and main effect (step 5) tests have not supported that result. Further experimentation is advised.
6. Compare the means:
   a. If the test for interaction (step 4) is significant, use a multiple comparisons procedure to compare any or all pairs of the treatment means.
   b. If the test for one or both main effects (step 5) is significant, use a multiple comparisons procedure to compare the pairs of means corresponding to the levels of the significant factor(s).

Tests Conducted in Analyses of Factorial Experiments: Completely Randomized Design, r Replicates per Treatment

TEST FOR TREATMENT MEANS

H_0: No difference among the ab treatment means

H_a: At least two treatment means differ

continued

---

*Some analysts prefer to proceed directly to test the interaction and main effect components, skipping the test of treatment means. We begin with this test to be consistent with our approach in the one-factor completely randomized design.

Test statistic: $F = \dfrac{\text{MST}}{\text{MSE}}$

Rejection region: $F \geq F_\alpha$, based on $(ab - 1)$ numerator and $(n - ab)$ denominator degrees of freedom [*Note: n = abr.*]

TEST FOR FACTOR INTERACTION

H_0: Factors A and B do not interact to affect the response mean

H_a: Factors A and B do interact to affect the response mean

Test statistic: $F = \dfrac{\text{MS}(AB)}{\text{MSE}}$

Rejection region: $F \geq F_\alpha$, based on $(a - 1)(b - 1)$ numerator and $(n - ab)$ denominator degrees of freedom

---

TEST FOR MAIN EFFECT OF FACTOR A

H_0: No difference among the a mean levels of factor A

H_a: At least two factor A mean levels differ

Test statistic: $F = \dfrac{\text{MS}(A)}{\text{MSE}}$

Rejection region: $F \geq F_\alpha$, based on $(a - 1)$ numerator and $(n - ab)$ denominator degrees of freedom

---

TEST FOR MAIN EFFECT OF FACTOR B

H_0: No difference among the b mean levels of factor B

H_a: At least two factor B mean levels differ

Test statistic: $F = \dfrac{\text{MS}(B)}{\text{MSE}}$

Rejection region: $F \geq F_\alpha$, based on $(b - 1)$ numerator and $(n - ab)$ denominator degrees of freedom

---

ASSUMPTIONS FOR ALL TESTS

1. The response distribution for each factor-level combination (treatment) is normal.
2. The response variance is constant for all treatments.
3. Random and independent samples of experimental units are associated with each treatment.

We assume the completely randomized design is a **balanced design**, meaning that the same number of observations are made for each treatment. That is, we assume that r experimental units are randomly and independently selected for each treatment. The numerical value of r must exceed 1 in order to have any degrees of freedom with which to measure the sampling variability. [Note that if $r = 1$, then $n = ab$, and the degrees of freedom associated with Error (Figure 10.20) is df $= n - ab = 0$.] The value of r is often referred to as the number of **replicates** of the factorial experiment since we assume that all ab treatments are repeated, or replicated, r times. Whatever approach is adopted in the analysis of a factorial experiment, several tests of hypotheses are usually conducted. The tests are summarized in the preceding box.

TABLE 10.11   Distance Data for 4 × 2 Factorial Golf Experiment

Club	BRAND			
	A	**B**	**C**	**D**
Driver	226.4	238.3	240.5	219.8
	232.6	231.7	246.9	228.7
	234.0	227.7	240.3	232.9
	220.7	237.2	244.7	237.6
Five-iron	163.8	184.4	179.0	157.8
	179.4	180.6	168.0	161.8
	168.6	179.5	165.2	162.1
	173.4	186.2	156.5	160.3

EXAMPLE 10.9

Suppose the USGA tests four different brands (A, B, C, D) of golf balls and two different clubs (driver, five-iron) in a completely randomized design. Each of the eight Brand–Club combinations (treatments) is randomly and independently assigned to four experimental units, each experimental unit consisting of a specific position in the sequence of hits by Iron Byron. The distance response is recorded for each of the 32 hits, and the results are shown in Table 10.11.

a. Use statistical software to partition the Total Sum of Squares into the components necessary to analyze this 4 × 2 factorial experiment.
b. Follow the steps for analyzing a two-factor factorial experiment and interpret the results of your analysis. Use $\alpha = .10$ for the tests you conduct.

Solution

a. The SAS printout that partitions the Total Sum of Squares for this factorial experiment is given in Figure 10.21. The partitioning takes place in two stages: First, the Total Sum of Squares is partitioned into the Model (Treatment) and Error Sums of Squares at the top of the printout. Note that SST is 33,659.8 with 7 degrees of freedom, and SSE is 822.2 with 24 degrees of freedom, adding to 34,482.0 and 31 degrees of freedom. In the second stage of partitioning, the

FIGURE 10.21

SAS printout for factorial golf experiment

```
                        Analysis of Variance Procedure

Dependent Variable: DISTANCE
                                    Sum of          Mean
Source                  DF         Squares        Square     F Value    Pr > F

Model                    7      33659.8087     4808.5441      140.35    0.0001
Error                   24        822.2400       34.2600
Corrected Total         31      34482.0487

                  R-Square           C.V.      Root MSE       DISTANCE Mean

                  0.976155      2.8964608       5.85320          202.081250

Source                  DF      Anova SS   Mean Square    F Value    Pr > F

CLUB                     1      32093.11      32093.11     936.75    0.0001
BRAND                    3        800.74        266.91       7.79    0.0008
CLUB*BRAND               3        765.96        255.32       7.45    0.0011
```

Treatment Sum of Squares is further divided into the Main Effect and Interaction Sums of Squares. At the bottom of the printout we see that SS(Club) is 32,093.1 with 1 degree of freedom, SS(Brand) is 800.7 with 3 degrees of freedom, and SS(Club × Brand) is 766.0 with 3 degrees of freedom, adding to 33,659.8 and 7 degrees of freedom.

b. Once partitioning is accomplished, our first test is

H_0: The eight treatment means are equal

H_a: At least two of the eight means differ

Test statistic: $F = \dfrac{\text{MST}}{\text{MSE}} = 140.35$ (top of printout)

Observed significance level: $p = .0001$ (top of printout)

Since $\alpha = .10$ exceeds p, we reject this null hypothesis and conclude that at least two of the Brand–Club combinations differ in mean distance.

After accepting the hypothesis that the treatment means differ, and therefore that the factors Brand and/or Club somehow affect the mean distance, we want to determine how the factors affect the mean response. We begin with a test of interaction between Brand and Club:

H_0: The factors Brand and Club do not interact to affect the mean response

H_a: Brand and Club interact to affect mean response

Test statistic: $F = \dfrac{\text{MS}(AB)}{\text{MSE}} = \dfrac{\text{MS}(\text{Brand} \times \text{Club})}{\text{MSE}}$

$$= \frac{255.32}{34.26} = 7.45 \quad \text{(bottom of printout)}$$

Observed significance level: $p = .0011$ (bottom of printout)

Since $\alpha = .10$ exceeds the p-value, we conclude that the factors Brand and Club interact to affect mean distance.

Because the factors interact, we do not test the main effects for Brand and Club. Instead, we compare the treatment means in an attempt to learn the nature of the interaction. Rather than compare all $8(7)/2 = 28$ pairs of treatment means, we test for differences only between pairs of brands within each club. That differences exist *between* clubs can be assumed. Therefore, only $4(3)/2 = 6$ pairs of means need to be compared for each club, or a total of 12 comparisons for the two clubs. The results of these comparisons using Tukey's method with an experimentwise error rate of $\alpha = .10$ for each club are displayed in the SAS printout, Figure 10.22. For each club, the brand means are listed in descending order in Figure 10.22, and those not significantly different are connected by the same letter in the *Tukey Grouping* column.

As shown in Figure 10.22, the picture is unclear with respect to Brand means. For the five-iron (top of Figure 10.22), the brand B mean significantly exceeds all other brands. However, when hit with a driver (bottom of Fig. 10.22), brand B's mean is not significantly different from any of the other brands. Note the nontransitive nature of the multiple comparisons. For example, for the driver the Brand C mean can be "the same" as the brand B mean, and the brand B mean can be "the same" as the brand D mean, and yet the brand C mean can significantly exceed the brand D mean. The reason lies in the definition of "the same"—we must be careful not to conclude two means are equal simply because they are connected by a vertical line. The line indicates only that *the connected means are not significantly different.* You should

FIGURE 10.22

SAS ranking of treatment means for factorial golf experiment

```
----------------------------CLUB=5IRON-----------------
                   Analysis of Variance Procedure

       Tukey's Studentized Range (HSD) Test for variable: DISTANCE

   Note: This test controls the type I experimentwise error rate, but
         generally has a higher type II error rate than REGWQ.

               Alpha= 0.1  df= 12  MSE= 36.10792
              Critical Value of Studentized Range= 3.621
               Minimum Significant Difference= 10.878

       Means with the same letter are not significantly different.

           Tukey Grouping          Mean      N   BRAND

                 A              182.675      4   B

                 D              171.300      4   A
                 B
                 B              167.175      4   C
                 B
                 B              160.500      4   D



----------------------------CLUB=DRIVER----------------------
                   Analysis of Variance Procedure

       Tukey's Studentized Range (HSD) Test for variable: DISTANCE

   Note: This test controls the type I experimentwise error rate, but
         generally has a higher type II error rate than REGWQ.

               Alpha= 0.1  df= 12  MSE= 32.41208
              Critical Value of Studentized Range= 3.621
               Minimum Significant Difference= 10.306

       Means with the same letter are not significantly different.

           Tukey Grouping          Mean      N   BRAND

                 A              243.100      4   C
                 A
           B     A              233.725      4   B
           B
           B                    229.750      4   D
           B
           B                    228.425      4   A
```

conclude (at the overall α level of significance) only that means *not* connected are different, while withholding judgment on those that are connected. The picture of which means differ and by how much will become clearer as we increase the number of replicates of the factorial experiment.

The Club × Brand interaction can be seen in the plot of means in Figure 10.23. Note that the difference between the mean distances of the two clubs

FIGURE 10.23

Sample mean plot for factorial golf experiment

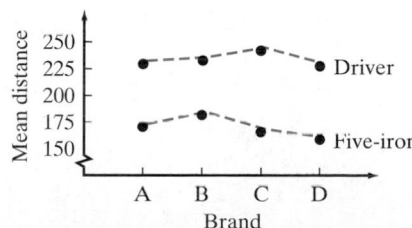

TABLE 10.12 **Distance Data for Second Factorial Golf Experiment**

Club	BRAND			
	E	**F**	**G**	**H**
Driver	238.6	261.4	264.7	235.4
	241.9	261.3	262.9	239.8
	236.6	254.0	253.5	236.2
	244.9	259.9	255.6	237.5
Five-iron	165.2	179.2	189.0	171.4
	156.9	171.0	191.2	159.3
	172.2	178.0	191.3	156.6
	163.2	182.7	180.5	157.4

(driver and five-iron) varies depending on brand. The biggest difference appears for Brand C while the smallest difference is for Brand B. ▲

EXAMPLE 10.10

Refer to Example 10.9. Suppose the same factorial experiment is performed on four other brands (E, F, G, and H), and the results are as shown in Table 10.12. Repeat the factorial analysis and interpret the results.

Solution

The printout for the second factorial experiment is shown in Figure 10.24. The F-ratio for Treatments is $F = 290.1$, which exceeds the tabled value of $F_{.10} = 1.98$ for 7 numerator and 24 denominator degrees of freedom. (Note that the same rejection regions will apply in this example as in Example 10.9 since the factors, treatments, and replicates are the same.) We conclude that at least two of the Brand–Club combinations have different mean distances.

We next test for interaction between Brand and Club:

$$F = \frac{MS(\text{Brand} \times \text{Club})}{MSE} = 1.42$$

FIGURE 10.24

SAS printout for second factorial golf experiment

```
                              Analysis of Variance Procedure
Dependent Variable: DISTANCE

                                     Sum of          Mean
Source                   DF          Squares         Square      F Value    Pr > F

Model                     7      49959.3747      7137.0535       290.12     0.0001
Error                    24        590.4075        24.6003
Corrected Total          31      50549.7822

                    R-Square            C.V.      Root MSE        DISTANCE Mean

                    0.988320       2.3515897       4.95987            210.915625


Source                   DF       Anova SS    Mean Square      F Value    Pr > F

CLUB                      1      46443.90        46443.90      1887.94     0.0001
BRAND                     3       3410.32         1136.77        46.21     0.0001
CLUB*BRAND                3        105.16           35.05         1.42     0.2600
```

Since this *F*-ratio does *not* exceed the tabled value of $F_{.10} = 2.33$ with 3 and 24 df, we cannot conclude at the .10 level of significance that the factors interact. In fact, note that the observed significance level (on the SAS printout) for the test of interaction is .26. Thus, at any level of significance lower than $\alpha = .26$, we could not conclude that the factors interact. We therefore test the main effects for Brand and Club.

We first test the Brand main effect:

H_0: No difference exists among the true Brand mean distances

H_a: At least two Brand mean distances differ

Test statistic: $F = \dfrac{\text{MS(Brand)}}{\text{MSE}} = \dfrac{1{,}136.77}{24.60} = 46.21$ (bottom of printout)

Observed significance level: $p = .0001$ (bottom of printout)

Since $\alpha = .10$ exceeds the *p*-value, we conclude that at least two of the brand means differ. We will subsequently determine which brand means differ using Tukey's multiple comparisons procedure. But first, we want to test the Club main effect:

H_0: No differences exist between the Club mean distances

H_a: The Club mean distances differ

Test statistic: $F = \dfrac{\text{MS(Club)}}{\text{MSE}} = \dfrac{46{,}443.9}{24.60} = 1{,}887.94$ (bottom of printout)

Observed significance level: $p = .0001$ (bottom of printout)

Since $\alpha = .10$ exceeds the *p*-value, we conclude that the two clubs are associated with different mean distances. Since only two levels of Club were utilized in the experiment, this *F*-test leads to the inference that the mean distance differs for the two clubs. It is no surprise (to golfers) that the mean distance for balls hit with the driver is significantly greater than the mean distance for those hit with the five-iron.

To determine which of the Brands' mean distances differ, we wish to compare the four Brand means using Tukey's method at $\alpha = .10$. The results of these multiple comparisons are displayed in the SAS printout, Figure 10.25. Again, the Brand means are shown in descending order in Figure 10.25, with a *Tukey Grouping* letter connecting the means that are not significantly different. Brands

FIGURE 10.25

SAS comparison of mean distances by brand: Tukey's method

```
                    Analysis of Variance Procedure
          Tukey's Studentized Range (HSD) Test for variable: DISTANCE

NOTE: This test controls the type I experimentwise error rate, but
      generally has a higher type II error rate than REGWQ.

             Alpha= 0.1  df= 12  MSE= 24.60031
           Critical Value of Studentized Range= 3.423
             Minimum Significant Difference= 6.003

   Means with the same letter are not significantly different.

        Tukey Grouping         Mean     N  BRAND

                    A        223.587     8  G
                    A
                    A        218.437     8  F

                    B        202.438     8  E
                    B
                    B        199.200     8  H
```

Anxiety Levels and Mathematical Achievement

CASE STUDY • 10.2 •

If you are very anxious when you take a mathematics examination, you may not be surprised to learn that the level of your anxiety and your achievement may be related. Writing in the *Journal for Research in Mathematics Education,* Pamela S. Clute (1984) describes her doctoral research on this subject. Clute's research involved a study of the effect of three factors on a student's mathematics examination score taken after completing a survey course in mathematics. One factor was the college where the course was taken. Students took the course at both the University of California at Riverside and California State College, San Bernardino. Therefore, the factor College was at two levels. The second factor was the anxiety level of the student. Students were tested to determine their anxiety level and placed into one of three anxiety level groups. Therefore, the factor Anxiety level was at three levels. The third factor, at two levels, was the manner in which the instructional material was presented. Two methods were employed; an expository direct instruction method and a direct instruction discovery method. This latter method developed the subject by a sequence of questions, eventually leading the student to the mathematical principle that was the object of the lesson.

Since all the students were taught by the same instructor (Clute), differences in research results between the two colleges would not be expected and in fact none were found in an analysis of the data. Consequently, we will combine the data for the two colleges and view Clute's experiment as a two-factor factorial experiment. The two factors are the mathematics anxiety level (three levels) and the method of instruction (two levels). The sample sizes, examination test score means, and standard deviations (in parentheses) for the six factor-level combinations are shown in Table 10.13.

Focus

a. Plot the six sample means in a graph similar to Figure 10.19.

b. Based on the plot, part a, give your opinion on whether or not the two factors, Anxiety level and Method of instruction, interact. Why is a statistical test preferred over the graph?

Clute's analysis of variance of the full three-factor factorial experiment is shown in Table 10.14. Although we have not explained how to compute the sums of squares for a three-factor factorial experiment, the interpretation of the ANOVA table is the same as the interpretation for a two-factor experiment. Sources of variation shown in Table 10.14 include the main effects for methods (M), anxiety (A), and college (C). They also include the three two-way interactions, $M \times A$, $M \times C$, and $A \times C$, and the three-way interaction $M \times A \times C$.

Typically, we conduct the F-test for the three-way interaction first. If the test is statistically significant, we perform no further F tests. This is because the three-way interaction implies that the difference between the means for any two levels of one factor depends on the levels of the other two factors. Usually, the analyst then compares all possible pairs of treatment means with a multiple comparisons procedure (e.g., Tukey or Bonferroni). When the three-way interaction test is nonsignificant, then F-tests for all two-way interactions are conducted and interpreted as outlined in this chapter. [*Note:* If all two-way interactions are nonsignificant, tests for main effects are conducted.]

Focus

c. Is there evidence of three-way interaction between Method, Anxiety, and College?

d. Conduct the tests for two-way interactions. (Why should we do this?) Interpret the results.

e. Clute concludes that "students with low mathematics anxiety tended to do better with the discovery method, whereas students with high mathematics anxiety tended to do better under the expository treatment." Which of the tests conducted above supports this statement statistically? Explain.

G and F are apparently associated with significantly greater mean distances than brands E and H, but we cannot distinguish between brands G and F or between brands E and H using these data. Since the interaction between Brand and Club was not significant, we conclude that this difference among brands applies to both clubs. The sample means for all Club–Brand combinations are shown in Figure 10.26 and appear to support the conclusions of the tests and comparisons. Note that the Brand means maintain their relative positions for each Club—brands F and G dominate brands E and H for both the driver and five-iron. ▲

FIGURE 10.26

Sample mean plot for second factorial golf experiment

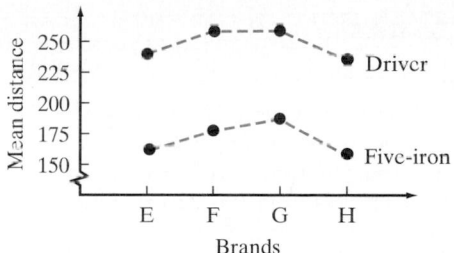

The analysis of factorial experiments can become complex if the number of factors is increased. Even the two-factor experiment becomes more difficult to analyze if some factor combinations have different numbers of observations than others. We have provided an introduction to these important experiments using two-factor factorials with equal numbers of observations for each treatment. Although similar principles apply to most factorial experiments, you should consult the references for this chapter at the end of the book if you need to design and analyze more complex factorials.

TABLE 10.13   Means and Standard Deviations on the Mathematics Achievement Test by Instructional Method and Level of Anxiety (Case Study 10.2)

| | MATHEMATICS ANXIETY LEVEL | | | | | | |
| | LOW | | MEDIUM | | HIGH | | |
Method	n	$\bar{x}$	n	$\bar{x}$	n	$\bar{x}$	**Total**
Discovery	14	352.50 (27.31)	14	299.71 (29.60)	13	225.38 (59.09)	294.17 (65.68)
Expository	14	288.86 (79.94)	13	265.69 (81.24)	13	260.69 (54.17)	272.17 (72.26)
Total		320.68 (66.98)		283.33 (61.52)		243.04 (58.38)	283.31 (69.46)

Note: Maximum score = 400.

TABLE 10.14   Analysis of Variance of Mathematics Achievement Scores (Case Study 10.2)

Source	df	MS	F
Method (M)	1	7,945	2.30
Anxiety (A)	2	34,935	10.11[a]
College (C)	1	2,239	.65
$M \times A$	2	17,125	4.96[a]
$M \times C$	1	516	.15
$A \times C$	2	7,914	2.29
$M \times A \times C$	2	1,985	.58
Residual	69	3,455	

[a]$p < .01$.

EXERCISES 10.50–10.63

Learning the Mechanics

10.50 Suppose you conduct a 3×5 factorial experiment.
   a. How many factors are used in the experiment?
   b. Can you determine the factor type(s)—qualitative or quantitative—from the information given? Explain.
   c. Can you determine the number of levels used for each factor? Explain.
   d. Describe a treatment for this experiment, and determine the number of treatments used.
   e. What problem is caused by using a single replicate of this experiment? How is the problem solved?

10.51 The partially completed ANOVA table for a 3×4 factorial experiment with two replications is shown here.

Source	df	SS	MS	F
A		.8		
B		5.3		
AB		9.6		
Error				
Total		18.1		

   a. Complete the ANOVA table.
   b. Which sums of squares are combined to find the Sum of Squares for Treatment? Do the data provide sufficient evidence to indicate that the treatment means differ? Use $\alpha = .05$.
   c. Does the result of the test in part **b** warrant further testing? Explain.
   d. What is meant by factor interaction, and what is the practical implication if it exists?
   e. Test to determine whether these factors interact to affect the response mean. Use $\alpha = .05$, and interpret the result.
   f. Does the result of the interaction test warrant further testing? Explain.

10.52 The partially complete ANOVA table given here is for a two-factor factorial experiment.

Source	df	SS	MS	F
A	3		.75	
B	1	.95		
AB			.30	
Error				
Total	23	6.5		

   a. Give the number of levels for each factor.
   b. How many observations were collected for each factor-level combination?
   c. Complete the ANOVA table.

   d. Test to determine whether the treatment means differ. Use $\alpha = .10$.
   e. Conduct the tests of factor interaction and mean effects, each at the $\alpha = .10$ level of significance. Which of the tests are warranted as part of the factorial analysis? Explain.

10.53 The following two-way table gives data for a 2×3 factorial experiment with two observations for each factor-level combination.

			FACTOR B	
	Level	**1**	**2**	**3**
Factor A	*1*	3.1, 4.0	4.6, 4.2	6.4, 7.1
	2	5.9, 5.3	2.9, 2.2	3.3, 2.5

   a. Identify the treatments for this experiment. Calculate and plot the treatment means, using the response variable as the y-axis and the levels of factor B as the x-axis. Use the levels of factor A as plotting symbols. Do the treatment means appear to differ? Do the factors appear to interact?
   b. The MINITAB ANOVA printout for this experiment is shown here. Test to determine whether the treatment means differ at the $\alpha = .05$ level of significance. Does the test support your visual interpretation from part **a**?

```
ANALYSIS OF VARIANCE ON RESPONSE

SOURCE            DF          SS          MS
A                  1       4.441       4.441
B                  2       4.127       2.063
INTERACTION        2      18.007       9.003
ERROR              6       1.475       0.246
TOTAL             11      28.049
```

   c. Does the result of the test in part **b** warrant a test for interaction between the two factors? If so, perform it using $\alpha = .05$.
   d. Do the results of the previous tests warrant tests of the two factor main effects? If so, perform them using $\alpha = .05$.
   e. Interpret the results of the tests. Do they support your visual interpretation from part **a**?

10.54 The accompanying two-way table gives data for a 2×2 factorial experiment with two observations per factor-level combination.
   a. Identify the treatments for this experiment. Calculate and plot the treatment means, using the response variable as the y-axis and the levels

	FACTOR B	
Level	**1**	**2**

Factor A	*1*	29.6, 35.2	47.3, 42.1
	2	12.9, 17.6	28.4, 22.7

of factor B as the x-axis. Use the levels of factor A as plotting symbols. Do the treatment means appear to differ? Do the factors appear to interact?

b. Use the computational formulas in Appendix C to create an ANOVA table for this experiment.

c. Test to determine whether the treatment means differ at the $\alpha = .05$ level of significance. Does the test support your visual interpretation from part **a**?

d. Does the result of the test in part **b** warrant a test for interaction between the two factors? If so, perform it using $\alpha = .05$.

e. Do the results of the previous tests warrant tests of the two factor main effects? If so, perform them using $\alpha = .05$.

f. Interpret the results of the tests. Do they support your visual interpretation from part **a**?

g. Given the results of your tests, which pairs of means, if any, should be compared?

10.55 Suppose a 3×3 factorial experiment is conducted with three replications. Assume that SS(Total) = 1,000. For each of the following scenarios, form an ANOVA table, conduct the appropriate tests, and interpret the results.

a. The Sum of Squares of factor A main effect [SS(A)] is 20% of SS(Total), the Sum of Squares for factor B main effect [SS(B)] is 10% of SS(Total), and the Sum of Squares for interaction [SS(AB)] is 10% of SS(Total).

b. SS(A) is 10%, SS(B) is 10%, and SS(AB) is 50% of SS(Total).

c. SS(A) is 40%, SS(B) is 10%, and SS(AB) is 20% of SS(Total).

d. SS(A) is 40%, SS(B) is 40%, and SS(AB) is 10% of SS(Total).

Applying the Concepts

10.56 The *American Journal of Psychology* (Winter 1991) reported on a study designed to investigate the way in which adolescents with low reading ability comprehend simple text-based problems. Fourteen-year-old students from New York City schools participated in the experiment. Based on expert evaluation, the students were divided into four groups: (1) learning-disabled, low socioeconomic status; (2) nondisabled, low socioeconomic status; (3) learning-disabled, high socioeconomic status;

and (4) nondisabled, high socio-economic status. Each student was asked to read and retell a "story problem"; however, the way in which the problem was presented was varied. Some students read a "no-priority" problem (i.e., a problem with no clear goals and/or objectives), while others read a "priority" problem (i.e., a problem with a clear statement of the character's priority). The experiment was designed as a 4×2 factorial, with Group at 4 levels and Problem type (priority or no-priority) at 2 levels. One of the dependent variables measured was proportion of ideas recalled correctly.

a. The test for Group by Problem type interaction was nonsignificant at $\alpha = .01$. Interpret this result.

b. The test for Problem type main effect was nonsignificant at $\alpha = .01$. Interpret this result.

c. The test for Group main effect was statistically significant at $\alpha = .01$. Interpret this result.

10.57 The cattle inhabiting the Biological Reserve of Doñana (Spain), live under free-range conditions, with virtually no human interference. The cattle population is organized into four herds (LGN, MTZ, PLC, and QMD). The *Journal of Zoology* (July 1995) investigated the ranging behavior of the four herds across the four seasons. Thus, a 4×4 factorial experiment was employed, with Herd and Season representing the two factors. Three animals from each herd during each season were sampled and the home range of each individual was measured (in square kilometers). The data were subjected to an ANOVA, with the results shown in the table below.

Source	df	F	p-Value
Herd (H)	3	17.2	$p < .001$
Season (S)	3	3.0	$p < .05$
$H \times S$	9	1.2	$p > .05$
Error	32		
Total	47		

a. Conduct the appropriate ANOVA F tests and interpret the results.

b. The researcher ranked the four herd means independently of season. Do you agree with this strategy? Explain.

c. Refer to part **b**. The Bonferroni rankings of the four herd means (at $\alpha = .05$) are shown below. Interpret the results.

Mean home range (km²)	.75	1.0	2.7	3.8
Herd	PLC	LGN	QMD	MTZ

10.58 Many temperate-zone animal species exhibit physiological and morphological changes when the hours of daylight begin to decrease during autumn months. A study was conducted to investigate the "short-day" traits of collared lemmings (*The Journal of Experimental Zoology,* Sept. 1993). A total of 124 lemmings were bred in a colony maintained with a photoperiod of 22 hours of light per day. At weaning (19 days of age), the lemmings were weighed and randomly assigned to live under one of two photoperiods: 16 hours or less of light per day and more than 16 hours light per day. (Each group was assigned the same number of males and females.) After 10 weeks, the lemmings were weighed again. The response variable of interest was the gain in body weight (measured in grams) over the 10-week experimental period. The researchers analyzed the data using an ANOVA for a 2×2 factorial design, where the two factors are Photoperiod (at two levels) and Gender (at two levels).

a. Construct an ANOVA table for the experiment, listing the sources of variation and associated degrees of freedom.

b. The *F* test for interaction was not significant. Interpret this result practically.

c. The *p*-values for testing for Photoperiod and Gender main effects were both smaller than .001. Interpret these results practically.

10.59 Most short-run supermarket strategies such as price reductions, media advertising, and in-store displays are designed to increase unit sales of particular products temporarily. Factorial designs have been employed in the *Journal of Marketing Research* (Feb. 1982) to evaluate the effectiveness of such strategies. Two of the factors were Price level (regular, reduced price, cost to supermarket) and Display level (normal display space, normal display space plus end-of-aisle display, twice the normal display space). A complete factorial design based on these two factors involves nine treatments. Suppose each treatment was applied three times to a particular product at a particular supermarket. Each application lasted a full week and the dependent variable of interest was unit sales for the week. To minimize treatment carryover effects, each treatment was preceded and followed by a week in which the product was priced at its regular price and was displayed in its normal manner. The table below reports the data collected.

a. The SAS ANOVA printout for this experiment is given below the table. Use the printout to complete the diagram on page 459 partitioning the Total Sum of Squares.

b. Do the data indicate that the mean sales differ among the nine treatments? Test using $\alpha = .10$.

	PRICE		
Display	**Regular**	**Reduced**	**Cost to Supermarket**
Normal	989, 1,025, 1,030	1,211, 1,215, 1,182	1,577, 1,559, 1,598
Normal Plus	1,191, 1,233, 1,221	1,860, 1,910, 1,926	2,492, 2,527, 2,511
Twice Normal	1,226, 1,202, 1,180	1,516, 1,501, 1,498	1,801, 1,833, 1,852

```
                    Analysis of Variance Procedure

Dependent Variable: SALES

                              Sum of         Mean
Source               DF      Squares       Square    F Value    Pr > F

Model                 8   5291151.19    661393.90    1336.85    0.0001
Error                18      8905.33       494.74
Corrected Total      26   5300056.52

                 R-Square         C.V.     Root MSE         SALES Mean

                 0.998320    1.4344689      22.2428         1550.59259

Source               DF    Anova SS   Mean Square    F Value    Pr > F

DISPLAY               2   1691392.5      845696.3    1709.37    0.0001
PRICE                 2   3089053.9     1544526.9    3121.89    0.0001
DISPLAY*PRICE         4    510704.8      127676.2     258.07    0.0001
```

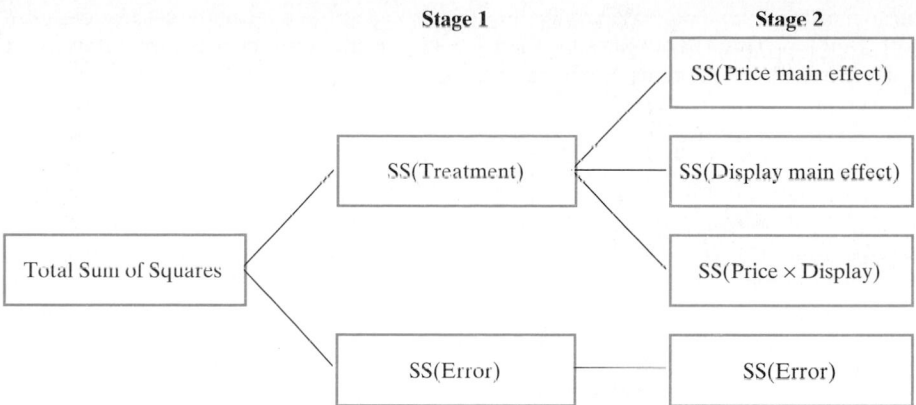

c. Is the test of interaction between the factors Price and Display warranted as a result of the test in part **b**? If so, conduct the test using $\alpha = .10$.

d. Are the tests of the main effects for Price and Display warranted as a result of the previous tests? If so, conduct them using $\alpha = .10$.

e. Which pairs of treatment means should be compared as a result of the tests in parts **b–d**?

f. The treatment means are graphically portrayed in the SAS graph printout below. Use the graph to interpret the results of your tests.

10.60 Traditionally, people protect themselves from mosquito bites by applying insect repellent to their skin and clothing. Recent research suggests that peremethrin, an insecticide with low toxicity to humans, can provide protection from mosquitoes. A study in the *Journal of the American Mosquito Control Association* (Mar. 1995) investigated whether a tent sprayed with a commercially available 1% peremethrin formulation would protect people, both inside and outside the tent, against biting mosquitoes. Two canvas tents—one treated with peremethrin, the other untreated—were positioned 25 meters apart on flat dry ground in an area infested with mosquitoes. Eight people participated in the experiment, with four randomly assigned to each tent. Of the four stationed at each tent, two were randomly assigned to stay inside the tent (at opposite corners) and two to stay outside the tent (at opposite corners). During a specified 20-minute period during the night, each person kept count of the number of mosquito bites re-

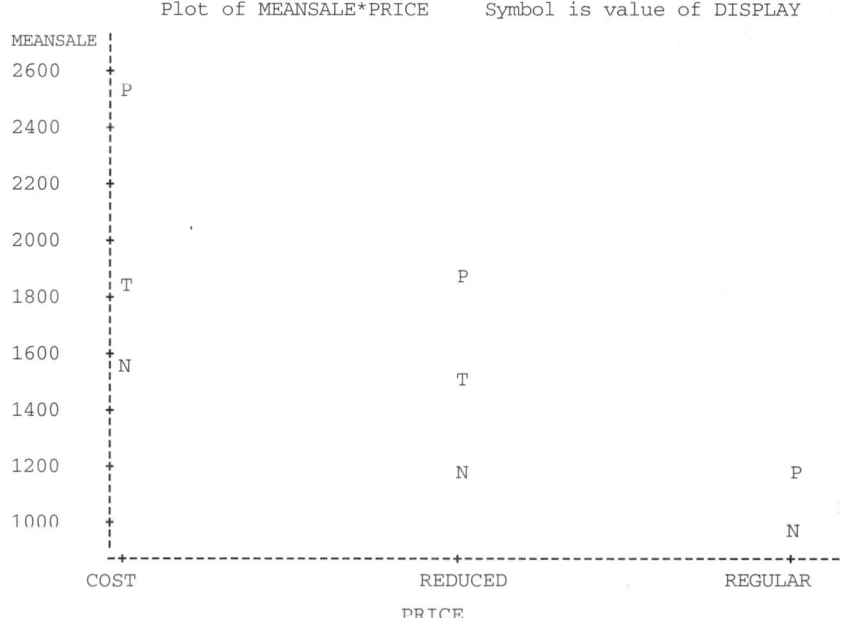

[*Note:* The means for the Normal plus (P) and Twice normal (T) displays are nearly equal for Regular price. Both are represented by the symbol P at this level of price.]

ceived. The goal of the study was to determine the effect of both Tent type (treated or untreated) and Location (inside or outside the tent) on the mean mosquito bite count.

a. What type of design was employed in the study?

b. Identify the factors and treatments.

c. Identify the response variable.

d. The study found statistical evidence of interaction between Tent type and Location. Give a practical interpretation of this result.

10.61 How do women compare with men in their ability to perform laborious tasks that require strength? Some information on this question is provided in a study of the firefighting ability of men and women (*Human Factors,* 1982). The researchers conducted a 2 × 2 factorial experiment to investigate the effect of the factor Sex (male or female) and the factor Weight (light or heavy) on the length of time required for a person to perform a particular firefighting task. Eight persons were selected for each of the 2 × 2 = 4 Sex–Weight categories of the 2 × 2 factorial experiment, and the length of time needed to complete the task was recorded for each of the 32 persons. The means and standard deviations of the four samples are shown in the following table.

	LIGHT		**HEAVY**	
	Mean	**Standard Deviation**	**Mean**	**Standard Deviation**
Female	18.30	6.81	14.50	2.93
Male	13.00	5.04	12.25	5.70

Source: M. D. Phillips and R. L. Pepper, "Shipboard firefighting performance of females and males." Reprinted with permission from *Human Factors,* Vol. 24, No. 3, 1982. Copyright 1982 by the Human Factors and Ergonomics Society. All rights reserved.

a. Calculate the total of the $n = 8$ time measurements for each of the four categories of the 2 × 2 factorial experiment.

b. Calculate the correction for mean, CM. (See Appendix C for computational formulas.)

c. Use the results of parts **a** and **b** to calculate the sums of squares for Sex, Weight, and for the Sex × Weight interaction.

d. Calculate each sample variance. Then calculate the sum of squares of deviations *within* each sample for each of the four samples.

e. Calculate SSE. [*Hint:* SSE is the pooled sum of squares for the deviations calculated in part **d**.]

f. Now that you know SS(Sex), SS(Weight), SS(Sex × Weight), and SSE, find SS(Total).

g. Summarize the calculations in an ANOVA table.

h. Conduct a complete analysis of these data. Use $\alpha = .05$ for any inferential techniques you employ. Interpret your conclusions graphically.

i. What assumptions are necessary to ensure the validity of the inferential techniques you utilized? State them in terms of this experiment.

10.62 The fact that most people are right-handed is attributed to the propensity of the left hemisphere of the brain to control sequential movement. Similarly, the fact that some tasks are performed better with the left hand is likely due to the superiority of the right hemisphere of the brain to process the necessary information. Does such cerebral specialization in spatial processing occur in adults with Down syndrome? A 2 × 2 factorial experiment was conducted to answer this question (*American Journal on Mental Retardation,* May 1995). A sample of adults with Down syndrome were compared to a control group of normal individuals of a similar age. Thus, one factor was Group at two levels (Down syndrome and control) and the second factor was the Handedness (left or right) of the subject. All the subjects performed a task that typically yields a left-hand advantage. The response variable was "laterality index," measured on a −100 to 100 point scale. (A large positive index indicates a right-hand advantage, while a large negative index indicates a left-hand advantage.)

a. Identify the treatments in this experiment.

b. Construct a graph that would support a finding of no interaction between the two factors.

c. Construct a graph that would support a finding of interaction between the two factors.

d. The *F*-test for factor interaction yielded an observed significance level of $p < .05$. Interpret this result.

e. Multiple comparisons of all pairs of treatment means yielded the rankings shown at the bottom of the page. Interpret the results.

f. The experimentwise error rate for the analysis, part **e**, was $\alpha = .05$. Interpret this value.

10.63 If you are overweight, how will your friends respond to a request for a favor? Researchers sought an answer to this question by experimentation (*Journal of General Psychology,* 1983). Each subject in the experiment contacted a second person (a confederate) who was either male or female and either overweight or normal weight. After following a standard procedure, the confederate requested a favor of the subject and a measure of the subject's compliance was recorded. The portion of the data

Mean laterality index	−30	−4	−.5	+5
Group/Handed	Down/Left	Control/Right	Control/Left	Down/Right

	CONFEDERATE'S WEIGHT			
Confederate's Sex	**Overweight**		**Normal**	
Male	32.0	(28.65)	34.67	(36.52)
Female	24.27	(12.94)	36.67	(14.48)

Source: Steinberg, C. L., and Birk, J. M. *Journal of General Psychology,* 1983, Vol. 109. A publication of the Helen Dwight Reid Educational Foundation.

summary that gives the measure of compliance of male subjects to the weight and sex of confederates is shown in the table above.

As the table suggests, the study was conducted as a 2 × 2 factorial experiment. The two factors were the confederate's sex (two levels) and the confederate's weight (two levels). Fifteen male subjects were assigned to each of the confederate Sex–Weight groups. The means and standard deviations of each of the four samples are shown in the table. (Standard deviations are in parentheses.)

a. Identify the following elements of this experiment: experimental units, response, factor(s) and factor type(s), and treatments.
b. What type of experimental design was employed?
c. Given that SS(Total) = 36,772.85, SS(Sex) = 123.12, SS(Weight) = 851.64, and SS(Sex × Weight) = 355.02, complete an ANOVA table for this experiment.
d. Perform a statistical analysis of the experiment. Plot the means to assist in your interpretation of the results of your analysis.

QUICK REVIEW

Key Terms

Analysis of variance (ANOVA)   413
Balanced design   448
Blocks   429
Bonferroni multiple comparisons
    procedure   422
Comparisonwise error rate   422
Complete factorial experiment   444
Completely randomized design   407
Dependent variable   402
Designed experiment   404
Experimental unit   403
Experimentwise error rate   422
F-test   411
Factor interaction   446
Factor levels   403
Factor main effect   446
Factorial experiments   444

Factors   402
Multiple comparisons of means   422
Observational experiment   404
Qualitative factors   402
Quantitative factors   402
Randomized block design   429
Replicates of the experiment   448
Response variable   402
Robust method   414
Scheffé multiple comparisons
    procedure   422
Single-factor experiment   444
Treatments   403
Tukey multiple comparisons
    procedure   422
Two-factor experiment   444

Key Formulas

Note: Computing formulas for sums of squares (SS) and mean squares (MS) in ANOVA are provided in Appendix C.

Completely randomized design:

$$F = \frac{MST}{MSE}$$

Testing treatments   411

Randomized block design:

$$F = \frac{MST}{MSE}$$ Testing treatments 432

$$F = \frac{MSB}{MSE}$$ Testing blocks 437

Factorial design with 2 factors:

$$F = \frac{MS(A)}{MSE}$$ Testing main effect A 448

$$F = \frac{MS(B)}{MSE}$$ Testing main effect B 448

$$F = \frac{MS(AB)}{MSE}$$ Testing $A \times B$ interaction 448

$$c = p(p-1)/2$$ Number of pairwise comparisons for p treatment means 422

LANGUAGE LAB

Symbol	Description
ANOVA	Analysis of variance
SST	Sum of Squares for Treatments (i.e., the variation among treatment means)
SSE	Sum of Squares for Error (i.e., the variability around the treatment means due to sampling error)
MST	Mean Square for Treatments
MSE	Mean Square for Error (an estimate of σ^2)
SSB	Sum of Squares for Blocks
MSB	Mean Square for Blocks
$a \times b$ factorial	Two-factor factorial experiment with one factor at a levels and the other at b levels (thus, there are $a \times b$ treatments in the experiment)
SS(A)	Sum of Squares for Factor A
MS(A)	Mean Square for Factor A
SS(B)	Sum of Squares for Factor B
MS(B)	Mean Square for Factor B
SS(AB)	Sum of Squares for $A \times B$ interaction
MS(AB)	Mean Square for $A \times B$ interaction

SUPPLEMENTARY EXERCISES 10.64–10.83

Note: List the assumptions necessary to ensure the validity of the procedure you use to solve these problems. Exercises marked with 💾 *require the use of a computer.*

Learning the Mechanics

10.64 Explain the difference between an experiment that utilizes a completely randomized design and one that utilizes a randomized block design.

10.65 What are the treatments in a two-factor experiment, with factor A at three levels and factor B at two levels?

10.66 Why does the experimentwise error rate of a multiple comparisons procedure differ from the significance level for each comparison (assuming the experiment has more than two treatments)?

10.67 A completely randomized design is utilized to compare four treatment means. The data are shown in the table below.

Treatment 1	Treatment 2	Treatment 3	Treatment 4
8	6	9	12
10	9	10	13
9	8	8	10
10	8	11	11
11	7	12	11

a. Given that SST = 36.95 and SS(Total) = 62.55, complete an ANOVA table for this experiment.

b. Is there evidence that the treatment means differ? Use $\alpha = .10$.

c. Place a 90% confidence interval on the mean response for treatment 4.

10.68 An experiment utilizing a randomized block design was conducted to compare the mean responses for four treatments, A, B, C, and D. The treatments were randomly assigned to the four experimental units in each of five blocks. The data are shown in the following table.

BLOCK

Treatment	1	2	3	4	5
A	8.6	7.5	8.7	9.8	7.4
B	7.3	6.3	7.3	8.4	6.3
C	9.1	8.3	9.0	9.9	8.2
D	9.3	8.2	9.2	10.0	8.4

a. Given that SS(Total) = 22.308, SS(Block) = 10.688, and SSE = .288, complete an ANOVA table for the experiment.

b. Do the data provide sufficient evidence to indicate a difference among treatment means? Test using $\alpha = .05$.

c. Does the result of the test in part **b** warrant further comparison of the treatment means? If so, how many pairwise comparisons need to be made?

d. Is there evidence that the block means differ? Use $\alpha = .05$.

10.69 The table shows a partially completed ANOVA table for a two-factor factorial experiment.

Source	df	SS	MS	F
A	3	2.6		
B	5	9.2		
$A \times B$				3.1
Error		18.7		
Total	47			

a. Complete the ANOVA table.

b. How many levels were used for each factor? How many treatments were used? How many replications were performed?

c. Find the value of the Sum of Squares for Treatments. Test to determine whether the data provide evidence that the treatment means differ. Use $\alpha = .05$.

d. Is further testing of the nature of factor effects warranted? If so, test to determine whether the factors interact. Use $\alpha = .05$. Interpret the result.

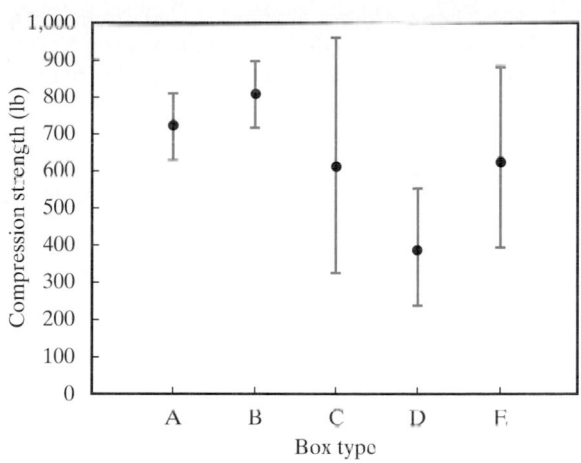

Source: Singh, S. P., *et al.* "Compression of single-wall corrugated shipping containers using fixed and floating test platens." *Journal of Testing and Evaluation*, Vol. 20, No. 4, July 1992, p. 319 (Figure 3). Copyright American Society for Testing and Materials. Reprinted with permission.

Applying the Concepts

10.70 The *Journal of Testing and Evaluation* (July 1992) published an investigation of the mean compression strength of corrugated fiberboard shipping containers. Comparisons were made for boxes of five different sizes: A, B, C, D, and E. Twenty identical boxes of each size were tested and the peak compression strength (pounds) was recorded for each box. The figure above shows the sample means for the five box types as well as the variation around each sample mean.

a. Explain why the data are collected as a completely randomized design.

b. Refer to box types B and D. Based on the graph, does it appear that the mean compressive strengths of these two box types are significantly different? Explain.

c. Based on the graph, does it appear that the mean compressive strengths of all five box types are significantly different? Explain.

10.71 In the past decade, many county-run local jails have assumed major responsibility for treating mentally ill patients. Historically, state mental hospitals have experienced conflicts between the correctional and mental health staffs. A study reported in *Criminology* was conducted to determine whether the conflict between correctional and mental health staffs found in state mental hospitals also exists in local jails. Staff members at 43 jails with mental health programs were mailed questionnaires asking about the conflict encountered in the day-to-day delivery of mental health services in the jails. A total of $n = 167$ questionnaires were returned. In one portion of the study, respondents were classified into one of four groups according to staff affiliation: (1) jail correctional, (2) jail mental health, (3) county mental health, and (4) other mental health. The variable in question was "perceived level of conflict in providing treatment to inmates/patients," measured on a scale of 1 (little or no conflict) to 7 (extreme conflict). The data were subjected to a one-way analysis of variance on staff affiliation with the results summarized in the following table.

Source	df	SS	MS	F
Staff affiliation	—	21.31	7.10	—
Error	163	——	——	
Total	—	120.62		

Source: Steadman, H. J., Morrissey, J. P., and Robbins, P. C. "Reevaluating the custody–therapy conflict paradigm in correctional mental health settings." *Criminology,* Vol. 23, No. 1, 1985, pp. 165–179.

a. Complete the ANOVA table.
b. Is there sufficient evidence of a difference among the mean levels of conflict perceived by the four groups of jail staff members? Test using $\alpha = .05$.
c. Bonferroni's method was used to rank the mean levels of conflict perceived by the four groups of jail staff members at $\alpha = .05$. The rankings are provided in the table below. Interpret the results.

Staff Affiliation	Sample Size	Mean Level of Conflict
Jail correctional	58	1.81
Jail mental health	29	2.32
County mental health	52	2.32
Other mental health	28	2.84

10.72 A large citrus products company is interested in purchasing several new orange juice extractors. To help them with their purchase decision, six different manufacturers of juice extractors have agreed to let the company conduct an experiment to compare the yields of juices for the six different brands of extractors. Because of the possibility of a variation in the amount of juice per orange from one truckload of oranges to another, equal weights of oranges from a single truckload were assigned to each extractor, and this process was repeated for

Source	df	SS	MS	F
Extractors		84.71		
Truckloads		159.29		
Error		95.33		
Total		339.33		

15 loads. The amount of juice recorded for each extractor for each truckload produced the sums of squares shown in the accompanying table.
a. Complete the ANOVA table.
b. Do the data provide sufficient evidence to indicate a difference among the mean amounts of juice extracted by the six extractors? Use $\alpha = .05$.

10.73 A large clothing manufacturer conducted an experiment to study the effect on productivity of increases in its employees' hourly wages. Four treatments were used in the experiment.

Treatment 1:   No increase in hourly wage

Treatment 2:   Increase hourly wage by $.50

Treatment 3:   Increase hourly wage by $1.00

Treatment 4:   Increase hourly wage by $1.50

Twelve employees were selected and grouped into three blocks of size four according to the length of time they had been with the company. The four treatments were randomly assigned to the four employees in each block. The employees were observed for 3 weeks, and their productivity was measured as the average number of nondefective garments each produced per hour. The resulting productivity measures appear in the table below.

	TREATMENT			
	1	2	3	4
Group 1 (less than 1 year)	2.4	3.0	3.1	3.2
Group 2 (1−5 years)	4.8	6.1	5.9	5.7
Group 3 (more than 5 years)	5.1	7.0	7.2	7.3

a. What type of experimental design was used in this study? Why do you think such a design was employed?
b. The SPSS printout for this experiment is shown at the top of page 465. Is there evidence that the mean productivity levels differ among the four pay programs? Use $\alpha = .05$.
c. What is the observed significance level for the test you conducted in part **b**?
d. The results of Bonferroni's technique to compare all the pairs of treatment means, with an overall significance level of $\alpha = .10$, are shown in the SAS printout on page 465. Interpret them.

10.74 A research psychologist wishes to investigate the difference in maze test scores for a strain of laboratory mice trained under different laboratory conditions. The experiment is conducted using 18 randomly selected mice of this strain, with six receiving no training at all (control group), six trained under condition 1, and six trained under condition 2. Then each of the mice is given a test

```
* * *  A N A L Y S I S   O F   V A R I A N C E * * *

               ·™™™··
       By   TREATMNT
            GROUP
                        Sum of              Mean              Signif
Source of Variation     Squares      DF     Square        F   of F

Main Effects            33.362       5      6.672    45.236    .000
   TREATMNT              3.740       3      1.247     8.452    .014
   GROUP                29.622       2     14.811   100.412    .000

Explained               33.362       5      6.672    45.236    .000

Residual                  .885       6       .148

Total                   34.247      11      3.113
```

[*Note:* SPSS shows the combined treatment and block effects in both the "Main Effects" and "Explained" rows of the printout. Also, SPSS uses "Residual" instead of "Error."]

```
                Analysis of Variance Procedure

          Bonferroni (Dunn) T tests for variable: NUMBER

NOTE: This test controls the type I experimentwise error rate but
      generally has a higher type II error rate than Tukey's for all
      pairwise comparisons.

          Alpha= 0.1  Confidence= 0.9  df= 6  MSE= 0.1475
                  Critical Value of T= 3.28746
              Minimum Significant Difference= 1.0309

   Comparisons significant at the 0.1 level are indicated by '***'.

                      Simultaneous           Simultaneous
                         Lower    Difference    Upper
            TREATMNT   Confidence   Between   Confidence
            Comparison    Limit      Means       Limit

         3   - 4        -1.031      0.000      1.031
         3   - 2        -0.998      0.033      1.064
         3   - 1         0.269      1.300      2.331      ***

         4   - 3        -1.031      0.000      1.031
         4   - 2        -0.998      0.033      1.064
         4   - 1         0.269      1.300      2.331      ***

         2   - 3        -1.064     -0.033      0.998
         2   - 4        -1.064     -0.033      0.998
         2   - 1         0.236      1.267      2.298      ***

         1   - 3        -2.331     -1.300     -0.269      ***
         1   - 4        -2.331     -1.300     -0.269      ***
         1   - 2        -2.298     -1.267     -0.236      ***
```

score between 0 and 100, depending on its performance in a test maze. The experiment produced the results shown in the following table.

a. Identify the following elements of this experiment: experimental units, response, factor(s) and factor type(s), and treatments.

b. Is this a designed or observational experiment?

Control	Condition 1	Condition 2
58	73	53
32	70	74
59	68	72
64	71	62
55	60	58
49	62	61

c. Is there sufficient evidence to indicate a difference among mean maze test scores for mice trained under the three different laboratory conditions? Use Appendix C or a computer program to perform the calculations, and test at $\alpha = .10$.

d. Use a compter package with a multiple comparisons routine to compare the control mean to each of the condition means. How many comparisons are involved? Use an overall significance level of $\alpha = .10$.

10.75 An evaluation of diffusion bonding of zircaloy components is performed. The main objective is to determine which of three elements—nickel, iron, or copper—is the best bonding agent. A series of zircaloy components are bonded with each of the possible bonding agents. Since there is a great deal of variation in components machined from different ingots, a randomized block design is used, blocking on the ingots. A pair of components from each ingot are bonded together using each of the three agents, and the pressure (in units of 1,000 pounds per square inch) required to separate the bonded components is measured. The data in the table are obtained.

BONDING AGENT

Ingot	Nickel	Iron	Copper
1	67.0	71.9	72.2
2	67.5	68.8	66.4
3	76.0	82.6	74.5
4	72.7	78.1	67.3
5	73.1	74.2	73.2
6	65.8	70.8	68.7
7	75.6	84.9	69.0

a. Identify the following elements of the experiment: experimental units, blocks, response, factor(s) and factor type(s), and treatments.

b. Is the experiment designed or observational? Explain.

c. The SAS printout for this experiment is shown below. Construct an ANOVA summary table using the printout.

d. Is there sufficient evidence that the mean pressure required to separate the components differs for the three bonding agents? Use $\alpha = .05$.

e. Interpret the results of the multiple comparisons of bonding agents shown on the printout.

f. What assumptions must hold to ensure the validity of the inferences in parts **d** and **e**?

```
                    Analysis of Variance Procedure

Dependent Variable: PRESSURE

                                Sum of          Mean
Source              DF         Squares        Square     F Value    Pr > F

Model                8     400.1904762    50.0238095       4.82     0.0076
Error               12     124.4590476    10.3715873
Corrected Total     20     524.6495238

                R-Square            C.V.      Root MSE       PRESSURE Mean
                0.762777        4.448490      3.220495          72.3952381

Source              DF       Anova SS   Mean Square     F Value    Pr > F

BONDING              2    131.9009524    65.9504762       6.36     0.0131
INGOT                6    268.2895238    44.7149206       4.31     0.0151

        Tukey's Studentized Range (HSD) Test for variable: PRESSURE

Note: This test controls the type I experimentwise error rate, but
      generally has a higher type II error rate than REGWQ.

            Alpha= 0.05  df= 12  MSE= 10.37159
        Critical Value of Studentized Range= 3.773
        Minimum Significant Difference= 4.5926

    Means with the same letter are not significantly different.

        Tukey Grouping         Mean     N   BONDING

                   A         75.900     7   IRON

                   B         71.100     7   NICKEL
                   B
                   B         70.186     7   COPPER
```

10.76 Sixteen workers were randomly selected to participate in an experiment to determine the effects of work scheduling and method of payment on attitude toward the job. Two types of scheduling were employed, the standard 8-to-5 workday and a modification in which the worker was permitted to start the day at either 7 or 8 A.M. and to vary the starting time as desired; in addition, the worker was allowed to choose, on a daily basis, either a ½-hour or 1-hour lunch period. The two methods of payment were a standard hourly rate and a reduced hourly rate with an added piece rate based on the worker's production. Four workers were randomly assigned to each of the four Scheduling–Payment combinations, and each completed an attitude test after 1 month on the job. The test scores are shown in the table below.

a. What type of experiment was performed? Identify the response, factor(s), factor type(s), treatments, and experimental units.

b. The SAS printout for this experiment is shown below the table. Is there evidence that the treatment means differ? Use $\alpha = .05$.

c. If the test in part **b** warrants further analysis, conduct the appropriate tests of interaction and main effects. Interpret your results.

d. What assumptions are necessary to ensure the validity of the inferences? State the assumptions in terms of this experiment.

10.77 In a nutrition experiment, an investigator studied the effects of different rations on the growth of young rats. Forty rats from the same inbred strain were divided at random into four groups of 10 and used for the experiment. A different ration was fed to each group and, after a specified length of time, the increase in growth of each rat was measured (in grams). The data are shown in the following table.

Ration A		Ration B		Ration C		Ration D	
10	6	13	9	12	10	15	21
8	6	15	10	16	12	13	18
12	9	14	8	13	10	15	20
11	5	13	10	11	9	10	19
9	6	17	8	15	9	12	22

a. Identify the experimental units, the response, the factor(s) and factor type(s), and the treatments for this experiment.

b. Is the experiment designed or observational? Explain.

c. Given that SS(Ration) = 348.675 and SSE = 342.300, complete an ANOVA table for this experiment.

d. Do the data provide sufficient evidence to conclude that a difference exists in the mean growth of rats receiving the different rations? Test using $\alpha = .05$.

e. If warranted, use a multiple comparisons procedure to compare the pairs of mean growths

	PAYMENT	
Scheduling	**Hourly Rate**	**Hourly and Piece Rate**
8–5	54, 68, 55, 63	89, 75, 71, 83
Worker-Modified Schedule	79, 65, 62, 74	83, 94, 91, 86

```
                   Analysis of Variance Procedure

Dependent Variable: SCORE

                           Sum of          Mean
Source            DF       Squares        Square      F Value    Pr > F

Model              3     1806.00000     602.00000      12.29     0.0006
Error             12      588.00000      49.00000
Corrected Total   15     2394.00000

                R-Square          C.V.       Root MSE          SCORE Mean
                0.754386     9.3959732       7.00000           74.5000000

Source            DF     Anova SS    Mean Square    F Value    Pr > F

SCHEDULE           1     361.0000       361.0000       7.37     0.0188
PAYMENT            1    1444.0000      1444.0000      29.47     0.0002
SCHEDULE*PAYMENT   1       1.0000         1.0000       0.02     0.8888
```

corresponding to the four rations. Use an over-all significance level of $\alpha = .10$.

f. Suppose that prior to conducting the experiment, the investigator expected the mean increase in weight for rats receiving ration B to exceed 10 grams. Construct a 95% confidence interval for the true mean increase for ration B. Does the interval support the investigator's theory?

10.78 A farmer wants to determine the effect of five different concentrations of lime on the pH (acidity) of the soil on a farm. Fifteen soil samples are to be used in the experiment, five from each of three different locations. The five soil samples from each location are then randomly assigned to the five concentrations of lime, and 1 week after the lime is applied the pH of the soil is measured. The data are shown in the following table.

LIME CONCENTRATION

Location	0	1	2	3	4
I	3.2	3.6	3.9	4.0	4.1
II	3.6	3.7	4.2	4.3	4.3
III	3.5	3.9	4.0	3.9	4.2

a. What type of experimental design was utilized? Explain.

b. Given that SS(Total) = 1.42933, SS(Lime) = 1.14267, and SSE = .11733, complete an ANOVA table for this experiment.

c. What is the estimated standard deviation for the experiment? Interpret it.

d. Is there evidence that the mean soil pH levels differ over the five lime concentrations? Use $\alpha = .05$.

e. Is there evidence that the mean pH levels vary over the three locations? Use $\alpha = .05$. What is the practical significance of the result of this test?

10.79 An agricultural research unit wished to study the yields of three strains of wheat, A, B, and C. Because of the direction of drainage, it was suspected that the field used for the experiment varied substantially in fertility along the north–south line. Consequently, to reduce data variation the three strains of wheat were planted in plots laid out in east–west blocks. One plot of each type of wheat was assigned to each block, and the strains were assigned in random order within each block. The layout of the design and the data on wheat yields for each strain within each block (in bushels) are shown in the accompanying figure.

a. Identify the following elements of this experiment: experimental units, response, factor(s) and factor type(s), blocks, treatments.

b. Use the appropriate formulas in Appendix C or a computer to construct an ANOVA summary table for this experiment.

A 53	B 62	C 47	A 56	B 58
B 60	C 44	A 49	B 55	C 45
C 49	A 57	B 61	C 43	A 51

c. Is there evidence that the mean yield differs among the three strains of wheat? Use $\alpha = .10$.

d. Can you use the results of this experiment to obtain a valid confidence interval for the mean yield for strain C? Explain.

10.80 Three anticoagulant drugs are studied to compare their effectiveness in dissolving blood clots. Each of five subjects receives the drugs at equally spaced time intervals and in random order. Time periods between drug applications permit a drug to be passed out of a subject's body before the subject receives the next drug. After each drug is in the bloodstream, the length of time (in seconds) required for a cut of specified size to stop bleeding is recorded. The results are shown in the following table.

DRUG

Person	A	B	C
1	127.5	129.0	135.5
2	130.6	129.1	138.0
3	118.3	111.7	110.1
4	155.5	144.3	162.3
5	180.7	174.4	181.8

a. What type of experimental design was used in this study? Identify the response, factor(s), factor type(s), treatments, and experimental units.

b. The SAS printout for this experiment is shown on page 469. Is there evidence of a difference in mean clotting time among the three drugs? Test using $\alpha = .01$.

c. What is the observed significance level of the test you conducted in part **a**? Interpret it.

d. Was blocking effective in reducing the variation among the data? That is, do the data support the contention that the mean clotting time varies from person to person?

e. If warranted, use a multiple comparisons technique to determine whether one of the drugs is most effective. Use an overall significance level of $\alpha = .10$.

10.81 In hope of attracting more riders, a city transit company plans to have express bus service from a suburban terminal to the downtown business district. These buses will travel along a major city street where there are numerous traffic lights that affect travel time. The city decides to perform a

```
                    Analysis of Variance Procedure

Dependent Variable. TIME

                              Sum of          Mean
Source                DF      Squares        Square    F Value    Pr > F

Model                  6   7802.1746667   1300.3624444   64.97    0.0001
Error                  8    160.1093333     20.0136667
Corrected Total       14   7962.2840000

            R-Square              C.V.      Root MSE        TIME Mean
            0.979892           3.1522433   4.4736637     141.92000000

Source                DF      Anova SS    Mean Square   F Value    Pr > F

DRUG                   2     156.36400      78.18200      3.91    0.0655
PERSON                 4    7645.81067    1911.45267     95.51    0.0001
```

study of the effect of four different plans (a special bus lane, traffic signal progression, etc.) on the travel times for the buses. Travel times (in minutes) are measured for several weekdays during a morning rush-hour trip while each plan is in effect. The results are recorded in the following table.

PLAN

1	2	3	4
27	25	34	30
25	28	29	33
29	30	32	31
26	27	31	
	24	36	

a. What experimental design was used?

b. Use the appropriate formulas in Appendix C or a computer program to construct an ANOVA summary table for this experiment.

c. Is there evidence of a difference between the mean travel times for the four plans? Use $\alpha = .01$.

d. If warranted, use a multiple comparisons procedure to compare the pairs of mean travel times corresponding to the four plans. Use $\alpha = .05$.

10.82 Refer to Case Study 10.2. Since there was no evidence to indicate any difference in mathematics scores between the two colleges, Clute (1984) combined the data for the two colleges and analyzed the portion of the mathematics examination scores that pertained to the low-level items on the test.

a. The sample sizes, mean scores, and standard deviations (in parentheses) for the six Method–Anxiety combinations are shown in the table below. Plot the mean score on the vertical axis against the anxiety level on the horizontal axis, using the first letter of the instructional method as a plotting symbol. Does it appear that teaching method and anxiety level interact to affect the mean test score?

b. Clute's ANOVA table is shown at the top left of page 470. Interpret the results and discuss whether they support your impressions from part **a.**

Means and Standard Deviations on the Low-Level Items by Instructional Method and Level of Anxiety

	MATHEMATICS ANXIETY LEVEL						
	LOW		MEDIUM		HIGH		
Method	n	$\bar{x}$	n	$\bar{x}$	n	$\bar{x}$	Total
Discovery	14	218.79 (20.81)	14	183.43 (27.17)	13	146.85 (36.93)	183.90 (40.77)
Expository	14	183.43 (52.27)	13	182.46 (41.64)	13	185.92 (33.72)	183.92 (42.38)
Total		201.11 (43.99)		182.96 (34.20)		166.38 (39.97)	183.91 (41.31)

Analysis of Variance for Low-Level Items

Source	df	MS	F
Method (M)	1	17	.01
Anxiety (A)	2	8,146	6.02[a]
$M \times A$	2	9,341	6.90[a]
Residual	75	1,354	

[a] $p < .01$.

Analysis of Variance for High-Level Items

Source	df	MS	F
Method (M)	1	5,724	5.65[a]
Anxiety (A)	2	12,573	12.39[b]
$M \times A$	2	1,176	1.16
Residual	75	1,015	

[a] $p < .05$.
[b] $p < .01$.

Means and Standard Deviations on the High-Level Items by Instructional Method and Level of Anxiety

	MATHEMATICS ANXIETY LEVEL						
	LOW		MEDIUM		HIGH		
Method	n	$\bar{x}$	n	$\bar{x}$	n	$\bar{x}$	Total
Discovery	14	134.43 (23.06)	14	109.14 (12.88)	13	78.08 (28.67)	107.93 (31.77)
Expository	14	104.71 (31.42)	13	91.69 (47.92)	13	74.77 (37.39)	90.75 (40.26)
Total		119.57 (30.99)		100.74 (34.95)		76.42 (32.69)	99.44 (37.01)

Note: Maximum score = 240.

10.83 Refer to Case Study 10.2 and Exercise 10.82. The table shown immediately above gives data for the portion of the mathematics examination scores that pertained to the high-level items on the test (again combining the two colleges). Clute's summary ANOVA table for this portion of the test is given at the top right of the page. Perform the same analysis described in parts **a** and **b** of Exercise 10.82 for the high-level item data. Compare the results to those obtained for the low-level items.

STUDENT PROJECTS

Owing to ever-increasing food costs, consumers are becoming more discerning in their choice of supermarkets. It is usually more convenient to shop at just one market, as opposed to buying different items at different markets. Thus, it would be useful to compare the mean food expenditure for a market basket of food items from store to store. Since there is a great deal of variability in the prices of products sold at any supermarket, we will consider an experiment that blocks on products.

Choose three (or more) supermarkets in your area that you want to compare. Then choose approximately 10 (or more) food products you typically purchase. For each food item, record the price each store charges in the following manner:

Food Item 1	Food Item 2	...	Food Item 10
Price store 1	Price store 1	...	Price store 1
Price store 2	Price store 2	...	Price store 2
Price store 3	Price store 3	...	Price store 3

Use the data you obtain to test

H_0: Mean expenditures at the stores are the same

H_a: Mean expenditures for at least two of the stores are different

Also, test to determine whether blocking on food items is advisable in this kind of experiment. Fully interpret the results of your analysis.

EXPLORING DATA WITH A COMPUTER

Do mean starting salaries of college graduates differ among majors? Treat the starting salary data set described in Appendix B as a collection of independent random samples of salaries from the different majors offered at the University of South Florida. Then select for analysis the data for four majors of your choice (e.g., Accounting, Electrical Engineering, Psychology, and Marketing).

a. Identify the following elements of this experiment: design, experimental units, response, factor(s) and factor levels, and treatments.

b. Construct a dot plot and stem-and-leaf display for each of the four samples. What do your graphs suggest about the differences in mean starting salaries of the four majors?

c. Conduct an analysis of variance to determine whether differences exist among the four majors' mean starting salaries.

d. Use a multiple comparisons procedure to compare the four major (treatment) means.

e. What assumptions must you make in order to employ the inferential procedures of parts b and c? Which of these assumptions are suspect in this exercise?

Chapter 11

SIMPLE LINEAR REGRESSION

Contents

Case Studies

WHERE WE'VE BEEN

We've learned how to estimate and test hypotheses about population parameters based on a random sample of observations from the population. We've also seen how to extend these methods to allow for a comparison of parameters from two or more populations.

WHERE WE'RE GOING

Suppose we want to predict the rainfall at a given location on a given day. We could select a single random sample of n daily rainfalls, use the methods of Chapter 7 to estimate the mean daily rainfall μ, and then use this quantity to predict the day's rainfall. A better method uses information that is available to any forecaster, e.g., barometric pressure and cloud cover. If we measure barometric pressure and cloud cover at the same time as daily rainfall, we establish the relationship between these variables—one that lets us use these variables for prediction. This chapter covers the simplest situation—relating two variables. The more complex problem of relating more than two variables is the topic of Chapter 12.

In Chapters 7–10 we described methods for making inferences about population means. The mean of a population has been treated as a *constant,* and we have shown how to use sample data to estimate or to test hypotheses about this constant mean. In many applications, the mean of a population is not viewed as a constant, but rather as a variable. For example, the mean sale price of residences in a large city during 1995 can be treated as a constant and might be equal to $150,000. But we might also treat the mean sale price as a variable that depends on the square feet of living space in the residence. For example, the relationship might be

$$\text{Mean sale price} = \$30,000 + \$60(\text{Square feet})$$

This formula implies that the mean sale price of 1,000-square-foot homes is $90,000, the mean sale price of 2,000-square-foot homes is $150,000, and the mean sale price of 3,000-square-foot homes is $210,000.

What do we gain by treating the mean as a variable rather than a constant? In many practical applications we will be dealing with highly variable data, data for which the standard deviation is so large that a constant mean is almost "lost" in a sea of variability. For example, if the mean residential sale price is $150,000 but the standard deviation is $75,000, then the actual sale prices will vary considerably, and the mean price is not a very meaningful or useful characterization of the price distribution. On the other hand, if the mean sale price is treated as a variable that depends on the square feet of living space, the standard deviation of sale prices for any given size of home might be only $10,000. In this case, the mean price will provide a much better characterization of sale prices when it is treated as a variable rather than a constant.

In this chapter we discuss situations in which the mean of the population is treated as a variable, dependent on the value of another variable. The dependence of residential sale price on the square feet of living space is one illustration. Other examples include the dependence of mean reaction time on the amount of a drug in the bloodstream, the dependence of mean starting salary of a college graduate on the student's GPA, and the dependence of mean number of years to which a criminal is sentenced on the number of previous convictions.

In this chapter we discuss the simplest of all models relating a population mean to another variable, the *straight-line model.* We show how to use the sample data to estimate the straight-line relationship between the mean value of one variable, *y,* as it relates to a second variable, *x.* The methodology of estimating and using a straight-line relationship is referred to as *simple linear regression analysis.*

11.1   PROBABILISTIC MODELS

An important consideration when taking a drug is how it may affect one's perception or general awareness. Suppose you want to model the length of time it takes to respond to a stimulus (a measure of awareness) as a function of the percentage of a certain drug in the bloodstream. The first question to be answered is this: "Do you think an exact relationship exists between these two variables?" That is, do you think it is possible to state the exact length of time it takes an individual (subject) to respond if the amount of the drug in the bloodstream is known? We think you will agree with us that this is *not* possible for several reasons. The reaction time depends on many variables other than the percentage of the drug in the bloodstream—for example, the time of day, the amount of sleep

FIGURE 11.1

Possible reaction times, *y*, for five different drug percentages, *x*

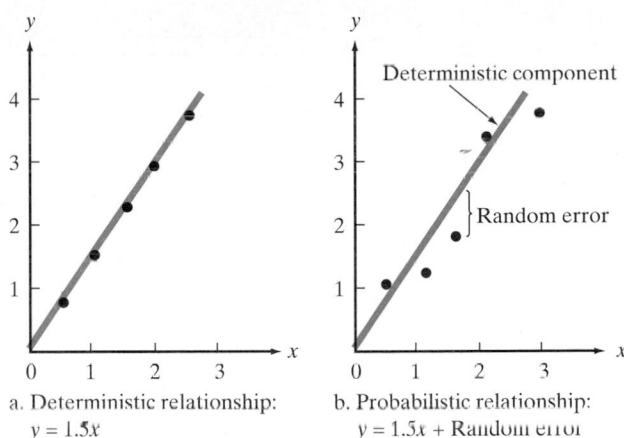

a. Deterministic relationship:
$y = 1.5x$

b. Probabilistic relationship:
$y = 1.5x + \text{Random error}$

the subject had the night before, the subject's visual acuity, the subject's general reaction time without the drug, and the subject's age would all probably affect reaction time. Even if many variables are included in a model (the topic of Chapter 12), it is still unlikely that we would be able to predict *exactly* the subject's reaction time. There will almost certainly be some variation in response times due strictly to *random phenomena* that cannot be modeled or explained.

If we were to construct a model that hypothesized an exact relationship between variables, it would be called a **deterministic model**. For example, if we believe that *y*, the reaction time (in seconds), will be exactly one and one-half times *x*, the amount of drug in the blood, we write

$$y = 1.5x$$

This represents a **deterministic relationship** between the variables *y* and *x*. It implies that *y* can always be determined exactly when the value of *x* is known. *There is no allowance for error in this prediction.*

If, on the other hand, we believe there will be unexplained variation in reaction times—perhaps caused by important but unincluded variables or by random phenomena—we discard the deterministic model and use a model that accounts for this **random error**. This **probabilistic model** includes both a deterministic component and a random error component. For example, if we hypothesize that the response time *y* is related to the percentage of drug *x* by

$$y = 1.5x + \text{Random error}$$

we are hypothesizing a **probabilistic relationship** between *y* and *x*. Note that the deterministic component of this probabilistic model is 1.5*x*.

Figure 11.1a shows the possible responses for five different values of *x*, the percentage of drug in the blood, when the model is deterministic. All the responses must fall exactly on the line because a deterministic model leaves no room for error.

Figure 11.1b shows a possible set of responses for the same values of *x* when we are using a probabilistic model. Note that the deterministic part of the model (the straight line itself) is the same. Now, however, the inclusion of a random error component allows the response times to vary from this line. Since we know that the response time does vary randomly for a given value of *x*, the probabilistic model provides a more realistic model for *y* than does the deterministic model.

General Form of Probabilistic Models

$$y = \text{Deterministic component} + \text{Random error}$$

where y is the variable of interest. We always assume that the mean value of the random error equals 0. This is equivalent to assuming that the mean value of y, $E(y)$, equals the deterministic component of the model; that is,

$$E(y) = \text{Deterministic component}$$

We begin with the simplest of probabilistic models—the **straight-line model**—which derives its name from the fact that the deterministic portion of the model graphs as a straight line. Fitting this model to a set of data is an example of **regression analysis**, or **regression modeling**. The elements of the straight-line model are summarized in the next box.

A First-Order (Straight-Line) Probabilistic Model

$$y = \beta_0 + \beta_1 x + \epsilon$$

where $y = $ *Dependent or response variable* (variable to be modeled)
$x = $ *Independent or predictor variable* (variable used as a predictor of y)*

$E(y) = \beta_0 + \beta_1 x = $ Deterministic component

ϵ (epsilon) $ = $ Random error component

β_0 (beta zero) $ = $ y-*intercept of the line,* that is, the point at which the line intercepts or cuts through the y-axis (see Figure 11.2)

β_1 (beta one) $ = $ *Slope of the line,* that is, the amount of increase (or decrease) in the deterministic component of y for every 1-unit increase in x. [As you can see in Figure 11.2, $E(y)$ increases by the amount β_1 as x increases from 2 to 3.]

FIGURE 11.2

The straight-line model

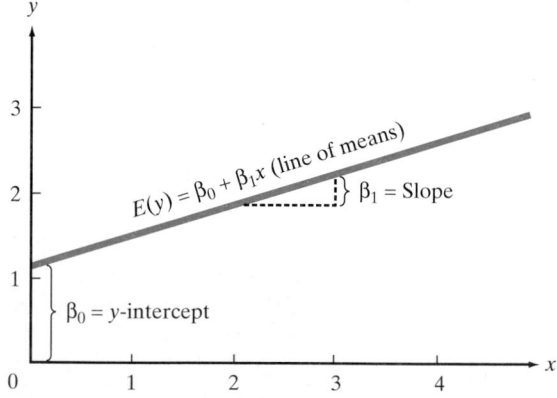

*The word *independent* should not be interpreted in a probabilistic sense, as defined in Chapter 3. The phrase *independent variable* is used in regression analysis to refer to a predictor variable for the response y.

In the probabilistic model, the deterministic component is referred to as the **line of means**, because the mean of y, $E(y)$, is equal to the straight-line component of the model. That is,

$$E(y) = \beta_0 + \beta_1 x$$

Note that the Greek symbols β_0 and β_1, respectively, represent the y-intercept and slope of the model. They are population parameters that will be known only if we have access to the entire population of (x, y) measurements. Together with a specific value of the independent variable x, they determine the mean value of y, which is just a specific point on the line of means (Figure 11.2).

The values of β_0 and β_1 will be unknown in almost all practical applications of regression analysis. The process of developing a model, estimating the unknown parameters, and using the model can be viewed as the five-step procedure shown in the next box.

> **Step 1**   Hypothesize the deterministic component of the model that relates the mean, $E(y)$, to the independent variable x (Section 11.1).
>
> **Step 2**   Use the sample data to estimate unknown parameters in the model (Section 11.2).
>
> **Step 3**   Specify the probability distribution of the random error term and estimate the standard deviation of this distribution (Sections 11.3 and 11.4).
>
> **Step 4**   Statistically evaluate the usefulness of the model (Sections 11.5, 11.6, and 11.7).
>
> **Step 5**   When satisfied that the model is useful, use it for prediction, estimation, and other purposes (Section 11.8).

In this chapter only the straight-line model is discussed; more complex models are addressed in Chapters 12 and 13.

EXERCISES 11.1–11.9

Learning the Mechanics

11.1 In each case, graph the line that passes through the given points.
  a. $(1, 1)$ and $(5, 5)$ **b.** $(0, 3)$ and $(3, 0)$
  c. $(-1, 1)$ and $(4, 2)$ **d.** $(-6, -3)$ and $(2, 6)$

11.2 Give the slope and y-intercept for each of the lines graphed in Exercise 11.1.

11.3 The equation for a straight line (deterministic) is

$$y = \beta_0 + \beta_1 x$$

If the line passes through the point $(-2, 4)$, then $x = -2, y = 4$ must satisfy the equation; that is,

$$4 = \beta_0 + \beta_1(-2)$$

Similarly, if the line passes through the point $(4, 6)$, then $x = 4, y = 6$ must satisfy the equation; that is,

$$6 = \beta_0 + \beta_1(4)$$

Use these two equations to solve for β_0 and β_1; then find the equation of the line that passes through the points $(-2, 4)$ and $(4, 6)$.

11.4 Refer to Exercise 11.3. Find the equations of the lines that pass through the points listed in Exercise 11.1.

11.5 Plot the following lines:
  a. $y = 4 + x$ **b.** $y = 5 - 2x$ **c.** $y = -4 + 3x$
  d. $y = -2x$ **e.** $y = x$ **f.** $y = .50 + 1.5x$

11.6 Give the slope and y-intercept for each of the lines defined in Exercise 11.5.

11.7 Why do we generally prefer a probabilistic model to a deterministic model? Give examples for which the two types of models might be appropriate.

11.8 What is the line of means?

11.9 If a straight-line probabilistic relationship relates the mean $E(y)$ to an independent variable x, does it imply that every value of the variable y will always fall exactly on the line of means? Why or why not?

11.2   FITTING THE MODEL: THE LEAST SQUARES APPROACH

After the straight-line model has been hypothesized to relate the mean $E(y)$ to the independent variable x, the next step is to collect data and to estimate the (unknown) population parameters, the y-intercept β_0 and the slope β_1.

To begin with a simple example, suppose an experiment involving five subjects is conducted to determine the relationship between the percentage of a certain drug in the bloodstream and the length of time it takes to react to a stimulus. The results are shown in Table 11.1. (The number of measurements and the measurements themselves are unrealistically simple in order to avoid arithmetic confusion in this introductory example.) This set of data will be used to demonstrate the five-step procedure of regression modeling given in Section 11.1. In this section we hypothesize the deterministic component of the model and estimate its unknown parameters (steps 1 and 2). The model assumptions and the random error component (step 3) are the subjects of Sections 11.3 and 11.4, whereas Sections 11.5–11.7 assess the utility of the model (step 4). Finally, we use the model for prediction and estimation (step 5) in Section 11.8.

Step 1 *Hypothesize the deterministic component of the probabilistic model.* As stated before, we will consider only straight-line models in this chapter. Thus the complete model to relate mean response time $E(y)$ to drug percentage x is given by

$$E(y) = \beta_0 + \beta_1 x$$

Step 2 *Use sample data to estimate unknown parameters in the model.* This step is the subject of this section—namely, how can we best use the information in the sample of five observations in Table 11.1 to estimate the unknown y-intercept β_0 and slope β_1?

To determine whether a linear relationship between y and x is plausible, it is helpful to plot the sample data. Such a plot, called a **scattergram**, locates each of the five data points on a graph, as shown in Figure 11.3. Note that the scattergram suggests a general tendency for y to increase as x increases. If you place a ruler on the scattergram, you will see that a line may be drawn through three of the five points, as shown in Figure 11.4. To obtain the equation of this visually fitted line, note that the line intersects the y-axis at $y = -1$, so the y-intercept is -1. Also, y increases exactly 1 unit for every 1-unit increase in x, indicating that the slope is $+1$. Therefore, the equation is

$$\tilde{y} = -1 + 1(x) = -1 + x$$

where $\tilde{y}$ is used to denote the predicted y from the visual model.

TABLE 11.1   Reaction Time Versus Drug Percentage

Subject	Amount of Drug x (%)	Reaction Time y (seconds)
1	1	1
2	2	1
3	3	2
4	4	2
5	5	4

FIGURE 11.3

Scattergram for data in
Table 11.1

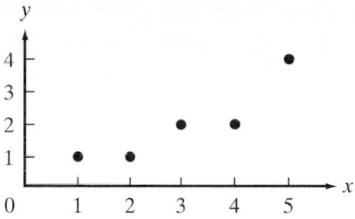

FIGURE 11.4

Visual straight line fitted to
the data in Figure 11.3

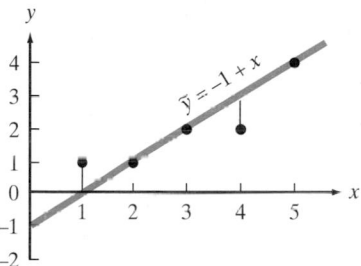

One way to decide quantitatively how well a straight line fits a set of data is to
note the extent to which the data points deviate from the line. For example, to
evaluate the model in Figure 11.4, we calculate the magnitude of the **deviations**,
i.e., the differences between the observed and the predicted values of y. These
deviations, or **errors**, are the vertical distances between observed and predicted
values (see Figure 11.4). The observed and predicted values of y, their differences, and their squared differences are shown in Table 11.2. Note that the *sum of
errors* equals 0 and the *sum of squares of the errors* (SSE), which gives greater
emphasis to large deviations of the points from the line, is equal to 2.

You can see by shifting the ruler around the graph that it is possible to find
many lines for which the sum of errors is equal to 0, but it can be shown that
there is one (and only one) line for which the SSE is a *minimum*. This line is
called the **least squares line**, the **regression line**, or the **least squares prediction
equation**.

To find the least squares prediction equation for a set of data, assume that we
have a sample of n data points consisting of pairs of values of x and y, say (x_1, y_1),
(x_2, y_2), ..., (x_n, y_n). For example, the $n = 5$ data points shown in Table 11.2 are
$(1, 1)$, $(2, 1)$, $(3, 2)$, $(4, 2)$, and $(5, 4)$. The fitted line, which we will calculate based
on the five data points, is written as

$$\hat{y} = \hat{\beta}_0 + \hat{\beta}_1 x$$

TABLE 11.2   Comparing Observed and Predicted Values for the Visual Model

x	y	$\tilde{y} = -1 + x$	$(y - \tilde{y})$	$(y - \tilde{y})^2$
1	1	0	$(1 - 0) =$ 1	1
2	1	1	$(1 - 1) =$ 0	0
3	2	2	$(2 - 2) =$ 0	0
4	2	3	$(2 - 3) = -1$	1
5	4	4	$(4 - 4) =$ 0	0
			Sum of errors = 0	Sum of squared errors (SSE) = 2

The "hats" indicate that the symbols below them are estimates: $\hat{y}$ (y-hat) is an estimator of the mean value of y, $E(y)$, and a predictor of some future value of y; and $\hat{\beta}_0$ and $\hat{\beta}_1$ are estimators of β_0 and β_1, respectively.

For a given data point, say the point (x_i, y_i), the observed value of y is y_i and the predicted value of y would be obtained by substituting x_i into the prediction equation:

$$\hat{y}_i = \hat{\beta}_0 + \hat{\beta}_1 x_i$$

And the deviation of the ith value of y from its predicted value is

$$(y_i - \hat{y}_i) = [y_i - (\hat{\beta}_0 + \hat{\beta}_1 x_i)]$$

Then the sum of squares of the deviations of the y-values about their predicted values for all the n data points is

$$\text{SSE} = \sum [y_i - (\hat{\beta}_0 + \hat{\beta}_1 x_i)]^2$$

The quantities $\hat{\beta}_0$ and $\hat{\beta}_1$ that make the SSE a minimum are called the **least squares estimates** of the population parameters β_0 and β_1, and the prediction equation $\hat{y} = \hat{\beta}_0 + \hat{\beta}_1 x$ is called the *least squares line.*

DEFINITION 11.1

The **least squares line** is one that has the following two properties:

1. the sum of the errors (SE) equals 0
2. the sum of squared errors (SSE) is smaller than that for any other straight-line model

---

The values of $\hat{\beta}_0$ and $\hat{\beta}_1$ that minimize the SSE are (proof omitted) given by the formulas in the next box.*

Formulas for the Least Squares Estimates

Slope: $\hat{\beta}_1 = \dfrac{\text{SS}_{xy}}{\text{SS}_{xx}}$

y-intercept: $\hat{\beta}_0 = \bar{y} - \hat{\beta}_1 \bar{x}$

where $\text{SS}_{xy} = \sum (x_i - \bar{x})(y_i - \bar{y}) = \sum x_i y_i - \dfrac{\left(\sum x_i\right)\left(\sum y_i\right)}{n}$

$\text{SS}_{xx} = \sum (x_i - \bar{x})^2 = \sum x_i^2 - \dfrac{\left(\sum x_i\right)^2}{n}$

n = Sample size

Preliminary computations for finding the least squares line for the drug reaction example are presented in Table 11.3. We can now calculate

---

*Students who are familiar with calculus should note that the values of β_0 and β_1 that minimize $\text{SSE} = \sum (y_i - \hat{y}_i)^2$ are obtained by setting the two partial derivatives $\partial \text{SSE}/\partial \beta_0$ and $\partial \text{SSE}/\partial \beta_1$ equal to 0. The solutions to these two equations yield the formulas shown in the box. Furthermore, we denote the *sample* solutions to the equations by $\hat{\beta}_0$ and $\hat{\beta}_1$, where the "hat" denotes that these are sample estimates of the true population intercept β_0 and β_1.

TABLE 11.3 **Preliminary Computations for the Drug Reaction Example**

	x_i	y_i	x_i^2	$x_i y_i$
	1	1	1	1
	2	1	4	2
	3	2	9	6
	4	2	16	8
	5	4	25	20
Totals	$\sum x_i = 15$	$\sum y_i = 10$	$\sum x_i^2 = 55$	$\sum x_i y_i = 37$

$$SS_{xy} = \sum x_i y_i - \frac{\left(\sum x_i\right)\left(\sum y_i\right)}{5} - 37 - \frac{(15)(10)}{5} = 37 \quad 30 - 7$$

$$SS_{xx} = \sum x_i^2 - \frac{\left(\sum x_i\right)^2}{5} - 55 - \frac{(15)^2}{5} = 55 \quad 45 = 10$$

Then the slope of the least squares line is

$$\hat{\beta}_1 = \frac{SS_{xy}}{SS_{xx}} = \frac{7}{10} = .7$$

and the *y*-intercept is

$$\hat{\beta}_0 = \bar{y} - \hat{\beta}_1 \bar{x} = \frac{\sum y_i}{5} - \hat{\beta}_1 \frac{\sum x_i}{5}$$

$$= \frac{10}{5} - (.7)\left(\frac{15}{5}\right) = 2 - (.7)(3) = 2 - 2.1 = -.1$$

The least squares line is thus

$$\hat{y} = \hat{\beta}_0 + \hat{\beta}_1 x = -.1 + .7x$$

The graph of this line is shown in Figure 11.5.

The predicted value of *y* for a given value of *x* can be obtained by substituting into the formula for the least squares line. Thus, when $x = 2$ we predict *y* to be

$$\hat{y} = -.1 + .7x = -.1 + .7(2) = 1.3$$

We show how to find a prediction interval for *y* in Section 11.8.

The observed and predicted values of *y*, the deviations of the *y* values about their predicted values, and the squares of these deviations are shown in Table 11.4. Note that the sum of squares of the deviations, SSE, is 1.10, and (as we would expect) this is less than the SSE = 2.0 obtained in Table 11.2 for the visually fitted line.

FIGURE 11.5

The line $\hat{y} = -.1 + .7x$
fit to the data

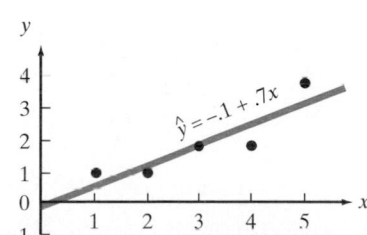

TABLE 11.4   Comparing Observed and Predicted Values for the Least Squares Prediction Equation

x	y	$\hat{y} = -.1 + .7x$	$(y - \hat{y})$	$(y - \hat{y})^2$
1	1	.6	$(1 - .6) = .4$	.16
2	1	1.3	$(1 - 1.3) = -.3$	.09
3	2	2.0	$(2 - 2.0) = 0$	.00
4	2	2.7	$(2 - 2.7) = -.7$	.49
5	4	3.4	$(4 - 3.4) = .6$	.36
			Sum of errors = 0	SSE = 1.10

The calculations required to obtain $\hat{\beta}_0$, $\hat{\beta}_1$, and SSE in simple linear regression, although straightforward, can become rather tedious. Even with the use of a pocket calculator, the process is laborious and susceptible to error, especially when the sample size is large. Fortunately, the use of a statistical software package can significantly reduce the labor involved in regression calculations. The SAS output for the simple linear regression of the data in Table 11.1 is displayed in Figure 11.6. The values of $\hat{\beta}_0$ and $\hat{\beta}_1$ are shaded on the SAS printout under the **Parameter Estimate** column in the rows labeled **INTERCEP** and **X**, respectively. These values, $\hat{\beta}_0 = -.1$ and $\hat{\beta}_1 = .7$, agree exactly with our hand-calculated values. The value of SSE = 1.10 is also shaded in Figure 11.6, under the **Sum of Squares** column in the row labeled **Error**.

Whether you use a hand calculator or a computer, it is important that you be able to interpret the intercept and slope in terms of the data being utilized to fit the model. In the drug reaction example, the estimated y-intercept, $\hat{\beta}_0 = -.1$, appears to imply that the estimated mean reaction time is equal to $-.1$ second when the amount of drug, x, is equal to 0%. Since negative reaction times are not possible, this seems to make the model nonsensical. However, *the model parameters should be interpreted only within the sampled range of the independent variable*—in this case, for amounts of drug in the bloodstream between 1% and 5%. Thus, the y-intercept—which is, by definition, at $x = 0$ (0% drug)—is not within the range of the sampled values of x and is not subject to meaningful interpretation.

FIGURE 11.6

SAS printout for reaction time–drug percentage regression

```
Dependent Variable: Y

                              Analysis of Variance

                              Sum of          Mean
        Source      DF        Squares         Square     F Value    Prob>F

        Model        1        4.90000        4.90000     13.364     0.0354
        Error        3        1.10000        0.36667
        C Total      4        6.00000

             Root MSE       0.60553      R-square     0.8167
             Dep Mean       2.00000      Adj R-sq     0.7556
             C.V.          30.27650

                              Parameter Estimates

                         Parameter      Standard      T for H0:
        Variable    DF    Estimate        Error     Parameter=0    Prob>|T|

        INTERCEP     1   -0.100000     0.63508530      -0.157       0.8849
        X            1    0.700000     0.19148542       3.656       0.0354
```

The slope of the least squares line, $\hat{\beta}_1 = .7$, implies that for every unit increase of x, the mean value of y is estimated to increase by .7 unit. In terms of this example, for every 1% increase in the amount of drug in the bloodstream, the mean reaction time is estimated to increase by .7 second *over the sampled range of drug amounts from 1% to 5%*. Thus, the model does not imply that increasing the drug amount from 5% to 10% will result in an increase in mean reaction time of 3.5 seconds, because the range of x in the sample does not extend to 10% ($x = 10$). In fact, 10% might be such a high concentration that the drug would kill the subject! Be careful to interpret the estimated parameters only within the sampled range of x.

Even when the interpretations of the estimated parameters are meaningful, we need to remember that they are only estimates based on the sample. As such, their values will typically change in repeated sampling. How much confidence do we have that the estimated slope, $\hat{\beta}_1$, accurately approximates the true slope, β_1? This requires statistical inference, in the form of confidence intervals and tests of hypotheses, which we address in Section 11.5.

To summarize, we defined the best-fitting straight line to be the one that minimizes the sum of squared errors around the line, and we called it the least squares line. We should interpret the least squares line only within the sampled range of the independent variable. In subsequent sections we show how to make statistical inferences about the model.

EXERCISES 11.10–11.22

Learning the Mechanics

11.10 The following table is similar to Table 11.3. It is used for making the preliminary computations for finding the least squares line for the given pairs of x and y values.

x_i	y_i	x_i^2	$x_i y_i$
7	2		
4	4		
6	2		
2	5		
1	7		
1	6		
3	5		
Totals $\Sigma x_i =$	$\Sigma y_i =$	$\Sigma x_i^2 =$	$\Sigma x_i y_i =$

a. Complete the table.   **b.** Find SS_{xy}.
c. Find SS_{xx}.   **d.** Find $\hat{\beta}_1$.   **e.** Find $\bar{x}$ and $\bar{y}$.
f. Find $\hat{\beta}_0$.   **g.** Find the least squares line.

11.11 Refer to Exercise 11.10. After the least squares line has been obtained, the table at the bottom of the page (which is similar to Table 11.4) can be used for (1) comparing the observed and the predicted values of y, and (2) computing SSE.

a. Complete the table.
b. Plot the least squares line on a scattergram of the data. Plot the following line on the same graph:

$$\hat{y} = 14 - 2.5x$$

c. Show that SSE is larger for the line in part **b** than it is for the least squares line.

x	y	$\hat{y} =$	$(y - \hat{y})$	$(y - \hat{y})^2$
7	2			
4	4			
6	2			
2	5			
1	7			
1	6			
3	5			
			$\Sigma(y - \hat{y}) =$	$SSE = \Sigma(y - \hat{y})^2 =$

11.12 Construct a scattergram for the data in the following table.

x	.5	1	1.5
y	2	1	3

a. Plot the following two lines on your scattergram:

$$y = 3 - x \quad \text{and} \quad y = 1 + x$$

b. Which of these lines would you choose to characterize the relationship between x and y? Explain.

c. Show that the sum of errors for both of these lines equals 0.

d. Which of these lines has the smaller SSE?

e. Find the least squares line for the data and compare it to the two lines described in part **a**.

11.13 Consider the following pairs of measurements:

x	5	3	−1	2	7	6	4
y	4	3	0	1	8	5	3

a. Construct a scattergram for these data.

b. What does the scattergram suggest about the relationship between x and y?

c. Given that $SS_{xx} = 43.4286$, $SS_{xy} = 39.8571$, $\bar{y} = 3.4286$, and $\bar{x} = 3.7143$, calculate the least squares estimates of β_0 and β_1.

d. Plot the least squares line on your scattergram. Does the line appear to fit the data well? Explain.

e. Interpret the y-intercept and slope of the least squares line. Over what range of x are these interpretations meaningful?

11.14 Suppose that $n = 100$ recent residential home sales in a city are used to fit a least squares straight-line model relating the sale price, y, to the square feet of living space, x. Homes in the sample range from 1,500 square feet to 4,000 square feet of living space, and the resulting least squares equation is

$$\hat{y} = -30,000 + 70x$$

a. What is the underlying hypothesized probabilistic model for this application? What does it imply about the relationship between the mean sale price and living space?

b. Identify the least squares estimates of the y-intercept and slope of the model.

c. Interpret the least squares estimate of the y-intercept. Is it meaningful for this application? Why?

d. Interpret the least squares estimate of the slope of the model. Over what range of x is the interpretation meaningful?

e. Use the least squares model to estimate the mean sale price of a 3,000-square-foot home. Is the estimate meaningful? Explain.

f. Use the least squares model to estimate the mean sale price of a 5,000-square-foot home. Is the estimate meaningful? Explain.

[*Note:* We show how to measure the statistical reliability of these least squares estimates in subsequent sections.]

Applying the Concepts

11.15 Neurologists have found that the hippocampus, a structure found in the brain, plays an important role in short-term memory. Research published in the *American Journal of Psychiatry* (July 1995) attempted to establish a link between hippocampal volume and short-term verbal memory of patients with combat-related posttraumatic stress disorder (PTSD). A sample of 21 Vietnam veterans with a history of combat-related PTSD participated in the study. Magnetic resonance imaging was used to measure the volume, x, of the right hippocampus (in cubic millimeters) of each subject. Verbal memory retention, y, of each subject was measured by the percent retention subscale of the Wechsler Memory Scale. The data for the 21 patients are plotted in the accompanying scattergram.

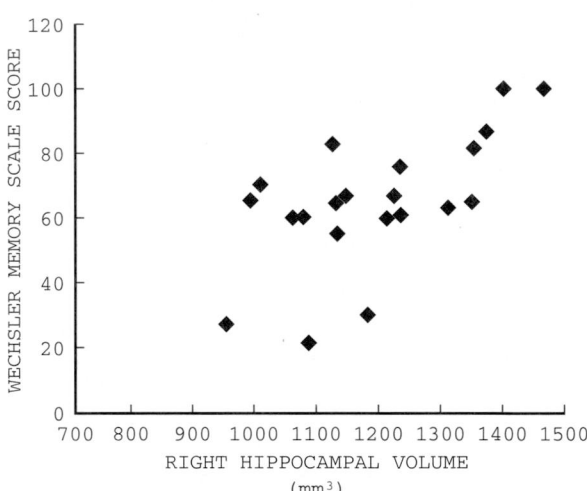

a. The researchers "hypothesized that smaller hippocampal volume would be associated with deficits in short-term verbal memory in patients with PTSD." Does the scattergram provide visual evidence to support this theory?

b. Propose a straight-line model relating verbal memory y with right hippocampal volume x.

c. Based on the scattergram, would you expect the slope of the line to be positive or negative? Explain.

11.16 The long jump is a track and field event in which a competitor attempts to jump a maximum distance into a sand pit after a running start. At the edge of the sand pit is a takeoff board. Jumpers usually try to plant their toes at the front edge of this board to maximize jumping distance. The absolute distance between the front edge of the takeoff board and the spot where the toe actually lands on the board prior to jumping is called "takeoff error." Is takeoff error in the long jump linearly related to best jumping distance? To answer this question, kinesiology researchers videotaped the performances of 18 novice long jumpers at a high school track meet (*Journal of Applied Biomechanics,* May 1995). The average takeoff error, x, and best jumping distance (out of three jumps), y, for each jumper are recorded in the table directly below.

Jumper	Best Jumping Distance y (meters)	Average Takeoff Error x (meters)
1	5.30	.09
2	5.55	.17
3	5.47	.19
4	5.45	.24
5	5.07	.16
6	5.32	.22
7	6.15	.09
8	4.70	.12
9	5.22	.09
10	5.77	.09
11	5.12	.13
12	5.77	.16
13	6.22	.03
14	5.82	.50
15	5.15	.13
16	4.92	.04
17	5.20	.07
18	5.42	.04

Source: Reprinted by permission from W. P. Berg and N. L. Greer, "A kinematic profile of the approach run of novice long jumpers." *Journal of Applied Biomechanics,* Vol. 11, No. 2, May 1995, p. 147 (Table 1).

a. Propose a straight-line model relating y to x.

b. Construct a scattergram for the data. Is there visual evidence of a linear relationship between y and x?

c. A MINITAB printout of the simple linear regression analysis is given below. Find the estimates of β_0 and β_1 on the printout.

d. Graph the least squares line on the scattergram. Does the least squares line seem to fit the points on the scattergram?

e. Interpret the least squares intercept and slope.

11.17 A study was conducted to investigate how people with a hearing impairment communicate with their conversational partners (*Journal of the Academy of Rehabilitative Audiology,* Vol. 27, 1994). Each of thirteen hearing-impaired subjects, all fitted with a cochlear implant, participated in a structured communication interaction with a familiar conversational partner (a family member) and with an unfamiliar conversational partner (who was instructed not to take the initiative to repair breakdowns in communication). The total number of words used by the subject in each of the two conversations is given in the table directly below.

Subject	TOTAL NUMBER OF WORDS IN CONVERSATION WITH	
	Familiar Partner, x	Unfamiliar Partner, y
1	65	47
2	160	78
3	55	90
4	83	75
5	0	6
6	140	101
7	49	40
8	164	215
9	62	29
10	56	75
11	207	121
12	207	139
13	93	83

Source: Tye-Murray, N., *et al.* "Communication breakdowns: Partner contingencies and partner reactions." *Journal of the Academy of Rehabilitative Audiology,* Vol. 27, 1994, pp. 116–117 (Tables 6 & 7).

```
The regression equation is
DISTANCE = 5.35 + 0.530 TAKOFFER

Predictor      Coef      Stdev     t-ratio       p
Constant     5.3480     0.1635      32.71    0.000
TAKOFFER     0.5299     0.9254       0.57    0.575

s = 0.4115     R-sq = 2.0%      R-sq(adj) = 0.0%

Analysis of Variance

SOURCE       DF         SS         MS        F       p
Regression    1     0.0555     0.0555     0.33   0.575
Error        16     2.7091     0.1693
Total        17     2.7646
```

a. Plot the data in a scattergram. Is there visual evidence of a linear relationship between x and y? If so, is it positive or negative?

b. Propose a straight-line model relating y to x.

c. An SPSS printout of the simple linear regression analysis is displayed below. Find the estimate of β_0 and β_1 on the printout.

d. Interpret the values of $\hat{\beta}_0$ and $\hat{\beta}_1$.

11.18 Chemists at Kyushu University (Japan) examined the linear relationship between the maximum absorption rate y (in nanomoles) and the Hammett substituent constant x for metacyclophane compounds (*Journal of Organic Chemistry,* July 1995). The data for variants of two compounds are given in the table directly below. The variants of compound 1 are labeled 1a, 1b, 1d, 1e, 1f, 1g, and 1h; the variants of compound 2 are 2a, 2b, 2c, and 2d.

Compound	Maximum Absorption, y	Hammett Constant, x
1a	298	0.00
1b	346	.75
1d	303	.06
1e	314	−.26
1f	302	.18
1g	332	.42
1h	302	−.19
2a	343	.52
2b	367	1.01
2c	325	.37
2d	331	.53

Source: Adapted from Tsuge, A., *et al.* "Preparation and spectral properties of disubstituted [2-2] metacyclophanes." *Journal of Organic Chemistry,* Vol. 60, No. 15, July 1995, pp. 4390–4391 (Table 1 and Figure 1).

a. Plot the data in a scattergram. Use two different plotting symbols for the two compounds. What do you observe?

b. Using only the data for compound 1, find $\hat{\beta}_0$ and $\hat{\beta}_1$. Interpret the results.

c. Using only the data for compound 2, find $\hat{\beta}_0$ and $\hat{\beta}_1$. Interpret the results.

11.19 Physicians have used the "diving reflex" to reduce abnormally rapid heartbeats in humans by briefly submerging the patient's face in cold water. The reflex, triggered by cold water temperatures, is an involuntary neural response that shuts off circulation to the skin, muscles, and internal organs to divert extra oxygen-carrying blood to the heart, lungs, and brain. A research physician conducted an experiment to investigate the effects of various cold water temperatures on the pulse rate of small children. The data for seven 6-year-old children are shown in the table directly below.

Child	Temperature of Water x (°F)	Decrease in Pulse Rate y (beats/minute)
1	68	2
2	65	5
3	70	1
4	62	10
5	60	9
6	55	13
7	58	10

a. Find the least squares line for the data.

b. Construct a scattergram for the data; then graph the least squares line as a check on your calculations.

c. If the water temperature is 60°F, predict the drop in pulse rate for a 6-year-old child. [*Note:* A measure of the reliability of such predictions is discussed in Section 11.8.]

11.20 Two species of predatory birds, collared flycatchers and tits, compete for nest holes during breeding season on the island of Gotland, Sweden. Frequently,

```
Multiple R            .76401
R Square              .58371
Adjusted R Square     .54587
Standard Error      36.12789

Analysis of Variance
                    DF      Sum of Squares      Mean Square
Regression           1         20131.76038      20131.76038
Residual            11         14357.47039       1305.22458

F =      15.42398     Signif F =  .0024

-------------------Variables in the Equation-------------------

Variable            B         SE B        Beta        T   Sig T

X              .624416     .158992     .764010     3.927   .0024
(Constant)   20.127511   19.219357                 1.047   .3174
```

dead flycatchers are found in nest boxes occupied by tits. A field study examined whether the risk of mortality to flycatchers is related to the degree of competition between the two bird species for nest sites (*The Condor*, May 1995). The table directly below gives data on the number y of flycatchers killed at each of 14 discrete locations (plots) on the island as well as the nest box tit occupancy x (that is, the percentage of nest boxes occupied by tits) at each plot. SAS was used to conduct a simple linear regression analysis for the model, $E(y) = \beta_0 + \beta_1 x$. The printout is shown below.

Plot	Number of Flycatchers Killed, y	Nest Box Tit Occupancy x (%)
1	0	24
2	0	33
3	0	34
4	0	43
5	0	50
6	1	35
7	1	35
8	1	38
9	1	40
10	2	31
11	2	43
12	3	55
13	4	57
14	5	64

Source: Merila, J., and Wiggins, D. A. "Interspecific competition for nest holes causes adult mortality in the collard flycatcher." *The Condor*, Vol. 97, No. 2, May 1995, p. 449 (Figure 2), Cooper Ornithological Society.

a. Plot the data in a scattergram. Does the frequency of flycatcher casualties per plot appear to increase linearly with increasing proportion of nest boxes occupied by tits?

b. Find the estimates of β_0 and β_1 in the SAS printout. Interpret their values.

11.21 In Exercise 9.40, we gave data on the mean daily air temperature and corresponding cocoon temperature of woolly-bear caterpillars of the High Arctic (*Arctic and Alpine Research*, Vol. 54, 1982). The data, collected over 12 days, are reproduced in the table directly below.

TEMPERATURE (°C)

Day	Air	Cocoon[a]
1	10.4	15.1
2	9.2	14.6
3	2.2	6.8
4	2.6	6.8
5	4.1	8.0
6	3.7	8.7
7	1.7	3.6
8	2.0	5.3
9	3.0	7.0
10	3.5	7.1
11	4.5	9.6
12	4.4	9.5

[a]Each cocoon temperature is the average of the temperatures of two cocoons.

Given that $SS_{xx} = 83.3425$, $SS_{xy} = 100.0825$, $\bar{x} = 4.2750$, and $\bar{y} = 8.5083$:

a. Fit a least squares line to relate the cocoon temperature y to the outside air temperature x.

b. Plot the data points and graph your least squares line on the same sheet of graph paper. Does your line provide a good fit to the data?

c. Interpret the least squares intercept and slope in terms of this application.

11.22 The sand lance is a small fish that can be found in the Northwest Atlantic from Cape Hatteras to

```
Dependent Variable: NOKILLED

                        Analysis of Variance

                            Sum of        Mean
        Source      DF     Squares       Square    F Value    Prob>F

        Model        1    19.11669     19.11669     16.029    0.0018
        Error       12    14.31188      1.19266
        C Total     13    33.42857

            Root MSE       1.09209     R-square     0.5719
            Dep Mean       1.42857     Adj R-sq     0.5362
            C.V.          76.44618

                        Parameter Estimates

                     Parameter     Standard     T for H0:
        Variable  DF   Estimate        Error  Parameter=0   Prob > |T|

        INTERCEP   1  -3.046856   1.15533214       -2.637       0.0217
        TITPCT     1   0.107656   0.02689001        4.004       0.0018
```

Greenland. In a 10-year study of the biological and demographic characteristics of the sand lance, data were collected on the mean length of specimens collected each year for sand lances whose ages ranged from 2 through 8 years (*Canadian Journal of Fisheries and Aquatic Sciences*, Vol. 40, 1983). The data are given below.

a. Find the least squares prediction equation relating the mean length *y* of the sand lance to its age *x*.

b. Plot the data points and graph your least squares lines on the same sheet of graph paper. Does your line appear to provide a good fit to the data?

c. Interpret the least squares intercept and slope in terms of this application. Is the intercept's interpretation meaningful in this case?

Mean Length (millimeters)	176	194	212	226	236	244	254
Age (years)	2	3	4	5	6	7	8

11.3   MODEL ASSUMPTIONS

In Section 11.2 we assumed that the probabilistic model relating drug reaction time *y* to the percentage of drug *x* in the bloodstream is

$$y = \beta_0 + \beta_1 x + \epsilon$$

We also recall that the least squares estimate of the deterministic component of the model, $\beta_0 + \beta_1 x$, is

$$\hat{y} = \hat{\beta}_0 + \hat{\beta}_1 x = -.1 + .7x$$

Now we turn our attention to the random component ϵ of the probabilistic model and its relation to the errors in estimating β_0 and β_1. We will use a probability distribution to characterize the behavior of ϵ. We will see how the probability distribution of ϵ determines how well the model describes the relationship between the dependent variable *y* and the independent variable *x*.

Step 3 in a regression analysis requires us to specify the probability distribution of the random error ϵ. We will make four basic assumptions about the general form of this probability distribution:

Assumption 1   The mean of the probability distribution of ϵ is 0. That is, the average of the values of ϵ over an infinitely long series of experiments is 0 for each setting of the independent variable *x*. This assumption implies that the mean value of *y*, $E(y)$, for a given value of *x* is $E(y) = \beta_0 + \beta_1 x$.

Assumption 2   The variance of the probability distribution of ϵ is constant for all settings of the independent variable *x*. For our straight-line model, this assumption means that the variance of ϵ is equal to a constant, say σ^2, for all values of *x*.

Assumption 3   The probability distribution of ϵ is normal.

Assumption 4   The values of ϵ associated with any two observed values of *y* are independent. That is, the value of ϵ associated with one value of *y* has no effect on the values of ϵ associated with other *y* values.

The implications of the first three assumptions can be seen in Figure 11.7, which shows distributions of errors for three values of *x*, namely, x_1, x_2, and x_3.

FIGURE 11.7

The probability distribution of ϵ

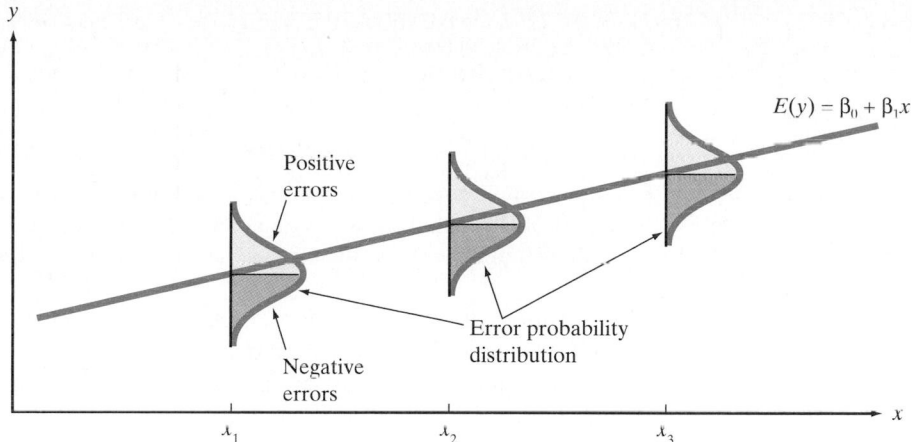

Note that the relative frequency distributions of the errors are normal with a mean of 0 and a constant variance σ^2. (All the distributions shown have the same amount of spread or variability.) The straight line shown in Figure 11.7 is the line of means. It indicates the mean value of y for a given value of x. We denote this mean value as $E(y)$. Then, the line of means is given by the equation

$$E(y) = \beta_0 + \beta_1 x$$

These assumptions make it possible for us to develop measures of reliability for the least squares estimators and to develop hypothesis tests for examining the usefulness of the least squares line. We have various techniques for checking the validity of these assumptions, and we have remedies to apply when they appear to be invalid. Several of these remedies are discussed in Chapters 12 and 13. Fortunately, the assumptions need not hold exactly in order for least squares estimators to be useful. The assumptions will be satisfied adequately for many applications encountered in practice.

11.4   AN ESTIMATOR OF σ^2

It seems reasonable to assume that the greater the variability of the random error ϵ (which is measured by its variance σ^2), the greater will be the errors in the estimation of the model parameters β_0 and β_1 and in the error of prediction when $\hat{y}$ is used to predict y for some value of x. Consequently, you should not be surprised, as we proceed through this chapter, to find that σ^2 appears in the formulas for all confidence intervals and test statistics that we will be using.

Estimation of σ^2 for a (First-Order) Straight-Line Model

$$s^2 = \frac{\text{SSE}}{\text{Degrees of freedom for error}} = \frac{\text{SSE}}{n-2}$$

where $\text{SSE} = \sum(y_i - \hat{y}_i)^2 = \text{SS}_{yy} - \hat{\beta}_1 \text{SS}_{xy}$

$\text{SS}_{yy} = \sum(y_i - \bar{y})^2 = \sum y_i^2 - \frac{(\sum y_i)^2}{n}$

Continued

To estimate the standard deviation σ of ϵ, we calculate

$$s = \sqrt{s^2} = \sqrt{\frac{SSE}{n-2}}$$

We will refer to s as the **estimated standard error of the regression model**.

Warning: When performing these calculations, you may be tempted to round the calculated values of SS_{yy}, $\hat{\beta}_1$, and SS_{xy}. Be certain to carry at least six significant figures for each of these quantities to avoid substantial errors in calculation of the SSE.

In most practical situations, σ^2 is unknown and we must use our data to estimate its value. The best estimate of σ^2, denoted by s^2, is obtained by dividing the sum of squares of the deviations of the y values from the prediction line,

$$SSE = \sum(y_i - \hat{y}_i)^2$$

by the number of degrees of freedom associated with this quantity. We use 2 df to estimate the two parameters β_0 and β_1 in the straight-line model, leaving $(n-2)$ df for the error variance estimation.

In the drug reaction example, we previously calculated $SSE = 1.10$ for the least squares line $\hat{y} = -.1 + .7x$. Recalling that there were $n = 5$ data points, we have $n - 2 = 5 - 2 = 3$ df for estimating σ^2. Thus,

$$s^2 = \frac{SSE}{n-2} = \frac{1.10}{3} = .367$$

is the estimated variance, and

$$s = \sqrt{.367} = .61$$

is the standard error of the regression model.

The values of s^2 and s can also be obtained from a simple linear regression printout. The SAS printout for the drug reaction example is reproduced in Figure 11.8. The value of s^2 is shaded on the printout in the **Mean Square** column in the row labeled **Error**. (In regression, the estimate of σ^2 is called Mean Square for Error, or MSE.) The value, $s^2 = .36667$, rounded to three decimal places, agrees with the one calculated by hand. The value of s is shaded in Figure 11.8 next to the heading **Root MSE**. This value, $s = .60553$, agrees (except for rounding) with our hand-calculated value.

FIGURE 11.8

SAS printout for reaction time – drug percentage example

Dependent Variable: Y

Analysis of Variance

Source	DF	Sum of Squares	Mean Square	F Value	Prob>F
Model	1	4.90000	4.90000	13.364	0.0354
Error	3	1.10000	0.36667		
C Total	4	6.00000			

Root MSE	0.60553	R-square	0.8167
Dep Mean	2.00000	Adj R-sq	0.7556
C.V.	30.27650		

Parameter Estimates

Variable	DF	Parameter Estimate	Standard Error	T for H0: Parameter=0	Prob > |T|
INTERCEP	1	-0.100000	0.63508530	-0.157	0.8849
X	1	0.700000	0.19148542	3.656	0.0354

You may be able to grasp s intuitively by recalling the interpretation of a standard deviation given in Chapter 2 and remembering that the least squares line estimates the mean value of y for a given value of x. Since s measures the spread of the distribution of y values about the least squares line, we should not be surprised to find that most of the observations lie within $2s$, or $2(.61) - 1.22$, of the least squares line. For this simple example (only five data points), all five data points fall within $2s$ of the least squares line. In Section 11.8, we use s to evaluate the error of prediction when the least squares line is used to predict a value of y to be observed for a given value of x.

Interpretation of s, the Estimated Standard Deviation of ϵ

We expect most of the observed y values to lie within $2s$ of their respective least squares predicted values, $\hat{y}$.

EXERCISES 11.23–11.31

Learning the Mechanics

11.23 Calculate SSE and s^2 for each of the following cases:
  a. $n = 20$, $SS_{yy} = 95$, $SS_{xy} = 50$, $\hat{\beta}_1 = .75$
  b. $n = 40$, $\Sigma y^2 = 860$, $\Sigma y = 50$, $SS_{xy} = 2,700$, $\hat{\beta}_1 - .2$
  c. $n = 10$, $\Sigma(y_i - y)^2 = 58$, $SS_{xy} = 91$, $SS_{xx} = 170$

11.24 Suppose you fit a least squares line to 12 data points and the calculated value of SSE is .429.
  a. Find s^2, the estimator of σ^2 (the variance of the random error term ϵ).
  b. What is the largest deviation that you might expect between any one of the 12 points and the least squares line?

11.25 Visually compare the scattergrams shown below. If a least squares line were determined for each data set, which do you think would have the smallest variance, s^2? Explain.

11.26 Refer to Exercises 11.10 and 11.13. Calculate SSE, s^2, and s for the least squares lines obtained in these exercises. Interpret the standard error of the regression model, s, for each.

Applying the Concepts

11.27 Refer to the simple linear regression relating best jumping distance y to average takeoff error x for novice long jumping, Exercise 11.16. The MINITAB printout is reproduced at the bottom of the page.

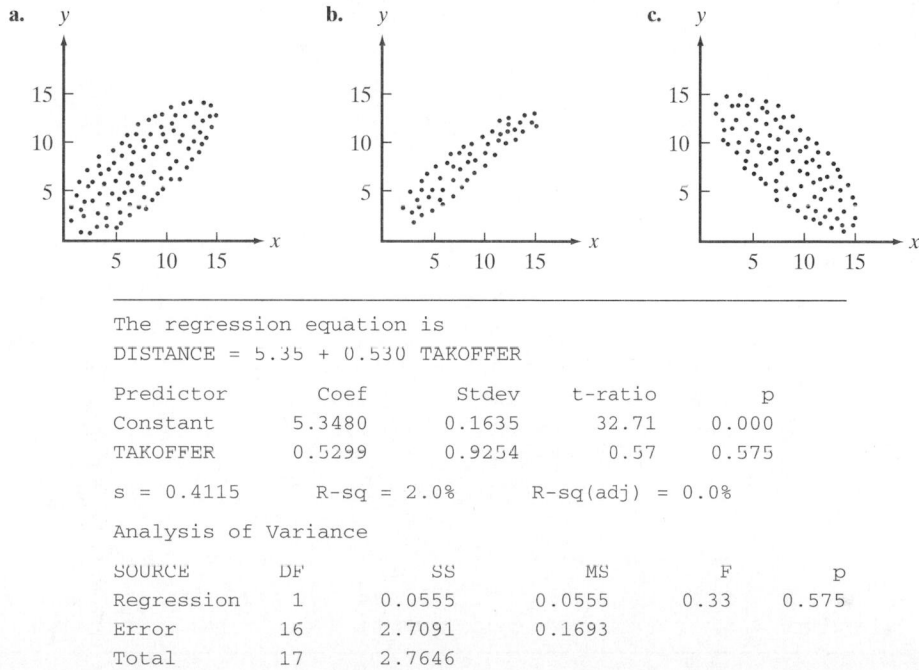

```
The regression equation is
DISTANCE = 5.35 + 0.530 TAKOFFER

Predictor       Coef       Stdev      t-ratio          p
Constant       5.3480     0.1635       32.71      0.000
TAKOFFER       0.5299     0.9254        0.57      0.575

s = 0.4115      R-sq = 2.0%      R-sq(adj) = 0.0%

Analysis of Variance

SOURCE         DF          SS          MS         F         p
Regression      1      0.0555      0.0555      0.33     0.575
Error          16      2.7091      0.1693
Total          17      2.7646
```

```
Multiple R              .76401
R Square                .58371
Adjusted R Square       .54587
Standard Error        36.12789

Analysis of Variance
                    DF      Sum of Squares      Mean Square
Regression           1         20131.76038      20131.76038
Residual            11         14357.47039       1305.22458

F =      15.42398      Signif F =  .0024

------------------Variables in the Equation------------------

Variable            B          SE B        Beta         T   Sig T

X                .624416    .158992    .764010      3.927  .0024
(Constant)     20.127511  19.219357                 1.047  .3174
```

a. Find SSE, s^2, and s on the printout.
b. Interpret the value of s.

11.28 Refer to the simple linear regression relating total number of words in a conversation with an unfamiliar partner y to total number of words with a familiar partner x for hearing-impaired subjects, Exercise 11.17. The SPSS printout is reproduced above.
a. Find SSE, s^2, and s on the printout.
b. Interpret the value of s.

11.29 Refer to the simple linear regression relating number of flycatchers killed y to nest box tit occupancy x for nest sites, Exercise 11.20. The SAS printout is reproduced below.
a. Find SSE, s^2, and s on the printout.
b. Interpret the value of s.

11.30 A much larger proportion of U.S. teenagers are working while attending high school than was the case a decade ago. These heavy workloads often result in underachievement in the classroom and lower grades. As a result, many states are imposing tighter child labor laws. (In fact, some business groups have mobilized to block these restrictions.) However, a study of high school students in California and Wisconsin showed that those who worked only a few hours per week had the highest grade point averages (*Newsweek,* Nov. 16, 1992). The following table shows grade point averages and number of hours worked per week for a sample of five students. A simple linear regression analysis of the data is displayed in the accompanying ASP printout on page 493.

Grade Point Average, y	2.93	3.00	2.86	3.04	2.66
Hours Worked per Week, x	12	0	17	5	21

```
Dependent Variable: NOKILLED

                      Analysis of Variance

                            Sum of      Mean
     Source        DF      Squares     Square    F Value    Prob>F

     Model          1     19.11669   19.11669     16.029    0.0018
     Error         12     14.31188    1.19266
     C Total       13     33.42857

          Root MSE        1.09209    R-square    0.5719
          Dep Mean        1.42857    Adj R-sq    0.5362
          C.V.           76.44618

                      Parameter Estimates

                   Parameter    Standard    T for H0:
     Variable  DF   Estimate      Error   Parameter=0   Prob > |T|

     INTERCEP   1  -3.046856   1.15533214     -2.637      0.0217
     TITPCT     1   0.107656   0.02689001      4.004      0.0018
```

```
                    SIMPLE LINEAR REGRESSION
Model: GPA(y) = -0.0154762HoursWork(x) + 3.06824CNST

                COEF.      SD. ER.      t(3)     P-VALUE  PT. R SQ.
HoursWork(x)  -0.0154762  4.67338E-3  -3.31156  0.0453392   0.785199
        CNST   3.06824    0.0626652   48.9624   1.87599E-5  0.99875

R SQ. = 0.785199,  ADJ. R SQ. = 0.713599,  D. W. = 2.39313
SQ. ROOT MSE = 0.0801318,  F(1/3) = 10.9664 (P-VALUE = 0.0453392)
```

a. Give the equation of the least squares line.

b. Plot the data and graph the least squares line.

c. Predict the grade point average of a high school student who works 10 hours per week. Repeat this for one who works 16 hours per week.

d. Find s on the printout.

e. Within what approximate distance do you expect your predictions in part **c** to fall from the student's true grade point average? [*Note:* A more precise measure of reliability for these predictions is discussed in Section 11.8.]

11.31 To improve the quality of the output of any production process, it is necessary first to understand the capabilities of the process (*Out of the Crisis,* Deming 1982). In a particular manufacturing process, the useful life of a cutting tool is related to the speed at which the tool is operated. It is necessary to understand this relationship in order to predict when the tool should be replaced and how many spare tools should be available. The data in the accompanying table were derived from life tests for the two different brands of cutting tools currently used in the production process.

a. Construct a scattergram for each brand of cutting tool.

b. For each brand, use the method of least squares to model the relationship between useful life and cutting speed.

c. Find SSE, s^2, and s for each least squares line.

d. For a cutting speed of 70 meters per minute, find $\hat{y} \pm 2s$ for each least squares line.

e. For which brand would you feel more confident in using the least squares line to predict useful life for a given cutting speed? Explain.

Cutting Speed (meters per minute)	USEFUL LIFE (HOURS)	
	Brand A	Brand B
30	4.5	6.0
30	3.5	6.5
30	5.2	5.0
40	5.2	6.0
40	4.0	4.5
40	2.5	5.0
50	4.4	4.5
50	2.8	4.0
50	1.0	3.7
60	4.0	3.8
60	2.0	3.0
60	1.1	2.4
70	1.1	1.5
70	.5	2.0
70	3.0	1.0

11.5   ASSESSING THE UTILITY OF THE MODEL: MAKING INFERENCES ABOUT THE SLOPE β_1

Now that we have specified the probability distribution of ϵ and found an estimate of the variance σ^2, we are ready to make statistical inferences about the model's usefulness for predicting the response y. This is step 4 in our regression modeling procedure.

Refer again to the data of Table 11.1 and suppose the reaction times are *completely unrelated* to the percentage of drug in the bloodstream. What could be said about the values of β_0 and β_1 in the hypothesized probabilistic model

$$y = \beta_0 + \beta_1 x + \epsilon$$

if x contributes no information for the prediction of y? The implication is that the mean of y—that is, the deterministic part of the model $E(y) = \beta_0 + \beta_1 x$—does not

FIGURE 11.9

Graphing the model
$y = \beta_0 + \epsilon \ (\beta_1 = 0)$

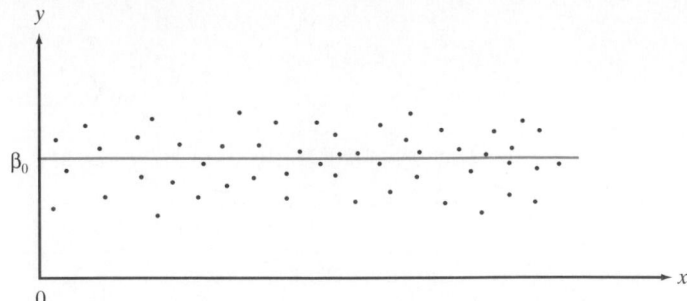

change as x changes. In the straight-line model, this means that the true slope, β_1, is equal to 0 (see Figure 11.9). Therefore, to test the null hypothesis that the linear model contributes no information for the prediction of y against the alternative hypothesis that the linear model is useful for predicting y, we test

$$H_0: \beta_1 = 0$$
$$H_a: \beta_1 \neq 0$$

If the data support the alternative hypothesis, we will conclude that x does contribute information for the prediction of y using the straight-line model [although the true relationship between $E(y)$ and x could be more complex than a straight line]. Thus, in effect, this is a test of the usefulness of the hypothesized model.

The appropriate test statistic is found by considering the sampling distribution of $\hat{\beta}_1$, the least squares estimator of the slope β_1, as shown in the following box.

Sampling Distribution of $\hat{\beta}_1$

If we make the four assumptions about ϵ (see Section 11.3), the sampling distribution of the least squares estimator $\hat{\beta}_1$ of the slope will be normal with mean β_1 (the true slope) and standard deviation

$$\sigma_{\hat{\beta}_1} = \frac{\sigma}{\sqrt{SS_{xx}}} \qquad \text{(see Figure 11.10)}$$

We estimate $\sigma_{\hat{\beta}_1}$ by $s_{\hat{\beta}_1} = \dfrac{s}{\sqrt{SS_{xx}}}$ and refer to this quantity as the estimated standard error of the least squares slope $\hat{\beta}_1$.

Since σ is usually unknown, the appropriate test statistic is a t statistic, formed as follows:

$$t = \frac{\hat{\beta}_1 - \text{Hypothesized value of } \beta_1}{s_{\hat{\beta}_1}} \qquad \text{where} \quad s_{\hat{\beta}_1} = \frac{s}{\sqrt{SS_{xx}}}$$

FIGURE 11.10

Sampling distribution of $\hat{\beta}_1$

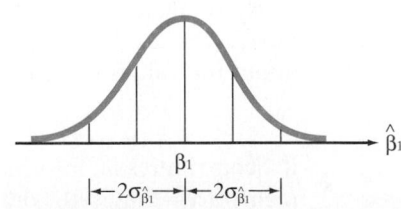

Thus,

$$t = \frac{\hat{\beta}_1 - 0}{s / \sqrt{SS_{xx}}}$$

Note that we have substituted the estimator s for σ and then formed the estimated standard error $s_{\hat{\beta}_1}$ by dividing s by $\sqrt{SS_{xx}}$. The number of degrees of freedom associated with this t statistic is the same as the number of degrees of freedom associated with s. Recall that this number is $(n - 2)$ df when the hypothesized model is a straight line (see Section 11.4). The setup of our test of the usefulness of the straight-line model is summarized in the next box.

A Test of Model Utility: Simple Linear Regression

ONE-TAILED TEST	TWO-TAILED TEST
H_0: $\beta_1 = 0$	H_0: $\beta_1 = 0$
H_a: $\beta_1 < 0$ (or H_a: $\beta_1 > 0$)	H_a: $\beta_1 \neq 0$

Test statistic:
$$t = \frac{\hat{\beta}_1}{s_{\hat{\beta}_1}} = \frac{\hat{\beta}_1}{s / \sqrt{SS_{xx}}}$$

Rejection region: $t < -t_\alpha$ *Rejection region:* $|t| > t_{\alpha/2}$
    (or $t > t_\alpha$ when H_a: $\beta_1 > 0$)

where t_α and $t_{\alpha/2}$ are based on $(n - 2)$ degrees of freedom

Assumptions: The four assumptions about ϵ listed in Section 11.3.

For the drug reaction example, we will choose $\alpha = .05$ and, since $n = 5$, t will be based on $n - 2 = 3$ df and the rejection region will be

$$|t| > t_{.025} = 3.182$$

We previously calculated $\hat{\beta}_1 = .7$, $s = .61$, and $SS_{xx} = 10$. Thus,

$$t = \frac{\hat{\beta}_1}{s / \sqrt{SS_{xx}}} = \frac{.7}{.61\sqrt{10}} = \frac{.7}{.19} = 3.7$$

Since this calculated t value falls in the upper-tail rejection region (see Figure 11.11), we reject the null hypothesis and conclude that the slope β_1 is not 0. The sample evidence indicates that amount of drug x in the blood stream contributes information for the prediction of reaction time y when a linear model is used.

We can reach the same conclusion by using the observed significance level (p-value) of the test from a computer printout. The SAS printout for the drug

FIGURE 11.11

Rejection region and calculated t value for testing H_0: $\beta_1 = 0$ versus H_a: $\beta_1 \neq 0$

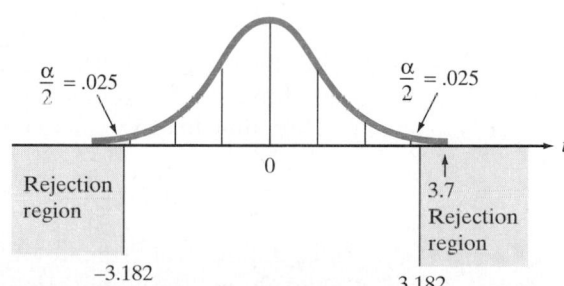

FIGURE 11.12

SAS printout for reaction time–drug percentage example

```
Dependent Variable: Y

                            Analysis of Variance

                            Sum of          Mean
   Source        DF        Squares         Square      F Value     Prob>F

   Model          1        4.90000        4.90000      13.364      0.0354
   Error          3        1.10000        0.36667
   C Total        4        6.00000

        Root MSE          0.60553      R-square      0.8167
        Dep Mean          2.00000      Adj R-sq      0.7556
        C.V.             30.27650

                          Parameter Estimates

                      Parameter       Standard      T for H0:
   Variable   DF       Estimate          Error    Parameter=0    Prob >|T|

   INTERCEP    1      -0.100000      0.63508530        -0.157       0.8849
   X           1       0.700000      0.19148542         3.656       0.0354
```

reaction example is reproduced in Figure 11.12. The test statistic is shaded on the printout under the **T for H0: Parameter=0** column in the row corresponding to **X**, while the two-tailed p-value is shaded under the column labeled **Prob > |T|**. Since the p-value $= .0354$ is smaller than $\alpha = .05$, we will reject H_0.

What conclusion can be drawn if the calculated t value does not fall in the rejection region or if the observed significance level of the test exceeds α? We know from previous discussions of the philosophy of hypothesis testing that such a t value does *not* lead us to accept the null hypothesis. That is, we do not conclude that $\beta_1 = 0$. Additional data might indicate that β_1 differs from 0, or a more complex relationship may exist between x and y, requiring the fitting of a model other than the straight-line model. We discuss several such models in Chapter 12.

Another way to make inferences about the slope β_1 is to estimate it using a confidence interval. This interval is formed as shown in the next box.

A 100(1 – α)% Confidence Interval for the Simple Linear Regression Slope β_1

$$\hat{\beta}_1 \pm t_{\alpha/2} s_{\hat{\beta}_1}$$

where the estimated standard error of $\hat{\beta}_1$ is calculated by

$$s_{\hat{\beta}_1} = \frac{s}{\sqrt{SS_{xx}}}$$

and $t_{\alpha/2}$ is based on $(n - 2)$ degrees of freedom.

Assumptions: The four assumptions about ϵ listed in Section 11.3.

For the drug reaction example, $t_{\alpha/2}$ is based on $(n - 2) = 3$ degrees of freedom. Therefore, a 95% confidence interval for the slope β_1, the expected change in reaction time for a 1% increase in the amount of drug in the bloodstream, is

$$\hat{\beta}_1 \pm t_{.025} s_{\hat{\beta}_1} = .7 \pm 3.182\left(\frac{s}{\sqrt{SS_{xx}}}\right) = .7 \pm 3.182\left(\frac{.61}{\sqrt{10}}\right) = .7 \pm .61$$

Thus, the interval estimate of the slope parameter β_1 is .09 to 1.31. In terms of this example, the implication is that we can be 95% confident that the *true* mean increase in reaction time per additional 1% of the drug is between .09 and

1.31 seconds. This inference is meaningful only over the sampled range of x— that is, from 1% to 5% of the drug in the bloodstream. Since all the values in this interval are positive, it appears that β_1 is positive and that the mean of y, $E(y)$, increases as x increases. However, the rather large width of the confidence interval reflects the small number of data points (and, consequently, a lack of information) in the experiment. We would expect a narrower interval if the sample size were increased.

EXERCISES 11.32–11.44

Learning the Mechanics

11.32 Construct both a 95% and a 90% confidence interval for β_1 for each of the following cases:
 a. $\hat{\beta}_1 = 31$, $s = 3$, $SS_{xx} = 35$, $n = 12$
 b. $\hat{\beta}_1 = 64$, $SSE = 1,960$, $SS_{xx} = 30$, $n = 18$
 c. $\hat{\beta}_1 = -8.4$, $SSE = 146$, $SS_{xx} = 64$, $n = 24$

11.33 Consider the following pairs of observations:

x	1	5	3	2	6	6	0
y	1	3	3	1	4	5	1

 a. Construct a scattergram for the data.
 b. Use the method of least squares to fit a straight line to the seven data points in the table.
 c. Plot the least squares line on your scattergram of part **a**.
 d. Specify the null and alternative hypotheses you would use to test whether the data provide sufficient evidence to indicate that x contributes information for the (linear) prediction of y.
 e. What is the test statistic that should be used in conducting the hypothesis test of part **d**? Specify the degrees of freedom associated with the test statistic.
 f. Conduct the hypothesis test of part **d** using $\alpha = .05$.

11.34 Refer to Exercise 11.33. Construct an 80% and a 98% confidence interval for β_1.

11.35 Do the accompanying data provide sufficient evidence to conclude that a straight line is useful for characterizing the relationship between x and y?

y	4	2	5	3	2	4
x	1	4	5	3	2	4

11.36 Suppose that $n = 100$ recent residential home sales in a city are used to fit a least squares straight-line model relating the sale price, y, to the square feet of living space, x. Homes in the sample range from 1,500 square feet to 4,000 square feet of living space, and the resulting least squares equation is

$$\hat{y} = -30,000 + 70x$$

 a. When the null hypothesis that the true slope is zero is tested, the result is a t value of 6.572. Give the appropriate p-value of this test, and interpret the result in the context of this application.
 b. The 95% confidence interval for the slope is calculated to be 49.1 to 90.9. Interpret this interval in the context of this application. What can be done to obtain a narrower confidence interval?

Applying the Concepts

11.37 Researchers at the University of North Carolina–Greensboro investigated a model for the rate of seed germination (*Journal of Experimental Botany*, Jan. 1993). In one experiment, alfalfa seeds were placed in a specially constructed germination chamber. Eleven hours later, the seeds were examined and the change in free energy (a measure of germination rate) was recorded. The results for seeds germinated at seven different temperatures are given in the following table. The data were used to fit a simple linear regression model, with y = change in free energy and x = temperature.

Change in Free Energy (kJ/mol) y	Temperature (K) x
7	295
6.2	297.5
9	291
9.5	289
8.5	301
7.8	293
11.2	286.5

Source: Hageseth, G. T., and Cody, A. L. "Energy-level model for isothermal seed germination." *Journal of Experimental Botany*, Vol. 44, No. 258, Jan. 1993, p. 123 (Figure 9). By permission of Oxford University Press.

 a. Plot the points in a scattergram.
 b. Find the least squares prediction equation.
 c. Plot the least squares line, part **b**, on the scattergram of part **a**.
 d. Conduct a test of model adequacy. Use $\alpha = .01$.
 e. Use the plot, part **c**, to locate any unusual data points (outliers).
 f. Eliminate the outlier, part **e**, from the data set and repeat parts **a–d**.

11.38 The U.S. Department of Agriculture has developed and adopted the Universal Soil Loss Equation (USLE) for predicting water erosion of soils. In geographic areas where runoff from melting snow is

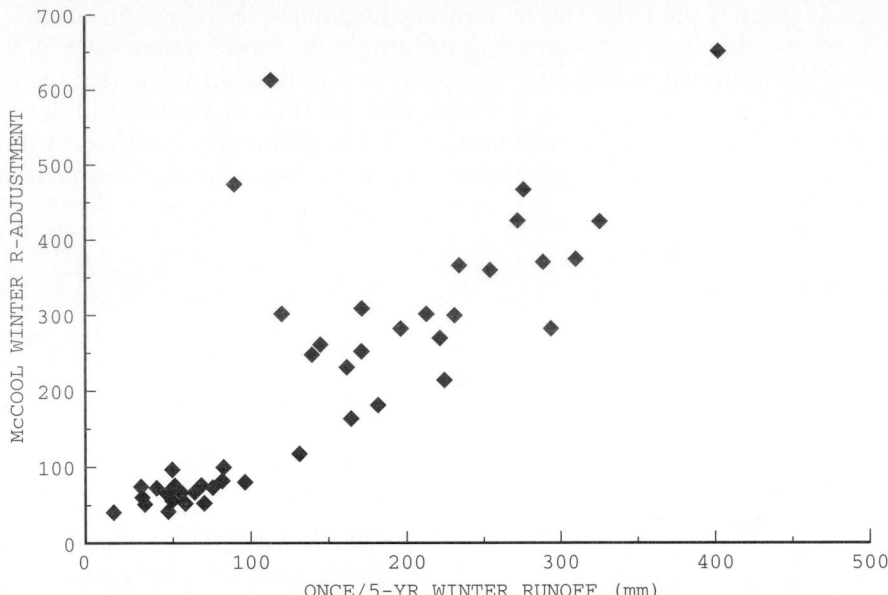

common, the USLE requires an accurate estimate of snowmelt runoff erosion. An article in the *Journal of Soil and Water Conservation* (Mar.–Apr. 1995) used simple linear regression to develop a snowmelt erosion index. Data for 54 climatological stations in Canada were used to model the McCool winter-adjusted rainfall erosivity index, y, as a straight-line function of the once-in-5-year snowmelt runoff amount, x (measured in millimeters).

a. The data points are plotted in the scattergram shown above. Is there visual evidence of a linear trend?

b. The data for seven stations were removed from the analysis due to lack of snowfall during the study period. Why is this strategy advisable?

c. The simple linear regression on the remaining $n = 47$ data points yielded the following results: $\hat{y} = -6.72 + 1.39x$, $s_{\hat{\beta}_1} = .06$. Use this information to construct a 90% confidence interval for β_1.

d. Interpret the interval, part **c**.

11.39 One of the most common types of "information retrieval" processes is document-database searching. An experiment was conducted to investigate the variables that influence search performance in the Medline database and retrieval system (*Journal of Information Science,* Vol. 21, 1995). Simple linear regression was used to model the fraction y of the set of potentially informative documents that are retrieved using Medline as a function of the number x of terms in the search query, based on a sample of $n = 124$ queries. The results are summarized below:

$$\hat{y} = .202 + .135x; \qquad t \text{ (for testing } H_0: \beta_1 = 0) = 4.98$$

Two-tailed p-value = .001

a. Is there sufficient evidence to indicate that x and y are linearly related? Test using $\alpha = .01$.

b. If appropriate, use the model to predict the fraction of documents retrieved for a search query with $x = 3$ terms.

11.40 Refer to Exercises 11.16 and 11.27. Reproduced below is the MINITAB simple linear regression

```
The regression equation is
DISTANCE = 5.35 + 0.530 TAKOFFER

Predictor      Coef       Stdev      t-ratio        p
Constant      5.3480     0.1635       32.71      0.000
TAKOFFER      0.5299     0.9254        0.57      0.575

s = 0.4115      R-sq = 2.0%      R-sq(adj) = 0.0%

Analysis of Variance

SOURCE       DF         SS          MS         F        p
Regression    1       0.0555      0.0555      0.33     0.575
Error        16       2.7091      0.1693
Total        17       2.7646
```

```
Dependent Variable: NOKILLED

                        Analysis of Variance

                               Sum of          Mean
         Source       DF       Squares        Square     F Value    Prob>F

         Model         1      19.11669       19.11669     16.029    0.0018
         Error        12      14.31188        1.19266
         C Total      13      33.42857

              Root MSE          1.09209     R-square     0.5719
              Dep Mean          1.42857     Adj R-sq     0.5362
              C.V.             76.44618

                           Parameter Estimates

                        Parameter    Standard     T for H0:
         Variable   DF    Estimate       Error   Parameter=0    Prob > |T|

         INTERCEP    1   -3.046856    1.15533214      -2.637       0.0217
         TITPCT      1    0.107656    0.02689001       4.004       0.0018
```

printout relating best jumping distance y to average takeoff error x for novice long jumpers.

a. Conduct a test of model adequacy using $\alpha = .05$. Interpret the results.

b. Construct a 95% confidence interval for the ratio of change of best jumping distance y with average takeoff error x. Interpret the interval.

11.41 Refer to Exercises 11.20 and 11.29. The SAS simple linear regression printout relating number of flycatchers killed y to nest box tit occupancy x is reproduced above.

a. Test to determine whether y is positively linearly related to x. Use $\alpha = .01$.

b. Construct a 99% confidence interval for β_1. Practically interpret the result.

11.42 A verbal test was given to a group of children, ranging from 10 to 12 years of age, in order to study the relationship between the number of words used and the silence interval before response. The tester believes that a linear relationship exists between the two variables. Each subject was asked a series of questions, and the total number of words used in answering was recorded. The time (in seconds) before the subject responded to each question was also recorded. The data for eight children are given in the table.

Subject	Total Words y	Total Silence Time x (seconds)
1	61	23
2	70	37
3	42	38
4	52	25
5	91	17
6	63	21
7	71	42
8	55	16

a. Write a simple linear probabilistic model relating total words to total silence time, and use the least squares method to estimate the deterministic part of the model.

b. Does x contribute information for the prediction of y? Test the null hypothesis, slope $\beta_1 = 0$, against the alternative hypothesis, $\beta_1 \neq 0$. Use $\alpha = .05$. Interpret the results of the test.

11.43 In Exercise 11.21, we fit a least squares line to relate the cocoon temperature y of a woolly-bear caterpillar to the outside air temperature x. The data are reproduced in the table.

TEMPERATURE (°C)

Day	Air	Cocoon[a]
1	10.4	15.1
2	9.2	14.6
3	2.2	6.8
4	2.6	6.8
5	4.1	8.0
6	3.7	8.7
7	1.7	3.6
8	2.0	5.3
9	3.0	7.0
10	3.5	7.1
11	4.5	9.6
12	4.4	9.5

[a]Each cocoon temperature is the average of the temperatures of two cocoons.

Given that $SS_{xx} = 83.3425$, $SS_{xy} = 100.0825$, $SS_{yy} = 127.5092$, $\bar{x} = 4.2750$, and $\bar{y} = 8.5083$:

a. Calculate SSE, s^2, and s.

b. Interpret the standard error of the regression model.

c. Do the data provide sufficient evidence to indicate that the outside air temperature provides information for predicting the woolly-bear caterpillar cocoon temperature? Test using $\alpha = .05$.

d. Find the observed significance level of the test.

```
                    SIMPLE LINEAR REGRESSION
MODEL:  Index(y) = 0.236617Interactions(x) + 44.1305CNST

                  COEF.   SD. ER.   t(17)    P-VALUE PT. R SQ.
Interactions(x)  0.236617 0.186501 1.26871 0.221641   0.0864948
        CNST 44.1305    9.36159  4.71399 2.00177E-4 0.566566

R SQ. = 0.0864948,  ADJ. R SQ. = 0.0327592,  D. W. = 2.52206
SQ. ROOT MSE = 19.4038,  F(1/17) = 1.60964 (P-VALUE = 0.221641)
```

11.44 In an observational study of 19 managers from a medium-sized manufacturing plant, regression analysis was used to investigate which recorded activities over a 2-week period were related to managerial success (*Journal of Applied Behavioral Science*, Aug. 1985). To measure success, the researchers devised an index based on the manager's length of time in the organization and his or her level within the firm; the higher the index, the more successful the manager. The table at the right presents data (which are representative of the data collected by the researchers) that can be used to determine whether managerial success is related to the extensiveness of a manager's network-building interactions with people outside the manager's work unit. Such interactions include engaging in phone and face-to-face meetings with customers and suppliers, attending outside meetings, and doing public relations work. An ASP printout of the simple linear regression is shown above.

a. Construct a scattergram for the data.

b. Find the prediction equation for managerial success.

c. Find s for your prediction equation. Interpret the standard deviation s in the context of this problem.

d. Plot the least squares line on your scattergram of part **a**. Does it appear that the number of inter-actions with outsiders contributes information for the prediction of managerial success? Explain.

e. Conduct a formal statistical hypothesis test to answer the question posed in part **d**. Use $\alpha = .05$.

f. Construct a 95% confidence interval for β_1. Interpret the interval in the context of the problem.

Manager	Manager Success Index y	Number of Interactions with Outsiders x
1	40	12
2	73	71
3	95	70
4	60	81
5	81	43
6	27	50
7	53	42
8	66	18
9	25	35
10	63	82
11	70	20
12	47	81
13	80	40
14	51	33
15	32	45
16	50	10
17	52	65
18	30	20
19	42	21

11.6 THE COEFFICIENT OF CORRELATION

The claim is often made that the number of cigarettes smoked and the incidence of lung cancer are "highly correlated." Another popular belief is that the crime rate and the unemployment rate are "correlated." Some people even believe that the Dow Jones Industrial Average and the lengths of fashionable skirts are "correlated." In this section we discuss the concept of **correlation** and show how it can be used to measure the linear relationship between two variables, x and y. A numerical descriptive measure of correlation is provided by the *Pearson product moment coefficient of correlation, r.*

DEFINITION 11.2

The **Pearson product moment coefficient of correlation**, *r*, is a measure of the strength of the *linear* relationship between two variables *x* and *y*. It is computed (for a sample of *n* measurements on *x* and *y*) as follows:

$$r = \frac{SS_{xy}}{\sqrt{SS_{xx}SS_{yy}}}$$

Note that the computational formula for the correlation coefficient *r* given in Definition 11.2 involves the same quantities that were used in computing the least squares prediction equation. In fact, since the numerators of the expressions for $\hat{\beta}_1$ and *r* are identical, you can see that $r = 0$ when $\hat{\beta}_1 = 0$ (the case where *x* contributes no information for the prediction of *y*) and that *r* is positive when the slope is positive and negative when the slope is negative. Unlike $\hat{\beta}_1$, the correlation coefficient *r* is *scaleless* and assumes a value between -1 and $+1$, regardless of the units of *x* and *y*.

A value of *r* near or equal to 0 implies little or no linear relationship between *y* and *x*. In contrast, the closer *r* comes to 1 or -1, the stronger the linear relationship between *y* and *x*. And if $r = 1$ or $r = -1$, all the sample points fall exactly on the least squares line. Positive values of *r* imply a positive linear relationship between *y* and *x;* that is, *y* increases as *x* increases. Negative values of *r* imply a negative linear relationship between *y* and *x;* that is, *y* decreases as *x* increases. Each of these situations is portrayed in Figure 11.13.

FIGURE 11.13

Values of *r* and their implications

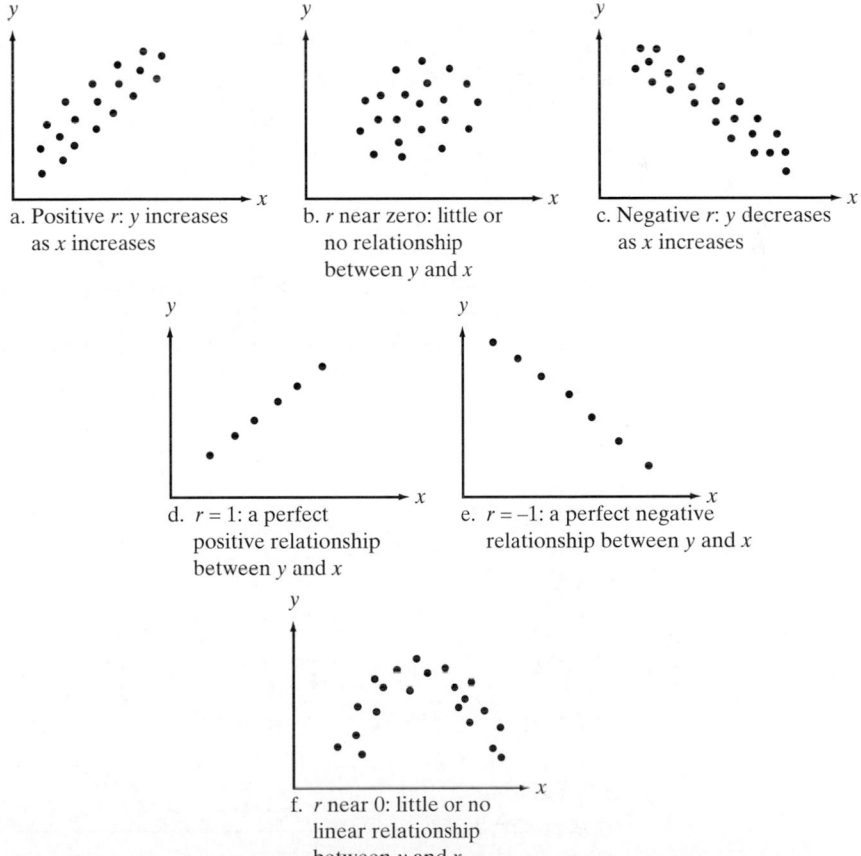

a. Positive *r*: *y* increases as *x* increases

b. *r* near zero: little or no relationship between *y* and *x*

c. Negative *r*: *y* decreases as *x* increases

d. $r = 1$: a perfect positive relationship between *y* and *x*

e. $r = -1$: a perfect negative relationship between *y* and *x*

f. *r* near 0: little or no linear relationship between *y* and *x*

We demonstrate how to calculate the coefficient of correlation r using the data in Table 11.1 for the drug reaction example. The quantities needed to calculate r are SS_{xy}, SS_{xx}, and SS_{yy}. The first two quantities have been calculated previously and are repeated here for convenience:

$$SS_{xy} = 7 \quad SS_{xx} = 10 \quad SS_{yy} = \sum y^2 - \frac{\left(\sum y\right)^2}{n}$$

$$= 26 - \frac{(10)^2}{5} = 26 - 20 = 6$$

We now find the coefficient of correlation:

$$r = \frac{SS_{xy}}{\sqrt{SS_{xx}SS_{yy}}} = \frac{7}{\sqrt{(10)(6)}} = \frac{7}{\sqrt{60}} = .904$$

The fact that r is positive and near 1 in value indicates that the reaction time tends to increase as the amount of drug in the bloodstream increases—*for this sample of five subjects.* This is the same conclusion we reached when we found the calculated value of the least squares slope to be positive.

EXAMPLE 11.1

Legalized gambling is available on several riverboat casinos operated by a city in Mississippi. The mayor of the city wants to know the correlation between the number of casino employees and yearly crime rate. The records for the past 10 years are examined, and the results listed in Table 11.5 are obtained. Calculate the coefficient of correlation r for the data.

Solution

Rather than use the computing formula given in Definition 11.2, we resort to a statistical software package. The data of Table 11.5 were entered into a computer and MINITAB was used to compute r. The MINITAB printout is shown in Figure 11.14.

TABLE 11.5   Data on Casino Employees and Crime Rate, Example 11.1

Year	Number of Casino Employees x (thousands)	Crime Rate y (number of crimes per 1,000 population)
1986	15	1.35
1987	18	1.63
1988	24	2.33
1989	22	2.41
1990	25	2.63
1991	29	2.93
1992	30	3.41
1993	32	3.26
1994	35	3.63
1995	38	4.15

FIGURE 11.14

MINITAB printout for Example 11.1

```
Correlation of NOEMPLOY and CRIMERAT = 0.987
```

FIGURE 11.15

Scattergram for
Example 11.1

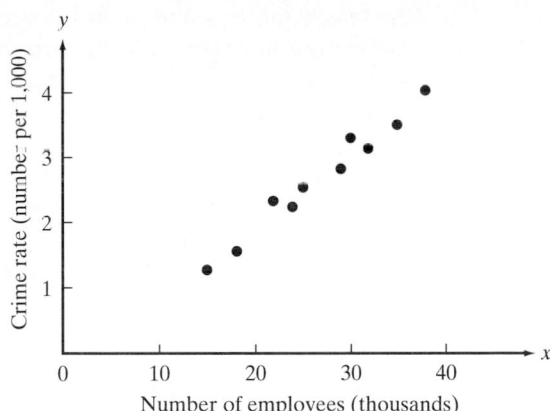

The coefficient of correlation, shaded on the printout, is $r = .987$. Thus, the size of the casino workforce and crime rate in this city are very highly correlated—at least over the past 10 years. The implication is that a strong positive linear relationship exists between these variables (see Figure 11.15). We must be careful, however, not to jump to any unwarranted conclusions. For instance, the mayor may be tempted to conclude that hiring more casino workers next year will increase the crime rate—that is, that there is a *causal relationship* between the two variables. However, high correlation does not imply causality. The fact is, many things have probably contributed both to the increase in the casino workforce and to the increase in crime rate. The city's tourist trade has undoubtedly grown since legalizing riverboat casinos and it is likely that the casinos have expanded both in services offered and in number. *We cannot infer a causal relationship on the basis of high sample correlation. When a high correlation is observed in the sample data, the only safe conclusion is that a linear trend may exist between x and y.* Another variable, such as the increase in tourism, may be the underlying cause of the high correlation between x and y.   ▲

Keep in mind that the correlation coefficient r measures the linear correlation between x values and y values in the sample, and a similar linear coefficient of correlation exists for the population from which the data points were selected. The **population correlation coefficient** is denoted by the symbol ρ (rho). As you might expect, ρ is estimated by the corresponding sample statistic, r. Or, instead of estimating ρ, we might want to test the null hypothesis H_0: $\rho = 0$ against H_a: $\rho \neq 0$ —that is, we can test the hypothesis that x contributes no information for the prediction of y by using the straight-line model against the alternative that the two variables are at least linearly related.

However, we already performed this *identical* test in Section 11.5 when we tested H_0: $\beta_1 = 0$ against H_a: $\beta_1 \neq 0$. That is, the null hypothesis H_0: $\rho = 0$ is equivalent to the hypothesis H_0: $\beta_1 = 0$.* When we tested the null hypothesis H_0: $\beta_1 = 0$ in connection with the drug reaction example, the data led to a rejection of the null hypothesis at the $\alpha = .05$ level. This rejection implies that the null hypothesis of a 0 linear correlation between the two variables (drug and reaction time) can also be rejected at the $\alpha = .05$ level. The only real difference between

---

*The correlation test statistic that is equivalent to $t = \hat{\beta}_1 / s_{\hat{\beta}_1}$ is $t = \dfrac{r}{\sqrt{(1 - r^2)/(n - 2)}}$

the least squares slope $\hat{\beta}_1$ and the coefficient of correlation r is the measurement scale. Therefore, the information they provide about the usefulness of the least squares model is to some extent redundant. For this reason, we will use the slope to make inferences about the existence of a positive or negative linear relationship between two variables.

11.7 THE COEFFICIENT OF DETERMINATION

Another way to measure the usefulness of the model is to measure the contribution of x in predicting y. To accomplish this, we calculate how much the errors of prediction of y were reduced by using the information provided by x. To illustrate, consider the sample shown in the scattergram of Figure 11.16a. If we assume that x contributes no information for the prediction of y, the best prediction for a value of y is the sample mean $\bar{y}$, which is shown as the horizontal line in Figure 11.16b. The vertical line segments in Figure 11.16b are the deviations of the points about the mean $\bar{y}$. Note that the sum of squares of deviations for the prediction equation $\hat{y} = \bar{y}$ is

$$SS_{yy} = \sum(y_i - \bar{y})^2$$

Now suppose you fit a least squares line to the same set of data and locate the deviations of the points about the line as shown in Figure 11.16c. Compare the deviations about the prediction lines in Figures 11.16b and 11.16c. You can see that

1. If x contributes little or no information for the prediction of y, the sums of squares of deviations for the two lines,

$$SS_{yy} = \sum(y_i - \bar{y})^2 \qquad \text{and} \qquad SSE = \sum(y_i - \hat{y}_i)^2$$

will be nearly equal.
2. If x does contribute information for the prediction of y, the SSE will be smaller than SS_{yy}. In fact, if all the points fall on the least squares line, then SSE = 0.

Then, the reduction in the sum of squares of deviations that can be attributed to x, expressed as a proportion of SS_{yy}, is

$$\frac{SS_{yy} - SSE}{SS_{yy}}$$

Note that SS_{yy} is the "total sample variation" of the observations around the mean $\bar{y}$ and that SSE is the remaining "unexplained sample variability" after fitting the line $\hat{y}$. Thus, the difference $(SS_{yy} - SSE)$ is the "explained sample variability" attributable to the linear relationship with x. Then a verbal description of the proportion is

$$\frac{SS_{yy} - SSE}{SS_{yy}} = \frac{\text{Explained sample variability}}{\text{Total sample variability}}$$
$$= \text{Proportion of total sample variability explained by the linear relationship}$$

In simple linear regression, it can be shown that this proportion is equal to the square of the simple linear coefficient of correlation r (the Pearson product moment coefficient of correlation).

FIGURE 11.16

A comparison of the sum of squares of deviations for two models

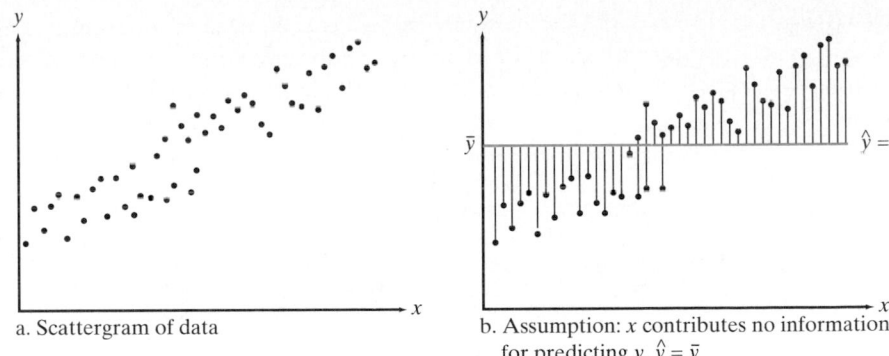

a. Scattergram of data

b. Assumption: x contributes no information for predicting y, $\hat{y} = \bar{y}$

c. Assumption: x contributes information for predicting y, $\hat{y} = \hat{\beta}_0 + \hat{\beta}_1 x$

TABLE 11.6

Amount of Drug x (%)	Reaction Time y (seconds)
1	1
2	1
3	2
4	2
5	4

DEFINITION 11.3

The **coefficient of determination** is

$$r^2 = \frac{SS_{yy} - SSE}{SS_{yy}} = 1 - \frac{SSE}{SS_{yy}}$$

It represents the proportion of the total sample variability around $\bar{y}$ that is explained by the linear relationship between y and x. (In simple linear regression, it may also be computed as the square of the coefficient of correlation r.)

Note that r^2 is always between 0 and 1, because r is between -1 and $+1$. Thus, an r^2 of .60 means that the sum of squares of deviations of the y values about their predicted values has been reduced 60% by the use of the least squares equation $\hat{y}$, instead of $\bar{y}$, to predict y.

EXAMPLE 11.2

Calculate the coefficient of determination for the drug reaction example. The data are repeated in Table 11.6 for convenience. Interpret the result.

Solution

From previous calculations,

$$SS_{yy} = 6 \quad \text{and} \quad SSE = \sum (y - \hat{y})^2 = 1.10$$

Then, from Definition 11.3, the coefficient of determination is given by

$$r^2 = \frac{SS_{yy} - SSE}{SS_{yy}} = \frac{6.0 - 1.1}{6.0} = \frac{4.9}{6.0} = .82$$

FIGURE 11.17

SAS printout for reaction time – drug percentage example

Dependent Variable: Y

Analysis of Variance

Source	DF	Sum of Squares	Mean Square	F Value	Prob>F
Model	1	4.90000	4.90000	13.364	0.0354
Error	3	1.10000	0.36667		
C Total	4	6.00000			

Root MSE	0.60553	R-square	0.8167
Dep Mean	2.00000	Adj R-sq	0.7556
C.V.	30.27650		

Parameter Estimates

Variable	DF	Parameter Estimate	Standard Error	T for H0: Parameter=0	Prob >|T|
INTERCEP	1	-0.100000	0.63508530	-0.157	0.8849
X	1	0.700000	0.19148542	3.656	0.0354

Another way to compute r^2 is to recall (Section 11.6) that $r = .904$. Then we have $r^2 = (.904)^2 = .82$. A third way to obtain r^2 is from a computer printout. This value is shaded on the SAS printout reproduced in Figure 11.17 next to the heading **R-square**. Our interpretation is as follows: We know that using the amount of drug in the blood, x, to predict y with the least squares line

$$\hat{y} = -.1 + .7x$$

accounts for 82% of the total sum of squares of deviations of the five sample y values about their mean. Or, stated another way, 82% of the sample variation in reaction time (y) can be "explained" by using amount (x) of drug in a straight-line model.   ▲

Practical Interpretation of the Coefficient of Determination, r^2

$100(r^2)\%$ of the sample variation in y (measured by the total sum of squares of deviations of the sample y values about their mean $\bar{y}$) can be explained by (or attributed to) using x to predict y in the straight-line model.

EXERCISES 11.45–11.57

Learning the Mechanics

11.45 Explain what each of the following sample correlation coefficients tells you about the relationship between the x and y values in the sample:
  a. $r = 1$ **b.** $r = -1$ **c.** $r = 0$
  d. $r = .90$ **e.** $r = .10$ **f.** $r = -.88$

11.46 Describe the slope of the least squares line if
  a. $r = .7$ **b.** $r = -.7$ **c.** $r = 0$ **d.** $r^2 = .64$

11.47 Construct a scattergram for each data set. Then calculate r and r^2 for each data set. Interpret their values.

a.

x	-2	-1	0	1	2
y	-2	1	2	5	6

b.

x	-2	-1	0	1	2
y	6	5	3	2	0

c.

x	1	2	2	3	3	3	4
y	2	1	3	1	2	3	2

d.

x	0	1	3	5	6
y	0	1	2	1	0

11.48 Calculate r^2 for the least squares line in each of the following exercises. Interpret their values.

a. Exercise 11.10   **b.** Exercise 11.13

Applying the Concepts

11.49 A robust and frequently adopted model of human movement is Fitts' Law. According to Fitts' Law, the time T required to move and select a target of width W that lies at a distance (or amplitude) A is

$$T = a + b \log_2(2A/W)$$

where a and b are constants estimated using simple linear regression. The quantity $\log_2(2A/W)$, which is termed the Index of Difficulty (ID), represents the independent variable (measured in "bits") in the model. Research reported in the *Special Interest Group on Computer–Human Interaction Bulletin* (July 1993) used Fitts' Law to model the time (in milliseconds) required to perform a certain task on a computer. Based on data collected for $n = 160$ trials (using different values of A and W), the following least squares prediction equation was obtained:

$$\hat{T} = 175.4 + 133.2(\text{ID})$$

a. Interpret the estimates 175.4 and 133.2.

b. The coefficient of correlation for the analysis is $r = .951$. Interpret this value.

c. Conduct a test to determine whether the Fitts' Law model is statistically adequate for predicting performance time. Use $\alpha = .05$. [*Hint:* See the last paragraph and accompanying footnote of Section 11.6.]

d. Calculate the coefficient of determination, r^2. Interpret the result.

11.50 Many high school students experience "math anxiety," which has been shown to have a negative effect on their learning achievement. Does such an attitude carry over to learning computer skills? A math and computer science researcher at Duquesne University investigated this question and published her results in *Educational Technology* (May–June 1995). A sample of 1,730 high school students—902 boys and 828 girls—from public schools in Pittsburgh, Pennsylvania, participated in the study. Using 5-point Likert scales, where 1 = "strongly disagree" and 5 = "strongly agree," the researcher measured the stu-

dents' interest and confidence in both mathematics and computers.

a. For boys, math confidence and computer interest were correlated at $r = .14$. Fully interpret this result.

b. For girls, math confidence and computer interest were correlated at $r = .33$. Fully interpret this result.

11.51 Is there a link between the loneliness of parents and that of their offspring? This question was examined in the *Journal of Marriage and the Family* (Aug. 1986). The participants in the study were 130 female college undergraduates and their parents. Each triad of daughter, mother, and father completed the UCLA Loneliness Scale, a 20-item questionnaire designed to assess loneliness and several variables theoretically related to loneliness, such as social accessibility to others, difficulty in making friends, and depression.

a. The correlation between daughter's loneliness score y and mother's loneliness score x was determined to be $r = .26$. Interpret this value.

b. The correlation between daughter's loneliness score y and father's loneliness score x was determined to be $r = .19$. Interpret this value.

c. The correlation between daughter's loneliness score y and mother's self-esteem score x was determined to be $r = .14$. Interpret this value.

d. The correlation between daughter's loneliness score y and father's assertiveness score x was determined to be $r = .01$. Interpret this value.

e. Calculate the coefficient of determination, r^2, for parts **a–d**. Interpret the results.

11.52 The fertility rate of a country is defined as the number of children a woman citizen bears, on average, in her lifetime. *Scientific American* (Dec. 1993) reported on the declining fertility rate in developing countries. The researchers found that family planning can have a great effect on fertility rate. The accompanying table on page 508 gives the fertility rate, y, and contraceptive prevalence, x, (measured as the percentage of married women who use contraception) for each of 27 developing countries. A SAS printout of the simple linear regression analysis follows the table.

a. According to the researchers, "the data reveal that differences in contraceptive prevalence explain about 90% of the variation in fertility rates." Do you concur?

b. The researchers also concluded that "if contraceptive use increases by 15 percent, women bear, on average, one fewer child." Is this statement supported by the data? Explain.

Country	Contraceptive Prevalence x	Fertility Rate y	Country	Contraceptive Prevalence x	Fertility Rate y
Mauritius	76	2.2	Egypt	40	4.5
Thailand	69	2.3	Bangladesh	40	5.5
Colombia	66	2.9	Botswana	35	4.8
Costa Rica	71	3.5	Jordan	35	5.5
Sri Lanka	63	2.7	Kenya	28	6.5
Turkey	62	3.4	Guatemala	24	5.5
Peru	60	3.5	Cameroon	16	5.8
Mexico	55	4.0	Ghana	14	6.0
Jamaica	55	2.9	Pakistan	13	5.0
Indonesia	50	3.1	Senegal	13	6.5
Tunisia	51	4.3	Sudan	10	4.8
El Salvador	48	4.5	Yemen	9	7.0
Morocco	42	4.0	Nigeria	7	5.7
Zimbabwe	46	5.4			

Source: Robey, B., *et al.* "The fertility decline in developing countries." *Scientific American*, Dec. 1993, p. 62. [*Note:* The data values are estimated from a scatterplot.]

```
Dependent Variable: FERTRATE

                    Analysis of Variance

                            Sum of         Mean
        Source      DF      Squares       Square    F Value    Prob>F

        Model        1     35.96633     35.96633    74.309     0.0001
        Error       25     12.10033      0.48401
        C Total     26     48.06667

            Root MSE        0.69571    R-square     0.7483
            Dep Mean        4.51111    Adj R-sq     0.7382
            C.V.           15.42216

                      Parameter Estimates

                      Parameter     Standard    T for H0:
        Variable  DF    Estimate       Error    Parameter=0   Prob > |T|

        INTERCEP   1    6.731929    0.29034252    23.186        0.0001
        CONTPREV   1   -0.054610    0.00633512    -8.620        0.0001
```

11.53 In cotherapy two or more therapists lead a group. An article in the *American Journal of Dance Therapy* (Spring/Summer 1995) examined the use of cotherapy in dance/movement therapy. Two of several variables measured on each of a sample of 136 professional dance/movement therapists were years of formal training x and reported success rate y (measured as a percentage) of coleading dance/movement therapy groups.

a. Propose a linear model relating y to x.

b. The researcher hypothesized that dance/movement therapists with more years in formal dance training will report higher perceived success rates in cotherapy relationships. State the hypothesis in terms of the parameter of the model, part **a**.

c. The correlation coefficient for the sample data was reported as $r = -.26$. Interpret this result.

d. Does the value of r in part **c** support the hypothesis in part **b**? Test using $\alpha = .05$. [*Hint:* See the last paragraph and accompanying footnote of Section 11.6.]

11.54 Refer to the *American Journal of Psychiatry* (July 1995) study of the linear relationship between hippocampal volume x (cubic millimeters) and verbal memory retention score y of patients suffering from posttraumatic stress, Exercise 11.15.

a. The coefficient of correlation between right hippocampal volume and verbal retention was $r = .64$. Interpret this value.

Trial Number	Static Weight of Truck x (thousand pounds)	Weigh-in-Motion Reading Prior to Calibration Adjustment y_1 (thousand pounds)	Weigh-in-Motion Reading After Calibration Adjustment y_2 (thousand pounds)
1	27.9	26.0	27.8
2	29.1	29.9	29.1
3	38.0	39.5	37.8
4	27.0	25.1	27.1
5	30.3	31.6	30.6
6	34.5	36.2	34.3
7	27.8	25.1	26.9
8	29.6	31.0	29.6
9	33.1	35.6	33.0
10	35.5	40.2	35.0

Source: Adapted from data in Wright, J. L., Owen, F., and Pena, D. "Status of MN/DOT's weigh-in-motion program." St. Paul: Minnesota Department of Transportation, January 1983.

b. A statistical test of H_0: $\beta_1 = 0$ versus H_a: $\beta_1 > 0$ resulted in a p-value smaller than .05. Interpret this result.

11.55 Refer to the *Journal of Information Science* study of the relationship between the fraction y of documents retrieved using Medline and the number x of terms in the search query, Exercise 11.39.

a. The value of r was reported in the article as $r = .679$. Interpret this result.

b. Calculate the coefficient of determination, r^2, and interpret the result.

11.56 The Minnesota Department of Transportation installed a state-of-the-art weigh-in-motion scale in the concrete surface of the eastbound lanes of Interstate 494 in Bloomington, Minnesota. After installation, a study was undertaken to determine whether the scale's readings correspond with the static weights of the vehicles being monitored. (Studies of this type are known as *calibration studies*.) After some preliminary comparisons using a two-axle, six-tire truck carrying different loads (see table above), calibration adjustments were made in the software of the weigh-in-motion system and the scales were reevaluated.

a. Construct two scattergrams, one of y_1 versus x and the other of y_2 versus x.

b. Use the scattergrams of part **a** to evaluate the performance of the weigh-in-motion scale both before and after the calibration adjustment.

c. Calculate the correlation coefficient for both sets of data and interpret their values. Explain how these correlation coefficients can be used to evaluate the weigh-in-motion scale.

d. Suppose the sample correlation coefficient for y_2 and x was 1. Could this happen if the static weights and the weigh-in-motion readings disagreed? Explain.

11.57 Is there a relationship between leisure activities and high school performance? This question was investigated in the *Journal of Leisure Research* (Vol. 24, 1992). A list of 43 leisure activities was presented to 159 high school students and the students were asked to identify the activities they participated in each week. The article reports that the correlation between high school GPA and the number of leisure activities is $r = .13$, which has a two-tailed p-value of .0512.

a. What are the appropriate null and alternative hypotheses to test whether the number of leisure activities and GPA are linearly related?

b. Interpret the p-value in terms of this test.

11.8 USING THE MODEL FOR ESTIMATION AND PREDICTION

If we are satisfied that a useful model has been found to describe the relationship between reaction time and amount of drug in the bloodstream, we are ready for step 5 in our regression modeling procedure: using the model for estimation and prediction.

FIGURE 11.18

Estimated mean value and predicted individual value of reaction time y for $x = 4$

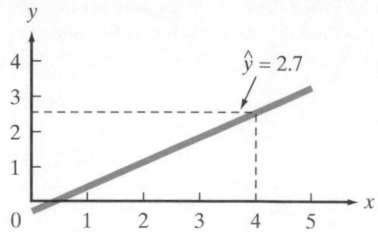

The most common uses of a probabilistic model for making inferences can be divided into two categories. The first is the use of the model for estimating the mean value of y, $E(y)$, for a specific value of x.

For our drug reaction example, we may want to estimate the mean response time for all people whose blood contains 4% of the drug.

The second use of the model entails predicting a new individual y value for a given x.

That is, we may want to predict the reaction time for a specific person who possesses 4% of the drug in the bloodstream.

In the first case, we are attempting to estimate the mean value of y for a very large number of experiments at the given x value. In the second case, we are trying to predict the outcome of a single experiment at the given x value. Which of these model uses—estimating the mean value of y or predicting an individual new value of y (for the same value of x)—can be accomplished with the greater accuracy?

Before answering this question, we first consider the problem of choosing an estimator (or predictor) of the mean (or a new individual) y value. We will use the least squares prediction equation

$$\hat{y} = \hat{\beta}_0 + \hat{\beta}_1 x$$

both to estimate the mean value of y and to predict a specific new value of y for a given value of x. For our example, we found

$$\hat{y} = -.1 + .7x$$

so that the estimated mean reaction time for all people when $x = 4$ (drug is 4% of blood content) is

$$\hat{y} = -.1 + .7(4) = 2.7 \text{ seconds}$$

The same value is used to predict a new y value when $x = 4$. That is, both the estimated mean and the predicted value of y are $\hat{y} = 2.7$ when $x = 4$, as shown in Figure 11.18.

The difference between these two model uses lies in the relative accuracy of the estimate and the prediction. These accuracies are best measured by using the sampling errors of the least squares line when it is used as an estimator and as a predictor, respectively. These errors are reflected in the standard deviations given in the next box.

Sampling Errors for the Estimator of the Mean of y and the Predictor of an Individual New Value of y

1. The standard deviation of the sampling distribution of the estimator $\hat{y}$ of the mean value of y at a specific value of x, say x_p, is

$$\sigma_{\hat{y}} = \sigma \sqrt{\frac{1}{n} + \frac{(x_p - \bar{x})^2}{SS_{xx}}} \, .$$

where σ is the standard deviation of the random error ϵ. We refer to $\sigma_{\hat{y}}$ as the standard error of $\hat{y}$.

2. The standard deviation of the prediction error for the predictor $\hat{y}$ of an individual new y value at a specific value of x is

$$\sigma_{(y-\hat{y})} = \sigma \sqrt{1 + \frac{1}{n} + \frac{(x_p - \bar{x})^2}{SS_{xx}}}$$

where σ is the standard deviation of the random error ϵ. We refer to $\sigma_{(y-\hat{y})}$ as the standard error of prediction.

The true value of σ is rarely known, so we estimate σ by s and calculate the estimation and prediction intervals as shown in the next two boxes.

A 100(1 − α)% Confidence Interval for the Mean Value of y at x = x_p

$$\hat{y} \pm t_{\alpha/2}(\text{Estimated standard error of } \hat{y})$$

or

$$\hat{y} \pm t_{\alpha/2}s \sqrt{\frac{1}{n} + \frac{(x_p - \bar{x})^2}{SS_{xx}}}$$

where $t_{\alpha/2}$ is based on $(n - 2)$ degrees of freedom.

A 100(1 − α)% Prediction Interval* for an Individual New Value of y at x = x_p

$$\hat{y} \pm t_{\alpha/2}(\text{Estimated standard error of prediction})$$

or

$$\hat{y} \pm t_{\alpha/2}s \sqrt{1 + \frac{1}{n} + \frac{(x_p - \bar{x})^2}{SS_{xx}}}$$

where $t_{\alpha/2}$ is based on $(n - 2)$ degrees of freedom.

EXAMPLE 11.3

Find a 95% confidence interval for the mean reaction time when the concentration of the drug in the bloodstream is 4%.

Solution

For a 4% concentration, $x = 4$ and the confidence interval for the mean value of y is

$$\hat{y} \pm t_{\alpha/2}s \sqrt{\frac{1}{n} + \frac{(x_p - \bar{x})^2}{SS_{xx}}} = \hat{y} \pm t_{.025}s \sqrt{\frac{1}{5} + \frac{(4 - \bar{x})^2}{SS_{xx}}}$$

where $t_{.025}$ is based on $n - 2 = 5 - 2 = 3$ degrees of freedom. Recall that $\hat{y} = 2.7$, $s = .61$, $\bar{x} = 3$, and $SS_{xx} = 10$. From Table VI in Appendix A, $t_{.025} = 3.182$. Thus, we have

---

*The term *prediction interval* is used when the interval formed is intended to enclose the value of a random variable. The term *confidence interval* is reserved for estimation of population parameters (such as the mean).

$$2.7 \pm (3.182)(.61)\sqrt{\frac{1}{5} + \frac{(4-3)^2}{10}} = 2.7 \pm (3.182)(.61)(.55)$$
$$= 2.7 \pm (3.182)(.34)$$
$$= 2.7 \pm 1.1$$

Therefore, when the percentage of drug in the bloodstream is 4%, the standard error of $\hat{y}$ is .34 and the corresponding 95% confidence interval for the mean reaction time for all possible subjects is 1.6 to 3.8 seconds. Note that we used a small amount of data (small sample size) for purposes of illustration in fitting the least squares line. The interval would probably be narrower if more information had been obtained from a larger sample. ▲

EXAMPLE 11.4

Predict the reaction time for the next performance of the experiment for a subject with a drug concentration of 4%. Use a 95% prediction interval.

Solution

To predict the response time for an individual new subject for whom $x = 4$, we calculate the 95% prediction interval as

$$\hat{y} \pm t_{\alpha/2}s\sqrt{1 + \frac{1}{n} + \frac{(x_p - \bar{x})^2}{SS_{xx}}} = 2.7 \pm (3.182)(.61)\sqrt{1 + \frac{1}{5} + \frac{(4-3)^2}{10}}$$
$$= 2.7 \pm (3.182)(.61)(1.14)$$
$$= 2.7 \pm (3.182)(.70)$$
$$= 2.7 \pm 2.2$$

Therefore, the standard error of prediction when the drug concentration is 4% is .70, and we predict with 95% confidence that the reaction time for this new individual will fall in the interval from .5 to 4.9 seconds. Like the confidence interval for the mean value of y, the prediction interval for y is quite large. This is because we have chosen a simple example (only five data points) to fit the least squares line. The width of the prediction interval could be reduced by using a larger number of data points. ▲

Both the confidence interval for $E(y)$ and the prediction interval for y can be obtained using a statistical software package. Figures 11.19 and 11.20 are SAS

FIGURE 11.19

SAS printout giving 95% confidence intervals for $E(y)$

Obs	X	Dep Var Y	Predict Value	Std Err Predict	Lower95% Mean	Upper95% Mean	Residual
1	1	1.0000	0.6000	0.469	-0.8927	2.0927	0.4000
2	2	1.0000	1.3000	0.332	0.2445	2.3555	-0.3000
3	3	2.0000	2.0000	0.271	1.1382	2.8618	0
4	4	2.0000	2.7000	0.332	1.6445	3.7555	-0.7000
5	5	4.0000	3.4000	0.469	1.9073	4.8927	0.6000

FIGURE 11.20

SAS printout giving 95% prediction intervals for y

Obs	X	Dep Var Y	Predict Value	Std Err Predict	Lower95% Predict	Upper95% Predict	Residual
1	1	1.0000	0.6000	0.469	-1.8376	3.0376	0.4000
2	2	1.0000	1.3000	0.332	-0.8972	3.4972	-0.3000
3	3	2.0000	2.0000	0.271	-0.1110	4.1110	0
4	4	2.0000	2.7000	0.332	0.5028	4.8972	-0.7000
5	5	4.0000	3.4000	0.469	0.9624	5.8376	0.6000

FIGURE 11.21

A 95% confidence interval for mean sales and a prediction interval for drug concentration when $x = 4$

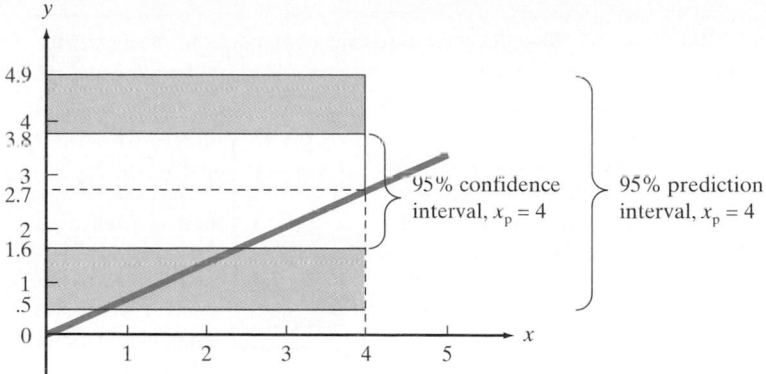

printouts showing confidence intervals and prediction intervals, respectively, for the data in the drug example.

The 95% confidence interval for $E(y)$ when $x = 4$ is shaded in Figure 11.19 in the row corresponding to **4** under the columns labeled **Lower95% Mean** and **Upper95% Mean**. The interval shown on the printout, (1.6445, 3.7555), agrees (except for rounding) with the interval calculated in Example 11.3. The 95% prediction interval for y when $x = 4$ is shaded in Figure 11.20 under the columns **Lower95% Predict** and **Upper95% Predict**. Again, except for rounding, the SAS interval (.5028, 4.8972) agrees with the one computed in Example 11.4.

A comparison of the confidence interval for the mean value of y and the prediction interval for a new value of y for 4% drug concentration ($x = 4$) is shown in Figure 11.21. Note that the prediction interval for an individual new value of y is *always* wider than the corresponding confidence interval for the mean value of y. You can see this by examining the formulas for the two intervals and by studying Figure 11.21.

The error in estimating the mean value of y, $E(y)$, for a given value of x, say x_p, is the distance between the least squares line and the true line of means, $E(y) = \beta_0 + \beta_1 x$. This error, $[\hat{y} - E(y)]$, is shown in Figure 11.22. In contrast, *the error* $(y_p - \hat{y})$ *in predicting some future value of y is the sum of two errors*—the error of estimating the mean of y, $E(y)$, shown in Figure 11.22, plus the random error

FIGURE 11.22

Error of estimating the mean value of y for a given value of x

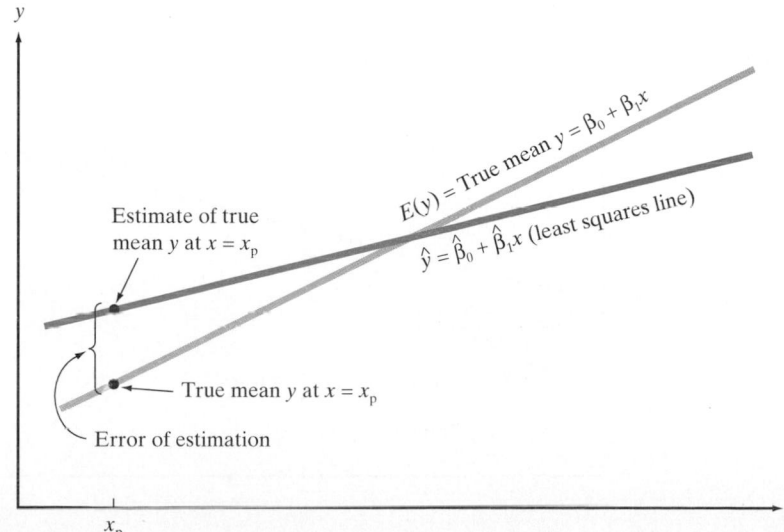

Statistical Assessment of Damage to Bronx Bricks

CASE STUDY

• 11.1 •

This case concerns an actual civil suit, described in *Chance* (Summer 1994), in which statistics played a key role in the jury's decision. The suit revolved around a five-building apartment complex located in the Bronx, New York. The buildings were constructed in the late 1970s from custom-designed jumbo (35-pound) bricks. Nearly three-quarters of a million bricks were used in the construction.

Over time, the bricks began to suffer *spalling* damage, i.e., separation of some portion of the face of a brick from its body. Experts agreed that the cause of the spalling was winter month freeze–thaw cycles in which water absorbed in the brick face alternates between freezing and thawing. The owner of the complex alleged that the bricks were defectively manufactured. The brick manufacturer countered that poor design and shoddy management of water runoff caused the water to be trapped and absorbed in the bricks, leading to the

damage. Ultimately, the suit required an estimate of the spall rate—the rate of damage per 1,000 bricks.

The owner estimated the spall rate using several *scaffold-drop* surveys. With this method, an engineer lowers a scaffold to selected places on building walls and counts the number of visible spalls for every 1,000 bricks in the observation area. The estimated spall rate is then multiplied by the total number of bricks (in thousands) in the entire complex to determine the total number of damaged bricks. When properly designed, the scaffold-drop survey, although extremely time-consuming and tedious to perform, is considered the "gold standard" for measuring spall damage. However, the owner did not drop the scaffolds at randomly selected wall areas. Instead, scaffolds were dropped primarily in areas of high spall concentration, leading to a substantially biased high estimate of total spall damage.

that is a component of the value of y to be predicted (see Figure 11.23). Consequently, the error of predicting a particular value of y will be larger than the error of estimating the mean value of y for a particular value of x. Note from their formulas that both the error of estimation and the error of prediction take their smallest values when $x_p = \bar{x}$. The farther x_p lies from $\bar{x}$, the larger will be the errors of estimation and prediction. You can see why this is true by noting the deviations for different values of x_p between the line of means $E(y) = \beta_0 + \beta_1 x$ and the predicted line of means $\hat{y} = \hat{\beta}_0 + \hat{\beta}_1 x$ shown in Figure 11.23. The

FIGURE 11.23

Error of predicting a future value of y for a given value of x

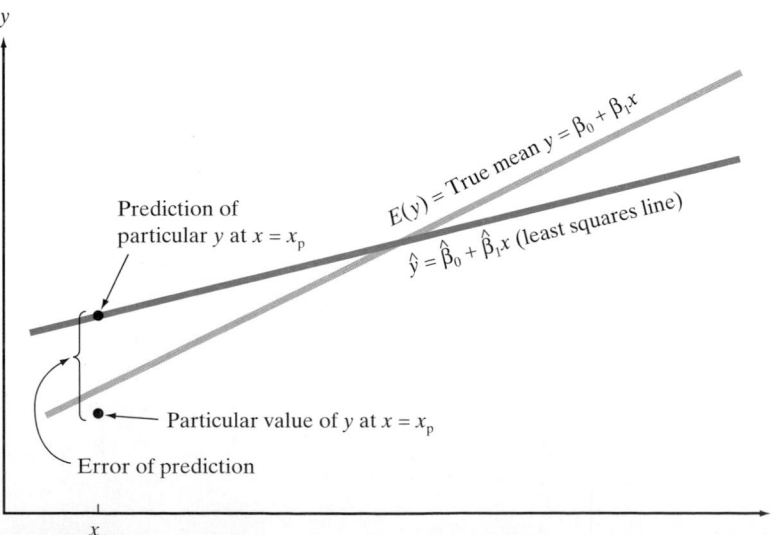

Case Study 11.1 continued

In an attempt to obtain an unbiased estimate of spall rate, the brick manufacturer conducted its own survey of the walls of the complex. The walls were divided into 83 wall segments and a photograph of each wall segment was taken. The number of spalled bricks that could be made out from each photo was recorded and the sum over all 83 wall segments was used as an estimate of total spall damage.

When the data from the two methods were compared, major discrepancies were discovered. At eleven locations that had been painstakingly surveyed by the scaffold drops, the spalls visible from the photos did not include all of the spalls identified on the drops. For these wall segments, the photo method provided a serious underestimate of the spall rate, as shown in Table 11.7. Consequently, the total spall damage estimated by the photo survey will also be underestimated.

In this court case, the jury was faced with the following dilemma: The scaffold-drop survey provided the most accurate estimate of spall rate in a given wall segment. Unfortunately, the drop areas were not selected at random from the entire complex; rather, drops were made at areas with high spall concentration, leading to an overestimate of the total damage. On the other hand, the photo survey was complete in that all 83 wall segments in the complex were checked for spall damage. But the spall rate estimated by the photos, at least in areas of high spall concentration, was biased low, leading to an underestimate of the total damage.

Use the data in Table 11.7, as did expert statisticians who testified in the case, to help the jury estimate the true spall rate at a given wall segment. Then explain how this information, coupled with the data (not given here) on all 83 wall segments, can provide a reasonable estimate of the total spall damage (i.e., total number of damaged bricks).

deviation is larger at the extremes of the interval where the largest and smallest values of x in the data set occur.

Both the confidence intervals for mean values and the prediction intervals for new values are depicted over the entire range of the regression line in Figure 11.24. You can see that the confidence interval is always narrower than the prediction interval, and that they are both narrowest at the mean $\bar{x}$, increasing steadily as the distance $|x - \bar{x}|$ increases. In fact, when x is selected far enough away from $\bar{x}$ so that it falls outside the range of the sample data, it is dangerous to make any inferences about $E(y)$ or y.

TABLE 11.7   Comparison of Spall Rates Estimated by Two Methods (Case Study 11.1)

Drop Location	Drop Spall Rate (per 1,000 bricks)	Photo Spall Rate (per 1,000 bricks)
1	0	0
2	5.1	0
3	6.6	0
4	1.1	.8
5	1.8	1.0
6	3.9	1.0
7	11.5	1.9
8	22.1	7.7
9	39.3	14.9
10	39.9	13.9
11	43.0	11.8

Source: Fairley, W. B., *et al.* "Bricks, buildings, and the Bronx: Estimating masonry deterioration." *Chance,* Vol. 7, No. 3, Summer 1994, p. 36 (Figure 3). [*Note:* The data points are estimated from the points shown on a scatterplot.]

FIGURE 11.24

Confidence intervals for mean values and prediction intervals for new values

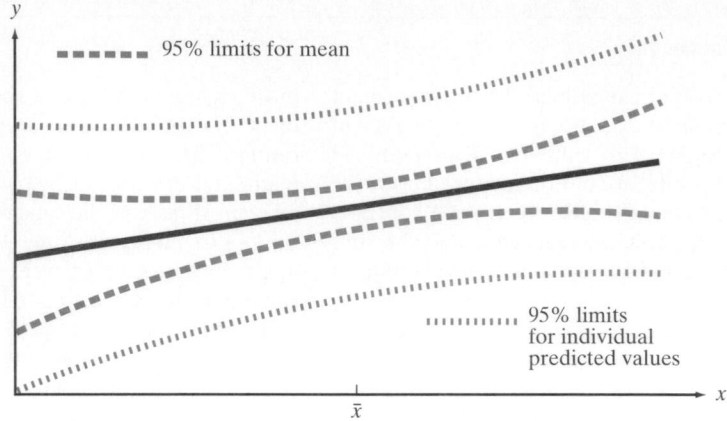

Warning

Using the least squares prediction equation to estimate the mean value of y or to predict a particular value of y for values of x that fall *outside the range* of the values of x contained in your sample data may lead to errors of estimation or prediction that are much larger than expected. Although the least squares model may provide a very good fit to the data over the range of x values contained in the sample, it could give a poor representation of the true model for values of x outside this region.

The confidence interval width grows smaller as n is increased; thus, in theory, you can obtain as precise an estimate of the mean value of y as desired (at any given x) by selecting a large enough sample. The prediction interval for a new value of y also grows smaller as n increases, but there is a lower limit on its width. If you examine the formula for the prediction interval, you will see that the interval can get no smaller than $\hat{y} \pm z_{\alpha/2}\sigma$.* Thus, the only way to obtain more accurate predictions for new values of y is to reduce the standard deviation of the regression model, σ. This can be accomplished only by improving the model, either by using a curvilinear (rather than linear) relationship with x or by adding new independent variables to the model, or both. Methods of improving the model are discussed in Chapters 12 and 13.

EXERCISES 11.58–11.69

Learning the Mechanics

11.58 Consider the following pairs of measurements:

x	1	2	3	4	5	6	7
y	3	5	4	6	7	7	10

  a. Construct a scattergram for these data.
  b. Find the least squares line, and plot it on your scattergram.
  c. Find s^2.

  d. Find a 90% confidence interval for the mean value of y when $x = 4$. Plot the upper and lower bounds of the confidence interval on your scattergram.
  e. Find a 90% prediction interval for a new value of y when $x = 4$. Plot the upper and lower bounds of the prediction interval on your scattergram.
  f. Compare the widths of the intervals you constructed in parts **d** and **e**. Which is wider and why?

*The result follows from the facts that, for large n, $t_{\alpha/2} \approx z_{\alpha/2}$, $s \approx \sigma$, and the last two terms under the radical in the standard error of the predictor are approximately 0.

11.59 Consider the following pairs of measurements:

x	y	x	y
1	−1	−2	−6
−1	−5	3	4
2	1	−1	−4
0	−3	5	4
4	7	1	0

Given that $SS_{xx} = 47.6$, $SS_{yy} = 168.1$, $SS_{xy} = 85.6$, and $\hat{y} = -2.458 + 1.7983x$:

a. Construct a scattergram for the data.

b. Plot the least squares line on your scattergram.

c. Use a 95% confidence interval to estimate the mean value of y when $x = 5$. Plot the upper and lower bounds of the interval on your scattergram.

d. Repeat part **c** for $x = 1.2$ and $x = -2$.

e. Compare the widths of the three confidence intervals you constructed in parts **c** and **d** and explain why they differ.

11.60 Refer to Exercise 11.59.

a. Using no information about x, estimate and calculate a 95% confidence interval for the mean value of y. [*Hint:* Use the one-sample t methodology of Section 7.3.]

b. Plot the estimated mean value and the confidence interval as horizontal lines on your scattergram.

c. Compare the confidence intervals you calculated in parts **c** and **d** of Exercise 11.59 with the one you calculated in part **a** of this exercise. Does x appear to contribute information about the mean value of y?

d. Check the answer you gave in part **c** with a statistical test of the null hypothesis $H_0: \beta_1 = 0$ against $H_a: \beta_1 \neq 0$. Use $\alpha = .05$.

11.61 Consider the pairs of measurements shown in the next table.

x	4	6	0	5	2	3	2	6	2	1
y	3	5	−1	4	3	2	0	4	1	1

For these data, $SS_{xx} = 38.900$, $SS_{yy} = 33.600$, $SS_{xy} = 32.8$, and $\hat{y} = -.414 + .843x$.

a. Construct a scattergram for these data.

b. Plot the least squares line on your scattergram.

c. Use a 95% confidence interval to estimate the mean value of y when $x_p = 6$. Plot the upper and lower bounds of the interval on your scattergram.

d. Repeat part **c** for $x_p = 3.2$ and $x_p = 0$.

e. Compare the widths of the three confidence intervals you constructed in parts **c** and **d** and explain why they differ.

11.62 In fitting a least squares line to $n = 10$ data points, the following quantities were computed:

$$SS_{xx} = 32 \quad \bar{x} = 3 \quad SS_{yy} = 26 \quad \bar{y} = 4 \quad SS_{xy} = 28$$

a. Find the least squares line.

b. Graph the least squares line.

c. Calculate SSE.   **d.** Calculate s^2.

e. Find a 95% confidence interval for the mean value of y when $x_p = 2.5$.

f. Find a 95% prediction interval for y when $x_p = 4$.

Applying the Concepts

11.63 Refer to the simple linear regression of best jumping distance y on average takeoff error x for $n = 18$ novice long jumpers, Exercises 11.16 and 11.27. The MINITAB printout of the analysis is reproduced below.

a. A 95% prediction interval for y when $x = .20$ is shown at the bottom of the printout under **95% P.I.** Interpret this interval.

b. A 95% confidence interval for $E(y)$ when $x = .20$ is also shown at the bottom of the printout under **95% C.I.** Interpret this interval.

```
The regression equation is
DISTANCE = 5.35 + 0.530 TAKOFFER

Predictor       Coef      Stdev    t-ratio        p
Constant      5.3480     0.1635      32.71    0.000
TAKOFFER      0.5299     0.9254       0.57    0.575

s = 0.4115      R-sq = 2.0%      R-sq(adj) = 0.0%

Analysis of Variance

SOURCE        DF         SS         MS        F        p
Regression     1     0.0555     0.0555     0.33    0.575
Error         16     2.7091     0.1693
Total         17     2.7646

    Fit   Stdev.Fit        95% C.I.          95% P.I.
 5.4539      0.1107   ( 5.2191, 5.6888)  ( 4.5504, 6.3575)
```

```
Dependent Variable: NOKILLED

                        Analysis of Variance

                              Sum of         Mean
          Source       DF     Squares       Square    F Value    Prob>F

          Model         1    19.11669      19.11669    16.029    0.0018
          Error        12    14.31188       1.19266
          C Total      13    33.42857

              Root MSE        1.09209     R-square     0.5719
              Dep Mean        1.42857     Adj R-sq     0.5362
              C.V.           76.44618

                        Parameter Estimates

                         Parameter    Standard    T for H0:
          Variable   DF   Estimate      Error    Parameter=0   Prob > |T|

          INTERCEP    1   -3.046856   1.15533214    -2.637        0.0217
          TITPCT      1    0.107656   0.02689001     4.004        0.0018

                   Dep Var   Predict   Std Err  Lower95%  Upper95%
     Obs  TITPCT  NOKILLED    Value    Predict   Predict   Predict   Residual
      1     24       0       -0.4631    0.555    -3.1326    2.2064    0.4631
      2     33       0        0.5058    0.372    -2.0078    3.0195   -0.5058
      3     34       0        0.6135    0.356    -1.8891    3.1161   -0.6135
      4     43       0        1.5824    0.294    -0.8820    4.0468   -1.5824
      5     50       0        2.3360    0.370    -0.1760    4.8480   -2.3360
      6     35    1.0000       0.7211    0.341    -1.7718    3.2140    0.2789
      7     35    1.0000       0.7211    0.341    -1.7718    3.2140    0.2789
      8     38    1.0000       1.0441    0.307    -1.4278    3.5159   -0.0441
      9     40    1.0000       1.2594    0.295    -1.2053    3.7241   -0.2594
     10     31    2.0000       0.2905    0.407    -2.2492    2.8301    1.7095
     11     43    2.0000       1.5824    0.294    -0.8820    4.0468    0.4176
     12     55    3.0000       2.8742    0.464     0.2887    5.4598    0.1258
     13     57    4.0000       3.0896    0.507     0.4659    5.7132    0.9104
     14     64    5.0000       3.8431    0.670     1.0516    6.6347    1.1569
```

c. Why would you recommend against computing and interpreting the intervals, parts **a** and **b**?

11.64 Refer to the simple linear regression of number of flycatchers killed y and nest box tit occupancy x for $n = 14$ nest sites, Exercises 11.20 and 11.29. The SAS printout of the analysis is reproduced above.

a. A 95% prediction interval for y when $x = 64$ is shown at the bottom of the printout. Interpret this interval.

b. How would the width of a 95% confidence interval for $E(y)$ when $x = 64$ compare to the interval, part **a**?

c. Would you recommend using the model to predict the number of flycatchers killed at a site with a nest box tit occupancy of 15%? Explain.

11.65 Certain dosages of a new drug developed to reduce a smoker's reliance on nicotine may reduce one's pulse rate to dangerously low levels. To investigate the drug's effect on pulse rate, different dosages of the drug were administered to six randomly selected patients, and 30 minutes later the decrease in each patient's pulse rate was recorded.

Patient	Dosage x (cubic centimeters)	Decrease in Pulse Rate y (beats/minute)
1	2.0	15
2	1.5	9
3	3.0	18
4	2.5	16
5	4.0	23
6	3.0	20

a. Is there evidence of a linear relationship between drug dosage and change in pulse rate? Test at $\alpha = .10$.

b. Find a 95% confidence interval for the slope β_1.

c. Find a 99% prediction interval for the decrease in pulse rate corresponding to a dosage of 3.5 cubic centimeters.

11.66 Refer to Exercises 11.21 and 11.43. The woolly-bear caterpillar data are shown in the table on page 519, preceded by a SAS printout of the simple linear regression analysis.

a. Find a 95% confidence interval for the mean cocoon temperature when the air temperature is 7°C on the SAS printout. Interpret the interval.

Dependent Variable: COCTEMP

Analysis of Variance

Source	DF	Sum of Squares	Mean Square	F Value	Prob>F
Model	1	120.18486	120.18486	164.090	0.0001
Error	10	7.32431	0.73243		
C Total	11	127.50917			

Root MSE	0.85582	R-square	0.9426	
Dep Mean	8.50833	Adj R-sq	0.9368	
C.V.	10.05863			

Parameter Estimates

Variable	DF	Parameter Estimate	Standard Error	T for H0: Parameter=0	Prob > |T|
INTERCEP	1	3.374666	0.47079265	7.168	0.0001
AIRTEMP	1	1.200858	0.09374540	12.810	0.0001

Obs	AIRTEMP	Dep Var COCTEMP	Predict Value	Std Err Predict	Lower95% Mean	Upper95% Mean	Residual
1	10.4	15.1000	15.8636	0.625	14.4708	17.2564	-0.7636
2	9.2	14.6000	14.4226	0.524	13.2558	15.5893	0.1774
3	2.2	6.8000	6.0166	0.314	5.3159	6.7172	0.7834
4	2.6	6.8000	6.4969	0.293	5.8446	7.1491	0.3031
5	4.1	8.0000	8.2982	0.248	7.7465	8.8499	-0.2982
6	3.7	8.7000	7.8178	0.253	7.2544	8.3813	0.8822
7	1.7	3.6000	5.4161	0.345	4.6465	6.1857	-1.8161
8	2	5.3000	5.7764	0.326	5.0492	6.5036	-0.4764
9	3	7.0000	6.9772	0.274	6.3657	7.5888	0.0228
10	3.5	7.1000	7.5777	0.258	7.0039	8.1515	-0.4777
11	4.5	9.6000	8.7785	0.248	8.2260	9.3310	0.8215
12	4.4	9.5000	8.6584	0.247	8.1073	9.2095	0.8416
13	7	.	11.7807	0.355	10.9888	12.5725	.

	TEMPERATURE (°C)	
Day	**Air**	**Cocoon**[a]
1	10.4	15.1
2	9.2	14.6
3	2.2	6.8
4	2.6	6.8
5	4.1	8.0
6	3.7	8.7
7	1.7	3.6
8	2.0	5.3
9	3.0	7.0
10	3.5	7.1
11	4.5	9.6
12	4.4	9.5

[a]Each cocoon temperature is the average of the temperatures of two cocoons.

b. Suppose you were to place a single woolly-bear caterpillar cocoon in a controlled environment of 7°C. Find a 95% prediction interval for the cocoon temperature. Interpret the interval.

11.67 The reasons given by workers for quitting their jobs generally fall into one of two categories: (1) worker quits to seek or take a different job, or (2) worker quits to withdraw from the labor force. Economic theory suggests that wages and quit rates are related. The table below lists quit rates (quits per 100 employees) and the average hourly wage in a sample of 15 manufacturing industries. A MINITAB printout of the simple linear regression of quit rate y on average wage x is shown on page 520.

Industry	Quit Rate y	Average Wage x
1	1.4	$ 8.20
2	.7	10.35
3	2.6	6.18
4	3.4	5.37
5	1.7	9.94
6	1.7	9.11
7	1.0	10.59
8	.5	13.29
9	2.0	7.99
10	3.8	5.54
11	2.3	7.50
12	1.9	6.43
13	1.4	8.83
14	1.8	10.93
15	2.0	8.80

```
The regression equation is
QuitRate = 4.86 - 0.347 AveWage

Predictor       Coef       Stdev      t-ratio       p
Constant      4.8615      0.5201        9.35    0.000
AveWage      -0.34655     0.05866      -5.91    0.000

s = 0.4862     R-sq = 72.9%     R-sq(adj) = 70.8%

Analysis of Variance

SOURCE         DF         SS         MS         F        p
Regression      1       8.2507     8.2507     34.90    0.000
Error          13       3.0733     0.2364
Total          14      11.3240

    Fit   Stdev.Fit       95% C.I.           95% P.I.
  1.743     0.128    ( 1.467,  2.018)   ( 0.656,  2.829)
```

a. Do the data present sufficient evidence to conclude that average hourly wage rate contributes useful information for the prediction of quit rates? What does your model suggest about the relationship between quit rates and wages?

b. A 95% prediction interval for the quit rate in an industry with an average hourly wage of $9.00 is given at the bottom of the MINITAB printout. Interpret the result.

c. A 95% confidence interval for the mean quit rate for industries with an average hourly wage of $9.00 is also shown on the printout. Interpret this result.

11.68 In Exercise 11.22, we found the least squares line relating the age x of a sand lance to its length y. The data are reproduced in the table below.

   a. Do the data provide sufficient evidence to indicate that age x contributes information for the prediction of the length y of a sand lance? Test using $\alpha = .05$.

   b. Find the p-value for the test and interpret it.

   c. Find a 95% confidence interval for the mean length of sand lances that are 4 years old. Interpret the interval.

   d. Find a 95% prediction interval for the length of a sand lance that is 4 years old. Interpret the interval. Explain the difference between this interval and the interval obtained in part **a**.

11.69 Refer to Exercise 11.31. The data are shown in the table at right.

   a. Use a 90% confidence interval to estimate the mean useful life of a brand A cutting tool when the cutting speed is 45 meters per minute. Repeat for brand B. Compare the widths of the two intervals and comment on the reasons for any difference.

Cutting Speed (meters per minute)	USEFUL LIFE (HOURS) Brand A	Brand B
30	4.5	6.0
30	3.5	6.5
30	5.2	5.0
40	5.2	6.0
40	4.0	4.5
40	2.5	5.0
50	4.4	4.5
50	2.8	4.0
50	1.0	3.7
60	4.0	3.8
60	2.0	3.0
60	1.1	2.4
70	1.1	1.5
70	.5	2.0
70	3.0	1.0

   b. Use a 90% prediction interval to predict the useful life of a brand A cutting tool when the cutting speed is 45 meters per minute. Repeat for brand B. Compare the widths of the two intervals to each other, and to the two intervals you calculated in part **a**. Comment on the reasons for any differences.

   c. Note that the estimation and prediction you performed in parts **a** and **b** were for a value of x that was not included in the original sample. That is, the value $x = 45$ was not part of the sample. However, the value is within the range of x values in the sample, so that the regression model spans the x value for which the estimation and prediction were made. In such situations, estimation and prediction represent **interpolations**.

Mean Length (millimeters)	176	194	212	226	236	244	254
Age (years)	2	3	4	5	6	7	8

Suppose you were asked to predict the useful life of a brand A cutting tool for a cutting speed of $x = 100$ meters per minute. Since the given value of x is outside the range of the sample x values, the prediction is an example of **extrapolation**. Predict the useful life of a brand A cutting tool that is operated at 100 meters per minute, and construct a 95% confidence interval for the actual useful life of the tool. What additional assumption do you have to make in order to ensure the validity of an extrapolation?

11.9   SIMPLE LINEAR REGRESSION: AN EXAMPLE

In the preceding sections we have presented the basic elements necessary to fit and use a straight-line regression model. In this section we will assemble these elements by applying them in an example with the aid of a computer.

Suppose a fire insurance company wants to relate the amount of fire damage in major residential fires to the distance between the burning house and the nearest fire station. The study is to be conducted in a large suburb of a major city; a sample of 15 recent fires in this suburb is selected. The amount of damage, y, and the distance between the fire and the nearest fire station, x, are recorded for each fire. The results are given in Table 11.8.

Step 1   First, we hypothesize a model to relate fire damage, y, to the distance from the nearest fire station, x. We hypothesize a straight-line probabilistic model:

$$y = \beta_0 + \beta_1 x + \epsilon$$

Step 2   Next, we enter the data of Table 11.8 into a computer and use a statistical software package to estimate the unknown parameters in the deterministic component of the hypothesized model. The SAS printout for the simple linear regression analysis is shown in Figure 11.25. The least squares estimate of the slope β_1 and intercept β_0, shaded on the printout, are

$$\hat{\beta}_1 = 4.919331$$
$$\hat{\beta}_0 = 10.277929$$

TABLE 11.8   Fire Damage Data

Distance from Fire Station x (miles)	Fire Damage y (thousands of dollars)
3.4	26.2
1.8	17.8
4.6	31.3
2.3	23.1
3.1	27.5
5.5	36.0
.7	14.1
3.0	22.3
2.6	19.6
4.3	31.3
2.1	24.0
1.1	17.3
6.1	43.2
4.8	36.4
3.8	26.1

FIGURE 11.25

SAS printout for fire
damage regression analysis

Dep Variable: Y

Analysis of Variance

Source	DF	Sum of Squares	Mean Square	F Value	Prob>F
Model	1	841.76636	841.76636	156.886	0.0001
Error	13	69.75098	5.36546		
C Total	14	911.51733			

Root MSE	2.31635	R-Square	0.9235	
Dep Mean	26.41333	Adj R-Sq	0.9176	
C.V.	8.76961			

Parameter Estimates

Variable	DF	Parameter Estimate	Standard Error	T for H0: Parameter=0	Prob >|T|
INTERCEP	1	10.277929	1.42027781	7.237	0.0001
X	1	4.919331	0.39274775	12.525	0.0001

Obs	X	Y	Predict Value	Residual	Lower95% Predict	Upper95% Predict
1	3.4	26.2000	27.0037	-0.8037	21.8344	32.1729
2	1.8	17.8000	19.1327	-1.3327	13.8141	24.4514
3	4.6	31.3000	32.9068	-1.6068	27.6186	38.1951
4	2.3	23.1000	21.5924	1.5076	16.3577	26.8271
5	3.1	27.5000	25.5279	1.9721	20.3573	30.6984
6	5.5	36.0000	37.3342	-1.3342	31.8334	42.8351
7	0.7	14.1000	13.7215	0.3785	8.1087	19.3342
8	3.0	22.3000	25.0359	-2.7359	19.8622	30.2097
9	2.6	19.6000	23.0682	-3.4682	17.8678	28.2686
10	4.3	31.3000	31.4311	-0.1311	26.1908	36.6713
11	2.1	24.0000	20.6085	3.3915	15.3442	25.8729
12	1.1	17.3000	15.6892	1.6108	10.1999	21.1785
13	6.1	43.2000	40.2858	2.9142	34.5906	45.9811
14	4.8	36.4000	33.8907	2.5093	28.5640	39.2175
15	3.8	26.1000	28.9714	-2.8714	23.7843	34.1585
16	3.5		27.4956		22.3239	32.6672

Sum of Residuals   -3.73035E-14
Sum of Squared Residuals   69.7510
Predicted Resid SS (Press)   93.2117

and the least squares equation is (rounded)

$$\hat{y} = 10.278 + 4.919x$$

This prediction equation is graphed in Figure 11.26 along with a plot of the
data points.

The least squares estimate of the slope, $\hat{\beta}_1 = 4.919$, implies that the estimated
mean damage increases by \$4,919 for each additional mile from the fire station.
This interpretation is valid over the range of x, or from .7 to 6.1 miles from the station. The estimated y-intercept, $\hat{\beta}_0 = 10.278$, has the interpretation that a fire 0
miles from the fire station has an estimated mean damage of \$10,278. Although
this would seem to apply to the fire station itself, remember that the y-intercept is
meaningfully interpretable only if $x = 0$ is within the sampled range of the independent variable.

FIGURE 11.26

Least squares model for the fire damage data

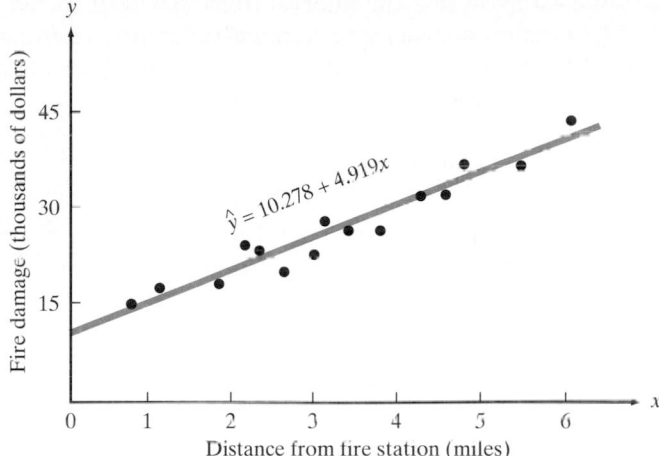

Step 3 Now we specify the probability distribution of the random error component ϵ. The assumptions about the distribution are identical to those listed in Section 11.3. Although we know that these assumptions are not completely satisfied (they rarely are for practical problems), we are willing to assume they are approximately satisfied for this example. The estimate of the standard deviation σ of ϵ, shaded on the printout, is

$$s = 2.31635$$

This implies that most of the observed fire damage (y) values will fall within approximately $2s = 4.64$ thousand dollars of their respective predicted values when using the least squares line.

Step 4 We can now check the usefulness of the hypothesized model—that is, whether x really contributes information for the prediction of y using the straight-line model. First, test the null hypothesis that the slope β_1 is 0, that is, that there is no linear relationship between fire damage and the distance from the nearest fire station, against the alternative hypothesis that fire damage increases as the distance increases. We test

$$H_0: \beta_1 = 0$$
$$H_a: \beta_1 > 0$$

The observed significance level for testing $H_a: \beta_1 \neq 0$, shaded on the printout, is .0001. Thus, the p-value for our one-tailed test is $p = .0001/2 = .00005$. This small p-value leaves little doubt that mean fire damage and distance between the fire and station are at least linearly related, with mean fire damage increasing as the distance increases.

We gain additional information about the relationship by forming a confidence interval for the slope β_1. A 95% confidence interval is

$$\hat{\beta}_1 \pm t_{.025} s_{\hat{\beta}_1}$$

where $\hat{\beta}_1 = 4.919$ and its standard error, $s_{\hat{\beta}_1} = .393$ are both obtained from the printout. The value $t_{.025}$, based on $n - 2 = 13$ df, is 2.160. Therefore, the 95% confidence interval is

$$\hat{\beta}_1 \pm t_{.025} s_{\hat{\beta}_1} = 4.919 \pm (2.160)(.393) = 4.919 \pm .849 = (4.070, 5.768)$$

We estimate that the interval from \$4,070 to \$5,768 encloses the mean increase (β_1) in fire damage per additional mile distance from the fire station.

Another measure of the utility of the model is the coefficient of determination, r^2. The value (shaded on the printout) is $r^2 = .9235$, which implies that about 92% of the sample variation in fire damage (y) is explained by the distance (x) between the fire and the fire station.

The coefficient of correlation, r, that measures the strength of the linear relationship between y and x is not shown on the SAS printout and must be calculated. Using the facts that $r = \sqrt{r^2}$ in simple linear regression and that r and $\hat{\beta}_1$ have the same sign, we find

$$r = +\sqrt{r^2} = \sqrt{.9235} = .96$$

The high correlation confirms our conclusion that β_1 is greater than 0; it appears that fire damage and distance from the fire station are positively correlated. All signs point to a strong linear relationship between y and x.

Step 5 We are now prepared to use the least squares model. Suppose the insurance company wants to predict the fire damage if a major residential fire were to occur 3.5 miles from the nearest fire station. The predicted value (shaded at the bottom of the printout) is $\hat{y} = 27.4956$, while the 95% prediction interval (also shaded) is (22.3239, 32.6672). Therefore, with 95% confidence we predict fire damage in a major residential fire 3.5 miles from the nearest station to be between \$22,324 and \$32,667.

One caution before closing: We would not use this prediction model to make predictions for homes less than .7 mile or more than 6.1 miles from the nearest fire station. A look at the data in Table 11.8 reveals that all the x values fall between .7 and 6.1. It is dangerous to use the model to make predictions outside the region in which the sample data fall. A straight line might not provide a good model for the relationship between the mean value of y and the value of x when stretched over a wider range of x values.

QUICK REVIEW

Key Terms

Coefficient of correlation   500
Coefficient of determination   505
Confidence interval for mean of y 511
Dependent variable   476
Deterministic model   475
Independent variable   476
Least squares line   480
Line of means   477
Method of least squares   478
Pearson product moment coefficient of
  correlation   501

Prediction interval for y 511
Predictor variable   476
Probabilistic model   475
Random error   475
Regression analysis   476
Response variable   476
Scattergrams   478
Slope   476
Straight-line (first-order) model   476
y-intercept   476

Key Formulas

$$\hat{\beta}_1 = \frac{SS_{xy}}{SS_{xx}}, \qquad \hat{\beta}_0 = \bar{y} - \hat{\beta}_1 \bar{x}$$

where $SS_{xy} = \sum xy - \dfrac{(\sum x)(\sum y)}{n}$

$SS_{xx} = \sum x^2 - \dfrac{(\sum x)^2}{n}$

Least squares estimates of β's   480

$$\hat{y} = \hat{\beta}_0 + \hat{\beta}_1 x$$

Least squares line   480

$$SSE = \sum(y_i - \hat{y}_i)^2 = SS_{yy} - \hat{\beta}_1 SS_{xy}$$

Sum of squared errors   490

$$\text{where } SS_{yy} - \sum y^2 - \frac{(\sum y)^2}{n}$$

$$s^2 = \frac{SSE}{n-2}$$

Estimated variance of σ^2 of ϵ 489

$$s_{\hat{\beta}_1} = \frac{s}{\sqrt{SS_{xx}}}$$

Estimated standard error of $\hat{\beta}_1$ 494

$$\hat{\beta}_1 \pm (t_{\alpha/2})s_{\hat{\beta}_1}$$

$(1-\alpha)100\%$ confidence interval for β_1 496

$$t = \frac{\hat{\beta}_1}{s_{\hat{\beta}_1}}$$

Test statistic for H_0: $\beta_1 = 0$ 495

$$r^2 = \frac{SS_{yy} - SSE}{SS_{yy}}$$

Coefficient of determination   505

$$r = \frac{SS_{xy}}{\sqrt{SS_{xx}SS_{yy}}} = \pm\sqrt{r^2} \quad (\text{same sign as } \hat{\beta}_1)$$

Coefficient of correlation   501

$$\hat{y} \pm (t_{\alpha/2})s\sqrt{\frac{1}{n} + \frac{(x_p - \bar{x})^2}{SS_{xx}}}$$

$(1-\alpha)100\%$ confidence interval for $E(y)$ when $x = x_p$ 511

$$\hat{y} + (t_{\alpha/2})s\sqrt{1 + \frac{1}{n} + \frac{(x_p - \bar{x})^2}{SS_{xx}}}$$

$(1-\alpha)100\%$ prediction interval for y when $x = x_p$ 511

LANGUAGE LAB

Symbol	Pronunciation	Description
y		Dependent variable (variable to be predicted or modeled)
x		Independent (predictor) variable
$E(y)$		Expected (mean) value of y
β_0	beta-zero	y-intercept of true line
β_1	beta-one	Slope of true line
$\hat{\beta}_0$	beta-zero hat	Least squares estimate of y-intercept
$\hat{\beta}_1$	beta-one hat	Least squares estimate of slope
ϵ	epsilon	Random error
$\hat{y}$	y-hat	Predicted value of y
$(y - \hat{y})$		Error of prediction
SE		Sum of errors (will equal zero with least squares line)
SSE		Sum of squared errors (will be smallest for least squares line)
SS_{xx}		Sum of squares of x-values
SS_{yy}		Sum of squares of y-values
SS_{xy}		Sum of squares of cross-products, $x \cdot y$
r		Coefficient of correlation
r^2	R-squared	Coefficient of determination
x_p		Value of x used to predict y

SUPPLEMENTARY EXERCISES 11.70–11.84

Learning the Mechanics

11.70 In fitting a least squares line to $n = 15$ data points, the following quantities were computed: $SS_{xx} = 55$, $SS_{yy} = 198$, $SS_{xy} = -88$, $\bar{x} = 1.3$, and $\bar{y} = 35$.
  a. Find the least squares line.
  b. Graph the least squares line.
  c. Calculate SSE.   **d.** Calculate s^2.
  e. Find a 90% confidence interval for β_1. Interpret this estimate.
  f. Find a 90% confidence interval for the mean value of y when $x = 15$.
  g. Find a 90% prediction interval for y when $x = 15$.

11.71 Consider the following sample data:

y	5	1	3
x	5	1	3

  a. Construct a scattergram for the data.
  b. It is possible to find many lines for which $\Sigma (y - \hat{y}) = 0$. For this reason, the criterion $\Sigma (y - \hat{y}) = 0$ is not used for identifying the "best-fitting" straight line. Find two lines that have $\Sigma (y - \hat{y}) = 0$.
  c. Find the least squares line.
  d. Compare the value of SSE for the least squares line to that of the two lines you found in part **b**. What principle of least squares is demonstrated by this comparison?

11.72 Consider the following 10 data points:

x	3	5	6	4	3	7	6	5	4	7
y	4	3	2	1	2	3	3	5	4	2

  a. Plot the data on a scattergram
  b. Calculate the values of r and r^2.

  c. Is there sufficient evidence to indicate that x and y are linearly correlated? Test at the $\alpha = .10$ level of significance.

Applying the Concepts

11.73 A study was conducted to determine whether a student's final grade in an introductory sociology course is linearly related to his or her performance on the verbal ability test administered before college entrance. The verbal test scores and final grades for a random sample of 10 students are shown in the table below. An SPSS printout of the simple linear regression follows the table.
  a. Find the least squares line relating y to x.
  b. Plot the data points and graph the least squares line.
  c. Do the data provide sufficient evidence to indicate that a positive correlation exists between verbal score and final grade? Use $\alpha = .01$.
  d. Find a 95% confidence interval for the slope β_1.

Student	Verbal Ability Test Score, x	Final Sociology Grade, y
1	39	65
2	43	78
3	21	52
4	64	82
5	57	92
6	47	89
7	28	73
8	75	98
9	34	56
10	52	75

```
Multiple R            .83979
R Square              .70524
Adjusted R Square     .66840
Standard Error       8.70363

Analysis of Variance
                 DF    Sum of Squares    Mean Square
Regression        1        1449.97413     1449.97413
Residual          8         606.02587       75.75323

F =     19.14076    Signif F =  .0024

------------------Variables in the Equation-------------------

Variable            B        SE B       Beta        T   Sig T

VERBAL        .765562     .174985     .839786    4.375   .0024
(Constant)  40.784155    8.506861                4.794   .0014
```

e. Predict a student's final grade in the introductory course when his or her verbal test score is 50. Use a 90% prediction interval.

f. Find a 95% confidence interval for the mean final grade for students scoring 35 on the college entrance verbal exam.

11.74 Common maize rust is a serious disease of sweet corn. Although fungicides are effective in controlling maize rust, the timing of the application is crucial. Researchers in New York state have developed an action threshold for initiation of fungicide applications based on a regression equation relating maize rust incidence to severity of the disease (*Phytopathology*, Vol. 80, 1990). In one particular field, data were collected on more than 100 plants of the sweet corn hybrid Jubilee. For each plant, incidence was measured as the percentage of leaves infected (x) and severity was calculated as the log (base 10) of the average number of infections per leaf (y). A simple linear regression analysis of the data produced the following results:

$$\hat{y} = -.939 + .020x$$
$$r^2 = .816$$
$$s = .288$$

a. Interpret the value of $\hat{\beta}_1$.

b. Interpret the value of r^2.

c. Interpret the value of s.

d. Calculate the value of r and interpret it.

e. Use the result, part **d**, to test the utility of the model. Use $\alpha = .05$. (Assume $n = 100$.)

f. Predict the severity of the disease when the incidence of maize rust for a plant is 80%. [*Note:* Take the antilog (base 10) of $\hat{y}$ to obtain the predicted average number of infections per leaf.]

11.75 At temperatures approaching absolute zero ($-273°C$), helium exhibits traits that seem to defy many laws of Newtonian physics. An experiment has been conducted with helium in solid form at various temperatures near absolute zero. The solid helium is placed in a dilution refrigerator along with a solid impure substance, and the fraction (in weight) of the impurity passing through the solid helium is recorded. (This phenomenon of solids passing directly through solids is known as *quantum tunneling*.) The data are given in the next table. The least squares printout for the simple linear regression model is on page 528.

a. Find the least squares estimates of the intercept and slope. Interpret them.

b. Use a 95% confidence interval to estimate the slope β_1. Interpret the interval in terms of this application. Does the interval support the hypothesis that temperature contributes information about the proportion of impurity passing through helium?

Temperature x (°C)	Proportion of Impurity Passing Through Helium y
−262.0	.315
−265.0	.202
−256.0	.204
−267.0	.620
−270.0	.715
−272.0	.935
−272.4	.957
−272.7	.906
−272.8	.985
−272.9	.987

c. Interpret the coefficient of determination for this model.

d. Find the 95% prediction interval for the percentage of impurity passing through solid helium at $-273°C$.

e. Note that the value of x in part **d** is outside the experimental region. Why might this lead to an unreliable prediction?

11.76 Is the maximum oxygen uptake, a measure often used by physiologists to indicate an individual's state of cardiovascular fitness, related to the performance of distance runners? Six long-distance runners submitted to treadmill tests for determination of their maximum oxygen uptake. The results, along with each runner's best mile time (in seconds), are shown in the table.

Athlete	Maximum Oxygen Uptake (milliliters/kilogram)	Mile Time (seconds)
1	63.3	241.5
2	60.1	249.8
3	53.6	246.1
4	58.8	232.4
5	67.5	237.2
6	62.5	238.4

a. Calculate r and r^2. Interpret the results.

b. Do the data provide sufficient evidence to indicate that mile time is negatively correlated with maximum oxygen uptake? Test $H_0: \rho = 0$ against the alternative hypothesis, $H_a: \rho < 0$, using $\alpha = .05$. [*Hint:* Recall (Section 11.6) that the test of $H_0: \rho = 0$ is equivalent to the test of $H_0: \beta_1 = 0$.]

11.77 A breeder of thoroughbred horses wishes to model the relationship between the gestation period and the life span of a horse. The breeder believes that the two variables may follow a linear trend. The information in the following table was supplied to the breeder from various thoroughbred stables across the state. (Note that the horse has the greatest variation of gestation period of any species owing to seasonal and nutritional factors.) A MINITAB printout of the simple linear regression analysis is given next to the table.

```
Dep Variable: IMPURITY
```

Analysis of Variance

Source	DF	Sum of Squares	Mean Square	F Value	Prob>F
Model	1	0.83089	0.83089	46.728	0.0001
Error	8	0.14225	0.01778		
C Total	9	0.97315			

Root MSE	0.13335	R-Square	0.8538	
Dep Mean	0.68260	Adj R-Sq	0.8356	
C.V.	19.53519			

Parameter Estimates

Variable	DF	Parameter Estimate	Standard Error	T for H0: Parameter=0	Prob >|T|
INTERCEP	1	-13.490347	2.07377160	-6.505	0.0002
TEMP	1	-0.052829	0.00772828	-6.836	0.0001

Obs	TEMP	IMPURITY	Predict Value	Residual	Lower95% Predict	Upper95% Predict
1	-262	0.3150	0.3508	-0.0358	0.00946	0.6922
2	-265	0.2020	0.5093	-0.3073	0.1816	0.8371
3	-256	0.2040	0.0339	0.1701	-0.3559	0.4236
4	-267	0.6200	0.6150	0.00502	0.2917	0.9383
5	-270	0.7150	0.7735	-0.0585	0.4495	1.0974
6	-272	0.9350	0.8791	0.0559	0.5499	1.2084
7	-272.4	0.9570	0.9003	0.0567	0.5695	1.2310
8	-272.7	0.9060	0.9161	-0.0101	0.5841	1.2481
9	-272.8	0.9850	0.9214	0.0636	0.5890	1.2538
10	-272.9	0.9870	0.9267	0.0603	0.5938	1.2595
11	-273	.	0.9320	.	0.5987	1.2653

```
Sum of Residuals           6.578071E-15
Sum of Squared Residuals        0.1423
Predicted Resid SS (Press)      0.3402
```

Horse	Gestation Period x (days)	Life Span y (years)
1	416	24
2	279	25.5
3	298	20
4	307	21.5
5	356	22
6	403	23.5
7	265	21

```
The regression equation is
LIFESPAN = 18.9 + 0.0109 GESTATE
```

Predictor	Coef	Stdev	t-ratio	p
Constant	18.890	4.499	4.20	0.009
GESTATE	0.01087	0.01337	0.81	0.453

```
s = 1.971      R-sq = 11.7%      R-sq(adj) = 0.0%
```

Analysis of Variance

SOURCE	DF	SS	MS	F	p
Regression	1	2.571	2.571	0.66	0.453
Error	5	19.429	3.886		
Total	6	22.000			

a. Do the data provide sufficient evidence to support the breeder's hypothesis? Test using $\alpha = .05$.

b. Find a 90% confidence interval for β_1. Interpret this interval.

11.78 Refer to Exercise 11.44, in which managerial success, *y*, was modeled as a function of the number of contacts a manager makes with people outside his or her work unit, *x*, during a specific period of time. The data are repeated in the accompanying table, with the ASP simple linear regression print-out shown at the bottom of the page.

Manager	Manager Success Index y	Number of Interactions with Outsiders x
1	40	12
2	73	71
3	95	70
4	60	81
5	81	43
6	27	50
7	53	42
8	66	18
9	25	35
10	63	82
11	70	20
12	47	81
13	80	40
14	51	33
15	32	45
16	50	10
17	52	65
18	30	20
19	42	21

a. A particular manager was observed for 2 weeks as in the *Journal of Applied Behavioral Science* (1985) study. She made 55 contacts with people outside her work unit. Predict the value of the manager's success index. Use a 90% prediction interval.

b. A second manager was observed for 2 weeks. This manager made 110 contacts with people outside his work unit. Give two reasons why caution should be exercised in using the least squares model developed from the given data set to construct a prediction interval for this manager's success index.

c. In the context of this problem, determine the value of x for which the associated prediction interval for y is the narrowest.

11.79 Firms planning to build new plants or make additions to existing facilities have become very conscious of the energy efficiency of proposed new structures, and are interested in the relation between yearly energy consumption and the number of square feet of building shell. The next table lists

the energy consumption in British thermal units (a BTU is the amount of heat required to raise 1 pound of water 1°F) for 22 buildings that were all subjected to the same climatic conditions. The SAS printout that fits the straight-line model relating BTU consumption, y, to building shell area, x, is given on page 530.

BTU/Year (thousands)	Shell Area (square feet)
3,870,000	30,001
1,371,000	13,530
2,422,000	26,060
672,200	6,355
233,100	4,576
218,900	24,680
354,000	2,621
3,135,000	23,350
1,470,000	18,770
1,408,000	12,220
2,201,000	25,490
2,680,000	23,680
337,500	5,650
567,500	8,001
555,300	6,147
239,400	2,660
2,629,000	19,240
1,102,000	10,700
423,500	9,125
423,500	6,510
1,691,000	13,530
1,870,000	18,860

a. Find the least squares estimates of the intercept β_0 and the slope β_1.

b. Investigate the usefulness of the model you developed in part **a**. Is yearly energy consumption positively linearly related to the shell area of the building? Test using $\alpha = .10$

c. Calculate the observed significance level of the test of part **b** using the printout. Interpret its value.

d. Find the coefficient of determination r^2 and interpret its value.

e. A company wishes to build a new warehouse that will contain 8,000 square feet of shell area. Find the predicted value of energy consumption

```
                    SIMPLE LINEAR REGRESSION

Model:  Index(y) = 0.236617Interactions(x) + 44.1305CNST

                    COEF.   SD. ER.   t(17)    P-VALUE PT. R SQ.
                    -------------------------------------------
Interactions(x)   0.236617 0.186501 1.26871 0.221641   0.0864948
           CNST   44.1305   9.36159  4.71399 2.00177E-4 0.566566

R SQ. = 0.0864948,  ADJ. R SQ. = 0.0327592,  D. W. = 2.52206
SQ. ROOT MSE = 19.4038,  F(1/17) = 1.60964 (P-VALUE = 0.221641)
```

```
Dep Variable: BTU

                        Analysis of Variance

                     Sum of          Mean
     Source      DF   Squares        Square      F Value    Prob>F

     Model        1   1.658498E+13   1.658498E+13   42.028    0.0001
     Error       20   7.89232E+12    394616010047
     C Total     21   2.44773E+13

         Root MSE     628184.69422     R-Square    0.6776
         Dep Mean    1357904.54545     Adj R-Sq    0.6614
         C.V.             46.26133

                        Parameter Estimates

                     Parameter      Standard      T for H0:
     Variable    DF   Estimate       Error      Parameter=0   Prob >|T|

     INTERCEP     1     -99045     261617.65980    -0.379       0.7090
     AREA         1    102.814048   15.85924082     6.483       0.0001

                          Predict                 Lower95%   Upper95%
     Obs  AREA        BTU    Value     Residual    Predict    Predict

       1  30001    3870000   2985479    884521    1546958    4424000
       2  13530    1371000   1292029    78971.2  -47949.3    2632007
       3  26060    2422000   2580289   -158289    1183940    3976637
       4   6355     672200    554338    117862    -810192    1918868
       5   4576     233100    371432   -138332   -1005463    1748327
       6  24680     218900   2438405  -2219505    1054223    3822588
       7   2621     354000    170430    183570   -1222796    1563657
       8  23350    3135000   2301663    833337     927871    3675455
       9  18770    1470000   1830774   -360774     482352    3179196
      10  12220    1408000   1157342    250658    -184021    2498706
      11  25490    2201000   2521685   -320685    1130530    3912840
      12  23680    2680000   2335591    344409     959345    3711838
      13   5650     337500    481854   -144354    -887287    1850995
      14   8001     567500    723570   -156070    -631698    2078838
      15   6147     555300    532953    22347.3   -832898    1898804
      16   2660     239400    174440    64959.9  -1218433    1567313
      17  19240    2629000   1879097    749903     528832    3229362
      18  10700    1102000   1001065    100935    -343656    2345786
      19   9125     423500    839133   -415633    -511035    2189301
      20   6510     423500    570274   -146774    -793294    1933842
      21  13530    1691000   1292029    398971   -47949.3    2632007
      22  18860    1870000   1840028    29972.3    491266    3188789
      23   8000        .      723467      .       -631806    2078740

     Sum of Residuals            1.6298145E-9
     Sum of Squared Residuals    7.89232E+12
     Predicted Resid SS (Press)  1.012747E+13
```

and a 95% prediction interval on the printout. Comment on the usefulness of this interval.

f. The application of the model you developed in part **a** to the warehouse problem of part **e** is appropriate only if certain assumptions can be made about the new warehouse. What are these assumptions?

11.80 Sometimes it is known from theoretical considerations that the straight-line relationship between two variables, x and y, passes through the origin of the xy-plane. Consider the relationship between the total weight of a shipment of 50-pound bags of flour, y, and the number of bags in the shipment, x. Since a shipment containing $x = 0$ bags (i.e., no shipment at all) has a total weight of $y = 0$, a straight-line model of the relationship between x and y should pass through the point $x = 0$, $y = 0$. In such a case you could assume $\beta_0 = 0$

and characterize the relationship between x and y with the following model:

$$y = \beta_1 x + \epsilon$$

The least squares estimate of β_1 for this model is

$$\hat{\beta}_1 = \frac{\sum x_i y_i}{\sum x_i^2}$$

From the records of past flour shipments, 15 shipments were randomly chosen and the data shown in the table were recorded.

Weight of Shipment	Number of 50-Pound Bags in Shipment
5,050	100
10,249	205
20,000	450
7,420	150
24,685	500
10,206	200
7,325	150
4,958	100
7,162	150
24,000	500
4,900	100
14,501	300
28,000	600
17,002	400
16,100	400

a. Find the least squares line for the given data under the assumption that $\beta_0 = 0$. Plot the least squares line on a scattergram of the data.

b. Find the least squares line for the given data using the model

$$y = \beta_0 + \beta_1 x + \epsilon$$

(i.e., do not restrict β_0 to equal 0). Plot this line on the same scatterplot you constructed in part **a**.

c. Refer to part **b**. Why might $\hat{\beta}_0$ be different from 0 even though the true value of β_0 is known to be 0?

d. The estimated standard error of $\hat{\beta}_0$ is equal to

$$s\sqrt{\frac{1}{n} + \frac{\bar{x}^2}{SS_{xx}}}$$

Use the t statistic

$$t = \frac{\hat{\beta}_0 - 0}{s\sqrt{(1/n) + (\bar{x}^2/SS_{xx})}}$$

to test the null hypothesis $H_0: \beta_0 = 0$ against the alternative $H_a: \beta_0 \neq 0$. Use $\alpha = .10$. Should you include β_0 in your model?

11.81 The data shown in the following table are part of a series of experiments conducted to investigate the effect of water temperature on the absorption by rainbow trout of sublethal levels of cyanide (*Canadian Journal of Fisheries and Aquatic Sciences,*

Vol. 39, 1982). This specific set of data is the result of a preliminary experiment to determine the relationship between mean weight gain (percentage of body weight) over a 20-day period as a function of the ration fed the fish (percentage of body weight). Twenty fish were included in the experiment for each ration level and water temperature combination. The mean percentage weight gain for each sample of 20 fish is shown in the table. (Note that, as expected, a 0 level of rations produced a negative weight gain, i.e., a loss.) Since all the sample sizes are equal, we can fit a simple linear regression model to the data for any water temperature level by fitting the model to the four sample means.

Ration (% body weight per day)	MEAN WET WEIGHT GAIN (%)		
	6°C	**12°C**	**18°C**
.0	−8.14	−10.33	−13.21
.8	12.31		
1.2		14.29	
1.5	28.19		15.51
2.5	29.12	38.05	
3.5			51.13
4.0		60.13	
4.5			65.49
Maintenance ration	.32	.48	.69

a. Find the least squares line relating mean weight gain to ration level for each of the three water temperature levels.

b. Plot the data points and graph the least squares lines on the same sheet of graph paper. Do the least squares lines appear to provide good fits to their respective data sets? Does the relationship between mean weight gain and ration appear to depend on the water temperature? (In Chapter 13 we test to determine whether differences exist.)

c. Find SSE and s^2 for the $n = 4$ data points.

d. Find a 95% confidence interval for the mean gain (percentage of body weight) for a 1% increase in ration level.

e. State the assumptions required for your inference in part **b** to be valid. Which assumption may not be satisfied?

f. Find a 95% confidence interval for the mean weight gain of rainbow trout (percentage of body weight) over a 20-day period when the ration level is 4% of body weight.

11.82 *Comparable worth* is a compensation plan designed to eliminate pay inequities among jobs of similar worth. A number of state and municipal governments have adopted comparable-worth plans, and some unions have attempted to negotiate comparable-worth clauses into their contracts. To develop such a plan, a sample of benchmark jobs are evaluated and assigned points, x, based on factors such as responsibility, skill, effort, and working conditions. A market survey is conducted to determine the market rates (or salaries), y, of the benchmark jobs. A regression analysis is then used to characterize the relationship between salary and job evaluation points (*Public Personnel Management,* Vol. 20, 1991). The following table gives the job evaluation points and salaries for a set of 21 benchmark jobs.

Job Evaluation

Points, x	Salary, y	
970	$15,704	Electrician
500	13,984	Semiskilled laborer
370	14,196	Motor equipment operator
220	13,380	Janitor
250	13,153	Laborer
1,350	18,472	Senior engineering technician
470	14,193	Senior janitor
2,040	20,642	Revenue agent
370	13,614	Engineering aide
1,200	16,869	Electrician supervisor
820	15,184	Senior maintenance technician
1,865	17,341	Registered nurse
1,065	15,194	Licensed practical nurse
880	13,614	Principal clerk typist
340	12,594	Clerk typist
540	13,126	Senior clerk stenographer
490	12,958	Senior clerk typist
940	13,894	Principal clerk stenographer
600	13,380	Institutional attendant
805	15,559	Eligibility technician
220	13,844	Cook's helper

a. Construct a scattergram for these data. What does it suggest about the relationship between salary and job evaluation points?

b. Using SAS, a straight-line model was fit to these data and the printout is given on page 533. Identify and interpret the least squares equation.

c. Interpret the value of r^2 for this least squares equation.

d. Is there sufficient evidence to conclude that a straight-line model provides useful information about the relationship in question? Interpret the p-value for this test.

e. A job outside the set of benchmark jobs is evaluated and receives a score of 800 points. Under the comparable-worth plan, what is a reasonable range within which a fair salary for this job should be found?

11.83 A problem of economic and social concern in the United States is the importation and sale of illicit drugs. The data shown in the table below are part of a larger body of data collected by the Florida attorney general's office in an attempt to relate the incidence of drug seizures and drug arrests to the characteristics of the Florida counties. Given are the number, y, of drug arrests per county in a particular year, the density, x_1, of the county (population per square mile), and the number, x_2, of law enforcement employees. In order to simplify the calculations, we show data for only 10 counties.

a. Fit a least squares line to relate the number, y, of drug arrests per county to the county population density, x_1.

b. We might expect the mean number of arrests to increase as the population density increases. Do the data support this theory? Test using $\alpha = .05$.

c. Calculate the coefficient of determination for this regression analysis and interpret its value.

11.84 Repeat parts **a**, **b**, and **c** of Exercise 11.83 using the number x_2 of county law enforcement employees as the independent variable. Then answer the following questions:

d. Which least squares line has the lower SSE?

e. Which independent variable explains more of the variation in y? Explain.

	COUNTY									
	1	2	3	4	5	6	7	8	9	10
Population Density, x_1	169	68	278	842	18	42	112	529	276	613
Number of Law Enforcement Employees, x_2	498	35	772	5,788	18	57	300	1,762	416	520
Number of Arrests, y	370	44	716	7,416	25	50	189	1,097	256	432

Dependent Variable: Y

Analysis of Variance

Source	DF	Sum of Squares	Mean Square	F Value	Prob>F
Model	1	66801750.334	66801750.334	74.670	0.0001
Error	19	16997968.904	894629.94232		
C Total	20	83799719.238			

Root MSE	945.84879	R-square	0.7972	
Dep Mean	14804.52381	Adj R-sq	0.7865	
C.V.	6.38892			

Parameter Estimates

Variable	DF	Parameter Estimate	Standard Error	T for H0: Parameter=0	Prob > |T|
INTERCEP	1	12024	382.31829064	31.449	0.0001
X	1	3.581616	0.41448305	8.641	0.0001

Obs	X	Dep Var Y	Predict Value	Std Err Predict	Lower95% Predict	Upper95% Predict	Residual
1	970	15704.0	15497.8	221.447	13464.6	17531.0	206.2
2	500	13984.0	13814.5	236.070	11774.1	15854.9	169.5
3	370	14196.0	13348.9	266.420	11292.1	15405.6	847.1
4	220	13380.0	12811.6	309.502	10728.6	14894.6	568.4
5	250	13153.0	12919.1	300.351	10842.0	14996.2	233.9
6	1350	18472.0	16858.8	314.833	14772.4	18945.3	1613.2
7	470	14193.0	13707.0	242.349	11663.4	15750.6	486.0
8	2040	20642.0	19330.2	562.933	17026.4	21633.9	1311.8
9	370	13614.0	13348.9	266.420	11292.1	15405.6	265.1
10	1200	16869.0	16321.6	270.968	14262.3	18380.9	547.4
11	820	15184.0	14960.6	207.190	12934.0	16987.2	223.4
12	1865	17341.0	18703.4	496.163	16467.8	20938.9	-1362.4
13	1065	15194.0	15838.1	238.553	13796.4	17879.8	-644.1
14	880	13614.0	15175.5	210.818	13147.2	17203.7	-1561.5
15	340	12594.0	13241.4	274.451	11180.1	15302.7	-647.4
16	540	13126.0	13957.7	228.483	11921.1	15994.4	-831.7
17	490	12958.0	13778.6	238.109	11737.2	15820.1	-820.6
18	940	13894.0	15390.4	217.251	13359.1	17421.6	-1496.4
19	600	13380.0	14172.6	218.972	12140.6	16204.7	-792.6
20	805	15559.0	14906.9	206.741	12880.4	16933.3	652.1
21	220	13844.0	12811.6	309.502	10728.6	14894.6	1032.4
22	800	.	14888.9	206.632	12862.6	16915.3	.

STUDENT PROJECTS

Many dependent variables in all areas of research serve as the subjects of regression modeling efforts. We list five such variables here:

1. Crime rate in various communities
2. Daily maximum temperature in your town
3. Grade-point average of students who have completed one academic year at your college
4. Gross Domestic Product of the United States
5. Points scored by your favorite football team in a single game

Choose one of these dependent variables or choose some other dependent variable for which you want to construct a prediction model. There may be a large number of independent variables that should be included in a prediction equation for the dependent variable you choose. List three potentially important independent variables, x_1, x_2, and x_3, that you think might be (individually) strongly related to your dependent variable. Next, obtain 10 data values, each of which consists of a measure of your dependent variable y and the corresponding values of x_1, x_2, and x_3.

a. Use the least squares formulas given in this chapter to fit three straight-line models—one for each independent variable—for predicting y.
b. Interpret the sign of the estimated slope coefficient $\hat{\beta}_1$ in each case, and test the utility of each model by testing H_0: $\beta_1 = 0$ against H_a: $\beta_1 \neq 0$. What assumptions must be satisfied to ensure the validity of these tests?
c. Calculate the coefficient of determination r^2 for each model. Which of the independent variables predicts y best for the 10 sampled sets of data? Is this variable necessarily best in general (that is, for the entire population)? Explain.

Be sure to keep the data and the results of your calculations, since you will need them for the Student Projects sections in Chapters 12 and 13.

EXPLORING DATA WITH A COMPUTER

Consider the Federal Trade Commission (FTC) data on tar, nicotine, and carbon monoxide content of cigarettes, Appendix B. Both tar and nicotine are easier to measure than carbon monoxide content. Thus, suppose the FTC wants to estimate the relationship between the carbon monoxide content y and tar content x of a cigarette using a straight-line model.

a. Draw a random sample of 100 cigarette brands from the 962 described in Appendix B, and extract y and x for each brand sampled. Use a statistical software package that includes a regression program to obtain the least squares fit of the straight-line model of interest.
   1. Graph the fitted model.
   2. Interpret the estimated intercept and slope.
   3. Interpret the estimated standard deviation of the error term.
   4. Evaluate the usefulness of the model.
   5. Predict the carbon monoxide content of a cigarette brand having a tar content of 10 milligrams using a 95% prediction interval.

b. Repeat part a using the entire set of 962 cigarette brands. Compare the results to those you obtained in part a.
c. Repeat part b, but use the nicotine content x as the independent variable.

Chapter 12

MULTIPLE REGRESSION

Contents

Case Studies

WHERE WE'VE BEEN

In Chapter 11 we saw how to model the relationship between a dependent variable y and an independent variable x using a straight line. We fitted the straight line to the data points, used r and r^2 to measure the strength of the relationship between y and x, and used the resulting prediction equation to estimate the mean value of y or to predict some future value of y for a given value of x.

WHERE WE'RE GOING

This chapter extends the basic concept of Chapter 11, converting it into a powerful estimation and prediction device by modeling the mean value of y as a function of two or more independent variables. The techniques developed will enable you to model a response, y, as a function of both quantitative and qualitative variables. As in the case of a simple linear regression, a multiple regression analysis involves fitting the model to a data set, testing the utility of the model, and using it for estimation and prediction.

12.1 MULTIPLE REGRESSION: THE MODEL AND THE PROCEDURE

Most practical applications of regression analysis utilize models that are more complex than the simple straight-line model. For example, a realistic probabilistic model for reaction time to a stimulus would include more than just the amount of a particular drug in the bloodstream. Factors such as age, a measure of visual perception, and sex of the subject are a few of the many variables that might be related to reaction time. Thus, we would want to incorporate these and other potentially important independent variables into the model in order to make accurate predictions.

Probabilistic models that include terms involving x^2, x^3 (or higher-order terms), or more than one independent variable are called **multiple regression models**. The general form of these models is

$$y = \beta_0 + \beta_1 x_1 + \beta_2 x_2 + \cdots + \beta_k x_k + \epsilon$$

The dependent variable y is now written as a function of k independent variables, $x_1, x_2, \ldots, x_k$. The random error term is added to make the model probabilistic rather than deterministic. The value of the coefficient β_i determines the contribution of the independent variable x_i, and β_0 is the y-intercept. The coefficients β_0, $\beta_1, \ldots, \beta_k$ are usually unknown because they represent population parameters.

At first glance it might appear that the regression model shown above would not allow for anything other than straight-line relationships between y and the independent variables, but this is not true. Actually, $x_1, x_2, \ldots, x_k$ can be functions of variables as long as the functions do not contain unknown parameters. For example, the dollar sales, y, in new housing in a region could be a function of the independent variables

$$x_1 = \text{Mortgage interest rate}$$
$$x_2 = (\text{Mortgage interest rate})^2 = x_1^2$$
$$x_3 = \text{Unemployment rate in the region}$$

and so on. You could even insert a cyclical term (if it would be useful) of the form $x_4 = \sin t$, where t is a time variable. The multiple regression model is quite versatile and can be made to model many different types of response variables.

The General Linear Model

$$y = \beta_0 + \beta_1 x_1 + \beta_2 x_2 + \cdots + \beta_k x_k + \epsilon$$

where

y is the dependent variable

$x_1, x_2, \ldots, x_k$ are the independent variables

$E(y) = \beta_0 + \beta_1 x_1 + \beta_2 x_2 + \cdots + \beta_k x_k$ is the deterministic portion of the model.

β_i determines the contribution of the independent variable x_i

Note: The symbols $x_1, x_2, \ldots, x_k$ may represent higher-order terms. For example, x_1 might represent the current interest rate, x_2 might represent x_1^2, and so forth.

As shown in the next box, we use the same steps to develop the multiple regression model that we used for the simple regression model.

Analyzing a Multiple Regression Model

Step 1   Hypothesize the deterministic component of the model. This component relates the mean, $E(y)$, to the independent variables $x_1, x_2, ..., x_k$. This involves the choice of the independent variables to be included in the model (Chapter 13).

Step 2   Use the sample data to estimate the unknown model parameters $\beta_0, \beta_1, \beta_2, ..., \beta_k$ in the model (Section 12.2).

Step 3   Specify the probability distribution of the random error term, ϵ, and estimate the standard deviation of this distribution, σ (Sections 12.3 and 12.8).

Step 4   Statistically evaluate the usefulness of the model (Sections 12.4 and 12.5).

Step 5   When satisfied that the model is useful, use it for prediction, estimation, and other purposes (Section 12.6).

Although we introduce several different types of models in this chapter, we defer formal discussion of model building (step 1) until Chapter 13.

12.2   FITTING THE MODEL: THE LEAST SQUARES APPROACH

The method of fitting multiple regression models is identical to that of fitting the simple straight-line model: the method of least squares. That is, we choose the estimated model

$$\hat{y} = \hat{\beta}_0 + \hat{\beta}_1 x_1 + \cdots + \hat{\beta}_k x_k$$

that minimizes

$$\text{SSE} = \sum(y - \hat{y})^2$$

As in the case of the simple linear model, the sample estimates $\hat{\beta}_0, \hat{\beta}_1, \hat{\beta}_2, ..., \hat{\beta}_k$ are obtained as a solution to a set of simultaneous linear equations.*

The primary difference between fitting the simple and multiple regression models is computational difficulty. The $(k + 1)$ simultaneous linear equations that must be solved to find the $(k + 1)$ estimated coefficients $\hat{\beta}_0, \hat{\beta}_1, \hat{\beta}_2, ..., \hat{\beta}_k$ are difficult (sometimes nearly impossible) to solve with a calculator. Consequently, we resort to the use of computers. Instead of presenting the tedious hand calculations required to fit the models, we present output from several statistical software packages, including SAS, SPSS, MINITAB, and ASP. Since these printouts are similar to those of most other packages, you should have little trouble interpreting regression output from other packages as you encounter them in the future. We demonstrate the SAS regression output with the following example.

In all-electric homes the amount of electricity expended is of interest to consumers, builders, and groups involved with energy conservation. Suppose we wish to investigate the monthly electrical usage, *y,* in all-electric homes and its relationship to the size, *x,* of the home. Moreover, suppose we think that

---

*Students who are familiar with calculus should note that $\hat{\beta}_0, \hat{\beta}_1, ..., \hat{\beta}_k$ are the solutions to the set of equations $\partial \text{SSE}/\partial \hat{\beta}_0 = 0, \partial \text{SSE}/\partial \hat{\beta}_1 = 0, ..., \partial \text{SSE}/\partial \hat{\beta}_k = 0$. The solution is usually given in matrix form, but we do not present the details here. See the references for details.

TABLE 12.1   Home Size—Electrical Usage Data

Size of Home x (sq. ft)	Monthly Usage y (kilowatt-hours)
1,290	1,182
1,350	1,172
1,470	1,264
1,600	1,493
1,710	1,571
1,840	1,711
1,980	1,804
2,230	1,840
2,400	1,956
2,930	1,954

monthly electrical usage in all-electric homes is related to the size of the home by the model

$$y = \beta_0 + \beta_1 x + \beta_2 x^2 + \epsilon$$

To estimate the unknown parameters β_0, β_1, and β_2, values of y and x are collected for 10 homes during a particular month. The data are shown in Table 12.1

Notice that we include a term involving x^2 in this model because we expect curvature in the graph of the response model relating y to x. The term involving x^2 is called a **quadratic term**. Figure 12.1 illustrates that the electrical usage appears to increase in a curvilinear manner with the size of the home. This provides some support for the inclusion of the quadratic term x^2 in the model.

Part of the output from the SAS multiple regression routine for the data in Table 12.1 is reproduced in Figure 12.2. The least squares estimates of the β parameters appear in the column labeled **Parameter Estimate**. You can see that $\hat{\beta}_0 = -1,216.1$, $\hat{\beta}_1 = 2.3989$, and $\hat{\beta}_2 = -.00045$. Therefore, the equation that minimizes the SSE for the data is

$$\hat{y} = -1,216.1 + 2.3989x - .00045x^2$$

FIGURE 12.1

Scattergram of the home size–electrical usage data

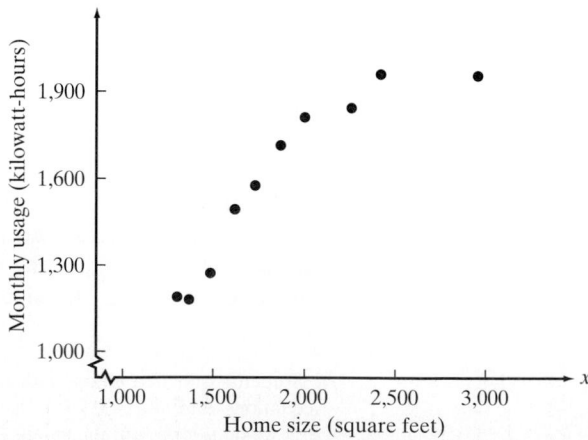

FIGURE 12.2

SAS output for the home size–electrical usage data

```
Dep Variable: Y

                        Analysis of Variance

                            Sum of          Mean
        Source      DF      Squares        Square      F Value    Prob>F

        Model        2   831069.54637   415534.77319   189.710    0.0001
        Error        7    15332.55363     2190.36480
        C Total      9   846402.10000

            Root MSE          46.80133     R-Square    0.9819
            Dep Mean        1594.70000     Adj R-Sq    0.9767
            C.V.               2.93480

                        Parameter Estimates

                        Parameter     Standard     T for H0:
        Variable   DF    Estimate       Error    Parameter=0   Prob > |T|

        INTERCEP    1  -1216.143887  242.80636850    -5.009       0.0016
        X           1      2.398930    0.24583560     9.758       0.0001
        XSQ         1     -0.000450    0.00005908    -7.618       0.0001
```

The minimum value of the SSE, 15,332.6, also appears (shaded) in the printout. [*Note:* Throughout this chapter we shade the aspects of the printout that are under discussion.]

Note that the graph of the multiple regression model (Figure 12.3, a response curve) provides a good fit to the data of Table 12.1. Furthermore, the small value of $\hat{\beta}_2$ does *not* imply that the curvature is insignificant, since the numerical value of $\hat{\beta}_2$ depends on the scale of the measurements. We will test the contribution of the quadratic coefficient β_2 in Section 12.4.

The ultimate goal of this multiple regression analysis is to use the fitted model to predict electrical usage y for a home of a specific size (area) x. And, of course, we will want to give a prediction interval for y so that we will know how much faith we can place in the prediction. That is, if the prediction model is used to predict electrical usage y for a given size of home, x, what will be the error of prediction? To answer this question, we need to estimate σ^2, the variance of ϵ.

The interpretation of the estimated coefficients in a quadratic model must be undertaken cautiously. First, the estimated y-intercept, $\hat{\beta}_0$, can be meaningfully interpreted only if the range of the independent variables includes zero—that is,

FIGURE 12.3

Least squares model for the home size–electrical usage data

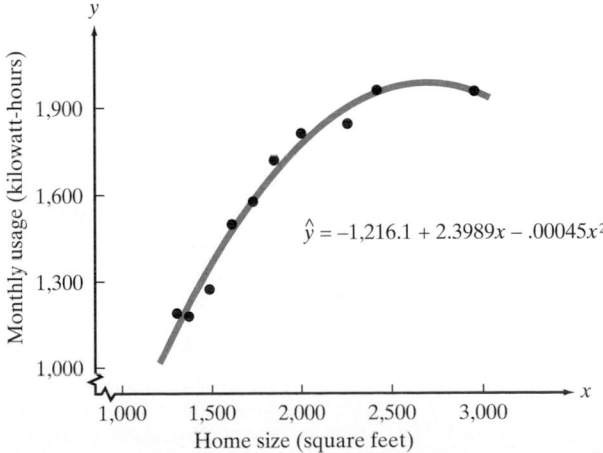

$$\hat{y} = -1,216.1 + 2.3989x - .00045x^2$$

FIGURE 12.4

Potential misuse of model

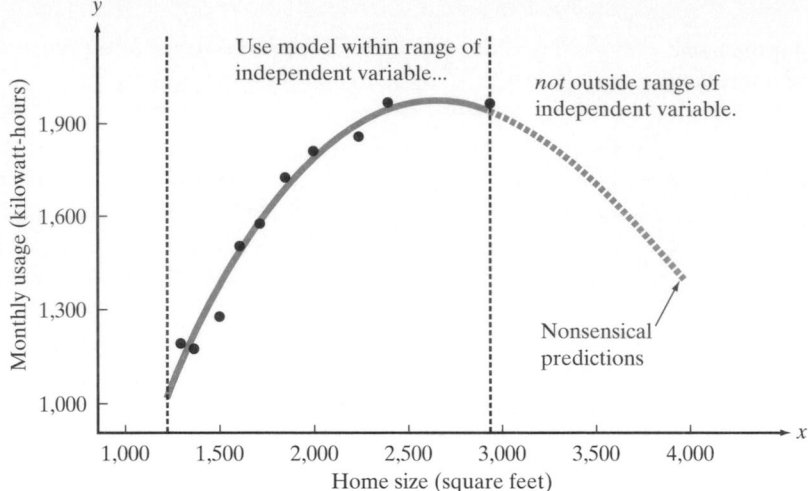

if $x = 0$ is included in the sampled range of x. In the electrical usage example, $\hat{\beta}_0 = -1{,}216.1$, which would seem to imply that the estimated electrical usage is negative when $x = 0$, but this zero point represents a home with 0 square feet. The zero point is, of course, not in the range of the sample (the lowest value of x is 1,290 square feet), and thus the interpretation of $\hat{\beta}_0$ is not meaningful.

The estimated coefficient of x is $\hat{\beta}_1$, but it no longer represents a slope in the presence of the quadratic term x^2.* The estimated coefficient of the linear term x will not, in general, have a meaningful interpretation in the quadratic model.

The sign of the coefficient, $\hat{\beta}_2$, of the quadratic term, x^2, is the indicator of whether the curve is concave downward (mound-shaped) or concave upward (bowl-shaped). A negative $\hat{\beta}_2$ implies downward concavity, as in the electrical usage example (Figure 12.3), and a positive $\hat{\beta}_2$ implies upward concavity. Rather than interpreting the numerical value of $\hat{\beta}_2$ itself, we utilize a graphical representation of the model, as in Figure 12.3, to describe the model.

Note that Figure 12.3 implies that the estimated electrical usage is leveling off as the home sizes increase beyond 2,500 square feet. In fact, the convexity of the model would lead to decreasing usage estimates if we were to display the model out to 4,000 square feet and beyond (see Figure 12.4). However, model interpretations are not meaningful outside the range of the independent variable, which has a maximum value of 2,930 square feet in this example. Thus, although the model appears to support the hypothesis that the *rate of increase* per square foot *decreases* for the home sizes near the high end of the sampled values, the conclusion that usage will actually begin to decrease for very large homes would be a *misuse* of the model, since no homes of 3,000 square feet or more were included in the sample.

All the interpretations of the electrical usage model were based on the sample estimates of unknown model parameters. As has been our theme throughout, we want to measure the reliability of these interpretations with regard to the real relationship between usage and home size. Additionally, we will want to use the model to estimate usage for homes not included in the sample, and to measure the

---

*For students with knowledge of calculus, note that the slope of the quadratic model is the first derivative $\partial y / \partial x = \beta_1 + 2\beta_2 x$. Thus, the slope varies as a function of x, rather than the constant slope associated with the straight-line model.

reliability of these estimates. All this requires an assessment of the error in the model—that is, of the probability distribution of the random error component, ϵ. This is the subject of the next section.

12.3 MODEL ASSUMPTIONS

We noted in Section 12.1 that the multiple regression model is of the form

$$y = \beta_0 + \beta_1 x_1 + \beta_2 x_2 + \cdots + \beta_k x_k + \epsilon$$

where y is the response variable that we wish to predict; $\beta_0, \beta_1, \ldots, \beta_k$ are parameters with unknown values; $x_1, x_2, \ldots, x_k$ are information-contributing variables that are measured without error; and ϵ is a random error component. Since $\beta_0, \beta_1, \ldots, \beta_k$, and $x_1, x_2, \ldots, x_k$ are nonrandom, the quantity

$$\beta_0 + \beta_1 x_1 + \beta_2 x_2 + \cdots + \beta_k x_k$$

represents the deterministic portion of the model. Therefore, y is composed of two components—one fixed and one random—and, consequently, y is a random variable.

$$y = \underbrace{\beta_0 + \beta_1 x_1 + \ldots + \beta_k x_k}_{\substack{\text{Deterministic} \\ \text{portion of model}}} + \underbrace{\epsilon}_{\substack{\text{Random} \\ \text{error}}}$$

We will assume (as in Chapter 11) that the random error can be positive or negative and that for any setting of the x values, $x_1, x_2, \ldots, x_k$, the random error ϵ has a normal probability distribution with mean equal to 0 and variance equal to σ^2. Further, we assume that the random errors associated with any (and every) pair of y values are probabilistically independent. That is, the error, ϵ, associated with any one y value is independent of the error associated with any other y value. These assumptions are summarized in the next box.

Assumptions for Random Error ϵ

1. For any given set of values of $x_1, x_2, \ldots, x_k$, the random error ϵ has a normal probability distribution with mean equal to 0 and variance equal to σ^2.
2. The random errors are independent (in a probabilistic sense).

Note that σ^2 represents the variance of the random error, ϵ. As such, σ^2 is an important measure of the usefulness of the model for the estimation of the mean and the prediction of actual values of y. If $\sigma^2 = 0$, all the random errors will equal 0 and the predicted values, $\hat{y}$, will be identical to $E(y)$; that is $E(y)$ will be estimated without error. In contrast, a large value of σ^2 implies large (absolute) values of ϵ and larger deviations between the predicted values, $\hat{y}$, and the mean value, $E(y)$. Consequently, the larger the value of σ^2, the greater will be the error in estimating the model parameters $\beta_0, \beta_1, \ldots, \beta_k$ and the error in predicting a value of y for a specific set of values of $x_1, x_2, \ldots, x_k$. Thus, σ^2 plays a major role in making inferences about $\beta_0, \beta_1, \ldots, \beta_k$, in estimating $E(y)$, and in predicting y for specific values of $x_1, x_2, \ldots, x_k$.

Since the variance, σ^2, of the random error, ϵ, will rarely be known, we must use the results of the regression analysis to estimate its value. Recall that σ^2 is the variance of the probability distribution of the random error, ϵ, for a given set of values for $x_1, x_2, \ldots, x_k$; hence it is the mean value of the squares of the deviations of the y values (for given values of $x_1, x_2, \ldots, x_k$) about the mean value $E(y)$.* Since the predicted value, $\hat{y}$, estimates $E(y)$ for each of the data points, it seems natural to use

$$\text{SSE} = \sum (y_i - \hat{y}_i)^2$$

to construct an estimator of σ^2.

For example, in the second-order (quadratic) model describing electrical usage as a function of home size, we found that SSE = 15,332.6. We now want to use this quantity to estimate the variance of ϵ. Recall that the estimator for the straight-line model is $s^2 = \text{SSE}/(n - 2)$ and note that the denominator is ($n -$ Number of estimated β parameters), which is $(n - 2)$ in the first-order (straight-line) model. Since we must estimate one more parameter, β_2, for the second-order model, the estimator of σ^2 is

$$s^2 = \frac{\text{SSE}}{n - 3}$$

That is, the denominator becomes $(n - 3)$ because there are now three β parameters in the model. The numerical estimate for this example is

$$s^2 = \frac{\text{SSE}}{10 - 3} = \frac{15,332.6}{7} = 2{,}190.36$$

In many computer printouts and textbooks, s^2 is called the **mean square for error (MSE)**. This estimate of σ^2 is shown in the column titled **Mean Square** in the SAS printout in Figure 12.2.

The units of the estimated variance are squared units of the dependent variable, y. In the electrical usage example, the units of s^2 are (kilowatt-hours)^2. This makes meaningful interpretation of s^2 difficult, so we use the standard deviation s to provide a more meaningful measure of variability. In the electrical usage example,

$$s = \sqrt{2{,}190.36} = 46.8,$$

which is given on the SAS printout in Figure 12.2 under **Root MSE**. One useful interpretation of the estimated standard deviation s is that the interval $\pm 2s$ will provide a rough approximation to the accuracy with which the model will predict future values of y for given values of x. Thus, in the electrical usage example, we expect the model to provide predictions of electrical usage to within about $\pm 2s = \pm 93.6$ kilowatt-hours.[†]

For the general multiple regression model

$$y = \beta_0 + \beta_1 x_1 + \beta_2 x_2 + \cdots + \beta_k x_k + \epsilon$$

we must estimate the $(k + 1)$ parameters $\beta_0, \beta_1, \beta_2, \ldots, \beta_k$. Thus, the estimator of σ^2 is SSE divided by the quantity ($n -$ Number of estimated β parameters).

We will use the estimator of σ^2 both to check the utility of the model (Sections 12.4 and 12.5) and to provide a measure of reliability of predictions and estimates when the model is used for those purposes (Section 12.6). Thus, you can see that the estimation of σ^2 plays an important part in the development of a regression model.

---

*Since $y = E(y) + \epsilon$, ϵ is equal to the deviation $y - E(y)$. Also, by definition, the variance of a random variable is the expected value of the square of the deviation of the random variable from its mean. According to our model, $E(\epsilon) = 0$. Therefore, $\sigma^2 = E(\epsilon^2)$.

[†]The $\pm 2s$ approximation will improve as the sample size is increased. We will provide more precise methodology for the construction of prediction intervals in Section 12.6.

> **Estimator of σ^2 for Multiple Regression Model with k Independent Variables**
>
> $$s^2 = MSE = \frac{SSE}{n - \text{Number of estimated } \beta \text{ parameters}} = \frac{SSE}{n - (k + 1)}$$

12.4   INFERENCES ABOUT THE β PARAMETERS

Sometimes the individual β parameters in a model have practical significance and we want to estimate their values or test hypotheses about them. For example, if electrical usage y is related to home size x by the straight-line relationship

$$y = \beta_0 + \beta_1 x + \epsilon$$

then β_1 has a very practical interpretation. That is, you saw in Chapter 11 that β_1 is the mean increase in electrical usage, y, for a 1-unit increase in home size x.

As proposed in the preceding sections, suppose that electrical usage y is related to home size x by the quadratic model

$$y = \beta_0 + \beta_1 x + \beta_2 x^2 + \epsilon$$

Then the mean value of y for a given value of x is

$$E(y) = \beta_0 + \beta_1 x + \beta_2 x^2$$

What is the practical interpretation of β_2? As noted earlier, the parameter β_2 measures the curvature of the response curve shown in Figure 12.3. If $\beta_2 > 0$, the slope of the curve will increase as x increases (upward concavity), as shown in Figure 12.5a. If $\beta_2 < 0$, the slope of the curve will decrease as x increases (downward concavity), as shown in Figure 12.5b.

Intuitively, we would expect the electrical usage, y, to rise almost proportionally to home size, x. Then, eventually, as the size of the home increases, the increase in electrical usage for a 1-unit increase in home size might begin to decrease. Thus, a forecaster of electrical usage would want to determine whether this type of curvature actually was present in the response curve or, equivalently, the forecaster would want to test the null hypothesis

$$H_0: \beta_2 = 0 \qquad \text{(No curvature in the response curve)}$$

against the alternative hypothesis

$$H_a: \beta_2 < 0 \qquad \text{(Downward concavity exists in the response curve)}$$

A test of this hypothesis can be performed using a t-test.

FIGURE 12.5

The interpretation of β_2 for a second-order model

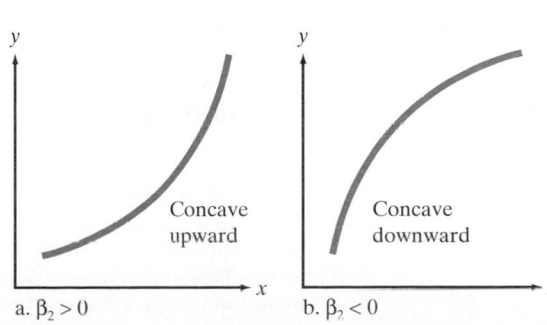

a. $\beta_2 > 0$ b. $\beta_2 < 0$

FIGURE 12.6

Rejection region for test of β_2

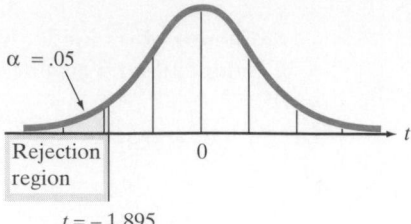

The t-test utilizes a test statistic analogous to that used to make inferences about the slope of the straight-line model (Section 11.5). The t statistic is formed by dividing the sample estimate, $\hat{\beta}_2$, of the parameter β_2 by the estimated standard deviation of the sampling distribution of $\hat{\beta}_2$:

$$Test\ statistic:\quad t = \frac{\hat{\beta}_2}{s_{\hat{\beta}_2}}$$

We use the symbol $s_{\hat{\beta}_2}$ to represent the estimated standard deviation of $\hat{\beta}_2$. The formula for computing $s_{\hat{\beta}_2}$ is very complex and its presentation is beyond the scope of this text,* but this omission will not cause difficulty. Most computer packages list the estimated standard deviation $s_{\hat{\beta}_i}$ for each of the estimated model coefficients $\hat{\beta}_i$. Moreover, they usually give the calculated t values for each coefficient in the model.

The rejection region for the test is found in exactly the same way as the rejection regions for the t-tests in previous chapters. That is, we consult Table VI in Appendix A to obtain an upper-tail value of t. This is a value t_α such that $P(t > t_\alpha) = \alpha$. We can then use this value to construct rejection regions for either one-tailed or two-tailed tests. To illustrate, in the electrical usage example the error degrees of freedom is $(n - 3) = 7$, the denominator of the estimate of σ^2. Then the rejection region (shown in Figure 12.6) for a one-tailed test with $\alpha = .05$ is

$$Rejection\ region:\quad t < -t_\alpha$$
$$t < -1.895$$

In Figure 12.7, we again show a portion of the SAS printout for the electrical usage example. The following quantities are shaded:

1. The estimated coefficients $\hat{\beta}_0$, $\hat{\beta}_1$, and $\hat{\beta}_2$.
2. The SSE and MSE (estimate of σ^2, variance of ϵ)
3. The t statistic, observed significance level, and standard error of $\hat{\beta}_2$ for testing $H_0: \beta_2 = 0$

The estimated standard deviations for the model coefficients appear under the column labeled **Standard Error**. The t statistics for testing the null hypothesis that the true coefficients are equal to 0 appear under the column headed **T for H0: Parameter = 0**. The t value corresponding to the test of the null hypothesis $H_0: \beta_2 = 0$ is the last one in the column, that is, $t = -7.618$. Since this value is less than -1.895, we conclude that the quadratic term $\beta_2 x^2$ makes a contribution to the prediction model of electrical usage.

---

*Because most of the formulas in a multiple regression analysis are so complex, the only reasonable way to present them is by using matrix algebra. We do not assume a prerequisite of matrix algebra for this text and, in any case, we think the formulas can be omitted in an introductory course without serious loss. They are programmed into almost all statistical software packages with multiple regression routines and are presented in some of the texts listed in the references.

FIGURE 12.7

SAS multiple regression output

Dep Variable: Y

Analysis of Variance

Source	DF	Sum of Squares	Mean Square	F Value	Prob>F
Model	2	831069.54637	415534.77319	189.710	0.0001
Error	7	15332.55363	2190.36480		
C Total	9	846402.10000			

Root MSE	46.80133	R-Square	0.9819	
Dep Mean	1594.70000	Adj R-Sq	0.9767	
C.V.	2.93480			

Parameter Estimates

Variable	DF	Parameter Estimate	Standard Error	T for H0: Parameter=0	Prob > |T|
INTERCEP	1	-1216.143887	242.80636850	-5.009	0.0016
X	1	2.398930	0.24583560	9.758	0.0001
XSQ	1	-0.000450	0.00005908	-7.618	0.0001

The SAS printout shown in Figure 12.7 also lists the two-tailed significance levels for each t value under the column headed **Prob** $>$ **|T|**. The significance level .0001 corresponds to the quadratic term, and this implies that we would reject H_0: $\beta_2 = 0$ in favor of H_a: $\beta_2 \neq 0$ at any α level larger than .0001. Since our alternative was one-sided, H_a: $\beta_2 < 0$, the significance level is half that given in the printout, that is, $\frac{1}{2}(.0001) = .00005$. Thus, there is very strong evidence that the mean electrical usage increases more slowly per square foot for large houses than for small houses.

We can also form a confidence interval for the parameter β_2 as follows:

$$\hat{\beta}_2 \pm t_{\alpha/2} s_{\hat{\beta}_2} = -.000450 \pm (2.365)(.0000591)$$

or $(-.000590, -.000310)$. Note that the t value 2.365 corresponds to $\alpha/2 = .025$ and $(n - 3) = 7$ df. This interval constitutes a 95% confidence interval for β_2 and can be used to estimate the rate of curvature in mean electrical usage as home size is increased. Note that all values in the interval are negative, reconfirming the conclusion of our test.

Note that the SAS printout in Figure 12.7 also provides the t-test statistic and corresponding two-tailed p-values for the tests of H_0: $\beta_0 = 0$ and H_0: $\beta_1 = 0$. Since the interpretation of these parameters is not meaningful for this model, the tests are not of interest.

Testing a hypothesis about a single β parameter that appears in any multiple regression model is accomplished in exactly the same manner as described for the quadratic electrical usage model. The t-test and a confidence interval for a β parameter are shown in the next two boxes.

Test of an Individual Parameter Coefficient in the Multiple Regression Model

ONE-TAILED TEST	**TWO-TAILED TEST**
H_0: $\beta_i = 0$	H_0: $\beta_i = 0$
H_a: $\beta_i < 0$ (or H_a: $\beta_i > 0$)	H_a: $\beta_i \neq 0$

Test statistic: $t = \dfrac{\hat{\beta}_i}{s_{\hat{\beta}_i}}$

continued

> *Rejection region:* $t < t_\alpha$ *Rejection region:* $|t| > t_{\alpha/2}$
>   (or $t > t_\alpha$ when H_a: $-\beta_i > 0$)
>
> where t_α and $t_{\alpha/2}$ are based on $n - (k + 1)$ degrees of freedom and
>
> n = Number of observations
>
> $k + 1$ = Number of β parameters in the model
>
> *Assumptions:*   See Section 12.3 for assumptions about the probability distribution for the random error component ϵ.

> ### A $100(1 - \alpha)$% Confidence Interval for a β Parameter
>
> $$\hat{\beta}_i \pm t_{\alpha/2} s_{\hat{\beta}_i}$$
>
> where $t_{\alpha/2}$ is based on $n - (k + 1)$ degrees of freedom and
>
> n = Number of observations
>
> $k + 1$ = Number of β parameters in the model

EXAMPLE 12.1

A collector of antique grandfather clocks knows that the price received for the clocks increases linearly with the age of the clocks. Moreover, the collector hypothesizes that the auction price of the clocks will increase linearly as the number of bidders increases. Thus, the following model is hypothesized:

$$y = \beta_0 + \beta_1 x_1 + \beta_2 x_2 + \epsilon$$

where

   y = Auction price

   x_1 = Age of clock (years)

   x_2 = Number of bidders

A sample of 32 auction prices of grandfather clocks, along with their age and the number of bidders, is given in Table 12.2. The model $y = \beta_0 + \beta_1 x_1 + \beta_2 x_2 + \epsilon$ is fitted to the data, and a portion of the MINITAB printout is shown in Figure 12.8.

a. Test the hypothesis that the mean auction price of a clock increases as the number of bidders increases when age is held constant, that is, $\beta_2 > 0$. Use $\alpha = .05$.

b. Interpret the estimates of the β coefficients in the model.

Solution

a. The hypotheses of interest concern the parameter β_2. Specifically,

$$H_0: \beta_2 = 0$$
$$H_a: \beta_2 > 0$$

Test statistic: $t = \dfrac{\hat{\beta}_2}{s_{\hat{\beta}_2}} = 9.85$

Rejection region: For $\alpha = .05$ and $n - (k + 1) = 32 - (2 + 1) = 29$ df, reject H_0 if $t > 1.699$ (see Figure 12.9).

     The calculated t-value, $t = 9.85$, is shaded on the MINITAB printout, Figure 12.8. This value falls in the rejection region shown in Figure 12.9. Thus,

TABLE 12.2   Auction Price Data

Age x_1	Number of Bidders x_2	Auction Price y	Age x_1	Number of Bidders x_2	Auction Price y
127	13	$1,235	170	14	$2,131
115	12	1,080	182	8	1,550
127	7	845	162	11	1,884
150	9	1,522	184	10	2,041
156	6	1,047	143	6	845
182	11	1,979	159	9	1,483
156	12	1,822	108	14	1,055
132	10	1,253	175	8	1,545
137	9	1,297	108	6	729
113	9	946	179	9	1,792
137	15	1,713	111	15	1,175
117	11	1,024	187	8	1,593
137	8	1,147	111	7	785
153	6	1,092	115	7	744
117	13	1,152	194	5	1,356
126	10	1,336	168	7	1,262

the collector can conclude that the mean auction price of a clock increases as the number of bidders increases, when age is held constant. Note that the observed significance level (p-value) of the test is 0. Therefore, any nonzero α will lead us to reject H_0.

b. The values $\hat{\beta}_1 - 12.74$ and $\hat{\beta}_2 = 85.95$ (shaded in Figure 12.8) are easily interpreted. We estimated that the mean auction price increases $12.74 per year of age of the clock (when the number of bidders is held constant), and that the mean price increases by $85.95 per additional bidder (when the age of the clock is held constant).

FIGURE 12.8

MINITAB printout for Example 12.1

```
The regression equation is
Y = -1339 + 12.7 X1 + 86.0 X2

Predictor        Coef      Stdev     t-ratio         p
Constant      -1339.0      173.8       -7.70     0.000
X1            12.7406     0.9047       14.08     0.000
X2             85.953      8.729        9.85     0.000

s = 133.5      R-sq = 89.2%      R-sq(adj) = 88.5%

Analysis of Variance

SOURCE        DF          SS          MS          F         p
Regression     2     4283063     2141532     120.19     0.000
Error         29      516727       17818
Total         31     4799789
```

FIGURE 12.9

Rejection region for H_0: $\beta_2 = 0$

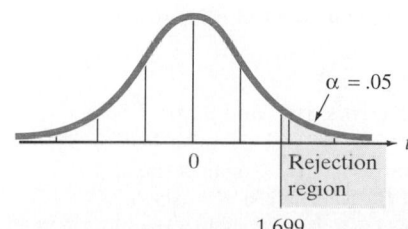

$\alpha = .05$

Rejection region

0

t

1.699

Be careful not to interpret the estimated intercept $\hat{\beta}_0 = -1,339$ in the same way $\hat{\beta}_1$ and $\hat{\beta}_2$ are interpreted. You might think that this negative value implies a negative price for clocks 0 years of age with 0 bidders. However, these zeros are meaningless numbers in this example, since the ages range from 108 to 194 and the number of bidders ranges from 5 to 15. Interpretations of predicted y values for values of the independent variables outside their sampled range can be very misleading.

Note: Some computer programs use an F statistic to conduct two-tailed tests concerning the individual β parameters, because the square of a Student t with v degrees of freedom is equal to an F statistic with 1 degree of freedom in the numerator and v degrees of freedom in the denominator. If you conduct a two-tailed t-test and reject the null hypothesis if $\mid t \mid > t_{\alpha/2}$, the corresponding F-test implies rejection if the computed value of F (which is equal to the square of the computed t statistic) is larger than F_α. Thus,

$$t^2_{\alpha/2} = F_\alpha$$

For example, when we tested a hypothesis about the curvature parameter β_2 in the quadratic model relating electrical usage to home size, the computed t value was -7.62 (see Figure 12.7). The equivalent F-test yields a value $F = t^2 = (-7.62)^2 = 58.06$. And the upper-tail rejection region for a two-tailed test with $\alpha = .05$, 1 df in the numerator, and 7 df in the denominator is

$$F > F_{.05} = 5.59$$

Note that the F value, 5.59, is equal to the square of 2.365, the value of t that corresponds to $t_{.025}$ based on 7 df. You can see that the conclusion is the same no matter which test statistic is used: There is very strong evidence that curvature is present in the model. ▲

EXERCISES 12.1–12.12

Note: Exercises marked with 💾 *require the use of a computer.*

Learning the Mechanics

12.1 SAS was used to fit the model $y = \beta_0 + \beta_1 x_1 + \beta_2 x_2 + \epsilon$ to $n = 20$ data points and the printout shown on page 549 was obtained.
  a. What are the sample estimates of β_0, β_1, and β_2?
  b. What is the least squares prediction equation?
  c. Find SSE, MSE, and s. Interpret the standard deviation in the context of the problem.
  d. Test H_0: $\beta_1 = 0$ against H_a: $\beta_1 \neq 0$. Use $\alpha = .05$.
  e. Use a 95% confidence interval to estimate β_2.

12.2 Suppose you fit the multiple regression model

$$y = \beta_0 + \beta_1 x_1 + \beta_2 x_2 + \beta_3 x_3 + \epsilon$$

to $n = 30$ data points and obtain the following result:

$$\hat{y} = 3.4 - 4.6x_1 + 2.7x_2 + 93x_3$$

The estimated standard errors of $\hat{\beta}_2$ and $\hat{\beta}_3$ are 1.86 and .29, respectively.
  a. Test the null hypothesis H_0: $\beta_2 = 0$ against the alternative hypothesis H_a: $\beta_2 \neq 0$. Use $\alpha = .05$.
  b. Test the null hypothesis H_0: $\beta_3 = 0$ against the alternative hypothesis H_a: $\beta_3 \neq 0$. Use $\alpha = .05$.

  c. The null hypothesis H_0: $\beta_2 = 0$ is not rejected. In contrast, the null hypothesis H_0: $\beta_3 = 0$ is rejected. Explain how this can happen even though $\hat{\beta}_2 > \hat{\beta}_3$.

12.3 Suppose you fit the second-order model

$$y = \beta_0 + \beta_1 x + \beta_2 x^2 + \epsilon$$

to $n = 25$ data points. Your estimate of β_2 is $\hat{\beta}_2 = .47$, and the estimated standard error of the estimate is $s_{\hat{\beta}_2} = .15$.
  a. Test the null hypothesis that the mean value of y is related to x by the (*first-order*) linear model

$$E(y) = \beta_0 + \beta_1 x$$

(H_0: $\beta_2 = 0$) against the alternative hypothesis that the true relationship is given by the (*second–order*) quadratic model

$$E(y) = \beta_0 + \beta_1 x + \beta_2 x^2$$

(H_a: $\beta_2 \neq 0$). Use $\alpha = .05$.
  b. Suppose you want to determine only whether the quadratic curve opens upward; that is, as x increases, the slope of the curve increases. Give the test statistic and the rejection region for the test for

```
Dep Variable: Y

                        Analysis of Variance

                     Sum of          Mean
      Source     DF   Squares        Square      F Value    Prob>F

      Model       2  128329.27624  64164.63812    7.223     0.0054
      Error      17  151015.72376   8883.27787
      C Total    19  279345.00000

              Root MSE     94.25114      R-Square    0.4594
              Dep Mean    360.50000      Adj R-Sq    0.3958
              C.V.         26.14456

                      Parameter Estimates

                      Parameter      Standard     T for H0:
      Variable    DF   Estimate        Error    Parameter=0   Prob > |T|

      INTERCEP     1   506.346067    45.16942487    11.210      0.0001
      X1           1  -941.900226   275.08555975    -3.424      0.0032
      X2           1  -429.060418   379.82566485    -1.130      0.2743
```

$\alpha = .05$. Do the data support the theory that the slope of the curve increases as x increases? Explain.

c. What is the value of the F statistic for testing the null hypothesis H_0: $\beta_2 = 0$ against the alternative hypothesis H_a: $\beta_2 \neq 0$?

d. Could the F statistic in part **c** be used to conduct the test in part **b**? Explain.

12.4 How is the number of degrees of freedom available for estimating σ^2 (the variance of ϵ) related to the number of independent variables in a regression model?

12.5 Use a computer to fit a second-order model to the following data:

x	0	1	2	3	4	5	6
y	1	2.7	3.8	4.5	5.0	5.3	5.2

a. Find SSE and s^2.

b. Do the data provide sufficient evidence to indicate that the second-order term provides information for the prediction of y? [*Hint:* Test H_0: $\beta_2 = 0$.]

c. Find the least squares prediction equation.

d. Plot the data points and graph $\hat{y}$. Does your prediction equation provide a good fit to the data?

Applying the Concepts

12.6 A disabled person's acceptance of a disability is critical to the rehabilitation process. The *Journal of Rehabilitation* (Sept. 1989) published a study that investigated the relationship between assertive behavior level and acceptance of disability in 160 disabled adults. The dependent variable, assertiveness (y), was measured using the Adult Self Expression Scale (ASES). Scores on the ASES range from 0

(no assertiveness) to 192 (extreme assertiveness). The model analyzed was $E(y) = \beta_0 + \beta_1 x_1 + \beta_2 x_2 + \beta_3 x_3$, where

x_1 = Acceptance of disability (AD) score

x_2 = Age (years)

x_3 = Length of disability (years)

The regression results are shown in the accompanying table.

Independent Variable	t	Two-Tailed p-Value
AD score (x_1)	5.96	.0001
Age (x_2)	0.01	.9620
Length (x_3)	1.91	.0576

a. Is there sufficient evidence to indicate that AD score is positively linearly related to assertiveness level, once age and length of disability are accounted for? Test using $\alpha = .05$.

b. Test the hypothesis H_0: $\beta_2 = 0$ against H_a: $\beta_2 \neq 0$. Use $\alpha = .05$. Give the conclusion in the words of the problem.

c. Test the hypothesis H_0: $\beta_3 = 0$ against H_a: $\beta_3 > 0$. Use $\alpha = .05$. Give the conclusion in the words of the problem.

12.7 Spinocerebellar ataxia type 1 (SCA1) is an inherited neurodegenerative disorder characterized by dysfunction of the brain. In the final stages of the disease, the patient suffers from frequent choking spells, poor cough reflex, and aspiration of food particles, leading to recurrent pneumonia and eventually death. From a DNA analysis of SCA1 chromosomes, researchers discovered the presence of repeat gene sequences (*Cell Biology*, Feb. 1995). In

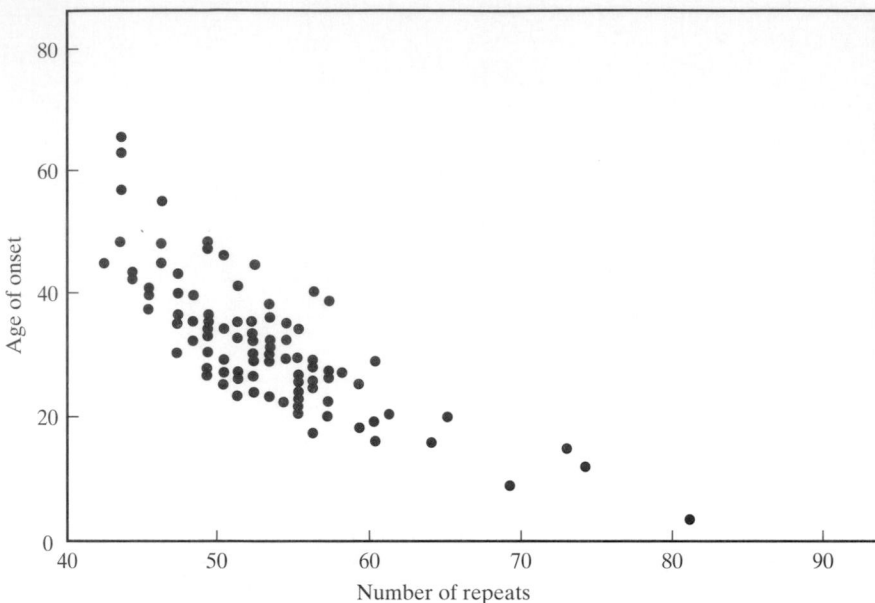

general, the more repeat sequences observed, the earlier the onset of the disease (in years of age). The scatterplot above shows this relationship for data collected on 113 individuals diagnosed with SCA1.

a. Suppose you want to model the age y of onset of the disease as a function of number x of repeat gene sequences in SCA1 chromosomes. Propose a quadratic model for y.

b. Will the sign of β_2 in the model, part **a**, be positive or negative? Base your decision on the results shown in the scatterplot.

12.8 Fish reared in captivity must be fed a diet containing an appropriate balance of nutrients to provide adequate energy to permit efficient growth. The amount of nitrogen excreted through the gills of a fish is one way to measure fish energy metabolism. *Fisheries Science* (Feb. 1995) reported on a study of the variables that affect endogenous nitrogen excretion (ENE) in carp raised in Japan. Carp were divided into groups of 2 to 15 fish each according to body weight and each group placed in a separate tank. The carp were then fed a protein-free diet three times daily for a period of 20 days. One day after terminating the feeding experiment, the amount of ENE in each tank was measured. The next table gives the mean body weight (in grams) and ENE amount (in milligrams per 100 grams of body weight per day) for each carp group.

a. Plot the data in a scattergram. Do you detect a pattern?

b. The quadratic model $E(y) = \beta_0 + \beta_1 x + \beta_2 x^2$ was fit to the data using MINITAB. The MINITAB printout is displayed on page 551. Use the printout information to test H_0: $\beta_2 = 0$ against H_a: $\beta_2 \neq 0$ using $\alpha = .10$. Give the conclusion in the words of the problem.

Tank	Body Weight, x	ENE, y
1	11.7	15.3
2	25.3	9.3
3	90.2	6.5
4	213.0	6.0
5	10.2	15.7
6	17.6	10.0
7	32.6	8.6
8	81.3	6.4
9	141.5	5.6
10	285.7	6.0

Source: Watanabe, T., and Ohta, M. "Endogenous nitrogen excretion and non-fecal energy losses in carp and rainbow trout." *Fisheries Science*, Vol. 61, No. 1, Feb. 1995, p. 56 (Table 5).

12.9 Does the length of time required to assemble a manufactured product depend on the assembler's experience? The data in the table on page 551 are the times to assemble and the months of experience for 15 assemblers in a manufacturing operation.

a. The SAS printout for fitting the model $y = \beta_0 + \beta_1 x + \beta_2 x^2 + \epsilon$ is shown on page 551. Find the least squares prediction equation.

b. Plot the fitted equation on a scattergram of the data. Is there sufficient evidence to support the inclusion of the quadratic term in the model? Explain.

c. Test the null hypothesis that $\beta_2 = 0$ against the alternative that $\beta_2 \neq 0$. Use $\alpha = .01$. Does the quadratic term make an important contribution to the model?

d. Your conclusion in part **c** should have been to drop the quadratic term from the model. Do so and fit the "reduced model," $y = \beta_0 + \beta_1 x + \epsilon$, to the data.

```
The regression equation is
ENE = 13.7 - 0.102 WEIGHT +0.000273 WGHTSQ

Predictor         Coef       Stdev      t-ratio        p
Constant        13.713       1.306       10.50      0.000
WEIGHT        -0.10184     0.02881       -3.53      0.010
WGHTSQ       0.0002735    0.0001016        2.69      0.031

s = 2.194       R-sq = 73.7%      R-sq(adj) = 66.2%

Analysis of Variance

SOURCE          DF          SS          MS          F          p
Regression       2      94.659      47.329       9.83      0.009
Error            7      33.705       4.815
Total            9     128.364
```

e. Define β_1 in the context of this exercise. Find a 90% confidence interval for β_1 in the reduced model of part **d.**

Time to Assemble y (minutes)	Months of Experience x
10	24
20	1
15	10
11	15
11	17
19	3
11	20
13	9
17	3
18	1
16	7
16	9
17	7
18	5
10	20

12.10 Empirical research was conducted to investigate the variables that affect the size distribution of manufacturing firms in international markets (*World Development*, Vol. 20, 1992). Data collected on $n = 54$ countries were used to model the country's size distribution, y, measured as the share of manufacturing firms in the country with 100 or more workers. The model studied was $E(y) = \beta_0 + \beta_1 x_1 + \beta_2 x_2 + \beta_3 x_3 + \beta_4 x_4 + \beta_5 x_5$, where

x_1 = natural logarithm of Gross National Product (LGNP)

x_2 = geographic area per capita (in thousands of square meters) (AREAC)

x_3 = share of heavy industry in manufacturing value added (SVA)

x_4 = Ratio of credit claims on the private sector to Gross Domestic Product (CREDIT)

x_5 = Ratio of stock equity shares to Gross Domestic Product (STOCK)

```
Dep Variable: TIME

                        Analysis of Variance

                             Sum of        Mean
        Source       DF      Squares       Square     F Value    Prob>F

        Model         2    156.11948     78.05974     65.594     0.0001
        Error        12     14.28052      1.19004
        C Total      14    170.40000

            Root MSE        1.09089     R-square      0.9162
            Dep Mean       14.80000     Adj R-sq      0.9022
            C.V.            7.37089

                        Parameter Estimates
                        Parameter     Standard     T for H0:
        Variable   DF    Estimate       Error     Parameter=0    Prob > |T|

        INTERCEP    1    20.091108    0.72470507     27.723       0.0001
        EXP         1    -0.670522    0.15470634     -4.334       0.0010
        EXPSQ       1     0.009535    0.00632580      1.507       0.1576
```

a. The researchers hypothesized that the higher the credit ratio of a country, the smaller the size distribution of manufacturing firms. Explain how to test this hypothesis.

b. The researchers hypothesized that the higher the stock ratio of a country, the larger the size distribution of manufacturing firms. Explain how to test this hypothesis.

12.11 The owner of an apartment building in Minneapolis believed that her 1990 property tax bill was too high because of an overassessment of the property's value by the city tax assessor. The owner hired an independent real estate appraiser to investigate the appropriateness of the city's assessment. The appraiser used regression analysis to explore the relationship between the sale prices of apartment buildings sold in Minneapolis during 1990 and various characteristics of the properties. Twenty-five apartment buildings were randomly sampled from all apartment buildings that were sold during 1990. The table lists the data collected by the appraiser. The real estate appraiser hypothesized that the sale price (that is, market value) of an apartment building is related to the other variables in the table according to the model $y = \beta_0 + \beta_1 x_1 + \beta_2 x_2 + \beta_3 x_3 + \beta_4 x_4 + \beta_5 x_5 + \epsilon$.

a. Fit the real estate appraiser's model to the data in the table. Report the least squares prediction equation.

b. Find the standard deviation of the regression model and interpret its value in the context of this problem.

c. Do the data provide sufficient evidence to conclude that value increases with the number of units in an apartment building? Report the observed significance level and reach a conclusion using $\alpha = .05$.

d. Interpret the value of $\hat{\beta}_1$ in terms of these data. Remember that your interpretation must recognize the presence of the other variables in the model.

e. Construct a scattergram of sale price versus age. What does your scattergram suggest about the relationship between these variables?

f. Test $H_0: \beta_2 = 0$ against $H_a: \beta_2 < 0$ using $\alpha = .01$. Interpret the result in the context of the problem. Does the result agree with your observation in part **e**? Why is it reasonable to conduct a one-tailed rather than a two-tailed test of this null hypothesis?

g. What is the observed significance level of the hypothesis test of part **f**?

Code No.	Sale Price y ($)	No. of Apartments x_1	Age of Structure x_2 (years)	Lot Size x_3 (sq. ft)	No. of On-Site Parking Spaces x_4	Gross Building Area x_5 (sq. ft)
0229	90,300	4	82	4,635	0	4,266
0094	384,000	20	13	17,798	0	14,391
0043	157,500	5	66	5,913	0	6,615
0079	676,200	26	64	7,750	6	34,144
0134	165,000	5	55	5,150	0	6,120
0179	300,000	10	65	12,506	0	14,552
0087	108,750	4	82	7,160	0	3,040
0120	276,538	11	23	5,120	0	7,881
0246	420,000	20	18	11,745	20	12,600
0025	950,000	62	71	21,000	3	39,448
0015	560,000	26	74	11,221	0	30,000
0131	268,000	13	56	7,818	13	8,088
0172	290,000	9	76	4,900	0	11,315
0095	173,200	6	21	5,424	6	4,461
0121	323,650	11	24	11,834	8	9,000
0077	162,500	5	19	5,246	5	3,828
0060	353,500	20	62	11,223	2	13,680
0174	134,400	4	70	5,834	0	4,680
0084	187,000	8	19	9,075	0	7,392
0031	155,700	4	57	5,280	0	6,030
0019	93,600	4	82	6,864	0	3,840
0074	110,000	4	50	4,510	0	3,092
0057	573,200	14	10	11,192	0	23,704
0104	79,300	4	82	7,425	0	3,876
0024	272,000	5	82	7,500	0	9,542

Source: Robinson Appraisal Co., Inc., Mankato, Minnesota.

12.12 *Sociology of Sport Journal* (Spring 1992) published research on Proposition 48, the NCAA regulation that requires a minimum of 700 on the Scholastic Assessment Test (SAT) for eligibility for college athletics. Critics have claimed that the SAT is biased against black student athletes and is not a valid predictor of academic success. To investigate the validity of the SAT as a predictor of academic success, high school GPA and study time were also considered as potential predictors of college GPA. The following variables are defined:

y = College GPA (the best measure of academic success)

x_1 = High school GPA

x_2 = Hours of studying in a season

x_3 = SAT score

The study conducted separate regression analyses for black and white student athletes. Some of the results are shown in the following table.

a. Write the model relating the mean value of the college GPA as a linear function of the three identified explanatory variables. Explain the meaning of each of the β parameters in your model.

b. Note that the estimate of the β parameter for SAT score for both models is .001. Interpret this estimate.

c. The reported *p*-value for testing H_0: $\beta_1 = 0$ is .734 for the black athletes' model and .000 for the white athletes' model. Interpret these values.

	Parameter Estimate	Standard Error
Blacks		
Intercept	2.245	.09
High School GPA	−.040	.01
Study Hours	.14	.10
SAT Score	.001	.0025
Whites		
Intercept	1.970	.16
High School GPA	−.089	.01
Study Hours	.20	.23
SAT Score	.001	.0002

12.5   CHECKING THE USEFULNESS OF A MODEL: R^2 AND THE ANALYSIS OF VARIANCE *F*-TEST

Conducting *t*-tests on each β parameter in a model is *not* a good way to determine whether a model is contributing information for the prediction of *y*. If we were to conduct a series of *t*-tests to determine whether the independent variables are contributing to the predictive relationship, we would be very likely to make one or more errors in deciding which terms to retain in the model and which to exclude. For example, even if all the β parameters (except β_0) are equal to 0, $100(\alpha)\%$ of the time you will reject the null hypothesis and conclude that some β parameter differs from 0. Thus, in multiple regression models for which a large number of independent variables are being considered, conducting a series of *t*-tests may include a large number of insignificant variables and exclude some useful ones. If we want to test the utility of a multiple regression model, we will need a global test (one that encompasses all the β parameters). We would also like to find some statistical quantity that measures how well the model fits the data.

We commence with the easier problem—finding a measure of how well a linear model fits a set of data. For this we use the multiple regression equivalent of r^2, the coefficient of determination for the straight-line model (Chapter 11). Thus, we define the **multiple coefficient of determination, R^2**, as

$$R^2 = 1 - \frac{\sum(y - \hat{y})^2}{\sum(y - \bar{y})^2} = 1 - \frac{\text{SSE}}{\text{SS}_{yy}} = \frac{\text{SS}_{yy} - \text{SSE}}{\text{SS}_{yy}} = \frac{\text{Explained variability}}{\text{Total variability}}$$

where $\hat{y}$ is the predicted value of *y* for the model. Just as for the simple linear model, R^2 represents the fraction of the sample variation of the *y* values (measured by SS_{yy}) that is explained by the least squares prediction equation. Thus,

FIGURE 12.10

SAS printout for electrical
usage example

```
Dep Variable: Y

                            Analysis of Variance

                              Sum of          Mean
        Source       DF       Squares         Square      F Value    Prob>F

        Model         2    831069.54637    415534.77319    189.710    0.0001
        Error         7     15332.55363      2190.36480
        C Total       9    846402.10000

            Root MSE           46.80133     R-Square     0.9819
            Dep Mean         1594.70000     Adj R-Sq     0.9767
            C.V.                2.93480

                            Parameter Estimates

                         Parameter      Standard      T for H0:
        Variable    DF    Estimate        Error     Parameter=0   Prob > |T|

        INTERCEP     1  -1216.143887   242.80636850    -5.009       0.0016
        X            1      2.398930     0.24583560     9.758       0.0001
        XSQ          1     -0.000450     0.00005908    -7.618       0.0001
```

$R^2 = 0$ implies a complete lack of fit of the model to the data and $R^2 = 1$ implies a perfect fit with the model passing through every data point. In general, the larger the value of R^2, the better the model fits the data.

To illustrate, the value $R^2 = .982$ for the electrical usage example is indicated in Figure 12.10. This very high value of R^2 implies that using the independent variable home size in a quadratic model explains 98.2% of the total **sample variation** (measured by SS_{yy}) of electrical usage y. Thus, R^2 is a sample statistic that tells how well the model fits the data and thereby represents a measure of the usefulness of the entire model.

The fact that R^2 is a sample statistic implies that it can be used to make inferences about the usefulness of the entire model for predicting the population of y values at each setting of the independent variables. In particular, for the electrical usage data, the test

$$H_0: \ \beta_1 = \beta_2 = 0$$

$$H_a: \ \text{At least one of the coefficients is nonzero}$$

would formally test the global usefulness of the model.

The test statistic used to test this hypothesis is an F statistic, and several equivalent versions of the formula can be used (although we will usually rely on the computer to calculate the F statistic):

$$\textit{Test statistic:} \quad F = \frac{(SS_{yy} - SSE)/k}{SSE/[n - (k + 1)]} = \frac{R^2/k}{(1 - R^2)/[n - (k + 1)]}$$

Both these formulas indicate that the F statistic is the ratio of the *explained* variability divided by the model degrees of freedom to the *unexplained* variability divided by the error degrees of freedom. Thus, the larger the proportion of the total variability accounted for by the model, the larger the F statistic.

To determine when the ratio becomes large enough that we can confidently reject the null hypothesis and conclude that the model is more useful than no model at all for predicting y, we compare the calculated F statistic to a tabulated F value with k df in the numerator and $[n - (k + 1)]$ df in the denominator.

Tabulations of the F-distribution for various values of α are given in Tables VIII, IX, X, and XI of Appendix A.

Rejection region: $F > F_{\alpha}$, where F is based on k numerator and $n - (k + 1)$ denominator degrees of freedom.

For the electrical usage example [$n = 10, k = 2, n - (k + 1) = 7$, and $\alpha = .05$], we will reject $H_0: \beta_1 = \beta_2 = 0$ if

$$F > F_{.05} = 4.74$$

From the SAS printout (Figure 12.10), we find that the computed F is 189.71. Since this value greatly exceeds the tabulated value of 4.74, we conclude that at least one of the model coefficients β_1 and β_2 is nonzero. Therefore, this global F-test indicates that the quadratic model $y = \beta_0 + \beta_1 x + \beta_2 x^2 + \epsilon$ is useful for predicting electrical usage.

The F statistic is also given as a part of most regression printouts, usually in a portion of the printout called the "Analysis of Variance." This is an appropriate descriptive term, since the F statistic relates the explained and unexplained portions of the total variance of y. For example, the elements of the SAS printout in Figure 12.10 that lead to the calculation of the F value are:

$$F\,\text{Value} = \frac{\text{Sum of Squares(Model)} / \text{df(Model)}}{\text{Sum of Squares(Error)} / \text{df(Error)}} = \frac{\text{Mean Square(Model)}}{\text{Mean Square(Error)}}$$

From Figure 12.10 we see that **F Value** = 189.71. Note, too, that the observed significance level for the F statistic is given under the heading **Prob>F** as .0001, which means that we would reject the null hypothesis $H_0: \beta_1 = \beta_2 = 0$ at any α value greater than .0001.

The analysis of variance F-test for testing the usefulness of the model is summarized in the next box.

Testing Global Usefulness of the Model: The Analysis of Variance *F*-Test

---

$H_0: \beta_1 = \beta_2 = \cdots = \beta_k = 0$ (All model terms are unimportant for predicting y)

$H_a:$ At least one $\beta_i \neq 0$ (At least one model term is useful for predicting y)

Test statistic: $F = \dfrac{(\text{SS}_{yy} - \text{SSE}) / k}{\text{SSE} / [n - (k + 1)]} = \dfrac{R^2 / k}{(1 - R^2) / [n - (k + 1)]}$

$\qquad\qquad = \dfrac{\text{Mean Square(Model)}}{\text{Mean Square(Error)}}$

where n is the sample size and k is the number of terms in the model.

Rejection region: $F > F_{\alpha}$, with k numerator degrees of freedom and $[n - (k + 1)]$ denominator degrees of freedom.

Assumptions: The standard regression assumptions about the random error component (Section 12.3).

Caution: A rejection of the null hypothesis leads to the conclusion [with $100(1 - \alpha)\%$ confidence] that the model is useful. However, "useful" does not necessarily mean "best." Another model may prove even more useful in terms of providing more reliable estimates and predictions. This global F-test is usually regarded as a test that the model *must* pass to merit further consideration.

EXAMPLE 12.2

Refer to Example 12.1, in which an antique collector modeled the auction price y of grandfather clocks as a function of the age of the clock, x_1, and the number of bidders, x_2. The hypothesized model is

$$y = \beta_0 + \beta_1 x_1 + \beta_2 x_2 + \epsilon$$

A sample of 32 observations is obtained, with the results summarized in the MINITAB printout repeated in Figure 12.11.

a. Find and interpret the coefficient of determination R^2 for this example.
b. Conduct the global F-test of model usefulness at the $\alpha = .05$ level of significance.

Solution

a. The R^2 value (shaded in Figure 12.11) is .892. This implies that the least squares model has explained about 89% of the total sample variation in y values (auction prices).

b. The elements of the global test of the model follow:

H_0: $\beta_1 = \beta_2 = 0$ [*Note: k = 2*]

H_a: At least one of the two model coefficients is nonzero

Test statistic: $F = \dfrac{R^2/k}{(1 - R^2)/[n - (k + 1)]} = 120.19$ (see Figure 12.11)

Rejection region: $F > F_\alpha$

For this example, $n = 32$, $k = 2$, and $n - (k + 1) = 32 - 3 = 29$. Then, for $\alpha = .05$, we will reject H_0: $\beta_1 = \beta_2 = 0$ if $F > F_{.05}$, where F is based on $k = 2$ numerator and $n - (k + 1) = 29$ denominator degrees of freedom—that is, if $F > 3.33$. Since the computed value of the F-test statistic falls in the rejection region ($F = 120.19$ greatly exceeds $F_{.05} = 3.33$, and $\alpha = .05$ exceeds $p = .000$), the data provide strong evidence that at least one of the model coefficients is nonzero. The model appears to be useful for predicting auction prices.   ▲

Can we be sure that the best prediction model has been found if the global F-test indicates that a model is useful? Unfortunately, we cannot. The addition of other independent variables may improve the usefulness of the model, as Example 12.3 demonstrates.

FIGURE 12.11

MINITAB printout for Example 12.2

```
The regression equation is
Y = -1339 + 12.7 X1 + 86.0 X2

Predictor       Coef       Stdev     t-ratio        p
Constant     -1339.0       173.8       -7.70    0.000
X1           12.7406      0.9047       14.08    0.000
X2            85.953       8.729        9.85    0.000

s = 133.5       R-sq = 89.2%       R-sq(adj) = 88.5%

Analysis of Variance

SOURCE        DF          SS          MS          F         p
Regression     2     4283063     2141532     120.19     0.000
Error         29      516727       17818
Total         31     4799789
```

EXAMPLE 12.3

Refer to Examples 12.1 and 12.2. Suppose the collector, having observed many auctions, believes that the *rate of increase* of the auction price with age will be driven upward by a large number of bidders. Thus, instead of a relationship like that shown in Figure 12.12a, in which the rate of increase in price with age is the same for any number of bidders, the collector believes the relationship is like that shown in Figure 12.12b. Note that as the number of bidders increases from 5 to 15, the slope of the price versus age line increases. When the slope of the relationship between y and one independent variable (x_1) depends on the value of a second independent variable (x_2), as is the case here, we say that x_1 and x_2 **interact.***

The **interaction model** is written

$$y = \beta_0 + \beta_1 x_1 + \beta_2 x_2 + \beta_3 x_1 x_2 + \epsilon$$

Note that the increase in the mean price, $E(y)$, for each 1-year increase in age, x_1, is no longer given by the constant β_1 but is now $\beta_1 + \beta_3 x_2$. *That is, the amount* $E(y)$ *increases for each 1-unit increase in* x_1 *is dependent on the number of bidders,* x_2. *Thus, the two variables* x_1 *and* x_2 *interact to affect* y.

The 32 data points listed in Table 12.2 were used to fit the model with interaction. A portion of the MINITAB printout is shown in Figure 12.13. Test the hypothesis that the price–age slope increases as the number of bidders increases—that is, that age and number of bidders, x_2, interact positively.

FIGURE 12.12

Examples of no-interaction and interaction models

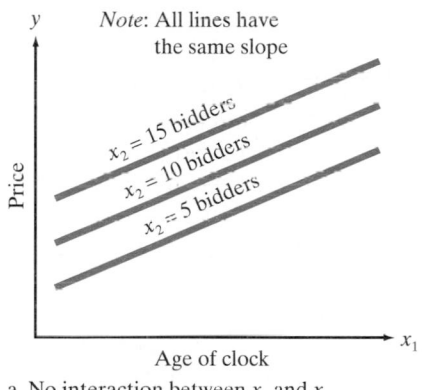

a. No interaction between x_1 and x_2

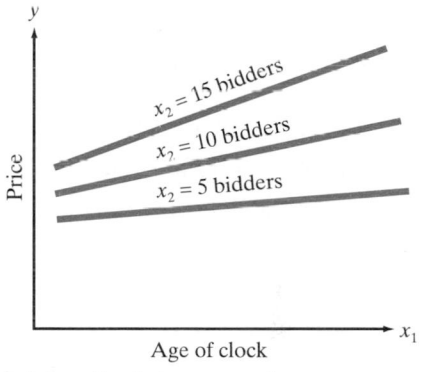

b. Interaction between x_1 and x_2

FIGURE 12.13

MINITAB printout for the model with interaction

```
The regression equation is
Y = 320 + 0.88 X1 - 93.3 X2 + 1.30 X1X2

Predictor        Coef        Stdev     t-ratio          p
Constant        320.5        295.1        1.09      0.287
X1              0.878        2.032        0.43      0.669
X2             -93.26        29.89       -3.12      0.004
X1X2           1.2978       0.2123        6.11      0.000

s = 88.91      R-sq = 95.4%      R-sq(adj) = 94.9%

Analysis of Variance

SOURCE       DF          SS          MS          F          p
Regression    3     4578428     1526142     193.04      0.000
Error        28      221362        7906
Total        31     4799789
```

*A detailed discussion of interaction is given in Chapter 13.

Solution

The model is

$$y = \beta_0 + \beta_1 x_1 + \beta_2 x_2 + \beta_3 x_1 x_2 + \epsilon$$

and the hypotheses of interest to the collector concern the parameter β_3. Specifically,

$$H_0: \beta_3 = 0$$
$$H_a: \beta_3 > 0$$

Test statistic: $\quad t = \dfrac{\hat{\beta}_3}{s_{\hat{\beta}_3}}$

Rejection region: $\quad$ For $\alpha = .05, t > t_{.05}$

where $n = 32$, $k = 3$, and $t_{.05} = 1.701$, based on $n - (k + 1) = 28$ df. [*Remember:* $(k + 1) = 4$ is the number of parameters in the regression model.]

The t value corresponding to $\hat{\beta}_3$ is shaded in Figure 12.13. The value $t = 6.11$ exceeds 1.701 and therefore falls in the rejection region. Thus, the collector can conclude that the rate of change of the mean price of the clocks with age increases as the number of bidders increases; that is, x_1 and x_2 interact. (The same conclusion can be reached by noting that the p-value of the test is 0.) Thus, it appears that the interaction term should be included in the model. ▲

One note of caution: Although the coefficient of x_2 is negative ($\hat{\beta}_2 = -93.26$) in Example 12.3, this does *not* imply that auction price decreases as the number of bidders increases. Since interaction is present, the rate of change (slope) of mean auction price with the number of bidders *depends* on x_1, the age of the clock. Thus, for example, the estimated rate of change of y for a unit increase in x_2 (one new bidder) for a 150-year-old clock is

$$\text{Estimated } x_2 \text{ slope} = \hat{\beta}_2 + \hat{\beta}_3 x_1 = -93.26 + 1.30(150) = 101.74$$

In other words, we estimate that the auction price of a 150-year-old clock will *increase* by about \$101.74 for every additional bidder. Although the rate of increase will vary as x_1 is changed, it will remain positive for the range of values of x_1 included in the sample. Extreme care is needed in interpreting the signs and sizes of coefficients in a multiple regression model.

To summarize the discussion in this section, the value of R^2 is an indicator of how well the prediction equation fits the data. More importantly, it can be used in the F statistic to determine whether the data provide sufficient evidence to indicate that the model contributes information for the prediction of y. Intuitive evaluations of the contribution of the model based on the computed value of R^2 must be examined with care. The value of R^2 increases as more and more variables are added to the model. Consequently, you could force R^2 to take a value very close to 1 even though the model contributes no information for the prediction of y. In fact, R^2 equals 1 when the number of terms in the model (including β_0) equals the number of data points. Therefore, you should not rely solely on the value of R^2 to tell you whether the model is useful for predicting y. Use the F-test.

After we have determined that the overall model is useful for predicting y by using the F-test, we may elect to conduct one or more t-tests on the individual β parameters (see Section 12.4). However, the test (or tests) to be conducted should be decided *a priori,* that is, prior to fitting the model. Also, we should limit the number of t-tests conducted to avoid the potential problem of making too many Type I errors. Generally, the regression analyst will conduct t-tests only on the "most important" β's. We provide insight in identifying the most important β's in a linear model in Chapter 13.

Recommendation for Checking the Utility of a Multiple Regression Model

1. Conduct a test of overall model adequacy using the F-test; that is, test

$$H_0: \beta_1 = \beta_2 = \cdots = \beta_k = 0$$

   If the model is deemed adequate (that is, if you reject H_0), proceed to step 2. Otherwise, you should hypothesize and fit another model. The new model may include more independent variables or higher-order terms.
2. Conduct t-tests on those β parameters in which you are particularly interested (that is, the "most important" β's). These usually involve only the β's associated with higher-order terms (x^2, $x_1 x_2$, etc.). However, it is a safe practice to limit the number of β's that are tested. Conducting a series of t-tests leads to a high overall Type I error rate α.

EXERCISES 12.13–12.28

Note: Exercises marked with 🖳 *require the use of a computer.*

Learning the Mechanics

12.13 Suppose you fit the model

$$y = \beta_0 + \beta_1 x_1 + \beta_2 x_2 + \beta_3 x_1 x_2 + \beta_4 x_1^2 + \beta_5 x_2^2 + \epsilon$$

to $n = 30$ data points and obtain

$$\text{SSE} = .33 \qquad R^2 = .92$$

a. Do the values of SSE and R^2 suggest that the model provides a good fit to the data? Explain.

b. Is the model of any use in predicting y? Test the null hypothesis that $E(y) = \beta_0$; that is, test

$$H_0: \beta_1 = \beta_2 = \cdots = \beta_5 = 0$$

against the alternative hypothesis

$$H_a: \text{At least one of the parameters } \beta_1, \beta_2, \ldots, \beta_5 \text{ is nonzero}$$

Use $\alpha = .05$.

12.14 The model $y = \beta_0 + \beta_1 x + \beta_2 x^2 + \epsilon$ was fit to $n = 19$ data points with the results shown in the printout below.

a. Find R^2 and interpret its value.

b. Test the null hypothesis that $\beta_1 = \beta_2 = 0$ against the alternative hypothesis that at least one of β_1 and β_2 is nonzero. Calculate the test statistic using the two formulas given in this section, and compare your results to each other and to that given on the printout. Use $\alpha = .05$ and interpret the result of your test.

```
Dep Variable: Y

                        Analysis of Variance

                     Sum of        Mean
    Source    DF     Squares       Square      F Value    Prob>F

    Model      2     24.22335     12.11167     65.478     0.0001
    Error     16      2.95955      0.18497
    C Total   18     27.18289

            Root MSE      0.43008     R-Square     0.8911
            Dep Mean      3.56053     Adj R-Sq     0.8775
            C.V.         12.07921

                        Parameter Estimates

                    Parameter     Standard     T for H0:
    Variable   DF    Estimate       Error     Parameter=0    Prob > |T|

    INTERCEP   1     0.734606     0.29313351      2.506       0.0234
    X          1     0.765179     0.08754136      8.741       0.0001
    XSQ        1    -0.030810     0.00452890     -6.803       0.0001
```

c. Find the observed significance level for this test on the printout and interpret it.

d. Test H_0: $\beta_2 = 0$ against H_a: $\beta_2 \neq 0$. Use $\alpha = .05$ and interpret the result of your test. Report and interpret the observed significance level of the test.

12.15 If the analysis of variance F-test leads to the conclusion that at least one of the model parameters is nonzero, can you conclude that the model is the best predictor for the dependent variable y? Can you conclude that all of the terms in the model are important for predicting y? What is the appropriate conclusion?

12.16 Suppose you fit the model

$$y = \beta_0 + \beta_1 x_1 + \beta_2 x_2 + \epsilon$$

to $n = 20$ data points and obtain

$$\sum (y_i - \hat{y}_i)^2 = 12.35 \qquad \sum (y_i - \bar{y})^2 = 24.44$$

a. Construct an analysis of variance table for this regression analysis, using the printout in Exercise 12.14 as a model. Be sure to include the sources of variability, the degrees of freedom, the sums of squares, the mean squares, and the F statistic. Calculate R^2 for the regression analysis.

b. Test the null hypothesis that $\beta_1 = \beta_2 = 0$ against the alternative hypothesis that at least one of the parameters differs from 0. Calculate the test statistic in two different ways and compare the results. Use $\alpha = .05$ to reach a conclusion about whether the model contributes information for the prediction of y.

Applying the Concepts

12.17 The *Journal of Quantitative Criminology* (Vol. 8, 1992) published a paper on the determinants of area property crime levels in the United Kingdom. Several multiple regression models for property crime prevalence, y, measured as the percentage of residents in a geographic area who were victims of at least one property crime, were examined. The results for one of the models, based on a sample of $n = 313$ responses collected for the British Crime Survey, are shown in the table below. [*Note:* All variables except Density are expressed as a percentage of the base area.]

a. Test the hypothesis that the density (x_1) of a region is positively linearly related to crime prevalence (y), holding the other independent variables constant.

b. Do you advise conducting t-tests on each of the 18 independent variables in the model to determine which variables are important predictors of crime prevalence? Explain.

c. The model yielded $R^2 = .411$. Use this information to conduct a test of the global utility of the model. Use $\alpha = .05$.

12.18 Refer to the *Cell Biology* study, Exercise 12.7, in which the age y of onset of a neurodegenerative disease is modeled as a function of the number x of repeat gene sequences in chromosomes. The researchers reported a correlation of $r = -.815$

Variable	$\hat{\beta}$	t	p-value
x_1 = Density (population per hectare)	.331	3.88	$p < .01$
x_2 = Unemployed male population	−.121	−1.17	$p > .10$
x_3 = Professional population	−.187	−1.90	$.01 < p < .10$
x_4 = Population aged less than 5	−.151	−1.51	$p > .10$
x_5 = Population aged between 5 and 15	.353	3.42	$p < .01$
x_6 = Female population	.095	1.31	$p > .10$
x_7 = 10-year change in population	.130	1.40	$p > .10$
x_8 = Minority population	−.122	−1.51	$p > .10$
x_9 = Young adult population	.163	5.62	$p < .01$
x_{10} = 1 if North region, 0 if not	.369	1.72	$.01 < p < .10$
x_{11} = 1 if Yorkshire region, 0 if not	−.210	−1.39	$p > .10$
x_{12} = 1 if East Midlands region, 0 if not	−.192	−0.78	$p > .10$
x_{13} = 1 if East Anglia region, 0 if not	−.548	−2.22	$.01 < p < .10$
x_{14} = 1 if South East region, 0 if not	.152	1.37	$p > .10$
x_{15} = 1 if South West region, 0 if not	−.151	−0.88	$p > .10$
x_{16} = 1 if West Midlands region, 0 if not	−.308	−1.93	$.01 < p < .10$
x_{17} = 1 if North West region, 0 if not	.311	2.13	$.01 < p < .10$
x_{18} = 1 if Wales region, 0 if not	−.019	−0.08	$p > .10$

Source: Osborn, D. R., Tickett, A., and Elder, R. "Area characteristics and regional variates as determinants of area property crime." *Journal of Quantitative Criminology,* Vol. 8, No. 3, 1992, Plenum Publishing Corp.

between age and number of repeats. Since $r^2 = (-.815)^2 = .664$, they concluded that about "66% of the variability in the age of onset can be accounted for by the number of repeats." Does this statement apply to the quadratic model, $E(y) = \beta_0 + \beta_1 x + \beta_2 x^2$? If not, give the equation of the model for which it does apply.

12.19 An important goal in occupational safety is "active caring." Employees demonstrate active caring (AC) about the safety of their co-workers when they identify environmental hazards and unsafe work practices and then implement appropriate corrective actions for these unsafe conditions or behaviors. Three factors hypothesized to increase the propensity for an employee to actively care for safety are (1) high self-esteem, (2) optimism, and (3) group cohesiveness. A study published in *Applied & Preventive Psychology* (Winter 1995) attempted to establish empirical support for the AC hypothesis by fitting the model $E(y) = \beta_0 + \beta_1 x_1 + \beta_2 x_2 + \beta_3 x_3$, where

y = AC score (measuring active caring on a 15-point scale)

x_1 = Self-esteem score

x_2 = Optimism score

x_3 = Group cohesion score

The regression analysis, based on data collected for $n = 31$ hourly workers at a large fiber-manufacturing plant, yielded a multiple coefficient of determination of $R^2 = .362$.
a. Interpret the value of R^2.
b. Use the R^2-value to test the global utility of the model. Use $\alpha = .05$.

12.20 *Trichuristrichiura*, a parasitic worm, affects millions of school-age children each year, especially children from developing countries. A study was conducted to determine the effects of treatment of the parasite on school achievement in 407 school-age Jamaican children infected with the disease (*Journal of Nutrition,* July 1995). About half the children in the sample received the treatment, while the others received a placebo. Multiple regression was used to model spelling test score y, measured as number correct, as a function of the following independent variables:

Treatment (T):

$$x_1 = \begin{cases} 1 & \text{if treatment} \\ 0 & \text{if placebo} \end{cases}$$

Disease Intensity (I):

$$x_2 = \begin{cases} 1 & \text{if more than 7,000 eggs per gram of stool} \\ 0 & \text{if not} \end{cases}$$

a. Propose a model for $E(y)$ that includes interaction between treatment and disease intensity.
b. The estimates of the β's in the model, part **a**, and the respective p-values for t-tests on the β's are given in the table below.* Is there sufficient evidence to indicate that the effect of the treatment on spelling score depends on disease intensity? Test using $\alpha = .05$.

Variable	β Estimate	*p*-Value
Treatment (x_1)	$-.1$	.62
Intensity (x_2)	$-.3$	.57
T × I $(x_1 x_2)$	1.6	.02

c. Based on the result, part **b**, explain why the analyst should avoid conducting t-tests for the treatment (x_1) and intensity (x_2) β's or interpreting these β's individually.

12.21 Regression analysis was employed to investigate the determinants of survival size of nonprofit hospitals (*Applied Economics,* Vol. 18, 1986). For a given sample of hospitals, survival size, y, is defined as the largest size hospital (in terms of number of beds) exhibiting growth in market share over a specific time interval. Suppose 10 states are randomly selected and the survival size for all nonprofit hospitals in each state is determined for two time periods five years apart, yielding two observations per state. The 20 survival sizes are listed in the table on page 562, along with the following data for each state, for the second year in each time interval:

x_1 = Percentage of beds that are in for-profit hospitals

x_2 = Ratio of the number of persons enrolled in health maintenance organizations (HMOs) to the number of persons covered by hospital insurance

x_3 = State population (in thousands)

x_4 = Percent of state that is urban

The article hypothesized that the following model characterizes the relationship between survival size and the four variables just listed:

$$y = \beta_0 + \beta_1 x_1 + \beta_2 x_2 + \beta_3 x_3 + \beta_4 x_4 + \epsilon$$

---

*The actual model fit in the study included several other variables such as age, gender, and socioeconomic status.

State	Time period	Survival size y	x_1	x_2	x_3	x_4
1	1	370	.13	.09	5,800	89
1	2	390	.15	.09	5,955	87
2	1	455	.08	.11	17,648	87
2	2	450	.10	.16	17,895	85
3	1	500	.03	.04	7,332	79
3	2	480	.07	.05	7,610	78
4	1	550	.06	.005	11,731	80
4	2	600	.10	.005	11,790	81
5	1	205	.30	.12	2,932	44
5	2	230	.25	.13	3,100	45
6	1	425	.04	.01	4,148	36
6	2	445	.07	.02	4,205	38
7	1	245	.20	.01	1,574	25
7	2	200	.30	.01	1,560	28
8	1	250	.07	.08	2,471	38
8	2	275	.08	.10	2,511	38
9	1	300	.09	.12	4,060	52
9	2	290	.12	.20	4,175	54
10	1	280	.10	.02	2,902	37
10	2	270	.11	.05	2,925	38

Source: Adapted from Bays, C. W. "The determinants of hospital size: A survivor analysis." *Applied Economics,* 1986, Vol. 18, pp. 359–377.

a. The model was fit to the data in the table using SAS, with the results given in the printout below. Report the least squares prediction equation.

b. Find the regression standard deviation s and interpret its value in the context of the problem.

c. Use an F-test to investigate the usefulness of the hypothesized model. Report the observed significance level, and use $\alpha = .025$ to reach your conclusion.

d. Prior to collecting the data it was hypothesized that increases in the number of for-profit hospital beds would decrease the survival size of non-profit hospitals. Do the data support this hypothesis? Test using $\alpha = .05$.

12.22 Because the coefficient of determination R^2 always increases when a new independent variable is added to the model, it is tempting to include many variables in a model to force R^2 to be near 1. However, doing so reduces the degrees of freedom

```
Dep Variable: Y

                         Analysis of Variance

                        Sum of          Mean
     Source     DF      Squares         Square      F Value    Prob>F

     Model       4  246537.05939    61634.26485     28.180     0.0001
     Error      15   32807.94061     2187.19604
     C Total    19  279345.00000

          Root MSE      46.76747     R-Square     0.8826
          Dep Mean     360.50000     Adj R-Sq     0.8512
          C.V.          12.97295

                         Parameter Estimates

                     Parameter      Standard      T for H0:
     Variable   DF    Estimate        Error     Parameter=0   Prob > |T|

     INTERCEP    1   295.327091    40.17888737      7.350       0.0001
     X1          1  -480.837576   150.39050364     -3.197       0.0060
     X2          1  -829.464955   196.47303539     -4.222       0.0007
     X3          1     0.007934     0.00355335      2.233       0.0412
     X4          1     2.360769     0.76150774      3.100       0.0073
```

available for estimating σ^2, which adversely affects our ability to make reliable inferences. Suppose you want to use 18 psychological and sociological factors to predict a student's Scholastic Assessment Test (SAT) score. You fit the model

$$y = \beta_0 + \beta_1 x_1 + \beta_2 x_2 + \cdots + \beta_{17} x_{17} + \beta_{18} x_{18} + \epsilon$$

where y = SAT score and $x_1, x_2, ..., x_{18}$ are the psychological and sociological factors. Only 20 years of data ($n = 20$) are used to fit the model, and you obtain $R^2 = .95$. Test to see whether this impressive-looking R^2 is large enough for you to infer that the model is useful, that is, that at least one term in the model is important for predicting SAT scores. Use $\alpha = .05$.

12.23 The *Journal of Applied Phycology* (Dec. 1994) published research on the seasonal growth activity of algae in an indoor environment. The daily growth rate y was regressed against temperature x using the quadratic model $E(y) = \beta_0 + \beta_1 x + \beta_2 x^2$. A particular algal strain was grown in a glass dish indoors at temperatures ranging from 10° to 32° Celsius. The data for $n = 33$ such experiments were used to fit the quadratic model, with the following results:

$$\hat{y} = -2.51 + .55x - .01x^2, \qquad R^2 = .67$$

a. Sketch the least squares prediction equation. Interpret the graph.
b. Interpret the value of R^2.
c. Is the model useful for predicting algae growth rate, y? Test using $\alpha = .05$.

12.24 In Exercise 11.22, we found the least squares line relating the age x of a sand lance to its length y. The data are shown in the table below.
a. Use a computer to fit a second-order model to the data.
b. Do the data provide sufficient evidence to indicate that the second-order term contributes information for the prediction of sand lance length? Test using $\alpha = .05$.
c. Find the p-value for the test in part **b**.
d. Find the coefficient of determination for the second-order model and interpret it.

12.25 To what degree do the attitudes of your peers influence your behavior? There is general agreement among sociologists and psychologists that your behavior is dependent on the attitudes of and social support from your friends, neighbors, etc. However, it is unclear whether the effects of atti-

tude and social support are additive or interactive. An attempt to resolve this attitude–behavior issue was presented in *Social Psychology Quarterly* (Vol. 30, 1987). The study included a sample of $n = 143$ adult drinkers in an urban setting characterized by high physical availability of alcoholic beverages. The goal of the study was to build a model relating frequency of drinking alcoholic beverages, y, to attitude toward drinking (x_1) and social support (x_2). Consider the interaction model

$$E(y) = \beta_0 + \beta_1 x_1 + \beta_2 x_2 + \beta_3 x_1 x_2$$

a. Interpret the phrase "x_1 and x_2 interact" in terms of the problem.
b. Write the null and alternative hypotheses for determining whether attitude (x_1) and social support (x_2) interact.
c. The reported p-value for the test, part **b**, was $p < .001$. Interpret this result.

12.26 Real estate appraisers make heavy use of regression analysis to model the relationship between the selling price of a house and its total living area. The table shown below contains the final selling prices and total living areas for a sample of 30 houses that were sold during 1995 in the same geographic area. The following model was fitted to the data:

$$y = \beta_0 + \beta_1 x + \beta_2 x^2 + \epsilon$$

where y = price and x = area. The computer printout for this model is shown on page 564.

Area (sq. ft)	Price	Area (sq. ft)	Price
1,600	$101,900	2,120	$113,600
1,980	108,700	1,800	106,500
2,130	114,500	1,775	105,000
1,990	110,300	1,650	99,000
1,710	100,900	2,250	116,100
2,357	121,100	2,390	123,300
2,335	118,700	2,560	133,000
2,650	137,200	2,370	125,000
2,070	109,300	2,420	123,300
1,930	108,400	2,300	119,200
1,910	104,000	2,320	120,000
1,810	104,500	2,470	126,800
2,035	112,000	2,460	127,100
3,000	210,000	3,650	225,100
2,500	160,000	2,750	205,600

a. What is the least squares equation for this model? Interpret the sign of the estimate of β_2.
b. What is the standard deviation for this model? Interpret it.

Mean Length (millimeters)	176	194	212	226	236	244	254
Age (years)	2	3	4	5	6	7	8

c. What is the value of R^2 for this model? Interpret it. Is there evidence that this model is useful for predicting price? Explain.

d. Do the data provide sufficient evidence to indicate that the second-order term (x^2) contributes to the prediction of price? Set up the appropriate null and alternative hypotheses, and interpret the p-value for this test.

12.27 Much research—and much litigation—has been conducted on the disparity between the salary levels of men and women. Research reported in *Work and Occupations* (Nov. 1992) analyzes the salaries for a sample of 191 Illinois managers using a regression analysis with the following independent variables:

$$x_1 = \text{Gender of manager} = \begin{cases} 1 & \text{if male} \\ 0 & \text{if not} \end{cases}$$

$$x_2 = \text{Race of manager} = \begin{cases} 1 & \text{if white} \\ 0 & \text{if not} \end{cases}$$

$x_3 = \text{Education level (in years)}$

$x_4 = \text{Tenure with firm (in years)}$

$x_5 = \text{Number of hours worked per week}$

The regression results are shown below as they were reported in the article.

Variable	$\hat{\beta}$	p-Value
x_1	12.774	<.05
x_2	.713	>.10
x_3	1.519	<.05
x_4	.320	<.05
x_5	.205	<.05
Constant	15.491	—

$$R^2 = .240 \qquad n = 191$$

a. Write the hypothesized model that was used, and interpret each of the β parameters in the model.

b. Write the least squares equation that estimates the model in part **a**, and interpret each of the β estimates.

c. Interpret the value of R^2. Test to determine whether the model is useful for predicting annual salary. Test using $\alpha = .05$.

d. Test to determine whether the gender variable indicates that male managers are paid more than female managers, even after adjusting for and holding constant the other four factors in the model. Test using $\alpha = .05$. [*Note:* The p-values given in the table are two-tailed.]

e. Why would one want to adjust for these other factors before conducting a test for salary discrimination?

f. Discuss how the interaction between gender (x_1) and tenure with the firm (x_4) might affect the results of the analysis. [*Note:* We will discuss the use and implications of this type of interaction term more fully in Chapter 13.]

12.28 Does extensive media coverage of a military crisis influence public opinion on how to respond to the crisis? Political scientists at UCLA researched this question and reported their results in *Communication Research* (June 1993). The military crisis of interest was the 1990 Persian Gulf War, precipitated by Iraqi leader Saddam Hussein's invasion of Kuwait. The researchers used multiple regression analysis to model the level y of U.S. public support for a military (rather than a diplomatic) response to the crisis. Values of y ranged from 0 (preference for a diplomatic response) to 4 (preference for a military response). The following independent variables were used in the model:

```
Dependent Variable: PRICE

                        Analysis of Variance

                         Sum of          Mean
        Source     DF    Squares         Square      F Value    Prob>F

        Model       2   24864565684   12432282842    63.644     0.0001
        Error      27   5274183982.3  195340147.49
        C Total    29   30138749667

            Root MSE      13976.41397    R-square    0.8250
            Dep Mean     126336.66667    Adj R-sq    0.8120
            C.V.             11.06283

                        Parameter Estimates

                      Parameter    Standard     T for H0:
        Variable  DF  Estimate     Error        Parameter=0    Prob > |T|

        INTERCEP   1     95521    51722.224045      1.847         0.0758
        AREA       1  -32.993909     42.27493672   -0.780         0.4419
        AREA_SQ    1    0.020089      0.00844736    2.378         0.0247
```

x_1 = Level of TV news exposure in a selected week (number of days)

x_2 = Knowledge of 7 political figures (1 point for each correct answer)

x_3 = Gender (1 if male, 0 if female)

x_4 = Race (1 if nonwhite, 0 if white)

x_5 = Partisanship (0–6 scale, where 0 = strong Democrat and 6 = strong Republican)

x_6 = Defense spending attitude (1–7 scale, where 1 = greatly decrease spending and 7 = greatly increase spending)

x_7 = Education level (1–7 scale, where 1 = less than eight grades and 7 = college)

Data from a survey of 1,763 U.S. citizens were used to fit the model

$$E(y) = \beta_0 + \beta_1 x_1 + \beta_2 x_2 + \beta_3 x_3 + \beta_4 x_4 + \beta_5 x_5 + \beta_6 x_6 + \beta_7 x_7 + \beta_8 x_2 x_3 + \beta_9 x_2 x_4$$

The regression results are shown in the accompanying table.

a. Interpret the β estimate for the variable x_1, TV news exposure.

b. Conduct a test to determine whether an increase in TV news exposure is associated with an increase in support for a military resolution of the crisis. Use $\alpha = .05$.

c. Is there sufficient evidence to indicate that the relationship between support for a military resolution (y) and gender (x_3) depends on political knowledge (x_2)? Test using $\alpha = .05$.

d. Is there sufficient evidence to indicate that the relationship between support for a military resolution (y) and race (x_4) depends on political knowledge (x_2)? Test using $\alpha = .05$.

e. The coefficient of determination for the model was $R^2 = .194$. Interpret this value.

f. Use the value of R^2, part **e**, to conduct a global test for model utility. Use $\alpha = .05$.

Variable	β Estimate	Standard Error	Two-Tailed *p*-Value
TV news exposure (x_1)	.02	.01	.03
Political knowledge (x_2)	.07	.03	.03
Gender (x_3)	.67	.11	<.001
Race (x_4)	−.76	.13	<.001
Partisanship (x_5)	.07	.01	<.001
Defense spending (x_6)	.20	.02	<.001
Education (x_7)	.07	.02	<.001
Knowledge × Gender ($x_2 x_3$)	−.09	.04	.02
Knowledge × Race ($x_2 x_4$)	.10	.06	.08

Source: Iyengar, S., and Simon, A. "News coverage of the Gulf Crisis and public opinion." *Communication Research*, Vol. 20, No. 3, June 1993, p. 380 (Table 2). Copyright 1993 by Sage Publications. Reprinted by permission of Sage Publications, Inc.

12.6   USING THE MODEL FOR ESTIMATION AND PREDICTION

In Section 11.8 we discussed the use of the least squares line for estimating the mean value of *y*, $E(y)$, for some particular value of *x*, say $x = x_p$. We also showed how to use the same fitted model to predict, when $x = x_p$, some new value of *y* to be observed in the future. Recall that the least squares line yielded the same value for both the estimate of $E(y)$ and the prediction of some future value of *y*. That is, both are the result of substituting x_p into the prediction equation, $\hat{y} = \hat{\beta}_0 + \hat{\beta}_1 x$ and calculating $\hat{y}_p$. There the equivalence ends. The confidence interval for the mean $E(y)$ is narrower than the prediction interval for *y* because of the additional uncertainty attributable to the random error ϵ when predicting some future value of *y*.

FIGURE 12.14

SAS printout for estimated mean values and corresponding confidence intervals

Obs	X	Dep Var Y	Predict Value	Std Err Predict	Lower95% Mean	Upper95% Mean	Residual
1	1290	1182.0	1129.6	30.072	1058.5	1200.7	52.4359
2	1350	1172.0	1202.2	25.851	1141.1	1263.3	-30.2136
3	1470	1264.0	1337.8	19.832	1290.9	1384.7	-73.7916
4	1600	1493.0	1470.0	17.392	1428.9	1511.2	22.9586
5	1710	1571.0	1570.1	17.976	1527.6	1612.6	0.9359
6	1840	1711.0	1674.2	19.919	1627.1	1721.3	36.7685
7	1980	1804.0	1769.4	21.887	1717.6	1821.2	34.5998
8	2230	1840.0	1895.5	23.348	1840.3	1950.7	-55.4654
9	2400	1956.0	1949.1	23.611	1893.2	2004.9	6.9431
10	2930	1954.0	1949.2	44.734	1843.4	2055.0	4.8287
11	1500	.	1369.7	18.892	1325.0	1414.3	.

These same concepts carry over to the multiple regression model. Suppose we want to estimate the mean electrical usage for a given home size, say $x_p = 1,500$ square feet. Assuming that the quadratic model represents the true relationship between electrical usage and home size, we want to estimate

$$E(y) = \beta_0 + \beta_1 x_p + \beta_2 x_p^2 = \beta_0 + \beta_1(1,500) + \beta_2(1,500)^2$$

Substituting into the least squares prediction equation, we find the estimate of $E(y)$ to be

$$\hat{y} = \hat{\beta}_0 + \hat{\beta}_1(1,500) + \hat{\beta}_2(1,500)^2$$
$$= -1,216.144 + 2.3989(1,500) - .00045004(1,500)^2 = 1,369.7$$

To form a confidence interval for the mean, we need to know the standard deviation of the sampling distribution for the estimator $\hat{y}$. For multiple regression models, the form of this standard deviation is rather complex. However, some regression packages allow us to obtain the confidence intervals for mean values of y for any given combination of values of the independent variables. A portion of the SAS output for the electrical usage example is shown in Figure 12.14. The mean value and corresponding 95% confidence interval for the x-values in the sample are shown in the columns labeled **Predict Value**, **Lower95% Mean**, and **Upper95% Mean**. In the last row, corresponding to $x_p = 1,500$, we observe that $\hat{y} = 1,369.7$, which agrees with our earlier calculation. The corresponding 95% confidence interval for the true mean of y, shaded on the printout, is 1,325.0 to 1,414.3. (This interval is shown graphically in Figure 12.15.)

FIGURE 12.15

Confidence interval for mean electrical usage when $x_p = 1,500$

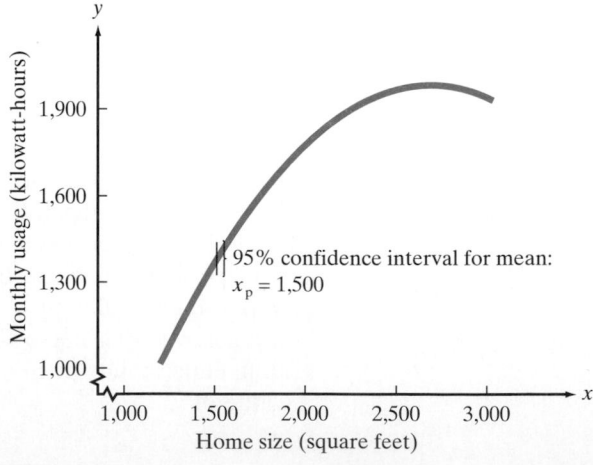

FIGURE 12.16

SAS printout for predicted values and corresponding prediction intervals

Obs	X	Dep Var Y	Predict Value	Std Err Predict	Lower95% Predict	Upper95% Predict	Residual
1	1290	1182.0	1129.6	30.072	998.0	1261.1	52.4359
2	1350	1172.0	1202.2	25.851	1075.8	1328.6	-30.2136
3	1470	1264.0	1337.8	19.832	1217.6	1458.0	-73.7916
4	1600	1493.0	1470.0	17.392	1352.0	1588.1	22.9586
5	1710	1571.0	1570.1	17.976	1451.5	1688.6	0.9359
6	1840	1711.0	1674.2	19.919	1554.0	1794.5	36.7685
7	1980	1804.0	1769.4	21.887	1647.2	1891.6	34.5998
8	2230	1840.0	1895.5	23.348	1771.8	2019.1	-55.4654
9	2400	1956.0	1949.1	23.611	1825.1	2073.0	6.9431
10	2930	1954.0	1949.2	44.734	1796.1	2102.3	4.8287
11	1500	.	1369.7	18.892	1250.3	1489.0	.

If we were interested in predicting the electrical usage for a particular 1,500-square-foot home, we would use $\hat{y} = 1,369.7$ as the predicted value. However, the prediction interval for a new value of y is wider than the confidence interval for the mean value. This is reflected by the SAS printout shown in Figure 12.16, which gives the predicted values of y and corresponding 95% prediction intervals under the columns **Lower95% Predict** and **Upper95% Predict**. Note that the prediction interval for $x_p = 1,500$ (shaded in the last row) is 1,250.3 to 1,489.0. (This interval is shown graphically in Figure 12.17.)

Unfortunately, not all statistical software packages have the capability to produce confidence intervals for means and prediction intervals for specific y values. This is a rather serious oversight, since the estimation of mean values and the prediction of specific values represent the culmination of our model-building efforts: using the model to make inferences about the dependent variable y.

12.7 MULTIPLE REGRESSION: AN EXAMPLE

Towers, Perrin, Forster & Crosby (TPF&C), an international management consulting firm, offers a unique service to its clients. Many firms are interested in evaluating their management salary structure, and TPF&C accomplishes this evaluation by using multiple regression models. The Compensation Management Service, as

FIGURE 12.17

Prediction interval for electrical usage when $x_p = 1,500$

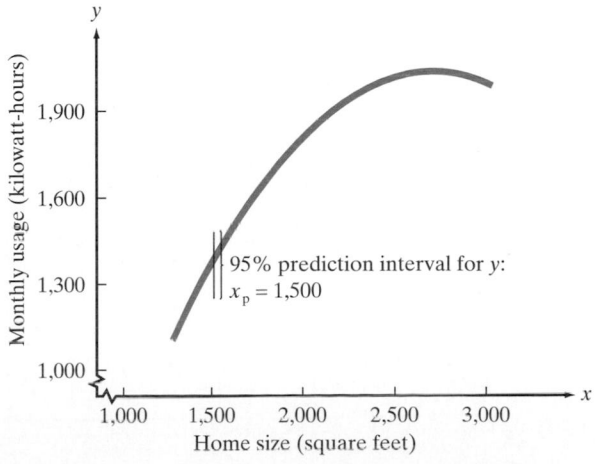

TPF&C calls it, measures both the internal and external consistency of a company's pay policies to determine whether they reflect the management's intent.

The dependent variable y used to measure executive compensation is annual salary. The independent variables used to explain salary structure include the executive's age, education, rank, and bonus eligibility; number of employees under the executive's direct supervision; as well as variables that describe the company for which the executive works, such as annual sales, profit, and total assets.

The initial step in developing models for executive compensation is to obtain a sample of executives from various client firms, which TPF&C calls the Compensation Data Bank. The data for these executives are used to estimate the model coefficients (the β parameters), and these estimates are then substituted into the linear model to form a prediction equation. To predict a particular executive's compensation, TPF&C substitutes into the prediction equation the values of the independent variables that pertain to the executive (age, rank, etc.).

We will use this application of multiple regression to demonstrate all the ideas we have introduced in this chapter. Suppose the list of independent variables given in Table 12.3 is to be used to build a model for the salaries of corporate executives.

Step 1 The first step is to hypothesize a model relating executive salary to the independent variables listed in Table 12.3. TPF&C has found that executive compensation models that use the logarithm of salary as the dependent variable provide better predictive power than those using the salary as the dependent variable. This is probably because salaries tend to be incremented in *percentages* rather than dollar values. When a response variable undergoes percentage changes as the independent variables are varied, the logarithm of the response variable will be more suitable as a dependent variable. The model we propose is

$$y = \beta_0 + \beta_1 x_1 + \beta_2 x_2 + \beta_3 x_3 + \beta_4 x_4 + \beta_5 x_5 + \beta_6 x_6 + \beta_7 x_7 + \epsilon$$

where $y = \log(\text{executive salary})$, $x_6 = x_1^2$ (quadratic term in years of experience), and $x_7 = x_3 x_4$ (cross product or interaction term between sex and number of employees supervised). The variable x_3 is a **dummy variable**; it is used to describe an independent variable that is not measured on a numerical scale but instead is qualitative (categorical) in nature. Sex is such a variable, since its values, male and female, are categories rather than numbers. Thus, we assign the value $x_3 = 1$ if the executive is male and $x_3 = 0$ if the executive is female. For more detail on the use and interpretation of dummy variables, see Chapter 13. The interaction term $x_3 x_4$ accounts for the fact that the relationship between corporate salary and the number of employees supervised, x_4, is dependent on sex, x_3. That is, as the

TABLE 12.3 List of Independent Variables for Executive Compensation Example

Independent Variable	Description
x_1	Years of experience
x_2	Years of education
x_3	1 if male; 0 if female
x_4	Number of employees supervised
x_5	Corporate assets (millions of dollars)
x_6	x_1^2
x_7	$x_3 x_4$

FIGURE 12.18

SAS printout for executive compensation example

Analysis of Variance

Source	DF	Sum of Squares	Mean Square	F Value	Prob>F
Model	7	27.06425	3.86632	1823.73	0.0001
Error	92	0.19551	0.00212		
C Total	99	27.25976			

Root MSE	0.0461	R-square	0.9928
Dep Mean	12.570	Adj R-sq	0.9923
C.V.	0.3660		

Parameter Estimates

Variable	DF	Parameter Estimate	Standard Error	T for H0: Parameter=0	Prob > |T|
INTERCEP	1	9.87878	0.04612	214.15	0.0001
X1	1	0.04460	0.00166	26.83	0.0001
X2	1	0.03326	0.00270	12.31	0.0001
X3	1	0.11892	0.01724	6.89	0.0001
X4	1	0.00033	0.00001	19.97	0.0001
X5	1	0.00201	0.00002	73.25	0.0001
X1X1	1	-0.00071	0.00004	-15.11	0.0001
X3X4	1	0.00031	0.00002	16.16	0.0001

Obs	X1	X2	X3	X4	X5	Dep Var Y	Predict Value	Std Err Predict	Lower95% Predict	Upper95% Predict
1	12	16	0	400	160.1	.	11.298	.048	11.203	11.392

number of supervised employees increases, a woman's salary (with all other factors being equal) might rise more (or less) rapidly than a man's. This concept (interaction), too, is explained in more detail in Chapter 13.

Step 2 Now we estimate the model coefficients $\beta_0, \beta_1, \ldots, \beta_7$. Suppose a sample of 100 executives is selected, and the variables y and $x_1, x_2, \ldots, x_7$ are recorded (or, in the case of x_6 and x_7, calculated). The sample is then used as input for the SAS regression routine; the output is shown in Figure 12.18. The least squares model is

$$\hat{y} = 9.88 + .045x_1 + .033x_2 + .119x_3 + .00033x_4 + .0020x_5 - .00071x_6 + .00031x_7$$

Step 3 The next step is to specify the probability distribution of ϵ, the random error component. We assume that ϵ is normally distributed with a mean of 0 and a constant variance σ^2. Furthermore, we assume that the errors are independent. The estimate of the variance σ^2, shaded on the SAS printout, is

$$s^2 = \text{MSE} = \frac{\text{SSE}}{n - (k + 1)} = \frac{\text{SSE}}{100 - (7 + 1)} = .00212$$

and the estimate of the standard deviation σ, also shaded on the printout, is $s = \sqrt{s^2} = .0461$. Our interpretation is that most of the observed y values (logarithms of salaries) lie within $2s = 2(.0461) = .0922$ of the least squares predicted values.

Step 4 We now want to see how well the model predicts salaries. First, note that $R^2 = .9928$. This implies that over 99% of the variation in y (the logarithm of salaries) for these 100 sampled executives is accounted for by the model. The significance of this can be tested:

$$H_0\text{: } \beta_1 = \beta_2 = \cdots = \beta_7 = 0$$

H_a: At least one of the model coefficients is nonzero

$$\text{Test statistic:} \quad F = \frac{\text{Mean Square(Model)}}{\text{MSE}} = 1{,}823.73$$
(shaded on Figure 12.18)

Rejection region: For $\alpha = .05$, $F > F_{.05}$

Here the tabulated value of F for $\alpha = .05$ based on $k = 7$ and $n - (k + 1) = 92$ df is $F_{.05} \approx 2.1$ (from Table IX of Appendix A). Since $F = 1{,}823.73$ exceeds the tabulated value of F and, in fact, has an observed significance level of .0001, we conclude that the model does contribute information for predicting executive salaries. It appears that at least one of the β parameters in the model differs from 0.

If the firm is discriminating (knowingly or unknowingly) against newly hired female executives, the mean salary for newly hired males will exceed the mean for newly hired females with the same qualifications. Therefore, we may be particularly interested in whether the data provide evidence that the mean salary of executives depends on sex when all other variables (experience, education, etc.) are held constant. In Chapter 13 we will learn that when the dummy variable x_3 is in the model and has a positive β coefficient, the mean salary of males exceeds the mean for females. Thus, one way to test the sex discrimination hypothesis is to test:*

$$H_0: \beta_3 = 0$$
$$H_a: \beta_3 > 0$$

$$\text{Test statistic:} \quad t = \frac{\hat{\beta}_3}{s_{\hat{\beta}_3}} = 6.89 \quad \text{(shaded on Figure 12.18)}$$

For $\alpha = .05$, $n = 100$, $k = 7$, and $n - (k + 1) = 92$, we will reject H_0 if $t > t_{.05}$, where (because the degrees of freedom, 92, of t is so large) $t_{.05} \approx z_{.05} = 1.645$. Thus, we reject H_0 if $t > 1.645$.

Since the test statistic exceeds 1.645 (and has an observed significance level of $.0001/2 = .00005$), we find evidence that the mean salaries of newly hired male executives do exceed the corresponding mean for females.

Step 5 The culmination of the modeling effort is to use it for estimation and/or prediction. Suppose a firm is trying to determine fair compensation for an executive with the characteristics shown in Table 12.4. The least squares model can be used to obtain a predicted value for the logarithm of salary. That is,

TABLE 12.4 **Values of Independent Variables for an Executive**

$x_1 = 12$ years of experience
$x_2 = 16$ years of education
$x_3 = 0$ (female)
$x_4 = 400$ employees supervised
$x_5 = \$160.1$ million (the firm's asset value)
$x_6 = x_1^2 = 144$
$x_7 = x_3 x_4 = 0$

*This test assumes that x_4, the number of employees supervised, is held fixed at 0. A better way to test the discrimination hypothesis is to test all β's associated with the dummy variable x_3 simultaneously, that is , test $H_0: \beta_3 = \beta_7 = 0$. This test on a partial set of β's is discussed in Chapter 13.

$$\hat{y} = \hat{\beta}_0 + \hat{\beta}_1(12) + \hat{\beta}_2(16) + \hat{\beta}_3(0) + \hat{\beta}_4(400) + \hat{\beta}_5(160.1) + \hat{\beta}_6(144) + \hat{\beta}_7(0)$$

This predicted value, $\hat{y} = 11.298$, is given at the bottom of Figure 12.18. The corresponding 95% prediction interval, shaded on the printout, is 11.203 to 11.392. To predict the salary of an executive with these characteristics we take the antilog of these values. That is, the predicted salary is $e^{11.298} = \$80,700$ (rounded to the nearest hundred) and the 95% prediction interval is from $e^{11.203}$ to $e^{11.392}$ (from \$73,400 to \$88,600). Thus, an executive with the characteristics in Table 12.4 should be paid between \$73,400 and \$88,600 to be consistent with the sample data.

Predicting the Price of Vintage Red Bordeaux Wine

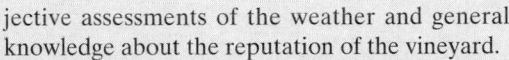

C A S E
S T U D Y
• 12.1 •

The vineyards in the Bordeaux region of France are known for producing excellent red wines. Wine experts agree that weather during the grape growing season is critical to producing good wines. The best vineyards in Europe are usually located near a large body of water (to delay spring frosts, prolong fall ripening, and reduce temperature fluctuations), on a slope with a southern exposure to the sun (to enhance spring ripening), and in a soil with good drainage (to overcome the dilution of the grapes that accompanies fall rains). The seaport city of Bordeaux, located on the Garonne River, meets all these desired conditions. Consequently, red Bordeaux wines are some of the most expensive wines in the world.

Usually, the more a wine ages, the more expensive it is. This is because young wines—even young wines produced from the best Bordeaux vineyards—are astringent and many wine drinkers find astringent wines unpalatable. As these wines age, they lose their astringency. Since older wines taste better than younger wines, they are deemed more valuable. Thus, both wine sellers and wine drinkers have an incentive to store young wines until they mature.

The combination of factors—the uncertainty of the weather during the growing season, the phenomenon that wine tastes better with age, and the fact that some vineyards produce better wines than others—encourages speculation concerning the value of a case of wine produced by a certain vineyard during a certain year (or vintage). As a result, many wine experts attempt to predict the auction price of a case of wine (on the London market). Trade magazines such as *Wine* and *The Wine Spectator* provide these opinions regularly, based on sub-jective assessments of the weather and general knowledge about the reputation of the vineyard.

Recently, a newsletter titled *Liquid Assets: The International Guide to Fine Wines* introduced a quantitative approach to predicting wine prices. The method, which employs statistics and multiple regression analysis to analyze wine prices, has sent the wine trade press into a frenzy. The front page headline in the *New York Times* (Mar. 4, 1995) read, "Wine Equation Puts Some Noses Out of Joint." In the article, the most influential wine critic in America called the approach "a Neanderthal way of looking at wine." According to Britain's *Wine* magazine, "the formula's self-evident silliness invited disrespect." Why has the use of multiple regression to predict wine prices engendered such controversy? One reason is simply that the theory behind the method is misunderstood. For example, the *Wine Spectator* condemned the multiple regression model because "the predictions come exactly true only 3 times in the 27 vintages since 1961 that were calculated, even though the formula was specifically designed to fit price data that already existed. The predicted prices are both under and over the actual prices." Obviously, the *Wine Spectator* does not understand that least squares regression proposes a probabilistic (not a deterministic) model that minimizes SSE and yields an average prediction error of 0.

The publishers of *Liquid Assets* discussed the multiple regression approach to predicting the London auction price of red Bordeaux wine in *Chance* (Fall 1995). The natural logarithm of the price y (in dollars) of a case containing a dozen bottles of red wine* was modeled as a function of weather during growing season

continued

---

*The price of a case is an index based on the wines of several Bordeaux vineyards. The bottles in the case were deliberately selected to represent the most expensive wines together with a selection of less expensive wines.

Case Study 12.1 continued

and age of vintage using data collected for the vintages of 1952–1980.* The natural log of price, denoted $\ln(y)$, was selected in order to measure increases (or decreases) in price as a percentage, as stated in Section 12.7. Three models were proposed:

Model 1: $\ln(y) = \beta_0 + \beta_1 x_1 + \epsilon$

where

  x_1 = Age of the vintage (year)

Model 2: $\ln(y) = \beta_0 + \beta_1 x_1 + \beta_2 x_2 + \beta_3 x_3 + \beta_4 x_4 + \epsilon$

where

  x_1 = Age of the vintage (year)

  x_2 = Average temperature (°C) over growing season (Apr.–Sept.)

  x_3 = Rainfall (cm) in September and August

  x_4 = Rainfall (cm) in the months preceding the vintage (Oct.–Mar.)

Model 3: $\ln(y) = \beta_0 + \beta_1 x_1 + \beta_2 x_2 + \beta_3 x_3 + \beta_4 x_4 + \beta_5 x_5 + \epsilon$

where

  x_1 = Age of the vintage (year)

  x_2 = Average temperature (°C) over growing season (Apr.–Sept.)

  x_3 = Rainfall (cm) in September and August

  x_4 = Rainfall (cm) in the months preceding the vintage (Oct.–Mar.)

  x_5 = Average temperature (°C) in September

The results of the regressions are summarized in Table 12.5.

Focus

a. Which of the three models would you use to predict red Bordeaux wine prices? Explain.
b. Interpret R^2 and s for the model you selected, part a.
c. Conduct a t-test for each of the β parameters in the model you selected, part a. Interpret the results.
d. When $\ln(y)$ is used as a dependent variable, the antilogarithm of a β coefficient minus 1, that is, $e^{\beta_i} - 1$, represents the percentage change in y for every 1-unit increase in the associated x-value.[†] Use this information to interpret the β-estimates of the model you selected in part a.

TABLE 12.5   Regressions of ln(Price) of Red Bordeaux Wine (Case Study 12.1)

Independent Variables in Model	BETA ESTIMATES (STANDARD ERRORS)		
	Model 1	Model 2	Model 3
Vintage (x_1)	.0354 (.0137)	.0238 (.00717)	.0240 (.00747)
Growing season temperature (x_2)	—	.616 (.0952)	.608 (.116)
Sept./Aug. rain (x_3)	—	−.00386 (.00081)	−.00380 (.00095)
Pre-rain (x_4)	—	.0001173 (.000482)	.00115 (.000505)
Sept. temperature (x_5)	—	—	.00765 (.0565)
R^2	.212	.828	.828
s	.575	.287	.293

Source: Ashenfelter, O., Ashmore, D., and LaLonde, R. "Bordeaux wine vintage quality and weather." *Chance*, Vol. 8, No. 4, Fall 1995, p. 116 (Table 2).

*The 1954 and 1956 vintages were excluded because they are now rarely sold.

[†]The result is derived by expressing the percentage change in price y as $(y_1 - y_0)/y_0$, where y_1 = the value of y when, say, $x = 1$, and y_0 = the value of y when $x = 0$. Now let $y^* = \ln(y)$ and assume the model is $y^* = \beta_0 + \beta_1 x$. Then

$$y = e^{y^*} = e^{\beta_0} e^{\beta_1 x} = \begin{cases} e^{\beta_0} & \text{when } x = 0 \\ e^{\beta_0} e^{\beta_1} & \text{when } x = 1 \end{cases}$$

Substituting, we have

$$\frac{y_1 - y_0}{y_0} = \frac{e^{\beta_0} e^{\beta_1} - e^{\beta_0}}{e^{\beta_0}} = e^{\beta_1} - 1$$

12.8 RESIDUAL ANALYSIS: CHECKING THE REGRESSION ASSUMPTIONS

When we apply regression analysis to a set of data, we never know for certain whether the assumptions of Section 12.3 are satisfied. How far can we deviate from the assumptions and still expect regression analysis to yield results that will have the reliability stated in this chapter? How can we detect departures (if they exist) from the assumptions of Section 12.3 and what can we do about them? We provide some partial answers to these questions in this section and direct you to further discussion in the next chapter.

Remember from Section 12.3 that

$$y = E(y) + \epsilon$$

where the expected value $E(y)$ of y for a given set of values of $x_1, x_2, ..., x_k$ is

$$E(y) = \beta_0 + \beta_1 x_1 + \beta_2 x_2 + \cdots + \beta_k x_k$$

and ϵ is a random error. The first assumption we made was that the mean value of the random error for *any* given set of values of $x_1, x_2, ..., x_k$ is $E(\epsilon) = 0$. One consequence of this assumption is that the mean $E(y)$ for a specific set of values of $x_1, x_2, ..., x_k$ is

$$E(y) = \beta_0 + \beta_1 x_1 + \beta_2 x_2 + \cdots + \beta_k x_k$$

That is,

$$
y \;\;=\;\; \underbrace{E(y)}_{\substack{\text{Mean value of } y \\ \text{for specific values} \\ \text{of } x_1, x_2, ..., x_k}} \;+\; \underbrace{\epsilon}_{\substack{\text{Random} \\ \text{error}}}
$$

The second consequence of the assumption is that the least squares estimators of the model parameters, $\beta_0, \beta_1, \beta_2, ..., \beta_k$, will be unbiased regardless of the remaining assumptions that we attribute to the random errors and their probability distributions.

The properties of the sampling distributions of the parameter estimators $\hat{\beta}_0, \hat{\beta}_1, ..., \hat{\beta}_k$ will depend on the remaining assumptions that we specify concerning the probability distributions of the random errors. Recall that we assumed that for any given set of values of $x_1, x_2, ..., x_k$, ϵ has a normal probability distribution with mean equal to 0 and variance equal to σ^2. Also, we assumed that the random errors are probabilistically independent.

It is unlikely that these assumptions are ever satisfied exactly in a practical application of regression analysis. Fortunately, experience has shown that least squares regression analysis produces reliable statistical tests, confidence intervals, and prediction intervals as long as the departures from the assumptions are not too great. In this section we present some methods for determining whether the data indicate significant departures from the assumptions.

Because the assumptions all concern the random error component, ϵ, of the model, the first step is to estimate the random error. Since the actual random error associated with a particular value of y is the difference between the actual y value and its unknown mean, we estimate the error by the difference between the actual y value and the *estimated* mean. This estimated error is called the **regression residual**, or simply the **residual**, and is denoted by $\hat{\epsilon}$. The actual error ϵ and residual $\hat{\epsilon}$ are shown in Figure 12.19.

FIGURE 12.19

Actual random error and
regression residual

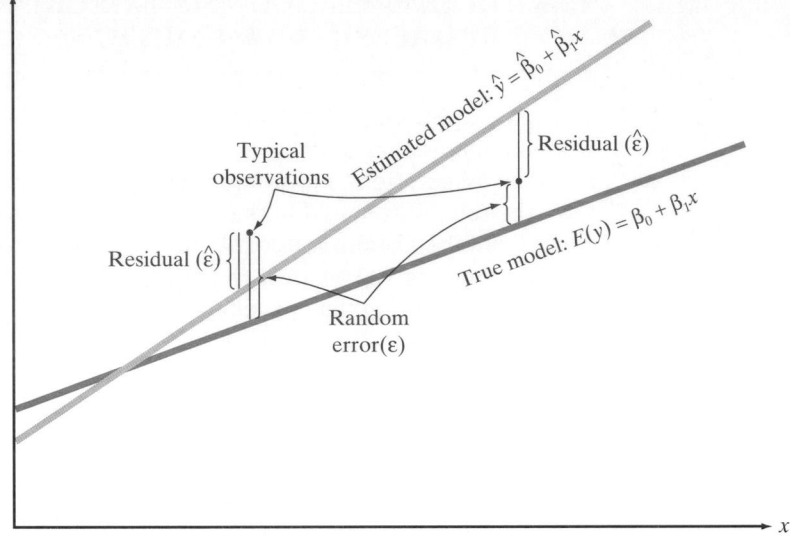

$$\text{Actual random error} = \epsilon$$
$$= (\text{Actual } y \text{ value}) - (\text{Mean of } y)$$
$$= y - E(y) = y - (\beta_0 + \beta_1 x_1 + \beta_2 x_2 + \cdots + \beta_k x_k)$$

$$\text{Estimated random error (residual)} = \hat{\epsilon}$$
$$= (\text{Actual } y \text{ value}) - (\text{Estimated mean of } y)$$
$$= y - \hat{y} = y - (\hat{\beta}_0 + \hat{\beta}_1 x_1 + \hat{\beta}_2 x_2 + \cdots + \hat{\beta}_k x_k)$$

Since the true mean of y (that is, the true regression model) is not known, the actual random error cannot be calculated. However, because the residual is based on the estimated mean (the least squares regression model), it can be calculated and used to estimate the random error and to check the regression assumptions. Such checks are generally referred to as **residual analyses**. Some useful properties of residuals are given in the next box.

Properties of Regression Residuals

1. A residual is equal to the difference between the observed y value and its estimated (regression) mean:

$$\text{Residual} = y - \hat{y}$$

2. The mean of the residuals is equal to 0. This property follows from the fact that the sum of the differences between the observed y values and their least squares predicted ($\hat{y}$) values is equal to 0.

$$\sum(\text{Residuals}) = \sum(y - \hat{y}) = 0$$

3. The standard deviation of the residuals is equal to the standard deviation of the fitted regression model, s. This property follows from the fact that the sum of the squared residuals is equal to SSE, which when divided by the error degrees of freedom is equal to the variance of the fitted regression model, s^2. The square root of the variance is both the standard deviation of the residuals and the standard deviation of the regression model.

$$\sum(\text{Residuals})^2 = \sum(y - \hat{y})^2 = \text{SSE}$$

$$s = \sqrt{\frac{\sum(\text{Residuals})^2}{n - (k + 1)}} = \sqrt{\frac{\text{SSE}}{n - (k + 1)}}$$

The following examples show how the analysis of regression residuals can be used to verify the assumptions associated with the model and to improve the model when the assumptions do not appear to be satisfied. Although the residuals can be calculated and plotted by hand, we rely on the computer for these tasks in the examples and exercises. Most statistical computer packages now include residual analyses as a standard component of their regression modeling programs.

EXAMPLE 12.4

The data for the home size–electrical usage example used throughout this chapter are repeated in Table 12.6. SAS printouts for a straight-line model and a quadratic model fitted to the data are shown in Figures 12.20a and 12.20b, respectively. The residuals from these models are shaded in the printouts. The residuals are then plotted on the vertical axis against the variable x, size of home, on the horizontal axis in Figures 12.21a and 12.21b, respectively.

a. Verify that each residual is equal to the difference between the observed y value and the estimated mean value, $\hat{y}$.
b. Analyze the residual plots.

Solution

a. For the straight-line model the residual is calculated for the first y value as follows:

$$\text{Residual} = (\text{Observed } y \text{ value}) - (\text{Estimated mean})$$
$$= y - \hat{y} = 1{,}182 - 1{,}275.9 = -93.9$$

where the estimated mean is the first number in the column labeled **Predict Value** (shaded) on the SAS printout in Figure 12.20a. Similarly, the residual for the first y value using the quadratic model (Figure 12.20b) is

$$\text{Residual} = 1{,}182 - 1{,}129.6 = 52.4$$

TABLE 12.6   Home Size—Electrical Usage Data

Size of Home x (sq. ft)	Monthly Usage y (kilowatt-hours)
1,290	1,182
1,350	1,172
1,470	1,264
1,600	1,493
1,710	1,571
1,840	1,711
1,980	1,804
2,230	1,840
2,400	1,956
2,930	1,954

FIGURE 12.20A

SAS printout for electrical usage example: Straight-line model

Dep Variable: Y

Analysis of Variance

Source	DF	Sum of Squares	Mean Square	F Value	Prob>F
Model	1	703957.18342	703957.18342	39.536	0.0002
Error	8	142444.91658	17805.61457		
C Total	9	846402.10000			

Root MSE	133.43766	R-Square	0.8317	
Dep Mean	1594.70000	Adj R-Sq	0.8107	
C. V.	8.36757			

Parameter Estimates

Variable	DF	Parameter Estimate	Standard Error	T for H0: Parameter=0	Prob > |T|
INTERCEP	1	578.927752	166.96805715	3.467	0.0085
X	1	0.540304	0.08592981	6.288	0.0002

Obs	Y	Predict Value	Residual
1	1182.0	1275.9	-93.9204
2	1172.0	1308.3	-136.3
3	1264.0	1373.2	-109.2
4	1493.0	1443.4	49.5852
5	1571.0	1502.8	68.1517
6	1711.0	1573.1	137.9
7	1804.0	1648.7	155.3
8	1840.0	1783.8	56.1935
9	1956.0	1875.7	80.3417
10	1954.0	2162.0	-208.0

Sum of Residuals   0
Sum of Squared Residuals   142444.9166

Both residuals agree (after rounding) with the first values given in the column labeled **Residual** in Figures 12.20a and 12.20b, respectively. Although the residuals both correspond to the same observed *y* value, 1,182, they differ because the estimated mean value changes depending on whether the straight-line model or quadratic model is used. Similar calculations produce the remaining residuals.

b. The plot of the residuals for the straight-line model (Figure 12.21a) reveals a nonrandom pattern. The residuals exhibit a curved shape, with the residuals for the small values of *x* below the horizontal 0 (mean of the residuals) line, the residuals corresponding to the middle values of *x* above the 0 line, and the residual for the largest value of *x* again below the 0 line. The indication is that the mean value of the random error ϵ *within* each of these ranges of *x* (small, medium, large) may not be equal to 0. Such a pattern usually indicates that curvature needs to be added to the model.

When the second-order term is added to the model, the nonrandom pattern disappears. In Figure 12.21b, the residuals appear to be randomly distributed around the 0 line, as expected. Note, too, that the ± 2 standard deviation lines are at about ± 95 on the quadratic residual plot, compared to (about) ± 275 on the straight-line plot. The implication is that the quadratic model provides a considerably better model for predicting electrical usage, verifying our conclusions from previous analyses in this chapter. ▲

FIGURE 12.20B

SAS printout for electrical usage example: Quadratic model

Dep Variable: Y

Analysis of Variance

Source	DF	Sum of Squares	Mean Square	F Value	Prob>F
Model	2	831069.54637	415534.77319	189.710	0.0001
Error	7	15332.55363	2190.36480		
C Total	9	846402.10000			

Root MSE	46.80133	R-Square	0.9819	
Dep Mean	1594.70000	Adj R-Sq	0.9767	
C. V.	2.93480			

Parameter Estimates

Variable	DF	Parameter Estimate	Standard Error	T for H0: Parameter=0	Prob > |T|
INTERCEP	1	-1216.143887	242.80636850	-5.009	0.0016
X	1	2.398930	0.24583560	9.758	0.0001
XSQ	1	-0.000450	0.00005908	-7.618	0.0001

Obs	Y	Predict Value	Residual
1	1182.0	1129.6	52.4359
2	1172.0	1202.2	-30.2136
3	1264.0	1337.8	-73.7916
4	1493.0	1470.0	22.9586
5	1571.0	1570.1	0.9359
6	1711.0	1674.2	36.7685
7	1804.0	1769.4	34.5998
8	1840.0	1895.5	-55.4654
9	1956.0	1949.1	6.9431
10	1954.0	1949.2	4.8287

Sum of Residuals   -2.27374E-12
Sum of Squared Residuals   15332.5536

FIGURE 12.21A

Residual plot for electrical usage example: Straight-line model

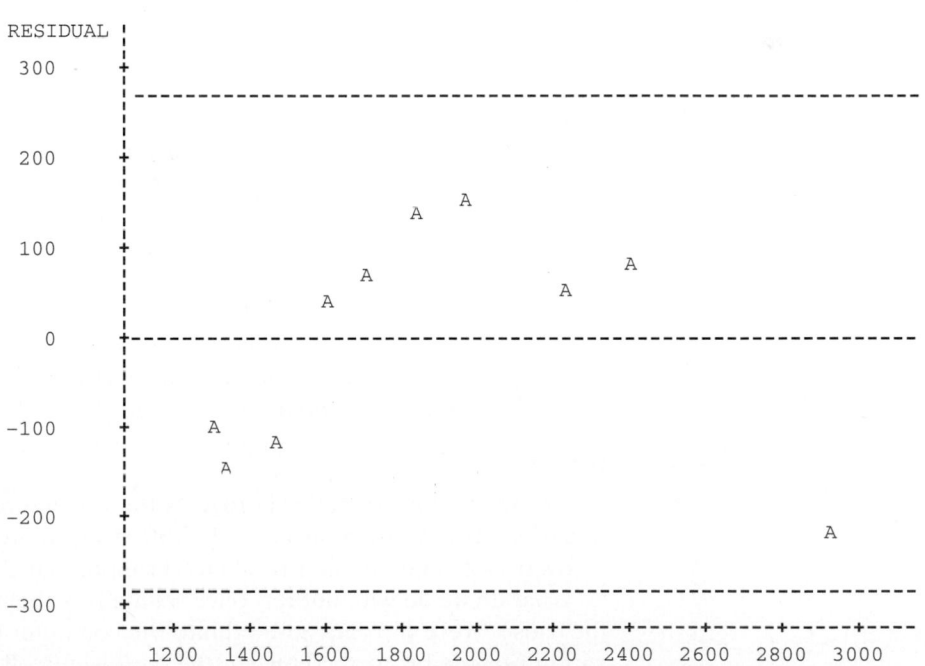

Plot of RESIDUAL*X   Legend: A = 1 obs, B = 2 obs, etc.

FIGURE 12.21B

Residual plot for electrical
usage example: Quadratic
model

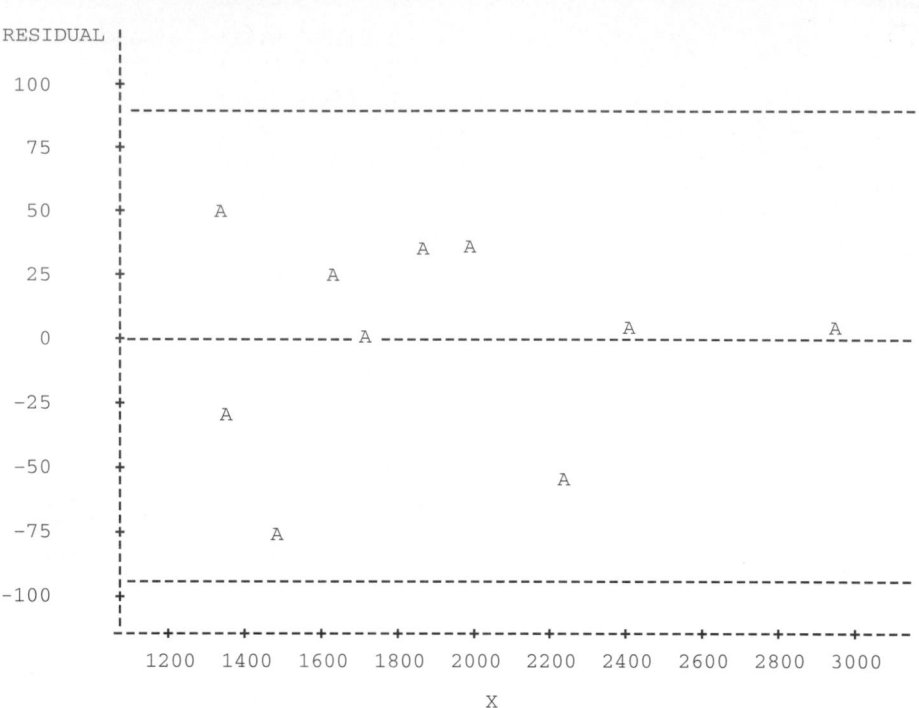

Residual analyses are also useful for detecting one or more observations that deviate significantly from the regression model. We expect approximately 95% of the residuals to fall within 2 standard deviations of the 0 line, and all or almost all of them to lie within 3 standard deviations of their mean of 0. Residuals that are extremely far from the 0 line, and disconnected from the bulk of the other residuals, are called **outliers**, which should receive special attention from the regression analyst.

EXAMPLE 12.5

The data for the grandfather clock example used throughout this chapter are repeated in Table 12.7, with one important difference: The auction price of the clock at the top of the second column has been changed from \$2,131 to \$1,131 (shaded in Table 12.7). The interaction model

$$E(y) = \beta_0 + \beta_1 x_1 + \beta_2 x_2 + \beta_3 x_1 x_2$$

is again fitted to these (modified) data, with the MINITAB printout shown in Figure 12.22. The residuals are shown shaded in the printout and then plotted against the number of bidders, x_2, in Figure 12.23. Analyze the residual plot.

Solution

The residual plot dramatically reveals the one altered measurement. Note that one of the two residuals at $x_2 = 14$ bidders falls more than 3 standard deviations below 0. Note that no other residual falls more than 2 standard deviations from 0.

What do we do with outliers once we identify them? First, we try to determine the cause. Were the data entered into the computer incorrectly? Was the observation recorded incorrectly when the data were collected? If so, we correct the

TABLE 12.7   Auction Price Data

Age x_1	Number of Bidders x_2	Auction Price y	Age x_1	Number of Bidders x_2	Auction Price y
127	13	$1,235	170	14	$1,131
115	12	1,080	182	8	1,550
127	7	845	162	11	1,884
150	9	1,522	184	10	2,041
156	6	1,047	143	6	845
182	11	1,979	159	9	1,483
156	12	1,822	108	14	1,055
132	10	1,253	175	8	1,545
137	9	1,297	108	6	729
113	9	946	179	9	1,792
137	15	1,713	111	15	1,175
117	11	1,024	187	8	1,593
137	8	1,147	111	7	785
153	6	1,092	115	7	744
117	13	1,152	194	5	1,356
126	10	1,336	168	7	1,262

observation and rerun the program. Another possibility is that the observation is not representative of the conditions we are trying to model. For example, in this case the low price may be attributable to extreme damage to the clock, or to a clock of inferior quality compared to the others. In these cases we probably would exclude the observation from the analysis. In many cases you may not be able to determine the cause of the outlier. Even so, you may want to rerun the regression analysis excluding the outlier in order to assess the effect of that observation on the results of the analysis.

Figure 12.24 shows the printout when the outlier observation is excluded from the grandfather clock analysis, and Figure 12.25 shows the new plot of the residuals against the number of bidders. Now only one of the residuals lies beyond 2 standard deviations from 0, and none of them lies beyond 3 standard deviations. Also, the model statistics indicate a much better model without the outlier. Most notably, the standard deviation (s) has decreased from 200.6 to 85.83, indicating a model that will provide more precise estimates and predictions (narrower confidence and prediction intervals) for clocks that are similar to those in the reduced sample. But remember that if the outlier is removed from the analysis when in fact it belongs to the same population as the rest of the sample, the resulting model may provide misleading estimates and predictions.   ▲

Outlier analysis is another example of testing the assumption that the expected (mean) value of the random error component is 0, since this assumption is in doubt for the error terms corresponding to the outliers. The last example in this section checks the assumption of the normality of the random error component.

EXAMPLE 12.6

Refer to Example 12.5. Use a stem-and-leaf display (Section 2.2) to plot the frequency distribution of the residuals in the grandfather clock example, both before and after the outlier residual is removed. Analyze the plots and determine whether the assumption of a normal distribution error is reasonable.

FIGURE 12.22

MINITAB printout for grandfather clock example with altered data

The regression equation is
PRICE = - 513 + 8.17 AGE = 19.9 BIDDERS + 0.320 AGE-BID

Predictor	Coef	Stdev	t-ratio	p
Constant	-512.8	665.9	-0.77	0.448
AGE	8.165	4.585	1.78	0.086
BIDDERS	19.89	67.44	0.29	0.770
AGE-BID	0.3196	0.4790	0.67	0.510

s = 200.6   R-sq = 72.9%   R-sq(adj) = 70.0%

Analysis of Variance

SOURCE	DF	SS	MS	F	p
Regression	3	3033587	1011196	25.13	0.000
Error	28	1126703	40239		
Total	31	4160289			

Obs.	AGE	PRICE	Fit	Stdev.Fit	Residual	St.Resid
1	127	1235.0	1310.4	59.3	-75.4	-0.39
2	115	1080.0	1105.9	62.1	-25.9	-0.14
3	127	845.0	947.5	61.1	-102.5	-0.54
4	150	1522.0	1322.5	37.1	199.5	1.01
5	156	1047.0	1179.5	60.3	-132.5	-0.69
6	182	1979.0	1831.9	82.9	147.1	0.81
7	156	1822.0	1598.0	61.9	224.0	1.17
8	132	1253.0	1185.8	39.7	67.2	0.34
9	137	1297.0	1178.9	39.0	118.1	0.60
10	113	946.0	913.9	58.6	32.1	0.17
11	137	1713.0	1561.0	78.4	152.0	0.82
12	117	1024.0	1072.6	53.1	-48.6	-0.25
13	137	1147.0	1115.2	44.3	31.8	0.16
14	153	1092.0	1149.2	59.0	-57.2	-0.30
15	117	1152.0	1187.2	69.7	-35.2	-0.19
16	126	1336.0	1117.6	43.4	218.4	1.12
17	170	1131.0	1914.4	116.7	-783.4	-4.80R
18	182	1550.0	1597.7	62.8	-47.7	-0.25
19	162	1884.0	1598.3	57.0	285.7	1.49
20	184	2041.0	1776.6	70.7	264.4	1.41
21	143	845.0	1048.4	58.9	-203.4	-1.06
22	159	1483.0	1421.8	40.6	61.2	0.31
23	108	1055.0	1130.7	97.9	-75.7	-0.43
24	175	1545.0	1522.7	55.4	22.3	0.12
25	108	729.0	695.5	99.6	33.5	0.19
26	179	1792.0	1642.7	57.6	149.3	0.78
27	111	1175.0	1224.0	107.2	-49.0	-0.29
28	187	1593.0	1651.3	68.6	-58.3	-0.31
29	111	785.0	781.1	80.9	3.9	0.02
30	115	744.0	822.7	75.5	-78.7	-0.42
31	194	1356.0	1480.7	133.6	-124.7	-0.83 X
32	168	1262.0	1374.0	57.7	-112.0	-0.58

R denotes an obs. with a large st. resid.
X denotes an obs. whose X value gives it large influence.

FIGURE 12.23

Residual plot against number of bidders

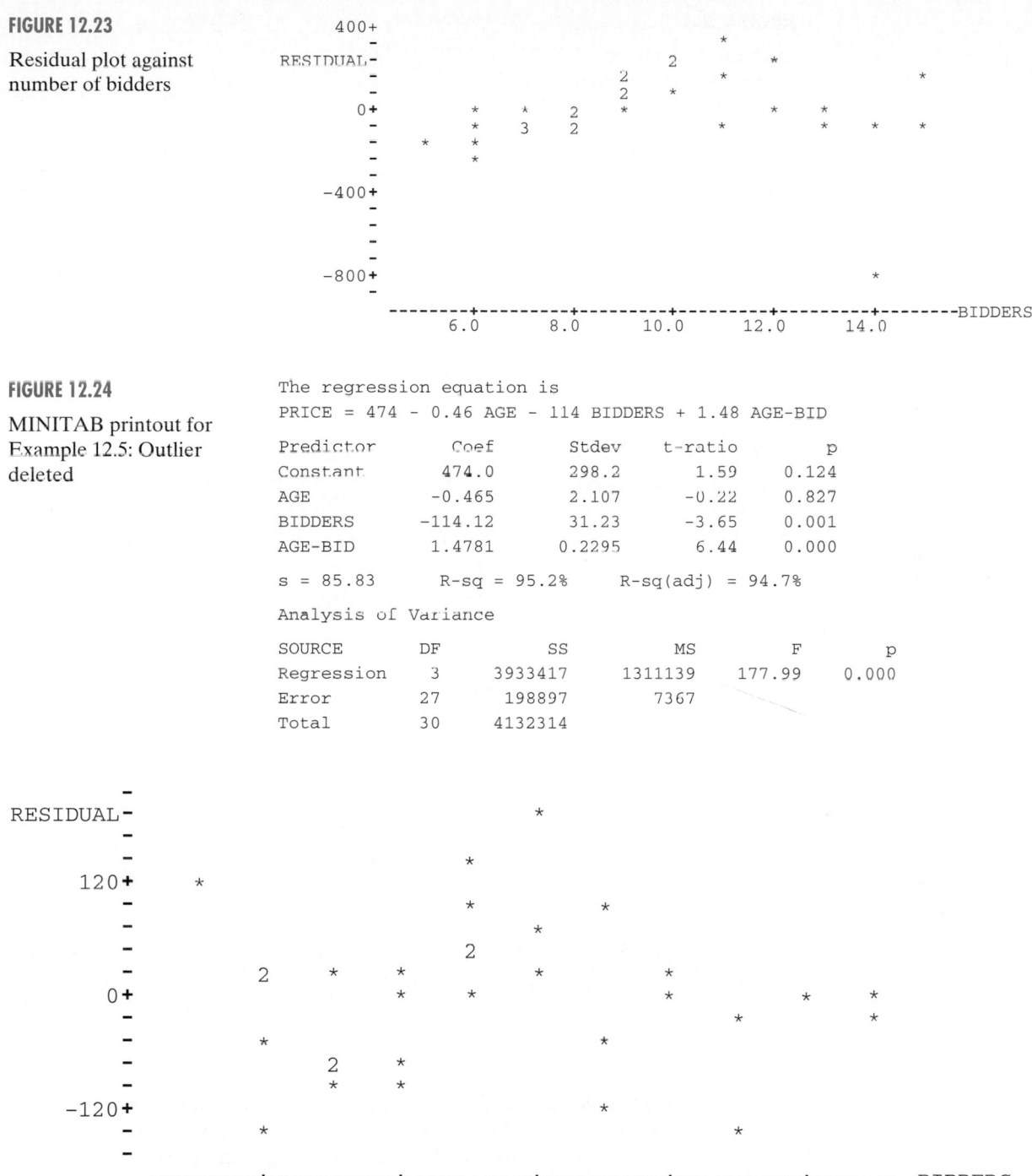

FIGURE 12.24

MINITAB printout for Example 12.5: Outlier deleted

The regression equation is
PRICE = 474 - 0.46 AGE - 114 BIDDERS + 1.48 AGE-BID

Predictor	Coef	Stdev	t-ratio	p
Constant	474.0	298.2	1.59	0.124
AGE	-0.465	2.107	-0.22	0.827
BIDDERS	-114.12	31.23	-3.65	0.001
AGE-BID	1.4781	0.2295	6.44	0.000

s = 85.83   R-sq = 95.2%   R-sq(adj) = 94.7%

Analysis of Variance

SOURCE	DF	SS	MS	F	p
Regression	3	3933417	1311139	177.99	0.000
Error	27	198897	7367		
Total	30	4132314			

FIGURE 12.25   MINITAB residual plot for Example 12.5: Outlier deleted

FIGURE 12.26A

Stem-and-leaf display for grandfather clock example: Outlier included

```
STEM-AND-LEAF DISPLAY OF RESIDUAL
LEAF DIGIT UNIT =  10.0000
1 2 REPRESENTS 120.

       STEM  LEAF
   1    -7   8
   1    -6
   1    -5
   1    -4
   1    -3
   1    -2
   6    -1   93210
  16    -0   7775544432
  16     0   0233366
   9     1   14459
   4     2   1268
```

FIGURE 12.26B

Stem-and-leaf display for grandfather clock example: Outlier excluded

```
STEM-AND-LEAF DISPLAY OF RESIDUAL
LEAF DIGIT UNIT =  10.0000
1 2 REPRESENTS 120.

   3    -1*  331
   9    -0.  987765
  (7)   -0*  4321000
  15    +0*  011223344
   6    +0.  79
   4     1*  004
   1     1.  9
```

Solution

The stem-and-leaf displays for the two sets of residuals are constructed using MINITAB and are shown in Figure 12.26.* Note that the outlier appears to skew the frequency distribution in Figure 12.26a, whereas the stem-and-leaf display in Figure 12.26b appears to be more mound-shaped. Although the displays do not provide formal statistical tests of normality, they do provide a descriptive display. Relative frequency histograms can also be used to check the normality assumption. In this example the normality assumption appears to be more plausible after the outlier is removed. Consult the references for methods to conduct statistical tests of normality using the residuals. ▲

Residual analysis is a useful tool for the regression analyst, not only to check the assumptions, but also to provide information about how the model can be improved. A summary of the residual analyses presented in this section to check the assumption that the random error ϵ is normally distributed with mean 0 is presented in the next box.

---

*Recall that the left column of the MINITAB printout shows the number of measurements at least as extreme as the stem. In Figure 12.26a, for example, the 6 corresponding to the STEM = −1 means that six measurements are less than or equal to −100. If one of the numbers in the leftmost column is enclosed in parentheses, the number in parentheses is the number of measurements in that row, and the median is contained in that row.

Steps in a Residual Analysis

1. Calculate and plot the residuals against each of the independent variables, preferably with the assistance of a statistical software program.

2. Analyze each plot, looking for curvature—either a mound or bowl shape. Both shapes are distinguished by groups of residuals at the low and high values of the x variable on one side of the 0 line, with the residuals for the medium x values on the opposite side of the 0 line. This shape signals the need for a curvature term in the model. Try a second-order term in the variable against which the residuals are plotted.

3. Examine the residual plots for outliers. Draw lines on the residual plots at 2- and 3-standard-deviation distances below and above the 0 line. Examine residuals outside the 3-standard-deviation lines as potential outliers, and check to see that approximately 5% of the residuals exceed the 2-standard-deviation lines. Determine whether each outlier can be explained as an error in data collection or transcription, or corresponds to a member of a population different from that of the remainder of the sample, or simply represents an unusual observation. If the observation is determined to be an error, fix it or remove it. Even if you can't determine the cause, you may want to rerun the regression analysis without the observation to determine its effect on the analysis.

4. Plot a frequency distribution of the residuals, using a stem-and-leaf display or a histogram. Check to see if obvious departures from normality exist. Extreme skewness of the frequency distribution may indicate the need for a transformation of the dependent variable. This topic is beyond the scope of this book, but you can find it in the references.

Analyzing Water/Oil Mixtures in High Electric Fields

In the oil industry, water mixes with crude oil during production and transportation. The organic properties of oil prevent it from dissolving in an inorganic medium; rather, tiny oil particles are suspended within the water. This water and oil (w/o) suspension is called an emulsion.

Chemists have found that the oil can be extracted from the w/o emulsion electrically. In a high electric field, the (lighter) emulsified droplets are enlarged while the (heavier) water settles out of the mix gravitationally. Researchers at the University of Bergen (Norway) conducted a series of experiments to study the factors that influence the voltage required to separate the water from the oil in w/o emulsions (*Journal of Colloid and Interface Science,* Aug. 1995). The seven independent variables investigated in the study are described below. Each variable was measured at two levels—a "low" level and a "high" level.

x_1: Volume fraction of disperse phase (as a percentage of weight); Low = 40%, High = 80%

C A S E
S T U D Y
• 12.2 •

x_2: Salinity of emulsion (as a percentage of weight); Low = 1%, High − 4%

x_3: Temperature of emulsion (in °C); Low = 4°, High = 23°

x_4: Time delay after emulsification (in hours); Low = .25 hour, High = 24 hours

x_5: Concentration of surface-active agent, or "surfactant" (as a percentage of weight); Low = 2%, High = 4%

x_6: Ratio of two chemicals (Span and Triton) used as surfactants; Low = .25, High = .75

x_7: Amount of solid particles added (as a percentage of weight); Low = .5%, High = 2%

Sixteen w/o emulsions were prepared using different combinations of the independent variables listed above; then each emulsion was exposed to a high electric field. In addition, three w/o emulsions were tested when all independent variables were set to 0. For all 19 emulsions, the amount of voltage (kilovolts per centimeter) where

continued

Case Study 12.2 continued

the first sign of macroscopic activity is observed was measured; this value represents the dependent variable, y. The data for the study are given in Table 12.8.

Focus

a. Propose a model for y as a function of all seven independent variables. Assume that a linear relationship exists between y and x_i, $i = 1, 2, ..., 7$.

b. Use a statistical software package to fit the model to the data in Table 12.8.

c. Fully interpret the results of the regression. Part of the analysis should include an interpretation of the β estimates.

d. According to the researchers, the model predicts a negative value for the voltage y for experiment #14. Verify this result.

e. The researchers state that the result, part d, "is physically not acceptable, and a model with interaction terms must be proposed." The model the researchers selected is $E(y) = \beta_0 + \beta_1 x_1 + \beta_2 x_2 + \beta_3 x_5 + \beta_4 x_1 x_2$ + $\beta_5 x_1 x_5$. Note that the model includes interaction between disperse phase volume (x_1) and salinity (x_2) as well as interaction between disperse phase volume (x_1) and surfactant concentration (x_5). Discuss how these interaction terms affect the hypothetical relationship between y and x_1. Draw a sketch to support your answer.

f. Fit the interaction model, part e, to the data. Does the model appear to fit the data better than the model, part a? Explain.

g. Interpret the β estimates of the interaction model, part e.

h. The researchers concluded that "in order to break an emulsion with the lowest possible voltage, the volume fraction of the disperse phase (x_1) should be high, while the salinity (x_2) and the amount of surfactant (x_5) should be low." Use this information and the interaction model to find a 95% prediction interval for this "low" voltage y. Interpret the interval.

TABLE 12.8   Data for W/O Emulsion Experiments (Case Study 12.2)

Experiment Number	Voltage y	Disperse Phase Volume x_1	Salinity x_2	Temperature x_3	Time Delay x_4	Surfactant Concentration x_5	S:T Ratio x_6	Solid Particles x_7
1	.64	40	1	4	.25	2	.25	.5
2	.80	80	1	4	.25	4	.25	2
3	3.20	40	4	4	.25	4	.75	.5
4	.48	80	4	4	.25	2	.75	2
5	1.72	40	1	23	.25	4	.75	2
6	.32	80	1	23	.25	2	.75	.5
7	.64	40	4	23	.25	2	.25	2
8	.68	80	4	23	.25	4	.25	.5
9	.12	40	1	4	24	2	.75	2
10	.88	80	1	4	24	4	.75	.5
11	2.32	40	4	4	24	4	.25	2
12	.40	80	4	4	24	2	.25	.5
13	1.04	40	1	23	24	4	.25	.5
14	.12	80	1	23	24	2	.25	2
15	1.28	40	4	23	24	2	.75	.5
16	.72	80	4	23	24	4	.75	2
17	1.08	0	0	0	0	0	0	0
18	1.08	0	0	0	0	0	0	0
19	1.04	0	0	0	0	0	0	0

Source: Førdedal, H., *et al.* "A multivariate analysis of W/O emulsions in high external electric fields as studied by means of dielectric time domain spectroscopy." *Journal of Colloid and Interface Science,* Vol. 173, No. 2, Aug. 1995, p. 398 (Table 2). Reprinted with permission of Academic Press, Inc.

12.9   SOME PITFALLS: ESTIMABILITY, MULTICOLLINEARITY, AND EXTRAPOLATION

You should be aware of several potential problems when constructing a prediction model for some response y. A few of the most important are discussed in this section.

Problem 1: Parameter Estimability

Suppose we want to fit a model relating annual crop yield y to the total expenditure for fertilizer, x. We propose the first-order model

$$E(y) = \beta_0 + \beta_1 x$$

Now suppose we have 3 years of data and \$1,000 is spent on fertilizer each year. The data are shown in Figure 12.27. You can see the problem: The parameters of the model cannot be estimated when all the data are concentrated at a single x value. Recall that it takes two points (x values) to fit a straight line. Thus, the parameters are not estimable when only one x is observed.

A similar problem would occur if we attempted to fit the quadratic model

$$E(y) = \beta_0 + \beta_1 x + \beta_2 x^2$$

to a set of data for which only one or two different x values were observed (see Figure 12.28). At least three different x values must be observed before a quadratic model can be fit to a set of data (that is, before all three parameters are estimable).

In general, the number of levels of observed x values must be one more than the order of the polynomial in x that you want to fit.

For controlled experiments, the researcher can select experimental designs that will permit estimation of the model parameters. Even when the values of the

FIGURE 12.27

Yield and fertilizer expenditure data: 3 years

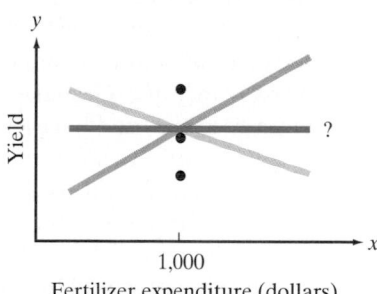

FIGURE 12.28

Only two x values observed—Quadratic model is not estimable

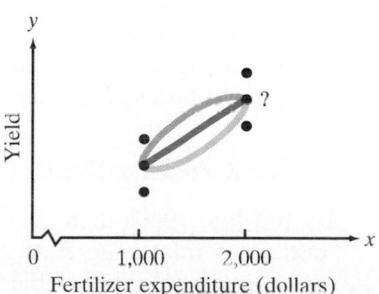

independent variables cannot be controlled by the researcher, the independent variables are almost always observed at a sufficient number of levels to permit estimation of the model parameters. When the computer program you use suddenly refuses to fit a model, however, the problem is likely to be inestimable parameters.

Problem 2: Multicollinearity

Often, two or more of the independent variables used in the model for $E(y)$ contribute redundant information. That is, the independent variables are correlated with each other. Suppose we want to construct a model to predict the gas mileage rating of a truck as a function of its load, x_1, and the horsepower, x_2, of its engine. In general, we would expect heavy loads to require greater horsepower and to result in lower mileage ratings. Thus, although both x_1 and x_2 contribute information for the prediction of mileage rating, some of the information is overlapping because x_1 and x_2 are correlated.

If the model

$$E(y) = \beta_0 + \beta_1 x_1 + \beta_2 x_2$$

were fit to a set of data, we might find that the t values for both $\hat{\beta}_1$ and $\hat{\beta}_2$ (the least squares estimates) are nonsignificant. However, the F-test for H_0: $\beta_1 = \beta_2 = 0$ would probably be highly significant. The tests may seem to produce contradictory conclusions, but really they do not. The t-tests indicate that the contribution of one variable, say $x_1 =$ Load, is not significant after the effect of $x_2 =$ Horsepower has been taken into account (because x_2 is also in the model). The significant F-test, on the other hand, tells us that at least one of the two variables is making a contribution to the prediction of y (that is, either β_1 or β_2, or both, differ from 0). In fact, both are probably contributing, but the contribution of one overlaps with that of the other.

When highly correlated independent variables are present in a regression model, the results are confusing. The researcher may want to include only one of the variables in the final model. One way of deciding which one to include is by using a technique called **stepwise regression** (Chapter 13). Generally, only one of a set of multicollinear independent variables is included in a stepwise regression model, since at each step every variable is tested in the presence of all the variables already in the model. For example, if at one step the variable Load is included as a significant variable in the prediction of the mileage rating, the variable Horsepower will probably never be added in a future step. Thus, if a set of independent variables is thought to be multicollinear, some screening by stepwise regression may be helpful.

Note that it would be fallacious to conclude that an independent variable x_1 is unimportant for predicting y *only* because it is not chosen by a stepwise regression procedure. The independent variable x_1 may be correlated with another one, x_2, that the stepwise procedure did select. The implication is that x_2 contributes *more* for predicting y (in the sample being analyzed), but it may still be true that x_1 alone contributes information for the prediction of y.

Problem 3: Prediction Outside the Experimental Region

By the late 1960s many research economists had developed highly technical models to relate the state of the economy to various economic indices and other independent variables. Many of these models were multiple regression

FIGURE 12.29

Using a regression model outside the experimental region

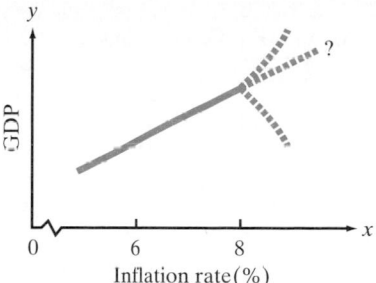

models, where, for example, the dependent variable y might be next year's growth in GDP and the independent variables might include this year's rate of inflation, this year's Consumer Price Index (CPI), etc. In other words, the model might be constructed to predict next year's economy using this year's knowledge.

Unfortunately, these models were almost all unsuccessful in predicting the recession in the early 1970s. What went wrong? One of the problems was that many of the regression models were used to predict y for values of the independent variables that were outside the region in which the model was developed. For example, the inflation rate in the late 1960s, when the models were developed, ranged from 6% to 8%. When the double-digit inflation of the early 1970s became a reality, some researchers attempted to use the same models to predict future growth in GDP. As you can see in Figure 12.29, the model may be very accurate for predicting y when x is in the range of experimentation, but the use of the model outside that range is a dangerous practice.

Problem 4: Correlated Errors

Another problem associated with using a regression model to predict a variable y based on independent variables $x_1, x_2, ..., x_k$ arises from the fact that the data are frequently **time series**. That is, the values of both the dependent and independent variables are observed sequentially over a period of time. The observations tend to be correlated over time, which in turn often causes the prediction errors of the regression model to be correlated. Thus, the assumption of independent errors is violated, and the model tests and prediction intervals are no longer valid. The solution to this problem is to construct a **time series model**; the interested reader should consult the references for details of time series analysis.

EXERCISES 12.29–12.34

Applying the Concepts

12.29   Chemical engineers at Tokyo Metropolitan University analyzed urban air specimens for the presence of low-molecular-weight dicarboxylic acid (*Environmental Science & Engineering,* Oct. 1993). The dicarboxylic acid (as a percentage of total carbon) and oxidant concentrations for 19 air speci- mens collected from urban Tokyo are listed in the table at the top of page 588. SAS printouts for the straight-line model relating dicarboxylic acid per- centage (y) to oxidant concentration (x) appear on pages 588–590. Conduct a complete residual analysis.

Dicarboxylic Acid (%)	Oxidant (ppm)	Dicarboxylic Acid (%)	Oxidant (ppm)
.85	78	.50	32
1.45	80	.38	28
1.80	74	.30	25
1.80	78	.70	45
1.60	60	.80	40
1.20	62	.90	45
1.30	57	1.22	41
.20	49	1.00	34
.22	34	1.00	25
.40	36		

Source: Kawamura, K., and Ikushima, K. "Seasonal changes in the distribution of dicarboxylic acids in the urban atmosphere." *Environmental Science & Technology,* Vol. 27, No. 10, Oct. 1993, p. 2232 (data extracted from Figure 4).

```
Dependent Variable: DICARBOX

                          Analysis of Variance

                           Sum of        Mean
        Source       DF    Squares      Square    F Value   Prob>F

        Model         1    2.41362     2.41362    17.080    0.0007
        Error        17    2.40234     0.14131
        C Total      18    4.81597

            Root MSE        0.37592    R-square    0.5012
            Dep Mean        0.92737    Adj R-sq    0.4718
            C.V.           40.53600

                          Parameter Estimates

                      Parameter     Standard     T for H0:
        Variable   DF   Estimate       Error   Parameter=0   Prob > |T|

        INTERCEP    1  -0.023737   0.24576577      -0.097       0.9242
        OXIDANT     1   0.019579   0.00473739       4.133       0.0007

                  Dep Var    Predict    Std Err                Std Err
     Obs  OXIDANT  DICARBOX    Value    Predict   Residual    Residual

       1      78    0.8500    1.5034     0.164    -0.6534       0.338
       2      80    1.4500    1.5425     0.172    -0.0925       0.334
       3      74    1.8000    1.4251     0.148     0.3749       0.346
       4      78    1.8000    1.5034     0.164     0.2966       0.338
       5      60    1.6000    1.1510     0.102     0.4490       0.362
       6      62    1.2000    1.1901     0.107     0.0099       0.360
       7      57    1.3000    1.0922     0.095     0.2078       0.364
       8      49    0.2000    0.9356     0.086    -0.7356       0.366
       9      34    0.2200    0.6419     0.110    -0.4219       0.359
      10      36    0.4000    0.6811     0.105    -0.2811       0.361
      11      32    0.5000    0.6028     0.117    -0.1028       0.357
      12      28    0.3800    0.5245     0.130    -0.1445       0.353
      13      25    0.3000    0.4657     0.141    -0.1657       0.348
      14      45    0.7000    0.8573     0.088    -0.1573       0.365
      15      40    0.8000    0.7594     0.095     0.0406       0.364
      16      45    0.9000    0.8573     0.088     0.0427       0.365
      17      41    1.2200    0.7790     0.093     0.4410       0.364
      18      34    1.0000    0.6419     0.110     0.3581       0.359
      19      25    1.0000    0.4657     0.141     0.5343       0.348
```

```
                         UNIVARIATE PROCEDURE
Variable=RESID              Residual

                              Moments

             N                   19   Sum Wgts              19
             Mean                 0   Sum                    0
             Std Dev       0.365327   Variance        0.133464
             Skewness      -0.41391   Kurtosis        -0.43294
             USS           2.402345   CSS             2.402345
             CV                   .   Std Mean        0.083812
             T:Mean=0             0   Prob>|T|          1.0000
             Sgn Rank             4   Prob>|S|          0.8906
             Num ^=0             19
             W:Normal      0.951334   Prob<W            0.4220

                          Quantiles(Def=5)

             100% Max      0.534273       99%        0.534273
              75% Q3       0.358066       95%        0.534273
              50% Med      0.009867       90%        0.449024
              25% Q1       -0.16573       10%        -0.65339
               0% Min      -0.73561        5%        -0.73561
                                          1%        -0.73561

             Range         1.269885
             Q3-Q1         0.523793
             Mode          -0.73561

                              Extremes

             Lowest        Obs         Highest        Obs
             -0.73561(        8)       0.358066(       18)
             -0.65339(        1)       0.374924(        3)
             -0.42193(        9)       0.441016(       17)
             -0.28109(       10)       0.449024(        5)
             -0.16573(       13)       0.534273(       19)


      Stem Leaf                      #              Boxplot
         4  453                      3                 |
         2  1067                     4              +-----+
         0  144                      3              *--+--*
        -0  76409                    5              +-----+
        -2  8                        1                 |
        -4  2                        1                 |
        -6  45                       2                 |
            ----+----+----+----+
      Multiply Stem.Leaf by 10**-1
```

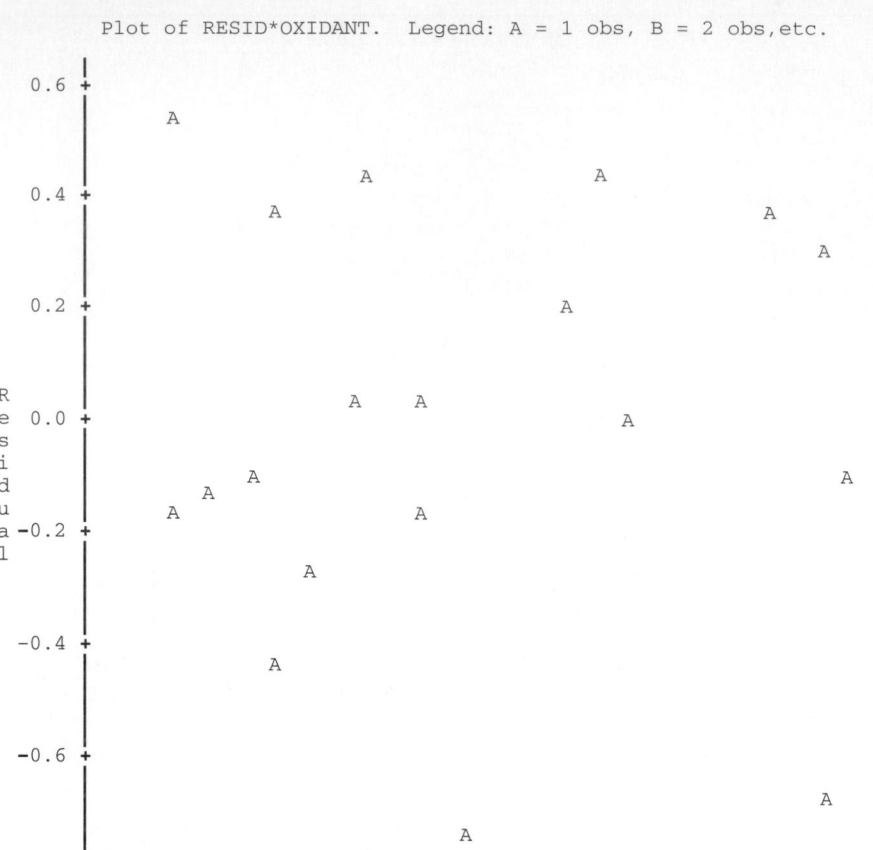

```
      Plot of RESID*OXIDANT.    Legend: A = 1 obs, B = 2 obs,etc.
```

Data for Exercise 12.31

Weight Percentile y	No. of Cigarettes Smoked per Day x	Weight Percentile y	No. of Cigarettes Smoked per Day x
6	0	43	0
6	15	49	0
2	40	50	0
8	23	49	22
11	20	46	30
17	7	54	0
24	3	58	0
25	0	62	0
17	25	66	0
25	20	66	23
25	15	83	0
31	23	87	44
35	10		

Source: Rubin, B. K. "Exposure of children with cystic fibrosis to environmental tobacco smoke." *The New England Journal of Medicine,* Sept. 20, 1990, Vol. 323, No. 12, p. 785 (data extracted from Figure 3).

```
The regression equation is
WTPCTILE = 41.2 - 0.262 SMOKED

Predictor        Coef       Stdev     t-ratio        p
Constant       41.153       6.843        6.01    0.000
SMOKED        -0.2619       0.3702       -0.71    0.486

s = 24.68      R-sq = 2.1%      R-sq(adj) = 0.0%

Analysis of Variance

SOURCE         DF          SS          MS          F         p
Regression      1       304.9       304.9       0.50     0.486
Error          23     14011.1       609.2
Total          24     14316.0

Obs.    SMOKED   WTPCTILE     Fit  Stdev.Fit  Residual   St.Resid
   1       0.0       6.00    41.15      6.04    -35.15      -1.48
   2      15.0       6.00    37.22      5.00    -31.22      -1.29
   3      40.0       2.00    30.68     11.22    -28.68      -1.30
   4      23.0       8.00    35.13      6.22    -27.13      -1.14
   5      20.0      11.00    35.91      5.61    -24.91      -1.04
   6       7.0      17.00    39.32      5.38    -22.32      -0.93
   7       3.0      24.00    40.37      6.13    -16.37      -0.68
   8       0.0      25.00    41.15      6.84    -16.15      -0.68
   9      25.0      17.00    34.60      6.69    -17.60      -0.74
  10      20.0      25.00    35.91      5.61    -10.91      -0.45
  11      15.0      25.00    37.22      5.00    -12.22      -0.51
  12      23.0      31.00    35.13      6.22     -4.13      -0.17
  13      10.0      35.00    38.53      5.04     -3.53      -0.15
  14       0.0      43.00    41.15      6.84      1.85       0.08
  15       0.0      49.00    41.15      6.84      7.85       0.33
  16       0.0      50.00    41.15      6.84      8.85       0.37
  17      22.0      49.00    35.39      6.00     13.61       0.57
  18      30.0      46.00    33.29      8.06     12.71       0.54
  19       0.0      54.00    41.15      6.84     12.85       0.54
  20       0.0      58.00    41.15      6.84     16.85       0.71
  21       0.0      62.00    41.15      6.84     20.85       0.88
  22       0.0      66.00    41.15      6.84     24.85       1.05
  23      23.0      66.00    35.13      6.22     30.87       1.29
  24       0.0      83.00    41.15      6.84     41.85       1.76
  25      44.0      87.00    29.63     12.56     57.37       2.70
```

12.30 Refer to the *World Development* study of the variables affecting the size distribution of manufacturing firms in international markets, Exercise 12.10. Five independent variables, LGNP, AREAC, SVA, CREDIT, and STOCK were used to model the share, *y*, of firms with 100 or more workers. The researchers detected a high correlation between pairs of the following independent variables: LGNP and SVA, LGNP and STOCK, and CREDIT and STOCK. Describe the problems that may arise if these high correlations are ignored in the multiple regression analysis of the model.

12.31 Passive exposure to environmental tobacco smoke has been associated with growth suppression and an increased frequency of respiratory tract infections in normal children. Is this association more pronounced in children with cystic fibrosis? To answer this question, a study was conducted on 43 children (18 girls and 25 boys) attending a 2-week summer camp for cystic fibrosis patients (*New England Journal of Medicine,* Sept. 20, 1990). Researchers investigated the correlation between a child's weight percentile (*y*) and the number of cigarettes smoked per day in the child's home (*x*). The table on page 590 lists the data for the 25 boys. A MINITAB printout (with residuals) for the straight-line model relating *y* to *x* is shown above. Examine the residuals. Do you detect any outliers?

Maximum Oxygen Uptake y	Age x_1 (years)	Height x_2 (centimeters)	Weight x_3 (kilograms)	Chest Depth x_4 (centimeters)
1.54	8.4	132.0	29.1	14.4
1.74	8.7	135.5	29.7	14.5
1.32	8.9	127.7	28.4	14.0
1.50	9.9	131.1	28.8	14.2
1.46	9.0	130.0	25.9	13.6
1.35	7.7	127.6	27.6	13.9
1.53	7.3	129.9	29.0	14.0
1.71	9.9	138.1	33.6	14.6
1.27	9.3	126.6	27.7	13.9
1.50	8.1	131.8	30.8	14.5

12.32 A physiologist wishes to investigate the relationship between the physical characteristics of preadolescent boys and their maximum oxygen uptake (measured in milliliters of oxygen per kilogram of body weight). The data shown in the table above are collected on a random sample of ten preadolescent boys. As a first step in the data analysis, the researcher fits the regression model

$$y = \beta_0 + \beta_1 x_1 + \beta_2 x_2 + \beta_3 x_3 + \beta_4 x_4 + \epsilon$$

to the data. The output for a SAS regression analysis is shown below.
a. Give the value of R^2 and interpret its value.
b. It seems reasonable to assume that the greater a child's weight, the greater should be the maximum oxygen uptake. But note that $\hat{\beta}_3$, the estimated coefficient of weight, x_3, is negative. Give an explanation for this result.
c. It would seem that the chest depth of a child should be positively correlated to lung volume and hence to maximum oxygen uptake. Can you explain the small t value associated with $\hat{\beta}_4$?

d. If you have access to the appropriate computer package, find a 95% prediction interval for maximum oxygen uptake for a boy with age = 8.8, height = 128.1, weight = 28.0, and chest depth = 13.7.

12.33 Refer to Exercise 12.32. Calculate the Pearson product moment correlation coefficients between all pairs of the independent variables x_1–x_4. Do these correlations provide an explanation for the confusing signs and small t values associated with the estimated regression coefficients of the model fit in Exercise 12.32?

12.34 *Teaching Sociology* (July 1995) developed a model for the professional socialization of graduate students working toward a Ph.D. in sociology. One of the dependent variables modeled was professional confidence, *y,* measured on a 5-point scale. The model included more than 20 independent variables and was fit to data collected for a sample of 309 sociology graduate students. One concern is whether multicollinearity exists in the data. A matrix of Pearson product moment correlations for

Source	DF	Sum of Squares	Mean Square	F Value	Prob>F
Model	4	0.20604	0.05151	37.204	0.0007
Error	5	0.00692	0.00138		
C Total	9	0.21296			

Root MSE	0.03721	R-Square	0.9675	
Dep Mean	1.49200	Adj R-Sq	0.9415	
C.V.	2.49391			

Parameter Estimates

Variable	DF	Parameter Estimate	Standard Error	T for H0: Parameter=0	Prob > |T|
INTERCEP	1	-4.774739	0.86281773	-5.534	0.0026
AGE	1	-0.035214	0.01538630	-2.289	0.0708
HEIGHT	1	0.051637	0.00621522	8.308	0.0004
WEIGHT	1	-0.023417	0.01342835	-1.744	0.1416
CHEST	1	0.034489	0.08523877	0.405	0.7025

Independent Variable	(1)	(2)	(3)	(4)	(5)	(6)	(7)	(8)	(9)	(10)
(1) Father's occupation	1.000	.363	.099	−.110	−.047	−.053	−.111	.178	.078	.049
(2) Mother's education	.363	1.000	.228	−.139	−.216	.084	−.118	.192	.125	.068
(3) Race	.099	.228	1.000	.036	−.515	.014	−.120	.112	.117	.337
(4) Sex	−.110	−.139	.036	1.000	.165	.256	.173	.106	−.117	.073
(5) Foreign status	−.047	−.216	−.515	.165	1.000	−.041	.159	−.130	−.165	−.171
(6) Undergraduate GPA	−.053	.084	.014	−.256	−.041	1.000	.032	.028	−.034	.092
(7) Year GRE taken	−.111	−.118	−.120	.173	.159	.032	1.000	−.086	−.602	.016
(8) Verbal GRE score	.178	.192	.112	−.106	−.130	.028	−.086	1.000	.132	.087
(9) Years in graduate program	.078	.125	.117	−.117	−.165	−.034	−.602	.132	1.000	−.071
(10) First-year graduate GPA	.049	.068	.337	.073	−.171	.092	.016	.087	−.071	1.000

Source: Keith, B., and Moore, H. A. "Training sociologists: An assessment of professional socialization and the emergence of career aspirations." *Teaching Sociology*, Vol, 23, No. 3, July 1995, p. 205 (Table 1).

ten of the independent variables is shown above. [*Note:* Each entry in the table is the correlation coefficient r between the variable in the corresponding row and corresponding column.]

a. Examine the correlation matrix and find the independent variables that are moderately or highly correlated.

b. What modeling problems may occur if the variables, part **a**, are left in the model? Explain.

QUICK REVIEW

Key Terms

Analysis of variance *F*-test   555
Correlated errors   587
Dummy variable   568
Extrapolation   585
Interaction   557
Interaction model   557
Mean square for error   542
Multicollinearity   586
Multiple coefficient of
  determination   553

Multiple regression model   536
Outlier   578
Parameter estimability   585
Quadratic model   539
Residual   573
Residual analysis   574
Time series model   587

Key Formulas

$$s^2 = \text{MSE} = \frac{\text{SSE}}{n - (k + 1)}$$

Estimator of σ^2 for a model with k independent variables   543

$$t = \frac{\hat{\beta}_i}{s_{\hat{\beta}_i}}$$

Test statistic for testing H_0: $\beta_i = 0$ 545

$\hat{\beta}_i \pm (t_{\alpha/2})s_{\hat{\beta}_i}$,
where t depends on $n - (k + 1)$ df

$(1 - \alpha)100\%$ confidence interval for β_i 546

$$F = \frac{\text{MS(Model)}}{\text{MSE}} = \frac{R^2 / k}{(1 - R^2) / [n - (k + 1)]}$$

Test statistic for testing H_0: $\beta_1 = \beta_2 = \cdots = \beta_k = 0$ 555

$$R^2 = \frac{\text{SS}_{yy} - \text{SSE}}{\text{SS}_{yy}}$$

Multiple coefficient of determination   553

$y - \hat{y}$

Regression residual   574

LANGUAGE LAB

Symbol	Pronunciation	Description
x_1^2	x-1 squared	Quadratic term that allows for curvature in the relationship between y and x
$x_1 x_2$	x-1 x-2	Interaction term
MSE	M-S-E	Mean square for error (estimates σ^2)
β_i	beta-i	Coefficient of x_i in the model
$\hat{\beta}_i$	beta-i-hat	Least squares estimate of β_i
$s_{\hat{\beta}_i}$	s of beta-i-hat	Estimated standard error of $\hat{\beta}_i$
R^2	R-squared	Multiple coefficient of determination
F		Test statistic for testing global usefulness of model
$\hat{\epsilon}$	epsilon-hat	Estimated random error, or residual

SUPPLEMENTARY EXERCISES 12.35–12.60

Note: Exercises marked with  *require the use of a computer.*

Learning the Mechanics

12.35 Suppose you fit the model

$$y = \beta_0 + \beta_1 x_1 + \beta_2 x_1^2 + \beta_3 x_2 + \beta_4 x_1 x_2 + \epsilon$$

to $n = 25$ data points and find that

$$\hat{\beta}_0 = 1.26 \qquad \hat{\beta}_1 = -2.43 \qquad \hat{\beta}_2 = .05 \qquad \hat{\beta}_3 = .62$$

$$\hat{\beta}_4 = 1.81 \qquad SSE = .41 \qquad R^2 = .83$$

$$s_{\hat{\beta}_1} = 1.21 \qquad s_{\hat{\beta}_2} = .16 \qquad s_{\hat{\beta}_3} = .26 \qquad s_{\hat{\beta}_4} = 1.49$$

a. Is there sufficient evidence to conclude that at least one of the parameters β_1, β_2, β_3, or β_4 is nonzero? Test using $\alpha = .05$.
b. Test H_0: $\beta_1 = 0$ against H_a: $\beta_1 < 0$. Use $\alpha = .05$.
c. Test H_0: $\beta_2 = 0$ against H_a: $\beta_2 > 0$. Use $\alpha = .05$.
d. Test H_0: $\beta_3 = 0$ against H_a: $\beta_3 \neq 0$. Use $\alpha = .05$.

12.36 When a multiple regression model is used for estimating the mean of the dependent variable and for predicting a new value of y, which will be narrower

—the confidence interval for the mean or the prediction interval for the new y value?

12.37 After a regression model is fit to a set of data, a confidence interval for the mean value of y at a given setting of the independent variables is *always* narrower than the corresponding prediction interval for a particular value of y at the same setting of the independent variables. Why?

12.38 Suppose you used MINITAB to fit the model

$$y = \beta_0 + \beta_1 x_1 + \beta_2 x_2 + \epsilon$$

to $n = 15$ data points and you obtained the printout shown below.

a. What is the least squares prediction equation?
b. Find R^2 and interpret its value.
c. Is there sufficient evidence to indicate that the model is useful for predicting y? Conduct an F-test using $\alpha = .05$.
d. Test the null hypothesis H_0: $\beta_1 = 0$ against the alternative hypothesis H_a: $\beta_1 \neq 0$. Test using $\alpha = .05$. Draw the appropriate conclusions.

```
The regression equation is
Y = 90.1 - 1.84 X1 + .285 X2

Predictor      Coef      Stdev    t-ratio        p
Constant      90.10      23.10       3.90    0.002
X1           -1.836      0.367      -5.01    0.001
X2            0.285      0.231       1.24    0.465

s = 10.68      R-sq = 91.6%     R-sq(adj) = 90.2%

Analysis of Variance

SOURCE       DF        SS         MS        F        p
Regression    2     14801       7400    64.91    0.001
Error        12      1364        114
Total        14     16165
```

y	-12	28	-15	-25	2	-11	25	-47	-28	12	29	-3	5	7	6
x_1	3	8	3	2	9	5	7	0	1	5	1	5	7	6	5
x_2	9	7	1	6	1	1	7	8	6	9	4	5	3	4	9

e. Find the standard deviation of the regression model and interpret it.

12.39 Suppose you have developed a regression model to explain the relationship between y and x_1, x_2, and x_3. A set of $n = 15$ data points is used to find the least squares prediction equation. The ranges of the variables you observed were as follows: $10 \le y \le 100$, $5 \le x_1 \le 55$, $.5 \le x_2 \le 1$, and $1,000 \le x_3 \le 2,000$. Will the error of prediction be smaller when you use the least squares equation to predict y when $x_1 = 30$, $x_2 = .6$, and $x_3 = 1,300$, or when $x_1 = 60$, $x_2 = .4$, and $x_3 = 900$? Why?

12.40 A response variable y was observed for $n = 15$ settings of two independent variables, x_1 and x_2. The data are shown in the table above.

  a. Fit the model $y = \beta_0 + \beta_1 x_1 + \beta_2 x_2 + \epsilon$ to the data.

  b. Fit the model $y = \beta_0 + \beta_1 x_1 + \beta_2 x_2 + \beta_3 x_1 x_2 + \epsilon$ to the data.

  c. Which of the models in parts **a** and **b** best describes the relationship between y and x_1 and x_2? Explain.

  d. Repeat part **a** using a different statistical program package. For example, if you originally used MINITAB, now use SAS, SPSS, or another available package. Compare these results with those found in part **a**.

12.41 Refer to Exercise 12.14, in which a quadratic model was fit to $n = 19$ data points. Two residual plots are shown here—one corresponding to the fitting of a straight-line model to the data below and the other to the quadratic model (on page 596) fit in Exercise 12.14. Analyze the two plots. Is the need for a quadratic term evident from the residual plot for the straight-line model? Does your conclusion agree with your test of the quadratic term in Exercise 12.14?

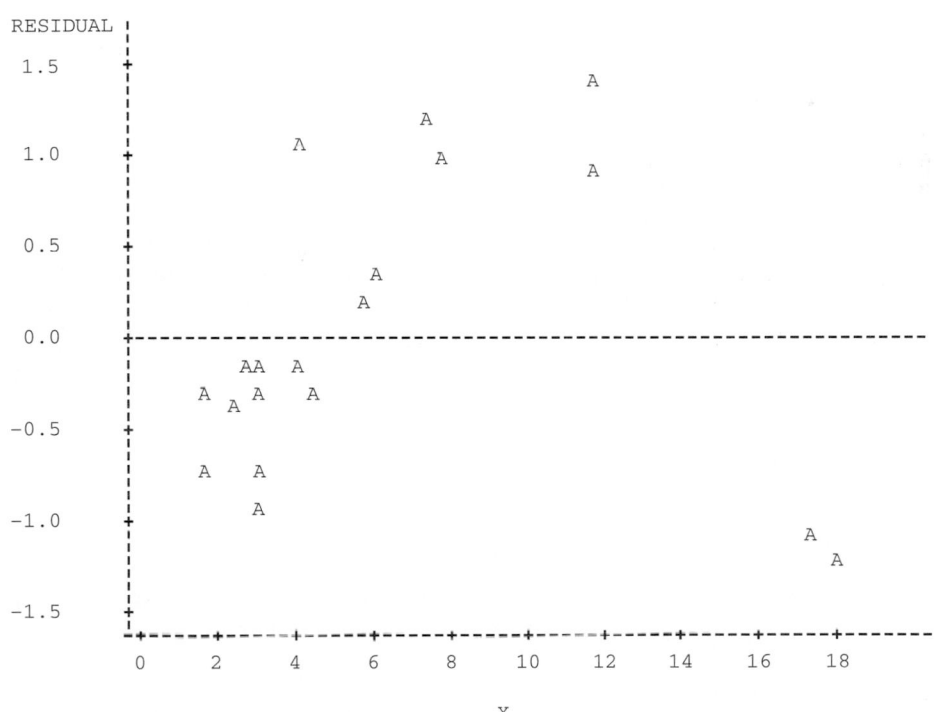

Residual plot for straight-line model

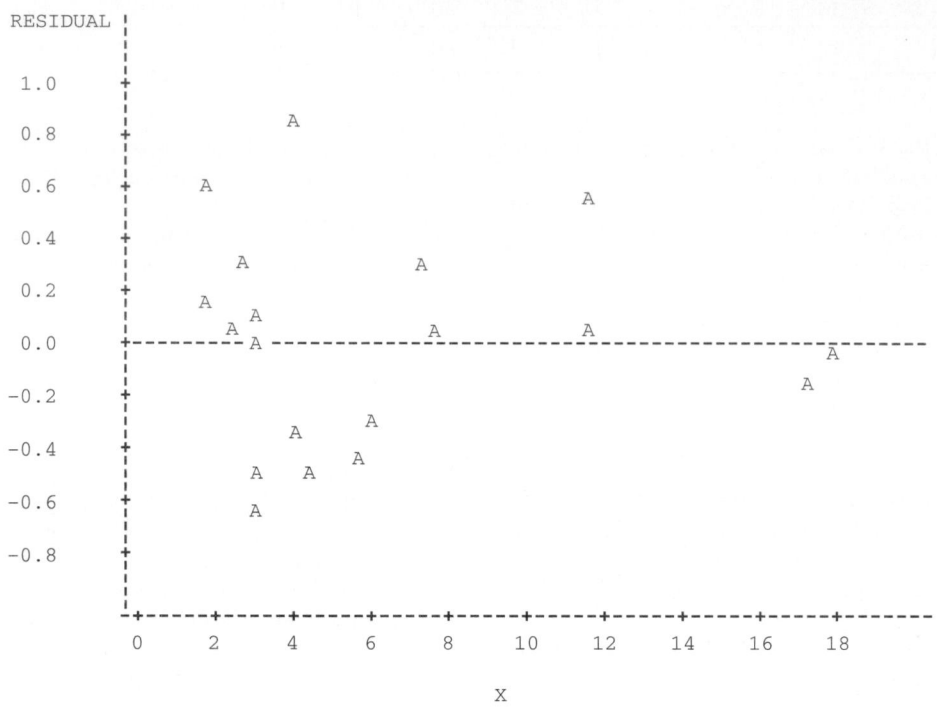

Residual plot for quadratic model

Applying the Concepts

12.42 Emergency services (EMS) personnel are constantly exposed to traumatic situations. However, few researchers have studied the psychological stress that EMS personnel may experience. *The Journal of Consulting and Clinical Psychology* (June 1995) reported on a study of EMS rescue workers who responded to the I-880 freeway collapse during the 1989 San Francisco earthquake. The goal of the study was to identify the predictors of symptomatic distress in the EMS workers. One of the distress variables studied was the Global Symptom Index (GSI). Several models for GSI, y, were considered based on the following independent variables:

x_1 = Critical Incident Exposure scale (CIE)

x_2 = Hogan Personality Inventory-Adjustment scale (HPI-A)

x_3 = Years of experience (EXP)

x_4 = Locus of Control scale (LOC)

x_5 = Social Support scale (SS)

x_6 = Dissociative Experiences scale (DES)

x_7 = Peritraumatic Dissociation Experiences Questionnaire, self-report (PDEQ-SR)

a. Write a first-order model for $E(y)$ as a function of the first five independent variables, x_1–x_5.

b. The model of part **a**, fitted to data collected for $n = 147$ EMS workers, yielded the following results: $R^2 = .469$, $F = 34.47$, p-value $< .001$. Interpret these results.

c. Write a first-order model for $E(y)$ as a function of all seven independent variables, x_1–x_7.

d. The model, part **c**, yielded $R^2 = .603$. Interpret this result.

e. The t-tests for testing the DES and PDEQ-SR variables both yielded a p-value of .001. Interpret these results.

12.43 Since the Great Depression of the 1930s, the link between the suicide rate and the state of the economy has been the subject of much research. Research exploring this link using regression analysis was reported in an article in the *Journal of Socio-Economics* (Spring, 1992). The researchers collected data from a 45-year period on the following variables:

y = Suicide rate

x_1 = Unemployment rate

x_2 = Percentage of females in the labor force

x_3 = Divorce rate

$$\hat{y} = .002 + .0204x_1 + (-.0231)x_2 + .0765x_3 + .2760x_4 + .0018x_5$$
$$(.002) \qquad (.02) \qquad (>.10) \quad (>.10) \quad (>.10)$$

$$R^2 = .45$$

x_4 = Logarithm of Gross National Product (GNP)

x_5 = Annual percent change in GNP

One of the models explored by the researchers was a multiple regression model relating y to linear terms in x_1 through x_5. The least squares model shown at the top of the page resulted (the observed significance levels of the β estimates are shown in parentheses beneath the estimates):

a. Interpret the value of R^2. Is there sufficient evidence to indicate that the model is useful for predicting the suicide rate? Use $\alpha = .05$.

b. Interpret each of the coefficients in the model, and each of the corresponding significance levels.

c. Is there sufficient evidence to indicate that the unemployment rate is a useful predictor of the suicide rate? Use $\alpha = .05$.

d. Discuss each of the following terms with respect to potential problems with the above model: curvature (second-order terms), interaction, and multicollinearity.

12.44 Perfectionists are persons who set themselves standards and goals that cannot be reasonably met or accomplished. One theory suggests that individuals who are depressed have a tendency toward perfectionism. To study this phenomenon, 76 members of an introductory psychology class completed four questionnaires: (1) the ASO scale, designed to measure self-acceptance; (2) the Burns scale, designed to measure perfectionism; (3) the Zung scale, designed to measure depression; and (4) the Rotter scale, designed to measure perceptions between actions and reinforcement (*The Journal of Adlerian Theory, Research, and Practice,* Mar. 1986).

a. Write a first-order model relating depression (Zung scale) to self-acceptance (ASO scale), perfectionism (Burns scale), and reinforcement (Rotter scale).

b. The model, part **a**, was fitted to the $n = 76$ points and resulted in a coefficient of determination of $R^2 = .70$. Interpret this value.

c. Is there sufficient evidence to indicate that the model is useful for predicting depression (Zung scale) score? Test using $\alpha = .05$.

d. A t-test for the perfectionism (Burns scale) variable resulted in a (two-tailed) p-value of .87. Interpret this value.

12.45 To meet the increasing demand for new software products, many systems development experts have adopted a prototyping methodology. The effects of prototyping on the system development life cycle (SDLC) were investigated in the *Journal of Computer Information Systems* (Spring 1993). A survey of 500 randomly selected corporate-level MIS managers was conducted. Three potential independent variables were (1) *importance* of prototyping to each phase of the SDLC; (2) degree of *support* prototyping provides for the SDLC; and (3) degree to which prototyping *replaces* each phase of the SDLC. The accompanying table gives the pairwise correlations of the three variables in the survey data for one particular phase of the SDLC. Use this information to assess the degree of multicollinearity in the survey data. Would you recommend using all three independent variables in a regression analysis? Explain.

Variable Pairs	Correlation Coefficient, r
Importance–Replace	.2682
Importance–Support	.6991
Replace–Support	−.0531

Source: Hardgrave, B. C., Doke, E. R., and Swanson, N. E. "Prototyping effects of the system development life cycle: An empirical study." *Journal of Computer Information Systems,* Vol. 33, No. 3, Spring 1993, p. 16 (Table 1).

12.46 The *Journal of Family Issues* (Mar. 1983) reported on a study of the effect of a wife's employment on the number of hours available for household tasks. Some 206 families were employed in the study. One dependent variable measured for each family was the total time y (in minutes per day) that the wife spent on household chores. Two independent variables were also recorded:

WEMP: x_1 = Wife's hours of employment per week

YAGE: x_2 = Age of the youngest child in the family

The researchers explain that they first attempted to fit the model

$$E(y) = \beta_0 + \beta_1 x_1 + \beta_2 x_2 + \beta_3 x_1 x_2$$

If the regression analysis did not indicate statistical significance for the interaction term, it was eliminated from the model. Looking at the accompanying computer printout on page 598, you can see that the interaction term was eliminated from their final regression analysis.

a. Find R^2 and give its practical implications.

DEPENDENT VARIABLE: WIFE'S HOUSEHOLD WORK		(MEAN=401.3[a])			
SOURCE OF VARIATION	DF	SUMS OF SQUARES	MEAN SQUARE	F	R^2
TOTAL	205	5,432,296		62.10***	0.38
MODEL	2	2,061,918	1,030,959		
ERROR	203	3,370,378	16,603		
SOURCE WITHIN MODEL	DF	SEQUENTIAL S.S.	PARTIAL S.S.	b-ESTIMATES	
WEMP	1	1,810,154***	1,110,288***	-4.08	
YAGE	1	251,765***	251,765***	-7.02	

DEPENDENT VARIABLE: HUSBAND'S WORK   (MEAN=105.2[a])

                REGRESSION ANALYSIS NOT SIGNIFICANT

[*Note:* In this particular printout "b-estimate" is our β estimate, or parameter estimate. Also, the asterisks indicate the level of significance of test statistics (or sum of squares to compute test statistics). Here, *** means $p < .001$.]

[a]Minutes per day.

Source: Fox, K. D., and Nickols, S. Y. "The time crunch." *Journal of Family Issues,* Mar. 1983, Vol. 4, No. 1, pp. 61–82. Copyright 1983 by Sage Publications. Reprinted by permission of Sage Publications, Inc.

b. Based on the computer printout, does the model contribute information for the prediction of y? Interpret the appropriate p-value.

c. In concluding, the researchers state that "for each additional hour of employment, wives decreased household work time 4 minutes per day." Do you agree?

12.47 Newspaper cartoons, although designed to be funny, often invoke hostility, pain, and/or aggression in readers, especially when those cartoons depict violence. A study was undertaken to determine how violence in cartoons is related to aggression or pain (*Motivation and Emotion*, Vol. 10, 1986). A group of volunteers (psychology students) rated each of 32 violent newspaper cartoons (16 "Herman" and 16 "Far Side" cartoons) on three dimensions:

y = Funniness (0 = not funny, ..., 9 = very funny)

x_1 = Pain (0 = none, ..., 9 = a very great deal)

x_2 = Aggression/hostility (0 = none, ..., 9 = a very great deal)

The ratings of the students on each dimension were averaged and the resulting $n = 32$ observations were subjected to a multiple regression analysis. Based on the underlying theory (called the *inverted-U theory*) that the funniness of a joke will increase at low levels of aggression or pain, level off, and then decrease at high levels of aggressiveness or pain, the following quadratic models were proposed:

Model 1: $E(y) = \beta_0 + \beta_1 x_1 + \beta_2 x_1^2,$ $R^2 = .099, F = 1.60$

Model 2: $E(y) = \beta_0 + \beta_1 x_2 + \beta_2 x_2^2,$ $R^2 = .100, F = 1.61$

a. According to the theory, what is the expected sign of β_2 in either model?

b. Is there sufficient evidence to indicate that the quadratic model relating pain to funniness rating is useful? Test at $\alpha = .05$.

c. Is there sufficient evidence to indicate that the quadratic model relating aggression/hostility to funniness rating is useful? Test at $\alpha = .05$.

12.48 A large government agency would like to predict the number of people it will hire within the next year to fill the 30 positions that are currently open. Historically, the agency has been unable to fill all its job openings. It has been decided to model the number of positions filled in a year, y, as a function of the number of positions open, x_1, and the recruiting budget (in dollars) for the year, x_2 (e.g., for advertising the positions, paying travel expenses, etc.). A random sample of 10 years of recruiting records was drawn from the agency's 30 years of records. The model

$$E(y) = \beta_0 + \beta_1 x_1 + \beta_2 x_2$$

was fitted to the data using MINITAB. The results shown in the computer printout on page 599 were obtained.

a. Identify the least squares prediction equation.

b. Is there sufficient evidence to indicate that the model contributes information for predicting the number y of positions that will be filled? Conduct an F-test using $\alpha = .05$.

c. Test the null hypothesis H_0: $\beta_2 = 0$ against the alternative hypothesis H_a: $\beta_2 \neq 0$ using $\alpha = .05$. Interpret the results of your test in the context of the problem.

d. Use the least squares prediction equation to predict how many of the 30 positions the agency will fill next year if the recruiting budget is $10,000.

e. Can you think of a situation for which the prediction in part **d** might possibly be inaccurate? Explain.

```
The regression equation is
Y  =   .056 + .273 X1 + .0006 X2

Predictor        Coef        Stdev      t-ratio        p
Constant        0.0562       0.902        0.06       0.998
X1              0.2733       0.0971       2.81       0.036
X2              0.000560     0.000129     4.34       0.007

s = 1.33        R-sq = 97.9%      R-sq(adj) = 97.3%

Analysis of Variance

SOURCE        DF         SS         MS         F          p
Regression     2       583.18     291.59     164.74     0.001
Error          7        12.42       1.77
Total          9       595.60
```

f. Which (if any) of the assumptions we make about ϵ in a regression analysis are likely to be violated in this problem? Explain.

12.49 Research was conducted to discover the factors in a person's education that determine future wages (*Southern Economic Journal*, Vol. 50, 1983). A first-order model was fit to a set of $n = 60$ data points and the prediction equation and t-test values were obtained. The t values used to test the individual model parameters are shown in parentheses below their respective estimates at the bottom of the page.

a. What are the interpretations of the coefficients?

b. Are they statistically significant at the $\alpha = .01$ level?

c. The standard error for the estimate of β_5 is .001225. Use this information to construct a 99% confidence interval for β_5.

d. Interpret the interval, part **c**.

12.50 A researcher wishes to investigate the effects of two independent variables on a teacher's attitude toward physically challenged students. A study is conducted involving 40 randomly selected teachers. The response y, a teacher's attitude toward physically challenged students, is measured with a standardized attitude scale. Independent variables in the study are

x_1 = Average number of physically challenged children taught per year

x_2 = Number of years of teaching experience

The researcher fits the model

$$y = \beta_0 + \beta_1 x_1 + \beta_2 x_2 + \beta_3 x_2^2 + \epsilon$$

to the data with the following results:

$$\hat{y} = 50 + 1.5x_1 + 5x_2 - .1x_2^2 \qquad s_{\hat{\beta}_3} = .03$$

a. Do these data provide sufficient evidence to indicate that the quadratic term in years of experience, x_2^2, is useful for predicting attitude score? Use $\alpha = .05$.

b. Sketch the predicted attitude score $\hat{y}$ as a function of the number of years of experience x_2 for $x_1 = 5$. Repeat this for $x_1 = 10$. [*Note:* For each value of x_1 ($x_1 = 5$ and $x_1 = 10$), plot $\hat{y}$ for $x_2 = 0, 2, 4, 6, 8$, and 10. Observe that (for this model) the vertical distance between the two prediction curves is the same for all values of x_2.]

$$y = 0 - .0945x_1 - .032x_2 + .009x_3 - .0028x_4 + .007x_5 + .105x_6 + .469x_7$$
$$\ (-2.61)\ \ (-.96)\ \ (2.74)\ \ (-2.23)\ \ (5.71)\ \ (4.02)\ \ (2.21)$$

where

y = Future wages

x_1 = Amount of business course work in high school

x_2 = Amount of college prep work in high school

x_3 = Math aptitude

x_4 = High school GPA

x_5 = Measure of socioeconomic status

x_6 = 1 if the individual is married, 0 if not

x_7 = Amount of on-the-job training

12.51 Refer to Exercise 12.50. Suppose the interaction terms x_1x_2 and $x_1x_2^2$ are added to the model to produce

$$y = \beta_0 + \beta_1 x_1 + \beta_2 x_2 + \beta_3 x_2^2 + \beta_4 x_1 x_2 + \beta_5 x_1 x_2^2 + \epsilon$$

This model is fit to the same 40 observations used in Exercise 12.50 with the result

$$\hat{y} = 50 + x_1 + 6x_2 - .2x_2^2 + x_1x_2 - .05x_1x_2^2$$

and $R^2 = .87$

a. Interpret the value of R^2.

b. Is there sufficient evidence to indicate that this model is useful for predicting attitude score? Test $H_0: \beta_1 = \beta_2 = \beta_3 = \beta_4 = \beta_5 = 0$ using $\alpha = .05$.

c. Sketch the predicted attitude score $\hat{y}$ as a function of the number of years of experience, x_2, for teachers who have taught an average of 10 physically challenged children per year ($x_1 = 10$). Repeat this for $x_1 = 5$. Compare these sketches with those obtained for the noninteraction model of Exercise 12.50. [*Note:* In Chapter 13 we test to determine whether this model provides more information for the prediction of y than the noninteraction model.]

12.52 The length of a mosquito's proboscis plays a large role in determining its feeding habits. An entomologist has proposed the following model for predicting the length of a mosquito's proboscis:

$$E(y) = \beta_0 + \beta_1 x_1 + \beta_2 x_2 + \beta_3 x_3$$

where

y = Length of proboscis (millimeters)

x_1 = Dry weight (milligrams)

x_2 = Length of wing (millimeters)

x_3 = Width of wing (millimeters)

An analysis of data obtained from a sample of 44 mosquitos of a certain species produces the least squares model

$$\hat{y} = .968 + .292x_1 + .614x_2 - .201x_3$$

Also,

$$s_{\hat{\beta}_1} = .248 \qquad s_{\hat{\beta}_2} = 1.31 \qquad s_{\hat{\beta}_3} = .267 \qquad R^2 = .536$$

a. Do these statistics indicate that the model is useful in predicting proboscis length? Use $\alpha = .10$.

b. Given the first-order model specified above, do the data provide sufficient evidence to indicate that x_3 (width of wing) is an important variable for predicting y? Test using $\alpha = .05$.

12.53 Refer to Exercise 12.11, in which regression analysis was used to explore the valuation of apartment buildings in Minneapolis.

a. Verify that number of apartment units, x_1, is highly correlated with gross building area, x_5.

b. Eliminate x_1 from the appraiser's model and refit the model to the data. Compare the standard deviations of the two models.

c. Use an F-test to investigate the usefulness of this model. Use $\alpha = .05$.

d. Use the least squares equation of part **b** to estimate the value of an apartment building that is 50 years old with gross area 20,000 square feet, lot size 15,000 square feet, and 10 parking spaces. Use a statistical software package to compute a 95% confidence interval for the model's estimate.

12.54 To determine whether extra personnel are needed for the day, the owners of a water adventure park would like to find a model that would allow them to predict the day's attendance each morning before opening based on the day of the week and weather conditions. The model is of the form

$$E(y) = \beta_0 + \beta_1 x_1 + \beta_2 x_2 + \beta_3 x_3$$

where

y = Daily admissions

$$x_1 = \begin{cases} 1 & \text{if weekend} \\ 0 & \text{otherwise} \end{cases} \quad \text{(dummy variable)}$$

$$x_2 = \begin{cases} 1 & \text{if sunny} \\ 0 & \text{if overcast} \end{cases} \quad \text{(dummy variable)}$$

x_3 = Predicted daily high temperature (°F)

These data were recorded for a random sample of 30 days, and a regression model was fit to the data. The least squares analysis produced the following results:

$$\hat{y} = -105 + 25x_1 + 100x_2 + 10x_3$$

with

$$s_{\hat{\beta}_1} = 10 \qquad s_{\hat{\beta}_2} = 30 \qquad s_{\hat{\beta}_3} = 4 \qquad R^2 = .65$$

a. Interpret the estimated model coefficients.

b. Is there sufficient evidence to conclude that this model is useful for the prediction of daily attendance? Use $\alpha = .05$.

c. Is there sufficient evidence to conclude that mean attendance increases on weekends? Use $\alpha = .10$.

d. Use the model to predict the attendance on a sunny weekday with a predicted high temperature of 95°F.

e. Suppose the 90% prediction interval for part **d** is (645, 1,245). Interpret this interval.

12.55 Refer to Exercise 12.54. The owners of the water adventure park are advised that the prediction model could probably be improved if interaction terms were added. In particular, it is thought that the *rate* at which mean attendance increases as predicted high temperature increases will be greater on weekends than on weekdays. The following model is therefore proposed:

$$E(y) = \beta_0 + \beta_1 x_1 + \beta_2 x_2 + \beta_3 x_3 + \beta_4 x_1 x_3$$

The same 30 days of data used in Exercise 12.54 are again used to obtain the least squares model

$$\hat{y} = 250 - 700x_1 + 100x_2 + 5x_3 + 15x_1 x_3$$

with

$$s_{\hat{\beta}_4} = 3.0 \qquad R^2 = .96$$

a. Graph the predicted day's attendance, y, against the day's predicted high temperature, x_3, for a sunny weekday and for a sunny weekend day. Plot both on the same paper for x_3 between 70 and 100°F. Note the increase in slope for the weekend day. Interpret this.

b. Do the data indicate that the interaction term is a useful addition to the model? Use $\alpha = .05$.

c. Use this model to predict the attendance for a sunny weekday with a predicted high temperature of 95°F.

d. Suppose the 90% prediction interval for part **c** is (800, 850). Compare this result with the prediction interval for the model without interaction in Exercise 12.54, part **e**. Do the relative widths of the confidence intervals support or refute your conclusion about the utility of the interaction term (part **b**)?

e. The owners, noting that the coefficient $\hat{\beta}_1 = -700$, conclude the model is ridiculous because it seems to imply that the mean attendance will be 700 less on weekends than on weekdays. Explain why this is *not* the case.

12.56 The data for Exercise 11.77 are reproduced at the top of the next column. The breeder of thoroughbred horses has been advised that the prediction model could probably be improved if a quadratic term were added. The following model is proposed:

$$y = \beta_0 + \beta_1 x + \beta_2 x^2 + \epsilon$$

where, as before,

y = Lifetime of horse (years)

x = Gestation period of horse (days)

Horse	Gestation Period x (days)	Life Span y (years)
1	416	24
2	279	25.5
3	298	20
4	307	21.5
5	356	22
6	403	23.5
7	265	21

a. Find the least squares prediction equation and test its adequacy.

b. Has the addition of the quadratic term contributed significant information for the prediction of a thoroughbred horse's lifetime? Test H_0: $\beta_2 = 0$ against the alternative H_a: $\beta_2 \neq 0$ using $\alpha = .05$.

c. Construct residual plots for the straight-line and quadratic models, displaying gestation period on the horizontal axis. Does the result of the residual analysis agree with the test you conducted in part **b**? Which is more reliable, and why?

12.57 Many colleges and universities develop regression models for predicting the GPA of incoming freshmen. This predicted GPA can then be used to make admission decisions. Although most models use many independent variables to predict GPA, we will illustrate by choosing two variables:

x_1 = Verbal score on college entrance examination (percentile)

x_2 = Mathematics score on college entrance examination (percentile)

Verbal x_1	Mathematics x_2	GPA y	Verbal x_1	Mathematics x_2	GPA y	Verbal x_1	Mathematics x_2	GPA y
81	87	3.49	83	76	3.75	97	80	3.27
68	99	2.89	64	66	2.70	77	90	3.47
57	86	2.73	83	72	3.15	49	54	1.30
100	49	1.54	93	54	2.28	39	81	1.22
54	83	2.56	74	59	2.92	87	69	3.23
82	86	3.43	51	75	2.48	70	95	3.82
75	74	3.59	79	75	3.45	57	89	2.93
58	98	2.86	81	62	2.76	74	67	2.83
55	54	1.46	50	69	1.90	87	93	3.84
49	81	2.11	72	70	3.01	90	65	3.01
64	76	2.69	54	52	1.48	81	76	3.33
66	59	2.16	65	79	2.98	84	69	3.06
80	61	2.60	56	78	2.58			
100	85	3.30	98	67	2.73			

```
Multiple R              .82527
R Square                .68106
Adjusted R Square       .66382
Standard Error          .40228

Analysis of Variance
                    DF      Sum of Squares      Mean Square
Regression           2           12.78595          6.39297
Residual            37            5.98755           .16183

F =     39.50530      Signif F =   .0000

------------------Variables in the Equation------------------

Variable             B          SE B        Beta         T    Sig T

X1             .02573  4.02357E-03        .59719     6.395   .0000
X2             .03361  4.92751E-03        .63702     6.822   .0000
(Constant)   -1.57054        .49375                 -3.181   .0030
```

The data in the table on page 601 are obtained for a random sample of 40 freshmen at one college.

The SPSS printout corresponding to the model

$$y = \beta_0 + \beta_1 x_1 + \beta_2 x_2 + \epsilon$$

is shown above.

a. Interpret the least squares estimates $\hat{\beta}_1$ and $\hat{\beta}_2$ in the context of this application.

b. Interpret the standard deviation and the coefficient of determination of the regression model in the context of this application.

c. Is this model useful for predicting GPA? Conduct a statistical test to justify your answer.

d. Sketch the relationship between predicted GPA, $\hat{y}$, and verbal score, x_1, for the following mathematics scores: $x_2 = 60, 75,$ and 90.

12.58 Refer to Exercise 12.57. The residuals from the first-order model are plotted against x_1 and x_2 below. Analyze the two plots, and determine whether visual evidence exists that curvature (a quadratic term) for either x_1 or x_2 should be added to the model.

12.59 Refer to Exercises 12.57 and 12.58. The complete second-order model

$$y = \beta_0 + \beta_1 x_1 + \beta_2 x_2 + \beta_3 x_1^2 + \beta_4 x_2^2 + \beta_5 x_1 x_2 + \epsilon$$

is fit to the data given in Exercise 12.57.

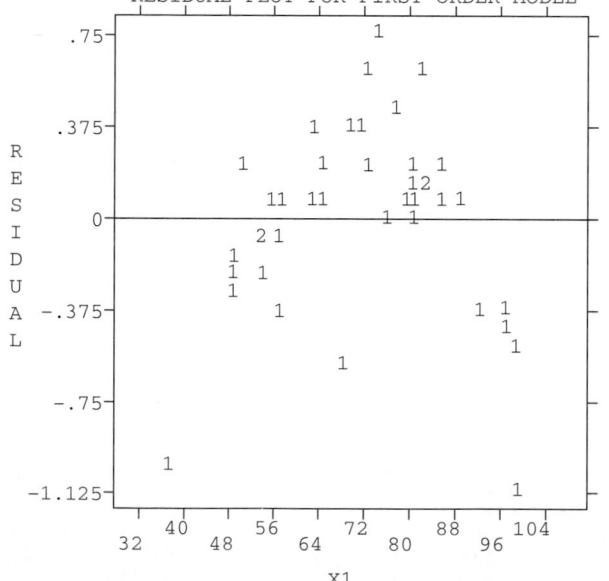

40 cases plotted.

Plot of residuals against x_1 for Exercise 12.58

RESIDUAL PLOT FOR FIRST-ORDER MODEL

40 cases plotted.

Plot of residuals against x_2 for Exercise 12.58

```
Multiple R                .96777
R Square                  .93657
Adjusted R Square         .92724
Standard Error            .18714

Analysis of Variance
                      DF        Sum of Squares        Mean Square
Regression             5              17.58274            3.51655
Residual              34               1.19076             .03502

F =      100.40901     Signif F =   .0000

-------------------Variables in the Equation-------------------

Variable                   B          SE B         Beta          T    Sig T
X1                    .16681       .02124      3.87132      7.852   .0000
X2                    .13760       .02673      2.60754      5.147   .0000
X1SQ            -1.10825E-03  1.17288E-04     -3.71359     -9.449   .0000
X2SQ            -8.43267E-04  1.59423E-04     -2.37284     -5.290   .0000
X1X2            2.410891E-04  1.43974E-04       .49600      1.675   .1032
(Constant)           -9.91676     1.35441                   -7.322   .0000
```

The resulting SPSS printout is shown above.

a. Compare the standard deviations of the first- and second-order regression models. With what relative precision will these two models predict GPA?

b. Test whether this model is useful for predicting GPA. Use $\alpha = .05$.

c. Test whether the interaction term, $\beta_5 x_1 x_2$, is important for the prediction of GPA. Use $\alpha = .10$.

12.60 Refer to Exercises 12.57–12.58. The residuals of the second-order model are plotted against x_1 and x_2 below. Compare the residual plots to those of the first-order model in Exercise 12.58. Given the analyses you have performed in Exercises 12.57–12.59, which of the two models do you think is preferable as a predictor of GPA: the first- or second-order model? [*Note:* In Chapter 13 we show how to conduct a statistical test to compare two models.]

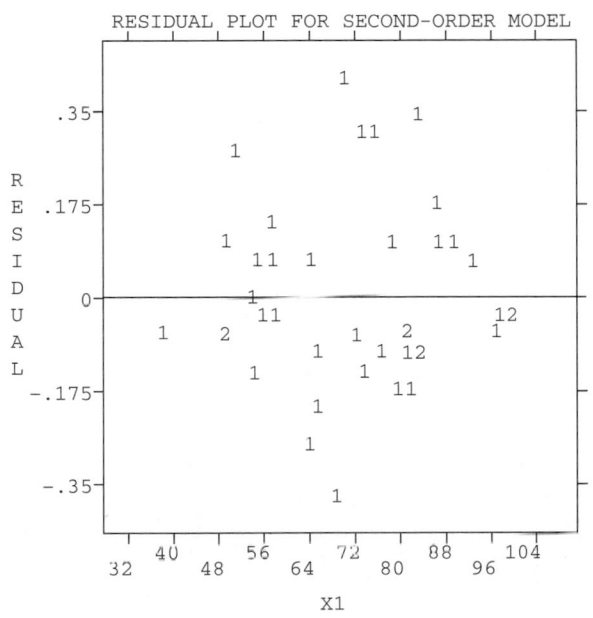

40 cases plotted.

Plot of residuals against x_1 for Exercise 12.60

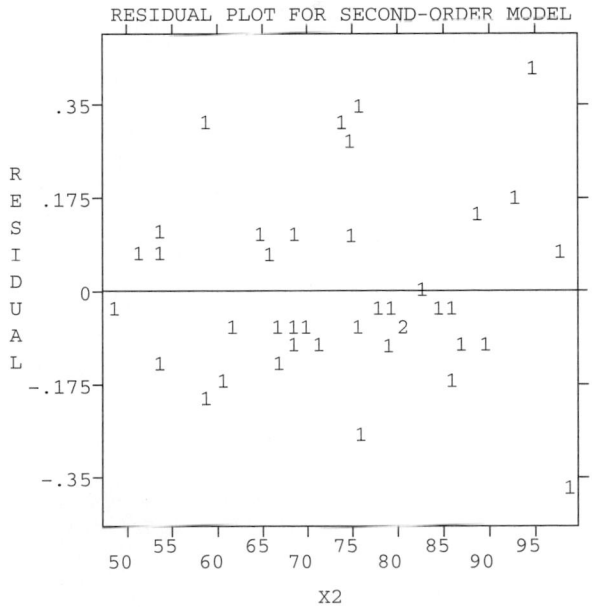

40 cases plotted.

Plot of residuals against x_2 for Exercise 12.60

STUDENT PROJECTS

Note: The use of a computer is required for this project.

This is a continuation of the Student Projects section in Chapter 11, in which you selected three independent variables as predictors of a dependent variable of your choice and obtained at least 10 data values. Now fit the multiple regression model using an available software package:

$$y = \beta_0 + \beta_1 x_1 + \beta_2 x_2 + \beta_3 x_3 + \epsilon$$

where

y = Dependent variable you chose

x_1 = First independent variable chosen

x_2 = Second independent variable chosen

x_3 = Third independent variable chosen

a. Compare the coefficients $\hat{\beta}_1$, $\hat{\beta}_2$, and $\hat{\beta}_3$ to their corresponding slope coefficients in the Chapter 11 Student Projects, where you fitted three separate straight-line models. How do you account for the differences?

b. Calculate the coefficient of determination, R^2, and conduct the F-test of the null hypothesis H_0: $\beta_1 = \beta_2 = \beta_3 = 0$. What is your conclusion?

c. Check the data for multicollinearity. If multicollinearity exists, how should you proceed?

EXPLORING DATA WITH A COMPUTER

Refer to the FTC cigarette brand data described in Appendix B. Suppose we wish to model the relationship of carbon monoxide content y to the tar content x_1, the nicotine content x_2, and the menthol type x_3, where $x_3 = 1$ if menthol and 0 if nonmenthol.

a. Drawing a random sample of 100 cigarette brands from the 962 described in Appendix B, extract y, x_1, x_2, and x_3 for each brand sampled. Use a software package to obtain the least squares fit of the model

$$E(y) = \beta_0 + \beta_1 x_1 + \beta_2 x_2 + \beta_3 x_3$$

1. Interpret the estimated β's of the model.
2. Interpret s and R^2 of the model.
3. Evaluate the usefulness of the model. Conduct a test of global model utility and tests on the individual β parameters.
4. Check for multicollinearity in the sample data. How will this affect the regression results?
5. Now fit the model $E(y) = \beta_0 + \beta_1 x_1 + \beta_2 x_3 + \beta_3 x_1 x_3$. Evaluate the usefulness of the model. Is there evidence of interaction between tar content and menthol type?

6. Which model is preferred, the interaction model or the model $E(y) = \beta_0 + \beta_1 x_1 + \beta_2 x_3$?
7. Use the preferred model to estimate the mean carbon monoxide content for all menthol brands with a tar content of 10 milligrams using a 95% confidence interval.
8. Use the same model to predict the carbon monoxide content for a particular menthol brand with a tar content of 10 milligrams using a 95% prediction interval.
9. Plot the residuals of the model against x_1. Check for evidence that curvature is present.
10. Plot a histogram or stem-and-leaf display of the residuals for the model. Comment on the assumption of normality for the random error component.
11. Use the residual plots to determine whether outliers exist.

b. Repeat part a using the entire set of 962 cigarette brands. Compare the results to those you obtained in part a.

Chapter 13

MODEL BUILDING

Contents

Case Studies

*W*HERE WE'VE BEEN

One of the most important topics in applied statistics—regression analysis—was presented in the last two chapters. Simple linear regression was the topic of Chapter 11, while multiple regression was the topic of Chapter 12. In both chapters, we learned how to fit regression models to a set of data and how to use the model to estimate the mean value of *y* or to predict a future value of *y* for a given value of *x*.

*W*HERE WE'RE GOING

Although we discussed, in general, how to use regression analysis to solve many practical problems, we skipped over one important problem—how to select a model that is appropriate for the given data. No matter how much you know about regression analysis, or how well you can fit a model to a set of data and interpret the results, the information will be of little value if you choose an inappropriate model. In Chapter 13, then, we take up the topic of model building, in which we choose a reasonable model and use the data to modify and improve it.

As indicated in Chapters 11 and 12, the first step in the construction of a regression model is to hypothesize the form of the deterministic portion of the probabilistic model. This *model building,* or model construction, is the key to the success (or failure) of the regression analysis. If the regression model does not reflect, at least approximately, the true nature of the relationship between the mean response $E(y)$ and the independent variables $x_1, x_2, ..., x_k$, the modeling effort will usually be unrewarded.

By **model building**, we mean developing a model that provides a good fit to a set of data, giving good estimates of the mean value of y and good predictions of future values of y for given values of the independent variables. Suppose, for example, you wish to relate a student's score y on an introductory sociology examination to amount of study time x and that (unknown to you) the second-order model

$$E(y) = \beta_0 + \beta_1 x + \beta_2 x^2$$

would permit you to predict y with a very small error of prediction (see Figure 13.1a).But suppose you erroneously choose the first-order model

$$E(y) = \beta_0 + \beta_1 x$$

to explain the relationship between y and x; see Figure 13.1b.

The consequence of choosing the wrong model is clearly demonstrated by comparing Figures 13.1a and 13.1b. The errors of prediction for the second-order model in Figure 13.1a are relatively small compared to those for the first-order model shown in Figure 13.1b. The lesson to be learned from this simple example is very clear. Choosing a good set of independent (predictor) variables $x_1, x_2, ..., x_k$, does not guarantee a good multivariable prediction equation. In addition to selecting independent variables that contain information about y, you must also develop an equation relating y to $x_1, x_2, ..., x_k$ that will provide a good fit to your data.

In the following sections, we present several useful models for relating a response y to one or more predictor variables. In addition, we provide tests and procedures for determining which model is appropriate.

FIGURE 13.1

Two models for relating examination score y to amount of study time x

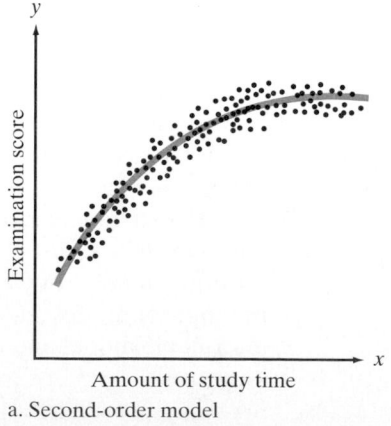

a. Second-order model

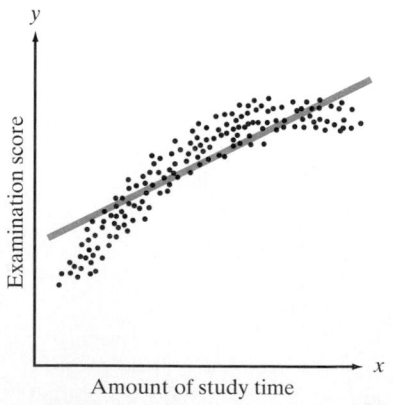

b. First-order model

13.1   THE TWO TYPES OF INDEPENDENT VARIABLES: QUANTITATIVE AND QUALITATIVE

In Chapter 1 we defined two types of data and, correspondingly, two types of variables that may be observed: quantitative and qualitative. In a regression analysis the dependent variable is always quantitative, but the independent variables may be either quantitative or qualitative. As you will see, the way in which an independent variable enters the model depends on its type.

Recall from Chapter 1 that quantitative variables are measured on a natural numerical scale, whereas qualitative variables are measured on a nonnumerical (categorical) scale. Thus, the number of murders committed in a large city per year, the dissolved oxygen content of a river, and the amount of fertilizer applied to an experimental agricultural plot are all examples of quantitative variables. On the other hand, the gender and race of a murder victim, the season of the year in which the water sample is taken, and the types of fertilizer applied are all examples of qualitative variables. Since both the type and the amount of fertilizer used are likely to affect crop yield, we would want to include both in a model describing yield. Thus, we want to be able to include both quantitative and qualitative independent variables in regression models.

Quantitative variables always assume numerical values, whereas qualitative variables usually assume nonnumerical values. The possible values of independent variables are also referred to as *levels.* For example, the type of fertilizer might have three possible levels: A, B, and C. And even if we designate these values arbitrarily as 1, 2, and 3, the numbers still represent categories and the variable is still qualitative. In addition, a variable we normally think of as quantitative, such as income, can sometimes be reported as a level, say low, medium, or high; this makes the variable qualitative for that application. Thus, the distinction between quantitative and qualitative variables is often a function of the type of information available.

DEFINITION 13.1

The possible values of an independent variable are called its **levels**.

EXAMPLE 13.1

In Chapter 12 we considered the problem of predicting executive salaries as a function of several independent variables. Consider the following four independent variables that may affect executive salaries:

a. Number of years of experience
b. Gender of the employee
c. Firm's net asset value
d. Rank of the employee

For each of these independent variables, give its type and describe the nature of the levels you would expect to observe.

Solution

a. The independent variable for the number of years of experience is quantitative since its values are numerical. We would expect to observe levels ranging from 0 to 40 (approximately) years.
b. The independent variable for gender is qualitative since its levels can be described only by the nonnumerical labels "female" and "male."

c. The independent variable for the firm's net asset value is quantitative, with a very large number of possible levels corresponding to various firms' net asset values.

d. Suppose the independent variable for the rank of the employee has three possible levels: supervisor, assistant vice president, and vice president. Since this is a categorical scale, rank is a qualitative independent variable. ▲

Quantitative independent variables are treated differently from qualitative variables in regression modeling. In the next section, we begin our discussion of how quantitative variables are used.

EXERCISES 13.1–13.6

Applying the Concepts

13.1 *New Scientist* (Apr. 3, 1993) published an article on strategies for foiling assassination attempts on politicians. The strategies are based on the findings of researchers at Middlesex University (United Kingdom), who used a multiple regression model for predicting the level *y* of assassination risk. Several of the variables used in the model are listed below. Identify each variable as quantitative or qualitative.
  a. Number of potential assailants
  b. Frequency of exposure (hours per day)
  c. Probability that the assailant will succeed in inflicting injury
  d. Economic status of country
  e. Political status of country

13.2 Multiple regression analysis was used to model the abundance *y* of an individual bird species in transects in the United Kingdom (*Journal of Applied Ecology,* Vol. 32, 1995). The independent variables used in the model, all field boundary attributes, are listed below. Identify each of the variables as quantitative or qualitative.
  a. Transect location (small pasture field, small arable field, or large arable field)
  b. Land use (pasture or arable) adjacent to the transect
  c. Average height of trees in transect
  d. Total number of trees in transect
  e. Length of hedgerow in transect
  f. Height of hedgerow in transect
  g. Width of hedgerow in transect
  h. Width of transect verge
  i. Depth of transect ditch
  j. Width of transect ditch
  k. Length of transect ditch

13.3 An Educational Testing Service (ETS) research scientist used multiple regression analysis to model the final grade point average (GPA) of business and management doctoral students (*Journal of Educational Statistics,* Spring 1993). A list of the potential independent variables measured for each doctoral student in the study is given below.

1. Quantitative Graduate Management Aptitude Test (GMAT) score
2. Verbal GMAT score
3. Undergraduate GPA
4. First-year graduate GPA
5. Student cohort (i.e., year in which student entered doctoral program: 1977, 1979, 1981, 1983, or 1985)
  a. Identify the variables as quantitative or qualitative.
  b. For each quantitative variable, give your opinion on whether the variable is positively or negatively related to final GPA.
  c. As we will learn (in Section 13.5), if a qualitative variable is an important predictor of *y*, each level of the variable yields a different value of $E(y)$. For each of the qualitative variables identified in part **a**, determine the number of different mean GPAs that are possible.

13.4 Over the years, Graduate Record Examination (GRE) scores have been used to aid college administrators in the graduate school admission process. Some educators argue, however, that the GRE is biased against minority students, especially African Americans, who are often unfairly denied admission to graduate study on the basis of test scores alone. The *Journal of Negro Education* (Jan. 1985) reported on a study that used regression to model the GPA, *y*, of graduate students as a function of the independent variables GRE score (x_1) and race (x_2). Classify the independent variables as quantitative or qualitative.

13.5 The *Journal of Human Stress* (Summer 1987) reported on a study of "psychological response of firefighters to chemical fire." The researchers used multiple regression to predict emotional distress as a function of the following independent variables. Identify each independent variable as quantitative or qualitative. For qualitative variables, suggest several levels that might be observed. For quantitative variables, give a range of values (levels) for which the variable might be observed.
  a. Number of preincident psychological symptoms
  b. Years of experience

c. Cigarette smoking behavior
d. Level of social support   **e.** Marital status
f. Age   **g.** Ethnic status
h. Exposure to a chemical fire   **i.** Education level
j. Distance lived from site of incident   **k.** Gender
13.6 *Environmental Science & Technology* (Oct. 1993) published an article that investigated the variables that affect the sorption of organic vapors on clay

minerals. The independent variables and levels considered in the study are listed below. Identify the type (quantitative or qualitative) of each.
a. Temperature ($50°, 60°, 75°, 90°$)
b. Relative humidity ($30\%, 50\%, 70\%$)
c. Organic compound (benzene, toluene, chloroform, methanol, anisole)

13.2   MODELS WITH A SINGLE QUANTITATIVE INDEPENDENT VARIABLE

The most common linear models relating y to a single quantitative independent variable x are those derived from a **pth-order polynomial** expression of the type shown in the accompanying box. Specific models, obtained by assigning values to p, are listed next.

Formula for a pth-Order Polynomial with One Independent Variable

$$E(y) = \beta_0 + \beta_1 x + \beta_2 x^2 + \beta_3 x^3 + \cdots + \beta_p x^p$$

where p is an integer and $\beta_0, \beta_1, \ldots, \beta_p$ are unknown parameters that must be estimated.

---

1. First-Order Model

$$E(y) = \beta_0 + \beta_1 x \qquad \text{(Figure 13.2)}$$

Comments on model parameters

β_0: y-intercept

β_1: Slope of the line

General comments   The first-order model is used when you expect the rate of change in y per unit change in x to remain fairly stable over the range of values of x for which you wish to predict y.* Most relationships between $E(y)$ and x are curvilinear, but the curvature over the range of values of x for which you wish to predict y may be very slight. When this occurs, a first-order (straight-line) model should provide a good fit to your data.

FIGURE 13.2

Graph of a first-order model

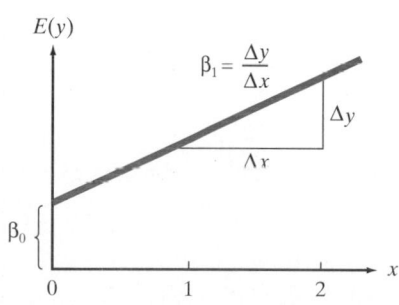

*The symbol Δy represents the change in y corresponding to a change in x equal to Δx.

FIGURE 13.3

Graphs for two second-order models

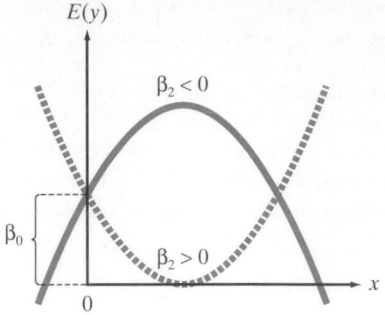

2. Second-Order Model

$$E(y) = \beta_0 + \beta_1 x + \beta_2 x^2 \qquad \text{(Figure 13.3)}$$

Comments on model parameters

β_0: y-intercept

β_1:   Shift parameter: Changing the value of β_1 shifts the parabola to the right or left (increasing the value of β_1 causes the parabola to shift to the left)

β_2:   Rate of curvature

General comments   A second-order model traces a parabola, one that opens either downward ($\beta_2 < 0$) or upward ($\beta_2 > 0$). Since most relationships possess some curvature, a second-order model is often a good choice to relate y to x.

3. Third-Order Model

$$E(y) = \beta_0 + \beta_1 x + \beta_2 x^2 + \beta_3 x^3 \qquad \text{(Figure 13.4)}$$

Comments on model parameters

β_0: y-intercept

β_3:   The magnitude of β_3 controls the rate of reversal of curvature for the curve

General comments   Reversals in curvature are not common, but such relationships can be modeled by third- and higher-order polynomials. As can be seen in Figure 13.3, a second-order model contains no reversal in curvature. The slope either continues to increase or continues to decrease as x increases and produces

FIGURE 13.4

Graphs for two third-order models

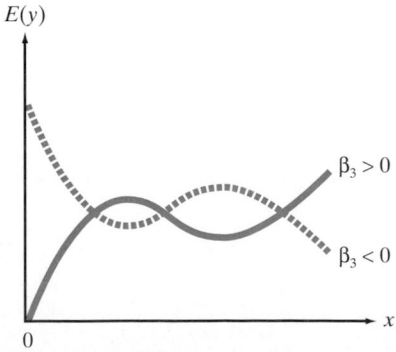

either a trough or a peak. A third-order model (see Figure 13.4) contains one reversal in curvature and produces one peak and one trough. In general, the graph of a kth-order polynomial contains a total of $(k - 1)$ peaks and troughs.

Most functional relationships in nature seem to be smooth (except for random error); that is, they are not subject to rapid and irregular reversals in direction. Consequently, the second-order polynomial model is perhaps the most useful of those described here. To develop a better understanding of how this model is employed, consider the following example.

EXAMPLE 13.2

Power companies have to be able to predict the peak power load at their various stations in order to operate efficiently. The peak power load is the maximum amount of power that must be generated each day in order to meet demand.

Suppose a power company in the southern part of the United States decides to model daily peak power load, y, as a function of the daily high temperature, x, and the model is to be constructed for the summer months when demand is greatest. Although we would expect the peak power load to increase as the high temperature increases, the *rate* of increase in $E(y)$ might also increase as x increases. That is, a 1-unit increase in high temperature from 100 to 101°F might result in a larger increase in power demand than would a 1-unit increase from 80 to 81°F. Therefore, we postulate the second-order model

$$E(y) = \beta_0 + \beta_1 x + \beta_2 x^2$$

and we expect β_2 to be positive.

A random sample of 25 summer days is selected, and the data are shown in Table 13.1. Fit a second-order model using these data, and test the hypothesis that the power load increases at an increasing *rate* with temperature—i.e., that $\beta_2 > 0$.

Solution

The ASP printout shown in Figure 13.5 gives the least squares fit of the second-order model using the data in Table 13.1. The prediction equation (shaded) is

$$\hat{y} = 385.048 - 8.293x + .05982x^2$$

TABLE 13.1   Power Load Data

Temperature (°F)	Peak Load (megawatts)	Temperature (°F)	Peak Load (megawatts)
94	136.0	100	151.9
96	131.7	79	106.2
95	140.7	97	153.2
108	189.3	98	150.1
67	96.5	87	114.7
88	116.4	76	100.9
89	118.5	68	96.3
84	113.4	92	135.1
90	132.0	100	143.6
106	178.2	85	111.4
67	101.6	89	116.5
71	92.5	74	103.9
		86	105.1

FIGURE 13.5

Portion of the ASP printout for the second-order model of Example 13.2

```
            QUADRATIC MODEL FOR PEAK POWER LOAD

MODEL:  Y = -8.29253X + 0.0598234XX + 385.048CNST

            COEF.       SD. ER.      t(22)      P-VALUE  PT. R SQ.
    X    -8.29253     1.29905     -6.38356  2.00975E-6   0.649401
    XX    0.0598234   7.54855E-3   7.92514  6.8979E-8    0.74059
CNST    385.048      55.1724       6.97899  5.2661E-7    0.688854

R SQ. = 0.959363,  ADJ. R SQ. = 0.955668,  D. W. = 2.20408
SD. ER. EST. = 5.3762,  F(2/22) = 259.687 (P-VALUE = 4.99085E-16)
```

A plot of this equation and the observed values is given in Figure 13.6.

We now test to determine whether the sample value, $\hat{\beta}_2 = .05982$, is large enough to conclude, *in general,* that the power load increases at an increasing rate with temperature:

$$H_0: \beta_2 = 0$$

$$H_a: \beta_2 > 0$$

Test statistic: $\dfrac{\hat{\beta}_2}{s_{\hat{\beta}_2}} = 7.93$ (shaded in Figure 13.5)

For $\alpha = .05$, $n = 25$, and $k = 2$, we will reject H_0 if

$$t > t_{.05}$$

where $t_{.05} = 1.717$ (from Table VI in Appendix A) possesses $n - (k + 1) = 22$ degrees of freedom. Since the test statistic exceeds $t_{.05} = 1.717$, we reject H_0 and conclude that the mean power load increases at an increasing rate with temperature. From Figure 13.5, the observed significance level (shaded) of the upper-tailed test is $p = .000000069/2 \approx 0$, which supports our conclusion. ▲

FIGURE 13.6

Plot of the observations and the second-order least squares fit, Example 13.2

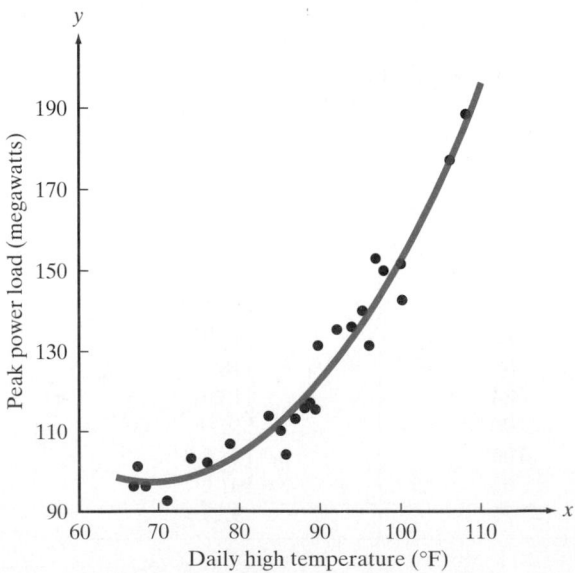

EXERCISES 13.7–13.22

Note: Exercises marked with 💻 *require the use of a computer.*

Learning the Mechanics

13.7 The accompanying graphs depict pth-order polynomials with one independent variable. For each graph, identify the order of the polynomial. Find the value of β_0 and β_1 for each.

a. $E(y)$

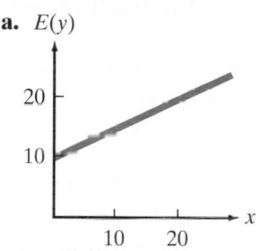

b. $E(y)$

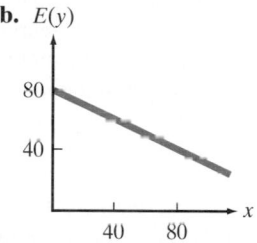

13.8 The accompanying graphs depict pth-order polynomials for one independent variable. For each graph, identify the order of the polynomial, the value of β_0, and the sign of β_2.

a. $E(y)$

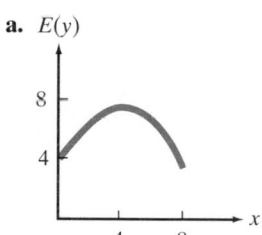

b. $E(y)$

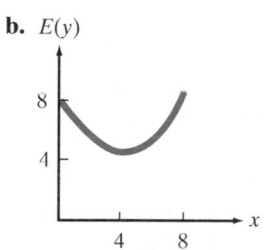

13.9 Graph the following polynomials and identify the order of each on your graph:
a. $E(y) = 2 + 3x$ **b.** $E(y) = 2 + 3x^2$
c. $E(y) = 1 + 2x + 2x^2 + x^3$
d. $E(y) = 2x + 2x^2 + x^3$
e. $E(y) = 2 - 3x^2$ **f.** $E(y) = -2 + 3x$

13.10 The following graphs depict pth-order polynomials for one independent variable:

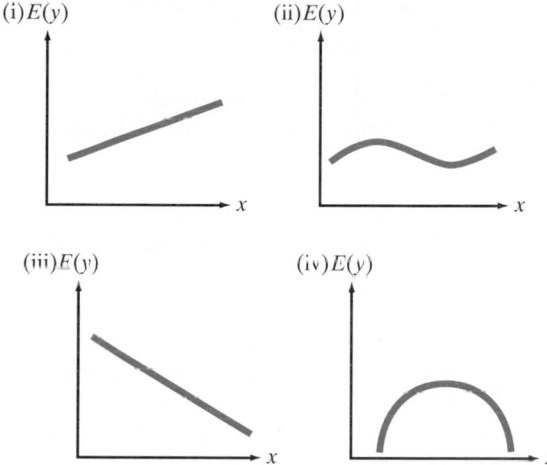

a. For each graph, identify the order of the polynomial.
b. Using the parameters β_0, β_1, β_2, etc., write an appropriate model relating $E(y)$ to x for each graph.
c. By examining the graphs, you can determine the signs ($+$ or $-$) of many of the parameters in the models of part **b**. Give the signs of those parameters that can be determined.

13.11 MINITAB was used to fit the model

$$y = \beta_0 + \beta_1 x + \beta_2 x^2 + \epsilon$$

to $n = 17$ data points. The printout is shown below.
a. Identify the least squares prediction equation in the printout.
b. Plot the prediction equation.
c. Specify the null and alternative hypotheses you would use to determine whether y decreases at an increasing rate as x increases.

```
The regression equation is
Y  = 63.1 + 1.17 X - 0.0374 XX

Predictor        Coef       Stdev     t-ratio        p
Constant        63.11       11.92        5.30    0.000
X              1.1683      0.8276        1.41    0.148
XX           -0.03742     0.01234       -3.03    0.005

s = 9.289      R-sq = 86.0%     R-sq(adj) = 84.0%

Analysis of Variance

SOURCE        DF          SS          MS         F        p
Regression     2      7434.4      3717.2     43.08    0.000
Error         14      1208.1        86.3
Total         16      8642.5
```

d. Conduct the hypothesis test of part **c** using $\alpha = .05$. Interpret your result.

13.12 Suppose $E(y)$ can best be modeled by a second-order polynomial in x, where x is a quantitative variable. Write the probabilistic model for y.

13.13 Consider the following polynomial model:

$$E(y) = 5 - 3x + x^2$$

a. Give the order of this polynomial.

b. Sketch the curve corresponding to the equation for $E(y)$.

c. How would the graph change if the coefficient of x^2 were negative rather than positive?

13.14 Consider the following polynomial model:

$$E(y) = 2 - 4x$$

a. Give the order of this polynomial.

b. Sketch the curve corresponding to the equation for $E(y)$.

c. How would the graph change if the coefficient of x were positive instead of negative?

Applying the Concepts

13.15 The optomotor responses of tree frogs were studied in the *Journal of Experimental Zoology* (Sept. 1993). Microspectrophotometry was used to measure the threshold quantal flux (the light intensity at which the optomotor response was first observed) of tree frogs tested at different spectral wavelengths. The data revealed the relationship between the log of quantal flux (y) and wavelength (x) shown in the accompanying graph. Hypothesize a model for $E(y)$ that corresponds to the graph.

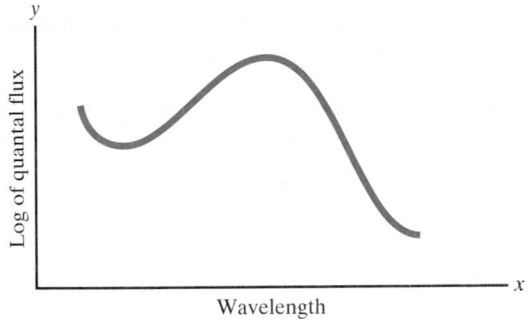

13.16 After spillage, light crude oil and refined oil can lose up to 75% of their volume to evaporation. Consequently, an understanding of the evaporation process is important for cleaning up oil spills. The *Journal of Hazardous Materials* (July 1995) presented a literature review of models designed to predict oil spill evaporation. One of the models presented is the polynomial model

$$E(y) = \beta_0 + \beta_1 x + \beta_2 x^2 + \beta_3 x^3$$

where y represents an oil spill variable and x represents the equivalent paraffin carbon number (that is, the number of carbon atoms characteristic of the type of oil).

a. What is the order of the polynomial model?

b. Give a rough sketch of the relationship between y and x hypothesized by the model.

c. Explain how you could determine whether the cubic term, x^3, is necessary in the model.

13.17 The amount of pressure used to produce a certain plastic is thought to be related to the strength of the plastic. Researchers hypothesize that, below a certain level, increases in pressure increase the strength of the plastic; at some point, however, additional increases in pressure will have a detrimental effect on its strength. Write a model in which the relationship of the plastic strength y to pressure x reflects this hypothesis. Sketch the model.

13.18 Underinflated or overinflated tires can increase tire wear. A new tire was tested for wear at different pressures with the results shown in the following table.

Pressure x (pounds per square inch)	Mileage y (thousands)
30	29
31	32
32	36
33	38
34	37
35	33
36	26

a. Plot the data on a scattergram.

b. If you were given only the information for $x = 30, 31, 32, 33$, what kind of model would you suggest? For $x = 33, 34, 35, 36$? For all the data?

13.19 An experiment was conducted to relate the number of visits per patient per year to a doctor's office to the age x of the patient. It is suspected that young and old patients tend to seek medical assistance more frequently than those of middle age. Write an appropriate model relating mean number of visits per patient per year to the patient's age.

13.20 A veterinarian who works for a large midwestern pig cooperative has developed a daily vitamin pellet that she believes will substantially increase the weight of mature pigs within 1 month. However, she is uncertain of the relationship between the daily dosage (amount of the vitamin) and the percentage gain in weight after 1 month. To understand this relationship better, she randomly selects 16 pigs of the same age and weight and feeds them different dosage levels for a 1-month trial period. The data are given in the next table.

```
The regression equation is
Y   = -6.17 + 2.04 X - 0.0323 XX

Predictor        Coef       Stdev      t-ratio         p
Constant       -6.173       1.666       -3.71      0.003
X               2.026       0.185       11.03      0.000
XX            -0.03231     0.00489      -6.60      0.000

s = 1.243      R-sq = 95.5%      R-sq(adj) = 95.1%

Analysis of Variance

SOURCE          DF          SS          MS          F          p
Regression       2      718.168     359.084     232.42     0.000
Error           22       33.992       1.545
Total           24      752.160
```

Weight Gain (% of original weight)	Daily Pellet Dose
82	0
78	0
80	1
87	1
87	2
95	2
97	3
90	3
90	4
95	4
89	5
93	5
90	6
85	6
84	7
90	7

a. Plot the data on a scattergram.

b. Fit the model $E(y) = \beta_0 + \beta_1 x + \beta_2 x^2$ to the data.

c. Is there evidence to support the inclusion of the quadratic (second-order) term in the model of part **b**? Test using $\alpha = .05$.

d. Plot the fitted model of part **b** on the scattergram of part **a**.

e. Use the fitted model to estimate the optimum daily dosage (i.e., the dosage that promotes the greatest weight gain). [*Note:* We would want to express this estimate as a confidence interval, but its computation is beyond the scope of this text. The procedure is described in the references. You may also find that it can be obtained by using your software program package.]

13.21 An economist has proposed the following model to describe the relationship between the number of items produced per day (output) and the number of work-hours expended per day (input) in a particular production process:

$$y = \beta_0 + \beta_1 x + \beta_2 x^2 + \epsilon$$

where

$y =$ Number of items produced per day

$x =$ Number of work-hours per day

A portion of the MINITAB computer printout that results from fitting this model to a sample of 25 weeks of production data is shown above. Do the data provide sufficient evidence to indicate that the *rate* of increase in output per unit increase of input decreases as the input increases? Test using $\alpha = .05$.

13.22 Does exercise improve the human immune system? An experiment was conducted by a physiologist at the University of Florida to determine whether such a relationship exists. Thirty subjects volunteered to participate in the study. The amount of immunoglobulin, IgG (an indicator of long-term immunity), and the maximum oxygen uptake (a measure of aerobic fitness level) were recorded for each subject. The resulting data are given in the table on page 616.

a. Construct a scattergram for the IgG–maximum oxygen uptake data.

b. Hypothesize a probabilistic model relating IgG to maximum oxygen uptake.

c. The SAS computer printout for fitting the model to the data is shown below the table. Is there sufficient evidence to indicate that the model provides information for the prediction of IgG, y? Test using $\alpha = .05$.

d. Give the observed significance level for the test of part **c** and interpret it.

e. Does the second-order term contribute information for the prediction of y? Test using $\alpha = .05$.

f. Give the observed significance level for the test of part **e** and interpret it.

Subject	IgG, y	Maximum Oxygen Uptake, x	Subject	IgG, y	Maximum Oxygen Uptake, x
1	881	34.6	16	1,660	52.5
2	1,290	45.0	17	2,121	69.9
3	2,147	62.3	18	1,382	38.8
4	1,909	58.9	19	1,714	50.6
5	1,282	42.5	20	1,959	69.4
6	1,530	44.3	21	1,158	37.4
7	2,067	67.9	22	965	35.1
8	1,982	58.5	23	1,456	43.0
9	1,019	35.6	24	1,273	44.1
10	1,651	49.6	25	1,418	49.8
11	752	33.0	26	1,743	54.4
12	1,687	52.0	27	1,997	68.5
13	1,782	61.4	28	2,177	69.5
14	1,529	50.2	29	1,965	63.0
15	969	34.1	30	1,264	43.2

```
Dependent Variable: Y

                          Analysis of Variance

                                 Sum of          Mean
           Source       DF       Squares        Square     F Value    Prob>F

           Model         2   4602145.2169  2301072.6085    203.113    0.0001
           Error        27   305883.74976   11329.02777
           C Total      29   4908028.9667

              Root MSE      106.43791     R-square    0.9377
              Dep Mean     1557.63333     Adj R-sq    0.9331
              C.V.            6.83331

                            Parameter Estimates

                          Parameter      Standard    T for H0:
           Variable   DF    Estimate        Error   Parameter=0    Prob > |T|

           INTERCEP    1  -1462.141458  411.48209286    -3.553       0.0014
           X           1     88.214044   16.47736086     5.354       0.0001
           XX          1     -0.535378    0.15822840    -3.384       0.0022
```

13.3 MODELS WITH TWO OR MORE QUANTITATIVE INDEPENDENT VARIABLES

The best way to understand how to write models relating a mean response $E(y)$ to a set of quantitative independent variables is to examine the different types of models in which only two independent variables are involved. If you can see geometrically (using three-dimensional figures) the relationships implied by these two-variable models, you will more easily understand the relationships in models that contain three or more quantitative independent variables.

1. First-Order Model with Two Independent Variables

$$E(y) = \beta_0 + \beta_1 x_1 + \beta_2 x_2$$

FIGURE 13.7

Graph of a first-order
model

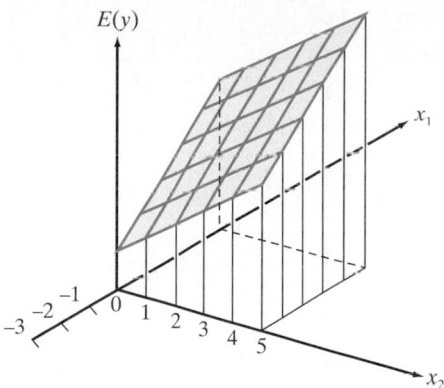

Comments on the parameters

β_0: y-intercept, the value of $E(y)$ when $x_1 = x_2 = 0$

β_1: Change in $E(y)$ for a 1-unit increase in x_1 when x_2 is held fixed

β_2: Change in $E(y)$ for a 1-unit increase in x_2 when x_1 is held fixed

General comments   The graph in Figure 13.7 traces a **response surface** (in contrast to the response *curve* that is used to relate $E(y)$ to a single quantitative variable). Particularly, a first-order model relating $E(y)$ to two independent quantitative variables, x_1 and x_2, graphs as a plane in a three-dimensional space. The plane traces the value of $E(y)$ for every combination of values (x_1, x_2) that correspond to points in the x_1, x_2-plane. Many response surfaces in the real world are well-behaved (i.e., smooth), and they possess curvature. Consequently, a first-order model is appropriate only if the response surface is fairly flat over the (x_1, x_2) region that is of interest.

The assumption that a first-order model will adequately characterize the relationship between $E(y)$ and the variables x_1 and x_2 is equivalent to assuming that x_1 and x_2 do not interact; i.e., you assume that the effect on $E(y)$ of changes in x_1 are the same, regardless of the value of x_2 (and vice versa). Thus, no interaction means that the effect of changes in one variable (say x_1) on $E(y)$ is *independent* of the value of the second variable (say x_2). For example, if we assign values to x_2 in a first-order model, the graph of $E(y)$ as a function of x_1 would produce parallel lines, as shown in Figure 13.8. These lines, called **contour lines**, show the contours

FIGURE 13.8

Graph indicating no
interaction between
x_1 and x_2

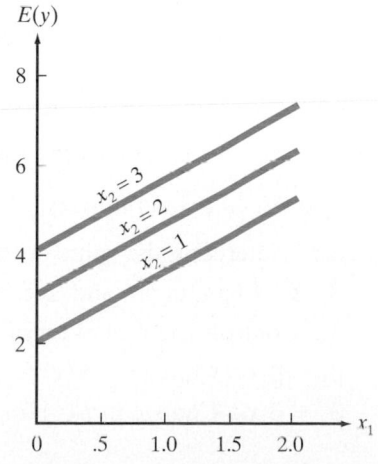

FIGURE 13.9

Graph for an interaction model (second-order)

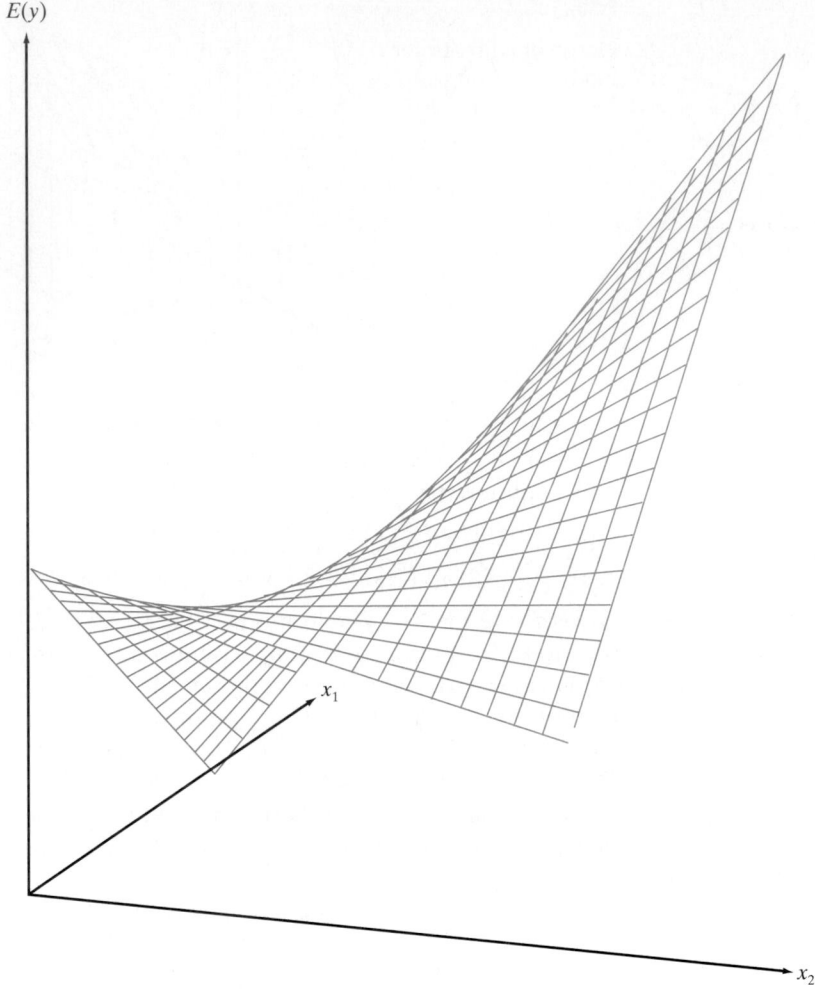

of the surface when it is sliced by three planes, each of which is parallel to the $E(y)$, x_1-plane, at distances $x_2 = 1, 2$, and 3 from the origin.

> **DEFINITION 13.2**
>
> Two variables x_1 and x_2 are said to **interact** if the change in $E(y)$ for a 1-unit increase in x_1 (when x_2 is held fixed) is dependent on the value of x_2.

2. Interaction Model (Second-Order) with Two Independent Variables

$$E(y) = \beta_0 + \beta_1 x_1 + \beta_2 x_2 + \beta_3 x_1 x_2$$

Comments on the parameters

β_0: y-intercept, the value of $E(y)$ when $x_1 = x_2 = 0$

β_1, β_2: Changing β_1 and β_2 causes the surface to shift along the x_1- and x_2-axes

β_3: Controls the rate of twist in the surface (Figure 13.9)

$\beta_1 + \beta_3 x_2$: Change in $E(y)$ for a 1-unit increase in x_1, when x_2 is held fixed

$\beta_2 + \beta_3 x_1$: Change in $E(y)$ for a 1-unit increase in x_2, when x_1 is held fixed

FIGURE 13.10

Graph indicating interaction between x_1 and x_2

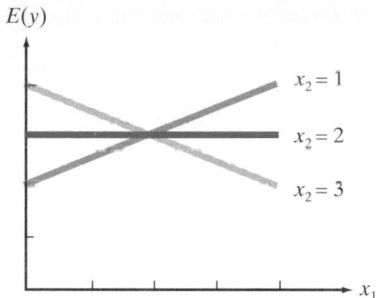

General comments   The model described here is said to be second-order because the highest-order term (x_1x_2) in x_1 and x_2 is of order 2; that is, the sum of the exponents of x_1 and x_2 equals 2.* This interaction model traces a twisted surface in a three-dimensional space (Figure 13.9). A graph of $E(y)$ as a function of x_1 for given values of x_2 (say $x_2 = 1, 2,$ and 3) produces nonparallel contour lines (see Figure 13.10), thus indicating that the change in $E(y)$ for a given change in x_1 is dependent on the value of x_2—that is, x_1 and x_2 interact.

It is extremely important to keep interaction in mind because it is too easy to get in the habit of fitting first-order models and individually examining the relationships between $E(y)$ and a set of independent variables, $x_1, x_2, …, x_k$. Because such a procedure is meaningless when interaction exists (which is, at least to some extent, almost always the case), it can lead to gross errors in interpretation. For example, suppose the relationship between $E(y)$ and x_1 and x_2 is as shown in Figure 13.10 and you have observed y for each of the $n = 9$ combinations of values of x_1 and x_2, $x_1 = 1, 2, 3$ and $x_2 = 1, 2, 3$. If you were to fit a first-order model in x_1 and x_2 to the data, the fitted plane would be (except for random error) approximately parallel to the x_1, x_2-plane, thus suggesting that x_1 and x_2 contribute very little information about $E(y)$. Figure 13.10 clearly indicates that this is not the case. Fitting a first-order model to the data would not allow for the twist in the true surface and would therefore give a false impression of the relationship between $E(y)$ and x_1 and x_2. The procedure for detecting interaction between two independent variables can be seen by examining the model. The interaction model differs from the noninteraction first-order model only in the inclusion of the $\beta_3 x_1 x_2$ term.

Interaction model: $E(y) = \beta_0 + \beta_1 x_1 + \beta_2 x_2 + \beta_3 x_1 x_2$
First-order model: $E(y) = \beta_0 + \beta_1 x_1 + \beta_2 x_2$

Therefore, to test for the presence of interaction, we test

$$H_0: \beta_3 = 0 \qquad \text{(No interaction)}$$

against the alternative hypothesis

$$H_a: \beta_3 \neq 0 \qquad \text{(Interaction)}$$

using the familiar Student t-test of Section 12.4.

---

*The *order of a term* is equal to the sum of the exponents of the variables included in the term, while the *order of a model* is determined by its highest-order term. Therefore, the interaction model is second-order.

3. Complete Second-Order Model with Two Independent Variables

$$E(y) = \beta_0 + \beta_1 x_1 + \beta_2 x_2 + \beta_3 x_1 x_2 + \beta_4 x_1^2 + \beta_5 x_2^2$$

Comments on the parameters

β_0: y-intercept, the value of $E(y)$ when $x_1 = x_2 = 0$

β_1, β_2: Changing β_1 and β_2 causes the surface to shift along the x_1- and x_2-axes

β_3: Controls the rotation of the surface

β_4, β_5: Signs and values of these parameters control the type of surface and the rates of curvature

Three types of surfaces are produced by a second-order model*: a **paraboloid** that opens upward (Figure 13.11a), a paraboloid that opens downward (Figure 13.11b), and a **saddle-shaped surface** (Figure 13.11c).

General comments   A complete second-order model is the three-dimensional equivalent of a second-order model in a single quantitative variable. Instead of tracing parabolas, it traces paraboloids and saddle surfaces. Since only a portion of the complete surface is used to fit the data, this model provides a very large variety of gently curving surfaces that can be used to fit data. It is a good choice for a model if you expect curvature in the response surface relating $E(y)$ to x_1 and x_2.

EXAMPLE 13.3

A social scientist would like to relate the number of hours worked per week (outside the home) by a married woman to the number of years of formal education she has completed and the number of children in her family.

a. Identify the dependent variable and the independent variables.
b. Write the first-order model for this example.
c. Modify the model in part b to include an interaction term.
d. Write a complete second-order model for $E(y)$.

Solution

a. The dependent variable is

$y = $ Number of hours worked per week by a married woman

FIGURE 13.11

Graphs for three second-order surfaces

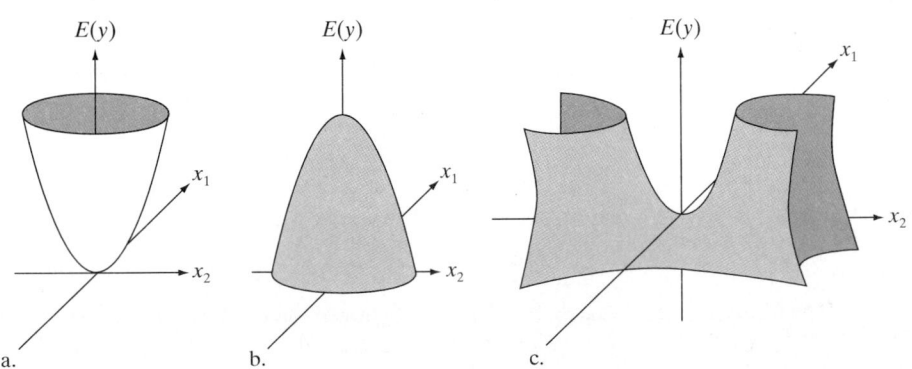

*The saddle-shaped surface (Figure 13.11c) is produced when $\beta_3^2 > 4\beta_4\beta_5$; the paraboloid opens upward (Figure 13.11a) when $\beta_4 + \beta_5 > 0$ and opens downward (Figure 13.11b) when $\beta_4 + \beta_5 < 0$.

The two independent variables, both quantitative in nature, are

x_1 = Number of years of formal education completed by the woman

x_2 = Number of children in the family

b. The first-order model is

$$E(y) = \beta_0 + \beta_1 x_1 + \beta_2 x_2$$

This model would probably not be appropriate in this situation because x_1 and x_2 may interact and/or second-order terms corresponding to x_1^2 and x_2^2 may be needed to obtain a good model for $E(y)$.

c. Adding the interaction term, we obtain

$$E(y) = \beta_0 + \beta_1 x_1 + \beta_2 x_2 + \beta_3 x_1 x_2$$

This model should be better than the model in part b, since we have now allowed for interaction between x_1 and x_2.

d. The complete second-order model is

$$E(y) = \beta_0 + \beta_1 x_1 + \beta_2 x_2 + \beta_3 x_1 x_2 + \beta_4 x_1^2 + \beta_5 x_2^2$$

Since it would not be surprising to find curvature in the response surface, the complete second-order model would be preferred to the models in parts b and c. How can we tell whether the complete second-order model really does provide better predictions of hours worked than the models in parts b and c? The answers to these and similar questions are examined in Section 13.4.   ▲

Models relating a mean response $E(y)$ to more than two quantitative variables are extensions of the corresponding models in two quantitative variables. For example, a first-order model for three quantitative independent variables, x_1, x_2, and x_3, is

> *First-order model:* $E(y) = \beta_0 + \beta_1 x_1 + \beta_2 x_2 + \beta_3 x_3$

As in the case of two independent variables, this model implies that the relationship between $E(y)$ and any one variable, say x_1, is linear when x_2 and x_3 are held constant. Further, it implies that there is no interaction among x_1, x_2, and x_3; that is, the slope of the linear relation is exactly the same, regardless of the values of x_2 and x_3. If one of the independent variables, say x_3, is held constant, the graph of $E(y)$ as a function of x_1 and x_2 is a plane. (See Figure 13.7.) Thus, for given values of β_0, β_1, β_2, and β_3, $E(y)$ will graph as a set of parallel planes, one for each different value of x_3.

Similarly, as in the case of two quantitative independent variables, we can add second-order cross-product terms, $x_1 x_2$, $x_1 x_3$, and $x_2 x_3$, to the first-order model and obtain a second-order model that allows interaction among x_1, x_2, and x_3.

> *Interaction model:* $E(y) = \beta_0 + \beta_1 x_1 + \beta_2 x_2 + \beta_3 x_3 + \beta_4 x_1 x_2 + \beta_5 x_1 x_3 + \beta_6 x_2 x_3$

If one of the independent variables, say x_3, is held constant, the graph of $E(y)$ as a function of x_1 and x_2 is a twisted plane of the type shown in Figure 13.9.

Finally, we could add second-order terms involving x_1^2, x_2^2, and x_3^2 to the interaction model and obtain the complete second-order model:

> *Complete second-order model:* $E(y) = \beta_0 + \beta_1 x_1 + \beta_2 x_2 + \beta_3 x_3 + \beta_4 x_1 x_2 +$
> $\beta_5 x_1 x_3 + \beta_6 x_2 x_3 + \beta_7 x_1^2 + \beta_8 x_2^2 + \beta_9 x_3^2$

If one of the independent variables, say x_3, is held constant, the graph of $E(y)$ as a function of x_1 and x_2 will trace a conic surface. (See Figure 13.11.)

Most relationships between $E(y)$ and two or more quantitative independent variables are second-order and require the use of either the interactive or the complete second-order model to obtain a good fit to a data set. As in the case of a single quantitative independent variable, however, the curvature in the response surface may be very slight over the range of values of the variables in the data set. When this happens, a first-order model may provide a good fit to the data.

How can you tell which model to use in a particular situation? We answer this question in Section 13.4.

EXERCISES 13.23–13.34

Note: Exercises marked with ⬛ *require the use of a computer.*

Learning the Mechanics

13.23 Consider an application in which you are trying to relate a response y to two independent variables, x_1 and x_2.
  a. Write a first-order model relating $E(y)$ to x_1 and x_2.
  b. Modify the model you constructed in part **a** to include an interaction term.
  c. Modify the model you constructed in part **b** to make it a complete second-order model.

13.24 Suppose the true relationship between $E(y)$ and the quantitative independent variables x_1 and x_2 is described by the first-order model

$$E(y) = 3 + x_1 - 2x_2$$

  a. Describe the corresponding response surface.
  b. Plot the contour lines of the response surface for $x_1 = 2, 3, 4$, where $0 \le x_2 \le 5$.
  c. Plot the contour lines of the response surface for $x_2 = 2, 3, 4$, where $0 \le x_1 \le 5$.
  d. Use the contour lines you plotted in parts **b** and **c** to explain how changes in the settings of x_1 and x_2 affect $E(y)$.

  e. Use your graph from part **b** to determine how much $E(y)$ changes when x_1 is changed from 4 to 2 and x_2 is simultaneously changed from 1 to 2.

13.25 Suppose the true relationship between $E(y)$ and the quantitative independent variables x_1 and x_2 is

$$E(y) = 3 + x_1 + 2x_2 - x_1x_2$$

  a. Describe the corresponding response surface.
  b. Plot the contour lines of the response surface for $x_1 = 0, 1, 2$, where $0 \le x_2 \le 5$.
  c. Explain why the contour lines you plotted in part **b** are not parallel.
  d. Use the contour lines you plotted in part **b** to explain how changes in the settings of x_1 and x_2 affect $E(y)$.
  e. Use your graph from part **b** to determine how much $E(y)$ changes when x_1 is changed from 2 to 0 and x_2 is simultaneously changed from 4 to 5.

13.26 If two variables, x_1 and x_2, do not interact, how would you describe their effect on the mean response $E(y)$?

13.27 MINITAB was used to fit the model

$$y = \beta_0 + \beta_1x_1 + \beta_2x_2 + \beta_3x_1x_2 + \epsilon$$

to $n = 15$ data points (printout below).

```
The regression equation is
Y  = -2.55 + 3.82 X1 + 2.63 X2 - 1.29 X1X2

Predictor      Coef      Stdev    t-ratio       p
Constant     -2.550      1.142      -2.23   0.043
X1            3.815      0.529       7.22   0.000
X2            2.630      0.344       7.64   0.000
X1X2         -1.285      0.159      -8.06   0.000

s = 0.713      R-sq = 85.6%     R-sq(adj) = 81.6%

Analysis of Variance

SOURCE      DF        SS         MS        F       p
Regression   3    33.149     11.050    21.75   0.000
Error       11     5.587      0.508
Total       14    38.736
```

a. What is the prediction equation for the response surface?

b. Describe the geometric form of the response surface of part **a**.

c. Plot the prediction equation for the case when $x_2 = 1$. Do this twice more on the same graph for the cases when $x_2 = 3$ and $x_2 = 5$.

d. Explain what it means to say that x_1 and x_2 interact. Explain why your graph of part **c** suggests that x_1 and x_2 interact.

e. Specify the null and alternative hypotheses you would use to test whether x_1 and x_2 interact.

f. Conduct the hypothesis test of part **e** using $\alpha = .01$.

Applying the Concepts

13.28 Refer to the *Environmental Science & Technology* study of sorption of organic vapors, Exercise 13.6. Consider modeling the retention coefficient y as a function of both

$$x_1 = \text{temperature (degrees)}$$

$$x_2 = \text{relative humidity (percent)}$$

a. Write a first-order model for $E(y)$.

b. Write a complete second-order model for $E(y)$.

c. Write a model for $E(y)$ that hypothesizes (i) that the relationships are linear and (ii) that the relationship between retention (y) and temperature (x_1) depends on relative humidity (x_2).

13.29 Refer to the *Journal of Hazardous Materials* study of oil spill evaporation, Exercise 13.16. Another model discussed in the article used boiling point (x_1) and API specific gravity (x_2) to predict the molecular weight (y) of the oil that is spilled. A complete second-order model for y was proposed.

a. Write the equation of the model.

b. Identify the terms in the model that allow for curvilinear relationships.

13.30 Baseball fans are involved in a never-ending search for variables (and a model) that will enable them to predict a professional baseball team's success for a given baseball season. In a *New York Times* review of the book *The Hidden Game of Baseball* (Doubleday, 1984), the reviewer wrote that the book's authors "measure players' offensive contributions by what they call a linear weights measure, which gives separate values to at-bats, singles, doubles, ..., home runs, ..., stolen bases, and times caught stealing."

a. Would regression analysis, using a first-order model, produce the *New York Times* description of a "linear weights measure"? Explain.

b. List some of the independent variables that you think might be related to a professional base-ball team's percentage of games won. Construct a first-order model to relate percentage of games won to these variables.

c. Can you propose a better model than the model in part **b**? Explain.

13.31 Refer to Exercise 12.12. Research in the *Sociology of Sport Journal* (Mar. 1992) examined the relationship of college athletes' grade point averages (GPAs, y) to their Scholastic Assessment Test (SAT, x_1) scores and their high school GPAs (x_2). Suppose you wish to model the expected college GPA as a function of the other two variables for college athletes.

a. Are these variables quantitative or qualitative? Explain.

b. Write the first-order model for $E(y)$. Interpret each of the β parameters in terms of this application.

c. Write the complete second-order model for $E(y)$.

d. For the second-order model of part **c**, specify the appropriate null and alternative hypotheses for testing whether the second-order terms contribute information for the prediction of the college athletes' GPA.

13.32 The dissolved oxygen content y in rivers and streams is related to the amount x_1 of nitrogen compounds per liter of water and the temperature x_2 of the water. Write the complete second-order model relating $E(y)$ to x_1 and x_2.

13.33 An economist is interested in modeling the relationship between quarterly sales of central air-conditioning systems (in $ thousands) for single-family homes in the United States and two quantitative independent variables, housing starts in the previous quarter (in thousands), x_1, and the Gross Domestic Product (in billions of 1972 dollars), x_2. Data were collected, and a second-order model was fit. The MINITAB printout (on page 624) describes the least squares prediction equation. In the prediction equation, $x_3 = x_1^2$ and $x_4 = x_2^2$.

a. Write the prediction equation for the response surface.

b. Describe the geometric form of the response surface of part **a**.

c. Do the data provide sufficient evidence to conclude that the model hypothesized by the economist is useful for predicting quarterly sales of central air-conditioning systems? Test using $\alpha = .01$.

d. Does it appear that the variation in air-conditioner sales could be adequately explained by a less complex regression model? Explain. [*Note:* In the next section we discuss a formal procedure for making this inference.]

```
The regression equation is
Y  = 149.5 + .47 X1 - .10 X2 - .0005 X3 + .0000 X4

Predictor        Coef       Stdev     t-ratio        p
Constant       148.50       224.5        0.66     0.662
X1              0.472       0.126        3.74     0.014
X2             -0.099       0.162       -0.61     0.708
X3          -0.000535    0.000359       -1.49     0.152
X4           0.0000153   0.0000294        0.52     0.887

s = 6.884      R-sq = 92.0%     R-sq(adj) = 89.1%

Analysis of Variance

SOURCE        DF          SS          MS        F        p
Regression     4     6002.08     1500.52    31.66    0.000
Error         11      521.36       47.40
Total         15     6523.44
```

13.34 A supermarket chain is interested in exploring the relationship between the sales of its store-brand canned vegetables (y), the amount spent on promotion of the vegetables in local newspapers (x_1), and the amount of shelf space allocated to the brand (x_2). One of the chain's supermarkets was randomly selected, and over a 20-week period x_1 and x_2 were varied, as reported in the next table.

a. Fit the following model to the data:

$$y = \beta_0 + \beta_1 x_1 + \beta_2 x_2 + \beta_3 x_1 x_2 + \epsilon$$

b. Conduct an F-test to investigate the overall usefulness of this model. Use $\alpha = .05$.

c. Test for the presence of interaction between advertising expenditure and shelf space. Use $\alpha = .05$.

d. Explain what it means to say that advertising expenditures and shelf space interact.

e. Explain how you could be misled by using a first-order model instead of an interaction model to explain how advertising expenditure and shelf space influence sales.

Week	Sales ($)	Advertising Expenditures ($)	Shelf Space (sq. ft)
1	2,010	201	75
2	1,850	205	50
3	2,400	355	75
4	1,575	208	30
5	3,550	590	75
6	2,015	397	50
7	3,908	820	75
8	1,870	400	30
9	4,877	997	75
10	2,190	515	30
11	5,005	996	75
12	2,500	625	50
13	3,005	860	50
14	3,480	1,012	50
15	5,500	1,135	75
16	1,995	635	30
17	2,390	837	30
18	4,390	1,200	50
19	2,785	990	30
20	2,989	1,205	30

13.4 TESTING PORTIONS OF A MODEL

The presentation of models with one and two quantitative independent variables raises a very general question. Do certain terms in the model contribute information for the prediction of y?

To illustrate, suppose you have collected data on a response y and two independent quantitative variables, x_1 and x_2, and you are considering the use of either a first-order or a second-order model to relate $E(y)$ to x_1 and x_2. Will the second-order model provide better predictions of y than the first-order model? To answer this question, examine the two models. Note that the second-order model contains all terms present in the first-order model plus three additional terms, those involving β_3, β_4, and β_5:

First-order model: $E(y) = \beta_0 + \beta_1 x_1 + \beta_2 x_2$

Second-order terms

Second-order model: $E(y) = \beta_0 + \beta_1 x_1 + \beta_2 x_2 + \overbrace{\beta_3 x_1 x_2 + \beta_4 x_1^2 + \beta_5 x_2^2}$

Therefore, asking whether the second-order model contributes more information for the prediction of y than the first-order model is equivalent to asking whether $\beta_3 = \beta_4 = \beta_5 = 0$; that is, whether the terms involving β_3, β_4, and β_5 should be retained in the model. To test whether the second-order terms should be included in the model, we test the null hypothesis

$$H_0: \beta_3 = \beta_4 = \beta_5 = 0$$

(i.e., the second-order terms do not contribute information for the prediction of y) against the alternative hypothesis

$$H_a: \text{At least one of the parameters, } \beta_3, \beta_4, \text{ or } \beta_5, \text{ differs from } 0$$

(i.e., at least one of the second-order terms contributes information for the prediction of y).

In Section 12.4 we presented the t-test for a single coefficient, and in Section 12.5 we gave the F-test for *all* the β parameters (except β_0) in the model. We now need a test for *some* of the β parameters in the model. The test procedure is intuitive: First, we use the method of least squares to fit the first-order model—called the **reduced model** because it is the smaller of the two models—and calculate the corresponding sum of squares for error, SSE_r (the sum of squares of the deviations between observed and predicted values). Next, we fit the second-order model—called the **complete model** because it is the larger of the two models—and calculate its sum of squares for error, SSE_c. Then, we compare SSE_r to SSE_c by calculating $SSE_r - SSE_c$. If the second-order terms contribute to the model, SSE_c should be much smaller than SSE_r and the difference $(SSE_r - SSE_c)$ will be large. The larger the difference, the greater the weight of evidence that the second-order (complete) model provides better predictions of y than does the first-order (reduced) model.

The sum of squares for error always decreases when new terms are added to the model. That is, the complete-model SSE_c is always smaller than a reduced-model SSE_r. The question is whether this decrease is large enough to conclude that it is due to more than just an increase in the number of model terms and to chance. To test the null hypothesis that the parameters of the second-order terms, β_3, β_4, and β_5, simultaneously equal 0, we use an F statistic calculated as follows:

$$\begin{aligned} F &= \frac{(SSE_r - SSE_c)/3}{SSE_c/[n - (5 + 1)]} \\ &= \frac{\text{Drop in SSE}/\text{Number of } \beta \text{ parameters being tested}}{s^2 \text{ for the larger model}} \end{aligned}$$

When the assumptions listed in Section 12.3 about the error term ϵ are satisfied and the β parameters for the second-order terms are all 0 (H_0 is true), this F statistic has an F-distribution with $\nu_1 = 3$ numerator df and $\nu_2 = n - 6$ denominator df. Note that ν_1 is the number of β parameters being tested and ν_2 is the number of degrees of freedom associated with s^2 in the complete second-order model.

If the second-order terms do contribute to the model (H_a is true), we expect the F statistic to be large. Thus, we use a one-tailed test and reject H_0 when F exceeds some critical value, F_α, as shown in Figure 13.12. The steps employed in testing the null hypothesis that a set of model parameters are all equal to 0 are summarized in the next box.

FIGURE 13.12

Rejection region for the
F-test of H_0: $\beta_3 = \beta_4 = \beta_5$

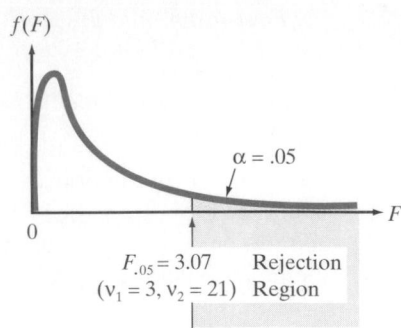

$f(F)$

$\alpha = .05$

0

$F_{.05} = 3.07$ Rejection
$(\nu_1 = 3, \nu_2 = 21)$ Region

F-Test for Testing the Null Hypothesis: Subset of β Parameters Equal to 0

Reduced model: $E(y) = \beta_0 + \beta_1 x_1 + \cdots + \beta_g x_g$

Complete model: $E(y) = \beta_0 + \beta_1 x_1 + \cdots + \beta_g x_g + \beta_{g+1} x_{g+1} + \cdots + \beta_k x_k$

H_0: $\beta_{g+1} = \beta_{g+2} = \cdots = \beta_k = 0$
H_a: At least one of the β parameters being tested is nonzero

Test statistic: $F = \dfrac{(SSE_r - SSE_c)/(k - g)}{SSE_c/[n - (k + 1)]}$

$= \dfrac{(SSE_r - SSE_c)/\text{Number of }\beta\text{'s tested}}{MSE_c}$

where

SSE_r = Sum of squared errors for the reduced model

SSE_c = Sum of squared errors for the complete model

MSE_c = Mean square error (or s^2) for the complete model

$k - g$ = Number of β parameters specified in H_0 (i.e., number of β's tested)

$k + 1$ = Number of β parameters in the complete model (including β_0)

n = Total sample size

Rejection region: $F > F_\alpha$

where F_α is based on $\nu_1 = (k - g)$ numerator df and $\nu_2 = [n - (k + 1)]$ denominator df.

EXAMPLE 13.4

Suppose you wish to study the growth of carnations as a function of the temperature x_1 (°F) in a greenhouse and the amount of fertilizer x_2 [kilograms (kg) per plot] applied to the soil. A total of 27 plots of equal size are treated with fertilizer in amounts varying between 50 and 60 kg per plot, and these plots are mechanically kept at constant temperatures between 80 and 100°F. Small carnation plants [approximately 15 centimeters (cm) in height] are planted in each plot, and their height y (cm) is measured after a 6-week growing period. The resulting data are shown in Table 13.2.

a. Fit a complete second-order model to the data.
b. Sketch the response surface.

TABLE 13.2   Temperature (x_1), Amount of Fertilizer (x_2), and Height of Carnations (y)

x_1	x_2	y	x_1	x_2	y	x_1	x_2	y
80	50	50.8	90	50	63.4	100	50	46.6
80	50	50.7	90	50	61.6	100	50	49.1
80	50	49.4	90	50	63.4	100	50	46.4
80	55	93.7	90	55	93.8	100	55	69.8
80	55	90.9	90	55	92.1	100	55	72.5
80	55	90.9	90	55	97.4	100	55	73.2
80	60	74.5	90	60	70.9	100	60	38.7
80	60	73.0	90	60	68.8	100	60	42.5
80	60	71.2	90	60	71.3	100	60	41.4

c. Do the data provide sufficient evidence to indicate that the second-order terms contribute information for the prediction of y?

Solution

a. The complete second-order model is

$$E(y) = \beta_0 + \beta_1 x_1 + \beta_2 x_2 + \beta_3 x_1 x_2 + \beta_4 x_1^2 + \beta_5 x_2^2$$

The data in Table 13.2 were used to fit this model, and a portion of the SAS output is shown in Figure 13.13.

The least squares prediction equation is

$$\hat{y} = -5{,}127.90 + 31.10x_1 + 139.75x_2 - .146x_1 x_2 - .133x_1^2 - 1.14x_2^2$$

b. A three-dimensional graph of this prediction model is shown in Figure 13.14. Note that the height seems to be greatest for temperatures of about 85–90°F

FIGURE 13.13

Portion of the SAS printout for Example 13.4

```
Dependent Variable: Y

                              Analysis of Variance

                                    Sum of          Mean
          Source        DF         Squares        Square      F Value    Prob>F

          Model          5      8402.26454     1680.45291     596.324    0.0001
          Error         21        59.17843        2.81802
          C Total       26      8461.44296

                 Root MSE        1.67870      R-square     0.9930
                 Dep Mean       66.96296      Adj R-sq     0.9913
                 C.V.            2.50690

                              Parameter Estimates

                         Parameter       Standard      T for H0:
          Variable   DF    Estimate          Error   Parameter=0    Prob > |T|

          INTERCEP    1  -5127.899074   110.29601493       -46.492       0.0001
          X1          1     31.096389     1.34441322        23.130       0.0001
          X2          1    139.747222     3.14005412        44.505       0.0001
          X1X2        1     -0.145500     0.00969196       -15.012       0.0001
          X1SQ        1     -0.133389     0.00685325       -19.464       0.0001
          X2SQ        1     -1.144222     0.02741299       -41.740       0.0001
```

FIGURE 13.14

Plot of second-order least squares model for Example 13.4

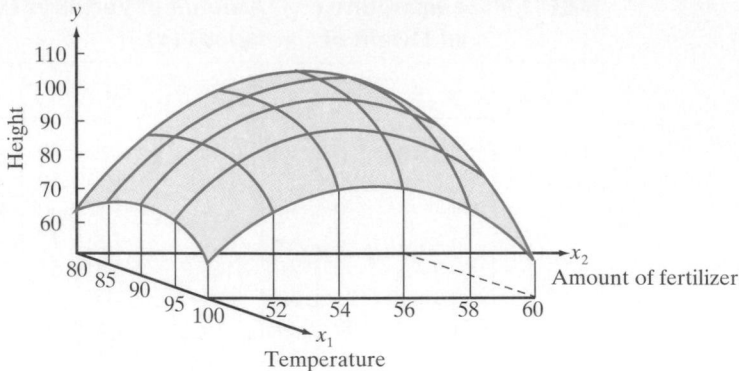

and for applications of about 55–57 kg of fertilizer per plot.* Further experimentation in these ranges might lead to a more precise determination of the optimal temperature–fertilizer combination.

c. To determine whether the data provide sufficient information to indicate that the second-order terms contribute information for the prediction of y, we wish to test

$$H_0: \beta_3 = \beta_4 = \beta_5 = 0$$

against the alternative hypothesis

H_a: At least one of the parameters, β_3, β_4, β_5, differs from 0

The first step in conducting the test is to drop the second-order terms out of the complete (second-order) model and fit the reduced model

$$E(y) = \beta_0 + \beta_1 x_1 + \beta_2 x_2$$

to the data. The SAS printout for this procedure is shown in Figure 13.15.

You can see that the sums of squares for error, given in Figures 13.13 and 13.15 for the complete and reduced models, respectively, are

$$SSE_c = 59.17843$$
$$SSE_r = 6{,}671.50852$$

and that s^2 for the complete model is

$$s^2 = MSE_c = 2.81802$$

Recall that $n = 27$, $k = 5$, and $g = 2$. Therefore, the calculated value of the F statistic, based on $\nu_1 = (k - g) = 3$ numerator df and $\nu_2 = [n - (k + 1)] = 21$ denominator df, is

$$F = \frac{(SSE_r - SSE_c)/(k - g)}{SSE_c/[n - (k + 1)]} = \frac{(SSE_r - SSE_c)/(k - g)}{MSE_c}$$

where $\nu_1 = (k - g)$ is equal to the number of parameters involved in H_0. Therefore,

*Students with knowledge of calculus should note that we can solve for the exact temperature and amount of fertilizer that maximize height in the least squares model by solving $\partial \hat{y}/\partial x_1 = 0$ and $\partial \hat{y}/\partial x_2 = 0$ for x_1 and x_2. Sample estimates of these estimated optimal values are $x_1 = 86.25°F$ and $x_2 = 55.58$ kg per plot.

FIGURE 13.15

SAS computer printout for
the reduced (first-order)
model, Example 13.4

Dependent Variable: Y

Analysis of Variance

Source	DF	Sum of Squares	Mean Square	F Value	Prob>F
Model	2	1789.93444	894.96722	3.22	0.0577
Error	24	6671.50852	277.97952		
C Total	26	8461.44296			

Root MSE	16.67272	R-square	0.2115	
Dep Mean	66.96296	Adj R-sq	0.1458	
C.V.	24.89840			

Parameter Estimates

Variable	DF	Parameter Estimate	Standard Error	T for H0: Parameter=0	Prob > |T|
INTERCEP	1	106.085185	55.94500427	1.90	0.0700
X1	1	-0.916111	0.39297973	-2.33	0.0285
X2	1	0.787778	0.78595946	1.00	0.3262

$$F = \frac{(6{,}671.50852 - 59.17843)/3}{2.81802} = 782.1$$

The final step in the test is to compare this computed value of F with the tabulated value based on $\nu_1 = 3$ and $\nu_2 = 21$ df. If we choose $\alpha = .05$, then $F_{.05} = 3.07$. Since the computed value of F falls in the rejection region (see Figure 13.12); i.e., it exceeds $F_{.05} = 3.07$, we reject H_0 and conclude that at least one of the second-order terms contributes information for the prediction of y. In other words, the data support the contention that the curvature we see in the response surface is not due simply to random variation in the data. The second-order model appears to provide better predictions of y than a first-order model. ▲

Example 13.4 demonstrates the motivation for testing a hypothesis that each one of a subset of β parameters equals 0; it also demonstrates the procedure. Other applications of this test appear in the following sections.

EXERCISES 13.35–13.46

Note: Exercises marked with 💾 *require the use of a computer.*

Learning the Mechanics

13.35 Suppose you fit the regression model

$$y = \beta_0 + \beta_1 x_1 + \beta_2 x_2 + \beta_3 x_1 x_2 + \beta_4 x_1^2 + \beta_5 x_2^2 + \epsilon$$

to $n = 30$ data points and you wish to test

$$H_0: \beta_3 = \beta_4 = \beta_5 = 0$$

a. State the alternative hypothesis H_a.
b. Explain in detail how you would find the quantities necessary to compute the F statistic for this test of hypothesis.

c. What are the numerator and denominator degrees of freedom associated with the F statistic?

13.36 Suppose you fit the complete and reduced models for the hypothesis test described in Exercise 13.35 and obtain $SSE_r = 1{,}250.2$ and $SSE_c = 1{,}125.2$. Conduct the hypothesis test and interpret the results of your test. Test using $\alpha = .05$.

13.37 Explain why the F-test used to compare complete and reduced models is a one-tailed, upper-tailed test.

13.38 MINITAB was used to fit the complete model

$$y = \beta_0 + \beta_1 x_1 + \beta_2 x_2 + \beta_3 x_3 + \beta_4 x_4 + \epsilon$$

to $n = 20$ data points (see top of page 630).

```
The regression equation is
Y = 14.6 - 0.611 X1 + 0.439 X2 - 0.080 X3 - 0.064 X4

Predictor        Coef       Stdev     t-ratio        p
Constant       14.575       4.887        2.98    0.010
X1            -0.6113      0.1775       -3.44    0.005
X2             0.4388      0.2199        2.00    0.062
X3            -0.0796      0.1083       -0.74    0.866
X4            -0.0636      0.1247       -0.51    0.921

s = 3.190      R-sq = 84.5%     R-sq(adj) = 80.3%

Analysis of Variance

SOURCE        DF          SS          MS          F          p
Regression     4      831.09      207.77      20.41      0.002
Error         15      152.66       10.18
Total         19      983.75
```

The independent variables x_3 and x_4 were dropped from the preceding model, and MINITAB was used to fit the resulting reduced model shown below.

a. Report the least squares prediction equations for the complete and reduced models.

b. Find SSE_r and SSE_c. Interpret each of the quantities.

c. How many β parameters are in the complete model? The reduced model?

d. Specify the null and alternative hypotheses you would use to investigate whether the complete model contributes more information for the prediction of y than the reduced model.

e. Conduct the hypothesis test of part **d**. Use $\alpha = .05$.

f. What is the approximate p-value of the test of part **e**?

Applying the Concepts

13.39 Refer to the *Journal of Family Issues* analysis of working wives' "time crunch" data, Exercise 12.46. The study investigated the impact of a wife's em-ployment on the number of hours available for household tasks. Some 206 families were included in the study. One independent variable measured for each family was the total time y (in minutes per day) that the wife spent on household chores. Two independent variables were also recorded:

WEMP: x_1 = Wife's hours of employment per week

YAGE: x_2 = Age of the youngest child in the family

The computer printout for the regression analysis is repeated on page 631. The sums of squares shown under **SEQUENTIAL S.S.** are those you would obtain by sequentially adding variables to the model. For example, the sum of squares opposite WEMP is for the model

$$E(y) = \beta_0 + \beta_1 x_1$$

The sum of squares opposite YAGE is the drop in SSE for the model obtained by adding x_2, that is, for

$$E(y) = \beta_0 + \beta_1 x_1 + \beta_2 x_2$$

```
The regression equation is
Y = 14.0 - 0.642 X1 + 0.396 X2

Predictor        Coef       Stdev     t-ratio        p
Constant       13.968       4.626        3.02    0.006
X1            -0.6422      0.1675       -3.84    0.001
X2             0.3959      0.2061        1.92    0.072

s = 3.072      R-sq = 83.7%     R-sq(adj) = 81.8%

Analysis of Variance

SOURCE        DF          SS          MS          F          p
Regression     2      823.31      411.66      43.61      0.000
Error         17      160.44        9.44
Total         19      983.75
```

```
DEPENDENT VARIABLE: WIFE'S HOUSEHOLD WORK    (MEAN=401.3ᵃ)

SOURCE OF VARIATION   DF    SUMS OF SQUARES    MEAN SQUARE        F         R²

TOTAL                 205    5,432,296                         62.10***     0.38
    MODEL             2      2,061,918          1,030,959
    ERROR             203    3,370,378             16,603

SOURCE WITHIN MODEL   DF    SEQUENTIAL S.S.    PARTIAL S.S.    b-ESTIMATES
WEMP                  1      1,810,154***       1,110,288***      -4.08
YAGE                  1        251,765***         251,765***      -7.02
--------------------------------------------------------------------------
DEPENDENT VARIABLE: HUSBAND'S WORK    (MEAN=105.2ᵃ)
               REGRESSION ANALYSIS NOT SIGNIFICANT
```

[*Note:* In this particular printout, "b-estimate" is our β estimate, or parameter estimate. Also, the asterisks indicate the level of significance of test statistics (or sum of squares to compute test statistics). Here, *** means $p < .001$.]

[a]Minutes per day.

Source: Fox, K. D., and Nickols, S. Y. "The time crunch." *Journal of Family Issues,* Mar. 1983, Vol. 4, No. 1, pp. 61–82. Copyright 1983 by Sage Publications. Reprinted by permission of Sage Publications, Inc.

Therefore, this sum of squares can be used in the F statistic to test $H_0: \beta_2 = 0$.

In contrast, the sums of squares shown under **PARTIAL S.S.** are the differences in the sums of squares between the reduced and complete models. These are the quantities needed to conduct individual F-tests on both of the parameters, β_1 and β_2. For example, 1,110,288 is the difference in the SSE values for the models

$$E(y) = \beta_0 + \beta_2 x_2$$

and

$$E(y) = \beta_0 + \beta_1 x_1 + \beta_2 x_2$$

Therefore, it can be used to compute the F statistic for testing $H_0: \beta_1 = 0$. [*Note:* This sum of squares differs substantially from the corresponding sum of squares, 1,810,154, shown under the **SEQUENTIAL S.S.** column.]

a. Use the information in the printout to test H_0: $\beta_1 = 0$. Test using $\alpha = .05$ and interpret your results.

b. State the alternative hypothesis implied in the test in part **a**.

c. Use the information in the printout to test H_0: $\beta_2 = 0$. Test using $\alpha = .05$ and interpret your results.

d. In concluding, the researchers state that "the wife's employment explained substantially more variation in the wife's housework time than age of the younger child." Do you agree? Explain.

13.40 A large hospital rates the performance of each member of its technical staff once a year. Each person is rated on a scale of 0 to 100 by his or her im-

mediate supervisor, and this merit rating is used to determine the size of the person's pay raise for the coming year. The hospital's personnel department is interested in developing a regression model to help them forecast the merit rating that an applicant for a technical position will receive after being employed 3 years. The hospital proposes to use the following model to forecast the merit ratings of applicants who have just completed their graduate studies and have no prior related job experience:

$$E(y) = \beta_0 + \beta_1 x_1 + \beta_2 x_2 + \beta_3 x_1 x_2 + \beta_4 x_1^2 + \beta_5 x_2^2$$

where

y = Applicant's merit rating after 3 years

x_1 = Applicant's GPA in graduate school

x_2 = Applicant's total score (verbal plus quantitative) on the Graduate Record Examination (GRE)

A random sample of $n = 40$ employees who have been on the technical staff of the hospital more than 3 years is selected. Each employee's merit rating after 3 years, graduate school GPA, and total score on the GRE are recorded. The model is fit to these data with the aid of a computer. A portion of the resulting computer printout is shown below.

SOURCE	DF	SUM OF SQUARES	MEAN SQUARE
MODEL	5	4911.56	982.31
ERROR	34	1830.44	53.84
TOTAL	39	6742.00	R-SQUARE
			0.73

The reduced model $E(y) = \beta_0 + \beta_1 x_1 + \beta_2 x_2$ is also fit to the same data. The resulting computer printout is partially reproduced here:

SOURCE	DF	SUM OF SQUARES	MEAN SQUARE
MODEL	2	3544.84	1772.42
ERROR	37	3197.16	86.41
TOTAL	39	6742.00	R-SQUARE
			0.53

a. Identify the appropriate null and alternative hypotheses to test whether the complete (second-order) model contributes information for the prediction of y.

b. Conduct the test of hypothesis given in part **a**. Test using $\alpha = .05$. Interpret the results in the context of this problem.

c. Identify the appropriate null and alternative hypotheses to test whether the complete model contributes more information than the reduced (first-order) model for the prediction of y.

d. Conduct the test of hypothesis given in part **c**. Test using $\alpha = .05$. Interpret the results in the context of this problem.

e. Which model, if either, would you use to predict y? Explain.

13.41 Refer to the *Motivation and Emotion* study on the relationship between cartoon funniness ratings (y) and pain (x_1) or aggression (x_2), Exercise 12.47. Since no evidence of an inverted-U relationship was found, the researchers fit the following two first-order models to the $n = 32$ data points:

Model 1: $E(y) = \beta_0 + \beta_1 x_1$,   SSE = 26.01

Model 2: $E(y) = \beta_0 + \beta_1 x_1 + \beta_2 x_2$,   SSE = 25.44

a. What hypothesis would you test to determine whether the addition of the aggression (x_2) rating improves the predictive ability of the model?

b. Use the F-test discussed in this section to test the hypothesis, part **a**. [Note that this test is equivalent to a t-test on β_2.]

13.42 The data in the next table were obtained from an experiment designed to investigate the relationship between the yield y of potatoes and the levels of three minerals in the soil, x_1, x_2, and x_3. [*Note:* The mineral levels have been coded by subtracting an appropriate constant from each of the x values.]

a. Use the method of least squares to fit the second-order polynomial,

$$y = \beta_0 + \beta_1 x_1 + \beta_2 x_2 + \beta_3 x_3 + \beta_4 x_1 x_2 + \beta_5 x_1 x_3 +$$
$$\beta_6 x_2 x_3 + \beta_7 x_1^2 + \beta_8 x_2^2 + \beta_9 x_3^2 + \epsilon$$

b. Test the hypothesis H_0: $\beta_4 = \beta_5 = \cdots = \beta_9 = 0$. That is, test whether the data provide sufficient evidence to indicate that a second-order model contributes more information for the prediction of yield than a first-order model. Test using $\alpha = .05$.

y	x_1	x_2	x_3
16.40	−1	−1	−1
13.51	1	−1	−1
14.41	−1	1	−1
9.38	1	1	−1
10.77	−1	−1	1
11.78	1	−1	1
4.11	−1	1	1
2.75	1	1	1
14.33	−1.682	0	0
5.44	1.682	0	0
19.80	0	−1.682	0
20.00	0	1.682	0
9.37	0	0	−1.682
10.03	0	0	1.682

13.43 An energy conservationist wants to develop a model that will estimate the mean annual gasoline consumption y in the United States (in millions of barrels) as a function of two independent variables:

x_1 = Number of cars (millions) in use during year

x_2 = Number of trucks (millions) in use during year

a. Identify the independent variables as quantitative or qualitative.

b. Write the first-order model for $E(y)$.

c. Write the complete second-order model for $E(y)$.

d. With respect to the model of part **c**, specify the null and alternative hypotheses you would employ in testing for the presence of interaction between x_1 and x_2.

e. Based on a sample of $n = 15$ years, the resulting values of SSE_r and SSE_c were 1,065.9 and 400.6, respectively. Conduct the test to determine whether the data present sufficient evidence to indicate interaction between x_1 and x_2. Test using $\alpha = .05$.

13.44 A firm would like to be able to forecast its yearly sales in each of its sales regions. The firm has decided to base its forecasts on regional population size and its yearly regional advertising expenditures. The population data in the table were obtained from the U.S. Bureau of the Census, and the advertising and sales data were obtained from the firm's internal records. The data are shown in the table on page 633.

Sales Region	Sales (thousands of units)	Population of Region (thousands)	Advertising Expenditures (thousands of dollars)
1	65	200	8
2	80	210	10
3	85	205	9
4	100	300	8.5
5	108	320	12
6	114	290	10
7	40	90	6
8	45	85	8
9	150	450	9
10	42	87	9
11	220	480	13
12	200	500	15

a. Fit a complete second-order model to these data.

b. Is the complete second-order model useful for forecasting sales? Test using $\alpha = .05$.

c. The firm is planning to market its product in a new sales region next year. The region has a population of 400,000, and the firm plans to spend \$12,000 on advertising. Use the fitted model you obtained in part **a** to forecast next year's sales in this new region. [*Note:* We would want to express this estimate as a prediction interval, but its computation is beyond the scope of this text. The procedure is described in the references. You may also find that it can be obtained using your computer program package.]

13.45 Refer to Exercise 13.44 in which a firm would like to develop a regression model to forecast its yearly sales in each of its sales regions. The model under consideration is a complete second-order model:

$$E(y) = \beta_0 + \beta_1 x_1 + \beta_2 x_2 + \beta_3 x_1 x_2 + \beta_4 x_1^2 + \beta_5 x_2^2$$

where

y = Yearly regional sales

x_1 = Population of sales region

x_2 = Yearly regional advertising expenditures

A portion of the computer printout that results from fitting this model to the $n = 12$ data points given in Exercise 13.44 is shown below.

SOURCE	DF	SUM OF SQUARES	MEAN SQUARE
MODEL	5	38638.97	7727.79
ERROR	6	159.94	26.66
TOTAL	11	38798.91	R-SQUARE
			0.996

The reduced first-order model

$$E(y) = \beta_0 + \beta_1 x_1 + \beta_2 x_2$$

was fit to the same data, and the resulting computer printout is partially reproduced below.

SOURCE	DF	SUM OF SQUARES	MEAN SQUARE
MODEL	2	36704.5	18352.2
ERROR	9	2094.4	232.7
TOTAL	11	38798.9	R-SQUARE
			0.946

Do these data present sufficient evidence to conclude that a second-order model contributes more information for the prediction of y than does a first-order model? Test using $\alpha = .05$.

13.46 In Exercise 12.11, a real estate appraiser used regression analysis to explore the relationship between the sale prices of apartments and various characteristics of the apartments. The data are repeated in the table on page 634.

a. Fit a first-order model to the data. (You may already have done this in Exercise 12.11.)

b. Do the data provide sufficient evidence to conclude that the model of part **a** is useful for predicting sale price? Test using $\alpha = .05$.

c. Drop x_3 and x_4 from the model of part **a** and refit the model to the data.

d. Do the data provide sufficient evidence to conclude that the model of part **c** is useful for predicting sale price? Test using $\alpha = .05$.

Code No.	Sale Price y ($)	No. of Apartments x₁	Age of Structure x₂ (years)	Lot Size x₃ (sq. ft)	No. of On-Site Parking Spaces x₄	Gross Building Area x₅ (sq. ft)
0229	90,300	4	82	4,635	0	4,266
0094	384,000	20	13	17,798	0	14,391
0043	157,500	5	66	5,913	0	6,615
0079	676,200	26	64	7,750	6	34,144
0134	165,000	5	55	5,150	0	6,120
0179	300,000	10	65	12,506	0	14,552
0087	108,750	4	82	7,160	0	3,040
0120	276,538	11	23	5,120	0	7,881
0246	420,000	20	18	11,745	20	12,600
0025	950,000	62	71	21,000	3	39,448
0015	560,000	26	74	11,221	0	30,000
0131	268,000	13	56	7,818	13	8,088
0172	290,000	9	76	4,900	0	11,315
0095	173,200	6	21	5,424	6	4,461
0121	323,650	11	24	11,834	8	9,000
0077	162,500	5	19	5,246	5	3,828
0060	353,500	20	62	11,223	2	13,680
0174	134,400	4	70	5,834	0	4,680
0084	187,000	8	19	9,075	0	7,392
0031	155,700	4	57	5,280	0	6,030
0019	93,600	4	82	6,864	0	3,840
0074	110,000	4	50	4,510	0	3,092
0057	573,200	14	10	11,192	0	23,704
0104	79,300	4	82	7,425	0	3,876
0024	272,000	5	82	7,500	0	9,542

Source: Robinson Appraisal Co., Inc., Mankato, Minnesota.

13.5 MODELS WITH ONE QUALITATIVE INDEPENDENT VARIABLE

Suppose we want to develop a model for the mean operating cost per mile, $E(y)$, of compact cars as a function of the car's manufacturer. (For the purpose of explanation, we ignore other independent variables that might affect the response.) Further suppose there are three manufacturers of interest, which we identify as A, B, and C. Then the automobile manufacturer is a single qualitative variable with three levels, A, B, and C. Note that with a qualitative independent variable, we cannot attach a quantitative measure to a given level. Even if we were to call the manufacturers 1, 2, and 3, the numbers would simply be identifiers of the manufacturers and would have no meaningful quantitative interpretation.

To simplify our notation, let μ_A be the mean cost per mile for compact cars produced by manufacturer A, and let μ_B and μ_C be the corresponding mean costs per mile for those produced by B and C. Our objective is to write a single equation that will give the mean value of y (cost per mile) for the three manufacturers. This can be done as follows:

$$E(y) = \beta_0 + \beta_1 x_1 + \beta_2 x_2$$

where

$$x_1 = \begin{cases} 1 & \text{if the car is manufactured by B} \\ 0 & \text{if the car is not manufactured by B} \end{cases}$$

$$x_2 = \begin{cases} 1 & \text{if the car is manufactured by C} \\ 0 & \text{if the car is not manufactured by C} \end{cases}$$

The variables x_1 and x_2 are not meaningful independent variables as they are in the case of the models with quantitative independent variables. Instead, they are **dummy** (or **indicator**) **variables** that make the model work. To see how they work, let $x_1 = 0$ and $x_2 = 0$. This condition will apply when we are seeking the mean response A. (Neither B nor C will be the manufacturer; hence, it must be A.) Then the mean cost per mile, $E(y)$, when A is the manufacturer is

$$\mu_A = E(y) = \beta_0 + \beta_1(0) + \beta_2(0) = \beta_0$$

This tells us that the mean cost per mile for A is β_0. Or, using our notation, it means that $\mu_A = \beta_0$.

Now suppose we want to represent the mean cost per mile, $E(y)$, for manufacturer B. Checking the dummy variable definitions, we see that we should let $x_1 = 1$ and $x_2 = 0$:

$$\mu_B = E(y) = \beta_0 + \beta_1 x_1 + \beta_2 x_2 = \beta_0 + \beta_1(1) + \beta_2(0) = \beta_0 + \beta_1$$

or, since $\beta_0 = \mu_A$,

$$\mu_B = \mu_A + \beta_1$$

Then it follows that the interpretation of β_1 is

$$\beta_1 = \mu_B - \mu_A$$

which is the difference between the mean costs per mile for manufacturers B and A.

Finally, if we want the mean value of y when C is the manufacturer, we let $x_1 = 0$ and $x_2 = 1$:

$$\mu_C = E(y) = \beta_0 + \beta_1(0) + \beta_2(1) = \beta_0 + \beta_2$$

or, since $\beta_0 = \mu_A$,

$$\mu_C = \mu_A + \beta_2$$

Then it follows that the interpretation of β_2 is

$$\beta_2 = \mu_C - \mu_A$$

Note that we were able to describe *three levels* of the qualitative variable with only *two dummy variables*. This is because the mean of the base level (manufacturer A, in this case) is accounted for by the intercept β_0.

Since the automobile manufacturer is a qualitative variable, we will use a bar graph to show the value of mean operating cost per mile $E(y)$ for the three levels of automobile manufacturer (see Figure 13.16). Particularly, note that the height of the bar, $E(y)$, for each level of automobile manufacturer is equal to the sum of the model parameters shown in the preceding equations. You can see that the height of the bar corresponding to manufacturer A is β_0, that is, $E(y) = \beta_0$.

FIGURE 13.16

Bar chart comparing $E(y)$ for three automobile manufacturers

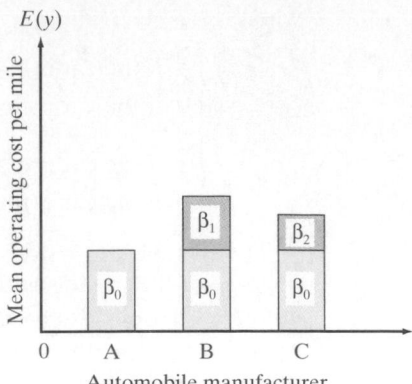

Similarly, the heights of the bars corresponding to manufacturers B and C are $E(y) = \beta_0 + \beta_1$ and $E(y) = \beta_0 + \beta_2$, respectively.*

Now carefully examine the model with a single qualitative independent variable at three levels, because we will use exactly the same pattern for any number of levels. Moreover, the interpretation of the parameters will always be the same.

One level is selected as the **base level**. (We used manufacturer A as the base level.) Then for the 1–0 system of coding† for the dummy variables,

$$\mu_A = \beta_0$$

Procedure for Writing a Model with One Qualitative Independent Variable with *k* Levels

Always use a number of dummy variables that is one less than the number of levels of the qualitative variable. Thus, for a qualitative variable with k levels, use $k - 1$ dummy variables:

$$y = \beta_0 + \beta_1 x_1 + \beta_2 x_2 + \cdots + \beta_{k-1} x_{k-1} + \epsilon$$

where x_i is the dummy variable for level $i + 1$ and

$$x_i = \begin{cases} 1 & \text{if } y \text{ is observed at level } i + 1 \\ 0 & \text{otherwise} \end{cases}$$

Then, for this system of coding

$$\mu_A = \beta_0 \qquad \text{and} \qquad \beta_1 = \mu_B - \mu_A$$
$$\mu_B = \beta_0 + \beta_1 \qquad\qquad\qquad \beta_2 = \mu_C - \mu_A$$
$$\mu_C = \beta_0 + \beta_2 \qquad\qquad\qquad \beta_3 = \mu_D - \mu_A$$
$$\mu_D = \beta_0 + \beta_3 \qquad\qquad\qquad\qquad \vdots$$
$$\vdots$$

---

*Note that either β_1 or β_2, or both, could be negative. If, for example, β_1 were negative, the height of the bar corresponding to manufacturer B would be *reduced* (rather than increased) from the height of the bar for manufacturer A by the amount β_1. Figure 13.16 is constructed assuming that β_1 and β_2 are positive quantities.

†You do not have to use a 1–0 system of coding for the dummy variables. Any two-value system will work, but the interpretation given to the model parameters will depend on the code. Using the 1–0 system makes the model parameters easy to interpret.

The coding for all dummy variables is as follows: To represent the mean value of y for a particular level, let that dummy variable equal 1; otherwise, the dummy variable is set equal to 0. Using this system of coding,

$$\mu_B = \beta_0 + \beta_1 \qquad \mu_C = \beta_0 + \beta_2$$

and so on. Because $\mu_A = \beta_0$, any other model parameter will represent the difference between means for that level and the base level.

$$\beta_1 = \mu_B - \mu_A \qquad \beta_2 = \mu_C - \mu_A$$

and so on.

EXAMPLE 13.5

Suppose a sociologist wants to compare the mean dollar amounts owed by delinquent credit-card customers in three different annual income groups: less than $15,000, $15,000–$30,000, and more than $30,000. A sample of 10 customers with delinquent accounts is selected from each group, and the amount owed by each is recorded, as shown in Table 13.3. Do the data provide sufficient evidence to indicate that the mean dollar amounts owed by customers differ for the three income groups?

Solution

Note that income is ordinarily a quantitative variable, but in this example only the group (low, medium, high) in which the customer's income falls is known. We therefore treat income as a qualitative variable (measured on an ordinal scale). For a three-level qualitative variable, we need two dummy variables in the regression model. The model relating $E(y)$ to this single qualitative variable, annual income group, is

$$E(y) = \beta_0 + \beta_1 x_1 + \beta_2 x_2$$

where

$$x_1 = \begin{cases} 1 & \text{if group 2} \\ 0 & \text{if not} \end{cases} \qquad x_2 = \begin{cases} 1 & \text{if group 3} \\ 0 & \text{if not} \end{cases}$$

and

$$\beta_1 = \mu_2 - \mu_1 \qquad \beta_2 = \mu_3 - \mu_1$$

TABLE 13.3   Dollars Owed, Example 13.5

Group 1 Less than $15,000	Group 2 $15,000–$30,000	Group 3 More than $30,000
$148	$513	$335
76	264	643
393	433	216
520	94	536
236	535	128
134	327	723
55	214	258
166	135	380
415	280	594
153	304	465

where μ_1, μ_2, and μ_3 are the mean responses for income groups 1, 2, and 3, respectively. Testing the null hypothesis that the means for the three groups are equal, that is, $\mu_1 = \mu_2 = \mu_3$, is equivalent to testing

$$H_0: \beta_1 = \beta_2 = 0$$

because if $\beta_1 = \mu_2 - \mu_1 = 0$ and $\beta_2 = \mu_3 - \mu_1 = 0$, then μ_1, μ_2, and μ_3 must be equal. The alternative hypothesis is

$$H_a: \text{At least one of the parameters, } \beta_1 \text{ or } \beta_2, \text{ differs from } 0$$

which implies that at least two of the three means (μ_1, μ_2, and μ_3) differ.

There are two ways to conduct this test. We can fit the complete model shown above and the reduced model (deleting the terms involving β_1 and β_2)

$$E(y) = \beta_0$$

and conduct the F-test described in the preceding section. (We leave this as an exercise for you.) Or we can use the F-test of the complete model (Section 12.5), which tests the null hypothesis that all parameters in the model, with the exception of β_0, equal 0. Either way you conduct the test, you will obtain the same computed value of F, the value shown on the SAS printout. The SAS printout for fitting the complete model,

$$E(y) = \beta_0 + \beta_1 x_1 + \beta_2 x_2$$

is shown in Figure 13.17; the value of the F statistic for testing the complete model, $F = 3.482$, is shaded. If we choose $\alpha = .05$, we will reject $H_0: \beta_1 = \beta_2 = 0$ because the computed value of F, $F = 3.482$, has an observed significance level of .0452 (also shaded). We therefore reject H_0 and conclude that at least one of the parameters, β_1 and β_2, differs from 0. Or, equivalently, we conclude that the data provide sufficient evidence to indicate that the mean indebtedness does vary from one income group to another. ▲

Note: The F-test in Example 13.5 can be obtained using two alternative methods. One way is to use the ANOVA (analysis of variance) procedure described in Chapter 10. Second, if you analyze the data by fitting complete and reduced

FIGURE 13.17

SAS computer printout for Example 13.5

Dependent Variable: Y

Analysis of Variance

Source	DF	Sum of Squares	Mean Square	F Value	Prob>F
Model	2	198772.46667	99386.23333	3.482	0.0452
Error	27	770670.90000	28543.36667		
C Total	29	969443.36667			

Root MSE	168.94782	R-square	0.2050	
Dep Mean	312.43333	Adj R-sq	0.1462	
C.V.	54.07484			

Parameter Estimates

Variable	DF	Parameter Estimate	Standard Error	T for H0: Parameter=0	Prob > |T|
INTERCEP	1	229.600000	53.42599243	4.30	0.0002
X1	1	80.300000	75.55576307	1.06	0.2973
X2	1	198.200000	75.55576307	2.62	0.0141

models (Section 13.4), you will find that the least squares estimate of β_0 in the reduced model

$$E(y) = \beta_0$$

is $\bar{y}$, the mean of all $n = 30$ observations, and the sum of squares for error for the reduced model is

$$SSE_r = \sum(y_i - \hat{y}_i)^2 = \sum(y_i - \bar{y})^2 = 969{,}443.367$$

This value is shown in the SAS printout (Figure 13.17) as the sum of squares corresponding to **C Total**. We leave the remaining steps—calculating the drop in SSE and the resulting F statistic—to you. You will find that the value you obtain is exactly the same as the value of F shown in the SAS printout.

EXERCISES 13.47–13.57

Learning the Mechanics

13.47 Write a regression model relating the mean value of y to a qualitative independent variable that can assume two levels. Interpret all the terms in the model.

13.48 Write a regression model relating $E(y)$ to a qualitative independent variable that can assume three levels. Interpret all the terms in the model.

13.49 The following model was used to relate $E(y)$ to a single qualitative variable with four levels:

$$E(y) = \beta_0 + \beta_1 x_1 + \beta_2 x_2 + \beta_3 x_3$$

where

$$x_1 = \begin{cases} 1 & \text{if level 2} \\ 0 & \text{if not} \end{cases}$$

$$x_2 = \begin{cases} 1 & \text{if level 3} \\ 0 & \text{if not} \end{cases}$$

$$x_3 = \begin{cases} 1 & \text{if level 4} \\ 0 & \text{if not} \end{cases}$$

This model was fit to $n = 30$ data points and the following result was obtained:

$$\hat{y} = 10.2 - 4x_1 + 12x_2 + 2x_3$$

a. Use the least squares prediction equation to find the estimate of $E(y)$ for each level of the qualitative independent variable.

b. Specify the null and alternative hypotheses you would use to test whether $E(y)$ is the same for all four levels of the independent variable.

13.50 MINITAB (see printout below) was used to fit the following model to $n = 15$ data points:

$$y = \beta_0 + \beta_1 x_1 + \beta_2 x_2 + \epsilon$$

where

$$x_1 = \begin{cases} 1 & \text{if level 2} \\ 0 & \text{if not} \end{cases} \qquad x_2 = \begin{cases} 1 & \text{if level 3} \\ 0 & \text{if not} \end{cases}$$

a. Report the least squares prediction equation.
b. Interpret the values of $\hat{\beta}_1$ and $\hat{\beta}_2$.
c. Interpret the following hypotheses in terms of μ_1, μ_2, and μ_3:

H_0: $\beta_1 = \beta_2 = 0$

H_a: At least one of the parameters β_1 and β_2 differs from 0

d. Conduct the hypothesis test of part **c.**

```
The regression equation is
Y  = 80.0 + 16.8 X1 + 40.4 X2

Predictor        Coef       Stdev     t-ratio         p
Constant       80.000       4.082       19.60     0.000
X1             16.800       5.774        2.91     0.013
X2             40.400       5.774        7.00     0.000

s = 9.129      R-sq = 80.5%      R-sq(adj) = 77.2%

Analysis of Variance

SOURCE         DF          SS          MS          F         p
Regression      2      4118.9      2059.5      24.72     0.000
Error          12      1000.0        83.3
Total          14      5118.9
```

Applying the Concepts

13.51 A field experiment was conducted to assess the effect of organic enrichment on the mean density of mosquito larvae (*Journal of the American Mosquito Control Association,* June 1995). Larval specimens were collected from a pond three days after the pond was flooded with canal water. A second sample of specimens was collected three weeks after flooding and enriching the pond with rabbit pellets. All specimens were returned to the laboratory and the number *y* of mosquito larvae was counted in each specimen.

    a. Write a model that will allow you to compare the mean number of mosquito larvae found in the enriched pond to the corresponding mean for the natural pond.

    b. Interpret the β coefficients in the model, part **a**.

    c. Set up the null and alternative hypotheses for testing whether the mean larval density for the enriched pond exceeds the mean for the natural pond.

    d. The *p*-value associated with the global *F*-test for the model, part **a**, was determined to be .004. Interpret this result.

13.52 Do grizzly bears segregate on the basis of sex? One hypothesis is that female grizzlies avoid male-occupied habitats because of competition for food and cannibalism. A competing theory is that females do not avoid males, but simply have different habitats available to them. These hypotheses were investigated in the *Journal of Wildlife Management* (July 1995). Grizzly bears were trapped, radio-collared, and released in the Highwood trapping zone (HTZ) in Alberta, Canada. The percentage of time, *y*, each bear used the HTZ as a habitat over a 4-year period was recorded. The researchers modeled $E(y)$ as a function of reproductive class at five levels: estrous adult females, adult females with offspring, independent subadult females, adult males, and independent subadult males. One goal was to compare the mean percent use of HTZ for the five classes of grizzly bears.

Grizzly Class	*n*	Mean Percent Use
Estrous adult females	5	38
Adult females with offspring	7	16
Independent subadult females	2	89
Adult males	7	43
Independent subadult males	8	58

Source: Wielgus, R. B., and Bunnell, F. L. "Tests of hypotheses for sexual segregation in grizzly bears." *Journal of Wildlife Management,* Vol. 59, No. 3, July 1995, p. 555 (Table 1).

    a. Write a model for $E(y)$ that will enable the researchers to carry out the comparison.

    b. The sample sizes and sample means for the five classes are shown in the table. Use this information to find estimates of the β's in the model, part **a**.

    c. Give the null hypothesis for a test to determine whether the mean percent use of HTZ differs among the grizzly bear classes.

    d. The *p*-value for the test, part **c**, was reported as .15. Interpret this result.

13.53 Five varieties of peas are currently being tested by a large agribusiness cooperative to determine which is best suited for production. A field was divided into 20 plots, with each variety of peas planted in four plots. The yields (in bushels of peas) produced from each plot are shown in the table.

VARIETY OF PEAS

A	B	C	D	E
26.2	29.2	29.1	21.3	20.1
24.3	28.1	30.8	22.4	19.3
21.8	27.3	33.9	24.3	19.9
28.1	31.2	32.8	21.8	22.1

SAS was used to fit the model

$$y = \beta_0 + \beta_1 x_1 + \beta_2 x_2 + \beta_3 x_3 + \beta_4 x_4 + \epsilon$$

to these data, using the coding $x_1 = 1$ for variety A, $x_2 = 1$ for variety B, $x_3 = 1$ for variety C, and $x_4 = 1$ for variety D. The resulting printout is shown at the top of page 641.

    a. Write the model relating mean yield to pea variety, and interpret all the parameters in the model.

    b. Report the least squares model from the SAS printout.

    c. What null and alternative hypotheses are tested by the global *F*-test for this model? Interpret the hypotheses both in terms of the β parameters and the mean yields for the five varieties of peas.

    d. Test the hypotheses of part **c** using $\alpha = .05$.

    e. Place a 95% confidence interval on the differences in the mean yields of varieties D and E.

13.54 Refer to Exercise 13.53. Note that this represents a completely randomized design for comparing the four varieties (treatments), as discussed in Section 10.2. The SAS analysis of variance (ANOVA) printout for this experiment is shown at the bottom of page 641.

    a. What are the null and alternative hypotheses tested by the ANOVA? How do they compare with those tested in parts **c** and **d** of Exercise 13.53?

    b. Find and interpret the test statistic and significance level on the ANOVA printout. Compare the results to those obtained in Exercise 13.53.

13.55 The amount of fructose in an athlete's bloodstream is critical to performance. A researcher ran an experiment to determine whether diet had any effect

```
Dep Variable: Y

                         Analysis of Variance

                           Sum of          Mean
       Source       DF     Squares        Square    F Value    Prob>F

       Model         4     342.0400      85.51000    23.966    0.0001
       Error        15      53.52000      3.56800
       C Total      19     395.56000

            Root MSE         1.88892    R-Square      0.8647
            Dep Mean        25.70000    Adj R-Sq      0.8286
            C.V.             7.34986

                         Parameter Estimates

                      Parameter      Standard     T for H0:
       Variable  DF    Estimate         Error    Parameter=0    Prob > |T|

       INTERCEP   1    20.350000    0.94445752      21.547       0.0001
       X1         1     4.750000    1.33566463       3.556       0.0029
       X2         1     8.600000    1.33566463       6.439       0.0001
       X3         1    11.300000    1.33566463       8.460       0.0001
       X4         1     2.100000    1.33566463       1.572       0.1367
```

on the level of fructose in the blood after 10 minutes of running on a treadmill. Twenty-one subjects were selected, and each subject's fructose level was measured after 10 minutes on a treadmill. Each subject was randomly assigned to one of three diets. One diet was high in protein, one high in carbohydrates, and one high in fruits. After 1 month on the diet, each subject was again asked to run on the treadmill for 10 minutes, and the fructose level in the bloodstream was measured. The dependent variable was the difference in the level of fructose in the blood between the second and the first runs on the treadmill.

a. Identify the independent variables in the experiment.

b. Write an appropriate regression model relating mean difference in fructose in the blood, $E(y)$,

to the independent variables. Identify and code all dummy variables.

13.56 A fisheries researcher believes that the growth rate of fish is related to the age of the fish, its feeding location, and other factors. Suppose the researcher wishes to collect fish of a certain species from four different locations.

a. Write a model relating the mean growth rate $E(y)$ to location, a qualitative variable.

b. Use your model to find the mean growth rate of the fish at location 1.

c. Use your model to find the mean growth rate of the fish at location 4.

d. In terms of your model, what is the difference in mean growth rates between locations 2 and 3?

13.57 The director of marketing of a company that sells business machines is interested in modeling mean

```
                   General Linear Models Procedure

Dependent Variable: Y

                         Sum of          Mean
Source           DF     Squares         Square     F Value    Pr > F

Model             4    342.0400000    85.5100000    23.97     0.0001
Error            15     53.5200000     3.5680000
Corrected Total  19    395.5600000

             R-Square         C.V.       Root MSE            Y Mean

             0.864698     7.3498639       1.88891         25.70000000

Source           DF    Type I SS    Mean Square    F Value    Pr > F

VARIETY           4    342.04000     85.51000       23.97     0.0001
```

monthly sales (in thousands of dollars) per salesperson, $E(y)$, as a function of the type of sales incentive plan that is in effect: commission only, straight salary, or salary plus commission on each sale. The director has proposed the following model:

$$E(y) = \beta_0 + \beta_1 x_1 + \beta_2 x_2$$

where

$$x_1 = \begin{cases} 1 & \text{if salesperson is paid a straight salary} \\ 0 & \text{otherwise} \end{cases}$$

$$x_2 = \begin{cases} 1 & \text{if salesperson is paid a salary plus commission} \\ 0 & \text{otherwise} \end{cases}$$

MINITAB was used to fit this model to the sales data collected from a sample of 15 salespersons

(five from each incentive plan). The printout is shown below.

a. Do the data provide sufficient evidence to conclude that there is a difference in mean monthly sales among the three incentive plans? Test using $\alpha = .05$.

b. Use the least squares prediction equation to estimate the mean sales for salespersons working on a straight-salary basis.

c. Use the least squares prediction equation to estimate the mean sales for salespersons working on a commission-only basis.

[*Note:* We would prefer to use confidence intervals for the estimates in parts **b** and **c**, but their calculation is beyond the scope of this text. This procedure can be found in the references at the end of the book.]

```
The regression equation is
Y  = 20.0 - 8.60 X1 + 3.80 X2

Predictor        Coef      Stdev     t-ratio        p
Constant        20.00      2.898        6.90    0.000
X1              -8.60      4.100       -2.10    0.055
X2               3.80      4.100        0.93    0.386

s = 6.481      R-sq = 44.5%    R-sq(adj) = 35.2%

Analyisis of Variance

SOURCE          DF         SS          MS         F        p
Regression       2      403.6      201.80      4.80    0.033
Error           12      504.0       42.00
Total           14      907.6
```

13.6   COMPARING THE SLOPES OF TWO OR MORE LINES

Suppose you wish to relate the mean monthly sales $E(y)$ of a company to monthly advertising expenditure x for three different advertising media (say, newspaper, radio, and television) and you wish to use first-order (straight-line) models to model the responses for all three media. Graphs of these three relationships might appear as shown in Figure 13.18.

Since the lines in Figure 13.18 are hypothetical, a number of practical questions arise. Is one advertising medium as effective as any other? That is, do the three mean sales lines differ for the three advertising media? Do the increases in mean sales per dollar input in advertising differ for the three advertising media? That is, do the slopes of the three lines differ? Note that the two practical questions have been rephrased into questions about the parameters that define the three lines of Figure 13.18. To answer them, we must write a single linear statistical model that will characterize the three lines of Figure 13.18 and that will allow us to test hypotheses about the lines.

FIGURE 13.18

Graphs of the relationship between mean sales $E(y)$ and advertising expenditure x

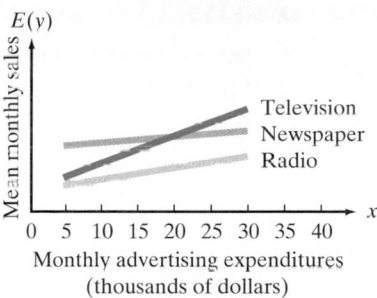

The response described previously, monthly sales, is a function of *two* independent variables, one quantitative (advertising expenditure x_1) and one qualitative (type of medium). We will proceed, in stages, to build a model relating $E(y)$ to these variables; then we will show graphically the interpretation we would give to the model at each stage. This will help you to see the contributions of the various terms in the model.

1. The straight-line relationship between mean sales $E(y)$ and advertising expenditure is the same for all three media—that is, a single line will describe the relationship between $E(y)$ and advertising expenditure x_1 for all the media (see Figure 13.19).

$$E(y) = \beta_0 + \beta_1 x_1 \qquad \text{where } x_1 = \text{Advertising expenditure}$$

2. The straight lines relating mean sales $E(y)$ to advertising expenditure x_1 differ from one medium to another, but the rate of increase in mean sales per increase in dollar advertising expenditure x_1 is the same for all media—that is, the lines are parallel but possess different y-intercepts (see Figure 13.20).

$$E(y) = \beta_0 + \beta_1 x_1 + \beta_2 x_2 + \beta_3 x_3$$

where

$x_1 = \text{Advertising expenditure}$

$$x_2 = \begin{cases} 1 & \text{if radio medium} \\ 0 & \text{if not} \end{cases}$$

$$x_3 = \begin{cases} 1 & \text{if television medium} \\ 0 & \text{if not} \end{cases}$$

Notice that this model is essentially a combination of a first-order model with a single quantitative variable and the model with a single qualitative variable:

FIGURE 13.19

The relationship between $E(y)$ and x_1 is the same for all media

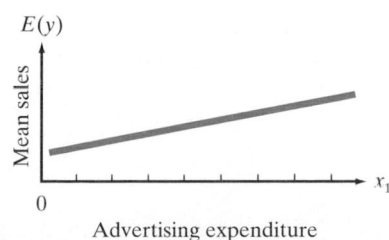

FIGURE 13.20

Parallel response lines for the three media

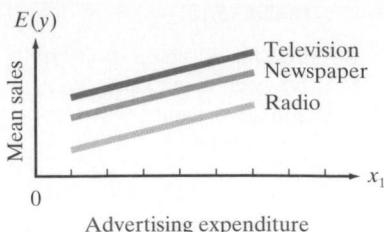

First-order model with a single quantitative variable:

Model with a single qualitative variable at three levels:

$$E(y) = \beta_0 + \beta_1 x_1$$

$$E(y) = \beta_0 + \beta_2 x_2 + \beta_3 x_3$$

where x_1, x_2, and x_3 are as just defined. The model described here implies no interaction between the two independent variables, advertising expenditure x_1, and the qualitative variable, type of advertising medium. The change in $E(y)$ for a 1-unit increase in x_1 is identical (the slopes of the lines are equal) for all three advertising media. The terms corresponding to each of the independent variables are called **main effect terms** because they imply no interaction.

3. The straight lines relating mean sales $E(y)$ to advertising expenditure x_1 differ for the three advertising media—that is, both the line intercepts and the slopes differ (see Figure 13.21). As you will see, this interaction model is obtained by adding terms involving the cross-product terms, one each from each of the two independent variables:

$$E(y) = \beta_0 + \underbrace{\beta_1 x_1}_{\substack{\text{Main effect,} \\ \text{advertising} \\ \text{expenditure}}} + \underbrace{\beta_2 x_2 + \beta_3 x_3}_{\substack{\text{Main effect,} \\ \text{type of} \\ \text{medium}}} + \underbrace{\beta_4 x_1 x_2 + \beta_5 x_1 x_3}_{\text{Interaction}}$$

Note that each of the preceding models is obtained by adding terms to model 1, the single first-order model used to model the responses for all three media. Model 2 is obtained by adding the main effect terms for type of medium, the qualitative variable. Model 3 is obtained by adding the interaction terms to model 2.

Will a single line (Figure 13.19) characterize the responses for all three media? Or do all three response lines differ as shown in Figure 13.21? A test of the null hypothesis that a single first-order model adequately describes the relationship between $E(y)$ and advertising expenditure x_1 for all three media is a test of the null hypothesis that the parameters of model 3, β_2, β_3, β_4, and β_5, equal 0; that is,

FIGURE 13.21

Different response lines for the three media

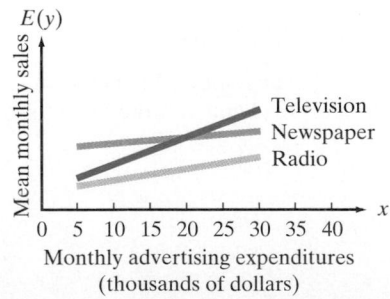

$$H_0: \beta_2 = \beta_3 = \beta_4 = \beta_5 = 0$$

This hypothesis can be tested by fitting the complete model (model 3) and the reduced model (model 1) and conducting an F-test.

Suppose we assume that the response lines for the three media will differ but wonder whether the data present sufficient evidence to indicate a difference among the slopes of the lines. To test the null hypothesis that model 2 adequately describes the relationship between $E(y)$ and advertising expenditure x_1, we wish to test

$$H_0: \beta_4 = \beta_5 = 0$$

that is, that the two independent variables, advertising expenditure x_1 and type of medium, do not interact. This test can be conducted by fitting the complete model (model 3) and the reduced model (model 2), calculating the drop in the sum of squares for error, and conducting an F-test as described in Section 13.4.

EXAMPLE 13.6

Substitute the appropriate values of the dummy variables in model 3 to obtain the equations of the three response lines in Figure 13.21.

Solution

The complete model that characterizes the three lines in Figure 13.21 is

$$E(y) = \beta_0 + \beta_1 x_1 + \beta_2 x_2 + \beta_3 x_3 + \beta_4 x_1 x_2 + \beta_5 x_1 x_3$$

where

$$x_1 = \text{Advertising expenditure}$$

$$x_2 = \begin{cases} 1 & \text{if radio medium} \\ 0 & \text{if not} \end{cases}$$

$$x_3 = \begin{cases} 1 & \text{if television medium} \\ 0 & \text{if not} \end{cases}$$

Examining the coding, you can see that $x_2 = x_3 = 0$ when the advertising medium is newspaper. Substituting these values into the expression for $E(y)$, we obtain the newspaper medium line:

$$E(y) = \beta_0 + \beta_1 x_1 + \beta_2(0) + \beta_3(0) + \beta_4 x_1(0) + \beta_5 x_1(0) = \beta_0 + \beta_1 x_1$$

Similarly, we substitute the appropriate values of x_2 and x_3 into the expression for $E(y)$ to obtain the radio medium line:

$$E(y) = \beta_0 + \beta_1 x_1 + \beta_2(1) + \beta_3(0) + \beta_4 x_1(1) + \beta_5 x_1(0)$$

$$= \underbrace{(\beta_0 + \beta_2)}_{y\text{-intercept}} + \underbrace{(\beta_1 + \beta_4)}_{\text{Slope}} x_1$$

and the television medium line:

$$E(y) = \beta_0 + \beta_1 x_1 + \beta_2(0) + \beta_3(1) + \beta_4 x_1(0) + \beta_5 x_1(1)$$

$$= \underbrace{(\beta_0 + \beta_3)}_{y\text{-intercept}} + \underbrace{(\beta_1 + \beta_5)}_{\text{Slope}} x_1$$

If you were to fit the general equation (model 3), obtain estimates of $\beta_0, \beta_1, \beta_2, \ldots,$ β_5, and substitute them into the equations for the three media lines shown, you would obtain exactly the same prediction equations as you would obtain if you were to fit three separate straight lines, one to each of the three sets of media data. You may ask why we would not fit the three lines separately. Why bother fitting a model that combines all three lines (model 3) into the same equation? The answer is that you need to use this procedure if you wish to use statistical tests to compare the three media lines. We need to be able to express a practical question about the lines in terms of a hypothesis that a set of parameters in the model equal 0. You could not do this if you were to perform three separate regression analyses and fit a line to each set of media data.   ▲

EXAMPLE 13.7

An industrial psychologist conducted an experiment to investigate the relationship between worker productivity and a measure of salary incentive for two manufacturing plants; one, plant A, had union representation and the other, plant B, had nonunion representation. The productivity y per worker was measured by recording the number of machined castings that a worker could produce in a 4-week period of 40 hours per week. The incentive was the amount x_1 of bonus (in cents per casting) paid for all castings produced in excess of 1,000 per worker for the 4-week period. Nine workers were selected from each plant, and three from each group of nine were assigned to receive a 20¢ bonus per casting, three a 30¢ bonus, and three a 40¢ bonus. The productivity data for the nine workers, three for each plant type and incentive combination, are shown in the table.

	INCENTIVE		
Type of Plant	**20¢/casting**	**30¢/casting**	**40¢/casting**
Union	1,435, 1,512, 1,491	1,583, 1,529, 1,610	1,601, 1,574, 1,636
Nonunion	1,575, 1,512, 1,488	1,635, 1,589, 1,661	1,645, 1,616, 1,689

a. Assume that the relationship between mean productivity and incentive is first-order. Plot the data points and graph the prediction equations for the two productivity lines.
b. Do the data provide sufficient evidence to indicate a difference in mean worker responses to incentives between the two plants?

Solution

If we assume that a first-order model* is adequate to detect a change in mean productivity as a function of incentive x_1, then the model that produces two productivity lines, one for each plant, is

$$E(y) = \beta_0 + \beta_1 x_1 + \beta_2 x_2 + \beta_3 x_1 x_2$$

where

$$x_1 = \text{Incentive} \qquad x_2 = \begin{cases} 1 & \text{if nonunion plant} \\ 0 & \text{if union plant} \end{cases}$$

a. The MINITAB printout for the regression analysis is shown in Figure 13.22. Reading the parameter estimates from the printout, you can see that

$$\hat{y} = 1{,}365.833 + 6.217 x_1 + 47.778 x_2 + .033 x_1 x_2$$

---

*Although the model contains a term involving $x_1 x_2$, it is first-order (graphs as a straight line) in the quantitative variable x_1. The variable x_2 is a dummy variable that introduces or deletes terms in the model. The order of a model is determined only by the quantitative variables that appear in the model.

FIGURE 13.22

MINITAB printout for the complete model, Example 13.7

```
The regression equation is
Y  = 1365 + 6.22 X1 + 47.8 X2 + .03 X1X2

Predictor        Coef        Stdev      t-ratio          p
Constant     1365.833       51.836        26.35      0.000
X1              6.217        1.667         3.73      0.002
X2             47.778       73.308         0.65      0.525
X1X2           0.0333        2.358         0.01      0.989

s = 40.839      R-sq = 71.1%       R-sq(adj) = 64.8%

Analysis of Variance

SOURCE         DF           SS           MS          F          p
Regression      3      57332.38     19110.80      11.46      0.001
Error          14      23349.23      1667.80
Total          17      80681.61
```

The prediction equation for the union plant can be obtained (see the coding) by substituting $x_2 = 0$ into the general prediction equation. Then

$$\hat{y} = \hat{\beta}_0 + \hat{\beta}_1 x_1 + \hat{\beta}_2(0) + \hat{\beta}_3 x_1(0) = \hat{\beta}_0 + \hat{\beta}_1 x_1$$
$$= 1{,}365.833 + 6.217 x_1$$

Similarly, the prediction equation for the nonunion plant is obtained by substituting $x_2 = 1$ into the general prediction equation. Then

$$\hat{y} = \hat{\beta}_0 + \hat{\beta}_1 x_1 + \hat{\beta}_2 x_2 + \hat{\beta}_3 x_1 x_2$$
$$= \hat{\beta}_0 + \hat{\beta}_1 x_1 + \hat{\beta}_2(1) + \hat{\beta}_3 x_1(1)$$
$$= \overbrace{(\hat{\beta}_0 + \hat{\beta}_2)}^{y\text{-intercept}} + \overbrace{(\hat{\beta}_1 + \hat{\beta}_3)}^{\text{Slope}} x_1$$
$$= (1{,}365.833 + 47.778) + (6.217 + .033)x_1$$
$$= 1{,}413.611 + 6.250 x_1$$

A graph of these prediction equations is shown in Figure 13.23. Note that the slopes of the two lines are nearly identical (6.217 for union and 6.250 for nonunion).

b. To determine whether the data provide sufficient evidence to indicate a difference in mean worker responses to incentives for the two plants, we test the null hypothesis that a *single* line characterizes the relationship between productivity per worker y and the amount of incentive x_1 against the alternative hypothesis that we need two separate lines to characterize the relationship—one for each plant. If there is no difference in mean response $E(y)$ to x_1

FIGURE 13.23

Graphs of the prediction equations for the two productivity lines, Example 13.7

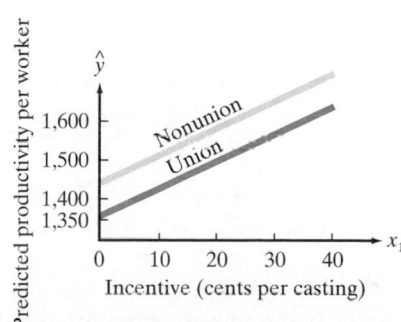

FIGURE 13.24

MINITAB printout for the reduced model, Example 13.7

```
The regression equation is
Y  = 1390 + 6.23 X1

Predictor        Coef        Stdev      t-ratio         p
Constant     1389.722       41.408        33.56     0.000
X1              6.233        1.332         4.68     0.000

s = 46.136      R-sq = 57.8%      R-sq(adj) = 55.2%

Analysis of Variance

SOURCE         DF          SS           MS        F        p
Regression      1     46625.33     46625.33    21.91    0.000
Error          16     34056.28      2128.52
Total          17     80681.61
```

between the two plants, then we do not need the variable Type of plant in the model; i.e., we do not need the terms involving x_2. Therefore, we wish to test

$$H_0:\ \beta_2 = \beta_3 = 0$$

against the alternative hypothesis

$H_a:$ At least one of the two parameters, β_2 or β_3, is not equal to 0

The MINITAB printout for fitting the reduced model

$$E(y) = \beta_0 + \beta_1 x_1$$

to the data is shown in Figure 13.24. Reading SSE_c and SSE_r from Figures 13.22 and 13.24, respectively, we obtain

Complete model: $SSE_c = 23{,}349.23$

Reduced model: $SSE_r = 34{,}056.28$

Drop in SSE: $SSE_r - SSE_c = 10{,}707.05$

The value of MSE, or s^2, for the complete model (Figure 13.22) is 1,667.80. Substituting these values, along with $k = 3$ and $g = 1$, into the formula for the F statistic yields

$$F = \frac{(SSE_r - SSE_c)/(k - g)}{MSE_c} = \frac{(10{,}707.05)/2}{1{,}667.80} = 3.21$$

The numerator degrees of freedom (the number of parameters involved in H_0) is $\nu_1 = 2$ and the denominator degrees of freedom (the degrees of freedom associated with s^2 in the complete model) is $\nu_2 = 14$. If we choose $\alpha = .05$, the tabulated value of $F_{.05}$ given in Table IX of Appendix A is 3.74. Since the computed value, $F = 3.21$, is less than the tabulated value, $F_{.05} = 3.74$, there is insufficient evidence (at the $\alpha = .05$ significance level) to indicate a difference in mean worker responses to incentives between the two plants. Therefore, there is no evidence to indicate that two different lines, one for each plant, are needed to describe the relationship between mean productivity per worker $E(y)$ and the amount of incentive x_1. ▲

EXAMPLE 13.8

Refer to Example 13.7 and explain how you would determine whether the data provide sufficient evidence to indicate that the incentive x_1 affects mean productivity.

Solution

If the variable incentive did not affect mean productivity, we would not need terms involving x_1 in the model. Therefore, we would test the null hypothesis

$$H_0: \beta_1 = \beta_3 = 0$$

against the alternative hypothesis

$$H_a: \text{ At least one of the parameters, } \beta_1 \text{ or } \beta_3, \text{ differs from } 0$$

We would fit the reduced model

$$E(y) = \beta_0 + \beta_2 x_2$$

to the data and find SSE_r. The values of SSE_c and s^2 for the complete model would be the same as those used in Example 13.7. Finally, you would calculate the value of the F statistic and compare it with a tabulated value of F based on $\nu_1 = 2$ numerator df and $\nu_2 = 14$ denominator df. If the test leads to rejection of H_0, you have evidence to indicate that the increase in mean productivity that appears to be present in the graphs in Figure 13.23 is due to random variation in the data. ▲

EXERCISES 13.58–13.67

Note: Exercises marked with 💾 *require the use of a computer.*

Learning the Mechanics

13.58 Suppose you are interested in an application with a response y, one quantitative independent variable x_1, and one qualitative variable at three levels.
  a. Write a first-order model that relates the mean response $E(y)$ to the quantitative independent variable.
  b. Add the main effect terms for the qualitative independent variable to the model of part **a**. Specify the coding scheme you use.
  c. Add terms to the model of part **b** to allow for interaction between the quantitative and qualitative independent variables.
  d. Under what circumstances will the response lines of the model in part **c** be parallel?

  e. Under what circumstances will the model in part **c** have only one response line?

13.59 SAS was used to fit the following model to $n = 15$ data points:

$$y = \beta_0 + \beta_1 x_1 + \beta_2 x_2 + \beta_3 x_3 + \epsilon$$

where x_1 is a quantitative variable and x_2 and x_3 are dummy variables describing a qualitative variable at three levels using the coding scheme

$$x_2 = \begin{cases} 1 & \text{if level 2} \\ 0 & \text{otherwise} \end{cases} \qquad x_3 = \begin{cases} 1 & \text{if level 3} \\ 0 & \text{otherwise} \end{cases}$$

The resulting printout is shown below.
  a. What is the response line (equation) for $E(y)$ when $x_2 = x_3 = 0$? When $x_2 = 1$ and $x_3 = 0$? When $x_2 = 0$ and $x_3 = 1$?

```
Dep Variable: Y

                        Analysis of Variance

                            Sum of          Mean
     Source        DF       Squares        Square      F Value    Prob>F

     Model          3     4747.70480    1582.56827     46.894     0.0001
     Error         11      371.22853      33.74805
     C Total       14     5118.93333

            Root MSE        5.80931      R-Square     0.9275
            Dep Mean       99.06667      Adj R-Sq     0.9077
            C.V.            5.86404

                        Parameter Estimates

                      Parameter      Standard     T for H0:
     Variable   DF     Estimate        Error     Parameter=0    Prob > |T|

     INTERCEP    1     44.802703     8.55817652      5.235         0.0003
     X1          1      2.172673     0.50335248      4.316         0.0012
     X2          1      9.412913     4.05315974      2.322         0.0404
     X3          1     15.631532     6.81368981      2.294         0.0425
```

```
Dep Variable: Y

                        Analysis of Variance

                        Sum of          Mean
        Source      DF  Squares         Square      F Value   Prob>F

        Model       1   4522.90741    4522.90741     98.650   0.0001
        Error      13    596.02592      45.84815
        C Total    14   5118.93333

            Root MSE       6.77113      R-Square    0.8836
            Dep Mean      99.06667      Adj R-Sq    0.8746
            C.V.           6.83492

                        Parameter Estimates

                    Parameter      Standard    T for H0:
        Variable  DF  Estimate       Error    Parameter=0   Prob > |T|

        INTERCEP   1   33.904568   6.78960378      4.994       0.0002
        X1         1    3.083380   0.31044102      9.932       0.0001
```

b. What is the general least squares prediction equation for the preceding model?

c. What is the least squares prediction equation associated with level 1? Level 2? Level 3? Plot these on the same graph.

d. Specify the null and alternative hypotheses that should be used to investigate whether a difference exists between the response lines of the qualitative variable levels 1, 2, and 3.

e. Use the printout above with only the quantitative variable x_1 in the model to conduct the hypothesis test of part **d**. Use $\alpha = .05$.

Applying the Concepts

13.60 The influence of cigarette smoking on resting energy expenditure (REE) in normal-weight and obese smokers was recently investigated (*Health Psychology,* Mar. 1995). The researchers hypothesized that the relationship between a smoker's REE and length of time since smoking differs for normal-weight and obese smokers. Consequently, the following interaction model was examined:

$$E(y) = \beta_0 + \beta_1 x_1 + \beta_2 x_2 + \beta_3 x_1 x_2$$

where

y = REE, measured in kilocalories per day

x_1 = Time, in minutes after smoking, of metabolic energy reading (levels = 10, 20, and 30 minutes)

$$x_2 = \begin{cases} 1 & \text{if normal weight} \\ 0 & \text{if obese} \end{cases}$$

a. Give the equation of the hypothesized line relating mean REE to time after smoking for obese smokers. What is the slope of the line?

b. Repeat part **a** for normal-weight smokers.

c. A test for interaction resulted in an observed significance level of .044. Interpret this value.

13.61 Excessive exposure to solar radiation is known to increase the risk of developing skin cancer, yet many people do not practice "sun safety." A group of University of Arizona researchers examined the feasibility of educating preschool (4- to 5-year-old) children about sun safety (*American Journal of Public Health,* July 1995). A sample of 122 preschool children was divided into two groups, the control group and the intervention group. Children in the intervention group received a *Be Sun Safe* curriculum in preschool, while the control group did not. All children were tested for their knowledge, comprehension, and application of sun safety at two times: prior to the implementation of the sun safety curriculum (pretest, x_1) and 7 weeks following the completion of the curriculum (posttest, y).

a. Write a first-order model for mean posttest score, $E(y)$, as a function of pretest score x_1 and group. Assume that no interaction exists between pretest score and group.

b. For the model, part **a**, show that the slope of the line relating posttest score to pretest score is the same for both groups of children.

c. Repeat part **a**, but assume that pretest score and group interact.

d. For the model, part **c**, show that the slope of the line relating posttest score to pretest score differs for the two groups of children.

e. Assuming interaction exists, give the reduced model for testing whether the mean posttest scores differ for the intervention and control groups.

f. When sun safety knowledge was used as the dependent variable, the test of part **e** resulted in a *p*-value of .03. Interpret this result.

g. When sun safety comprehension was used as the dependent variable, the test of part **e** resulted in a *p*-value of .033. Interpret this result.

h. When sun safety application was used as the dependent variable, the test of part **e** resulted in a *p*-value of .322. Interpret this result.

13.62 Researchers for a dog food company have developed a new puppy food they hope will compete with the major brands. One premarketing test involved the comparison of the new food with that of two competitors in terms of weight gain. Fifteen 8-week-old German shepherd puppies, each from a different litter, were divided into three groups of five puppies each. Each group was fed one of the three brands of food.

a. Set up a model that assumes the final weight *y* is linearly related to initial weight *x* but does not allow for differences among the three brands; that is, assume the response curve is the same for the three brands of dog food. Sketch the response curve as it might appear.

b. Set up a model that assumes the final weight is linearly related to initial weight and allows the intercepts of the lines to differ for the three brands. In other words, assume that the initial weight and brand both affect final weight, but the two variables do not interact. Sketch typical response curves.

c. Now write the main effects with interaction model. For this model we assume the final weight is linearly related to initial weight, but both the slopes and the intercepts of the lines depend on the brand. Sketch typical response curves.

13.63 In Exercises 12.11 and 13.46, a real estate appraiser used regression analysis to explore the relationship between the sale prices of apartments and various characteristics of the apartments. Some of the data from those exercises are reproduced in the next table, which also contains data on the physical condition of each apartment building (E: excellent; G: good; and F: fair).

a. Write a model that describes the relationship between sale price and number of apartment units as three parallel lines, one for each level of physical condition. Be sure to specify the dummy variable coding scheme you use.

b. Plot *y* against x_1 (number of apartment units) for all buildings in excellent condition. On the same graph, plot *y* against x_1 for all buildings in good condition. Do this again for all buildings in fair condition. Does it appear that the model you specified in part **a** is appropriate? Explain.

c. Fit the model from part **a** to the data. Report the least squares prediction equation for each of the three building condition levels.

d. Plot the three prediction equations of part **c** on a scattergram of the data.

Code No.	Sale Price	No. of Apartments	Condition of Apartment Bldg.
0229	$ 90,300	4	F
0094	384,000	20	G
0043	157,500	5	G
0079	676,200	26	E
0134	165,000	5	G
0179	300,000	10	G
0087	108,750	4	G
0120	276,538	11	G
0246	420,000	20	G
0025	950,000	62	G
0015	560,000	26	G
0131	268,000	13	F
0172	290,000	9	E
0095	173,200	6	G
0121	323,650	11	G
0077	162,500	5	G
0060	353,500	20	F
0174	134,400	4	E
0084	187,000	8	G
0031	155,700	4	E
0019	93,600	4	F
0074	110,000	4	G
0057	573,200	14	E
0104	79,300	4	F
0024	272,000	5	E

Source: Robinson Appraisal Co., Inc., Mankato, Minnesota.

e. Do the data provide sufficient evidence to conclude that the relationship between sale price and number of units differs depending on the physical condition of the apartments? Test using $\alpha = .05$.

13.64 Many women suffer from anemia. A female physician, who is also an avid jogger, wanted to know if women who exercise regularly have a different mean red blood cell count than women who do not. She also wanted to know if the amount of a particular iron supplement a woman takes has any effect and whether the effect is the same for both groups. Write a model that will reflect the relationship between red blood cell count and the two independent variables described previously, assuming that

a. The effect of the iron supplement on mean blood cell count is the same regardless of whether a woman exercises regularly.

b. The effect of the iron supplement on mean blood cell count depends on whether a woman exercises regularly.

13.65 A research physician wishes to find a model that will predict the mean time to relief after the administration of a certain drug. Two important independent variables are considered to be good predictors of relief time. They are age of the patient and method of administration. Physicians administer a standard dose of a drug to a patient in one of three different ways: orally in liquid form (method 1), orally in pill form (method 2), and intravenously (method 3). Consequently, we might use the model

$$E(y) = \beta_0 + \beta_1 x_1 + \beta_2 x_2 + \beta_3 x_3 + \beta_4 x_1 x_2 + \beta_5 x_1 x_3$$

where

y = Time to relief (in minutes)

x_1 = Age of patient (in years)

$$x_2 = \begin{cases} 1 & \text{if drug is administered orally in pill form} \\ 0 & \text{if not} \end{cases}$$

$$x_3 = \begin{cases} 1 & \text{if drug is administered intravenously} \\ 0 & \text{if not} \end{cases}$$

The data in the following table, obtained on 12 patients, were used to fit this model.

METHOD 1		METHOD 2		METHOD 3	
Age	Time to Relief	Age	Time to Relief	Age	Time to Relief
51	22	46	28	37	19
36	25	40	24	60	17
31	20	26	23	25	21
20	25	32	25	38	20

a. A multiple regression analysis produced the results shown below (upper printout). Test whether the model is useful in predicting mean time to relief. Use $\alpha = .05$.

b. What hypothesis would you use to test whether the age–drug method interaction terms contribute to the prediction of mean time to relief?

c. The reduced model

$$E(y) = \beta_0 + \beta_1 x_1 + \beta_2 x_2 + \beta_3 x_3$$

was fit to the data and produced an SSE of 38.289. Does this result provide sufficient evidence at the $\alpha = .05$ level of significance to indicate that the interaction terms should be kept in the model?

d. Use an available computer program to check the results given in parts **a** and **c**.

13.66 An economist is interested in modeling the mean monthly demand $E(y)$ (in thousands of units) for a certain product as a function of its price (in dollars) and the season of the year. The following model has been proposed:

$$E(y) = \beta_0 + \beta_1 x_1 + \beta_2 x_2 + \beta_3 x_3 + \beta_4 x_4$$

where

x_1 = Price

$$x_2 = \begin{cases} 1 & \text{if spring} \\ 0 & \text{otherwise} \end{cases}$$

$$x_3 = \begin{cases} 1 & \text{if summer} \\ 0 & \text{otherwise} \end{cases}$$

$$x_4 = \begin{cases} 1 & \text{if fall} \\ 0 & \text{otherwise} \end{cases}$$

A portion of the MINITAB computer printout that results from fitting this model to a sample of 16 months of sales data selected from among the last 2 years is shown below (lower printout).

SOURCE	DF	SUM OF SQUARES	MEAN SQUARE	F	PR > F
MODEL	5	87.473	17.495	4.90	0.0394
ERROR	6	21.443	3.574		
CORRECTED TOTAL	11	108.916		R-SQUARE	0.803120

```
The regression equation is
Y   = 12.1 - 1.11 X1 + 3.94 X2 + 7.17 X3 + 3.74 X4

Predictor      Coef      Stdev     t-ratio        p
Constant      12.067     1.473        8.19    0.000
X1            -1.113     0.187       -5.93    0.000
X2             3.942     0.858        4.59    0.001
X3             7.166     0.815        8.80    0.000
X4             3.724     0.819        4.55    0.001

s = 1.135      R-sq = 93.1%     R-sq(adj) = 90.6%

Analysis of Variance

SOURCE       DF         SS          MS        F        p
Regression    4     191.59      47.897    37.22    0.000
Error        11      14.16       1.287
Total        15     205.75
```

```
The regression equation is
Y =  17.4 - 1.35 X1

Predictor       Coef       Stdev      t-ratio        p
Constant       17.394      2.833        6.08      0.000
X1             -1.348      0.403       -3.34      0.006

s = 2.859      R-sq = 44.4%      R-sq(adj) = 40.4%

Analysis of Variance

SOURCE         DF          SS          MS        F        p
Regression      1        91.35       91.35     11.18    0.006
Error          14       114.40        8.17
Total          15       205.75
```

The reduced model $E(y) = \beta_0 + \beta_1 x_1$ was also fit to the same data, and the resulting computer printout is partially reproduced above.

a. Do these data present sufficient evidence to conclude that mean monthly demand depends on the season of the year? Test using $\alpha = .05$.

b. Find an estimate for $E(y)$ when the price is $8 and it is summer. Interpret your result. [*Note:* We prefer to use a confidence interval for this estimate, but its calculation is beyond the scope of this text. This procedure can be found in the references.]

13.67 In Exercise 11.81, we fit a regression line to relate the mean gain in weight of a rainbow trout to the trout's ration level. Twenty fish were fed at each of four ration levels while living in water maintained at 6°C. We also fit regression lines to similar data for fish fed in water at 12°C and at 18°C. The table below gives the means and standard deviations (in parentheses) for each sample of 20 fish.

a. Write a single model to relate the mean weight gain to ration level and water temperature (°C). A graphic representation of your model should depict three straight lines with differing intercepts and equal slopes.

b. Use a computer to fit the model in part **a** to the sample means.

c. Write a single model to relate the mean weight gain to ration level and water temperature (°C). A graphical representation of your model should depict three straight lines with differing intercepts and slopes.

d. Use a computer to fit the model in part **c** to the sample means.

e. Do the data provide sufficient evidence to indicate differences in mean gain in weight for a 1% change in the ration level (i.e., the slopes of the lines) for the three levels of water temperatures? Test using $\alpha = .05$

	MEAN WET WEIGHT GAIN (%)		
Ration (% body weight per day)	**6°C**	**12°C**	**18°C**
.0	–8.14 (3.27)	–10.33 (2.35)	–13.21 (1.96)
.8	12.31 (3.37)		
1.2		14.29 (5.32)	
1.5	28.19 (5.39)		15.51 (4.48)
2.5	29.12 (5.67)	38.05 (6.99)	
3.5			51.13 (7.14)
4.0		60.13 (12.71)	
4.5			65.49 (13.04)
Maintenance ration	.32	.48	.69

Source: T. G. Kovacs and G. Leduc, "Sublethal toxicity to rainbow trout (*Salmo gairdneri*) at different temperatures," *Canadian Journal of Fisheries and Aquatic Sciences,* 1982, Volume 39, p. 1391. Reproduced with permission of the Minister of Supply and Services Canada, 1996.

13.7   COMPARING TWO OR MORE RESPONSE CURVES

Suppose we think that the relationship between mean monthly sales $E(y)$ and advertising expenditure x_1 (Section 13.6) is second-order. We will construct, stage by stage, a model relating $E(y)$ to the quantitative variable x_1 and the qualitative variable, Type of medium. This will enable you to compare the procedure with the stage-by-stage construction of the first-order model (Section 13.6). The graphical interpretations will help you understand the contributions of the model terms.

1. The mean sales curves are identical for all three advertising media, that is, a single second-order curve will suffice to describe the relationship between $E(y)$ and x_1 for all the media (see Figure 13.25):

$$E(y) = \beta_0 + \beta_1 x_1 + \beta_2 x_1^2$$

where x_1 = Advertising expenditure

2. The response curves possess the same shapes but different y-intercepts (see Figure 13.26):

$$E(y) = \beta_0 + \beta_1 x_1 + \beta_2 x_1^2 + \beta_3 x_2 + \beta_4 x_3$$

where

x_1 = Advertising expenditure

$$x_2 = \begin{cases} 1 & \text{if radio medium} \\ 0 & \text{if not} \end{cases}$$

$$x_3 = \begin{cases} 1 & \text{if television medium} \\ 0 & \text{if not} \end{cases}$$

FIGURE 13.25

The relationship between $E(y)$ and x_1 is the same for all media

FIGURE 13.26

The response curves have the same shapes but different y-intercepts

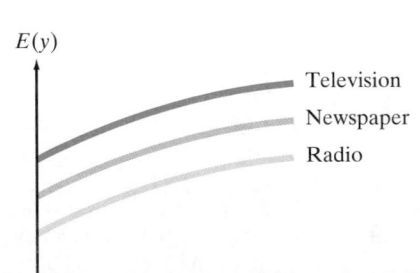

FIGURE 13.27

The response curves for the three media differ

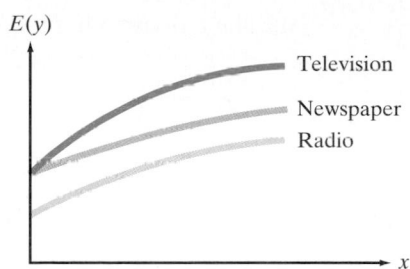

3. The response curves for the three advertising media are different (i.e., Advertising expenditure and Type of medium interact), as shown in Figure 13.27:

$$E(y) = \beta_0 + \beta_1 x_1 + \beta_2 x_1^2 + \beta_3 x_2 + \beta_4 x_3 + \beta_5 x_1 x_2 + \beta_6 x_1 x_3 + \beta_7 x_1^2 x_2 + \beta_8 x_1^2 x_3$$

Now that you know how to write a model with two independent variables—one qualitative and one quantitative—we ask a question. Why do it? Why not write a separate second-order model for each type of medium where $E(y)$ is a function only of advertising expenditure? One reason we write the single model representing all three response curves is so that we can test to determine whether the curves are different. We illustrate this procedure in Examples 13.10 and 13.11. A second reason for writing a single model is that we obtain a pooled estimate of σ^2, the variance of the random error component ϵ. If the variance of ϵ is truly the same for each type of medium, the pooled estimate is superior to three separate estimates calculated by fitting a separate model for each type of medium.

EXAMPLE 13.9   Give the equation of the second-order model for the radio advertising medium.

Solution

Model 3 characterizes the relationship between $E(y)$ and x_1 for the radio advertising medium (see the coding) when $x_2 = 1$ and $x_3 = 0$. Substituting these values into model 3, we obtain

$$
\begin{aligned}
E(y) &= \beta_0 + \beta_1 x_1 + \beta_2 x_1^2 + \beta_3 x_2 + \beta_4 x_3 + \beta_5 x_1 x_2 + \beta_6 x_1 x_3 + \beta_7 x_1^2 x_2 + \beta_8 x_1^2 x_3 \\
&= \beta_0 + \beta_1 x_1 + \beta_2 x_1^2 + \beta_3(1) + \beta_4(0) + \beta_5 x_1(1) \\
&\quad + \beta_6 x_1(0) + \beta_7 x_1^2(1) + \beta_8 x_1^2(0) \\
&= (\beta_0 + \beta_3) + (\beta_1 + \beta_5)x_1 + (\beta_2 + \beta_7)x_1^2
\end{aligned}
$$

Note: The y-intercept of the response curve is represented by $(\beta_0 + \beta_3)$, the shift parameter by $(\beta_1 + \beta_5)$, and the rate of curvature by $(\beta_2 + \beta_7)$.   ▲

EXAMPLE 13.10   What null hypothesis about the parameters of model 3 would you test if you wished to determine whether the second-order curves for the three media are identical?

Solution

If the curves were identical, we would not need the independent variable Type of medium in the model; that is, we would delete all terms involving x_2 and x_3. This would produce model 1

$$E(y) = \beta_0 + \beta_1 x_1 + \beta_2 x_1^2$$

and the null hypothesis would be

$$H_0: \beta_3 = \beta_4 = \beta_5 = \beta_6 = \beta_7 = \beta_8 = 0 \qquad \blacktriangle$$

EXAMPLE 13.11

Suppose we assume that the response curves for the three media differ, but we want to know whether the second-order terms contribute information for the prediction of y. Or, equivalently, will a second-order model give better predictions than a first-order model?

Solution

The only difference between model 3 and a first-order model are those terms involving x_1^2. Therefore, the null hypothesis, "the second-order terms contribute no information for the prediction of y," is equivalent to

$$H_0: \beta_2 = \beta_7 = \beta_8 = 0 \qquad \blacktriangle$$

Examples 13.10 and 13.11 identify two tests that answer practical questions concerning a collection of second-order models. Other comparisons between the curves can be made by testing appropriate sets of model parameters (see the Exercises).

The models described in the preceding sections provide merely an introduction to statistical modeling. Models can be constructed to relate $E(y)$ to any number of quantitative and qualitative independent variables. You can compare response curves and surfaces for different levels of a qualitative variable or for different combinations of levels of two or more qualitative independent variables. A general explanation of how to write linear statistical models can be found in Mendenhall and Sinich (1996, Chapter 5).

Building a Model for Condo Sale Prices

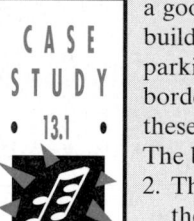

**C A S E
S T U D Y
• 13.1 •**

This case involves an investigation of the factors that affect the sale price of oceanside condominium units. It represents an extension of an analysis of the same data by Herman Kelting (1979). Although condo sale prices have increased dramatically over the past 20 years, the relationships between these factors and sale price remain about the same. Consequently, the data provide valuable insight into (if not current dollar values of) today's condominium sales market.

The sales data were obtained for a newly built oceanside condominium complex consisting of two adjacent and connected eight-floor buildings. The complex contains 200 units of equal size (approximately 500 square feet each). The locations of the buildings relative to the ocean, the swimming pool, the parking lot, etc., are shown in Figure 13.28. There are several features of the complex that you should note:

1. The units facing south, called *ocean-view,* face the beach and ocean. In addition, units in building 1 have a good view of the pool. Units to the rear of the building, optimistically called *bay-view,* face the parking lot and an area of land that, ultimately, borders a bay. The view from the upper floors of these units is primarily of wooded, sandy terrain. The bay is very distant and barely visible.

2. The only elevator in the complex is located at the east end of building 1, next to the office and the game room. People moving to or from the higher floor units in building 2 would probably use the elevator, then move through the passages to their units. Thus, units on the higher floors and at a greater distance from the elevator would be less convenient for their occupants; they would have to expend a greater effort in moving baggage, groceries, etc., and would be farther away from the game room, the office, and the swimming pool. These units also possess an advantage: Because traffic through the hallways in the area would be minimal, these units would be the most private.

Case Study 13.1 continued

3. Lower-floor oceanside units are most suited to active people; they open onto the beach, ocean, and pool. They are within easy reach of the game room and have ready access to the parking area.
4. Checking Figure 13.28, you will see that the views in some of the units at the center of the complex, units numbered _11 and _14, are partially blocked.
5. The condominium complex was completed at the time of the 1975 recession; sales were slow and the developer was forced to sell most of the units at auction approximately 18 months after opening. Many unsold units were furnished by the developer and rented prior to the auction.

This condominium complex is particularly suited to our study. Because the single elevator is located at one end of the complex, it is the source of a remarkably high level of both inconvenience and privacy for the people occupying units on the top floors in building 2. Consequently, the data provide a good opportunity to investigate the relationship that might exist between sale price, height of the unit (floor number), distance of the unit from the elevator, and presence or absence of an ocean view. In addition, the presence or absence of furniture in each of the units permits an investigation of the effect of the availability of furniture on sale price.

Finally, the auction data are completely buyer-specified and hence consumer-oriented, in contrast to most other real estate sales data which are, to a high degree, seller-oriented and broker-specified.

In addition to the sale price, the following data were recorded for each of the 106 units sold at the auction:

1. *Floor height* (x_1): The floor location of the unit; the variable levels are 1, 2, …, 8.
2. *Distance from elevator* (x_2): This distance, measured along the length of the complex, is expressed in number of condominium units. An additional two units of distance was added to the units in building 2 to account for the walking distance in the connecting area between the two buildings. Thus, the distance of unit 105 from the elevator would be 3, and the distance between unit 113 and the elevator would be 9. The variable levels are 1, 2, …, 15.
3. *View of ocean* (x_3): The presence or absence of an ocean view is recorded for each unit and specified with a dummy variable, where $x_3 = 1$ if the unit offered an ocean view and $x_3 = 0$ if not. Note that units not offering an ocean view would face the parking lot.
4. *End unit* (x_4): We expect the partial reduction of view of end units on the ocean side (numbers ending in
continued

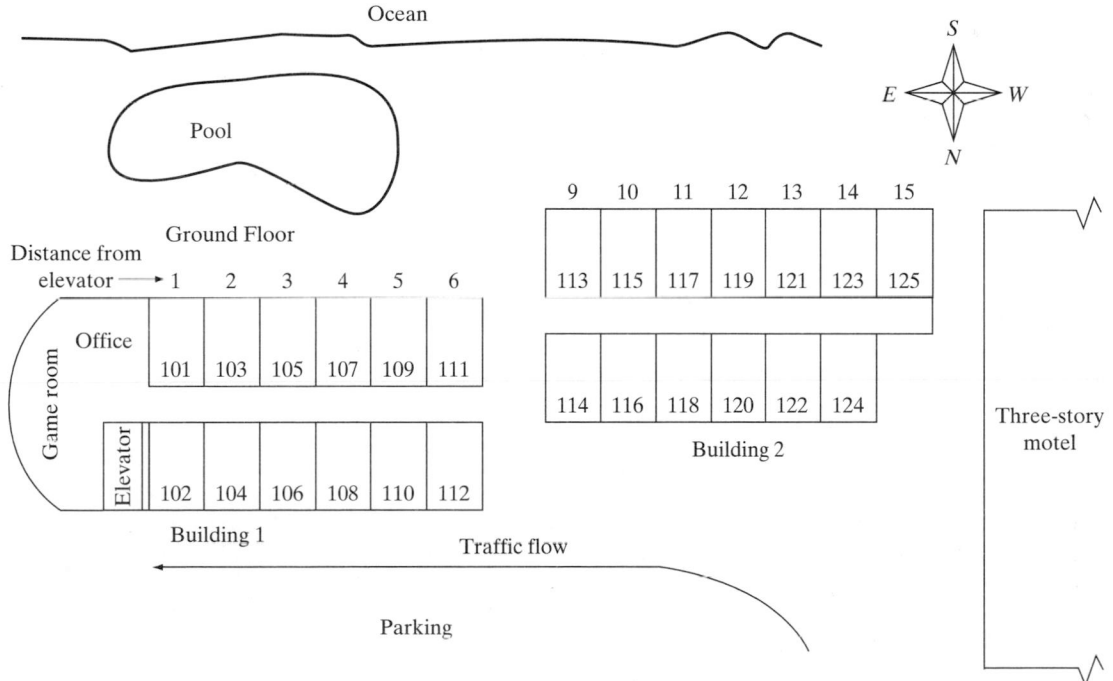

FIGURE 13.28   Layout of condominium complex

Case Study 13.1 continued

11) to reduce their sale price. The ocean view of these end units is partially blocked by building 2. This qualitative variable is specified with a dummy variable, where $x_4 = 1$ if the unit has a unit number ending in 11 and $x_4 = 0$ if not.

5. *Furniture* (x_5): The presence or absence of furniture is recorded for each unit. This qualitative variable is represented with a single dummy variable, where $x_5 = 1$ if the unit is furnished and $x_5 = 0$ if not.

Three models relating mean sale price to these five factors are postulated. The models, numbered 1–3, are given below:

Model 1

$$E(y) = \beta_0 + \beta_1 x_1 + \beta_2 x_2 + \beta_3 x_3 + \beta_4 x_4 + \beta_5 x_5$$

Model 2

$$E(y) = \beta_0 + \overbrace{\beta_1 x_1 + \beta_2 x_2 + \beta_3 x_1 x_2 + \beta_4 x_1^2 + \beta_5 x_2^2}^{\text{Second-order model in } x_1 \text{ and } x_2}$$

$$+ \underbrace{\beta_6 x_3}_{\text{View of ocean}} + \underbrace{\beta_7 x_4}_{\text{End unit}} + \underbrace{\beta_8 x_5}_{\text{Furniture}}$$

Model 3

$$E(y) = \beta_0 + \overbrace{\beta_1 x_1 + \beta_2 x_2 + \beta_3 x_1 x_2 + \beta_4 x_1^2 + \beta_5 x_2^2}^{\text{Second-order model in } x_1 \text{ and } x_2}$$

$$+ \underbrace{\beta_6 x_3}_{\text{View of ocean}} + \underbrace{\beta_7 x_4}_{\text{End unit}} + \underbrace{\beta_8 x_5}_{\text{Furniture}}$$

$$+ \overbrace{\beta_9 x_1 x_3 + \beta_{10} x_2 x_3 + \beta_{11} x_1 x_2 x_3}^{\text{Interaction of the second-order model with view of ocean}}$$
$$+ \beta_{12} x_1^2 x_3 + \beta_{13} x_2^2 x_3$$

Focus

a. Verbally describe the theoretical relationships hypothesized by each of the three models.

b. For each model, sketch the theoretical relationship between sale price y and floor height x_1 for each level of view of ocean.

The SAS printouts for the three models are displayed in Figures 13.29–13.31.

Focus

c. In Model 1, interpret the *p*-value for the variable floor height x_1. Why is it dangerous to conclude, at this stage, that floor height is *not* an important predictor of sale price? [*Note:* The failure of floor height x_1 to reveal itself as an important information contributor goes against our intuition, thereby demonstrating the pitfalls inherent in attempting to interpret the results of *t*-tests in a regression analysis. Intuitively, we would expect floor height to be an important factor. You might argue that units on the higher floors offer a better view and hence should command a higher mean sale price. Or, you might

continued

FIGURE 13.29

SAS regression analysis: Model 1

ANALYSIS OF VARIANCE

SOURCE	DF	SUM OF SQUARES	MEAN SQUARE	F VALUE	PROB>F
MODEL	5	236620761	47324152.11	48.419	0.0001
ERROR	100	97737975.28	977379.75		
C TOTAL	105	334358736			

ROOT MSE	988.6252	R-SQUARE	0.7077	
DEP MEAN	19176.04	ADJ R-SQ	0.6931	
C.V.	5.155524			

PARAMETER ESTIMATES

VARIABLE	DF	PARAMETER ESTIMATE	STANDARD ERROR	T FOR H0: PARAMETER=0	PROB > |T|
INTERCEP	1	17770.00813	416.07044	42.709	0.0001
X1	1	-73.68311471	52.97836340	-1.391	0.1674
X2	1	-86.41176445	24.44923524	-3.534	0.0006
X3	1	3136.59235	222.70788	14.084	0.0001
X4	1	-1781.05526	397.45816	-4.481	0.0001
X5	1	986.99033	204.77016	4.820	0.0001

FIGURE 13.30

SAS regression analysis: Model 2

ANALYSIS OF VARIANCE

SOURCE	DF	SUM OF SQUARES	MEAN SQUARE	F VALUE	PROB>F
MODEL	8	244325938	30540742.24	32.904	0.0001
ERROR	97	90032797.92	928173.17		
C TOTAL	105	334358736			

ROOT MSE	963.4174	R-SQUARE	0.7307	
DEP MEAN	19176.04	ADJ R-SQ	0.7085	
C.V.	5.024069			

PARAMETER ESTIMATES

VARIABLE	DF	PARAMETER ESTIMATE	STANDARD ERROR	T FOR H0: PARAMETER=0	PROB > |T|
INTERCEP	1	19461.81838	764.78661	25.447	0.0001
X1	1	-683.94788	245.08261	2.791	0.0063
X2	1	-264.54977	122.58866	-2.158	0.0334
X1X2	1	4.81613443	13.54099144	0.356	0.7229
X1X1	1	57.93750913	23.74605653	2.440	0.0165
X2X2	1	11.33982388	7.70152793	1.472	0.1441
X3	1	3051.55998	219.25992	13.918	0.0001
X4	1	-1680.93326	409.72999	-4.103	0.0001
X5	1	1114.78991	205.04676	5.437	0.0001

FIGURE 13.31

SAS regression analysis: Model 3

ANALYSIS OF VARIANCE

SOURCE	DF	SUM OF SQUARES	MEAN SQUARE	F VALUE	PROB>F
MODEL	13	255930191	19686937.74	23.094	0.0001
ERROR	92	78428545.24	852484.19		
C TOTAL	105	334358736			

ROOT MSE	923.3007	R-SQUARE	0.7654	
DEP MEAN	19176.04	ADJ R-SQ	0.7323	
C.V.	4.814867			

PARAMETER ESTIMATES

VARIABLE	DF	PARAMETER ESTIMATE	STANDARD ERROR	T FOR H0: PARAMETER=0	PROB > |T|
INTERCEP	1	14412.04197	2795.75178	5.155	0.0001
X1	1	819.81446	941.47799	0.871	0.3861
X2	1	-413.28219	303.20570	-1.363	0.1762
X1X2	1	33.53465808	34.14413805	0.982	0.3286
X1X1	1	-54.41406576	80.65633152	-0.675	0.5016
X2X2	1	11.87166612	16.59695367	0.715	0.4762
X3	1	8247.71451	2885.53387	2.858	0.0053
X4	1	-1632.24166	400.11602	-4.079	0.0001
X5	1	1242.52585	204.36346	6.080	0.0001
X1X3	1	-1350.92966	974.75335	-1.386	0.1691
X2X3	1	50.74158830	334.14998	0.152	0.8796
X1X2X3	1	-29.98215571	37.25553940	-0.805	0.4230
X1X1X3	1	91.37334655	84.49650785	1.081	0.2824
X2X2X3	1	5.55440578	19.06625192	0.291	0.7715

Case Study 13.1 continued

argue that units on the lower floors have readier access to the pool and ocean and, consequently, should be in greater demand. Why, then, is the t-test for floor height not statistically significant? The answer is that both of the preceding arguments are correct, one for the oceanside and one for the bayside. Thus, you will subsequently see that there is an interaction between floor height and view of ocean. Oceanview units on the lower floors sell at higher prices than oceanview units on the higher floors. In contrast, bayview units on the higher floors command higher prices than bayview units on the lower floors. These two contrasting effects tend to cancel (because we have not included interaction terms in the model), thus giving the false impression that floor height is not an important variable for predicting sale price.]

d. Conduct tests to determine which of the three models is "best" for predicting sale price.

e. Compare the value of s for the preferred model to the s-values for the other models. Do the results support your choice of models?

f. A plot of the residuals for Model 3 is shown in Figure 13.32. Do you detect any outliers? If these outliers are not found to have strong influence on the analysis, would you recommend they be deleted?

g. The outliers identified in part f were not found to be very influential. Consequently, Model 3 was refitted with these outliers deleted. The SAS printout for the refitted model is shown in Figure 13.33. Interpret the new model's standard deviation, s.

h. A histogram of the residuals for the refitted model is shown in Figure 13.34 and a residual plot in Figure 13.35. Interpret these graphs.

i. Write the prediction equation for Model 3, with outliers deleted.

The effect of floor height x_1 and distance from elevator x_2 can be determined by plotting $\hat{y}$ as a function of one of the variables for given values of the other. For example, suppose we wish to examine the relationship between $\hat{y}$, x_1, and x_2 for bayview ($x_3 = 0$), non-end units ($x_4 = 0$) with no furniture ($x_5 = 0$). The prediction curve relating $\hat{y}$ to distance from elevator x_2 can be graphed for each floor by first setting $x_1 = 1$ then $x_1 = 2, ..., x_1 = 8$. The graphs of these bayview curves are shown in Figure 13.36; the 1's indicate the curve for the first floor, the 2's the curve for the second floor, and so forth. A similar set of curves for the oceanview units are shown in Figure 13.37.

Focus

j. Describe the relationships predicted by Model 3 for the bayview units.

k. Describe the relationships predicted by Model 3 for the oceanview units.

l. Use the patterns depicted in Figures 13.36 and 13.37 to explain why the t-test for floor height x_1 is nonsignificant in Model 1.

FIGURE 13.32

Plot of residuals versus $\hat{y}$
for Model 3

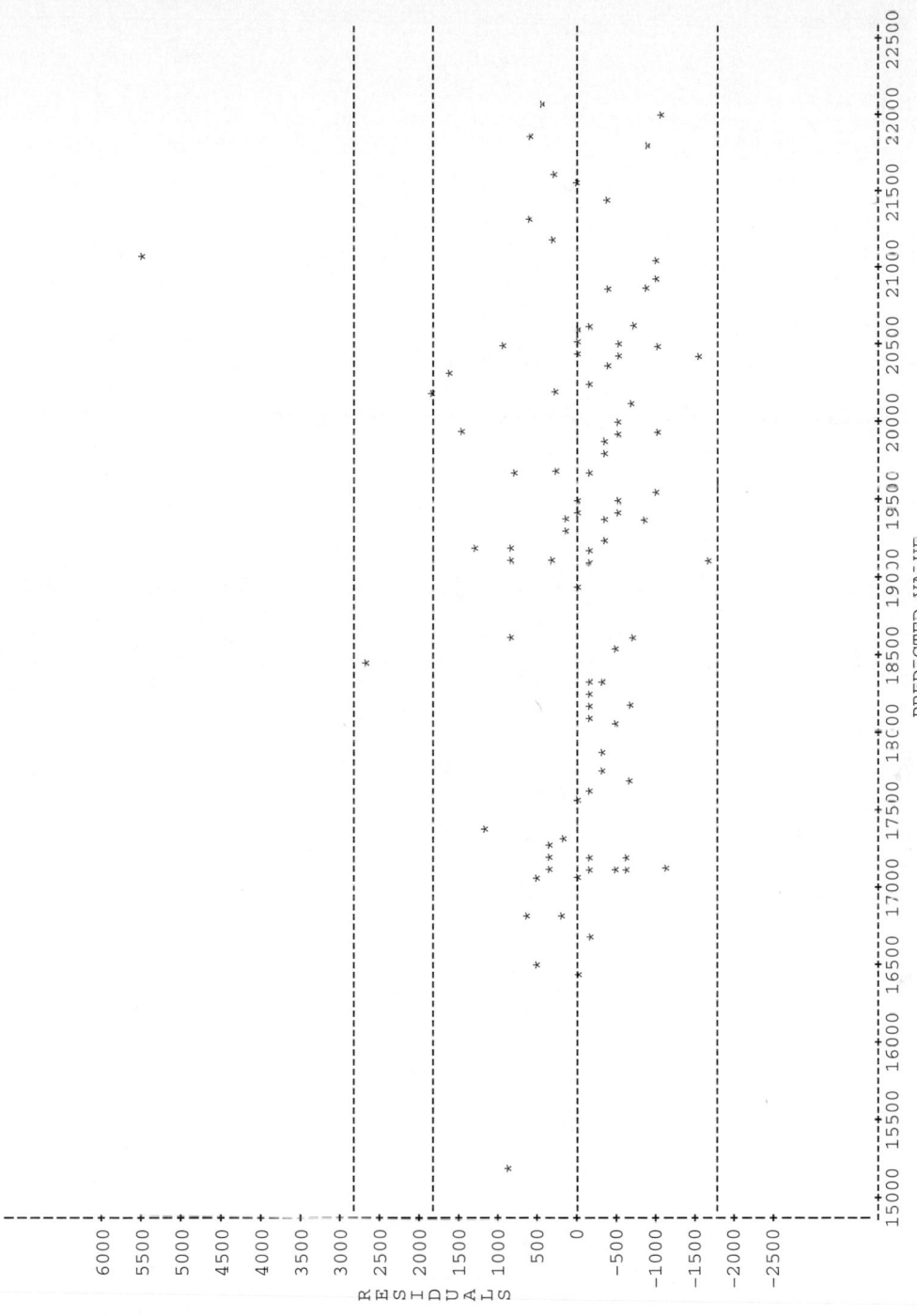

NOTE: 14 OBS HIDDEN

FIGURE 13.33

SAS printout for Model 3
with outliers deleted

ANALYSIS OF VARIANCE

SOURCE	DF	SUM OF SQUARES	MEAN SQUARE	F VALUE	PROB>F
MODEL	13	237649572	18280736.29	42.254	0.0001
ERROR	90	38937243.63	432636.04		
C TOTAL	103	276586815			

ROOT MSE	657.7507	R-SQUARE	0.8592	
DEP MEAN	19088.08	ADJ R-SQ	0.8389	
C.V.	3.445872			

PARAMETER ESTIMATES

VARIABLE	DF	PARAMETER ESTIMATE	STANDARD ERROR	T FOR H0: PARAMETER=0	PROB > |T|
INTERCEP	1	15390.19258	2002.11309	7.687	0.0001
X1	1	310.30678	682.48668	0.455	0.6504
X2	1	-203.35236	222.11810	-0.916	0.3624
X1X2	1	26.08674126	24.39118628	1.070	0.2877
X1X1	1	-10.85790288	58.55677607	-0.185	0.8533
X2X2	1	1.09206281	12.11658757	0.090	0.9284
X3	1	7537.23222	2065.74801	3.649	0.0004
X4	1	-1500.60982	285.47441	-5.257	0.0001
X5	1	1075.38112	146.64993	7.333	0.0001
X1X3	1	-1024.61067	706.31066	-1.451	0.1504
X2X3	1	-167.24636	243.44976	-0.687	0.4939
X1X2X3	1	-25.43791062	26.60757746	-0.956	0.3416
X1X1X3	1	67.67866117	61.32668383	1.104	0.2727
X2X2X3	1	19.21642228	13.82949599	1.390	0.1681

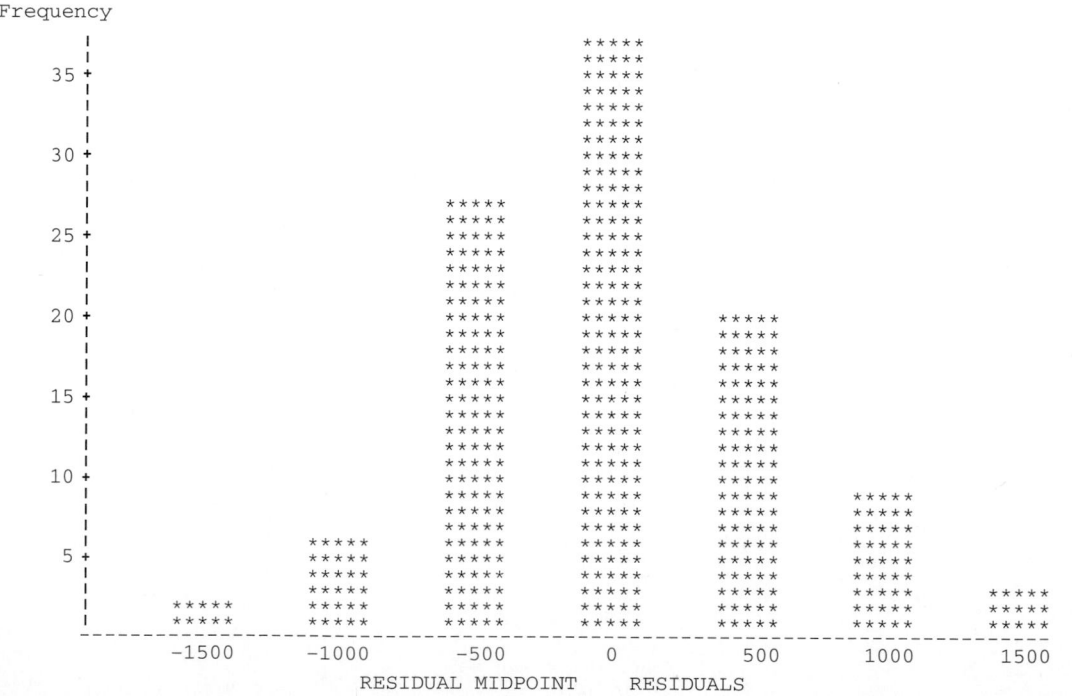

FIGURE 13.35   Histogram of residuals from Model 3 with outliers deleted

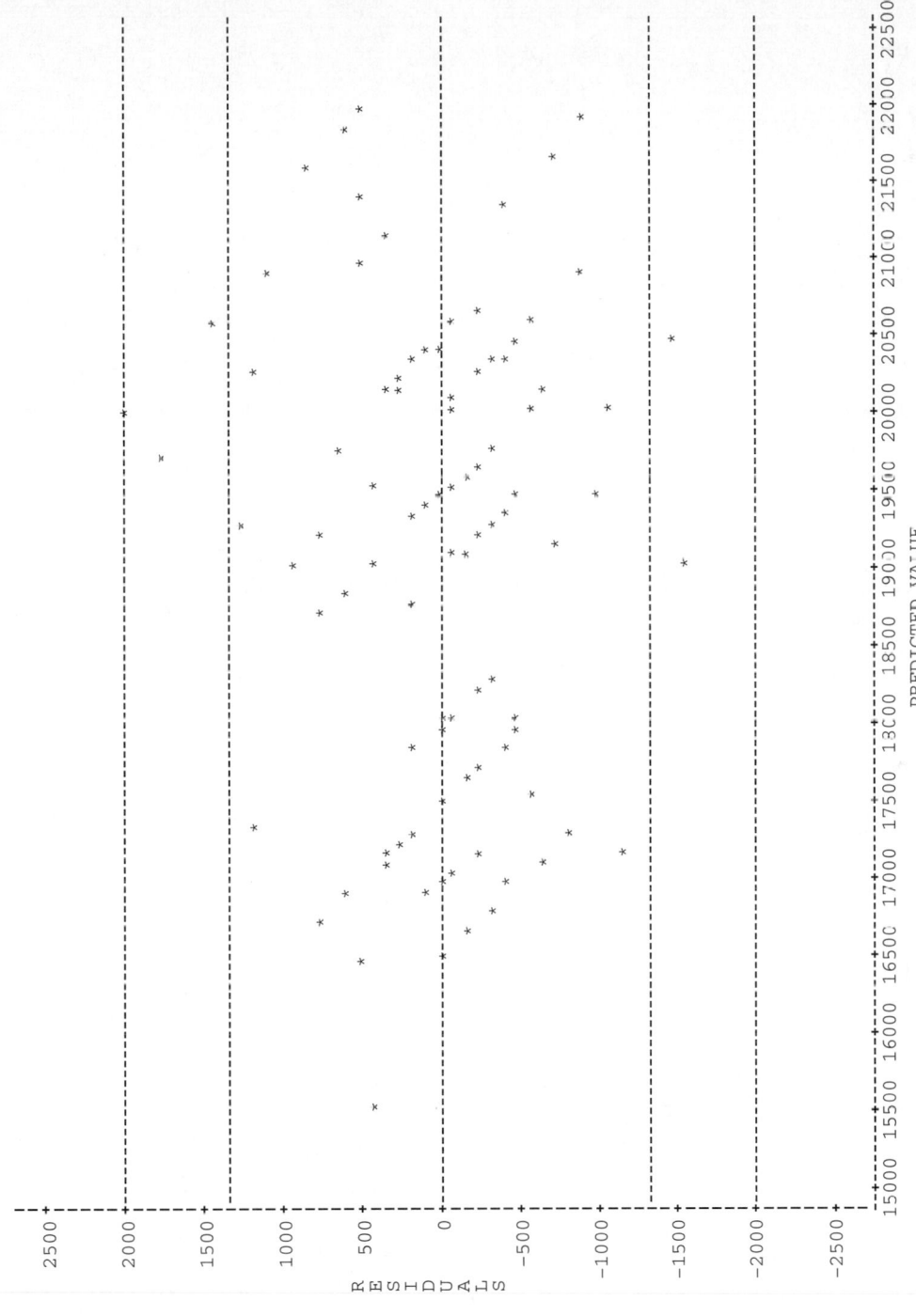

663

FIGURE 13.36

Plot of predicted price versus distance from elevator: Bayview units

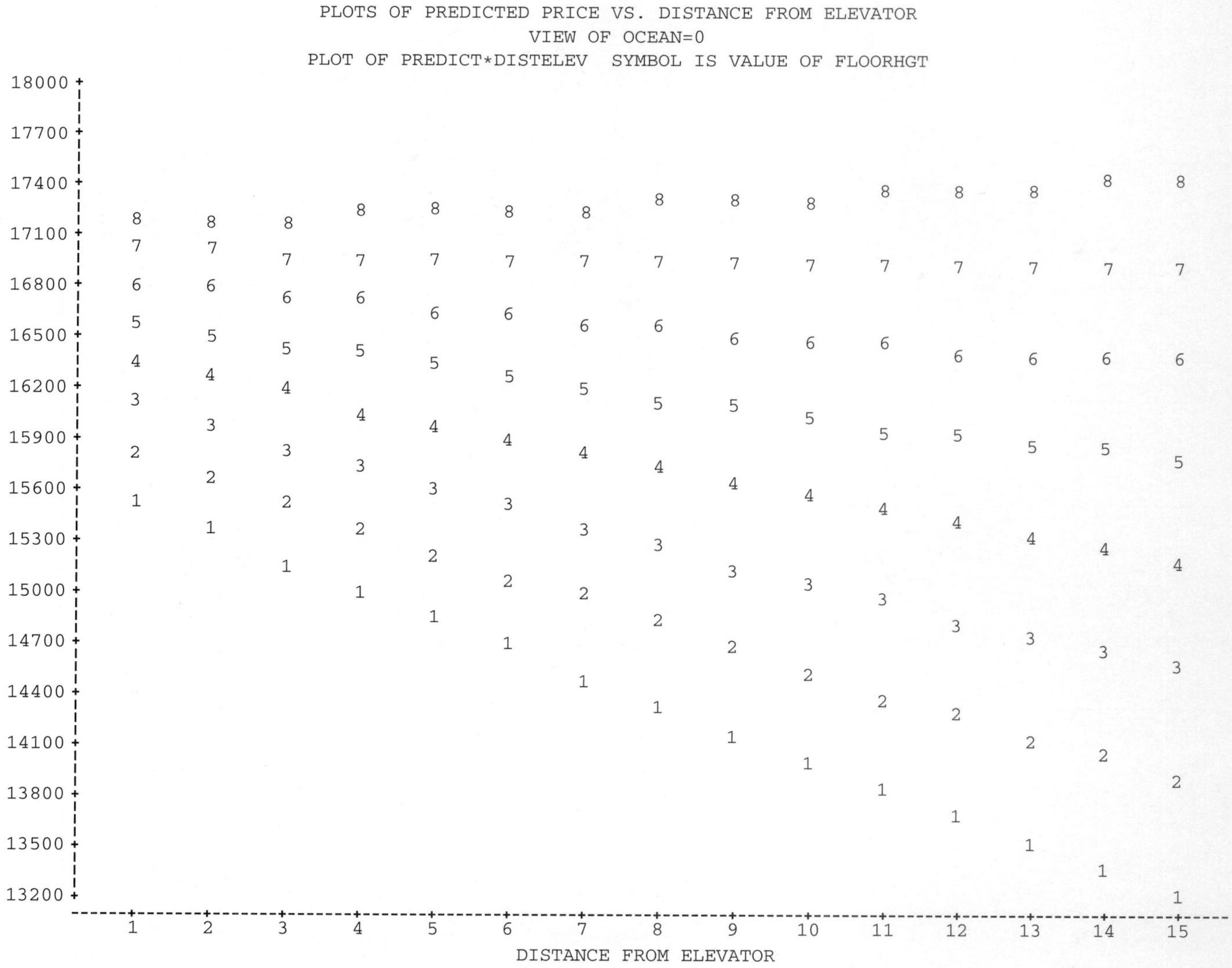

PLOTS OF PREDICTED PRICE VS. DISTANCE FROM ELEVATOR
VIEW OF OCEAN=0
PLOT OF PREDICT*DISTELEV   SYMBOL IS VALUE OF FLOORHGT

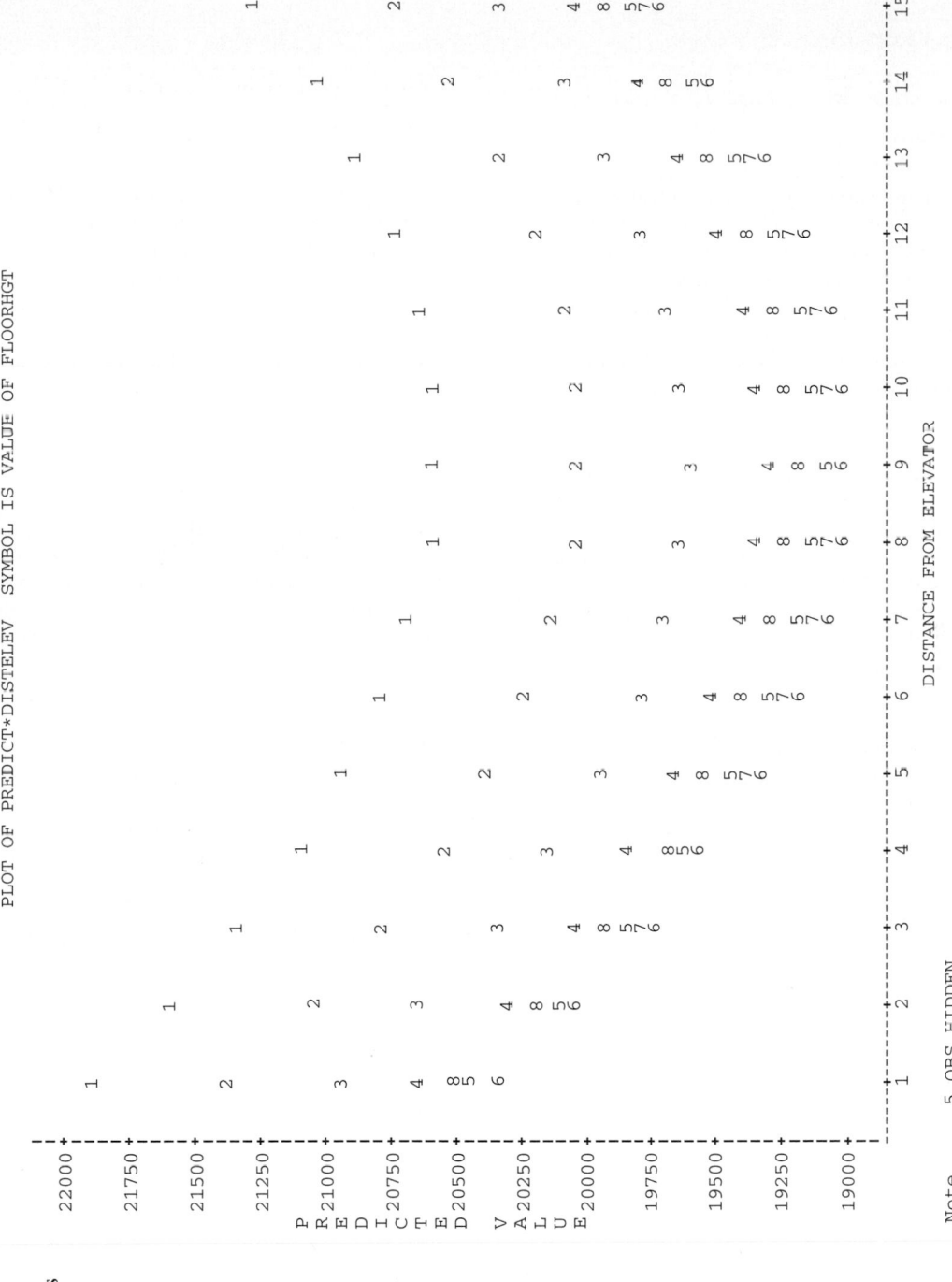

FIGURE 13.37

Plot of predicted price versus distance from elevator: Oceanview units

EXERCISES 13.68–13.78

Note: Exercises marked with 🖥 *require the use of a computer.*

Learning the Mechanics

13.68 Consider an application for which you want to relate the response variable y to one quantitative variable and one qualitative variable at three levels.
  a. Write a complete second-order model that relates $E(y)$ to the quantitative variable.
  b. Add the main effect terms for the qualitative variable (at three levels) to the model of part **a**.
  c. Add terms to the model of part **b** to allow for interaction between the quantitative and qualitative independent variables.

13.69 Refer to Exercise 13.68, part **c**.
  a. Under what circumstances will the response curves of the model have the same shape but different y-intercepts?
  b. Under what circumstances will the response curves of the model be parallel lines?
  c. Under what circumstances will the response curves of the model be identical?

13.70 Write a model that relates $E(y)$ to two independent variables—one quantitative and one qualitative at four levels. Construct a model that allows the associated response curves to be second-order but does not allow for interaction between the two independent variables.

13.71 MINITAB was used to fit the following model to $n = 25$ data points:

$$y = \beta_0 + \beta_1 x_1 + \beta_2 x_1^2 + \beta_3 x_2 + \beta_4 x_3 + \beta_5 x_1 x_2 + \beta_6 x_1 x_3 + \beta_7 x_1^2 x_2 + \beta_8 x_1^2 x_3 + \epsilon$$

where x_1 is a quantitative variable and

$$x_2 = \begin{cases} 1 & \text{if level 2} \\ 0 & \text{otherwise} \end{cases} \qquad x_3 = \begin{cases} 1 & \text{if level 3} \\ 0 & \text{otherwise} \end{cases}$$

as shown in the printout below. MINITAB was also used to fit the reduced model $y = \beta_0 + \beta_1 x_1 + \beta_2 x_1^2 + \epsilon$ to the same data set, as shown in the printout at the top of page 667.
  a. What is the general least squares prediction equation for the complete model?
  b. What is the equation of the response curve for $E(y)$ when $x_2 = 0$ and $x_3 = 0$? When $x_2 = 1$ and $x_3 = 0$? When $x_2 = 0$ and $x_3 = 1$?
  c. On the same graph, plot the least squares prediction equation associated with level 1, with level 2, and with level 3.
  d. Specify the null and alternative hypotheses that should be used to investigate whether the second-order response curves for the three levels differ.
  e. Conduct the hypothesis test of part **d**. Use $\alpha = .05$.

Applying the Concepts

13.72 An equal rights group has charged that women are being discriminated against in terms of the salary structure in a state university system. It is thought that a complete second-order model will be adequate to describe the relationship between salary and years of experience for both men and women. A sample is to be taken from the records for faculty

```
The regression equation is
Y  = 48.8 - 3.36 X1 + 0.075 X1SQ - 2.36 X2 - 7.60 X3
       + 3.71 X1X2 + 2.66 X1X3 - 0.018 X1SQX2 - 0.037 X1SQX3

Predictor      Coef      Stdev     t-ratio        p
Constant     48.784      4.186      11.65     0.000
X1           -3.362      0.471      -7.14     0.000
X1SQ          0.075      0.011       6.89     0.000
X2           -2.364      6.922      -0.34     0.779
X3           -7.595      6.102      -1.24     0.226
X1X2          3.714      0.830       4.48     0.001
X1X3          2.659      0.647       4.11     0.002
X1SQX2       -0.018      0.022      -0.84     0.455
X1SQX3       -0.037      0.015      -2.56     0.022

s = 3.190      R-sq = 98.9%      R-sq(adj) = 98.4%

Analysis of Variance

SOURCE        DF          SS          MS         F        p
Regression     8      15256.9      1907.1    186.97    0.000
Error         16        162.9        10.2
Total         24      15419.8
```

```
The regression equation is
Y  = 33.9 + 0.86 X1 - 0.011 X1SQ

Predictor          Coef         Stdev      t-ratio         p
Constant         33.930         20.74         1.64     0.083
X1                0.862         2.218         0.39     0.705
X1SQ           -0.01102       0.05103        -0.22     0.844

s = 26.10          R-sq = 2.8%        R-sq(adj) = .0%

Analysis of Variance

SOURCE         DF            SS          MS         F         p
Regression      2         433.7       216.9      0.32     0.789
Error          22       14986.1       681.2
Total          24       15419.8
```

members (all of equal rank) within the system and the following model is to be fit.*

$$E(y) = \beta_0 + \beta_1 x_1 + \beta_2 x_1^2 + \beta_3 x_2 + \beta_4 x_1 x_2 + \beta_5 x_1^2 x_2$$

where

y = Annual salary (in thousands of dollars)

x_1 = Experience (years) $x_2 = \begin{cases} 1 & \text{if female} \\ 0 & \text{if male} \end{cases}$

a. What hypothesis would you test to determine whether the *rate* of increase of mean salary with experience is different for males and females?

b. What hypothesis would you test to determine whether there are differences in mean salaries that are attributable to gender?

13.73 Refer to Exercise 13.72 and the model that was proposed to describe the relationship between salary and years of experience for both men and women. A portion of the computer printout that results from fitting this model to a sample of 200 faculty members in the university system is shown below (below left). The reduced model $E(y) = \beta_0 + \beta_1 x_1 + \beta_2 x_1^2$ is fit to the same data, and the resulting computer printout is partially reproduced (below right). Do the data provide sufficient evidence to support the claim that the mean salary of faculty members is dependent on gender? Use $\alpha = .05$.

13.74 Refer to the *Journal of Human Stress* study of firefighters, Exercise 13.5. It is thought that the following complete second-order model will be adequate to describe the relationship between emotional distress and years of experience for two groups of firefighters—those exposed to a chemical fire and those not exposed.[†]

$$E(y) = \beta_0 + \beta_1 x_1 + \beta_2 x_1^2 + \beta_3 x_2 + \beta_4 x_1 x_2 + \beta_5 x_1^2 x_2$$

where

y = Emotional distress

x_1 = Experience (years)

$x_2 = \begin{cases} 1 & \text{if exposed to chemical fire} \\ 0 & \text{if not} \end{cases}$

a. What hypothesis would you test to determine whether the *rate* of increase of emotional distress with experience is different for the two groups of firefighters?

b. What hypothesis would you test to determine whether there are differences in mean emotional distress levels that are attributable to exposure group?

c. A portion of the SAS printout that results from fitting the second-order model to a sample of 200 firefighters is shown at the top of page 668. The reduced model is fit to the same data, and the

SOURCE	DF	SUM OF SQUARES	MEAN SQUARE
MODEL	5	2351.70	470.34
ERROR	194	783.90	.04
TOTAL	199	3135.60	R-SQUARE
			0.750

SOURCE	DF	SUM OF SQUARES	MEAN SQUARE
MODEL	2	2340.37	1170.185
ERROR	197	795.23	4.04
TOTAL	199	3135.60	R-SQUARE
			0.746

---

*In practice, we would include other variables in the model. We include only two here to simplify the exercise.
[†]In practice, we would include other variables in the model.

		Analysis of Variance			
Source	DF	Sum of Squares	Mean Square	F Value	Prob>F
Model	5	2351.70	470.34	116.42	0.0001
Error	194	783.90	4.04		
C Total	199	3135.60			

Root MSE	2.0102	R-square	0.7500	
Dep Mean	24.221	Adj R-sq	0.7436	
C.V.	8.299			

$$E(y) = \beta_0 + \beta_1 x_1 + \beta_2 x_1^2$$

resulting computer printout is reproduced below. Is there sufficient evidence to support the claim that the mean emotional distress levels differ for the two groups of firefighters? Use $\alpha = .05$.

13.75 Since glass is not subject to radiation damage, encapsulation of waste in glass is considered to be one of the most promising solutions to the problem of storing low-level nuclear waste. However, glass undergoes chemical changes when exposed to extreme environmental conditions, and certain of its constituents, such as silicon salts, can leach into the surroundings. In addition, these chemical reactions tend to weaken the glass. These concerns led to a study undertaken jointly by the Department of Materials Science and Engineering at the University of Florida and the U.S. Department of Energy to assess the utility of glass as a waste encapsulant.* Corrosive chemical solutions (called corrosion baths) were prepared and applied directly to glass samples containing one of three types of waste (TDS-3A, FE, and AL); the chemical reactions were observed over time. A few of the key variables measured were

y = Amount of silicon (in parts per million) found in solution at end of experiment. (This is both a

measure of the degree of breakdown in the glass and a proxy for the amount of radioactive species released into the environment.)

x_1 = Temperature (°C) of the corrosion bath

$$x_2 = \begin{cases} 1 & \text{if waste type TDS-3A} \\ 0 & \text{if not} \end{cases}$$

$$x_3 = \begin{cases} 1 & \text{if waste type FE} \\ 0 & \text{if not} \end{cases}$$

Waste type AL is the base level. Suppose we want to model amount y of silicon as a function of temperature (x_1) and type of waste (x_2, x_3).

a. Write a model that proposes parallel straight-line relationships between amount of silicon and temperature, one line for each of the three waste types.

b. Add terms for the interaction between temperature and waste type to the model of part **a**.

c. Refer to the model of part **b**. For each waste type, give the slope of the line relating amount of silicon to temperature.

d. Explain how you could test for the presence of temperature–waste type interaction.

13.76 A medical researcher wants to model the extent of lung damage in emphysema patients as a function of two variables: the number of years the patient

		Analysis of Variance			
Source	DF	Sum of Squares	Mean Square	F Value	Prob>F
Model	2	2340.37	1170.185	289.87	0.0001
Error	197	795.23	4.037		
C Total	199	3135.60			

Root MSE	2.0092	R-square	0.7464	
Dep Mean	24.221	Adj R-sq	0.7438	
C.V.	8.295			

*The background information for this exercise was provided by Dr. David Clark, Department of Materials Science and Engineering, University of Florida.

has smoked (x_1) and the sex of the patient (x_2). Fifty emphysema patients are used in the study, and the response y for each patient is a subjective score ranging from 0 to 50. (High scores indicate extensive lung damage.)

a. Identify the independent variables as qualitative or quantitative.
b. Write the main effects (first-order) model for predicting the mean lung damage score, $E(y)$, from the two variables identified in part **a**.
c. Using a female patient as the base level, the researcher obtained the prediction equation

$$\hat{y} = -.85 + .65x_1 + .09x_2$$

What is the estimated difference between the mean lung damage scores for a male and female who have both smoked for 20 years?
d. Find the estimated mean lung damage score for a female patient who has smoked for 15 years. [*Note:* With the original data and an appropriate computer program package, you would be able to produce confidence intervals for the estimates in parts **c** and **d**.]
e. It would not be surprising if the functional relation between lung damage and years of smoking were second-order. Further, it is conceivable that years of smoking might affect men differently than women. Write a model that allows for these conditions.

13.77 An operations manager is interested in modeling $E(y)$, the expected length of time per month (in hours) that a machine will be shut down for repairs, as a function of the type of machine (001 or 002) and the age of the machine (in years). The manager has proposed the following model:

$$E(y) = \beta_0 + \beta_1 x_1 + \beta_2 x_1^2 + \beta_3 x_2$$

where

$x_1 = $ Age of machine

$x_2 = \begin{cases} 1 & \text{if machine type 001} \\ 0 & \text{if machine type 002} \end{cases}$

Data obtained on $n = 20$ machine breakdowns were used to estimate the parameters of this model. A portion of the regression analysis computer printout is shown in the first printout in the next column.

The reduced model $E(y) = \beta_0 + \beta_1 x_1 + \beta_3 x_2$ was fit to the same data. The regression analysis computer printout is partially reproduced in the second printout in the next column. Do these data provide sufficient evidence to conclude that the

second-order (x_1^2) term in the model proposed by the operations manager is necessary? Test using $\alpha = .05$.

SOURCE	DF	SUM OF SQUARES	MEAN SQUARE
MODEL	3	2396.364	798.788
ERROR	16	128.586	8.037
TOTAL	19	2524.950	R-SQUARE 0.949

SOURCE	DF	SUM OF SQUARES	MEAN SQUARE
MODEL	2	2342.42	1171.21
ERROR	17	182.53	10.74
TOTAL	19	2524.95	R-SQUARE 0.928

13.78 Refer to Exercise 13.77. The data used to fit the operations manager's complete and reduced models are displayed in the following table.

Downtime (hours per month)	Machine Age x_1 (years)	Machine Type	x_2
10	1.0	001	1
20	2.0	001	1
30	2.7	001	1
40	4.1	001	1
9	1.2	001	1
25	2.5	001	1
19	1.9	001	1
41	5.0	001	1
22	2.1	001	1
12	1.1	001	1
10	2.0	002	0
20	4.0	002	0
30	5.0	002	0
44	8.0	002	0
9	2.4	002	0
25	5.1	002	0
20	3.5	002	0
42	7.0	002	0
20	4.0	002	0
13	2.1	002	0

a. Use these data to test the null hypothesis that $\beta_1 = \beta_2 = 0$. Test using $\alpha = .10$.
b. Interpret the results of the test in the context of the problem.

13.8   STEPWISE REGRESSION

The problem of predicting executive salaries was discussed in Chapter 12. Perhaps the biggest problem in building a model to describe executive salaries is choosing the important independent variables to be included in the model. The list of potentially important independent variables is extremely long, and we need some objective method of screening out those that are not important.

The problem of deciding which of a large set of independent variables to include in a model is a common one. Trying to determine which variables influence the profit of a firm, affect blood pressure of humans, or are related to a student's performance in college are only a few examples.

A systematic approach to building a model with a large number of independent variables is difficult because the interpretation of multivariable interactions and higher-order polynomials is tedious. We therefore turn to a screening procedure known as **stepwise regression**.

The most commonly used stepwise regression procedure, available in most popular software packages, works as follows. The user first identifies the response, *y*, and the set of potentially important independent variables, $x_1, x_2, ..., x_k$, where *k* is generally large. [*Note:* This set of variables could include both first-order and higher-order terms. However, we may often include only the main effects of both quantitative variables (first-order terms) and qualitative variables (dummy variables), since the inclusion of second-order terms greatly increases the number of independent variables.] The response and independent variables are then entered into the computer, and the stepwise procedure begins.

Step 1   The computer fits all possible one-variable models of the form

$$E(y) = \beta_0 + \beta_1 x_i$$

to the data, where x_i is the *i*th independent variable, $i = 1, 2, ..., k$. For each model, the test of the null hypothesis

$$H_0: \beta_1 = 0$$

against the alternative hypothesis

$$H_a: \beta_1 \neq 0$$

is conducted using the *t*-test (or the equivalent *F*-test) for a single β parameter. The independent variable that produces the largest (absolute) *t* value is declared the best one-variable predictor of *y*.* Call this independent variable x_1.

Step 2   The stepwise program now begins to search through the remaining $(k - 1)$ independent variables for the best two-variable model of the form

$$E(y) = \beta_0 + \beta_1 x_1 + \beta_2 x_i$$

This is done by fitting all two-variable models containing x_1 and each of the other $(k - 1)$ options for the second variable x_i. The *t* values for the test $H_0: \beta_2 = 0$ are computed for each of the $(k - 1)$ models (corresponding to the remaining independent variables x_i, $i = 2, 3, ..., k$), and the variable having the largest *t* is retained. Call this variable x_2.

---

*Note that the variable with the largest *t* value is also the one with the largest (absolute) Pearson product moment correlation, *r* (Section 11.6), with *y*.

At this point, some software packages diverge in methodology. The better packages now go back and check the t value of $\hat{\beta}_1$ after $\hat{\beta}_2 x_2$ has been added to the model. If the t value has become nonsignificant at some specified α level (say $\alpha = .10$), the variable x_1 is removed and a search is made for the independent variable with a β parameter that will yield the most significant t value in the presence of $\hat{\beta}_2 x_2$. Other packages do not recheck the significance of $\hat{\beta}_1$ but proceed directly to step 3.

The reason the t value for x_1 may change from step 1 to step 2 is that the meaning of the coefficient β_1 changes. In step 2, we are approximating a complex response surface in two variables with a plane. The best-fitting plane may yield a different value for $\hat{\beta}_1$ than that obtained in step 1. Thus, both the value of $\hat{\beta}_1$ and its significance usually changes from step 1 to step 2. For this reason, the software packages that recheck the t values at each step are preferred.

Step 3 The stepwise procedure now checks for a third independent variable to include in the model with x_1 and x_2. That is, we seek the best model of the form

$$E(y) = \beta_0 + \beta_1 x_1 + \beta_2 x_2 + \beta_3 x_i$$

To do this, we fit all the $(k - 2)$ models using x_1, x_2, and each of the $(k - 2)$ remaining variables, x_i, as a possible x_3. The criterion is again to include the independent variable with the largest t value. Call this best third variable x_3.

The better programs now recheck the t values corresponding to the x_1 and x_2 coefficients, replacing the variables with t values that have become nonsignificant. This procedure is continued until no further independent variables can be found that yield significant t values (at the specified α level) in the presence of the variables already in the model.

The result of the stepwise procedure is a model containing only those terms with t values that are significant at the specified α level. Thus, in most practical situations only several of the large number of independent variables remain. However, it is very important *not* to jump to the conclusion that all the independent variables important for predicting y have been identified or that the unimportant independent variables have been eliminated. Remember, the stepwise procedure is using only *sample estimates* of the true model coefficients (β's) to select the important variables. An extremely large number of single β parameter t-tests have been conducted, and the probability is very high that one or more errors have been made in including or excluding variables. That is, we have very probably included some unimportant independent variables in the model (Type I errors) and eliminated some important ones (Type II errors).

There is a second reason why we might not have arrived at a good model. When we choose the variables to be included in the stepwise regression, we may often omit high-order terms (to keep the number of variables manageable). Consequently, we may have initially omitted several important terms from the model. Thus, we should recognize stepwise regression for what it is: an objective screening procedure.

Now, we will consider second-order terms (for quantitative variables) and other interactions among variables screened by the stepwise procedure. It would be best to develop this response surface model with a second set of data independent of that used for the screening, so the results of the stepwise procedure can be partially verified with new data. This is not always possible, however, because in many modeling situations only a small amount of data is available.

Do not be deceived by the impressive-looking t values that result from the stepwise procedure—it has retained only the independent variables with the largest *t* values. Also, be certain to consider second-order terms in systematically developing the prediction model. Finally, if you have used a first-order model for your stepwise procedure, remember that it may be greatly improved by the addition of higher-order terms.

Warning

Be cautious when using the results of stepwise regression to make inferences about the relationship between $E(y)$ and the independent variables in the resulting first-order model. First, an extremely large number of *t*-tests have been conducted, leading to a high probability of making either one or more Type I or Type II errors. Second, the stepwise model does not include any higher-order or interaction terms. Stepwise regression should be used only when necessary, that is, when you want to determine which of a large number of potentially important independent variables should be used in the model-building process.

EXAMPLE 13.12

In Section 12.7 we fit a multiple regression model for executive salaries as a function of experience, education, sex, etc. A preliminary step in the construction of this model was the determination of the most important independent variables. Ten independent variables (seven quantitative and three qualitative) were considered, as shown in Table 13.4. It would be very difficult to construct a complete second-order model with ten independent variables. Therefore, use the sample of 100 executives from Section 12.7 to decide which of the 10 variables should be included in the construction of the final model for executive salaries.

Solution

We will use stepwise regression with the main effects of the 10 independent variables to identify the most important variables. The dependent variable y is the natural logarithm of the executive salaries. The SAS stepwise regression printout is shown in Figure 13.38.* Note that the first variable included in the model is x_4, the number of employees supervised by the executive. At the second step, x_5,

TABLE 13.4 **Independent Variables in the Executive Salary Example**

Independent Variable	Description
x_1	Experience (years)—quantitative
x_2	Education (years)—quantitative
x_3	Sex (1 if male, 0 if female)—qualitative
x_4	Number of employees supervised—quantitative
x_5	Corporate assets (millions of dollars)—quantitative
x_6	Board member (1 if yes, 0 if no)—qualitative
x_7	Age (years)—quantitative
x_8	Company profits (past 12 months, millions of dollars)—quantitative
x_9	Has international responsibility (1 if yes, 0 if no)—qualitative
x_{10}	Company's total sales (past 12 months, millions of dollars)—quantitative

*Note that there are several slight changes in the SAS printout labels for the stepwise procedure. For example, the $\hat{\beta}$ values, labeled **ESTIMATE** in the multiple regression procedure, are labeled **B VALUE** in the stepwise procedure.

```
STEP 1
    Variable X4 Entered        R-Square - 0.42071677      C(P) = 1274.7576

                      DF    Sum of Squares   Mean Square      F    Prob > F

    Regression         1       11.46854285   11.46854285   71.17    0.0001
    Error             98       15.79113802    0.16113696
    Total             99       27.25977087

                              B Value       Std Error        F    Prob > F

    Intercept             10.20077500
    X4                     0.00057284      0.00006790    71.17      0.0001
-----------------------------------------------------------------------

STEP 2
    Variable X5 Entered        R-Square = 0.78299675      C(P) = 419.4947

                      DF    Sum of Squares   Mean Square      F    Prob > F

    Regression         2       21.34431198   10.67215599  175.00     0.0001
    Error             97        5.91545889    0.06098411
    Total             99       27.25977087

                              B Value       Std Error        F    Prob > F

    Intercept              9.87702903
    X4                     0.00058353      0.00004178   195.06      0.0001
    X5                     0.00183730      0.00014438   161.94      0.0001
-----------------------------------------------------------------------

STEP 3
    Variable X1 Entered        R-Square = 0.89667614      C(P) = 152.4952

                      DF    Sum of Squares   Mean Square      F    Prob > F

    Regression         3       24.44318616    8.14772872  277.71     0.0001
    Error             96        2.81658471    0.02933942
    Total             99       27.25977087

                              B Value       Std Error        F    Prob > F

    Intercept              9.66449288
    X4                     0.00055251      0.00002914   359.59      0.0001
    X5                     0.00191195      0.00010041   362.60      0.0001
    X1                     0.01870784      0.00182032   105.62      0.0001
-----------------------------------------------------------------------

STEP 4
    Variable X3 Entered        R-Square = 0.94815717      C(P) = 32.6757

                      DF    Sum of Squares   Mean Square      F    Prob > F

    Regression         4       25.84654710    8.46163678  434.37     0.0001
    Error             95        1.41322377    0.01487604
    Total             99       27.25977087

                              B Value       Std Error        F    Prob > F

    Intercept              9.40077349
    X4                     0.00055288      0.00002075   710.15      0.0001
    X5                     0.00190876      0.00007150   712.74      0.0001
    X1                     0.02074868      0.00131310   249.68      0.0001
    X3                     0.30011726      0.03089939    94.34      0.0001
-----------------------------------------------------------------------
```

FIGURE 13.38 Stepwise regression printout for Example 13.12

STEP 5
  Variable X2 Entered   R-Square = 0.96039323   C(P) = 5.7215

	DF	Sum of Squares	Mean Square	F	Prob > F
Regression	5	26.18009940	5.23601988	455.87	0.0001
Error	94	1.07967147	0.01148587		
Total	99	27.25977087			

		B Value	Std Error	F	Prob > F
Intercept		8.85387930			
X4		0.00056061	0.00001829	939.84	0.0001
X5		0.00193684	0.00006304	943.98	0.0001
X1		0.02141724	0.00116047	340.61	0.0001
X3		0.31927842	0.02738298	135.95	0.0001
X2		0.03315807	0.00615303	29.04	0.0001

STEP 6
  Variable X6 Entered   R-Square = 0.96100666   C(P) = 6.2699

	DF	Sum of Squares	Mean Square	F	Prob > F
Regression	6	26.19682148	4.36613691	382.00	0.0001
Error	93	1.06294939	0.01142956		
Total	99	27.25977087			

		B Value	Std Error	F	Prob > F
Intercept		8.87509152			
X4		0.00055820	0.00001835	925.32	0.0001
X5		0.00193764	0.00006289	949.31	0.0001
X1		0.02133460	0.00115963	338.48	0.0001
X3		0.31093801	0.02817264	121.81	0.0001
X2		0.03272195	0.00614851	28.32	0.0001
X6		0.03866226	0.03196369	1.46	0.2295

STEP 7
  Variable X6 Removed   R-Square = 0.96039323   C(P) = 5.7215

	DF	Sum of Squares	Mean Square	F	Prob > F
Regression	5	26.18009940	5.23601988	455.87	0.0001
Error	94	1.07967147	0.01148587		
Total	99	27.25977087			

		B Value	Std Error	F	Prob > F
Intercept		8.85387930			
X4		0.00056061	0.00001829	939.84	0.0001
X5		0.00193684	0.00006304	943.98	0.0001
X1		0.02141724	0.00116047	340.61	0.0001
X3		0.31927842	0.02738298	135.95	0.0001
X2		0.03315807	0.00615303	29.04	0.0001

FIGURE 13.38   Continued

corporate assets, enters the model. At the sixth step, x_6, a dummy variable for the qualitative variable Board member or not, is brought into the model. However, because the significance (.2295) of the F statistic (SAS uses the $F = t^2$ statistic rather than the t statistic in the stepwise procedure) for x_6 is above the preassigned $\alpha = .10$, x_6 is then removed from the model. Thus, at step 7 the procedure indicates that the five-variable model including x_1, x_2, x_3, x_4, and x_5 is best. That is, none of the other independent variables can meet the $\alpha = .10$ criterion for admission to the model.

Thus, in our final modeling effort (Section 12.7) we concentrated on these five independent variables and determined that several second-order terms were important in the prediction of executive salaries. ▲

EXERCISES 13.79–13.82

Note: Exercises marked with 🖫 *require the use of a computer.*

Learning the Mechanics

13.79 There are six independent variables, x_1, x_2, x_3, x_4, x_5, and x_6, that might be useful in predicting a response y. A total of $n = 50$ observations are available, and it is decided to employ stepwise regression to help in selecting the independent variables that appear to be useful. The computer fits all possible one-variable models of the form

$$E(y) = \beta_0 + \beta_1 x_i$$

where x_i is the ith independent variable, $i = 1, 2, \ldots, 6$. The information in the following table is provided from the computer printout.

Independent Variable	$\hat{\beta}_i$	$s_{\hat{\beta}_i}$
x_1	1.6	.42
x_2	−.9	.01
x_3	3.4	1.14
x_4	2.5	2.06
x_5	−4.4	.73
x_6	.3	.35

a. Which independent variable is declared the best one-variable predictor of y? Explain.
b. Would this variable be included in the model at this stage? Explain.
c. Describe the next phase that a stepwise procedure would execute.

Applying the Concepts

13.80 The *Journal of College Student Personnel* (1984) provided an analysis of data collected on 230 male and 72 female freshmen attending a midwestern university specializing in engineering and technology. Data collected on each student included a score on a self-concept of ability test (SCPT); scores on the SAT, ACT, and various other assessment and achievement tests; a standardized measure (called a *T*-score) of the student's rank in high school class; and the student's first-semester GPA. One aspect of the study included fitting a stepwise regression model for GPA involving first-order terms for the test scores shown in the first column of the table. The variables are listed in the order in which they entered the model. Column 2 gives the

value of R^2 at each stage of the stepwise regression; column 3 gives the simple coefficient r of correlation between each variable and GPA; and column 4 gives the F value at each stage of the stepwise regression for testing the contribution of the least variable included in the stepwise model for the prediction of GPA.

Variable	R^2	r	F
T-score	.14	.37	18.44
Mathematics test	.18	.34	10.93
SAT	.24	.18	29.57
ACT	.44	.27	14.15
Verbal test	.50	.30	3.93
SCPT	.52	.29	1.89
Trigonometry test	.52	.28	.16

Source: Robinson, D. A. G., and Cooper, S. E. "The influence of self-concept on academic success in technological careers." *Journal of College Student Personnel*, 1984. Copyright 1984 AACD. Reproduced with permission. No further reproduction is authorized without further permission of AACD.

a. Write the models used in the first, second, third, and fourth stages of the fitting process.
b. Does the self-concept of ability (SCPT) score come into the model?
c. Compare the values of r between GPA and the individual variables. How do you explain your answer to part **b** given that the correlation coefficient for the SCPT score is larger than that for the SAT score?
d. Use the value of R^2 given in stage 4 of the stepwise regression to calculate the value of the F statistic for testing the utility of the model.
e. Calculate the approximate p-value for the F statistic in part **d** and interpret it.

13.81 Many power plants dump hot waste water into the surrounding rivers, streams, and oceans, an action that may adversely affect the marine life in the dumping areas. A marine biologist was hired by the EPA to determine whether the hot-water runoff from a particular power plant located near a large gulf is having an adverse effect on the marine life in the area. In the initial phase of the study, the biologist's goal is to acquire a prediction equation for the number of marine animals located at certain designated areas, or stations, in the gulf. Based on past experience, the biologist considered

Country	Newspaper Copies per 1,000	Radios per 1,000	Television Sets per 1,000	Literacy Rate
Czechoslovakia	280	266	228	.98
Italy	142	230	201	.93
Kenya	10	114	2	.25
Norway	391	313	227	.99
Panama	86	329	82	.79
Philippines	17	42	11	.72
Tunisia	21	49	16	.32
USA	314	1,695	472	.99
Former USSR	333	430	185	.99
Venezuela	91	182	89	.82

the following environmental factors as predictors for the number of animals at a particular station:

x_1 = Temperature of water (TEMP)

x_2 = Salinity of water (SAL)

x_3 = Dissolved oxygen content of water (DO)

x_4 = Turbidity index, a measure of the turbidity of the water (TI)

x_5 = Depth of the water at the station (ST_DEPTH)

x_6 = Total weight of sea grasses in sampled area (TGRSWT)

As a preliminary step in the construction of this model, the biologist used a stepwise regression procedure to identify the most important of these six variables. A total of 716 samples were taken at different stations in the gulf, producing the SAS printout shown below and on page 677. (The response measured was y, the logarithm of the number of marine animals found in the sampled area.)

a. According to the SAS printout, which of the six independent variables should be used in the model? (Use α = .10.)

b. Are we able to assume that the marine biologist has identified all the important independent variables for the prediction of y? Why?

c. Using the variables identified in part **a**, write the first-order model with interaction that may be used to predict y.

d. How would the marine biologist determine whether the model specified in part **c** is better than the first-order model?

e. Note the small value of R^2. What action might the biologist take to improve the model?

13.82 Literacy rate is a reflection of the educational facilities and quality of education available in a country, and mass communication plays a large part in the educational process. In an effort to relate the literacy rate of a country to various mass communication outlets, a demographer has proposed to relate the response

$$y = \text{Literacy rate}$$

to the independent variables

x_1 = Number of daily newspaper copies (per 1,000 population)

x_2 = Number of radios (per 1,000 population)

x_3 = Number of television sets (per 1,000 population)

a. Apply stepwise regression to the data in the table above to find the variables that are most suitable for modeling y.

b. Interpret the β estimates in the resulting stepwise model.

c. What are the dangers associated with drawing inferences from the stepwise model?

```
STEP 1
    Variable ST_DEPTH Entered   R-Square = 0.1223

                    DF      Sum of Squares   Mean Square       F    Prob > F

    Regression       1             57.44         57.44     99.47      0.0001
    Error          714            412.33          0.58
    Total          715            469.77

                            B Value        Std Error         F    Prob > F

    Intercept               8.38559
    ST_DEPTH               -0.43678          0.04379     99.47      0.0001
-----------------------------------------------------------------------------
```

STEP 2

    Variable TGRSWT Entered   R-Square = 0.1821

	DF	Sum of Squares	Mean Square	F	Prob > F
Regression	2	85.55	42.78	79.38	0.0001
Error	713	384.22	0.54		
Total	715	469.77			

	B Value	Std Error	F	Prob > F
Intercept	8.07682			
ST_DEPTH	-0.35355	0.04385	65.02	0.0001
TGRSWT	0.00271	0.00038	52.16	0.0001

STEP 3

    Variable TI Entered   R-Square = 0.1870

	DF	Sum of Squares	Mean Square	F	Prob > F
Regression	3	87.85	29.28	54.59	0.0001
Error	712	381.92	0.54		
Total	715	469.77			

	B Value	Std Error	F	Prob > F
Intercept	7.38864			
TI	0.65774	0.31783	4.28	0.0389
ST_DEPTH	-0.31451	0.47641	43.58	0.0001
TGRSWT	0.00261	0.00038	47.73	0.0001

STEP 4

    Variable DO Entered   R-Square = 0.1889

	DF	Sum of Squares	Mean Square	F	Prob > F
Regression	4	88.75	22.19	41.40	0.0001
Error	711	381.02	0.54		
Total	715	469.77			

	B Value	Std Error	F	Prob > F
Intercept	7.22576			
DO	0.01769	0.01363	1.69	0.1946
TI	0.67347	0.31791	4.49	0.0345
ST_DEPTH	-0.30417	0.04828	39.69	0.0001
TGRSWT	0.00267	0.00038	49.23	0.0001

STEP 5

    Variable DO Removed   R-Square = 0.1870

	DF	Sum of Squares	Mean Square	F	Prob > F
Regression	3	87.85	29.28	54.59	0.0001
Error	712	381.92	0.54		
Total	715	469.77			

	B Value	Std Error	F	Prob > F
Intercept	7.38864			
TI	0.65774	0.31783	4.28	0.0389
ST_DEPTH	-0.31451	0.04764	43.58	0.0001
TGRSWT	0.00261	0.00038	47.73	0.0001

QUICK REVIEW

Key Terms

Base level   636
Complete model   625
Contour lines   617
Dummy variables   635
First-order model   609
Interaction model   618
Level of a variable   607
Model building   606

pth-order polynomial   609
Paraboloid   620
Reduced model   625
Response surface   617
Saddle-shaped surface   620
Second-order model   610
Stepwise regression   670

Key Formulas

$E(y) = \beta_0 + \beta_1 x + \beta_2 x^2 + \beta_3 x^3 + \cdots + \beta_p x^p$
pth-order polynomial with one quantitative independent variable   609

$E(y) = \beta_0 + \beta_1 x_1 + \beta_2 x_2$
First-order model with two quantitative independent variables   616

$E(y) = \beta_0 + \beta_1 x_1 + \beta_2 x_2 + \beta_3 x_1 x_2$
Interaction model with two quantitative independent variables   618

$E(y) = \beta_0 + \beta_1 x_1 + \beta_2 x_2 + \beta_3 x_1 x_2 + \beta_4 x_1^2 + \beta_5 x_2^2$
Complete second-order model with two quantitative independent variables   620

$E(y) = \beta_0 + \beta_1 x_1 + \beta_2 x_2 + \cdots + \beta_{k-1} x_{k-1},$
Model with one qualitative variable at k levels   636

$$\text{where } x_i = \begin{cases} 1 & \text{if level } i + 1 \\ 0 & \text{if not} \end{cases}$$

$F = \dfrac{(\text{SSE}_r - \text{SSE}_c)/\text{number of } \beta\text{'s tested}}{\text{MSE}_c}$
Test statistic for comparing reduced and complete models   626

LANGUAGE LAB

Symbol	Pronunciation	Description
x^p	x to the pth power	$x^p = \underbrace{(x)(x)(x) \cdots (x)}_{\text{multiply } p \text{ times}}$
SSE$_r$		Sum of squared errors for reduced model
SSE$_c$		Sum of squared errors for complete model
MSE$_c$		Mean square error for complete model

SUPPLEMENTARY EXERCISES 13.83–13.99

Note: Exercises marked with  *require the use of a computer.*

Learning the Mechanics

13.83 Why is the model-building step the key to the success or failure of a regression analysis?

13.84 Suppose you fit the regression model

$E(y) = \beta_0 + \beta_1 x_1 + \beta_2 x_2 + \beta_3 x_2^2 + \beta_4 x_1 x_2 + \beta_5 x_1 x_2^2$

to $n = 35$ data points and wish to test the null hypothesis

$$H_0\text{:}\ \beta_4 = \beta_5 = 0$$

a. State the alternative hypothesis.

b. Explain in detail how to compute the F statistic needed to test the null hypothesis.

c. What are the numerator and denominator degrees of freedom associated with the F statistic in part **b**?

d. Give the rejection region for the test if $\alpha = .05$.

13.85 Graph each of the following polynomials, and give the order of each:

a. $E(y) = 6 + 3x - 2x^2$ **b.** $E(y) = 4x$

c. $E(y) = x^2$ **d.** $E(y) = 2 - 4x$

e. $E(y) = 1 + x - x^2 + x^3$ **f.** $E(y) = 1 - 4x$

13.86 Write a model relating $E(y)$ to one qualitative independent variable that is at four levels. Define all the terms in your model.

13.87 Explain the difference between qualitative and quantitative variables.

13.88 It is desired to relate $E(y)$ to a quantitative variable x_1 and a qualitative variable at three levels.

a. Write a first-order model.

b. Write a model that will graph as three different second-order curves—one for each level of the qualitative variable.

13.89 Explain why stepwise regression is used. What is its value in the model-building process?

13.90 a. Write a first-order model relating $E(y)$ to two quantitative independent variables, x_1 and x_2.

b. Write a complete second-order model.

13.91 To model the relationship between y, a dependent variable, and x, an independent variable, a researcher has taken one measurement on y at each of three different x values. Drawing on his mathematical expertise, the researcher realizes that he can fit the second-order polynomial model

$$E(y) = \beta_0 + \beta_1 x + \beta_2 x^2$$

and it will pass exactly through all three points, yielding SSE = 0. The researcher, delighted with the "excellent" fit of the model, eagerly sets out to use it to make inferences. What problems will he encounter in attempting to make inferences?

Applying the Concepts

13.92 The number of practicing attorneys more than doubled in the decade 1973–1983, with the percentage of female attorneys increasing from 5% to 15% of the legal profession. A random sample of 400 female attorneys and 200 male attorneys, from among the approximately 606,000 total number of attorneys in the United States, revealed that 25% of the women finished in the top 10% of their classes versus 18% of the males. The survey also found that women tended to enter law school at a later age than men. (Nearly 33% of women began practicing after age 30, compared with 14% for

men.) For older women and men beginning practice, women's starting salaries tended to be higher than those of men of the same age. Finally, the median salary for different age groups increased with age, peaking for men at ages 51–55 and leveling off thereafter (*American Bar Association Journal*, Oct. 1983). Use this information to construct a model that estimates attorneys' salaries as a function of age, sex, years of experience, and rank in class. Justify each term in the model.

13.93 The *American Sociological Review* (Summer 1993) published a study of the variables that affect academic achievement in high school. Two variables hypothesized to have an impact on achievement were track (college program of courses or not) and sector (Catholic school or public school). One of the models considered in the study was

$$E(y) = \beta_0 + \beta_1 x_1 + \beta_2 x_2 + \beta_3 x_1 x_2$$

where

y = Verbal achievement score

$$x_1 = \begin{cases} 1 & \text{if college track} \\ 0 & \text{if not} \end{cases}$$

$$x_2 = \begin{cases} 1 & \text{if Catholic school} \\ 0 & \text{if not} \end{cases}$$

How would you determine whether the impact of college track (x_1) on verbal achievement score (y) depends on sector (x_2)? Set up the null and alternative hypotheses for the test.

13.94 Refer to Exercise 13.93. The impact of interaction between two qualitative independent variables can be seen by performing the following:

a. In terms of the β's, give the mean verbal achievement score, $E(y)$, for public school students not on a college program.

b. Repeat part **a** for public school students on a college program.

c. Subtract the results, parts **a** and **b**, to obtain the difference between the means of college-track and regular-track students in public schools.

d. In terms of the β's, give the mean verbal achievement score, $E(y)$, for Catholic school students not on a college program.

e. Repeat part **a** for Catholic school students on a college program.

f. Subtract the results, parts **d** and **e**, to obtain the difference between the means of college-track and regular-track students in Catholic schools.

13.95 An experiment was conducted to compare three weight-reducing programs. Ten people were assigned to each of the three diets, and their weights were measured at the beginning and end of a

DIET A		DIET B		DIET C	
x_1 (weight before)	y (weight loss)	x_1 (weight before)	y (weight loss)	x_1 (weight before)	y (weight loss)
227	14	255	19	206	7
286	16	193	8	222	9
180	−2	186	4	168	2
176	8	145	15	132	0
204	15	219	16	173	−3
155	5	273	19	210	8
303	17	289	25	269	10
146	7	168	6	275	15
215	15	194	12	241	8
187	6	248	21	219	5

1-month period; the results are recorded (in pounds) in the table above.

a. Construct a regression model to relate the weight loss y at the end of the 1-month period to

x_1 = Weight before program

$$x_2 = \begin{cases} 1 & \text{if diet B} \\ 0 & \text{otherwise} \end{cases} \qquad x_3 = \begin{cases} 1 & \text{if diet C} \\ 0 & \text{otherwise} \end{cases}$$

Assume that the effect of initial weight on weight loss is identical for each of the three diets. Sketch the type of response curve you would expect to observe, showing y as a function of x_1 for each of the three diets.

b. Suppose the effect of initial weight on weight loss varies from diet to diet. Write the appropriate regression model for this case. Sketch typical response curves depicting this situation.

c. Fit the models in parts **a** and **b** to the data given in the table.

d. Do the data provide sufficient evidence to indicate an interaction between initial weight and type of diet? That is, is the model of part **b** preferable to the model of part **a**?

e. If you wish to test the null hypothesis of no difference among diets, which model parameters would be included in the hypothesis? [*Note:* Assume you are using the model from part **b**.]

f. Conduct the test in part **e** using $\alpha = .05$.

13.96 To make a product more appealing to the consumer, an automobile manufacturer is experimenting with a new type of paint that is supposed to preserve a new-car look. The durability of this paint depends on the length of time the car body is in the oven after it has been painted. In the initial experiment, three groups of 10 car bodies each are baked for three different lengths of time—12, 24, and 36 hours—at the standard temperature setting. Then the paint finish of each of the 30 cars is analyzed to determine a durability rating, y.

a. Write a second-order model relating the mean durability $E(y)$ to the length of baking.

b. Could a third-order model be fitted to the data? Explain.

13.97 Many companies must accurately estimate their costs before a job is begun in order to acquire a contract and make a profit. For example, a heating and plumbing contractor may estimate costs for new homes based on the total area of the house and whether central air conditioning is to be installed.

a. Write a main effects model relating the mean cost of material and labor, $E(y)$, to the area and central–air-conditioning variables.

b. Write a complete second-order model for the mean cost as a function of the same two variables.

c. The contractor samples 25 recent jobs, fitting both the complete second-order model (part **b**) and the reduced main effects model (part **a**), so that a test can be conducted to determine whether the additional complexity of the second-order model is necessary. The resulting SSE and R^2 values are shown in the table below. Is there sufficient evidence to conclude that the second-order terms are important for predicting the mean cost? Use $\alpha = .05$.

	SSE	R^2
Main effects	8.548	.950
Second-order	6.133	.964

13.98 A firm that has developed a new type of light bulb is interested in evaluating its performance in order to decide whether to market it. It is known that the light output of the bulb depends on the cleanliness of its surface area and the length of time the bulb has been in operation. Use the data in the table on page 681 and the procedures you learned in this chapter to build a regression model that relates drop in light output to bulb surface cleanliness and length of operation.

Drop in Light Output (% original output)	Bulb Surface (C = clean, D = dirty)	Length of Operation (hours)
0	C	0
16	C	400
22	C	800
27	C	1,200
32	C	1,600
36	C	2,000
38	C	2,400
0	D	0
4	D	400
6	D	800
8	D	1,200
9	D	1,600
11	D	2,000
12	D	2,400

City	Traffic Flow (thousands of cars)	Weekly Sales y ($ thousands)
1	59.3	6.3
1	60.3	6.6
1	82.1	7.6
1	32.3	3.0
1	98.0	9.5
1	54.1	5.9
1	54.4	6.1
1	51.3	5.0
1	36.7	3.6
2	23.6	2.8
2	57.6	6.7
2	44.6	5.2
3	75.8	8.2
3	48.3	5.0
3	41.4	3.9
3	52.5	5.4
3	41.0	4.1
3	29.6	3.1
3	49.5	5.4
4	73.1	8.4
4	81.3	9.5
4	72.4	8.7
4	88.4	10.6
4	23.2	3.3

13.99 A fast-food restaurant chain is interested in modeling the mean weekly sales of a restaurant, $E(y)$, as a function of the weekly traffic flow on the street where the restaurant is located and the city in which the restaurant is located. The table at right contains data collected on 24 restaurants in four cities. The model that has been proposed is

$$E(y) = \beta_0 + \beta_1 x_1 + \beta_2 x_2 + \beta_3 x_3 + \beta_4 x_4$$

where

x_1 = Traffic flow

$$x_2 = \begin{cases} 1 & \text{if city 1} \\ 0 & \text{otherwise} \end{cases}$$

$$x_3 = \begin{cases} 1 & \text{if city 2} \\ 0 & \text{otherwise} \end{cases}$$

$$x_4 = \begin{cases} 1 & \text{if city 3} \\ 0 & \text{otherwise} \end{cases}$$

A portion of the computer printout that results from fitting the model to the data in the table is shown below. The reduced model, $E(y) = \beta_0 + \beta_1 x_1$, was then fit to the same data and the resulting printout is partially reproduced on page 682.

a. Test the null hypothesis that $\beta_1 = \beta_2 = \beta_3 = \beta_4 = 0$ using $\alpha = .05$. Interpret the results of your test.

b. Is mean weekly sales, $E(y)$, dependent on the city where a restaurant is located? Test using $\alpha = .05$. Interpret the results of your test.

c. Describe the nature of the response lines that the complete model

$$E(y) = \beta_0 + \beta_1 x_1 + \beta_2 x_2 + \beta_3 x_3 + \beta_4 x_4$$

would generate. Does the model imply interaction between city and traffic flow?

d. Use the prediction equation based on the complete model to graph the response lines that relate predicted weekly sales, $\hat{y}$ to traffic flow, x_1 (for each of the four cities). Do the graphed response lines suggest an interaction between city and traffic flow?

e. Write a model that includes interaction between city and traffic flow.

f. Fit the model of part **e** to the data.

g. Do the data provide sufficient evidence to indicate that the slopes of the lines differ for at least two of the four cities? Test using $\alpha = .05$.

SOURCE	DF	SUM OF SQUARES	MEAN SQUARE
MODEL	1	111.3423	111.3423
ERROR	22	7.8073	0.3549
CORRECTED TOTAL	23	119.1496	R-SQUARE
			0.934

```
Dep Variable: Y

                        Analysis of Variance

                      Sum of          Mean
        Source    DF   Squares        Square      F Value    Prob>F

        Model      4   116.65552      29.16388    222.173    0.0001
        Error     19     2.49407       0.13127
        C Total   23   119.14958

              Root MSE      0.36231      R-Square     0.9791
              Dep Mean      5.99583      Adj R-Sq     0.9747
              C.V.          6.04265

                        Parameter Estimates

                      Parameter      Standard     T for H0:
        Variable  DF   Estimate        Error     Parameter=0   Prob >|T|

        INTERCEP   1    1.083388     0.32100795      3.375      0.0032
        X1         1    0.103673     0.00409449     25.320      0.0001
        X2         1   -1.215762     0.20538681     -5.919      0.0001
        X3         1   -0.530757     0.28481946     -1.863      0.0779
        X4         1   -1.076525     0.22650014     -4.753      0.0001
```

STUDENT PROJECTS

Here, we continue the Student Projects theme from Chapters 11 and 12. Remember that you selected three independent variables related to a dependent variable of your choice. Now increase your list of three variables to include approximately ten that you think would be useful in predicting the dependent variable. Obtain data for as many years as possible for the entire new list of variables. With the aid of statistical software, employ a stepwise regression program to choose the important variables among those you have listed. To test your intuition, list the variables in the order you think they will be selected before you conduct the analysis. How does your list compare with the stepwise regression results?

After the group of ten variables has been narrowed to a smaller group of variables by the stepwise analysis, try to improve the model by including interactions and quadratic terms. Be sure to consider the meaning of each interaction or quadratic term before adding it to the model—a quick sketch can be very helpful. See if you can systematically construct a useful model for prediction. You might want to hold out the last several years of data to test the predictive ability of your model after it is constructed. (As noted in Section 13.8, using the same data to construct *and* to evaluate predictive ability can lead to invalid statistical tests and a false sense of security.)

EXPLORING DATA WITH A COMPUTER

Consider modeling the carbon monoxide content y of a cigarette brand using the variables listed in Appendix B (i.e., tar, nicotine, weight, filter, menthol, pack, length). Randomly select 100 brands from the data in Appendix B.

a. Select the best predictors of y using stepwise regression.

b. Fit each of the following models to your sample, using only the variables you selected in part a.
   (1) Complete second-order model
   (2) Main effects plus interaction (complete first-order) model
   (3) First-order main effects model (no interaction)

c. Test the complete second-order model (1) against each of the reduced models, (2) and (3). Be sure to write the hypotheses you are testing in terms of this exercise. Which of the models is preferred as a result of your testing?

d. Interpret the preferred model. Discuss the value of R^2 and the standard deviation of the model. Plot the predicted value of carbon monoxide content against one of the quantitative predictor variables, using a different plotting symbol for each level of one of the qualitative predictors in the model. Interpret your plot.

e. Use a prediction interval to predict the carbon monoxide content for a particular cigarette brand of your choice.

f. What assumptions are necessary to ensure the validity of the inferences in parts b–e? Perform a residual analysis to determine whether outliers exist. If any are found, remove them and refit the preferred model. Compare the models with and without the outliers to determine their sensitivity to the extreme observations.

g. Repeat this exercise using all the measurements in the data base, rather than your sample. Compare the results to those based on your sample.

Chapter 14

THE CHI-SQUARE TEST AND THE ANALYSIS OF CONTINGENCY TABLES

Contents

Case Studies

*W*HERE WE'VE BEEN

In the preceding chapters we've examined several statistical methods for analyzing many types of data. Except for those that pertain to binomial proportions, most of the techniques presented in Chapters 7–10 are appropriate for populations of data on quantitative random variables. Regression models were presented in Chapters 11–13 for the purpose of predicting a quantitative dependent variable from the values of both quantitative and qualitative independent variables.

*W*HERE WE'RE GOING

The methods we discuss in this chapter are appropriate for a type of data commonly found in many disciplines, a type exemplified by the binomial data we looked at in Chapters 7–9. We refer to *count,* or *enumerative,* data. Recall that the binomial experiment (Chapter 4) consists of n trials, each of which results in *two* outcomes. Thus, we have two classifications into which all the data fall. In this chapter, we consider the analysis of data that fall into *two or more* categories.

Many useful experiments consist of *enumerating* the number of occurrences of some event. For example, we may *count* the number of people who recover from leukemia when receiving a newly developed treatment, or the number of consumers who choose each of three brands of coffee, or the number of students who major in each of five different academic disciplines. In some instances, the objective of collecting the **count data** is to analyze the distribution of the counts in the various **classes**, or **cells**. For example, we may want to estimate the proportion of smokers who prefer each of three different brands of cigarettes by counting the number in a sample of smokers who buy each brand. We say that count data classified on a single scale have a **one-dimensional classification**. The analysis of one-dimensional count data is discussed in Section 14.1.

In other instances, the objective of collecting the count data is to study the relationship between two different categorical factors. For example, we may be interested in investigating whether the style of package purchased is related to the sex of the buyer. Or the relationship between socioeconomic status and political party affiliation could be of interest. When count data are classified in a *two-dimensional table,* we call the result a *contingency table.* The analysis of contingency tables is discussed in Section 14.2.

14.1   ONE-DIMENSIONAL COUNT DATA: THE MULTINOMIAL DISTRIBUTION

We first consider experiments that result in classification according to a single criterion, that is, classification on a one-dimensional scale. Suppose we wish to compare the percentage of voters favoring each of three political candidates running for the same elective position. The voting preferences of a random sample of 150 eligible voters are obtained, and the resulting count data are classified according to a single criterion: candidate preference. The data are shown in Table 14.1. Do you think these data indicate a voter preference for any of the candidates?

To answer this question with a valid statistical analysis, we need to know the underlying probability distribution of these count data. This distribution, called the **multinomial probability distribution**, is an extension of the binomial distribution (Section 4.4). The properties of a multinomial experiment are given in the following box.

Properties of the Multinomial Experiment

1. The experiment consists of n identical trials.
2. There are k possible outcomes to each trial.
3. The probabilities of the k outcomes, denoted by $p_1, p_2, ..., p_k$, remain the same from trial to trial, where $p_1 + p_2 + \cdots + p_k = 1$.
4. The trials are independent.
5. The random variables of interest are the counts $n_1, n_2, ..., n_k$ in each of the k cells.

TABLE 14.1   Voter-Preference Survey

CANDIDATE		
1	**2**	**3**
61	53	36

Note that our voter-preference survey satisfies the properties of a multinomial experiment. The experiment consists of randomly sampling $n = 150$ voters from a large population of voters containing an unknown proportion p_1 who favor candidate 1, a proportion p_2 who favor candidate 2, and a proportion p_3 who favor candidate 3. Each voter sampled represents a single trial that can result in one of three outcomes: The voter will favor candidate 1, 2, or 3 with probabilities $p_1, p_2,$

and p_3, respectively. (Assume that all voters will have a preference.) The voting preference of any single voter in the sample does not affect the preference of another; consequently, the trials are independent. And, finally, you can see that the recorded data are the numbers of voters in each of the three voter-preference categories. Thus, the voter-preference survey satisfies the five properties of a multinomial experiment. You can see that the properties of the multinomial experiment closely resemble those of the binomial experiment and that, in fact, a binomial experiment is a multinomial experiment for the special case where $k = 2$.

In the voter-preference survey, and in most practical applications of the multinomial experiment, the k outcome probabilities $p_1, p_2, ..., p_k$ are unknown and we want to use the survey data to make inferences about their values. The unknown probabilities in the voter-preference survey are

$$p_1 = \text{Proportion of all voters who favor candidate 1}$$
$$p_2 = \text{Proportion of all voters who favor candidate 2}$$
$$p_3 = \text{Proportion of all voters who favor candidate 3}$$

To decide whether the voters have a preference for any of the candidates, we will want to test the null hypothesis that the candidates are equally preferred (that is, $p_1 = p_2 = p_3 = \frac{1}{3}$) against the alternative hypothesis that one candidate is preferred (that is, at least one of the probabilities $p_1, p_2,$ and p_3 exceeds $\frac{1}{3}$). Thus, we want to test

H_0: $p_1 = p_2 = p_3 - \frac{1}{3}$ (no preference)

H_a: At least one of the proportions exceeds $\frac{1}{3}$ (a preference exists)

If the null hypothesis is true and $p_1 = p_2 = p_3 - \frac{1}{3}$, the expected value (mean value) of the number of voters who prefer candidate 1 is given by

$$E(n_1) = np_1 = (n)\frac{1}{3} = (150)\frac{1}{3} = 50$$

Similarly, $E(n_2) = E(n_3) = 50$ if the null hypothesis is true and no preference exists.

The following test statistic—the **chi-square test**—measures the degree of disagreement between the data and the null hypothesis:

$$\chi^2 = \frac{[n_1 - E(n_1)]^2}{E(n_1)} + \frac{[n_2 - E(n_2)]^2}{E(n_2)} + \frac{[n_3 - E(n_3)]^2}{E(n_3)}$$
$$= \frac{(n_1 - 50)^2}{50} + \frac{(n_2 - 50)^2}{50} + \frac{(n_3 - 50)^2}{50}$$

Note that the farther the observed numbers $n_1, n_2,$ and n_3 are from their expected value (50), the larger χ^2 will become. That is, large values of χ^2 imply that the null hypothesis is false.

We have to know the distribution of χ^2 in repeated sampling before we can decide whether the data indicate that a preference exists. When H_0 is true, χ^2 can be shown to have (approximately) the familiar chi-square distribution of Sections 8.7 and 11.4. For this one-way classification, the χ^2 distribution has $(k - 1)$ degrees of freedom.* The rejection region for the voter-preference survey for $\alpha = .05$ and $k - 1 = 3 - 1 - 2$ df is

---

*The derivation of the degrees of freedom for χ^2 involves the number of linear restrictions imposed on the count data. In the present case, the only constraint is that $\Sigma n_i = n$, where n (the sample size) is fixed in advance. Therefore, df $= k - 1$. For other cases, we will give the degrees of freedom for each usage of χ^2 and refer the interested reader to the references for more detail.

FIGURE 14.1

Rejection region for voter-preference survey

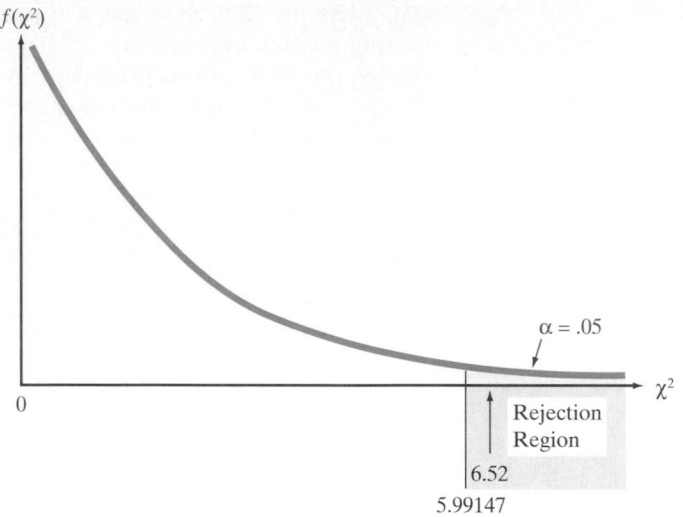

$$\textit{Rejection region:}\quad \chi^2 > \chi^2_{.05}$$

This value of $\chi^2_{.05}$ (found in Table VII) is 5.99147. (See Figure 14.1.) The computed value of the test statistic is

$$\chi^2 = \frac{(n_1 - 50)^2}{50} + \frac{(n_2 - 50)^2}{50} + \frac{(n_3 - 50)^2}{50}$$

$$= \frac{(61 - 50)^2}{50} + \frac{(53 - 50)^2}{50} + \frac{(36 - 50)^2}{50} = 6.52$$

Since the computed $\chi^2 = 6.52$ exceeds the critical value of 5.99147, we conclude at the $\alpha = .05$ level of significance that there does exist a voter preference for one or more of the candidates.

Now that we have evidence to indicate that the proportions p_1, p_2, and p_3 are unequal, we can make inferences concerning their individual values using the methods of Section 7.3. [*Note:* We cannot use the methods of Section 9.3 to compare two proportions because the cell counts are dependent random variables.] The general form for a test of a hypothesis concerning multinomial probabilities is shown in the next box.

A Test of a Hypothesis About Multinomial Probabilities: One-Way Table

H_0: $p_1 = p_{1,0}$, $p_2 = p_{2,0}$,   ..., $p_k = p_{k,0}$

where $p_{1,0}, p_{2,0}, ..., p_{k,0}$ represent the hypothesized values of the multinomial probabilities

H_a: At least one of the multinomial probabilities does not equal its hypothesized value

Test statistic: $\chi^2 = \sum \dfrac{[n_i - E(n_i)]^2}{E(n_i)}$

where $E(n_i) = np_{i,0}$ is the **expected cell count**, that is, the expected number of outcomes of type i assuming that H_0 is true. The total sample size is n.

Rejection region: $\chi^2 > \chi^2_\alpha$,   where χ^2_α has $(k - 1)$ df

> *Assumptions:* 1. A multinomial experiment has been conducted. This is generally satisfied by taking a random sample from the population of interest.
> 2. The sample size n will be large enough so that for every cell, the expected cell count $E(n_i)$ will be equal to 5 or more.*

EXAMPLE 14.1

Suppose an educational television station has broadcast a series of programs on the physiological and psychological effects of smoking marijuana. Now that the series is finished, the station wants to see whether the citizens within the viewing area have changed their minds about how possession of marijuana should be considered legally. Before the series was shown, it was determined that 7% of the citizens favored legalization, 18% favored decriminalization, 65% favored the existing law (an offender could be fined or imprisoned), and 10% had no opinion.

A summary of the opinions (after the series was shown) of a random sample of 500 people in the viewing area is given in Table 14.2. Test at the $\alpha = .01$ level to see whether these data indicate that the distribution of opinions differs significantly from the proportions that existed before the educational series was aired.

Solution

Define the proportions after the airing to be

$$p_1 = \text{Proportion of citizens favoring legalization}$$
$$p_2 = \text{Proportion of citizens favoring decriminalization}$$
$$p_3 = \text{Proportion of citizens favoring existing laws}$$
$$p_4 = \text{Proportion of citizens with no opinion}$$

Then the null hypothesis representing no change in the distribution of percentages is

$$H_0: \quad p_1 = .07, \quad p_2 = .18, \quad p_3 = .65, \quad p_4 = .10$$

and the alternative is

H_a: At least one of the proportions differs from its null hypothesized value

Test statistic: $\chi^2 = \sum \dfrac{[n_i - E(n_i)]^2}{E(n_i)}$

where

$$E(n_1) = np_{1,0} = 500(.07) = 35$$
$$E(n_2) = np_{2,0} = 500(.18) = 90$$
$$E(n_3) = np_{3,0} = 500(.65) = 325$$
$$E(n_4) = np_{4,0} = 500(.10) = 50$$

TABLE 14.2 **Distribution of Opinions About Marijuana Possession**

Legalization	Decriminalization	Existing Laws	No Opinion
39	99	336	26

*The assumption that all expected cell counts are at least 5 is necessary in order to ensure that the χ^2 approximation is appropriate. Exact methods for conducting the test of a hypothesis exist and may be used for small expected cell counts, but these methods are beyond the scope of this text.

Since all these values are larger than 5, the χ^2 approximation is appropriate. Also, if the citizens in the sample were randomly selected, the properties of the multinomial probability distribution are satisfied.

Rejection region: For $\alpha = .01$ and df $= k - 1 = 3$, reject H_0 if $\chi^2 > \chi^2_{.01}$, where (from Table VII in Appendix A) $\chi^2_{.01} = 11.3449$.

We now calculate the test statistic:

$$\chi^2 = \frac{(39 - 35)^2}{35} + \frac{(99 - 90)^2}{90} + \frac{(336 - 325)^2}{325} + \frac{(26 - 50)^2}{50} = 13.249$$

Since this value exceeds the table value of χ^2 (11.3449), the data provide sufficient evidence ($\alpha = .01$) that the opinions on legalization of marijuana have changed since the series was aired.

The χ^2 test can also be conducted using an available statistical software package. Figure 14.2 is an ASP printout of the analysis of the data in Table 14.2. The test statistic and *p*-value of the test are shaded on the printout. Since $\alpha = .01$ exceeds $p = .00412706$, there is sufficient evidence to reject H_0.   ▲

If we focus on one particular outcome of a multinomial experiment, we can use the methods developed in Section 7.4 for a binomial proportion to establish a confidence interval for any one of the multinomial probabilities.* For example, if we want a 95% confidence interval for the proportion of citizens in the viewing area who have no opinion about the issue, we calculate

$$\hat{p}_4 \pm 1.96\sigma_{\hat{p}_4}$$

where

$$\hat{p}_4 = \frac{n_4}{n} = \frac{26}{500} = .052 \qquad \text{and} \qquad \sigma_{\hat{p}_4} \approx \sqrt{\frac{\hat{p}_4(1 - \hat{p}_4)}{n}}$$

Thus, we get

$$.052 \pm 1.96\sqrt{\frac{(.052)(.948)}{500}} = .052 \pm .019$$

or (.033, .071). Thus, we estimate that between 3.3% and 7.1% of the citizens now have no opinion on the issue of marijuana legalization. The series of programs

FIGURE 14.2

ASP analysis of data in Table 14.2

```
ONE-WAY TABLE: GROUP 1=LEGAL, 2=DECRIM, 3=LAWS, 4=NO OPINION

              OBSERVED  EXPECTED
              --------  --------
    GROUP 1       39        35
    GROUP 2       99        90
    GROUP 3      336       325
    GROUP 4       26        50
    ------------------------------
    TOTAL        500       500

NUMBER OF ESTIMATED PARAMETERS = 0

        CHI SQUARE   =   13.2495
             D. F.   =   3
         P-VALUE     =   4.12706E-3
```

*Note that focusing on one outcome has the effect of lumping the other $(k - 1)$ outcomes into a single group. Thus, we obtain, in effect, two outcomes—or a binomial experiment.

may have helped citizens who formerly had no opinion on the issue to form an opinion, since it appears that the proportion of "no opinions" is now less than 10%.

EXERCISES 14.1–14.17

Learning the Mechanics

14.1 Use Table VII of Appendix A to find each of the following χ^2 values:
  a. $\chi^2_{.05}$ for df = 10 **b.** $\chi^2_{.990}$ for df = 50
  c. $\chi^2_{.10}$ for df = 16 **d.** $\chi^2_{.005}$ for df = 50

14.2 Use Table VII of Appendix A to find the following probabilities:
  a. $P(\chi^2 \leq 1.063623)$ for df = 4
  b. $P(\chi^2 > 30.5779)$ for df = 15
  c. $P(\chi^2 \geq 82.3581)$ for df = 100
  d. $P(\chi^2 < 18.4926)$ for df = 30

14.3 Find the rejection region for a one-dimensional χ^2-test of a null hypothesis concerning $p_1, p_2, ..., p_k$ if
  a. $k = 3; \alpha = .05$ **b.** $k = 5; \alpha = .10$
  c. $k = 4; \alpha = .01$

14.4 What are the characteristics of a multinomial experiment? Compare the characteristics to those of a binomial experiment.

14.5 What conditions must n satisfy to make the χ^2-test valid?

14.6 A multinomial experiment with $k = 3$ cells and $n = 320$ produced the data shown in the following table. Do these data provide sufficient evidence to contradict the null hypothesis that $p_1 = .25, p_2 = .25$, and $p_3 = .50$? Test using $\alpha = .05$.

	CELL		
	1	2	3
n_i	78	60	182

14.7 A multinomial experiment with $k = 4$ cells and $n = 205$ produced the data shown in the table.

	CELL			
	1	2	3	4
n_i	43	56	59	47

  a. Do these data provide sufficient evidence to conclude that the multinomial probabilities differ? Test using $\alpha = .05$.
  b. What are the Type I and Type II errors associated with the test of part **a**?

14.8 Refer to Exercise 14.7. Construct a 95% confidence interval for the multinomial probability associated with cell 3.

14.9 A multinomial experiment with $k = 4$ cells and $n = 400$ produced the data shown in the following table. Do these data provide sufficient evidence to contradict the null hypothesis that $p_1 = .2, p_2 = .4$, $p_3 = .1$, and $p_4 = .3$? Test using $\alpha = .05$.

	CELL			
	1	2	3	4
n_i	70	196	46	88

Applying the Concepts

14.10 Each year, approximately 1.3 million people in the United States suffer adverse drug effects (ADEs), that is, unintended injuries caused by prescribed medication. A study in the *Journal of the American Medical Association* (July 5, 1995) identified the cause of 247 ADEs that occurred at two Boston hospitals. The researchers found that dosing errors (that is, wrong dosage prescribed and/or dispensed) were the most common. The table summarizes the proximate cause of 95 ADEs that resulted from a dosing error. Conduct a test (at $\alpha = .10$) to determine whether the true percentages of ADEs in the five "cause" categories are different. Use the ASP printout on page 692 to arrive at your decision.

Wrong Dosage Cause	Number of ADEs
(1) Lack of knowledge of drug	29
(2) Rule violation	17
(3) Faulty dose checking	13
(4) Slips	9
(5) Other	27

14.11 In education, the term *instructional technology* refers to products such as computers, spreadsheets, CD-ROMs, videos, and presentation software. How frequently do professors use instructional technology in the classroom? To answer this question, researchers at Western Michigan University surveyed 306 of their fellow faculty (*Educational Technology,* Mar.–Apr. 1995). Responses to the frequency-of-technology use in teaching were recorded as "weekly to every class," "once a semester to monthly," or "never." The faculty responses

```
              ONE-WAY TABLE: GROUP 1=LACK 2=RULE 3=DOSE 4=SLIP 5=OTHER

                      OBSERVED  EXPECTED
                      --------  --------
            GROUP 1       29        19
            GROUP 2       17        19
            GROUP 3       13        19
            GROUP 4        9        19
            GROUP 5       27        19
            --------------------------
            TOTAL         95        95

   NUMBER OF ESTIMATED PARAMETERS = 0

              CHI SQUARE  =   16
                   D. F.  =    4
                 P-VALUE  =    3.01916E-3
```

(number in each response category) for the three technologies are summarized in the table below.

Technology	Weekly	Once a Semester/ Monthly	Never
Computer spreadsheets	58	67	181
Word processing	168	61	77
Statistical software	37	82	187

a. Determine whether the percentages in the three frequency-of-use response categories differ for computer spreadsheets. Use $\alpha = .01$.

b. Repeat part **a** for word processing.

c. Repeat part **a** for statistical software.

d. Construct a 99% confidence interval for the true percentage of faculty who never use computer spreadsheets in the classroom. Interpret the interval.

14.12 A Gallup survey portrays U.S. entrepreneurs as "… the mavericks, dreamers, and loners whose rough edges and uncompromising need to do it their own way set them in sharp contrast to senior executives in major American corporations" (*Wall Street Journal,* May 1985). One of the many questions put to a sample of $n = 100$ entrepreneurs about their job characteristics, work habits, social activities, etc., concerned the origin of the car they personally drive most frequently. The responses given in the following table were obtained.

United States	Europe	Japan
45	46	9

a. Do these data provide evidence of a difference in the preference of entrepreneurs for the cars made in the United States, Europe, and Japan? Test using $\alpha = .05$.

b. Do these data provide evidence of a difference in the preference of entrepreneurs for domestic versus foreign cars? Test using $\alpha = .05$.

c. What assumptions must you make in order to ensure that your inferences of parts **a** and **b** are valid?

14.13 The *Journal of Intellectual Disability Research* (Feb. 1995) published a longitudinal study of hearing impairment in a group of elderly patients with intellectual disability. The hearing function of each patient was screened each year over a 10-year period. At the study's conclusion, the hearing loss of each patient was categorized as severe, moderate, mild, or none. The classifications of the 28 surviving patients are summarized in the following table. Conduct a test to determine whether the true proportions of intellectually disabled elderly patients in each of the hearing-loss categories differ. Use $\alpha = .05$.

Hearing Loss	Number of Patients
None	7
Mild	7
Moderate	9
Severe	5
Total	28

14.14 Refer to Exercise **14.12**. Use a 90% confidence interval to estimate the proportion of U.S. entrepreneurs who drive foreign cars.

14.15 *Human Factors* (Dec. 1988) published a study of color brightness as a body orientation clue. Ninety college students, reclining on their backs in the dark, were disoriented when positioned on a rotating platform under a slowly rotating disk that blocked their field of vision. The subjects were asked to say "Stop" when they felt as if they were right-side up. The position of the brightness pattern on the disk in relation to each student's body orientation was then recorded. Subjects selected only three disk brightness patterns as subjective vertical clues: (1) brighter side up, (2) darker side up, and (3) brighter and darker sides aligned on either side of the subjects' heads. The frequency counts for the experiment are given in the accompanying table. Conduct a test to compare the proportions of subjects that fall in the three disk orientation categories. Assume you want to determine whether the three proportions differ. Use $\alpha = .05$.

DISK ORIENTATION

Brighter Side Up	Darker Side Up	Bright and Dark Sides Aligned
58	15	17

14.16 In chess, the first few moves often play a determining role in the final outcome. Five different opening strategies are highly favored by chess experts. To determine whether one or more of these strategies is most preferred by grand masters in international competition, a random sample of 100 grand masters is taken, and each is asked which of the strategies he or she would prefer to employ. A summary of their responses is shown below. Do these data present sufficient evidence to indicate a preference for one or more of the strategies? Use $\alpha = .05$.

Strategy	A	B	C	D	E
Frequency	17	27	22	15	19

14.17 *Nature* (Sept. 1993) reported on a study of animal and plant species "hotspots" in Great Britain. A hotspot is defined as a 10-km^2 area that is species-rich, that is, is heavily populated by the species of interest. Analogously, a coldspot is a 10-km^2 area that is species-poor. The table below gives the number of butterfly hotspots and the number of butterfly coldspots in a sample of 2,588 10-km^2 areas. In theory, 5% of the areas should be butterfly hotspots and 5% should be butterfly coldspots, while the remaining areas (90%) are neutral. Test the theory using $\alpha = .01$.

Butterfly Hotspots	123
Butterfly Coldspots	147
Neutral Areas	2,318
Total	2,588

Source: Prendergast, J.R., *et al.* "Rare species, the coincidence of diversity hotspots and conservation strategies." *Nature,* Vol. 365, No. 6444, Sept. 23, 1993, p. 335 (Table 1).

14.2 CONTINGENCY TABLES

In Section 14.1, we introduced the multinomial probability distribution and considered data classified according to a single criterion. We now consider multinomial experiments in which the data are classified according to two criteria, that is, *classification with respect to two factors.*

For example, high gasoline prices have made many consumers more aware of the size of the automobiles they purchase. Suppose an automobile manufacturer is interested in determining the relationship between the size and manufacturer of newly purchased automobiles. One thousand recent buyers of U.S.-made cars are randomly sampled, and each purchase is classified with respect to the size and manufacturer of the automobile. The data are summarized in the **two-way table** shown in Table 14.3. This table is called a **contingency table**; it presents

TABLE 14.3 Contingency Table for Automobile Size Example

	MANUFACTURER				
	A	**B**	**C**	**D**	**Totals**
Small	157	65	181	10	413
Intermediate	126	82	142	46	396
Large	58	45	60	28	191
Totals	341	192	383	84	1,000

TABLE 14.4A   Observed Counts for Contingency Table 14.3

	MANUFACTURER				
	A	**B**	**C**	**D**	**Totals**
Small	n_{11}	n_{12}	n_{13}	n_{14}	r_1
Intermediate	n_{21}	n_{22}	n_{23}	n_{24}	r_2
Large	n_{31}	n_{32}	n_{33}	n_{34}	r_3
Totals	c_1	c_2	c_3	c_4	n

TABLE 14.4B   Probabilities for Contingency Table 14.3

	MANUFACTURER				
	A	**B**	**C**	**D**	**Totals**
Small	p_{11}	p_{12}	p_{13}	p_{14}	p_{r1}
Intermediate	p_{21}	p_{22}	p_{23}	p_{24}	p_{r2}
Large	p_{31}	p_{32}	p_{33}	p_{34}	p_{r3}
Totals	p_{c1}	p_{c2}	p_{c3}	p_{c4}	1

multinomial count data classified on two scales, or **dimensions**, of classification—namely, automobile size and manufacturer.

The symbols representing the cell counts for the multinomial experiment in Table 14.3 are shown in Table 14.4A; and the corresponding cell, row, and column probabilities are shown in Table 14.4B. Thus, n_{11} represents the number of buyers who purchase a small car of manufacturer A and p_{11} represents the corresponding cell probability. Note the symbols for the row and column totals and also the symbols for the probability totals. The latter are called **marginal probabilities** for each row and column. The marginal probability p_{r1} is the probability that a small car is purchased; the marginal probability p_{c1} is the probability that a car by manufacturer A is purchased. Thus,

$$p_{r1} = p_{11} + p_{12} + p_{13} + p_{14} \qquad \text{and} \qquad p_{c1} = p_{11} + p_{21} + p_{31}.$$

Thus, we can see that this really is a multinomial experiment with a total of 1,000 trials, $(3)(4) = 12$ cells or possible outcomes, and probabilities for each cell as shown in Table 14.4B. If the 1,000 recent buyers are randomly chosen, the trials are considered independent and the probabilities are viewed as remaining constant from trial to trial.

Suppose we want to know whether the two classifications, manufacturer and size, are dependent. That is, if we know which size car a buyer will choose, does that information give us a clue about the manufacturer of the car the buyer will choose? In a probabilistic sense we know (Chapter 3) that independence of events A and B implies $P(AB) = P(A)P(B)$. Similarly, in the contingency table analysis, if the **two classifications are independent**, the probability that an item is classified in any particular cell of the table is the product of the corresponding marginal probabilities. Thus, under the hypothesis of independence, in Table 14.4B, we must have

$$p_{11} = p_{r1}p_{c1} \qquad p_{12} = p_{r1}p_{c2}$$

and so forth.

To test the hypothesis of independence, we use the same reasoning employed in the one-dimensional tests of Section 14.1. First, we calculate the *expected,* or *mean, count in each cell* assuming that the null hypothesis of independence is true. We do this by noting that the expected count in a cell of the table is just the total number of multinomial trials, n, times the cell probability. Recall that n_{ij} represents the **observed count** in the cell located in the ith row and jth column. Then the expected cell count for the upper-left-hand cell (first row, first column) is

$$E(n_{11}) = np_{11}$$

or, when the null hypothesis (the classifications are independent) is true,

$$E(n_{11}) = np_{r1}p_{c1}$$

Since these true probabilities are not known, we estimate p_{r1} and p_{c1} by the same proportions $\hat{p}_{r1} = r_1/n$ and $\hat{p}_{c1} = c_1/n$. Thus, the estimate of the expected value $E(n_{11})$ is

$$\hat{E}(n_{11}) = n\left(\frac{r_1}{n}\right)\left(\frac{c_1}{n}\right) = \frac{r_1 c_1}{n}$$

Similarly, for each i, j,

$$\hat{E}(n_{ij}) = \frac{(\text{Row total})(\text{Column total})}{\text{Total sample size}}$$

Thus,

$$\hat{E}(n_{12}) = \frac{r_1 c_2}{n}$$

$$\vdots \qquad \vdots$$

$$\hat{E}(n_{34}) = \frac{r_3 c_4}{n}$$

Using the data in Table 14.3, we find

$$\hat{E}(n_{11}) = \frac{r_1 c_1}{n} = \frac{(413)(341)}{1{,}000} = 140.833$$

$$\hat{E}(n_{12}) = \frac{r_1 c_2}{n} = \frac{(413)(192)}{1{,}000} = 79.296$$

$$\vdots \qquad \vdots \qquad \vdots \qquad \vdots$$

$$\hat{E}(n_{34}) = \frac{r_3 c_4}{n} = \frac{(191)(84)}{1{,}000} = 16.044$$

The observed data and the estimated expected values (in parentheses) are shown in Table 14.5.

TABLE 14.5 Observed and Estimated Expected (in Parentheses) Counts

| | MANUFACTURER | | | | |
	A	B	C	D	Totals
Small	157	65	181	10	
	(140.833)	(79.296)	(158.179)	(34.692)	413
Intermediate	126	82	142	46	
	(135.036)	(76.032)	(151.668)	(33.264)	396
Large	58	45	60	28	
	(65.131)	(36.672)	(73.153)	(16.044)	191
Totals	341	192	383	84	1,000

We now use the χ^2 statistic to compare the observed and expected (estimated) counts in each cell of the contingency table:

$$\chi^2 = \frac{[n_{11} - \hat{E}(n_{11})]^2}{\hat{E}(n_{11})} + \frac{[n_{12} - \hat{E}(n_{12})]^2}{\hat{E}(n_{12})} + \cdots + \frac{[n_{34} - \hat{E}(n_{34})]^2}{\hat{E}(n_{34})}$$

$$= \sum \frac{[n_{ij} - \hat{E}(n_{ij})]^2}{\hat{E}(n_{ij})}$$

Note: The use of Σ in the context of a contingency table analysis refers to a sum over all cells in the table.

Substituting the data of Table 14.5 into this expression, we get

$$\chi^2 = \frac{(157 - 140.833)^2}{140.833} + \frac{(65 - 79.296)^2}{79.296} + \cdots + \frac{(28 - 16.044)^2}{16.044} = 45.81$$

Large values of χ^2 imply that the observed counts do not closely agree and hence that the hypothesis of independence is false. To determine how large χ^2 must be before it is too large to be attributed to chance, we make use of the fact that the sampling distribution of χ^2 is approximately a χ^2 probability distribution when the classifications are independent.

When testing the null hypothesis of independence in a two-way contingency table, the appropriate degrees of freedom will be $(r - 1)(c - 1)$, where r is the number of rows and c is the number of columns in the table.

For the size and make of automobiles example, the degrees of freedom for χ^2 is $(r - 1)(c - 1) = (3 - 1)(4 - 1) = 6$. Then, for $\alpha = .05$, we reject the hypothesis of independence when

$$\chi^2 > \chi^2_{.05} = 12.5916$$

Since the computed $\chi^2 = 45.81$ exceeds the value 12.5916, we conclude that the size and manufacturer of a car selected by a purchaser are dependent events.

The pattern of **dependence** can be seen more clearly by expressing the data as percentages. We first select one of the two classifications to be used as the base variable. In the automobile size preference example, suppose we select manufacturer as the classificatory variable to be the base. Next, we represent the responses for each level of the second categorical variable (size of automobile in our example) as a percentage of the subtotal for the base variable. For example, from Table 14.5 we convert the response for small car sales for manufacturer A (157) to a percentage of the total sales for manufacturer A (341). That is,

$$(^{157}/_{341})100\% = 46\%$$

The conversions of all Table 14.5 entries are similarly computed, and the values are shown in Table 14.6. The value shown at the right of each row is the row's total expressed as a percentage of the total number of responses in the entire table. Thus, the small car percentage is $^{413}/_{1,000}(100\%) = 41\%$ (rounded to the nearest percent).

If the size and manufacturer variables are independent, then the percentages in the cells of the table are expected to be approximately equal to the corresponding row percentages. Thus, we would expect the small car percentages for each of the four manufacturers to be approximately 41% if size and manufacturer are independent. The extent to which each manufacturer's percentage departs from this value determines the dependence of the two classifications, with greater variability of the row percentages meaning a greater degree of dependence. A plot of the percentages helps summarize the observed pattern. In Figure 14.3 we show the

TABLE 14.6   Percentage of Car Sizes by Manufacturer

	MANUFACTURER				
	A	**B**	**C**	**D**	**All**
Small	46	34	47	12	41
Intermediate	37	43	37	55	40
Large	17	23	16	33	19
Totals	100	100	100	100	100

manufacturer (the base variable) on the horizontal axis, and the size percentages on the vertical axis. The "expected" percentages under the assumption of independence are shown as horizontal lines, and each observed value is represented by a symbol indicating the size category.

Figure 14.3 clearly indicates the reason that the test resulted in the conclusion that the two classifications in the contingency table are dependent. Note that the sales of manufacturers A, B, and C fall relatively close to the expected percentages under the assumption of independence. However, the sales of manufacturer D deviate significantly from the expected values, with much higher percentages for large and intermediate cars and a much smaller percentage for small cars than expected under independence. Also, manufacturer B deviates slightly from the expected pattern, with a greater percentage of intermediate than small car sales. Statistical measures of the degree of dependence and procedures for making comparisons of pairs of levels for classifications are available. They are beyond the scope of this text, but can be found in the references. We will, however, utilize descriptive summaries such as Figure 14.3 to examine the degree of dependence exhibited by the sample data.

The general form of a two-way contingency table containing r rows and c columns (called an $r \times c$ contingency table) is shown in Table 14.7. Note that the observed count in the (ij) cell is denoted by n_{ij}, the ith row total is r_i, the jth column total is c_j, and the total sample size is n. Using this notation, we give the general form of the contingency table test for independent classifications in the next box.

FIGURE 14.3

Size as a percentage of manufacturer subtotals

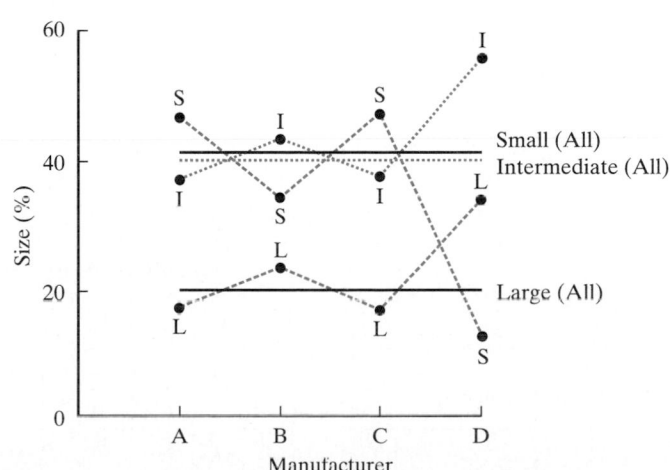

TABLE 14.7   General $r \times c$ Contingency Table

		COLUMN				Row Totals
		1	**2**	$\cdots$	**c**	
Row	*1*	n_{11}	n_{12}	$\cdots$	n_{1c}	r_1
	2	n_{21}	n_{22}	$\cdots$	n_{2c}	r_2
	$\vdots$	$\vdots$	$\vdots$		$\vdots$	$\vdots$
	r	n_{r1}	n_{r2}	$\cdots$	n_{rc}	r_r
Column Totals		c_1	c_2	$\cdots$	c_c	n

General Form of a Contingency Table Analysis: A Test for Independence

H_0: The two classifications are independent

H_a: The two classifications are dependent

Test statistic: $\chi^2 = \sum \dfrac{[n_{ij} - \hat{E}(n_{ij})]^2}{\hat{E}(n_{ij})}$

where $\hat{E}(n_{ij}) = \dfrac{r_i c_j}{n}$

Rejection region: $\chi^2 > \chi_\alpha^2$, where χ_α^2 has $(r-1)(c-1)$ df.

Assumptions: 1. The n observed counts are a random sample from the population of interest. We may then consider this to be a multinomial experiment with $r \times c$ possible outcomes.
2. The sample size, n, will be large enough so that, for every cell, the expected count, $E(n_{ij})$, will be equal to 5 or more.

EXAMPLE 14.2

A social scientist wants to determine whether the marital status (divorced or not divorced) of U.S. men is independent of their religious affiliation (or lack thereof). A sample of 500 U.S. men is surveyed and the results are tabulated as shown in Table 14.8.

a. Test to see whether there is sufficient evidence to indicate that the marital status of men who have been or are currently married is dependent on religious affiliation. Test using $\alpha = .01$.

TABLE 14.8   Survey Results (Observed Counts), Example 14.2

		RELIGIOUS AFFILIATION					
		A	**B**	**C**	**D**	**None**	**Totals**
Marital Status	*Divorced*	39	19	12	28	18	116
	Never divorced	172	61	44	70	37	384
	Totals	211	80	56	98	55	500

FIGURE 14.4

SAS contingency table
printout

```
                      TABLE OF MARITAL BY RELIGION

MARITAL       RELIGION

Frequency|        |        |        |        |        |
Expected |A       |B       |C       |D       | NONE   |   Total
---------+--------+--------+--------+--------+--------+
DIVORCED |     39 |     19 |     12 |     28 |     18 |    116
         | 48.952 |  18.56 | 12.992 | 22.736 |  12.76 |
         |        |        |        |        |        |
NEVER    |    172 |     61 |     44 |     70 |     37 |    384
         | 162.05 |  61.44 | 43.008 | 75.264 |  42.24 |
---------+--------+--------+--------+--------+--------+
Total         211       80       56       98       55       500


            STATISTICS FOR TABLE OF MARITAL BY RELIGION

    Statistic                           DF    Value          Prob
    -----------------------------------------------------------------
    Chi-Square                           4    7.135          0.129
    Likelihood Ratio Chi-Square          4    6.985          0.137
    Mantel-Haenszel Chi-Square           1    6.494          0.011
    Phi Coefficient                           0.119
    Contingency Coefficient                   0.119
    Cramer's V                                0.119

    Sample Size = 500
```

b. Plot the data and describe the patterns revealed. Is the result of the test supported by the plot?

Solution

a. The first step is to calculate estimated expected cell frequencies under the assumption that the classifications are independent. Rather than compute these values by hand, we resort to a computer. The SAS printout of the analysis of Table 14.8 is displayed in Figure 14.4. Each cell in Figure 14.4 contains the observed (top) and expected (bottom) frequency in that cell. Note that $\hat{E}(n_{11})$, the estimated expected count for the Divorced, A cell, is 48.952. Similarly, the estimated expected count for the Divorced, B cell is $\hat{E}(n_{12}) = 18.56$. Since all the estimated expected cell frequencies are greater than 5, the χ^2 approximation for the test statistic is appropriate. Assuming the men chosen were randomly selected from all married or previously married American men, the characteristics of the multinomial probability distribution are satisfied.

The null and alternative hypotheses we want to test are

H_0: The marital status of U.S. men and their religious affiliation are independent

H_a: The marital status of U.S. men and their religious affiliation are dependent

The test statistic, $\chi^2 = 7.135$, is shaded at the bottom of the printout, as is the observed significance level (*p*-value) of the test. Since $\alpha = .01$ is less than $p = .129$, we fail to reject H_0; that is, we cannot conclude that the marital status of U.S. men depends on their religious affiliation. (Note that we could not reject H_0 even with $\alpha = .10$.)

b. The marital status frequencies are expressed as percentages of the number of men in each religious affiliation category in Table 14.9. The expected percentages under the assumption of independence are shown at the right of each

Lifestyles of the Married (and Not Famous)

CASE STUDY • 14.1 •

According to Adlerian psychology theory, the selection of a spouse is not an accident but a purposeful choice, one that reflects a person's lifestyle. This theory was tested in a psychological study reported in the *Journal of Adlerian Theory, Research and Practice* (Mar. 1986). The main purpose of the study was to investigate the role lifestyles play in forming a marriage.

The data were obtained by distributing questionnaires to 202 married couples living in family housing at the University of Georgia. Each spouse in the sample completed the Langenfield Inventory of Personality Profiles (LIPP), a 75-item paper-and-pencil test designed to assess "personality priorities." In Adlerian theory, a personality priority is a *dominant* behavior pattern consisting of a set of convictions that a person uses to gain a sense of belonging. Adlerian psychologists use personality priorities to group people into one of the following five lifestyle types:

1. *Pleasing priority*—a strong desire to make others happy; seeks to win the approval of other people
2. *Achieving priority*—industrious, responsible, orderly; maintains an active approach to life (particularly in the workplace)
3. *Outdoing priority*—a strong desire to be on top, in a position of superiority over others; likes to be considered the best at whatever he/she does
4. *Suppressing (control) priority*—maintains control over his/her emotions; discloses very little of himself/herself

5. *Avoiding priority*—easily hurt; avoids any potentially painful (physical or emotional) situation

First, the personality priority of each respondent was determined on the basis of his/her answers to the LIPP. The researchers then classified the 202 couples according to both the wife's personality priority and the husband's personality priority. The results are summarized in Table 14.10.

Focus

a. The key question of the study was: "Do the personality profiles contribute to marriage pairings?" In other words, is there a relationship between the personality priorities of the husband and wife pairs? Analyze the data for the researchers.
b. Assuming no relationship is found between the personality priorities of the husband and wife pairs, part a, analyze the data for husbands and wives separately. Are there differences in the proportions of husbands who fall in the five personality priorities? Are there differences in the proportions of wives who fall in the five personality profiles?
c. Assuming a relationship exists between the personality priorities of the husband and wife pairs, illustrate the relationship with a graph similar to Figure 14.3.

row. The plot of the percentage data is shown in Figure 14.5, where horizontal lines represent the expected percentages assuming independence. Note that the response percentages deviate only slightly from those expected under the assumption of independence, supporting the result of the test in part a. That is, neither the descriptive plot nor the statistical test provides evidence that the male divorce rate depends on (varies with) religious affiliation. ▲

TABLE 14.9 Marital Status as Percentage of Religious Affiliation

| | | RELIGIOUS AFFILIATION | | | | |
		A	B	C	D	All
Marital Status	*Divorced*	18	24	21	33	23
	Never divorced	82	76	79	67	77
	Totals	100	100	100	100	100

FIGURE 14.5

Plot of marital status–religious affiliation contingency table

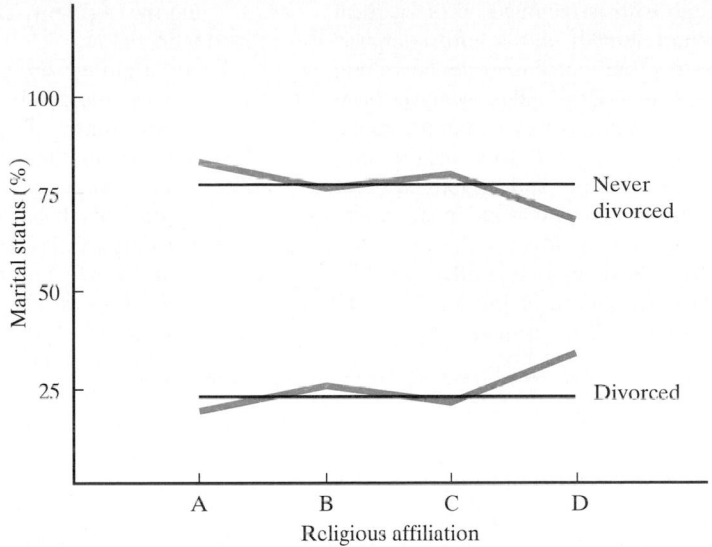

TABLE 14.10   Contingency Table for Study of Married Couples (Case Study 14.1)

		WIFE'S PERSONALITY PRIORITY					
		Pleasing	**Outdoing**	**Avoiding**	**Control**	**Achieving**	**Totals**
	Pleasing	9	6	6	3	9	33
Husband's	*Outdoing*	7	11	12	11	5	46
Personality	*Avoiding*	8	8	6	11	5	38
Priority	*Control*	8	10	7	15	11	51
	Achieving	4	6	7	10	7	34
	Totals	36	41	38	50	37	202

Source: Evans, T. D. and Bozarth, J. "Pairing of personality profiles in marriage." *Journal of Adlerian Theory, Research and Practice,* Vol. 42, No. 1, March 1986, pp. 59–64.

EXERCISES 14.18–14.32

Learning the Mechanics

14.18 Find the rejection region for a test of independence of two classifications where the contingency table contains r rows and c columns and
a. $r = 5$, $c = 5$, $\alpha = .05$
b. $r = 3$, $c = 6$, $\alpha = .10$
c. $r = 2$, $c = 3$, $\alpha = .01$

14.19 Consider the accompanying 2×3 (i.e., $r = 2$ and $c = 3$) contingency table.

		COLUMN		
		1	**2**	**3**
Row	*1*	9	34	53
	2	16	30	25

a. Specify the null and alternative hypotheses that should be used in testing the independence of the row and column classifications.
b. Specify the test statistic and the rejection region that should be used in conducting the hypothesis test of part **a**. Use $\alpha = .01$.
c. Assuming the row classification and the column classification are independent, find estimates for the expected cell counts.
d. Conduct the hypothesis test of part **a**. Interpret your result.

14.20 Refer to Exercise 14.19.
a. Convert the frequency responses to percentages by calculating the percentage of each column total falling in each row. Also convert the row totals to percentages of the total number of responses. Display the percentages in a table.

b. Create a graph with percentage on the vertical axis and column number on the horizontal axis. Showing the row total percentages as horizontal lines on the graph, plot the cell percentages from part **a** using the row number as a plotting symbol.

c. What pattern do you expect to see if the rows and columns are independent? Does the plot support the result of the test of independence in Exercise 14.19?

14.21 Test the null hypothesis of independence of the two classifications, A and B, of the 3 × 3 contingency table shown here. Test using $\alpha = .05$.

		B		
		B_1	B_2	B_3
	A_1	40	72	42
A	A_2	63	53	70
	A_3	31	38	30

14.22 Refer to Exercise 14.21. Convert the responses to percentages by calculating the percentage of each B class total falling into each A classification. Also, calculate the percentage of the total number of responses that constitute each of the A classification totals. Create a graph with percentage on the vertical axis and B classification on the horizontal axis. Show the percentages corresponding to the A classification totals as horizontal lines on the graph, and plot the individual cell percentages using the A class number as a plotting symbol. Does the graph support the result of the test of hypothesis in Exercise 14.21? Explain.

Applying the Concepts

14.23 The *American Journal of Public Health* (July 1995) reported on a population-based study of trauma in Hispanic children. One of the objectives of the study was to compare the use of protective devices in motor vehicles used to transport Hispanic and non-Hispanic white children. On the basis of data collected from the San Diego County Regionalized Trauma System, 792 children treated for injuries sustained in vehicular accidents were classified according to ethnic status (Hispanic or non-Hispanic white) and seatbelt usage (worn or not worn) dur-

ing the accident. The data are summarized in the table below.

a. Calculate the sample proportion of injured Hispanic children who were not wearing seatbelts during the accident.

b. Calculate the sample proportion of injured non-Hispanic white children who were not wearing seatbelts during the accident.

c. Compare the two sample proportions, parts **a** and **b**. Do you think the true population proportions differ?

d. Conduct a test to determine whether seatbelt usage in motor vehicle accidents depends on ethnic status in the San Diego County Regionalized Trauma System. Use $\alpha = .01$.

e. Construct a 99% confidence interval for the difference between the proportions, parts **a** and **b**. Interpret the interval.

14.24 A standardized procedure for determining a person's susceptibility to hypnosis is the Stanford Hypnotic Susceptibility Scale, Form C (SHSS:C). Recently, a new method called the Computer-Assisted Hypnosis Scale (CAHS), which uses a computer as a facilitator of hypnosis, has been developed. Each scale classifies a person's hypnotic susceptibility as low, medium, high, and very high. Researchers at the University of Tennessee compared the two scales by administering both tests to each of 130 undergraduate volunteers (*Psychological Assessment*, Mar. 1995). The hypnotic classifications are summarized on page 703. A chi-square analysis of the data in the contingency table is shown in the SAS printout also on page 703. Note that the expected cell counts are provided in the computer printout.

a. Check to see if the assumption of expected cell counts of 5 or more is satisfied. Should you proceed with the analysis? Explain.

b. One way to satisfy the assumption, part **a**, is to combine the data for two or more categories in the contingency table (e.g., high and very high). Form a new contingency table by combining the data for the high and very high categories in both the rows and columns.

c. Calculate the expected cell counts in the new contingency table, part **c**. Is the assumption now satisfied?

	Hispanic	Non-Hispanic White	Totals
Seatbelts worn	31	148	179
Seatbelts not worn	283	330	613
Totals	314	478	792

Source: Matteneci, R.M. *et al.* "Trauma among Hispanic children: A population-based study in a regionalized system of trauma care." *American Journal of Public Health*, Vol. 85, No. 7, July 1995, p. 1007 (Table 2). Copyright 1995 American Public Health Association.

		CAHS LEVEL				
		Low	**Medium**	**High**	**Very High**	**Totals**
	Low	32	14	2	0	48
SHSS:C Level	*Medium*	11	14	6	0	31
	High	6	14	19	3	42
	Very High	0	2	4	3	9
	Totals	49	44	31	6	130

Source: Grant, C. D., and Nash, M. R. "The Computer-Assisted Hypnosis Scale: Standardization and norming of a computer-administered measure of hypnotic ability." *Psychological Assessment,* Vol. 7, No. 1, Mar. 1995, p. 53 (Table 4). Copyright 1995 American Psychological Association. Reprinted with permission.

```
                    TABLE OF SHSSC BY CAHS

     SHSSC      CAHS

     Frequency|       |        |        |        |
     Expected |High   |Low     |Medium  |VeryHi  |  Total
     ---------+-------+--------+--------+--------+
     High     |    19 |     6  |    14  |     3  |    42
              | 10.015| 15.831 | 14.215 | 1.9385 |
     ---------+-------+--------+--------+--------+
     Low      |     2 |    32  |    14  |     0  |    48
              | 11.446| 18.092 | 16.246 | 2.2154 |
     ---------+-------+--------+--------+--------+
     Medium   |     6 |    11  |    14  |     0  |    31
              | 7.3923| 11.685 | 10.492 | 1.4308 |
     ---------+-------+--------+--------+--------+
     VeryHi   |     4 |     0  |     2  |     3  |     9
              | 2.1462| 3.3923 | 3.0462 | 0.4154 |
     ---------+-------+--------+--------+--------+
     Total         31      49       44        6     130


          STATISTICS FOR TABLE OF SHSSC BY CAHS

     Statistic                  DF     Value      Prob
     ------------------------------------------------------
     Chi-Square                  9     60.103     0.000
     Likelihood Ratio Chi-Square 9     59.638     0.000
     Mantel-Haenszel Chi-Square  1      2.351     0.125
     Phi Coefficient                    0.680
     Contingency Coefficient            0.562
     Cramer's V                         0.393

     Sample Size = 130
     WARNING:  44% of the cells have expected counts less
               than 5. Chi-Square may not be a valid test.
```

d. Perform the chi-square test on the new contingency table. Use $\alpha = .05$. Interpret the results.

14.25 Many companies use well known celebrities in their ads, while other companies create their own spokesperson (such as the Maytag repairman). A study in the *Journal of Marketing* (Fall, 1992) investigated the relationship between the gender of the spokesperson and gender of the viewer in order to see how this relationship affected brand awareness. Three hundred television viewers were asked to identify the products advertised by celebrity spokespersons. See the next two tables.

Male Spokesperson

	AUDIENCE GENDER		
	Male	**Female**	**Total**
Identified Product	95	41	136
Could Not Identify Product	55	109	164
Total	150	150	300

Female Spokesperson

	AUDIENCE GENDER		
	Male	**Female**	**Total**
Identified Product	47	61	108
Could Not Identify Product	103	89	192
Total	150	150	300

a. For the products advertised by male spokespersons, conduct a test to determine whether audience gender and product identification are dependent factors. Test using $\alpha = .05$.

b. Repeat part **a** for the products advertised by female spokespersons.

c. How would you interpret these results?

14.26 A field study was conducted to identify the natural predators of the gypsy moth (*Environmental Entomology,* June 1995). For one part of the study, 24 black-capped chickadees (common wintering birds) were captured in mist nets and individually caged. Each bird was offered a mass of gypsy moth eggs attached to a piece of bark. Half the birds were offered no other food (no choice) and half were offered a variety of other naturally occurring foods such as spruce and pine seeds (choice). The numbers of birds that did and did not feed on the gypsy moth egg mass are given in the table at the top of the next column. Analyze the data in the table to determine if a relationship exists between food choice and whether or not the chickadees fed on gypsy moth eggs. Use $\alpha = .10$.

	FED ON EGG MASS	
	Yes	**No**
Choice of foods	2	10
No choice	8	4

14.27 In order to evaluate their situational awareness, fighter aircraft pilots participate in battle simulations. At a random point in the trial, the simulator is frozen and data on situation awareness are immediately collected. The simulation is then continued until, ultimately, performance (e.g., number of kills) is measured. A study reported in *Human Factors* (Mar. 1995) investigated whether temporarily stopping the simulation results in any change in pilot performance. Trials were designed so that some simulations were stopped to collect situation awareness data while others were not stopped. Each trial was then classified according to the number of kills made by the pilot. The data for 180 trials are summarized in the contingency table below, and an SPSS printout of the analysis follows the table. Fully interpret the results.

	NUMBER OF KILLS					
	0	**1**	**2**	**3**	**4**	**Totals**
Stops	32	33	19	5	2	91
No Stops	24	36	18	8	3	89
Totals	56	69	37	13	5	180

```
STOPS   by  KILLS

                    KILLS
            Count
            Exp Val

  STOPS                                                       Row
                     .00|   1.00|   2.00|   3.00|   4.00| Total
                 ----------------------------------------
         no          24 |    36 |    18 |     8 |     3 |    89
                   27.7 |  34.1 |  18.3 |   6.4 |   2.5 | 49.4%
                 ----------------------------------------
         yes         32 |    33 |    19 |     5 |     2 |    91
                   28.3 |  34.9 |  18.7 |   6.6 |   2.5 | 50.6%
                 ----------------------------------------
        Column       56      69      37      13       5     180
         Total     31.1%   38.3%   20.6%    7.2%    2.8%  100.0%


      Chi-Square              Value          DF        Significance
   --------------------     ----------      ----       ------------

   Pearson                   2.17067          4           .70440
   Likelihood Ratio          2.18200          4           .70233

   Minimum Expected Frequency -    2.472
   Cells with Expected Frequency < 5 -     2 OF    10 (20.0%)
```

14.28 Research has indicated that the stress produced by today's lifestyles results in health problems for a large proportion of society. An article in the *International Journal of Sports Psychology* (July–Sept. 1990) evaluated the relationship between physical fitness and stress. Five hundred forty-nine employees of companies participating in the Health Examination Program offered by Health Advancement Services (HAS) were classified into three groups of fitness levels: good, average, and poor. Each person was tested for signs of stress. The following table reports the results for the three groups. [*Note:* The proportions given are the proportions of the entire group that show signs of stress and fall into each particular fitness level.] Do the data provide evidence to indicate that the likelihood for stress is dependent on an employee's fitness level?

Fitness Level	Sample Size	Proportions with Signs of Stress
Poor	242	.155
Average	212	.133
Good	95	.108

14.29 In the United States, people over age 50 represent 25% of the population, yet they control 70% of the wealth. Research indicates that the highest priority of retirees is travel. A study in the *Annals of Tourism Research* (Vol. 19, 1992) investigates the relationship of retirement status (pre- and post-retirement) to various items related to the travel industry. One part of the study investigated the differences in the length of stay of a trip for pre- and postretirees. A sample of 703 travelers were asked how long they stayed on a typical trip.

Number of Nights	Preretirement	Postretirement
4–7	247	172
8–13	82	67
14–21	35	52
22 or more	16	32
Total	380	323

The results are shown in the accompanying table. Use the information in the table to determine whether the retirement status of a traveler and the duration of a typical trip are dependent. Test using $\alpha = .05$.

14.30 Researchers conducted a study of 58 children admitted to the child psychiatry ward of the University of Iowa for aggressive conduct (*Journal of Child Psychology and Psychiatry,* Vol. 25, 1984). The purpose of the study was to determine the influence of social class, sex, and age on the clinical characteristics of the children. A small portion of their data is shown in the following table, which lists the numbers of children exhibiting antisocial behavior for two categories of social class. Classes I–III represent children from middle-class families. Those in Classes IV and V were from poor families.

	Classes I—III	Classes IV—V
Number Exhibiting Antisocial Behavior	24	17
Sample Size	28	30

Source: Behar, D., and Stewart, M. A. "Aggressive conduct disorder: The influence of social class, sex, and age on the clinical picture." *Journal of Child Psychology and Psychiatry*, 1984, p. 25.

a. Do the data provide sufficient evidence to indicate a dependence between antisocial behavior and social class? Test using $\alpha = .05$.

b. The authors give the approximate *p*-value for the test as .015. Calculate the approximate *p*-value for the test and compare it with the authors' value. Interpret the *p*-value.

14.31 According to research reported in the *Journal of the National Cancer Institute* (Apr. 1991), eating foods high in fiber may help protect against breast cancer. The researchers randomly divided 120 laboratory rats into four groups of 30 each. All rats were injected with a drug that causes breast cancer, then each rat was fed a diet of fat and fiber for 15 weeks. However, the levels of fat and fiber varied from group to group. At the end of the feeding period, the number of rats with cancer tumors was determined for each group. The data are summarized in the contingency table below.

		DIET				
		High-Fat/No-Fiber	High-Fat/Fiber	Low-Fat/No-Fiber	Low-Fat/Fiber	Totals
Cancer Tumors	*Yes*	27	20	19	14	80
	No	3	10	11	16	40
	Totals	30	30	30	30	120

Source: *Tampa Tribune*, Apr. 3, 1991.

a. Does the sampling appear to satisfy the assumptions for a multinomial experiment (see Section 14.1)? Explain.

b. Calculate the expected cell counts for the contingency table.

c. Calculate the χ^2 statistic.

d. Is there evidence to indicate that diet and presence/absence of cancer are independent? Test using $\alpha = .05$.

e. Compare the percentage of rats on a high-fat/no-fiber diet with cancer to the percentage of rats on a high-fat/fiber diet with cancer using a 95% confidence interval. Interpret the result.

14.32 The *American Journal on Mental Retardation* (Jan. 1992) published a study of the social interactions of two groups of children. Independent random samples of 15 children who did and 15 children who did not display developmental delays (i.e., mild mental retardation) were taken in the experiment. After observing the children during "freeplay," the number of children who exhibited disruptive behavior (e.g., ignoring or rejecting other children, taking toys from another child) was recorded for each group. The data are summarized in the following two-way table. Analyze the data and interpret the results.

	Disruptive Behavior	**Nondisruptive Behavior**	**Totals**
With developmental delays	12	3	15
Without developmental delays	5	10	15
Totals	17	13	30

Source: Kopp, C. B., Baker, B., and Brown, K. W. "Social skills and their correlates: Preschoolers with developmental delays." *American Journal on Mental Retardation*, Vol. 96, No. 4, Jan. 1992.

14.3 A WORD OF CAUTION ABOUT CHI-SQUARE TESTS

Because the χ^2 statistic for testing hypotheses about multinomial probabilities is one of the most widely applied statistical tools, it is also one of the most abused statistical procedures. Consequently, the user should always be certain that the experiment satisfies the assumptions given with each procedure. Furthermore, the user should be certain that the sample is drawn from the correct population—that is, from the population about which the inference is to be made.

The use of the χ^2 probability distribution as an approximation to the sampling distribution for χ^2 should be avoided when the expected counts are very small. The approximation can become very poor when these expected counts are small, and thus the true α level may be quite different from the tabled value. As a rule of thumb, an expected cell count of at least 5 means that the χ^2 probability distribution can be used to determine an approximate critical value.

If the χ^2 value does not exceed the established critical value of χ^2, *do not accept the hypothesis of independence.* You would be risking a Type II error (accepting H_0 if it is false), and the probability β of committing such an error is unknown. The usual alternative hypothesis is that the classifications are dependent. Because the number of ways in which two classifications can be dependent is virtually infinite, it is difficult to calculate one or even several values of β to represent such a broad alternative hypothesis. Therefore, we avoid concluding that two classifications are independent, even when χ^2 is small.

Finally, if a contingency table χ^2 value does exceed the critical value, we must be careful to avoid inferring that a *causal* relationship exists between the classifications. Our alternative hypothesis states that the two classifications are statistically dependent—and a statistical dependence does not imply causality. Therefore, *the existence of a causal relationship cannot be established by a contingency table analysis.*

QUICK REVIEW

Key Terms

Chi-square test   687
Contingency table   693
Count data   686
Dependence   696
Expected cell count   688
Independence of two classifications   694

Marginal probabilities   694
Multinomial experiment   686
Multinomial probability distribution   686
Observed cell count   695
One-dimensional (one-way) table   688
Two-dimensional (two-way) table   693

Key Formulas

$$\chi^2 = \sum \frac{[n_i - E(n_i)]^2}{E(n_i)}$$ One-way table   688

where n_i = count for cell i

$$E(n_i) = np_{i,0}$$

$p_{i,0}$ = hypothesized value of p_i in H_0

$$\chi^2 = \sum \frac{[n_{ij} - \hat{E}(n_{ij})]^2}{\hat{E}(n_{ij})}$$ Two-way table   698

where n_{ij} = count for cell in row i, column j

$$\hat{E}(n_{ij}) = r_i c_j / n$$

r_i = total for row i

c_j = total for column j

n = total sample size

LANGUAGE LAB

Symbol	Pronunciation	Description
$p_{i,0}$	p-i-zero	Value of multinomial probability p_i hypothesized in H_0
χ^2	Chi-square	Test statistic used in analysis of count data
n_i	n-i	Number of observed outcomes in cell i of one-way table
$E(n_i)$	Expected value of n_i	Expected number of outcomes in cell i of one-way table when H_0 is true
p_{ij}	p-i-j	Probability of an outcome in row i and column j of a two-way contingency table
n_{ij}	n-i-j	Number of observed outcomes in row i and column j of a two-way contingency table
$\hat{E}(n_{ij})$	Estimated expected value of n_{ij}	Estimated expected number of outcomes in row i and column j of a two-way contingency table
r_i	r-i	Total number of outcomes in row i of a contingency table
c_j	c-j	Total number of outcomes in column j of a contingency table

SUPPLEMENTARY EXERCISES 14.33–14.50

Learning the Mechanics

14.33 A random sample of 250 observations was classified according to the row and column categories shown in the following table.

		COLUMN		
		1	**2**	**3**
	1	20	20	10
Row	*2*	10	20	70
	3	20	50	30

a. Do the data provide sufficient evidence to conclude that the rows and columns are dependent? Test using $\alpha = .05$.

b. Would the analysis change if the row totals were fixed before the data were collected?

c. Do the assumptions required for the analysis to be valid differ according to whether the row (or column) totals are fixed? Explain.

d. Convert the table entries to percentages by using each column total as a base and calculating each row response as a percentage of the corresponding column total. In addition, calculate the row totals and convert them to percentages of all 250 observations.

e. Plot the row percentages on the vertical axis against the column number on the horizontal axis. Draw horizontal lines corresponding to the row total percentages. Does the deviation (or lack thereof) of the individual row percentages from the row total percentages support the result of the test conducted in part **a**?

14.34 A random sample of 150 observations was classified into the categories shown in the table below.

	CATEGORY				
	1	**2**	**3**	**4**	**5**
n_i	28	35	33	25	29

a. Do the data provide sufficient evidence that the categories are not equally likely? Use $\alpha = .10$.

b. Form a 90% confidence interval for p_2, the probability that an observation will fall in category 2.

Applying the Concepts

14.35 An article in the *Annals of the Association of American Geographers* (June, 1992) investigated many factors that are considered when purchasing earthquake insurance. One factor investigated was the proximity to a major earthquake fault. Surveys were mailed to residents in four California counties. The data collected are shown in the table at the bottom of the page.

a. State the hypothesis necessary to test the theory that the proportion of earthquake-insured residents differs for the four sampled counties.

b. Do the data provide evidence of a difference in the proportion of earthquake-insured residents for the four sampled counties? Test using $\alpha = .05$.

14.36 Despite a good winning percentage, a certain major league baseball team has not drawn as many fans as expected. In hopes of finding ways to increase attendance, the management plans to interview fans who come to the games to find out why they come. One thing that the management might want to know is whether there are differences in support for the team among various age groups. Suppose the information in the following table was collected during interviews with fans selected at random. Can you conclude that there is a relationship between age and number of games attended per year? Use $\alpha = .05$.

		NUMBER OF GAMES ATTENDED PER YEAR		
		1 or 2	**3–5**	**Over 5**
	Under 20	78	107	17
	21–30	147	87	13
Age of Fan	*31–40*	129	86	19
	41–55	55	103	40
	Over 55	23	74	22

	COUNTIES			
	Contra Costa	**Santa Clara**	**Los Angeles**	**San Bernardino**
Sample Size	521	556	337	372
Number with Earthquake Insurance	117	222	133	109

14.37 If a company can identify times of day when accidents are most likely to occur, extra precautions can be instituted during those times. A random sampling of the accident reports over the last year at a plant gives the frequency of occurrence of accidents during the different hours of the workday. Can it be concluded from the data in the table that the proportions of accidents are different for at least two of the four time periods?

Hours	1–2	3–4	5–6	7–8
Number of Accidents	31	28	45	47

14.38 The classification of solder joints as acceptable or rejectable is a particularly difficult inspection task owing to its subjective nature. Westinghouse Electric Company has experimented with different means of evaluating the performance of solder inspectors. One approach involves comparing an individual inspector's classifications with those of the group of experts that comprise Westinghouse's Work Standards Committee. In an experiment reported by Joseph J. Meagher and Joseph A. Scazzero (1985) of Westinghouse, 153 solder connections were evaluated by the committee and 111 were classified as acceptable. An inspector evaluated the same 153 connections and classified 124 as acceptable. Of the items rejected by the inspector, the committee agreed with 19.

a. Construct a contingency table that summarizes the classifications of the committee and the inspector.

b. Based on a visual examination of the table you constructed in part **a**, does it appear that there is a relationship between the inspector's classifications and the committee's? Explain. (A plot of the percentage rejected by committee and inspector will aid your examination.)

c. Conduct a chi-square test of independence for these data. Use $\alpha = .05$. Carefully interpret the results of your test in the context of the problem.

14.39 A sociologist was interested in knowing whether sons have a tendency to choose the same occupation as their fathers. To investigate this question, 500 males were polled and questioned concerning their occupations and those of their fathers. A summary of the numbers of father–son pairs falling in each occupational category is shown in the table on page 710. Do the data provide sufficient evidence at $\alpha = .05$ to indicate a dependence between a son's choice of occupation and his father's occupation? Use the SAS printout below to conduct the analysis.

```
                  TABLE OF FATHER BY SON

FATHER        SON

Frequency|         |         |       |         |
Expected |Farmer   |Prof/Bus |Skill  |Unskill  |  Total
---------+---------+---------+-------+---------+
Farmer   |     32  |     15  |   23  |     10  |    80
         |   6.72  |  27.36  | 33.12 |   12.8  |
---------+---------+---------+-------+---------+
Prof/Bus |      0  |     55  |   38  |      7  |   100
         |    8.4  |   34.2  | 41.4  |     16  |
---------+---------+---------+-------+---------+
Skill    |      0  |     79  |   71  |     25  |   175
         |   14.7  |  59.85  | 72.45 |     28  |
---------+---------+---------+-------+---------+
Unskill  |     10  |     22  |   75  |     38  |   145
         |  12.18  |  49.59  | 60.03 |   23.2  |
---------+---------+---------+-------+---------+
Total          42       171     207       80       500


      STATISTICS FOR TABLE OF FATHER BY SON

Statistic                     DF    Value     Prob
----------------------------------------------------
Chi-Square                     9   180.874    0.000
Likelihood Ratio Chi-Square    9   160.832    0.000
Mantel-Haenszel Chi-Square     1    52.040    0.000
Phi Coefficient                     0.601
Contingency Coefficient             0.515
Cramer's V                          0.347

Sample Size = 500
```

		SON			
		Professional or Business	Skilled	Unskilled	Farmer
Father	Professional or Business	55	38	7	0
	Skilled	79	71	25	0
	Unskilled	22	75	38	10
	Farmer	15	23	10	32

14.40 A study was done on the accuracy of newspaper advertisements by the five types of food stores in a southeastern city. On each of 4 days, items were randomly selected from the advertisements for each type of store and the actual price was compared to the advertised price. Each of the stores in the city was classified as one of the following types: national, regional chain A, regional chain B, regional chain C, or independent. Values in the table represent the number of items that were correctly and incorrectly priced.

Type of Store	Number Correctly Priced	Number Incorrectly Priced
National chain	89	10
Regional chain A	53	14
Regional chain B	43	12
Regional chain C	32	13
Independent	41	7

    a. Determine whether these data provide sufficient evidence to conclude that the proportion of correctly priced items differs for at least two types of stores. Use $\alpha = .10$.

    b. Use a 95% confidence interval to estimate the proportion of correctly priced items in the stores in the national chain category.

14.41 Teenage alcoholism is a big problem in the United States. To discover why teenagers are turning to alcohol, a survey was conducted to find out whether a teenager's family status has any relationship to the amount of alcohol he or she consumes. A random sample of 200 teenagers between the ages of 15 and 19 were questioned concerning their use of alcohol. A summary of the responses is shown in the table below.

    a. Do the data provide sufficient evidence to indicate a relationship between family status and the use of alcohol? Test using $\alpha = .05$.

    b. What assumptions are necessary to ensure the validity of the inferential procedure you applied in part **a**? Do you think they are satisfied in this application?

    c. Calculate the percentages of the total number of teenagers in each family status classification falling into each alcohol use classification. Graph these percentages on the vertical axis against the family status on the horizontal axis. Show the overall percentages for each alcohol use classification as horizontal lines on the graph. Does the graph support the result of the test in part **a**? Explain.

14.42 Five candidates have just entered the race for mayor of a large city. To determine whether any of the candidates has an early lead in popularity, a poll is conducted. Some 2,000 voters are asked to indicate the candidate they prefer. A summary of their responses is shown in the table below. Do the data provide sufficient evidence to indicate a preference for at least one of the five candidates? Test using $\alpha = .01$.

	CANDIDATE				
	I	II	III	IV	V
Responses	385	493	628	235	259

14.43 An experimenter wishes to determine whether there is a relationship between hair color and eye color. One hundred people are randomly sampled and the eyes and hair of each person are judged to

		ALCOHOL		
		None	Occasional	Frequent
Family Status	Upper Class	4	16	10
	Upper Middle Class	11	40	24
	Lower Middle Class	9	47	9
	Lower Class	6	17	7

```
Expected counts are printed below observed counts

         LiteHair  DarkHair    Total
    1         31        21        52
           23.40     28.60

    2         14        34        48
           21.60     26.40

  Total       45        55       100

ChiSq =   2.468 +  2.020 +
          2.674 +  2.188 = 9.350

df = 1
```

be light or dark. A summary of the number of people in each of the four categories is shown in the table below. Do the data provide sufficient evidence at $\alpha = .10$ to indicate a relationship between eye and hair color? Use the MINITAB printout above to perform the analysis.

	Light Hair	**Dark Hair**	**Totals**
Light Eyes	31	21	52
Dark Eyes	14	34	48
Totals	45	55	100

14.44 According to the Magnuson–Moss Warranty Act (enacted by Congress in 1977 to reform consumer product warranty practices), all warranties must be designated as "full" or "limited" and must be clearly written in readily understood language. A study was undertaken to investigate consumer satisfaction with warranty practices since the implementation of the Magnuson–Moss Warranty Act (*Baylor Business Studies,* Nov.–Dec. 1983). By means of a mailed questionnaire, 237 midwestern consumers who had purchased a major appliance within the past 6 to 18 months were sampled. One of the questions asked of the consumers was, "Do most retailers and dealers make a conscientious effort to satisfy their customers' warranty claims?" One hundred fifty-six answered yes, 61 were uncertain, and 20 said no. The population of consumers from which this sample was drawn had also been investigated two years prior to the passage of the Magnuson–Moss Warranty Act. At that time,

37.0% of the population answered yes to the same question, 53.3% were uncertain, and 9.7% said no.

a. As reflected in the answers to the above question, have consumer attitudes toward warranties changed since the pre–Magnuson–Moss Act study? Test using $\alpha = .05$.

b. Compare the pre– and post Magnuson–Moss Warranty Act responses and describe the changes that have occurred.

14.45 The typical method of treating moderately advanced cancer of the larynx is surgical removal. Attempts have been made to treat cancer of the larynx by radiation therapy alone, thereby saving the patient's voice box. Data on a group of patients treated by this method at the University of Florida's Shands Hospital were reported in the *International Journal of Radiation Oncology and Biological Physics* (Vol. 10, 1984). Eighteen patients with laryngeal cancer were treated by radiation alone and 23 were treated by surgery alone. Of those treated by radiation, 15 cancers were ultimately controlled at the primary site (the larynx). Ultimate control of the cancer was achieved for 21 of the 23 patients treated by surgery alone. These data are summarized in the table below. The objective of the study was to determine whether the rate of controlling laryngeal cancer is independent of type of treatment (surgery or radiation therapy).

a. Assuming the null hypothesis of independence is true, how many patients are expected to fall in each cell of the table?

b. The researchers were unable to use the chi-square test to analyze the data. Show why.

	Surgery	**Radiation Therapy**
Number of Cancers Ultimately Controlled	21	15
Number Not Ultimately Controlled	2	3
Totals	23	18

Source: Reprinted by permission of the publisher from Mendenhall, W. M., Million, R. R., Sharkey, D. E., and Cassisi, N. J. "Stage T3 squamous cell carcinoma of the larynx treated with surgery and/or radiation therapy." *International Journal of Radiation Oncology Biology Physics,* 1984, p. 10. Copyright 1984 by Elsevier Science, Inc.

c. A test of the null hypothesis can be conducted by using a small-sample method known as *Fisher's exact test.* This method calculates the exact probability (*p*-value) of observing sample results at least as contradictory to the hypothesis of independence as those observed for the researchers' data. The researchers report the *p*-value for this test as .187. Interpret this result.

d. Calculate the χ^2 statistic for the data using the formula provided in this chapter, then find its associated *p*-value. Compare the result to the *p*-value given in part **c.** Do the tests lead to the same conclusion?

14.46 According to one geneticist's theory, a crossing of red and white snapdragons should produce offspring that are 25% red, 50% pink, and 25% white. An experiment conducted to test the theory produced 30 red, 78 pink, and 36 white offspring in 144 crossings. Do the data provide sufficient evidence to contradict the geneticist's theory?

14.47 Refer to Exercise 14.30 and the *Journal of Child Psychology and Psychiatry* study on aggressive behavior in children. Another comparison given in their paper was the numbers of male and female children for whom physical aggressiveness was a presenting complaint. The data are shown in the following table.

	Female	Male
Number for which physical aggressiveness is a presenting complaint	7	40
Sample size	12	46

a. Do the data provide sufficient evidence to indicate a dependence between sex of the child and the proportion for which physical aggressiveness is a presenting complaint? Test using $\alpha = .05$.

b. The researchers give the *p*-value for the test as .025. Find the *p*-value for your test and compare it with the researchers' value. Interpret the *p*-value.

14.48 A computer used by a 24-hour banking service is supposed to assign each transaction to one of five memory locations at random. A check at the end of a day's transactions gives the following counts to each of the five memory locations:

1	2	3	4	5
90	78	100	72	85

Is there evidence to indicate a difference among the proportions of transactions assigned to the five memory locations? Test using $\alpha = .025$.

14.49 A statistical analysis is to be done on a set of data consisting of 1,000 monthly salaries. The analysis requires the assumption that the sample was drawn from a normal distribution. A preliminary test, called the χ^2 *goodness-of-fit test,* can be used to help determine whether it is reasonable to assume that the sample is from a normal distribution. Suppose the mean and standard deviation of the 1,000 salaries are hypothesized to be $1,200 and $200, respectively. Using the standard normal table, we can approximate the probability of a salary being in the intervals listed in the table below. The third column represents the expected number of the 1,000 salaries to be found in each interval if the sample was drawn from a normal distribution with $\mu = \$1,200$ and $\sigma = \$200$. Suppose the last column contains the actual observed frequencies in the sample. Large differences between the observed and expected frequencies cast doubt on the normality assumption.

Interval	Probability	Expected Frequency	Observed Frequency
Less than $800	.023	23	26
$800 < $1,000	.136	136	146
$1,000 < $1,200	.341	341	361
$1,200 < $1,400	.341	341	311
$1,400 < $1,600	.136	136	143
$1,600 or above	.023	23	13

a. Compute the χ^2 statistic based on the observed and expected frequencies—just as you did in Section 14.1.

b. Find the tabulated χ^2 value when $\alpha = .05$ and there are 5 degrees of freedom. (There are $k - 1 = 5$ df associated with this χ^2 statistic.)

c. Based on the χ^2 statistic and the tabulated χ^2 value, is there evidence that the salary distribution is nonnormal?

d. Find the approximate observed significance level for the test in part **c.**

14.50 Suppose a random variable is hypothesized to be normally distributed with mean 0 and standard deviation 1. A random sample of 200 observations on the variable yields frequencies in the intervals listed in the following table. Do the data provide sufficient evidence to contradict the hypothesis that *x* is normally distributed with $\mu = 0$ and $\sigma = 1$? Use the technique developed in Exercise 14.49.

Interval	$x < -2$	$-2 \leq x < -1$	$-1 \leq x < 0$	$0 \leq x < 1$	$1 \leq x < 2$	$x \geq 2$
Frequency	7	20	61	77	26	9

STUDENT PROJECTS

Many researchers rely on surveys to estimate the proportions of experimental units in populations that possess certain specified characteristics. A political scientist may want to estimate the proportion of an electorate in favor of a certain legislative bill. A social scientist may be interested in the proportions of people in a geographical region who fall in certain socioeconomic classifications. A psychologist might want to compare the proportions of patients who have different psychological disorders.

Choose a specific topic, similar to those described above, that interests you. Clearly define the population of interest, identify data categories of specific interest, and identify the proportions associated with them. Now *guesstimate* the proportions of the population that you

think fall in each of the categories. For example, you might guess that all the proportions are equal, or that the first proportion is twice as large as the second but equal to the third, etc.

You are now ready to collect the data by obtaining a random sample from your population of interest. Select a sample size so that all expected cell counts are at least 5 (preferably larger), then collect the data.

Use the count data you have obtained to test the null hypothesis that the true proportions in the population are equal to your presampling guesstimates. Would failure to reject this null hypothesis imply that your guesstimates are correct?

EXPLORING DATA WITH A COMPUTER

Refer to the data on cardiac surgery patients described in Appendix B. Recall that some patients received a new drug, designed to reduce blood loss, prior to surgery and some did not. Is type of complication (if any) experienced by the patient dependent on whether or not they received the drug?

a. Form the appropriate contingency table for the analysis.
b. Find the χ^2 test statistic and p-value. Interpret the results.

c. Calculate the percentage of patients in each complication type using the total number of patients in each drug group as the base. Then calculate the percentage in each complication type for all patients in this study. Construct a table like Table 14.9 to summarize the results. Display the percentages in a graph like Figure 14.5. Does the graphical display support the results of the test in part b?

NONPARAMETRIC STATISTICS

Contents

Case Studies

*W*HERE WE'VE BEEN

In Chapters 7–10, we explored techniques for making inferences about the mean of a single population and for comparing the means of two or more populations. Most of the techniques discussed in those chapters are based on the assumption that the sampled populations have probability distributions that are approximately normal with equal variances. But how can you analyze data that evolve from populations that do not satisfy these assumptions? How can you make comparisons between populations when you cannot assign specific numerical values to your observations?

*W*HERE WE'RE GOING

In this chapter, we present statistical techniques for comparing two or more populations that are based on an ordering of the sample measurements according to their relative magnitudes. These techniques, which require fewer or less stringent assumptions concerning the nature of the probability distributions of the populations, are called *nonparametric statistical methods*.

The *t*- and *F*-tests for making inferences about means presented in Chapters 7–10 are unsuitable for analyzing some types of data. These data fall into two categories. The first includes data from populations that do not satisfy the assumptions upon which the *t*- and *F*-tests are based. For both tests, we assume that the random variables being measured have normal probability distributions with equal variances. In practice, however, the observations from one population may exhibit much greater variability than those from another, or the probability distributions may be decidedly nonnormal. For example, the distribution might be very flat, peaked, or strongly skewed to the right or left. When any of the assumptions required for the *t*- and *F*-tests is seriously violated, the computed *t* and *F* statistics may not follow the standard *t*- and *F*-distributions. In this case, the tabulated values of *t* and *F* (Tables VI, VIII, IX, X, and XI in Appendix A) are not applicable, the correct value of α for the test is unknown, and the *t*- and *F*-tests are of dubious value.

The second type of data for which *t*- and *F*- tests are inappropriate are responses that are not susceptible to numerical measurement but can be *ranked in order of magnitude*.* For example, if we want to compare the teaching ability of two college instructors based on subjective evaluations of students, we cannot assign a number that exactly measures the teaching ability of a single instructor. But we can compare two instructors by ranking them according to a subjective evaluation of their teaching abilities. If instructors A and B are evaluated by each of 10 students, we have the standard problem of comparing the probability distributions for two populations of ratings—one for instructor A and one for B. But the *t*-test of Chapter 9 would be inappropriate because the only data that can be recorded are preference statements of 10 students—that is, each student decides that either A is better than B or vice versa.

Nonparametric tests—the counterparts of the *t*- and *F*-tests—compare the probability distributions of the sampled populations rather than specific parameters of these populations (such as the means or variances). For example, nonparametric tests can be used to compare the probability distribution of the strengths of preferences for a new product to the probability distributions of the strengths of preferences for the currently popular brands. If it can be inferred that the distribution for the new product lies above (to the right of) the others (see Figure 15.1), the implication is that the new product tends to be more preferred than the currently popular products. Such an inference might lead to a decision to market the product nationally.

Most nonparametric methods use the **relative ranks** of the sample observations rather than their actual numerical values. These tests are particularly valuable when we are unable to obtain numerical measurements of some phenomena but are able to rank them in comparison to each other. Statistics based on ranks of measurements are called **rank statistics**. In Section 15.1, we develop a test to

FIGURE 15.1

Probability distributions of strengths of preference measurements: New product is preferred

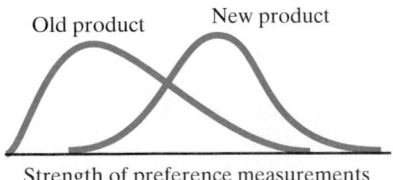

*Qualitative data that can be ranked in order of magnitude are called **ordinal data**.

make inferences about the central tendency of a single population. In Sections 15.2 and 15.4, we present rank statistics for comparing two probability distributions using independent samples. In Sections 15.3 and 15.5, the matched-pairs and randomized block designs are used to make nonparametric comparisons of populations. Finally, in Section 15.6, we present a nonparametric measure of correlation between two variables—*Spearman's rank correlation coefficient*.

15.1   SINGLE POPULATION INFERENCES: THE SIGN TEST

In Chapter 8 we utilized the z and t statistics for testing hypotheses about a population mean. The z statistic is appropriate for large random samples selected from "general" populations—that is, with few limitations on the probability distribution of the underlying population. The t statistic was developed for small-sample tests in which the sample is selected at random from a *normal* distribution. The question is: How can we conduct a test of hypothesis when we have a small sample from a *nonnormal* distribution?

The **sign test** is a relatively simple nonparametric procedure for testing hypotheses about the central tendency of a nonnormal probability distribution. Note that we used the phrase *central tendency* rather than *population mean*. This is because the sign test, like many nonparametric procedures, provides inferences about the population *median* rather than the population mean μ. Denoting the population median by the Greek letter η, we know (Chapter 2) that η is the 50th percentile of the distribution (Figure 15.2) and as such is less affected by the skewness of the distribution and the presence of outliers (extreme observations). Since the nonparametric test must be suitable for all distributions, not just the normal, it is reasonable for nonparametric tests to focus on the more robust (less sensitive to extreme values) measure of central tendency, the median.

For example, increasing numbers of both private and public agencies are requiring their employees to submit to tests for substance abuse. One laboratory that conducts such testing has developed a system with a normalized measurement scale, in which values less than 1.00 indicate "normal" ranges and values equal to or greater than 1.00 are indicative of potential substance abuse. The lab reports a normal result as long as the median level for an individual is less than 1.00. Eight independent measurements of each individual's sample are made. Suppose, then, that one individual's results were as follows:

.78   .51   3.79   .23   .77   .98   .96   .89

If the objective is to determine whether the *population* median (that is, the true median level if an indefinitely large number of measurements were made on the same individual sample) is less than 1.00, we establish that as our alternative hypothesis and test

FIGURE 15.2

Location of the population median, η

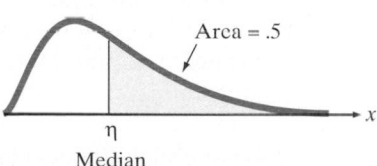

FIGURE 15.3

MINITAB printout of
sign test

SIGN TEST OF MEDIAN = 1.000 VERSUS L.T. 1.000

	N	BELOW	EQUAL	ABOVE	P-VALUE	MEDIAN
READING	8	7	0	1	0.0352	0.8350

$$H_0: \eta = 1.00$$
$$H_a: \eta < 1.00$$

The one-tailed sign test is conducted by counting the number of sample measurements that "favor" the alternative hypothesis—in this case, the number that are less than 1.00. If the null hypothesis is true, we expect approximately half of the measurements to fall on each side of the hypothesized median and if the alternative is true, we expect significantly more than half to favor the alternative—that is, to be less than 1.00. Thus,

Test statistic: S = Number of measurements less than 1.00,
the null hypothesized median

If we wish to conduct the test at the $\alpha = .05$ level of significance, the rejection region can be expressed in terms of the observed significance level, or *p*-value, of the test:

Rejection region: *p*-value $\leq .05$

In this example, $S = 7$ of the 8 measurements are less than 1.00. To determine the observed significance level associated with this outcome, we note that the number of measurements less than 1.00 is a binomial random variable (check the binomial characteristics presented in Chapter 4), and *if H_0 is true,* the binomial probability p that a measurement lies below (or above) the median 1.00 is equal to .5 (Figure 15.2). What is the probability that a result is *as contrary to or more contrary to H_0* than the one observed? That is, what is the probability that 7 *or more* of 8 binomial measurements will result in Success (be less than 1.00) if the probability of Success is .5? Binomial Table II in Appendix A (using $n = 8$ and $p = .5$) indicates that

$$P(x \geq 7) = 1 - P(x \leq 6) = 1 - .965 = .035$$

Thus, the probability that at least 7 of 8 measurements would be less than 1.00 *if the true median were* 1.00 is only .035. The *p*-value of the test is therefore .035.

This *p*-value can also be obtained by using a statistical software package. The MINITAB printout of the analysis is shown in Figure 15.3, with the *p*-value shaded on the printout. Since $p = .035$ is less than $\alpha = .05$, we conclude that this sample provides sufficient evidence to reject the null hypothesis. The implication of this rejection is that the laboratory can conclude at the $\alpha = .05$ level of significance that the true median level for the tested individual is less than 1.00. However, we note that one of the measurements greatly exceeds the others, with a value of 3.79, and deserves special attention. Note that this large measurement is an outlier that would make the use of a *t*-test and its concomitant assumption of normality dubious. The only assumption necessary to ensure the validity of the sign test is that the probability distribution of measurements is continuous.

The use of the sign test for testing hypotheses about population medians is summarized in the next box.

Sign Test for a Population Median η

ONE-TAILED TEST	TWO-TAILED TEST
H_0: $\eta = \eta_0$	H_0: $\eta = \eta_0$
H_a: $\eta > \eta_0$ [or H_a: $\eta < \eta_0$]	H_a: $\eta \neq \eta_0$

Test statistic:

S = Number of sample measurements greater than η_0 [or S = number of measurements less than η_0]

S = Larger of S_1 and S_2, where S_1 is the number of measurements less than η_0 and S_2 is the number of measurements greater than η_0

Observed significance level:

p-value = $P(x \geq S)$

p-value = $2P(x \geq S)$

where x has a binomial distribution with parameters n and $p = .5$. (Use Table II, Appendix A.)

Rejection region: Reject H_0 if p-value $\leq .05$.

Assumption: The sample is selected randomly from a continuous probability distribution. [*Note:* No assumptions need to be made about the shape of the probability distribution.]

Recall that the normal probability distribution provides a good approximation for the binomial distribution when the sample size is large. For tests about the median of a distribution, the null hypothesis implies that $p = .5$, and the normal distribution provides a good approximation if $n \geq 10$. (Samples with $n \geq 10$ satisfy the condition that $np \pm 3\sqrt{npq}$ is contained in the interval 0 to n.) Thus, we can use the standard normal z-distribution to conduct the sign test for large samples. The large-sample sign test is summarized in the next box.

Large-Sample Sign Test for a Population Median η

ONE-TAILED TEST	TWO-TAILED TEST
H_0: $\eta = \eta_0$	H_0: $\eta = \eta_0$
H_a: $\eta > \eta_0$	H_a: $\eta \neq \eta_0$
[or H_a: $\eta < \eta_0$]	

Test statistic: $z = \dfrac{(S - .5) - .5n}{.5\sqrt{n}}$

[*Note:* S is calculated as shown in the previous box. We subtract .5 from S as the "correction for continuity." The null hypothesized mean value is $np = .5n$, and the standard deviation is

$$\sqrt{npq} = \sqrt{n(.5)(.5)} = .5\sqrt{n}$$

See Chapter 5 for details on the normal approximation to the binomial distribution.]

Rejection region: $z > z_\alpha$

Rejection region: $z > z_{\alpha/2}$

where tabulated z values can be found inside the front cover.

EXAMPLE 15.1

A manufacturer of compact disk (CD) players has established that the median time to failure for its players is 5,250 hours of utilization. A sample of 20 CDs from a competitor is obtained, and they are continuously tested until each fails. The 20 failure times range from 5 hours (a "defective" player) to 6,575 hours, and 14 of the 20 exceed 5,250 hours. Is there evidence that the median failure time of the competitor differs from 5,250 hours? Use $\alpha = .10$.

Solution

The null and alternative hypotheses of interest are

H_0: $\eta = 5,250$ hours

H_a: $\eta \neq 5,250$ hours

Test statistic: Since $n \geq 10$, we use the standard normal z statistic:

$$z = \frac{(S - .5) - .5n}{.5\sqrt{n}}$$

where S is the maximum of S_1, the number of measurements greater than 5,250, and S_2, the number of measurements less than 5,250.

Rejection region: $z > 1.645$, where $z_{\alpha/2} = z_{.05} = 1.645$

Assumptions: The distribution of the failure times is continuous (time is a continuous variable), but nothing is assumed about the shape of its probability distribution.

Since the number of measurements exceeding 5,250 is $S_2 = 14$ and thus the number of measurements less than 5,250 is $S_1 = 6$, then $S = 14$, the greater of S_1 and S_2. The calculated z statistic is therefore

$$z = \frac{(S - .5) - .5n}{.5\sqrt{n}} = \frac{13.5 - 10}{.5\sqrt{20}} = \frac{3.5}{2.236} = 1.565$$

The value of z is not in the rejection region, so we cannot reject the null hypothesis at the $\alpha = .10$ level of significance. Thus, the CD manufacturer should not conclude, on the basis of this sample, that its competitor's CDs have a median failure time that differs from 5,250 hours.   ▲

The one-sample nonparametric sign test for a median provides an alternative to the *t*-test for small samples from nonnormal distributions. However, if the distribution is approximately normal, the *t*-test provides a more powerful test about the central tendency of the distribution.

EXERCISES 15.1–15.10

Learning the Mechanics

15.1 Under what circumstances is the sign test preferred to the *t*-test for making inferences about the central tendency of a population?

15.2 What is the probability that a randomly selected observation exceeds the
  a. Mean of a normal distribution?
  b. Median of a normal distribution?
  c. Mean of a nonnormal distribution?
  d. Median of a nonnormal distribution?

15.3 Use Table II of Appendix A to calculate the following binomial probabilities:

  a. $P(x \geq 6)$ when $n = 7$ and $p = .5$
  b. $P(x \geq 5)$ when $n = 9$ and $p = .5$
  c. $P(x \geq 8)$ when $n = 8$ and $p = .5$
  d. $P(x \geq 10)$ when $n = 15$ and $p = .5$. Also use the normal approximation to calculate this probability, then compare the approximation with the exact value.
  e. $P(x \geq 15)$ when $n = 25$ and $p = .5$. Also use the normal approximation to calculate this probability, then compare the approximation with the exact value.

15.4 Consider the following sample of 10 measurements:

8.4   16.9   15.8   12.5   10.3   4.9   12.9   9.8   23.7   7.3

Use these data to conduct each of the following sign tests using the binomial tables (Table II, Appendix A) and $\alpha = .05$:

a. H_0: $\eta = 9$ versus H_a: $\eta > 9$
b. H_0: $\eta = 9$ versus H_a: $\eta \neq 9$
c. H_0: $\eta = 20$ versus H_a: $\eta < 20$
d. H_0: $\eta = 20$ versus H_a: $\eta \neq 20$
e. Repeat each of the preceding tests using the normal approximation to the binomial probabilities. Compare the results.
f. What assumptions are necessary to ensure the validity of each of the preceding tests?

15.5 Suppose you wish to conduct a test of the research hypothesis that the median of a population is greater than 80. You randomly sample 25 measurements from the population and determine that 16 of them exceed 80. Set up and conduct the appropriate test of hypothesis at the .10 level of significance. Be sure to specify all necessary assumptions.

Applying the Concepts

15.6 Certain species of fish may move excessively from one confined area to another; in zoology, this phenomenon is known as excessive transitory migration (ETM). In a study of the ETM of guppy populations, 40 adult female guppies were placed in the left compartment of an experimental aquarium tank which was divided in half by a glass plate. After the plate was removed, the numbers of fish passing through the slit from the left compartment to the right one, and vice versa, were monitored every minute for 30 minutes (*Zoological Science,* Vol. 6, 1989). If an equilibrium is reached, the researchers would expect the median number of fish remaining in the left compartment to be 20. The data for the 30 observations (that is, numbers of fish in the left compartment at the end of each one-minute interval) are shown below. Use the large sample sign test to determine whether the median is less than 20. Test using $\alpha = .05$.

15.7 The biting rate of a particular species of fly was investigated in a study reported in the *Journal of the American Mosquito Control Association* (Mar. 1995). Biting rate was defined as the number of flies biting a volunteer during 15 minutes of exposure. This species of fly is known to have a median biting rate of 5 bites per 15 minutes on Stanbury Island, Utah. However, it is theorized that the median biting rate is higher in bright, sunny weather. To test this theory, 122 volunteers were exposed to the flies during a sunny day on Stanbury Island. Of these volunteers, 95 experienced biting rates greater than 5.

a. Set up the null and alternative hypotheses for the test.
b. Calculate the approximate *p*-value of the test. [*Hint:* Use the normal approximation for a binomial probability.]
c. Make the appropriate conclusion at $\alpha = .01$.

15.8 The American Cancer Society funds medical research focused on finding treatments for various forms of cancer. One new treatment is being tested to determine whether the average remission time for a particularly aggressive form of cancer is extended by the treatment. Assume that current treatments have been shown to provide a median remission time of 4.5 years. Seven patients with this cancer are given the new treatment, and their remission times (in years) are as follows:

5.3   7.3   3.6   5.2   6.1   4.8   8.4

Do these data provide sufficient evidence to indicate that the median remission time is increased by the new treatment?

15.9 Many states now require that mathematics teachers pass a basic math literacy test in order to qualify to teach math in the state. An important part of the process is the development of a fair test. Suppose that one state board of education has developed a test and that a sample of 20 math teachers are asked to take it. The test will be considered too hard if at least 50% of *all* math teachers in the state score less than 60 (on a scale of 0 to 100).

a. Set up the appropriate null and alternative hypotheses to test whether the test is too hard.
b. Establish the appropriate test statistic and rejection region for the test using $\alpha = .05$.
c. What assumptions are necessary to ensure the validity of the test?
d. Suppose that 14 of the 20 sampled teachers score less than 60. What is the appropriate conclusion?
e. Calculate the observed significance level of the test. Interpret it.
f. Considering your answers to parts **d** and **e**, what would your recommendation be concerning the adoption of the test statewide?

16	11	12	15	14	16	18	15	13	15
14	14	16	13	17	17	14	22	18	19
17	17	20	23	18	19	21	17	21	17

Source: Terami, H., and Watanabe, M. "Excessive transitory migration of guppy populations. III. Analysis of perception of swimming space and a mirror effect." *Zoological Science,* Vol. 6, 1989, p. 977 (Figure 2).

15.10 The following table lists the lymphocyte (LYMPHO) count results from hematology tests administered to a sample of 50 West Indian or African workers. Test (at $\alpha = .05$) the hypothesis that the median lymphocyte count of all West Indian/African workers exceeds 20. Use the accompanying MINITAB printout to make your conclusion.

Case Number	LYMPHO	Case Number	LYMPHO	Case Number	LYMPHO
1	14	18	28	35	11
2	15	19	17	36	25
3	19	20	14	37	30
4	23	21	8	38	32
5	17	22	25	39	17
6	20	23	37	40	22
7	21	24	20	41	20
8	16	25	15	42	20
9	27	26	9	43	20
10	34	27	16	44	26
11	26	28	18	45	40
12	28	29	17	46	22
13	24	30	23	47	61
14	26	31	43	48	12
15	23	32	17	49	20
16	9	33	23	50	35
17	18	34	31		

Source: Royston, J. P. "Some techniques for assessing multivariate normality based on the Shapiro–Wilk W." *Applied Statistics,* Vol. 32, No. 2, pp. 121–133.

```
SIGN TEST OF MEDIAN = 20.00  VERSUS  G.T.   20.00

               N  BELOW  EQUAL  ABOVE  P-VALUE     MEDIAN
LYMPHO        50    19     6     25    0.2257      20.50
```

15.2 COMPARING TWO POPULATIONS: THE WILCOXON RANK SUM TEST FOR INDEPENDENT SAMPLES

Suppose two independent random samples are to be used to compare two populations and the *t*-test of Chapter 9 is inappropriate for making the comparison. We may be unwilling to make assumptions about the form of the underlying population probability distributions or we may be unable to obtain exact values of the sample measurements. If the data can be ranked in order of magnitude for either of these situations, the **Wilcoxon rank sum test** (developed by Frank Wilcoxon) can be used to test the hypothesis that the probability distributions associated with the two populations are equivalent.

Suppose, for example, an experimental psychologist wants to compare reaction times for adult males under the influence of drug A to those under the influence of drug B. Experience has shown that the populations of reaction time measurements often possess probability distributions that are skewed to the right, as shown in Figure 15.4. Consequently, a *t*-test should not be used to compare the mean reaction times for the two drugs because the normality assumption that is required for the *t*-test may not be valid.

Suppose the psychologist randomly assigns seven subjects to each of two groups, one group to receive drug A and the other to receive drug B. The reaction

FIGURE 15.4

Typical probability distribution of reaction times

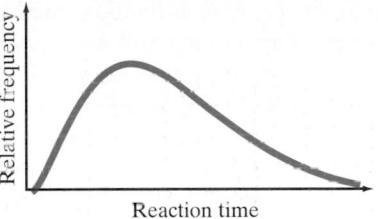

time for each subject is measured at the completion of the experiment. These data (with the exception of the measurement for one subject in group A who was eliminated from the experiment for personal reasons) are shown in Table 15.1.

The population of reaction times for either of the drugs, say drug A, is that which could conceptually be obtained by giving drug A to all adult males. To compare the probability distributions for populations A and B, *we first rank the sample observations as though they were all drawn from the same population.* That is, we pool the measurements from both samples and then rank all the measurements from the smallest (a rank of 1) to the largest (a rank of 13). The results of this ranking process are also shown in Table 15.1.

If the two populations were identical, we would expect the ranks to be *randomly mixed* between the two samples. If, on the other hand, one population tends to have longer reaction times than the other, we would expect the larger ranks to be mostly in one sample and the smaller ranks mostly in the other. Thus, the test statistic for the Wilcoxon test is based on the totals of the ranks for each of the two samples—that is, on the **rank sums**. When the sample sizes are equal, for example, the greater the difference in the rank sums, the greater will be the weight of evidence to indicate a difference between the probability distributions for populations A and B. In the reaction times example, we denote the rank sum for drug A by T_A and that for drug B by T_B. Then

$$T_A = 4 + 7 + 2 + 8 + 1 + 3 = 25$$
$$T_B = 6 + 9 + 5 + 11 + 10 + 12 + 13 = 66$$

The sum of T_A and T_B will always equal $n(n + 1)/2$, where $n = n_1 + n_2$. So, for this example, $n_1 = 6, n_2 = 7$, and

$$T_A + T_B = \frac{13(13 + 1)}{2} = 91$$

TABLE 15.1   Reaction Times of Subjects Under the Influence of Drug A or B

DRUG A		DRUG B	
Reaction Time (seconds)	**Rank**	**Reaction Time (seconds)**	**Rank**
1.96	4	2.11	6
2.24	7	2.43	9
1.71	2	2.07	5
2.41	8	2.71	11
1.62	1	2.50	10
1.93	3	2.84	12
		2.88	13

Since $T_A + T_B$ is fixed, a small value for T_A implies a large value for T_B (and vice versa) and a large difference between T_A and T_B. Therefore, the smaller the value of one of the rank sums, the greater the evidence to indicate that the samples were selected from different populations.

The test statistic for this test is the rank sum for the smaller sample; or, in the case where $n_1 = n_2$, either rank sum can be used. Values that locate the rejection region for this rank sum are given in Table XII of Appendix A. A partial reproduction of this table is shown in Table 15.2. The columns of the table represent n_1, the first sample size, and the rows represent n_2, the second sample size. *The T_L and T_U entries in the table are the boundaries of the lower and upper regions, respectively, for the rank sum associated with the sample that has fewer measurements.* If the sample sizes n_1 and n_2 are the same, either rank sum may be used as the test statistic. To illustrate, suppose $n_1 = 8$ and $n_2 = 10$. For a two-tailed test with $\alpha = .05$, we consult part **a** of the table and find that the null hypothesis will be rejected if the rank sum of sample 1 (the sample with fewer measurements), T, is less than or equal to $T_L = 54$ *or* greater than or equal to $T_U = 98$. The Wilcoxon rank sum test is summarized in the next box.

Wilcoxon Rank Sum Test: Independent Samples*

ONE-TAILED TEST	**TWO-TAILED TEST**

H_0: Two sampled populations have identical probability distributions

H_a: The probability distribution for population A is shifted to the right of that for B

Test statistic: The rank sum T associated with the sample with fewer measurements (if sample sizes are equal, either rank sum can be used)

Rejection region: Assuming the smaller sample size is associated with distribution A (if sample sizes are equal, we use the rank sum T_A), we reject the null hypothesis if

$$T_A \geq T_U$$

where T_U is the upper value given by Table XII in Appendix A for the chosen *one-tailed* α value.

H_0: Two sampled populations have identical probability distributions

H_a: The probability distribution for population A is shifted to the left *or* to the right of that for B

Test statistic: The rank sum T associated with the sample with fewer measurements (if sample sizes are equal, either rank sum can be used)

Rejection region: $T \leq T_L$
$\qquad$ or $T \geq T_U$

where T_L is the lower value given by Table XII in Appendix A for the chosen *two-tailed* α value and T_U is the upper value from Table XII.

[*Note:* If the one-sided alternative is that the probability distribution for A is shifted to the *left* of B (and T_A is the test statistic), we reject the null hypothesis if $T_A \leq T_L$.]

Assumptions: 1. The two samples are random and independent.
2. The two probability distributions from which the samples are drawn are continuous.

Ties: Assign tied measurements the average of the ranks they would receive if they were unequal but occurred in successive order. For example, if the third-ranked and fourth-ranked measurements are tied, assign each a rank of $(3 + 4)/2 = 3.5$.

---

*Another statistic used for comparing two populations based on independent random samples is the **Mann–Whitney U statistic**. The U statistic is a simple function of the rank sums. It can be shown that the Wilcoxon rank sum test and the Mann–Whitney U-test are equivalent.

**TABLE 15.2   Reproduction of Part of Table XII in Appendix A:
Critical Values for the Wilcoxon Rank Sum Test**

$\alpha = .025$ one-tailed; $\alpha = .05$ two-tailed

n_2 \ n_1	3		4		5		6		7		8		9		10	
	T_L	T_U	T_L	T_U	T_L	T_U	T_L	T_U	T_L	T_U	T_L	T_U	T_L	T_U	T_L	T_U
3	5	16	6	18	6	21	7	23	7	26	8	28	8	31	9	33
4	6	18	11	25	12	28	12	32	13	35	14	38	15	41	16	44
5	6	21	12	28	18	37	19	41	20	45	21	49	22	53	24	56
6	7	23	12	32	19	41	26	52	28	56	29	61	31	65	32	70
7	7	26	13	35	20	45	28	56	37	68	39	73	41	78	43	83
8	8	28	14	38	21	49	29	61	39	73	49	87	51	93	54	98
9	8	31	15	41	22	53	31	65	41	78	51	93	63	108	66	114
10	9	33	16	44	24	56	32	70	43	83	54	98	66	114	79	131

Note that the assumptions necessary for the validity of the Wilcoxon rank sum test do not specify the shape or type of probability distribution. However, the distributions are assumed to be continuous so that the probability of tied measurements is 0 (see Chapter 5), and each measurement can be assigned a unique rank. In practice, however, rounding of continuous measurements will sometimes produce ties. As long as the number of ties is small relative to the sample sizes, the Wilcoxon test procedure will still have approximate significance level α. The test is not recommended to compare discrete distributions for which many ties are expected.

EXAMPLE 15.2

Do the data given in Table 15.1 provide sufficient evidence to indicate a **shift in the probability distributions** for drugs A and B, that is, that the probability distribution corresponding to drug A lies either to the right or left of the probability distribution corresponding to drug B? Test at the .05 level of significance.

Solution

H_0: The two populations of reaction times corresponding to drug A and drug B have the same probability distribution

H_a: The probability distribution for drug A is shifted to the right or left of the probability distribution corresponding to drug B*

Test statistic: Since drug A has fewer subjects than drug B, the test statistic is T_A, the rank sum of drug A's reaction times.

Rejection region: Since the test is two-sided, we consult part **a** of Table XII for the rejection region corresponding to $\alpha = .05$. We will reject H_0 for $T_A \leq T_L$ or $T_A \geq T_U$. Thus, we will reject H_0 if $T_A \leq 28$ or $T_A \geq 56$.

---

*The alternative hypotheses in this chapter will be stated in terms of a difference in the *location* of the distributions. However, since the shapes of the distributions may also differ under H_a, some of the figures (e.g., Figure 15.5) depicting the alternative hypothesis will show probability distributions with different shapes.

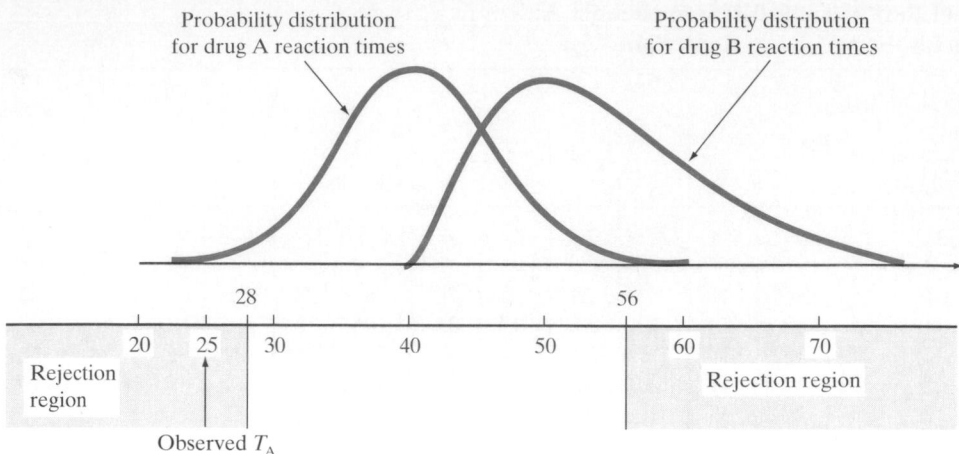

FIGURE 15.5 Alternative hypothesis and rejection region for Example 15.2

Since T_A, the rank sum of drug A's reaction times in Table 15.1, is 25, it is in the rejection region (see Figure 15.5).* Therefore, there is sufficient evidence to reject H_0. This same conclusion can be reached using a statistical software package. The SAS printout of the analysis is shown in Figure 15.6. Both the test statistic ($T_A = 25$) and two-tailed *p*-value ($p = .0184$) are shaded on the printout. The one-tailed *p*-value, $p = .0184/2 = .0092$, is less than $\alpha = .05$, leading us to reject H_0.

Our conclusion is that the probability distributions for drugs A and B are not identical. In fact, it appears that drug B tends to be associated with reaction times that are larger than those associated with drug A (because T_A falls in the lower tail of the rejection region). ▲

Table XII in Appendix A gives values of T_L and T_U for values of n_1 and n_2 less than or equal to 10. When both sample sizes, n_1 and n_2, are 10 or larger, the

FIGURE 15.6

SAS printout for
Example 15.2

```
              N P A R 1 W A Y   P R O C E D U R E

         Wilcoxon Scores (Rank Sums) for Variable TIME
                  Classified by Variable DRUG

                        Sum of      Expected      Std Dev          Mean
     Drug      N        Scores      Under H0      Under H0        Score

      A        6         25.0         42.0          7.0       4.16666667
      B        7         66.0         49.0          7.0       9.42857143

     Wilcoxon 2-Sample Test (Normal Approximation)
     (with Continuity Correction of .5)

     S= 25.0000      Z= -2.35714     Prob > |Z| =   0.0184

     T-Test approx. Significance =     0.0362

     Kruskal-Wallis Test (Chi-Square Approximation)
     CHISQ=  5.8980      DF= 1      Prob > CHISQ=     0.0152
```

*Figure 15.5 depicts only one side of the two-sided alternative hypothesis. The other would show distribution A shifted to the right of distribution B.

sampling distribution of T_A can be approximated by a normal distribution with mean and variance

$$E(T_A) = \frac{n_1(n_1 + n_2 + 1)}{2} \quad \text{and} \quad \sigma^2_{T_A} = \frac{n_1 n_2(n_1 + n_2 + 1)}{12}$$

Therefore, for $n_1 \geq 10$ and $n_2 \geq 10$ we can conduct the Wilcoxon rank sum test using the familiar z-test of Chapters 9 and 10. The test is summarized in the next box.

Wilcoxon Rank Sum Test: Large Independent Samples

ONE-TAILED TEST	**TWO-TAILED TEST**
H_0: Two sampled populations have identical probability distributions	H_0: Two sampled populations have identical probability distributions
H_a: The probability distribution for population A is shifted to the right of that for B	H_a: The probability distribution for population A is shifted to the left *or* to the right of that for B

$$\text{Test statistic:} \quad z = \frac{T_A - \dfrac{n_1(n_1 + n_2 + 1)}{2}}{\sqrt{\dfrac{n_1 n_2(n_1 + n_2 + 1)}{12}}}$$

Rejection region: $z > z_\alpha$	*Rejection region:* $|z| > z_{\alpha/2}$
Assumptions: $n_1 \geq 10$ and $n_2 \geq 10$	*Assumptions:* $n_1 \geq 10$ and $n_2 \geq 10$

EXERCISES 15.11–15.23

Learning the Mechanics

15.11 Specify the test statistic and the rejection region for the Wilcoxon rank sum test for independent samples in each of the following situations:

a. H_0: Two probability distributions, A and B, are identical

H_a: Probability distribution for population A is shifted to the right or left of the probability distribution for population B

$n_A = 7, n_B = 8, \alpha = .10$

b. H_0: Two probability distributions, A and B, are identical

H_a: Probability distribution for population A is shifted to the right of the probability distribution for population B

$n_A = 6, n_B = 6, \alpha = .05$

c. H_0: Two probability distributions, A and B, are identical

H_a: Probability distribution for population A is shifted to the left of the probability distribution for population B

$n_A = 7, n_B = 10, \alpha = .025$

d. H_0: Two probability distributions, A and B, are identical

H_a: Probability distribution for population A is shifted to the right or left of the probability distribution for population B

$n_A = 20, n_B = 20, \alpha = .05$

15.12 Suppose you want to compare two treatments, A and B. In particular, you wish to determine whether the distribution for population B is shifted to the right of the distribution for population A. You plan to use the Wilcoxon rank sum test.

a. Specify the null and alternative hypotheses you would test.

b. Suppose you obtained the following independent random samples of observations on experimental units subjected to the two treatments:

Sample A: 37, 40, 33, 29, 42, 33, 35, 28, 34
Sample B: 65, 35, 47, 52

Conduct a test of the hypotheses described in part **a**. Test using $\alpha = .05$.

15.13 Explain the difference between the one-tailed and two-tailed versions of the Wilcoxon rank sum test for independent random samples.

15.14 Random samples of sizes $n_1 = 16$ and $n_2 = 12$ were drawn from populations 1 and 2, respectively.

The measurements obtained are listed in the following table.

Population 1				Population 2		
9.0	15.6	25.6	31.1	10.1	11.1	13.5
21.1	26.9	24.6	20.0	12.0	18.2	10.3
24.8	16.5	26.0	25.1	9.2	7.0	14.2
17.2	30.1	18.7	26.1	15.8	13.6	13.2

a. Conduct a hypothesis test to determine whether the probability distribution for population 2 is shifted to the left of the probability distribution for population 1. Use $\alpha = .05$.

b. What is the approximate p-value of the test of part **a**?

15.15 Independent random samples are selected from two populations. The data are shown in the table.

Population 1		Population 2		
15	16	5	9	5
10	13	12	8	10
12	8	9	4	

a. Use the Wilcoxon rank sum test to determine whether the data provide sufficient evidence to indicate a shift in the locations of the probability distributions of the sampled populations. Test using $\alpha = .05$.

b. Do the data provide sufficient evidence to indicate that the probability distribution for population 1 is shifted to the right of the probability distribution for population 2? Use the Wilcoxon rank sum test with $\alpha = .05$.

15.16 Suppose you wish to compare two treatments, A and B, based on independent random samples of 15 observations selected from each of the two populations. If $T_A = 173$, do the data provide sufficient evidence to indicate that distribution A is shifted to the left of distribution B? Test using $\alpha = .05$.

Applying the Concepts

15.17 Are tax assessments fair? One measure of "fairness" is the ratio of a property's assessed value to its market value—or a proxy for market value such as a recent sale price (*Journal of Business & Economic Statistics,* Jan. 1985). The accompanying table lists assessment ratios for random samples of 10 properties in neighborhood A and 8 properties in neighborhood B.

Neighborhood A		Neighborhood B	
.850	.880	.911	.835
1.060	.895	.770	.800
.910	.844	.815	.793
.813	.965	.748	.796
.737	.875		

a. Use the Wilcoxon rank sum test to investigate the fairness of the assessments between the two neighborhoods. Use $\alpha = .05$ and interpret your findings in the context of the problem.

b. Under what circumstances could the two-sample *t*-test of Chapter 9 be used to investigate the fairness issue of part **a**?

c. What assumptions are necessary to ensure the validity of the test you conducted in part **a**?

15.18 The term *brood-parasitic intruder* is used to describe birds that search for and lay eggs in the nests built by another bird species. For example, the brown-headed cowbird is known to be a brood parasite of the smaller-sized willow flycatcher. Ornithologists theorize that those flycatchers that recognize but do not vocally react to cowbird calls are more apt to defend their nests and less likely to be found and parasitized. In a study published in *The Condor* (May 1995), each of 13 active flycatcher nests was categorized as parasitized (if at least one cowbird egg was present) or nonparasitized. Cowbird songs were taped and played back while the flycatcher pairs were sitting in the nest prior to incubation. The vocalization rate (number of calls per minute) of each flycatcher pair was recorded. The data for the two groups of flycatchers are given in the following table. Do the data suggest (at $\alpha = .05$) that the vocalization rates of parasitized flycatchers are higher than those of nonparasitized flycatchers? Use the SAS printout on page 729 to form your conclusion.

Parasitized	Not Parasitized
2.00	1.00
1.25	1.00
8.50	0
1.10	3.25
1.25	1.00
3.75	.25
5.50	

Source: Uyehara, J. C., and Narins, P. M. "Nest defense by Willow Flycatchers to brood-parasitic intruders." *The Condor,* Vol. 97, No. 2, May 1995, p. 364 (Figure 1).

15.19 An educational psychologist claims that the order in which test questions are asked affects a student's ability to answer correctly. To investigate this assertion, a professor randomly divides a class of 13 students into two groups—7 in one group and 6 in the other. The professor prepares one set of test questions but arranges the questions in two different orders. On test A the questions are arranged in order of increasing difficulty (that is, from easiest to most difficult), while on test B the order is reversed. One group of students is given test A, the

```
              N P A R 1 W A Y   P R O C E D U R E

        Wilcoxon Scores (Rank Sums) for Variable VOCRATE
                Classified by Variable PARGROUP

                        Sum of    Expected     Std Dev      Mean
        PARGROUP    N   Scores    Under H0     Under H0     Score

        PARASITE    7    66.0       49.0      6.95175683   9.42857143
        NOTPAR      6    25.0       42.0      6.95175683   4.16666667
                    Average Scores were used for Ties
        Wilcoxon 2-Sample Test (Normal Approximation)
        (with Continuity Correction of .5)

        S=  25.0000     Z= -2.37350    Prob > |Z| =   0.0176

        T-Test approx. Significance =     0.0352

        Kruskal-Wallis Test (Chi-Square Approximation)
        CHISQ=  5.9001     DF=  1     Prob > CHISQ=     0.0145
```

other test B, and the test score is recorded for each student. The results are as follows:

Test A: 90   71   83   82   75   91   65
Test B: 66   78   50   68   80   60

Do the data provide sufficient evidence to indicate a difference (a shift in location) in the probability distributions of student scores on the two tests? Test using $\alpha = .05$.

15.20 The data in the table below, extracted from *Technometrics* (Feb. 1986), represent daily accumulated stream flow and precipitation (in inches) for two U.S. Geological Survey stations in Colorado. Conduct a test to determine whether the distributions of daily accumulated stream flow and precipitation for the two stations differ in location. Use $\alpha = .10$. Why is a nonparametric test appropriate for these data?

Station 1			Station 2		
127.96	108.91	100.85	114.79	85.54	280.55
210.07	178.21	85.89	109.11	117.64	145.11
203.24	285.37		330.33	302.74	95.36

Source: Gastwirth, J. L., and Mahmoud, H. "An efficient robust non-parametric test for scale change for data from a gamma distribution." *Technometrics*, Vol. 28, No. 1, Feb. 1986, p. 83 (Table 2).

15.21 Fourteen rats were used in an experiment aimed at comparing two deprivation schedules, A and B, for their effect on hoarding behavior. An independent sampling design was used, with seven rats randomly assigned to each schedule. At the end of the deprivation period, the rats were permitted free access to food pellets and the number of pellets hoarded

(taken but not eaten) during a given time period was recorded. The data are given in the following table. Is there sufficient evidence to indicate that rats on one of the deprivation schedules have a greater tendency to hoard than those on the other schedule? Test using $\alpha = .05$.

Schedule A		Schedule B	
15	4	5	2
10	9	1	6
5	7	2	3
7		8	

15.22 In a comparison of visual acuity of deaf and hearing children, eye movement rates are taken on 10 deaf and 10 hearing children. (See the table below.) A clinical psychologist believes that deaf children have greater visual acuity than hearing children. Test the psychologist's claim by using the data in the table. (The larger a child's eye movement rate, the more visual acuity the child possesses.) Use $\alpha = .05$.

Deaf Children		Hearing Children	
2.75	1.95	1.15	1.23
3.14	2.17	1.65	2.03
3.23	2.45	1.43	1.64
2.30	1.83	1.83	1.96
2.64	2.23	1.75	1.37

15.23 Conduct the test in Exercise 15.22 by using the large-sample approximation for the Wilcoxon rank sum test. Compare the results with those found in Exercise 15.22.

15.3   COMPARING TWO POPULATIONS: THE WILCOXON SIGNED RANK TEST FOR THE PAIRED DIFFERENCE EXPERIMENT

Nonparametric techniques may also be employed to compare two probability distributions when a paired difference design is used. For example, consumer preferences for two competing products are often compared by having each of a sample of consumers rate both products. Thus, the ratings have been paired on each consumer. Here is an example of this type of experiment.

For some paper products, softness is an important consideration in determining consumer acceptance. One method of determining softness is to have judges give a sample of the products a softness rating. Suppose each of ten judges is given a sample of two products that a company wants to compare. Each judge rates the softness of each product on a scale from 1 to 10, with higher ratings implying a softer product. The results of the experiment are shown in Table 15.3.

Since this is a paired difference experiment, we analyze the differences between the measurements (see Section 9.2). However, the nonparametric approach requires that we calculate the ranks of the absolute values of the differences between the measurements, that is, the ranks of the differences after removing any minus signs. *Note that tied absolute differences are assigned the average of the ranks they would receive if they were unequal but successive measurements.* After the absolute differences are ranked, the sum of the ranks of the positive differences of the original measurements, T_+, and the sum of the ranks of the negative differences of the original measurements, T_-, are computed.

We are now prepared to test the nonparametric hypotheses:

H_0: The probability distributions of the ratings for products A and B are identical.

H_a: The probability distributions of the ratings differ (in location) for the two products. (Note that this is a two-sided alternative and that it implies a two-tailed test.)

TABLE 15.3   Softness Ratings of Paper

	PRODUCT		DIFFERENCE		
Judge	A	B	(A − B)	Absolute Value of Difference	Rank of Absolute Value
1	6	4	2	2	5
2	8	5	3	3	7.5
3	4	5	−1	1	2
4	9	8	1	1	2
5	4	1	3	3	7.5
6	7	9	−2	2	5
7	6	2	4	4	9
8	5	3	2	2	5
9	6	7	−1	1	2
10	8	2	6	6	10

$$T_+ = \text{Sum of positive ranks} = 46$$
$$T_- = \text{Sum of negative ranks} = 9$$

TABLE 15.4   **Reproduction of Part of Table XIII of Appendix A: Critical Values for the Wilcoxon Paired Difference Signed Rank Test**

One-Tailed	Two-Tailed	$n = 5$	$n = 6$	$n = 7$	$n = 8$	$n = 9$	$n = 10$
$\alpha = .05$	$\alpha = .10$	1	2	4	6	8	11
$\alpha = .025$	$\alpha = .05$		1	2	4	6	8
$\alpha = .01$	$\alpha = .02$			0	2	3	5
$\alpha = .005$	$\alpha = .01$				0	2	3
		$n = 11$	$n = 12$	$n = 13$	$n = 14$	$n = 15$	$n = 16$
$\alpha = .05$	$\alpha = .10$	14	17	21	26	30	36
$\alpha = .025$	$\alpha = .05$	11	14	17	21	25	30
$\alpha = .01$	$\alpha = .02$	7	10	13	16	20	24
$\alpha = .005$	$\alpha = .01$	5	7	10	13	16	19
		$n = 17$	$n = 18$	$n = 19$	$n = 20$	$n = 21$	$n = 22$
$\alpha = .05$	$\alpha = .10$	41	47	54	60	68	75
$\alpha = .025$	$\alpha = .05$	35	40	46	52	59	66
$\alpha = .01$	$\alpha = .02$	28	33	38	43	49	56
$\alpha = .005$	$\alpha = .01$	23	28	32	37	43	49
		$n = 23$	$n = 24$	$n = 25$	$n = 26$	$n = 27$	$n = 28$
$\alpha = .05$	$\alpha = .10$	83	92	101	110	120	130
$\alpha = .025$	$\alpha = .05$	73	81	90	98	107	117
$\alpha = .01$	$\alpha = .02$	62	69	77	85	93	102
$\alpha = .005$	$\alpha = .01$	55	61	68	76	84	92

Test statistic: T = Smaller of the positive and negative rank sums T_+ and T_-

The smaller the value of T, the greater the evidence to indicate that the two probability distributions differ in location. The rejection region for T can be determined by consulting Table XIII in Appendix A (part of the table is shown in Table 15.4). This table gives a value T_0 for both one-tailed and two-tailed tests for each value of n, the number of matched pairs. For a two-tailed test with $\alpha = .05$, we will reject H_0 if $T \leq T_0$. You can see in Table 15.4 that the value of T_0 that locates the boundary of the rejection region for the judges' ratings for $\alpha = .05$ and $n = 10$ pairs of observations is 8. Thus, the rejection region for the test (see Figure 15.7) is

Rejection region: $T \leq 8$ for $\alpha = .05$

Since the smaller rank sum for the paper data, $T_- = 9$, does not fall within the rejection region, the experiment has not provided sufficient evidence to indicate that the two paper products differ with respect to their softness ratings at the $\alpha = .05$ level.

Note that if a significance level of $\alpha = .10$ had been used, the rejection region would have been $T \leq 11$ and we would have rejected H_0. In other words, the

FIGURE 15.7

Rejection region for paired difference experiment

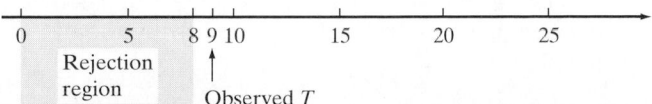

samples do provide evidence that the probability distributions of the softness ratings differ at the $\alpha = .10$ significance level.

The Wilcoxon signed rank test is summarized in the following box. Note that the difference measurements are assumed to have a continuous probability distribution so that the absolute differences will have unique ranks. Although tied (absolute) differences can be assigned ranks by averaging, the number of ties should be small relative to the number of observations to ensure the validity of the test.

Wilcoxon Signed Rank Test for a Paired Difference Experiment

ONE-TAILED TEST	TWO-TAILED TEST
H_0: Two sampled populations have identical probability distributions	H_0: Two sampled populations have identical probability distributions
H_a: The probability distribution for population A is shifted to the right of that for population B	H_a: The probability distribution for population A is shifted to the right *or* to the left of that for population B

Test statistic: T_-, the negative rank sum (we assume the differences are computed by subtracting each paired B measurement from the corresponding A measurement)

Test statistic: T, the smaller of the positive and negative rank sums, T_+ and T_-

Rejection region: $T_- \leq T_0$, where T_0 is found in Table XIII (in Appendix A) for the one-tailed significance level α and the number of untied pairs, n.

Rejection region: $T \leq T_0$, where T_0 is found in Table XIII (in Appendix A) for the two-tailed significance level α and the number of untied pairs, n.

[*Note:* If the alternative hypothesis is that the probability distribution for A is shifted to the left of B, we use T_+ as the test statistic and reject H_0 if $T_+ \leq T_0$.]

Assumptions: 1. The sample of differences is randomly selected from the population of differences.
2. The probability distribution from which the sample of paired differences is drawn is continuous.

Ties: Assign tied absolute differences the average of the ranks they would receive if they were unequal but occurred in successive order. For example, if the third-ranked and fourth-ranked differences are tied, assign both a rank of $(3 + 4)/2 = 3.5$.

EXAMPLE 15.3

Suppose the police commissioner in a small community must chose between two plans for patrolling the town's streets. Plan A, the less expensive plan, uses voluntary citizen groups to patrol certain high-risk neighborhoods. In contrast, plan B would utilize police patrols. As an aid in reaching a decision, both plans are examined by 10 trained criminologists, each of whom is asked to rate the plans on a scale from 1 to 10. (High ratings imply a more effective crime prevention plan.) The city will adopt plan B (and hire extra police) only if the data provide sufficient evidence that criminologists tend to rate plan B more effective than plan A.

The results of the survey are shown in Table 15.5. Do the data provide evidence at the $\alpha = .05$ level that the distribution of ratings for plan B lies above that for plan A?

TABLE 15.5   Effectiveness Ratings by 10 Qualified Crime Prevention Experts

	PLAN		DIFFERENCE	
Crime Prevention Expert	**A**	**B**	**(A − B)**	**Rank of Absolute Difference**
1	7	9	−2	4.5
2	4	5	−1	2
3	8	8	0	(Eliminated)
4	9	8	1	2
5	3	6	−3	6
6	6	10	−4	7.5
7	8	9	−1	2
8	10	8	2	4.5
9	9	4	5	9
10	5	9	−4	7.5

Positive rank sum $= T_+ = 15.5$

Solution

The null and alternative hypotheses are

H_0: The two probability distributions of effectiveness ratings are identical

H_a: The effectiveness ratings of the more expensive plan (B) tend to exceed those of plan A

Observe that the alternative hypothesis is one-sided (that is, we only wish to detect a shift in the distribution of the B ratings to the right of the distribution of A ratings) and therefore it implies a one-tailed test of the null hypothesis (see Figure 15.8). If the alternative hypothesis is true, the B ratings will tend to be larger than the paired A ratings, more negative differences in pairs will occur, T_- will be large, and T_+ will be small. Because Table XIII is constructed to give lower-tail values of T_0, we will use T_+ as the test statistic and reject H_0 for $T_+ \leq T_0$.

The differences in ratings for the pairs (A − B) are shown in Table 15.5. Note that one of the differences equals 0. Consequently, we eliminate this pair from the ranking and reduce the number of pairs to $n = 9$. Looking in Table XIII, we have $T_0 = 8$ for a one-tailed test with $\alpha = .05$ and $n = 9$. Therefore, the test statistic and rejection region for the test are

Test statistic: T_+, the positive rank sum

Rejection region: $T_+ \leq 8$

FIGURE 15.8

The alternative hypothesis for Example 15.3

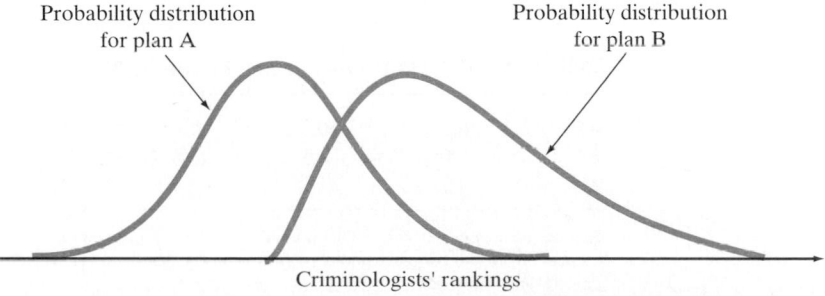

Probability distribution for plan A

Probability distribution for plan B

Criminologists' rankings

FIGURE 15.9

MINITAB printout for
Example 15.3

```
TEST OF MEDIAN = 0.000000 VERSUS MEDIAN L.T. 0.000000

              N FOR    WILCOXON              ESTIMATED
         N    TEST     STATISTIC   P-VALUE     MEDIAN
AminusB  10    9         15.5       0.221      -1.000
```

Summing the ranks of the positive differences from Table 15.5, we find $T_+ = 15.5$. Since this value exceeds the critical value, $T_0 = 8$, we conclude that this sample provides insufficient evidence at the $\alpha = .05$ level to support the alternative hypothesis. The commissioner *cannot* conclude that the plan utilizing police patrols tends to be rated higher than the plan using citizen volunteers. That is, on the basis of this study, extra police will not be hired. A MINITAB printout of the analysis, shown in Figure 15.9, confirms this conclusion. The *p*-value of the test (shaded) is .221, which exceeds $\alpha = .05$. ▲

As is the case for the rank sum test for independent samples, the sampling distribution of the signed rank statistic can be approximated by a normal distribution when the number *n* of paired observations is large (say $n \geq 25$). The large-sample *z*-test is summarized in the next box.

Wilcoxon Signed Rank Test for a Paired Difference Experiment: Large Sample

ONE-TAILED TEST	**TWO-TAILED TEST**
H_0: Two sampled populations have identical probability distributions	H_0: Two sampled populations have identical probability distributions
H_a: The probability distribution for population A is shifted to the right of that for population B	H_a: The probability distribution for population A is shifted to the right *or* to the left of that for population B

Test statistic: $$z = \frac{T_+ - \dfrac{n(n+1)}{4}}{\sqrt{\dfrac{n(n+1)(2n+1)}{24}}}$$

Rejection region: $z > z_\alpha$	Rejection region: $\mid z \mid > z_{\alpha/2}$
Assumptions: $n \geq 25$	Assumptions: $n \geq 25$

TABLE 15.6 **TCDD Levels in Fat Tissue of Vietnam Veterans (Case Study 15.1)**

4.9	6.9	10.0	4.4	4.6	1.1	2.3
5.9	7.0	5.5	7.0	1.4	11.0	2.5
4.4	4.2	41.0	2.9	7.7	2.5	

Source: Schechter, A., *et al.* "Partitioning of 2,3,7,8-chlorinated dibenzo-*p*-dioxins and dibenzofurans between adipose tissue and plasma liquid of 20 Massachusetts Vietnam veterans." *Chemosphere,* Vol. 20, Nos. 7–9, 1990, pp. 954–955 (Table II).

Deadly Exposure: Agent Orange and Vietnam Vets

CASE STUDY • 15.1 •

uring the Vietnam War, American soldiers were exposed to Agent Orange, a defoliant used by the U.S. Armed Forces to destroy the dense plant and tree cover of the Asian jungle. As a result of this exposure, many Vietnam veterans have dangerously high 2,3,7,8-tetrachlorodibenzo-p-levels of the dioxin (TCDD) in their blood and adipose (fatty) tissue.

A study published in *Chemosphere* (Vol. 20, 1990) reported on the TCDD levels of 20 Massachusetts Vietnam vets who were possibly exposed to Agent Orange. The amounts of TCDD (measured in parts per trillion) in fat tissue drawn from each vet are shown in Table 15.6 on page 734. The data provide an opportunity to apply the nonparametric tests of the previous sections.

Focus

a. Suppose we want to use this small sample ($n = 20$) to test whether the median TCDD level in fat tissue of Vietnam vets exceeds 3 parts per trillion. Set up the null and alternative hypotheses of this test.

b. If the population of TCDD levels is normally distributed, how are the median η and the mean μ related? Why is the parametric small-sample t-test of Section 8.4 appropriate in this case?

c. A MINITAB stem-and-leaf plot of the sample TCDD levels is displayed in Figure 15.10. Comment on the validity of the normality assumption required to conduct the t-test of part b.

d. Test the hypothesis of part a using the nonparametric sign test discussed in this chapter. Use $\alpha = .05$. Interpret the result.

e. Repeat part d, using $\alpha = .10$.

f. Although the sign test of parts d and e is easy to apply, it is not the most powerful nonparametric test available. Recall (Section 8.6) that the "power" of a test is the probability that the test rejects the null hypothesis when it is false. Define "power" in practical terms of this case.

A nonparametric test with higher power than the sign test is the Wilcoxon signed ranks test. Although we presented the signed ranks procedure as a test to compare two populations in a matched-pairs design, it can also be used to test the location (median) of a single population. To conduct the signed ranks test, we calculate the differences ($y_i - 3$) for the sample, where y_i is the TCDD level for veteran i and 3 is the hypothesized value of the median η. The absolute values of the differences are ranked, and the rank sum of the negative differences represents the test statistic.

Focus

g. Compute the differences ($y_i - 3$) for the sample data in Table 15.6 and rank the absolute values of these differences.

h. Use the results of part g to calculate the rank sum of the negative differences, T_-. This is the Wilcoxon signed rank test statistic.

i. The p-value of the test, part h, is shown on the MINITAB printout, Figure 15.11. Interpret this value.

j. Compare the results, parts d, e, and i. Comment on the power of the signed rank test as compared to the sign test.

continued

FIGURE 15.10

MINITAB stem-and-leaf display for the data of Table 15.6

```
Stem-and-leaf of TCDD      N = 20
Leaf Unit = 1.0
 (11)  0  11222244444
   9   0  556777
   3   1  01
   1   1
   1   2
   1   2
   1   3
   1   3
   1   4  1
```

FIGURE 15.11

MINITAB printout of signed rank analysis

```
TEST OF MEDIAN = 0.000000 VERSUS MEDIAN N.E.  0.000000

                N FOR   WILCOXON              ESTIMATED
          N     TEST    STATISTIC   P-VALUE    MEDIAN
TCDD_3   20      20       181.0      0.005      2.100
```

Case Study 15.1 continued

The researchers also measured the TCDD levels in blood plasma drawn from the 20 Vietnam veterans. These values (also measured in parts per trillion) are given in Table 15.7.

Focus

k. Conduct an appropriate test to determine whether the true median TCDD level in plasma of Vietnam vets exceeds 3 parts per trillion. Justify your choice of a test and fully interpret the results.

TABLE 15.7   TCDD Levels in Plasma of Vietnam Veterans (Case Study 15.1)

2.5	6.9	3.1	3.3	2.1	4.6	3.5
7.2	1.8	1.8	6.0	20.0	3.0	2.0
4.7	4.1	3.1	2.5	1.6	36.0	

Source: Schecter, A., *et al.* "Partitioning of 2,3,7,8-chlorinated dibenzo-*p*-dioxins and dibenzofurans between adipose tissue and plasma lipid of 20 Massachusetts Vietnam veterans." *Chemosphere*, Vol. 20, Nos. 7–9, 1990, pp. 954–955 (Table I).

EXERCISES 15.24–15.36

Learning the Mechanics

15.24 Specify the test statistic and the rejection region for the Wilcoxon signed rank test for the paired difference design in each of the following situations:

a. H_0: Two probability distributions, A and B, are identical

H_a: Probability distribution for population A is shifted to the right or left of probability distribution for population B

$n = 20$, $\alpha = .10$

b. H_0: Two probability distributions, A and B, are identical

H_a: Probability distribution for population A is shifted to the right of the probability distribution for population B

$n = 39$, $\alpha = .05$

c. H_0: Two probability distributions, A and B, are identical

H_a: Probability distribution for population A is shifted to the left of the probability distribution for population B

$n = 7$, $\alpha = .005$

15.25 Suppose you want to test a hypothesis that two treatments, A and B, are equivalent against the alternative hypothesis that the responses for A tend to be larger than those for B. You plan to use a paired difference experiment and to analyze the resulting data using the Wilcoxon signed rank test.

a. Specify the null and alternative hypotheses you would test.

b. Suppose the paired difference experiment yielded the data in the following table. Conduct the test of part **a**. Test using $\alpha = .025$.

	TREATMENT			TREATMENT	
Pair	A	B	Pair	A	B
1	54	45	6	77	75
2	60	45	7	74	63
3	98	87	8	29	30
4	43	31	9	63	59
5	82	71	10	80	82

15.26 Explain the difference between the one- and two-tailed versions of the Wilcoxon signed rank test for the paired difference experiment.

15.27 In order to conduct the Wilcoxon signed rank test, why do we need to assume the probability distribution of differences is continuous?

15.28 Suppose you wish to test a hypothesis that two treatments, A and B, are equivalent against the alternative that the responses for A tend to be larger than those for B.

a. If the number of pairs equals 25, give the rejection region for the large-sample Wilcoxon signed rank test for $\alpha = .05$.

b. Suppose that $T_+ = 273$. State your test conclusions.

c. Find the p-value for the test and interpret it.

15.29 A paired difference experiment with $n = 30$ pairs yielded $T_+ = 354$.

a. Specify the null and alternative hypotheses that should be used in conducting a hypothesis test to determine whether the probability distribution for population A is located to the right of that for population B.

b. Conduct the test of part **a** using $\alpha = .05$.

c. What is the approximate *p*-value of the test of part **b**?

d. What assumptions are necessary to ensure the validity of the test you performed in part **b**?

15.30 A random sample of nine pairs of measurements is shown in the following table.

Pair	Sample Data from Population 1	Sample Data from Population 2
1	8	7
2	10	1
3	6	4
4	10	10
5	7	4
6	8	3
7	4	6
8	9	2
9	8	4

a. Use the Wilcoxon signed rank test to determine whether the data provide sufficient evidence to indicate that the probability distribution for population 1 is shifted to the right of the probability distribution for population 2. Test using $\alpha = .05$.

b. Use the Wilcoxon signed rank test to determine whether the data provide sufficient evidence to indicate that the probability distribution for population 1 is shifted either to the right or to the left of the probability distribution for population 2. Test using $\alpha = .05$.

Applying the Concepts

15.31 The regional atlas is an important educational resource that is updated on a periodic basis. One of the most critical aspects of a new atlas design is its thematic content. In a survey of atlas users

Theme	RANKINGS	
	High School Teachers	**Geography Alumni**
Tourism	10	2
Physical	2	1
Transportation	7	3
People	1	6
History	2	5
Climate	6	4
Forestry	5	8
Agriculture	7	10
Fishing	9	7
Energy	2	8
Mining	10	11
Manufacturing	12	12

Source: Keller, C. P., *et al.* "Planning the next generation of regional atlases: Input from educators." *Journal of Geography*, Vol. 94, No. 3, May/June 1995, p. 413 (Table 1).

(*Journal of Geography,* May/June 1995), a large sample of high school teachers in British Columbia ranked 12 thematic atlas topics for usefulness. The consensus rankings of the teachers (based on the percentage of teachers who responded they "would definitely use" the topic) are given in the accompanying table. These teacher rankings were compared to the rankings of a group university geography alumni made three years earlier. Compare the distributions of theme rankings for the two groups with an appropriate nonparametric test. Use $\alpha = .05$. Interpret the results practically.

15.32 A study published in the *Journal of Business Communications* (Fall 1985) considered the question: "Which is the more effective means of dealing with complex group problem-solving tasks—face-to-face meetings or video teleconferencing?" In an experiment similar to the one described in this exercise, two Stetson University researchers concluded that video teleconferencing may be the more effective method. Ten groups of four people each were randomly assigned both to a specific communication setting (face-to-face or video teleconferencing) and to one of two specific complex problems. Upon completion of the problem-solving task, the same groups were placed in the alternative communication setting and asked to complete the second problem-solving task. The percentage of each problem task correctly completed was recorded for each group, with the results given in the accompanying table.

Group	Face-to-Face	Video Teleconferencing
1	65%	75%
2	82	80
3	54	60
4	69	65
5	40	55
6	85	90
7	98	98
8	35	40
9	85	89
10	70	80

a. What type of experimental design was used in this study?

b. Specify the null and alternative hypotheses that should be used in determining whether the data provide sufficient evidence to conclude that the problem-solving performance of video teleconferencing groups is superior to that of groups that interact face-to-face.

c. Conduct the hypothesis test of part **b**. Use $\alpha = .05$. Interpret the results of your test in the context of the problem.

d. What is the *p*-value of the test in part **c**?

15.33 Hypoglycemia is a condition in which blood sugar is below normal limits. To compare the effectiveness of two compounds, X and Y, for treating hypoglycemia, each compound is applied to half the diaphragms of each of seven white mice. Blood glucose uptake in milligrams per gram of tissue is measured for each half, producing the results listed in the table below. Do the data provide sufficient evidence to indicate that one of the compounds tends to produce higher blood sugar uptake readings than the other? Test using $\alpha = .10$.

Mouse	COMPOUND X	COMPOUND Y	Mouse	COMPOUND X	COMPOUND Y
1	4.7	5.1	5	7.0	6.1
2	3.3	4.6	6	4.7	4.1
3	8.5	8.7	7	5.2	5.1
4	3.9	3.6			

15.34 Children completing the sixth grade at a school located in a large city have the choice of going to one of two junior high schools, A or B. Members of the school board want to compare the academic effectiveness of the two schools. The parents of six sets of identical twins agree to send one child to school A and the other to school B. Since each set of twins is in the same class at each grade level through the sixth grade, a paired difference design could be employed. Near the end of the ninth grade, an achievement test is given to each child in the experiment. The results are given in the following table. Test to determine whether there is evidence of a difference (shift in location) in the probability distributions of achievement test scores at the two schools. Use $\alpha = .10$.

Twin Pair	SCHOOL A	SCHOOL B	Twin Pair	SCHOOL A	SCHOOL B
1	65	69	4	50	52
2	72	72	5	60	47
3	86	74	6	81	72

15.35 Twelve sets of identical twins are given psychological tests to determine whether the firstborn of the twins tends to be more aggressive than the secondborn. The results are shown in the following table, where the higher score indicates greater aggressiveness. Do the data provide sufficient evidence (at $\alpha = .05$) to indicate that the firstborn of a pair of twins is more aggressive than the other? Use the SPSS printout at the bottom of the page to make your conclusion.

Set	Firstborn	Secondborn
1	86	88
2	71	77
3	77	76
4	68	64
5	91	96
6	72	72
7	77	65
8	91	90
9	70	65
10	71	80
11	88	81
12	87	72

15.36 In Exercise 9.98 we compared matched pairs of measurements on the radon exhalation rate (a measure of radiation) of 15 soil samples from waste gypsum and phosphate mounds in Polk County, Florida. Each soil sample was measured for exhalation rate by the Polk County Health Department (PCHD) and the Eastern Environmental Radiation Facility (EERF). The data are reproduced in the table on page 739. Do the data provide sufficient evidence to indicate that one of the measuring facilities, PCHD or EERF, tends to read higher or lower than the other? Test using the Wilcoxon signed rank test with $\alpha = .05$.

```
- - - - - Wilcoxon Matched-pairs Signed-ranks Test

     FRSTBORN
with SECBORN

   Mean Rank    Cases

       5.93        7  - Ranks (SECBORN Lt FRSTBORN)
       6.13        4  + Ranks (SECBORN Gt FRSTBORN)
                   1    Ties (SECBORN Eq FRSTBORN)
                  --
                  12    Total

     Z =   -.7557          2-tailed P =   .4498
```

Charcoal Canister No.	PCHD	EERF	Charcoal Canister No.	PCHD	EERF
71	1,709.79	1,479.0	85	393.55	187.7
58	357.17	257.8	46	880.84	630.4
84	1,150.94	1,287.0	4	2,996.49	3,707.0
91	1,572.69	1,395.0	20	2,367.40	2,791.0
44	558.33	416.5	36	599.84	706.8
43	4,132.28	3,993.0	42	538.37	618.5
79	1,489.86	1,351.0	55	2,770.23	2,639.0
61	3,017.48	1,813.0			

Source: Horton, T. R. "Preliminary radiological assessment of radon exhalation from phosphate gypsum piles and inactive uranium mill tailings piles." EPA-520/5-79-004. Washington, D.C.: Environmental Protection Agency, 1979.

15.4 THE KRUSKAL–WALLIS *H*-TEST FOR A COMPLETELY RANDOMIZED DESIGN

In Chapter 10 we used an analysis of variance and the *F*-test to compare the means of *p* populations (treatments) based on random sampling from populations that were normally distributed with a common variance σ^2. We now present a nonparametric technique for comparing the populations that requires no assumptions concerning the population probability distributions.

Suppose a health administrator wants to compare the unoccupied bed space for three hospitals located in the same city. She randomly selects 10 different days from the records of each hospital and lists the number of unoccupied beds for each day (see Table 15.8). Because the number of unoccupied beds per day may occasionally be quite large, it is conceivable that the population distributions of data may be skewed to the right and that this type of data may not satisfy the assumptions necessary for a parametric comparison of the population means. We therefore use a nonparametric analysis and base our comparison on the rank sums for the three sets of sample data. Just as with two independent samples

TABLE 15.8 Number of Available Beds

HOSPITAL 1		HOSPITAL 2		HOSPITAL 3	
Beds	Rank	Beds	Rank	Beds	Rank
6	5	34	25	13	9.5
38	27	28	19	35	26
3	2	42	30	19	15
17	13	13	9.5	4	3
11	8	40	29	29	20
30	21	31	22	0	1
15	11	9	7	7	6
16	12	32	23	33	24
25	17	39	28	18	14
5	4	27	18	24	16
$R_1 = 120$		$R_2 = 210.5$		$R_3 = 134.5$	

(Section 15.2), the ranks are computed for each observation according to the relative magnitude of the measurements *when the data for all the samples are combined* (see Table 15.8). Ties are treated as they were for the Wilcoxon rank sum and signed rank tests by assigning the average value of the ranks to each of the tied observations.

We test

H_0: The probability distributions of the number of unoccupied beds are the same for all three hospitals

H_a: At least two of the three hospitals have probability distributions of number of unoccupied beds that differ in location

If we denote the three sample rank sums by R_1, R_2, and R_3, the test statistic is given by

$$H = \frac{12}{n(n + 1)}\sum\frac{R_j^2}{n_j} - 3(n + 1)$$

where n_j is the number of measurements in the jth sample and n is the total sample size ($n = n_1 + n_2 + \cdots + n_p$). For the data in Table 15.8, we have $n_1 = n_2 = n_3 = 10$ and $n = 30$. The rank sums are $R_1 = 120$, $R_2 = 210.5$, and $R_3 = 134.5$. Thus,

$$H = \frac{12}{30(31)}\left[\frac{(120)^2}{10} + \frac{(210.5)^2}{10} + \frac{(134.5)^2}{10}\right] - 3(31)$$

$$= 99.097 - 93 = 6.097$$

The H statistic measures the extent to which the p samples differ with respect to their relative ranks. This is more easily seen by writing H in an alternative but equivalent form:

$$H = \frac{12}{n(n + 1)}\sum n_j(\overline{R}_j - \overline{R})^2$$

where $\overline{R}_j$ is the mean rank corresponding to sample j and $\overline{R}$ is the mean of all the ranks [that is, $\overline{R} = \frac{1}{2}(n + 1)$]. Thus, the H statistic is 0 if all samples have the same mean rank and becomes increasingly large as the distance between the sample mean ranks grows.

If the null hypothesis is true, the distribution of H in repeated sampling is approximately a χ^2 (chi-square) distribution. This approximation for the sampling distribution of H is adequate as long as one of the p sample sizes exceeds 5. (See the references for more detail.) The degrees of freedom corresponding to the approximate sampling distribution of H will always be $(p - 1)$—one less than the number of probability distributions being compared. Because large values of H support the alternative hypothesis that the populations have different probability distributions, the rejection region for the test is located in the upper tail of the χ^2 distribution.

For the data of Table 15.8, the approximate distribution of the test statistic H is χ^2 with $(p - 1) = 2$ df. To determine how large H must be before we will reject the null hypothesis, we consult Table VII in Appendix A. For $\alpha = .05$ and df = 2, $\chi^2_{.05} = 5.99147$. Therefore, we can reject the null hypothesis that the three probability distributions are the same if

$$H > 5.99147$$

FIGURE 15.12

Rejection region for the comparison of three probability distributions

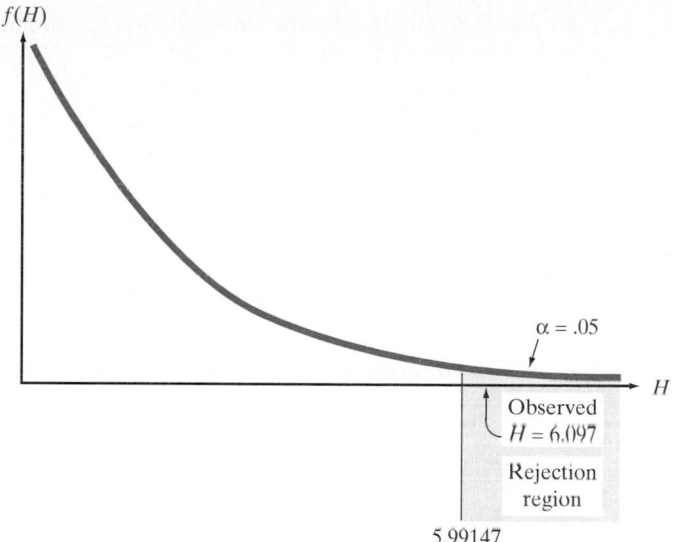

The rejection region is pictured in Figure 15.12. Since the calculated $H = 6.097$ exceeds the critical value of 5.99147, we conclude that at least one of the three hospitals tends to have a larger number of unoccupied beds than the others.

The same conclusion can be reached from a computer printout of the analysis. The rank sums, test statistic, and p-value of the nonparametric test are shaded on the SAS printout in Figure 15.13. Since $\alpha = .05$ exceeds p-value $= .0474$, there is sufficient evidence to reject H_0.

The **Kruskal–Wallis H-test** for comparing more than two probability distributions is summarized in the next box. Note that we can use the Wilcoxon rank sum test of Section 15.2 to compare the separate pairs of populations if the Kruskal–Wallis H-test supports the alternative hypothesis that at least two of the probability distributions differ.*

FIGURE 15.13

SAS Kruskal–Wallis printout

```
               N P A R 1 W A Y   P R O C E D U R E

          Wilcoxon Scores (Rank Sums) for Variable BEDS
                  Classified by Variable HOSPITAL

                        Sum of      Expected      Std Dev         Mean
     HOSPITAL    N      Scores      Under H0      Under H0        Score

        1       10    120.000000      155.0     22.7277743    12.0000000
        2       10    210.500000      155.0     22.7277743    21.0500000
        3       10    134.500000      155.0     22.7277743    13.4500000
                 Average Scores were used for Ties

          Kruskal-Wallis Test (Chi-Square Approximation)
          CHISQ=  6.0988      DF=  2      Prob > CHISQ=      0.0474
```

*The multiple comparisons procedure of Chapter 10 can be used to determine the appropriate significance level of each test such that the overall level of significance is less than or equal to a prespecified α. If c Wilcoxon rank sum tests are to be performed with an overall level of significance α, then conduct each test at $\alpha^* = \alpha/c$.

Kruskal–Wallis *H*-Test for Comparing *p* Probability Distributions

H_0: The *p* probability distributions are identical

H_a: At least two of the *p* probability distributions differ in location

Test statistic: $H = \dfrac{12}{n(n + 1)}\sum\dfrac{R_j^2}{n_j} - 3(n + 1)$

where

n_j = Number of measurements in sample *j*

R_j = Rank sum for sample *j*, where the rank of each measurement is computed according to its relative magnitude in the totality of data for the *p* samples

n = Total sample size = $n_1 + n_2 + \cdots + n_p$

Rejection region: $H > \chi_\alpha^2$ with $(p - 1)$ degrees of freedom

Assumptions: 1. The *p* samples are random and independent.
2. There are 5 or more measurements in each sample.
3. The *p* probability distributions from which the samples are drawn are continuous.

Ties: Assign tied measurements the average of the ranks they would receive if they were unequal but occurred in successive order. For example, if the third-ranked and fourth-ranked measurements are tied, assign both a rank of $(3 + 4)/2 = 3.5$. The number of ties should be small relative to the total number of observations.

EXERCISES 15.37–15.46

Learning the Mechanics

15.37 Data were collected from three populations, A, B, and C, using a completely randomized design. The following describes the sample data:

$$n_A = n_B = n_C = 15$$
$$R_A = 235 \qquad R_B = 439 \qquad R_C = 361$$

a. Specify the null and alternative hypotheses that should be used in conducting a test of hypothesis to determine whether the probability distributions of populations A, B, and C differ in location.

b. Conduct the test of part **a**. Use $\alpha = .05$.

c. What is the approximate *p*-value of the test of part **b**?

d. Calculate the mean rank for each sample and compute *H* according to the formula that utilizes these means. Verify that this formula yields the same value of *H* that you obtained in part **b**.

15.38 Suppose you want to use the Kruskal–Wallis *H*-test to compare the probability distributions of three populations. The following are independent random samples selected from the three populations:

I:	34, 56, 65, 59, 82, 70, 45
II:	24, 18, 27, 41, 34, 42, 33
III:	72, 101, 91, 76, 80, 75

a. What experimental design was used?

b. Specify the null and alternative hypotheses you would test.

c. Specify the rejection region you would use for your hypothesis test at $\alpha = .01$.

d. Conduct the test at $\alpha = .01$.

15.39 Under what circumstances does the χ^2 distribution provide an appropriate characterization of the sampling distribution of the Kruskal–Wallis *H* statistic?

Applying the Concepts

15.40 Two University of Georgia researchers studied the effect of social reinforcement on exercise duration in adolescents with moderate mental retardation (*Clinical Kinesiology*, Spring 1995). Eleven adolescents with IQs ranging from 32 to 61 were divided into two groups. All participated in a 6-week exercise program. Group A (4 subjects) received verbal and social reinforcement during the program, while Group B (7 subjects) received verbal and social reinforcement and kept a self-record of their individual performances. The researchers theorized that Group B subjects would exercise for longer periods than Group A. Upon completion of the exercise program, all 11 subjects participated in a run/walk "race" in which the goal was to complete as many laps as possible during a 15-minute period. The number of laps completed (to the nearest quarter lap) was used as a measure of exercise duration.

a. Specify the null and alternative hypotheses for a nonparametric analysis of the data.

b. The Kruskal–Wallis H test was applied to the data. The researchers reported the test statistic as $H = 5.1429$ and the observed significance level of the test as $p = .0233$. Interpret these results.

c. Are the assumptions for the test, part **b**, satisfied? If not, propose an alternative nonparametric method for the analysis.

15.41 Sixth graders in an elementary school are taught reading by three different methods. Students were randomly assigned to three classes. One class used programmed instruction, a second used standard memorization techniques, and the third used an open classroom approach. The increases in reading levels attained by five students randomly selected from each of the three classes are shown in the following table. Do the data provide sufficient evidence to indicate that the probability distributions of increases in reading level differ for at least two of the methods? Use $\alpha = .05$.

Programmed	Standard	Open
.9	1.0	1.7
1.5	.8	.5
.7	.9	1.6
1.1	1.2	1.4
.5	1.4	1.0

15.42 Refer to the *Journal of the American Mosquito Control Association* study of biting flies, Exercise 15.7. The effect of wind speeds in kilometers per hour (kph) on the biting rate of the fly on Stanbury Island, Utah, was investigated by exposing samples of volunteers to one of six wind speed conditions.

Wind Speed (kph)	Number of Volunteers (n_j)	Rank Sum of Biting Rates (R_j)
< 1	11	1,804
1 – 2.9	49	6,398
3 – 4.9	62	7,328
5 – 6.9	39	4,075
7 – 8.9	35	2,660
9 – 20	21	1,388
Totals	217	23,653

Source: Strickman, D., *et al.* "Meteorological effects on the biting activity of *Leptoconops americanus* (Diptera: Ceratopogonidae)." *Journal of the American Mosquito Control Association*, Vol. II, No. 1, Mar. 1995, p. 17 (Table 1).

The distributions of the biting rates for the six wind speeds were compared using the Kruskal–Wallis

test. The rank sums of the biting rates for the six conditions are shown in the accompanying table.

a. The researchers reported the test statistic as $H = 35.2$. Verify this value.

b. Find the rejection region for the test using $\alpha = .01$.

c. Make the proper conclusions.

d. The researchers reported the p-value of the test as $p < .01$. Does this value support your inference in part **c**? Explain.

15.43 An experiment was conducted to compare the length of time it takes a human to recover from each of three types of influenza—Victoria A, Texas, and Russian. Twenty-one human subjects were selected at random from a group of volunteers and divided into three groups of seven each. Each group was randomly assigned a strain of the virus, and the influenza was induced in the subjects. All the subjects were then cared for under identical conditions, and the recovery time (in days) was recorded. The results are shown in the accompanying table.

Victoria A	Texas	Russian
12	9	7
6	10	3
13	5	7
10	4	5
8	9	6
11	8	4
7	11	8

a. Do the data provide sufficient evidence to indicate that the recovery times for one or more types of influenza tend to be longer than for the other types? Test using $\alpha = .05$.

b. Do the data provide sufficient evidence to indicate a difference in locations of the distributions of recovery times for the Victoria A and Russian types? Test using $\alpha = .05$.

15.44 Three lists of words, representing three levels of abstractness, are randomly assigned to 21 experimental subjects so that seven subjects receive each list. The subjects are asked to respond to each word on their list with as many associated words as possible within a given period of time. A subject's score is the total number of word associates, summing over all words in the list. Scores for each list are given in the accompanying table. Do the data provide sufficient evidence to indicate a difference (shift in location) between at least two of the probability distributions of the numbers of word associates that subjects can name for the three lists? Use $\alpha = .05$.

List 1	List 2	List 3
48	41	18
43	36	42
39	29	28
57	40	38
21	35	15
47	45	33
58	32	31

15.45 An experiment was conducted to determine whether a test designed to identify a certain form of mental disturbance could be easily interpreted by a person with little psychological training. Thirty judges were selected to review the results of 100 tests, half of which were given to disturbed patients and half to nondisturbed people. Of the 30 judges chosen, 10 were staff members of a mental hospital, 10 were trainees at the hospital, and 10 were undergraduate psychology majors. The results in the table below give the number of the 100 tests correctly classified by each judge. Do the data provide sufficient evidence (at $\alpha = .05$) of a difference in the probability distributions of the number of correct identifications among the three types of judges? Use the SAS printout at the bottom of the page to form your conclusion.

15.46 The EPA wants to determine whether ocean temperature changes caused by a nuclear power plant will have a significant effect on marine life in the region. Recently hatched specimens of a certain species of fish are randomly divided into four groups. The groups are placed in separate simulated ocean environments that are identical in every way except for water temperature. Six months later, the specimens are weighed. The results (in ounces) are given in the table. Do the data provide sufficient evidence to indicate that one or more of the temperatures tend to produce larger weight increases than the other temperatures? Test using $\alpha = .10$.

38°F	42°F	46°F	50°F
22	15	14	17
24	21	28	18
16	26	21	13
18	16	19	20
19	25	24	21
	17	23	

Staff		Trainees		Undergraduates	
78	76	80	69	65	74
79	86	75	81	70	80
85	88	72	76	74	73
93	84	68	72	78	75
90	81	75	76	68	73

```
              N P A R 1 W A Y   P R O C E D U R E

            Wilcoxon Scores (Rank Sums) for Variable CORRECT
                    Classified by Variable JUDGE

                         Sum of      Expected      Std Dev         Mean
         JUDGE     N     Scores      Under H0      Under H0        Score

         STAFF    10   243.000000      155.0     22.6923452   24.3000000
         TRAIN    10   122.500000      155.0     22.6923452   12.2500000
         UGRAD    10    99.500000      155.0     22.6923452    9.9500000
                 Average Scores were used for Ties

            Kruskal-Wallis Test (Chi-Square Approximation)
              CHISQ= 15.381    DF= 2    Prob > CHISQ=      0.0005
```

15.5 THE FRIEDMAN F_r-TEST FOR A RANDOMIZED BLOCK DESIGN

In Section 10.4 we employed an analysis of variance to compare p population (treatment) means when the data were collected using a randomized block design. The **Friedman F_r-test** provides another method for testing to detect a shift in

TABLE 15.9 **Reaction Times for Three Drugs**

Subject	Drug A	Rank	Drug B	Rank	Drug C	Rank
1	1.21	1	1.48	2	1.56	3
2	1.63	1	1.85	2	2.01	3
3	1.42	1	2.06	3	1.70	2
4	2.43	2	1.98	1	2.64	3
5	1.16	1	1.27	2	1.48	3
6	1.94	1	2.44	2	2.81	3
	$R_1 = 7$		$R_2 = 12$		$R_3 = 17$	

location of a set of p populations.* Like other nonparametric tests, it requires no assumptions concerning the nature of the populations other than the capacity of individual observations to be ranked.

In Section 15.2, we gave an example in which a completely randomized design was used to compare the reaction times of subjects under the influence of one of two drugs. When the effect of a drug is short-lived (there is no carryover effect) and when the drug effect varies greatly from person to person, it may be useful to employ a *randomized block design.* Using the subjects as blocks, we would hope to eliminate the variability among subjects and thereby increase the amount of information in the experiment. Suppose that three drugs, A, B, and C, are to be compared using a randomized block design. Each of the three drugs is administered to the *same subject* with suitable time lags between the three doses. The order in which the drugs are administered is randomly determined for each subject. Thus, one drug would be administered to a subject and its reaction time would be noted; then after a sufficient length of time, the second drug administered; etc.

Suppose six subjects are chosen and that the reaction times for each drug are as shown in Table 15.9. To compare the three drugs, we rank the observations within each subject (block) and then compute the rank sums for each of the drugs (treatments). Tied observations within blocks are handled in the usual manner by assigning the average value of the ranks to each of the tied observations.

The null and alternative hypotheses are

H_0: The populations of reaction times are identically distributed for all three drugs

H_a: At least two of the drugs have probability distributions of reaction times that differ in location

The **Friedman F_r-test statistic**, which is based on the rank sums for each treatment, is

$$F_r = \frac{12}{bp(p+1)}\sum R_j^2 - 3b(p+1)$$

where b is the number of blocks, p is the number of treatments, and R_j is the jth rank sum. For the data in Table 15.9,

$$F_r = \frac{12}{(6)(3)(4)}\left[(7)^2 + (12)^2 + (17)^2\right] - 3(6)(4) = 80.33 - 72 = 8.33$$

*The Friedman F_r-test was developed by the Nobel prize–winning economist Milton Friedman.

The Friedman F_r statistic measures the extent to which the p samples differ with respect to their relative ranks within the blocks. This is more easily seen by writing F_r in an alternative, but equivalent, form:

$$F_r = \frac{12}{bp(p+1)}\sum b(\overline{R}_j - \overline{R})^2$$

where $\overline{R}_j$ is the mean rank corresponding to treatment j and $\overline{R}$ is the mean of all the ranks (i.e., $\overline{R} = \frac{1}{2}(p+1)$]. Thus, the F_r statistic is 0 if all treatments have the same mean rank and becomes increasingly large as the distance between the sample mean ranks grows.

As for the Kruskal–Wallis H statistic, the Friedman F_r statistic has approximately a χ^2 sampling distribution with $(p-1)$ degrees of freedom. Empirical results show the approximation to be adequate if either b (the number of blocks) or p (the number of treatments) exceeds 5. The Friedman F_r-test for a randomized block design is summarized in the next box.

Friedman F_r-Test for a Randomized Block Design

H_0: The probability distributions for the p treatments are identical

H_a: At least two of the probability distributions differ in location

Test statistic: $F_r = \dfrac{12}{bp(p+1)}\sum R_j^2 - 3b(p+1)$

where

     b = Number of blocks

     p = Number of treatments

     R_j = Rank sum of the jth treatment, where the rank of each measurement is computed relative to its position *within its own block*

Rejection region: $F_r > \chi_\alpha^2$ with $(p-1)$ degrees of freedom

Assumptions: 1. The treatments are randomly assigned to experimental units within the blocks.
           2. The measurements can be ranked within blocks.
           3. The p probability distributions from which the samples within each block are drawn are continuous.

Ties: Assign tied measurements within a block the average of the ranks they would receive if they were unequal but occurred in successive order. For example, if the third-ranked and fourth-ranked measurements are tied, assign each a rank of $(3+4)/2 = 3.5$. The number of ties should be small relative to the total number of observations.

For the drug example, we will use $\alpha = .05$ to form the rejection region:

$$F_r > \chi_{.05}^2 = 5.99147 \qquad \text{(see Figure 15.14)}$$

where $\chi_{.05}^2$ is based on $(p-1) = 2$ degrees of freedom. Consequently, because the observed value, $F_r = 8.33$, exceeds 5.99147, we conclude that at least two of the three drugs have probability distributions of reaction times that differ in location.

An SPSS printout of the nonparametric analysis, shown in Figure 15.15, confirms our inference. Both the test statistic and p-value are shaded on the printout. Since $p = .0155$ is less than our selected $\alpha = .05$, there is evidence to reject H_0.

FIGURE 15.14

Rejection region for reaction time example

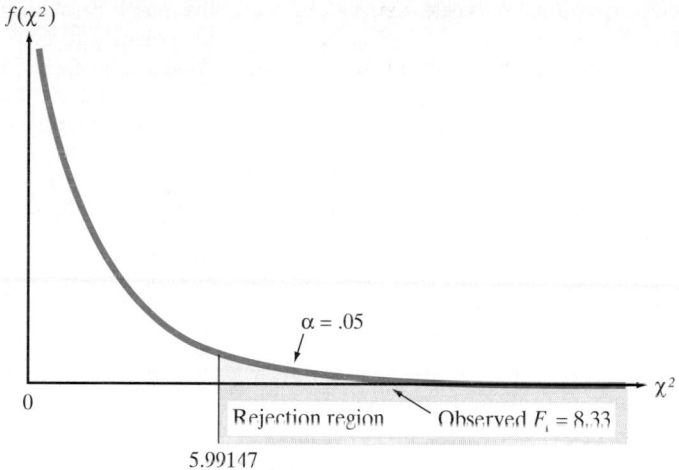

FIGURE 15.15

SPSS Friedman printout

```
- - - - - Friedman Two-way ANOVA

     Mean Rank    Variable

        1.17      DRUGA
        2.00      DRUGB
        2.03      DRUGC

        Cases       Chi-Square        D.F.      Significance
          6           8.3333            2           .0155
```

Clearly, the assumptions for this test—that the measurements are ranked within blocks and that the number of blocks (subjects) is greater than 5—are satisfied. However, we must be sure that the treatments are randomly assigned to blocks. For the procedure to be valid, we assume that the three drugs are administered in a random order to each subject. If this were not true, the difference in the reaction times for the three drugs might be due to the order in which the drugs are given.

EXERCISES 15.47–15.55

Learning the Mechanics

15.47 Data were collected using a randomized block design with four treatments (A, B, C, and D) and $b = 6$. The following rank sums were obtained:

$$R_A = 11 \quad R_B = 21 \quad R_C = 21 \quad R_D = 7$$

a. How many blocks were used in the experimental design?

b. Specify the null and alternative hypotheses that should be used in conducting a hypothesis test to determine whether the probability distributions for at least two of the treatments differ in location.

c. Conduct the test of part **b**. Use $\alpha = .10$.

d. What is the approximate p-value of the test of part **c**?

e. Calculate the mean rank for each of the four treatments and compute the value of the F_r-test statistic according to the formula that utilizes those means. Verify that the test statistic is the same as that you obtained in part **c**.

15.48 Suppose you have used a randomized block design to help you compare the effectiveness of three different treatments, A, B, and C. You obtained the data given in the following table and plan to conduct a Friedman F_r-test.

a. Specify the null and alternative hypotheses you will test.

b. Specify the rejection region for the test. Use α = .10.

c. Conduct the test and interpret the results.

TREATMENT

Block	A	B	C
1	9	11	18
2	13	13	13
3	11	12	12
4	10	15	16
5	9	8	10
6	14	12	16
7	10	12	15

15.49 An experiment was conducted using a randomized block design with four treatments and six blocks. The ranks of the measurements within each block are shown in the table. Use the Friedman F_r-test for a randomized block design to determine whether the data provide sufficient evidence to indicate that at least two of the treatment probability distributions differ in location. Test using α = .05.

BLOCK

Treatment	1	2	3	4	5	6
1	3	3	2	3	2	3
2	1	1	1	2	1	1
3	4	4	3	4	4	4
4	2	2	4	1	3	2

Applying the Concepts

15.50 Refer to the Kansas State study designed to investigate the effects of plants on human stress levels, Exercise 10.45. The data are given as finger temperatures for each of ten students in a dimly lit room under three experimental conditions: presence of a live plant, presence of a plant photo, and absence of a plant (either live or photo; see table below). Analyze the data using a nonparametric procedure. Do the students' finger temperatures depend on the experimental condition?

Student	Live Plant	Plant Photo	No Plant (control)
1	91.4	93.5	96.6
2	94.9	96.6	90.5
3	97.0	95.8	95.4
4	93.7	96.2	96.7
5	96.0	96.6	93.5
6	96.7	95.5	94.8
7	95.2	94.6	95.7
8	96.0	97.2	96.2
9	95.6	94.8	96.0
10	95.6	92.6	96.6

Source: Elizabeth Schreiber, Department of Statistics, Kansas State University, Manhattan, Kansas

15.51 Refer to the *Journal of Geography*'s published rankings of regional atlas theme topics, Exercise 15.31. In addition to high school teachers and university geography alumni, university geography students and representatives of the general public also ranked the 12 thematic topics. The rankings of all four groups are shown in the table below. Use the SPSS printout on page 749 to compare (at α = .01) the atlas theme ranking distributions of the four groups. Give the conclusion in the words of the problem.

15.52 An *optical mark reader* (OMR) is a machine that is able to "read" pencil marks entered on a scannable form. OMRs are used by schools to grade exams

Theme	RANKINGS			
	High School Teachers	**Geography Alumni**	**Geography Students**	**General Public**
Tourism	10	2	5	1
Physical	2	1	1	5
Transportation	7	3	7	2
People	1	6	2	3
History	2	5	9	4
Climate	6	4	4	8
Forestry	5	8	2	7
Agriculture	7	10	6	9
Fishing	9	7	10	6
Energy	2	8	7	10
Mining	10	11	11	11
Manufacturing	12	12	12	12

Source: Keller, C. P., *et al.* "Planning the next generation of regional atlases: Input from educators." *Journal of Geography,* Vol. 94, No. 3, May/June 1995, p. 413 (Table 1).

```
- - - - - Friedman Two-way ANOVA

    Mean Rank    Variable

        2.25     HSTEACH
        2.71     GEOALUM
        2.42     GEOSTUD
        2.63     GENPUB

        Cases         Chi-Square        D.F.    Significance
         12              .9250            3         .8194
```

and by survey research organizations to compile data from questionnaires. A manufacturer of OMRs believes its product can operate equally well in a variety of temperature and humidity environments. To determine whether operating data contradict this belief, the manufacturer asks a well-known industrial testing laboratory to test its product. Five recently manufactured OMRs were randomly selected and each was operated in five different environments. The number of forms each was able to process in an hour was recorded and used as a measure of the OMR's operating efficiency. These data are given in the following table. Use the Friedman F_r-test to determine whether evidence exists to indicate that the probability distributions for the number of forms processed per hour differ in location for at least two of the environments. Test using $\alpha = .10$.

ENVIRONMENT

Machine Number	1	2	3	4	5
1	8,001	8,025	8,100	8,055	7,991
2	7,910	7,932	7,900	7,990	7,892
3	8,111	8,101	8,201	8,175	8,102
4	7,802	7,820	7,904	7,850	7,819
5	7,500	7,601	7,702	7,633	7,600

15.53 Suppose we want to compare the mean flood crests on the Suwannee and Santa Fe rivers for the flood level years 1984, 1983, 1973, and 1948. The data (in feet), matched by year at six locations on the rivers, are shown in the table below. Do the data indicate that the flood crests on the rivers were higher or lower in any one of the flood level years than any other? Test using the Friedman F_r-test with $\alpha = .05$.

15.54 Corrosion of different metals is a problem in many mechanical devices. Three sealers used to help retard the corrosion of metals were tested to see whether there were any differences among them. Samples of 10 different metal compositions were treated with each of the three sealers, and the amount of corrosion was measured after exposure to the same environmental conditions for 1 month. The data are given in the following table. Is there any evidence of a difference in the probability distributions of the amounts of corrosion among the three types of sealer? Use $\alpha = .05$.

SEALER

Metal	1	2	3
1	4.6	4.2	4.9
2	7.2	6.4	7.0
3	3.4	3.5	3.4
4	6.2	5.3	5.9
5	8.4	6.8	7.8
6	5.6	4.8	5.7
7	3.7	3.7	4.1
8	6.1	6.2	6.4
9	4.9	4.1	4.2
10	5.2	5.0	5.1

15.55 A serious drought-related problem for farmers is the spread of aflatoxin, a highly toxic substance

Station	Flood Stage	1984	1983	1973	1948
Suwannee River					
White Springs	77.0	79.9	74.4	88.5	85.2
Ellaville	54.0	59.8	52.8	65.0	68.1
Branford	29.0	29.3	29.2	35.6	38.9
Wilcox	14.0	11.6	13.0	18.6	22.3
Santa Fe River					
Three Rivers	19.0	24.1	24.3	27.2	30.6
U.S. 129 Bridge	21.0	23.8	23.8	30.7	34.2

Source: *Gainesville Sun,* March 15, 1984.

generated by mold, which contaminates field corn. In higher levels of contamination, aflatoxin is potentially hazardous to animal and possibly human health. (Officials of the Food and Drug Administration have set a maximum limit of 20 parts per billion aflatoxin as safe for interstate marketing). Three sprays, A, B, and C, have been developed to control aflatoxin in field corn. To determine whether differences exist among the sprays, 10 ears of corn are randomly chosen from a contaminated corn field and each is divided into three pieces of equal size. The sprays are then randomly assigned to the pieces for each ear of corn, thus setting up a randomized block design. The following table gives the amount (in parts per billion) of aflatoxin present in the corn samples after spraying.

a. Use the Friedman F_r-test to determine whether there is evidence that the distributions of the levels of aflatoxin in corn differ for at least two of the three sprays. Test at $\alpha = .05$.

b. Do the results of the test warrant further comparisons of the pairs of sprays? If so, conduct the comparison of the pairs of sprays using the Wilcoxon signed rank test and $\alpha = .01$ for each comparison. Can you conclude that one spray appears to best control the level of aflatoxin?

c. What assumptions are necessary to ensure the validity of the procedures used in parts **a** and **b**?

	SPRAY		
Ear	A	B	C
1	21	23	15
2	29	30	21
3	16	19	18
4	20	19	18
5	13	10	14
6	5	12	6
7	18	18	12
8	26	32	21
9	17	20	9
10	4	10	2

15.6 SPEARMAN'S RANK CORRELATION COEFFICIENT

Suppose 10 new paintings are shown to two art critics and each critic ranks the paintings from 1 (best) to 10 (worst). We want to determine whether the critics' ranks are related. Does a correspondence exist between their ratings? If a painting is ranked high by critic 1, is it likely to be ranked high by critic 2? Or do high rankings by one critic correspond to low rankings by the other? That is, are the rankings of the critics **correlated**?

If the rankings are as shown in the "Perfect Agreement" columns of Table 15.10, we immediately notice that the critics agree on the rank of every painting.

TABLE 15.10 **Rankings of 10 Paintings by Two Critics**

	PERFECT AGREEMENT		PERFECT DISAGREEMENT	
Painting	Critic 1	Critic 2	Critic 1	Critic 2
1	4	4	9	2
2	1	1	3	8
3	7	7	5	6
4	5	5	1	10
5	2	2	2	9
6	6	6	10	1
7	8	8	6	5
8	3	3	4	7
9	10	10	8	3
10	9	9	7	4

High ranks correspond to high ranks and low ranks to low ranks. This is an example of **perfect positive correlation** between the ranks. In contrast, if the rankings appear as shown in the "Perfect Disagreement" columns of Table 15.10, high ranks for one critic correspond to low ranks for the other. This is an example of **perfect negative correlation**.

In practice, you will rarely see perfect positive or negative correlation between the ranks. In fact, it is quite possible for the critics' ranks to appear as shown in Table 15.11. You will note that these rankings indicate some agreement between the critics, but not perfect agreement, thus indicating a need for a measure of rank correlation.

Spearman's rank correlation coefficient, r_s, provides a measure of correlation between ranks. The formula for this measure of correlation is given in the next box. We also give a formula that is identical to r_s when there are no ties in rankings; this provides a good approximation to r_0 when the number of ties is small relative to the number of pairs.

Note that if the ranks for the two critics are identical, as in the second and third columns of Table 15.10, the differences between the ranks, *d,* will all be 0. Thus,

$$r_s = 1 - \frac{6\sum d^2}{n(n^2 - 1)} = 1 - \frac{6(0)}{10(99)} = 1$$

That is, *perfect positive correlation* between the pairs of ranks is characterized by a Spearman correlation coefficient of $r_s = 1$. When the ranks indicate perfect disagreement, as in the fourth and fifth columns of Table 15.10, $\sum d_i^2 = 330$ and

$$r_s = 1 - \frac{6(330)}{10(99)} = -1$$

Thus, *perfect negative correlation* is indicated by $r_s = -1$.

TABLE 15.11   Rankings of Paintings: Less Than Perfect Agreement

Painting	CRITIC 1	2	DIFFERENCE BETWEEN RANK 1 AND RANK 2 d	d^2
1	4	5	−1	1
2	1	2	−1	1
3	9	10	−1	1
4	5	6	−1	1
5	2	1	1	1
6	10	9	1	1
7	7	7	0	0
8	3	3	0	0
9	6	4	2	4
10	8	8	0	0
				$\sum d^2 = 10$

Spearman's Rank Correlation Coefficient

$$r_s = \frac{SS_{uv}}{\sqrt{SS_{uu}SS_{vv}}}$$

where

$$SS_{uv} = \sum(u_i - \bar{u})(v_i - \bar{v}) = \sum u_i v_i - \frac{(\sum u_i)(\sum v_i)}{n}$$

$$SS_{uu} = \sum(u_i - \bar{u})^2 = \sum u_i^2 - \frac{(\sum u_i)^2}{n}$$

$$SS_{vv} = \sum(v_i - \bar{v})^2 = \sum v_i^2 - \frac{(\sum v_i)^2}{n}$$

u_i = Rank of the ith observation in sample 1

v_i = Rank of the ith observation in sample 2

n = Number of pairs of observations (number of observations in each sample)

Shortcut Formula for r_s*

$$r_s = 1 - \frac{6\sum d_i^2}{n(n^2 - 1)}$$

where

$d_i = u_i - v_i$ (difference in the ranks of the ith observations for samples 1 and 2)

For the data of Table 15.11,

$$r_s = 1 - \frac{6\sum d^2}{n(n^2 - 1)} = 1 - \frac{6(10)}{10(99)} = 1 - \frac{6}{99} = .94$$

The fact that r_s is *close* to 1 indicates that the critics tend to agree, but the agreement is not perfect.

The value of r_s *always falls between* −1 *and* +1, *with* +1 *indicating perfect positive correlation and* −1 *indicating perfect negative correlation.* The closer r_s falls to +1 or −1, the greater the correlation between the ranks. Conversely, the nearer r_s is to 0, the less the correlation.

Note that the concept of correlation implies that two responses are obtained for each experimental unit. In the art critics example, each painting received two ranks (one for each critic) and the objective of the study was to determine the degree of positive correlation between the two rankings. Rank correlation methods can be used to measure the correlation between any pair of variables. If two variables are measured on each of n experimental units, we rank the measurements associated with each variable separately. Ties receive the average of the ranks of the tied observations. Then we calculate the value of r_s for the two rankings. This value measures the rank correlation between the two variables. We illustrate the procedure in Example 15.4.

EXAMPLE 15.4

A study is conducted to investigate the relationship between cigarette smoking during pregnancy and the weights of newborn infants. A sample of 15 women

*The shortcut formula is not exact when there are tied measurements, but it is a good approximation when the total number of ties is not large relative to n.

TABLE 15.12   Data and Calculations for Example 15.4

Woman	Cigarettes per Day	Rank	Baby's Weight (pounds)	Rank	d	d^2
1	12	1	7.7	5	−4	16
2	15	2	8.1	9	−7	49
3	35	13	6.9	4	9	81
4	21	7	8.2	10	−3	9
5	20	5.5	8.6	13.5	−8	64
6	17	3	8.3	11.5	−8.5	72.25
7	19	4	9.4	15	−11	121
8	46	15	7.8	6	9	81
9	20	5.5	8.3	11.5	−6	36
10	25	8.5	5.2	1	7.5	56.25
11	39	14	6.4	3	11	121
12	25	8.5	7.9	7	1.5	2.25
13	30	12	8.0	8	4	16
14	27	10	6.1	2	8	64
15	29	11	8.6	13.5	−2.5	6.25
					Total = 795	

smokers kept accurate records of the number of cigarettes smoked during their pregnancies, and the weights of their children were recorded at birth. The data are given in Table 15.12. Calculate and interpret Spearman's rank correlation coefficient for the data.

Solution

We first rank the number of cigarettes smoked per day, assigning a 1 to the smallest number (12) and a 15 to the largest (46). Note that the two ties receive the averages of their respective ranks. Similarly, we assign ranks to the 15 babies' weights. Since the number of ties is relatively small, we will use the shortcut formula to calculate r_s. The differences d between the ranks of the babies' weights and the ranks of the number of cigarettes smoked per day are shown in Table 15.12. The squares of the differences, d^2, are also given. Thus,

$$r_s = 1 - \frac{6\sum d_i^2}{n(n^2 - 1)} = 1 - \frac{6(795)}{15(15^2 - 1)} = 1 - 1.42 = -.42$$

The value of r_s can also be obtained using a computer. An ASP printout of the analysis is shown in Figure 15.16. The value of r_s, shaded on the printout, agrees (except for rounding) with our hand-calculated value of $-.42$.

This negative correlation coefficient indicates that in this sample an increase in the number of cigarettes smoked per day is *associated with* (but is not necessarily the *cause of*) a decrease in the weight of the newborn infant. Can this conclusion be generalized from the sample to the population? That is, can we conclude that

FIGURE 15.16

ASP printout for
Example 15.4

```
                        RANK ORDER CORRELATION MATRIX

                   Cigarettes  Weight
                   ----------  -------
        Cigarettes      1      0.4247
          Weight    -0.4247    1
```

TABLE 15.13 **Reproduction of Part of Table XIV in Appendix A: Critical Values of Spearman's Rank Correlation Coefficient**

n	$\alpha = .05$	$\alpha = .025$	$\alpha = .01$	$\alpha = .005$
5	.900	—	—	—
6	.829	.886	.943	—
7	.714	.786	.893	—
8	.643	.738	.833	.881
9	.600	.683	.783	.833
10	.564	.648	.745	.794
11	.523	.623	.736	.818
12	.497	.591	.703	.780
13	.475	.566	.673	.745
14	.457	.545	.646	.716
15	.441	.525	.623	.689
16	.425	.507	.601	.666
17	.412	.490	.582	.645
18	.399	.476	.564	.625
19	.388	.462	.549	.608
20	.377	.450	.534	.591

weights of newborns and the number of cigarettes smoked per day are negatively correlated for the populations of observations for *all* smoking mothers?

If we define ρ as the **population rank correlation coefficient** [i.e., the rank correlation coefficient that could be calculated from all (x, y) values in the population], this question can be answered by conducting the test

H_0: $\rho = 0$ (no population correlation between ranks)

H_a: $\rho < 0$ (negative population correlation between ranks)

Test statistic: r_s (the *sample* Spearman rank correlation coefficient)

To determine a rejection region, we consult Table XIV in Appendix A, which is partially reproduced in Table 15.13. Note that the left-hand column gives values of n, the number of pairs of observations. The entries in the table are values for an upper-tail rejection region, since only positive values are given. Thus, for $n = 15$ and $\alpha = .05$, the value .441 is the boundary of the upper-tailed rejection region, so that $P(r_s > .441) = .05$ if H_0: $\rho = 0$ is true. Similarly, for negative values of r_s, we have $P(r_s < -.441) = .05$ if $\rho = 0$. That is, we expect to see $r_s < -.441$ only 5% of the time if there is really no relationship between the ranks of the variables. The lower-tailed rejection region is therefore

Rejection region ($\alpha = .05$): $r_s < -.441$

Since the calculated $r_s = -.42$ is not less than $-.441$, we cannot reject H_0 at the $\alpha = .05$ level of significance. That is, this sample of 15 smoking mothers provides insufficient evidence to conclude that a negative correlation exists between number of cigarettes smoked and the weight of newborns for the populations of measurements corresponding to all smoking mothers. This does not, of course, mean that no relationship exists. A study using a larger sample of smokers and taking other factors into account (father's weight, sex of newborn child, etc.) would be more likely to reveal whether smoking and the weight of a newborn child are related. ▲

A summary of Spearman's nonparametric test for correlation is given in the next box.

Spearman's Nonparametric Test for Rank Correlation

ONE-TAILED TEST	TWO-TAILED TEST
H_0: $\rho = 0$	H_0: $\rho = 0$
H_a: $\rho > 0$	H_a: $\rho \neq 0$
(or H_a: $\rho < 0$)	

Test statistic: r_s, the sample rank correlation (see the formulas for calculating r_s)

Rejection region: $r_s > r_{s,\alpha}$	*Rejection region:* $\mid r_s \mid > r_{s,\alpha/2}$
(or $r_s < -r_{s,\alpha}$ when H_a: $\rho_s < 0$)	

where $r_{s,\alpha}$ is the value from Table XIV corresponding to the upper-tail area α and n pairs of observations

where $r_{s,\alpha/2}$ is the value from Table XIV corresponding to the upper tail area $\alpha/2$ and n pairs of observations

Assumptions: 1. The sample of experimental units on which the two variables are measured is randomly selected.
2. The probability distributions of the two variables are continuous.

Ties: Assign tied measurements the average of the ranks they would receive if they were unequal but occurred in successive order. For example, if the third-ranked and fourth ranked measurements are tied, assign each a rank of $(3 + 4)/2 = 3.5$. The number of ties should be small relative to the total number of observations.

EXERCISES 15.56–15.66

Learning the Mechanics

15.56 Use Table XIV of Appendix A to find each of the following probabilities:
a. $P(r_s > .508)$ when $n = 22$
b. $P(r_s > .448)$ when $n = 28$
c. $P(r_s \leq .648)$ when $n = 10$
d. $P(r_s < -.738$ or $r_s > .738)$ when $n = 8$

15.57 Specify the rejection region for Spearman's nonparametric test for rank correlation in each of the following situations:
a. H_0: $\rho = 0$; H_a: $\rho \neq 0, n = 10, \alpha = .05$
b. H_0: $\rho = 0$; H_a: $\rho > 0, n = 20, \alpha = .025$
c. H_0: $\rho = 0$; H_a: $\rho < 0, n = 30, \alpha = .01$

15.58 Compute Spearman's rank correlation coefficient for each of the following pairs of sample observations:

a.

x	33	61	20	19	40
y	26	36	65	25	35

b.

x	89	102	120	137	41
y	81	94	75	52	136

c.

x	2	15	4	10
y	11	2	15	21

d.

x	5	20	15	10	3
y	80	83	91	82	87

15.59 The following sample data were collected on variables x and y:

x	0	3	0	−4	3	0	4
y	0	2	2	0	3	1	2

a. Specify the null and alternative hypotheses that should be used in conducting a hypothesis test to determine whether the variables x and y are correlated.
b. Conduct the test of part **a** using $\alpha = .05$.
c. What is the approximate p-value of the test of part **b**?
d. What assumptions are necessary to ensure the validity of the test of part **b**?

Applying the Concepts

15.60 The Milan Overall Dementia Assessment (MODA) is a neuropsychologically oriented test that provides an overall age- and education-adjusted measure of an individual's cognitive deterioration. MODA scores range from 0 to 100, with higher scores indicating a greater degree of deterioration. A team of psychologists and neurologists administered the MODA to a sample of 30 patients with Alzheimer's disease (*Neuropsychologia,* June 1995). In addition, all patients were given a "face matching" test in which they were requested to match photographs of unknown faces. Scores on this test were recorded as the number of matching errors, with a maximum score of 27.

a. The researchers reported Spearman's rank correlation between the MODA and face-matching scores as $r_s = .48$. Interpret this result.

b. Test the hypothesis of a positive correlation between MODA score and face-matching score in Alzheimer's patients. Use $\alpha = .05$.

15.61 Two expert wine tasters were asked to rank six brands of wine. Their rankings are shown below. Do the data present sufficient evidence to indicate a positive correlation in the rankings of the two experts?

Brand	Expert 1	Expert 2
A	6	5
B	5	6
C	1	2
D	3	1
E	2	4
F	4	3

15.62 The metabolic and cardiopulmonary responses during maximal effort exercise in persons with multiple sclerosis (MS) was studied (*Clinical Kinesiology,* Spring 1995). The following variables were measured for each of 10 MS patients:

1. Expanded Disability Status Scale (EDSS)—Ratings range from 1 to 4.5
2. Peak oxygen uptake (liters per minute) during an arm cranking exercise (ARM)
3. Peak oxygen uptake (liters per minute) during a leg cycling exercise (LEG)
4. Peak oxygen uptake (liters per minute) during a combined leg cycling and arm cranking exercise (LEG/ARM)

Spearman Rank Correlation Coefficients

	EDSS	ARM	LEG/ARM	LEG
EDSS	1.000	−.439	−.655	−.512
ARM		1.000	.634	.722
LEG/ARM			1.000	.890
LEG				1.000

Source: Ponichtera-Mulcare, J. A., *et al.* "Maximal aerobic exercise of individuals with multiple sclerosis using three modes of energy." *Clinical Kinesiology,* Vol. 49, No. 1, Spring 1995, p. 10 (Table 2).

Spearman's rank correlation was calculated for each pair of variables. The accompanying table gives the correlation r_s between the variables in the corresponding row and column.

a. Explain why the table shows rank correlations of 1.000 in the diagonal.

b. Practically interpret each of the other rank correlations in the table.

c. Is there sufficient evidence (at $\alpha = .01$) to conclude that EDSS and peak oxygen uptake are negatively correlated? Perform the test for each of the three exercises.

15.63 It is frequently conjectured that income is one of the primary determinants of social status for an adult male. To investigate this theory, 15 adult males are chosen at random from a community comprising primarily professional people, and their annual gross incomes are noted. Each subject is then asked to complete a questionnaire designed to measure social status within the community. The social status scores (higher scores correspond to higher social status) and gross incomes (in thousands of dollars) are given in the following table for the 15 adult males.

Subject	Social Status	Income
1	92	29.9
2	51	18.7
3	88	32.0
4	65	15.0
5	80	26.0
6	31	9.0
7	38	11.3
8	75	22.1
9	45	16.0
10	72	25.0
11	53	17.2
12	43	9.7
13	87	20.1
14	30	15.5
15	74	16.5

a. Compute Spearman's rank correlation coefficient for these data.

b. Is there evidence that social status and income are positively correlated? Use $\alpha = .05$.

15.64 An experiment was designed to study whether eye pupil size is related to a person's attempt at deception. Eight students were asked to respond verbally to a series of questions. Before the questioning began, the pupil size of each student was noted and the students were instructed to answer some of the questions dishonestly. (The number of questions answered dishonestly was left to individual choice.) During questioning, the percentage increase in pupil size was recorded. Each student was then given a deception score based on the proportion of questions answered dishonestly. (High scores indicate a large number of deceptive responses.) The results are given in the table below. Can you conclude that the percentage increase in eye pupil size is positively correlated with deception score? Use $\alpha = .05$.

Student	Deception Score	Percentage Increase in Pupil Size
1	87	10
2	63	6
3	95	11
4	50	7
5	43	0
6	89	15
7	33	4
8	55	5

15.65 Recreation therapy is currently being used for the treatment of the mentally retarded. Particularly, psychologists theorize that certain types of recreation have a tranquilizing effect on highly excitable patients. In one experiment, nine mentally retarded patients were monitored for 1 week, and the total number of hours each spent in a specially designed recreation room was recorded. The patients were permitted to spend as much time as they wanted in the room. Also recorded was the number of times during the week that some type of tranquilizing medication had to be given to a patient in a highly excited state. Using the data in the table (right), determine whether there is sufficient evidence to indicate that recreation is negatively correlated with the number of times a tranquilizer has to be given. Test using $\alpha = .05$.

Patient	Recreation Time (hours)	Tranquilizers
1	16	3
2	22	1
3	10	4
4	8	9
5	14	5
6	34	2
7	26	0
8	13	10
9	5	9

15.66 Health maintenance organizations (HMOs) provide and monitor a variety of short-term outpatient services, including crisis intervention mental health services. To investigate the standards used by HMOs in interpreting crisis intervention, two researchers (Cheifetz and Salloway, 1985) conducted a comprehensive questionnaire survey of 145 national HMOs. Each HMO was asked to "write a brief, descriptive definition of situations or states you would include as qualifying for 'crisis intervention.'" The researchers sorted these situations into 10 categories and then asked three experienced clinicians to rate each category for two criteria: validity of crisis intervention (i.e., is the situation defined really a "crisis") and clarity of guidelines for offering service. A 4-point rating scale was provided for both criteria. The mean ratings for the 10 categories on both the crisis intervention and clarity scales are given in the following table. Is there evidence of a positive relationship between the mean crisis intervention and mean clarity ratings? Test using $\alpha = .05$.

Category (Situation)	Crisis Intervention Rating (1 = definitely a crisis, 4 = definitely not a crisis)	Clarity Rating (1 = very clear guideline, 4 = very unclear guideline)
Psychosis	1.31	1.33
Drug/alcohol abuse	1.33	1.29
Depression/anxiety	1.48	1.59
Emphasis on acuteness	1.76	2.50
Insistence on "short-term" response	2.48	3.22
Suicide	1.13	1.32
Family problems	2.59	2.30
Violence/harm	1.06	1.86
Miscellaneous	2.60	2.33
Nondefinition	3.57	3.57

Source: Cheifetz, D. I. and Salloway, J. C. "Crisis intervention: Interpretation and practice by HMO." *Medical Care,* Vol. 23, No. 1, Jan. 1985, pp. 89–93.

QUICK REVIEW

Key Terms

Correlation   750
Friedman F_r statistic   745
Kruskal–Wallis H statistic   742
Nonparametric tests   716
Population correlation coefficient   754
Rank statistics   716
Rank sum   723
Relative ranks   716

Shift in probability distributions   725
Sign test   717
Spearman's nonparametric test for
   rank correlation   755
Spearman's rank correlation
   coefficient   751
Wilcoxon rank sum test   722
Wilcoxon signed rank test   732

Key Formulas

Test	Test Statistic	Large Sample Approximation
Sign	S = number of sample measurements greater than (or less than) hypothesized mean, η_0	$z = \dfrac{(S - .5) - .5n}{.5\sqrt{n}}$
Wilcoxon rank sum	T_A = rank sum of sample A or T_B = rank sum of sample B	$z = \dfrac{T_A - \dfrac{n_1(n_1 + n_2 + 1)}{2}}{\sqrt{\dfrac{n_1 n_2(n_1 + n_2 + 1)}{12}}}$
Wilcoxon signed ranks	T_- = negative rank sum or T_+ = positive rank sum	$z = \dfrac{T_+ - \dfrac{n(n + 1)}{4}}{\sqrt{\dfrac{n(n + 1)(2n + 1)}{24}}}$
Kruskal–Wallis	$H = \dfrac{12}{n(n + 1)}\sum\dfrac{R_j^2}{n_j} - 3(n + 1)$	
Friedman	$F_r = \dfrac{12}{bp(p + 1)}\sum R_j^2 - 3b(p + 1)$	
Spearman rank correlation (shortcut formula)	$r_s = 1 - \dfrac{6\sum d_i^2}{n(n^2 - 1)}$ where d_i = difference in ranks of ith observation for samples 1 and 2	

LANGUAGE LAB

Symbol	Description
η (eta)	Population median
S	Test statistic for sign test (see Key Formulas)
T_A	Sum of ranks of observations in sample A
T_B	Sum of ranks of observations in sample B
T_L	Critical lower Wilcoxon rank sum value
T_U	Critical upper Wilcoxon rank sum value
T_+	Sum of ranks of positive differences of paired observations
T_-	Sum of ranks of negative differences of paired observations
T_0	Critical value of Wilcoxon signed rank test
R_j	Rank sum of observations in sample j

H	Test statistic for Kruskal–Wallis test (see Key Formulas)
F_r	Test statistic for Friedman test (see Key Formulas)
r_s	Spearman's rank correlation coefficient (see Key Formulas)
ρ (rho)	Population correlation coefficient

SUPPLEMENTARY EXERCISES 15.67–15.91

Learning the Mechanics

15.67 When is it appropriate to use the *t*- and *F*-tests of Chapters 9 and 10 for comparing two or more population means?

15.68 The data for three independent random samples are shown in the table below. It is known that the sampled populations are not normally distributed. Use an appropriate test to determine whether the data provide sufficient evidence to indicate that at least two of the populations differ in location. Test using $\alpha = .05$.

Sample from Population 1		Sample from Population 2		Sample from Population 3	
18	15	12	34	87	50
32	63	33	18	53	64
43		10		65	77

15.69 A random sample of nine pairs of observations are recorded on two variables, *x* and *y*. The data are shown in the following table.

Pair	*x*	*y*	Pair	*x*	*y*
1	19	12	6	29	10
2	27	19	7	16	16
3	15	7	8	22	10
4	35	25	9	16	18
5	13	11			

    a. Do the data provide sufficient evidence to indicate that ρ, the rank correlation between *x* and *y*, differs from 0? Test using $\alpha = .05$.

    b. Do the data provide sufficient evidence to indicate that the probability distribution for *x* is shifted to the right of that for *y*? Test using $\alpha = .05$.

15.70 Two independent random samples produced the measurements listed in the table below. Do the data provide sufficient evidence to conclude that there is a difference between the locations of the probability distributions for the sampled populations? Test using $\alpha = .05$.

Sample from Population 1		Sample from Population 2	
1.2	1.0	1.5	1.9
1.9	1.8	1.3	2.7
.7	1.1	2.9	3.5
2.5			

15.71 An experiment was conducted using a randomized block design with five treatments and four blocks.

The data are shown in the following table. Do the data provide sufficient evidence to conclude that at least two of the treatment probability distributions differ in location? Test using $\alpha = .05$.

Treatment	BLOCK			
	1	2	3	4
1	75	77	70	80
2	65	69	63	69
3	74	78	69	80
4	80	80	75	86
5	69	72	63	77

Applying the Concepts

15.72 A major razor blade manufacturer advertises that its twin-blade disposable razor will "get you more shaves" than any single-blade disposable razor on the market. A rival blade company that has been very successful in selling single-blade razors wishes to test this claim. Marketing managers in this company randomly sampled eight single-blade shavers and eight twin-blade shavers, and counted the number of shaves that each got before a change of blades was indicated. The results are shown in the table below.

Twin Blades		Single Blade	
8	15	10	13
17	10	6	14
9	6	3	5
11	12	7	7

    a. Do the data support the twin-blade manufacturer's claim? Use $\alpha = .05$.

    b. Do you think this experiment was designed in the best possible way? If not, what design might have been better?

    c. What assumptions are necessary for the test performed in part **a** to be valid? Do the assumptions seem reasonable for this application?

15.73 An experiment was conducted to compare two print types, A and B, to determine whether type A is easier to read. Ten subjects were randomly divided into two groups of five. Each subject was given the same material to read, one group receiving the material printed with type A, the other group receiving print type B. The times necessary for

Person	Lawyer	Politician	Physician	Corporate President	College Professor
1	3	5	1	4	2
2	4	5	1	2	3
3	1	4	2	5	3
4	3	5	2	4	1
5	4	5	1	3	2
6	5	4	3	2	1
7	1	5	3	4	2
8	4	5	1	3	2
9	3	5	1	4	2
10	4	5	2	3	1
11	5	4	2	3	1
12	3	5	1	4	2
13	4	5	2	3	1
14	3	4	1	5	2
15	4	5	1	2	3

each subject to read the material (in seconds) are shown below:

Type A:	95	122	101	99	108
Type B:	110	102	115	112	120

Do the data provide sufficient evidence to indicate that print type A is easier to read? Test using $\alpha = .05$.

15.74 Refer to Exercise 15.73. Test the research hypothesis that the median of the type A probability distribution exceeds 100 seconds. Repeat the test for the type B probability distribution. Use $\alpha = .05$ for both tests.

15.75 A sociologist conducted an experiment to investigate the general public's perception of certain occupations. Each of a random sample of 15 people was asked to rank, in order of prestige, five leading professions: lawyer, politician, physician, corporate president, and college professor. Do the data shown in the table at the top of the page provide sufficient evidence to indicate a difference in the amount of prestige the public attaches to these five professions? Test using $\alpha = .025$.

15.76 The length of time required for a human subject to respond to a new painkiller was tested in the following manner. Seven randomly selected subjects were assigned to receive both aspirin and the new drug. The two treatments were spaced in time and assigned in random order. The length of time (in minutes) required for a subject to indicate that he or she could feel pain relief was recorded for both the aspirin and the drug. The data are shown in the next table. Do the data provide sufficient evidence to indicate that the probability distribution of the times required to obtain relief with aspirin is shifted to the right of the probability distribution of the times required to obtain relief with the new drug? Test using $\alpha = .05$.

Subject	Aspirin	New Drug
1	15	7
2	20	14
3	12	13
4	20	11
5	17	10
6	14	16
7	17	11

15.77 A 1974 Supreme Court decision (Milliken *v.* Bradley) held that desegregation could not extend beyond the boundary of the school systems that were found to be segregated. One dissenting justice feared that the decision would trigger "white flight," the migration of white families out of the inner cities to the suburbs. In a discussion of this decision, the *Journal of Legal Studies* (Vol. 5, 1976) gave the percentage change in number of white students in the school systems of 12 large cities over two time periods, 1968–1970 and 1970–1972 (see table at top of page 761). Was the percentage change in white student population larger in the time period 1970–1972 than for the comparable period, 1968–1970? Test using $\alpha = .05$.

15.78 A union wants to determine the preferences of its members before negotiating with management. Ten union members are randomly selected, and an extensive questionnaire is completed by each member. The responses to the various aspects of the questionnaire will enable the union to rank in order of importance the items to be negotiated. The rankings are shown in the table on page 761.

a. What type of experimental design was employed? Identify the key elements of the experiment: response, factor(s), factor type(s), treatments, and experimental units. [*Hint:* Be careful not to confuse the ranking of the response with the response itself.]

City	PERCENTAGE CHANGE IN WHITE STUDENTS[a]	
	1968–1970	1970–1972
Boston	−3.9	−7.4
Philadelphia	−5.7	−2.2
Baltimore	−5.6	−9.3
St. Louis	−9.4	−13.8
Chicago	−9.0	−14.7
Detroit	−15.6	−14.0
Atlanta	−22.3	−34.4
Charlotte	−3.1	−5.6
Jacksonville	−1.8	−11.4
Houston	−9.1	−17.5
San Francisco	−13.5	−22.4
San Diego	−1.1	−5.5

[a]Percentages based on beginning years.

Source: U.S. Dept. of Health, Education, and Welfare, Office for Civil Rights, Directory of Public Elementary and Secondary Schools in Selected Districts, Fall 1968, Fall 1970, and Fall 1972.

Person	More Pay	Job Stability	Fringe Benefits	Shorter Hours
1	2	1	3	4
2	1	2	3	4
3	4	3	2	1
4	1	4	2	3
5	1	2	3	4
6	1	3	4	2
7	2.5	1	2.5	4
8	3	1	4	2
9	1.5	1.5	3	4
10	2	3	1	4

b. Test to determine whether evidence exists that the probability distributions of ratings differ for at least two of the four negotiable items. Use $\alpha = .05$.

c. What assumptions are necessary to ensure the validity of the test?

d. Do the results of the test warrant further comparisons of the pairs of negotiable items? If so, compare all pairs of probability distributions using the Wilcoxon signed rank test. To determine the significance level of each test, use the technique of Section 10.3 and an overall $\alpha = .05$. Does one item appear to be most important?

15.79 A state highway patrol was interested in knowing whether frequent patrolling of highways substantially reduces the number of speeders. Two similar interstate highways were selected for the study—one heavily patrolled and the other only occasionally patrolled. After 1 month, random samples of 100 cars were chosen on each highway, and the number of cars exceeding the speed limit was recorded. This process was repeated on 5 randomly selected days. The data are shown in the following table.

Day	Highway 1 Heavily patrolled	Highway 2 Occasionally patrolled
1	35	60
2	40	36
3	25	48
4	38	54
5	47	63

a. Do the data provide evidence to indicate that the heavily patrolled highway tends to have fewer speeders per 100 cars than the occasionally patrolled highway? Test using $\alpha = .05$.

b. Use the paired t-test with $\alpha = .05$ to compare the population mean number of speeders per 100 cars for the two highways. What assumptions are necessary for this procedure to be valid?

15.80 A drug company has synthesized two new compounds to be used in sleeping pills. The data in the table below represent the additional hours of sleep gained by ten patients through the use of the two drugs. Do the data present sufficient evidence to indicate that the probability distributions of additional hours of sleep differ for the two drugs? Test using $\alpha = .10$.

Patient	Drug A	Drug B	Patient	Drug A	Drug B
1	.4	.7	6	2.9	3.4
2	−.7	−1.6	7	4.0	3.7
3	−.4	−.2	8	.1	.8
4	−1.4	−1.4	9	3.1	.0
5	−1.6	−.2	10	1.9	2.0

15.81 Weevils cause millions of dollars worth of damage each year to cotton crops. Three chemicals (A, B, and C) designed to control weevil populations were applied, one to each of three fields of cotton. After 3 months, 10 plots of equal size were randomly selected within each field and the percentage of cotton plants with weevil damage was recorded for each. Do the data in the accompanying table provide sufficient evidence to indicate a difference in location among the distributions of damage rates corresponding to the three treatments? Use $\alpha = .05$.

A		B		C	
10.8	9.8	22.3	20.4	9.8	10.8
15.6	16.7	19.5	23.6	12.3	12.2
19.2	19.0	18.6	21.2	16.2	17.3
17.9	20.3	24.3	19.8	14.1	15.1
18.3	19.4	19.9	22.6	15.3	11.3

15.82 Three new traps were tested to compare their ability to trap mosquitoes. Three traps, A, B, and C, were placed side-by-side at each of five different locations. After a specified length of time, the number of mosquitoes in each trap was recorded, as shown in the table below.

Location	Trap A	Trap B	Trap C
1	3	5	0
2	23	17	15
3	11	5	7
4	19	11	5
5	8	4	2

a. What type of design was employed?
b. Identify the experimental units, response, factor(s) and factor type(s), treatments, and blocks.
c. Is there evidence that the three probability distributions associated with the number of mosquitoes trapped by the devices differ? Use $\alpha = .05$.
d. If warranted, use the Wilcoxon signed rank test to compare the pairs of traps. Use a significance level of $\alpha = .10$ for each comparison.
e. What assumptions are necessary to ensure the validity of the inferential procedures used in parts **c** and **d**?

15.83 Many water treatment facilities supplement the natural fluoride concentration with hydrofluosilicic acid in order to reach a target concentration of fluoride in drinking water. Certain levels are thought to enhance dental health, but very high concentrations can be dangerous. Suppose that one such treatment plant targets .75 milligrams per liter (mg/L) for their water. The plant tests 25 samples each day to determine whether the median level differs from the target.
a. Set up the null and alternative hypotheses.

b. Set up the test statistic and rejection region using $\alpha = .10$.
c. Explain the implication of a Type I error in the context of this application. A Type II error.
d. Suppose that one day's samples result in 18 values that exceed .75 mg/L. Conduct the test and state the appropriate conclusion in the context of this application.
e. When it was suggested to the plant's supervisor that a t-test should be used to conduct the daily test, she replied that the probability distribution of the fluoride concentrations was "heavily skewed to the right." Show graphically what she meant by this, and explain why this is a reason to prefer the sign test to the t-test.

15.84 The trend among doctors in some areas of the country is to form a group practice. By combining treatment resources, doctors hope to be more efficient. One doctor who recently joined a group family practice wants to compare the distribution of the number of patients he treated during a day when he was an individual practitioner with the distribution after joining the group practice. Records were checked for 8 days before and 8 days after, with the results listed in the following table. Use a nonparametric test to determine whether these samples indicate that the doctor tends to treat more patients per day now than before. Use $\alpha = .05$.

Before		After	
26	24	28	29
25	20	30	23
27	22	27	28
26	21	31	29

15.85 For many years, the Girl Scouts of America have sold cookies using various sales techniques. One troop experimented with several techniques and reported the number of boxes of cookies sold per scout, as listed in the table below.

Door-to-Door	Telephone	Grocery Store Stand	Department Store Stand
47	63	113	25
93	19	50	36
58	29	68	21
37	24	37	27
62	33	39	18
		77	31

a. Is there sufficient evidence to indicate that at least one sales method tends to produce a larger number of sales per scout than the others? Use $\alpha = .10$.
b. Does the door-to-door technique tend to produce a different number of sales per scout than the grocery store stand? Test using $\alpha = .05$.

15.86 Each applicant to a certain university is judged by his or her high school grade-point average and a score on a standard aptitude test. The GPAs and aptitude scores for eight randomly selected applicants are shown in the following table. Do these data provide sufficient evidence to indicate a positive correlation between high school GPA and aptitude test score? Use $\alpha = .05$.

Applicant	GPA	Aptitude Test
1	3.25	1,200
2	2.85	890
3	3.01	980
4	4.00	1,150
5	3.10	1,510
6	2.90	950
7	2.75	1,010
8	3.35	1,080

15.87 Refer to Exercise 15.86. Test whether this sample indicates that the median GPA of all applicants exceeds 3.0. Use $\alpha = .05$. Calculate and interpret the observed significance level of this test.

15.88 The coach of a mediocre basketball team has one all-star player who attempts the vast majority of the team's shots. For each of the last 10 games, the star's number of shots and the team's winning margin (negative number implies a loss) are recorded:

Game	Star's Shots	Winning Margin
1	19	-3
2	15	13
3	21	-8
4	17	-5
5	25	-2
6	22	16
7	14	4
8	11	1
9	18	-5
10	18	-12

Compute Spearman's rank correlation coefficient. Is there evidence of a relationship between the team's performance and the star's number of shots?

15.89 A manufacturer wants to determine whether the number of defectives produced by its employees tends to increase as the day progresses. Unknown to the employees, a complete inspection is made of every item that was produced on one day, and the hourly fraction defective is recorded. The resulting data are given in the table (top right). Do they provide evidence that the fraction defective increases as the day progresses? Test at the $\alpha = .05$ level.

Hour	Fraction Defective
1	.02
2	.05
3	.03
4	.08
5	.06
6	.09
7	.11
8	.10

15.90 Refer to Exercise 15.89. Assume that the eight measurements of the fraction defective constitute a random sample from the probability distribution of all fraction defective measurements. The manufacturer wants to determine whether the median hourly fraction defective exceeds .05.

  a. Conduct the appropriate test to make this determination at the .10 level of significance.

  b. Why might the manufacturer wish to conduct the test with this relatively high level of significance?

  c. What assumptions are necessary to ensure the validity of this test? How do they differ from the assumptions that would be necessary to conduct a *t*-test about the mean fraction defective?

15.91 A psychologist ranked a random sample of 10 children on two subjective scales according to the amount of paranoid behavior and the amount of aggressiveness they exhibit. The rankings according to these two criteria are given in the following table. Compute Spearman's rank correlation coefficient. Is there evidence of a relationship between aggression and paranoia in children as judged by this psychologist?

Child	Paranoia	Aggression
1	7	5
2	3	1
3	6	4
4	1	2
5	2	8
6	4	7
7	10	9
8	8	3
9	5	6
10	9	10

STUDENT PROJECTS

In Chapters 10 and 15 we have discussed two methods of analyzing a randomized block design. When the populations have normal probability distributions and their variances are equal, we can employ the analysis of variance described in Chapter 10. Otherwise, we can use the Friedman F_r-test.

In the Student Projects section of Chapter 10, we asked you to conduct a randomized block design to compare supermarket prices, and to use an analysis of variance to interpret the data. Now use the Friedman F_r-test to compare the supermarket prices.

How do the results of the two analyses compare? Explain the similarity (or lack of similarity) between the two results.

EXPLORING DATA WITH A COMPUTER

In the Exploring Data with a Computer sections of Chapters 8–11, you conducted various parametric tests on several data sets described in Appendix B. These tests are listed below. Now use an appropriate nonparametric procedure to conduct each test. Compare and contrast the results.

Chapter	Data Set	Parametric Test
8	FTC cigarette data	H_0: $\mu = \mu_0$, where μ is the true mean of either tar, nicotine, or carbon monoxide content
9	Cardiac patient blood loss data	H_0: $\mu_{Drug} - \mu_{No\ drug}$, where μ is true mean blood loss
10	College graduate salary data	H_0: $\mu_1 = \mu_2 = \mu_3 = \mu_4$, where μ_i is the true mean starting salary of graduates with major i
11	FTC cigarette data	H_0: $\beta_1 = 0$, where β_1 is the true slope of a straight line relating tar (x) to carbon monoxide content (y) [*Note:* This test is equivalent to H_0: $\rho = 0$]

✦ APPENDIX A
TABLES

Contents

TABLE I **Random Numbers**

Row \ Column	1	2	3	4	5	6	7	8	9	10	11	12	13	14
1	10480	15011	01536	02011	81647	91646	69179	14194	62590	36207	20969	99570	91291	90700
2	22368	46573	25595	85393	30995	89198	27982	53402	93965	34095	52666	19174	39615	99505
3	24130	48360	22527	97265	76393	64809	15179	24830	49340	32081	30680	19655	63348	58629
4	42167	93093	06243	61680	07856	16376	39440	53537	71341	57004	00849	74917	97758	16379
5	37570	39975	81837	16656	06121	91782	60468	81305	49684	60672	14110	06927	01263	54613
6	77921	06907	11008	42751	27756	53498	18602	70659	90655	15053	21916	81825	44394	42880
7	99562	72905	56420	69994	98872	31016	71194	18738	44013	48840	63213	21069	10634	12952
8	96301	91977	05463	07972	18876	20922	94595	56869	59014	60045	18425	84903	42508	32307
9	89579	14342	63661	10281	17453	18103	57740	84378	25331	12566	58678	44947	05585	56941
10	85475	36857	53342	53988	53060	59533	38867	62300	08158	17983	16439	11458	18593	64952
11	28918	69578	88231	33276	70997	79936	56865	05859	90106	31595	01547	85590	91610	78188
12	63553	40961	48235	03427	49626	69445	18663	72695	52180	20847	12234	90511	33703	90322
13	09429	93969	52636	92737	88974	33488	36320	17617	30015	08272	84115	27156	30613	74952
14	10365	61129	87529	85689	48237	52267	67689	93394	01511	26358	85104	20285	29975	89868
15	07119	97336	71048	08178	77233	13916	47564	81056	97735	85977	29372	74461	28551	90707
16	51085	12765	51821	51259	77452	16308	60756	92144	49442	53900	70960	63990	75601	40719
17	02368	21382	52404	60268	89368	19885	55322	44819	01188	65255	64835	44919	05944	55157
18	01011	54092	33362	94904	31273	04146	18594	29852	71585	85030	51132	01915	92747	64951
19	52162	53916	46369	58586	23216	14513	83149	98736	23495	64350	94738	17752	35156	35749
20	07056	97628	33787	09998	42698	06691	76988	13602	51851	46104	88916	19509	25625	58104
21	48663	91245	85828	14346	09172	30168	90229	04734	59193	22178	30421	61656	99904	32812
22	54164	58492	22421	74103	47070	25306	76468	26384	58151	06646	21524	15227	96909	44592
23	32639	32363	05597	24200	13363	38005	94342	28728	35806	06912	17012	64161	18296	22851
24	29334	27001	87637	87308	58731	00256	45834	15398	46557	41135	10367	07684	36188	18510
25	02488	33062	28834	07351	19731	92420	60952	61280	50001	67658	32586	86679	50720	94953
26	81525	72295	04839	96423	24878	82651	66566	14778	76797	14780	13300	87074	79666	95725
27	29676	20591	68086	26432	46901	20849	89768	81536	86645	12659	92259	57102	80428	25280
28	00742	57392	39064	66432	84673	40027	32832	61362	98947	96067	64760	64584	96096	98253
29	05366	04213	25669	26422	44407	44048	37937	63904	45766	66134	75470	66520	34693	90449
30	91921	26418	64117	94305	26766	25940	39972	22209	71500	64568	91402	42416	07844	69618
31	00582	04711	87917	77341	42206	35126	74087	99547	81817	42607	43808	76655	62028	76630
32	00725	69884	62797	56170	86324	88072	76222	36086	84637	93161	76038	65855	77919	88006
33	69011	65795	95876	55293	18988	27354	26575	08625	40801	59920	29841	80150	12777	48501
34	25976	57948	29888	88604	57917	48708	18912	82271	65424	69774	33611	54262	85963	03547
35	09763	83473	73577	12908	30883	18317	28290	35797	05998	41688	34952	37888	38917	88050

continued

TABLE I *Continued*

Row	1	2	3	4	5	6	7	8	9	10	11	12	13	14
36	91576	42595	27958	30134	04024	86385	29880	99730	55536	84855	29080	09250	79656	73211
37	17955	56349	90999	49127	20044	59931	06115	20542	18059	02008	73708	83517	36103	42791
38	46503	18584	18845	49618	02304	51038	20655	58727	28168	15475	56942	53389	20562	87338
39	92157	89634	94824	78171	84610	82834	09922	25417	44137	48413	25555	21246	35509	20468
40	14577	62765	35605	81263	39667	47358	56873	56307	61607	49518	89656	20103	77490	18062
41	98427	07523	33362	64270	01638	92477	66969	98420	04880	45585	46565	04102	46880	45709
42	34914	63976	88720	82765	34476	17032	87589	40836	32427	70002	70663	88863	77775	69348
43	70060	28277	39475	46473	23219	53416	94970	25832	69975	94884	19661	72828	00102	66794
44	53976	54914	06990	67245	68350	82948	11398	42878	80287	88267	47363	46634	06541	97809
45	76072	29515	40980	07391	58745	25774	22987	80059	39911	96189	41151	14222	60697	59583
46	90725	52210	83974	29992	65831	38857	50490	83765	55657	14361	31720	57375	56228	41546
47	64364	67412	33339	31926	14883	24413	59744	92351	97473	89286	35931	04110	23726	51900
48	08962	00358	31662	25388	61642	34072	81249	35648	56891	69352	48373	45578	78547	81788
49	95012	68379	93526	70765	10592	04542	76463	54328	02349	17247	28865	14777	62730	92277
50	15664	10493	20492	38391	91132	21999	59516	81652	27195	48223	46751	22923	32261	85653
51	16408	81899	04153	53381	79401	21438	83035	92350	36693	31238	59649	91754	72772	02338
52	18629	81953	05520	91962	04739	13092	97662	24822	94730	06496	35090	04822	86774	98289
53	73115	35101	47498	87637	99016	71060	88824	71013	18735	20286	23153	72924	35165	43040
54	57491	16703	23167	49323	45021	33132	12544	41035	80780	45393	44812	12512	98931	91202
55	30405	83946	23792	14422	15059	45799	22716	19792	09983	74353	68668	30429	70735	25499
56	16631	35006	85900	98275	32388	52390	16815	69290	82732	38480	73817	32523	41961	44437
57	96773	20206	42559	78985	05300	22164	24369	54224	35083	19687	11052	91491	60383	19746
58	38935	64202	14349	82674	66523	44133	00697	35552	35970	19124	63318	29686	03387	59846
59	31624	76384	17403	53363	44167	64486	64758	75366	76554	31601	12614	33072	60332	92325
60	78919	19474	23632	27889	47914	02584	37680	20801	72152	39339	34806	08930	85001	87820
61	03931	33309	57047	74211	63445	17361	62825	39908	05607	91284	68833	25570	38818	46920
62	74426	33278	43972	10110	89917	15665	52872	73823	73144	88662	88970	74492	51805	99378
63	09066	00903	20795	95452	92648	45454	09552	88815	16553	51125	79375	97596	16296	66092
64	42238	12426	87025	14267	20979	04508	64535	31355	86064	29472	47689	05974	52468	16834
65	16153	08002	26504	41744	81959	65642	74240	56302	00033	67107	77510	70625	28725	34191
66	21457	40742	29820	96783	29400	21840	15035	34537	33310	06116	95240	15957	16572	06004
67	21581	57802	02050	89728	17937	37621	47075	42080	97403	48626	68995	43805	33386	21597
68	55612	78095	83197	33732	05810	24813	86902	60397	16489	03264	88525	42786	05269	92532
69	44657	66999	99324	51281	84463	60563	79312	93454	68876	25471	93911	25650	12682	73572
70	91340	84979	46949	81973	37949	61023	43997	15263	80644	43942	89203	71795	99533	50501

TABLE I *Continued*

Row	1	2	3	4	5	6	7	8	9	10	11	12	13	14
71	91227	21199	31935	27022	84067	05462	35216	14486	29891	68607	41867	14951	91696	85065
72	50001	38140	66321	19924	72163	09538	12151	06878	91903	18749	34405	56037	82790	70925
73	65390	05224	72958	28609	81406	39147	25549	48542	42627	45233	57202	94617	23772	07896
74	27504	96131	83944	41575	10573	08619	64482	73923	36152	05184	94142	25299	84387	34925
75	37169	94851	39117	89632	00959	16487	65536	49071	39782	17095	02330	74301	00275	48280
76	11508	70225	51111	38351	19444	66499	71945	05422	13442	78675	84081	66938	93654	59894
77	37449	30362	06694	54690	04052	53115	62757	95348	78662	11163	81651	50245	34971	52924
78	46515	70331	85922	38329	57015	15765	97161	17869	45349	61796	66345	81073	49106	79860
79	30986	81223	42416	58353	21532	30502	32305	86482	05174	07901	54339	58861	74818	46942
80	63798	64995	46583	09785	44160	78128	83991	42865	92520	83531	80377	35909	81250	54238
81	82486	84846	99254	67632	43218	50076	21361	64816	51202	88124	41870	52689	51275	83556
82	21885	32906	92431	09060	64297	51674	64126	62570	26123	05155	59194	52799	28225	85762
83	60336	98782	07408	53458	13564	59089	26445	29789	85205	41001	12535	12133	14645	23541
84	43937	46891	24010	25560	86355	33941	25786	54990	71899	15475	95434	98227	21824	19585
85	97656	63175	89303	16275	07100	92063	21942	18611	47348	20203	18534	03862	78095	50136
86	03299	01221	05418	38982	55758	92237	26759	86367	21216	98442	08303	56613	91511	75928
87	79626	06486	03574	17668	07785	76020	79924	25651	83325	88428	85076	72811	22717	50585
88	85636	68335	47539	03129	65651	11977	02510	26113	99447	68645	34327	15152	55230	93448
89	18039	14367	64337	06177	12143	46609	32989	74014	64708	00533	35398	58408	13261	47908
90	08362	15656	60627	36478	65648	16764	53412	09013	07832	41574	17639	82163	60859	75567
91	79556	29068	04142	16268	15387	12856	66227	38358	22478	73373	88732	09443	82558	05250
92	92608	82674	27072	32534	17075	27698	98204	63863	11951	34648	88022	56148	34925	57031
93	23982	25835	40055	67006	12293	02753	14827	23235	35071	99704	37543	11601	35503	85171
94	09915	96306	05908	97901	28395	14186	00821	80703	70426	75647	76310	88717	37890	40129
95	59037	33300	26695	62247	69927	76123	50842	43834	86654	70959	79725	93872	28117	19233
96	42488	78077	69882	61657	34136	79180	97526	43092	04098	73571	80799	76536	71255	64239
97	46764	86273	63003	93017	31204	36692	40202	35275	57306	55543	53203	18098	47625	88684
98	03237	45430	55417	63282	90816	17349	88298	90183	36600	78406	06216	95787	42579	90730
99	86591	81482	52667	61582	14972	90053	89534	76036	49199	43716	97548	04379	46370	28672
100	38534	01715	94964	87288	65680	43772	39560	12918	86537	62738	19636	51132	25739	56947

Source: Abridged from W. H. Beyer (ed.), *CRC Standard Mathematical Tables*, 24th edition. (Cleveland: The Chemical Rubber Company), 1976. Reproduced by permission of the publisher.

TABLE II Binomial Probabilities

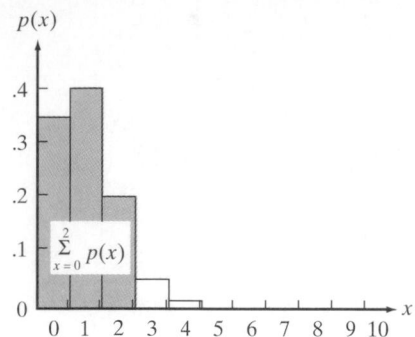

Tabulated values are $\sum_{x=0}^{k} p(x)$. *(Computations are rounded at the third decimal place.)*

a. n = 5

k \ p	.01	.05	.10	.20	.30	.40	.50	.60	.70	.80	.90	.95	.99
0	.951	.774	.590	.328	.168	.078	.031	.010	.002	.000	.000	.000	.000
1	.999	.977	.919	.737	.528	.337	.188	.087	.031	.007	.000	.000	.000
2	1.000	.999	.991	.942	.837	.683	.500	.317	.163	.058	.009	.001	.000
3	1.000	1.000	1.000	.993	.969	.913	.812	.663	.472	.263	.081	.023	.001
4	1.000	1.000	1.000	1.000	.998	.990	.969	.922	.832	.672	.410	.226	.049

b. n = 6

k \ p	.01	.05	.10	.20	.30	.40	.50	.60	.70	.80	.90	.95	.99
0	.941	.735	.531	.262	.118	.047	.016	.004	.001	.000	.000	.000	.000
1	.999	.967	.886	.655	.420	.233	.109	.041	.011	.002	.000	.000	.000
2	1.000	.998	.984	.901	.744	.544	.344	.179	.070	.017	.001	.000	.000
3	1.000	1.000	.999	.983	.930	.821	.656	.456	.256	.099	.016	.002	.000
4	1.000	1.000	1.000	.998	.989	.959	.891	.767	.580	.345	.114	.033	.001
5	1.000	1.000	1.000	1.000	.999	.996	.984	.953	.882	.738	.469	.265	.059

c. n = 7

k \ p	.01	.05	.10	.20	.30	.40	.50	.60	.70	.80	.90	.95	.99
0	.932	.698	.478	.210	.082	.028	.008	.002	.000	.000	.000	.000	.000
1	.998	.956	.850	.577	.329	.159	.063	.019	.004	.000	.000	.000	.000
2	1.000	.996	.974	.852	.647	.420	.227	.096	.029	.005	.000	.000	.000
3	1.000	1.000	.997	.967	.874	.710	.500	.290	.126	.033	.003	.000	.000
4	1.000	1.000	1.000	.995	.971	.904	.773	.580	.353	.148	.026	.004	.000
5	1.000	1.000	1.000	1.000	.996	.981	.937	.841	.671	.423	.150	.044	.002
6	1.000	1.000	1.000	1.000	1.000	.998	.992	.972	.918	.790	.522	.302	.068

TABLE II *Continued*

d. $n = 8$

k \ p	.01	.05	.10	.20	.30	.40	.50	.60	.70	.80	.90	.95	.99
0	.923	.663	.430	.168	.058	.017	.004	.001	.000	.000	.000	.000	.000
1	.997	.943	.813	.503	.255	.106	.035	.009	.001	.000	.000	.000	.000
2	1.000	.994	.962	.797	.552	.315	.145	.050	.011	.001	.000	.000	.000
3	1.000	1.000	.995	.944	.806	.594	.363	.174	.058	.010	.000	.000	.000
4	1.000	1.000	1.000	.990	.942	.826	.637	.406	.194	.056	.005	.000	.000
5	1.000	1.000	1.000	.999	.989	.950	.855	.685	.448	.203	.038	.006	.000
6	1.000	1.000	1.000	1.000	.999	.991	.965	.894	.745	.497	.187	.057	.003
7	1.000	1.000	1.000	1.000	1.000	.999	.996	.983	.942	.832	.570	.337	.077

e. $n = 9$

k \ p	.01	.05	.10	.20	.30	.40	.50	.60	.70	.80	.90	.95	.99
0	.914	.630	.387	.134	.040	.010	.002	.000	.000	.000	.000	.000	.000
1	.997	.929	.775	.436	.196	.071	.020	.004	.000	.000	.000	.000	.000
2	1.000	.992	.947	.738	.463	.232	.090	.025	.004	.000	.000	.000	.000
3	1.000	.999	.992	.914	.730	.483	.254	.099	.025	.003	.000	.000	.000
4	1.000	1.000	.999	.980	.901	.733	.500	.267	.099	.020	.001	.000	.000
5	1.000	1.000	1.000	.997	.975	.901	.746	.517	.270	.086	.008	.001	.000
6	1.000	1.000	1.000	1.000	.996	.975	.910	.768	.537	.262	.053	.008	.000
7	1.000	1.000	1.000	1.000	1.000	.996	.980	.929	.804	.564	.225	.071	.003
8	1.000	1.000	1.000	1.000	1.000	1.000	.998	.990	.960	.866	.613	.370	.086

f. $n = 10$

k \ p	.01	.05	.10	.20	.30	.40	.50	.60	.70	.80	.90	.95	.99
0	.904	.599	.349	.107	.028	.006	.001	.000	.000	.000	.000	.000	.000
1	.996	.914	.376	.376	.149	.046	.011	.002	.000	.000	.000	.000	.000
2	1.000	.988	.930	.678	.383	.167	.055	.012	.002	.000	.000	.000	.000
3	1.000	.999	.987	.879	.650	.382	.172	.055	.011	.001	.000	.000	.000
4	1.000	1.000	.998	.967	.850	.633	.377	.166	.047	.006	.000	.000	.000
5	1.000	1.000	1.000	.999	.953	.834	.623	.367	.150	.033	.002	.000	.000
6	1.000	1.000	1.000	.999	.989	.945	.828	.618	.350	.121	.013	.001	.000
7	1.000	1.000	1.000	1.000	.998	.988	.945	.833	.617	.322	.070	.012	.000
8	1.000	1.000	1.000	1.000	1.000	.998	.989	.954	.851	.624	.264	.086	.004
9	1.000	1.000	1.000	1.000	1.000	1.000	.999	.994	.972	.893	.651	.401	.096

continued

TABLE II *Continued*

g. *n* = 15

k \ p	.01	.05	.10	.20	.30	.40	.50	.60	.70	.80	.90	.95	.99
0	.860	.463	.206	.035	.005	.000	.000	.000	.000	.000	.000	.000	.000
1	.990	.829	.549	.167	.035	.005	.000	.000	.000	.000	.000	.000	.000
2	1.000	.964	.816	.398	.127	.027	.004	.000	.000	.000	.000	.000	.000
3	1.000	.995	.944	.648	.297	.091	.018	.002	.000	.000	.000	.000	.000
4	1.000	.999	.987	.838	.515	.217	.059	.009	.001	.000	.000	.000	.000
5	1.000	1.000	.998	.939	.722	.403	.151	.034	.004	.000	.000	.000	.000
6	1.000	1.000	1.000	.982	.869	.610	.304	.095	.015	.001	.000	.000	.000
7	1.000	1.000	1.000	.996	.950	.787	.500	.213	.050	.004	.000	.000	.000
8	1.000	1.000	1.000	.999	.985	.905	.696	.390	.131	.018	.000	.000	.000
9	1.000	1.000	1.000	1.000	.996	.966	.849	.597	.278	.061	.002	.000	.000
10	1.000	1.000	1.000	1.000	.999	.991	.941	.783	.485	.164	.013	.001	.000
11	1.000	1.000	1.000	1.000	1.000	.998	.982	.909	.703	.352	.056	.005	.000
12	1.000	1.000	1.000	1.000	1.000	1.000	.996	.973	.873	.602	.184	.036	.000
13	1.000	1.000	1.000	1.000	1.000	1.000	1.000	.995	.965	.833	.451	.171	.010
14	1.000	1.000	1.000	1.000	1.000	1.000	1.000	1.000	.995	.965	.794	.537	.140

h. *n* = 20

k \ p	.01	.05	.10	.20	.30	.40	.50	.60	.70	.80	.90	.95	.99
0	.818	.358	.122	.012	.001	.000	.000	.000	.000	.000	.000	.000	.000
1	.983	.736	.392	.069	.008	.001	.000	.000	.000	.000	.000	.000	.000
2	.999	.925	.677	.206	.035	.004	.000	.000	.000	.000	.000	.000	.000
3	1.000	.984	.867	.411	.107	.016	.001	.000	.000	.000	.000	.000	.000
4	1.000	.997	.957	.630	.238	.051	.006	.000	.000	.000	.000	.000	.000
5	1.000	1.000	.989	.804	.416	.126	.021	.002	.000	.000	.000	.000	.000
6	1.000	1.000	.998	.913	.608	.250	.058	.006	.000	.000	.000	.000	.000
7	1.000	1.000	1.000	.968	.772	.416	.132	.021	.001	.000	.000	.000	.000
8	1.000	1.000	1.000	.990	.887	.596	.252	.057	.005	.000	.000	.000	.000
9	1.000	1.000	1.000	.997	.952	.755	.412	.128	.017	.001	.000	.000	.000
10	1.000	1.000	1.000	.999	.983	.872	.588	.245	.048	.003	.000	.000	.000
11	1.000	1.000	1.000	1.000	.995	.943	.748	.404	.113	.010	.000	.000	.000
12	1.000	1.000	1.000	1.000	.999	.979	.868	.584	.228	.032	.000	.000	.000
13	1.000	1.000	1.000	1.000	1.000	.994	.942	.750	.392	.087	.002	.000	.000
14	1.000	1.000	1.000	1.000	1.000	.998	.979	.874	.584	.196	.011	.000	.000
15	1.000	1.000	1.000	1.000	1.000	1.000	.994	.949	.762	.370	.043	.003	.000
16	1.000	1.000	1.000	1.000	1.000	1.000	.999	.984	.893	.589	.133	.016	.000
17	1.000	1.000	1.000	1.000	1.000	1.000	1.000	.996	.965	.794	.323	.075	.001
18	1.000	1.000	1.000	1.000	1.000	1.000	1.000	.999	.992	.931	.608	.264	.017
19	1.000	1.000	1.000	1.000	1.000	1.000	1.000	1.000	.999	.988	.878	.642	.182

TABLE II *Continued*

i. $n = 25$

k \ p	.01	.05	.10	.20	.30	.40	.50	.60	.70	.80	.90	.95	.99
0	.778	.277	.072	.004	.000	.000	.000	.000	.000	.000	.000	.000	.000
1	.974	.642	.271	.027	.002	.000	.000	.000	.000	.000	.000	.000	.000
2	.998	.873	.537	.098	.009	.000	.000	.000	.000	.000	.000	.000	.000
3	1.000	.966	.764	.234	.033	.002	.000	.000	.000	.000	.000	.000	.000
4	1.000	.993	.902	.421	.090	.009	.000	.000	.000	.000	.000	.000	.000
5	1.000	.999	.967	.617	.193	.029	.002	.000	.000	.000	.000	.000	.000
6	1.000	1.000	.991	.780	.341	.074	.007	.000	.000	.000	.000	.000	.000
7	1.000	1.000	.998	.891	.512	.154	.022	.001	.000	.000	.000	.000	.000
8	1.000	1.000	1.000	.953	.677	.274	.054	.004	.000	.000	.000	.000	.000
9	1.000	1.000	1.000	.983	.811	.425	.115	.013	.000	.000	.000	.000	.000
10	1.000	1.000	1.000	.994	.902	.586	.212	.034	.002	.000	.000	.000	.000
11	1.000	1.000	1.000	.998	.956	.732	.345	.078	.006	.000	.000	.000	.000
12	1.000	1.000	1.000	1.000	.983	.846	.500	.154	.017	.000	.000	.000	.000
13	1.000	1.000	1.000	1.000	.994	.922	.655	.268	.044	.002	.000	.000	.000
14	1.000	1.000	1.000	1.000	.998	.966	.788	.414	.098	.006	.000	.000	.000
15	1.000	1.000	1.000	1.000	1.000	.987	.885	.575	.189	.017	.000	.000	.000
16	1.000	1.000	1.000	1.000	1.000	.996	.946	.726	.323	.047	.000	.000	.000
17	1.000	1.000	1.000	1.000	1.000	.999	.978	.846	.488	.109	.002	.000	.000
18	1.000	1.000	1.000	1.000	1.000	1.000	.993	.926	.659	.220	.009	.000	.000
19	1.000	1.000	1.000	1.000	1.000	1.000	.998	.971	.807	.383	.033	.001	.000
20	1.000	1.000	1.000	1.000	1.000	1.000	1.000	.991	.910	.579	.098	.007	.000
21	1.000	1.000	1.000	1.000	1.000	1.000	1.000	.998	.967	.766	.236	.034	.000
22	1.000	1.000	1.000	1.000	1.000	1.000	1.000	1.000	.991	.902	.463	.127	.002
23	1.000	1.000	1.000	1.000	1.000	1.000	1.000	1.000	.998	.973	.729	.358	.026
24	1.000	1.000	1.000	1.000	1.000	1.000	1.000	1.000	1.000	.996	.928	.723	.222

TABLE III **Poisson Probabilities**

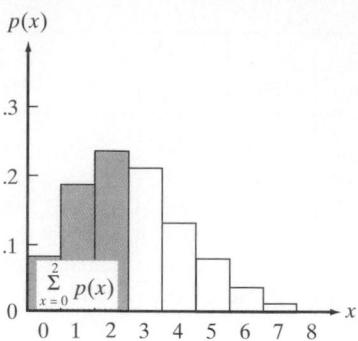

Tabulated values are $\sum_{x=0}^{k} p(x)$. (Computations are rounded at the third decimal place.)

λ \ x	0	1	2	3	4	5	6	7	8	9
.02	.980	1.000								
.04	.961	.999	1.000							
.06	.942	.998	1.000							
.08	.923	.997	1.000							
.10	.905	.995	1.000							
.15	.861	.990	.999	1.000						
.20	.819	.982	.999	1.000						
.25	.779	.974	.998	1.000						
.30	.741	.963	.996	1.000						
.35	.705	.951	.994	1.000						
.40	.670	.938	.992	.999	1.000					
.45	.638	.925	.989	.999	1.000					
.50	.607	.910	.986	.998	1.000					
.55	.577	.894	.982	.998	1.000					
.60	.549	.878	.977	.997	1.000					
.65	.522	.861	.972	.996	.999	1.000				
.70	.497	.844	.966	.994	.999	1.000				
.75	.472	.827	.959	.993	.999	1.000				
.80	.449	.809	.953	.991	.999	1.000				
.85	.427	.791	.945	.989	.998	1.000				
.90	.407	.772	.937	.987	.998	1.000				
.95	.387	.754	.929	.981	.997	1.000				
1.00	.368	.736	.920	.981	.996	.999	1.000			
1.1	.333	.699	.900	.974	.995	.999	1.000			
1.2	.301	.663	.879	.966	.992	.998	1.000			
1.3	.273	.627	.857	.957	.989	.998	1.000			
1.4	.247	.592	.833	.946	.986	.997	.999	1.000		
1.5	.223	.558	.809	.934	.981	.996	.999	1.000		

TABLE III *Continued*

λ \ x	0	1	2	3	4	5	6	7	8	9
1.6	.202	.525	.783	.921	.976	.994	.999	1.000		
1.7	.183	.493	.757	.907	.970	.992	.998	1.000		
1.8	.165	.463	.731	.891	.964	.990	.997	.999	1.000	
1.9	.150	.434	.704	.875	.956	.987	.997	.999	1.000	
2.0	.135	.406	.677	.857	.947	.983	.995	.999	1.000	
2.2	.111	.355	.623	.819	.928	.975	.993	.998	1.000	
2.4	.091	.308	.570	.779	.904	.964	.988	.997	.999	1.000
2.6	.074	.267	.518	.736	.877	.951	.983	.995	.999	1.000
2.8	.061	.231	.469	.692	.848	.935	.976	.992	.998	.999
3.0	.050	.199	.423	.647	.815	.916	.966	.988	.996	.999
3.2	.041	.171	.380	.603	.781	.895	.955	.983	.994	.998
3.4	.033	.147	.340	.558	.744	.871	.942	.977	.992	.997
3.6	.027	.126	.303	.515	.706	.844	.927	.969	.988	.996
3.8	.022	.107	.269	.473	.668	.816	.909	.960	.984	.994
4.0	.018	.092	.238	.433	.629	.785	.889	.949	.979	.992
4.2	.015	.078	.210	.395	.590	.753	.867	.936	.972	.989
4.4	.012	.066	.185	.359	.551	.720	.844	.921	.964	.985
4.6	.010	.056	.163	.326	.513	.686	.818	.905	.955	.980
4.8	.008	.048	.143	.294	.476	.651	.791	.887	.944	.975
5.0	.007	.040	.125	.265	.440	.616	.762	.867	.932	.968
5.2	.006	.034	.109	.238	.406	.581	.732	.845	.918	.960
5.4	.005	.029	.095	.213	.373	.546	.702	.822	.903	.951
5.6	.004	.024	.082	.191	.342	.512	.670	.797	.886	.941
5.8	.003	.021	.072	.170	.313	.478	.638	.771	.867	.929
6.0	.002	.017	.062	.151	.285	.446	.606	.744	.847	.916

λ	10	11	12	13	14	15	16			
2.8	1.000									
3.0	1.000									
3.2	1.000									
3.4	.999	1.000								
3.6	.999	1.000								
3.8	.998	.999	1.000							
4.0	.997	.999	1.000							
4.2	.996	.999	1.000							
4.4	.994	.998	.999	1.000						
4.6	.992	.997	.999	1.000						
4.8	.990	.996	.999	1.000						
5.0	.986	.995	.998	.999	1.000					
5.2	.982	.993	.997	.999	1.000					
5.4	.977	.990	.996	.999	1.000					
5.6	.972	.988	.995	.998	.999	1.000				
5.8	.965	.984	.993	.997	.999	1.000				
6.0	.957	.980	.991	.996	.999	.999	1.000			

continued

TABLE III *Continued*

λ＼x	0	1	2	3	4	5	6	7	8	9
6.2	.002	.015	.054	.134	.259	.414	.574	.716	.826	.902
6.4	.002	.012	.046	.119	.235	.384	.542	.687	.803	.886
6.6	.001	.010	.040	.105	.213	.355	.511	.658	.780	.869
6.8	.001	.009	.034	.093	.192	.327	.480	.628	.755	.850
7.0	.001	.007	.030	.082	.173	.301	.450	.599	.729	.830
7.2	.001	.006	.025	.072	.156	.276	.420	.569	.703	.810
7.4	.001	.005	.022	.063	.140	.253	.392	.539	.676	.788
7.6	.001	.004	.019	.055	.125	.231	.365	.510	.648	.765
7.8	.000	.004	.016	.048	.112	.210	.338	.481	.620	.741
8.0	.000	.003	.014	.042	.100	.191	.313	.453	.593	.717
8.5	.000	.002	.009	.030	.074	.150	.256	.386	.523	.653
9.0	.000	.001	.006	.021	.055	.116	.207	.324	.456	.587
9.5	.000	.001	.004	.015	.040	.089	.165	.269	.392	.522
10.0	.000	.000	.003	.010	.029	.067	.130	.220	.333	.458

λ	10	11	12	13	14	15	16	17	18	19
6.2	.949	.975	.989	.995	.998	.999	1.000			
6.4	.939	.969	.986	.994	.997	.999	1.000			
6.6	.927	.963	.982	.992	.997	.999	.999	1.000		
6.8	.915	.955	.978	.990	.996	.998	.999	1.000		
7.0	.901	.947	.973	.987	.994	.998	.999	1.000		
7.2	.887	.937	.967	.984	.993	.997	.999	.999	1.000	
7.4	.871	.926	.961	.980	.991	.996	.998	.999	1.000	
7.6	.854	.915	.954	.976	.989	.995	.998	.999	1.000	
7.8	.835	.902	.945	.971	.986	.993	.997	.999	1.000	
8.0	.816	.888	.936	.966	.983	.992	.996	.998	.999	1.000
8.5	.763	.849	.909	.949	.973	.986	.993	.997	.999	.999
9.0	.706	.803	.876	.926	.959	.978	.989	.995	.998	.999
9.5	.645	.752	.836	.898	.940	.967	.982	.991	.996	.998
10.0	.583	.697	.792	.864	.917	.951	.973	.986	.993	.997

λ	20	21	22
8.5	1.000		
9.0	1.000		
9.5	.999	1.000	
10.0	.998	.999	1.000

TABLE III *Continued*

λ \ x	0	1	2	3	4	5	6	7	8	9
10.5	.000	.000	.002	.007	.021	.050	.102	.179	.279	.397
11.0	.000	.000	.001	.005	.015	.038	.079	.143	.232	.341
11.5	.000	.000	.001	.003	.011	.028	.060	.114	.191	.289
12.0	.000	.000	.001	.002	.008	.020	.046	.090	.155	.242
12.5	.000	.000	.000	.002	.005	.015	.035	.070	.125	.201
13.0	.000	.000	.000	.001	.004	.011	.026	.054	.100	.166
13.5	.000	.000	.000	.001	.003	.008	.019	.041	.079	.135
14.0	.000	.000	.000	.000	.002	.006	.014	.032	.062	.109
14.5	.000	.000	.000	.000	.001	.004	.010	.024	.048	.088
15.0	.000	.000	.000	.000	.001	.003	.008	.018	.037	.070

λ	10	11	12	13	14	15	16	17	18	19
10.5	.521	.639	.742	.825	.888	.932	.960	.978	.988	.994
11.0	.460	.579	.689	.781	.854	.907	.944	.968	.982	.991
11.5	.402	.520	.633	.733	.815	.878	.924	.954	.974	.986
12.0	.347	.462	.576	.682	.772	.844	.899	.937	.963	.979
12.5	.297	.406	.519	.628	.725	.806	.869	.916	.948	.969
13.0	.252	.353	.463	.573	.675	.764	.835	.890	.930	.957
13.5	.211	.304	.409	.518	.623	.718	.798	.861	.908	.942
14.0	.176	.260	.358	.464	.570	.669	.756	.827	.883	.923
14.5	.145	.220	.311	.413	.518	.619	.711	.790	.853	.901
15.0	.118	.185	.268	.363	.466	.568	.664	.749	.819	.875

λ	20	21	22	23	24	25	26	27	28	29
10.5	.997	.999	.999	1.000						
11.0	.995	.998	.999	1.000						
11.5	.992	.996	.998	.999	1.000					
12.0	.988	.994	.987	.999	.999	1.000				
12.5	.983	.991	.995	.998	.999	.999	1.000			
13.0	.975	.986	.992	.996	.998	.999	1.000			
13.5	.965	.980	.989	.994	.997	.998	.999	1.000		
14.0	.952	.971	.983	.991	.995	.997	.999	.999	1.000	
14.5	.936	.960	.976	.986	.992	.996	.998	.999	.999	1.000
15.0	.917	.947	.967	.981	.989	.994	.997	.998	.999	1.000

continued

TABLE III   *Continued*

λ＼x	4	5	6	7	8	9	10	11	12	13
16	.000	.001	.004	.010	.022	.043	.077	.127	.193	.275
17	.000	.001	.002	.005	.013	.026	.049	.085	.135	.201
18	.000	.000	.001	.003	.007	.015	.030	.055	.092	.143
19	.000	.000	.001	.002	.004	.009	.018	.035	.061	.098
20	.000	.000	.000	.001	.002	.005	.011	.021	.039	.066
21	.000	.000	.000	.000	.001	.003	.006	.013	.025	.043
22	.000	.000	.000	.000	.001	.002	.004	.008	.015	.028
23	.000	.000	.000	.000	.000	.001	.002	.004	.009	.017
24	.000	.000	.000	.000	.000	.000	.001	.003	.005	.011
25	.000	.000	.000	.000	.000	.000	.001	.001	.003	.006

	14	15	16	17	18	19	20	21	22	23
16	.368	.467	.566	.659	.742	.812	.868	.911	.942	.963
17	.281	.371	.468	.564	.655	.736	.805	.861	.905	.937
18	.208	.287	.375	.469	.562	.651	.731	.799	.855	.899
19	.150	.215	.292	.378	.469	.561	.647	.725	.793	.849
20	.105	.157	.221	.297	.381	.470	.559	.644	.721	.787
21	.072	.111	.163	.227	.302	.384	.471	.558	.640	.716
22	.048	.077	.117	.169	.232	.306	.387	.472	.556	.637
23	.031	.052	.082	.123	.175	.238	.310	.389	.472	.555
24	.020	.034	.056	.087	.128	.180	.243	.314	.392	.473
25	.012	.022	.038	.060	.092	.134	.185	.247	.318	.394

	24	25	26	27	28	29	30	31	32	33
16	.978	.987	.993	.996	.998	.999	.999	1.000		
17	.959	.975	.985	.991	.995	.997	.999	.999	1.000	
18	.932	.955	.972	.983	.990	.994	.997	.998	.999	1.000
19	.893	.927	.951	.969	.980	.988	.993	.996	.998	.999
20	.843	.888	.922	.948	.966	.978	.987	.992	.995	.997
21	.782	.838	.883	.917	.944	.963	.976	.985	.991	.994
22	.712	.777	.832	.877	.913	.940	.959	.973	.983	.989
23	.635	.708	.772	.827	.873	.908	.936	.956	.971	.981
24	.554	.632	.704	.768	.823	.868	.904	.932	.953	.969
25	.473	.553	.629	.700	.763	.818	.863	.900	.929	.950

	34	35	36	37	38	39	40	41	42	43
19	.999	1.000								
20	.999	.999	1.000							
21	.997	.998	.999	.999	1.000					
22	.994	.996	.998	.999	.999	1.000				
23	.988	.993	.996	.997	.999	.999	1.000			
24	.979	.987	.992	.995	.997	.998	.999	.999	1.000	
25	.966	.978	.985	.991	.991	.997	.998	.999	.999	1.000

TABLE IV **Normal Curve Areas**

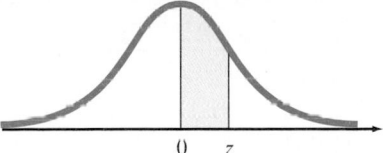

z	.00	.01	.02	.03	.04	.05	.06	.07	.08	.09
.0	.0000	.0040	.0080	.0120	.0160	.0199	.0239	.0279	.0319	.0359
.1	.0398	.0438	.0478	.0517	.0557	.0596	.0636	.0675	.0714	.0753
.2	.0793	.0832	.0871	.0910	.0948	.0987	.1026	.1064	.1103	.1141
.3	.1179	.1217	.1255	.1293	.1331	.1368	.1406	.1443	.1480	.1517
.4	.1554	.1591	.1628	.1664	.1700	.1736	.1772	.1808	.1844	.1879
.5	.1915	.1950	.1985	.2019	.2054	.2088	.2123	.2157	.2190	.2224
.6	.2257	.2291	.2324	.2357	.2389	.2422	.2454	.2486	.2517	.2549
.7	.2580	.2611	.2642	.2673	.2704	.2734	.2764	.2794	.2823	.2852
.8	.2881	.2910	.2939	.2967	.2995	.3023	.3051	.3078	.3106	.3133
.9	.3159	.3186	.3212	.3238	.3264	.3289	.3315	.3340	.3365	.3389
1.0	.3413	.3438	.3461	.3485	.3508	.3531	.3554	.3577	.3599	.3621
1.1	.3643	.3665	.3686	.3708	.3729	.3749	.3770	.3790	.3810	.3830
1.2	.3849	.3869	.3888	.3907	.3925	.3944	.3962	.3980	.3997	.4015
1.3	.4032	.4049	.4066	.4082	.4099	.4115	.4131	.4147	.4162	.4177
1.4	.4192	.4207	.4222	.4236	.4251	.4265	.4279	.4292	.4306	.4319
1.5	.4332	.4345	.4357	.4370	.4382	.4394	.4406	.4418	.4429	.4441
1.6	.4452	.4463	.4474	.4484	.4495	.4505	.4515	.4525	.4535	.4545
1.7	.4554	.4564	.4573	.4582	.4591	.4599	.4608	.4616	.4625	.4633
1.8	.4641	.4649	.4656	.4664	.4671	.4678	.4686	.4693	.4699	.4706
1.9	.4713	.4719	.4726	.4732	.4738	.4744	.4750	.4756	.4761	.4767
2.0	.4772	.4778	.4783	.4788	.4793	.4798	.4803	.4808	.4812	.4817
2.1	.4821	.4826	.4830	.4834	.4838	.4842	.4846	.4850	.4854	.4857
2.2	.4861	.4864	.4868	.4871	.4875	.4878	.4881	.4884	.4887	.4890
2.3	.4893	.4896	.4898	.4901	.4904	.4906	.4909	.4911	.4913	.4916
2.4	.4918	.4920	.4922	.4925	.4927	.4929	.4931	.4932	.4934	.4936
2.5	.4938	.4940	.4941	.4943	.4945	.4946	.4948	.4949	.4951	.4952
2.6	.4953	.4955	.4956	.4957	.4959	.4960	.4961	.4962	.4963	.4964
2.7	.4965	.4966	.4967	.4968	.4969	.4970	.4971	.4972	.4973	.4974
2.8	.4974	.4975	.4976	.4977	.4977	.4978	.4979	.4979	.4980	.4981
2.9	.4981	.4982	.4982	.4983	.4984	.4984	.4985	.4985	.4986	.4986
3.0	.4987	.4987	.4987	.4988	.4988	.4989	.4989	.4989	.4990	.4990

Source: Abridged from Table I of A. Hald, *Statistical Tables and Formulas* (New York: Wiley), 1952. Reproduced by permission of A. Hald.

TABLE V   Exponentials

λ	$e^{-\lambda}$	λ	$e^{-\lambda}$	λ	$e^{-\lambda}$	λ	$e^{-\lambda}$	λ	$e^{-\lambda}$
.00	1.000000	2.05	.128735	4.05	.017422	6.05	.002358	8.05	.000319
.05	.951229	2.10	.122456	4.10	.016573	6.10	.002243	8.10	.000304
.10	.904837	2.15	.116484	4.15	.015764	6.15	.002133	8.15	.000289
.15	.860708	2.20	.110803	4.20	.014996	6.20	.002029	8.20	.000275
.20	.818731	2.25	.105399	4.25	.014264	6.25	.001930	8.25	.000261
.25	.778801	2.30	.100259	4.30	.013569	6.30	.001836	8.30	.000249
.30	.740818	2.35	.095369	4.35	.012907	6.35	.001747	8.35	.000236
.35	.704688	2.40	.090718	4.40	.012277	6.40	.001661	8.40	.000225
.40	.670320	2.45	.086294	4.45	.011679	6.45	.001581	8.45	.000214
.45	.637628	2.50	.082085	4.50	.011109	6.50	.001503	8.50	.000204
.50	.606531	2.55	.078082	4.55	.010567	6.55	.001430	8.55	.000194
.55	.576950	2.60	.074274	4.60	.010052	6.60	.001360	8.60	.000184
.60	.548812	2.65	.070651	4.65	.009562	6.65	.001294	8.65	.000175
.65	.522046	2.70	.067206	4.70	.009095	6.70	.001231	8.70	.000167
.70	.496585	2.75	.063928	4.75	.008652	6.75	.001171	8.75	.000158
.75	.472367	2.80	.060810	4.80	.008230	6.80	.001114	8.80	.000151
.80	.449329	2.85	.057844	4.85	.007828	6.85	.001059	8.85	.000143
.85	.427415	2.90	.055023	4.90	.007447	6.90	.001008	8.90	.000136
.90	.406570	2.95	.052340	4.95	.007083	6.95	.000959	8.95	.000130
.95	.386741	3.00	.049787	5.00	.006738	7.00	.000912	9.00	.000123
1.00	.367879	3.05	.047359	5.05	.006409	7.05	.000867	9.05	.000117
1.05	.349938	3.10	.045049	5.10	.006097	7.10	.000825	9.10	.000112
1.10	.332871	3.15	.042852	5.15	.005799	7.15	.000785	9.15	.000106
1.15	.316637	3.20	.040762	5.20	.005517	7.20	.000747	9.20	.000101
1.20	.301194	3.25	.038774	5.25	.005248	7.25	.000710	9.25	.000096
1.25	.286505	3.30	.036883	5.30	.004992	7.30	.000676	9.30	.000091
1.30	.272532	3.35	.035084	5.35	.004748	7.35	.000643	9.35	.000087
1.35	.259240	3.40	.033373	5.40	.004517	7.40	.000611	9.40	.000083
1.40	.246597	3.45	.031746	5.45	.004296	7.45	.000581	9.45	.000079
1.45	.234570	3.50	.030197	5.50	.004087	7.50	.000553	9.50	.000075
1.50	.223130	3.55	.028725	5.55	.003887	7.55	.000526	9.55	.000071
1.55	.212248	3.60	.027324	5.60	.003698	7.60	.000501	9.60	.000068
1.60	.201897	3.65	.025991	5.65	.003518	7.65	.000476	9.65	.000064
1.65	.192050	3.70	.024724	5.70	.003346	7.70	.000453	9.70	.000061
1.70	.182684	3.75	.023518	5.75	.003183	7.75	.000431	9.75	.000058
1.75	.173774	3.80	.022371	5.80	.003028	7.80	.000410	9.80	.000056
1.80	.165299	3.85	.021280	5.85	.002880	7.85	.000390	9.85	.000053
1.85	.157237	3.90	.020242	5.90	.002739	7.90	.000371	9.90	.000050
1.90	.149569	3.95	.019255	5.95	.002606	7.95	.000353	9.95	.000048
1.95	.142274	4.00	.018316	6.00	.002479	8.00	.000336	10.00	.000045
2.00	.135335								

TABLE VI **Critical Values of *t***

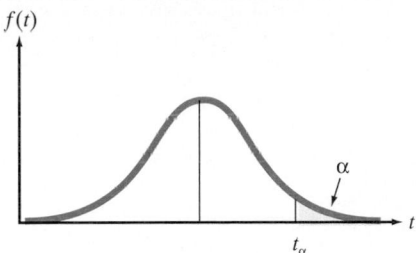

ν	$t_{.100}$	$t_{.050}$	$t_{.025}$	$t_{.010}$	$t_{.005}$	$t_{.001}$	$t_{.0005}$
1	3.078	6.314	12.706	31.821	63.657	318.31	636.62
2	1.886	2.920	4.303	6.965	9.925	22.326	31.598
3	1.638	2.353	3.182	4.541	5.841	10.213	12.924
4	1.533	2.132	2.776	3.747	4.604	7.173	8.610
5	1.476	2.015	2.571	3.365	4.032	5.893	6.869
6	1.440	1.943	2.447	3.143	3.707	5.208	5.959
7	1.415	1.895	2.365	2.998	3.499	4.785	5.408
8	1.397	1.860	2.306	2.896	3.355	4.501	5.041
9	1.383	1.833	2.262	2.821	3.250	4.297	4.781
10	1.372	1.812	2.228	2.764	3.169	4.144	4.587
11	1.363	1.796	2.201	2.718	3.106	4.025	4.437
12	1.356	1.782	2.179	2.681	3.055	3.930	4.318
13	1.350	1.771	2.160	2.650	3.012	3.852	4.221
14	1.345	1.761	2.145	2.624	2.977	3.787	4.140
15	1.341	1.753	2.131	2.602	2.947	3.733	4.073
16	1.337	1.746	2.120	2.583	2.921	3.686	4.015
17	1.333	1.740	2.110	2.567	2.898	3.646	3.965
18	1.330	1.734	2.101	2.552	2.878	3.610	3.922
19	1.328	1.729	2.093	2.539	2.861	3.579	3.883
20	1.325	1.725	2.086	2.528	2.845	3.552	3.850
21	1.323	1.721	2.080	2.518	2.831	3.527	3.819
22	1.321	1.717	2.074	2.508	2.819	3.505	3.792
23	1.319	1.714	2.069	2.500	2.807	3.485	3.767
24	1.318	1.711	2.064	2.492	2.797	3.467	3.745
25	1.316	1.708	2.060	2.485	2.787	3.450	3.725
26	1.315	1.706	2.056	2.479	2.779	3.435	3.707
27	1.314	1.703	2.052	2.473	2.771	3.421	3.690
28	1.313	1.701	2.048	2.467	2.763	3.408	3.674
29	1.311	1.699	2.045	2.462	2.756	3.396	3.659
30	1.310	1.697	2.042	2.457	2.750	3.385	3.646
40	1.303	1.684	2.021	2.423	2.704	3.307	3.551
60	1.296	1.671	2.000	2.390	2.660	3.232	3.460
120	1.289	1.658	1.980	2.358	2.617	3.160	3.373
∞	1.282	1.645	1.960	2.326	2.576	3.090	3.291

Source: This table is reproduced with the kind permission of the Trustees of Biometrika from E. S. Pearson and H. O. Hartley (eds.), *The Biometrika Tables for Statisticians*, Vol. 1, 3d ed., Biometrika, 1966.

TABLE VII   Critical Values of χ^2

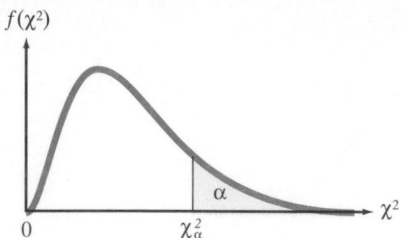

Degrees of Freedom	$\chi^2_{.995}$	$\chi^2_{.990}$	$\chi^2_{.975}$	$\chi^2_{.950}$	$\chi^2_{.900}$
1	.0000393	.0001571	.0009821	.0039321	.0157908
2	.0100251	.0201007	.0506356	.102587	.210720
3	.0717212	.114832	.215795	.351846	.584375
4	.206990	.297110	.484419	.710721	1.063623
5	.411740	.554300	.831211	1.145476	1.61031
6	.675727	.872085	1.237347	1.63539	2.20413
7	.989265	1.239043	1.68987	2.16735	2.83311
8	1.344419	1.646482	2.17973	2.73264	3.48954
9	1.734926	2.087912	2.70039	3.32511	4.16816
10	2.15585	2.55821	3.24697	3.94030	4.86518
11	2.60321	3.05347	3.81575	4.57481	5.57779
12	3.07382	3.57056	4.40379	5.22603	6.30380
13	3.56503	4.10691	5.00874	5.89186	7.04150
14	4.07468	4.66043	5.62872	6.57063	7.78953
15	4.60094	5.22935	6.26214	7.26094	8.54675
16	5.14224	5.81221	6.90766	7.96164	9.31223
17	5.69724	6.40776	7.56418	8.67176	10.0852
18	6.26481	7.01491	8.23075	9.39046	10.8649
19	6.84398	7.63273	8.90655	10.1170	11.6509
20	7.43386	8.26040	9.59083	10.8508	12.4426
21	8.03366	8.89720	10.28293	11.5913	13.2396
22	8.64272	9.54249	10.9823	12.3380	14.0415
23	9.26042	10.19567	11.6885	13.0905	14.8479
24	9.88623	10.8564	12.4011	13.8484	15.6587
25	10.5197	11.5240	13.1197	14.6114	16.4734
26	11.1603	12.1981	13.8439	15.3791	17.2919
27	11.8076	12.8786	14.5733	16.1513	18.1138
28	12.4613	13.5648	15.3079	16.9279	18.9392
29	13.1211	14.2565	16.0471	17.7083	19.7677
30	13.7867	14.9535	16.7908	18.4926	20.5992
40	20.7065	22.1643	24.4331	26.5093	29.0505
50	27.9907	29.7067	32.3574	34.7642	37.6886
60	35.5346	37.4848	40.4817	43.1879	46.4589
70	43.2752	45.4418	48.7576	51.7393	55.3290
80	51.1720	53.5400	57.1532	60.3915	64.2778
90	59.1963	61.7541	65.6466	69.1260	73.2912
100	67.3276	70.0648	74.2219	77.9295	82.3581

Source: From C. M. Thompson, "Tables of the Percentage Points of the χ^2-Distribution," *Biometrika*, 1941, 32, 188–189. Reproduced by permission of the *Biometrika* Trustees.

TABLE VII *Continued*

Degrees of Freedom	$\chi^2_{.100}$	$\chi^2_{.050}$	$\chi^2_{.025}$	$\chi^2_{.010}$	$\chi^2_{.005}$
1	2.70554	3.84146	5.02389	6.63490	7.87944
2	4.60517	5.99147	7.37776	9.21034	10.5966
3	6.25139	7.81473	9.34840	11.3449	12.8381
4	7.77944	9.48773	11.1433	13.2767	14.8602
5	9.23635	11.0705	12.8325	15.0863	16.7496
6	10.6446	12.5916	14.4494	16.8119	18.5476
7	12.0170	14.0671	16.0128	18.4753	20.2777
8	13.3616	15.5073	17.5346	20.0902	21.9550
9	14.6837	16.9190	19.0228	21.6660	23.5893
10	15.9871	18.3070	20.4831	23.2093	25.1882
11	17.2750	19.6751	21.9200	24.7250	26.7569
12	18.5494	21.0261	23.3367	26.2170	28.2995
13	19.8119	22.3621	24.7356	27.6883	29.8194
14	21.0642	23.6848	26.1190	29.1413	31.3193
15	22.3072	24.9958	27.4884	30.5779	32.8013
16	23.5418	26.2962	28.8454	31.9999	34.2672
17	24.7690	27.5871	30.1910	33.4087	35.7185
18	25.9894	28.8693	31.5264	34.8053	37.1564
19	27.2036	30.1435	32.8523	36.1908	38.5822
20	28.4120	31.4104	34.1696	37.5662	39.9968
21	29.6151	32.6705	35.4789	38.9321	41.4010
22	30.8133	33.9244	36.7807	40.2894	42.7956
23	32.0069	35.1725	38.0757	41.6384	44.1813
24	33.1963	36.4151	39.3641	42.9798	45.5585
25	34.3816	37.6525	40.6465	44.3141	46.9278
26	35.5631	38.8852	41.9232	45.6417	48.2899
27	36.7412	40.1133	43.1944	46.9630	49.6449
28	37.9159	41.3372	44.4607	48.2782	50.9933
29	39.0875	42.5569	45.7222	49.5879	52.3356
30	40.2560	43.7729	46.9792	50.8922	53.6720
40	51.8050	55.7585	59.3417	63.6907	66.7659
50	63.1671	67.5048	71.4202	76.1539	79.4900
60	74.3970	79.0819	83.2976	88.3794	91.9517
70	85.5271	90.5312	95.0231	100.425	104.215
80	96.5782	101.879	106.629	112.329	116.321
90	107.565	113.145	118.136	124.116	128.299
100	118.498	124.342	129.561	135.807	140.169

TABLE VIII Percentage Points of the *F*-distribution, $\alpha = .10$

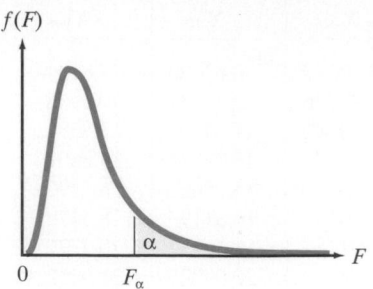

v_1	NUMERATOR DEGREES OF FREEDOM								
v_2	1	2	3	4	5	6	7	8	9
1	39.86	49.50	53.59	55.83	57.24	58.20	58.91	59.44	59.86
2	8.53	9.00	9.16	9.24	9.29	9.33	9.35	9.37	9.38
3	5.54	5.46	5.39	5.34	5.31	5.28	5.27	5.25	5.24
4	4.54	4.32	4.19	4.11	4.05	4.01	3.98	3.95	3.94
5	4.06	3.78	3.62	3.52	3.45	3.40	3.37	3.34	3.32
6	3.78	3.46	3.29	3.18	3.11	3.05	3.01	2.98	2.96
7	3.59	3.26	3.07	2.96	2.88	2.83	2.78	2.75	2.72
8	3.46	3.11	2.92	2.81	2.73	2.67	2.62	2.59	2.56
9	3.36	3.01	2.81	2.69	2.61	2.55	2.51	2.47	2.44
10	3.29	2.92	2.73	2.61	2.52	2.46	2.41	2.38	2.35
11	3.23	2.86	2.66	2.54	2.45	2.39	2.34	2.30	2.27
12	3.18	2.81	2.61	2.48	2.39	2.33	2.28	2.24	2.21
13	3.14	2.76	2.56	2.43	2.35	2.28	2.23	2.20	2.16
14	3.10	2.73	2.52	2.39	2.31	2.24	2.19	2.15	2.12
15	3.07	2.70	2.49	2.36	2.27	2.21	2.16	2.12	2.09
16	3.05	2.67	2.46	2.33	2.24	2.18	2.13	2.09	2.06
17	3.03	2.64	2.44	2.31	2.22	2.15	2.10	2.06	2.03
18	3.01	2.62	2.42	2.29	2.20	2.13	2.08	2.04	2.00
19	2.99	2.61	2.40	2.27	2.18	2.11	2.06	2.02	1.98
20	2.97	2.59	2.38	2.25	2.16	2.09	2.04	2.00	1.96
21	2.96	2.57	2.36	2.23	2.14	2.08	2.02	1.98	1.95
22	2.95	2.56	2.35	2.22	2.13	2.06	2.01	1.97	1.93
23	2.94	2.55	2.34	2.21	2.11	2.05	1.99	1.95	1.92
24	2.93	2.54	2.33	2.19	2.10	2.04	1.98	1.94	1.91
25	2.92	2.53	2.32	2.18	2.09	2.02	1.97	1.93	1.89
26	2.91	2.52	2.31	2.17	2.08	2.01	1.96	1.92	1.88
27	2.90	2.51	2.30	2.17	2.07	2.00	1.95	1.91	1.87
28	2.89	2.50	2.29	2.16	2.06	2.00	1.94	1.90	1.87
29	2.89	2.50	2.28	2.15	2.06	1.99	1.93	1.89	1.86
30	2.88	2.49	2.28	2.14	2.05	1.98	1.93	1.88	1.85
40	2.84	2.44	2.23	2.09	2.00	1.93	1.87	1.83	1.79
60	2.79	2.39	2.18	2.04	1.95	1.87	1.82	1.77	1.74
120	2.75	2.35	2.13	1.99	1.90	1.82	1.77	1.72	1.68
∞	2.71	2.30	2.08	1.94	1.85	1.77	1.72	1.67	1.63

Denominator Degrees of Freedom

Source: From M. Merrington and C. M. Thompson, "Tables of Percentage Points of the Inverted Beta (*F*)-Distribution," *Biometrika*, 1943, 33, 73–88. Reproduced by permission of the *Biometrika* Trustees.

TABLE VIII *Continued*

ν_1 / ν_2	\multicolumn{10}{c}{NUMERATOR DEGREES OF FREEDOM}
	10	**12**	**15**	**20**	**24**	**30**	**40**	**60**	**120**	**∞**
1	60.19	60.71	61.22	61.74	62.00	62.26	62.53	62.79	63.06	63.33
2	9.39	9.41	9.42	9.44	9.45	9.46	9.47	9.47	9.48	9.49
3	5.23	5.22	5.20	5.18	5.18	5.17	5.16	5.15	5.14	5.13
4	3.92	3.90	3.87	3.84	3.83	3.82	3.80	3.79	3.78	3.76
5	3.30	3.27	3.24	3.21	3.19	3.17	3.16	3.14	3.12	3.10
6	2.94	2.90	2.87	2.84	2.82	2.80	2.78	2.76	2.74	2.72
7	2.70	2.67	2.63	2.59	2.58	2.56	2.54	2.51	2.49	2.47
8	2.54	2.50	2.46	2.42	2.40	2.38	2.36	2.34	2.32	2.29
9	2.42	2.38	2.34	2.30	2.28	2.25	2.23	2.21	2.18	2.16
10	2.32	2.28	2.24	2.20	2.18	2.16	2.13	2.11	2.08	2.06
11	2.25	2.21	2.17	2.12	2.10	2.08	2.05	2.03	2.00	1.97
12	2.19	2.15	2.10	2.06	2.04	2.01	1.99	1.96	1.93	1.90
13	2.14	2.10	2.05	2.01	1.98	1.96	1.93	1.90	1.88	1.85
14	2.10	2.05	2.01	1.96	1.94	1.91	1.89	1.86	1.83	1.80
15	2.06	2.02	1.97	1.92	1.90	1.87	1.85	1.82	1.79	1.76
16	2.03	1.99	1.94	1.89	1.87	1.84	1.81	1.78	1.75	1.72
17	2.00	1.96	1.91	1.86	1.84	1.81	1.78	1.75	1.72	1.69
18	1.98	1.93	1.89	1.84	1.81	1.78	1.75	1.72	1.69	1.66
19	1.96	1.91	1.86	1.81	1.79	1.76	1.73	1.70	1.67	1.63
20	1.94	1.89	1.84	1.79	1.77	1.74	1.71	1.68	1.64	1.61
21	1.92	1.87	1.83	1.78	1.75	1.72	1.69	1.66	1.62	1.59
22	1.90	1.86	1.81	1.76	1.73	1.70	1.67	1.64	1.60	1.57
23	1.89	1.84	1.80	1.74	1.72	1.69	1.66	1.62	1.59	1.55
24	1.88	1.83	1.78	1.73	1.70	1.67	1.64	1.61	1.57	1.53
25	1.87	1.82	1.77	1.72	1.69	1.66	1.63	1.59	1.56	1.52
26	1.86	1.81	1.76	1.71	1.68	1.65	1.61	1.58	1.54	1.50
27	1.85	1.80	1.75	1.70	1.67	1.64	1.60	1.57	1.53	1.49
28	1.84	1.79	1.74	1.69	1.66	1.63	1.59	1.56	1.52	1.48
29	1.83	1.78	1.73	1.68	1.65	1.62	1.58	1.55	1.51	1.47
30	1.82	1.77	1.72	1.67	1.64	1.61	1.57	1.54	1.50	1.46
40	1.76	1.71	1.66	1.61	1.57	1.54	1.51	1.47	1.42	1.38
60	1.71	1.66	1.60	1.54	1.51	1.48	1.44	1.40	1.35	1.29
120	1.65	1.60	1.55	1.48	1.45	1.41	1.37	1.32	1.26	1.19
∞	1.60	1.55	1.49	1.42	1.38	1.34	1.30	1.24	1.17	1.00

Denominator Degrees of Freedom

TABLE IX   Percentage Points of the *F*-distribution, $\alpha = .05$

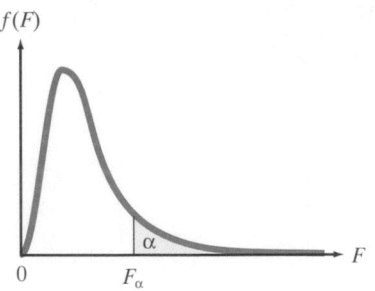

ν_1				NUMERATOR DEGREES OF FREEDOM					
ν_2	1	2	3	4	5	6	7	8	9
1	161.4	199.5	215.7	224.6	230.2	234.0	236.8	238.9	240.5
2	18.51	19.00	19.16	19.25	19.30	19.33	19.35	19.37	19.38
3	10.13	9.55	9.28	9.12	9.01	8.94	8.89	8.85	8.81
4	7.71	6.94	6.59	6.39	6.26	6.16	6.09	6.04	6.00
5	6.61	5.79	5.41	5.19	5.05	4.95	4.88	4.82	4.77
6	5.99	5.14	4.76	4.53	4.39	4.28	4.21	4.15	4.10
7	5.59	4.74	4.35	4.12	3.97	3.87	3.79	3.73	3.68
8	5.32	4.46	4.07	3.84	3.69	3.58	3.50	3.44	3.39
9	5.12	4.26	3.86	3.63	3.48	3.37	3.29	3.23	3.18
10	4.96	4.10	3.71	3.48	3.33	3.22	3.14	3.07	3.02
11	4.84	3.98	3.59	3.36	3.20	3.09	3.01	2.95	2.90
12	4.75	3.89	3.49	3.26	3.11	3.00	2.91	2.85	2.80
13	4.67	3.81	3.41	3.18	3.03	2.92	2.83	2.77	2.71
14	4.60	3.74	3.34	3.11	2.96	2.85	2.76	2.70	2.65
15	4.54	3.68	3.29	3.06	2.90	2.79	2.71	2.64	2.59
16	4.49	3.63	3.24	3.01	2.85	2.74	2.66	2.59	2.54
17	4.45	3.59	3.20	2.96	2.81	2.70	2.61	2.55	2.49
18	4.41	3.55	3.16	2.93	2.77	2.66	2.58	2.51	2.46
19	4.38	3.52	3.13	2.90	2.74	2.63	2.54	2.48	2.42
20	4.35	3.49	3.10	2.87	2.71	2.60	2.51	2.45	2.39
21	4.32	3.47	3.07	2.84	2.68	2.57	2.49	2.42	2.37
22	4.30	3.44	3.05	2.82	2.66	2.55	2.46	2.40	2.34
23	4.28	3.42	3.03	2.80	2.64	2.53	2.44	2.37	2.32
24	4.26	3.40	3.01	2.78	2.62	2.51	2.42	2.36	2.30
25	4.24	3.39	2.99	2.76	2.60	2.49	2.40	2.34	2.28
26	4.23	3.37	2.98	2.74	2.59	2.47	2.39	2.32	2.77
27	4.21	3.35	2.96	2.73	2.57	2.46	2.37	2.31	2.25
28	4.20	3.34	2.95	2.71	2.56	2.45	2.36	2.29	2.24
29	4.18	3.33	2.93	2.70	2.55	2.43	2.35	2.28	2.22
30	4.17	3.32	2.92	2.69	2.53	2.42	2.33	2.27	2.21
40	4.08	3.23	2.84	2.61	2.45	2.34	2.25	2.18	2.12
60	4.00	3.15	2.76	2.53	2.37	2.25	2.17	2.10	2.04
120	3.92	3.07	2.68	2.45	2.29	2.17	2.09	2.02	1.96
∞	3.84	3.00	2.60	2.37	2.21	2.10	2.01	1.94	1.88

Denominator Degrees of Freedom

Source: From M. Merrington and C. M. Thompson, "Tables of Percentage Points of the Inverted Beta (*F*)-Distribution." *Biometrika,* 1943, 33, 73–88. Reproduced by permission of the *Biometrika* Trustees.

TABLE IX   *Continued*

ν_1	10	12	15	20	24	30	40	60	120	∞
ν_2	\multicolumn{10}{c}{**NUMERATOR DEGREES OF FREEDOM**}									
1	241.9	243.9	245.9	248.0	249.1	250.1	251.1	252.2	253.3	254.3
2	19.40	19.41	19.43	19.45	19.45	19.46	19.47	19.48	19.49	19.50
3	8.79	8.74	8.70	8.66	8.64	8.62	8.59	8.57	8.55	8.53
4	5.96	5.91	5.86	5.80	5.77	5.75	5.72	5.69	5.66	5.63
5	4.74	4.68	4.62	4.56	4.53	4.50	4.46	4.43	4.40	4.36
6	4.06	4.00	3.94	3.87	3.84	3.81	3.77	3.74	3.70	3.67
7	3.64	3.57	3.51	3.44	3.41	3.38	3.34	3.30	3.27	3.23
8	3.35	3.28	3.22	3.15	3.12	3.08	3.04	3.01	2.97	2.93
9	3.14	3.07	3.01	2.94	2.90	2.86	2.83	2.79	2.75	2.71
10	2.98	2.91	2.85	2.77	2.74	2.70	2.66	2.62	2.58	2.54
11	2.85	2.79	2.72	2.65	2.61	2.57	2.53	2.49	2.45	2.40
12	2.75	2.69	2.62	2.54	2.51	2.47	2.43	2.38	2.34	2.30
13	2.67	2.60	2.53	2.46	2.42	2.38	2.34	2.30	2.25	2.21
14	2.60	2.53	2.46	2.39	2.35	2.31	2.27	2.22	2.18	2.13
15	2.54	2.48	2.40	2.33	2.29	2.25	2.20	2.16	2.11	2.07
16	2.49	2.42	2.35	2.28	2.24	2.19	2.15	2.11	2.06	2.01
17	2.45	2.38	2.31	2.23	2.19	2.15	2.10	2.06	2.01	1.96
18	2.41	2.34	2.27	2.19	2.15	2.11	2.06	2.02	1.97	1.92
19	2.38	2.31	2.23	2.16	2.11	2.07	2.03	1.98	1.93	1.88
20	2.35	2.28	2.20	2.12	2.08	2.04	1.99	1.95	1.90	1.84
21	2.32	2.25	2.18	2.10	2.05	2.01	1.96	1.92	1.87	1.81
22	2.30	2.23	2.15	2.07	2.03	1.98	1.94	1.89	1.84	1.78
23	2.27	2.20	2.13	2.05	2.01	1.96	1.91	1.86	1.81	1.76
24	2.25	2.18	2.11	2.03	1.98	1.94	1.89	1.84	1.79	1.73
25	2.24	2.16	2.09	2.01	1.96	1.92	1.87	1.82	1.77	1.71
26	2.22	2.15	2.07	1.99	1.95	1.90	1.85	1.80	1.75	1.69
27	2.20	2.13	2.06	1.97	1.93	1.88	1.84	1.79	1.73	1.67
28	2.19	2.12	2.04	1.96	1.91	1.87	1.82	1.77	1.71	1.65
29	2.18	2.10	2.03	1.94	1.90	1.85	1.81	1.75	1.70	1.64
30	2.16	2.09	2.01	1.93	1.89	1.84	1.79	1.74	1.68	1.62
40	2.08	2.00	1.92	1.84	1.79	1.74	1.69	1.64	1.58	1.51
60	1.99	1.92	1.84	1.75	1.70	1.65	1.59	1.53	1.47	1.39
120	1.91	1.83	1.75	1.66	1.61	1.55	1.50	1.43	1.35	1.25
∞	1.83	1.75	1.67	1.57	1.52	1.46	1.39	1.32	1.22	1.00

Denominator Degrees of Freedom

TABLE X   Percentage Points of the *F*-distribution, $\alpha = .025$

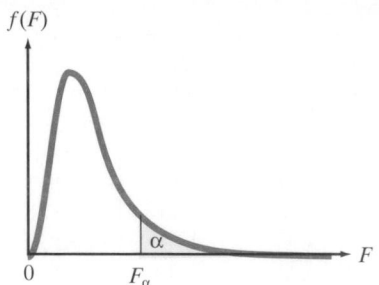

ν_1	NUMERATOR DEGREES OF FREEDOM								
ν_2	1	2	3	4	5	6	7	8	9
1	647.8	799.5	864.2	899.6	921.8	937.1	948.2	956.7	963.3
2	38.51	39.00	39.17	39.25	39.30	39.33	39.36	39.37	39.39
3	17.44	16.04	15.44	15.10	14.88	14.73	14.62	14.54	14.47
4	12.22	10.65	9.98	9.60	9.36	9.20	9.07	8.98	8.90
5	10.01	8.43	7.76	7.39	7.15	6.98	6.85	6.76	6.68
6	8.81	7.26	6.60	6.23	5.99	5.82	5.70	5.60	5.52
7	8.07	6.54	5.89	5.52	5.29	5.12	4.99	4.90	4.82
8	7.57	6.06	5.42	5.05	4.82	4.65	4.53	4.43	4.36
9	7.21	5.71	5.08	4.72	4.48	4.32	4.20	4.10	4.03
10	6.94	5.46	4.83	4.47	4.24	4.07	3.95	3.85	3.78
11	6.72	5.26	4.63	4.28	4.04	3.88	3.76	3.66	3.59
12	6.55	5.10	4.47	4.12	3.89	3.73	3.61	3.51	3.44
13	6.41	4.97	4.35	4.00	3.77	3.60	3.48	3.39	3.31
14	6.30	4.86	4.24	3.89	3.66	3.50	3.38	3.29	3.21
15	6.20	4.77	4.15	3.80	3.58	3.41	3.29	3.20	3.12
16	6.12	4.69	4.08	3.73	3.50	3.34	3.22	3.12	3.05
17	6.04	4.62	4.01	3.66	3.44	3.28	3.16	3.06	2.98
18	5.98	4.56	3.95	3.61	3.38	3.22	3.10	3.01	2.93
19	5.92	4.51	3.90	3.56	3.33	3.17	3.05	2.96	2.88
20	5.87	4.46	3.86	3.51	3.29	3.13	3.01	2.91	2.84
21	5.83	4.42	3.82	3.48	3.25	3.09	2.97	2.87	2.80
22	5.79	4.38	3.78	3.44	3.22	3.05	2.93	2.84	2.76
23	5.75	4.35	3.75	3.41	3.18	3.02	2.90	2.81	2.73
24	5.72	4.32	3.72	3.38	3.15	2.99	2.87	2.78	2.70
25	5.69	4.29	3.69	3.35	3.13	2.97	2.85	2.75	2.68
26	5.66	4.27	3.67	3.33	3.10	2.94	2.82	2.73	2.65
27	5.63	4.24	3.65	3.31	3.08	2.92	2.80	2.71	2.63
28	5.61	4.22	3.63	3.29	3.06	2.90	2.78	2.69	2.61
29	5.59	4.20	3.61	3.27	3.04	2.88	2.76	2.67	2.59
30	5.57	4.18	3.59	3.25	3.03	2.87	2.75	2.65	2.57
40	5.42	4.05	3.46	3.13	2.90	2.74	2.62	2.53	2.45
60	5.29	3.93	3.34	3.01	2.79	2.63	2.51	2.41	2.33
120	5.15	3.80	3.23	2.89	2.67	2.52	2.39	2.30	2.22
∞	5.02	3.69	3.12	2.79	2.57	2.41	2.29	2.19	2.11

Denominator Degrees of Freedom

Source: From M. Merrington and C. M. Thompson, "Tables of Percentage Points of the Inverted Beta (*F*)-Distribution," *Biometrika*, 1943, 33, 73–88. Reproduced by permission of the *Biometrika* Trustees.

TABLE X *Continued*

ν_1 / ν_2	**NUMERATOR DEGREES OF FREEDOM**									
	10	**12**	**15**	**20**	**24**	**30**	**40**	**60**	**120**	**∞**
1	968.6	976.7	984.9	993.1	997.2	1,001	1,006	1,010	1,014	1,018
2	39.40	39.41	39.43	39.45	39.46	39.46	39.47	39.48	39.49	39.50
3	14.42	14.34	14.25	14.17	14.12	14.08	14.04	13.99	13.95	13.90
4	8.84	8.75	8.66	8.56	8.51	8.46	8.41	8.36	8.31	8.26
5	6.62	6.52	6.43	6.33	6.28	6.23	6.18	6.12	6.07	6.02
6	5.46	5.37	5.27	5.17	5.12	5.07	5.01	4.96	4.90	4.85
7	4.76	4.67	4.57	4.47	4.42	4.36	4.31	4.25	4.20	4.14
8	4.30	4.20	4.10	4.00	3.95	3.89	3.84	3.78	3.73	3.67
9	3.96	3.87	3.77	3.67	3.61	3.56	3.51	3.45	3.39	3.33
10	3.72	3.62	3.52	3.42	3.37	3.31	3.26	3.20	3.14	3.08
11	3.53	3.43	3.33	3.23	3.17	3.12	3.06	3.00	2.94	2.88
12	3.37	3.28	3.18	3.07	3.02	2.96	2.91	2.85	2.79	2.72
13	3.25	3.15	3.05	2.95	2.89	2.84	2.78	2.72	2.66	2.60
14	3.15	3.05	2.95	2.84	2.79	2.73	2.67	2.61	2.55	2.49
15	3.06	2.96	2.86	2.76	2.70	2.64	2.59	2.52	2.46	2.40
16	2.99	2.89	2.79	2.68	2.63	2.57	2.51	2.45	2.38	2.32
17	2.92	2.82	2.72	2.62	2.56	2.50	2.44	2.38	2.32	2.25
18	2.87	2.77	2.67	2.56	2.50	2.44	2.38	2.32	2.26	2.19
19	2.82	2.72	2.62	2.51	2.45	2.39	2.33	2.27	2.20	2.13
20	2.77	2.68	2.57	2.46	2.41	2.35	2.29	2.22	2.16	2.09
21	2.73	2.64	2.53	2.42	2.37	2.31	2.25	2.18	2.11	2.04
22	2.70	2.60	2.50	2.39	2.33	2.27	2.21	2.14	2.08	2.00
23	2.67	2.57	2.47	2.36	2.30	2.24	2.18	2.11	2.04	1.97
24	2.64	2.54	2.44	2.33	2.27	2.21	2.15	2.08	2.01	1.94
25	2.61	2.51	2.41	2.30	2.24	2.18	2.12	2.05	1.98	1.91
26	2.59	2.49	2.39	2.28	2.22	2.16	2.09	2.03	1.95	1.88
27	2.57	2.47	2.36	2.25	2.19	2.13	2.07	2.00	1.93	1.85
28	2.55	2.45	2.34	2.23	2.17	2.11	2.05	1.98	1.91	1.83
29	2.53	2.43	2.32	2.21	2.15	2.09	2.03	1.96	1.89	1.81
30	2.51	2.41	2.31	2.20	2.14	2.07	2.01	1.94	1.87	1.79
40	2.39	2.29	2.18	2.07	2.01	1.94	1.88	1.80	1.72	1.64
60	2.27	2.17	2.06	1.94	1.88	1.82	1.74	1.67	1.58	1.48
120	2.16	2.05	1.94	1.82	1.76	1.69	1.61	1.53	1.43	1.31
∞	2.05	1.94	1.83	1.71	1.64	1.57	1.48	1.39	1.27	1.00

Denominator Degrees of Freedom

TABLE XI   Percentage Points of the *F*-distribution, α = .01

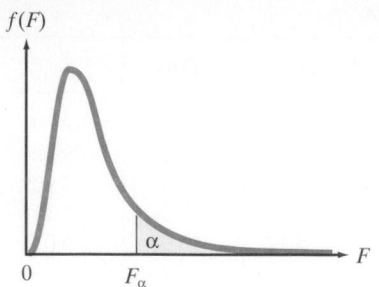

$f(F)$

0 F_α F

α

				NUMERATOR DEGREES OF FREEDOM					
ν_1									
ν_2	1	2	3	4	5	6	7	8	9
1	4,052	4,999.5	5,403	5,625	5,764	5,859	5,928	5,982	6,022
2	98.50	99.00	99.17	99.25	99.30	99.33	99.36	99.37	99.39
3	34.12	30.82	29.46	28.71	28.24	27.91	27.67	27.49	27.35
4	21.20	18.00	16.69	15.98	15.52	15.21	14.98	14.80	14.66
5	16.26	13.27	12.06	11.39	10.97	10.67	10.46	10.29	10.16
6	13.75	10.92	9.78	9.15	8.75	8.47	8.26	8.10	7.98
7	12.25	9.55	8.45	7.85	7.46	7.19	6.99	6.84	6.72
8	11.26	8.65	7.59	7.01	6.63	6.37	6.18	6.03	5.91
9	10.56	8.02	6.99	6.42	6.06	5.80	5.61	5.47	5.35
10	10.04	7.56	6.55	5.99	5.64	5.39	5.20	5.06	4.94
11	9.65	7.21	6.22	5.67	5.32	5.07	4.89	4.74	4.63
12	9.33	6.93	5.95	5.41	5.06	4.82	4.64	4.50	4.39
13	9.07	6.70	5.74	5.21	4.86	4.62	4.44	4.30	4.19
14	8.86	6.51	5.56	5.04	4.69	4.46	4.28	4.14	4.03
15	8.68	6.36	5.42	4.89	4.56	4.32	4.14	4.00	3.89
16	8.53	6.23	5.29	4.77	4.44	4.20	4.03	3.89	3.78
17	8.40	6.11	5.18	4.67	4.34	4.10	3.93	3.79	3.68
18	8.29	6.01	5.09	4.58	4.25	4.01	3.84	3.71	3.60
19	8.18	5.93	5.01	4.50	4.17	3.94	3.77	3.63	3.52
20	8.10	5.85	4.94	4.43	4.10	3.87	3.70	3.56	3.46
21	8.02	5.78	4.87	4.37	4.04	3.81	3.64	3.51	3.40
22	7.95	5.72	4.82	4.31	3.99	3.76	3.59	3.45	3.35
23	7.88	5.66	4.76	4.26	3.94	3.71	3.54	3.41	3.30
24	7.82	5.61	4.72	4.22	3.90	3.67	3.50	3.36	3.26
25	7.77	5.57	4.68	4.18	3.85	3.63	3.46	3.32	3.22
26	7.72	5.53	4.64	4.14	3.82	3.59	3.42	3.29	3.18
27	7.68	5.49	4.60	4.11	3.78	3.56	3.39	3.26	3.15
28	7.64	5.45	4.57	4.07	3.75	3.53	3.36	3.23	3.12
29	7.60	5.42	4.54	4.04	3.73	3.50	3.33	3.20	3.09
30	7.56	5.39	4.51	4.02	3.70	3.47	3.30	3.17	3.07
40	7.31	5.18	4.31	3.83	3.51	3.29	3.12	2.99	2.89
60	7.08	4.98	4.13	3.65	3.34	3.12	2.95	2.82	2.72
120	6.85	4.79	3.95	3.48	3.17	2.96	2.79	2.66	2.56
∞	6.63	4.61	3.78	3.32	3.02	2.80	2.64	2.51	2.41

Denominator Degrees of Freedom

TABLE XI *Continued*

v_2 \ v_1	NUMERATOR DEGREES OF FREEDOM									
	10	**12**	**15**	**20**	**24**	**30**	**40**	**60**	**120**	**∞**
1	6,056	6,106	6,157	6,209	6,235	6,261	6,287	6,313	6,339	6,366
2	99.40	99.42	99.43	99.45	99.46	99.47	99.47	99.48	99.49	99.50
3	27.23	27.05	26.87	26.69	26.60	26.50	26.41	26.32	26.22	26.13
4	14.55	14.37	14.20	14.02	13.93	13.84	13.75	13.65	13.56	13.46
5	10.05	9.89	9.72	9.55	9.47	9.38	9.29	9.20	9.11	9.02
6	7.87	7.72	7.56	7.40	7.31	7.23	7.14	7.06	6.97	6.88
7	6.62	6.47	6.31	6.16	6.07	5.99	5.91	5.82	5.74	5.65
8	5.81	5.67	5.52	5.36	5.28	5.20	5.12	5.03	4.95	4.86
9	5.26	5.11	4.96	4.81	4.73	4.65	4.57	4.48	4.40	4.31
10	4.85	4.71	4.56	4.41	4.33	4.25	4.17	4.08	4.00	3.91
11	4.54	4.40	4.25	4.10	4.02	3.94	3.86	3.78	3.69	3.60
12	4.30	4.16	4.01	3.86	3.78	3.70	3.62	3.54	3.45	3.36
13	4.10	3.96	3.82	3.66	3.59	3.51	3.43	3.34	3.25	3.17
14	3.94	3.80	3.66	3.51	3.43	3.35	3.27	3.18	3.09	3.00
15	3.80	3.67	3.52	3.37	3.29	3.21	3.13	3.05	2.96	2.87
16	3.69	3.55	3.41	3.26	3.18	3.10	3.02	2.93	2.84	2.75
17	3.59	3.46	3.31	3.16	3.08	3.00	2.92	2.83	2.75	2.65
18	3.51	3.37	3.23	3.08	3.00	2.92	2.84	2.75	2.66	2.57
19	3.43	3.30	3.15	3.00	2.92	2.84	2.76	2.67	2.58	2.49
20	3.37	3.23	3.09	2.94	2.86	2.78	2.69	2.61	2.52	2.42
21	3.31	3.17	3.03	2.88	2.80	2.72	2.64	2.55	2.46	2.36
22	3.26	3.12	2.98	2.83	2.75	2.67	2.58	2.50	2.40	2.31
23	3.21	3.07	2.93	2.78	2.70	2.62	2.54	2.45	2.35	2.26
24	3.17	3.03	2.89	2.74	2.66	2.58	2.49	2.40	2.31	2.21
25	3.13	2.99	2.85	2.70	2.62	2.54	2.45	2.36	2.27	2.17
26	3.09	2.96	2.81	2.66	2.58	2.50	2.42	2.33	2.23	2.13
27	3.06	2.93	2.78	2.63	2.55	2.47	2.38	2.29	2.20	2.10
28	3.03	2.90	2.75	2.60	2.52	2.44	2.35	2.26	2.17	2.06
29	3.00	2.87	2.73	2.57	2.49	2.41	2.33	2.23	2.14	2.03
30	2.98	2.84	2.70	2.55	2.47	2.39	2.30	2.21	2.11	2.01
40	2.80	2.66	2.52	2.37	2.29	2.20	2.11	2.02	1.92	1.80
60	2.63	2.50	2.35	2.20	2.12	2.03	1.94	1.84	1.73	1.60
120	2.47	2.34	2.19	2.03	1.95	1.86	1.76	1.66	1.53	1.38
∞	2.32	2.18	2.04	1.88	1.79	1.70	1.59	1.47	1.32	1.00

Denominator Degrees of Freedom

TABLE XII   **Critical Values of T_L and T_U for the Wilcoxon Rank Sum Test: Independent Samples**

Test statistic is the rank sum associated with the smaller sample (if equal sample sizes, either rank sum can be used).

a. $\alpha = .025$ one-tailed; $\alpha = .05$ two-tailed

n_2	n_1 3 T_L	T_U	4 T_L	T_U	5 T_L	T_U	6 T_L	T_U	7 T_L	T_U	8 T_L	T_U	9 T_L	T_U	10 T_L	T_U
3	5	16	6	18	6	21	7	23	7	26	8	28	8	31	9	33
4	6	18	11	25	12	28	12	32	13	35	14	38	15	41	16	44
5	6	21	12	28	18	37	19	41	20	45	21	49	22	53	24	56
6	7	23	12	32	19	41	26	52	28	56	29	61	31	65	32	70
7	7	26	13	35	20	45	28	56	37	68	39	73	41	78	43	83
8	8	28	14	38	21	49	29	61	39	73	49	87	51	93	54	98
9	8	31	15	41	22	53	31	65	41	78	51	93	63	108	66	114
10	9	33	16	44	24	56	32	70	43	83	54	98	66	114	79	131

b. $\alpha = .05$ one-tailed; $\alpha = .10$ two-tailed

n_2	n_1 3 T_L	T_U	4 T_L	T_U	5 T_L	T_U	6 T_L	T_U	7 T_L	T_U	8 T_L	T_U	9 T_L	T_U	10 T_L	T_U
3	6	15	7	17	7	20	8	22	9	24	9	27	10	29	11	31
4	7	17	12	24	13	27	14	30	15	33	16	36	17	39	18	42
5	7	20	13	27	19	36	20	40	22	43	24	46	25	50	26	54
6	8	22	14	30	20	40	28	50	30	54	32	58	33	63	35	67
7	9	24	15	33	22	43	30	54	39	66	41	71	43	76	46	80
8	9	27	16	36	24	46	32	58	41	71	52	84	54	90	57	95
9	10	29	17	39	25	50	33	63	43	76	54	90	66	105	69	111
10	11	31	18	42	26	54	35	67	46	80	57	95	69	111	83	127

Source: From F. Wilcoxon and R. A. Wilcox, "Some Rapid Approximate Statistical Procedures," 1964, 20–23. Courtesy of Lederle Laboratories Division of American Cyanamid Company, Madison, NJ.

TABLE XIII **Critical Values of T_0 in the Wilcoxon Paired Difference Signed Rank Test**

One-Tailed	Two-Tailed	$n = 5$	$n = 6$	$n = 7$	$n = 8$	$n = 9$	$n = 10$
$\alpha = .05$	$\alpha = .10$	1	2	4	6	8	11
$\alpha = .025$	$\alpha = .05$		1	2	4	6	8
$\alpha = .01$	$\alpha = .02$			0	2	3	5
$\alpha = .005$	$\alpha = .01$				0	2	3
		$n = 11$	$n = 12$	$n = 13$	$n = 14$	$n = 15$	$n = 16$
$\alpha = .05$	$\alpha = .10$	14	17	21	26	30	36
$\alpha = .025$	$\alpha = .05$	11	14	17	21	25	30
$\alpha = .01$	$\alpha = .02$	7	10	13	16	20	24
$\alpha = .005$	$\alpha = .01$	5	7	10	13	16	19
		$n = 17$	$n = 18$	$n = 19$	$n = 20$	$n = 21$	$n = 22$
$\alpha = .05$	$\alpha = .10$	41	47	54	60	68	75
$\alpha = .025$	$\alpha = .05$	35	40	46	52	59	66
$\alpha = .01$	$\alpha = .02$	28	33	38	43	49	56
$\alpha = .005$	$\alpha = .01$	23	28	32	37	43	49
		$n = 23$	$n = 24$	$n = 25$	$n = 26$	$n = 27$	$n = 28$
$\alpha = .05$	$\alpha = .10$	83	92	101	110	120	130
$\alpha = .025$	$\alpha = .05$	73	81	90	98	107	117
$\alpha = .01$	$\alpha = .02$	62	69	77	85	93	102
$\alpha = .005$	$\alpha = .01$	55	61	68	76	84	92
		$n = 29$	$n = 30$	$n = 31$	$n = 32$	$n = 33$	$n = 34$
$\alpha = .05$	$\alpha = .10$	141	152	163	175	188	201
$\alpha = .025$	$\alpha = .05$	127	137	148	159	171	183
$\alpha = .01$	$\alpha = .02$	111	120	130	141	151	162
$\alpha = .005$	$\alpha = .01$	100	109	118	128	138	149
		$n = 35$	$n = 36$	$n = 37$	$n = 38$	$n = 39$	
$\alpha = .05$	$\alpha = .10$	214	228	242	256	271	
$\alpha = .025$	$\alpha = .05$	195	208	222	235	250	
$\alpha = .01$	$\alpha = .02$	174	186	198	211	224	
$\alpha = .005$	$\alpha = .01$	160	171	183	195	208	
		$n = 40$	$n = 41$	$n = 42$	$n = 43$	$n = 44$	$n = 45$
$\alpha = .05$	$\alpha = .10$	287	303	319	336	353	371
$\alpha = .025$	$\alpha = .05$	264	279	295	311	327	344
$\alpha = .01$	$\alpha = .02$	238	252	267	281	297	313
$\alpha = .005$	$\alpha = .01$	221	234	248	262	277	292
		$n = 46$	$n = 47$	$n = 48$	$n = 49$	$n = 50$	
$\alpha = .05$	$\alpha = .10$	389	408	427	446	466	
$\alpha = .025$	$\alpha = .05$	361	379	397	415	434	
$\alpha = .01$	$\alpha = .02$	329	345	362	380	398	
$\alpha = .005$	$\alpha = .01$	307	323	339	356	373	

Source: From F. Wilcoxon and R. A. Wilcox, "Some Rapid Approximate Statistical Procedures," 1964, p. 28. Courtesy of Lederle Laboratories Division of American Cyanamid Company, Madison, NJ.

TABLE XIV   Critical Values of Spearman's Rank Correlation Coefficient

The α values correspond to a one-tailed test of H_0: ρ = 0. The value should be doubled for two-tailed tests.

n	α = .05	α = .025	α = .01	α = .005	n	α = .05	α = .025	α = .01	α = .005
5	.900	—	—	—	18	.399	.476	.564	.625
6	.829	.886	.943	—	19	.388	.462	.549	.608
7	.714	.786	.893	—	20	.377	.450	.534	.591
8	.643	.738	.833	.881	21	.368	.438	.521	.576
9	.600	.683	.783	.833	22	.359	.428	.508	.562
10	.564	.648	.745	.794	23	.351	.418	.496	.549
11	.523	.623	.736	.818	24	.343	.409	.485	.537
12	.497	.591	.703	.780	25	.336	.400	.475	.526
13	.475	.566	.673	.745	26	.329	.392	.465	.515
14	.457	.545	.646	.716	27	.323	.385	.456	.505
15	.441	.525	.623	.689	28	.317	.377	.448	.496
16	.425	.507	.601	.666	29	.311	.370	.440	.487
17	.412	.490	.582	.645	30	.305	.364	.432	.478

Source: From E. G. Olds, "Distribution of Sums of Squares of Rank Differences for Small Samples," *Annals of Mathematical Statistics,* 1938, 9. Reproduced with the permission of the Editor, *Annals of Mathematical Statistics.*

✦ APPENDIX B
DATA SETS

Contents

B.1   CORONARY ARTERY PATIENTS' BLOOD LOSS DATA

ASCII file name: B1BLOOD.DAT

Number of Observations: 114

Variable	Column(s)	Type	Description
PATIENT	1–3	QL	Patient identification number
DRUG	5–7	QL	Drug status (YES, or NO)
CABG	9	QN	Coronary artery bypass graft number
BYPASS	11–13	QN	Length of time on bypass machine (minutes)
XCLAMP	15–16	QN	Length of time aorta is clamped (minutes)
TEMP	18–19	QN	Body temperature (degrees Celsius)
DIABETE	21–24	QL	Diabetes type (I, II, or NONE)
COMP	26–31	QL	Complications (REDO, INFECT, BOTH, or NONE)
LOSS	33–36	QN	Amount of blood loss (cubic centimeters)
PLATELET	38–43	QN	Platelet count on day of surgery
BLTIME	45–48	QN	Bleeding time (minutes)

Data for the first 25 observations are listed below:

PATIENT	DRUG	CABG	BYPASS	XCLAMP	TEMP	DIABETE	COMP	LOSS	PLATELET	BLTIME
1A	NO	5	182	86	26	II	REDO	1045	99000	5.5
2A	NO	4	87	49	25	NONE	NONE	1170	327000	8.0
3A	NO	4	106	48	20	NONE	NONE	1020	154000	8.0
4A	NO	5	132	69	28	NONE	NONE	1145	145000	.
5A	NO	5	110	67	27	I	NONE	2015	227000	4.5
6A	NO	5	136	83	28	NONE	NONE	1275	61000	3.5
7A	NO	3	99	43	26	NONE	NONE	1160	131000	5.5
8A	NO	3	104	67	28	NONE	NONE	810	145000	4.5
9A	NO	4	88	47	28	NONE	NONE	1030	158000	4.5
10A	NO	1	30	11	35	NONE	INFECT	1585	111000	5.0
11A	NO	3	97	60	28	II	REDO	750	300000	7.0
12A	NO	5	114	79	20	NONE	NONE	1180	116000	8.5
13A	NO	4	73	45	28	NONE	NONE	580	241000	4.5
14A	NO	4	116	48	25	NONE	NONE	675	125000	.
15A	NO	2	47	20	36	NONE	NONE	1015	153000	3.5
16A	NO	4	93	54	28	II	INFECT	485	235000	4.0
17A	NO	5	114	74	21	NONE	NONE	905	172000	8.5
18A	NO	5	126	68	27	NONE	NONE	705	90000	4.5
19A	NO	2	86	40	28	II	NONE	485	175000	2.5
20A	NO	4	104	54	29	II	NONE	970	194000	8.5
21A	NO	2	72	31	27	NONE	BOTH	3370	226000	4.0
22A	NO	5	111	64	27	NONE	NONE	570	123000	5.0
23A	NO	2	60	23	25	NONE	NONE	1190	182000	4.5
24A	NO	3	64	35	28	NONE	NONE	1220	141000	4.5
25A	NO	1	49	16	28	II	NONE	525	110000	3.5

B.2 *CAR & DRIVER* DATA

ASCII file name: B2CAR.DAT

Number of observations: 129

Variable	Column(s)	Type	Description
CAR	1–34	QL	Make/model of car
PRICE	36–41	QN	New car dealer price (dollars)
TIME1	44–47	QN	Acceleration time (seconds) from 0 to 60 miles per hour
TIME2	50–53	QN	Acceleration time (seconds) required to travel 1/4 mile
MAXSPEED	56–58	QN	Maximum speed attained (miles per hour)
BRAKDIST	61–63	QN	Braking distance (feet) from 70 miles per hour
MPG	66–67	QN	EPA estimated fuel economy (miles per gallon)
GRIP	70–73	QN	Road-holding "grip" (g's) during cornering

Data for the first 25 observations are listed below:

CAR	PRICE	TIME1	TIME2	MAXSPEED	BRAKDIST	MPG	GRIP
Acura Integra GS-R	20015	7.0	15.6	133	185	25	0.84
Acura Integra LS	18560	7.5	15.9	125	189	25	0.82
Acura NSX-T	86642	5.2	13.8	162	173	18	0.95
AM General Hummer	71760	18.1	21.2	83	253	10	0.62
Audi A6 Quattro	36802	8.3	16.5	127	191	18	0.79
Audi Cabriolet	40960	9.7	17.2	126	189	18	0.73
BMW M3	38782	5.3	14.0	138	165	19	0.88
BMW 318ti	22770	7.8	16.1	116	175	22	.
BMW 325i	36015	7.7	16.0	127	182	20	0.82
BMW 540i Six-speed	54044	5.7	14.3	129	186	14	0.78
BMW 740i	66837	8.4	16.6	127	181	16	0.79
BMW 840Ci	74790	7.1	15.5	155	170	15	0.83
BMW 850CSi	108395	5.3	13.9	158	167	12	0.85
Buick Park Avenue Ultra	33797	7.0	15.5	107	216	17	0.74
Buick Regal Gran Sport	24305	8.3	16.5	110	202	19	0.79
Buick Riviera	30643	8.2	16.3	109	199	19	0.75
Cadillac Eldorado Touring Coupe	45626	6.4	14.8	148	184	16	0.80
Cadillac SLS	46501	6.7	15.0	115	195	16	0.77
Chevrolet Blazer LT	26969	9.1	17.0	103	218	16	0.67
Chevrolet Camaro	20509	7.7	15.9	114	192	19	0.83
Chevrolet Camaro Convertible	22484	9.3	17.0	110	195	19	0.84
Chevrolet Camaro Z28	22686	5.5	14.1	156	162	17	0.86
Chevrolet Cavalier	13417	8.8	16.7	113	209	25	0.76
Chevrolet Corvette	42614	5.1	13.7	161	166	17	0.85
Chevrolet Impala SS	23355	6.5	15.0	142	179	17	0.86

B.3   STARTING SALARIES OF USF GRADUATES

ASCII file name: B3SALARY.DAT

Number of observations: 159

Variable	Column(s)	Type	Description
DEGREE	1	QL	Degree type (B=bachelors, M=masters)
DATE	9–13	QL	Graduation date (mm/yy)
MAJOR	20–22	QL	Major abbreviation
GENDER	27	QL	Gender (M=male, F=female)
JOB	35–63	QL	Job title
SALARY	65–69	QN	Starting salary (dollars)

Data for the first 25 observations are listed below:

DEGREE	DATE	MAJOR	GENDER	JOB	SALARY
B	05/94	ACC	M	Staff Accountant	36000
B	12/93	ISM	M	Application Developer	28000
B	05/94	EEL	M	Electrical Engineer	42500
B	12/93	HTY	F	Marketing Specialist	31800
B	12/93	ACC	M	Staff Accountant	32000
B	12/93	ACC	M	Administrative Manager	20000
B	12/93	ECP	M	Software Engineer	29000
B	12/93	ISM	F	Staff Consultant	28500
M	12/93	EEL	M	Engineer	35000
B	05/94	ISM	F	Team Member	26500
B	12/93	FIN	M	Assistant Manager	23000
B	12/93	ISM	F	Computer Program Analyst	22000
B	12/93	MKT	M	Management Trainee	21000
B	12/93	MKT	M	Account Manager	29472
B	12/93	FIN	M	Lender	21000
B	12/93	EME	M	PGS Engineer	37000
B	12/93	FIN	M	Sales Manager	23000
B	12/93	FIN	M	Budget Analyst	25000
B	12/93	MKT	F	Sales Representative	20000
B	05/94	EEL	M	Staff Consultant	30500
B	05/94	ACC	M	Staff Auditor	31500
B	12/93	ACC	M	Staff Accountant Auditor	29000
B	12/93	ISM	M	Sr. Technical Engineer	25000
B	12/93	FIN	F	International Group	19600
B	08/94	FIN	M	Audit Training	26000

B.4 SEALED MILK BIDS DATA

ASCII file name: B4MILK.DAT

Number of observations: 395

Variable	Column(s)	Type	Description
YEAR	7–10	QN	Year in which milk contract awarded
MARKET	15–24	QL	Kentucky market (TRI-COUNTY or SURROUND)
WINNER	29–44	QL	Name of winning dairy
WWBID	47–52	QN	Winning bid price of whole white milk (dollars per half-pint)
LFWBID	57–62	QN	Winning bid price of low fat white milk (dollars per half-pint)
LFCBID	67–72	QN	Winning bid price of low fat chocolate milk (dollars per half-pint)

Data for the first 25 observations are listed below:

YEAR	MARKET	WINNER	WWBID	LFWBID	LFCBID
1983	TRI-COUNTY	MEYER	0.1260	0.1260	.
1983	TRI-COUNTY	TRAUTH	0.1495	0.1395	.
1983	SURROUND	HILL'S DAIRY	0.1400	0.1350	0.1430
1983	TRI-COUNTY	TRAUTH	0.1193	0.1113	.
1983	TRI-COUNTY	MEYER	0.1094	0.1094	0.1094
1983	SURROUND	CLOVERLEAF	0.1200	.	.
1983	SURROUND	CLOVERLEAF	0.1200	0.1100	0.1100
1983	SURROUND	BORDEN	0.1200	0.1190	0.1190
1983	SURROUND	TRAUTH	.	0.1143	0.1143
1983	SURROUND	BORDEN	0.1300	0.1300	.
1983	SURROUND	CLOVERLEAF	0.1200	0.1100	0.1100
1983	SURROUND	FLAV-O-RICH	0.1225	0.1225	0.1225
1983	SURROUND	CLOVERLEAF	0.1200	.	0.1100
1983	SURROUND	FLAV-O-RICH	0.1290	0.1220	0.1220
1983	TRI-COUNTY	TRAUTH	0.1064	0.1064	.
1983	TRI-COUNTY	MEYER	0.1074	0.1074	0.1074
1983	TRI-COUNTY	MEYER	.	0.0987	.
1983	TRI-COUNTY	TRAUTH	0.1110	0.1010	0.1010
1983	TRI-COUNTY	MEYER	.	0.1026	0.1026
1983	TRI-COUNTY	MEYER	0.1095	0.1095	0.1095
1983	SURROUND	WILSON DIST.	0.1375	0.1330	0.1425
1983	SURROUND	ROYAL CREST	0.1370	0.1116	0.1325
1983	SURROUND	CLOVERLEAF	0.1250	0.1250	0.1250
1983	SURROUND	THOMPSON-GLASS	0.1300	0.1200	0.1250
1983	TRI-COUNTY	TRAUTH	0.1293	0.1215	.

B.5   FEDERAL TRADE COMMISSION RANKINGS OF DOMESTIC CIGARETTE BRANDS

ASCII file name: B5FTC.DAT

Number of observations: 962

Variable	Column(s)	Type	Description
BRAND	1–23	QL	Cigarette brand name
LENGTH	25–27	QN	Length (millimeters)
MENTHOL	29–30	QL	Menthol type (M=menthol, NM=non-menthol)
FILTER	32–33	QL	Filter type (F=filter, NF=non-filter)
LIGHT	35–36	QL	Light type (FF=full-flavor, LT=light, UL=ultra-light)
PACK	38–39	QL	Pack type (SP=soft pack, HP=hard pack)
TAR	41–42	QN	Tar content (milligrams)
NICOTINE	44–46	QN	Nicotine content (milligrams)
CO	48–49	QN	Carbon monoxide content (milligrams)

Data for the first 25 observations are listed below:

BRAND	LENGTH	MENTHOL	FILTER	LIGHT	PACK	TAR	NICOTINE	CO
ALPINE	100	M	F	FF	SP	15	1.0	15
ALPINE	100	M	F	LT	SP	9	0.7	10
ALPINE	120	M	F	FF	HP	15	1.0	15
ALPINE	120	M	F	LT	HP	9	0.7	9
ALPINE	120	M	F	FF	SP	15	1.0	15
ALPINE	120	M	F	LT	SP	9	0.7	10
AMERICAN FILTER	100	NM	F	FF	SP	17	1.3	16
AMERICAN FILTER	120	NM	F	FF	SP	16	1.2	14
AMERICAN LIGHTS	100	NM	F	LT	SP	11	0.9	12
AMERICAN LIGHTS	100	M	F	LT	SP	11	0.9	11
AMERICAN LIGHTS	120	NM	F	LT	SP	11	0.9	12
AUSTIN	100	NM	F	FF	SP	14	0.9	18
AUSTIN	100	M	F	FF	SP	17	1.0	18
AUSTIN	100	NM	F	LT	SP	8	0.7	11
AUSTIN	100	M	F	LT	SP	9	0.7	12
AUSTIN	100	NM	F	UL	SP	5	0.4	7
AUSTIN	120	NM	F	FF	HP	13	0.8	15
AUSTIN	120	NM	F	LT	HP	9	0.6	12
AUSTIN	120	NM	F	FF	SP	14	0.9	16
AUSTIN	120	M	F	FF	SP	14	0.9	16
AUSTIN	120	NM	F	LT	SP	9	0.6	12
AUSTIN	120	M	F	LT	SP	9	0.7	13
AUSTIN	120	NM	F	UL	SP	5	0.4	6
BARCLAY	100	NM	F	FF	SP	5	0.4	4
BARCLAY	120	NM	F	FF	HP	4	0.3	3

APPENDIX C
CALCULATION FORMULAS
FOR ANALYSIS OF VARIANCE

Contents

C.1   FORMULAS FOR THE CALCULATIONS IN THE COMPLETELY RANDOMIZED DESIGN

$$CM = \text{Correction for mean}$$

$$= \frac{(\text{Total of all observations})^2}{\text{Total number of observations}} = \frac{\left(\sum y_i\right)^2}{n}$$

$$SS(\text{Total}) = \text{Total sum of squares}$$

$$= (\text{Sum of squares of all observations}) - CM = \sum y_i^2 - CM$$

$$SST = \text{Sum of squares for treatments}$$

$$= \left(\begin{array}{c} \text{Sum of squares of treatment totals with} \\ \text{each square divided by the number of} \\ \text{observations for that treatment} \end{array} \right) - CM$$

$$= \frac{T_1^2}{n_1} + \frac{T_2^2}{n_2} + \cdots + \frac{T_p^2}{n_p} - CM$$

$$SSE = \text{Sum of squares for error} = SS(\text{Total}) - SST$$

$$MST = \text{Mean square for treatments} = \frac{SST}{p-1}$$

$$MSE = \text{Mean square for error} = \frac{SSE}{n-p}$$

$$F = \text{Test statistic} = \frac{MST}{MSE}$$

where

n = Total number of observations

p = Number of treatments

T_i = Total for treatment i $(i = 1, 2, \ldots, p)$

C.2 FORMULAS FOR THE CALCULATIONS IN THE RANDOMIZED BLOCK DESIGN

CM = Correction for mean

$$= \frac{(\text{Total of all observations})^2}{\text{Total number of observations}} = \frac{\left(\sum y_i\right)^2}{n}$$

SS(Total) = Total sum of squares

$$= (\text{Sum of squares of all observations}) - \text{CM} = \sum y_i^2 - \text{CM}$$

SST = Sum of squares for treatments

$$= \left(\begin{array}{c} \text{Sum of squares of treatment totals with} \\ \text{each square divided by } b \text{, the number of} \\ \text{observations for that treatment} \end{array} \right) - \text{CM}$$

$$= \frac{T_1^2}{b} + \frac{T_2^2}{b} + \cdots + \frac{T_p^2}{b} - \text{CM}$$

SST = Sum of squares for blocks

$$= \left(\begin{array}{c} \text{Sum of squares of block totals with} \\ \text{each square divided by } p \text{, the number} \\ \text{of observations in that block} \end{array} \right) - \text{CM}$$

$$= \frac{B_1^2}{p} + \frac{B_2^2}{p} + \cdots + \frac{B_b^2}{p} - \text{CM}$$

SSE = Sum of squares for error = SS(Total) − SST − SSB

$$\text{MST} = \text{Mean square for treatments} = \frac{\text{SST}}{p - 1}$$

$$\text{MSB} = \text{Mean square for blocks} = \frac{\text{SSB}}{b - 1}$$

$$\text{MSE} = \text{Mean square for error} = \frac{\text{SSE}}{n - p - b + 1}$$

$$F = \text{Test statistic} = \frac{\text{MST}}{\text{MSE}}$$

where

n = Total number of observations

b = Number of blocks

p = Number of treatments

T_i = Total for treatment i ($i = 1, 2, \ldots, p$)

B_i = Total for block i ($i = 1, 2, \ldots, b$)

C.3 FORMULAS FOR THE CALCULATIONS FOR A TWO-FACTOR FACTORIAL EXPERIMENT

CM = Correction for mean

$$= \frac{(\text{Total of all } n \text{ measurements})^2}{n} = \frac{\left(\sum\limits_{i=1}^{n} y_i\right)^2}{n}$$

SS(Total) = Total sum of squares

$$= (\text{Sum of squares of all } n \text{ measurements}) - \text{CM} = \sum_{i=1}^{n} y_i^2 - \text{CM}$$

SS(A) = Sum of squares for main effects, factor A

$$= \left(\begin{array}{c} \text{Sum of squares of the totals } A_1, A_2, ..., A_a \\ \text{divided by the number of measurements} \\ \text{in a single total, namely } br \end{array} \right) - \text{CM}$$

$$= \frac{\sum\limits_{i=1}^{a} A_i^2}{br} - \text{CM}$$

SS(B) = Sum of squares for main effects, factor B

$$= \left(\begin{array}{c} \text{Sum of squares of the totals } B_1, B_2, ..., B_b \\ \text{divided by the number of measurements} \\ \text{in a single total, namely } ar \end{array} \right) - \text{CM}$$

$$= \frac{\sum\limits_{i=1}^{b} B_i^2}{ar} - \text{CM}$$

SS(AB) = Sum of squares for AB interaction

$$= \left(\begin{array}{c} \text{Sum of squares of the cell totals} \\ AB_{11}, AB_{12}, ..., AB_{ab} \text{ divided by} \\ \text{the number of measurements in} \\ \text{a single total, namely } r \end{array} \right) - \text{SS}(A) - \text{SS}(B) - \text{CM}$$

$$= \frac{\sum\limits_{j=1}^{b} \sum\limits_{i=1}^{a} AB_{ij}^2}{r} - \text{SS}(A) - \text{SS}(B) - \text{CM}$$

where

a = Number of levels of factor A

b = Number of levels of factor B

r = Number of replicates (observations per treatment)

A_i = Total for level i of factor A ($i = 1, 2, ..., a$)

B_i = Total for level i of factor B ($i = 1, 2, ..., b$)

AB_{ij} = Total for treatment (i, j) i.e. for ith level of factor A and jth level of factor B

ANSWERS TO SELECTED EXERCISES

CHAPTER 1

1.11 Qualitative; qualitative **1.13b.** Qualitative **c.** Survey data **1.15a.** quantitative **b.** qualitative **c.** qualitative **d.** quantitative **1.17a.** quantitative **b.** quantitative **c.** qualitative **d.** quantitative **e.** qualitative **f.** quantitative **g.** qualitative **1.19b.** Job satisfaction and Machiavellian rating **d.** Survey data **1.21b.** Number of headers per game and IQ **c.** Both quantitative **1.23b.** Amount of acid required; quantitative **d.** Designed experiment **1.25b.** Answer to question posed; qualitative **d.** Survey data

CHAPTER 2

2.9a. 195 **b.** 90 **2.13a.** 23 **b.** Stem = 0; leaves = 0, 1, 2; values = 0, 1, 2 **2.21b.** 100, AL **2.23a.**

Stem	Leaf
0	0 11234559
1	1 123
2	2 00
3	3 9
4	4 ⑥
5	5 ⑥
6	6 ①⑤
7	7
8	8
9	9
10	10 ⓪

c. Eclipse **2.25a.** 12 **b.** 40 **c.** 7 **d.** 21 **e.** 144 **2.27a.** 11.2 **b.** 12 **c.** 30 **2.29** Mode = 15; median = 15; mean = 14.55 **2.33a.** 8.5 **b.** 25 **c.** .778 **d.** 13.444 **2.35a.** Mean < median **b.** Mean > median **c.** Mean = median **2.37a.** Mean = $-.1544$; median = $-.11$ **b.** Outlier = -1.07 **c.** Mean = $-.116$; median = $-.10$; mean is most affected **2.41a.** Skewed right **b.** Skewed left **c.** Skewed right **d.** Symmetric **e.** Skewed right **f.** Skewed left **2.43a.** No **b.** Yes, 110 **2.47a.** $s^2 = 4.89$, $s = 2.2$ **b.** $s^2 = 3.33$, $s = 1.83$ **c.** $s^2 = .187$, $s = .432$ **2.49a.** 5, 3.7, 1.92 **b.** 99, 1,949.25, 44.15 **c.** 98, 1,307.84, 36.16 **2.53a.** 3, 1.3, 1.14 **b.** 3, 1.3, 1.14 **c.** 3, 1.3, 1.14 **2.55** First professor **2.59b.** At least $\frac{3}{4}$; at least $\frac{8}{9}$; nothing; nothing **2.61a.** $\approx 68\%$ **b.** $\approx 95\%$ **c.** $\approx 100\%$ **2.63** $s = 104.17$; no **2.65b.** $\bar{x} \pm 2s \Rightarrow (-.68, 5.24)$ **2.67a.** Adolescents: $x + s \Rightarrow (8.41, 13.37)$, $\bar{x} \pm 2s \Rightarrow (5.93, 15.85)$; Adults: $\bar{x} \pm s \Rightarrow (3.56, 9.34)$, $\bar{x} \pm 2s \Rightarrow (0.67, 12.23)$ **b.** $\bar{x} \pm 1s$, $\approx 68\%$; $\bar{x} \pm 2s$, $\approx 95\%$ **c.** No **2.69a.** At most 25% **b.** $\approx 2.5\%$ **2.71a.** $\approx .025$ **b.** Yes **2.73a.** $z = 2$ **b.** $z = .5$ **c.** $z = 0$ **d.** $z = -2.5$ **2.75** Median **2.77a.** $z = 2.0$: 3.7; $z = 1.0$: 2.2; $z = .5$: 2.95; $z = -2.5$: 1.45 **b.** GPA = 1.9 **c.** Mound-shaped and symmetric **2.79a.** Skewed right distribution **b.** 90% of the scores are below 660. **c.** Only

6% of the scores exceeded yours.   **2.81a.** $z = -3$ **b.** Yes   **c.** No   **d.** Yes   **2.85a.** 39
b. $Q_L = 31.5$, $Q_U = 45$ **c.** IQR $= 13.5$ **d.** Skewed to the left   **e.** 50%, 75%
2.89b. Customers 238, 268, 269, and 264   **c.** $z = 1.92, 2.06, 2.13,$ and 3.14   **2.91a.** Median
higher in 1975   **b.** Variability is greater in 1975   **c.** No   **2.95a.** $-1, 1, 2$ **b.** $-2, 2, 4$
c. 1, 3, 4   **d.** .1, .3, .4   **2.97a.** $s^2 = 3.1234$ **b.** $s^2 = 9.0233$ **c.** $s^2 = 9.7857$ **2.99a.** $\bar{x} =$
5.67, $s^2 = 1.0667$, $s = 1.03$ **b.** $\bar{x} = -\$1.5$, $s^2 = 11.5$, $s = \$3.39$ **c.** $\bar{x} = .4125\%$, $s^2 =$
.0883, $s = .30\%$ **d.** (a) 3   (b) \$10   (c) .7375%   **2.101a.** $\bar{x} = 1$, $s^2 = .9524$, $s = .976$, 9%
b. $\bar{x} = 1$, $s^2 = .2439$, $s = .494$, 76%   **c.** $\bar{x} = 1$, $s^2 = .1091$, $s = .330$, 89%   **2.105a.** Skewed
to the right   **b.** Chebyshev's Rule is appropriate.   **c.** ≈ 38 homes   **d.** $z = 3.333$; no
2.107c. For A1775A, $\bar{x} = 19,462.2$, $s = 532.29$; For A1775B, $\bar{x} = 22,838.5$, $s = 560.98$
d. Cluster A1775A   **2.109a.** Quantitative, qualitative   **2.111b.** $\bar{x} = 16.48$, median $= 15$,
mode $= 14$ **c.** $s \approx 5.5$ **d.** $s^2 = 33.93$, $s = 5.82$ **e.** 100%   **2.113** 52 minutes   **2.115a.** At
most 25%   **b.** $\approx 2.5\%$

CHAPTER 3

3.1a. .5   **b.** .3   **c.** 6   **3.3** $P(A) = .55$; $P(B) = .50$, $P(C) = .70$ **3.5a.** $P(A) = \frac{4}{9}$ **b.** $P(B)$
$= \frac{1}{3}$ **c.** 0   **3.7** A: $P(A) = \frac{1}{12}$; B: $P(B) = \frac{1}{4}$; C: $P(C) = \frac{1}{2}$; $P(D) = \frac{1}{2}$ **3.11d.** .202
3.13a. .286   **3.15** Probabilities are the same.   **3.17a.** 1 to 2   **b.** .5   **c.** .4   **3.19** $^{1,000}/_{1,250} = .8$
3.21

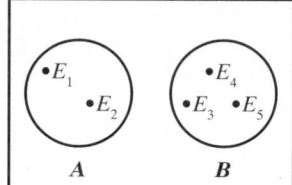

3.23a. $P(A) = \frac{3}{4}$ **b.** $P(B) = \frac{13}{20}$ **c.** $P(A \cup B) = 1$ **d.** $P(A \cap B) = \frac{2}{5}$ **e.** $P(A') = \frac{1}{4}$
f. $P(B') = \frac{7}{20}$ **g.** $P(A \cup A') = 1$ **h.** $P(A' \cap B) = \frac{1}{4}$ **3.25a.** .5   **b.** .31875   **c.** .125
d. .75   **e.** .75   **f.** .3125   **g.** $(A, B), (A, C), (B, C), (D, E), (D, F), (E, F)$ **3.27a.** $\frac{7}{8}$
b. $\frac{7}{8}$ **3.29a.** $B \cap A$ **b.** A' **c.** $B \cup C$ **d.** $B' \cap C'$ **3.31b.** Sample space   **c.** $P(C) = .21$
d. $P(G) = .55$ **e.** $P(A) = .32$ **f.** $P(D) = .08$ **g.** $P(E) = .18$ **3.33a.** Yes   **b.** $P(A) =$
.26; $P(B) = .35$; $P(C) = .72$; $P(D) = .28$; $P(E) = .05$ **c.** $P(A \cup B) = .56$;
$P(A \cap B) = .05$; $P(A \cup C) = .77$ **d.** $P(A') = .74$ **e.** $(C, D), (D, E)$ **3.35** $P(A \mid B) = .25$;
$P(B \mid A) = .50$
3.37a.

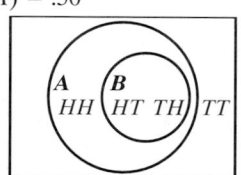

b. $P(A) = \frac{3}{4}$; $P(B) = \frac{1}{2}$; $P(A \cap B) = \frac{1}{2}$ **c.** $P(A \mid B) = 1$ **d.** $P(B \mid A) = \frac{2}{3}$ **3.39a.** .557
b. .075   **c.** .028   **d.** .504   **3.41a.** .553   **c.** .856   **d.** .552   **3.43a.** .543   **b.** .221   **c.** .052
d. .914   **3.45a.** .105   **b.** .638   **c.** .517   **3.47a.** .37   **b.** .68   **c.** .15   **d.** .221   **e.** 0
f. 0   **g.** None are independent.   **3.49a.** $(A, C), (B, C)$ **b.** None are independent.
c. $P(A \cup B) = .65$ **3.51a.** $P(A \cap B) = .08$; $P(A \mid B) = .4$; $P(A \cup B) = .52$ **b.** $P(A \cap B)$
$= .12$; $P(B \mid A) = .30$ **3.53c.** .891   **3.55a.** $(^{1}/_{2},780)^4$; assume independence   **b.** $(^{2,779}/_{2,780})^4$
c. $1 - (^{2,779}/_{2,780})^4$ **d.** $(^{1}/_{2,920})^4$, $(^{2,919}/_{2,920})^4$, $1 - (^{2,919}/_{2,920})^4$ **3.57a.** .02   **b.** .08   **3.59a.** $\frac{2}{9}$
b. $\frac{1}{36}$ **c.** $\frac{11}{36}$ **3.61a.** $\frac{27}{64}$ **b.** $\frac{1}{64}$ **c.** $\frac{27}{64}$ **3.63a.** $n = 20$ **b.** $\frac{1}{20}$ **3.69a.** 20   **b.** 20
c. 12   **d.** 4950   **e.** 1   **f.** 1   **g.** 60   **h.** 1   **3.71a.** 6   **b.** 36   **c.** 1,296   **d.** 6^n **3.73** 6
choices   **3.75a.** 6 assignments   **b.** All are stable.   **c.** 1   **3.77** 2,598,960   **3.79a.** 3,003
b. $^{56}/_{3,003}$ **c.** $^{1,722}/_{3,003}$ **3.81a.** 24   **b.** 12   **3.83a.** 56   **b.** 1,680   **c.** 6,720   **3.85a.** $^{21}/_{252}$
b. $^{21}/_{252}$ **c.** $^{105}/_{252}$ **3.87c.** $P(A) = \frac{1}{4}$; $P(B) = \frac{1}{2}$ **e.** $P(A') = \frac{3}{4}$; $P(B') = \frac{1}{2}$; $P(A \cap B) =$
$\frac{1}{4}$; $P(A \cup B) = \frac{1}{2}$; $P(A \mid B) = \frac{1}{2}$; $P(B \mid A) = 1$ **f.** No, no   **3.89a.** No   **b.** Yes   **c.** No
3.91a. No   **b.** .3, .1   **c.** .37   **3.93a.** 720   **b.** 10   **c.** 10   **d.** 30   **e.** 20   **f.** 1   **g.** 5,040
h. 2,450   **3.97a.** $\frac{1}{3}$ **b.** $\frac{1}{3}$ **c.** $\frac{1}{6}$ **d.** $\frac{1}{6}$ **3.99a.** Independent   **b.** Independent
c. Dependent   **d.** Dependent   **e.** Independent   **f.** Independent   **3.101** $^{15}/_{36}$, $^{315}/_{1,296}$,

$^{6,615}/_{46,656}$, $(^{21}/_{36})^{n-1}(^{15}/_{36})$ **3.103a.** .13 **b.** .52 **c.** .2917 **d.** No, $P(A \cap B)$ 0 **e.** No, $P(A \cap C)$ $P(A)P(C)$ **3.105a.** .4096 **b.** .0016 **c.** .9984 **3.107** .993 **3.109a.** .7127 **b.** .2873 **3.111a.** .3886 **b.** .3049 **c.** .1720 **d.** No; $P(A \cap B)$ $P(A)P(B)$ **3.113b.** All $^1/_{20}$ **c.** $^1/_{20}$ **d.** $^1/_2$ **3.115a.** 142,506 **b.** .236 **c.** .671 **3.117** 4.4739×10^{28}

CHAPTER 4

4.3a. Discrete **b.** Continuous **c.** Continuous **d.** Discrete **e.** Continuous **f.** Continuous **4.5a.** Continuous **b.** Discrete **c.** Discrete **d.** Discrete **e.** Discrete **f.** Continuous **4.7a.** .25 **b.** .40 **c.** .75 **4.9a.** .7 **b.** .3 **c.** 1 **d.** .2 **e.** .8 **4.11b.** $P(0) = ^1/_8$, $P(1) = ^3/_8$, $P(2) = ^3/_8$, $P(3) = ^1/_8$ **d.** $^1/_2$ **4.13a.** Yes **b.** .051 **c.** .757 **d.** .192 **4.15** $^7/_8$ **4.17b.** .85 **c.** a_2: .6, a_3: 0, a_4: 0, a_5: 0, a_6: .7 **d.** .51 **4.19a.** $\mu = 3.8$ **b.** $\sigma^2 = 10.56$ **c.** $\sigma = 3.2496$ **e.** No **f.** Yes **4.21a.** $\mu_x = 1, \mu_y = 1$ **b.** Distribution of x. **c.** $\mu_x = 1$, $\sigma_x^2 = .6$; $\mu_y = 1$, $\sigma_y^2 = .2$ **4.23a.** $E(x) = 3.678$ **b.** $\sigma = .983$ **c.** .949 **4.25a.** Mean = 2.9; median = 3 **b.** 3, 4 **c.** 3, 3 **4.27a.** $\mu = \$7,200$ **b.** $\mu = \$11,000$ **c.** $\mu = \$4,800$ **4.29** $\mu = -\$.50$ **4.31a.** Discrete **b.** Binomial **d.** $\mu = 3.5, \sigma = 1.0247$ **4.33a.** $P(0) = .343, P(1) = .441, P(2) = .189, P(3) = .027$ **4.35a.** .121 **b.** .034 **c.** .081 **d.** 0 **e.** .998 **f.** .137 **4.37a.** $p = .5$ **b.** $p < .5$ **c.** $p > .5$ **e.** See above **4.39a.** .99998 **b.** .0000 **c.** Yes, $z = -1.66$ **4.41a.** No **b.** Yes **c.** No **d.** $\mu = 4, \sigma^2 = 3.2, \sigma = 1.789$ from part b. **4.43a.** .7908 **b.** .0555 **4.45** $p = .0005$ **4.47a.** .20 **b.** $\mu = 4$ **c.** .196 **d.** No. **4.49a.** $\mu = 520, \sigma = 13.491$ **b.** No, $z = -8.895$ **4.51a.** .920 **b.** .677 **c.** .423 **d.** Increases **4.53b.** $\mu = 1, \sigma = 1$ **c.** .920 **4.57a.** $\sigma = 2$ **b.** No, $p = .003$ **4.59** $P(x \le 1) = .040$ **4.61a.** .0000075 **b.** $\mu = 11.8, \sigma = 3.435$ **4.63a.** .010 **b.** Yes **4.67a.** .3 **b.** .119 **c.** .167 **d.** .167 **4.69b.** $\mu = 4, \sigma = .853$ **d.** .939 **4.71a.** Hypergeometric **b.** Binomial **4.73** $P(x = 1) = .491$

4.75

x	$p(x)$
0	.091
1	.485
2	.424

4.77a. $\mu = 113.24, \sigma = 4.194$ **b.** $z = 6.38$ **4.79a.** Poisson **b.** Binomial **c.** Binomial **4.81a.** .243 **b.** .131 **c.** .36 **d.** .157 **e.** .128 **f.** .121 **4.83a.** $\mu = 15.4, \sigma^2 = 18.44, \sigma = 4.294$ **b.** .5 **d.** 1 **4.85a.** .180 **b.** .015 **c.** .076 **4.87a.** $P(0) = .0016, P(1) = .0256$, $P(2) = .1536, P(3) = .4096, P(4) = .4096$ **c.** .0272 **4.89** $\mu = 22,250$ **4.91a.** $\mu = 1.25, \sigma = 1.09$ **b.** .007 **c.** The percentage is larger than 5%. **4.93a.** .4019 **b.** .1608 **c.** Independent calls **d.** \$50,000 **4.95a.** .006 **b.** The insecticide is less effective than claimed. **4.97a.** .230 **b.** .143 **c.** .100

CHAPTER 5

5.1b. $\mu = 20, \sigma = 5.774$ **c.** (8.452, 31.548) **5.3b.** $\mu = 3, \sigma = .577$ **c.** .577 **d.** .61 **e.** .65 **f.** 0 **5.5a.** 0 **b.** 1 **c.** 1 **5.7a.** Continuous **c.** $\mu = 7, \sigma = .2887$ **d.** .5 **e.** 0 **f.** .75 **g.** .0002 **5.9** $p = .3125$ **5.11** $p = .4444$ **5.13a.** .4772 **b.** .3413 **c.** .4987 **d.** .2190 **e.** .4772 **f.** .3413 **g.** .4545 **h.** .2190 **5.15a.** .0721 **b.** .0594 **c.** .2434 **d.** .3457 **e.** .5 **f.** .9233 **5.17a.** .6826 **b.** .9500 **c.** .90 **d.** .9544 **5.19a.** $z_0 = -1.75$ **b.** $z_0 = 1.96$ **c.** $z_0 = 1.645$ **d.** $z_0 = 1.53$ **e.** $z_0 = -.83$ **f.** $z_0 = 2.50$ **5.21a.** $z = 1$ **b.** $z = -1$ **c.** $z = 0$ **d.** $z = -2.5$ **e.** $z = 3$ **5.23a.** .0456, .0026 **b.** .6826, .9544 **c.** $x_0 = 261.6$ **5.25a.** .8413 **b.** .3300 **c.** .6450 **d.** .2047 **e.** .9292 **f.** .0708 **5.27a.** .9544 **b.** .0228 **c.** .1587 **d.** .8185 **e.** .1498 **f.** .9974 **5.29a.** .5124 **b.** Yes, $p = .2912$ **c.** No, $p = .0027$ **5.31a.** .2514 **b.** $x_0 = 90.375$ **5.33a.** .0301 **5.35a.** .3745 **b.** .2447 **5.37** $\mu = 5.068$ **5.39** $d = 56.24$ **5.41a.** $z_L = -.675, z_U = .675$ **b.** $-2.68, 2.68$ **c.** $-4.69, 4.69$ **d.** .0074, 0 **5.45a.** Yes **b.** $\mu = 10, \sigma^2 = 6$ **c.** .726 **d.** .7291 **5.47a.** .1788 **b.** .5236 **c.** .6950 **5.49a.** 0 **b.** .4641 **c.** No, $p \approx 0$ **5.51** ≈ 0 **5.53a.** No **b.** .6026 **c.** No for $n = 100$; yes for $n = 500$; yes for $n = 1000$ **d.** .7190 **5.55a.** $\mu = 6,000, \sigma^2 = 2,400$ **b.** .0202 **c.** No, $p \approx 0$ **5.57** $P = .2676$; no **5.59a.** .0559 **b.** Yes, $p < .0559$

5.61a. .367879 **b.** .082085 **c.** .000553 **d.** .223130 **5.63a.** .798103 **b.** .818731 **c.** .550671 **d.** .301194 **5.65a.** .018316 **b.** .950213 **c.** .383401 **5.67a.** $a \approx .35$ hour **b.** .534985 **5.69a.** $\mu = 20$ **b.** .223130 **c.** .9502 **5.71** $p = .181269$ **5.73a.** .550671 **b.** .263597 **5.75b.** $\mu = 55, \sigma = 8.660$ **d.** .167 **e.** .4 **f.** 1 **g.** .577 **h.** .333 **5.77a.** $z_0 = 1.13$ **b.** $z_0 = 1.62$ **c.** $z_0 = 0$ **d.** $z_0 = 1.33$ **e.** $z_0 = -.85$ **f.** $z_0 = 2.64$ **5.79a.** $x_0 = 40$ **b.** $x_0 = 54.22$ **c.** $x_0 = 23.38$ **d.** $x_0 = 52$ **e.** $x_0 = 32.32$ **f.** $x_0 = 34.96$ **5.81a.** .283469 **b.** .716531 **c.** 0 **d.** .864665 **e.** .477743 **5.83** $p = .0154$, very unlikely **5.85** $p = .8315$ **5.87** $p = .982$ **5.89a.** .8861 **b.** .0301 **5.91a.** .0901 **b.** .0384 **c.** .9573 **5.93a. i.** .384 **ii.** .49 **iii.** .212 **iv.** .84 **b. i.** .9869 **ii.** .4938 **iii.** .2119 **iv.** .8413 **c. i.** .0029 **ii.** .0038 **iii.** .0001 **iv.** .0013 **5.95** At $p = .2$, $P(x \geq 400) \approx 0$ **5.97a.** .135335 **b.** .594

CHAPTER 6

6.1c. $\frac{1}{16}$ **6.3c.** .05 **d.** No **6.9a.** $\mu = 5$, **b.** $E(\bar{x}) = 5$ **c.** $E(m) = 4.778$ **d.** Use $\bar{x}$. **6.13b.** $\sigma^2 = 1.61$ **c.** $E(s^2) = 1.61 = \sigma^2$ **e.** $E(s) = 1.004 \neq \sigma$ **6.15a.** $\mu_{\bar{x}} = 100, \sigma_{\bar{x}} = 5$ **b.** $\mu_{\bar{x}} = 100, \sigma_{\bar{x}} = 2$ **c.** $\mu_{\bar{x}} = 100, \sigma_{\bar{x}} = 1$ **d.** $\mu_{\bar{x}} = 100, \sigma_{\bar{x}} = 1.414$ **e.** $\mu_{\bar{x}} = 100, \sigma_{\bar{x}} = .447$ **f.** $\mu_{\bar{x}} = 100, \sigma_{\bar{x}} = .316$ **6.17a.** $\mu = 2.9, \sigma^2 = 3.29, \sigma = 1.814$ **b.** $\mu_{\bar{x}} = 2.9 = \mu, \sigma_{\bar{x}} = 1.283 = \sigma/\sqrt{n}$ **6.19a.** $\mu_{\bar{x}} = 20, \sigma_{\bar{x}} = 2$ **b.** Approximately normal, yes **c.** $z = -2.25$ **d.** $z = 1.50$ **6.21a.** .8944 **b.** .0228 **c.** .1292 **d.** .9699 **6.25a.** $\mu_{\bar{x}} = 5.1, \sigma_{\bar{x}} = .4981$ **b.** Yes **c.** .2119 **d.** .4071 **6.27a.** $\mu_{\bar{x}} = 89.34, \sigma_{\bar{x}} = 1.3083$ **c.** .8461 **d.** .0367 **6.29a.** .0031 **b.** More likely if $\mu = 156$; less likely if $\mu = 158$ **c.** Less likely if $\sigma = 2$; more likely if $\sigma = 6$. **6.33a.** .5 **b.** .0606 **c.** .0985 **d.** .8436 **6.39a.** Discrete **b.** Approximately normal with $\mu_{\bar{x}} = 1.5, \sigma_{\bar{x}} = .01764$ **c.** .0023 **d.** No, $p \approx 0$ **6.41a.** Approximately normal with $\mu_{\bar{x}} = 107, \sigma_{\bar{x}} = 1.637$ **b.** .0336 **c.** No, $p = .0336$ **6.43a.** $\mu_{\bar{x}} = 1.3, \sigma_{\bar{x}} = .240$ **b.** Yes **c.** .1056 **d.** .0062 **6.45** $p = .9772$ **6.47a.** Approximately normal with $\mu_{\bar{x}} = 38, \sigma_{\bar{x}} = .5$ **b.** .0139 **c.** No, $p = .0139$ **6.49** $p = .9332$

CHAPTER 7

7.1a. 1.645 **b.** 2.58 **c.** 1.96 **d.** 1.28 **7.3a.** 10.9375 ± 2.0094 **7.5a.** 83.2 ± 1.25 **c.** 83.2 ± 1.65 **d.** Increases **e.** Yes **7.9** Yes **7.11b.** No assumptions needed **c.** 95% confident claim is false **7.13a.** $3.39 \pm .0466$; no assumptions **7.15** $.72 \pm .05$ **7.17a.** $.044 \pm .162$ **7.19a.** μ_Y: $4.17 \pm .095$; μ_{MA}: $4.04 \pm .057$; μ_O: $4.31 \pm .062$ **b.** More likely **7.23a.** $t_0 = 2.228$ **b.** $t_0 = 2.567$ **c.** $t_0 = -3.707$ **d.** $t_0 = -1.771$ **7.25a.** 5 ± 1.876 **b.** 5 ± 2.394 **c.** 5 ± 3.754 **d.** (a) $5 \pm .780$ (b) $5 \pm .941$ (c) 5 ± 1.276; width decreases **7.27** $1.3731 \pm .7618$ **7.29a.** $3.8 \pm .464$ **7.31a.** 49.3 ± 8.60 **c.** Population is normal. **7.33a.** Population must be normal. **b.** $\$67,648.583 \pm \$9,939.553$ **7.35** Approximately normal with $\mu_{\hat{p}} = p$ and $\sigma_{\hat{p}} = \sqrt{\dfrac{pq}{n}}$ **7.37a.** Yes **b.** $.64 \pm .067$ **7.39a.** Yes **b.** No **c.** Yes **d.** No **7.41** $.556 \pm .044$ **7.43a.** $.78 \pm .056$ **b.** Appears $p > .25$. **7.45a.** $.644 \pm .099$ **b.** Yes **7.47a.** $.524 \pm .095$ **7.49** .047, .036 **7.51** $n = 519$ **7.53a.** $n = 482$ **b.** $n = 214$ **7.55** $n = 34$ **7.57a.** $n = 16$: $W = .98$; $n = 25$: $W = .784$; $n = 49$: $W = .56$; $n = 100$: $W = .392$; $n = 400$: $W = .196$ **7.59** No. n needed to be 984. **7.61** $n = 1,351$ **7.63** $n = 271$ **7.65** No. $n = 457$, $457(\$25) = \$11,425$ **7.67** $n = 3,701$ **7.69a.** $t = 2.086$ **b.** $z = 1.96$ **c.** $z = 1.96$ **d.** $z = 1.96$ **e.** Neither t nor z. **7.73** (1) p (2) μ (3) μ (4) μ **7.75** $.812 \pm .409$ **7.77a.** $.257 \pm .003$ **b.** $.241 \pm .003$ **7.79** $.5325 \pm .0457$ **7.81** $49.70 \pm .1498$ **7.83a.** $.623 \pm .086$ **d.** Probably not. **7.85** $2 \pm .47$ **7.87a.** $\bar{x} = 42,250$ **b.** $42,250 \pm 1,683.713$ **c.** Interval estimation gives a measure of reliability. **7.89a.** $\hat{p} = .094$ **b.** Yes **c.** $.094 \pm .037$ **7.91a.** 34.76 ± 6.344 **b.** Population is normal.

CHAPTER 8

8.1 Null hypothesis **8.3** α **8.7** No **8.9c.** α **8.13a.** $z = 1.67$, reject H_0 **b.** $z = 1.67$, fail to reject H_0 **8.15a.** H_0: $\mu = 16, H_a$: $\mu < 16$ **b.** $z = -4.31$, reject H_0 **8.17a.** $z = 4.03$, reject H_0 **8.19** Yes. $z = -4.34$, reject H_0 **8.21** $z = -1.70$, fail to reject H_0 **8.23a.** $z = 0.70$, fail to reject H_0 **b.** Yes **8.25a.** Fail to reject H_0 **b.** Reject H_0 **c.** Reject H_0 **d.** Fail to reject H_0 **e.** Fail to reject H_0 **8.27** $p = .0150$ **8.29** $p = .9279$, fail to reject H_0 **8.31a.** Fail to reject H_0 **b.** Fail to reject H_0 **c.** Reject H_0 **d.** Fail to reject H_0 **8.33a.** No, $z = -1.29$. Fail to reject H_0 **b.** $p = .0985$ **c.** Small values **8.35b.** $p \approx 0$ **c.** Reject H_0 **8.37a.** $H_0: \mu = 250, H_a: \mu > 250$ **b.** $p = .0455$. At $\alpha = .05$, reject H_0 **c.** $p = .0910$. At $\alpha = .05$, fail to reject **8.39** Small n, normal population **8.43a.** $t < -2.160$ or $t > 2.160$ **b.** $t > 2.500$ **c.** $t > 1.397$ **d.** $t < -2.718$ **e.** $t < -1.729$ or $t > 1.729$ **f.** $t < -2.353$ **8.45a.** $t = -2.064$; fail to reject H_0 **b.** $t = -2.064$; fail to reject H_0 **c.** **(a)** $.05 < p < .10$ **(b)** $.10 < p < .20$ **8.47** Yes. $t = 8.75$. Reject H_0 **8.49a.** $H_0: \mu = 15, H_a: \mu < 15$ **b.** $p = .0556$ **c.** At $\alpha = .05$, reject H_0 **8.51** $t = 2.97$, fail to reject H_0 **8.53a.** $H_0: \mu = 10, H_a: \mu > 10$ **b.** $.05 < p < .10$ **8.55b.** $z = -1.39$, fail to reject H_0 **c.** $p = .0823$ **8.57a.** $z = 1.89$, fail to reject H_0 **b.** $z = 1.89$, reject H_0 **c.** $z = -5.33$, reject H_0 **d.** $.74 \pm .09$ **e.** $.74 \pm .11$ **8.59a.** No **b.** $p = .6600$ **8.61** $z = 33.47$, reject H_0 **8.63a.** $z = 15.61$, reject H_0 **b.** $p \approx 0$ **8.65** $z = 21.66$, reject H_0 **8.69a.** $\beta = .3594$ **b.** Power $= .6406$ **8.71a.** 49: $\beta = .8106$, 47: $\beta = .5319$, 45: $\beta = .2358$, 43: $\beta = .0643$, 41: $\beta = .0102$ **c.** $\beta \approx .67$ **d.** 49, power $= .1884$; 47, power $= .4681$; 45, power $= .7642$; 43, power $= .9357$; 41, power $= .9898$ **8.73a.** 15.5, power $= .1660$; 15.0, power $= .2514$; 14.5, power $= .3594$; 14.0, power $= .4801$; 13.5, power $= .6026$ **c.** Power $\approx .42$; power $= .4207$ **d.** Power $\approx .85$, $\beta = .0078$ **8.75a.** $\beta = .1949$, Type II **b.** $\alpha = .05$, Type I **c.** Power $= .8051$ **8.77a.** $\chi_0^2 = 17.2750$ **b.** $\chi_0^2 = 15.5073$ **c.** $\chi_0^2 = 11.1433$ **8.79a.** $\chi_{.025}^2 = 28.8454$, $\chi_{.975}^2 = 6.90766$ **b.** $\chi_{.05}^2 = 26.2962$, $\chi_{.95}^2 = 7.96164$ **c.** $\chi_{.005}^2 = 34.2672$, $\chi_{.995}^2 = 5.14224$ **8.81a.** $\chi^2 = 479.16$, reject H_0 **c.** $(3.854, 6.149)$ **8.83a.** $H_0: \sigma^2 = .54, H_a: \sigma^2 > .54$ **b.** $s \approx .7425$ **c.** $\chi^2 = 40.8375$, fail to reject H_0 **8.85** $\chi^2 = 4.5$, fail to reject H_0 **8.87** $\chi^2 = 133.90$, $.01 < p < .025$. At $\alpha = .05$, reject H_0. **8.89a.** $(.163, 29.141)$ **b.** Normal population **8.93** Small **8.95a.** 72.6 ± 1.70 **b.** $t = -7.51$, reject H_0 **c.** $t = -7.51$, reject H_0 **d.** 72.6 ± 2.82 **e.** $n = 75$ **8.97a.** $8.2 \pm .12$ **b.** $z = -1.67$, fail to reject H_0 **c.** $z = -3.35$, reject H_0 **8.99a.** $p = .1288$ **b.** Normal population **c.** $p = .2576$ **8.101a.** $t = -3.46$, reject H_0 **b.** Normal population **8.103a.** $t = 3.512$; $.025 < p < .05$ **b.** $p = .0362$ **c.** Normal population **d.** Type I **8.105a.** $z = 6.11$, reject H_0 **b.** $p \approx 0$ **c.** $\beta = .4247$ **8.107a.** No, $z = 1.41$. Fail to reject H_0 **b.** Small α **8.109a.** No **b.** $\alpha = .05$; power $= .4090$ **c.** Power increases **8.111b.** $\beta = .8264$ **c.** $\beta = .7389$ **8.113** $p = .1423$ **8.115a.** $H_0: \mu = 1, H_a: \mu > 1$ **b.** $t = 2.41$ reject if $t > 1.345$ **c.** Normal population **d.** $t = 2.41$, reject H_0 **e.** $.01 < p < .025$ **8.117** No, $z = -1.45$, fail to reject H_0 **8.119** Power $= .2611$ **8.121** No. $\chi^2 = 23.66$, fail to reject H_0

CHAPTER 9

9.1a. $.35 \pm 24.5$ **b.** $p = .0052$, reject H_0 **c.** $p = .0026$ **d.** $p = .4238$, fail to reject H_0 **e.** Independent random samples **9.3a.** 14, .4 **b.** 10, .3 **c.** 4, .5 **d.** Yes **9.5a.** No **b.** No **c.** No **d.** Yes **e.** No **9.7a.** 190 **b.** 29.8214 **c.** .2611 **d.** 2,138.7097 **9.9a.** 1.2806 **c.** Yes **d.** $z < -1.96$ or $z > 1.96$ **e.** $z = 11.71$, reject H_0 **9.11a.** $p = .1150$, fail to reject H_0 at $\alpha = .10$. **b.** Yes, $p = .0575$. Reject H_0 at $\alpha = .10$ **9.13a.** $p = .1114$, fail to reject H_0 at $\alpha = .05$. **b.** -2.50 ± 3.12 **9.15a.** $-.52 \pm 3.546$ **b.** Both populations normal and $\sigma_1^2 \neq \sigma_2^2$ **c.** 8.59 ± 3.152 **9.17** $z = 2.19$, reject H_0 at $\alpha = .05$. **9.19** Activity score: $z = 13.31$, reject H_0; Impulsive-sensation: $z = -.36$, fail to reject H_0; Sociability: $z = -8.04$, reject H_0; Neuroticism-anxiety: $z = -1.63$, fail to reject H_0; Aggression-hostility: $z = 5.21$, reject H_0 **9.21a.** t test **b.** Yes **c.** No **d.** Fail to reject H_0 **e.** Reject H_0 at $\alpha = .05$ **9.23a.** Beginning: 6.09 ± 41.835; First: -52.24 ± 40.842; Second: -48.41 ± 40.643; Third: -37.68 ± 39.784; Fourth: -38.54 ± 42.958 **9.25b.** $H_0: \mu_1 - \mu_2 = 0, H_a: \mu_1 - \mu_2 > 0$ **c.** $p = .078$, fail to reject H_0 at $\alpha = .05$ **9.27a.** $H_0: \mu_1 - \mu_2 = 0, H_a: \mu_1 - \mu_2 < 0$ **b.** Fail to reject H_0 **c.** Independent samples **9.29** $\mu_D = \mu_1 - \mu_2$ **9.31a.** $\bar{x}_D = 2, s_D^2 = 2$ **b.** $\mu_D = \mu_1 - \mu_2$ **c.** 2 ± 1.484 **d.** $t = 3.46$, reject H_0 **9.33a.** $t < -1.341$ **b.** $t = -3.5$, reject H_0 **d.** -7 ± 3.506 **e.** Confidence interval **9.35a.** Time needed **b.** Climbers **c.** Paired experiment **9.37** $t = -2.31$; reject H_0 at $\alpha = .05$ **9.39** $t = 1.77$; fail to reject H_0 at $\alpha = .05$ **9.41a.** $H_0: \mu_D = 0, H_a: \mu_D > 0$ **b.** Paired difference test **d.** Leadership: fail to reject H_0 at $\alpha = .05$; Popularity: reject H_0 at $\alpha = .05$; Intellectual self-confidence: reject at $\alpha = .05$ **9.43a.** No, $t = -3.08$; fail to reject H_0 **b.** -31.13 ± 32.17 **9.47a.** $z < -2.33$ **b.** $z < -1.96$ **c.** $z < -1.645$ **d.** $z < -1.28$ **9.49a.** $.07 \pm .067$ **b.** $.06 \pm .086$ **c.** $-.15 \pm .131$ **9.51** $z = 1.16$, fail to reject H_0 **9.53a.** Yes, $z = 13.02$, reject H_0 **9.55** $.148 \pm .069$ **9.57a.** $H_0: p_1 - p_2 = 0$, $H_a: p_1 - p_2 \neq 0$ **c.** right tail **d.** $p < .10$, reject H_0 if $\alpha = .10$ **9.59** $z = -2.23$, fail to reject H_0 at $\alpha = .01$ **9.61** $z = 1.21$, fail to reject H_0 at $\alpha = .01$ **9.63** $n_1 = n_2 = 24$ **9.65a.** $n_1 = n_2 = 193$ **b.** $n_1 = n_2 = 47$ **c.** $n_1 = n_2 = 144$ **9.67** $n_1 = n_2 = 125$ **9.69a.** $n_1 = n_2 = 911$ **b.** $n = 1692$ **9.71** $n_1 = n_2 = 4,802$ **9.73** $n_1 = n_2 = 136$ **9.77a.** $.025$ **b.** $.90$ **c.** $.99$ **d.** $.05$ **9.79a.** $F > 2.36$ **b.** $F > 3.04$ **c.** $F > 3.84$ **d.** $F > 5.12$ **9.81a.** $F_{L,.025} = 3.77$, $F_{U,.025} = 3.12$ **b.** $F_{L,.025} = 3.28$, $F_{U,.025} = 3.28$ **c.** $F_{L,.025} = 2.07$, $F_{U,.025} = 2.29$ **d.** $F_{L,.025} = 2.62$, $F_{U,.025} = 2.78$ **9.83a.** $(.02, .99)$ **b.** $F = 5.47$, fail to reject H_0 **c.** $F = 5.47$, reject H_0 **9.85a.** No, $F = 2.26$, fail to reject H_0 **b.** $.05 < p < .10$ **9.87** $F = 1.045$, fail to reject H_0 **9.89a.** $t = .78$, fail to reject H_0 **b.** 2.5 ± 8.99 **c.** $n_1 = n_2 = 225$ **9.91a.** $3.90 \pm .31$ **b.** $z = 20.60$, reject H_0 **c.** $n_1 = n_2 = 346$ **9.93a.** $t = 5.73$, reject H_0 **b.** 3.8 ± 1.84 **9.95a.** H_0: $\mu_1 - \mu_2 = 0$, H_a: $\mu_1 - \mu_2 \neq 0$ **b.** $z = -7.69$, reject H_0 **9.97b.** $t = 1.18$, fail to reject H_0 **9.99** No, $z = 1.93$, fail to reject H_0 at $\alpha = .05$. **9.101a.** H_0: $\mu_1 - \mu_2 = 0$, H_a: $\mu_1 - \mu_2 \neq 0$ **b.** $p = .0072$, reject H_0 at $\alpha = .01$ **9.103a.** H_0: $\mu_D = 0$, H_a: $\mu_D > 0$ **b.** $p = .015$, reject H_0 at $\alpha = .025$ **c.** $(1.39, 19.01)$ **9.105** $n_1 = n_2 = 193$ **9.107a.** H_0: $\mu_1 - \mu_2 = 0$, H_a: $\mu_1 - \mu_2 \neq 0$ **b.** Hours of work: fail to reject H_0; Free time: Reject H_0 at $\alpha = .05$; Breaks: reject H_0 at $\alpha = .01$ **9.109** Yes, $z = -2.61$, reject H_0 **9.111** $n_1 = n_2 = 1,905$ **9.113a.** Reject H_0 **b.** Fail to reject H_0 **c.** Reject H_0 **d.** Reject H_0 **e.** Fail to reject H_0 **f.** Fail to reject H_0 **9.115a.** H_0: $\mu_D = 0$, H_a: $\mu_D > 0$ **b.** Paired difference **c.** Aad: $p = .091$, reject H_0 at $\alpha = .10$; Ab: $p = .032$, reject H_0 at $\alpha = .05$; Interaction: $p = .050$, reject H_0 at $\alpha = .10$ **9.117a.** -6.58 ± 16.88 **b.** No **c.** $t = -.85$, fail to reject H_0 **9.119** Yes, $F = 3.17$, reject H_0

CHAPTER 10

10.1 A, B, C, and D **10.3a.** College students **b.** Households **c.** You **d.** Patients **10.5a.** Cholesterol levels **b.** Socioeconomic class; qualitative **c.** Poverty, low, middle, and high **d.** Adults **10.7a.** Observational **b.** Designed **c.** Designed **d.** Observational **e.** Observational **f.** Observational **10.9a.** $.95$ **b.** $.01$ **c.** $.05$ **d.** $.90$ **10.11a.** 9, 14; 9, 14 **b.** 75, 75 **c.** 20, 144 **d.** 95, 78.95%; 219, 34.25% **e.** 37.5, 5.21 **f.** Reject H_0; reject H_0 **10.13** $F = 37.5, 5.21$ **10.15a.** $F = 4.01$ **b.** 7 **c.** Yes; $F = 4.01$ **d.** $p < .01$ **e.** No, $t = -.79$ **f.** $-.4 \pm .8573$ **g.** $3.7 \pm .6062$ **10.17a.** $F = 3.39$; $F = 3.39$ **b.** Fail to reject H_0 **10.19a.** Reject H_0 for each variable **b.** No **10.21a.** Completely randomized design **b.** $p = .0041$, reject H_0 **c.** $p = .0041$ **10.23a.** The 3 groups of women **b.** H_0: $\mu_1 = \mu_2 = \mu_3$ **c.** df Group = 2; df Error = 80; df Total = 82 **d.** Reject H_0 **10.25a.** $F = 5.29$ **b.** Yes, reject H_0 **c.** $p < .01$ **10.27a.** 3 **b.** 10 **c.** 6 **d.** 45 **10.31** $p = .0016$; reject H_0; $\mu_{OND} < \mu_{SLI}$, μ_{YND} **10.33a.** $\alpha = .05$ **b.** $(.3340, .6122)$ **c.** $\mu_E < \mu_C$, μ_A **10.35a.** Yes, $F = 3.15$ **b.** $\mu_T > \mu_F$, μ_{GY}, μ_V, μ_{TF}, μ_S; $\mu_{GO} > \mu_{TF}$, μ_S; $\mu_B > \mu_S$; $\mu_F > \mu_S$; $\mu_{GY} > \mu_S$ **10.37** μ_M, $\mu_F < \mu_A$ **10.39a.** $F = 5.54, .23$ **b.** H_0: $\mu_1 = \mu_2 = \mu_3$; H_a: At least two means differ **c.** $F = 5.54$ **e.** Fail to reject H_0 **10.41a.** $F = 4.13, 6.07$ **b.** No, $F = 4.13$ **c.** Yes, $F = 6.07$ **d.** -2.4 ± 1.526 **10.43a.** The 3 positions **b.** The 10 students **c.** Medial rotation **10.45** Treatments: $F = .0019$; fail to reject H_0 **10.47a.** Yes, $F = 5.06$ **b.** $F = 2.28$, fail to reject H_0 **10.49a.** Randomized block design **b.** No **c.** $F = .672$; fail to reject H_0 **10.51a.** $F = 2.00, 8.83, 8.00$ **b.** $F = 7.14$, reject H_0 **c.** Yes **e.** $F = 8.00$; reject H_0 **f.** No **10.53a.** $(1, 1), (1, 2), (1, 3), (2, 1), (2, 2), (2, 3)$ **b.** $F = 21.61$; reject H_0 **c.** Yes, $F = 36.60$; reject H_0 **d.** No **e.** Yes **10.55a.** $F = 1.50$, fail to reject H_0 **b.** $F = 5.25$, reject H_0 **c.** $F = 5.25$, reject H_0 **d.** $F = 20.25$, reject H_0 **10.57a.** Interaction: $F = 1.2$, fail to reject H_0; Herds: $F = 17.2$, reject H_0; Seasons: $F = 3.0$, fail to reject H_0 **b.** Yes **c.** μ_{PLC}, $\mu_{LGN} < \mu_{QMD}$, μ_{MTZ} **10.59b.** $F = 1,336.85$, reject H_0 **c.** Yes, $F = 258.07$, reject H_0 **d.** No **e.** The nine treatments **10.61a.** 464.4 **b.** $6,739.605$ **c.** $SS(S) = 114.005$ $SS(W) = 41.405$, $SS(S \times W) = 18.605$ **e.** $SSE = 789.9682$ **f.** $SS(T) = 963.9832$ **h.** $F = 2.06$, fail to reject H_0 **10.63b.** 2×2 factorial design **c.** $F = .194, 1.35, .560$ **d.** $F = .70$, fail to reject H_0 **10.65** $(1, 1), (1, 2), (2, 1), (2, 2), (3, 1), (3, 2)$ **10.67a.** $F = 7.70$ **b.** Yes, $F =$

7.70, reject H_0 **c.** $11.4 \pm .99$ **10.69a.** $F = 1.11, 2.36, 3.98$ **b.** $4, 6, 4 \times 6 = 24, 2$ **c.** SST $= 58.3; F = 3.25$, reject H_0 **10.71a.** $F = 11.65$ **b.** Yes, $F = 11.65$ **c.** $\mu_{JC} < \mu_{JM}, \mu_{CM}, \mu_{OM}$ **10.73a.** Randomized block design **b.** Yes, $F = 8.452$ **c.** $p = .014$ **d.** $\mu_1 < \mu_2, \mu_3, \mu_4$ **10.75b.** Designed **c.** $F = 6.36, 4.31$ **d.** Yes, $p = .0131$ **10.77b.** Designed **c.** $F = 12.22$ **d.** Yes, $F = 12.22$ **e.** $\mu_D > \mu_C, \mu_B > \mu_A$ **f.** 11.7 ± 1.98 **10.79b.** $F = 25.06, .67$ **c.** Yes, $F = 25.06$ **d.** No **10.81a.** Completely randomized design **b.** $F = 7.79$ **c.** Yes, $F = 7.79$ **d.** $\mu_3 > \mu_1, \mu_2$ **10.83a.** No **b.** $F = 6.55$, reject H_0

CHAPTER 11

11.3 $\beta_0 = 14/3; \beta_1 = 1/3$ **11.9** No **11.11c.** SSE $= 108$ **11.13b.** Positive **c.** $\hat{y} = .020 + .918x$ **d.** Yes **e.** $-1 \le x \le 7$ **11.15a.** Yes **b.** $y = \beta_0 + \beta_1 x + \epsilon$ **c.** Positive **11.17a.** Positive relationship **b.** $y = \beta_0 + \beta_1 x + \epsilon$ **c.** $\hat{\beta}_0 = 20.127511; \hat{\beta}_1 = .624416$ **11.19a.** $\hat{y} = 57.9122 - .8114x$ **c.** $\hat{y} = 9.2282$ **11.21a.** $\hat{y} = 3.375 + 1.201x$ **b.** It is a good fit. **11.23a.** SSE $= 57.5, s^2 = 3.19444$ **b.** SSE $= 257.5, s^2 = 6.7763$ **c.** SSE $= 9.288, s^2 = 1.1610$ **11.25** Graph b **11.27a.** SSE $= 2.7091, s^2 = .1693, s = .4115$ **11.29a.** SSE $= 14.31188, s^2 = 1.19266, s = 1.09209$ **11.31b.** $\hat{y}_A = 6.62 - .0727x; \hat{y}_B = 9.31 - .1077x$ **c.** A: SSE $= 19.056, s^2 = 1.466, s = 1.211$; B: SSE $= 4.833, s^2 = .3718, s = .610$ **d.** A: 1.531 ± 2.422; B: 1.771 ± 1.22 **e.** Brand B **11.33b.** $\hat{y} = .544 + .617x$ **d.** $H_0: \beta_1 = 0; H_a: \beta_1 \ne 0$ **e.** $t = 5.50$, df $= 5$ **f.** Reject H_0 **11.35** $t = .627$, fail to reject H_0 **11.37b.** $\hat{y} = 78.516 - .2389x$ **d.** $t = -2.31$, fail to reject H_0 **e.** One outlier **f.** $\hat{y} = 139.759 - .4497x; t = -15.35$, reject H_0 **11.39a.** $t = 4.98$, reject H_0 **b.** $\hat{y} = .607$ **11.41a.** $t = 4.004$, reject H_0 **b.** $.1077 \pm .0822$ **11.43a.** SSE $= 7.324, s^2 = .7324, s = .8558$ **c.** $t = 12.81$, reject H_0 **d.** $p < .001$ **11.47a.** $r = .9853, r^2 = .9709$ **b.** $r = -.9934, r^2 = .9868$ **c.** $r = 0, r^2 = 0$ **d.** $r = 0, r^2 = 0$ **11.49c.** $t = 38.66$, reject H_0 **d.** $r^2 = .904$ **11.51e.** **(a)** $r^2 = .0676$ **(b)** $r^2 = .0361$ **(c)** $r^2 = .0196$ **(d)** $r^2 = .0001$ **11.53a.** $y = \beta_0 + \beta_1 x + \epsilon$ **b.** $\beta_1 > 0$ **d.** $t = -3.12$, do not reject H_0 **11.55b.** $r^2 = .461$ **11.57a.** $H_0: \rho = 0; H_a: \rho \ne 0$ **b.** $p = .0512$ **11.59c.** 6.5335 ± 1.9487 **d.** -6.0546 ± 1.7225 **11.61c.** 4.644 ± 1.118 **d.** $-.414 \pm 1.717$ **11.63c.** No **11.65a.** $t = 6.97$, reject H_0 **b.** 5.261 ± 2.096 **c.** 21.22 ± 4.01 **11.67a.** $t = -5.91$, reject H_0 **11.69a.** A: $3.349 \pm .587$; B: $4.464 \pm .296$ **b.** A: 3.349 ± 2.224; B: 4.464 ± 1.120 **c.** $-.65 \pm 3.606$ **11.71b.** $\hat{y} = x, \hat{y} = 3$ **c.** $\hat{y} = x$ **d.** SSE $= 0$; SSE $= 8$ **11.73a.** $\hat{y} = 40.7842 + .7656x$ **c.** $t = 4.375$, reject H_0 **d.** $.7656 \pm .4035$ **e.** 79.06 ± 17.03 **f.** 67.58 ± 7.74 **11.75a.** $\hat{y} = -13.49 - .0528x$ **b.** $-.0528 \pm .0178$ **c.** $r^2 = .8538$ **d.** $(.5987, 1.2653)$ **11.77a.** $t = .81$, fail to reject H_0 **b.** $.011 \pm .027$ **11.79a.** $\hat{y} = -99,045 + 102.814x$ **b.** $t = 6.483$, reject H_0 **c.** $p = .00005$, reject H_0 **d.** $r^2 = .6776$ **e.** $(-631,806, 2,078,740)$ **11.81a.** $\hat{y}_6 = -2.792 + 15.135x; \hat{y}_{12} = -8.368 + 17.612x; \hat{y}_{18} = -12.068 + 17.599x$ **c.** $6°$: SSE $= 141.227, s^2 = 70.613$; $12°$: SSE $= 15.676, s^2 = 7.838$; $18°$: SSE $= 7.942, s^2 = 3.971$ **d.** $6°$: 15.135 ± 19.668; $12°$: 17.612 ± 4.045; $18°$: 17.599 ± 2.456 **f.** $6°$: 57.748 ± 54.994; $12°$: 62.08 ± 10.332; $18°$: 58.328 ± 5.858 **11.83a.** $\hat{y} = -768.6 + 6.2033x_1$ **b.** $t = 3.35$, reject H_0 **c.** $r^2 = .584$

CHAPTER 12

12.1a. $\hat{\beta}_0 = 506.346, \hat{\beta}_1 = -941.900, \hat{\beta}_2 = -429.060$ **b.** $\hat{y} = 506.346 - 941.900x_1 - 429.060x_2$ **c.** SSE $= 151,015.72$, MSE $= 8,883.28, s = 94.251$ **d.** $t = -3.424$, reject H_0 **e.** -429.060 ± 801.4322 **12.3a.** $t = 3.133$ **b.** $t = 3.133$, yes **c.** $F = 9.816$ **d.** No **12.5a.** SSE $= .0438, s^2 = .01095$ **b.** $t = -13.97$, reject H_0 **c.** $\hat{y} = 1.095 + 1.636x - .1595x^2$ **d.** Yes **12.7a.** $E(y) = \beta_0 + \beta_1 x + \beta_2 x^2$ **b.** Positive **12.9a.** $\hat{y} = 20.0911 - .6705x + .0095x^2$ **c.** $t = 1.51$, fail to reject H_0 **d.** $\hat{y} = 19.279 - .445x$ **e.** $-.445 \pm .0727$ **12.11a.** $\hat{y} = 93,074 + .4152x_1 - 855x_2 + .924x_3 + 2,692x_4 + 15.5x_5$ **b.** $s = 33,225.9$ **c.** $t = 2.78$, reject H_0 **e.** Very little relationship **f.** $t = -2.86$, reject H_0 **g.** $p = .00495$ **12.13a.** Yes **b.** $F = 55.2$, reject H_0 **12.15** No **12.17a.** $t = 3.88$, reject H_0 **b.** No **c.** $F = 11.397$, reject H_0 **12.19a.** $R^2 = .362$ **b.** $F = 5.11$, reject H_0 **12.21a.** $\hat{y} = 295.33 - 480.84x_1 - 829.46x_2 + .0079x_3 + 2.3608x_4$ **b.** $s = 46.76747$ **c.** $F = 28.18$, reject H_0 **d.** $t = -3.197$, reject H_0 **12.23b.** $R^2 = .67$ **c.** $F = 30.45$, reject H_0 **12.25b.** $H_0: \beta_3 = 0; H_a: \beta_3$

$\neq 0$ **c.** Reject H_0 for $\alpha > .001$ **12.27a.** $E(y) = \beta_0 + \beta_1 x_1 + \beta_2 x_2 + \beta_3 x_3 + \beta_4 x_4 + \beta_5 x_5$ **b.** $\hat{y} = 15.491 + 12.774 x_1 + .713 x_2 + 1.519 x_3 + .32 x_4 + .205 x_5$ **c.** $R^2 = .240$ **d.** $p = .025$, reject H_0 **12.31** One possible outlier **12.35a.** $F = 24.41$, reject H_0 **b.** $t = -2.01$, reject H_0 **c.** $t = .31$, fail to reject H_0 **d.** $t = 2.38$, reject H_0 **12.39** $x_1 = 30$, $x_2 = .6$, $x_3 = 1300$ **12.41** Yes **12.43a.** $R^2 = .45$, $F = 6.38$, reject H_0 **c.** $p = .002$, reject H_0 **12.45** Use only two of the variables **12.47a.** $\beta_2 < 0$ **b.** $F = 1.60$, fail to reject H_0 **c.** $F = 1.61$, fail to reject H_0 **12.49b.** β_3, β_5, and β_6 are significant **c.** $.007 \pm .0033$ **12.51a.** $R^2 = .87$ **b.** $F = 45.51$, reject H_0 **12.53a.** $r = .878$ **b.** $s_1 = 33,225.9$, $s_2 = 38,425.1$ **c.** $F = 176.829$, reject H_0 **d.** $(482,786, 559,593)$ **12.55b.** $t = 5$, reject H_0 **c.** $\hat{y} = 825$ **d.** Support **12.57a.** $\hat{\beta}_1 = .02573$, $\hat{\beta}_2 = .03361$ **b.** $s = .40228$, $R^2 = .68106$ **c.** $F = 39.505$, reject H_0 **d.** $\hat{y}_{60} = .47 + .026 x_1$; $\hat{y}_{75} = .98 + .026 x_1$; $\hat{y}_{90} = 1.49 + .026 x_1$ **12.59a.** $s_1 = .4023$, $s_2 = .1871$ **b.** $F = 100.409$, reject H_0 **c.** $t = 1.675$, fail to reject H_0

CHAPTER 13

13.1a. Quantitative **b.** Quantitative **c.** Quantitative **d.** Qualitative **e.** Qualitative **13.3a.** **1.** Quantitative **2.** Quantitative **3.** Quantitative **4.** Quantitative **5.** Quantitative **b.** **1.** Positive **2.** Positive **3.** Positive **4.** Positive **5.** Negative **13.5a.** Quantitative **b.** Quantitative **c.** Qualitative **d.** Qualitative **e.** Qualitative **f.** Quantitative **g.** Qualitative **h.** Qualitative **i.** Qualitative **j.** Quantitative **k.** Qualitative **13.7a.** $\beta_0 = 10$, $\beta_1 = .5$ **b.** $\beta_0 = 80$, $\beta_1 = -.5$ **13.9a.** First **b.** Second **c.** Third **d.** Third **e.** Second **f.** First **13.11a.** $\hat{y} = 63.1 + 1.17x - .0374 x^2$ **c.** H_0: $\beta_2 = 0$, H_a: $\beta_2 < 0$ **d.** $t = -3.03$, reject H_0 **13.13a.** Second **c.** Open downward **13.15** $E(y) = \beta_0 + \beta_1 x_1 + \beta_2 x^2 + \beta_3 x^3$ **13.17** $E(y) = \beta_0 + \beta_1 x + \beta_2 x^2$ **13.19** $E(y) = \beta_0 + \beta_1 x + \beta_2 x^2$ **13.21** $t = -6.60$, reject H_0 **13.23a.** $E(y) = \beta_0 + \beta_1 x_1 + \beta_2 x_2$ **b.** $E(y) = \beta_0 + \beta_1 x_1 + \beta_2 x_2 + \beta_3 x_1 x_2$ **c.** $E(y) = \beta_0 + \beta_1 x_1 + \beta_2 x_2 + \beta_3 x_1 x_2 + \beta_4 x_1^2 + \beta_5 x_2^2$ **13.25e.** From 5 to 13 **13.27a.** $\hat{y} = -2.55 + 3.82 x_1 + 2.63 x_2 - 1.29 x_1 x_2$ **e.** H_0: $\beta_3 = 0$, H_a: $\beta_3 \neq 0$ **f.** $t = -8.06$, reject H_0 **13.29a.** $E(y) = \beta_0 + \beta_1 x_1 + \beta_2 x_2 + \beta_3 x_1 x_2 + \beta_4 x_1^2 + \beta_5 x_2^2$ **b.** $\beta_4 x_1^2$ and $\beta_5 x_2^2$ **13.31a.** Both are quantitative **b.** $E(y) = \beta_0 + \beta_1 x_1 + \beta_2 x_2$ **c.** $E(y) = \beta_0 + \beta_1 x_1 + \beta_2 x_2 + \beta_3 x_1 x_2 + \beta_4 x_1^2 + \beta x_2^2$ **d.** H_0: $\beta_3 = \beta_4 = \beta_5 = 0$, H_a: At least one $\beta_i \neq 0$ **13.33a.** $\hat{y} = 148.5 + .472 x_1 - .099 x_2 - .00054 x_1^2 + .000015 x_2^2$ **c.** $F = 31.66$, reject H_0 **d.** Yes **13.35a.** H_a: At least one $\beta_i \neq 0$ **c.** num: 3; den: 24 **13.39a.** $F = 66.87$, reject H_0 **b.** H_a: $\beta_1 \neq 0$ **c.** $F = 15.16$, reject H_0 **d.** Yes **13.41a.** H_0: $\beta_2 = 0$; H_a: $\beta_2 \neq 0$ **b.** $F = .65$, fail to reject H_0 **13.43a.** Both are quantitative **b.** $E(y) = \beta_0 + \beta_1 x_1 + \beta_2 x_2$ **c.** $E(y) = \beta_0 + \beta_1 x_1 + \beta_2 x_2 + \beta_3 x_1 x_2 + \beta_4 x_1^2 + \beta_5 x_2^2$ **d.** H_0: $\beta_3 = 0$, H_a: $\beta_3 \neq 0$ **e.** $F = 14.95$, reject H_0 **13.45** $F = 24.19$, reject H_0 **13.47** $E(y) = \beta_0 + \beta_1 x_1$ **13.49a.** 1: $y = 10.2$; 2: $\hat{y} = 6.2$; 3: $\hat{y} = 22.2$; 4: $\hat{y} = 12.2$ **b.** H_0: $\beta_1 = \beta_2 = \beta_3 = 0$, H_a: At least one $\beta_i \neq 0$ **13.51a.** $E(y) = \beta_0 + \beta_1 x_1$ **c.** H_0: $\beta_1 = 0$; H_a: $\beta_1 > 0$ **13.53a.** $E(y) = \beta_0 + \beta_1 x_1 + \beta_2 x_2 + \beta_3 x_3 + \beta_4 x_4$ **b.** $\hat{y} = 20.35 + 4.75 x_1 + 8.6 x_2 + 11.3 x_3 + 2.1 x_4$ **c.** H_0: $\beta_1 = \beta_2 = \beta_3 = \beta_4 = 0$; H_a: At least one $\beta_i \neq 0$ **d.** $F = 23.966$, reject H_0 **e.** 2.1 ± 2.846 **13.55a.** Diet **b.** $E(y) = \beta_0 + \beta_1 x_1 + \beta_2 x_2$ **13.57a.** $F = 4.80$, reject H_0 **b.** \$11,400 **c.** \$20,000 **13.59a.** $E(y) = \beta_0 + \beta_1 x_1$; $E(y) = \beta_0 + \beta_1 x_1 + \beta_2$; $E(y) = \beta_0 + \beta_1 x_1 + \beta_3$ **b.** $\hat{y} = 44.803 + 2.173 x_1 + 9.413 x_2 + 15.632 x_3$ **c.** $\hat{y} = 44.803 + 2.173 x_1$; $\hat{y} = 54.216 + 2.173 x_1$; $\hat{y} = 60.435 + 2.173 x_1$ **d.** H_0: $\beta_2 = \beta_3 = 0$; H_a: At least one $\beta_i \neq 0$ **e.** $F = 3.33$, fail to reject H_0 **13.61a.** $E(y) = \beta_0 + \beta_1 x_1 + \beta_2 x_2$ **b.** Slope $= \beta_1$ **c.** $E(y) = \beta_0 + \beta_1 x_1 + \beta_2 x_2 + \beta_3 x_1 x_2$ **d.** Slope $= \beta_1$; slope $= \beta_1 + \beta_3$ **e.** $E(y) = \beta_0 + \beta_1 x_1$ **f.** Reject H_0 at $\alpha = .05$ **g.** Reject H_0 at $\alpha = .05$ **h.** Fail to reject H_0 at $\alpha = .05$ **13.63a.** $E(y) = \beta_0 + \beta_1 x_1 + \beta_2 x_2 + \beta_3 x_3$ **c.** $\hat{y} = 36,388 + 15,617 x_1 + 152,487 x_2 + 49,441 x_3$ **e.** $F = 8.43$, reject H_0 **13.65a.** $F = 4.90$, reject H_0 **b.** H_0: $\beta_4 = \beta_5 = 0$ **c.** $F = 2.36$, fail to reject H_0 **13.67a.** $E(y) = \beta_0 + \beta_1 x_1 + \beta_2 x_2 + \beta_3 x_3$ **b.** $\hat{y} = -5.346 + 17.263 x_1 - 2.351 x_2 - 5.924 x_3$ **c.** $E(y) = \beta_0 + \beta_1 x_1 + \beta_2 x_2 + \beta_3 x_3 + \beta_4 x_1 x_2 + \beta_5 x_1 x_3$ **d.** $\hat{y} = -2.792 + 15.135 x_1 - 5.576 x_2 - 9.276 x_3 + 2.477 x_1 x_2 + 2.464 x_1 x_3$ **e.** $F = .32$, fail to reject H_0 **13.69a.** If $\beta_5 = \beta_6 = \beta_7 = \beta_8 = 0$ **b.** If $\beta_2 = \beta_5 = \beta_6 = \beta_7 = \beta_8 = 0$ **c.** If $\beta_3 = \beta_4 = \beta_5 = \beta_6 = \beta_7 = \beta_8 = 0$ **13.71a.** $\hat{y} = 48.8 - 3.36 x_1 + .0749 x_1^2 - 2.36 x_2 - 7.60 x_3 + 3.71 x_1 x_2 + 2.66 x_1 x_3 - .0183 x_1^2 x_2 - .0372 x_1^2 x_3$ **b.** $\hat{y} = 46.44 + .35 x_1 + .0566 x_1^2$; $\hat{y} = 41.2 - .7 x_1 + .0377 x_1^2$ **d.** H_0: $\beta_3 = \beta_4 = \beta_5 = \beta_6 = \beta_7 = \beta_8 = 0$; H_a: At least one $\beta_i \neq 0$ **e.** $F = 242.65$, reject H_0 **13.73** $F = .93$, fail to reject H_0

13.75a. $E(y) = \beta_0 + \beta_1 x_1 + \beta_2 x_2 + \beta_3 x_3$ **b.** $E(y) = \beta_0 + \beta_1 x_1 + \beta_2 x_2 + \beta_3 x_3 + \beta_4 x_1 x_2 + \beta_5 x_1 x_3$ **c.** AL: β_1; TDS-3A: $\beta_1 + \beta_4$; FE: $\beta_1 + \beta_5$ **d.** H_0: $\beta_4 = \beta_5 = 0$ **13.77** $F = 6.71$, reject H_0 **13.79a.** x_2 **b.** Yes **13.81a.** x_4, x_5, x_6 **b.** No **c.** $E(y) = \beta_0 + \beta_1 x_4 + \beta_2 x_5 + \beta_3 x_6 + \beta_4 x_4 x_5 + \beta_5 x_4 x_6 = \beta_6 x_5 x_6$ **d.** H_0: $\beta_4 = \beta_5 = \beta_6 = 0$ **13.85a.** Second **b.** First **c.** Second **d.** First **e.** Third **f.** First **13.93** Test H_0: $\beta_3 = 0$; H_a: $\beta_3 \neq 0$ **13.95a.** $E(y) = \beta_0 + \beta_1 x_1 + \beta_2 x_2 + \beta_3 x_3$ **b.** $E(y) = \beta_0 + \beta_1 x_1 + \beta_2 x_2 + \beta_3 x_3 + \beta_4 x_1 x_2 + \beta_5 x_1 x_3$ **d.** $F = .22$, fail to reject H_0 **e.** H_0: $\beta_2 = \beta_3 = \beta_4 = \beta_5 = 0$ **f.** $F = 4.99$, reject H_0 **13.97a.** $E(y) = \beta_0 + \beta_1 x_1 + \beta_2 x_2$ **b.** $E(y) = \beta_0 + \beta_1 x_1 + \beta_2 x_2 + \beta_3 x_1^2 + \beta_4 x_1 x_2 + \beta_5 x_1^2 x_2$ **c.** $F = 2.49$, fail to reject H_0 **13.99a.** $F = 222.173$, reject H_0 **b.** $F = 13.49$, reject H_0 **e.** $E(y) = \beta_0 + \beta_1 x_1 + \beta_2 x_2 + \beta_3 x_3 + \beta_4 x_4 + \beta_5 x_1 x_2 + \beta_6 x_1 x_3 + \beta_7 x_1 x_4$ **f.** $\hat{y} = .709 + .109 x_1 - .252 x_2 - .618 x_3 - 1.205 x_4 - .016 x_1 x_2 + .0055 x_1 x_3 + .0049 x_1 x_4$ **g.** $F = 1.78$, fail to reject H_0

CHAPTER 14

14.1a. 18.3070 **b.** 29.7067 **c.** 23.5418 **d.** 79.4900 **14.3a.** $\chi^2 > 5.99147$ **b.** $\chi^2 > 7.77944$ **c.** $\chi^2 > 11.3449$ **14.7a.** $\chi^2 = 3.293$, fail to reject H_0 **14.9** $\chi^2 = 18.78$, reject H_0 **14.11a.** $\chi^2 = 92.176$, reject H_0 **b.** $\chi^2 = 65.314$, reject H_0 **c.** $\chi^2 = 116.176$, reject H_0 **d.** $.59 \pm .07$ **14.13** $\chi^2 = 1.143$, fail to reject H_0 **14.15** $\chi^2 = 39.267$, reject H_0 **14.17** $\chi^2 = 2.764$, fail to reject H_0 **14.19a.** H_0: classifications independent; H_a: classifications dependent **b.** $\chi^2 > 9.21034$ **d.** $\chi^2 = 8.71$, fail to reject H_0 **14.21** $\chi^2 = 12.36$, reject H_0 **14.23a.** .901 **b.** .690 **c.** Probably **d.** $\chi^2 = 48.191$, reject H_0 **e.** $.211 \pm .070$ **14.25a.** $\chi^2 = 39.22$, reject H_0 **b.** $\chi^2 = 2.84$, fail to reject H_0 **14.27** $\chi^2 = 2.17067$, fail to reject H_0 **14.29** $\chi^2 = 19.10$, reject H_0 **14.31a.** No **c.** $\chi^2 = 12.9$ **d.** Reject H_0 **e.** $.233 \pm 200$ **14.33a.** $\chi^2 = 54.14$, reject H_0 **b.** No **c.** Yes **14.35a.** H_0: classifications independent; H_a: classifications dependent **b.** $\chi^2 = 47.11$, reject H_0 **14.37** $\chi^2 = 7.384$, fail to reject H_0 **14.39** $\chi^2 = 180.874$, reject H_0 **14.41a.** $\chi^2 = 8.664$, fail to reject H_0 **b.** No **c.** Yes **14.43** $\chi^2 = 9.35$, reject H_0 **14.45c.** $p = .187$ **d.** $\chi^2 = .592$, $p > .10$ **14.49a.** $\chi^2 = 9.647$ **b.** $\chi^2 = 11.0705$ **c.** No **d.** $.05 < p < .10$

CHAPTER 15

15.1 Population not normal **15.3a.** .063 **b.** .500 **c.** .004 **d.** .1515 **e.** .2119 **15.5** $p = .115$, fail to reject H_0 **15.7a.** H_0: $\eta = 5$, H_a: $\eta > 5$ **b.** $p = 0$ **c.** Reject H_0 **15.9a.** H_0: $\eta = 60$; H_a: $\eta < 60$ **b.** Reject if $p < .05$ **d.** $p = .058$, fail to reject H_0 **e.** $p = .058$ **15.11a.** $T_A \leq 41$ or $T_A \geq 71$ **b.** $T_A \geq 50$ **c.** $T_A \leq 43$ **d.** $z < -1.96$ or $z > 1.96$ **15.15a.** $T_A = 62.5$, reject H_0 **b.** $T_A = 62.5$, reject H_0 **15.17a.** $T_B = 53$, reject H_0 **15.19** $T_B = 29$, fail to reject H_0 **15.21** $T_A = 69.5$, reject H_0 **15.23** $z = 3.44$, reject H_0 **15.25b.** $T_- = 3.5$, reject H_0 **15.29b.** $z = 2.499$, reject H_0 **c.** $p = .0062$ **15.31** $T_+ = 27.5$ fail to reject H_0 **15.33** $T_- = 13$, fail to reject H_0 **15.35** $p = .2249$, fail to reject H_0 **15.37a.** $H = 8.190$, reject H_0 **c.** $.01 < p < .025$ **15.41** $H = 1.565$, fail to reject H_0 **15.43a.** $H = 6.66$, reject H_0 **b.** $T_A = 71$, reject H_0 **15.45** $H = 15.381$, $p = .0005$, reject H_0 **15.47a.** $b = 6$, **c.** $F_r = 15.2$, reject H_0 **d.** $p < .005$ **15.49** $F_r = 13$, reject H_0 **15.51** $F_r = .9250$, fail to reject H_0 **15.53** $F_r = 15.25$, reject H_0 **15.55a.** $F_r = 7.85$, reject H_0 **b.** Yes. C is better than B. **15.57a.** $r_s > .648$ or $r_s < -.648$ **b.** $r_s > .450$ **c.** $r_s < -.432$ **15.59a.** H_0: $\rho_s = 0$; H_a: $\rho_s \neq 0$ **b.** $r_s = .745$, fail to reject H_0 **c.** $.05 < p < .10$ **15.61** $r_s = .657$, fail to reject H_0 **15.63a.** $r_s = .861$ **b.** Reject H_0 **15.65** $r_s = -.812$; reject H_0 **15.69a.** $r_s = .40$, fail to reject H_0 **b.** $T_- = 1.5$, reject H_0 **15.71** $F_r = 14.9$, reject H_0 **15.73** $T_A = 21$, fail to reject H_0 **15.75** $F_r = 39.413$, reject H_0 **15.77** $T_- = 4.5$, reject H_0 **15.79a.** $T_+ = 1$, reject H_0 **b.** $t = -2.96$, reject H_0 **15.81** $H = 19.466$, reject H_0 **15.83a.** H_0: $\eta = .57$; H_a: $\eta \neq .75$ **b.** Reject if $p < .10$ **d.** $p = .044$, reject H_0 **15.85a.** $H = 12.7927$, reject H_0 **b.** $T_A = 28.5$, fail to reject H_0 **15.87** $p = .363$, fail to reject H_0 **15.89** $r_s = .929$, reject H_0 **15.91** $r_s = .479$, fail to reject H_0

REFERENCES

CHAPTER 1

Careers in Statistics. American Statistical Association, Biometric Society, Institute of Mathematical Statistics and Statistical Society of Canada, 1995.

Ethical Guidelines for Statistical Practice. American Statistical Association, 1995.

Tanur, J. M., Mosteller, F., Kruskal, W. H., Link, R. F., Pieters, R. S., and Rising, G. R. *Statistics: A Guide to the Unknown* (E. L. Lehmann, special editor). San Francisco: Holden-Day, 1989.

What Is a Survey? Section on Survey Research Methods, American Statistical Association, 1995.

CHAPTER 2

Hoff, D. *How to Lie with Statistics.* New York: Norton, 1954.

Mendenhall, W. *Introduction to Probability and Statistics,* 9th ed. North Scituate, Mass.: Duxbury, 1994.

Wasserman, P., and Bernero, J. *Statistics Sources,* 5th ed. Detroit: Gale Research Company, 1978.

CHAPTER 3

Feller, W. *An Introduction to Probability Theory and Its Applications,* 3d ed., Vol. 1. New York: Wiley, 1968.

Kuehn, A. A. "An analysis of the dynamics of consumer behavior and its implications for marketing management." Unpublished doctoral dissertation, Graduate School of Industrial Administration, Carnegie Institute of Technology, 1958.

Lindley, D.V. *Making Decisions,* 2d ed. London: Wiley, 1985.

Parzen, E. *Modern Probability Theory and Its Applications.* New York: Wiley, 1960.

Scheaffer, R. L., and Mendenhall, W. *Introduction to Probability: Theory and Applications.* North Scituate, Mass.: Duxbury, 1975.

Williams, B. *A Sampler on Sampling.* New York: Wiley, 1978.

Winkler, R. L. *An Introduction to Bayesian Inference and Decision.* New York: Holt, Rinehart and Winston, 1972.

CHAPTER 4

Hogg, R. V., and Craig, A. *Introduction to Mathematical Statistics,* 4th ed. New York: Macmillan, 1978.

Mendenhall, W., Wackerly, D., and Scheaffer, R. L. *Mathematical Statistics with Applications,* 4th ed. Boston: PWS-Kent, 1990.

CHAPTER 5

Hogg, R. V., and Craig, A. T. *Introduction to Mathematical Statistics,* 4th ed. New York: Macmillan, 1978.

Mendenhall, W., Wackerly, D., and Scheaffer, R. *Mathematical Statistics with Applications,* 4th ed. Boston: PWS-Kent, 1990.

CHAPTER 6

Hogg, R. V., and Craig, A. T. *Introduction to Mathematical Statistics,* 4th ed., New York: Macmillan, 1978.

Lindgren, B. W. *Statistical Theory,* 3d ed. New York: Macmillan, 1976.

Mendenhall, W., Wackerly, D., and Scheaffer, R. L. *Mathematical Statistics with Applications,* 4th ed. Boston: PWS-Kent, 1990.

CHAPTER 7

Deming, W. E. *Out of the Crisis.* Cambridge, Mass.: M.I.T. Center for Advanced Study of Engineering, 1986.

Mendenhall, W., and Beaver, B. *Introduction to Probability and Statistics,* 8th ed. Boston: PWS-Kent, 1991.

CHAPTER 8

Snedecor, G. W., and Cochran, W. G. *Statistical Methods,* 7th ed. Ames: Iowa State University Press, 1980.

CHAPTER 9

Freedman, D., Pisani, R., and Purves, R. *Statistics.* New York: W. W. Norton and Co., 1978.

Mendenhall, W. *Introduction to Probability and Statistics,* 8th ed. Boston: PWS-Kent, 1991.

Satterthwaite, F. W. "An approximate distribution of estimates of variance components." *Biometrics Bulletin,* Vol. 2, 1946, pp. 110–114.

Snedecor, G. W., and Cochran, W. *Statistical Methods,* 7th ed. Ames: Iowa State University Press, 1980.

Steel, R. G. D., and Torrie, J. H. *Principles and Procedures of Statistics,* 2nd ed. New York: McGraw-Hill, 1980.

CHAPTER 10

Kramer, C. Y. "Extension of multiple range tests to group means with unequal number of replications." *Biometrics,* Vol. 12, 1956, pp. 307–310.

Mendenhall, W. *Introduction to Linear Models and the Design and Analysis of Experiments.* Belmont, Calif.: Wadsworth, 1968.

Mendenhall, W. and Sincich, T. *A Second Course in Statistics: Regression Analysis,* 5th ed. Upper Saddle River, N.J.: Prentice-Hall, 1996.

Miller, R. G. Jr. *Simultaneous Statistical Inference.* New York: Springer-Verlag, 1981.

Scheffé, H. "A method for judging all contrasts in the analysis of variance." *Biometrika,* Vol. 40, 1953, pp. 87–104.

Scheffé, H. *The Analysis of Variance.* New York: Wiley, 1959.

Snedecor, G. W., and Cochran, W. G. *Statistical Methods,* 7th ed. Ames: Iowa State University Press, 1980.

Steele, R. G. D., and Torrie, J. H. *Principles and Procedures of Statistics: A Biometrical Approach,* 2d ed. New York: McGraw-Hill, 1980.

Tukey, J. W. "Comparing individual means in the analysis of variance." *Biometrics,* Vol. 5, 1949, pp. 99–114.

Winer, B. J. *Statistical Principles in Experimental Design,* 2d ed. New York: McGraw-Hill, 1971.

CHAPTER 11

Graybill, F. *Theory and Application of the Linear Model.* North Scituate, Mass.: Duxbury, 1976.

Mendenhall, W. *Introduction to Linear Models and the Design and Analysis of Experiments.* Belmont, Ca.: Wadsworth, 1968.

Younger, M. S. *A First Course in Linear Regression,* 2d ed. Boston, Mass.: Duxbury, 1985.

CHAPTER 12

Draper, N., and Smith, H. *Applied Regression Analysis.* New York: Wiley, 1966.

Graybill, F. *Theory and Application of the Linear Model.* North Scituate, Mass.: Duxbury, 1976.

Kleinbaum, D., and Kupper, L. *Applied Regression Analysis and Other Multivariable Methods.* North Scituate, Mass.: Duxbury, 1978.

Mendenhall, W. *Introduction to Linear Models and the Design and Analysis of Experiments.* Belmont, Ca.: Wadsworth, 1968.

Mendenhall, W., and Sincich, T. *A Second Course in Statistics: Regression Analysis,* 5th ed. Saddle Lake, N.J.: Prentice-Hall, 1996.

Neter, J., Wasserman, W., and Kutner, M. *Applied Linear Statistical Models,* 3d ed. Homewood, Ill.: Richard Irwin, 1992.

Younger, M. S. *A First Course in Linear Regression,* 2d ed. Boston: Duxbury, 1985.

CHAPTER 13

Draper, N., and Smith, H. *Applied Regression Analysis.* New York: Wiley, 1966.

Graybill, F. A. *Theory and Application of the Linear Model.* North Scituate, Mass.: Duxbury, 1976.

Kelting, H. "Investigation of condominium sale prices in three market scenarios: Utility of stepwise, interactive, multiple regression analysis and implications for design and appraisal methodology." Unpublished paper, University of Florida, Gainesville, 1979.

Mendenhall, W. *Introduction to Linear Models and the Design and Analysis of Experiments.* Belmont, Cal.: Wadsworth, 1968.

Mendenhall, W., and Sincich, T. *A Second Course in Statistics: Regression Analysis,* 5th ed. Saddle River, N.J.: Prentice Hall, 1996.

CHAPTER 14

Agresti, A., and Agresti, B. F. *Statistical Methods for the Social Sciences,* 2d edition. San Francisco: Dellen, 1986.

Cochran, W. G. "The χ^2 test of goodness of fit." *Annals of Mathematical Statistics,* 1952, p. 23.

Savage, I. R. "Bibliography of nonparametric statistics and related topics." *Journal of the American Statistical Association,* 1953, p. 48.

Siegel, S. *Nonparametric Statistics for the Behavioral Sciences.* New York: McGraw-Hill, 1956.

CHAPTER 15

Agresti, A., and Agresti, B. F. *Statistical Methods for the Social Sciences,* 2d ed. San Francisco: Dellen, 1986.

Gibbons, J. D. *Nonparametric Statistical Inference,* 2d ed. New York: McGraw-Hill, 1985.

Hollander, M., and Wolfe, D. A. *Nonparametric Statistical Methods.* New York: Wiley, 1973.

Lehmann, E. L. *Nonparametrics: Statistical Methods Based on Ranks.* San Francisco: Holden-Day, 1975.

Noether, G. E. *Elements of Nonparametric Statistics.* New York: Wiley, 1967.

Siegel, S. *Nonparametric Statistics for the Behavioral Sciences.* New York: McGraw-Hill, 1956.

INDEX

Symbol	Description
$P(A)$	Probability of event A 98
$P(A \mid B)$	Probability of event A given that event B occurs 115
P_n^N	Permutation of N elements taken n at a time 138
q	Probability of Failure for the binomial distribution 170
Q_L	Lower quartile of a data set 70
Q_U	Upper quartile of a data set 70
r	Pearson product moment coefficient of correlation between two samples 501
r^2	Coefficient of determination in simple linear regression 505
r_i	Total count in row i of a contingency table 694
r_s	Spearman's rank correlation coefficient between two samples 752
R_j	Rank sum for the jth sample in a nonparametric analysis of variance 742
R^2	Coefficient of determination in multiple regression 553
ρ (rho)	Pearson product moment coefficient of correlation between two populations 503
ρ_s	Spearman's rank correlation coefficient between two populations 754
s	Sample standard deviation 54
s^2	Sample variance 53
$s_{\hat{\beta}_i}$	Estimated standard deviation of the sampling distribution of $\hat{\beta}_i$ 544
$s_{\hat{\beta}_i}$	Sample estimator of $\sigma_{\hat{\beta}_i}$ 494
s_D	Sample standard deviation of differences in a paired difference experiment 361
s_p^2	Pooled sample variance in two-sample t test 345
S	(1) Sample space 94 (2) Test statistic for sign test for population median 719
SS(Total)	Total sum of squares in an analysis of variance 413
SS_{xx}	Sum of squares of the distances between x measurements and their mean, i.e., $\sum (x - \bar{x})^2$ 480
SS_{xy}	Sum of products of distances of x and y measurements from their means, i.e., $\sum (x - \bar{x})(y - \bar{y})$ 480
SS_{yy}	Sum of squares of the distances between y measurements and their mean, i. e., $\sum (y - \bar{y})^2$ 489
SS(A)	Sum of squares for factor A in factorial experiment 446
SSB	Sum of squares for blocks in a randomized block design 430
SSE	(1) Sum of squared distances between observed and predicted values in a regression model, i.e., $\sum (y - \hat{y})^2$ 489 (2) Difference between total sum of squares and treatment sum of squares (and block sum of squares in a randomized block design) in analysis of variance 408
SST	Sum of squares for treatments in an analysis of variance 408

Symbol	Description
σ (lowercase sigma)	Population standard deviation 54
σ^2	Population variance 54
$\sigma_{\hat{\beta}_1}$	Standard deviation of the sampling distribution of $\hat{\beta}_1$ 494
$\sigma_{\hat{p}}$	Standard deviation of the sampling distribution of $\hat{p}$ 270
$\sigma_{(\hat{p}_1 - \hat{p}_2)}$	Standard deviation of the sampling distribution of $(\hat{p}_1 - \hat{p}_2)$ 372
$\sigma_{\bar{x}}$	Standard deviation of the sampling distribution of $\bar{x}$ 240
$\sigma_{(x_1 - x_2)}$	Standard deviation of the sampling distribution of $(\bar{x}_1 - \bar{x}_2)$ 341
$\sigma_{\hat{y}}$	Standard deviation of the sampling distribution of $\hat{y}$ 511
$\sigma_{(y - \hat{y})}$	Standard deviation of prediction error when $\hat{y}$ is used to predict a particular value of y 511
$\sum$	Symbol for summation 42
t	Statistic used for small-sample tests of hypotheses 262
t_α	Value of t distribution with an area α to its right 263
T_A, T_B	Rank sums corresponding to the two samples in a Wilcoxon rank sum test (independent samples) 723
T_i	Total of observations for treatment i in an analysis of variance 802
T_L, T_U	Lower and upper rejection region values in a Wilcoxon rank sum test 724
T_0	Rejection region value in a Wilcoxon signed rank test for a paired difference experiment 732
T_-, T_+	Rank sums corresponding to the negative and positive rank sums in a Wilcoxon signed rank test for a paired difference experiment 732
θ (theta)	(1) Mean of exponential distribution 218 (2) General symbol for any population parameter 235
x	(1) Sample measurement 42 (2) Random variable 156 (3) Independent variable in a regression model 476
$\bar{x}$	(1) Sample mean 43 (2) Mean of all observations in an analysis of variance 408
$\bar{x}_i$	Sample mean for ith treatment in an analysis of variance 409
$\bar{x}_D$	Sample mean of differences in a paired difference experiment 361
X^2	Test statistic for multinomial and contingency table tests 688
$\hat{y}$	Least squares prediction of y using a regression model 480
z	(1) z-score 65 (2) Standard normal random variable 204 (3) Test statistic 290
z_α	Value of standard normal random variable with an area α to its right 257